Standard Normal Distribution *(continued)*

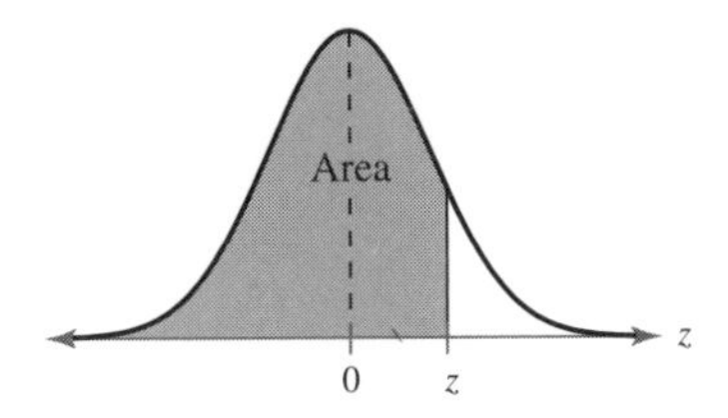

z	.00	.01	.02	.03	.04	.05	.06	.07	.08	.09
0.0	.5000	.5040	.5080	.5120	.5160	.5199	.5239	.5279	.5319	.5359
0.1	.5398	.5438	.5478	.5517	.5557	.5596	.5636	.5675	.5714	.5753
0.2	.5793	.5832	.5871	.5910	.5948	.5987	.6026	.6064	.6103	.6141
0.3	.6179	.6217	.6255	.6293	.6331	.6368	.6406	.6443	.6480	.6517
0.4	.6554	.6591	.6628	.6664	.6700	.6736	.6772	.6808	.6844	.6879
0.5	.6915	.6950	.6985	.7019	.7054	.7088	.7123	.7157	.7190	.7224
0.6	.7257	.7291	.7324	.7357	.7389	.7422	.7454	.7486	.7517	.7549
0.7	.7580	.7611	.7642	.7673	.7704	.7734	.7764	.7794	.7823	.7852
0.8	.7881	.7910	.7939	.7967	.7995	.8023	.8051	.8078	.8106	.8133
0.9	.8159	.8186	.8212	.8238	.8264	.8289	.8315	.8340	.8365	.8389
1.0	.8413	.8438	.8461	.8485	.8508	.8531	.8554	.8577	.8599	.8621
1.1	.8643	.8665	.8686	.8708	.8729	.8749	.8770	.8790	.8810	.8830
1.2	.8849	.8869	.8888	.8907	.8925	.8944	.8962	.8980	.8997	.9015
1.3	.9032	.9049	.9066	.9082	.9099	.9115	.9131	.9147	.9162	.9177
1.4	.9192	.9207	.9222	.9236	.9251	.9265	.9278	.9292	.9306	.9319
1.5	.9332	.9345	.9357	.9370	.9382	.9394	.9406	.9418	.9429	.9441
1.6	.9452	.9463	.9474	.9484	.9495	.9505	.9515	.9525	.9535	.9545
1.7	.9554	.9564	.9573	.9582	.9591	.9599	.9608	.9616	.9625	.9633
1.8	.9641	.9649	.9656	.9664	.9671	.9678	.9686	.9693	.9699	.9706
1.9	.9713	.9719	.9726	.9732	.9738	.9744	.9750	.9756	.9761	.9767
2.0	.9772	.9778	.9783	.9788	.9793	.9798	.9803	.9808	.9812	.9817
2.1	.9821	.9826	.9830	.9834	.9838	.9842	.9846	.9850	.9854	.9857
2.2	.9861	.9864	.9868	.9871	.9875	.9878	.9881	.9884	.9887	.9890
2.3	.9893	.9896	.9898	.9901	.9904	.9906	.9909	.9911	.9913	.9916
2.4	.9918	.9920	.9922	.9925	.9927	.9929	.9931	.9932	.9934	.9936
2.5	.9938	.9940	.9941	.9943	.9945	.9946	.9948	.9949	.9951	.9952
2.6	.9953	.9955	.9956	.9957	.9959	.9960	.9961	.9962	.9963	.9964
2.7	.9965	.9966	.9967	.9968	.9969	.9970	.9971	.9972	.9973	.9974
2.8	.9974	.9975	.9976	.9977	.9977	.9978	.9979	.9979	.9980	.9981
2.9	.9981	.9982	.9982	.9983	.9984	.9984	.9985	.9985	.9986	.9986
3.0	.9987	.9987	.9987	.9988	.9988	.9989	.9989	.9989	.9990	.9990
3.1	.9990	.9991	.9991	.9991	.9992	.9992	.9992	.9992	.9993	.9993
3.2	.9993	.9993	.9994	.9994	.9994	.9994	.9994	.9995	.9995	.9995
3.3	.9995	.9995	.9995	.9996	.9996	.9996	.9996	.9996	.9996	.9997
3.4	.9997	.9997	.9997	.9997	.9997	.9997	.9997	.9997	.9997	.9998

t-Distribution

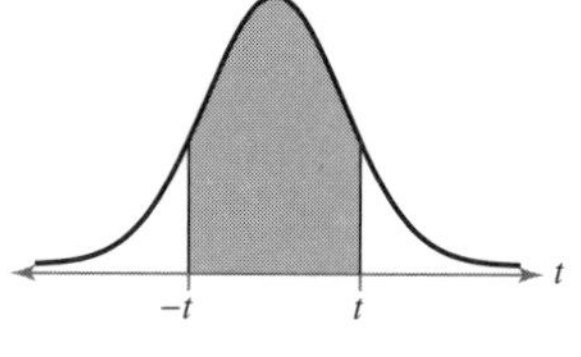

c-confidence interval

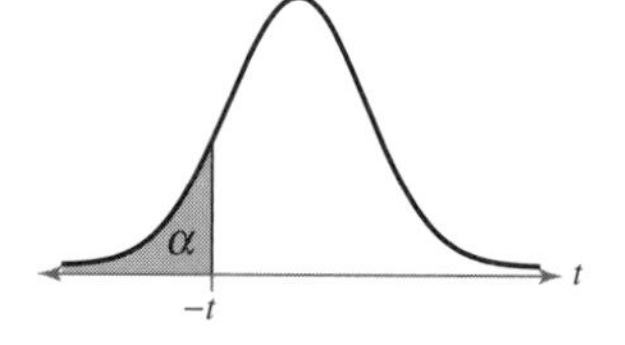

Left-tailed test

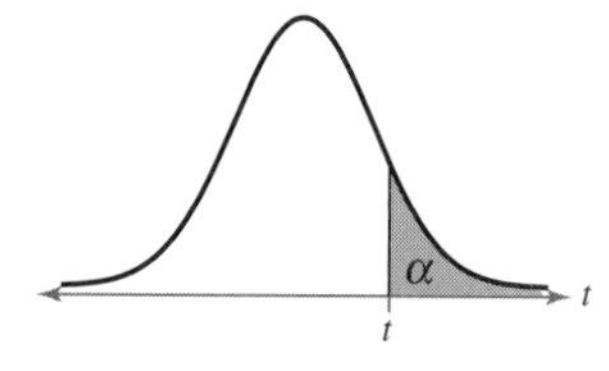

Right-tailed test

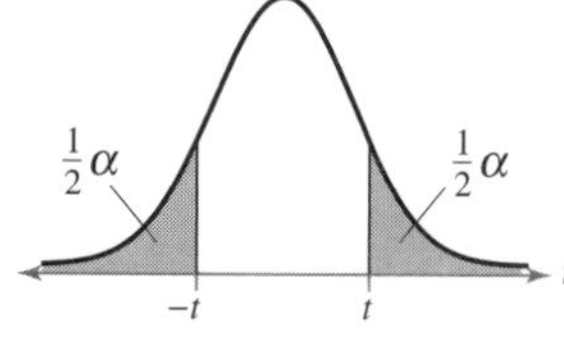

Two-tailed test

	Level of confidence, c	**0.50**	**0.80**	**0.90**	**0.95**	**0.98**	**0.99**
	One tail, α	**0.25**	**0.10**	**0.05**	**0.025**	**0.01**	**0.005**
d.f.	**Two tails, α**	**0.50**	**0.20**	**0.10**	**0.05**	**0.02**	**0.01**
1		1.000	3.078	6.314	12.706	31.821	63.657
2		.816	1.886	2.920	4.303	6.965	9.925
3		.765	1.638	2.353	3.182	4.541	5.841
4		.741	1.533	2.132	2.776	3.747	4.604
5		.727	1.476	2.015	2.571	3.365	4.032
6		.718	1.440	1.943	2.447	3.143	3.707
7		.711	1.415	1.895	2.365	2.998	3.499
8		.706	1.397	1.860	2.306	2.896	3.355
9		.703	1.383	1.833	2.262	2.821	3.250
10		.700	1.372	1.812	2.228	2.764	3.169
11		.697	1.363	1.796	2.201	2.718	3.106
12		.695	1.356	1.782	2.179	2.681	3.055
13		.694	1.350	1.771	2.160	2.650	3.012
14		.692	1.345	1.761	2.145	2.624	2.977
15		.691	1.341	1.753	2.131	2.602	2.947
16		.690	1.337	1.746	2.120	2.583	2.921
17		.689	1.333	1.740	2.110	2.567	2.898
18		.688	1.330	1.734	2.101	2.552	2.878
19		.688	1.328	1.729	2.093	2.539	2.861
20		.687	1.325	1.725	2.086	2.528	2.845
21		.686	1.323	1.721	2.080	2.518	2.831
22		.686	1.321	1.717	2.074	2.508	2.819
23		.685	1.319	1.714	2.069	2.500	2.807
24		.685	1.318	1.711	2.064	2.492	2.797
25		.684	1.316	1.708	2.060	2.485	2.787
26		.684	1.315	1.706	2.056	2.479	2.779
27		.684	1.314	1.703	2.052	2.473	2.771
28		.683	1.313	1.701	2.048	2.467	2.763
29		.683	1.311	1.699	2.045	2.462	2.756
∞		.674	1.282	1.645	1.960	2.326	2.576

Elementary Statistics

Picturing the World

Second Edition

Ron Larson

Penn State University at Erie

Betsy Farber

Bucks County Community College

Prentice Hall, Upper Saddle River, NJ 07458

Library of Congress Cataloging-in-Publication Data

Larson, Ron
Elementary statistics : picturing the world / Ron Larson, Betsy Farber.—2nd ed.
p. cm.
Includes index.
ISBN 0-13-065595-3
1. Statistics. I. Farber, Elizabeth. II. Title.

QA276.12.L373 2003
519.5—de21 2002016911

Editor-in-Chief: ***Sally Yagan***
Acquisition Editor: ***Quincy McDonald***
Editorial/Production Supervision: ***Bayani Mendoza de Leon***
Vice President/Director of Production and Manufacturing: ***David W. Riccardi***
Senior Managing Editor: ***Linda Mihatov Behrens***
Executive Managing Editor: ***Kathleen Schiaparelli***
Manufacturing Buyer: ***Alan Fischer***
Manufacturing Manager: ***Trudy Pisciotti***
Assistant Editor of Media: ***Vince Jansen***
Marketing Manager: ***Angela Battle***
Marketing Assistant: ***Rachel Beckman***
Creative Director: ***Carole Anson***
Director of Creative Services: ***Paul Belfanti***
Art Director: ***Maureen Eide***
Assistant to the Art Director: ***John Christiana***
Managing Editor, Audio/Video Assets: ***Grace Hazeldine***
Art Editor: ***Thomas Benfatti***
Photo Editor: ***Beth Boyd***
Photo Researcher: ***Kathy Ringrose***
Logo Permissions Specialist: ***Shirley Webster***
Cover Designer: ***Kiwi Design***
Interior Design: ***Meridian Creative Group***
Composition and Art: ***Meridian Creative Group***
Editorial Assistant/Supplement Editor: ***Joanne Wendelken***
Cover photos: Body Scan—Elscint/Photo Researchers, Inc.; Baseball–Corbis Digital Stock; Newsstand/Belgium—Craig Aurness/Corbis; Businesspeople/Cafe—Jules Frazier/Getty Images, Inc./PhotoDisc, Inc.

The authors and publisher of this book have used their best efforts in preparing this book. These efforts include the development, research, and testing of the theories and programs to determine their effectiveness. The author and publisher make no warranty of any kind, expressed or implied, with regard to these programs or the documentation contained in this book. The author and publisher shall not be liable in any event for incidental or consequential damages in connection with, or arising out of, the furnishing, performance, or use of these programs.

Printed in the United States of America

10 9 8 7 6 5 4 3

ISBN 0-13-065595-3 (Higher Education)
ISBN 0-13-048885-2 (School version)

Pearson Education LTD., *London*
Pearson Education Australia PTY, Limited, *Sydney*
Pearson Education Singapore, Pte. Ltd
Pearson Education North Asia Ltd, *Hong Kong*
Pearson Education Canada, Ltd., *Toronto*
Pearson Educación de Mexico, S.A. de C.V.
Pearson Education—Japan, *Tokyo*
Pearson Education Malaysia, Pte. Ltd

About the Authors

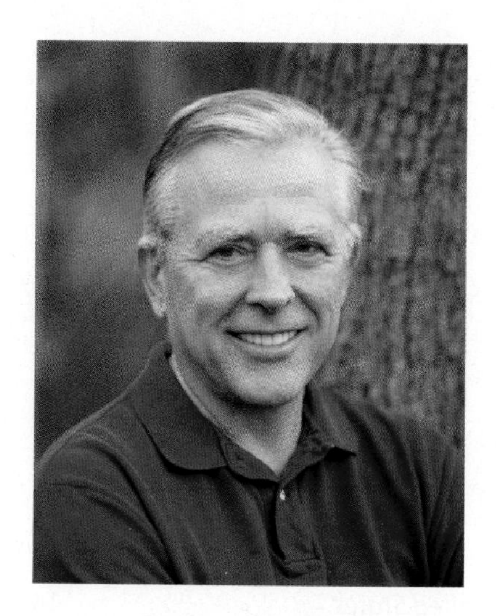

Ron Larson
Penn State University at Erie

Ron Larson received his Ph.D. in mathematics from the University of Colorado in 1970. At that time he accepted a position with Penn State University, and he currently holds the rank of professor of mathematics at the university. Larson is the lead author of more than two dozen mathematics textbooks that range from sixth grade through calculus levels. Many of his texts, such as the seventh edition of his calculus text, are leaders in their markets. Larson is also one of the pioneers in the use of multimedia and the Internet to enhance the learning of mathematics. He has authored multimedia programs, extending from the elementary school through calculus levels. Larson is a member of several professional groups and is a frequent speaker at national and regional mathematics meetings.

Betsy Farber
Bucks County Community College

Betsy Farber received her Bachelor's degree in mathematics from Penn State University and Master's degree in mathematics from the College of New Jersey. Since 1976, she has been teaching all levels of mathematics at Bucks County Community College in Newtown, Pennsylvania, where she currently holds the rank of professor. She is particularly interested in developing new ways to make statistics relevant and interesting to her students, and has been teaching statistics in many different modes—with *TI-83*, with *MINITAB*, and by distance learning as well as in the traditional classroom. A member of the American Mathematical Association of Two-Year Colleges (AMATYC), she is an author of *The Student Edition to MINITAB* and *A Guide to MINITAB.* She served as consulting editor for *Statistics, A First Course* and has written computer tutorials for the CD-ROM correlating to the texts in the Streeter Series in mathematics.

Contents

◆ Chapter 5 Normal Probability Distributions 204

◆ Chapter 6 Confidence Intervals 268

Chapter 9 Correlation and Regression 440

Keeping cars longer.
The median age of vehicles on U.S. roads for seven different years:

Median age in years	
Cars, x	Trucks, y
8.1	7.8
7.7	7.6
6.5	6.5
6.9	7.6
6.0	6.3
5.4	5.8
4.9	5.9

Chapter 10 Chi-Square Tests and the *F*-Distribution 492

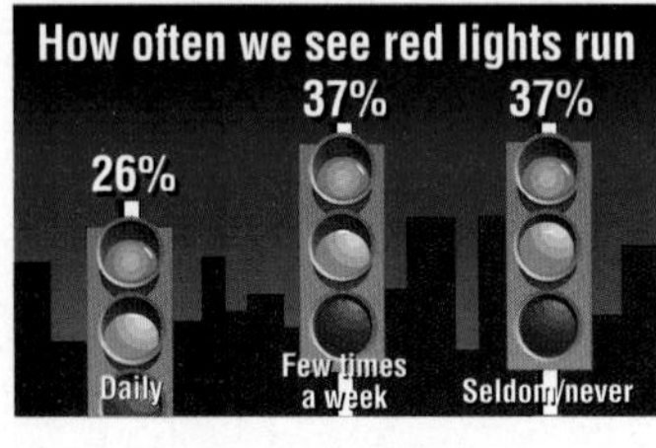

Chapter 11 Nonparametric Tests 544

Appendices

Preface

Welcome to *Elementary Statistics: Picturing the World*, Second Edition. We are grateful for the overwhelming acceptance and support of the First Edition. It is gratifying to know that our vision of combining theory, pedagogy, and design to exemplify how statistics is used to picture and describe the world has helped students learn about statistics and make informed decisions. This message—picturing the world—begins with the cover and is carefully integrated into every feature of the text.

New to the Second Edition

Two features in the Second Edition help students apply statistics to real-life situations and practice making decision about statistics.

Uses and Abuses Each chapter now has a full page summarizing the uses of concepts in the chapter, as well as a description of common misuses. Each "abuse" is accompanied by one or more exercises.

Real Statistics–Real Decisions—Putting It All Together Following the **Review Exercises** in each chapter, we have added a full-page real-life situation accompanied by exercises that ask students to use the concepts in the chapter to make decisions.

The exercise sets in the Second Edition include approximately 200 new exercises, giving the students more practice in performing calculations, making decisions, providing explanations, and applying results to a real-life setting.

In response to suggestions from statistics instructors, the coverage of topics in Chapters 2, 5, 7, and 9 is revised in the Second Edition.

- In **Chapter 2,** the z-score is now introduced in Section 2.5, **Measures of Position.**
- In **Chapter 5,** we added two new sections—Section 5.3, **Normal Distributions: Finding Probabilities,** and Section 5.4, **Normal Distributions: Finding Values.** These sections replace Section 5.3, **Applications of Normal Distributions** in the First Edition. Changing these sections allows the instructor to cover applications of normal distributions in greater detail and from two perspectives.
- In **Chapter 7,** Section 7.1, **Introduction to Hypothesis Testing,** we now introduce the concept of hypothesis testing using probability values, or P-values. The concept of using rejection regions is now introduced in Section 7.2, **Hypothesis Testing for the Mean (Large Samples).**
- In **Chapter 9,** for instructors who prefer to cover Section 9.1, **Correlation,** immediately after covering graphing paired data in Chapter 2, we added a method for testing a population correlation coefficient that does not involve hypothesis testing. The method is simple and can easily be covered after Chapters 1 and 2.

General Features

Versatile Course Coverage The table of contents of the text was developed to give instructors **many options.** For instance, by assigning the **Extending the Basics** exercises and spending time on the chapter projects, there is sufficient content to use the text in a two-semester course. More commonly, we expect the text to be used in a three-credit semester course or a four-semester course that includes a lab component. In such cases, instructors will have to pare down the text's 46 sections. If you want more information on sample syllabi, check the Web site that accompanies the text, *www.prenhall.com/larson*.

Choice of Tables Our experience has shown that students find a **cumulative density function** (CDF) table easier to use than a "0-to-z" table. Using the CDF table to find the area under a normal curve is the topic of Section 5.2 on pages 214–218. Because we realize that many teachers prefer to use the "0-to-z" table, we have provided an alternative presentation of Section 5.2 using the "0-to-z" table in Appendix A of the book.

Graphical Approach As with most introductory statistics texts, we begin the descriptive statistics chapter with a survey of different ways to display data graphically. A difference between this text and many others is that **we continue to incorporate the graphical display of data throughout the text.** For example, see the use of stem-and-leaf plots to display data on pages 348 and 349. In all, the text has over 900 graphs—surpassing all other introductory statistics texts.

Variety of Real-Life Applications We have chosen real-life applications that are representative of the majors of the students taking introductory statistics courses. These include business, psychology, health sciences, sports, computer science, political science, and many others. Choosing meaningful applications for such a diverse audience is difficult. We wanted the applications to be **authentic**—but they also need to be **accessible.**

Data and Source Lines The data sets in the book were chosen for interest, variety, and their ability to illustrate concepts. Most of the **over 200 data sets** contain actual data with source lines. The remaining data sets contain simulated data that, though not actual, are representative of real-life situations. All data sets containing 20 or more entries are available in a variety of electronic forms, including disk and Internet. In the exercise sets, the data sets that are available electronically are indicated by the icon DATA.

Accuracy Every effort was made to **ensure the mathematical accuracy** of the examples and exercise solutions. The examples and exercises were solved by two people independently. A third person compared the independent solutions and resolved differences. If you encounter errors that we missed, please contact us so that we can correct the problem in a subsequent printing.

Balanced Approach The text strikes a **balance between computation, decision making, and conceptual understanding.** We have provided many Examples, Exercises, and Try It problems that go beyond mere computation. For instance, look at Exercises 31 and 32 on page 45. Students are not just asked to construct a relative frequency histogram for the given data, they are asked to go a step further and use the histogram to make a decision.

Prerequisites Statistics contains many formulas and variables, including radicals, summation notation, Greek letters, and subscripts. So, **some familiarity with algebra and evaluation of algebraic expressions** is a prerequisite. Nevertheless, we have made every effort to keep algebraic manipulations to a minimum—often we display informal versions of formulas using words in place of or in addition to variables. For instance, see the definitions of midpoint and relative frequency on page 34.

Flexible Technology Although most formulas in the book are illustrated with tabular "hand" calculations, we assume that most students who take this course have access to some form of technology tool, such as *MINITAB*, *Excel*, the *TI-83*, or *SPSS*. Because the use of technology varies widely, we have made the text flexible. **It can be used in courses with no more technology than a scientific calculator—or it can be used in courses that require frequent use of sophisti-cated technology tools.** For those who want specific instructions on particular technology tools, separate technology manuals are available to augment the text. Whatever your use of technology, we are sure that you agree with us that the goal of this course is not computation. Rather, it is to gain an understanding of the basic concepts and uses of statistics.

Page Layout We believe that statistics is more accessible to students when it is carefully formatted on each page with a consistent open layout. This text is the **first college-level statistics book to be written to design,** which means that its features (Examples, Try It problems, Definitions, or Guidelines) are not split from one page to the next. Although this process requires extra planning and work in the development stage, the result is a presentation that is clean and clear.

MAA, AMATYC, NCTM Standards This text answers the call for a **student-friendly text that emphasizes the uses of statistics** and not just the computation of its myriad of formulas. Our experience indicates that our job as instructors of an introductory course in statistics is not to produce statisticians but to produce informed consumers of statistical reports. For this reason, we have included many exercises that require students to provide written explanations, find patterns, and make decisions.

Features

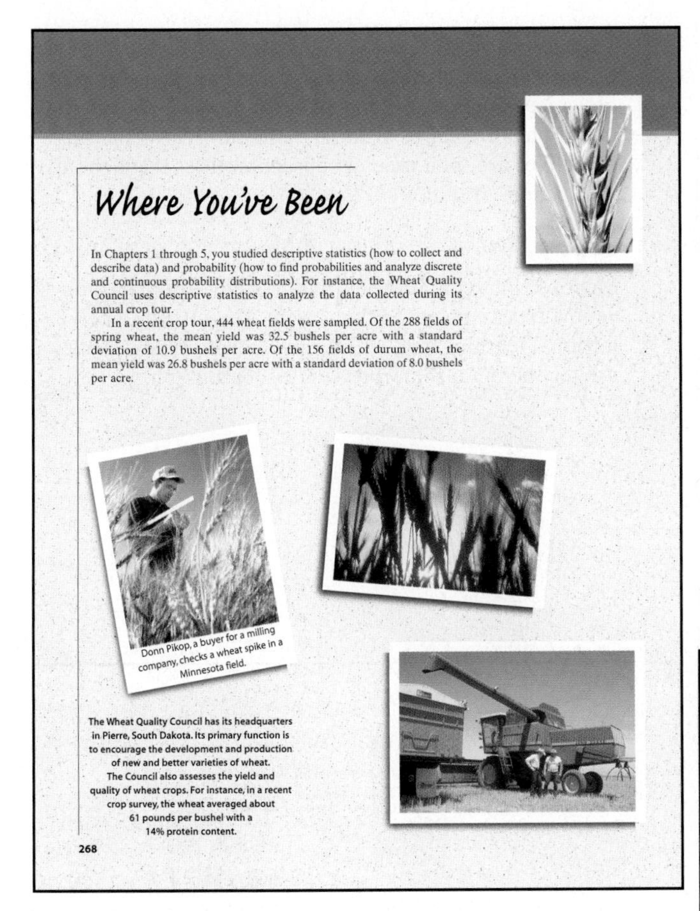

Where You've Been

In Chapters 1 through 5, you studied descriptive statistics (how to collect and describe data) and probability (how to find probabilities and analyze discrete and continuous probability distributions). For instance, the Wheat Quality Council uses descriptive statistics to analyze the data collected during its annual crop tour.

In a recent crop tour, 444 wheat fields were sampled. Of the 288 fields of spring wheat, the mean yield was 32.5 bushels per acre with a standard deviation of 10.9 bushels per acre. Of the 156 fields of durum wheat, the mean yield was 26.8 bushels per acre with a standard deviation of 8.0 bushels per acre.

Donn Pikop, a buyer for a milling company, checks a wheat spike in a Minnesota field.

The Wheat Quality Council has its headquarters in Pierre, South Dakota. Its primary function is to encourage the development and production of new and better varieties of wheat. The Council also assesses the yield and quality of wheat crops. For instance, in a recent crop survey, the wheat averaged about 61 pounds per bushel with a 14% protein content.

268

Chapter Openers

Where You've Been

Each chapter begins with a two-page photographic description of a real-life problem. The first page has a feature called *Where You've Been*. It shows students how the chapter fits into the bigger picture of statistics, by connecting it to topics learned in earlier chapters.

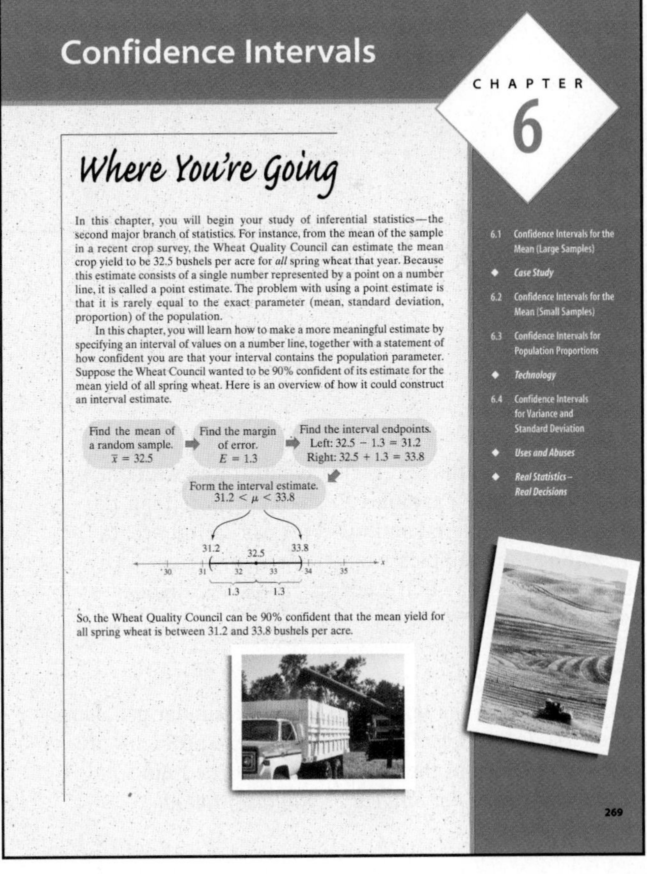

Confidence Intervals

CHAPTER 6

Where You're Going

In this chapter, you will begin your study of inferential statistics—the second major branch of statistics. For instance, from the mean of the sample in a recent crop survey, the Wheat Quality Council can estimate the mean crop yield to be 32.5 bushels per acre for *all* spring wheat that year. Because this estimate consists of a single number represented by a point on a number line, it is called a point estimate. The problem with using a point estimate is that it is rarely equal to the exact parameter (mean, standard deviation, proportion) of the population.

In this chapter, you will learn how to make a more meaningful estimate by specifying an interval of values on a number line, together with a statement of how confident you are that your interval contains the population parameter. Suppose the Wheat Council wanted to be 90% confident of its estimate for the mean yield of all spring wheat. Here is an overview of how it could construct an interval estimate.

Find the mean of a random sample. $\bar{x} = 32.5$ → Find the margin of error. $E = 1.3$ → Find the interval endpoints. Left: $32.5 - 1.3 = 31.2$ Right: $32.5 + 1.3 = 33.8$ → Form the interval estimate. $31.2 < \mu < 33.8$

So, the Wheat Quality Council can be 90% confident that the mean yield for all spring wheat is between 31.2 and 33.8 bushels per acre.

269

Chapter Openers

Where You're Going

The second page of the chapter opener has a feature called *Where You're Going*. It gives students an overview of the chapter, exploring concepts in the context of real-world settings.

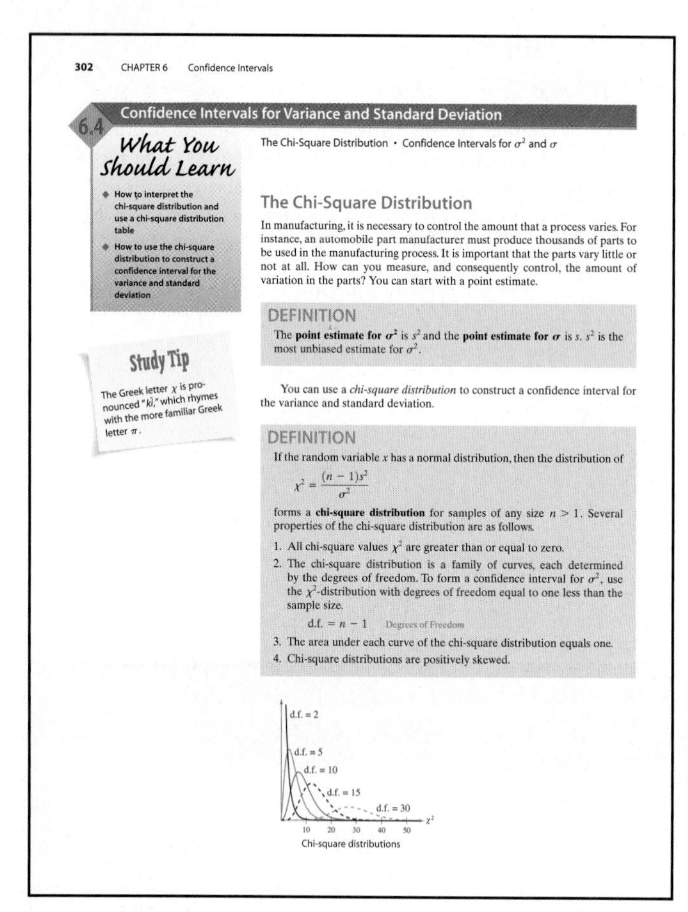

302 CHAPTER 6 Confidence Intervals

6.4 Confidence Intervals for Variance and Standard Deviation

The Chi-Square Distribution • Confidence Intervals for σ^2 and σ

What You Should Learn

- How to interpret the chi-square distribution and use a chi-square distribution table
- How to use the chi-square distribution to construct a confidence interval for the variance and standard deviation

The Chi-Square Distribution

In manufacturing, it is necessary to control the amount that a process varies. For instance, an automobile part manufacturer must produce thousands of parts to be used in the manufacturing process. It is important that the parts vary little or not at all. How can you measure, and consequently control, the amount of variation in the parts? You can start with a point estimate.

DEFINITION

The **point estimate for σ^2** is s^2 and the **point estimate for σ** is s. s^2 is the most unbiased estimate for σ^2.

Study Tip

The Greek letter χ is pronounced "kī," which rhymes with the more familiar Greek letter π.

You can use a *chi-square distribution* to construct a confidence interval for the variance and standard deviation.

DEFINITION

If the random variable x has a normal distribution, then the distribution of

$$\chi^2 = \frac{(n-1)s^2}{\sigma^2}$$

forms a **chi-square distribution** for samples of any size $n > 1$. Several properties of the chi-square distribution are as follows.

1. All chi-square values χ^2 are greater than or equal to zero.
2. The chi-square distribution is a family of curves, each determined by the degrees of freedom. To form a confidence interval for σ^2, use the χ^2-distribution with degrees of freedom equal to one less than the sample size.
   d.f. $= n - 1$ Degrees of Freedom
3. The area under each curve of the chi-square distribution equals one.
4. Chi-square distributions are positively skewed.

Chi-square distributions

Section Organization

Each section is organized by learning objectives. These objectives are presented in everyday language in a margin feature called *What You Should Learn*. The same objectives are then used as subsection titles throughout the section.

Study Tips

Most sections contain one or more study tips placed on yellow "sticky notes" in the margin. These tend to be informal learning aids, which show how to read a table, use technology, or interpret a result or a graph.

Titled Examples

Every concept in the text is clearly illustrated with one or more step-by-step examples. Each of the more than 200 examples is numbered and titled for easy reference. In presenting the examples, we used an open format with a step-by-step display that students can use as a model when solving the exercises.

Try Its

Each example in the text is followed by a similar problem called *Try It Yourself*. The answers to these problems are given in the back of the book, and the worked-out solutions are given in the *Student's Solutions Manual*.

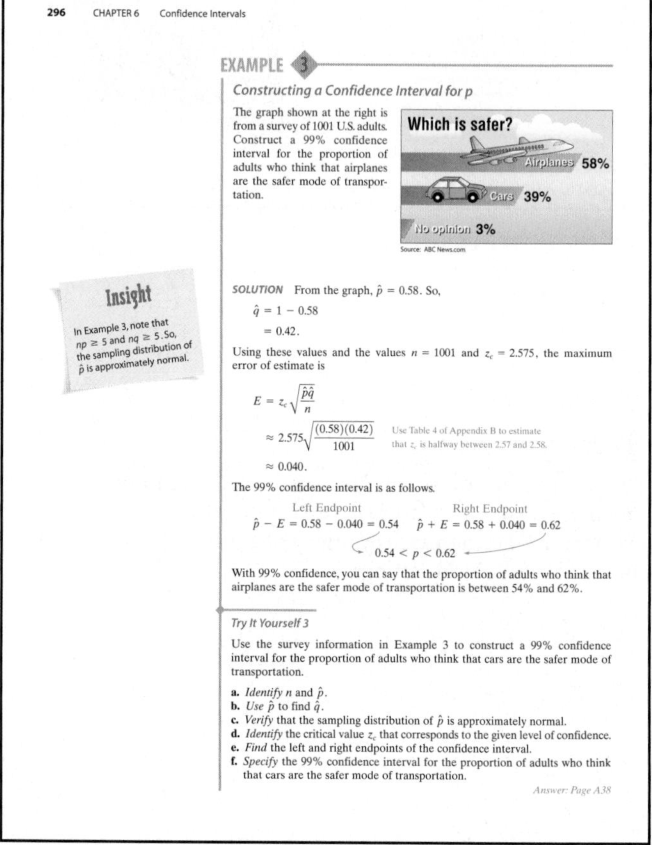

296 CHAPTER 6 Confidence Intervals

EXAMPLE 3

Constructing a Confidence Interval for p

The graph shown at the right is from a survey of 1001 U.S. adults. Construct a 99% confidence interval for the proportion of adults who think that airplanes are the safer mode of transportation.

Source: ABC News.com

Insight

In Example 3, note that $np \geq 5$ and $nq \geq 5$. So, the sampling distribution of $\hat{p}$ is approximately normal.

SOLUTION From the graph, $\hat{p} = 0.58$. So,

$$\hat{q} = 1 - 0.58 = 0.42.$$

Using these values and the values $n = 1001$ and $z_c = 2.575$, the maximum error of estimate is

$$E = z_c\sqrt{\frac{\hat{p}\hat{q}}{n}} \approx 2.575\sqrt{\frac{(0.58)(0.42)}{1001}} \approx 0.040.$$

Use Table 4 of Appendix B to estimate that z_c is halfway between 2.57 and 2.58.

The 99% confidence interval is as follows.

Left Endpoint: $\hat{p} - E = 0.58 - 0.040 = 0.54$
Right Endpoint: $\hat{p} + E = 0.58 + 0.040 = 0.62$

$$0.54 < p < 0.62$$

With 99% confidence, you can say that the proportion of adults who think that airplanes are the safer mode of transportation is between 54% and 62%.

Try It Yourself 3

Use the survey information in Example 3 to construct a 99% confidence interval for the proportion of adults who think that cars are the safer mode of transportation.

a. *Identify* n and $\hat{p}$.
b. *Use* $\hat{p}$ to find $\hat{q}$.
c. *Verify* that the sampling distribution of $\hat{p}$ is approximately normal.
d. *Identify* the critical value z_c that corresponds to the given level of confidence.
e. *Find* the left and right endpoints of the confidence interval.
f. *Specify* the 99% confidence interval for the proportion of adults who think that cars are the safer mode of transportation.

Answer: Page A38

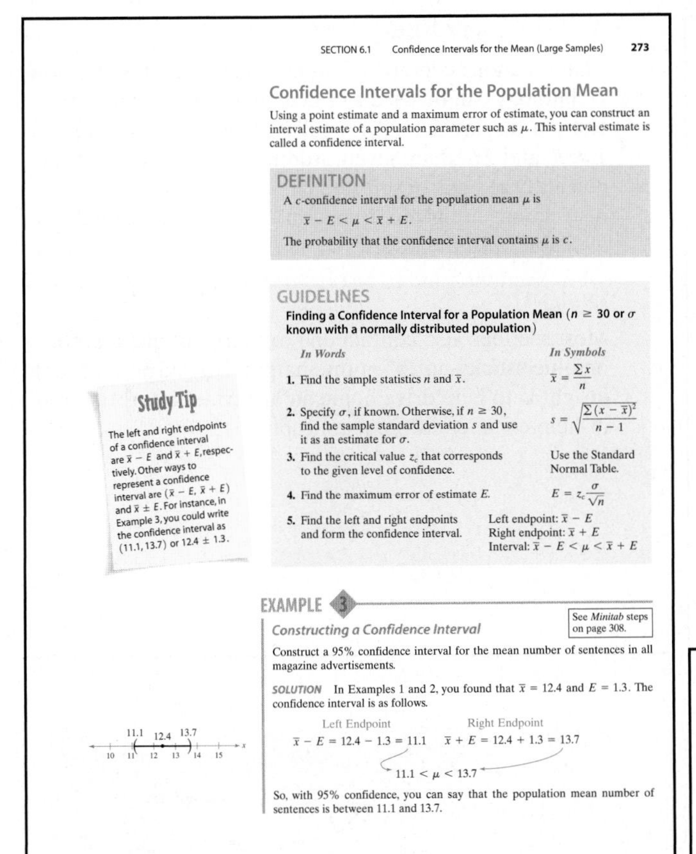

Confidence Intervals for the Population Mean

Using a point estimate and a maximum error of estimate, you can construct an interval estimate of a population parameter such as μ. This interval estimate is called a confidence interval.

DEFINITION

A c-confidence interval for the population mean μ is

$\bar{x} - E < \mu < \bar{x} + E.$

The probability that the confidence interval contains μ is c.

GUIDELINES

Finding a Confidence Interval for a Population Mean ($n \geq 30$ or σ known with a normally distributed population)

In Words	*In Symbols*
1. Find the sample statistics n and $\bar{x}$.	$\bar{x} = \dfrac{\Sigma x}{n}$
2. Specify σ, if known. Otherwise, if $n \geq 30$, find the sample standard deviation s and use it as an estimate for σ.	$s = \sqrt{\dfrac{\Sigma(x - \bar{x})^2}{n-1}}$
3. Find the critical value z_c that corresponds to the given level of confidence.	Use the Standard Normal Table.
4. Find the maximum error of estimate E.	$E = z_c \dfrac{\sigma}{\sqrt{n}}$
5. Find the left and right endpoints and form the confidence interval.	Left endpoint: $\bar{x} - E$ Right endpoint: $\bar{x} + E$ Interval: $\bar{x} - E < \mu < \bar{x} + E$

Study Tip

The left and right endpoints of a confidence interval are $\bar{x} - E$ and $\bar{x} + E$, respectively. Other ways to represent a confidence interval are $(\bar{x} - E, \bar{x} + E)$ and $\bar{x} \pm E$. For instance, in Example 3, you could write the confidence interval as (11.1, 13.7) or 12.4 ± 1.3.

EXAMPLE 3

Constructing a Confidence Interval

See *Minitab* steps on page 308.

Construct a 95% confidence interval for the mean number of sentences in all magazine advertisements.

SOLUTION In Examples 1 and 2, you found that $\bar{x} = 12.4$ and $E = 1.3$. The confidence interval is as follows.

Left Endpoint: $\bar{x} - E = 12.4 - 1.3 = 11.1$

Right Endpoint: $\bar{x} + E = 12.4 + 1.3 = 13.7$

$11.1 < \mu < 13.7$

So, with 95% confidence, you can say that the population mean number of sentences is between 11.1 and 13.7.

Definitions

The critical statistics definitions are set off with gold screens. Formal definitions are often followed by guidelines that explain, in everyday English, how to apply the definition.

Guidelines

Throughout the book, the presentation of a statistical formula is followed by a set of step-by-step guidelines for applying the formula. The guidelines are divided into two columns titled *In Words* and *In Symbols*.

Picturing the World

Each section contains a real-life "mini case study" that illustrates the important concept or concepts of the section. Each *Picturing the World* concludes with a question.

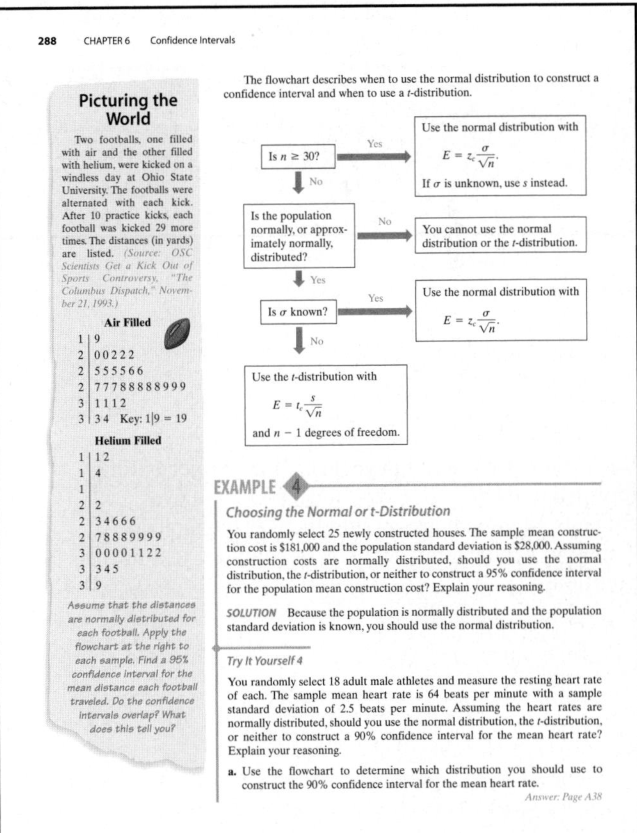

Picturing the World

Two footballs, one filled with air and the other filled with helium, were kicked on a windless day at Ohio State University. The footballs were alternated with each kick. After 10 practice kicks, each football was kicked 29 more times. The distances (in yards) are listed. *(Source: OSC Scientists Get a Kick Out of Sports Controversy, "The Columbus Dispatch," November 21, 1993.)*

Air Filled

```
1 | 9
2 | 0 0 2 2 2
2 | 5 5 5 5 6 6
2 | 7 7 7 8 8 8 8 8 9 9 9
3 | 1 1 1 2
3 | 3 4   Key: 1|9 = 19
```

Helium Filled

```
1 | 1 2
1 | 4
1 |
2 | 2
2 | 3 4 6 6 6
2 | 7 8 8 8 9 9 9 9
3 | 0 0 0 0 1 1 2 2
3 | 3 4 5
3 | 9
```

Assume that the distances are normally distributed for each football. Apply the flowchart at the right to each sample. Find a 95% confidence interval for the mean distance each football traveled. Do the confidence intervals overlap? What does this tell you?

The flowchart describes when to use the normal distribution to construct a confidence interval and when to use a t-distribution.

EXAMPLE 4

Choosing the Normal or t-Distribution

You randomly select 25 newly constructed houses. The sample mean construction cost is \$181,000 and the population standard deviation is \$28,000. Assuming construction costs are normally distributed, should you use the normal distribution, the t-distribution, or neither to construct a 95% confidence interval for the population mean construction cost? Explain your reasoning.

SOLUTION Because the population is normally distributed and the population standard deviation is known, you should use the normal distribution.

Try It Yourself 4

You randomly select 18 adult male athletes and measure the resting heart rate of each. The sample mean heart rate is 64 beats per minute with a sample standard deviation of 2.5 beats per minute. Assuming the heart rates are normally distributed, should you use the normal distribution, the t-distribution, or neither to construct a 90% confidence interval for the mean heart rate? Explain your reasoning.

a. Use the flowchart to determine which distribution you should use to construct the 90% confidence interval for the mean heart rate.

Answer: Page A38

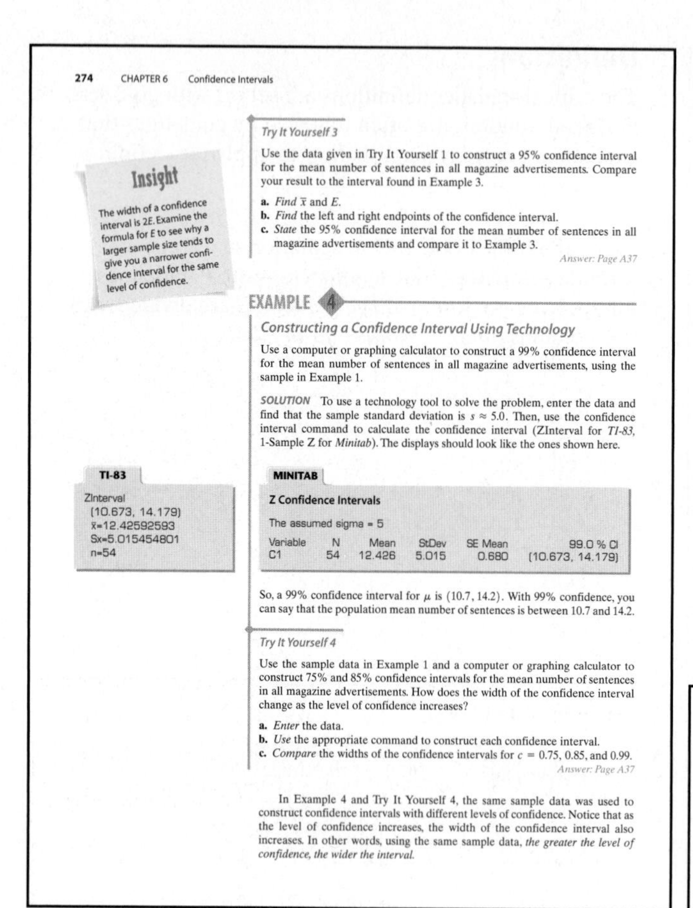

274 CHAPTER 6 Confidence Intervals

Insight

The width of a confidence interval is 2E. Examine the formula for E to see why a larger sample size tends to give you a narrower confidence interval for the same level of confidence.

Try It Yourself 3

Use the data given in Try It Yourself 1 to construct a 95% confidence interval for the mean number of sentences in all magazine advertisements. Compare your result to the interval found in Example 3.

a. *Find* $\bar{x}$ and E.
b. *Find* the left and right endpoints of the confidence interval.
c. *State* the 95% confidence interval for the mean number of sentences in all magazine advertisements and compare it to Example 3.

Answer: Page A37

EXAMPLE 4

Constructing a Confidence Interval Using Technology

Use a computer or graphing calculator to construct a 99% confidence interval for the mean number of sentences in all magazine advertisements, using the sample in Example 1.

SOLUTION To use a technology tool to solve the problem, enter the data and find that the sample standard deviation is $s \approx 5.0$. Then, use the confidence interval command to calculate the confidence interval (ZInterval for *TI-83*, 1-Sample Z for *Minitab*). The displays should look like the ones shown here.

TI-83

ZInterval
(10.673, 14.179)
$\bar{x}$=12.42592593
Sx=5.015454801
n=54

MINITAB

Z Confidence Intervals

The assumed sigma = 5

Variable	N	Mean	StDev	SE Mean	99.0 % CI
C1	54	12.426	5.015	0.680	(10.673, 14.179)

So, a 99% confidence interval for μ is (10.7, 14.2). With 99% confidence, you can say that the population mean number of sentences is between 10.7 and 14.2.

Try It Yourself 4

Use the sample data in Example 1 and a computer or graphing calculator to construct 75% and 85% confidence intervals for the mean number of sentences in all magazine advertisements. How does the width of the confidence interval change as the level of confidence increases?

a. *Enter* the data.
b. *Use* the appropriate command to construct each confidence interval.
c. *Compare* the widths of the confidence intervals for $c = 0.75$, 0.85, and 0.99.

Answer: Page A37

In Example 4 and Try It Yourself 4, the same sample data was used to construct confidence intervals with different levels of confidence. Notice that as the level of confidence increases, the width of the confidence interval also increases. In other words, using the same sample data, *the greater the level of confidence, the wider the interval.*

Technology Examples

Many sections contain a worked example that shows how technology can be used to calculate formulas, perform tests, or display data. Screen displays from *MINITAB*, *Excel*, and *TI-83* are given. Additional screen displays are given at the ends of selected chapters, and detailed instructions are given in separate technology manuals available with the book.

Insights

Most sections also contain one or more insights placed on blue "sticky notes" in the margin. The purpose of each insight is to help drive home an important interpretation or help connect different concepts.

Chapter Technology Project

Each chapter has a full-page technology projects using tools from *MINITAB*, *Excel*, and *TI-83*, that gives students additional insight into the way technology is used to handle large data sets or complex, real-life questions.

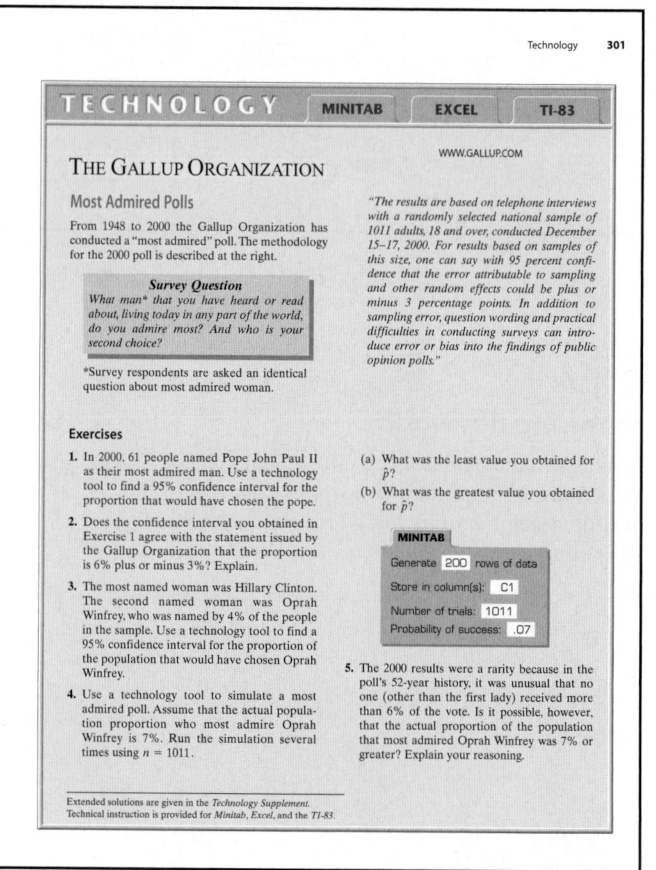

Technology 301

TECHNOLOGY MINITAB EXCEL TI-83

WWW.GALLUP.COM

THE GALLUP ORGANIZATION

Most Admired Polls

From 1948 to 2000 the Gallup Organization has conducted a "most admired" poll. The methodology for the 2000 poll is described at the right.

Survey Question

What man that you have heard or read about, living today in any part of the world, do you admire most? And who is your second choice?*

*Survey respondents are asked an identical question about most admired woman.

"The results are based on telephone interviews with a randomly selected national sample of 1011 adults, 18 and over, conducted December 15–17, 2000. For results based on samples of this size, one can say with 95 percent confidence that the error attributable to sampling and other random effects could be plus or minus 3 percentage points. In addition to sampling error, question wording and practical difficulties in conducting surveys can introduce error or bias into the findings of public opinion polls."

Exercises

1. In 2000, 61 people named Pope John Paul II as their most admired man. Use a technology tool to find a 95% confidence interval for the proportion that would have chosen the pope.
2. Does the confidence interval you obtained in Exercise 1 agree with the statement issued by the Gallup Organization that the proportion is 6% plus or minus 3%? Explain.
3. The most named woman was Hillary Clinton. The second named woman was Oprah Winfrey, who was named by 4% of the people in the sample. Use a technology tool to find a 95% confidence interval for the proportion of the population that would have chosen Oprah Winfrey.
4. Use a technology tool to simulate a most admired poll. Assume that the actual population proportion who most admire Oprah Winfrey is 7%. Run the simulation several times using $n = 1011$.

(a) What was the least value you obtained for $\hat{p}$?

(b) What was the greatest value you obtained for $\hat{p}$?

MINITAB

Generate 200 rows of data
Store in column(s): C1
Number of trials: 1011
Probability of success: .07

5. The 2000 results were a rarity because in the poll's 52-year history, it was unusual that no one (other than the first lady) received more than 6% of the vote. Is it possible, however, that the actual proportion of the population that most admired Oprah Winfrey was 7% or greater? Explain your reasoning.

Extended solutions are given in the *Technology Supplement*.
Technical instruction is provided for *Minitab*, *Excel*, and the *TI-83*.

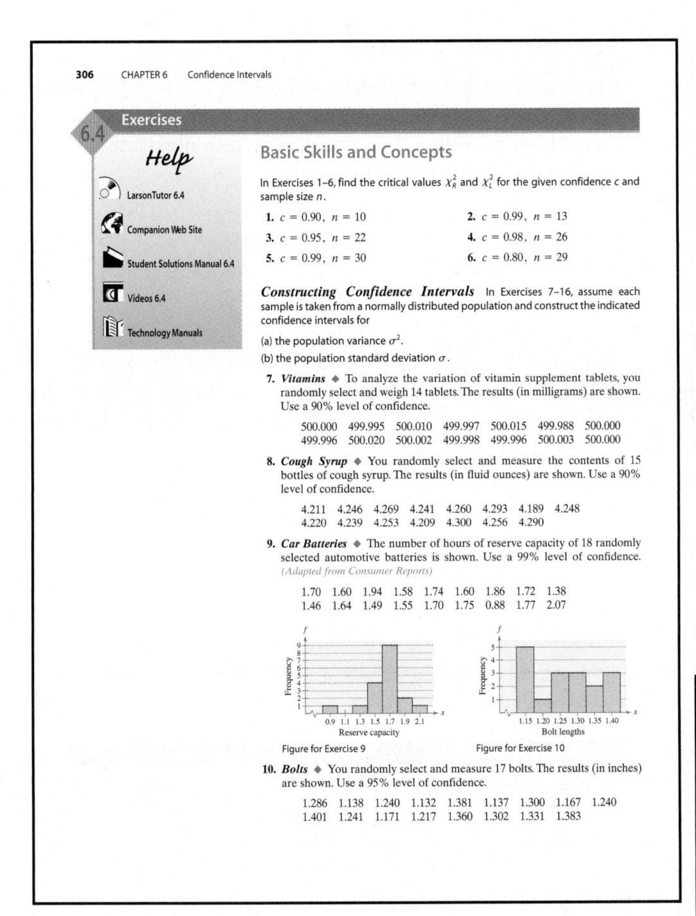
306 CHAPTER 6 Confidence Intervals

6.4 Exercises

Help

LarsonTutor 6.4

Companion Web Site

Student Solutions Manual 6.4

Videos 6.4

Technology Manuals

Basic Skills and Concepts

In Exercises 1–6, find the critical values χ^2_R and χ^2_L for the given confidence c and sample size n.

1. $c = 0.90,\ n = 10$
2. $c = 0.99,\ n = 13$
3. $c = 0.95,\ n = 22$
4. $c = 0.98,\ n = 26$
5. $c = 0.99,\ n = 30$
6. $c = 0.80,\ n = 29$

Constructing Confidence Intervals In Exercises 7–16, assume each sample is taken from a normally distributed population and construct the indicated confidence intervals for

(a) the population variance σ^2.

(b) the population standard deviation σ.

7. *Vitamins* ◆ To analyze the variation of vitamin supplement tablets, you randomly select and weigh 14 tablets. The results (in milligrams) are shown. Use a 90% level of confidence.

500.000 499.995 500.010 499.997 500.015 499.988 500.000
499.996 500.020 500.002 499.998 499.996 500.003 500.000

8. *Cough Syrup* ◆ You randomly select and measure the contents of 15 bottles of cough syrup. The results (in fluid ounces) are shown. Use a 90% level of confidence.

4.211 4.246 4.269 4.241 4.260 4.293 4.189 4.248
4.220 4.239 4.253 4.209 4.300 4.256 4.290

9. *Car Batteries* ◆ The number of hours of reserve capacity of 18 randomly selected automotive batteries is shown. Use a 99% level of confidence. (Adapted from Consumer Reports)

1.70 1.60 1.94 1.58 1.74 1.60 1.86 1.72 1.38
1.46 1.64 1.49 1.55 1.70 1.75 0.88 1.77 2.07

Figure for Exercise 9 (Frequency vs. Reserve capacity)

Figure for Exercise 10 (Frequency vs. Bolt lengths)

10. *Bolts* ◆ You randomly select and measure 17 bolts. The results (in inches) are shown. Use a 95% level of confidence.

1.286 1.138 1.240 1.132 1.381 1.137 1.300 1.167 1.240
1.401 1.241 1.171 1.217 1.360 1.302 1.331 1.383

Section Exercise Sets

Each section concludes with a set of exercises carefully written to nurture student understanding and proficiency. They move from basic concepts and skill development to more challenging and interpretive problems.

Labeled Exercises

Most exercises are labeled for easy reference. For instance, exercises labeled Graphical Analysis ask students to use the graphs provided to answer the questions.

Paired Format

Almost all exercises are given in "paired format" so that the odd-numbered exercise, whose answer is given in the back of the book, is paired with an even-numbered exercise, whose answer is not given. This paired format is commonly used in mathematics texts, but is less common in statistics texts.

SECTION 6.4 Confidence Intervals for Variance and Standard Deviation 307

11. *Lawn Mower* ◆ A lawn mower manufacturer is trying to determine the standard deviation of the life of one of its lawn mower models. To do this, it randomly selects 12 lawn mowers that were sold several years ago and finds that the sample standard deviation is 3.25 years. Use a 99% level of confidence. (Adapted from Consumer Reports)

12. *CD Players* ◆ A magazine includes a report on the prices of compact disc players. The article states that 26 randomly selected CD players had a standard deviation of \$150. Use a 95% level of confidence. (Adapted from Consumer Reports)

13. *Hotels* ◆ As part of your vacation planning, you randomly contact 10 hotels in your destination area and record the room rate of each. The results are shown in the stem-and-leaf plot. Use $c = 0.90$. (Adapted from Smith Travel Research)

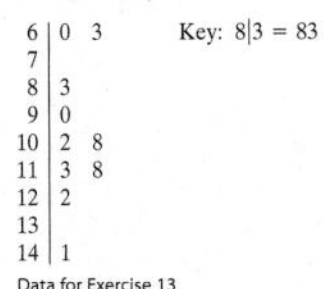

Data for Exercise 13

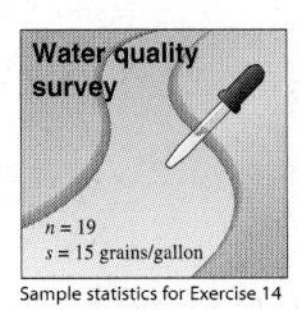

Sample statistics for Exercise 14

14. *Water Quality* ◆ As part of a water quality survey, you test the water hardness in several randomly selected streams. The results are shown above. Use $c = 0.95$.

15. *Monthly Income* ◆ The monthly incomes of 20 randomly selected individuals who have recently graduated with a bachelor's degree in social science have a sample standard deviation of \$107. Use a 95% level of confidence. (Adapted from U.S. Bureau of the Census)

16. *Sodium Chloride Concentration* ◆ The sodium chloride concentrations of 13 randomly selected seawater samples have a standard deviation of 6.7 cc/cubic meter. Use a 98% level of confidence. (Adapted from Dorling Kindersley Visual Encyclopedia)

Extending the Basics

17. *Vitamin Tablet Weights* ◆ You are analyzing the sample of vitamin supplement tablets in Exercise 7. The population standard deviation of the tablet's weights should be less than 0.015 milligram. Does the confidence interval you constructed for σ suggest that the variation in the tablet's weights is at an acceptable level? Explain your reasoning.

18. *Cough Syrup Bottle Contents* ◆ You are analyzing the sample of cough syrup bottles in Exercise 8. The population standard deviation of the bottle's contents should be less than 0.025 fluid ounce. Does the confidence interval you constructed for σ suggest that the variation in the bottle's contents is at an acceptable level? Explain your reasoning.

Extending the Basics

Each exercise set ends with a group of exercises called *Extending the Basics*. These exercises go beyond the material presented in the section (they tend to be more challenging and are not required as prerequisites of subsequent sections).

Answers and Solutions

The answers to all odd-numbered exercises are given in the back of the book, and the worked-out solutions are available in the *Student's Solutions Manual.*

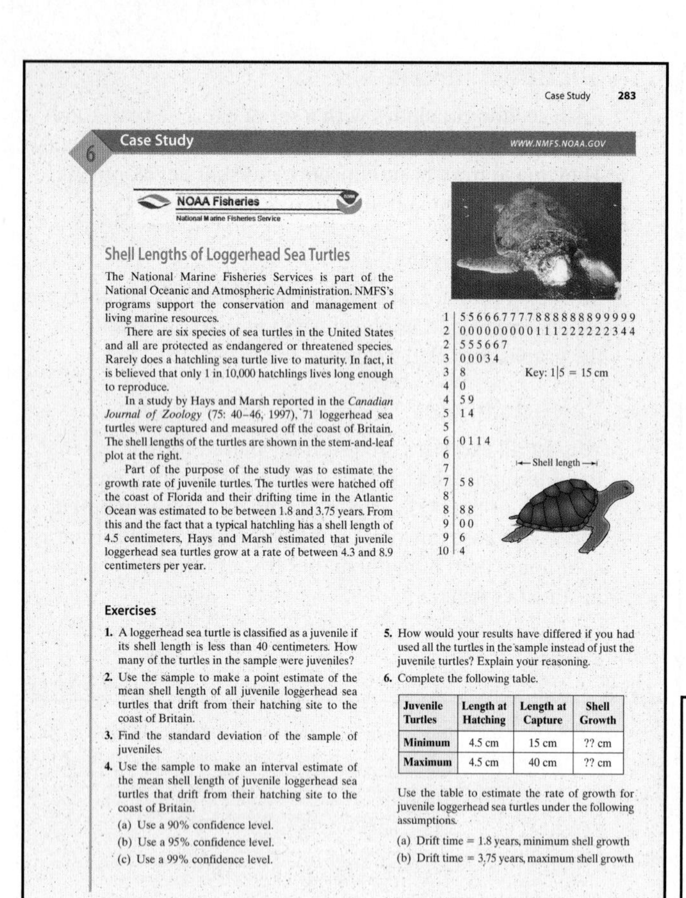

Case Study 283

Case Study 6 WWW.NMFS.NOAA.GOV

NOAA Fisheries
National Marine Fisheries Service

Shell Lengths of Loggerhead Sea Turtles

The National Marine Fisheries Services is part of the National Oceanic and Atmospheric Administration. NMFS's programs support the conservation and management of living marine resources.

There are six species of sea turtles in the United States and all are protected as endangered or threatened species. Rarely does a hatchling sea turtle live to maturity. In fact, it is believed that only 1 in 10,000 hatchlings lives long enough to reproduce.

In a study by Hays and Marsh reported in the *Canadian Journal of Zoology* (75: 40–46, 1997), 71 loggerhead sea turtles were captured and measured off the coast of Britain. The shell lengths of the turtles are shown in the stem-and-leaf plot at the right.

Part of the purpose of the study was to estimate the growth rate of juvenile turtles. The turtles were hatched off the coast of Florida and their drifting time in the Atlantic Ocean was estimated to be between 1.8 and 3.75 years. From this and the fact that a typical hatchling has a shell length of 4.5 centimeters, Hays and Marsh estimated that juvenile loggerhead sea turtles grow at a rate of between 4.3 and 8.9 centimeters per year.

1 | 5 5 6 6 6 7 7 7 7 8 8 8 8 8 8 8 9 9 9 9 9
2 | 0 0 0 0 0 0 0 0 0 1 1 1 2 2 2 2 2 2 3 4 4
2 | 5 5 5 6 6 7
3 | 0 0 0 3 4
3 | 8
4 | 0
4 | 5 9
5 | 1 4
5 |
6 | 0 1 1 4
6 |
7 |
7 | 5 8
8 |
8 | 8 8
9 | 0 0
9 | 6
10 | 4

Key: 1|5 = 15 cm

Exercises

1. A loggerhead sea turtle is classified as a juvenile if its shell length is less than 40 centimeters. How many of the turtles in the sample were juveniles?
2. Use the sample to make a point estimate of the mean shell length of all juvenile loggerhead sea turtles that drift from their hatching site to the coast of Britain.
3. Find the standard deviation of the sample of juveniles.
4. Use the sample to make an interval estimate of the mean shell length of juvenile loggerhead sea turtles that drift from their hatching site to the coast of Britain.
   (a) Use a 90% confidence level.
   (b) Use a 95% confidence level.
   (c) Use a 99% confidence level.
5. How would your results have differed if you had used all the turtles in the sample instead of just the juvenile turtles? Explain your reasoning.
6. Complete the following table.

Juvenile Turtles	Length at Hatching	Length at Capture	Shell Growth
Minimum	4.5 cm	15 cm	?? cm
Maximum	4.5 cm	40 cm	?? cm

Use the table to estimate the rate of growth for juvenile loggerhead sea turtles under the following assumptions.

(a) Drift time = 1.8 years, minimum shell growth
(b) Drift time = 3.75 years, maximum shell growth

Chapter Case Study

Each chapter has a full-page case study featuring actual data from a real-world context and a series of thought-provoking questions that are designed to illustrate the important concepts of the chapter.

310 CHAPTER 6 Confidence Intervals

6 Chapter Summary

What did you learn?	*Review Exercises*
Section 6.1	
◆ How to find a point estimate and a maximum error of estimate	*1, 2*
◆ How to construct and interpret confidence intervals for the population mean	*3, 4*
◆ How to determine the required minimum sample size when estimating μ	*5, 6*
Section 6.2	
◆ How to interpret the t-distribution and use a t-distribution table	*7, 8*
◆ How to construct confidence intervals when $n < 30$ and σ is unknown	*9–14*
Section 6.3	
◆ How to find a point estimate for the population proportion	*15–20*
◆ How to construct a confidence interval for a population proportion	*21–26*
◆ How to determine the required minimum sample size when estimating a population proportion	*27, 28*
Section 6.4	
◆ How to interpret the chi-square distribution and use a chi-square distribution table	*29–32*
◆ How to use the chi-square distribution to construct a confidence interval for the variance and standard deviation	*33, 34*

Chapter Summary

Each chapter concludes with a Chapter Summary that answers the questions *What did you learn?* This can be used as a study aid in conjunction with the Chapter Review exercises.

Statistics: Uses and Abuses

Each chapter features an expanded discussion on how statistical techniques should be used, while cautioning students about common abuses. New exercises help students to apply their knowledge.

STATISTICS

Uses and Abuses

Uses

By now, you know that complete information about population parameters is often not available. The techniques of this chapter can be used to make interval estimates of these parameters so that you can make informed decisions.

From what you learned in this chapter, you know that point estimates of population parameters are rarely exact. Remembering this can help you make good decisions in your career and in everyday life. For instance, suppose the results of a survey tell you that 52% of the population plans to vote for a certain item on a ballot. You know that this is only a point estimate of the actual proportion that will vote for the item. If the interval estimate is $0.49 < p < 0.55$, then you know this means it is possible that the item will not receive a majority vote.

Abuses

Unrepresentative Samples There are many ways that surveys can result in incorrect predictions. When you read the results of a survey, remember to question the sample size, the sampling technique, and the questions asked. For example, suppose you want to know how a group of people will vote in an election. From the diagram at the right, you can see that even if your sample is large enough, it may not consist of actual voters.

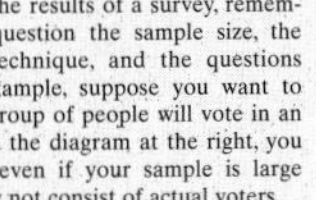

Biased Survey Questions In surveys, it is also important to analyze the wording of the questions. For example, a question about rezoning might be presented as: "Knowing that rezoning will result in more businesses contributing to school taxes, would you support the rezoning ?"

Exercises

1. ***Unrepresentative Samples*** Find an example of a survey that is reported in a newspaper or magazine. Describe different ways that the sample could have been unrepresentative of the population.
2. ***Biased Survey Questions*** Find an example of a survey that is reported in a newspaper or magazine. Describe different ways that the survey questions could have been biased.

Real Statistics–Real Decisions

This new feature encourages students to think critically and make informed decisions about real-world data. Exercises guide students from interpretation to drawing conclusions.

Real Statistics–Real Decisions 315

PUTTING IT ALL TOGETHER

Real Statistics Real Decisions

As part of the U.S. Environmental Protection Agency's (EPA) efforts to "protect human health and safeguard the natural environment," the EPA conducts the Urban Air Toxics Monitoring Program (UATMP). The program has gathered thousands of air samples and analyzed them for concentrations of over 50 different organic compounds, such as acetylene (used for cutting and welding metals, and in the production of neoprene rubber, plastics, and resins). The results from UATMP are used to gain insight into the effects of air pollution and if efforts to clean up the air are working.

For instance, using air samples from a major city, the EPA can analyze the results and estimate the mean concentration of acetylene in the air using a 95% confidence interval. They can then compare the interval to previous years' results to see if there are any trends and if there has been a significant change in the amount of acetylene in the air.

You work for the EPA and are asked to interpret the results shown in the graph at the right. The graph shows the point estimate for the population mean concentration and the 95% confidence interval for μ for acetylene over a three-year period. The data is based on air samples taken at one major city.

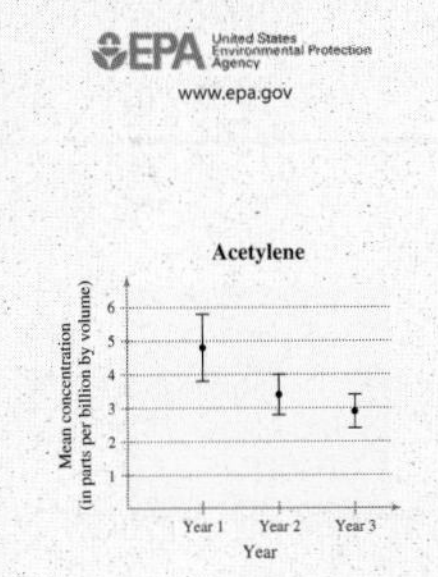

Exercises

1. *Interpreting the Results*

   Consider the graph of the mean concentration levels of acetylene. For the following years, decide if there has been a change in the mean concentration level of acetylene. Explain your reasoning.

   (a) From Year 1 to Year 2
   (b) From Year 2 to Year 3
   (c) From Year 1 to Year 3

2. *What Can You Conclude?*

   Using the results of Exercise 1, what can you conclude about the efforts to reduce the concentration of acetylene in the air?

3. *How Do You Think They Did It?*

   How do you think the EPA constructed the 95% confidence interval for the population mean concentration of the organic compounds in the air? Do the following to answer the question (you do not need to make any calculations).

   (a) What sampling distribution do you think they used? Why?
   (b) Do you think they used the population standard deviation in calculating the maximum error of estimate? Why or why not? If not, what could they have used?

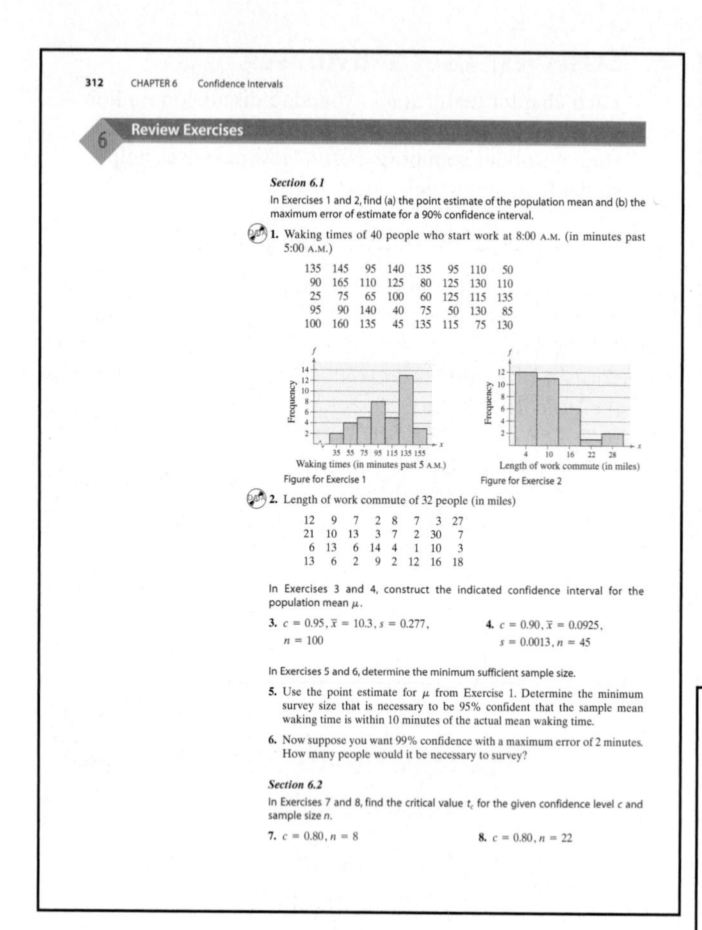

312 CHAPTER 6 Confidence Intervals

6 Review Exercises

Section 6.1

In Exercises 1 and 2, find (a) the point estimate of the population mean and (b) the maximum error of estimate for a 90% confidence interval.

1. Waking times of 40 people who start work at 8:00 A.M. (in minutes past 5:00 A.M.)

135	145	95	140	135	95	110	50
90	165	110	125	80	125	130	110
25	75	65	100	60	125	115	135
95	90	140	40	75	50	130	85
100	160	135	45	135	115	75	130

Figure for Exercise 1

Figure for Exercise 2

2. Length of work commute of 32 people (in miles)

12	9	7	2	8	7	3	27
21	10	13	3	7	2	30	7
6	13	6	14	4	1	10	3
13	6	2	9	2	12	16	18

In Exercises 3 and 4, construct the indicated confidence interval for the population mean μ.

3. $c = 0.95, \bar{x} = 10.3, s = 0.277, n = 100$

4. $c = 0.90, \bar{x} = 0.0925, s = 0.0013, n = 45$

In Exercises 5 and 6, determine the minimum sufficient sample size.

5. Use the point estimate for μ from Exercise 1. Determine the minimum survey size that is necessary to be 95% confident that the sample mean waking time is within 10 minutes of the actual mean waking time.

6. Now suppose you want 99% confidence with a maximum error of 2 minutes. How many people would it be necessary to survey?

Section 6.2

In Exercises 7 and 8, find the critical value t_c for the given confidence level c and sample size n.

7. $c = 0.80, n = 8$

8. $c = 0.80, n = 22$

Chapter Review Exercises

A set of Review Exercises follows each Chapter Summary. The order of the exercises follows the chapter organization. Answers to all odd-numbered exercises are given in the back of the book. Worked-out solutions are available in the *Student's Solutions Manual.*

316 CHAPTER 6 Confidence Intervals

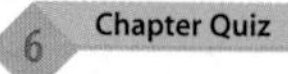

Chapter Quiz

Take this quiz as you would take a quiz in class. After you are done, check your work against the answers given in the back of the book.

1. The following data set represents the repair costs (in dollars) for a random sample of 30 dishwashers. (Adapted from Consumer Reports)

41.82	52.81	57.80	68.16	73.48	78.88	88.13	88.79
90.07	90.35	91.68	91.72	93.01	95.21	95.34	96.50
100.05	101.32	103.59	104.19	105.62	111.32	117.14	118.42
118.77	119.01	120.70	140.52	141.84	147.06		

(a) Find the point estimate of the population mean.

(b) Find the maximum error of estimate for a 95% level of confidence.

(c) Construct a 95% confidence interval for the population mean and interpret the results.

2. You want to estimate the mean repair cost for dishwashers. The estimate must be within $10 of the population mean. Determine the required sample size to construct a 99% confidence interval for the population mean. Assume the population standard deviation is $22.50. (Adapted from Consumer Reports)

3. The following data set represents the time (in minutes) for a random sample of phone calls made by employees at a company.

7.5 2.0 12.1 8.8 9.4 7.3 1.9 2.8 7.0 7.3

(a) Find the sample mean.

(b) Find the sample standard deviation.

(c) Use the t-distribution to construct a 90% confidence interval for the population mean and interpret the results. Assume the population of the data set is normally distributed.

(d) Repeat part (c), assuming $\sigma = 3.5$ minutes. Compare the results.

4. In a random sample of eight people with advanced degrees in biology, the mean monthly income was $3705 and the standard deviation was $566. Assume the monthly incomes are normally distributed and use a t-distribution to construct a 95% confidence interval for the population mean monthly income for people with advanced degrees in biology. (Adapted from U.S. Bureau of the Census)

5. In a survey of 2000 adults from the U.S. age 65 and over, 1320 received a flu shot. (Adapted from The Centers for Disease Control and Prevention)

(a) Find a point estimate for the population proportion p of those receiving flu shots.

(b) Construct a 90% confidence interval for the population proportion.

(c) Find the minimum sample size needed to estimate the population proportion at the 99% confidence level in order to ensure that the estimate is accurate within 4% of the population proportion.

6. Refer to the data set in Exercise 1. Assume the population of dishwasher repair costs is normally distributed.

(a) Construct a 95% confidence interval for the population variance.

(b) Construct a 95% confidence interval for the population standard deviation.

Chapter Quizzes and Cumulative Tests

Each chapter ends with a Chapter Quiz. Additionally, Cumulative Tests help students check their mastery of content from multiple chapters. The answers to all quiz and test questions are provided in the back of the book.

Supplements

Elementary Statistics: Picturing the World, Second Edition and its ancillary package have been developed in tandem to provide instructors and students with the most comprehensive, supportive package available.

Resources for the Instructor

Annotated Instructor's Edition

Betsy Farber, Bucks County Community College

ISBN 0-13-065938-X

Notes to Instructors appear in the margin of the text to suggest activities that correspond to the example or concept, additional ways to present the material, common pitfalls students encounter, alternate formulas or approaches that may be used, and other helpful teaching tips for instructors. Short answers (numerical, tabular, and/or graphical) to the section and review exercises appear in the margin next to the exercise.

Instructor's Solutions Manual

Jay Schaffer, University of Northern Colorado

ISBN 0-13-065930-4

Complete solutions to all of the exercises, *Try It Yourself* problems, case studies, technology pages, Uses and Abuses exercises, and Real Statistics–Real Decisions exercises are found in a single convenient volume.

Printed Test Bank

ISBN 0-13-065942-8

Includes more than 1100 additional questions—75% multiple choice and 25% open ended—with an answer key. Computerized versions are available.

TestGen-EQ with QuizMaster-EQ

ISBN 0-13-065952-5

Available on CD-ROM for IBM and Macintosh formats, this easy-to-use cross-platform test generator contains all the questions from the printed test bank. This testing program

- is algorithmically driven and text specific.
- is networkable for administering tests and capturing grades online.
- is easy to edit—instructors can edit or add their own questions to create a nearly unlimited number of tests and work sheets.
- can be used to create graphs with the new "Function Plotter."
- produces tests that can be easily exported to HTML so they can be posted to the Web.

PowerPoint Presentation

Betsy Farber, Bucks County Community College

Designed to enhance an instructor's classroom presentation, a mini-lecture that corresponds to each chapter of the text has been developed using PowerPoint. Featuring sound, graphics, and numerous additional examples, the PowerPoint presentation is ideal for part-timers or other teachers new to statistics. Most slides include notes offering suggestions for how the material may effectively be presented in class. Each presentation may be edited by the user, as desired, to reflect his or her individual teaching style. In addition, the PowerPoint slides may be printed to use as transparency masters or handed out as note-taking tools for students.

Instructor's Resource CD

ISBN 0-13-065943-6

This CD-ROM contains an electronic version of the Instructor's Solutions Manual, the TestGen-EQ computerized test bank, the data sets from the text, and the PowerPoint Presentation.

Resources for the Student

Student's Solutions Manual

Jay Schaffer, University of Northern Colorado

ISBN 0-13-065941-X

Complete worked-out solutions to all of the *Try It Yourself* problems, the odd- numbered exercises, and all of the Chapter Quiz and Cumulative Test exercises are included.

Technology Companion Guides

TI-83/TI-83 Plus

Dorothy Wakefield & Kate McLaughlin, University of Connecticut & Manchester Community College

ISBN 0-13-065949-5

MINITAB Manual

Dorothy Wakefield & Kate McLaughlin, University of Connecticut & Manchester Community College

ISBN 0-13-065940-1

Excel Manual

Beverly Dretzke, University of Wisconsin, Eau Clair

ISBN 0-13-065948-7

Each spiral-bound companion manual works hand-in-glove with the text. Step-by-step keystroke level instructions, with screen captures, provide detailed help for using the technology to work pertinent examples and all of the technology projects in the text. A cross-reference chart indicates which text examples are included and the exact page reference in both the text and technology manual. Output with brief instruction is provided for selected odd-numbered exercises to reinforce the examples.

The Excel Manual includes ***PHStat2,*** a statistics add-in for Microsoft *Excel* (CD-ROM) featuring a custom menu of choices that lead to dialog boxes to help perform statistical analyses more quickly and easily than off-the-shelf *Excel* permits.

Text and Technology Packages

Text and MINITAB Rel. 12.0 (Student Version)

ISBN 0-13-077509-6

Order the above ISBN for a package of the text with the student version of this popular software program at a significant discount.

Text and SPSS 11.0 (Student Version)

ISBN 0-13-078054-5

Offering exceptional value to the student, a CD-ROM of MINITAB Rel. 12.0 Student Version or of SPSS 11.0 Student Version may be packaged with the text for a small additional cost. Consult the publisher for more information.

Media Resources

Data CD

Packaged free with every copy of the text. Data files for the technology projects, all exercises identified with a (DATA) symbol, and selected examples in the book are saved on the data CD as *ASCII*, *MINITAB*, and *Excel* files. The data files may also be downloaded from the Web site that accompanies the text located at *www.prenhall.com/larson*.

LarsonTutor

Student CD: ISBN 0-13-065946-0
Instructor CD: ISBN 0-13-047371-5

LarsonTutor is a new CD-ROM tutorial statistics program modeled on Prentice Hall's celebrated MathPro. LarsonTutor features three main components. The first is keyed to the *Try It Yourself* exercises and provides an example that models the exercise, interactive guided instructions for solving the exercise, and additional problems for practice. Students may use the tutorial section in both a practice or graded-quiz mode. The second component is a course management system, complete with a reports wizard, customizable syllabus, and messaging system between students and instructors. The third component features explorations which help the student's conceptual understanding of the material and promote discovery learning.

Video Lecture Series

Betsy Farber, Bucks County CC

ISBN 0-13-008808-0 (CD-ROM)
ISBN 0-13-065945-2 (VHS tape)

Available on CD and VHS tape, this comprehensive set of videos provides a short lecture and worked examples for almost every section in the book. These videos provide excellent support for students who missed class or need to study the material at their own pace. They are also an invaluable resource for distance learning or self-paced study programs.

Companion Web Site (*www.prenhall.com/larson*)

The free accompanying companion Web site is a helpful teaching and learning resource and includes chapter-specific objectives and quizzes, PowerPoint presentations available for download, data files in multiple formats, technology correlation guides, destinations, a customizable syllabus builder, technology projects, case studies, and new animated Java applets (StatLets, developed by StatPoint, LLC) for each text section.

Blackboard, Web CT, and Course Compass Solutions

by Bobbi Allen, Delta College

Course Compass Access Card ISBN 0-13-067097-9
WebCT Access Card ISBN 0-13-067099-5
Blackboard Access Card ISBN 0-13-067095-2

Combining Blackboard and WebCT course management with Prentice Hall's award-winning content, Prentice Hall is pleased to present three premium options for your course management needs. Featured content includes Quizzes, Introduction to each Chapter, Learning Outcomes, guided textbook reading, suggested textbook homework, suggested LarsonTutor exercises, and StatLet problems. (StatLets, designed by StatPoint, LLC, are Java applets scripted to animate and display specific statistics concepts or problems.)

Acknowledgments

We owe a debt of gratitude to the many reviewers who helped us shape and refine *Elementary Statistics: Picturing the World*, Second Edition. In particular we must thank our content advisory panel for their invaluable contribution in carefully reviewing each chapter of the book and making detailed suggestions for improvement.

Content Advisory Panel

Mike McGann, Ventura Community College
Vicki McMillian, Ocean County College
Lindsay Packer, College of Charleston
Carol Shapero, Oakton Community College

Pedagogy and Design Survey Reviewers

John Bernard, University of Texas—Pan American
G. Andy Chang, Youngstown State University
Gary Egan, Monroe Community College
Charles Ehler, Anne Arundel Community College
Douglas Frank, Indiana University of Pennsylvania
Rita Kolb, Catonsville Community College
Jeffrey Linek, St. Petersburg Junior College
Rowan Lindley, Westchester Community College
Diane Long, College of DuPage
Rhonda Magel, North Dakota State University
Aileen Solomon, Trident Technical College
Agnes Tuska, California State University—Fresno
Dex Whittinghill, Rowan University

Reviewers

Michelle Strager-McCarney, Penn State Erie, The Behrend College
Sandra L. Spain, Thomas Nelson Community College
John J. Avioli, Christopher Newport University
David Kay, Moorpark College
Hyune-Ju Kim, Syracuse University
Frieda Ganter, California State University–Fresno
Neal Rogness, Grand Valley State University
Jill Fanter, Walters State Community College
Sonja Hensler, St. Petersbury Jr. College
Ting-Xiu Wang, Oakton Community College
Keith J. Craswell, Western Washington University
Lynn Onken, San Juan College
Julie Norton, California State University-Hayward
Jean Sells, Sacred Heart University

Accuracy Checkers

Aileen Solomon, Trident Technical College
Sarah Streett

We also want to give special thanks to the people at Prentice Hall who worked with us in the development of *Elementary Statistics: Picturing the World*, Second Edition, especially Quincy McDonald—who worked with us on every aspect of the book, Sally Yagan, Joanne Wendelken, Angela Battle, Bayani Mendoza deLeon, Donna Crilly, Linda Behrens, Maureen Eide, and Alan Fischer.

We would also like to thank the staff of Larson Texts, Inc., who assisted with the development and production of the book.

On a personal level, we are grateful to our spouses, Deanna Gilbert Larson and Richard Farber for their love, patience, and support. Also, a special thanks goes to R. Scott O'Neil.

We have worked hard to make *Elementary Statistics: Picturing the World*, Second Edition, a clean, clear, and enjoyable text from which to teach and learn statistics. Despite our best efforts to ensure accuracy and ease of use, many users will undoubtedly have suggestions for improvement. We welcome your suggestions.

Ron Larson

Betsy Farber

How to Study Statistics

Studying Statistics Congratulations! You are about to begin your study of statistics. As you progress through the course, you should discover how to use statistics in your everyday life and in your career. The prerequisites for this course are two years of algebra, an open mind, and a willingness to study. When studying statistics, the material you learn each day builds on material you learned previously. There are no shortcuts—you must keep up with your studies every day. Before you begin, read through the following hints that will help you succeed.

Making a Plan Make your own course plan right now! A good rule of thumb is to study at least two hours for every hour in class. After your first major exam, you will know if your efforts were sufficient. If you did not get the grade you wanted, then you should increase your study time, improve your study efficiency, or both.

Preparing for Class Before every class, review your notes from the previous class and read the portion of the text that is to be covered. Pay special attention to the definitions and rules that are highlighted. Read the examples and work through the Try Its that accompany each example. Use LarsonTutor for additional practice problems and detailed Try It explanations. These steps take self-discipline, but they pay off because you will benefit much more from your instructor's presentation.

Attending Class Attend every class. Arrive on time with your text, materials for taking notes, and your calculator. If you must miss a class, get the notes from another student, go to a tutor for help, or view the appropriate statistics videotape. Try to learn the material that was covered in the missed class before attending the next class.

Participating in Class When reading the text before class, reviewing your notes from a previous class, or working on your homework, write down any questions you have about the material. Ask your instructor these questions during class. Doing so will help you (and others in your class) understand the material better.

Taking Notes During class, be sure to take notes on definitions, examples, concepts, and rules. Focus on the instructor's cues to identify important material. Then, as soon as possible after class, review your notes and add any explanations that will help to make your notes more understandable to you.

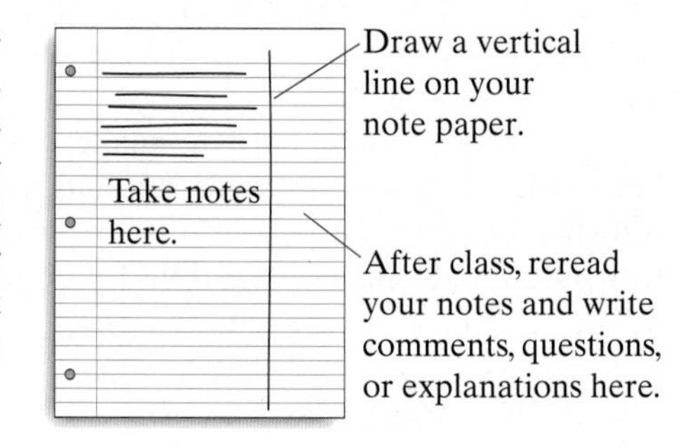

Doing the Homework Learning statistics is like learning to play the piano or basketball. You cannot develop skills just by watching someone do it; you must do it yourself. The best time to do your homework is right after class, when the concepts are still fresh in your mind. Doing homework at this time increases your chances of retaining the information in long-term memory.

Finding a Study Partner When you get stuck on a problem, you may find that it helps to work with a partner. Even if you feel you are giving more help than you are getting, you will find that teaching others is an excellent way to learn.

Keeping Up with the Work Don't let yourself fall behind in this course. If you are having trouble, seek help immediately—from your instructor, a statistics tutor, your study partner, or additional study aids such as videotapes and software tutorials. Remember: If you have trouble with one section of your statistics text, there's a good chance that you will have trouble with later sections unless you take steps to improve your understanding.

Getting Stuck Every statistics student has had this experience: You work a problem and cannot solve it, or the answer you get does not agree with the one given in the text. When this happens, consider asking for help or taking a break to clear your thoughts. You might even want to sleep on it, or rework the problem, or reread the section in the text. Avoid getting frustrated or spending too much time on a single problem.

Preparing for Tests Cramming for a statistics test seldom works. If you keep up with the work and follow the suggestions given here, you should be almost ready for the test. To prepare for the chapter test, review the Chapter Summary and work the Review Exercises. Then set aside some time to take the sample Chapter Quiz and the sample Cumulative Test. Analyze the results of your Chapter Quiz and Cumulative Test to locate and correct test-taking errors.

Taking a Test Most instructors do not recommend studying right up to the minute the test begins. Doing so tends to make people anxious. The best cure for test-taking anxiety is to prepare well in advance. Once the test begins, read the directions carefully and work at a reasonable pace. (You might want to read the entire test first, then work the problems in the order in which you feel most comfortable.) Don't rush! People who hurry tend to make careless errors. If you finish early, take a few moments to clear your thoughts and then go over your work.

Learning from Mistakes After your test is returned to you, go over any errors you might have made. This will help you avoid repeating some systematic or conceptual errors. Don't dismiss any error as just a "dumb mistake." Take advantage of any mistakes by hunting for ways to improve your test-taking skills.

Where You've Been

You are already familiar with many of the practices of statistics, such as taking surveys, collecting data, and describing populations. What you may not know is that collecting accurate statistical data is often difficult and costly. Consider, for instance, the monumental task of counting and describing the entire population of the United States. If you were in charge of such a census, how would you do it? How would you ensure that your results are accurate? These and many more concerns are the responsibility of the United States Census Bureau.

New York City, with a population of over 8 million, has more than twice the population of any other American city. The next four most populated cities in the United States are Los Angeles, Chicago, Houston, and Philadelphia.

Introduction to Statistics

CHAPTER 1

Where You're Going

In Chapter 1, you will be introduced to the basic concepts and goals of statistics. For instance, statistics was used to obtain the following data, which show the population (in millions) of the United States at the 10-year census intervals. When conducting the census, the Bureau sends short forms to 83% of the population. The short form asks only a few questions, such as gender, age, race, and home ownership. A long form, which covers many additional topics, is sent to the other 17% of the population. This 17% forms a sample. In this course, you will learn how the data collected from a sample are used to infer characteristics about the entire population.

1790	**3.9**	1850	**23.2**	1910	**92.0**	1970	**203.3**
1800	**5.3**	1860	**31.4**	1920	**105.7**	1980	**226.5**
1810	**7.2**	1870	**39.8**	1930	**122.8**	1990	**248.7**
1820	**9.6**	1880	**50.2**	1940	**131.7**	2000	**281.4**
1830	**12.9**	1890	**62.9**	1950	**151.3**		
1840	**17.1**	1900	**76.0**	1960	**179.3**		

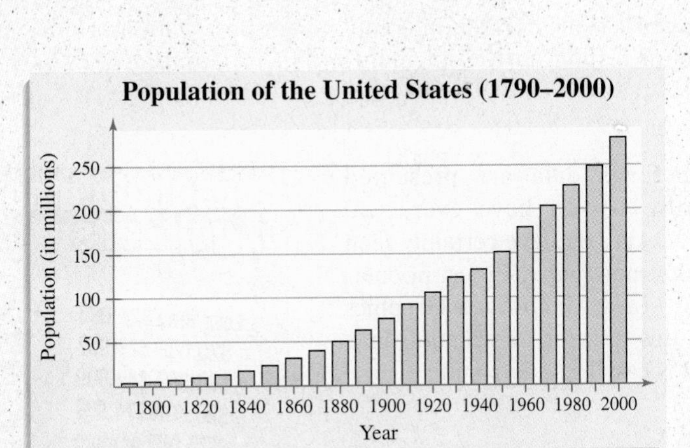

1.1 An Overview of Statistics

What You Should Learn

- The definition of statistics
- How to distinguish between a population and a sample and between a parameter and a statistic
- How to distinguish between descriptive statistics and inferential statistics

A Definition of Statistics • Data Sets • Branches of Statistics

A Definition of Statistics

What is statistics? Why should I study statistics? How can studying statistics help me in my profession? These are questions that you may have asked yourself when choosing this course. Almost every day you are exposed to statistics. For example, consider the following excerpts from recent newspapers and journals.

- "A . . . survey of traffic deaths during this past Memorial Day weekend shows a 36% decrease in fatalities compared with last year." *(Source: National Safety Council)*
- " . . . men who eat just two servings of tomatoes a week in raw, sauce, or pizza form, have a 34% less risk of developing prostate cancer." *(Source: Journal of the National Cancer Institute)*
- " . . . more than three fourths of all college seniors in the United States complete at least one internship by graduation and 55% participate in two or more." *(Source: UPI)*

The three statements you just read are based on the collection of **data.**

> **DEFINITION**
>
> **Data** consist of information coming from observations, counts, measurements, or responses. The singular for data is datum.

Sometimes data are presented graphically. If you have ever read *USA TODAY,* you have certainly seen one of that newspaper's most popular features—*USA Snapshots.* Graphics such as this present information in a way that is easy to understand.

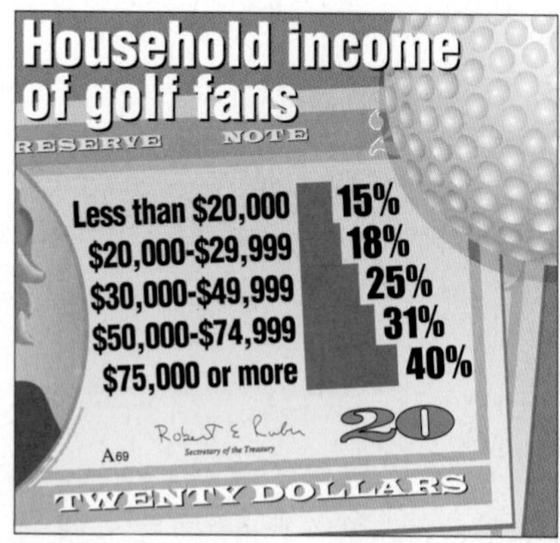

The use of statistics dates back to census taking in ancient Babylonia, Egypt, and later in the Roman Empire, when data were collected about matters concerning the state, such as births and deaths. In fact, the word *statistics* is derived from the Latin word *status*, meaning "state."

But statistics today involves more than collecting data, presenting facts, calculating averages, and drawing graphs. So, what is statistics?

DEFINITION

Statistics is the science of collecting, organizing, analyzing, *and* interpreting data in order to make decisions.

Insight

Unless a population is small, it is usually impractical to obtain all the population data. In most studies, information must be obtained from a sample.

Data Sets

There are two types of data sets you will use when studying statistics. These data sets are called *populations* and *samples.*

DEFINITION

A **population** is the collection of *all* outcomes, responses, measurements, or counts that are of interest.

A **sample** is a subset of a population.

EXAMPLE 

Identifying Data Sets

In a recent survey, 3002 adults in the United States were asked if they read news on the Internet at least once a week. Six hundred of the adults said yes. Identify the population and the sample. Describe the data set. *(Source: Pew Research Center)*

SOLUTION The population consists of the responses of all adults in the United States and the sample consists of the responses of the 3002 adults in the United States in the survey. The sample is a subset of the responses of all adults in the U.S. The data set consists of 600 yes's and 2402 no's.

Responses of all adults in the United States (population)

Responses of adults in survey (sample)

Try It Yourself 1

The U.S. Department of Energy conducts weekly surveys of 800 gasoline stations to determine the average price per gallon of regular gasoline. On May 14, 2001, the average price was $1.713 per gallon. Identify the population and the sample. *(Source: U.S. Department of Energy)*

a. Identify the *population.*
b. Identify the *sample.*
c. What does the data set consist of?

Answer: Page A29

Whether a data set is a population or a sample usually depends on the context of the real-life situation. For instance, in Example 1, the population was the set of responses of all adults in the United States. Depending on the purpose of the survey, the population could have been the set of responses of all adults who live in California or who have telephones or who read a particular newspaper.

Two important terms that are used throughout this course are *parameter* and *statistic.*

DEFINITION

A **parameter** is a numerical description of a *population* characteristic.

A **statistic** is a numerical description of a *sample* characteristic.

Study Tip

The terms *parameter* and *statistic* are easy to remember if you use the mnemonic device of matching the first letters in ***p****opulation* ***p****arameter* and the first letters in ***s****ample* ***s****tatistic.*

EXAMPLE 

Distinguishing Between a Parameter and a Statistic

Decide whether the numerical value describes a population parameter or a sample statistic. Explain your reasoning.

1. A recent survey of a sample of MBAs reported that the average starting salary for an MBA is less than $65,000. *(Source: The Washington Post Company)*
2. Starting salaries for the 667 MBA graduates from the University of Chicago Graduate School of Business increased 8.5% from the previous year.
3. In a random check of a sample of retail stores, the Food and Drug Administration found that 34% of the stores were not storing fish at the proper temperature.

SOLUTION

1. Because the numerical measure of $65,000 is based on a subset of the population, it is a sample statistic.
2. Because the numerical measure of 8.5% is based on all 667 graduates' starting salaries, it is a population parameter.
3. Because the numerical measure of 34% is based on a subset of the population, it is sample statistic.

Try It Yourself 2

In 2001, major league baseball teams spent a total of $1,968,088,814 on players' salaries. Does this numerical value describe a population parameter or a sample statistic? *(Source: USA TODAY)*

a. Decide whether the numerical value is from a *population* or a *sample.*
b. Specify whether the numerical value is a *parameter* or a *statistic.*

Answer: Page A29

Picturing the World

How accurate is the U.S. census? According to a post-census evaluation conducted by the Census Bureau, the 1990 census undercounted the U.S. population by an estimated 4.7 million people. The 1990 census was the first census since at least 1940 to be less accurate than its predecessor.

U.S. Census Undercount

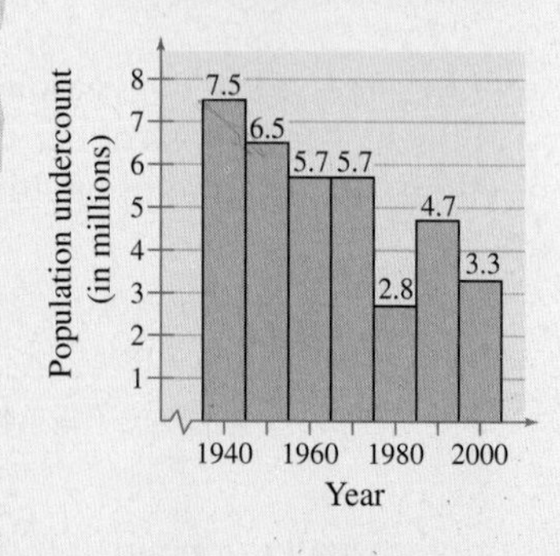

What are some difficulties in collecting population data?

In this course, you will see how the use of statistics can help you make informed decisions that affect your life. Consider the census that the U.S. government takes every decade. When taking the census, the Census Bureau attempts to contact everyone living in the United States. This is an impossible task. It is important that the census is accurate, because public officials make many decisions based on the census information. Data collected in the 2010 census will determine how to assign congressional seats and how to distribute public funds.

Branches of Statistics

The study of statistics has two major branches—**descriptive statistics** and **inferential statistics.**

DEFINITION

Descriptive statistics is the branch of statistics that involves the organization, summarization, and display of data.

Inferential statistics is the branch of statistics that involves using a sample to draw conclusions about a population. A basic tool in the study of inferential statistics is probability.

EXAMPLE 3

Descriptive and Inferential Statistics

Decide which part of the study represents the descriptive branch of statistics. What conclusions might be drawn from the study using inferential statistics?

1. A large sample of men, aged 48, was studied for 18 years. For unmarried men, 60% to 70% were alive at age 65. For married men, 90% were alive at age 65. Which part of the study represents the descriptive branch of statistics? What conclusions might be drawn from this study using inferential statistics? *(Source: The Journal of Family Issues)*
2. In a sample of Wall Street analysts, the percent who incorrectly forecasted high-tech earnings in a recent year was 44%. *(Source: Bloomberg News)*

SOLUTION

1. Descriptive statistics involves statements such as "For unmarried men, 60% to 70% were alive at age 65" and "For married men, 90% were alive at 65." A possible inference drawn from the study is that being married is associated with a longer life for men.
2. The part of this study which represents the descriptive branch of statistics involves the statement "the percent [of Wall Street analysts] who incorrectly forecasted high-tech earnings in a recent year was 44%." A possible inference drawn from the study is that the stock market is difficult to forecast, even for professionals.

Try It Yourself 3

A survey conducted among 1017 men and women by Opinion Research Corporation International found that 76% of women and 60% of men had a physical examination within the previous year. *(Source: Men's Health)*

a. Identify the descriptive aspect of the survey.
b. What inferences could be drawn from this survey? *Answer: Page A29*

Throughout this course you will see applications of both branches. A major theme in this course will be how to use sample statistics to make inferences about unknown population parameters.

1.1 Exercises

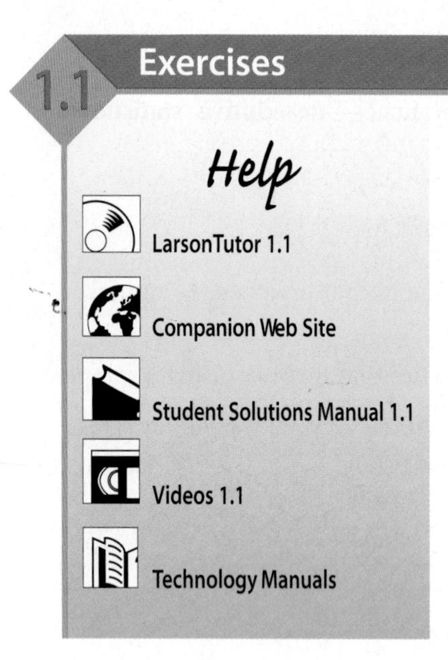

Basic Skills and Concepts

1. How is a sample related to a population?
2. Why is a sample used more often than a population?

True or False In Exercises 3–8, determine whether the statement is true or false. If it is false, rewrite it as a true statement.

3. A statistic is a measure that describes a population characteristic.
4. A sample is a subset of a population.
5. It is impossible for the Census Bureau to obtain all the census data about the population of the United States.
6. Inferential statistics involves using a population to draw a conclusion about a corresponding sample.
7. A population is the collection of some outcomes, responses, measurements, or counts that are of interest.
8. The word statistics is derived from the Latin word status, meaning "state."

Classifying a Data Set In Exercises 9–12, determine whether the data set is a population or a sample. Explain your reasoning.

9. The age of each state governor
10. The speed of every fifth car passing a police speed trap
11. A survey of 500 students from a university with 2000 students
12. The annual salary for each employee at a company

Graphical Analysis In Exercises 13–16, use the Venn diagram to identify the population and the sample.

13.
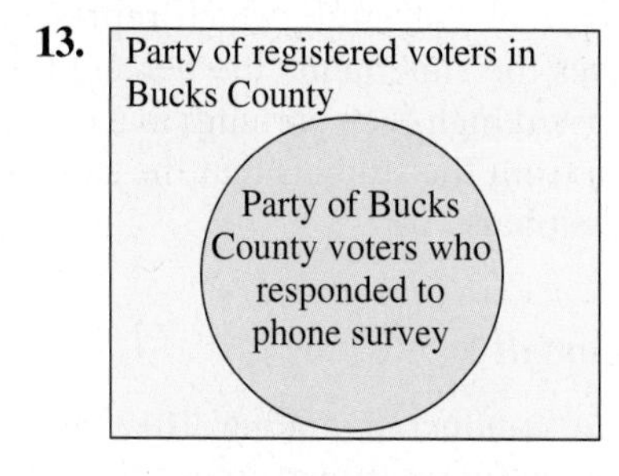

14.
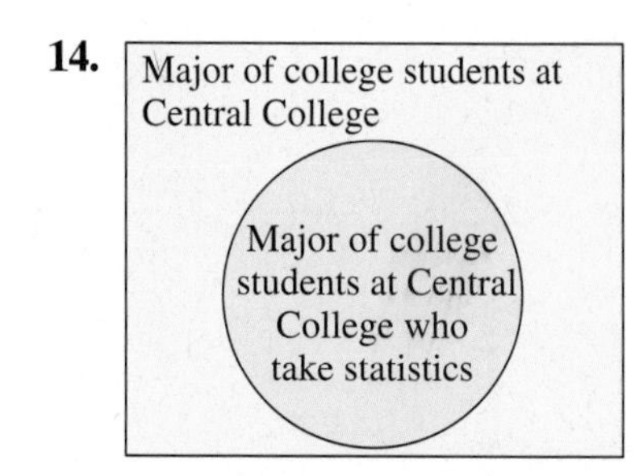

15.
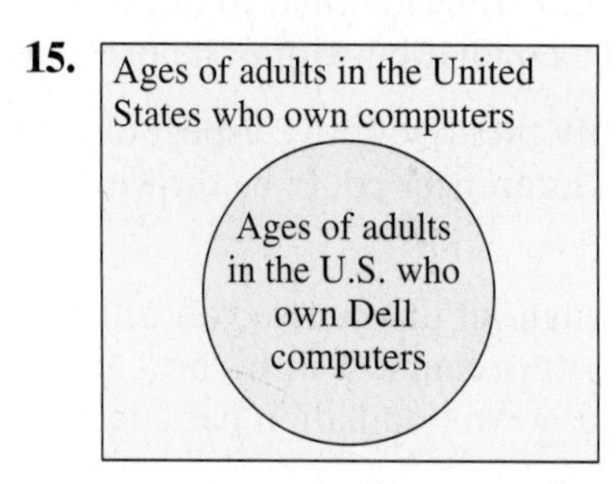

16.
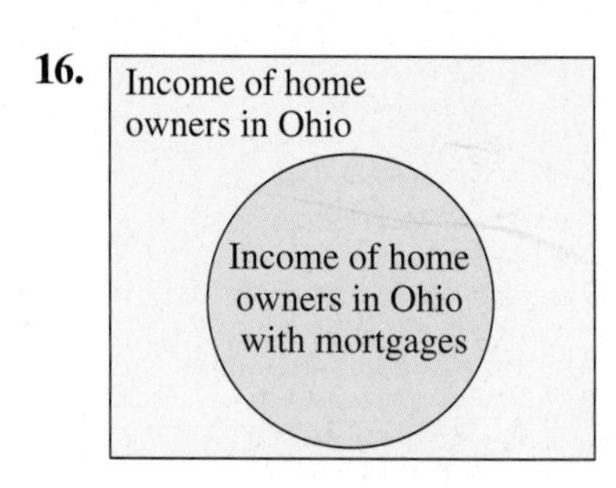

Identifying Data Sets In Exercises 17–20, identify the population and the sample.

17. A study of 33,043 infants in Italy was conducted to find a link between a heart rhythm abnormality and sudden infant death syndrome. *(Source: New England Journal of Medicine)*

18. A survey of 1023 households in the U.S. found that 65% subscribe to cable television.

19. A survey of 546 women found that more than 56% are the primary investor in their household. *(Adapted from: Roper Starch Worldwide for Intuit)*

20. A survey of 872 vacationers from the U.S. found that they planned on spending at least $1800 for their next vacation.

Distinguishing between a Parameter and a Statistic In Exercises 21–26, determine whether the numerical value is a parameter or a statistic. Explain your reasoning.

21. The average annual salary for 35 of a company's 1200 accountants is $57,000.

22. In a survey of a sample of high school students, 43% said that their mother has taught them the most about managing money. *(Source: Harris Poll for Girls Incorporated)*

23. In a survey of a sample of computer users, 10% said their computer had a malfunction that needed to be repaired by a service technician.

24. In a recent year, the interest category for 12% of all new magazines was sports. *(Source: Oxbridge Communications)*

25. In a recent survey of 1000 adults from the U.S., 47% said using a cell phone while driving should be illegal. *(Source: Rasmussen Research)*

26. In a recent year, the average Math scores for all graduates on the ACT was 20.7. *(Source: ACT, Inc.)*

27. Which part of the survey described in Exercise 19 represents the descriptive branch of statistics? Make an inference based on the results of the survey.

28. Which part of the survey described in Exercise 20 represents the descriptive branch of statistics? Make an inference based on the results of the survey.

Extending the Basics

29. ***Identifying Data Sets in Articles*** ◆ Find a newspaper or magazine article that describes a survey. Identify the sample used in the survey. What is the sample's population?

30. ***Writing*** ◆ Write an essay about the importance of statistics for one of the following.

(a) A study on the effectiveness of a new drug

(b) An analysis of a manufacturing process

(c) Making conclusions about voter opinions using surveys

1.2 Data Classification

What You Should Learn

- How to distinguish between qualitative data and quantitative data
- How to classify data with respect to the four levels of measurement: nominal, ordinal, interval, and ratio

Types of Data • Levels of Measurement

Types of Data

In this section, you will learn how to classify data by type and by level of measurement. Data sets can consist of two types of data: *qualitative data* and *quantitative data.*

DEFINITION

Qualitative data consist of attributes, labels, or nonnumerical entries.

Quantitative data consist of numerical measurements or counts.

EXAMPLE 

Classifying Data by Type

The base prices of several vehicles are shown in the table. Which data are qualitative data and which are quantitative data? Explain your reasoning. *(Source: Ford Motor Company)*

Model	Base Price
Ranger XL	$12,595
ZX2	$12,730
Focus LX	$13,120
Taurus LX	$19,075
Explorer Sport-Trac	$22,510
Windstar LX 3-Door	$22,740
Crown Victoria	$22,835
Expedition XLT	$30,855

SOLUTION The information shown in the table can be separated into two data sets. One data set contains the names of vehicle models and the other contains the base prices of vehicle models. The names are nonnumerical entries, so these are qualitative data. The base prices are numerical entries, so these are quantitative data.

Try It Yourself 1

The populations of several U.S. cities are shown in the table. Which data are qualitative data and which are quantitative data? *(Source: U.S. Census Bureau)*

City	Population
Baltimore, MD	651,154
Boston, MA	589,141
Dallas, TX	1,188,580
Las Vegas, NV	478,434
Lincoln, NE	225,581
Seattle, WA	563,374

a. Identify the contents of each data set.
b. Decide whether each data set consists of numerical or nonnumerical entries.
c. Specify the qualitative data and the quantitative data.

Answer: Page A29

Levels of Measurement

Another data characteristic is the data's level of measurement. The level of measurement determines which statistical calculations are meaningful. The four levels of measurement, in order from lowest to highest, are *nominal, ordinal, interval,* and *ratio.*

Insight

When numbers are at the nominal level of measurement, they simply represent a label. Examples of numbers used as labels include social security numbers and numbers on sports jerseys.

DEFINITION

Data at the **nominal level of measurement** are qualitative only. Data at this level are categorized using names, labels, or qualities. No mathematical computations can be made at this level.

Data at the **ordinal level of measurement** are qualitative or quantitative. Data at this level can be arranged in order, but differences between data entries are not meaningful.

EXAMPLE 

Classifying Data by Level

Two data sets are shown. Which data set consists of data at the nominal level? Which data set consists of data at the ordinal level? Explain your reasoning. *(Source: Nielsen Media Research)*

Top 5 TV Programs (from 5/14/01 to 5/20/01)
1. *E.R.*
2. *Friends*
3. *Law and Order*
4. *West Wing*
5. *Will & Grace*

Network Affiliates in Portland, Oregon	
KATU	(ABC)
KGW	(NBC)
KOIN	(CBS)
KPDX	(FOX)

SOLUTION The first data set lists the rank of five TV programs. The data consists of the ranks 1, 2, 3, 4, and 5. Because the rankings can be listed in order, these data are at the ordinal level. Note that the difference between a rank of 1 and 5 has no mathematical meaning. The second data set consists of the call letters of each network affiliate in Portland. The call letters are simply the names of network affiliates, so these data are at the nominal level.

Try It Yourself 2

Consider the following data sets. For each data set, decide whether the data are at the nominal level or at the ordinal level.

1. The final standings for the Northeast Division of the National Hockey League
2. A collection of phone numbers

a. *Identify* what each data set represents.
b. Specify the *level of measurement.*

Answer: Page A29

Picturing the World

In 1998, the American Film Institute chose the 100 greatest American movies. The institute started with a population of more than 40,000 movies and chose a sample of 400 to put on a ballot. The ballot was mailed to over 1500 film artists and executives. The films were judged on popularity over time, historical significance, cultural impact, and other factors.

The American Film Institute's Top Five American Films
1. *Citizen Kane*
2. *Casablanca*
3. *The Godfather*
4. *Gone With the Wind*
5. *Lawrence of Arabia*

In this list, what is the level of measurement?

DEFINITION

Data at the **interval level of measurement** are quantitative. The data can be ordered and you can calculate meaningful differences between data entries. At the interval level, a zero entry simply represents a position on a scale; the entry is not an inherent zero.

Data at the **ratio level of measurement** are similar to data at the interval level, with the added property that a zero entry is an inherent zero. A ratio of two data values can be formed so one data value can be expressed as a multiple of another.

An *inherent* zero is a zero that implies "none." For instance, the amount of money you have in a savings account could be zero dollars. In this case, the zero represents no money—it is an inherent zero. On the other hand, a temperature of 0°C does not represent a condition where no heat is present. The 0°C temperature is simply a position on the Celsius scale; it is not an inherent zero.

To distinguish between data at the interval level and at the ratio level, determine whether the expression "twice as much" has any meaning in the context of the data. For instance, $100 is twice as much as $50, so these data are at the ratio level. On the other hand, 100°C is not twice as warm as 50°C, so these data are at the interval level.

New York Yankees' World Series Victories (Years)
1923, 1927, 1928, 1932, 1936, 1937, 1938, 1939, 1941, 1943, 1947, 1949, 1950, 1951, 1952, 1953, 1956, 1958, 1961, 1962, 1977, 1978, 1996, 1998, 1999, 2000

2000 American League Home Run Totals (by team)	
Anaheim	236
Baltimore	184
Boston	167
Chicago	216
Cleveland	221
Detroit	177
Kansas City	150
Minnesota	116
New York	205
Oakland	239
Seattle	198
Tampa Bay	162
Texas	173
Toronto	244

EXAMPLE 

Classifying Data by Level

Two data sets are shown at the left. Which data set consists of data at the interval level? Which data set consists of data at the ratio level? Explain your reasoning. *(Source: Major League Baseball)*

SOLUTION Both of these data sets contain quantitative data. Consider the dates of the Yankees' World Series victories. It makes sense to find differences between specific dates. For instance, the time between the Yankees' first and last World Series victories is

$$2000 - 1923 = 77 \text{ years.}$$

But it does not make sense to write a ratio using these dates. So, these data are at the interval level. Using the home run totals, you can find differences *and* write ratios. From the data, you can see that New York hit fifty-five more home runs than Kansas City hit, and that Anaheim hit about twice as many home runs as Minnesota hit. So, these data are at the ratio level.

Try It Yourself 3

Decide whether the data are at the interval level or at the ratio level.

1. The body temperatures (°F) of an athlete who is exercising
2. The heart rates, in beats per minute, of an athlete who is exercising

a. *Identify* what each data set represents.
b. Specify the *level of measurement.*

Answer: Page A29

The following tables summarize meaningful operations at the four levels of measurement.

Level of measurement	Put data in categories	Arrange data in order	Subtract data values	Determine if one data value is a multiple of another
Nominal	Yes	No	No	No
Ordinal	Yes	Yes	No	No
Interval	Yes	Yes	Yes	No
Ratio	Yes	Yes	Yes	Yes

Summary of Four Levels of Measurement

	Example of a Data Set	Meaningful Calculations
Nominal Level (Qualitative data)	*Major PGA Tournaments* The Masters The U.S. Open The British Open The PGA Championship	*Put in a category.* For instance, these are four categories of major PGA tournaments.
Ordinal Level (Qualitative or quantitative data)	*Motion Picture Association of America Ratings Description* G General Audiences PG Parental Guidance Suggested PG-13 Parents Strongly Cautioned R Restricted NC-17 17 and Under Not Permitted	Put in a category and *put in order.* For instance, a PG rating has a stronger restriction than a G rating.
Interval Level (Quantitative data)	*Average Monthly Temp. (°F) for Sacramento, CA* Jan 45.3 Jul 75.6 Feb 50.3 Aug 74.7 Mar 53.2 Sep 71.7 Apr 58.2 Oct 63.9 May 64.9 Nov 53.0 Jun 71.2 Dec 45.6 *(Source: PC USA)*	Put in a category, put in order, and *find difference in values.* For instance, $50.3 - 45.3 = 5$°F. So, February was 5° warmer than January.
Ratio Level (Quantitative data)	*Average Monthly Precipitation (in inches) for Sacramento, CA* Jan 4.0 Jul 0.1 Feb 2.9 Aug 0.1 Mar 2.1 Sep 0.3 Apr 1.3 Oct 0.9 May 0.3 Nov 2.2 Jun 0.1 Dec 2.9 *(Source: PC USA)*	Put in a category, put in order, find difference in values, and *find ratios of values.* For instance, $\frac{0.3}{0.1} = 3$. So, there was three times as much rain in May as in June.

1.2 Exercises

Help

LarsonTutor 1.2

Companion Web Site

Student Solutions Manual 1.2

Videos 1.2

Technology Manuals

Basic Skills and Concepts

1. Name each level of measurement for which data can be qualitative.
2. Name each level of measurement for which data can be quantitative.

True or False In Exercises 3–6, determine whether the statement is true or false. If it is false, rewrite it as a true statement.

3. Data at the nominal level are qualitative only.
4. For data at the ratio level, zero entries represent position only and are not inherent zeros.
5. Data at the ordinal level are quantitative only.
6. For data at the interval level, you cannot calculate meaningful differences between data entries.

Classifying Data by Type In Exercises 7–10, determine whether the data are qualitative or quantitative.

7. The telephone numbers in a telephone directory
8. The daily high temperatures for the month of June
9. The percentage scores of a class in an exam
10. The player numbers in a baseball team

Classifying Data by Level In Exercises 11–16, identify the data set's level of measurement. Explain your reasoning.

11. ***Football*** ◆ The top five teams in the final college football poll released on January 3, 2001 are listed. *(Source: Associated Press)*

    1. Oklahoma 2. Miami (Florida) 3. Washington
    4. Oregon State 5. Florida State

12. ***Tennis*** ◆ The four major professional tennis tournaments are listed below.

    Australian Open
    French Open
    The Lawn Tennis Championships (Wimbledon)
    U.S. Open

13. ***Fish Catches*** ◆ The total catches (in thousands) for different species of fish in Alabama waters are listed. *(Adapted from National Marine Fisheries Service, Fisheries Statistics and Economics Division)*

    36.3 14.7 48.8 87.3 48.5 17.2
    359.0 0.6 32.3 545.9 1.9

14. ***Sales Staff*** ◆ A corporation's U.S. sales staff is split among the following regions.

    Northwest Southwest Southeast Northeast

15. ***Monthly Temperatures*** ◆ The average monthly temperatures (in degrees Celsius) at Gulkana Glacier basin in Alaska are listed. *(Source: U.S. Geological Survey, a bureau of the U.S. Department of the Interior)*

−4.6	−11.4	−13.0	−16.7	−13.1	−9.8
−4.6	−0.1	4.7	6.1	2.8	−0.7

16. ***Fish Lengths*** ◆ The lengths (in inches) of a sample of striped bass caught in Maryland waters are listed. *(Adapted from National Marine Fisheries Service, Fisheries Statistics and Economics Division)*

16 17.25 19 18.75 21 20.3 19.8 24 21.82

Graphical Analysis In Exercises 17–20, identify the level of measurement of the data listed on the horizontal axis in the graph.

17. **Does Global Warming Contribute to More Severe El Niños?**

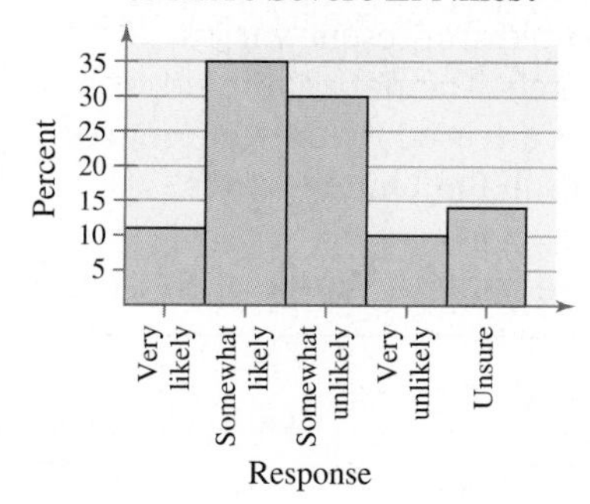

(Source: Yankelovich for the National Representatives Science Foundation, American Meteorological Society)

18. **Average January Snowfall for 15 Cities**

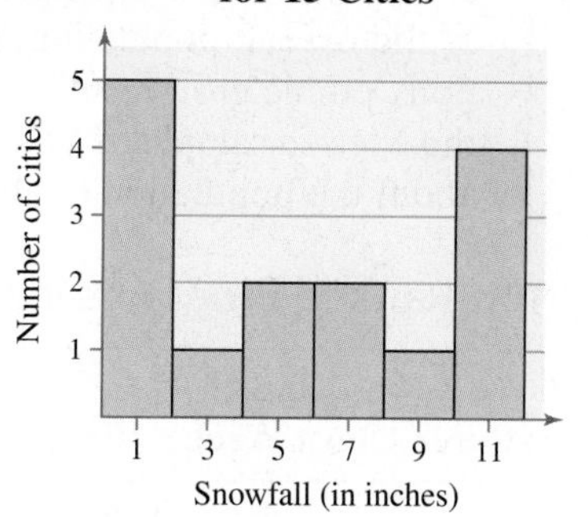

(Source: National Climatic Data Center)

19. **Gender Profile of the 107th Congress**

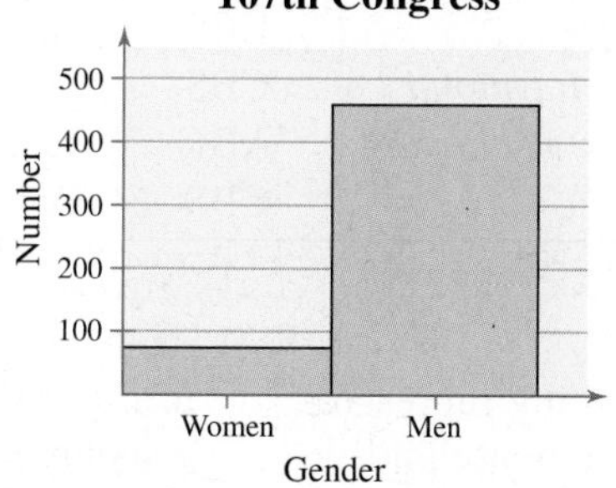

(Source: U.S. House of Representatives, Office of the Clerk)

20. **State Government Tax Collections by Year**

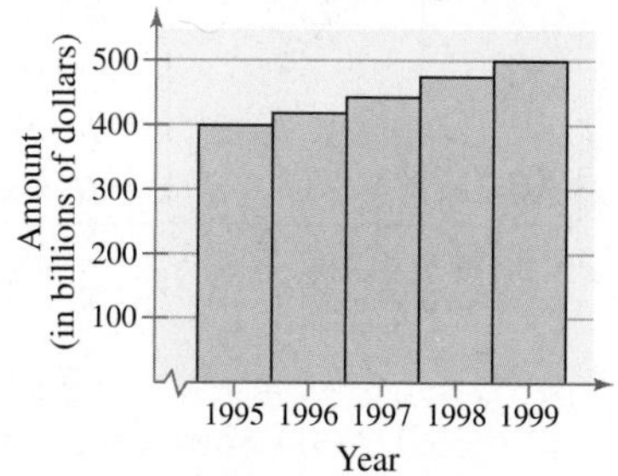

(Source: U.S. Census Bureau)

Extending the Basics

21. ***Writing*** ◆ Explain how to distinguish between the interval level and the ratio level. Give examples of each.

22. ***Writing*** ◆ What is an inherent zero? Describe three examples of data sets that have an inherent zero and three that do not.

Case Study 1

WWW.NIELSENMEDIA.COM

Rating Television Shows in the United States

Nielsen Media Research has been rating television programs for nearly 50 years. Nielsen uses several sampling procedures, but its main one is to track the viewing patterns of 5000 households. These contain over 13,000 people and are chosen to form a cross section of the overall population. The households represent various locations, ethnic groups, and income brackets. The data gathered from the Nielsen sample of 5000 households are used to draw inferences about the population of all households in the United States.

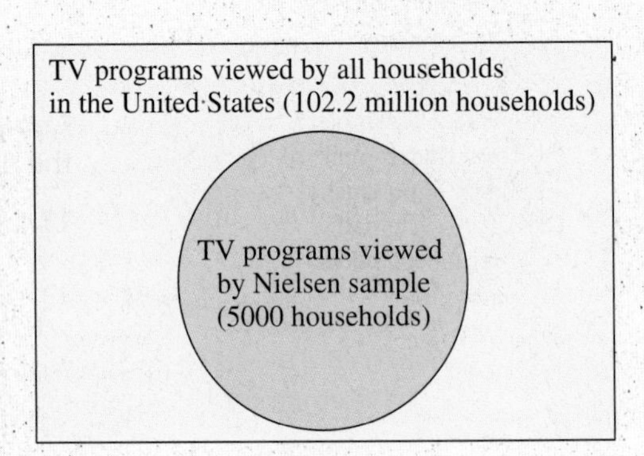

Top-Ranked Programs in Prime Time for the Week of 5/14/01 – 5/20/01

Rank	Rank Last Week	Program Name	Network	Day, Time	Rating	Share	Audience
1	1	*E.R.*	NBC	Thu., 10:00 P.M.	20.1	32	20,495,000
2	9	*Friends*	NBC	Thu., 8:00 P.M.	18.7	31	19,157,000
3	3	*Law and Order*	NBC	Wed.,10:00 P.M.	14.5	24	14,856,000
4	6	*West Wing*	NBC	Wed., 9:00 P.M.	14.1	22	14,414,000
5	19	*Will & Grace*	NBC	Thu., 9:00 P.M.	13.4	20	13,710,000
6	4	*CSI*	CBS	Thu., 9:00 P.M.	12.4	19	12,684,000
7	–	*CBS Sunday Movie*	CBS	Sun., 9:00 P.M.	11.3	18	11,557,000
8	5	*Everybody Loves Raymond*	CBS	Mon., 9:00 P.M.	11.2	17	11,494,000
9	7	*Who Wants to be a Millionaire*	ABC	Tue., 8:00 P.M.	11.1	19	11,323,000
10	13	*Judging Amy*	CBS	Tue., 10:00 P.M.	10.8	18	11,027,000

Exercises

1. ***Rating Points*** Each rating point represents 1,022,000 households or 1% of the households in the United States. Does a program with a rating of 8.4 have twice the number of households as a program with a rating of 4.2? Explain your reasoning.

2. ***Sampling Percent*** What percent of the total number of U.S. households is used in the Nielsen sample?

3. ***Nominal Level of Measurement*** Which columns in the table contain data at the nominal level?

4. ***Ordinal Level of Measurement*** Which columns in the table contain data at the ordinal level? Describe two ways that the data can be ordered.

5. ***Interval Level of Measurement*** Which column in the table contains data at the interval level? How can these data be ordered? What is the unit of measure for the difference of two entries in the data set?

6. ***Ratio Level of Measurement*** Which three columns contain data at the ratio level?

7. ***Share*** The column listed as "Share" gives the percent of televisions in use at a given time. Does the Nielsen rating rank shows by rating or by share? Explain your reasoning.

8. ***Inferences*** What decisions (inferences) can be made based on the Nielsen ratings?

Experimental Design

What You Should Learn

- How to design a statistical study
- How to collect data by taking a census using a sampling, using a simulation, or performing an experiment
- How to create a sample using random sampling, simple random sampling, stratified sampling, cluster sampling, and systematic sampling, and how to identify a biased sample

Experimental Design • Data Collection • Sampling Techniques

Experimental Design

The goal of every statistical study is to collect data and then use the data to make a decision. Any decision you make using the results of a statistical study is only as good as the process used to obtain the data. If the process is flawed, then the resulting decision is questionable.

While you may never have to develop a statistical study, it is likely that you will have to interpret the results of one. And before you interpret the results of a study, you should determine whether or not the results are valid. In other words, you should be familiar with how to design a statistical study.

GUIDELINES

Designing a Statistical Study

1. Identify the variable(s) of interest (the focus) and the population of the study.
2. Develop a detailed plan for collecting data. If you use a sample, make sure the sample is representative of the population.
3. Collect the data.
4. Describe the data using descriptive statistics techniques.
5. Interpret the data and make decisions about the population using inferential statistics.
6. Identify any possible errors.

Data Collection

There are several ways you can collect data. Often, the focus of the study dictates the best way to collect data. The following is a brief summary of four methods of data collection.

- ***Take a census*** A **census** is a count or measure of an *entire* population. Taking a census provides complete information, but it is often costly and difficult to perform.
- ***Use sampling*** A **sampling** is a count or measure of part of a population. The statistics calculated from a sample are used to predict various population parameters. For instance, every year the U.S. Census Bureau samples the U.S. population to update the most recent census data. Using sampling is often more practical than taking a census.
- ***Use a simulation*** A **simulation** is the use of a mathematical or physical model to reproduce the conditions of a situation or process. Collecting data often involves the use of computers. Simulations allow you to study situations that are impractical or even dangerous to create in real life and often save time and money. For instance, automobile manufacturers use simulations with dummies to study the effects of crashes on humans.

- ***Perform an experiment*** When performing an **experiment,** a treatment is applied to part of a population and responses are observed. A second part of the population is often used as a control group. This group receives no treatment or is given a placebo. After responses from both groups are observed, results are compared. For instance, to test the effect of imposing a new marketing strategy, you could perform an experiment by using the new marketing strategy in a certain region. Each experimental unit is called a *block*. Care must be taken to ensure that the blocks are similar.

Once you determine which method you will use to collect data, you might decide that a survey can help you. Surveys can be used to take a census or a sampling. A **survey** is an investigation of one or more characteristics of a population. Most often, surveys are carried out on *people* by asking them questions. A disadvantage of using a survey to collect data is that the wording of the questions can lead to biased results.

Picturing the World

The Gallup Organization conducts many polls (or surveys) regarding the president, Congress, and political and nonpolitical issues. A commonly cited Gallup poll is the public approval rating of the president. For example, the approval ratings for President Bill Clinton from 1995 to 2000 are shown in the following graph. (The rating is from the first poll conducted in June of each year.)

President's Approval Ratings, 1995–2000

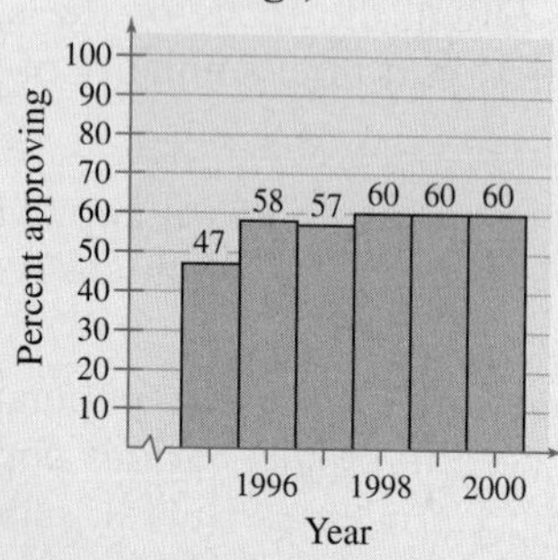

Discuss some ways that Gallup could select a biased sample to conduct a poll. How could Gallup select a sample that is unbiased?

EXAMPLE 

Deciding Upon Methods of Data Collection

Consider the following statistical studies. Which method of data collection would you use to collect data for each study? Explain your reasoning.

1. A study of the effect of changing flight patterns on the number of airplane accidents
2. A study of the effect of aspirin on preventing heart attacks
3. A study of the weights of all linemen in the National Football League
4. A study of U.S. residents' approval rating of the U.S. president

SOLUTION

1. Because it is impractical to create this situation, you would want to use a simulation.
2. In this study, you want to measure the effect a treatment (taking aspirin) has on patients. So, you would want to perform an experiment.
3. Because National Football League teams keep accurate physical records of *all* players, you could take a census.
4. It would be nearly impossible to ask every person in the United States whether or not he or she approves of the president's job performance. So, you should use sampling to collect these data.

Try It Yourself 1

Consider the following statistical studies. Which method of data collection would you use to collect data for each study?

1. A study of the effect of exercise on senior citizens
2. A study of the effect of radiation fallout on senior citizens

a. Identify the *focus* of the study.
b. Identify the *population* of the study.
c. Choose an appropriate *method of data collection*.

Answer: Page A29

Sampling Techniques

To collect unbiased data, it is important that the sample be representative of the population. Appropriate sampling techniques must be used to ensure that inferences about the population are valid. Remember that when a study is done with faulty data, the results are questionable.

A **random sample** is one in which every member of the population has an equal chance of being selected. A **simple random sample** is a sample in which every possible sample of the same size has the same chance of being selected. One way to collect a simple random sample is to assign a different number to each member of the population and then use a random number table like the one in Appendix B. Responses, counts, or measures from members of the population whose numbers correspond to those generated using the table would be in the sample. Calculators and computer software programs are also used to generate random numbers (see page 22).

Insight

A **biased sample** is one that is not representative of the population from which it is drawn. For instance, a sample consisting of only 18 to 22-year-old college students would not be representative of the entire 18 to 22-year-old population in the country.

```
randInt(1,731,8)
{537 33 249 728...
```

Random Numbers Generated using TI-83

Table 1—Random Numbers

92630	78240	19267	95457	53497	23894	37708	79862
79445	78735	71549	44843	26104	67318	00701	34986
59654	71966	27386	50004	05358	94031	29281	18544
31524	49587	76612	39789	13537	48086	59483	60680
06348	76938	90379	51392	55887	71015	09209	79157

Portion of Table 1 found in Appendix B

For instance, to use a simple random sample to measure the number of people who live in Dade County households, you could assign a different number to each household, use a computer or table of random numbers to generate a sample of numbers, and then count the number of people living in each selected household.

EXAMPLE 

Using a Simple Random Sample

There are 731 students currently enrolled in statistics at your school. You wish to form a sample of eight students to answer some survey questions. Select the students who will belong to the simple random sample.

SOLUTION Assign numbers 1 to 731 to each student in the course. On the table of random numbers, choose a starting place at random and read the digits in the first column in groups of three (because 731 is a three-digit number). For example, if you start in the third row of the table at the beginning of the second column, you would group the numbers as follows.

719|66 2|738|6 50|004| 053|58 9|403|1 29|281| 185|44

Ignoring numbers that are greater than 731, the first eight numbers are 719, 662, 650, 4, 53, 589, 403, and 129. Students who were assigned these numbers will make up the sample.

Try It Yourself 2

A company employs 79 people. Choose a simple random sample of five to survey.

a. On the table, randomly choose a *starting place.*
b. *Read the digits* in groups of two.
c. Write the five random numbers.

Answer: Page A29

When you choose members of a sample, you should decide whether it is acceptable to have the same population member selected more than once. If it is acceptable, then the sampling process is said to be *with replacement*. If it is not acceptable, then the sampling process is said to be *without replacement*.

There are several other commonly used sampling techniques. Each has advantages and disadvantages.

- ***Stratified Sample*** When it is important for the sample to have members from each segment of the population, you should use a stratified sample. Depending on the focus of the study, members of the population are divided into two or more different subsets, called *strata*, that share a similar characteristic such as age, gender, ethnicity, or even political preference. A sample is then randomly selected from each of the strata. Using a stratified sample ensures that each segment of the population is represented. For example, to collect a stratified sample of the number of people who live in Dade County households, you could divide the households into socioeconomic levels, and then randomly select households from each level.

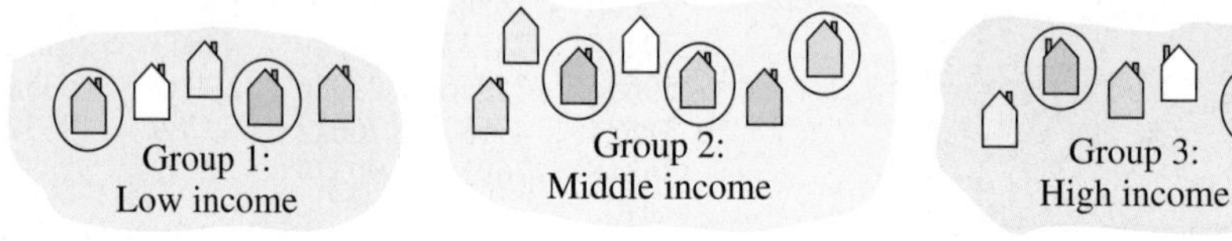

Stratified Sampling

- ***Cluster Sample*** When the population falls into naturally occurring subgroups, each having similar characteristics, a cluster sample may be the most appropriate. To select a cluster sample, divide the population into groups, called *clusters*, and select all of the members in one or more (but not all) of the clusters. Examples of clusters could be different sections of the same course or different branches of a bank. For instance, to collect a cluster sample of the number of people who live in Dade County households, divide the households into groups according to zip codes, then select all the households in one or more, but not all, zip codes and count the number of people living in each household. When using a cluster sample, care must be taken to ensure that all clusters have similar characteristics. For example, if the West cluster has a greater proportion of high-income people, the data might not be representative of the population.

Insight

For stratified sampling, each of the strata contains members with a certain characteristic (for example, a particular age group). In contrast, clusters consist of geographic groupings, and each cluster should consist of members with all of the characteristics (for example, all age groups). With stratified samples, some of the members of each group are used. In a cluster sampling, all of the members of one or more groups are used.

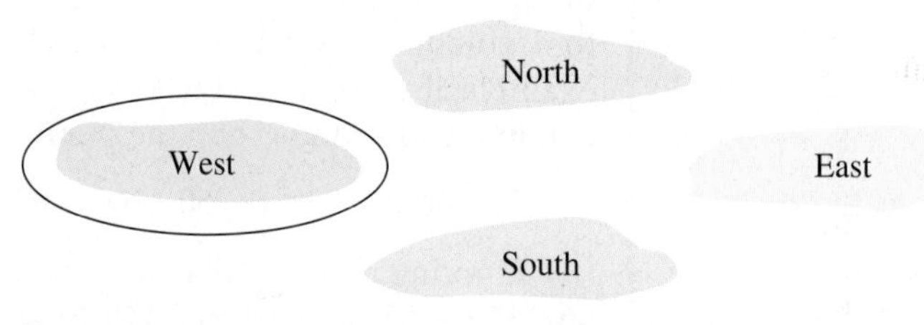

Cluster Sampling

- ***Systematic Sample*** A systematic sample is a sample in which each member of the population is assigned a number. The members of the population are ordered in some way, a starting number is randomly selected, and then sample members are selected at regular intervals from the starting number. (For instance, every 3rd, 5th, or 100th member is selected.)

For example, to collect a systematic sample of the number of people who live in Dade County households, you could assign a different number to each household, randomly choose a starting number, select every hundredth household, and count the number of people living in each.

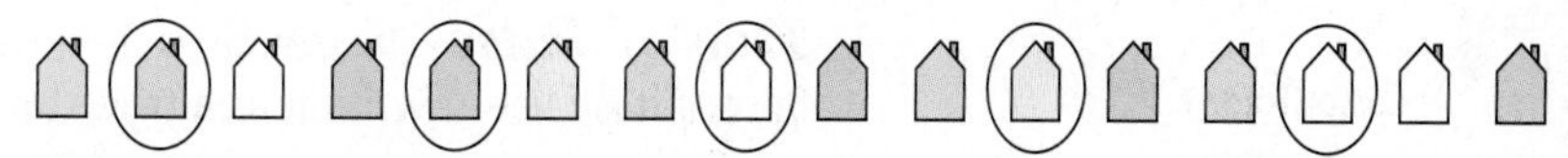

Systematic Sampling

An advantage of systematic sampling is that it is easy to use. In the case of any regularly occurring pattern in the data, however, this type of sampling should be avoided.

A type of sample that often leads to biased studies (and so it is not recommended) is a **convenience sample.** A convenience sample consists only of available people.

EXAMPLE 

Identifying Sampling Techniques

You are doing a study to determine the opinion of students at your school regarding gun control. Identify the sampling technique you are using if you select the samples listed.

1. You select a class at random and question each student in the class.
2. You divide the student population with respect to majors and randomly select and question some students in each major.
3. You assign each student a number and generate random numbers. You then question each student whose number is randomly selected.

SOLUTION

1. Because each class is a naturally occurring subgroup (a cluster) and you question each student in the class, this is a cluster sample.
2. Because students are divided into strata (majors) and a sample is selected from each major, this is a stratified sample.
3. Each sample of the same size has an equal chance of being selected and each student has an equal chance of being selected, so this is a simple random sample.

Try It Yourself 3

You want to determine the opinion of students at your school regarding gun control. Identify the sampling technique you are using if you select the samples listed.

1. You select students who are in your statistics class.
2. You assign each student a number, and after choosing a starting number, question every 25th student.

a. Determine *how* the sample is *selected.*
b. Identify the corresponding *sampling technique.*

Answer: Page A29

1.3 Exercises

Help

 LarsonTutor 1.3

 Companion Web Site

 Student Solutions Manual 1.3

 Videos 1.3

 Technology Manuals

Basic Skills and Concepts

True or False In Exercises 1–4, determine whether the statement is true or false. If it is false, rewrite it as a true statement.

1. Using a systematic sample guarantees that members of each group within a population will be sampled.
2. A census is a count of part of a population.
3. To select a stratified sample, a population is ordered in some way and then members of the population are selected at regular intervals.
4. To select a cluster sample, divide a population into groups and then select all of the members in at least one (but not all) of the groups.

Deciding Upon the Method of Data Collection In Exercises 5–8, decide which method of data collection you would use. Explain.

5. A study of the effect on the human digestive system of potato chips made with a fat substitute
6. A study of the effect of a product's warning label to determine whether consumers still buy the product
7. A study of how fast a virus would spread in a metropolitan area
8. A study of the salaries of the 535 members of the U.S. Congress

Identifying Sampling Techniques In Exercises 9–18, identify the sampling technique used and discuss potential sources of bias (if any). Explain.

9. Using random digit dialing, 1599 people were called and asked what obstacles (such as childcare) kept them from exercising. *(Source: Yankelovich Partners, Inc. for Shape Up America!)*
10. Chosen at random, 200 rural and 200 urban persons age 65 or older were asked about their health and experience with prescription drugs.
11. Questioning students as they left a university library, a researcher asked 358 students about their drinking habits.
12. After a hurricane, a disaster area is divided into 200 equal grids. Thirty of the grids are selected and every occupied household in the grid is interviewed to help focus relief efforts on what residents require the most.
13. Chosen at random, 1819 hospital outpatients were contacted and asked their opinion of the care they received.
14. For quality assurance, every twelfth engine part is selected from an assembly line and tested for durability.
15. Soybeans are planted on a 48-acre field. The field is divided into one-acre subplots. A sample of plants is taken from each subplot to estimate the harvest.

PS **16.** Questioning teachers as they left a faculty lounge, a researcher asked 56 teachers about their teaching styles and grading methods.

17. A list of managers is compiled and ordered. After randomly choosing a starting number, every twentieth name is selected until 1000 managers are selected. The managers are questioned about the use of digital media.

18. From calls made with randomly generated telephone numbers, 1012 respondents were asked if they rented or owned their residence.

Recognizing a Biased Question In Exercises 19 and 20, determine whether the survey question is biased. If the question is biased, suggest a better wording.

19. Why is drinking fruit juice good for you?

20. Why are drivers who change lanes several times dangerous?

21. ***Writing*** ◆ Television program ratings by Nielsen Media Research are described in the case study on page 14. Discuss the strata used in the sample.

22. ***Writing*** ◆ Television program ratings by Nielsen Media Research are described in the case study on page 14. Why is it important to have a stratified sample for these ratings?

Extending the Basics

23. ***Open and Closed Questions*** ◆ Two types of survey questions are open questions and closed questions. An open question allows for any kind of response, while a closed question allows only for a fixed response. For example, an open question and a closed question are given.

Open Question

What can be done to get students to eat healthier foods?

Closed Question

How would you get students to eat healthier foods?

1. Mandatory nutrition course
2. Offer only healthy foods in the cafeteria and remove unhealthy foods
3. Offer more healthy foods in the cafeteria and raise the prices on unhealthy foods

(a) List an advantage and a disadvantage of an open question.

(b) List an advantage and a disadvantage of a closed question.

24. ***Who Picked These People?*** ◆ Sometimes, instead of selecting a sample to survey, some polling agencies ask people to call a telephone number and give their response to a question.

(a) List an advantage and a disadvantage of a survey conducted in this manner.

(b) What sampling technique is used in such a survey?

TECHNOLOGY

Using Technology in Statistics

With large data sets, you will find that calculators or computer software programs can help perform calculations and create graphics. Of the many calculators and statistical software programs that are available, we have chosen to incorporate the Minitab and Excel software and TI-83 graphing calculator into this text.

The following example shows how to use these three technologies to generate a list of random numbers.

EXAMPLE

Generating a List of Random Numbers

A quality control department inspects a random sample of 15 of the 167 cars that are assembled at an auto plant. How should the cars be chosen?

SOLUTION

One way to choose the sample is to number the cars from 1 to 167. To form a list of random numbers from 1 to 167, you can use technology. Each of the technology tools shown requires different steps to generate the list. Each, however, does require that you identify the minimum value as 1 and the maximum value as 167. Check your user's manual for specific instructions.

TI-83

randInt(1, 167, 15)
{17 42 152 59 5 116
125 64 122 55 58 60
82 152 105}

MINITAB

↓	C1
1	167
2	11
3	74
4	160
5	18
6	70
7	80
8	56
9	37
10	6
11	82
12	126
13	98
14	104
15	137

EXCEL

	A
1	41
2	16
3	91
4	58
5	151
6	36
7	96
8	154
9	2
10	113
11	157
12	103
13	64
14	135
15	90

MINITAB EXCEL TI-83

Recall that when you generate a list of random numbers, you should decide whether it is acceptable to have numbers that repeat. If it is acceptable, then the sampling process is said to be *with replacement.* If it is not acceptable, then the sampling process is said to be *without replacement.*

With each of the three technology tools shown on page 22, you have the capability of sorting the list so that the numbers appear in order. Doing this helps you see whether any of the numbers in the list repeat. If it is not acceptable to have repeats, you should specify that the tool generate more random numbers than you need.

Exercises

1. A CPA is examining the records of a business. The business has 74 major accounts, and the CPA decides to audit a random sample of 8 of the accounts. Describe how this could be done. Then use technology to generate a list of 8 random numbers from 1 to 74 and order the list.

2. A quality control department is testing 20 batteries from a shipment of 200 batteries. Describe how this could be done. Then use technology to generate a list of 20 random numbers from 1 to 200 and order the list.

3. Consider the population of ten digits: 0, 1, 2, 3, 4, 5, 6, 7, 8, and 9. Select three random samples of five digits from this list. Find the average of each sample. Compare your results to the average of the entire population. Comment on your results. (*Hint:* To find the average, sum the data entries and divide the sum by the number of entries.)

4. Consider the population of 41 whole numbers from 0 to 40. What is the average of these numbers? Select three random samples of seven numbers from this list. Find the average of each sample. Compare your results to the average of the entire population. Comment on your results. (*Hint:* To find the average, sum the data entries and divide the sum by the number of entries.)

5. Use random numbers to simulate rolling a six-sided die 60 times. Make a tally of your results. How many times did you obtain each number from 1 to 6? Are the results what you expected?

6. Suppose you rolled a six-sided die 60 times and got the following tally.

20 ones	20 twos	15 threes
3 fours	2 fives	0 sixes

Does this seem like a reasonable result? What inference might you draw from the result?

7. Use random numbers to simulate tossing a coin 100 times. Let 0 represent heads and let 1 represent tails. Make a tally of your results. How many times did you obtain each number? Are the results what you expected?

8. Suppose you tossed a coin 100 times and got 77 heads and 23 tails. Does this seem like a reasonable result? What inference might you draw from the result?

Extended solutions are given in the *Technology Supplement.*
Technical instruction is provided for *Minitab*, *Excel*, and the *TI-83*.

1

Chapter Summary

What did you learn?	*Review Exercises*
Section 1.1	
◆ How to distinguish between a population and a sample	*1–4*
◆ How to distinguish between a parameter and a statistic	*5–8*
Section 1.2	
◆ How to distinguish between qualitative and quantitative data	*9–12*
◆ How to classify data with respect to the four levels of measurement: nominal, ordinal, interval, and ratio	*13–16*
Section 1.3	
◆ How data are collected: by taking a census, using a sampling, using a simulation, or performing an experiment.	*17–20*
◆ How to create a sample using random sampling, simple random sampling, stratified sampling, cluster sampling, and systematic sampling, and how to identify a biased sample.	*21–30*

STATISTICS

Uses and Abuses

Uses

Surveys Surveys can be valuable in determining the attitude of a population about a candidate, product, or issue.

If you are working for a political candidate, it is important that you know how the voting population views your candidate. By knowing this information, you might be able to address voter concerns and increase your candidate's chance of winning the election.

If you are working in the market research department of a manufacturing company, it is important that you know how the public will react to a proposed new product *before* the product is produced. Perhaps you will be able to alter the product's design to make it more appealing and capture a greater market share.

If you are working for an activist organization, it is important that you know how the population feels about your organization's issues. If you discover that the population does not support an issue, perhaps you will be able to change this through an advertising campaign.

Abuses

Biased Samples The most common abuse (or misuse) of statistics is using a sample that does not represent the entire population of the study. Consider a phone survey of opinions about a candidate for a local school board. The survey consisted of phone calls to numbers listed in the local phone directory. Of the 930 numbers dialed, 543 were answered. Of the calls that were answered, 162 people agreed to take the survey. Of those surveyed, 62% plan on voting for the candidate. It should be clear that you cannot conclude from this survey that 62% of the voters in the locality plan on voting for the candidate. The survey did not include voters with unlisted phone numbers, or who did not happen to be home, or who did not agree to participate in the phone survey. In such cases, we say that the survey is biased toward people who have listed phone numbers, tend to be at home in the evening, and are willing to participate in surveys.

Biased Survey Questions Another common abuse is using survey questions that encourage respondents, either intentionally or unintentionally, to answer in a certain way. For instance, consider a survey about gun control. It seems clear that the following questions would produce very different results. (a) Do you agree with the statement: "People have the right to own a gun to protect themselves and their families." (b) Do you agree with the statement: "People have the right to possess loaded guns in their homes."

Exercises

1. ***Biased Survey*** Find an example of a survey that used a biased sample. Try hunting in magazines, newspapers, or on the Internet. Explain why the sample is biased. What could have been done to make the sample unbiased?
2. ***Biased Survey Question*** Find an example of a survey that used a biased survey question. Try hunting in magazines, newspapers, or on the Internet. Explain why the question is biased. Rewrite the survey question so that it is unbiased, or at least less biased.

1 Review Exercises

Section 1.1

In Exercises 1–4, identify the population and the sample.

1. A survey of 898 U.S. adult VCR owners found that 16% had VCR clocks that were currently blinking "12:00." *(Source: Wirthlin Worldwide)*

2. Thirty-eight nurses working in the San Francisco area were surveyed concerning their opinions of managed health care.

3. A study of 860 U.S. ATMs (automated teller machines) found that the average surcharge for withdrawals from a competing bank was $1.15. *(Source: Public Interest Research Groups)*

4. A survey of 1420 U.S. undergraduate English majors asked which Shakespearean play was most relevant in the year 2000.

In Exercises 5–8, determine whether the numerical value describes a parameter or a statistic.

5. The 2001 team payroll of the Baltimore Orioles was $74,279,540. *(Source: Major League Baseball)*

6. In a survey of a sample of U.S. adults, 22% owned a portable cellular phone. *(Source: Wirthlin Worldwide)*

7. In a recent survey at the University of Arizona, 89 students were majoring in astronomy. *(Source: University of Arizona Student Research Office)*

8. Nineteen percent of a sample of Indiana ninth graders surveyed smoked cigarettes daily. *(Source: Indiana University)*

Section 1.2

In Exercises 9–12, determine which data are qualitative data and which are quantitative data. Explain your reasoning.

9. The monthly salaries of the employees at an accounting firm

10. The social security numbers of the employees at an accounting firm

11. The ages of a sample of 350 residents of nursing homes

12. The zip codes of a sample of 350 residents of nursing homes

In Exercises 13–16, identify the data set's level of measurement. Explain your reasoning.

13. The daily high temperatures (in degrees Fahrenheit) for Mohave, Arizona, for a week in June are listed. *(Source: Arizona Meteorological Network)*

93 91 86 94 103 104 103

14. The EPA size classes for automobiles are listed. *(Source: Carspec)*

subcompact compact midsize fullsize

15. The four teams in the American League West Division are listed.

Anaheim Texas Oakland Seattle

16. The heights (in inches) of the 2001–2002 Chicago Bulls are listed. *(Source: National Basketball Association)*

79 85 78 84 77 83 75 81 82 73
77 77 79 79 84 80 81 83 83

Section 1.3

In Exercises 17–20, decide which method of data collection you would use to gather data for each study. Explain your reasoning.

17. A study of charitable donations of the judges in Sioux Falls, South Dakota

18. A study of the effect of kangaroos on the Florida Everglades ecosystem

19. A study of the effects of a plant hormone on chrysanthemums

20. A study of the awareness of college students of the ozone layer

In Exercises 21–26, identify which sampling technique was used in the study. Explain your reasoning.

21. Calling randomly generated telephone numbers, a study asked 1001 U.S. adults which medical conditions could be prevented by their diet. *(Adapted from Wirthlin Worldwide)*

22. A student asks 12 friends to participate in a psychology experiment.

23. A pregnancy study in Cebu, Philippines, randomly selected 33 communities from the Cebu metropolitan area, then interviewed all available pregnant women in these communities. *(Adapted from Cebu Longitudinal Health and Nutrition Survey)*

24. Law enforcement officials use a radar gun to measure the speed of every tenth vehicle on an interstate.

25. Twenty-five students are randomly selected from each grade level at a high school and surveyed about their study habits.

26. A journalist interviews 123 people after they leave a restaurant and asks them how confident they are that the food is safe.

In Exercises 27–30, identify a possible bias or error that might occur in the indicated survey or study.

27. The phone survey in Exercise 21

28. The psychology experiment in Exercise 22

29. The pregnancy study in Exercise 23

30. The vehicle speed sampling in Exercise 24

PUTTING IT ALL TOGETHER

Real Statistics Real Decisions

You are a researcher for a professional research firm. Your firm has won a contract on doing a study for an automobile industry publication. The publication would like to get its readers' (engineers, manufacturers, researchers and developers) thoughts on the future of automobiles, such as what type of fuel they think will be used in the future. The publication would like to get input from those who work for automakers and from those who work for automaker suppliers.

The publication has given you their readership database and the 20 questions they would like to ask (two sample questions from a previous study are given at the right). It is too expensive to contact all of the readers, so you need to determine a way to contact a representative sample of the entire readership population.

Exercises

1. *How Would You Do It?*

(a) What sampling technique would you use to select the sample for the study? Why?

(b) Will the technique you chose in part (a) give you a sample that is representative of the population?

(c) Describe the method for collecting data.

(d) Identify possible flaws or biases in your study.

2. *Data Classification*

(a) What type of data do you expect to collect: qualitative, quantitative, or both? Why?

(b) What levels of measurement do you think the data in the study will be? Why?

(c) Will the data collected for the study represent a population or a sample?

(d) Will the numerical descriptions of the data be parameters or statistics?

3. *How They Did It*

When *Ward's AutoWorld* did a similar study, they hired a professional research firm. The firm mailed out 2400 surveys to a broad cross section of *Ward's AutoWorld* readers. About six weeks later, the firm had received 446 complete surveys. (a) Describe some possible errors in collecting data by mailed surveys. (b) Compare your method for collecting the data in Exercise 1 to this method.

www.wardsauto.com

How will the internal combustion engine of the future be fueled?

Fuel	Percent Responding
Gasoline	42.6%
Hydrogen	24.8%
Diesel fuel	18.3%
Natural gas	9.6%
Other	4.3%
No responses	0.4%

Source: Wardsauto.com

When will affordable fuel-cell vehicles be on the market?

Time	Percent Responding
5 years or less	6.5%
More than 5 years to at most 10	35.0%
More than 10 years	49.6%
Not likely	8.3%
No response	0.5%

Source: Wardsauto.com

1 Chapter Quiz

Take this quiz as you would take a quiz in class. After you are done, check your work against the answers given in the back of the book.

1. Identify the population and the sample in the following study.

A study of 163 patients with sleep disorders was conducted to find a link between obesity and sleep disorders. *(Source: Archives of Internal Medicine 1998)*

2. Determine whether the numerical value is a parameter or a statistic.

(a) In a survey of a sample of parents, 53% said they protect their children from sun exposure using sunscreen. *(Source: Morbidity and Mortality Weekly Report)*

(b) In a union's vote, 67% of all union members voted to ratify a contract proposal.

3. Determine whether the data are qualitative or quantitative.

(a) A database of student identification numbers

(b) The test scores in a statistics class

4. Identify each data set's level of measurement. Explain your reasoning.

(a) A list of the uniform numbers retired by each major league baseball team

(b) The number of products sold by a toy manufacturer each quarter for the current fiscal year

(c) The finishing order of the runners in the Boston marathon

5. Decide which method of data collection you would use to gather data for each study. Explain your reasoning.

(a) A study on the effect of low dietary intake of vitamin C and iron on lead levels in adults

(b) The ages of people living within 500 miles of your home

6. Identify which sampling technique was used in each study. Explain your reasoning.

(a) A journalist goes to a beach to ask people how they feel about water pollution.

(b) For quality assurance, every fifth engine part is selected from an assembly line and tested for durability.

(c) A study on attitudes about smoking is conducted at a college. The students are divided by class (freshman, sophomore, junior, and senior). Then a random sample is selected from each class and interviewed.

7. Which sampling technique used in Exercise 6 could lead to a biased study?

8. Determine whether each statement is true or false. If it is false, rewrite it as a true statement.

(a) A parameter is a numerical measure that describes a sample characteristic.

(b) Ordinal data represent the highest level of measurement.

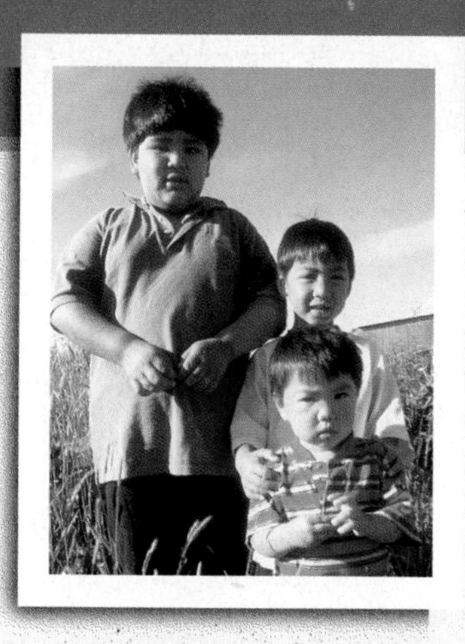

Where You've Been

In Chapter 1, you learned that there are many ways to collect data. Usually, researchers must work with sample data in order to analyze populations, but occasionally it is possible to collect all the data for a given population. For instance, the following represents census data reporting the ages of the entire population of the 77 residents of Akhiok, Alaska.

28, 6, 17, 48, 63, 47, 27, 21, 3, 7, 12, 39, 50, 54, 33, 45, 15, 24, 1, 7, 36, 53, 46, 27, 5, 10, 32, 50, 52, 11, 42, 22, 3, 17, 34, 56, 25, 2, 30, 10, 33, 1, 49, 13, 16, 8, 31, 21, 6, 9, 2, 11, 32, 25, 0, 55, 23, 41, 29, 4, 51, 1, 6, 31, 5, 5, 11, 4, 10, 26, 12, 6, 16, 8, 2, 4, 28

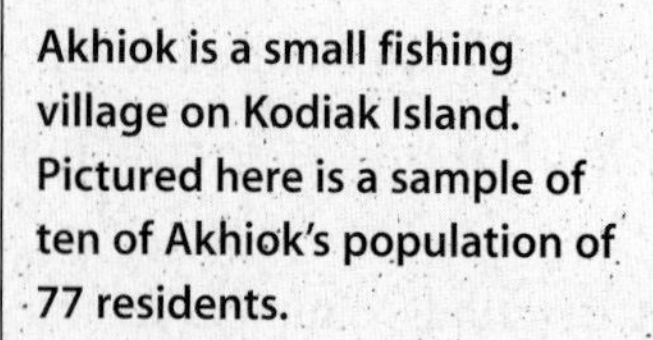

Akhiok is a small fishing village on Kodiak Island. Pictured here is a sample of ten of Akhiok's population of 77 residents.

Photographs © Roy Corral

Descriptive Statistics

CHAPTER 2

Where You're Going

In Chapter 2, you will learn ways to organize and describe data sets. The goal is to make the data easier to understand and make it easier to see trends, averages, and variations. For instance, in the raw data showing the ages of the residents of Akhiok, it is not easy to see any patterns or special characteristics. Here are some ways you can organize and describe the data.

0, 1, 1, 1, 2, 2, 2, 3, 3, 4, 4,
4, 5, 5, 5, 6, 6, 6, 6, 7, 7, 8,
8, 9, 10, 10, 10, 11, 11, 11, 12, 12, 13,
15, 16, 16, 17, 17, 21, 21, 22, 23, 24, 25,
25, 26, 27, 27, 28, 28, 29, 30, 31, 31, 32,
32, 33, 33, 34, 36, 39, 41, 42, 45, 46, 47,
48, 49, 50, 50, 51, 52, 53, 54, 55, 56, 63

Order the data.

Class	Frequency, f
0–12	32
13–25	13
26–38	15
39–51	11
52–64	6

Make a frequency distribution table.

Draw a histogram.

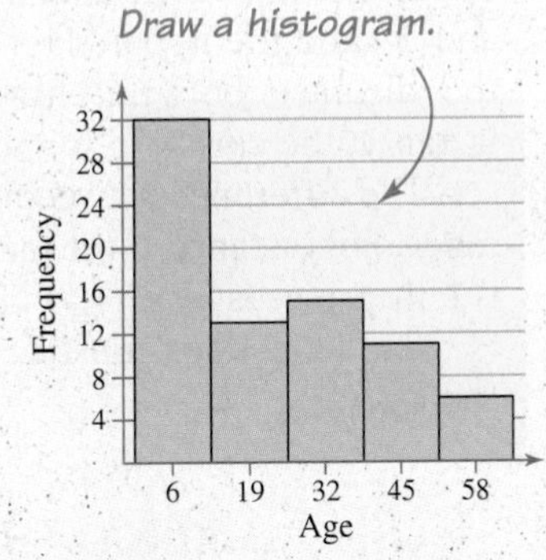

$$\text{Mean} = \frac{0 + 1 + 1 + 1 + 2 + \cdots + 54 + 55 + 56 + 63}{77}$$

$$= \frac{1745}{77}$$

$$\approx 22.7 \text{ years}$$

Find an average.

2.1 Frequency Distributions and Their Graphs

Frequency Distributions • Graphs of Frequency Distributions

What You Should Learn

- How to construct a frequency distribution including limits, boundaries, midpoints, relative frequencies, and cumulative frequencies
- How to construct frequency histograms, frequency polygons, relative frequency histograms, and ogives

Frequency Distributions

When a data set has many entries, it can be difficult to see patterns. In this section, you will learn how to organize data sets by grouping the data into intervals called classes and forming a frequency distribution. You will also learn how to use frequency distributions to construct graphs.

DEFINITION

A **frequency distribution** is a table that shows **classes** or **intervals** of data entries with a count of the number of entries in each class. The **frequency** f of a class is the number of data entries in the class.

Example of a Frequency Distribution

Class	Frequency, f
1–5	5
6–10	8
11–15	6
16–20	8
21–25	5
26–30	4

In the frequency distribution at the left there are six classes. The frequencies for each of the six classes are 5, 8, 6, 8, 5, and 4, respectively. Each class has a **lower class limit,** which is the least number that can belong to the class, and an **upper class limit,** which is the greatest number that can belong to the class. In the frequency distribution at the left, the lower class limits are 1, 6, 11, 16, 21, and 26, and the upper class limits are 5, 10, 15, 20, 25, and 30. The **class width** is the distance between lower (or upper) limits of consecutive classes. For instance, the class width in the frequency distribution at the left is $6 - 1 = 5$.

The difference between the maximum and minimum data entries is called the **range.** For instance, if the maximum data entry is 29, and the minimum data entry is 1, the range is $29 - 1 = 28$. You will learn more about the range in Section 2.4.

Guidelines for constructing a frequency distribution from a data set are as follows.

GUIDELINES

Constructing a Frequency Distribution from a Data Set

1. Decide on the number of classes to include in the frequency distribution. The number of classes should be between five and twenty; otherwise, it may be difficult to detect any patterns.
2. Find the class width as follows. Determine the range of the data, divide the range by the number of classes, and *round up to the next convenient number.*
3. Find the class limits. You can use the minimum data entry as the lower limit of the first class. To find the remaining lower limits, add the class width to the lower limit of the preceding class. Then find the upper limit of the first class. Remember that classes cannot overlap. Find the remaining upper class limits.
4. Make a tally mark for each data entry in the row of the appropriate class.
5. Count the tally marks to find the total frequency f for each class.

Study Tip

In a frequency distribution, it is best if each class has the same width. Answers shown will use the minimum data value for the lower limit of the first class. Sometimes it may be more convenient to choose a value that is slightly lower than the minimum value. The frequency distribution produced will vary slightly.

EXAMPLE 

Constructing a Frequency Distribution from a Data Set

The following sample data set lists the number of minutes 50 Internet subscribers spent on the Internet during their most recent session. Construct a frequency distribution that has seven classes.

50 40 41 17 11 7 22 44 28 21 19 23 37 51 54 42 88
41 78 56 72 56 17 7 69 30 80 56 29 33 46 31 39 20
18 29 34 59 73 77 36 39 30 62 54 67 39 31 53 44

Lower limit	Upper limit
7	18
19	30
31	42
43	54
55	66
67	78
79	90

SOLUTION

1. The number of classes (7) is stated in the problem.
2. The minimum data entry is 7 and the maximum data entry is 88, so the range is 81. Divide the range by the number of classes and round up to find that the class width is 12.

$$\text{Class width} = \frac{88 - 7}{7} \qquad \frac{\text{maximum entry} - \text{minimum entry}}{\text{number of classes}}$$

$$= \frac{81}{7} \qquad \frac{\text{range}}{\text{number of classes}}$$

$$\approx 11.57 \qquad \text{Round up to 12.}$$

3. The minimum data entry is a convenient lower limit for the first class. To find the lower limits of the remaining six classes, add the class width of 12 to the lower limit of each previous class. The upper limit of the first class is 18, which is one less than the lower limit of the second class. The upper limits of the other classes are $18 + 12 = 30$, $30 + 12 = 42$, and so on. The lower and upper limits for all seven classes are shown at the left.
4. Tally the entries for each class.
5. The number of tally marks for a class is the frequency for that class.

Study Tip

The uppercase Greek letter sigma (Σ) is used throughout statistics to indicate a summation of values.

The frequency distribution is shown in the following table. The first class, 7–18, has six tally marks. So the frequency for this class is 6. Notice that the sum of the frequencies is 50, which is the number of entries in the sample data set. The sum is denoted by Σf, where Σ is the uppercase Greek letter **sigma.**

Frequency Distribution for Internet Usage (in minutes)

Number of minutes → Class; Number of subscribers → Frequency, f

Class	Tally	Frequency, f
7–18	卌 I	6
19–30	卌 卌	10
31–42	卌 卌 III	13
43–54	卌 III	8
55–66	卌	5
67–78	卌 I	6
79–90	II	2
		$\Sigma f = 50$

Check that the sum of the frequencies equals the number in the sample.

Try It Yourself 1

Construct a frequency distribution using the ages of the residents of Akhiok given in the chapter opener on page 30. Use six classes.

a. State the *number of classes.*
b. Find the minimum and maximum values and the *class width.*
c. Find the *class limits.*
d. *Tally* the data entries.
e. Write the *frequency f* for each class.

Answer: Page A29

After constructing a standard frequency distribution such as the one in Example 1, there are several additional features you can include that will help provide a better understanding of the data. These features are the midpoint, relative frequency, and cumulative frequency of each class, and can be included as additional columns in your table.

DEFINITION

The **midpoint** of a class is the sum of the lower and upper limits of the class divided by two. The midpoint is sometimes called the *class mark.*

$$\text{Midpoint} = \frac{(\text{Lower class limit}) + (\text{Upper class limit})}{2}$$

The **relative frequency** of a class is the portion or percent of the data that falls in that class. To find the relative frequency of a class, divide the frequency f by the sample size n.

$$\text{Relative frequency} = \frac{\text{Class frequency}}{\text{Sample size}} = \frac{f}{n}$$

The **cumulative frequency** of a class is the sum of the frequency for that class and all previous classes. The cumulative frequency of the last class is equal to the sample size n.

After finding the first midpoint, you can find the remaining midpoints by adding the class width to the previous midpoint. For instance, if the first midpoint is 12.5 and the class width is 12, then the remaining midpoints are

$$12.5 + 12 = 24.5$$

$$24.5 + 12 = 36.5$$

$$36.5 + 12 = 48.5$$

$$48.5 + 12 = 60.5$$

and so on.

Also, you can write the relative frequency as a decimal or as a percent. The sum of the relative frequencies of all the classes must equal one or 100%.

49

EXAMPLE 2 

Midpoints, Relative and Cumulative Frequencies

Using the frequency distribution constructed in Example 1, find the midpoint, relative frequency, and cumulative frequency for each class. Identify any patterns.

SOLUTION The midpoint, relative frequency, and cumulative frequency for the first three classes are calculated as follows.

Class	f	Midpoint	Relative frequency	Cumulative frequency
7–18	6	$\frac{7+18}{2} = 12.5$	$\frac{6}{50} = 0.12$	6
19–30	10	$\frac{19+30}{2} = 24.5$	$\frac{10}{50} = 0.2$	$6 + 10 = 16$
31–42	13	$\frac{31+42}{2} = 36.5$	$\frac{13}{50} = 0.26$	$16 + 13 = 29$

The remaining midpoints, relative frequencies, and cumulative frequencies are shown in the following expanded frequency distribution.

Frequency Distribution for Internet Usage (in minutes)

Minutes online → Class; Number of subscribers → Frequency, f; Portion of subscribers → Relative frequency

Class	Frequency, f	Midpoint	Relative frequency	Cumulative frequency
7–18	6	12.5	0.12	6
19–30	10	24.5	0.2	16
31–42	13	36.5	0.26	29
43–54	8	48.5	0.16	37
55–66	5	60.5	0.1	42
67–78	6	72.5	0.12	48
79–90	2	84.5	0.04	50
	$\Sigma f = 50$		$\Sigma \frac{f}{n} = 1$	

There are several patterns in the data set. For instance, the most common time span that users spent online was 31 to 42 minutes.

Try It Yourself 2

Using the frequency distribution constructed in Try It Yourself 1, find the midpoint, relative frequency, and cumulative frequency for each class. Identify any patterns.

a. Use the formulas to *find each midpoint, relative frequency, and cumulative frequency.*
b. *Organize your results* in a frequency distribution.
c. *Identify* patterns that emerge from the data.

Answer: Page A29

Graphs of Frequency Distributions

Sometimes it is easier to identify patterns of a data set by looking at a graph of the frequency distribution. One such graph is a frequency histogram.

DEFINITION

A **frequency histogram** is a bar graph that represents the frequency distribution of a data set. A histogram has the following properties.

1. The horizontal scale is quantitative and measures the data values.
2. The vertical scale measures the frequencies of the classes.
3. Consecutive bars must touch.

Study Tip

If data entries are integers, subtract 0.5 from each lower limit to find the lower class boundaries. To find the upper class boundaries, add 0.5 to each upper limit. The upper boundary of a class will equal the lower boundary of the next higher class.

Because consecutive bars of a histogram must touch, bars must begin and end at class boundaries instead of class limits. **Class boundaries** are the numbers that separate classes *without* forming gaps between them. You can mark the horizontal scale at either the midpoints or at the class boundaries as shown in Example 3.

EXAMPLE 3

Constructing a Frequency Histogram

Draw a frequency histogram for the frequency distribution in Example 2. Describe any patterns.

Class	Class Boundaries	Frequency, f
7–18	6.5–18.5	6
19–30	18.5–30.5	10
31–42	30.5–42.5	13
43–54	42.5–54.5	8
55–66	54.5–66.5	5
67–78	66.5–78.5	6
79–90	78.5–90.5	2

SOLUTION First, find the class boundaries. The distance from the upper limit of the first class to the lower limit of the second class is $19 - 18 = 1$. Half this distance is 0.5. So, the lower and upper boundaries of the first class are as follows.

$$\text{First class lower boundary} = 7 - 0.5 = 6.5$$

$$\text{First class upper boundary} = 18 + 0.5 = 18.5.$$

The boundaries of the remaining classes are shown in the table at the left. Using the class midpoints or class boundaries for the horizontal scale and choosing possible frequency values for the vertical scale, you can construct the histogram. From either histogram, you can see that more than 50% of the subscribers spent between 19 and 54 minutes on the Internet during their most recent session.

Internet Usage (labeled with class midpoints)

Internet Usage (labeled with class boundaries)

Try It Yourself 3

Use the frequency distribution from Try It Yourself 1 to construct a frequency histogram that represents the ages of the residents of Akhiok. Describe any patterns.

a. Find the *class boundaries*.
b. Choose appropriate *horizontal and vertical scales*.
c. Use the frequency distribution to *find the height of each bar*.
d. *Describe* any patterns for the data.

Answer: Page A30

Another way to graph a frequency distribution is to use a frequency polygon. A **frequency polygon** is a line graph that emphasizes the continuous change in frequencies.

EXAMPLE 4

Constructing a Frequency Polygon

Draw a frequency polygon for the frequency distribution in Example 2.

SOLUTION To construct the frequency polygon, use the same horizontal and vertical scales that were used in the histogram labeled with class midpoints in Example 3. Then plot points that represent the midpoint and frequency of each class and connect the points in order from left to right. Because the graph should begin and end on the horizontal axis, extend the left side to one class width before the first class midpoint and extend the right side to one class width after the last class midpoint. You can see that the frequency of subscribers increases up to 36.5 minutes and then decreases.

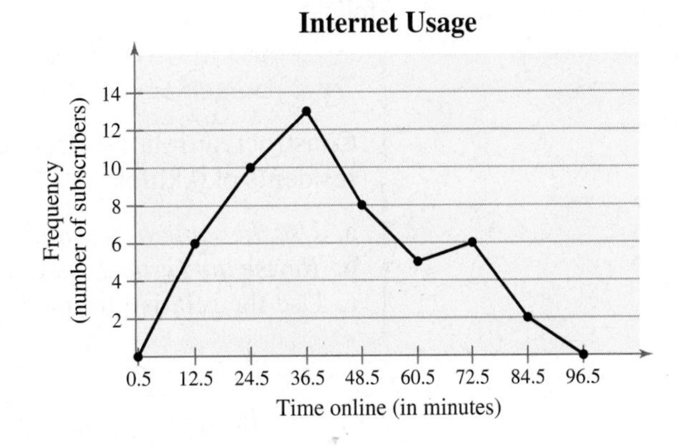

Study Tip

A histogram and its corresponding frequency polygon are often drawn together. If you have not already constructed the histogram, begin constructing the frequency polygon by choosing appropriate horizontal and vertical scales. The horizontal scale should consist of the class midpoints and the vertical scale should consist of appropriate frequency values.

Try It Yourself 4

Construct a frequency polygon that represents the ages of the residents of Akhiok. Describe any patterns.

a. Choose appropriate *horizontal and vertical scales.*
b. *Plot points* that represent the midpoint and frequency for each class.
c. *Connect the points* and extend the sides as necessary.
d. *Describe* any patterns for the data.

Answer: Page A30

Picturing the World

Old Faithful, a geyser at Yellowstone National Park, erupts on a regular basis. The time spans of a sample of eruptions are given in the relative frequency histogram. *(Source: Yellowstone National Park)*

Old Faithful Eruptions

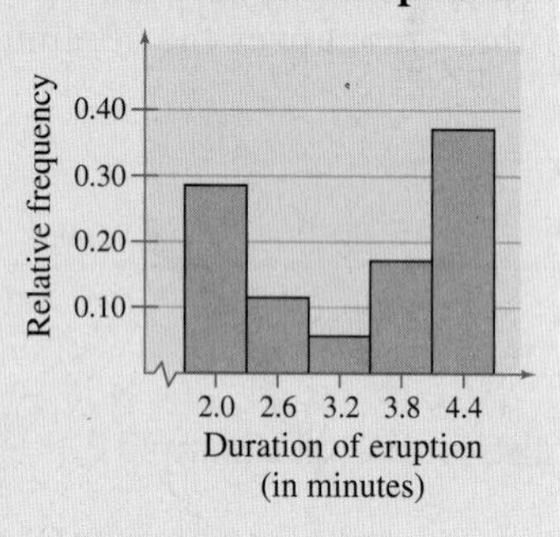

Fifty percent of the eruptions last less than how many minutes?

A **relative frequency histogram** has the same shape and the same horizontal scale as the corresponding frequency histogram. The difference is that the vertical scale measures the *relative* frequencies, not frequencies.

EXAMPLE 5

Constructing a Relative Frequency Histogram

Draw a relative frequency histogram for the frequency distribution in Example 2.

SOLUTION The relative frequency histogram is shown. Notice that the shape of the histogram is the same as the frequency histogram constructed in Example 3. The only difference is the vertical scale measures the relative frequencies. From this graph, you can quickly see that 0.20 or 20% of the Internet subscribers spent between 18.5 minutes and 30.5 minutes online, which is not as immediately obvious from the frequency histogram.

Internet Usage

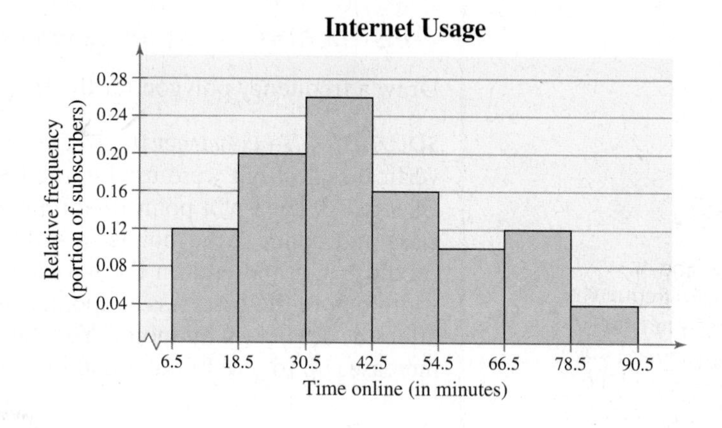

Try It Yourself 5

Construct a relative frequency histogram that represents the ages of the residents of Akhiok.

a. *Use the same horizontal scale* as used in the frequency histogram.
b. *Revise the vertical scale* to reflect relative frequencies.
c. Use the relative frequencies to *find the height of each bar.*

Answer: Page A30

If you want to describe the number of data entries that are above or below a certain value, then you can easily do so by constructing a cumulative frequency graph.

DEFINITION

A **cumulative frequency graph,** or **ogive** (pronounced ō′jīve), is a line graph that displays the cumulative frequency of each class at its upper class boundary. The upper boundaries are marked on the horizontal axis and the cumulative frequencies are marked on the vertical axis.

GUIDELINES

Constructing an Ogive (Cumulative Frequency Graph)

1. Construct a frequency distribution that includes cumulative frequencies as one of the columns.
2. Specify the horizontal and vertical scales. The horizontal scale consists of upper class boundaries and the vertical scale measures cumulative frequencies.
3. Plot points that represent the upper class boundaries and their corresponding cumulative frequencies.
4. Connect the points in order from left to right.
5. The graph should start at the lower boundary of the first class (cumulative frequency is zero) and should end at the upper boundary of the last class (cumulative frequency is equal to the sample size).

EXAMPLE 

Constructing an Ogive

Draw an ogive for the frequency distribution in Example 2. Estimate how many subscribers spent less than 60 minutes during their last session. Also, use the graph to estimate when the greatest increase in usage occurs.

SOLUTION Using the frequency distribution, you can construct the ogive shown. The upper class boundaries, frequencies, and the cumulative frequencies are listed in the table. Notice that the graph starts at 6.5, where the cumulative frequency is 0, and the graph ends at 90.5, where the cumulative frequency is 50.

Upper Class Boundaries	f	Cumulative Frequencies
18.5	6	6
30.5	10	16
42.5	13	29
54.5	8	37
66.5	5	42
78.5	6	48
90.5	2	50

Internet Usage

Cumulative frequency (number of subscribers)
50
40
30
20
10
6.5 18.5 30.5 42.5 54.5 66.5 78.5 90.5
Time online (in minutes)

From the ogive, you can see that about 40 subscribers spent less than 60 minutes online during their last session. The greatest increase in usage occurs between 30.5 minutes and 42.5 minutes because the line segment is steepest between these two class boundaries.

Another type of ogive uses percent as the vertical axis instead of frequency (see Example 5 in Section 2.5).

Try It Yourself 6

Construct an ogive that represents the ages of the residents of Akhiok. Estimate the number of residents who are less than 45 years old.

a. Specify the *horizontal* and *vertical scales.*
b. *Plot* the points given by the upper class boundaries and the cumulative frequencies.
c. *Construct* the graph.
d. *Estimate* the number of residents who are less than 45 years old.

Answer: Page A30

EXAMPLE 

Using Technology to Construct Histograms

Use a calculator or a computer to construct a histogram for the frequency distribution in Example 2.

SOLUTION *Minitab*, *Excel*, and the *TI-83* each have features for graphing histograms. Try using this technology to draw the histograms as shown.

MINITAB

Frequency
10
5
0
12.5 24.5 36.5 48.5 60.5 72.5 84.5
Minutes

EXCEL

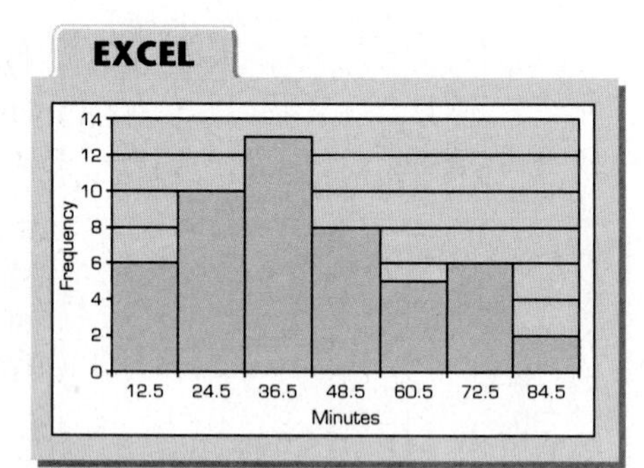

TI-83

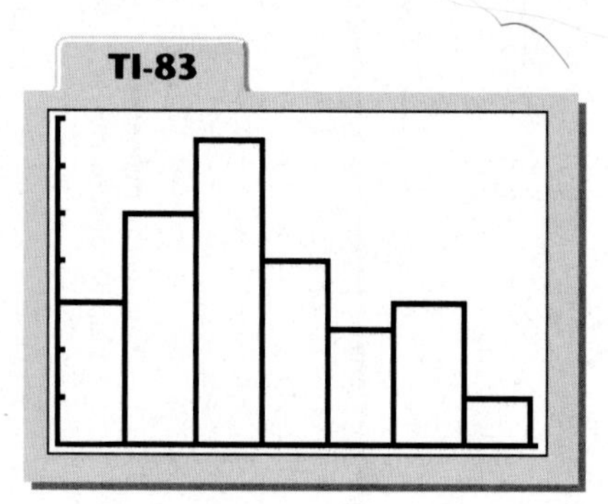

Study Tip

Detailed instructions for using *Minitab*, *Excel*, and the *TI-83* are shown in the Technology Guide that accompanies this text. For instance, here are instructions for creating a histogram on a *TI-83*.

STAT ENTER

Enter midpoints in L1.
Enter frequencies in L2.

2nd STATPLOT

Turn on Plot 1.
Highlight Histogram.

Xlist: L1
Freq: L2

ZOOM 9
WINDOW

Xscl=12

GRAPH

Try It Yourself 7

Use a calculator or a computer to construct a frequency histogram that represents the ages of the residents of Akhiok. Use six classes.

a. *Enter* the data.
b. *Construct* the histogram.

Answer: Page A30

2.1 Exercises

Help

LarsonTutor 2.1

Companion Web Site

Student Solutions Manual 2.1

Videos 2.1

Technology Manuals

Basic Skills and Concepts

1. What are some benefits of representing data sets using frequency distributions?

2. What are some benefits of representing data sets using graphs of frequency distributions?

True or False In Exercises 3 and 4, determine whether the statement is true or false. If it is false, rewrite it as a true statement.

3. The midpoint of a class is the sum of its lower and upper limits.

4. The relative frequency of a class is the sample size divided by the frequency of the class.

Reading a Frequency Distribution In Exercises 5 and 6, use the given frequency distribution to find the midpoint, relative frequency, and cumulative frequency for each class. Write your results as shown in the expanded frequency distribution in Example 2.

5. **Employee Age**

Class	Frequency, f
20–29	10
30–39	132
40–49	284
50–59	300
60–69	175
70–79	65
80–89	25

6. **Tree Height**

Class	Frequency, f
16–20	100
21–25	122
26–30	900
31–35	207
36–40	795
41–45	568
46–50	322

Graphical Analysis In Exercises 7 and 8, use the frequency histogram to

(a) estimate the frequency of the class with the least frequency.

(b) estimate the frequency of the class with the greatest frequency.

(c) determine the class width.

7. 

8. 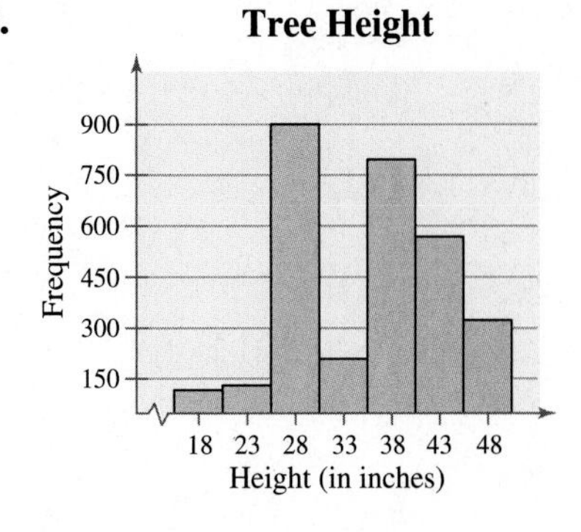

9. ***Weights of Monkeys*** ◆ Use the ogive to approximate (a) the number in the sample, (b) the location of the greatest increase in frequency, (c) the cumulative frequency for a weight of 14.5 pounds, and (d) the weight for which the cumulative frequency is 45.

Adult Male Rhesus Monkeys

Cumulative frequency

Weight (in pounds)

10. ***Heights of Males*** ◆ Use the ogive to approximate (a) the number in the sample, (b) the location of the greatest increase in frequency, (c) the cumulative frequency for a height of 74 inches, and (d) the height for which the cumulative frequency is 25.

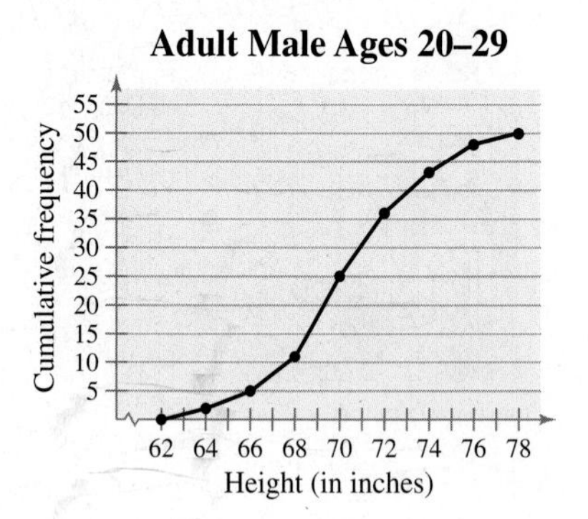

Graphical Analysis In Exercises 11 and 12, use the relative frequency histogram to

(a) identify the class with the greatest and the least relative frequency.

(b) approximate the greatest and least relative frequency.

(c) approximate the relative frequency of the second class.

11.

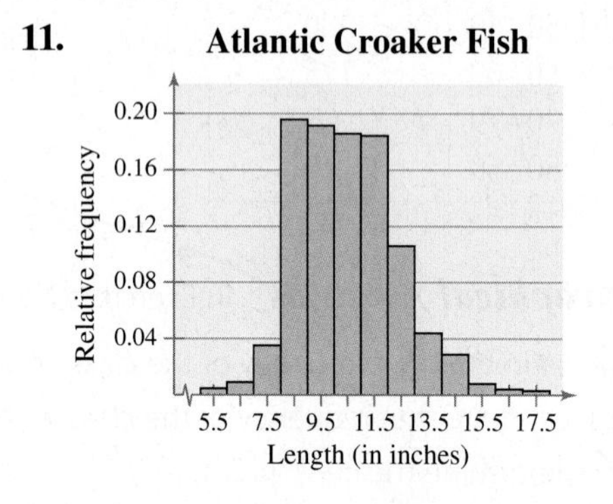

12.

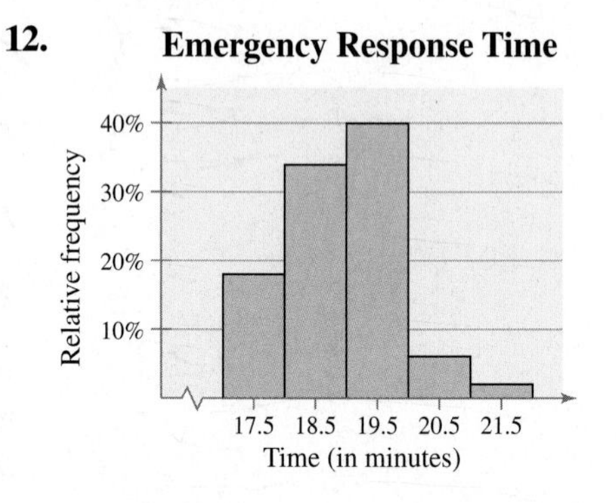

Graphical Analysis In Exercises 13 and 14, use the frequency polygon to identify the class with the greatest and the least frequency.

13.

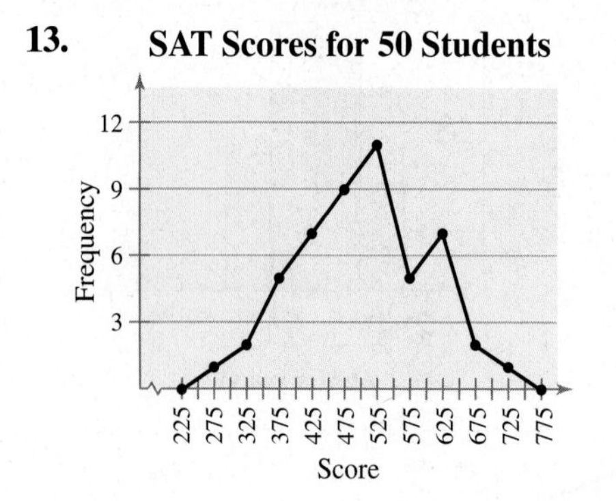

14.

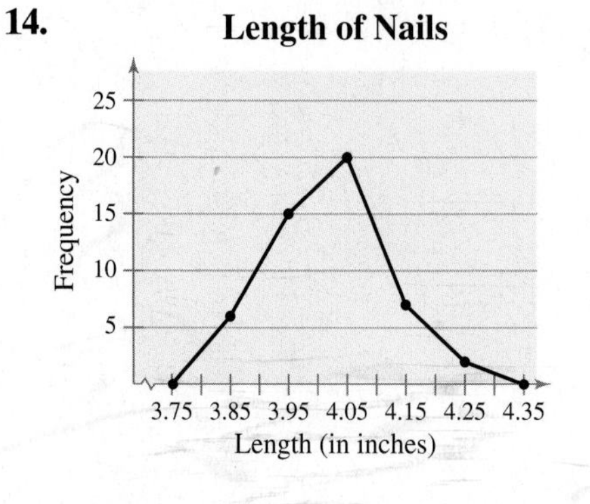

Constructing a Frequency Distribution In Exercises 15 and 16, construct a frequency distribution for the data set using the indicated number of classes. In the table, include the midpoints, relative frequencies, and cumulative frequencies.

 **15.** ***Newspaper Reading Times*** ◆

Number of classes: 5

Data set: Time (in minutes) spent reading the newspaper in a day

7 39 13 9 25 8 22 0 2 18 2 30 7
35 12 15 8 6 5 29 0 11 39 16 15

 **16.** ***Book Spending*** ◆

Number of classes: 6

Data set: Amount (in dollars) spent on books for a semester

91 472 279 249 530 376 188 341 266 199
142 273 189 130 489 266 248 101 375 486
190 398 188 269 43 30 127 354 84

Constructing a Frequency Distribution and a Frequency Histogram In Exercises 17–20, construct a frequency distribution and a frequency histogram for the data set using the indicated number of classes. Which class has the greatest frequency and which has the least frequency?

 **17.** ***Sales*** ◆

Number of classes: 6

Data set: July sales (in dollars) for all sales representatives at a company

2114 2468 7119 1876 4105 3183 1932 1355
4278 1030 2000 1077 5835 1512 1697 2478
3981 1643 1858 1500 4608 1000

18. ***Pepper Pungencies*** ◆

Number of classes: 5

Data set: Pungencies (in 1000s of Scoville units) of sixteen tabasco peppers

35 51 44 42 37 38 36 39 44 43 40 40 32 39 41 38

 **19.** ***Reaction Times*** ◆

Number of classes: 8

Data set: Reaction times (in milliseconds) of a sample of 22 adult females to an auditory stimulus

507 389 305 291 336 310 514 442 307 337 373
428 387 454 323 441 388 426 469 351 411 382

 **20.** ***Fracture Times*** ◆

Number of classes: 5

Data set: Amount of pressure (in pounds per square inch) at fracture time for 25 samples of brick mortar

2750 2862 2885 2490 2512 2456 2554 2532 2885
2872 2601 2877 2721 2692 2888 2755 2853 2517
2867 2718 2641 2834 2466 2596 2519

 indicates that the data set for this exercise is available electronically.

Constructing a Frequency Distribution and a Relative Frequency Histogram In Exercises 21–24, construct a frequency distribution and a relative frequency histogram for the data set using five classes. Which class has the greatest relative frequency and which has the least relative frequency?

21. *Bowling Scores* ◆

Data set: Bowling scores of a sample of league members

154 257 195 220 182 240 177 228 235
146 174 192 165 207 185 180 264 169
225 239 148 190 182 205 148 188

22. *ATM Withdrawals* ◆

Data set: A sample of ATM withdrawals (in dollars)

35 10 30 25 75 10 30 20
20 10 40 50 40 30 60 70
25 40 10 60 20 80 40 25
20 10 20 25 30 50 80 20

23. *Tree Heights* ◆

Data set: Heights (in feet) of a sample of Douglas fir trees

40 44 35 49 35 43 35 36 39 37 41 41 48
52 37 45 40 36 35 50 42 51 33 34 51 39

24. *Farm Acreage* ◆

Data set: Number of acres on a sample of small farms

12 7 9 8 9 8 12 10 9 16 8 13 12
10 11 7 14 12 9 8 10 9 11 13 8

Constructing a Cumulative Frequency Distribution and an Ogive
In Exercises 25–28, construct a cumulative frequency distribution and an ogive for the data set using six classes. Then describe the location of the greatest increase in frequency.

25. *Retirement Ages* ◆

Data set: Retirement ages for a sample of engineers

60 65 68 63 66 67 69 67 58 65 67 61
63 65 62 64 73 50 61 71 62 69 72 63

26. *Saturated Fat Intakes* ◆

Data set: Daily saturated fat intakes (in grams) of a sample of people

38 32 34 39 40 54 32 17 29 33
57 40 25 36 33 24 42 16 31 33

27. *Gasoline Purchases* ◆

Data set: Gasoline (in gallons) purchased by a sample of drivers during one fill-up

7 4 18 4 9 8 8 7 6 2 9 5 9 12
4 14 15 7 10 2 3 11 4 4 9 12 5 3

28. *Long-Distance Phone Calls* ◆

Data set: Lengths (in minutes) of a sample of long-distance phone calls

1 20 10 20 13 23 3 7 18 7 4 5
15 7 29 10 18 10 10 23 4 12 8 6

Constructing a Frequency Distribution and a Frequency Polygon

In Exercises 29 and 30, construct a frequency distribution and a frequency polygon for the data set. Then determine which class has the greatest frequency and which has the least frequency.

29. Number of classes: 5

Data set: Exam scores for all students in a statistics class

83 92 94 82 73 98 78 85 72 90 89 92 96 89 75 85 63 47 75 82

30. Number of classes: 6

Data set: Number of children of the U.S. presidents *(Source: The World Almanac and Book of Facts 1998 and infoplease.com)*

0 5 6 0 2 4 0 4 10 14 0 6 2 3 0 4 5 4 8 5
3 5 3 2 6 3 3 0 2 2 5 1 2 2 4 4 4 6 1 2

31. ***What Would You Do?*** ◆ You work at a bank and are asked to recommend the amount of cash to put in an ATM each day. You don't want to put in too much (security) or too little (customer irritation). Here are the daily withdrawals (in 100s of dollars) for a period of 30 days.

72 84 61 76 104 76 86 92 80 88 98 76 97 82 84
67 70 81 82 89 74 73 86 81 85 78 82 80 91 83

(a) Construct a relative frequency histogram for the data, using eight classes.

(b) If you are willing to run out of cash for 10% of the days, how much cash, in hundreds of dollars, should you put in the ATM each day? Explain your reasoning.

(c) If you put $9000 in the ATM each day, what percent of the days in a month should you expect to run out of cash? Explain your reasoning.

32. ***What Would You Do?*** ◆ You work in the admissions department for a college and are asked to recommend the minimum SAT scores that the college will accept for a position as a full-time student. Here are the SAT scores for a sample of 50 applicants.

1325 1072 982 996 872 849 785 706 669 1049 885 1367 935
980 1188 869 1006 1127 979 1034 1052 1165 1359 667 1264 727
808 955 544 1202 1051 1173 410 1148 1195 1141 1193 768 812
887 1211 1266 830 672 917 988 791 1035 688 700

(a) Construct a relative frequency histogram for the data, using 10 classes.

(b) If you want to accept the top 88% of the applicants, what should the minimum score be?

(c) If you set the minimum score at 986, what percent of the applicants will you be accepting?

Extending the Basics

33. ***Writing*** ◆ What happens when the number of classes is increased for a frequency histogram? Use the data set listed and create frequency histograms with 5, 10, and 20 classes. Which graph displays the data best?

2 7 3 2 11 3 15 8 4 9 10 13 9
7 11 10 1 2 12 5 6 4 2 9 15

2.2 More Graphs and Displays

What You Should Learn

- How to graph and interpret quantitative data sets using stem-and-leaf plots and dot plots
- How to graph and interpret qualitative data sets using pie charts and Pareto charts
- How to graph and interpret paired data sets using scatter plots and time series charts

Graphing Quantitative Data Sets • Graphing Qualitative Data Sets • Graphing Paired Data Sets

Graphing Quantitative Data Sets

In Section 2.1, you learned several traditional ways to display quantitative data graphically. In this section, you will learn a newer way to display quantitative data, called a **stem-and-leaf plot.** Stem-and-leaf plots are examples of **exploratory data analysis (EDA),** which was developed by John Tukey in 1977.

In a stem-and-leaf plot, each number is separated into a **stem** (for instance, the entry's leftmost digits) and a **leaf** (for instance, the rightmost digit). A stem-and-leaf plot is similar to a histogram but has the advantage that the graph still contains the original data values. Another advantage of a stem-and-leaf plot is that it provides an easy way to sort data.

EXAMPLE 

Constructing a Stem-and-Leaf Plot

The following are the numbers of league-leading runs batted in (RBIs) for baseball's American League during a recent 50-year period. Display the data in a stem-and-leaf plot. What can you conclude? *(Source: Major League Baseball)*

155 159 144 129 105 145 126 116 130 114 122 112 112 142 126
118 118 108 122 121 109 140 126 119 113 117 118 109 109 119
139 139 122 78 133 126 123 145 121 134 124 119 132 133 124
129 112 126 148 147

SOLUTION Because the data entries go from a low of 78 to a high of 159, you should use stem values from 7 to 15. To construct the plot, list these stems to the left of a vertical line. For each data entry, list a leaf to the right of its stem. For instance, the entry 155 has a stem of 15 and a leaf of 5. The resulting stem-and-leaf plot will be unordered. To obtain an ordered stem-and-leaf plot, rewrite the plot with the leaves in increasing order from left to right. It is important to include a key for the display to identify the values of the data.

RBIs for American League Leaders

Key: 15|5 = 155

Stem	Leaves
7	8
8	
9	
10	5 8 9 9 9
11	6 4 2 2 8 8 9 3 7 8 9 9 2
12	9 6 2 6 2 1 6 2 6 3 1 4 4 9 6
13	0 9 9 3 4 2 3
14	4 5 2 0 5 8 7
15	5 9

Unordered Stem-and-Leaf Plot

RBIs for American League Leaders

Key: 15|5 = 155

Stem	Leaves
7	8
8	
9	
10	5 8 9 9 9
11	2 2 2 3 4 6 7 8 8 8 9 9 9
12	1 1 2 2 2 3 4 4 6 6 6 6 6 9 9
13	0 2 3 3 4 9 9
14	0 2 4 5 5 7 8
15	5 9

Ordered Stem-and-Leaf Plot

From the ordered stem-and-leaf plot, you can conclude that more than 50% of the RBI leaders had between 110 and 130 RBIs.

Study Tip

In a stem-and-leaf plot, you should have as many leaves as there are entries in the original data set.

Try It Yourself 1

Use a stem-and-leaf plot to organize the Akhiok population data set listed on page 30. What can you conclude?

a. List all possible *stems.*
b. List the leaf of each data entry to the right of its stem and include a key.
c. Rewrite the stem-and-leaf plot so that the leaves are ordered.
d. Use the plot to make a conclusion.

Answer: Page A30

EXAMPLE 2

Constructing Variations of Stem-and-Leaf Plots

Organize the data given in Example 1 using a stem-and-leaf plot that has two lines for each stem. What can you conclude?

SOLUTION Construct the stem-and-leaf plot as described in Example 1, except now list each stem twice. Use the leaves 0, 1, 2, 3, and 4 in the first stem row and the leaves 5, 6, 7, 8, and 9 in the second stem row. The revised stem-and-leaf plot is shown. From the display, you can conclude that most of the RBI leaders had between 105 and 135 RBIs.

RBIs for American League Leaders

Key: 15|5 = 155

Stem	Leaves
7	
7	8
8	
8	
9	
9	
10	
10	5 8 9 9 9
11	4 2 2 3 2
11	6 8 8 9 7 8 9 9
12	2 2 1 2 3 1 4 4
12	9 6 6 6 6 9 6
13	0 3 4 2 3
13	9 9
14	4 2 0
14	5 5 8 7
15	
15	5 9

Unordered Stem-and-Leaf Plot

RBIs for American League Leaders

Key: 15|5 = 155

Stem	Leaves
7	
7	8
8	
8	
9	
9	
10	
10	5 8 9 9 9
11	2 2 2 3 4
11	6 7 8 8 8 9 9 9
12	1 1 2 2 2 3 4 4
12	6 6 6 6 6 9 9
13	0 2 3 3 4
13	9 9
14	0 2 4
14	5 5 7 8
15	
15	5 9

Ordered Stem-and-Leaf Plot

Insight

Compare Examples 1 and 2. Notice that by using two lines per stem, you obtain a more detailed picture of the data.

Try It Yourself 2

Using two rows for each stem, revise the stem-and-leaf plot you constructed in Try It Yourself 1.

a. List each stem *twice.*
b. List all leaves *using the appropriate stem row.*

Answer: Page A30

You can also use a dot plot to graph quantitative data. In a **dot plot,** each data entry is plotted, using a point, above a horizontal axis. Like a stem-and-leaf plot, a dot plot allows you to see how data are distributed and to determine specific data entries.

EXAMPLE 3

Constructing a Dot Plot

Use a dot plot to organize the RBI data given in Example 1.

155 159 144 129 105 145 126 116 130
114 122 112 112 142 126 118 118 108
122 121 109 140 126 119 113 117 118
109 109 119 139 139 122 78 133 126
123 145 121 134 124 119 132 133 124
129 112 126 148 147

SOLUTION So that each data entry is included in the dot plot, the horizontal axis should include numbers between 70 and 160. To represent a data entry, plot a point above the entry's position on the axis. If an entry is repeated, plot another point above the previous point.

RBIs for American League Leaders

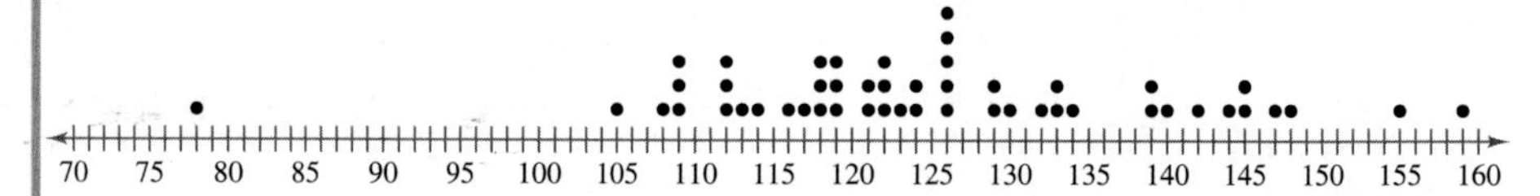

From the dot plot, you can see that most values cluster between 105 and 148 and the value that occurs the most is 126.

Try It Yourself 3

Use a dot plot to organize the data listed in the chapter opener on page 30. What can you conclude from the graph?

a. Choose an appropriate scale for the *horizontal axis.*
b. Represent each data entry by *plotting a point.*
c. *Describe* any patterns for the data.

Answer: Page A30

Technology can be used to construct stem-and-leaf plots and dot plots. For instance, a *Minitab* dot plot for the RBI data is shown.

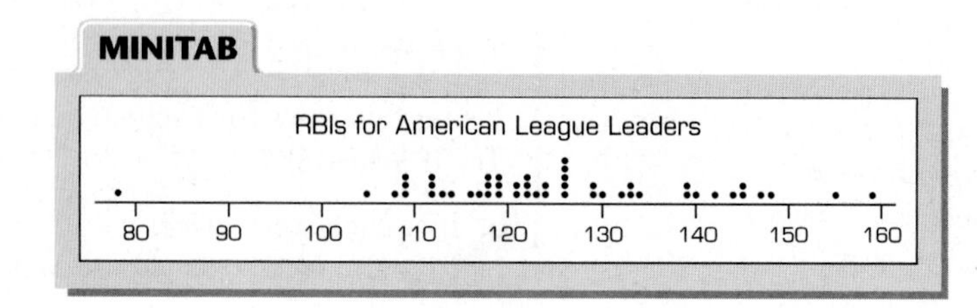

Graphing Paired Data Sets

If two data sets have the same number of entries, and each entry in the first data set corresponds to one entry in the second data set, the sets are called **paired data sets.** For instance, suppose a data set contains the costs of an item and a second data set contains sales amounts for the item at each cost. Because each cost corresponds to a sales amount, the data sets are paired. One way to graph paired data sets is to use a **scatter plot,** where the ordered pairs are graphed as points in a coordinate plane.

EXAMPLE 

Interpreting a Scatter Plot

The British statistician Ronald Fisher (see page 292) introduced a famous data set called Fisher's Iris data set. This data set describes various physical characteristics, such as petal length and petal width, for three species of iris. The petal lengths and petal widths for each species are graphed in the scatter plot. As the petal length increases, what tends to happen to the petal width? *(Source: Fisher, R. A., 1936)*

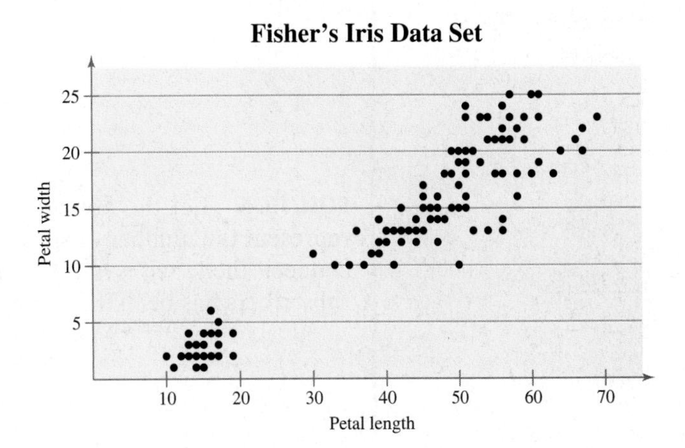

SOLUTION The horizontal axis represents the petal length and the vertical axis represents the petal width. Each point in the scatter plot represents the petal length and petal width of one flower. From the scatter plot, you can see that for each of the three species, as the petal length increases, the petal width also tends to increase.

Try It Yourself 6

The lengths of employment and the salaries of 10 employees are listed in the table at the left. Graph the data using a scatter plot. What can you conclude?

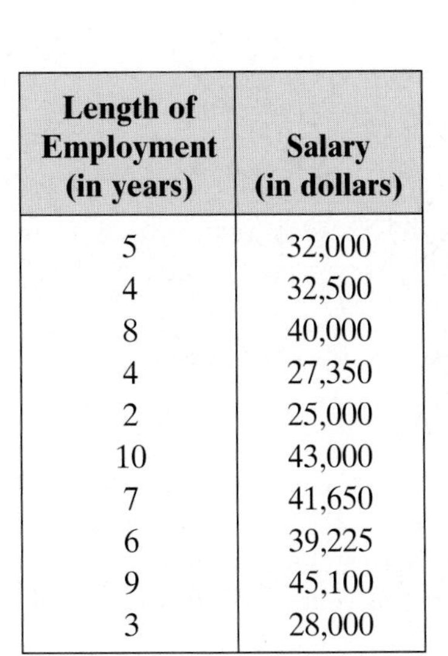

Length of Employment (in years)	Salary (in dollars)
5	32,000
4	32,500
8	40,000
4	27,350
2	25,000
10	43,000
7	41,650
6	39,225
9	45,100
3	28,000

a. Label the *horizontal and vertical axes.*
b. *Plot* the paired data.
c. Describe any trends.

Answer: Page A31

You will learn more about scatter plots and how to analyze them in Chapter 9.

A data set that is composed of entries taken at regular intervals over a period of time is a **time series.** For instance, the amount of precipitation measured each day for one month is an example of a time series. You can use a **time series chart** to graph a time series.

See *Minitab* and *TI-83* steps on pages 98, 99.

EXAMPLE 

Constructing a Time Series Chart

The table lists the number of cellular telephone subscribers, in millions, and a subscriber's average local monthly bill for service, in dollars, for the years 1987 through 1999. Construct a time series chart for the number of cellular subscribers. What can you conclude? *(Source: Cellular Telecommunications Industry Association)*

Year	Subscribers (in millions)	Average Bill (in dollars)
1987	1.2	96.83
1988	2.1	98.02
1989	3.5	89.30
1990	5.3	80.90
1991	7.6	72.74
1992	11.0	68.68
1993	16.0	61.48
1994	24.1	56.21
1995	33.8	51.00
1996	44.0	47.70
1997	55.3	42.78
1998	69.2	39.43
1999	86.0	41.24

SOLUTION Let the horizontal axis represent the years and the vertical axis represent the number of subscribers, in millions. Then plot the paired data and connect them with line segments. The graph shows that the number of subscribers has been increasing since 1987, with greater increases recently.

Cellular Telephone Subscribers

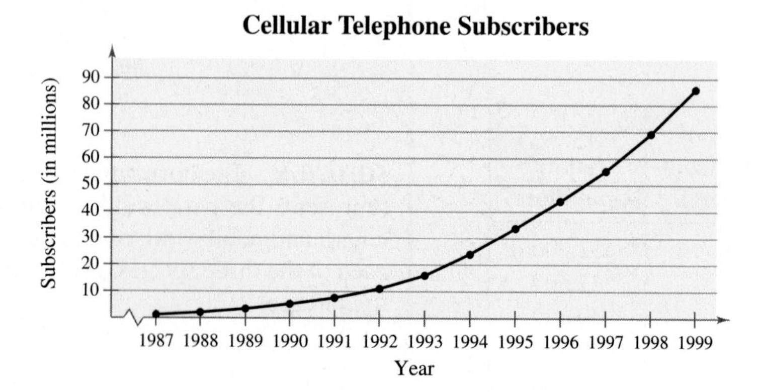

Try It Yourself 7

Use the table in Example 7 to construct a time series chart for a subscriber's average local monthly cellular telephone bill for the years 1987 through 1999. What can you conclude?

a. Label the *horizontal and vertical axes.*
b. *Plot* the paired data and *connect* them with line segments.
c. *Describe* any patterns you see.

Answer: Page A31

2.2 Exercises

Help

LarsonTutor 2.2

Companion Web Site

Student Solutions Manual 2.2

Videos 2.2

Technology Manuals

Basic Skills and Concepts

1. Name some ways to display quantitative data graphically. Name some ways to display qualitative graphically.

2. What is an advantage of using a stem-and-leaf plot instead of a histogram? What is a disadvantage?

Putting Graphs in Context In Exercises 3–6, match the plot with the description of the sample.

3. Key: 2|8 = 28

Stem	Leaves
2	8 9
3	2 2 2 3 4 5 7 7 8 9
4	0 2 4 5
5	1
6	5 6
7	2

4. Key: 6|7 = 67

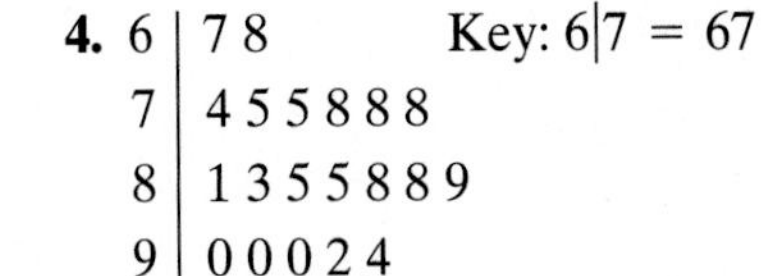

Stem	Leaves
6	7 8
7	4 5 5 8 8 8
8	1 3 5 5 8 8 9
9	0 0 0 2 4

5.

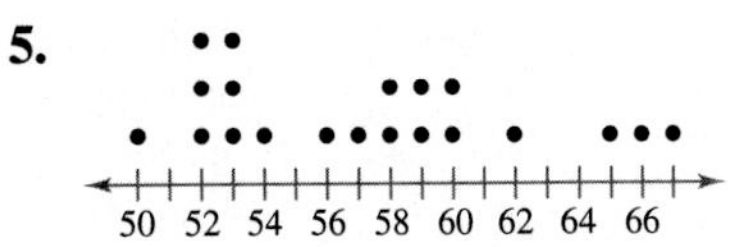

6.

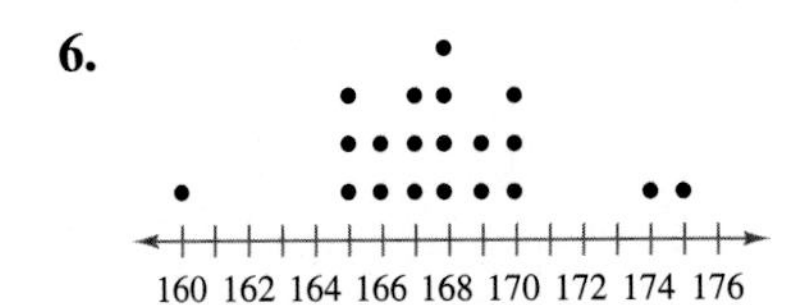

(a) Prices (in dollars) of a sample of 20 brands of jeans
(b) Weights (in pounds) of a sample of 20 first grade students
(c) Volumes (in cubic centimeters) of a sample of 20 oranges
(d) Ages (in years) of a sample of 20 residents of a retirement home

Graphical Analysis In Exercises 7–10, use the stem-and-leaf plot or dot plot to list the actual data entries. What is the maximum data entry? What is the minimum data entry?

7. Key: 2|7 = 27

Stem	Leaves
2	7
3	2
4	1 3 3 4 7 7 8
5	0 1 1 2 3 3 3 4 4 4 4 5 6 6 8 9
6	8 8 8
7	3 8 8
8	5

8. Key: 12|9 = 12.9

Stem	Leaves
12	
12	9
13	3
13	6 7 7
14	1 1 1 1 3 4 4
14	6 9 9
15	0 0 0 1 2 4
15	6 7 8 8 8 9
16	1
16	6 7

9.

13 14 15 16 17 18 19

10.

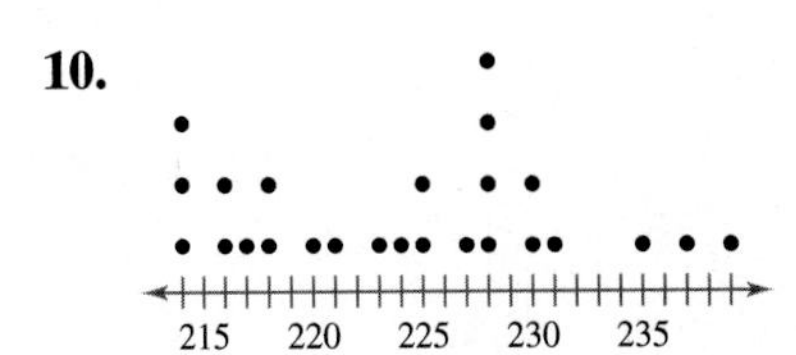

Graphical Analysis In Exercises 11–14, what can you conclude from the graph?

11. **Top 5 Sports Advertisers**

Advertising (in millions of dollars)

40
30
20
10

Anheuser-Busch
Nike
Coors
Chevrolet
Ford

Company

(Source: Nielsen Media Research)

12. **Stock Portfolio**

Value (in dollars)

30,000
20,000
10,000

1996 1997 1998 1999 2000

Year

13. **How Other Drivers Irk Us**

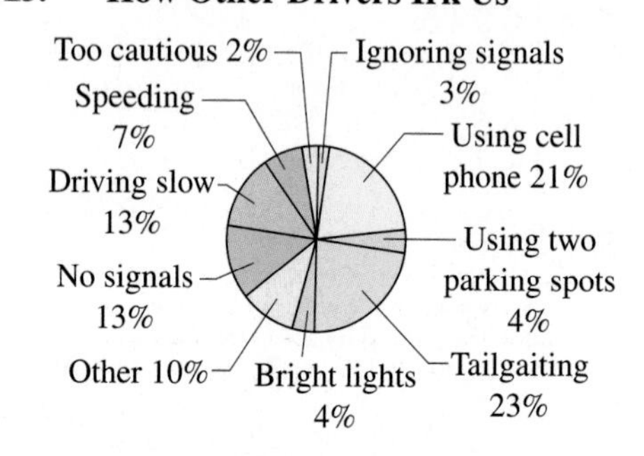

(Adapted from Reuters/Zogby)

14. **Driving and Cell Phone Use**

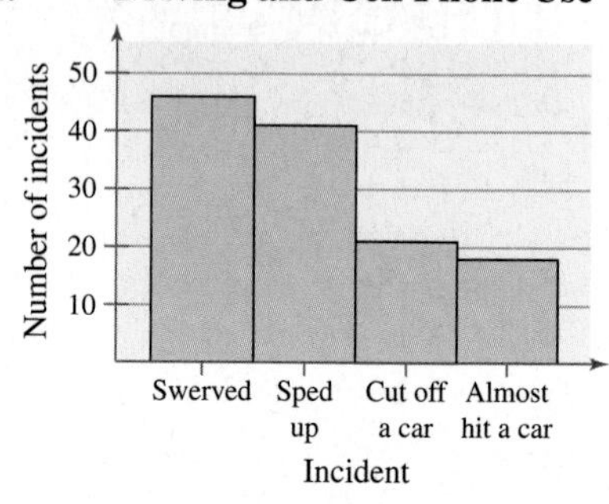

(Adapted from USA TODAY)

Graphing Data Sets In Exercises 15–28, organize the data using the indicated type of graph. What can you conclude about the data?

15. ***Elephants: Water Consumed*** ◆ Use a stem-and-leaf plot to display the data. The data represent the amount of water (in gallons) consumed by 24 elephants in one day.

33 45 34 47 43 48 35 69 45 60 46 51
41 60 66 41 32 40 44 39 46 33 53 53

16. ***Elephants: Hay Eaten*** ◆ Use a stem-and-leaf plot to display the data. The data represent the amount of hay (in pounds) eaten daily by 24 elephants.

449 450 419 448 479 410 446 465 415 455 345 305
491 479 390 393 403 298 503 327 460 351 409 319

17. ***Apple Prices*** ◆ Use a stem-and-leaf plot to display the data. The data represent the price (in cents per pound) paid to 28 farmers for apples.

19.2 19.6 16.4 17.1 19.0 17.4 17.3 20.1 19.0 17.5
17.6 18.6 18.4 17.7 19.5 18.4 18.9 17.5 19.3 20.8
19.3 18.6 18.6 18.3 17.1 18.1 16.8 17.9

18. ***Advertisements*** ◆ Use a dot plot to display the data. The data represent the number of advertisements seen or heard in one week by a sample of 30 people from the U.S.

598 494 441 595 728 690 684 486 735 808 734 590 673 545 702
481 298 135 846 764 317 649 732 582 637 588 540 727 486 703

19. *Life Spans of Houseflies* ◆ Use a dot plot to display the data. The data represent the life span (in days) of 40 houseflies.

9	9	4	4	8	11	10	5	8	13
9	6	7	11	13	11	6	9	8	14
10	6	10	10	8	7	14	11	7	8
6	11	13	10	14	14	8	13	14	10

20. *Nobel Prize* ◆ Use a pie chart to display the data. The data represent the number of Nobel Prize laureates by country during the years 1901–1993.

United States	170	France	24	Germany	59
United Kingdom	69	USSR	10	Other	88

21. *NASA Budget* ◆ Use a pie chart to display the data. The data represent the 2001 NASA budget (in millions of dollars) divided among four categories. *(Source: NASA)*

Human space flight	5499.9
Science, aeronautics, and technology	5929.4
Mission support	2584.0
Inspector General	22.0

22. *NASA Expenditures* ◆ Use a Pareto chart to display the data. The data represent the 2001 NASA space shuttle operations expenditures (in millions of dollars). *(Source: NASA)*

Orbiter and integration	724.5	Reusable solid rocket motor	418.3
External tank	349.7	Solid rocket booster	137.5
Main engine	261.9	Mission/launch operations	780.9

23. *UV Index* ◆ Use a Pareto chart to display the data. The data represent the ultraviolet index for five cities at noon on June 14, 2001. *(Source: National Oceanic and Atmospheric Administration)*

Atlanta, GA	Boise, ID	Concord, NH	Denver, CO	Miami, FL
9	7	8	7	10

24. *Hourly Wages* ◆ Use a scatter plot to display the data in the table. The data represent the hours worked and the hourly wage (in dollars) for a sample of 12 production workers. Describe any trends shown.

Hours	Hourly Wage
33	12.16
37	9.98
34	10.79
40	11.71
35	11.80
33	11.51
40	13.65
33	12.05
28	10.54
45	10.33
37	11.57
28	10.17

Number of Students per Teacher	Average Teacher's Salary
17.1	28.7
17.5	47.5
18.9	31.8
17.1	28.1
20.0	40.3
18.6	33.8
14.4	49.8
16.5	37.5
13.3	42.5
18.4	31.9

Table for Exercise 25

25. ***Salaries*** ◆ Use a scatter plot to display the data given in the table. The data represent the number of students per teacher and the average teacher salary (in thousands of dollars) for a sample of 10 school districts. Describe any trends shown.

26. ***UV Index*** ◆ Use a time series chart to display the data. The data represent the ultraviolet index for Memphis, TN on June 14–23, 2001. *(Source: Weather Services International)*

June 14	June 15	June 16	June 17	June 18
9	4	10	10	10

June 19	June 20	June 21	June 22	June 23
10	10	10	9	9

27. ***Egg Prices*** ◆ Use a time series chart to display the data. The data represent the prices of Grade A eggs (in dollars per dozen) for the indicated years. *(Source: U.S. Bureau of Labor Statistics)*

1990	1991	1992	1993	1994
1.00	1.01	0.93	0.87	0.87

1995	1996	1997	1998	1999
1.16	1.31	1.17	1.09	0.92

28. ***T-Bone Steak Prices*** ◆ Use a time series chart to display the data. The data represent the prices of T-bone steak (in dollars per pound) for the indicated years. *(Source: U.S. Bureau of Labor Statistics)*

1990	1991	1992	1993	1994
5.45	5.21	5.39	5.77	5.86

1995	1996	1997	1998	1999
5.92	5.87	6.07	6.40	6.71

Extending the Basics

A Misleading Graph? In Exercises 29 and 30,

(a) explain why the graph is misleading.

(b) redraw the graph so that it is not misleading.

29.

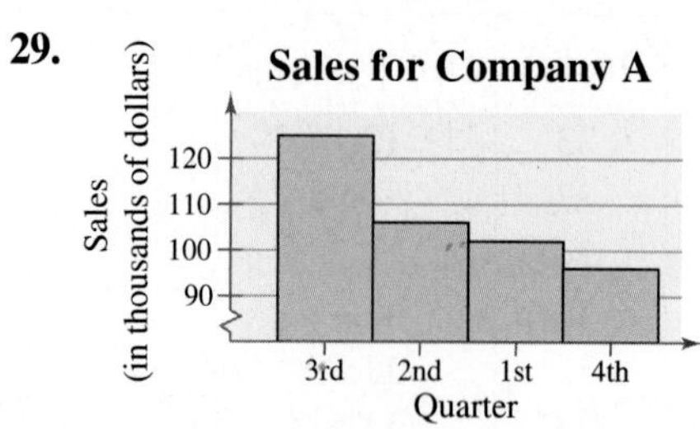

30.

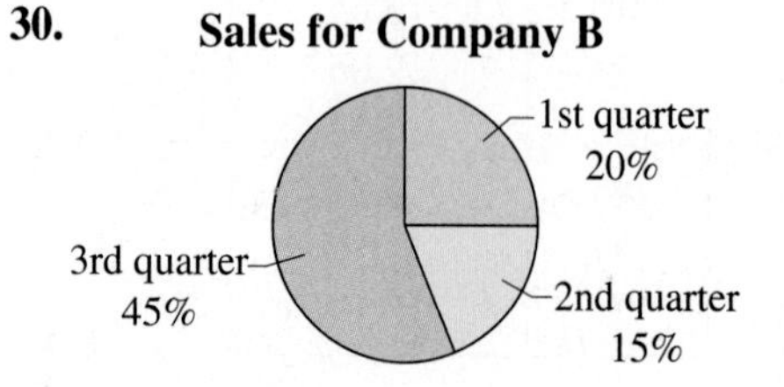

1st quarter	2nd quarter	3rd quarter	4th quarter
20%	15%	45%	20%

2.3 Measures of Central Tendency

Mean, Median, and Mode • Weighted Mean and Mean of Grouped Data • The Shape of Distributions

What You Should Learn

- How to find the mean, median, and mode of a population and a sample
- How to find a weighted mean and the mean of a frequency distribution
- How to describe the shape of a distribution as symmetric, uniform, or skewed and how to compare the mean and median for each

Mean, Median, and Mode

A **measure of central tendency** is a value that represents a typical, or central, entry of a data set. The three most commonly used measures of central tendency are the mean, the median, and the mode.

DEFINITION

The **mean** of a data set is the sum of the data entries divided by the number of entries. To find the mean of a data set, use one of the following formulas.

Population Mean: $\mu = \dfrac{\Sigma x}{N}$ Sample Mean: $\bar{x} = \dfrac{\Sigma x}{n}$

EXAMPLE 1

Finding a Sample Mean

The prices, in dollars, for a sample of room air conditioners (10,000 Btu/hr) are listed. What is the mean price of the air conditioners? *(Source: Consumer Reports)*

500 840 470 480 420 440 440

SOLUTION The sum of the air conditioner prices is

$$\Sigma x = 500 + 840 + 470 + 480 + 420 + 440 + 440$$
$$= 3590.$$

To find the mean price, divide the sum of the prices by the number of prices in the sample.

$$\bar{x} = \frac{\Sigma x}{n}$$
$$= \frac{3590}{7}$$
$$\approx 512.86$$

So, the mean price for one of the air conditioners is about \$513.

Try It Yourself 1

Find the mean age of the residents of Akhiok. Use the population data set given in the chapter opener on page 30.

a. *Find the sum* of the data entries.
b. *Divide the sum* by the number of data entries.
c. *Interpret the results* in the context of the data.

Answer: Page A31

Study Tip

Symbol	Description
Σ	The uppercase Greek letter sigma; indicates a summation of values
x	A variable that represents quantitative data entries
N	Number of entries in a population
n	Number of entries in a sample
μ	The lowercase Greek letter mu; the population mean
$\bar{x}$	Read as "x bar;" the sample mean

DEFINITION

The **median** of a data set is the middle data entry when the data set is sorted in ascending or descending order. If the data set has an even number of entries, the median is the mean of the two middle data entries.

EXAMPLE 

Finding the Median

Find the median of the air conditioner prices given in Example 1.

SOLUTION To find the median price, first order the data.

420 440 440 470 480 500 840

Because there are seven entries (an odd number), the median is the middle, or fourth, data entry. So, the median air conditioner price is \$470.

Study Tip

In a data set, there are an equal number of data values above the median as there are below the median. For instance, in Example 2, three of the prices are below \$470 and three are above \$470.

Try It Yourself 2

Find the median age of the residents of Akhiok.

a. *Order* the data entries.
b. Find the *middle data entry.*
c. *Interpret the results* in the context of the data.

Answer: Page A31

EXAMPLE 

Finding the Median

The air conditioner priced at \$480 is discontinued. What is the median price of the remaining air conditioners?

SOLUTION The remaining prices, in order, are

420, 440, 440, 470, 500, and 840.

Because there are six entries (an even number), the median is the mean of the two middle entries.

$$\text{Median} = \frac{440 + 470}{2} = 455$$

So, the median price of the remaining air conditioners is \$455.

Akhiok, Alaska, is a fishing village on Kodiak Island.
(Photograph © Roy Corral.)

Try It Yourself 3

Suppose that one of the families of Akhiok relocates to another city. The ages of the family members are 33, 34, 11, 6, and 56. What is the median age of the remaining residents?

a. *Order* the data entries.
b. *Find the mean* of the two middle data entries.

Answer: Page A31

DEFINITION

The **mode** of a data set is the data entry that occurs with the greatest frequency. If no entry is repeated, the data set has no mode. If two entries occur with the same greatest frequency, each entry is a mode and the data set is called **bimodal.**

EXAMPLE 4

Finding the Mode

Find the mode of the air conditioner prices given in Example 1.

SOLUTION To find the mode, it helps to order the data.

420 440 440 470 480 500 840

From the ordered data, you can see that the entry of 440 occurs twice while the other data entries occur only once. So, the mode of the air conditioner prices is $440.

Insight

The mode is the only measure of central tendency that can be used to describe data at the nominal level of measurement.

Try It Yourself 4

Find the mode of the ages of the Akhiok residents. The data is given below.

28, 6, 17, 48, 63, 47, 27, 21, 3, 7, 12, 39, 50, 54, 33, 45, 15, 24, 1,
7, 36, 53, 46, 27, 5, 10, 32, 50, 52, 11, 42, 22, 3, 17, 34, 56, 25, 2,
30, 10, 33, 1, 49, 13, 16, 8, 31, 21, 6, 9, 2, 11, 32, 25, 0, 55, 23, 41,
29, 4, 51, 1, 6, 31, 5, 5, 11, 4, 10, 26, 12, 6, 16, 8, 2, 4, 28

a. Find the *frequency* of each data entry.
b. Identify the entry, or entries, that occur with the *greatest frequency.*
c. *Interpret the results* in the context of the data.

Answer: Page A31

EXAMPLE 5

Finding the Mode

At a political debate a sample of audience members was asked to name the political party to which they belong. Their responses are summarized in the table at the left. What is the mode of the responses?

Political Party	Frequency, f
Democrat	34
Republican	56
Other	21
Did not respond	9

SOLUTION The response occurring with the greatest frequency is "Republican." So, the mode is "Republican," which means that, in this sample, there were more Republicans than people of any other single affiliation.

Try It Yourself 5

In a survey, 250 baseball fans were asked if Barry Bonds's home run record would ever be broken. One hundred sixty-nine of the fans responded "yes," 54 responded "no," and 27 "didn't know." What is the mode of the responses?

a. Identify the entry that occurs with the *greatest frequency.*
b. *Interpret the results* in the context of the data.

Answer: Page A31

Insight

The sample mean is the most commonly used measure of central tendency for inferential statistics.

While the mean, the median, and the mode each describe a typical entry of a data set, there are advantages and disadvantages of using each, especially when the data set contains outliers. An **outlier** is a data entry that is far removed from the other entries in the data set.

EXAMPLE 6

Comparing the Mean, the Median, and the Mode

Find the mean, the median, and the mode of the following sample ages of a class. Which measure of central tendency best describes a typical entry?

20 20 20 20 20 20 21 21 21 21
22 22 22 23 23 23 23 24 24 65 ← Outlier

SOLUTION

Mean: $\bar{x} = \frac{\Sigma x}{n} = \frac{475}{20} = 23.75$ years

Median: $\text{Median} = \frac{21 + 22}{2} = 21.5$ years

Mode: The entry occurring with the greatest frequency is 20 years.

There is no right answer to the question "Which measure of central tendency best describes a typical data entry?" The mean takes every entry into account but is influenced by the outlier of 65. The median also takes every entry into account and it is not affected by the outlier. In this case the mode exists, but it doesn't appear to represent a typical entry. Sometimes a graphical comparison can help you decide which measure of central tendency best represents a data set. The histogram shows the distribution of the data and the location of the mean, the median, and the mode.

Ages of Students in a Class

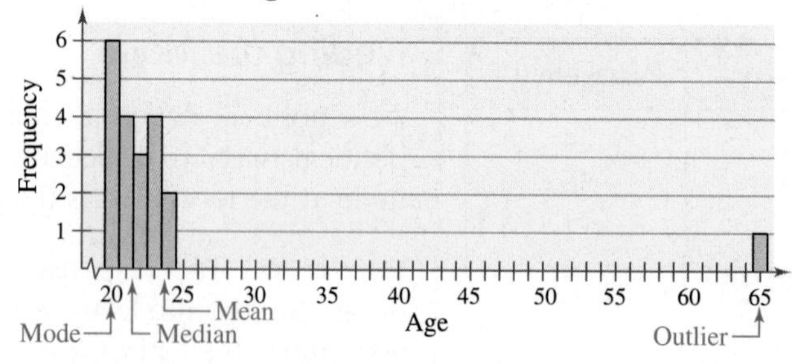

Try It Yourself 6

Remove the data entry of 65 from the preceding data set. Then rework the example. How does the absence of this outlier change the measures?

a. Find the *mean,* the *median,* and the *mode.*
b. Compare these measures of central tendency with those found in Example 6.

Answer: Page A31

Picturing the World

The National Association of Realtors keeps a databank of existing-home sales. One list uses the *median* price of existing homes sold and another uses the *mean* price of existing homes sold. The sales for the first quarter of 2001 are shown in the following graph. *(Source: National Association of Realtors)*

2001 U.S. Existing-Home Sales

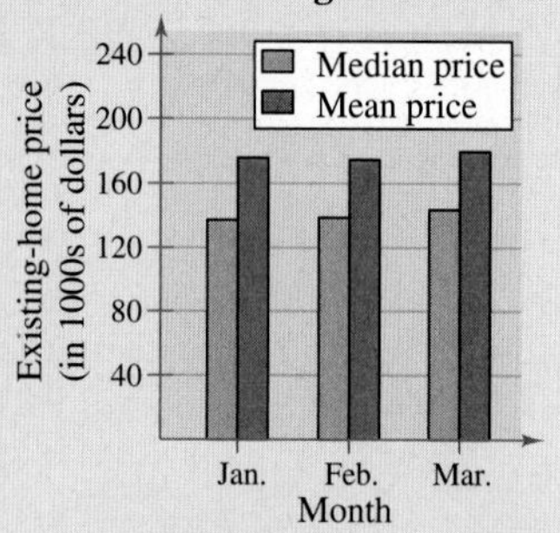

Notice in the graph that each month the mean price is about $35,000 more than the median price. What factors would cause the mean price to be greater than the median price?

Weighted Mean and Mean of Grouped Data

Sometimes data sets contain entries that have a greater effect on the mean than do other entries. To find the mean of such data sets, you must find the weighted mean.

DEFINITION

A **weighted mean** is the mean of a data set whose entries have varying weights. A weighted mean is given by

$$\bar{x} = \frac{\Sigma(x \cdot w)}{\Sigma w}$$

where w is the weight of each entry x.

EXAMPLE 7

Finding a Weighted Mean

You are taking a class in which your grade is determined from five sources: 50% from your test mean, 15% from your midterm, 20% from your final exam, 10% from your computer lab work, and 5% from your homework. Your scores are 86 (test mean), 96 (midterm), 82 (final exam), 98 (computer lab), and 100 (homework). What is the weighted mean of your scores?

SOLUTION Begin by organizing the scores and the weights in a table.

Source	**Score, x**	**Weight, w**	xw
Test Mean	86	0.5	43
Midterm	96	0.15	14.4
Final	82	0.2	16.4
Computer Lab	98	0.1	9.8
Homework	100	0.05	5
		$\Sigma w = 1$	$\Sigma(x \cdot w) = 88.6$

$$\bar{x} = \frac{\Sigma(x \cdot w)}{\Sigma w} = \frac{88.6}{1} = 88.6$$

So, your weighted mean for the course is 88.6.

Try It Yourself 7

An error was made in grading your final exam. Instead of getting 82, you scored 98. What is your new weighted mean?

a. Multiply each score by its weight and *find the sum of these products.*
b. Find the *sum of the weights.*
c. Find the *weighted mean.*
d. *Interpret the results* in the context of the data.

Answer: Page A31

To find the mean of grouped data, use the following.

DEFINITION

The **mean of a frequency distribution** for a sample is approximated by

$$\bar{x} = \frac{\Sigma(x \cdot f)}{n} \qquad \text{Note that } n = \Sigma f$$

where x and f are the midpoints and frequencies of a class, respectively.

Study Tip

If the frequency distribution represents a population, then the mean of the frequency distribution is approximated by

$$\mu = \frac{\Sigma(x \cdot f)}{N}$$

where $N = \Sigma f$.

GUIDELINES

Finding the Mean of a Frequency Distribution

In Words	*In Symbols*
1. Find the midpoint of each class.	$x = \dfrac{(\text{Lower limit}) + (\text{Upper limit})}{2}$
2. Find the sum of the products of the midpoints and the frequencies.	$\Sigma(x \cdot f)$
3. Find the sum of the frequencies.	$n = \Sigma f$
4. Find the mean of the frequency distribution.	$\bar{x} = \dfrac{\Sigma(x \cdot f)}{n}$

EXAMPLE 

Finding the Mean of a Frequency Distribution

Use the frequency distribution at the left to approximate the mean number of minutes that a sample of Internet subscribers spent online during their most recent session.

Class Midpoint

x	Frequency, f	$(x \cdot f)$
12.5	6	75.0
24.5	10	245.0
36.5	13	474.5
48.5	8	388.0
60.5	5	302.5
72.5	6	435.0
84.5	2	169.0
	$n = 50$	$\Sigma = 2089.0$

SOLUTION

$$\bar{x} = \frac{\Sigma(x \cdot f)}{n}$$

$$= \frac{2089}{50} \approx 41.8$$

So, the mean time spent online was approximately 41.8 minutes.

Try It Yourself 8

Use a frequency distribution to approximate the mean age of the residents of Akhiok. (See Try It Yourself 2 on page 35.)

a. Find the *midpoint* of each class.
b. Find the *sum of the products* of each midpoint and corresponding frequency.
c. Find the *sum of the frequencies.*
d. Find the *mean of the frequency distribution.*

Answer: Page A32

The Shape of Distributions

A graph reveals several characteristics of a frequency distribution. One such characteristic is the shape of the distribution.

DEFINITION

A frequency distribution is **symmetric** when a vertical line can be drawn through the middle of a graph of the distribution and the resulting halves are approximately mirror images.

A frequency distribution is **uniform** (or **rectangular**) when all entries, or classes, in the distribution have equal frequencies. A uniform distribution is also symmetric.

A frequency distribution is skewed if the "tail" of the graph elongates more to one side than to the other. A distribution is **skewed left (negatively skewed)** if its tail extends to the left. A distribution is **skewed right (positively skewed)** if its tail extends to the right.

When a distribution is symmetric, the mean, median, and mode are equal. If a distribution is skewed left, the mean is less than the median and the median is usually less than the mode. If a distribution is skewed right, the mean is greater than the median and the median is usually greater than the mode. Examples of these commonly occurring distributions are shown.

Insight

The mean will always fall in the direction the distribution is skewed. For instance, when a distribution is skewed left, the mean is to the left of the median.

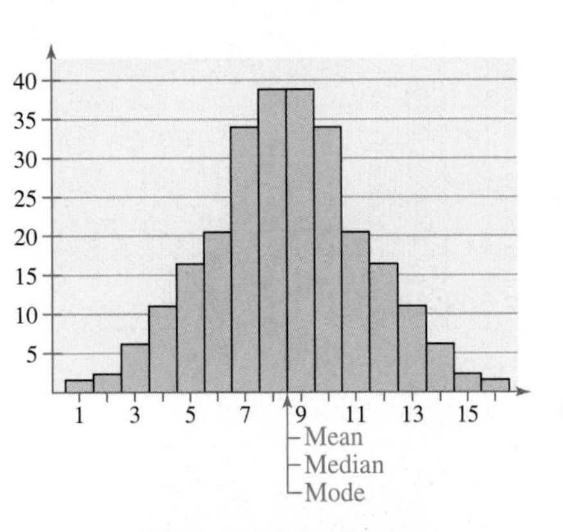

Symmetric Distribution

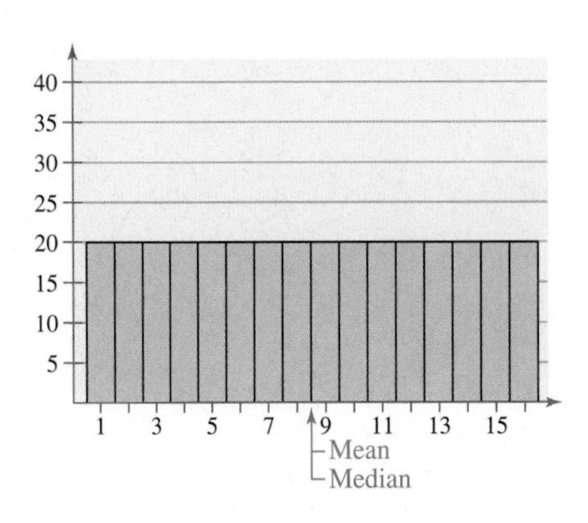

Uniform Distribution

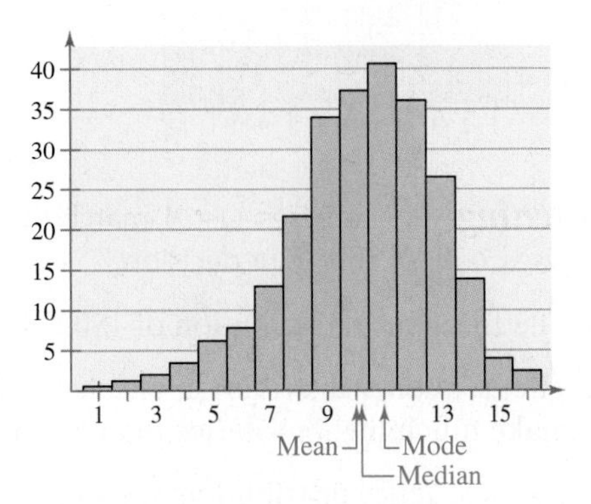

Skewed-Left Distribution

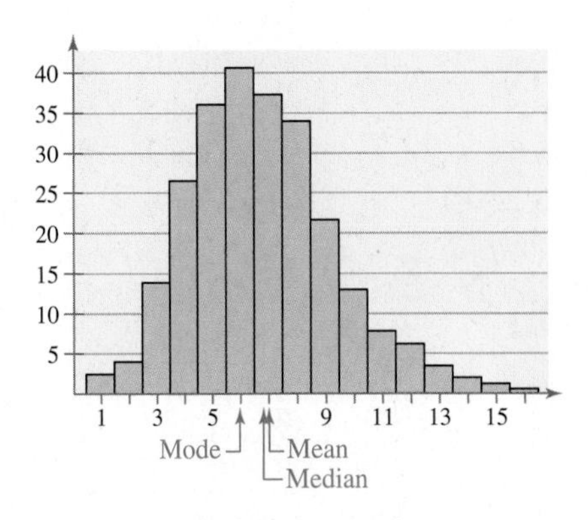

Skewed-Right Distribution

2.3 Exercises

Basic Skills and Concepts

True or False In Exercises 1–4, determine whether the statement is true or false. If it is false, rewrite it so it is a true statement.

1. The median is the measure of central tendency most likely to be affected by an extreme value (an outlier).

2. Every data set must have a mode.

3. Some quantitative data sets do not have a median.

4. The mean is the only measure of central tendency that can be used for data at the nominal level of measurement.

5. Give an example in which the mean of a data set is *not* representative of a typical number in the data set.

6. Give an example in which the median and the mode of a data set are the same.

Graphical Analysis In Exercises 7–10, determine whether the shape of the distribution in the histogram is symmetric, uniform, skewed left, skewed right, or none of these. Justify your answer.

7.

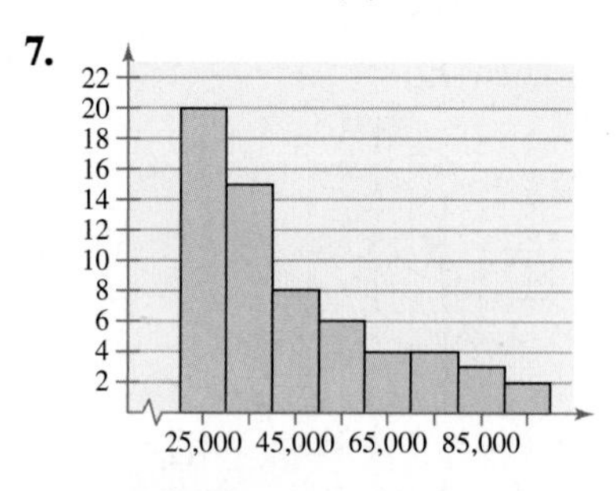

8.

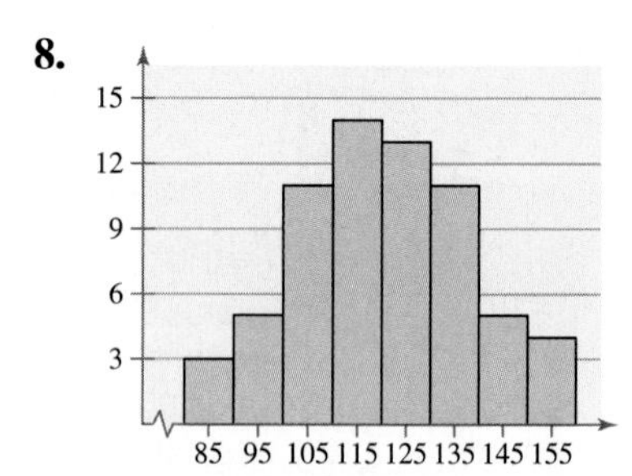

9.

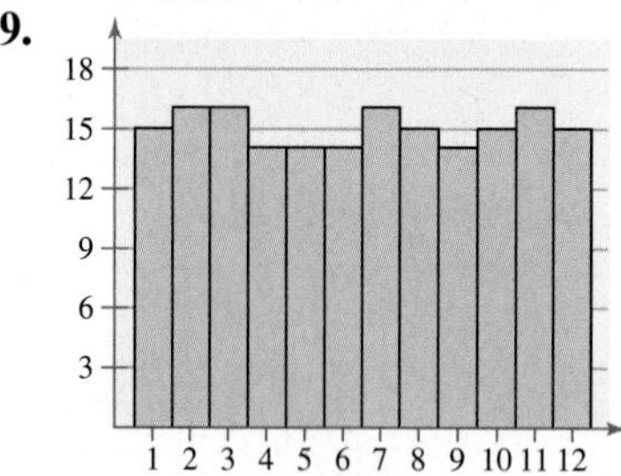

10.

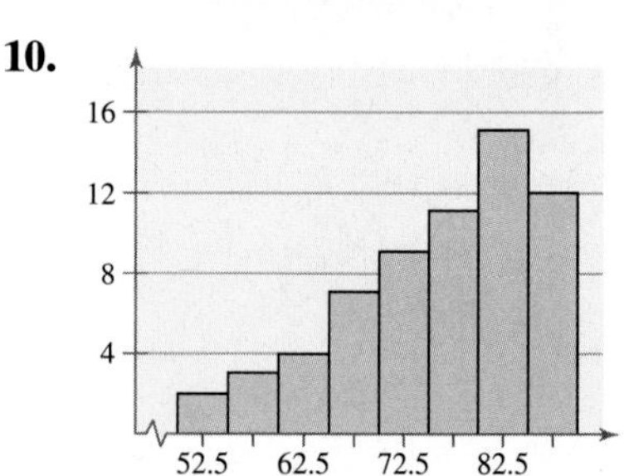

Matching In Exercises 11–14, match the distribution with one of the graphs in Exercises 7–10. Justify your decision.

11. The frequency distribution of 180 rolls of a dodecagon (a 12-sided die)

12. The frequency distribution of salaries at a company where a few executives make much higher salaries than the majority of employees

13. The frequency distribution of scores on a 90-point test where a few students scored much lower than the majority of students

14. The frequency distribution of weights for a sample of seventh-grade boys

Finding and Discussing the Mean, Median, and Mode In Exercises 15–28, do the following.

(a) Find the mean, median and mode of the data, if possible. If it is not possible, explain why the measure of central tendency cannot be found.

(b) Which measure of central tendency best represents the data? Explain your reasoning.

15. ***SUVs*** ◆ The maximum number of seats in a sample of sport utility vehicles *(Source: Consumer Reports)*

6 6 9 9 6 5 5 5 7 5 5 5 8

16. ***Education*** ◆ The education cost per student (in thousands of dollars) from a sample of ten liberal arts colleges *(Source: U.S. News and World Report)*

22 26 19 20 20 18 21 17 19 14

17. ***Sports Cars*** ◆ The time (in seconds) for a sample of seven sports cars to go from 0 to 60 miles per hour *(Source: Motor Trend)*

3.7 4.0 4.8 4.8 4.8 4.8 5.1

18. ***Cholesterol*** ◆ The cholesterol level of a sample of 10 female employees

154 216 171 188 229 203 184 173 181 147

19. ***NBA*** ◆ The points per game scored by each NBA team during a recent season *(Source: NBA)*

94.8 100.6 98.9 103.1 87.5 90.6 97.8
94.2 99.6 100.5 95.4 97.2 100.0 94.8
95.3 96.1 97.2 95.4 100.2 102.8 99.0
96.4 90.5 100.9 95.5 103.1 89.2 99.4

20. ***Power Failures*** ◆ The duration (in minutes) of every power failure at a residence in the last 10 years

18 26 45 75 125 80 33 40 44 49
89 80 96 125 12 61 31 63 103 28

21. ***Air Quality*** ◆ The responses of a sample of 1040 people who were asked if the air quality in their community is better or worse than it was 10 years ago

Better: 346 Worse: 450 Same: 244

22. ***Crime*** ◆ The responses of a sample of 1019 people who were asked how they felt when they thought about crime

Unconcerned: 34 Watchful: 672 Nervous: 125 Afraid: 188

23. ***Top Speeds*** ◆ The top speed (in miles per hour) for a sample of seven sports cars *(Source: Motor Trend)*

187.3 181.8 180.0 169.3 162.2 158.1 155.7

24. ***Purchase Preference*** ◆ The responses of a sample of 1001 people who were asked if their next vehicle purchase will be foreign or domestic

Domestic: 704 Foreign: 253 Don't know: 44

25. ***Stocks*** ◆ The recommended prices for several stocks that analysts predict should produce at least 10% annual returns *(Source: Money)*

41 20 22 14 15 25 18 40 17 14

26. ***Weight Loss*** ◆ The number of weeks it took to reach a target weight for a sample of five patients with eating disorders treated by psychodynamic psychotherapy *(Source: The Journal of Consulting and Clinical Psychology)*

15.0 31.5 10.0 25.5 1.0

27. ***Eating Disorders*** ◆ The number of weeks it took to reach a target weight for a sample of 14 patients with eating disorders treated by psychodynamic psychotherapy and cognitive behavior techniques *(Source: The Journal of Consulting and Clinical Psychology)*

2.5	20.0	11.0	10.5	17.5	16.5	13.0
15.5	26.5	2.5	27.0	28.5	1.5	5.0

28. ***Aircraft*** ◆ The number of aircraft 11 airlines have in their fleets *(Source: Airline Transport Association)*

563	667	443	544	358	178
289	182	105	82	66	

Graphical Analysis In Exercises 29 and 30, the letters A, B, and C are marked on the horizontal axis. From these, determine which is the mean, which is the median, and which is the mode. Justify your answers.

29. **Sick Days Used by Employees**

Frequency
16
14
12
10
8
6
4
2
10 14 16 18 20 22 24 26 28
A B C
Days

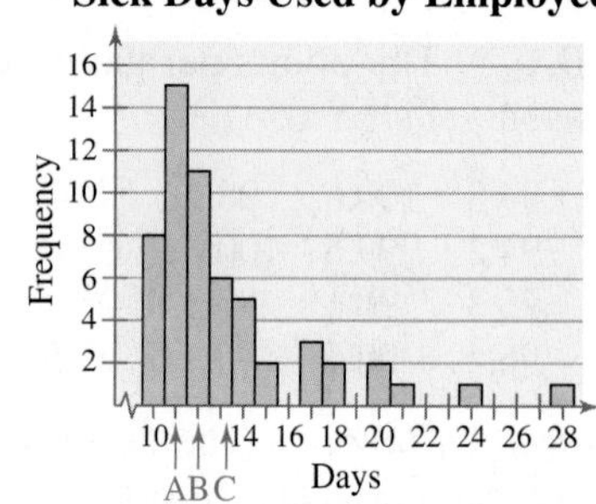

30. **Hourly Wages of Employees**

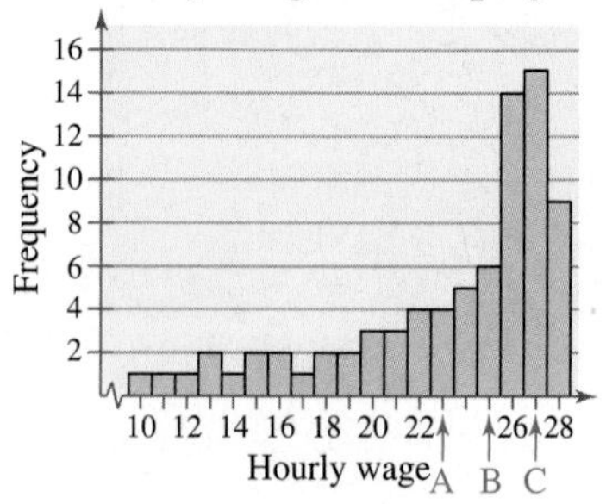

Frequency
16
14
12
10
8
6
4
2
10 12 14 16 18 20 22 26 28
Hourly wage
A B C

Finding the Weighted Mean In Exercises 31–34, find the weighted mean of the data.

31. ***Final Grade*** ◆ The scores and their percent of the final grade for a statistics student are given. What is the student's mean score?

	Score	Percent of final grade
Homework	85	15%
Quiz	80	10%
Quiz	92	10%
Quiz	76	10%
Project	100	15%
Speech	90	15%
Final Exam	93	25%

32. ***Salaries*** ◆ The average starting salaries (by degree attained) for 25 employees at a company are given. What is the mean starting salary for these employees?

8 with MBAs: $42,500

17 with BAs in business: $28,000

33. ***Grades*** ◆ A student receives the following grades, with an A worth 4 points, a B worth 3 points, a C worth 2 points, and a D worth 1 point. What is the student's mean grade point score?

B in two 3-credit classes	D in one 2-credit class
A in one 4-credit class	C in one 3-credit class

34. ***Scores*** ◆ The mean scores for a statistics course (by major) are given. What is the mean score for the class?

8 engineering majors: 83	5 math majors: 87
11 business majors: 79	

Finding the Mean of Grouped Data

In Exercises 35–38, approximate the mean of the grouped data.

35. ***Heights of Females*** ◆ The heights (in inches) of 16 female students in a physical education class

Height (in inches)	Frequency
60–62	3
63–65	4
66–68	7
69–71	2

36. ***Heights of Males*** ◆ The heights (in inches) of 21 male students in a physical education class

Height (in inches)	Frequency
63–65	2
66–68	4
69–71	8
72–74	5
75–77	2

37. ***Ages*** ◆ The ages of residents of Medicine Bow, Wyoming *(Source: U.S. Bureau of the Census)*

Age	Frequency
0–9	57
10–19	68
20–29	36
30–39	55
40–49	71
50–59	44
60–69	36
70–79	14
80–89	8

38. ***Phone Calls*** ◆ The lengths of long-distance calls (in minutes) made by one person in one year

Length of call	Number of calls
1–5	12
6–10	26
11–15	20
16–20	7
21–25	11
26–30	7
31–35	4
36–40	4
41–45	1

Identifying the Shape of a Distribution

In Exercises 39–42, construct a frequency distribution and a frequency histogram of the data using the indicated number of classes. Describe the shape of the histogram as symmetric, uniform, negatively skewed, positively skewed, or none of these.

39. ***Hospitalization*** ◆

Number of classes: 6

Data set: The number of days 20 patients remained hospitalized

6	9	7	14	4	5	6	8	4	11
10	6	8	6	5	7	6	6	3	11

40. *Hospital Beds* ◆

Number of classes: 5

Data set: The number of beds in a sample of 24 hospitals

149	167	162	127	130	180	160	167
221	145	137	194	207	150	254	262
244	297	137	204	166	174	180	151

41. *Height of Males* ◆

Number of classes: 5

Data set: The heights (to the nearest inch) of 30 males

67	76	69	68	72	68	65	63	75	69
66	72	67	66	69	73	64	62	71	73
68	72	71	65	69	66	74	72	68	69

42. *Six-Sided Die* ◆

Number of classes: 6

Data set: The results of rolling a six-sided die 30 times

1	4	6	1	5	3	2	5	4	6
1	2	4	3	5	6	3	2	1	1
5	6	2	4	4	3	1	6	2	4

43. *Coffee Content* ◆ During a quality assurance check, the actual coffee content of six jars of instant coffee was recorded as 6.03, 5.59, 6.40, 6.00, 5.99 and 6.02.

(a) Find the mean and the median of the coffee content.

(b) Suppose the third value had been incorrectly measured and was actually 6.04. Find the mean and median of the coffee content again.

(c) Which measure of central tendency, the mean or the median, was affected more by the data entry error?

44. *U.S. Exports* ◆ The following data are the U.S. exports (in billions of dollars) to 19 countries for a recent year. *(Source: U.S. Department of Commerce)*

Canada	133.7	Japan	67.5
Mexico	56.8	United Kingdom	30.9
Germany	23.5	South Korea	26.6
Taiwan	18.4	Singapore	16.7
Netherlands	16.6	France	14.4
China	12.0	Brazil	12.7
Australia	12.0	Belgium	12.5
Malaysia	8.5	Italy	8.8
Switzerland	8.4	Thailand	7.2
Saudi Arabia	7.3		

(a) Find the mean and median.

(b) Find the mean and median without the U.S. exports to Canada.

(c) Which measure of central tendency, the mean or the median, was affected more by the elimination of the Canadian export data?

Extending the Basics

45. ***Data Analysis*** ◆ A consumer testing service obtained the following miles per gallon in five test runs performed with three types of compact cars.

	Run 1	Run 2	Run 3	Run 4	Run 5
Car A:	28	32	28	30	34
Car B:	31	29	31	29	31
Car C:	29	32	28	32	30

(a) If the manufacturer of Car A wants to advertise that their car performed best in this test, which measure of central tendency—mean, median, or mode—should be used for their claim? Explain your reasoning.

(b) If the manufacturer of Car B wants to advertise that their car performed best in this test, which measure of central tendency—mean, median, or mode—should be used for their claim? Explain your reasoning.

(c) If the manufacturer of Car C wants to advertise that their car performed best in this test, which measure of central tendency—mean, median, or mode—should be used for their claim? Explain your reasoning.

46. ***Midrange*** ◆ The midrange is

$$\frac{(\text{Maximum data entry}) + (\text{Minimum data entry})}{2}.$$

Which of the dealers in Exercise 45 would prefer to use the midrange statistic in their ads? Explain your reasoning.

DATA **47.** ***Data Analysis*** ◆ Students in an experimental psychology class did research on depression as a sign of stress. A test was administered to a sample of 30 students. The scores are given.

44 51 11 90 76 36 64 37 43 72 53 62 36 74 51
72 37 28 38 61 47 63 36 41 22 37 51 46 85 13

(a) Find the mean of the data.

(b) Find the median of the data.

(c) Draw a stem-and-leaf display for the data using one line per stem. Locate the mean and median on the display.

(d) Describe the shape of the distribution.

48. ***Trimmed Mean*** ◆ To find the 10% trimmed mean of a data set, order the data, delete the lowest 10% of the entries and the highest 10% of the entries, and find the mean of the remaining entries.

(a) Find the 10% trimmed mean for the data in the previous problem.

(b) Compare the four measures of central tendency.

(c) What is the benefit of using a trimmed mean versus using a mean found using all data entries? Explain your reasoning.

49. ***Writing*** ◆ The population mean μ and the sample mean $\bar{x}$ have essentially the same formulas. Explain why it is necessary to have two different symbols.

2.4 Measures of Variation

What You Should Learn

- How to find the range of a data set
- How to find the variance and standard deviation of a population and of a sample
- How to use the Empirical Rule and Chebychev's Theorem to interpret standard deviation
- How to approximate the sample standard deviation for grouped data

Range • Deviation, Variance, and Standard Deviation • Interpreting Standard Deviation • Standard Deviation for Grouped Data

Range

In this section, you will learn different ways to measure the variation of a data set. The simplest measure is the range of the set.

DEFINITION

The **range** of a data set is the difference between the maximum and minimum data entries in the set.

Range = (Maximum data entry) − (Minimum data entry)

EXAMPLE 1

Finding the Range of a Data Set

Two corporations each hired 10 graduates. The starting salaries for each are as follows.

Starting Salaries for Corporation A (1000s of dollars)

Salary	41	38	39	45	47	41	44	41	37	42

Starting Salaries for Corporation B (1000s of dollars)

Salary	40	23	41	50	49	32	41	29	52	58

Find the range of the starting salaries for Corporation A.

SOLUTION To find the least and greatest salaries, it helps to order the data.

37 38 39 41 41 41 42 44 45 47

Minimum (37) ... Maximum (47)

Range = (Maximum salary) − (Minimum salary)

= 47 − 37

= 10

So, the range of the starting salaries for Corporation A is 10 or $10,000.

Insight

Both data sets in Example 1 have a mean of 41.5, a median of 41, and a mode of 41. And yet the two sets differ significantly.

The difference is that the entries in the second set have greater variation. Your goal in this section is to learn how to measure the variation of a data set.

Try It Yourself 1

Find the range of the starting salaries for Corporation B.

a. Identify the *minimum* and *maximum* salaries.
b. Find the *range*.
c. Compare your answer with that of Example 1.

Answer: Page A32

Deviation, Variance, and Standard Deviation

As a measure of variation, the range has the advantage of being easy to compute. Its disadvantage, however, is that it uses only two entries from the data set. Two measures of variation that use all the entries in a data set are the variance and the standard deviation. However, before you learn about these measures of variation, you need to know what is meant by the deviation of an entry in a data set.

> **DEFINITION**
>
> The **deviation** of an entry x in a population data set is the difference between the entry and the mean μ of the data set.
>
> Deviation of $x = x - \mu$

EXAMPLE 

Finding the Deviations of a Data Set

Find the deviation of each starting salary for Corporation A.

SOLUTION The mean starting salary is $\mu = 415/10 = 41.5$. To find out how much each salary deviates from the mean, subtract 41.5 from the salary. For instance, the deviation of 41 (or \$41,000) is

$$41 - 41.5 = -0.5 \text{ (or } -\$500\text{)}.$$

Deviation of $x = x - \mu$

(x → 41, μ → 41.5)

The table at the left lists the deviations of each of the 10 starting salaries.

Salary (1000s of dollars) x	Deviation (1000s of dollars) $x - \mu$
41	−0.5
38	−3.5
39	−2.5
45	3.5
47	5.5
41	−0.5
44	2.5
41	−0.5
37	−4.5
42	0.5
$\Sigma x = 415$	$\Sigma(x - \mu) = 0$

Deviations of starting salaries for Corporation A

Try It Yourself 2

Find the deviation of each starting salary for Corporation B in Example 1.

a. Find the *mean* of the data set.
b. *Subtract* the mean from each salary.
c. *Organize* your results in a table.

Answer: Page A32

In Example 2, notice that the sum of the deviations is zero. Because this is true for any data set, it doesn't make sense to find the average of the deviations. To overcome this problem, you can square each deviation. In a population data set, the mean of the squares of the deviations is called the **population variance.**

> **Study Tip**
>
> When you add the squares of the deviations, you compute a quantity called the sum of squares, denoted SS_x.

> **DEFINITION**
>
> The **population variance** of a population data set of N entries is
>
> $$\text{Population variance} = \sigma^2 = \frac{\Sigma(x - \mu)^2}{N}$$
>
> The symbol σ is the lowercase Greek letter sigma.

Insight

The disadvantage with the variance is that its units are usually meaningless. For instance, the variance for the starting salaries is measured in "square dollars." You'll be able to return to the original unit of the data by using the standard deviation.

DEFINITION

The **population standard deviation** of a population data set of N entries is the square root of the population variance.

$$\text{Population standard deviation} = \sigma = \sqrt{\sigma^2} = \sqrt{\frac{\Sigma(x-\mu)^2}{N}}$$

GUIDELINES

Finding the Population Variance and Standard Deviation

In Words	*In Symbols*
1. Find the mean of the population data set.	$\mu = \frac{\Sigma x}{N}$
2. Find the deviation of each entry.	$x - \mu$
3. Square each deviation.	$(x-\mu)^2$
4. Add to get the **sum of squares.**	$SS_x = \Sigma(x-\mu)^2$
5. Divide by N to get the **population variance.**	$\sigma^2 = \frac{\Sigma(x-\mu)^2}{N}$
6. Find the square root of the variance to get the **population standard deviation.**	$\sigma = \sqrt{\frac{\Sigma(x-\mu)^2}{N}}$

EXAMPLE 3

Finding the Population Standard Deviation

Find the population variance and standard deviation of the starting salaries for Corporation A.

SOLUTION The table at the left summarizes the steps used to find SS_x.

$$SS_x = 88.5, \quad N = 10, \quad \sigma^2 = \frac{88.5}{10} = 8.85, \quad \sigma = \sqrt{8.85} \approx 2.97$$

So, the population variance is 8.85 and the population standard deviation is about 3.0 or $3000.

Salary x	Deviation $x-\mu$	Squares $(x-\mu)^2$
41	−0.5	0.25
38	−3.5	12.25
39	−2.5	6.25
45	3.5	12.25
47	5.5	30.25
41	−0.5	0.25
44	2.5	6.25
41	−0.5	0.25
37	−4.5	20.25
42	0.5	0.25
	$\Sigma = 0$	$SS_x = 88.5$

Sum of squares of starting salaries for Corporation A

Try It Yourself 3

Find the population standard deviation of the starting salaries for Corporation B given in Example 1.

a. Find the *mean* and each *deviation,* as you did in Try It Yourself 2.
b. *Square* each deviation and *add* to get the sum of squares.
c. *Divide* by N to get the population variance.
d. Find the *square root* of the population variance.
e. *Interpret* the results by giving the population standard deviation in dollars.

Answer: Page A32

Study Tip

Note that when you find the *population* variance, you divide by *N*, the number of entries, but when you find the *sample* variance, you divide by $n - 1$, one less than the number of entries.

DEFINITION

The **sample variance** and **sample standard deviation** of a sample data set of n entries are listed below.

$$\text{Sample variance} = s^2 = \frac{\Sigma(x - \bar{x})^2}{n - 1}$$

$$\text{Sample standard deviation} = s = \sqrt{s^2} = \sqrt{\frac{\Sigma(x - \bar{x})^2}{n - 1}}$$

Symbols in Variance and Standard Deviation Formulas

	Population	Sample
Variance	σ^2	s^2
Standard deviation	σ	s
Mean	μ	$\bar{x}$
Number of entries	N	n
Deviation	$x - \mu$	$x - \bar{x}$
Sum of squares	$\Sigma(x - \mu)^2$	$\Sigma(x - \bar{x})^2$

GUIDELINES

Finding the Sample Variance and Standard Deviation

In Words	*In Symbols*
1. Find the mean of the sample data set.	$\bar{x} = \frac{\Sigma x}{n}$
2. Find the deviation of each entry.	$x - \bar{x}$
3. Square each deviation.	$(x - \bar{x})^2$
4. Add to get the **sum of squares.**	$SS_x = \Sigma(x - \bar{x})^2$
5. Divide by $n - 1$ to get the **sample variance.**	$s^2 = \frac{\Sigma(x - \bar{x})^2}{n - 1}$
6. Find the square root of the variance to get the **sample standard deviation.**	$s = \sqrt{\frac{\Sigma(x - \bar{x})^2}{n - 1}}$

See *Minitab* and *TI-83* steps on pages 98, 99

EXAMPLE  4

Finding the Sample Standard Deviation

Suppose that the starting salaries given in Example 1 are for the Chicago branches of Corporations A and B. Each corporation has several other branches, and you plan to use the starting salaries of the Chicago branches to estimate the starting salaries for the larger populations. Find the *sample* standard deviation of the starting salaries for the Chicago branch of Corporation A.

SOLUTION

$$SS_x = 88.5, \quad n = 10, \quad s^2 = \frac{88.5}{9} \approx 9.83, \quad s = \sqrt{9.83} \approx 3.14$$

So, the sample variance is about 9.83 and the sample standard deviation is about 3.1 or \$3100.

Insight

Notice that in Example 4, all graduates' starting salaries for the Chicago branch of Corporation A constitute a cluster sample.

Try It Yourself 4

Find the sample standard deviation of the starting salaries for the Chicago branch of Corporation B.

a. Find the *sum of squares,* as you did in Try It Yourself 3.
b. *Divide* by $n - 1$ to get the sample variance.
c. Find the *square root* of the sample variance.

Answer: Page A32

EXAMPLE 5

Using Technology to Find the Standard Deviation

Office Rental Rates		
35.00	33.50	37.00
23.75	26.50	31.25
36.50	40.00	32.00
39.25	37.50	34.75
37.75	37.25	36.75
27.00	35.75	26.00
37.00	29.00	40.50
24.50	33.00	38.00

Sample office rental rates, in dollars per square foot per year, for Miami's central business district are listed at the left. Use a calculator or a computer to find the mean rental rate and the sample standard deviation. *(Adapted from Cushman & Wakefield Inc.)*

SOLUTION *Minitab, Excel,* and the *TI-83* each have features that automatically calculate the mean and the standard deviation of data sets. Try using this technology to find the mean and the standard deviation of the office rental rates. From the displays, you can see that $\bar{x} \approx 33.73$ and $s \approx 5.09$.

MINITAB

Descriptive Statistics

Variable	N	Mean	Median	TrMean	StDev
Rental Rates	24	33.73	35.38	33.88	5.09

Variable	SE Mean	Minimum	Maximum	Q1	Q3
Rental Rates	1.04	23.75	40.50	29.56	37.44

EXCEL

	A	B
1	Mean	33.72917
2	Standard Error	1.038864
3	Median	35.375
4	Mode	37
5	Standard Deviation	5.089373
6	Sample Variance	25.90172
7	Kurtosis	-0.74282
8	Skewness	-0.70345
9	Range	16.75
10	Minimum	23.75
11	Maximum	40.5
12	Sum	809.5
13	Count	24

TI-83

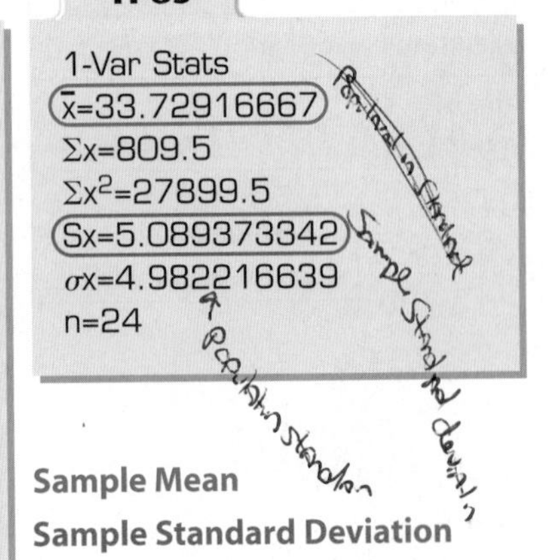

Try It Yourself 5

Sample office rental rates, in dollars per square foot per year, for Seattle's central business district are listed. Use a calculator or a computer to find the mean rental rate and the sample standard deviation. *(Adapted from Cushman & Wakefield Inc.)*

40.00	43.00	46.00	40.50	35.75	39.75	32.75
36.75	35.75	38.75	38.75	36.75	38.75	39.00
29.00	35.00	42.75	32.75	40.75	35.25	

a. *Enter* the data.

b. *Calculate* the sample mean and the sample standard deviation.

Answer: Page A32

Interpreting Standard Deviation

When interpreting the standard deviation, remember that it is a measure of the typical amount an entry deviates from the mean. The more the entries are spread out, the greater the standard deviation.

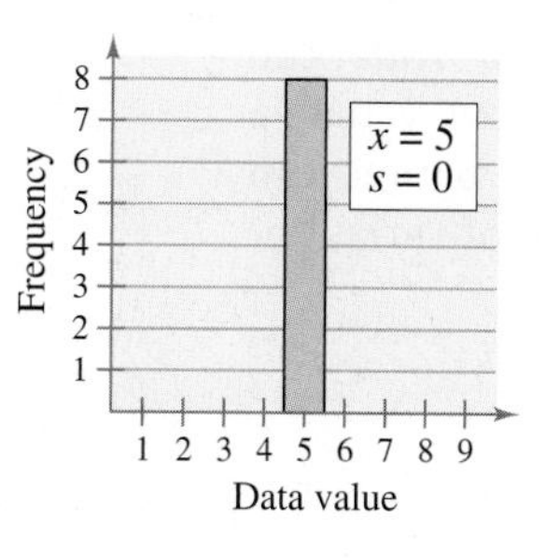

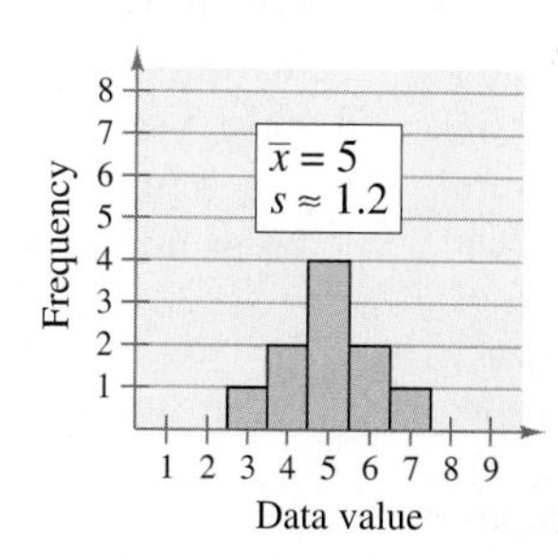

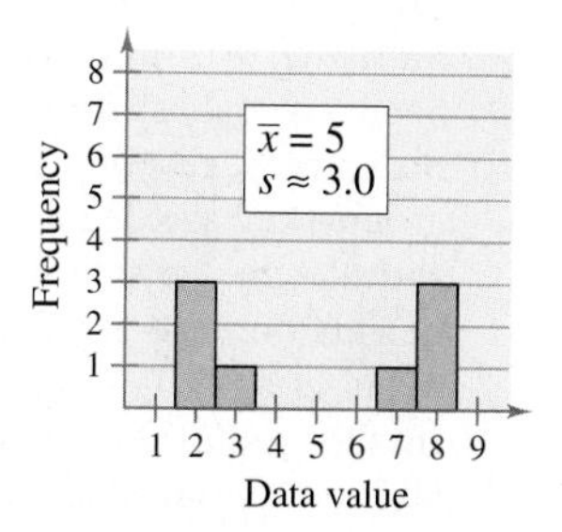

EXAMPLE 6

Estimating Standard Deviation

Without calculating, estimate the population standard deviation of each data set.

1. 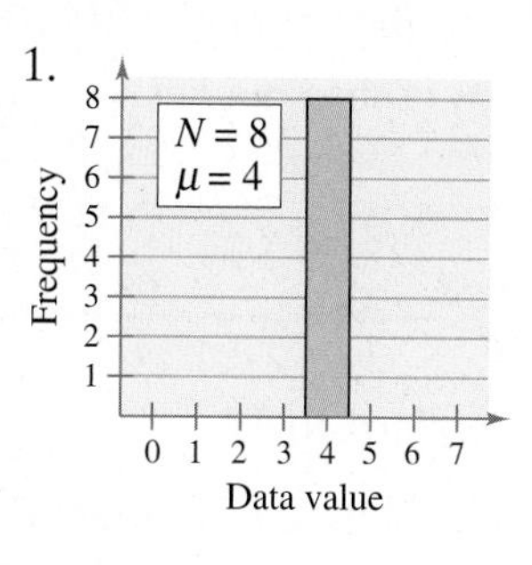

2. 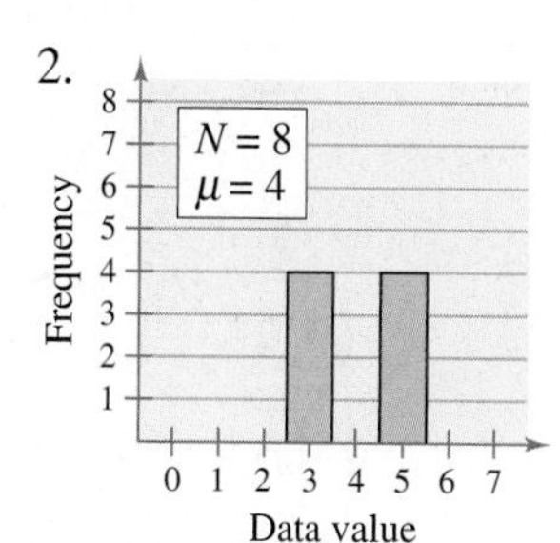

3. 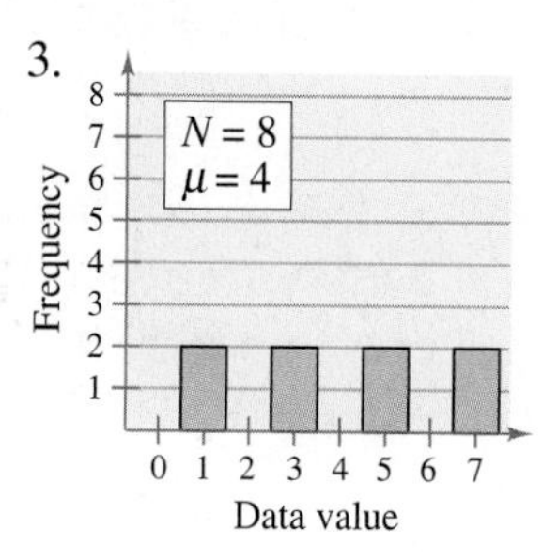

SOLUTION

1. Each of the eight entries is 4. So, each deviation is 0, which implies that

   $\sigma = 0.$

2. Each of the eight entries has a deviation of ± 1. So, the population standard deviation should be 1. By calculating, you can see that

   $\sigma = 1.$

3. Each of the eight entries has a deviation of ± 1 or ± 3. So, the population standard deviation should be about 2. By calculating, you can see that

   $\sigma \approx 2.24.$

Try It Yourself 6

Write a data set that has 10 entries, a mean of 10, and a population standard deviation that is approximately 3. (There are many correct answers.)

a. *Write* a data set that has five entries that are 3 units less than 10 and five entries that are 3 units more than 10.

b. *Calculate* the population standard deviation to check that σ is approximately 3.

Answer: Page A32

Picturing the World

A survey was conducted by the National Center for Health Statistics to find the mean height of males in the U.S. In the survey, respondents were grouped by age. The histogram shows the distribution of heights for the 2485 respondents in the 20–29 age group. In this group, the mean was 69.2 inches and the standard deviation was 2.9 inches.

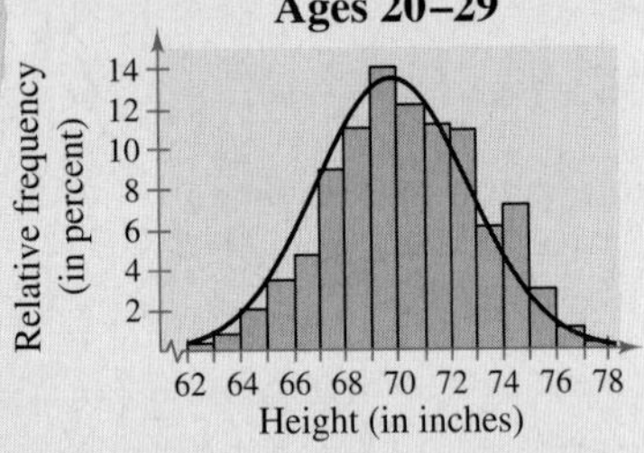

About what percent of the heights lie within two standard deviations of the mean?

Many real-life data sets have distributions that are approximately bell shaped. Later in the text, you will study this type of distribution in detail. For now, however, the following *Empirical Rule* can help you see how valuable the standard deviation can be as a measure of variation.

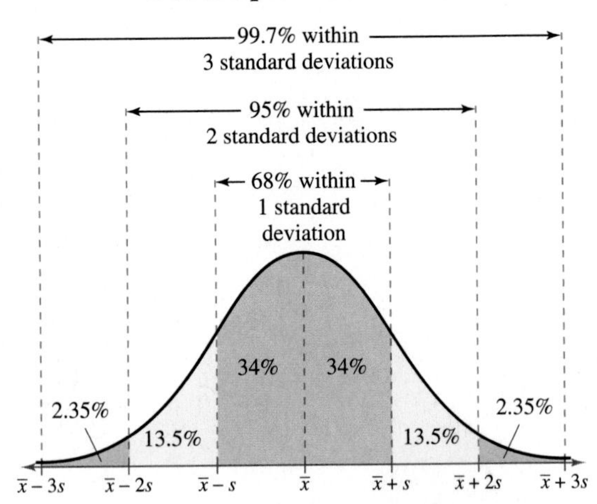

Empirical Rule (or 68-95-99.7 Rule)

For data with a (symmetric) bell-shaped distribution, the standard deviation has the following characteristics.

1. About 68% of the data lies within 1 standard deviation of the mean.
2. About 95% of the data lies within 2 standard deviations of the mean.
3. About 99.7% of the data lies within 3 standard deviations of the mean.

EXAMPLE 7

Using the Empirical Rule

In a survey conducted by the National Center for Health Statistics, the sample mean height of women in the U.S. (ages 20–29) was 64 inches with a sample standard deviation of 2.75 inches. Estimate the percent of the women whose heights are between 64 inches and 69.5 inches.

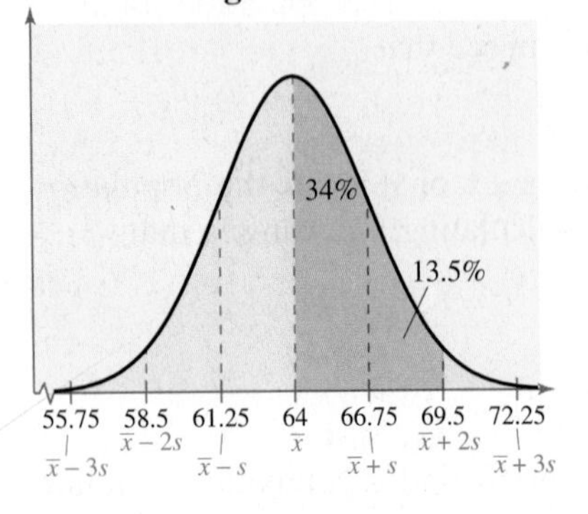

SOLUTION The distribution of the women's heights is shown at the left. Because the distribution is bell shaped, you can use the Empirical Rule. The mean height is 64, so when you add two standard deviations to the mean height, you get

$$\bar{x} + 2s = 64 + 2(2.75) = 69.5.$$

Because 69.5 is two standard deviations above the mean height, the percent of the heights between 64 inches and 69.5 inches is

$$34\% + 13.5\% = 47.5\%.$$

Try It Yourself 7

Estimate the percent of the heights that are between 61.25 and 64 inches.

a. How many standard deviations is 61.25 to the left of 64?

b. Use the Empirical Rule to estimate the percent of the data between $\bar{x} - s$ and $\bar{x}$.

c. *Interpret* the result in the context of the data.

Answer: Page A32

The following theorem applies to *all* distributions. It is named after the Russian statistician Pafnuti Chebychev (1821–1894).

Chebychev's Theorem

The portion of any data set lying within k standard deviations ($k > 1$) of the mean is at least

$$1 - \frac{1}{k^2}.$$

- $k = 2$: In any data set, at least $1 - \frac{1}{2^2} = \frac{3}{4}$, or 75%, of the data lie within 2 standard deviations of the mean.
- $k = 3$: In any data set, at least $1 - \frac{1}{3^2} = \frac{8}{9}$, or 88.9%, of the data lie within 3 standard deviations of the mean.

EXAMPLE 

Using Chebychev's Theorem

The age distributions for Alaska and Florida are shown in the histograms. Decide which is which. Apply Chebychev's Theorem to the data for Florida using $k = 2$. What can you conclude?

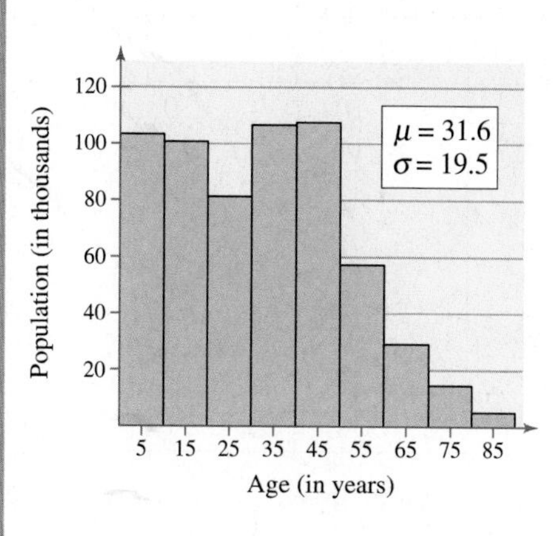

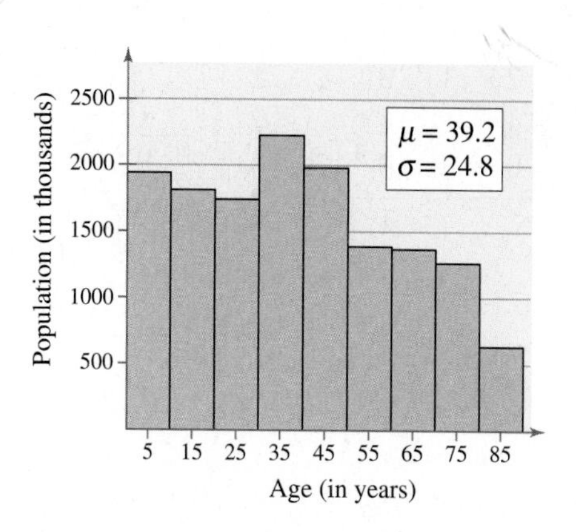

SOLUTION The histogram on the right shows Florida's age distribution. You can tell this because the population is greater and older. Moving two standard deviations to the left of the mean puts you below 0, because $\mu - 2\sigma = 39.2 - 2(24.8) = -10.4$. Moving two standard deviations to the right of the mean puts you at $\mu + 2\sigma = 39.2 + 2(24.8) = 88.8$. By Chebychev's Theorem, you can say that at least 75% of the population of Florida is between 0 and 88.8 years old.

Insight

In Example 8, Chebychev's Theorem tells you that at least 75% of the population of Florida is under the age of 88.8. This is a true statement, but it is not nearly as strong a statement as could be made from reading the histogram.

In general, Chebychev's Theorem gives cautious estimates of the percent lying within k standard deviations of the mean. Remember, the theorem applies to all distributions.

Try It Yourself 8

Apply Chebychev's Theorem to the data for Alaska using $k = 2$.

a. *Subtract* two standard deviations from the mean.
b. *Add* two standard deviations to the mean.
c. *Apply* Chebychev's Theorem for $k = 2$.
d. *Interpret* the result in the context of the data.

Answer: Page A32

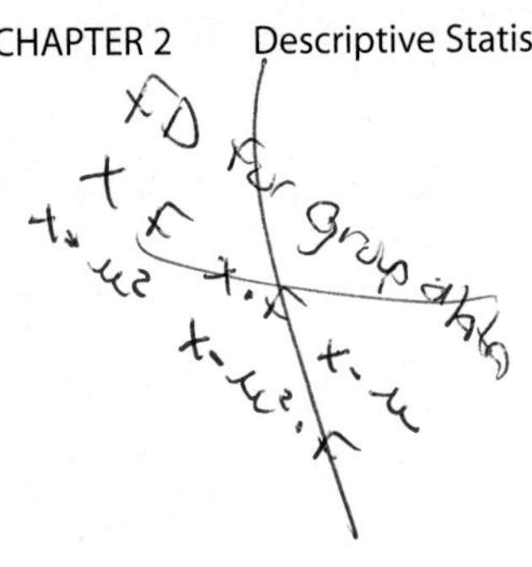

Standard Deviation for Grouped Data

In Section 2.1, you learned that large data sets are usually best represented by a frequency distribution. The formula for the sample standard deviation for a frequency distribution is

$$\text{Sample standard deviation} = s = \sqrt{\frac{\Sigma(x - \bar{x})^2 f}{n - 1}}$$

where $n = \Sigma f$ is the number of entries in the data set.

EXAMPLE 9

Finding the Standard Deviation for Grouped Data

You collect a random sample of the number of children per household in a region. The results are shown at the left. Find the sample mean and the sample standard deviation of the data set.

Number of Children in 50 Households

1	3	1	1	1
1	2	2	1	0
1	1	0	0	0
1	5	0	3	6
3	0	3	1	1
1	1	6	0	1
3	6	6	1	2
2	3	0	1	1
4	1	1	2	2
0	3	0	2	4

SOLUTION These data could be treated as 50 individual entries, and you could use the formulas for mean and standard deviation. Because there are so many repeated numbers, however, it is easier to use a frequency distribution.

x	f	xf
0	10	0
1	19	19
2	7	14
3	7	21
4	2	8
5	1	5
6	4	24
	$\Sigma = 50$	$\Sigma = 91$

$x - \bar{x}$	$(x - \bar{x})^2$	$(x - \bar{x})^2 f$
−1.8	3.24	32.40
−0.8	0.64	12.16
0.2	0.04	0.28
1.2	1.44	10.08
2.2	4.84	9.68
3.2	10.24	10.24
4.2	17.64	70.56
		$\Sigma = 145.40$

$$\bar{x} = \frac{\Sigma xf}{n} = \frac{91}{50} \approx 1.8 \quad \text{Sample mean}$$

Use the sum of squares to find the sample standard deviation.

$$s = \sqrt{\frac{\Sigma(x - \bar{x})^2 f}{n - 1}} = \sqrt{\frac{145.4}{49}} \approx 1.7 \quad \text{Sample standard deviation}$$

So, the sample mean is 1.8 children and the standard deviation is 1.7 children.

Study Tip

Remember that formulas for grouped data require you to multiply by the frequencies.

Try It Yourself 9

Change three of the 6s in the data set to 4s. How does this change the sample mean and sample standard deviation?

a. Write the first three columns of a *frequency distribution.*
b. Find the *sample mean.*
c. Complete the *last three columns* of the frequency distribution.
d. Find the *sample standard deviation.*

Answer: Page A32

When a frequency distribution has classes, you can estimate the sample mean and standard deviation by using the midpoint of each class.

EXAMPLE 

Using Midpoints of Classes

The circle graph at the right shows the results of a survey in which 1000 adults were asked how much they spend in preparation for personal travel each year. Make a frequency distribution for the data. Then use the table to estimate the sample mean and the sample standard deviation of the data set. *(Adapted from Travel Industry Association of America)*

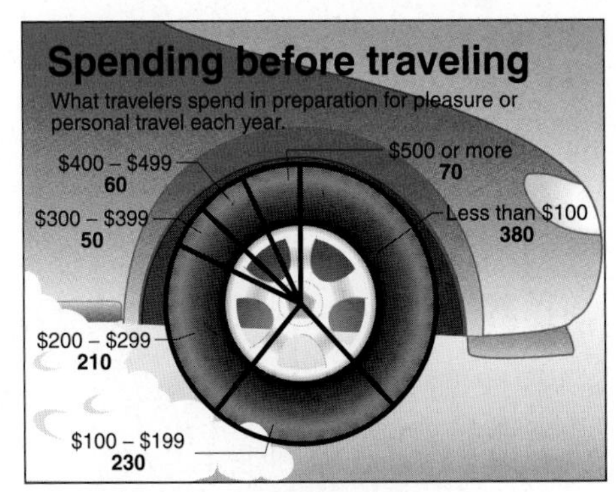

SOLUTION Begin by using a frequency distribution to organize the data.

Class	x	f	xf	$x - \bar{x}$	$(x - \bar{x})^2$	$(x - \bar{x})^2 f$
0–99	49.5	380	18,810	−142.5	20,306.25	7,716,375
100–199	149.5	230	34,385	−42.5	1,806.25	415,437.5
200–299	249.5	210	52,395	57.5	3,306.25	694,312.5
300–399	349.5	50	17,475	157.5	24,806.25	1,240,312.5
400–499	449.5	60	26,970	257.5	66,306.25	3,978,375
500+	599.5	70	41,965	407.5	166,056.25	11,623,937.5
		$\Sigma = 1{,}000$	$\Sigma = 192{,}000$			$\Sigma = 25{,}668{,}750$

Study Tip

When a class is open, as in the last class, you must assign a single value to represent the midpoint. For this example, we selected 599.5.

$$\bar{x} = \frac{\Sigma xf}{n} = \frac{192{,}000}{1{,}000} = 192$$ Sample mean

Use the sum of squares to find the sample standard deviation.

$$s = \sqrt{\frac{\Sigma(x - \bar{x})^2 f}{n - 1}} = \sqrt{\frac{25{,}668{,}750}{999}} \approx 160.3$$ Sample standard deviation

So, the sample mean is 192 dollars per year and the sample standard deviation is about 160.3 dollars per year.

Try It Yourself 10

In the frequency distribution, 599.5 was chosen to represent the class of $500 or more. How would the sample mean and standard deviation change if you used 650 to represent this class?

a. Write the first four columns of a *frequency distribution.*
b. Find the *sample mean.*
c. Complete the *last three columns* of the frequency distribution.
d. Find the *sample standard deviation.*

Answer: Page A32

Exercises 2.4

Help

LarsonTutor 2.4

Companion Web Site

Student Solutions Manual 2.4

Videos 2.4

Technology Manuals

Basic Skills and Concepts

In Exercises 1 and 2, find the range, mean, variance, and standard deviation of the population data set.

1. 11 10 8 4 6 7 11 6 11 7

2. 13 23 15 13 18 13 15
14 20 20 18 17 20 13

In Exercises 3 and 4, find the range, mean, variance, and standard deviation of the sample data set.

3. 15 8 12 5 19 14 8 6 13

4. 24 26 27 23 9 14 8 8 26 15 15 27 11

Graphical Reasoning In Exercises 5 and 6, find the range of the data set represented by the display or graph.

5.

2	3 9	Key: 2|3 = 23
3	0 0 2 3 6 7	
4	0 1 2 3 3 8	
5	0 1 1 9	
6	1 2 9 9	
7	5 9	
8	4 8	
9	0 2 5 6	

6. **Bride's Age at First Marriage**

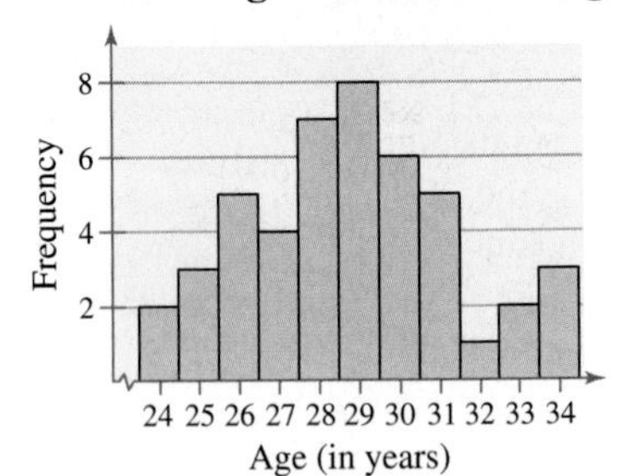

7. Explain how to find the range of a data set. What is an advantage of using the range as a measure of variation? What is a disadvantage?

8. Explain how to find the deviation of an entry in a data set. What is the sum of all the deviations in any data set?

9. Why is the standard deviation used more frequently than the variance? (*Hint:* Consider the units of the variance.)

10. Explain the relationship between variance and standard deviation. Can either of these measures be negative? Explain. Find a data set for which $n = 5$, $\bar{x} = 7$, and $s = 0$.

11. ***Marriage Ages*** ◆ The ages of 10 grooms at their first marriage are given below.

24.3 46.6 41.6 32.9 26.8 39.8 21.5 45.7 33.9 35.1

(a) Find the range of the data set.

(b) Change 46.6 to 66.6 and find the range of the new data set.

(c) Compare your answer to parts (a) and (b).

12. Find a population data set that contains six entries, has a mean of 5, and has a standard deviation of 2.

13. ***Graphical Reasoning*** ◆ Both data sets represented below have a mean of 50. One has a standard deviation of 2.4 and the other has a standard deviation of 5. Which is which? Explain your reasoning.

(a)

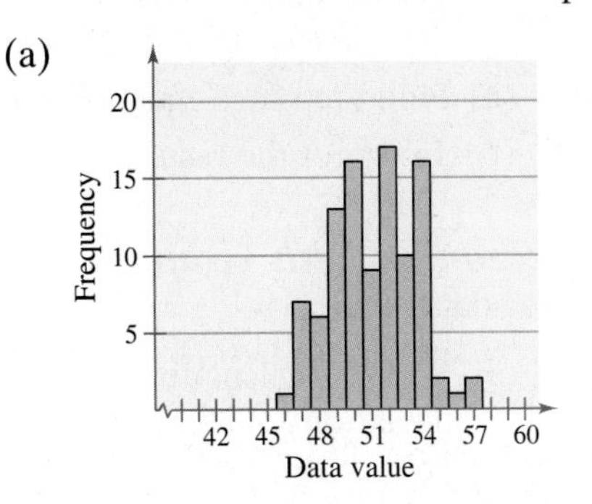

(b)

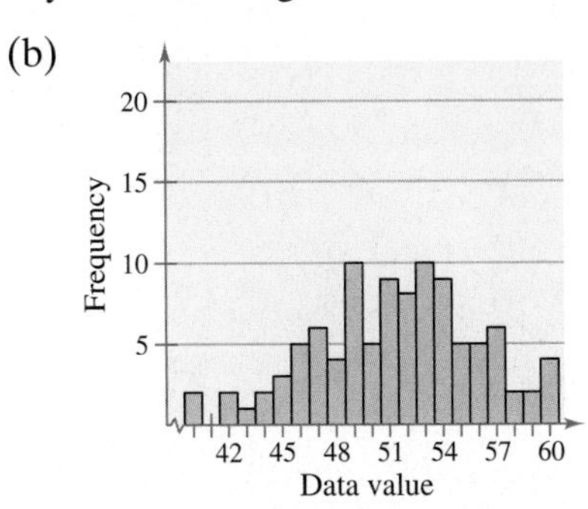

14. ***Writing*** ◆ Describe the difference between the calculation of population standard deviation and sample standard deviation. Given a data set, how do you know whether to calculate σ or s?

15. ***Salary Offers*** ◆ You are applying for a job at two companies. Company A offers starting salaries with $\mu = \$31,000$ and $\sigma = \$1000$. Company B offers starting salaries with $\mu = \$31,000$ and $\sigma = \$5000$. From which company are you more likely to get an offer of \$33,000 or more?

16. ***Golf Strokes*** ◆ An Internet site compares the strokes per round of two professional golfers. Which golfer is more consistent: Player A with $\mu -$ 71.5 strokes and $\sigma = 2.3$ strokes, or Player B with $\mu = 70.1$ strokes and $\sigma =$ 1.2 strokes?

Comparing Two Data Sets In Exercises 17–20, you are asked to compare two data sets and interpret the results.

17. ***Annual Salaries*** ◆ Sample annual salaries, in thousands of dollars, for municipal employees in Los Angeles and Long Beach are listed.

Los Angeles:	20.2	26.1	20.9	32.1	35.9	23.0	28.2	31.6	18.3
Long Beach:	20.9	18.2	20.8	21.1	26.5	26.9	24.2	25.1	22.2

(a) Find the range, variance, and standard deviation of each data set.

(b) Interpret the results in the context of the real-life setting.

18. ***Annual Salaries*** ◆ Sample annual salaries, in thousands of dollars, for municipal employees in Dallas and Houston are listed.

Dallas:	34.9	25.7	17.3	16.8	26.8	24.7	29.4	32.7	25.5
Houston:	25.6	23.2	26.7	27.7	25.4	26.4	18.3	26.1	31.3

(a) Find the range, variance, and standard deviation of each data set.

(b) Interpret the results in the context of the real-life setting.

19. ***SAT Scores*** ◆ Sample SAT scores for eight males and eight females are listed.

Male SAT scores:	1059	1328	1175	1123	923	1017	1214	1042
Female SAT scores:	1226	965	841	1053	1056	1393	1312	1222

(a) Find the range, variance, and standard deviation of each data set.

(b) Interpret the results in the context of the real-life setting.

20. *Annual Salaries* ◆ Sample annual salaries, in thousands of dollars, for public and private elementary school teachers are listed.

Public teachers:	38.6	38.1	38.7	36.8	34.8	35.9	39.9	36.2
Private teachers:	21.8	18.4	20.3	17.6	19.7	18.3	19.4	20.8

(a) Find the range, variance, and standard deviation of each data set.

(b) Interpret the results in the context of the real-life setting.

Reasoning with Graphs In Exercises 21 and 22, you are asked to compare three data sets.

21. (a) Without calculating, which data set has the greatest sample standard deviation? Which has the least sample standard deviation? Explain your reasoning.

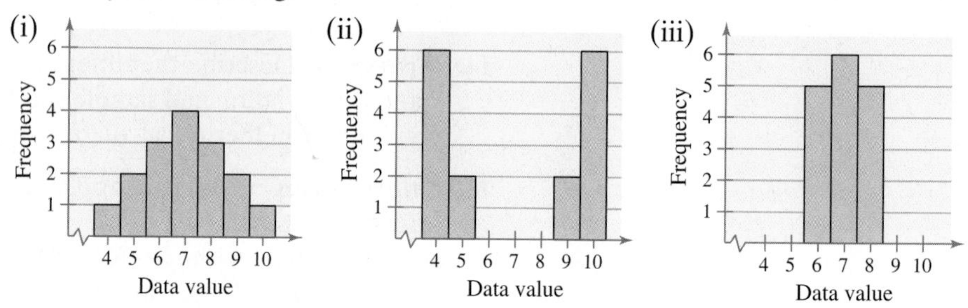

(b) How are the data sets the same? How do they differ?

22. (a) Without calculating, which data set has the greatest sample standard deviation? Which has the least sample standard deviation? Explain your reasoning.

(i)

Stem	Leaves
0	9
1	5 8
2	3 3 7 7
3	2 5
4	1

Key: 4|1 = 41

(ii)

Stem	Leaves
0	9
1	5
2	3 3 3 7 7 7
3	5
4	1

Key: 4|1 = 41

(iii)

Stem	Leaves
0	
1	5
2	3 3 3 3 7 7 7 7
3	5
4	

Key: 4|1 = 41

(b) How are the data sets the same? How do they differ?

23. *Writing* ◆ Discuss the similarities and the differences between the Empirical Rule and Chebychev's Theorem.

24. *Writing* ◆ What must you know about a data set before you can use the Empirical Rule?

Using the Empirical Rule In Exercises 25–28, you are asked to use the Empirical Rule.

25. The mean value of land and buildings per acre from a sample of farms is \$1000 with a standard deviation of \$200. The data set has a bell-shaped distribution. Estimate the percent of farms whose land and building values per acre are between \$800 and \$1200.

26. The mean value of land and buildings per acre from a sample of farms is \$1200 with a standard deviation of \$350. Between what two values does about 95% of the data lie? (Assume the data set has a bell-shaped distribution.)

27. Using the sample statistics from Exercise 25, do the following. (Assume the number of farms in the sample is 75.)

(a) Use the Empirical Rule to estimate the number of farms whose land and building values per acre are between \$800 and \$1200.

(b) If 25 additional farms were sampled, about how many of these farms would you expect to have land and building values between \$800 per acre and \$1200 per acre?

28. Using the sample statistics from Exercise 26, do the following. (Assume the number of farms in the sample is 40.)

(a) Using the Empirical Rule, estimate the number of farms whose land and building values per acre are between \$500 and \$1900.

(b) If 20 additional farms were sampled, about how many of these farms would you expect to have land and building values between \$500 per acre and \$1900 per acre?

29. ***Chebychev's Theorem*** ◆ Old Faithful is a famous geyser at Yellowstone National Park. From a sample with $n = 32$, the mean duration of Old Faithful's eruptions is 3.32 minutes and the standard deviation is 1.09 minutes. Using Chebychev's Theorem, at least how many of the eruptions lasted between 1.14 minutes and 5.5 minutes? *(Source: Yellowstone National Park)*

30. ***Chebychev's Theorem*** ◆ The mean time in a women's 400-meter dash is 52.37 seconds with a standard deviation of 2.15. Apply Chebychev's Theorem to the data using $k = 2$. Interpret the results.

Calculating Using Grouped Data In Exercises 31–38, use the grouped data formulas to find the indicated mean and standard deviation.

31. ***Pets Per Household*** ◆ The results of a random sample of the number of pets per household in a region are shown in the histogram. Estimate the sample mean and the sample standard deviation of the data set.

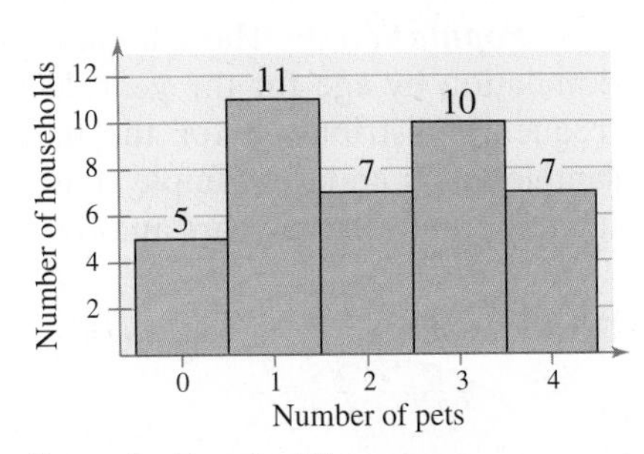

Figure for Exercise 31

Figure for Exercise 32

32. ***Cars Per Household*** ◆ A random sample of households in a region and the number of cars per household are shown in the histogram. Estimate the sample mean and the sample deviation of the data set.

33. ***Football Wins*** ◆ The number of wins for each National Football League team in 2000 are listed. Make a frequency distribution (using five classes) for the data set. Then approximate the population mean and the population standard deviation of the data set. *(Source: National Football League)*

11 10 9 8 5 13 12 9 7 4 3
12 11 7 6 1 12 11 8 5 3 11
10 9 9 5 10 10 7 6 4

34. *Water Consumption* ◆ The number of gallons of water consumed per day by a small village are listed. Make a frequency distribution (using five classes) for the data set. Then approximate the population mean and the population standard deviation of the data set.

167	180	192	173	145	151	174	175	178	160
195	224	244	146	162	146	177	163	149	188

35. *Amount of Caffeine* ◆ The amount of caffeine in a sample of 5-ounce servings of brewed coffee is given in the bar graph. Make a frequency distribution for the data. Then use the table to estimate the sample mean and the sample standard deviation of the data set. *(Adapted from the American Dietetic Association's Complete Food and Nurtrition Guide)*

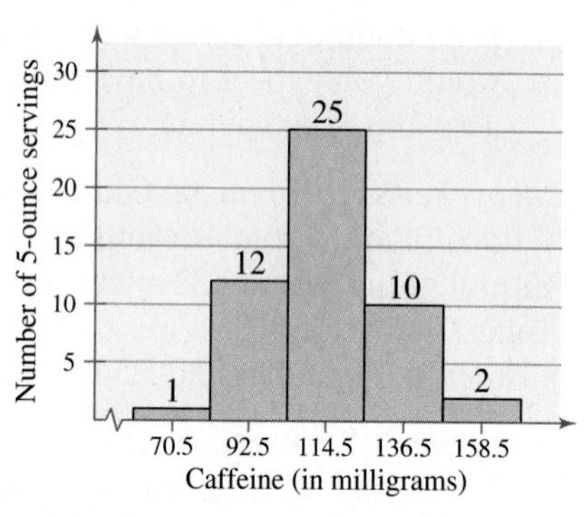

Figure for Exercise 35

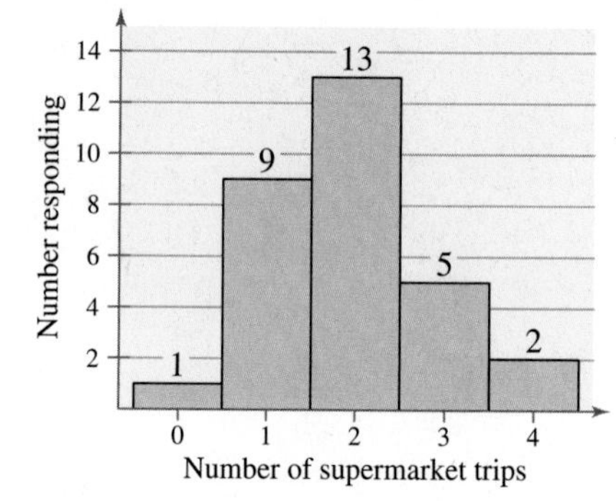

Figure for Exercise 36

36. *Supermarket Trips* ◆ Thirty people were randomly selected and asked how many trips to the supermarket they made in the past week. The responses are given in the bar graph. Make a frequency distribution for the data. Then use the table to estimate the sample mean and the sample standard deviation of the data set. *(Adapted from the Food Marketing Institute)*

37. *U.S. Population* ◆ The estimated distribution (in millions) of the U.S. population by age for the year 2006 is shown in the circle graph. Make a frequency distribution for the data. Then use the table to estimate the sample mean and the sample standard deviation of the data set. Use 70 as the midpoint for "65 years and over." *(Source: U.S. Census Bureau)*

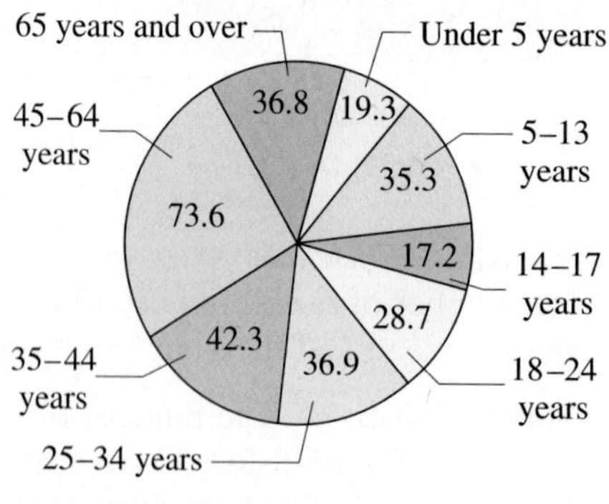

Figure for Exercise 37

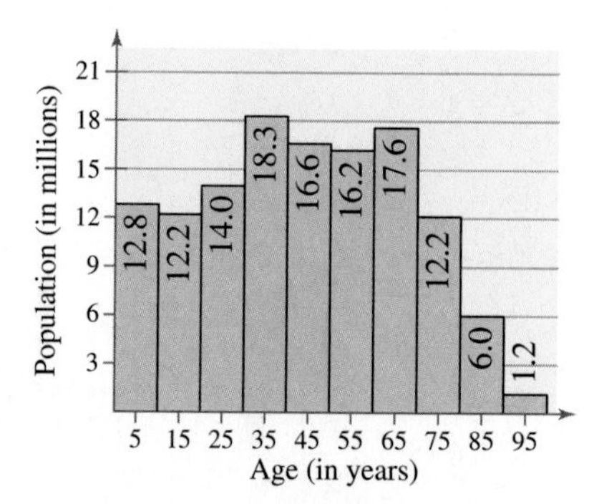

Figure for Exercise 38

38. *Japan's Population* ◆ Japan's estimated population for the year 2010 is given in the bar graph. Make a frequency distribution for the data. Then use the table to estimate the sample mean and the sample standard deviation of the data set. *(Source: U.S. Census Bureau, International Data Base)*

Extending the Basics

39. *Coefficient of Variation* ◆ The coefficient of variation CV describes the standard deviation as a percent of the mean. Because it has no units, you can use the coefficient of variation to compare data with different units.

$$CV = \frac{\text{Standard deviation}}{\text{mean}} \times 100\%$$

The following data show the heights (in inches) and weights (in pounds) of the members of a basketball team. Find the coefficient of variation for each data set. What can you conclude?

Heights:	72	74	68	76	74	69	72	79	70	69	77	73
Weights:	180	168	225	201	189	192	197	162	174	171	185	210

40. *Shortcut Formula* ◆ You used $SS_x = \Sigma(x - \bar{x})^2$ when calculating variance and standard deviation. An alternate formula that is sometimes more convenient for hand calculations is

$$SS_x = \Sigma x^2 - \frac{(\Sigma x)^2}{n}.$$

You can find the sample variance by dividing the sum of squares by $n - 1$, and the sample standard deviation by finding the square root of the sample variance.

(a) Use the shortcut formula to calculate the sample standard deviation for the data set given in Exercise 19.

(b) Compare your results to those obtained in Exercise 19.

41. *Team Project: Scaling Data* ◆ Consider the following sample data set.

100 200 300 400 500 600 700 800 900 1000

(a) Find $\bar{x}$ and s.

(b) Multiply each entry by 10. Find $\bar{x}$ and s for the revised data.

(c) Divide the original data by 10. Find $\bar{x}$ and s for the revised data.

(d) What can you conclude from the results of (a), (b), and (c)?

42. *Team Project: Shifting Data* ◆ Consider the following sample data set.

100 200 300 400 500 600 700 800 900 1000

(a) Find $\bar{x}$ and s.

(b) Add 10 to each entry. Find $\bar{x}$ and s for the revised data.

(c) Subtract 10 from the original data. Find $\bar{x}$ and s for the revised data.

(d) What can you conclude from the results of (a), (b), and (c)?

43. *Pearson's Index of Skewness* ◆ The English statistician Karl Pearson (1857–1936) introduced a formula for the skewness of a distribution.

$$P = \frac{3(\bar{x} - \text{median})}{s}$$ Pearson's index of skewness

Most distributions have an index of skewness between -3 and 3. When $P > 0$, the data are skewed right. When $P < 0$, the data are skewed left. When $P = 0$, the data are symmetric. Calculate the coefficient of skewness for each distribution. Describe the shape of each.

(a) $\bar{x} = 17, s = 2.3$, median $= 19$

(b) $\bar{x} = 32, s = 5.1$, median $= 25$

2 **Case Study** WWW.SUNGLASSASSOCIATION.COM

Sunglass Sales in the United States

The Sunglass Association of America is a not-for-profit association of manufacturers and distributors of sunglasses. Part of the association's mission is to gather and distribute marketing information about the sale of sunglasses. The data presented here are based on surveys administered by Jobson Optical Research International.

Outlet Type	Number of Locations
Optical Store	34,043
Sunglass Specialty	2,060
Dept. Store	6,866
Discount Dept. Store	10,376
Catalog Showroom	887
General Merchandise	11,868
Supermarket	21,613
Convenience Store	83,613
Chain Drug Store	31,127
Indep. Drug Store	7,034
Chain Apparel Store	26,831
Chain Sports Store	5,760
Indep. Sports Store	14,683

Number (in 1000s) of Pairs of Sunglasses Sold

Price	\$0–\$10	\$11–\$30	\$31–\$50	\$51–\$75	\$76–\$100	\$101–\$150	\$151+
Optical Store	0	290	3,164	1,240	3,654	842	478
Sunglass Specialty	192	708	2,515	1,697	1,145	805	378
Dept. Store	1,224	1,464	1,527	488	38	16	5
Discount Dept. Store	8,793	5,284	147	67	16	8	0
Catalog Showroom	153	100	65	35	29	9	0
General Merchandise	6,147	495	0	0	0	0	0
Supermarket	14,108	316	0	0	0	0	0
Convenience Store	19,726	2,985	0	0	0	0	0
Chain Drug Store	17,883	3,432	50	0	0	0	0
Indep. Drug Store	1,352	1,110	12	0	0	0	0
Chain Apparel Store	3,464	1,804	186	112	40	17	7
Chain Sports Store	672	526	430	72	45	18	4
Indep. Sports Store	875	1,997	1,320	528	206	85	11

Exercises

1. ***Mean Price*** Estimate the mean price of a pair of sunglasses sold at (a) an optical store, (b) a sunglass specialty store, and (c) a department store. Use \$200 as the midpoint for \$151+.

2. ***Revenue*** Which type of outlet had the greatest total revenue? Explain your reasoning.

3. ***Revenue*** Which type of outlet had the greatest revenue per location? Explain your reasoning.

4. ***Standard Deviation*** Estimate the standard deviation for the number of pairs of sunglasses sold at (a) optical stores, (b) sunglass specialty stores, and (c) department stores.

5. ***Standard Deviation*** Of the 13 distributions, which has the greatest standard deviation? Explain your reasoning.

6. ***Bell-Shaped Distribution*** Of the 13 distributions, which is more bell shaped? Explain.

2.5 Measures of Position

What You Should Learn

- How to find the first, second, and third quartiles of a data set
- How to find the interquartile range of a data set
- How to represent a data set graphically using a box-and-whisker plot
- How to interpret other fractiles such as percentiles
- How to find and interpret the standard score (*z*-score)

Quartiles • Percentiles and Other Fractiles • The Standard Score

Quartiles

In this section, you will learn how to use fractiles to specify the position of a data entry within a data set. **Fractiles** are numbers that partition, or divide, an ordered data set into equal parts. For instance, the median is a fractile because it divides an ordered data set into two equal parts.

DEFINITION

The three **quartiles** Q_1, Q_2, and Q_3 approximately divide an ordered data set into four equal parts. About one quarter of the data falls on or below the **first quartile** Q_1. About one half the data falls on or below the **second quartile** Q_2 (the second quartile is the same as the median of the data set). About three quarters of the data falls on or below the **third quartile** Q_3.

EXAMPLE 1

Finding the Quartiles of a Data Set

The test scores of 15 employees enrolled in a CPR training course are listed. Find the first, second, and third quartiles of the test scores.

13 9 18 15 14 21 7 10 11 20 5 18 37 16 17

SOLUTION First, order the data set and find the median Q_2. Once you find Q_2, you can divide the data set into two halves. The first and third quartiles are the medians of the lower and upper halves of the data set.

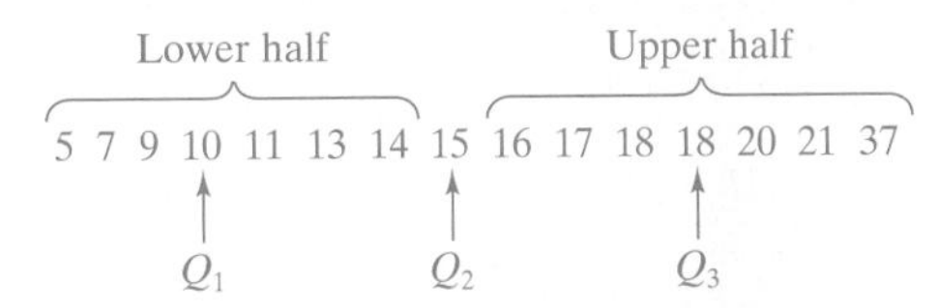

So about one fourth of the employees scored 10 or less, about one half scored 15 or less, and about three fourths scored 18 or less.

Try It Yourself 1

Find the first, second, and third quartiles for the ages of the Akhiok residents. Use the data set given in the chapter opener on page 30.

a. *Order* the data set.
b. Find the median Q_2.
c. Find the first and third quartiles Q_1 and Q_3.

Answer: Page A33

EXAMPLE 

Using Technology to Find Quartiles

The tuition costs, in thousands of dollars, for 25 liberal arts colleges are listed. Use a calculator or a computer to find the first, second, and third quartiles. *(Source: U.S. News and World Report)*

23 25 30 23 20 22 21 15 25 24 30 25 30
20 23 29 20 19 22 23 29 23 28 22 28

SOLUTION *Minitab*, *Excel*, and the *TI-83* each have features that automatically calculate quartiles. Try using this technology to find the first, second, and third quartiles of the tuition data. From the displays, you can see that $Q_1 = 21.5$, $Q_2 = 23$, and $Q_3 = 28$, which means that about one quarter of these colleges charge tuition of $21,500 or less, one half charge $23,000 or less, and about three quarters charge $28,000 or less.

Study Tip

There are several ways to find the quartiles of a data set. Regardless of how you find the quartiles, the results are rarely off by more than one data entry. For instance, in Example 2, the first quartile, as determined by *Excel*, is 22 instead of 21.5.

EXCEL

	A	B	C	D
1	23			
2	25		Quartile(A1:A25,1)	
3	30		22	
4	23			
5	20		Quartile(A1:A25,2)	
6	22		23	
7	21			
8	15		Quartile(A1:A25,3)	
9	25		28	
10	24			
11	30			
12	25			
13	30			
14	20			
15	23			
16	29			
17	20			
18	19			
19	22			
20	23			
21	29			
22	23			
23	28			
24	22			
25	28			

TI-83

1-Var Stats
↑n=25
minX=15
Q_1=21.5
Med=23
Q_3=28
maxX=30

MINITAB

Descriptive Statistics

Variable	N	Mean	Median	TrMean	StDev
Tuition	25	23.960	23.000	24.087	3.942

Variable	SE Mean	Minimum	Maximum	Q1	Q3
Tuition	0.788	15.000	30.000	21.500	28.000

Try It Yourself 2

The tuition costs, in thousands of dollars, for 25 universities are listed. Use a calculator or a computer to find the first, second, and third quartiles. *(Source: U.S. News and World Report)*

20 26 28 25 31 14 23 15 12 26 29 24 31
19 31 17 15 17 20 31 32 16 21 22 28

a. *Enter* the data.
b. *Calculate* the first, second, and third quartiles.
c. What can you conclude?

Answer: Page A33

After finding the quartiles of a data set, you can find the interquartile range.

DEFINITION

The **interquartile range (IQR)** of a data set is the difference between the first and third quartiles.

$$\text{Interquartile range (IQR)} = Q_3 - Q_1$$

Insight

The IQR is a measure of variation that gives you an idea of how much the middle 50% of the data varies. It can also be used to identify outliers. Any data value that lies more than 1.5 IQRs to the left of Q_1 or to the right of Q_3 is an outlier. For instance, 37 is an outlier of the 15 test scores in Example 1.

EXAMPLE 

Finding the Interquartile Range

Find the interquartile range of the 15 test scores given in Example 1. What can you conclude from the result?

SOLUTION From Example 1, you know that $Q_1 = 10$ and $Q_3 = 18$. So, the interquartile range is

$$\begin{aligned} \text{IQR} &= Q_3 - Q_1 \\ &= 18 - 10 \\ &= 8. \end{aligned}$$

This means that the test scores in the middle half of the data set vary by at most eight points.

Try It Yourself 3

Find the interquartile range for the ages of the Akhiok residents. Use the data set given in the chapter opener on page 30.

a. Find the first and third quartiles, Q_1 and Q_3.
b. *Subtract* Q_1 from Q_3.
c. *Interpret* the result in the context of the data.

Answer: Page A33

Another important application of quartiles is to represent data sets using box-and-whisker plots. A **box-and-whisker plot** is an exploratory data analysis tool that highlights the important features of a data set. To graph a box-and-whisker plot, you must know the following values.

Picturing the World

Of the first 43 U.S. presidents, Theodore Roosevelt was the youngest at the time of inauguration at the age of 42. Ronald Reagan was the oldest president inaugurated at the age of 69. The box-and-whisker plot summarizes the ages of the first 43 U.S. presidents at inauguration. *(Source: infoplease.com)*

Ages of U.S. Presidents at Inauguration

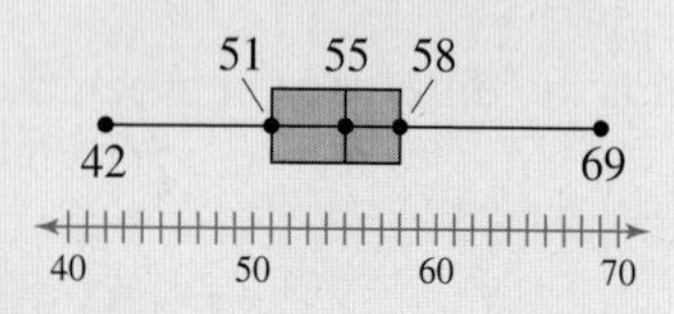

How many U.S. presidents' ages are represented by the box?

1. The minimum entry
2. The first quartile Q_1
3. The median Q_2
4. The third quartile Q_3
5. The maximum entry

These five numbers are called the **five-number summary** of the data set.

GUIDELINES

Drawing a Box-and-Whisker Plot

1. Find the five-number summary of the data set.
2. Construct a horizontal scale that spans the range of the data.
3. Plot the five numbers above the horizontal scale.
4. Draw a box above the horizontal scale from Q_1 to Q_3 and draw a vertical line in the box at Q_2.
5. Draw whiskers from the box to the minimum and maximum entries.

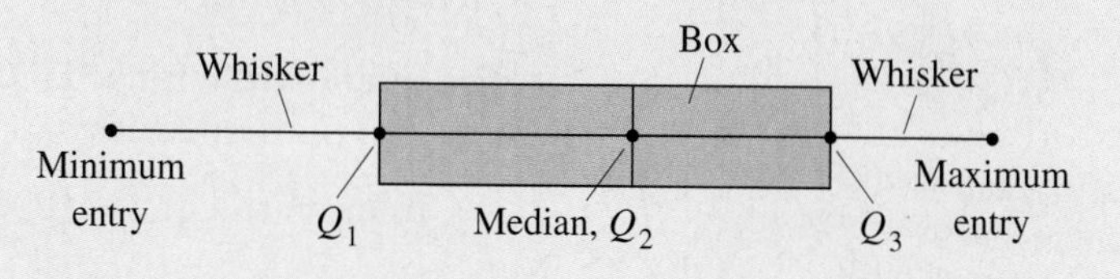

EXAMPLE 4

Drawing a Box-and-Whisker Plot

See *Minitab* and *TI-83* steps on pages 98, 99.

Draw a box-and-whisker plot that represents the 15 test scores given in Example 1. What can you conclude from the display?

SOLUTION The five-number summary of the test scores is as follows.

Min = 5 $\quad Q_1 = 10 \quad Q_2 = 15 \quad Q_3 = 18 \quad$ Max = 37

Using these five numbers, you can construct the box-and-whisker plot shown. You can make several conclusions from the display. One is that about half the scores are between 10 and 18.

Test Scores in CPR Class

5 10 15 18 37

5 6 7 8 9 10 11 12 13 14 15 16 17 18 19 20 21 22 23 24 25 26 27 28 29 30 31 32 33 34 35 36 37

Try It Yourself 4

Draw a box-and-whisker plot that represents the ages of the residents of Akhiok. Use the data set given in the chapter opener on page 30.

a. Find the *five-number summary* of the data set.
b. Construct a *horizontal scale* and *plot* the five numbers above it.
c. Draw the *box*, the *vertical line*, and the *whiskers.*
d. Make some conclusions.

Answer: Page A33

Percentiles and Other Fractiles

In addition to using quartiles to specify a measure of position, you can also use percentiles and deciles. These common fractiles are summarized as follows.

Fractiles	Summary	Symbols
Quartiles	Divide a data set into four equal parts.	Q_1, Q_2, Q_3
Deciles	Divide a data set into ten equal parts.	$D_1, D_2, D_3, \ldots, D_9$
Percentiles	Divide a data set into one hundred equal parts.	$P_1, P_2, P_3, \ldots, P_{99}$

Percentiles are often used in education and health-related fields to indicate how one individual compares with others in a group. For instance, test scores and children's growth measurements are often expressed in percentiles.

Study Tip

It is important that you understand what a percentile means. For instance, if the weight of a six-month-old infant is at the 78th percentile, that means that the infant's weight is greater than 78% of all six-month-old infants. It does not mean that the infant weighs 78% of some ideal weight.

EXAMPLE 5

Interpreting Percentiles

The ogive represents the cumulative frequency distribution for SAT test scores of college-bound students in the year 2000. What test score represents the 64th percentile? How should you interpret this? *(Source: College Board Online)*

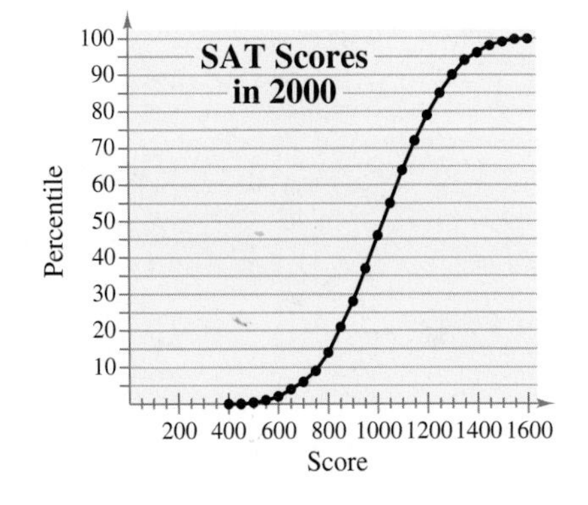

SOLUTION From the ogive, you can see that the 64th percentile corresponds to a test score of 1100. This means that 64% of the students had an SAT score of 1100 or less.

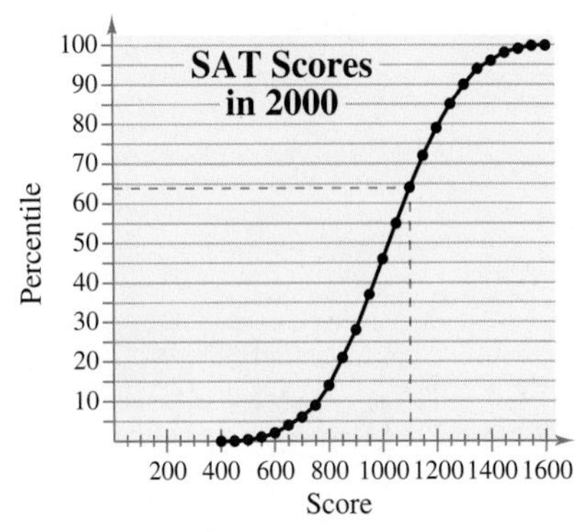

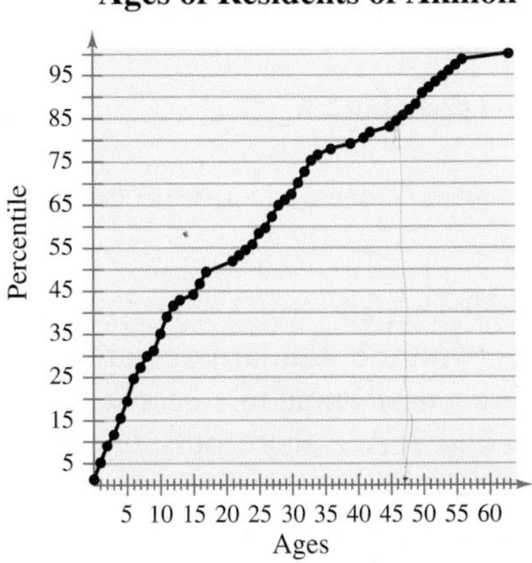

Try It Yourself 5

The ages of the residents of Akhiok are represented in the cumulative frequency graph at the left. At what percentile is a resident whose age is 47? How should you interpret this?

a. *Use the graph* to find the percentile that corresponds to the given age.
b. *Interpret* the results in the context of the data.

Answer: Page A33

The Standard Score

Another measure of position is the standard score, or z-score.

DEFINITION

The **standard score,** or **z-score,** represents the number of standard deviations a given value x falls from the mean μ. To find the z-score for a given value, use the following formula.

$$z = \frac{\text{value} - \text{mean}}{\text{standard deviation}} = \frac{x - \mu}{\sigma}$$

A z-score can be negative, positive, or zero. If z is negative, the corresponding x-value is below the mean. If z is positive, the corresponding x-value is above the mean. And if $z = 0$, the corresponding x-value is equal to the mean.

EXAMPLE 

Finding z-Scores

The mean speed of vehicles along a stretch of highway is 56 mph with a standard deviation of 4 mph. You measure the speed of three cars traveling along this stretch of highway as 62 mph, 47 mph, and 56 mph. Find the z-score that corresponds to each speed. What can you conclude?

SOLUTION The z-score that corresponds to each speed is calculated below.

$x = 62$ **mph**

$$z = \frac{62 - 56}{4} = 1.5$$

$x = 47$ **mph**

$$z = \frac{47 - 56}{4} = -2.25$$

$x = 56$ **mph**

$$z = \frac{56 - 56}{4} = 0$$

From the z-scores, you can conclude that a speed of 62 mph is 1.5 standard deviations above the mean, a speed of 47 mph is 2.25 standard deviations below the mean, and a speed of 56 mph is equal to the mean.

Try It Yourself 6

The monthly utility bills in a city have a mean of \$70 and a standard deviation of \$8. Find the z-scores that correspond to utility bills of \$60, \$71, and \$92. What can you conclude?

a. *Identify* μ and σ of the nonstandard normal distribution.
b. *Transform* each value to a z-score.
c. *Interpret* the results.

Answer: Page A33

Insight

Notice that if the distribution of the speeds in Example 6 is approximately bell shaped, the car going 47 mph would be traveling at an unusually slow speed because the speed corresponds to a z-score of -2.25.

When a distribution is approximately bell shaped, you know from the Empirical Rule that about 95% of the data lie within 2 standard deviations of the mean. So, when this distribution's values are transformed to z-scores, about 95% of the z-scores should fall between -2 and 2. A z-score outside of this range will occur about 5% of the time and would be considered unusual. So, according to the Empirical Rule, a z-score less than -3 or greater than 3 would be *very* unusual, with such a score occurring about 0.3% of the time.

Exercises

Help

LarsonTutor 2.5

Companion Web Site

Student Solutions Manual 2.5

Videos 2.5

Technology Manuals

Basic Skills and Concepts

In Exercises 1 and 2, (a) find the three quartiles and (b) draw a box-and-whisker plot of the data.

1. 4 7 7 5 2 9 7 6 8 5 8 4 1 5 2 8 7 6 6 9

2. 2 7 1 3 1 2 8 9 9 2 5 4 7 3 7 5 4 7
2 3 5 9 5 6 3 9 3 4 9 8 8 2 3 9 5

3. The points scored per game by a basketball team represent the third quartile for all teams in a league. What can you conclude about the team's points scored per game?

4. A salesperson at a company sold $6,903,435 of hardware equipment last year, a figure that represented the eighth decile of sales performance at the company. What can you conclude about the salesperson's performance?

5. A student's score on the ACT placement test for college algebra is in the 63rd percentile. What can you conclude about the student's test score?

6. A doctor tells a child's parents that their child's height is in the 87th percentile for the child's age group. What can you conclude about the child's height?

True or False In Exercises 7 and 8, determine whether the statement is true or false. If it is false, rewrite it as a true statement.

7. The second quartile is the median of an ordered data set.

8. The five numbers you need to graph a box-and-whisker plot are the minimum, the maximum, Q_1, Q_3, and the mean.

Graphical Analysis In Exercises 9–14, use the box-and-whisker plot to identify

(a) the minimum entry.
(b) the maximum entry.
(c) the first quartile.
(d) the second quartile.
(e) the third quartile.
(f) the interquartile range.

9.

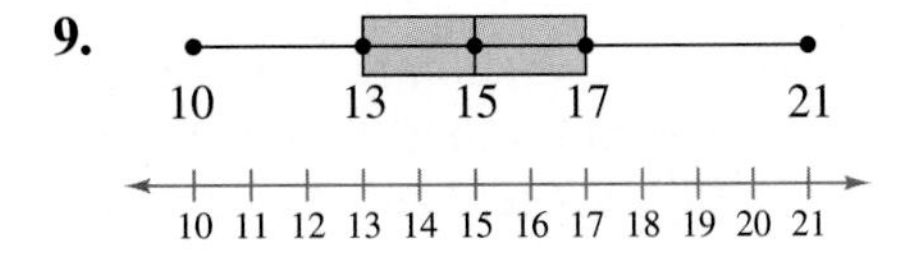

10.

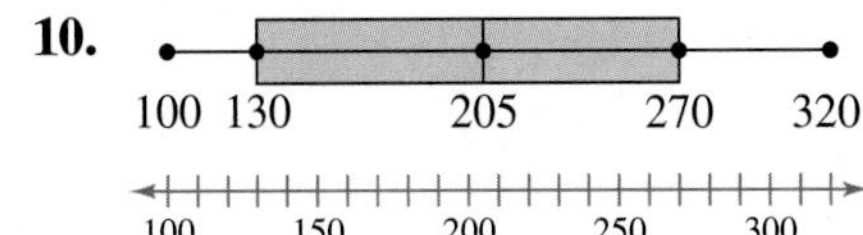

11.

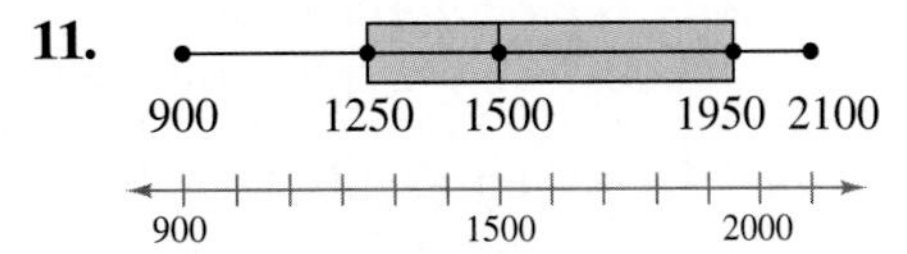

12.

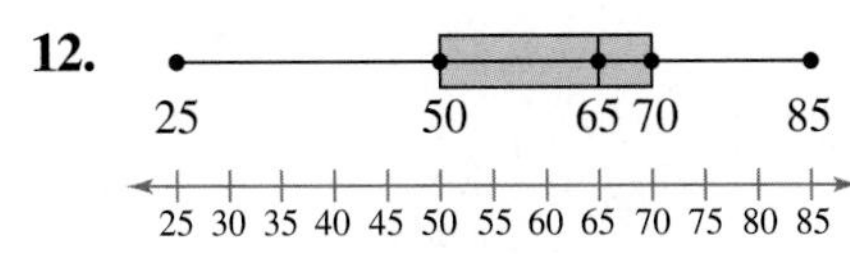

13.

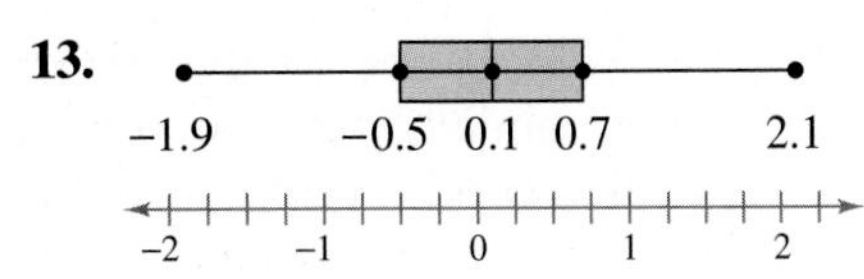

14.

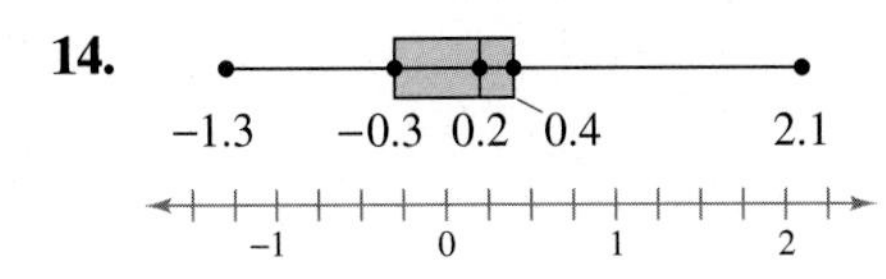

15. *Graphical Analysis* ◆ The letters A, B, and C are marked on the histogram. Match them to Q_1, Q_2, (the median), and Q_3. Justify your answer.

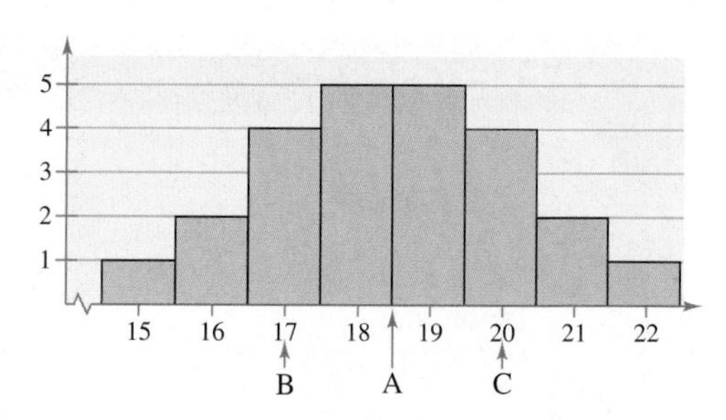

16. *Graphical Analysis* ◆ The letters R, S, and T are marked on the histogram. Match them to P_{10}, P_{50}, and P_{80}. Justify your answer.

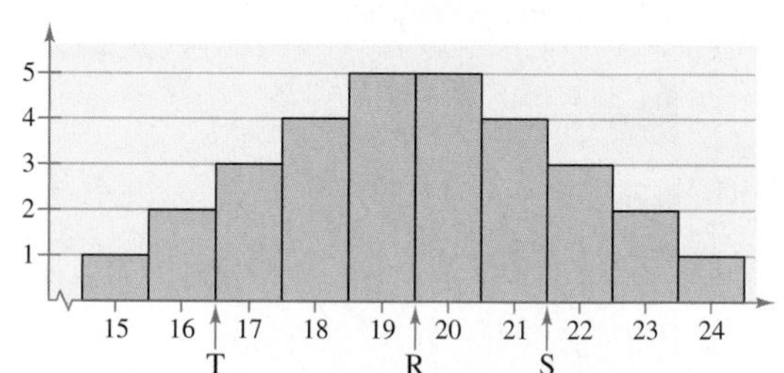

Using Technology to Find Quartiles and Draw Graphs In Exercises 17 and 18, use a calculator or a computer to (a) find the data set's first, second, and third quartiles, and (b) draw a box-and-whisker plot that represents the data set.

17. *TV Viewing* ◆ The number of hours of television watched per day by a sample of 28 people

2 4 1 5 7 2 5 4 4 2 3 6 4 3
5 2 0 3 5 9 4 5 2 1 3 6 7 2

18. *Hourly Earnings* ◆ The hourly earnings (in dollars) of a sample of 25 railroad equipment manufacturers

15.60 18.75 14.60 15.80 14.35 13.90 17.50 17.55 13.80
14.20 19.05 15.35 15.20 19.45 15.95 16.50 16.30 15.25
15.05 19.10 15.20 16.22 17.75 18.40 15.25

19. *TV Viewing* ◆ Refer to the data set given in Exercise 17 and the box-and-whisker plot you drew that represents the data set.

(a) About 75% of the people watched no more than how many hours of television per day?

(b) What percent of the people watched more than 4 hours of television per day?

(c) If you randomly selected one person from the sample, what is the likelihood that the person watched less than 2 hours of television per day? Write your answer as a percent.

20. *Manufacturer Earnings* ◆ Refer to the data set given in Exercise 18 and the box-and-whisker plot you drew that represents the data set.

(a) About 75% of the manufacturers made less than what amount per hour?

(b) What percent of the manufacturers made more than $15.80 per hour?

(c) If you randomly selected one manufacturer from the sample, what is the likelihood that the manufacturer made less than $15.80 per hour? Write your answer as a percent.

Graphical Analysis In Exercises 21 and 22, the midpoints A, B, and C are marked on the histogram. Match them to the indicated *z*-scores. Which *z*-scores, if any, would be considered unusual?

21. $z = 0$

$z = 2.14$

$z = -1.43$

22. $z = 0.77$

$z = 1.54$

$z = -1.54$

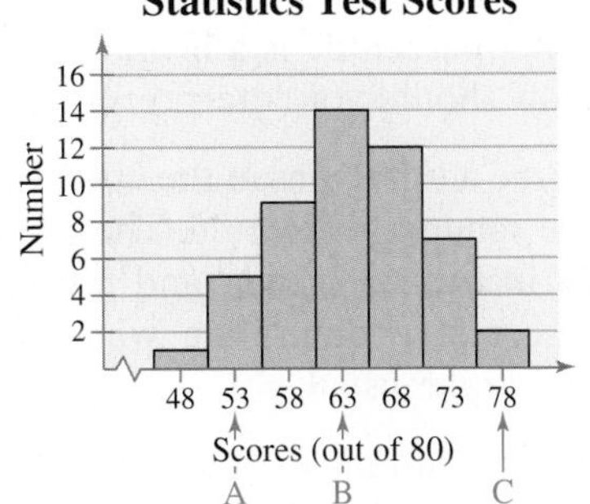

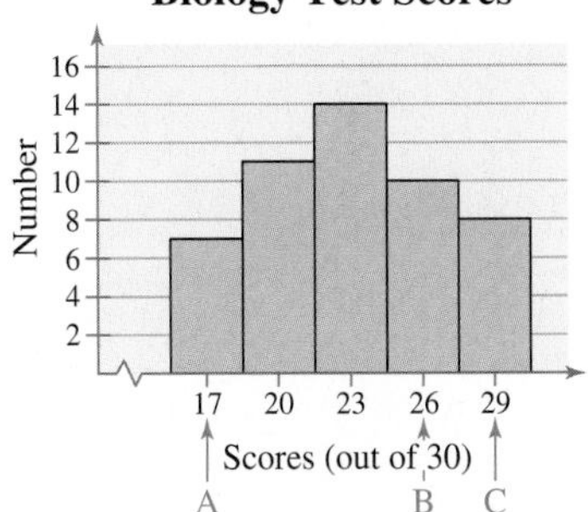

Comparing Test Scores For the statistics test scores in Exercise 21, the mean is 63 and the standard deviation is 7.0, and for the biology test scores in Exercise 22 the mean is 23 and the standard deviation is 3.9. In Exercises 23–26, you are given the test scores of a student who took both tests.

(a) Transform each test score to a *z*-score.

(b) Determine on which test the student had a better score.

23. A student gets a 73 on the statistics test and a 26 on the biology test.

24. A student gets a 60 on the statistics test and a 20 on the biology test.

25. A student gets a 78 on the statistics test and a 29 on the biology test.

26. A student gets a 63 on the statistics test and a 23 on the biology test.

27. ***Tires*** ◆ A certain brand of automobile tire has a mean life of 35,000 miles and a standard deviation of 2250 miles. (Assume the life spans of the tires have a bell-shaped distribution.)

(a) The life spans of three randomly selected tires are 34,000 miles, 37,000 miles, and 31,000 miles. Find the *z*-score that corresponds to each life span. According to the *z*-scores, would the life spans of any of these tires be considered unusual?

(b) The life spans of three randomly selected tires are 30,500 miles, 37,250 miles, and 35,000 miles. Using the Empirical Rule, find the percentile that corresponds to each life span.

28. ***Fruit Flies*** ◆ The life spans of a species of fruit fly have a bell-shaped distribution, with a mean of 33 days and a standard deviation of 4 days.

(a) The life spans of three randomly selected fruit flies are 34 days, 30 days, and 42 days. Find the *z*-score that corresponds to each life span and determine if any of these life spans are unusual.

(b) The life spans of three randomly selected fruit flies are 29 days, 41 days, and 25 days. Using the Empirical Rule, find the percentile that corresponds to each life span.

Interpreting Percentiles In Exercises 29–32, use the cumulative frequency distribution to answer the questions. The cumulative frequency distribution represents the heights of males in the U.S. in the 20–29 age group. The heights have a bell-shaped distribution (see Picturing the World, page 76) with a mean of 69.2 inches and a standard deviation of 2.9 inches. *(Source: National Center for Health Statistics)*

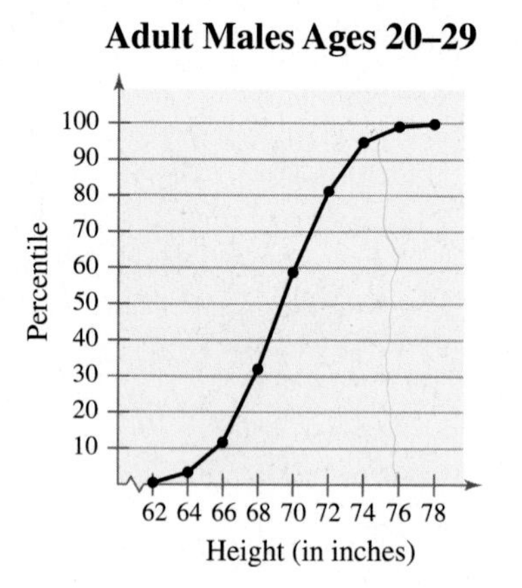

29. What height represents the 20th percentile? How should you interpret this?

30. What percentile is a height of 76 inches? How should you interpret this?

31. Three adult males in the 20–29 age group are randomly selected. Their heights are 74 inches, 62 inches, and 80 inches. Use z-scores to determine which heights, if any, are unusual.

32. Three adult males in the 20–29 age group are randomly selected. Their heights are 70 inches, 66 inches, and 68 inches. Use z-scores to determine which heights, if any, are unusual.

Extending the Basics

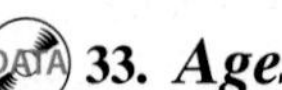

33. ***Ages of Executives*** The ages of a sample of 100 executives are listed.

31 62 51 44 61 47 49 45 40 52 60 51 67 47 63 54 59 43 63 52
50 54 61 41 48 49 51 54 39 54 47 52 36 53 74 33 53 68 44 40
60 42 50 48 42 42 36 57 42 48 56 51 54 42 27 43 43 41 54 49
49 47 51 28 54 36 36 41 60 55 42 59 35 65 48 56 82 39 54 49
61 56 57 32 38 48 64 51 45 46 62 63 59 63 32 47 40 37 49 57

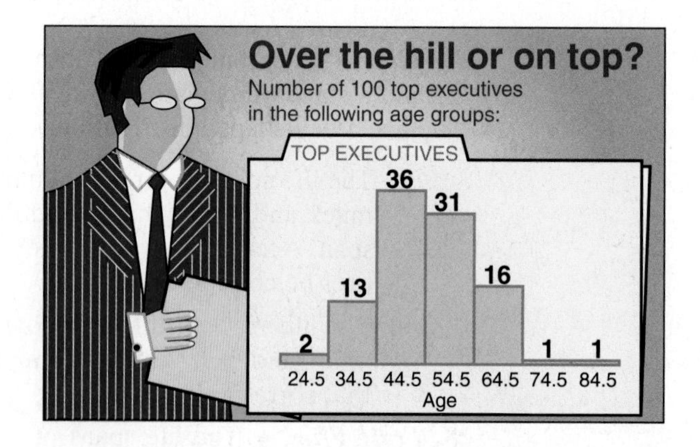

(a) Order the data and find the first, second, and third quartiles.

(b) Draw a box-and-whisker plot that represents the data set.

(c) Interpret the results in the context of the data.

(d) Based on this sample, at what age would you expect to be an executive? Explain your reasoning.

TECHNOLOGY MINITAB EXCEL TI-83

The Dairy Farmers of America is an association that provides help to dairy farmers. Part of this help is gathering and distributing statistics on milk production.

www.dfamilk.com

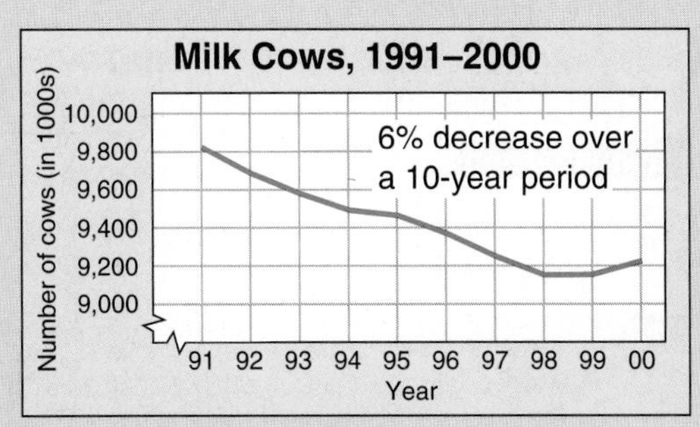

Source: National Agricultural Statistics Service

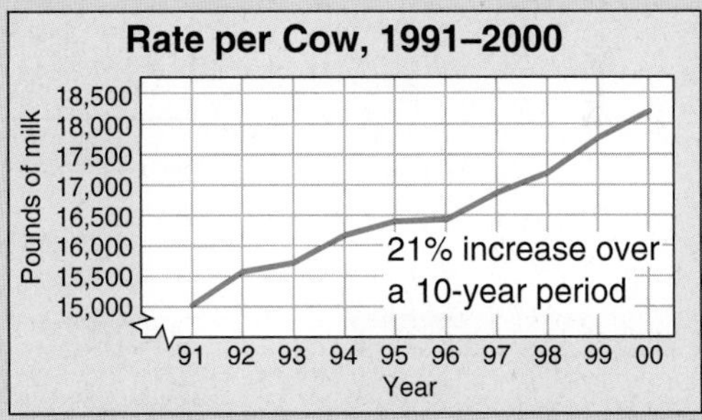

Source: National Agricultural Statistics Service

From 1991 to 2000, the number of dairy cows in the United States decreased and the yearly milk production increased.

Monthly Milk Production

The following data set was supplied by a dairy farmer. It lists the monthly milk production (in pounds) for 50 different Holstein dairy cows. *(Source: Matlink Dairy, Clymer, NY)*

2825	2072	2733	2069	2484
4285	2862	3353	1449	2029
1258	2982	2045	1677	1619
2597	3512	2444	1773	2284
1884	2359	2046	2364	2669
3109	2804	1658	2207	2159
2207	2882	1647	2051	2202
3223	2383	1732	2230	1147
2711	1874	1979	1319	2923
2281	1230	1665	1294	2936

Exercises

In Exercises 1–4, use a computer or calculator. If possible, print your results.

1. Find the sample mean of the data.
2. Find the sample standard deviation of the data.
3. Make a frequency distribution for the data. Use a class width of 500.
4. Draw a histogram for the data. Does the distribution appear to be bell shaped?
5. What percent of the distribution lies within one standard deviation of the mean? Within two standard deviations of the mean? How do these results agree with the Empirical Rule?

In Exercises 6–8, use the frequency distribution found in Exercise 3.

6. Use the frequency distribution to estimate the sample mean of the data. Compare your results to Exercise 1.
7. Use the frequency distribution to find the sample standard deviation for the data. Compare your results to Exercise 2.
8. ***Writing*** Use the results of Exercises 6 and 7 to write a general statement about the mean and standard deviation for grouped data. Do the formulas for grouped data give results that are as accurate as the individual entry formulas?

Extended solutions are given in the *Technology Supplement*.
Technical instruction is provided for *Minitab*, *Excel*, and the *TI-83*.

Using Technology to Determine Descriptive Statistics

Here are some *Minitab* and *TI-83* printouts for three examples in this chapter.

(See Example 7, page 52)

Layout...
Plot...
Time Series Plot...
Chart...
Histogram...
Boxplot...
Matrix Plot...
Draftsman Plot...
Contour Plot...

MINITAB

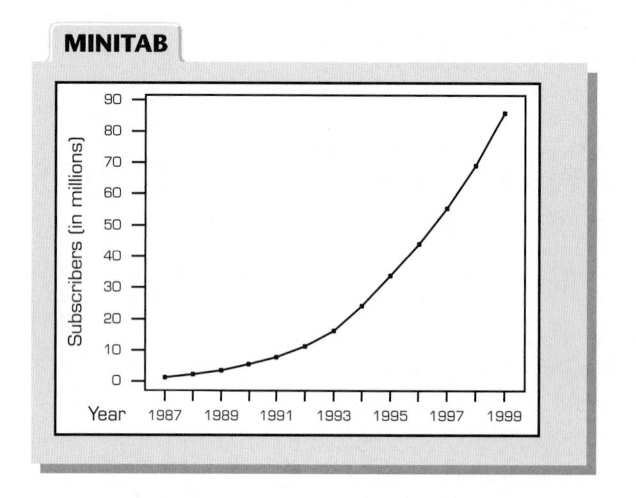

(See Example 4, page 73)

Display Descriptive Statistics...
Store Descriptive Statistics...
1-Sample Z...
1-Sample t...
2-Sample t...
Paired t...
1 Proportion...
2 Proportions...
2 Variances...
Correlation...
Covariance...
Normality Test...

MINITAB

Descriptive Statistics

Variable	N	Mean	Median	TrMean	StDev	SE Mean
C1	10	41.500	41.000	41.375	3.136	0.992

Variable	Minimum	Maximum	Q1	Q3
C1	37.000	47.000	38.750	44.250

(See Example 4, page 90)

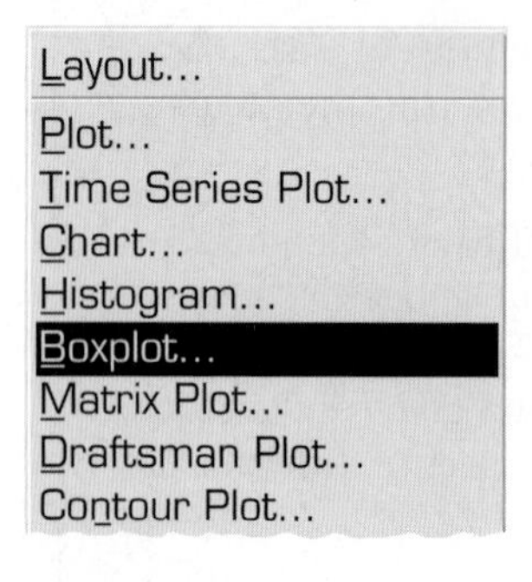

MINITAB

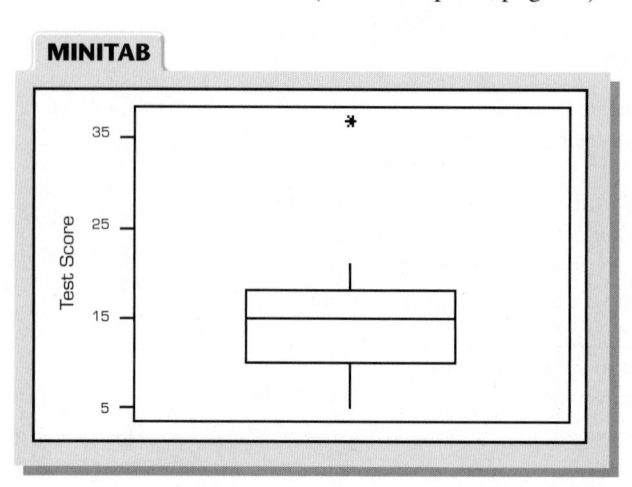

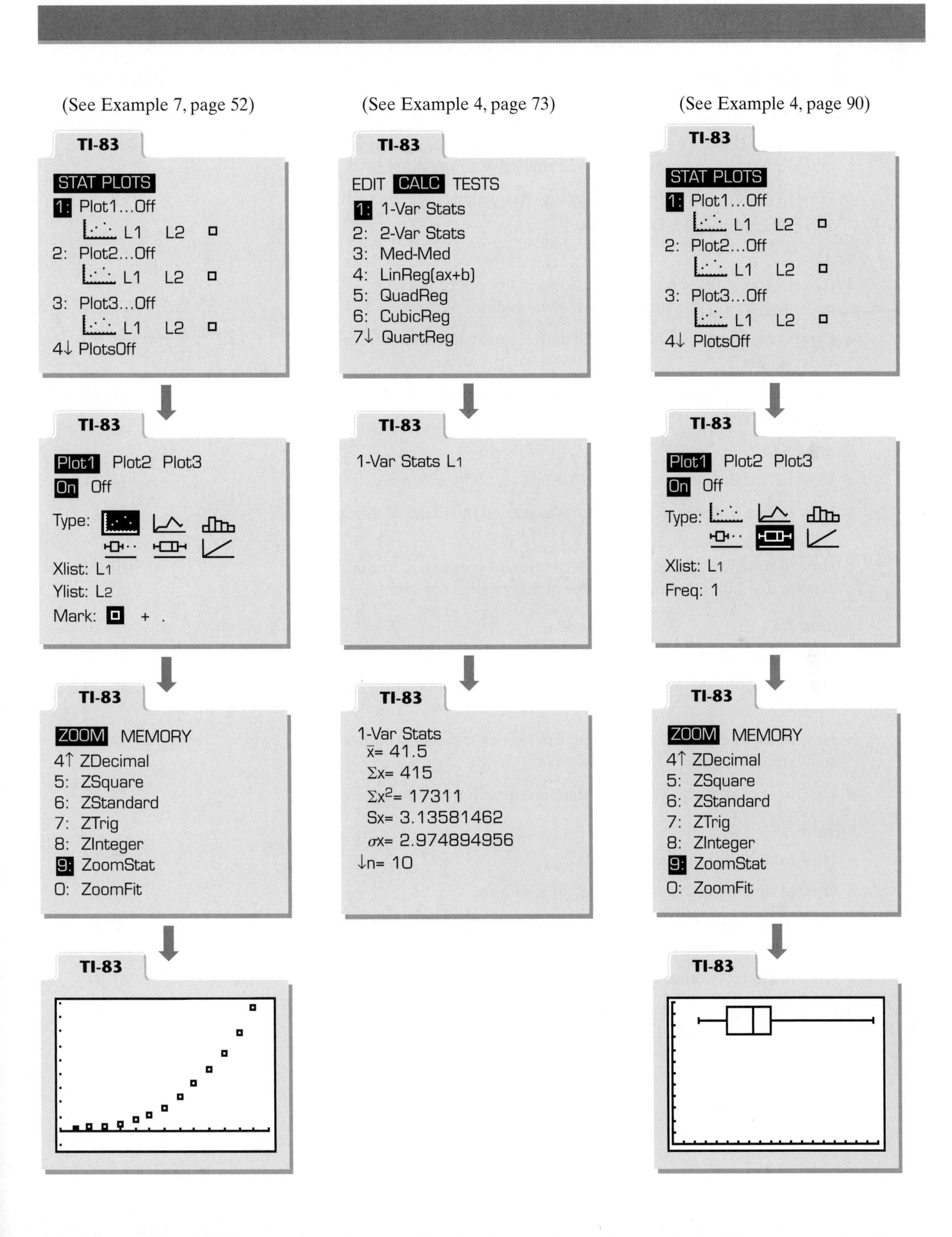
(See Example 7, page 52)
TI-83
STAT PLOTS
1: Plot1...Off
L1 L2
2: Plot2...Off
L1 L2
3: Plot3...Off
L1 L2
4↓ PlotsOff
TI-83
Plot1 Plot2 Plot3
On Off
Type:
Xlist: L1
Ylist: L2
Mark: + .
TI-83
ZOOM MEMORY
4↑ ZDecimal
5: ZSquare
6: ZStandard
7: ZTrig
8: ZInteger
9: ZoomStat
0: ZoomFit
TI-83
(See Example 4, page 73)
TI-83
EDIT CALC TESTS
1: 1-Var Stats
2: 2-Var Stats
3: Med-Med
4: LinReg(ax+b)
5: QuadReg
6: CubicReg
7↓ QuartReg
TI-83
1-Var Stats L1
TI-83
1-Var Stats
x̄= 41.5
Σx= 415
Σx²= 17311
Sx= 3.13581462
σx= 2.974894956
↓n= 10
(See Example 4, page 90)
TI-83
STAT PLOTS
1: Plot1...Off
L1 L2
2: Plot2...Off
L1 L2
3: Plot3...Off
L1 L2
4↓ PlotsOff
TI-83
Plot1 Plot2 Plot3
On Off
Type:
Xlist: L1
Freq: 1
TI-83
ZOOM MEMORY
4↑ ZDecimal
5: ZSquare
6: ZStandard
7: ZTrig
8: ZInteger
9: ZoomStat
0: ZoomFit
TI-83

2

Chapter Summary

What did you learn?	*Review Exercises*
Section 2.1	
◆ How to construct a frequency distribution including limits, boundaries, midpoints, relative frequencies, and cumulative frequencies	*1*
◆ How to construct frequency histograms, frequency polygons, relative frequency histograms, and ogives	*2–6*
Section 2.2	
◆ How to graph quantitative data sets using the exploratory data analysis tools of stem-and-leaf plots and dot plots	*7, 8*
◆ How to graph and interpret paired data sets using scatter plots and time series charts	*9, 10*
◆ How to graph qualitative data sets using pie charts and Pareto charts	*11, 12*
Section 2.3	
◆ How to find the mean, median, and mode of a population and a sample	*13, 14*
◆ How to find a weighted mean of a data set and the mean of a frequency distribution	*15–18*
◆ How to describe the shape of a distribution as symmetric, uniform, or skewed and how to compare the mean and median for each	*19–24*
Section 2.4	
◆ How to find the range of a data set	*25, 26*
◆ How to find the mean and standard deviation of a population and of a sample	*27–30*
◆ How to use the Empirical Rule and Chebychev's Theorem to interpret standard deviation	*31–34*
◆ How to approximate the sample standard deviation for grouped data	*35, 36*
Section 2.5	
◆ How to find the quartiles of a data set	*37, 38*
◆ How to find the interquartile range of a data set	*39, 41*
◆ How to use the exploratory data analysis tool of describing a data set using a box-and-whisker plot	*40, 42*
◆ How to interpret other fractiles such as percentiles	*43, 44*
◆ How to find and interpret the standard score (z-score)	*45–48*

STATISTICS

Uses and Abuses

Uses

It can be difficult to see trends or patterns from a set of raw data. Descriptive statistics helps you do so. A good description of a data set consists of three features: (1) the shape of the data, (2) a measure of the center of the data, and (3) a measure of how much variability there is in the data. When you read reports, news items, or advertisements prepared by other people, you are seldom given raw data sets. Instead, you are given graphs, measures of central tendency, and measures of variation. To be a discerning reader, you need to understand the terms and techniques of descriptive statistics.

Abuses

Cropped Vertical Axis Misleading statistical graphs are common in newspapers and magazines. Compare the two time series charts below. The data are the same for each. However, the first graph has a cropped vertical axis, which makes it appear that the stock price has increased greatly over the 10-year period. In the second graph, the scale on the vertical axis begins at zero. This graph correctly shows that stock prices increased only modestly during the 10-year period.

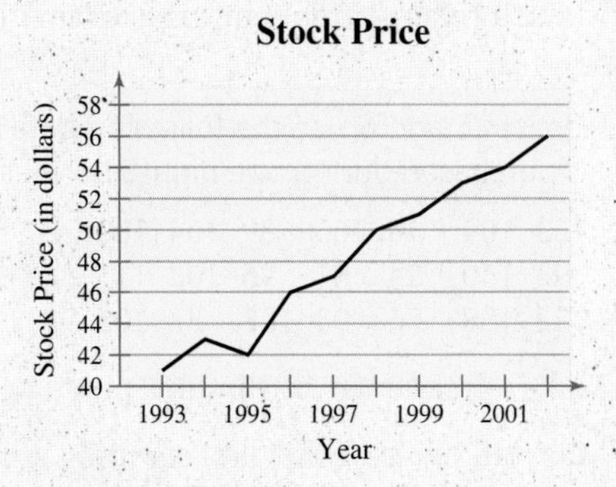

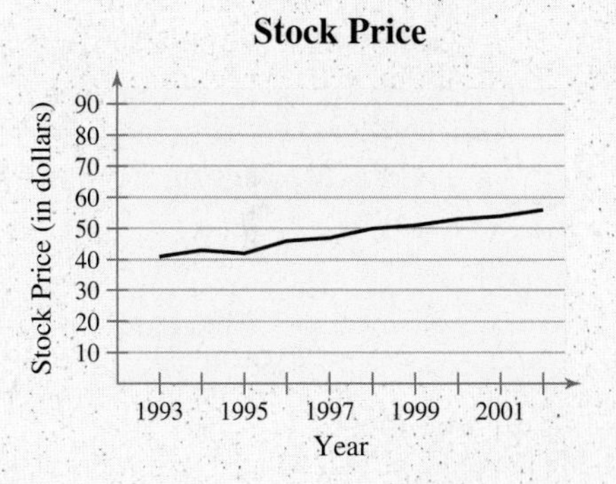

Effect of Outliers on the Mean Outliers or extreme values can have significant effects on the mean. Suppose, for example, that in recruiting information, a company stated that the average commission earned by the five people in its sales force was \$60,000 last year. This would be misleading if four of the five earned \$25,000 and the fifth person earned \$200,000.

Exercises

1. ***Cropped Vertical Axis*** In a newspaper or magazine, find an example of a graph that has a cropped vertical axis. Is the graph misleading? Do you think this was intentional? Redraw the graph so that it is not misleading.

2. ***Effect of Outliers on the Mean*** Describe a situation in which an outlier can make the mean misleading. Is the median also affected significantly by outliers? Explain your reasoning.

2 Review Exercises

Section 2.1

In Exercises 1–3, use the following data set. The data set represents the income (in thousands of dollars) of 20 employees at a small business.

30 28 26 39 34 33 20 39 28 33
26 39 32 28 31 39 33 31 33 32

1. Make a frequency distribution of the data set using five classes. Include the class midpoints, limits, boundaries, frequencies, relative frequencies, and cumulative frequencies.

2. Make a relative frequency histogram using the frequency distribution in Exercise 1. Then determine which class has the greatest relative frequency and which has the least relative frequency.

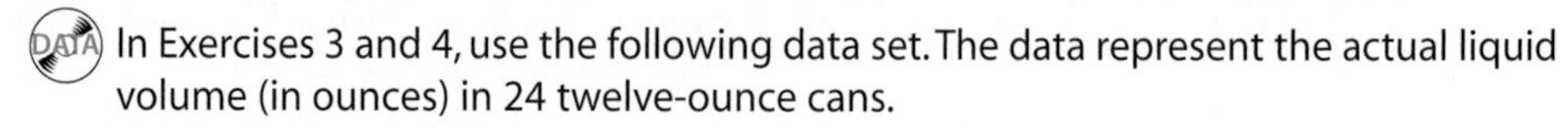

In Exercises 3 and 4, use the following data set. The data represent the actual liquid volume (in ounces) in 24 twelve-ounce cans.

11.95 11.91 11.86 11.94 12.00 11.93 12.00 11.94
12.10 11.95 11.99 11.94 11.89 12.01 11.99 11.94
11.92 11.98 11.88 11.94 11.98 11.92 11.95 11.93

3. Make a frequency histogram using seven classes.

4. Make a relative frequency histogram of the data set using seven classes.

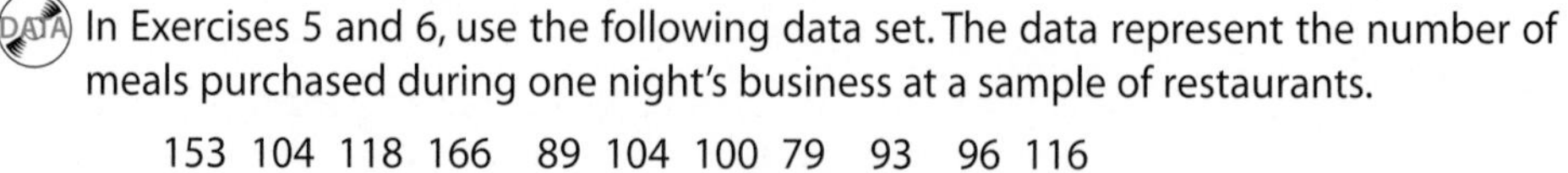

In Exercises 5 and 6, use the following data set. The data represent the number of meals purchased during one night's business at a sample of restaurants.

153 104 118 166 89 104 100 79 93 96 116
94 140 84 81 96 108 111 87 126 101 111
122 108 126 93 108 87 103 95 129 93

5. Make a frequency distribution with six classes and draw a frequency polygon.

6. Make an ogive of the data set using six classes.

Section 2.2

In Exercises 7 and 8, use the following data set. The data represent the average daily high temperature (in degrees Fahrenheit) during the month of January for Chicago, Illinois. *(Source: National Oceanic and Atmospheric Administration)*

33 31 25 22 38 51 32 23 23 34 44 43 47 37 29 25
28 35 21 24 20 19 23 27 24 13 18 28 17 25 31

7. Make a stem-and-leaf plot of the data set. Use one line per stem.

8. Make a dot plot of the data set.

9. The following are the height (in feet) and the number of stories of nine notable buildings in Miami. Use the data to construct a scatter plot. What type of pattern is shown in the scatter plot? *(Source: Skyscrapers.com)*

Height (in feet)	764	625	520	510	484	480	450	430	410
Number of stories	55	47	51	28	35	40	33	31	40

10. The U.S. unemployment rate over a 12-year period is given. Use the data to construct a time series chart. *(Source: U.S. Bureau of Labor Statistics)*

Year	1989	1990	1991	1992	1993	1994
Unemployment rate	5.3	5.6	6.8	7.5	6.9	6.1
Year	1995	1996	1997	1998	1999	2000
Unemployment rate	5.6	5.4	4.9	4.5	4.2	4.0

In Exercises 11 and 12, use the following data set. The data set represents the top seven American Kennel Club registrations (in thousands) in 2000. *(Source: American Kennel Club, New York, NY)*

Breed	Labrador Retriever	Golden Retriever	German Shepherd	Dachshund	Beagle	Poodle	Yorkshire Terrier
Number registered (in thousands)	173	66	58	55	52	46	44

11. Make a Pareto chart of the data set.

12. Make a pie chart of the data set.

Section 2.3

13. Find the mean, median, and mode of the data set.

9 7 8 6 9 12 11 5 9 10

14. Find the mean, median, and mode of the data set.

28 35 29 29 33 32 29 33 31 29

15. Estimate the mean of the frequency distribution you made in Exercise 1.

16. The following frequency distribution shows the number of magazine subscriptions per household for a sample of 60 households. Find the mean number of subscriptions per household.

Number of magazines	0	1	2	3	4	5	6
Frequency	13	9	19	8	5	2	4

17. Six test scores are given. The first five test scores are 15% of the final grade, and the last test score is 25% of the final grade. Find the weighted mean of the test scores.

75 67 86 77 79 88

18. Four test scores are given. The first three test scores are 10% of the final grade, and the last test score is 70% of the final grade. Find the weighted mean of the test scores.

81 95 89 87

19. Describe the shape of the distribution in the histogram you made in Exercise 3. Is the distribution symmetric, uniform, or skewed?

20. Describe the shape of the distribution in the histogram you made in Exercise 4. Is the distribution symmetric, uniform, or skewed?

In Exercises 21 and 22, decide if the distribution is skewed right, skewed left, or symmetric.

21.

22.

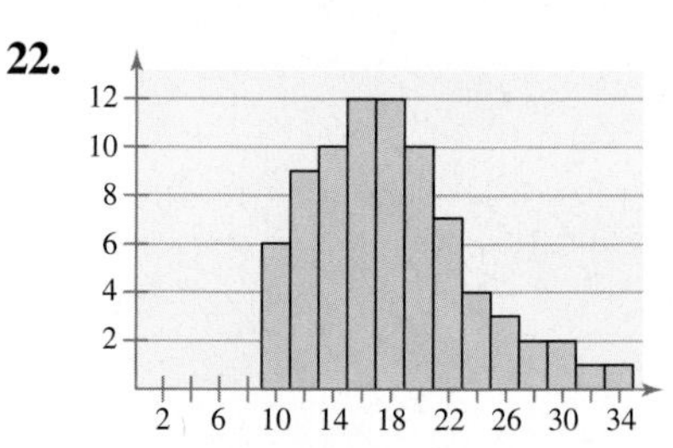

23. For the histogram in Exercise 21, which is greater, the mean or the median?

24. For the histogram in Exercise 22, which is greater, the mean or the median?

Section 2.4

25. The data set represents the mean price of a movie ticket (in U.S. dollars) for a sample of 12 U.S. cities. Find the range of the data set.

7.82 7.38 6.42 6.76 6.34 7.44 6.15 5.46 7.92 6.58 8.26 7.17

26. The data set represents the mean price of a movie ticket (in U.S. dollars) for a sample of 12 Japanese cities. Find the range of the data set.

19.73 16.48 19.10 18.56 17.68 17.19
16.63 15.99 16.66 19.59 15.89 16.49

27. The mileage (in thousands) for a rental car company's fleet is listed. Find the population mean and standard deviation of the data.

6 14 3 7 11 13 8 5 10 9 12 10

28. The age of each Supreme Court justice as of April 30, 2001 is listed. Find the population mean and standard deviation of the data. *(Source: Supreme Court of the United States)*

76 81 71 65 64 61 52 68 62

29. Dormitory room prices (in dollars for one school year) for a sample of four-year universities are listed. Find the sample mean and the sample standard deviation of the data.

2445 2940 2399 1960 2421 2940 2657 2153
2430 2278 1947 2383 2710 2761 2377

30. Sample salaries (in dollars) of public school teachers are listed. Find the sample mean and standard deviation of the data.

46,098 36,259 35,084 38,617 42,690 26,202 47,169 37,109

31. The mean rate for cable television from a sample of households was \$29.00 per month with a standard deviation of \$2.50 per month. Between what two values does 99.7% of the data lie? (Assume a bell-shaped distribution.)

32. The mean rate for cable television from a sample of households was \$29.50 per month with a standard deviation of \$2.75 per month. Estimate the percent of cable television rates between \$26.75 and \$32.25. (Assume that the data set has a bell-shaped distribution.)

33. The mean sale per customer for 40 customers at a grocery store is \$23.00 with a standard deviation of \$6.00. Using Chebychev's Theorem, at least how many of the customers spent between \$11.00 and \$35.00?

34. The mean length of the first 20 space shuttle flights was about 7 days and the standard deviation was about 2 days. Using Chebychev's Theorem, at least how many of the flights lasted between 3 days and 11 days? *(Source: NASA)*

35. From a random sample of households, the number of television sets are listed. Find the sample mean and standard deviation of the data.

Number of televisions	0	1	2	3	4	5
Number of households	1	8	13	10	5	3

36. From a random sample of airplanes, the number of defects found in their fuselages are listed. Find the sample mean and standard deviation of the data.

Number of defects	0	1	2	3	4	5	6
Number of planes	4	5	2	9	1	3	1

Section 2.5

In Exercises 37–40, use the following data set. The data represent the heights (in inches) of students in a statistics class.

50 51 54 54 56 59 60 61 61 63
64 65 68 69 70 70 71 71 75

37. Find the height that corresponds to the first quartile.

38. Find the height that corresponds to the third quartile.

39. Find the interquartile range.

40. Make a box-and-whisker plot of the data.

41. Find the interquartile range of the data from Exercise 14.

42. The weights (in pounds) of the defensive players on a high school football team are given. Make a box-and-whisker plot of the data.

173 145 205 192 197 227 156 240 172 185
208 185 190 167 212 228 190 184 195

43. A student's test grade of 68 represents the 77th percentile of the grades. What percent of students scored higher than 68?

44. In 2001 there were 718 "oldies" radio stations in the United States. If one station finds that 84 stations have a larger daily audience than it does, what percentile does this station come closest to in the daily audience rankings? *(Source: Radioinfo.com)*

In Exercises 45–48, use the following information. The weights of 19 high school football players have a bell-shaped distribution with a mean of 192 pounds and a standard deviation of 24 pounds. Use *z*-scores to determine if the weights of the following randomly selected football players are unusual.

45. 251 pounds **46.** 162 pounds **47.** 219 pounds **48.** 178 pounds

PUTTING IT ALL TOGETHER

Real Statistics Real Decisions

You are a consumer journalist for a newspaper. You have received several letters and e-mails from readers who are concerned about the price of gasoline. One of the readers wrote the following.

"I think, on the average, a driver in our city pays more for a gallon of regular gasoline than a driver in other cities like ours in this state."

Your editor asks you to investigate the gasoline prices and write an article about it. You have gathered the data given at the right (your city is City B). The data represent the price paid for one gallon of regular gasoline from a random sample of gas stations in your city and the other cities of similar size in your state.

City A	City B	City C	City D
1.69	1.75	1.72	1.79
1.64	1.74	1.71	1.76
1.63	1.72	1.69	1.73
1.68	1.74	1.77	1.72
1.71	1.77	1.71	1.75
1.72	1.78	1.72	1.78
1.68	1.75	1.72	1.71
1.69	1.79	1.76	1.78
1.68	1.77	1.71	1.72
1.63	1.75	1.75	1.74

The retail prices for one gallon of regular gasoline at 10 randomly selected stations in four cities.

Adapted from Energy Information Administration

Exercises

1. *How Would You Do It?*
   (a) How would you investigate the statement about the price of gas?
   (b) What statistical measures in this chapter would you use?
2. *Displaying The Data*
   (a) What type of graphs would you choose to display the data? Why?
   (b) Make the graphs from part (a).
   (c) Using what you did in part (b), does it appear that the average price for a gallon of gas in your city, City B, is higher than any of the other cities?
3. *Measuring The Data*
   (a) What statistical measures discussed in this chapter would you use to analyze the gas price data?
   (b) Calculate the measures from part (a).
   (c) Compare the measurements from part (b) to the graphs you made in Exercise 2. Do the measurements support your conclusion in Exercise 2?
4. *Discussing The Data*
   (a) What would you tell your readers? Is the average price for a gallon of gasoline more than the other cities?
   (b) What reasons might you give to your readers as to why the price of gasoline varies from city to city?

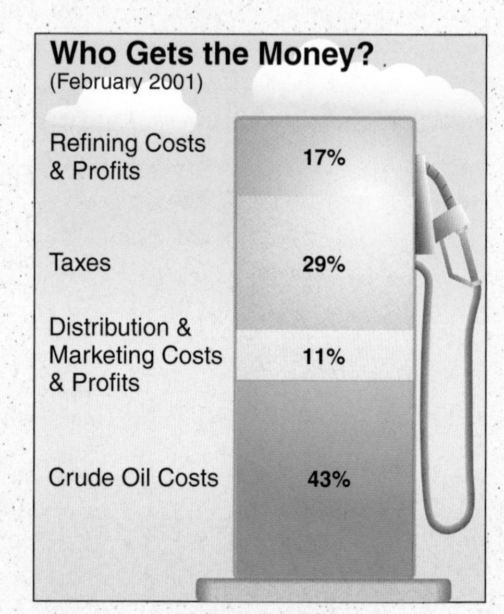

Adapted from Energy Information Administration

2

Chapter Quiz

Take this quiz as you would take a quiz in class. After you are done, check your work against the answers given in the back of the book.

1. The data set is the number of minutes a sample of 25 people exercises each week.

108 139 120 123 120 132 123 131 131 157 150 124 111
101 135 119 116 117 127 128 139 119 118 114 127

(a) Make a frequency distribution of the data set using five classes. Include class limits, midpoints, frequencies, boundaries, relative frequencies, and cumulative frequencies.

(b) Display the data using a frequency histogram and a frequency polygon on the same axes.

(c) Display the data using a relative frequency histogram.

(d) Describe the distribution's shape as symmetric, uniform, or skewed.

(e) Display the data using a box-and-whisker plot.

(f) Display the data using an ogive.

2. Use frequency distribution formulas to approximate the sample mean and standard deviation of the data set in Exercise 1.

3. U.S. sporting goods sales (in billions of dollars) can be classified in four areas: clothing (9.6), footwear (5.9), equipment (13.5), and recreational transport (15.1). Display the data using (a) a pie chart and (b) a Pareto chart. *(Source: National Sporting Goods Association)*

4. Weekly salaries (in dollars) for a sample of registered nurses are listed.

774 446 1019 795 908 667 444 960

(a) Find the mean, the median, and the mode of the salaries. Which best describes a typical salary?

(b) Find the range, variance, and standard deviation of the data set. Interpret the results in the context of the real-life setting.

5. The mean price of new homes from a sample of houses is $155,000 with a standard deviation of $15,000. The data set has a bell-shaped distribution. Between what two prices do 95% of the houses fall?

6. Refer to the sample statistics of Exercise 5 and use z-scores to determine which, if any, of the following house prices is unusual.

(a) $200,000 (b) $55,000 (c) $175,000 (d) $122,000

7. The number of wins for each Major League Baseball team in 2000 are listed. *(Source: Major League Baseball)*

87 85 83 74 69 95 90 79 77 69 91 91 82 71 95
94 79 67 65 95 85 73 72 69 65 97 86 85 82 76

(a) Find the quartiles of the data set.

(b) Find the interquartile range.

(c) Construct a box-and-whisker plot.

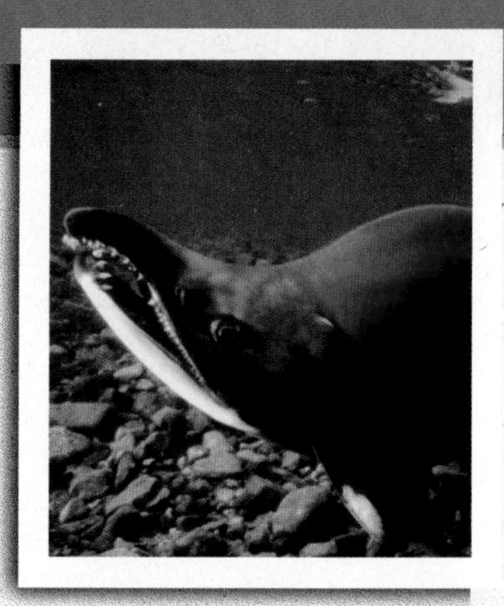

Where You've Been

In Chapters 1 and 2, you learned how to collect and describe data. Once the data are collected and described, you can use the results to write summaries, form conclusions, and make decisions. For instance, the Army Corp of Engineers maintains several dams on the Columbia and Snake Rivers. From studies at these dams, researchers have estimated that a hydroelectric turbine in the dams kills about 15% of all fish that pass through it during their downstream migration as juveniles. There are many proposals for solving this problem: placing bypass channels near the turbine entrances, spilling water over the dams during downstream migrations, and capturing and transporting juvenile fish downstream.

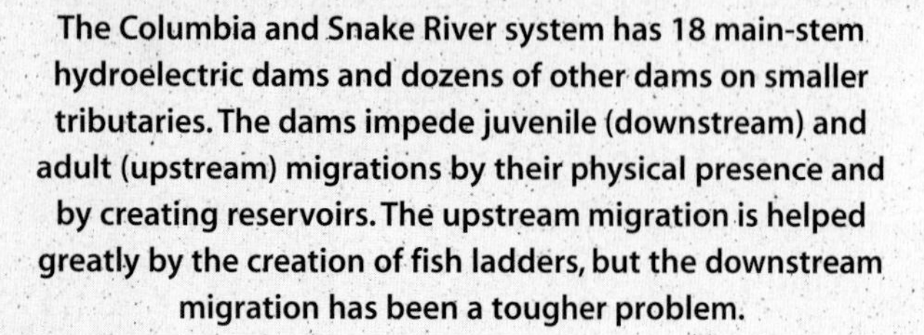

The Columbia and Snake River system has 18 main-stem hydroelectric dams and dozens of other dams on smaller tributaries. The dams impede juvenile (downstream) and adult (upstream) migrations by their physical presence and by creating reservoirs. The upstream migration is helped greatly by the creation of fish ladders, but the downstream migration has been a tougher problem.

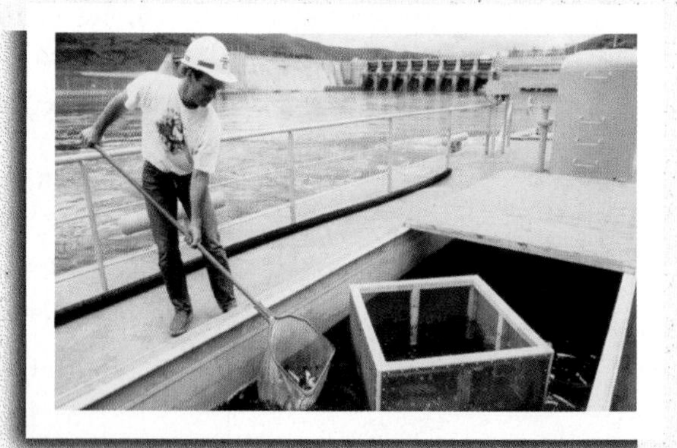

Probability

CHAPTER 3

Where You're Going

In Chapter 3, you will learn how to use data to determine the probability that an event will occur. For instance, suppose that you wanted to find the probability that three juvenile salmons will pass unharmed through a turbine of one of the hydroelectric dams on the Columbia and Snake Rivers. After many samplings, you determine that 85 out of 100 juveniles can pass safely through a turbine. So, you can say that the probability of one, two, or three fish passing through a turbine is as follows.

Probability of one fish passing safely through a turbine $= \frac{85}{100} = 0.85$

Probability of two fish passing safely through a turbine $= \frac{85}{100} \cdot \frac{85}{100} \approx 0.723$

Probability of three fish passing safely through a turbine $= \frac{85}{100} \cdot \frac{85}{100} \cdot \frac{85}{100} \approx 0.614$

This process can be continued, and you will learn to calculate the probability of n fish passing safely through a turbine as

Probability of n fish passing safely through a turbine $= \left(\frac{85}{100}\right)^n$.

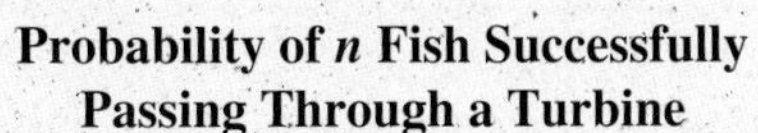

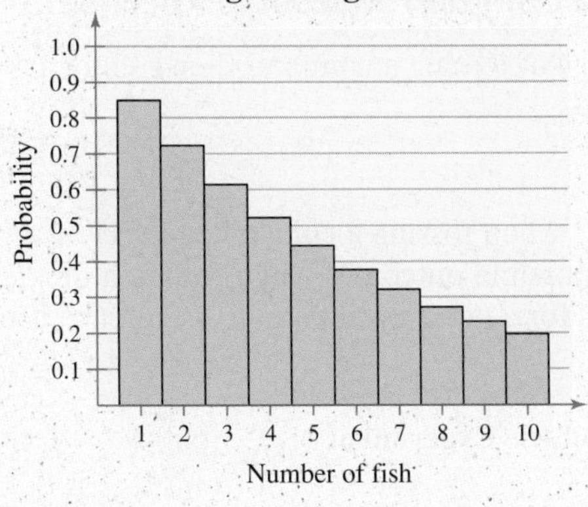

3.1 Basic Concepts of Probability

Probability Experiments • Types of Probability • Properties of Probability

What You Should Learn

- How to identify the sample space of a probability experiment and to identify simple events
- How to distinguish among classical probability, empirical probability, and subjective probability
- How to identify and use properties of probability

Probability Experiments

When weather forecasters say that there is a 90% chance of rain or a physician says there is a 35% chance for a successful surgery, they are stating the chance, or *probability,* that a specific event will occur. Decisions such as "should you wash your car" or "should you proceed with surgery" are often based on these probabilities. In the previous chapter, you learned about the role of the descriptive branch of statistics. Since probability is the foundation of inferential statistics, it is necessary to learn about probability before proceeding to the second branch—inferential statistics.

DEFINITION

A **probability experiment** is an action, or trial, through which specific results (counts, measurements, or responses) are obtained. The result of a single trial in a probability experiment is an **outcome.** The set of all possible outcomes of a probability experiment is the **sample space.** An **event** consists of one or more outcomes and is a subset of the sample space.

Study Tip

Here is a simple example of the use of the terms *probability experiment, sample space, event,* and *outcome.*

Probability Experiment:
Roll a six-sided die.
Sample Space:
{1, 2, 3,4, 5, 6}
Event:
Roll an even number, {2, 4, 6}
Outcome:
Roll a 2, {2}

EXAMPLE 1

Identifying the Sample Space of a Probability Experiment

A probability experiment consists of tossing a coin and then rolling a six-sided die. Describe the sample space.

SOLUTION

There are two possible outcomes when tossing a coin, a head (H) or a tail (T). For each of these, there are six possible outcomes when rolling a die: 1, 2, 3, 4, 5, or 6. One way to list outcomes for actions occurring in a sequence is to use a **tree diagram.**

Tree Diagram for Coin and Die Experiment

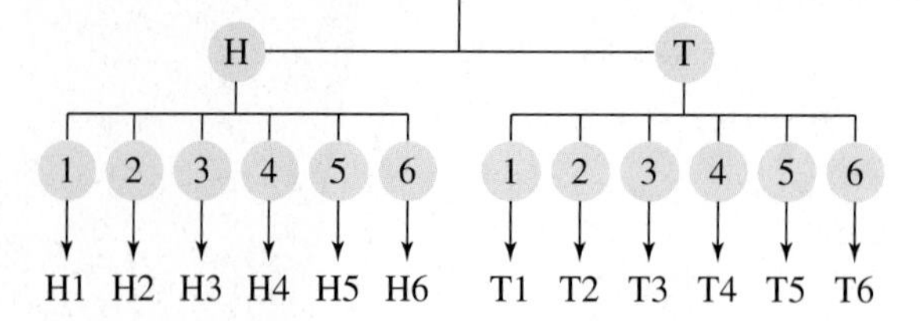

From the tree diagram, the sample space has 12 outcomes.

{H1, H2, H3, H4, H5, H6, T1, T2, T3, T4, T5, T6}

SURVEY

There should be a limit to the number of terms a U.S. senator can serve.

Check one response:

☐ Agree
☐ Disagree
☐ No opinion

Try It Yourself 1

For each probability experiment, identify the sample space.

1. A probability experiment consists of recording a response to the survey statement at the left *and* the gender of the respondent.
2. A probability experiment consists of recording a response to the survey statement at the left *and* the political party (Democrat, Republican, or Other) of the respondent.

a. Start a tree diagram by forming a branch for each possible response to the survey.
b. At the end of each survey response branch, draw a new branch for each possible outcome.
c. Find the *number of outcomes* in the sample space.
d. List the *sample space.*

Answer: Page A33

In the rest of this chapter, you will learn how to calculate the probability or likelihood of an event. Events are often represented by uppercase letters, such as A, B, and C. An event that consists of a single outcome is called a **simple event.** For instance, if you determine the blood type of a sample, then a simple event A is "the blood is type A." In contrast, the event E is "the blood is not type A" and is not simple because it consists of three possible outcomes, {B, AB, O}.

EXAMPLE 

Identifying Simple Events

Decide whether each event is simple or not. Explain your reasoning.

1. For quality control, you randomly select a computer chip from a batch that has been manufactured that day. Event A is selecting a specific defective chip.
2. You roll a six-sided die. Event B is rolling at least a 4.

SOLUTION

1. Event A has only one outcome: choosing the specific defective chip. So, the event is a simple event.
2. B has three outcomes: rolling a 4, a 5, or a 6. Because the event has more than one outcome, it is not simple.

Try It Yourself 2

You ask for a student's age at his or her last birthday. Decide whether each event is simple or not.

1. Event C: The student's age is between 18 and 23, inclusive.
2. Event D: The student's age is 20.

a. Decide how many outcomes are in the event.
b. State whether the event is *simple* or not.

Answer: Page A33

Types of Probability

There are three types of probability: classical probability, empirical probability, and subjective probability. The probability that event E will occur is written as $P(E)$ and is read "the probability of event E."

DEFINITION

Classical (or **theoretical**) **probability** is used when each outcome in a sample space is equally likely to occur. The classical probability for an event E is given by

$$P(E) = \frac{\text{Number of outcomes in } E}{\text{Total number of outcomes in sample space}}.$$

Study Tip

Probabilities can be expressed as fractions, decimals, or percents. In this text, probabilities will usually be expressed as fractions or decimals, rounded if necessary to three places.

EXAMPLE 

Finding Classical Probabilities

You roll a six-sided die. Find the probability of the following events.

1. Event A: rolling a 3
2. Event B: rolling a 7
3. Event C: rolling a number less than 5

SOLUTION When rolling a six-sided die, the sample space consists of six outcomes: $\{1, 2, 3, 4, 5, 6\}$.

1. There is one outcome in event $A = \{3\}$. So,

$$P(3) = \frac{1}{6} \approx 0.167.$$

2. Because 7 is not in the sample space, there are no outcomes in event B. So,

$$P(7) = \frac{0}{6} = 0.$$

3. There are four outcomes in event $C = \{1, 2, 3, 4\}$. So,

$$P(\text{number less than 5}) = \frac{4}{6} = \frac{2}{3} \approx 0.667.$$

Standard Deck of Playing Cards

Hearts	Diamonds	Spades	Clubs
A ♥	A ♦	A ♠	A ♣
K ♥	K ♦	K ♠	K ♣
Q ♥	Q ♦	Q ♠	Q ♣
J ♥	J ♦	J ♠	J ♣
10♥	10♦	10♠	10♣
9 ♥	9 ♦	9 ♠	9 ♣
8 ♥	8 ♦	8 ♠	8 ♣
7 ♥	7 ♦	7 ♠	7 ♣
6 ♥	6 ♦	6 ♠	6 ♣
5 ♥	5 ♦	5 ♠	5 ♣
4 ♥	4 ♦	4 ♠	4 ♣
3 ♥	3 ♦	3 ♠	3 ♣
2 ♥	2 ♦	2 ♠	2 ♣

Try It Yourself 3

You select a card from a standard deck. Find the probability of the following.

1. Event D: Selecting a seven of diamonds
2. Event E: Selecting a diamond
3. Event F: Selecting a diamond, heart, club, or spade

a. Identify the *total number of outcomes* in the sample space.
b. Find the *number of outcomes* in the event.
c. Use the *classical probability formula*.

Answer: Page A33

A second type of probability is empirical probability. Empirical probability can be used even if each outcome is not equally likely to occur.

DEFINITION

Empirical (or **statistical**) **probability** is based on observations obtained from probability experiments. The empirical probability of an event E is the relative frequency of event E.

$$P(E) = \frac{\text{Frequency of event } E}{\text{Total frequency}} = \frac{f}{n}.$$

EXAMPLE 

Finding Empirical Probabilities

A pond contains three types of fish: bluegills, redgills, and crappies. Each fish in the pond is equally likely to be caught. You catch 40 fish and record the type. Each time, you release the fish back into the pond. The following frequency distribution shows your results.

Fish Type	Number of Times Caught, f
Bluegill	13
Redgill	17
Crappy	10
	$\Sigma f = 40$

If you catch another fish, what is the probability that it is a bluegill?

SOLUTION The event is "catching a bluegill." In your experiment, the frequency of this event is 13. Because the total of the frequencies is 40, the empirical probability of catching a bluegill is

$$P(\text{bluegill}) = \frac{13}{40} = 0.325.$$

Try It Yourself 4

An insurance company determines that in every 100 claims, 4 are fraudulent. What is the probability that the next claim the company processes is fraudulent?

a. *Identify* the event. Find the *frequency* of the event.
b. *Find the total frequency for the experiment.*
c. Find the *relative frequency* of the event.

Answer: Page A33

Picturing the World

It seems as if no matter how strange an event is, somebody wants to know the probability that it will occur. The following table lists the probability that some intriguing events will happen. *(Source: What Are The Chances)*

Event	Probability
Appearing on *The Tonight Show*	1 in 490,000
Being a victim of serious crime	5%
Writing a best-selling novel	0.00205
Congress will override a veto	4%
Earning a Ph.D.	0.008

Which of these events is most likely to occur? Least likely?

As you increase the number of times a probability experiment is repeated, the empirical probability (relative frequency) of an event approaches the theoretical probability of the event. This is known as the **law of large numbers.**

Law of Large Numbers

As an experiment is repeated over and over, the empirical probability of an event approaches the theoretical (actual) probability of the event.

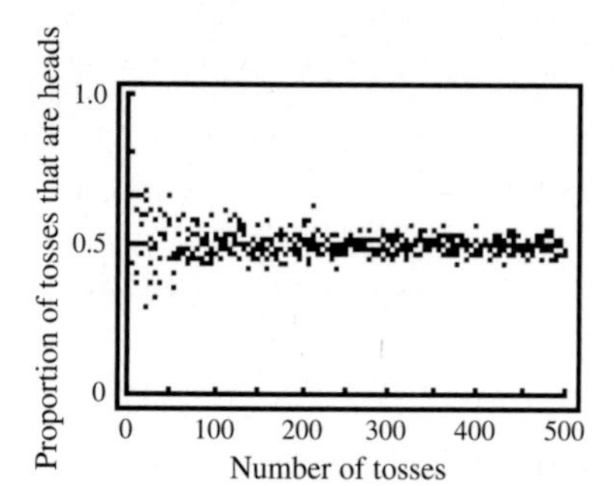

As an example of this law, suppose you want to determine the probability of tossing a head with a fair coin. If you toss the coin 10 times and get only 3 heads, you obtain an empirical probability of $\frac{3}{10}$. Because you tossed the coin only a few times, your empirical probability is not representative of the theoretical probability, which is $\frac{1}{2}$. If, however, you toss the coin several thousand times, then the law of large numbers tells you that the empirical probability will be very close to the theoretical or actual probability.

The scatter plot at the left shows the results of simulating a coin toss 500 times. Notice that as the number of tosses increases, the probability of tossing a head gets closer and closer to the theoretical probability of 0.5.

EXAMPLE 

Using Frequency Distributions to Find Probabilities

You survey a sample of 1000 employees at a company and record the age of each. The results are shown at the left in the frequency distribution. If you randomly select another employee, what is the probability that the employee is between 25 and 34 years old?

Employee Ages	Frequency, f
15 to 24	54
25 to 34	366
35 to 44	233
45 to 54	180
55 to 64	125
65 and over	42
	$\Sigma f = 1000$

SOLUTION The event is selecting an employee who is between 25 and 34 years old. In your survey, the frequency of this event is 366. Because the total of the frequencies is 1000, the probability of selecting an employee between the ages of 25 and 34 years old is

$$P(\text{age 25 to 34}) = \frac{366}{1000} = 0.366.$$

Try It Yourself 5

Find the probability that an employee chosen at random is between 15 and 24 years old.

a. Find the *frequency* of the event.
b. Find the *total of the frequencies.*
c. Find the *relative frequency* of the event.

Answer: Page A33

The third type of probability is **subjective probability.** Subjective probabilities result from intuition, educated guesses, and estimates. For instance, given a patient's health and extent of injuries, a doctor may feel that the patient has a 90% chance of a full recovery. Or a business analyst may predict that the chance of the employees of a certain company going on strike is 0.25.

Type	Summary	Formula
Classical (Theoretical) Probability	The number of outcomes in the sample space is known and each outcome is equally likely to occur.	$P(E) = \dfrac{\text{Number of outcomes in event } E}{\text{Number of outcomes in sample space}}$
Empirical (Statistical) Probability	The frequency of outcomes in the sample space is estimated from experimentation.	$P(E) = \dfrac{\text{Frequency of event } E}{\text{Total frequency}} = \dfrac{f}{n}$
Subjective Probability	Probabilities result from intuition, educated guesses, and estimates.	None

EXAMPLE 6

Classifying Types of Probability

Classify each statement as an example of classical probability, empirical probability, or subjective probability. Explain your reasoning.

1. The probability of your phone ringing during dinner is 0.5.
2. The probability that a voter chosen at random will vote Republican is 0.45.
3. The probability of winning a 1000-ticket raffle with one ticket is $\frac{1}{1000}$.

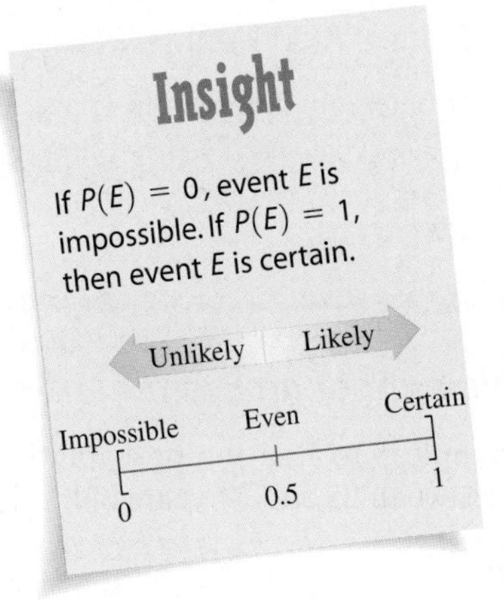

SOLUTION

1. This probability is most likely based on an educated guess. It is an example of subjective probability.
2. This statement is most likely based on a survey of a sample of voters, so it is an example of empirical probability.
3. Because you know the number of outcomes and each is equally likely, this is an example of classical probability.

Try It Yourself 6

Based on previous counts, the probability of a salmon successfully passing through a dam on the Columbia River is 0.85. Is this statement an example of classical probability, empirical probability, or subjective probability?

a. Identify the *event.*
b. Decide whether the probability is *determined* by knowing all possible outcomes, whether the probability is *estimated* from the results of an experiment, or whether the probability is an *educated guess.*
c. Make a *conclusion.*

Answer: Page A33

A probability cannot be negative or greater than 1. So, the probability of an event E is between 0 and 1, inclusive. That is,

$$0 \leq P(E) \leq 1.$$

Properties of Probability

The sum of the probabilities of all outcomes in a sample space is 1 or 100%. An important result of this fact is that if you know the probability of an event E, you can find the probability of the *complement of event E*.

E'
E
6
5
1
2
3
4

The area of the rectangle represents the total probability of the sample space ($1 = 100\%$). The area of the circle represents the probability of event E, while the area outside the circle represents the probability of the complement of event E.

DEFINITION

The **complement of event E** is the set of all outcomes in a sample space that are not included in event E. The complement of event E is denoted by E' and is read as "E prime."

For instance, if you roll a die and let E be the event "the number is at least 5," then the complement of E is the event "the number is less than 5." In other words, $E = \{5, 6\}$ and $E' = \{1, 2, 3, 4\}$.

Using the definition of the complement of an event and the fact that the sum of the probabilities of all outcomes is 1, you can determine the following formulas:

$$P(E) + P(E') = 1 \qquad P(E) = 1 - P(E') \qquad P(E') = 1 - P(E)$$

The Venn diagram illustrates the relationship between the sample space, an event E, and its complement E'.

EXAMPLE 7

Finding the Probability of the Complement of an Event

Use the frequency distribution given in Example 5 to find the probability of randomly choosing an employee who is not between 25 and 34 years old.

SOLUTION From Example 5, you know that

$$P(\text{age 25 to 34}) = \frac{366}{1000} = 0.366.$$

So, the probability that an employee is not between 25 and 34 years old is

$$P(\text{age is not 25 to 34}) = 1 - \frac{366}{1000} = \frac{634}{1000} = 0.634.$$

Try It Yourself 7

Use the frequency distribution in Example 4 to find the probability that a fish that is caught is *not* a redgill.

a. *Find the probability* that the fish is a redgill.
b. *Subtract* the resulting probability from 1.
c. State the probability as a fraction and as a decimal.

Answer: Page A33

3.1 Exercises

Help

LarsonTutor 3.1

Companion Web Site

Student Solutions Manual 3.1

Videos 3.1

Technology Manuals

Basic Skills and Concepts

1. Determine which of the numbers (a) 0, (b) 1.5, (c) -1, (d) 50%, and (e) $\frac{2}{3}$ could represent the probability of an event. Explain your reasoning.

2. Explain why the following statement is incorrect:
The probability of rain tomorrow is 120%.

Identifying a Sample Space In Exercises 3–6, identify the sample space of each probability experiment. Draw a tree diagram if it is appropriate.

3. Guessing the last digit in a telephone number

4. Tossing three coins

5. Determining a person's blood type (A, B, AB, O) and Rh-factor (positive, negative)

6. Rolling two dice

Recognizing Simple Events In Exercises 7 and 8, decide whether the event is a simple event or not. Explain your reasoning.

7. A computer is used to select randomly a number between 1 and 2000. Event A is selecting 359.

8. A computer is used to select randomly a number between 1 and 2000. Event B is selecting a number less than 200.

Classifying Types of Probability In Exercises 9 and 10, classify the statement as an example of classical probability, empirical probability, or subjective probability. Explain your reasoning.

9. According to company records, the probability that a washing machine will need repairs during a six-year period is 0.10.

10. The probability of choosing six numbers from 1 to 40 that match the six numbers drawn by a state lottery is $1/3{,}838{,}380 \approx 0.00000026$.

Graphical Analysis In Exercises 11 and 12, use the diagram to answer the question. *(Source: Federal Election Commission)*

11. What is the probability that a voter chosen at random did not vote for George W. Bush in the 2000 election?

About 50,456,141 voted for George W. Bush

About 54,948,405 voted for another candidate

12. What is the probability that a voter chosen at random did not vote for a Democratic representative in the 1998 election?

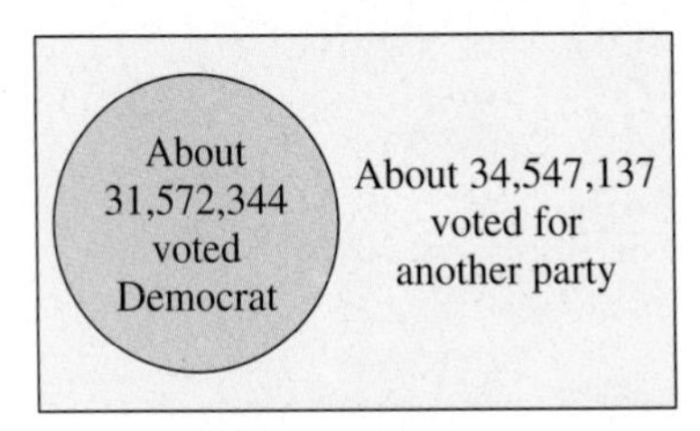

Finding Probabilities In Exercises 13–16, consider a company that selects employees for random drug tests. The company uses a computer to select randomly employee numbers that range from 1 to 6296.

13. Find the probability of selecting a number less than 1000.

14. Find the probability of selecting a number greater than 1000.

15. Find the probability of selecting a number divisible by 1000.

16. Find the probability of selecting a number that is not divisible by 1000.

Using a Frequency Distribution to Find Probabilities In Exercises 17–20, use the following frequency distribution. The distribution shows the number of American voters (in millions) according to age. *(Source: U.S. Bureau of the Census)*

Ages of Voters	Frequency (in millions)
18 to 20 years old	10.8
21 to 24 years old	13.9
25 to 34 years old	40.1
35 to 44 years old	43.3
45 to 64 years old	53.7
65 years old and over	31.9

Find the probability that a voter chosen at random is

17. between 21 and 24 years old.

18. between 35 and 44 years old.

19. not between 18 and 20 years old.

20. not between 25 and 34 years old.

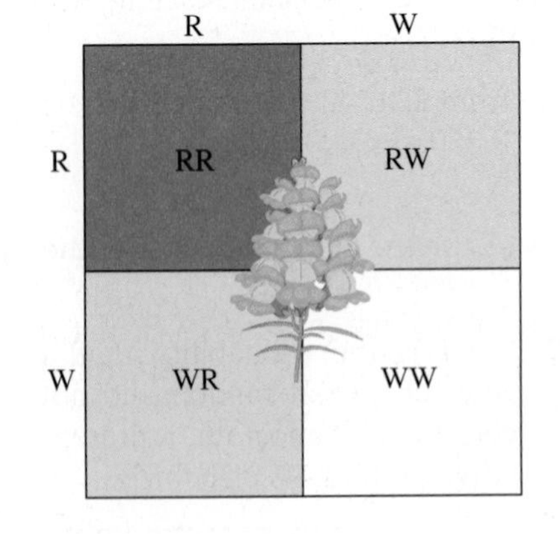

21. ***Genetics*** ◆ When two pink snapdragon flowers (RW) are crossed, there are four equally likely possible outcomes for the genetic makeup of the offspring: red (RR), pink (RW), pink (WR), and white (WW). If two pink snapdragons are crossed, what is the probability that the offspring is (a) pink?, (b) red?, and (c) white?

22. ***Genetics*** ◆ There are six basic types of coloring in registered collies: sable (SSmm), tricolor (ssmm), trifactored sable (Ssmm), blue merle (ssMm), sable merle (SSMm), and trifactored sable merle (SsMm). The *Punnett square* shows the possible coloring of the offspring of a trifactored sable merle collie and a trifactored sable collie. What is the probability that the offspring has the same coloring as one of its parents?

	SM	Sm	sM	sm
Sm	SSMm	SSmm	SsMm	Ssmm
Sm	SSMm	SSmm	SsMm	Ssmm
sm	SsMm	Ssmm	ssMm	ssmm
sm	SsMm	Ssmm	ssMm	ssmm

Parents
Ssmm and SsMm

Using a Pie Chart to Find Probabilities

In Exercises 23–26, use the following pie chart. The pie chart shows the number of workers (in thousands) by industry for the United States. *(Source: U.S. Department of Labor)*

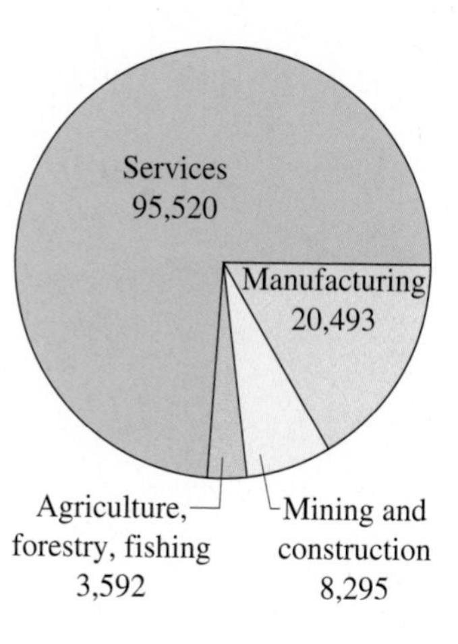

23. Find the probability that a worker chosen at random was employed in the services industry.

24. Find the probability that a worker chosen at random was employed in the manufacturing industry.

25. Find the probability that a worker chosen at random was not employed in the services industry.

26. Find the probability that a worker chosen at random was not employed in the agriculture, forestry, or fishing industry.

Writing In Exercises 27 and 28, write a statement that represents the complement of the given probability.

27. The probability of randomly choosing a tea drinker who has a college degree (Assume that you are choosing from the population of all tea drinkers.)

28. The probability of randomly choosing a smoker whose mother also smoked (Assume that you are choosing from the population of all smokers.)

Extending the Basics

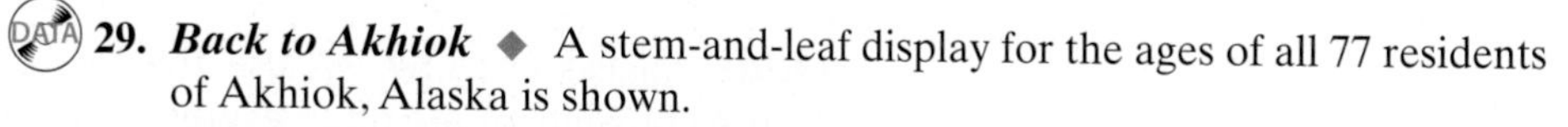

29. ***Back to Akhiok*** ◆ A stem-and-leaf display for the ages of all 77 residents of Akhiok, Alaska is shown.

Stem	Leaves
0	0 1 1 1 2 2 2 3 3 4 4 4 5 5 5 6 6 6 6 7 7 8 8 9
1	0 0 0 1 1 1 2 2 3 5 6 6 7 7
2	1 1 2 3 4 5 5 6 7 7 8 8 9
3	0 1 1 2 2 3 3 4 6 9
4	1 2 5 6 7 8 9
5	0 0 1 2 3 4 5 6
6	3

Key: $1|0 = 10$

If a resident is selected at random, find the probability he or she will be (a) at least 21 years old, (b) between 40 and 50 years old inclusive, and (c) older than 65.

30. ***Individual Stock Price*** ◆ An individual stock is selected at random from the portfolio represented by the box-and-whisker plot shown.

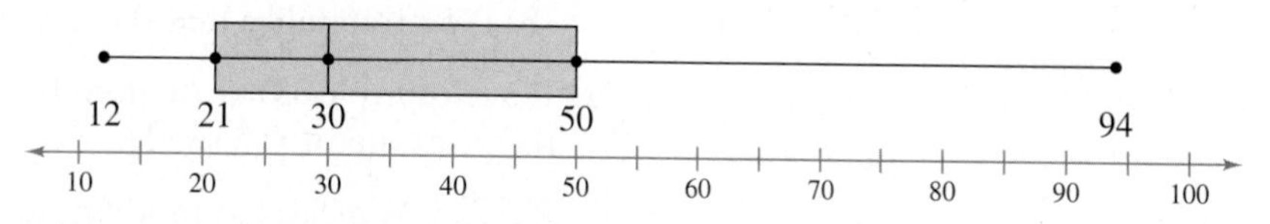

Find the probability that the stock price is (a) less than \$21, (b) between \$21 and \$50, and (c) \$30 or more.

31. ***Wet or Dry?*** ◆ You are planning a three-day trip to Seattle, Washington, in October. Use the following tree diagram to answer the questions.

(a) List the sample space.

(b) List the outcome(s) of the event "It rains all three days."

(c) List the outcome(s) of the event "It rains on exactly one day."

(d) List the outcome(s) of the event "It rains on at least one day."

Tree Diagram for Rainy Days

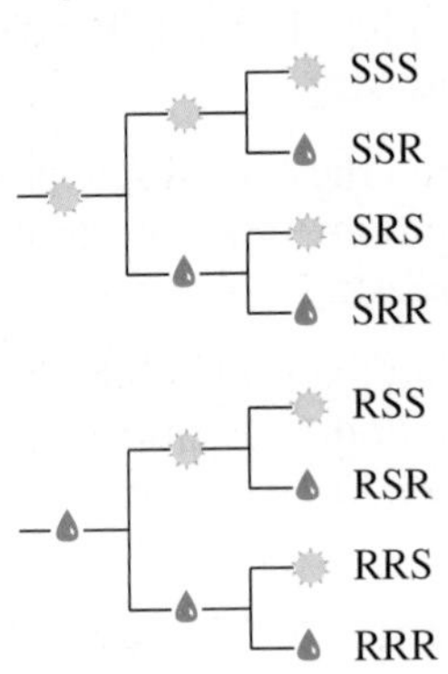

32. ***Sunny and Rainy Days*** ◆ You are planning a four-day trip to Seattle, Washington, in October. Make a sunny day/rainy day tree diagram (see Exercise 31) for your trip.

33. ***Machine Part Suppliers*** ◆ Your company buys machine parts from three different suppliers. Make a tree diagram that shows the three suppliers and whether the parts they supply are defective.

Odds In Exercises 34–37, use the following information. In gambling, the chances of winning are often expressed in terms of odds rather than probabilities. The odds of winning is the ratio of the number of successful outcomes to the number of unsuccessful outcomes. The odds of losing is the ratio of the number of unsuccessful outcomes to the number of successful outcomes. For example, if the number of successful outcomes is 2 and the number of unsuccessful outcomes is 3, the odds of winning are $2:3$ (read "2 to 3") or $\frac{2}{3}$. (*Note:* The *probability* of success is $\frac{2}{5}$.)

34. A beverage company puts game pieces under the caps of its drinks and claims that one in six game pieces wins a prize. The official rules of the contest state that the odds of winning a prize are $1:6$. Is the claim "one in six game pieces wins a prize" correct? Why or why not?

35. The odds of an event occurring are $4:5$. Find

(a) the probability that the event will occur.

(b) the probability that the event will not occur.

36. If a card is picked at random from a standard deck of 52 playing cards, find the odds that it is a spade.

37. If a card is picked at random from a standard deck of 52 playing cards, find the odds that it is not a spade.

3.2 Conditional Probability and the Multiplication Rule

What You Should Learn

- How to find the probability of an event given that another event has occurred
- How to distinguish between independent and dependent events
- How to use the Multiplication Rule to find the probability of two events occurring in sequence
- How to use the Multiplication Rule to find conditional probabilities

Conditional Probability • Independent and Dependent Events • The Multiplication Rule

Conditional Probability

In this section, you will learn how to find the probability that two events occur in sequence. Before you can find this probability, however, you must know how to find conditional probabilities.

DEFINITION

A **conditional probability** is the probability of an event occurring, given that another event has already occurred. The conditional probability of event B occurring, given that event A has occurred, is denoted by $P(B|A)$ and is read as "probability of B, given A."

EXAMPLE 1

Finding Conditional Probabilities

1. Two cards are selected in sequence from a standard deck. Find the probability that the second card is a queen, given that the first card is a king. (Assume that the king is not replaced.)
2. The table at left shows the results of a study in which researchers examined a child's IQ and the presence of a specific gene in the child. Find the probability that a child has a high IQ, given that the child has the gene. *(Source: Psychological Science)*

	Gene present	Gene not present	Total
High IQ	33	19	52
Normal IQ	39	11	50
Total	72	30	102

SOLUTION

1. Because the first card is a king and is not replaced, the remaining deck has 51 cards, 4 of which are queens. So,

$$P(B|A) = \frac{4}{51} \approx 0.078.$$

2. There are 72 children who have the gene. So, the sample space consists of these 72 children. Of these, 33 have a high IQ. So,

$$P(B|A) = \frac{33}{72} \approx 0.458.$$

Try It Yourself 1

1. Find the probability that a child does not have the gene.
2. Find the probability that a child does not have the gene, given that the child has a normal IQ.

a. Find the *number of outcomes* in the event and in the sample space.
b. *Divide* the number of outcomes in the event by the number of outcomes in the sample space.

Answer: Page A33

Independent and Dependent Events

The question of the independence of two or more events is important to researchers in fields such as marketing, medicine, and psychology. You can use conditional probabilities to determine whether events are independent or dependent.

DEFINITION

Two events are **independent** if the occurrence of one of the events does not affect the probability of the occurrence of the other event. Two events A and B are independent if

$$P(B|A) = P(B) \quad \text{or if} \quad P(A|B) = P(A).$$

Events that are not independent are **dependent.**

Often it is important to determine whether two events are independent. To determine if A and B are independent, calculate $P(B)$ and $P(B|A)$. If the values are equal, the events are independent. If $P(B) \neq P(B|A)$, then A and B are dependent events.

Picturing the World

Truman Collins, a probability and statistics enthusiast, wrote a program that finds the probability of landing on each square of a Monopoly board during a game. Collins explored various scenarios, including the effects of the Chance and Community Chest cards and the various ways of landing in or getting out of jail. Interestingly, Collins discovered that the length of each jail term affects the probabilities.

Monopoly Square	Probability Given Short Jail Term	Probability Given Long Jail Term
Go	0.0310	0.0291
Chance	0.0087	0.0082
In Jail	0.0395	0.0946
Free Parking	0.0288	0.0283
Park Place	0.0218	0.0206
B&O RR	0.0307	0.0289
Water Works	0.0281	0.0265

Why do the probabilities depend on how long you stay in jail?

EXAMPLE 

Classifying Events as Independent or Dependent

Decide whether the events are independent or dependent.

1. Selecting a king from a standard deck (A), not replacing it, and then selecting a queen from the deck (B)
2. Tossing a coin and getting a head (A), and then rolling a six-sided die and obtaining a 6 (B)
3. Practicing the piano (A), and then becoming a concert pianist (B)

SOLUTION

1. $P(B|A) = \frac{4}{51}$ and $P(B) = \frac{4}{52}$. The occurrence of A changes the probability of the occurrence of B, so the events are dependent.
2. $P(B|A) = \frac{1}{6}$ and $P(B) = \frac{1}{6}$. The occurrence of A does not change the probability of the occurrence of B, so the events are independent.
3. If you practice the piano, the chances of becoming a concert pianist are greatly increased, so these events are dependent.

Try It Yourself 2

Decide whether the events are independent or dependent events. Explain.

1. A salmon swims successfully through a dam (A) and another salmon swims successfully through the same dam (B).
2. Exercising frequently (A) and having a low resting heart rate (B)

a. *Decide* whether the occurrence of the first event affects the probability of the second event.
b. State if the events are *independent* or *dependent.*
c. *Explain* your reasoning.

Answer: Page A33

The Multiplication Rule

To find the probability of two events occurring in sequence, you can use the Multiplication Rule.

The Multiplication Rule for the Probability of *A* and *B*

The probability that two events *A* and *B* will occur in sequence is

$$P(A \text{ and } B) = P(A) \cdot P(B|A).$$

If events *A and B* are independent, then the rule can be simplified to $P(A \text{ and } B) = P(A) \cdot P(B)$. This simplified rule can be extended for any number of independent events.

EXAMPLE 

Using the Multiplication Rule to Find Probabilities

1. Two cards are selected, without replacing the first card, from a standard deck. Find the probability of selecting a king and then selecting a queen.
2. A coin is tossed and a die is rolled. Find the probability of getting a head and then rolling a 6.

Study Tip

If you are not sure if the events are independent, use the formula for dependent events.

SOLUTION

1. Because the first card is not replaced, the events are dependent.

$$\begin{aligned} P(K \text{ and } Q) &= P(K) \cdot P(Q|K) \\ &= \frac{4}{52} \cdot \frac{4}{51} \\ &= \frac{16}{2652} \\ &\approx 0.006 \end{aligned}$$

So, the probability of selecting a king and then a queen is about 0.006.

2. The events are independent.

$$\begin{aligned} P(H \text{ and } 6) &= P(H) \cdot P(6) \\ &= \frac{1}{2} \cdot \frac{1}{6} \\ &= \frac{1}{12} \\ &\approx 0.083 \end{aligned}$$

So, the probability of tossing a head and then rolling a 6 is about 0.083.

Try It Yourself 3

1. The probability that a salmon swims successfully through a dam is 0.85. Find the probability that two salmon successfully swim through the dam.
2. Consider the table shown in Example 1. Find the probability that a child has a normal IQ but does not have the gene.

a. Decide if the events are *independent* or *dependent.*
b. Use the *Multiplication Rule* to find the probability.

Answer: Page A33

EXAMPLE 

Using the Multiplication Rule to Find Probabilities

1. A coin is tossed and a die is rolled. Find the probability of getting a head and then rolling a 2.
2. The probability that a salmon swims successfully through a dam is 0.85. Find the probability that three salmon swim successfully through the dam.
3. Find the probability that none of the three salmon is successful.
4. Find the probability that at least one of the three salmon is successful in swimming through the dam.

SOLUTION

1. $P(H) = \frac{1}{2}$. Whether or not the coin is a head, $P(2) = \frac{1}{6}$. The events are independent.

$$P(H \text{ and } 2) = P(H) \cdot P(2) = \tfrac{1}{2} \cdot \tfrac{1}{6} = \tfrac{1}{12} \approx 0.083$$

So, the probability of tossing a head and then rolling a 2 is about 0.083.

2. The probability that each salmon is successful is 0.85. One salmon's chance of success is independent of the others.

$$P(3 \text{ salmon are successful}) = (0.85)(0.85)(0.85) \approx 0.614$$

So, the probability that all three are successful is about 0.614.

3. Because the probability of success for one salmon is 0.85, the probability of failure for one salmon is $1 - 0.85 = 0.15$.

$$P(\text{none of the three is successful}) = (0.15)(0.15)(0.15) \approx 0.003$$

So, the probability that none are successful is about 0.003.

4. The phrase "at least one" means one or more. The complement to the event "at least one is successful" is the event "none are successful." Using the rule of complements,

$$\begin{aligned} P(\text{at least 1 is successful}) &= 1 - P(\text{none are successful}) \\ &\approx 1 - 0.003 \\ &= 0.997. \end{aligned}$$

There is about a 0.997 probability that at least one of the three salmon is successful.

Try It Yourself 4

Suppose engineers can increase the probability of a salmon successfully swimming through a dam to 0.90.

1. Find the probability that three salmon swim successfully through the dam.
2. Find the probability that at least one of three salmon swims successfully through the dam.

a. Determine whether to find the probability of the event or its complement.
b. Use the *Multiplication Rule* to find the probability. If necessary, use the *Complement Rule*.

Answer: Page A33

3.2 Exercises

Help

LarsonTutor 3.2

Companion Web Site

Student Solutions Manual 3.2

Videos 3.2

Technology Manuals

Basic Skills and Concepts

1. What is the difference between independent and dependent events?

2. List examples of the following types of events.

(a) Two events that are independent

(b) Two events that are dependent

True or False In Exercises 3 and 4, determine whether the statement is true or false. If it is false, rewrite it so that it is a true statement.

3. If two events are not independent, $P(A|B) = P(B)$.

4. If events A and B are dependent, then $P(A \text{ and } B) = P(A) \cdot P(B)$.

Classifying Events In Exercises 5–8, decide whether the events are independent or dependent. Explain your reasoning.

5. Selecting a king from a standard deck, *replacing it,* and then selecting a queen from the deck

6. Parking beside a fire hydrant on Tuesday and getting a parking ticket on Tuesday

7. A numbered ball between 1 and 40 is selected from a bin, and then a second numbered ball is selected from the remaining balls in the bin.

8. A numbered ball between 1 and 52 is selected from a bin, *replaced,* and then a second numbered ball is selected from the bin.

9. ***BRCA Gene*** ◆ In the general population, one woman in nine will develop breast cancer. Research has shown that one woman in 250 carries a mutation of the BRCA gene. Eight out of 10 women with this mutation develop breast cancer. *(Source: Journal of National Cancer Institute)*

(a) Find the probability that a randomly selected woman will develop breast cancer, given that she has a mutation of the BRCA gene.

(b) Find the probability that a randomly selected woman will carry the mutation of the BRCA gene and will develop breast cancer.

(c) Are the events of carrying this mutation and developing breast cancer independent or dependent? Explain.

Breast Cancer and the BRCA Gene

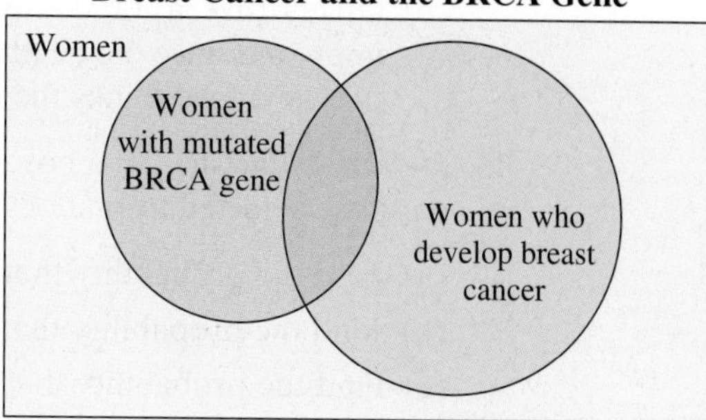

10. ***Summer Vacation*** ◆ The following table shows the results of a survey in which 104 families were asked if they own a computer and if they will be taking a summer vacation this year.

		Summer Vacation This Year		
		Yes	**No**	**Total**
Own a Computer	**Yes**	37	8	45
	No	40	19	59
	Total	77	27	104

(a) Find the probability a randomly selected family is taking a summer vacation this year.

(b) Find the probability a randomly selected family is taking a summer vacation this year, given that they own a computer.

(c) Are the events of owning a computer and taking a summer vacation this year independent or dependent events? Explain.

11. ***Assisted Reproductive Technology*** ◆ A study found that 24% of the assisted reproductive technology (ART) cycles resulted in a pregnancy. Seven percent of the ART pregnancies resulted in multiple births. *(Source: U.S. Department of Health and Human Services)*

Pregnancies

ART cycles

Pregnancies

Multiple births

(a) Find the probability that a randomly selected ART cycle resulted in a pregnancy *and* produced a multiple birth.

(b) Find the probability that a randomly selected ART cycle that resulted in a pregnancy did *not* produce a multiple birth.

12. ***Race Relations*** ◆ In a survey, 60% of adults in the U.S. think race relations have improved since the death of Martin Luther King, Jr. Of these 60%, 4 out of 10 said the rate of civil rights progress is too slow. *(Source: Marist Institute for Public Opinion)*

Race Relations

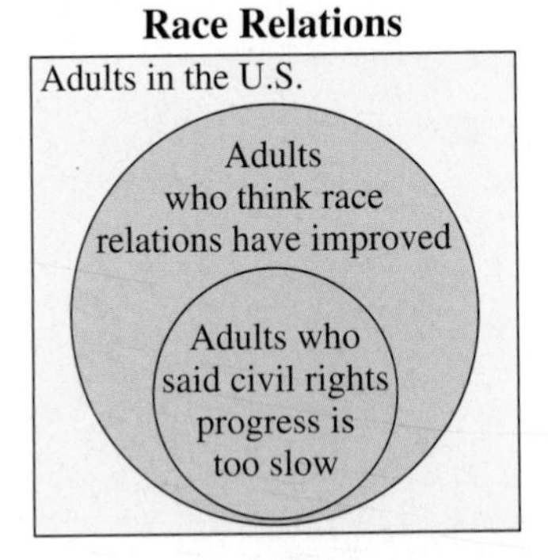

(a) Find the probability that a randomly selected adult thinks race relations have improved since the death of Martin Luther King, Jr. *and* thinks the rate of civil rights progress is too slow.

(b) Given that a randomly selected adult thinks race relations have improved since the death of Martin Luther King, Jr., find the probability that he or she thinks the rate of civil rights progress is *not* too slow.

13. ***Defective Parts*** ◆ In a box of 11 parts, four of the parts are defective. Two parts are selected at random without replacement.

(a) Find the probability that both parts are defective.

(b) Find the probability that both parts are *not* defective.

(c) Find the probability that at least one part is defective.

14. ***Light Bulbs*** ◆ Twelve light bulbs are tested to see if they last as long as the manufacturer claims they do. Three light bulbs fail the test. Two light bulbs are selected at random without replacement.

(a) Find the probability that both light bulbs failed the test.

(b) Find the probability that both light bulbs passed the test.

(c) Find the probability that at least one light bulb failed the test.

15. ***Emergency Savings*** ◆ The following table shows the results of a survey in which 102 men and 103 women workers ages 25 to 64 were asked if they have at least one month's income set aside for emergencies. *(Adapted from Merrill Lynch)*

	Men	Women	Total
Less than one month's income	47	59	106
One month's income or more	55	44	99
Total	102	103	205

(a) Find the probability that a randomly selected worker has one month's income or more set aside for emergencies.

(b) Given that a randomly selected worker is a male, find the probability that the worker has less than one month's income.

(c) Given that a randomly selected worker has one month's income or more, find the probability that the worker is a female.

(d) Are the events of having less than one month's income or more saved and being male independent or dependent? Explain.

16. ***Health Care for Dogs*** ◆ The following table shows the results of a survey in which 90 dog owners were asked

- how much they have spent in the last year for their dog's health care.
- whether their dogs were purebred or mixed breeds.

		Type of Dog		
		Purebred	**Mixed Breed**	**Total**
Health Care	**Less than $100**	19	21	40
	$100 or more	35	15	50
	Total	54	36	90

(a) Find the probability that $100 or more was spent on a randomly selected dog's health care in the last year.

(b) Given that a randomly selected dog owner spent less than $100, find the probability that the dog was a mixed breed.

(c) Find the probability that a randomly selected dog owner spent $100 or more on health care and the dog was a mixed breed.

(d) Are the events "spending $100 or more on health care" and "having a mixed breed dog" independent or dependent? Explain.

17. ***Blood*** ◆ The probability that a person in the United States has type AB^+ blood is 3%. Five unrelated people in the U.S. are selected at random. *(Source: American Association of Blood Banks)*

(a) Find the probability that all five have type AB^+ blood.

(b) Find the probability that none of the five has type AB^+ blood.

(c) Find the probability that at least one of the five has type AB^+ blood.

18. ***Blood*** ◆ The probability that a person in the United States has type O^+ blood is 38%. Three unrelated people in the U.S. are selected at random. *(Source: American Association of Blood Banks)*

(a) Find the probability that all three have type O^+ blood.

(b) Find the probability that none of the three has type O^+ blood.

(c) Find the probability that at least one of the three has type O^+ blood.

19. ***Guessing*** ◆ A multiple-choice quiz has three questions, each with five answer choices. Only one of the choices is correct. You have no idea what the answer is to any question, and have to guess each answer.

(a) Find the probability of answering the first question correctly.

(b) Find the probability of answering the first two questions correctly.

(c) Find the probability of answering all three questions correctly.

(d) Find the probability of answering none of the questions correctly.

(e) Find the probability of answering at least one of the questions correctly.

20. ***Bookbinding Defects*** ◆ A printing company's bookbinding machine has a probability of 0.005 of producing a defective book. If this machine is used to bind three books, find the probability that (a) none of the books are defective, (b) at least one of the books is defective, and (c) all of the books are defective.

21. ***Warehouses*** ◆ A distribution center receives shipments of a certain product from three different factories in the following quantities: 5000, 3500, and 2500. Three times a product is selected at random, each time without a replacement. Find the probability that all three products came from the third factory.

22. ***Birthdays*** ◆ Three people are selected at random. Find the probability that (a) all three share the same birthday and (b) none of the three share the same birthday. Assume 365 days in a year.

Extending the Basics

According to **Bayes's Theorem,** the probability of event A, given that event B has occurred, is

$$P(A|B) = \frac{P(A) \cdot P(B|A)}{P(A) \cdot P(B|A) + P(A') \cdot P(B|A')}.$$

In Exercises 23 and 24, use Bayes's Theorem to find $P(A|B)$.

23. $P(A) = \frac{2}{3}$, $P(A') = \frac{1}{3}$, $P(B|A) = \frac{1}{5}$, and $P(B|A') = \frac{1}{2}$

24. $P(A) = \frac{3}{8}$, $P(A') = \frac{5}{8}$, $P(B|A) = \frac{2}{3}$, and $P(B|A') = \frac{3}{5}$

25. ***Reliability of Testing*** ◆ A certain virus infects one in every 200 people. A test used to detect the virus in a person is positive 80% of the time if the person has the virus and 5% of the time if the person does not have the virus. (This 5% result is called a *false positive*.) Let A be the event "the person is infected" and B be the event "the person tests positive."

(a) If a person tests positive, what is the probability that the person is infected?

(b) If a person tests negative, what is the probability that the person is *not* infected?

26. ***Birthday Problem*** ◆ You are in a class that has 24 students. You want to find the probability that at least two of the students share the same birthday.

(a) First, find the probability that each student has a different birthday.

$$P(\text{different birthdays}) = \overbrace{\frac{365}{365} \cdot \frac{364}{365} \cdot \frac{363}{365} \cdot \frac{362}{365} \cdots \frac{343}{365} \cdot \frac{342}{365}}^{24 \text{ factors}}$$

(b) The probability that at least two students have the same birthday is the complement of the probability in part (a). What is this probability?

(c) We used a technology tool to generate 24 random numbers between 1 and 365. Each number represents a birthday. Did we get at least two people with the same birthday?

228	348	181	317	81	183
52	346	177	118	315	273
252	168	281	266	285	13
118	360	8	193	57	107

(d) Use a technology tool to simulate the "Birthday Problem." Repeat the simulation 10 times. How many times did you get "at least two people" with the same birthday?

The Multiplication Rule and Conditional Probability By rewriting the formula for the Multiplication Rule, you can write a formula for finding conditional probabilities. The conditional probability of event B occurring, given that event A has occurred, is

$$P(B|A) = \frac{P(A \text{ and } B)}{P(A)}.$$

In Exercises 27 and 28, use the following information.

- The probability that an airplane flight departs on time is 0.89.
- The probability that a flight arrives on time is 0.87.
- The probability that a flight departs and arrives on time is 0.83.

27. Find the probability that a flight departed on time given that it arrives on time.

28. Find the probability that a flight arrives on time given that it departed on time.

3.3 The Addition Rule

Mutually Exclusive Events • The Addition Rule • A Summary of Probability

What You Should Learn

- How to determine if two events are mutually exclusive
- How to use the Addition Rule to find the probability of two events

Mutually Exclusive Events

In Section 3.2, you learned how to find the probability of two events, A and B, occurring in sequence. Such probabilities are denoted by $P(A \text{ and } B)$. In this section, you will learn how to find the probability that at least one of two events will occur. Probabilities such as these are denoted by $P(A \text{ or } B)$ and depend on whether the events are mutually exclusive.

DEFINITION

Two events A and B are **mutually exclusive** if A and B cannot occur at the same time.

The Venn diagrams below show the relationship between events that are mutually exclusive and events that are not mutually exclusive.

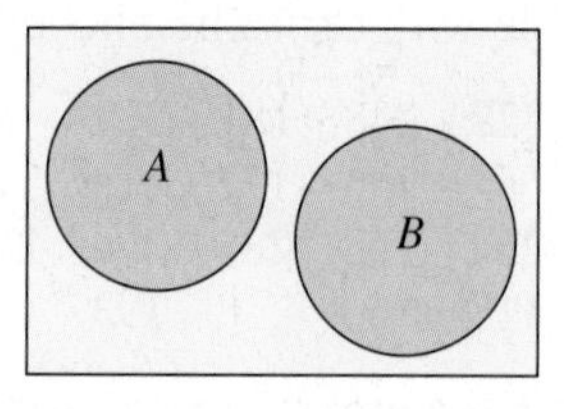

A and *B* are mutually exclusive.

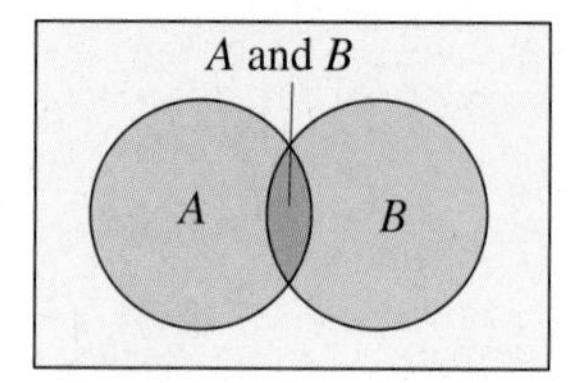

A and *B* are not mutually exclusive.

Study Tip

In probability and statistics the word *or* is usually used as an "inclusive or" rather than an "exclusive or." For instance, there are three ways for "Event *A* or *B*" to occur.

(1) *A* occurs and *B* does not occur.

(2) *B* occurs and *A* does not occur.

(3) *A* and *B* both occur.

EXAMPLE 1

Mutually Exclusive Events

Decide if the events are mutually exclusive. Explain your reasoning.

1. Roll a die. A: Roll a 3. B: Roll a 4.
2. Select a student. A: Select a male student. B: Select a nursing major.
3. Select a blood donor. A: The donor's blood is type O. B: The donor is a female.

SOLUTION

1. The first event has one outcome, a 3. The second event also has one outcome, a 4. These outcomes cannot occur at the same time, so the events are mutually exclusive.
2. Because the student can be a male nursing major, the events are not mutually exclusive.
3. Because the donor can be a female with type O blood, the events are not mutually exclusive.

Try It Yourself 1

Decide if the events are mutually exclusive.

1. *Select a card from a standard deck.*
   A: The card is a jack. B: The card is a face card.
2. *Select a student.*
   A: The student is 20 years old. B: The student has blue eyes.
3. *Select a registered vehicle.*
   A: The vehicle is a Ford. B: The vehicle is a Toyota.

a. Decide if one of the following statements is true.
- Events A and B cannot occur at the same time.
- Events A and B have no outcomes in common.
- $P(A \text{ and } B) = 0$

b. Make a conclusion.

Answer: Page A33

The Addition Rule

The Addition Rule for the Probability of *A* or *B*

The probability that events A *or* B will occur $P(A \text{ or } B)$ is given by

$$P(A \text{ or } B) = P(A) + P(B) - P(A \text{ and } B).$$

If events A and B are mutually exclusive, then the rule can be simplified to $P(A \text{ or } B) = P(A) + P(B)$. This simplified rule can be extended to any number of mutually exclusive events.

Study Tip

By subtracting $P(A \text{ and } B)$ you avoid double counting the probability of outcomes that occur in both A and B.

EXAMPLE 2

Using the Addition Rule to Find Probabilities

1. You select a card from a standard deck. Find the probability that the card is a 4 or an ace.
2. You roll a die. Find the probability of rolling a number less than three or rolling an odd number.

SOLUTION

1. If the card is a 4, it cannot be an ace. So, the events are mutually exclusive. The probability of selecting a 4 or an ace is

$$P(4 \text{ or ace}) = P(4) + P(\text{ace}) = \frac{4}{52} + \frac{4}{52} = \frac{8}{52} = \frac{2}{13} \approx 0.154.$$

2. The events are not mutually exclusive because 1 is an outcome of both events. So, the probability of rolling a number less than 3 or an odd number is

$$P(\text{less than 3 or odd}) = P(\text{less than 3}) + P(\text{odd}) - P(\text{less than 3 and odd})$$
$$= \frac{2}{6} + \frac{3}{6} - \frac{1}{6} = \frac{4}{6} = \frac{2}{3} \approx 0.667.$$

Deck of 52 Cards

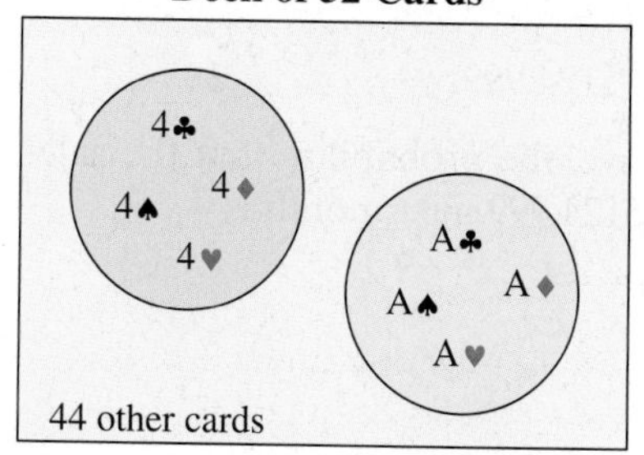

Roll a Die

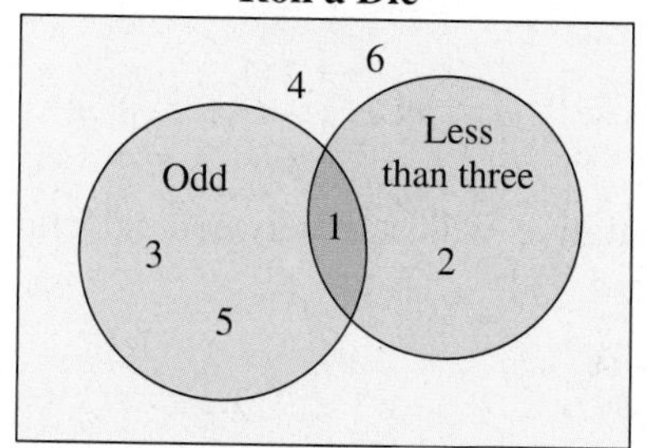

Try It Yourself 2

1. A die is rolled. Find the probability of rolling a 6 or an odd number.
2. A card is selected from a standard deck. Find the probability that the card is a face card or a heart.

a. Decide whether the events are *mutually exclusive.*
b. Find $P(A)$, $P(B)$, and, if necessary, $P(A \text{ and } B)$.
c. Use the *Addition Rule* to find the probability.

Answer: Page A33

EXAMPLE

Finding Probabilities of Mutually Exclusive Events

The following frequency distribution shows the volume of sales, in dollars, and the number of months a sales representative reached each sales level during the past three years. If this sales pattern continues, what is the probability that the sales representative will sell between \$75,000 and \$124,999 next month?

Sales Volume (\$)	Months
0–24,999	3
25,000–49,999	5
50,000–74,999	6
75,000–99,999	7
100,000–124,999	9
125,000–149,999	2
150,000–174,999	3
175,000–199,999	1

SOLUTION To solve this problem, define events A and B as follows.

A = monthly sales between \$75,000 and \$99,999

B = monthly sales between \$100,000 and \$124,999

Because events A and B are mutually exclusive, the probability that the sales representative will sell between \$75,000 and \$124,999 next month is

$$P(A \text{ or } B) = P(A) + P(B)$$
$$= \frac{7}{36} + \frac{9}{36}$$
$$= \frac{16}{36} = \frac{4}{9} \approx 0.444.$$

Try It Yourself 3

Find the probability that the sales representative will sell between \$0 and \$49,999.

a. *Identify* events A and B.
b. Verify that A and B are *mutually exclusive.*
c. Find the *probability* of each event.
d. Use the *Addition Rule* to find the probability.

Answer: Page A34

EXAMPLE 4

Using the Addition Rule to Find Probabilities

A blood bank catalogs the types of blood, including positive or negative Rh-factor, given by donors during the last five days. The number of donors who gave each blood type is listed in the following table. A donor is selected at random.

1. Find the probability that the donor has type O or type A blood.
2. Find the probability that the donor has type B blood or is Rh-negative.

		Blood Type				
		O	**A**	**B**	**AB**	**Total**
Rh-factor	**Positive**	156	139	37	12	344
	Negative	28	25	8	4	65
	Total	184	164	45	16	409

Picturing the World

In a survey conducted by the National Family Organization, new mothers were asked to rate the difficulty of delivering their first child compared with what they expected.

How Difficult was the Delivery?

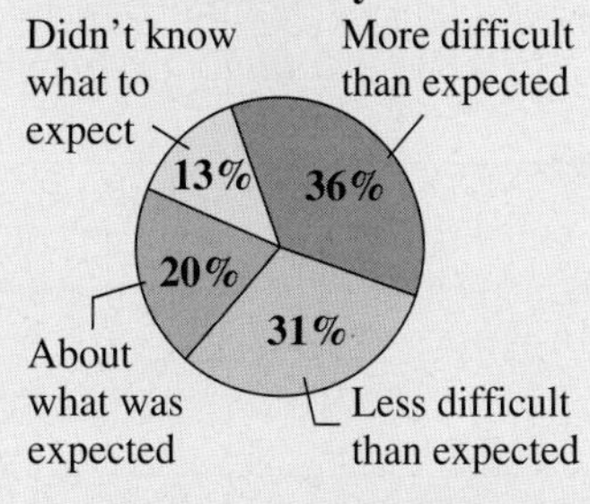

If you selected at random a new mother and asked her to compare the difficulty of her delivery to what she expected, what is the probability that she would say that it was the same or more difficult than what she expected?

SOLUTION

1. Because a donor cannot have type O blood and type A blood, these events are mutually exclusive. So, using the Addition Rule, the probability that a randomly chosen donor has type O or type A blood is

$$\begin{aligned} P(\text{type O or type A}) &= P(\text{type O}) + P(\text{type A}) \\ &= \frac{184}{409} + \frac{164}{409} \\ &= \frac{348}{409} \\ &\approx 0.851. \end{aligned}$$

2. Because a donor can have type B blood and be Rh-negative, these events are not mutually exclusive. So, using the Addition Rule, the probability that a randomly chosen donor has type B blood or is Rh-negative is

$$\begin{aligned} P(\text{type B or Rh-neg}) &= P(\text{type B}) + P(\text{Rh-neg}) \\ &\quad - P(\text{type B and Rh-neg}) \\ &= \frac{45}{409} + \frac{65}{409} - \frac{8}{409} \\ &= \frac{102}{409} \\ &\approx 0.249. \end{aligned}$$

Try It Yourself 4

1. Find the probability that the donor has type B or type AB blood.
2. Find the probability that the donor has type O blood or is Rh-positive.

a. Decide if the events are *mutually exclusive.*
b. Use the *Addition Rule.*

Answer: Page A34

A Summary of Probability

Type of Probability and Probability Rules	Formula
Classical Probability	$P(E) = \dfrac{\text{Number of outcomes in event } E}{\text{Number of outcomes in sample space}}$
Empirical Probability	$P(E) = \dfrac{\text{Frequency of event } E}{\text{Total frequency}} = \dfrac{f}{n}$
Complementary Events	$P(E) + P(E') = 1,\ P(E) = 1 - P(E')$ $P(E') = 1 - P(E)$
Multiplication Rule	$P(A \text{ and } B) = P(A) \cdot P(B \mid A)$ $P(A \text{ and } B) = P(A) \cdot P(B)$ *Independent events*
Addition Rule	$P(A \text{ or } B) = P(A) + P(B) - P(A \text{ and } B)$ $P(A \text{ or } B) = P(A) + P(B)$ *Mutually exclusive events*

EXAMPLE 

Finding Probabilities

Use the graph at the right to find the probability that a randomly selected draft pick is not a running back or a wide receiver.

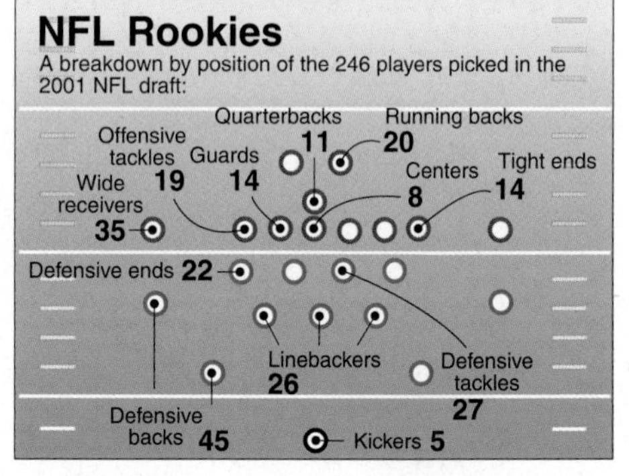

Source: NFL.com

SOLUTION Define events A and B.

A: Draft pick is a running back.
B: Draft pick is a wide receiver.

These events are mutually exclusive, so the probability that the draft pick is a running back or wide receiver is

$$P(A \text{ or } B) = P(A) + P(B) = \tfrac{20}{246} + \tfrac{35}{246} = \tfrac{55}{246}.$$

By taking the complement of $P(A \text{ or } B)$, you can determine that the probability of randomly selecting a draft pick who is not a running back or wide receiver is

$$1 - P(A \text{ or } B) = 1 - \tfrac{55}{246} = \tfrac{191}{246} \approx 0.776.$$

Try It Yourself 5

Find the probability that a randomly selected draft pick is not a linebacker or a quarterback.

a. Find the *probability* that the draft pick is a linebacker or a quarterback.
b. Find the *complement* of the event.

Answer: Page A34

3.3 Exercises

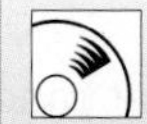
LarsonTutor 3.3

Companion Web Site

Student Solutions Manual 3.3

Videos 3.3

Technology Manuals

Basic Skills and Concepts

1. If two events are mutually exclusive, why is $P(A \text{ and } B) = 0$?

2. List examples of

(a) two events that are mutually exclusive.

(b) two events that are not mutually exclusive.

True or False In Exercises 3 and 4, determine whether the statement is true or false. If it is false, explain why.

3. If two events are mutually exclusive, they have no outcomes in common.

4. The probability that event A or event B will occur is

$$P(A \text{ or } B) = P(A) + P(B) - P(A \text{ or } B).$$

Graphical Analysis In Exercises 5 and 6, decide if the events shown in the Venn diagram are mutually exclusive. Explain your reasoning.

5.

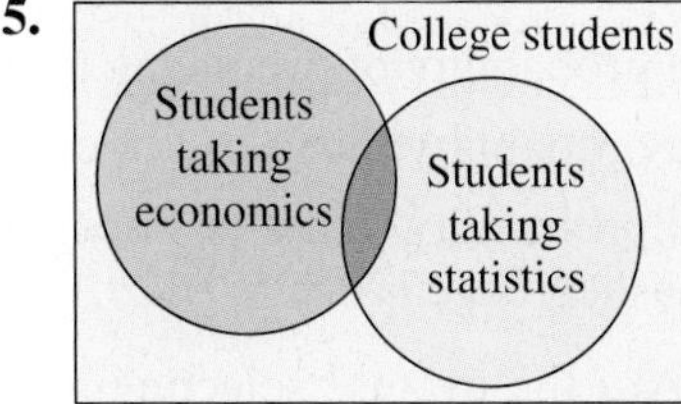

6.

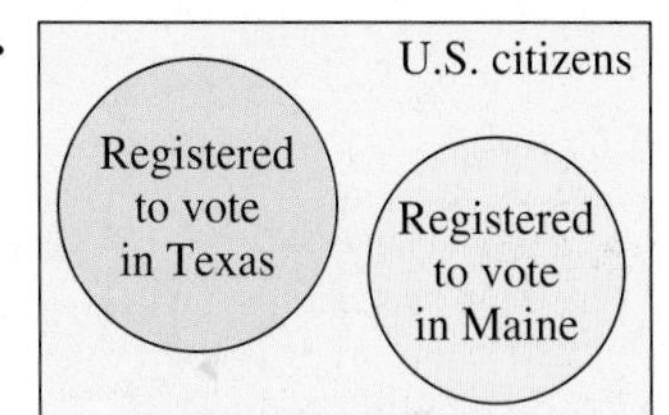

Recognizing Mutually Exclusive Events In Exercises 7–10, decide if the events are mutually exclusive. Explain your reasoning.

7. Event A: Randomly select a female worker.
Event B: Randomly select a worker with a college degree.

8. Event A: Randomly select a male worker.
Event B: Randomly select a worker employed part time.

9. Event A: Randomly select a person between 18 and 24 years old.
Event B: Randomly select a person between 25 and 34 years old.

10. Event A: Randomly select a person between 18 and 24 years old.
Event B: Randomly select a person earning between \$20,000 and \$29,999.

11. ***Audit*** ◆ During a 52-week period, a company paid overtime wages for 18 weeks and hired temporary help for 9 weeks. During 5 weeks, the company paid overtime *and* hired temporary help.

(a) Are the events "selecting a week that contained overtime wages" and "selecting a week that contained temporary help wages" mutually exclusive? Explain.

(b) If an auditor randomly examined the payroll records for only one week, what is the probability that the payroll for that week contained overtime wages or temporary help wages?

12. ***Newspaper Survey*** ◆ A college has an undergraduate enrollment of 3500. Of these, 860 are business majors and 1800 are women. Of the business majors, 425 are women.

(a) Are the events "selecting a woman student" and "selecting a business major" mutually exclusive? Explain.

(b) If a college newspaper conducts a poll and selects students at random to answer a survey, find the probability that a selected student is a woman or a business major.

13. ***Carton Defects*** ◆ A company that makes cartons finds

- the probability of producing a carton with a puncture is 0.05.
- the probability that a carton has a smashed corner is 0.08.
- the probability that a carton has a puncture and has a smashed corner is 0.004.

(a) Are the events "selecting a carton with a puncture" and "selecting a carton with a smashed corner" mutually exclusive? Explain.

(b) If a quality inspector randomly selects a carton, find the probability that the carton has a puncture or has a smashed corner.

14. ***Can Defects*** ◆ A company that makes soda pop cans finds

- the probability of producing a can without a puncture is 0.96.
- the probability that a can does not have a smashed edge is 0.93.
- the probability that a can does not have a puncture and does not have a smashed edge is 0.893.

(a) Are the events "selecting a can without a puncture" and "selecting a can without a smashed edge" mutually exclusive? Explain.

(b) If a quality inspector randomly selects a can, find the probability that the can does not have a puncture or does not have a smashed edge.

15. ***U.S. Age Distribution*** ◆ The estimated percent distribution of the U.S. population for 2006 is shown in the pie chart. Find the following probabilities. *(Source: U.S. Census Bureau)*

(a) Randomly selecting someone under five years old

(b) Randomly selecting someone who is not 65 years or over

(c) Randomly selecting someone who is between 18 and 34 years old

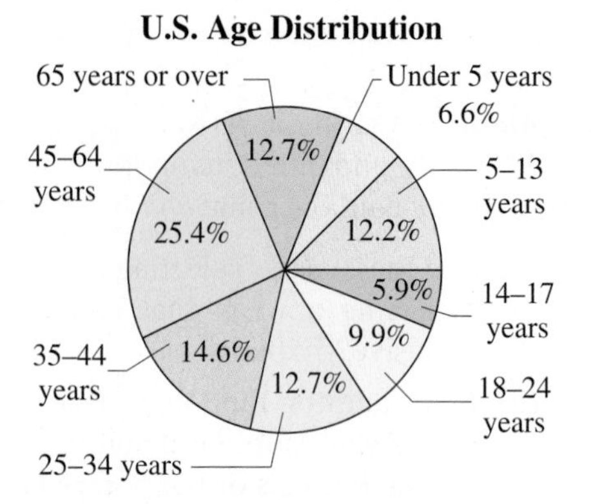

Figure for 15

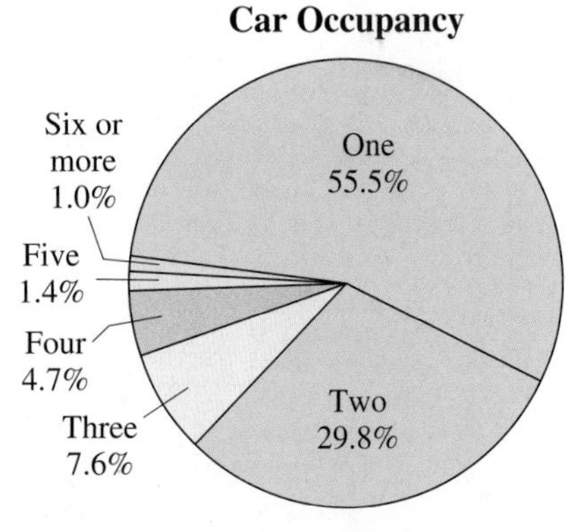

Figure for 16

16. ***Tacoma Narrows Bridge*** ◆ The percent distribution of the number of occupants in vehicles crossing the Tacoma Narrows Bridge in Washington is shown in the pie chart. Find the following probabilities. *(Source: Washington State Department of Transportation)*

(a) Randomly selecting a car with two occupants

(b) Randomly selecting a car with two or more occupants

(c) Randomly selecting a car with between two and five occupants, inclusive

17. ***Air Travel*** ◆ The number of responses to a survey are given in the Pareto chart. The survey asked 1018 U.S. adults how they feel about the safety standards of the major U.S. commercial airlines. Each person gave one response. Find the following probabilities. *(Source: Gallup/CNN/USA Today)*

(a) Randomly selecting a person from the sample who is not confident about air travel safety

(b) Randomly selecting a person from the sample who is somewhat confident or very confident about air travel safety

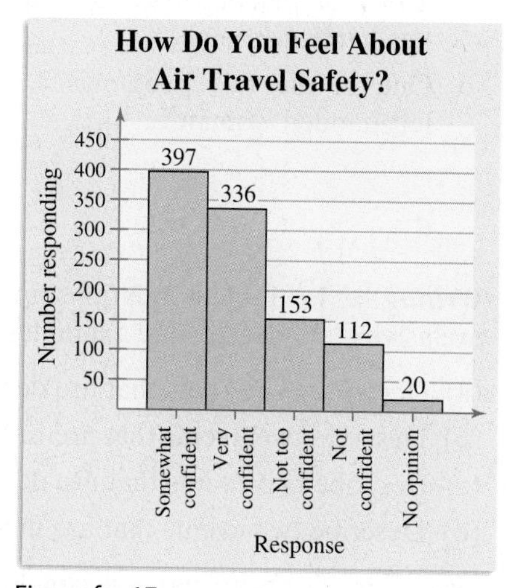

Figure for 17

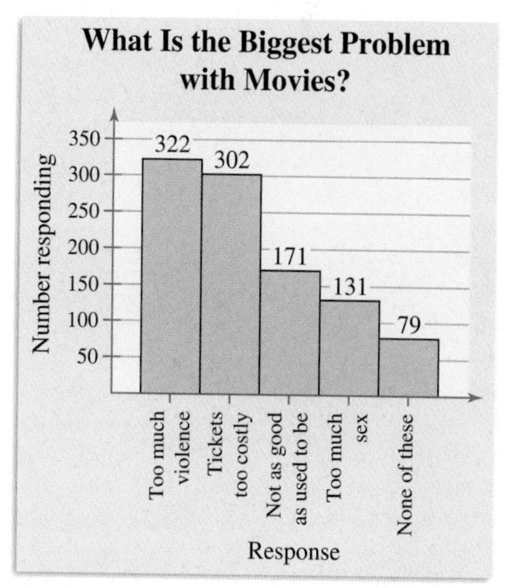

Figure for 18

18. ***Movies*** ◆ The number of responses to a survey are given in the Pareto chart. The survey asked 1005 U.S. adults what they feel is the biggest problem with movies. Each person gave one response. Find the following probabilities. *(Source: Associated Press)*

(a) Randomly selecting a person from the sample who feels the biggest problem with movies is that movies are not as good as they used to be

(b) Randomly selecting a person from the sample who feels the biggest problem with movies is that movies have too much violence or that the tickets cost too much

19. ***Left-Handed People*** ◆ In a sample of 1000 people, 120 are left handed. If two unrelated people are selected at random from the sample, find the probability of the following.

(a) Both people are left handed.

(b) At least one of the two people is left handed.

(c) Neither person is left handed.

(d) Two of the events from (a), (b), and (c) are complementary events. Which two are they? Explain.

20. ***Left-Handed People*** ◆ In a sample of 1000 people (500 men and 500 women), 113 are left handed (63 men and 50 women). The results of the sample are given in the table. If two unrelated people are selected at random from the sample, find the probability of the following.

		Gender		
		Men	***Women***	***Total***
Dominant Hand	***Left***	63	50	113
	Right	437	450	887
	Total	500	500	1000

(a) Both people are left-handed men.

(b) Both people are left-handed women.

(c) At least one of the two people is left handed.

(d) The first person is a left-handed man and the second person is a left-handed woman.

(e) One of the two people is a left-handed man and the other is a left-handed woman.

Extending the Basics

21. ***Writing*** ◆ Is there a relationship between independence and mutual exclusivity? To decide, find examples of the following, *if possible.*

(a) Describe two events that are dependent and mutually exclusive.

(b) Describe two events that are independent and mutually exclusive.

(c) Describe two events that are dependent and not mutually exclusive.

(d) Describe two events that are independent and not mutually exclusive.

Use your results to write a conclusion about the relationship between independence and mutual exclusivity.

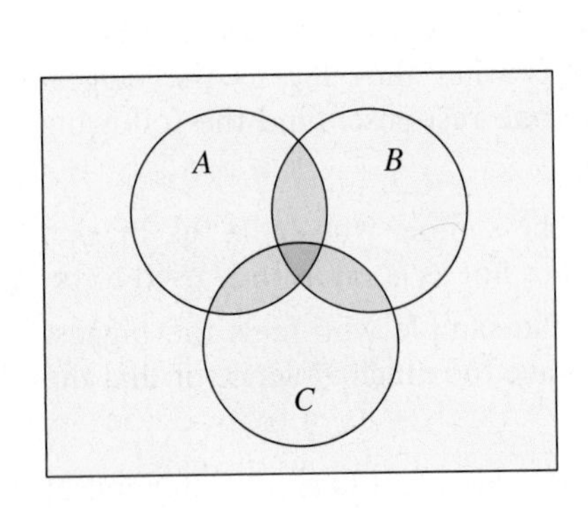

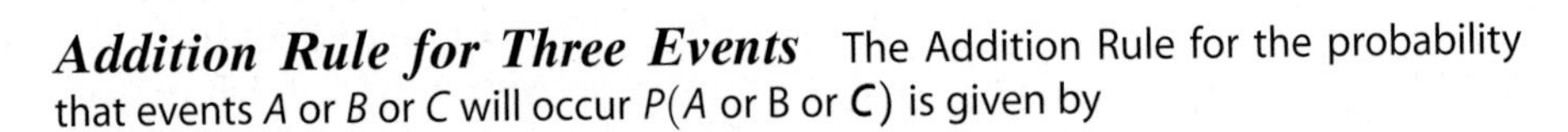

Addition Rule for Three Events The Addition Rule for the probability that events A or B or C will occur $P(A \text{ or } B \text{ or } C)$ is given by

$$P(A \text{ or } B \text{ or } C) = P(A) + P(B) + P(C) - P(A \text{ and } B) - P(A \text{ and } C) - P(B \text{ and } C) + P(A \text{ and } B \text{ and } C).$$

In the Venn diagram at the left, $P(A \text{ or } B \text{ or } C)$ is represented by the blue areas.

In Exercises 22 and 23, find $P(A \text{ or } B \text{ or } C)$ for the given probabilities.

22. $P(A) = 0.20$, $P(B) = 0.50$, $P(C) = 0.30$,
$P(A \text{ and } B) = 0.15$, $P(A \text{ and } C) = 0.13$, $P(B \text{ and } C) = 0.20$,
$P(A \text{ and } B \text{ and } C) = 0.10$

23. $P(A) = 0.25$, $P(B) = 0.25$, $P(C) = 0.50$,
$P(A \text{ and } B) = 0.15$, $P(A \text{ and } C) = 0.25$, $P(B \text{ and } C) = 0.19$,
$P(A \text{ and } B \text{ and } C) = 0.13$

Case Study 3

WWW.INFORMS.ORG

Probability and Parking Lot Strategies

The Institute for Operations Research and the Management Sciences (INFORMS) is an international scientific society with over 12,000 members. It is dedicated to the application of scientific methods to improve decision making, management, and operations. Members of the institute work primarily in business, government, and education. They represent fields as diverse as airlines, health care, law enforcement, the military, the stock market, and telecommunications.

One study published by INFORMS was the result of research conducted by Richard Cassady of Mississippi State University and John Kobza of Virginia Polytechnic Institute. The parking space study was conducted at a mall that has four entrances, seven rows with 72 spaces each, and directional restrictions. The researchers compared several parking lot strategies to see which strategy saves the most time. The two best strategies are called *Pick a Lane* and *Cycling*. The results are listed below.

Pick a Lane
Choose a lane. Enter it and select the closest available space.

Cycling
Enter the closest lane. Park in any of the 20 closest spaces. If all are full, cycle to next row.

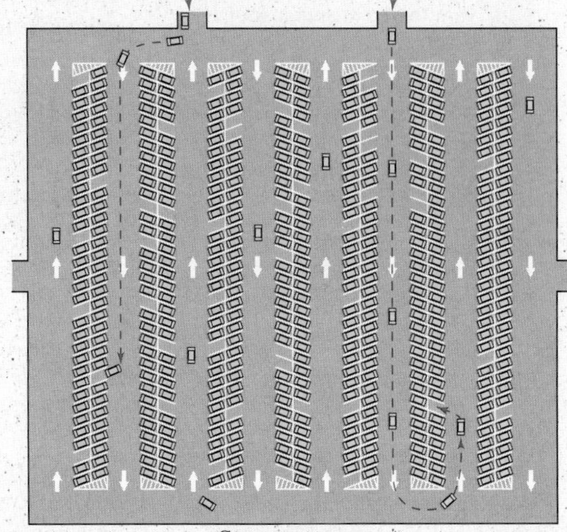

Store entrance

Time or Distance	Pick a Lane	Cycling
Time from lot entrance to parking space	37.7 sec	52.5 sec
Time from lot entrance to store's door	61.3 sec	70.7 sec
Average walking distance to store	257 feet	208 feet

Exercises

1. In a parking lot study, is each parking space equally likely to be empty? Explain your reasoning.
2. According to the results of the study, are you more likely to spend less time using the Pick-a-Lane strategy or the Cycling strategy? Explain.
3. According to the results of the study, are you more likely to walk less using the Pick-a-Lane strategy or the Cycling strategy? Explain.
4. A key assumption in the study was that the drivers can see which spaces are available as soon as they enter a lane. Why is that important?
5. The parking lot is completely full, and one car leaves. What is the probability that the car was in the first row? Explain your reasoning.
6. A person is leaving from a row that is full. What is the probability that the person was parked in one of the 20 spaces that are closest to the store?
7. Draw a diagram of the parking lot. Color code the parking spaces into three categories of 168 spaces each: most desirable, moderately desirable, and least desirable. Assume that the parking lot is half full. Estimate the probability that you can find a parking space in the most desirable category. Explain your reasoning.

3.4 Counting Principles

What You Should Learn

- How to use the Fundamental Counting Principle to find the number of ways two or more events can occur
- How to find the number of ways a group of objects can be arranged in order
- How to find the number of ways to choose several objects from a group without regard to order
- How to use counting principles to find probabilities

The Fundamental Counting Principle • Permutations • Combinations • Applications of Counting Principles

The Fundamental Counting Principle

In this section, you will study several techniques for counting the number of ways an event can occur. One is the Fundamental Counting Principle. You can use this principle to find the number of ways two or more events can occur in sequence.

The Fundamental Counting Principle

If one event can occur in m ways and a second event can occur in n ways, the number of ways the two events can occur in sequence is $m \cdot n$. This rule can be extended for any number of events occurring in sequence.

EXAMPLE 1

Using the Fundamental Counting Principle

You are purchasing a new car. Using the following manufacturers, car sizes, and colors, how many different ways can you select one manufacturer, one car size, and one color?

Manufacturer:	Ford, GM, Chrysler
Car size:	small, medium
Color:	white (W), red (R), black (B), green (G)

SOLUTION There are three choices of manufacturers, two car sizes, and four colors. So, the number of ways to select one manufacturer, one car size, and one color is

$$3 \cdot 2 \cdot 4 = 24 \text{ ways.}$$

A tree diagram can help you see why there are 24 options.

Tree Diagram for Car Selections

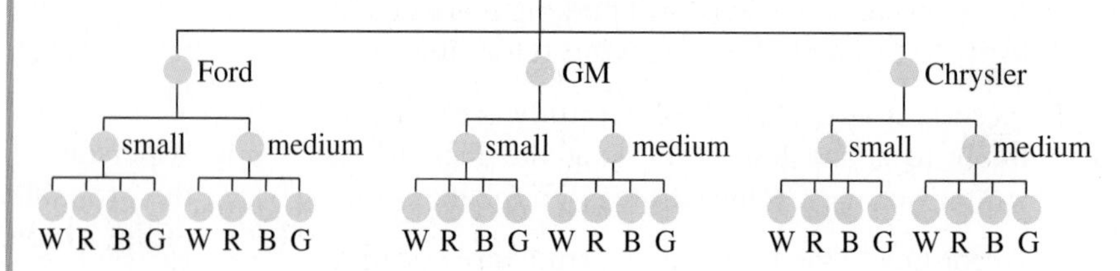

Try It Yourself 1

You increase your choices to include a Toyota, a large car, or a tan or gray car. How many different ways can you select one manufacturer, one car size, and one color now?

a. Find the *number of ways* each event can occur.
b. Use the *Fundamental Counting Principle*.

Answer: Page A34

EXAMPLE 2

Using the Fundamental Counting Principle

The access code for a car's security system consists of four digits. Each digit can be 0 through 9. How many access codes are possible if

1. each digit can be used only once and not repeated?
2. each digit can be repeated?

SOLUTION

1. Because each digit can be used only once, there are 10 choices for the first digit, 9 choices left for the second digit, 8 choices left for the third digit, and 7 for the fourth digit. Using the Fundamental Counting Principle, you can conclude that there are

   $$10 \cdot 9 \cdot 8 \cdot 7 = 5040$$

   possible access codes.
2. Because each digit can be repeated, there are 10 choices for each of the four digits. So, there are

   $$10 \cdot 10 \cdot 10 \cdot 10 = 10^4 = 10{,}000$$

   possible access codes.

Try It Yourself 2

How many license plates can you make if a license plate consists of

1. six (out of 26) alphabetical letters each of which can be repeated?
2. six (out of 26) alphabetical letters each of which cannot be repeated?

a. *Identify* each event and the *number of ways* each event can occur.
b. Use the *Fundamental Counting Principle.*

Answer: Page A34

Permutations

An important application of the Fundamental Counting Principle is determining the number of ways that n objects can be arranged in order or in a permutation.

Study Tip

The six permutations for the letters *A*, *B*, and *C* are

ABC, *ACB*, *BAC*
BCA, *CAB*, *CBA*.

DEFINITION

A **permutation** is an ordered arrangement of objects. The number of different permutations of n distinct objects is $n!$.

The expression $\boldsymbol{n!}$ is read as ***n* factorial** and is defined as follows.

$$n! = n \cdot (n-1) \cdot (n-2) \cdot (n-3) \cdots 3 \cdot 2 \cdot 1$$

As a special case,

$$0! = 1.$$

> **Study Tip**
>
> Notice at the right that as n increases, $n!$ becomes very large. Take some time now to learn how to use the factorial key on your calculator.

Here are several other values of $n!$.

$1! = 1$
$2! = 2 \cdot 1 = 2$
$3! = 3 \cdot 2 \cdot 1 = 6$
$4! = 4 \cdot 3 \cdot 2 \cdot 1 = 24$
$5! = 5 \cdot 4 \cdot 3 \cdot 2 \cdot 1 = 120$

EXAMPLE 3

Finding the Number of Permutations of n Objects

The starting lineup for a baseball team consists of nine players. How many different batting orders are possible using the starting lineup?

SOLUTION

The number of permutations is $9! = 9 \cdot 8 \cdot 7 \cdot 6 \cdot 5 \cdot 4 \cdot 3 \cdot 2 \cdot 1 = 362{,}880$. So, there are 362,880 different batting lineups.

Try It Yourself 3

The teams in the National League Central Division are listed at the right. How many different final standings are possible?

National League Central Division	
Chicago Cubs	Cincinnati Reds
Houston Astros	Milwaukee Brewers
Pittsburgh Pirates	St. Louis Cardinals

a. Determine *how many teams, n,* are in the Central Division.
b. Evaluate $n!$.

Answer: Page A34

Suppose you want to choose some of the objects in a group and put them in order. Such an ordering is called a **permutation of *n* objects taken *r* at a time.**

> **Permutations of *n* Objects Taken *r* at a Time**
>
> The number of permutations of n distinct objects taken r at a time is
>
> $${}_nP_r = \frac{n!}{(n-r)!} \quad \text{where } r \le n.$$

EXAMPLE 

Finding ${}_nP_r$

Find the number of ways of forming three-digit codes in which no digit is repeated.

SOLUTION

To form a three-digit code with no repeating digits, you need to select 3 digits from a group of 10, so $n = 10$ and $r = 3$.

$${}_nP_r = {}_{10}P_3 = \frac{10!}{(10-3)!} = \frac{10!}{7!} = \frac{10 \cdot 9 \cdot 8 \cdot \not{7} \cdot \not{6} \cdot \not{5} \cdot \not{4} \cdot \not{3} \cdot \not{2} \cdot \not{1}}{\not{7} \cdot \not{6} \cdot \not{5} \cdot \not{4} \cdot \not{3} \cdot \not{2} \cdot \not{1}} = 720$$

So, there are 720 possible three-digit codes that do not have repeating digits.

Try It Yourself 4

In a race with eight horses, how many ways can three of the horses finish in first, second, and third place? Assume that there are no ties.

a. Find the *quotient* of $n!$ and $(n - r)!$. (List the factors and cancel.)
b. *Write* the result as a sentence.

Answer: Page A34

EXAMPLE 5

Permutations of n Objects Taken r at a Time

Forty-three race cars started the 2001 Daytona 500. How many ways can the cars finish first, second, and third? *(Source: NASCAR.com)*

SOLUTION Because there are 43 race cars and order is important, the number of ways the cars can finish first, second, and third is

$$
\begin{aligned}
{}_{43}P_3 &= \frac{43!}{(43-3)!} \\
&= \frac{43!}{40!} \\
&= 43 \cdot 42 \cdot 41 \\
&= 74{,}046.
\end{aligned}
$$

Try It Yourself 5

The board of directors for a company has twelve members. One member is the president, another is the vice-president, another is the secretary, and another is the treasurer. How many ways can can these positions be assigned?

a. *Identify* the total number of objects n and the number of objects r being chosen in order.
b. Evaluate ${}_nP_r$.

Answer: Page A34

Suppose you want to order a group of n objects where some of the objects are the same. For instance, consider a group of letters consisting of four A's, two B's, and one C. How many ways can you order such a group? Using the previous formula, you might conclude that there are ${}_7P_7 = 7!$ possible orders. However, because some of the objects are the same, not all of these permutations are *distinguishable.* How many distinguishable permutations are possible? The answer can be found using the following formula.

Study Tip

The letters *AAAABBC* can be rearranged in 7! orders, but many of these are not distinguishable. The number of distinguishable orders is

$$\frac{7!}{4! \cdot 2! \cdot 1!} = \frac{7 \cdot 6 \cdot 5}{2} = 105.$$

Distinguishable Permutations

The number of **distinguishable permutations** of n objects where n_1 are one type, n_2 are of another type, and so on is

$$\frac{n!}{n_1! \cdot n_2! \cdot n_3! \cdots n_k!}, \text{ where } n_1 + n_2 + n_3 + \cdots + n_k = n.$$

EXAMPLE 6

Distinguishable Permutations

A building contractor is planning to develop a subdivision. The subdivision is to consist of six one-story houses, four two-story houses, and two split-level houses. In how many distinguishable ways can the houses be arranged?

SOLUTION There are to be twelve houses in the subdivision, six of which are of one type (one-story), four of another type (two-story), and two of a third type (split-level). So, there are

$$\frac{12!}{6! \cdot 4! \cdot 2!} = \frac{12 \cdot 11 \cdot 10 \cdot 9 \cdot 8 \cdot 7 \cdot \cancel{6!}}{\cancel{6!} \cdot 4! \cdot 2!}$$
$$= 13{,}860$$

distinguishable ways to arrange the houses in the subdivision.

Try It Yourself 6

The contractor wants to plant six oak trees, nine maple trees, and five poplar trees along the subdivision street. If the trees are spaced evenly apart, in how many distinguishable ways can they be planted?

a. *Identify* the total number of objects, n, and the number of each type of object in the group, $n_1, n_2,$ and n_3.

b. *Evaluate* $\frac{n!}{n_1! \cdot n_2! \cdots n_k!}$.

Answer: Page A34

Combinations

Suppose you want to buy three CDs from a selection of five CDs. There are 10 ways to make your selections.

ABC, ABD, ABE,
ACD, ACE,
ADE,
BCD, BCE,
BDE,
CDE

In each selection, order does not matter (*ABC* is the same set as *BAC*). The number of ways to choose r objects from n objects without regard to order is called the number of **combinations of n objects taken r at a time.**

Combination of *n* Objects taken *r* at a Time

A **combination** is a selection of r objects from a group of n objects without regard to order and is denoted by ${}_nC_r$. The number of combinations of r objects selected from a group of n objects is

$${}_nC_r = \frac{n!}{(n-r)!r!}.$$

EXAMPLE 7

Finding the Number of Combinations

A state's department of transportation plans to develop a new section of interstate highway and receives 16 bids for the project. The state plans to hire four of the bidding companies. How many different combinations of four companies can be selected from the 16 bidding companies?

SOLUTION Because order is not important, there are

$$
\begin{aligned}
{}_{16}C_4 &= \frac{16!}{(16-4)!4!} \\
&= \frac{16!}{12!4!} \\
&= \frac{16 \cdot 15 \cdot 14 \cdot 13 \cdot \cancel{12!}}{\cancel{12!} \cdot 4!} \\
&= 1820
\end{aligned}
$$

different combinations.

Try It Yourself 7

The manager of an accounting department wants to form a three-person advisory committee from the 16 employees in the department. In how many ways can the manager do this?

a. *Identify* the number of objects in the group n and the number of objects to be selected r.
b. *Evaluate* ${}_nC_r$.
c. *Write* the results as a sentence.

Answer: Page A34

The following table summarizes the counting principles.

Principle	Description	Formula
Fundamental Counting Principle	If one event can occur in m ways and a second event can occur in n ways, the number of ways the two events can occur in sequence is $m \cdot n$.	$m \cdot n$
Permutations	The number of different ordered arrangements of n distinct objects	$n!$
	The number of permutations of n distinct objects taken r at a time, where $r \le n$	${}_nP_r = \frac{n!}{(n-r)!}$
	The number of distinguishable permutations of n objects where n_1 are of one type, n_2 are of another type, and so on	$\frac{n!}{n_1! \cdot n_2! \cdots n_k!}$
Combinations	The number of combinations of r objects selected from a group of n objects without regard to order	${}_nC_r = \frac{n!}{(n-r)!r!}$

Picturing the World

On May 9, 2000, the largest lottery jackpot ever, \$363 million, was won in The Big Game lottery. The Big Game is a lottery game in which five numbers are chosen from 1 to 50 and one number, the Big Money Ball, is chosen from 1 to 36. The winning numbers are shown below.

If you buy one ticket, what is the probability that you will win The Big Game lottery?

Applications of Counting Principles

EXAMPLE 8

Finding Probabilities

A word consists of one M, four I's, four S's, and two P's. If the letters are randomly arranged in order, what is the probability that the arrangement spells the word *Mississippi*?

SOLUTION There is one favorable outcome and there are

$$\frac{11!}{1! \cdot 4! \cdot 4! \cdot 2!} = 34{,}650 \qquad \text{11 letters with 1, 4, 4, and 2 like letters}$$

distinguishable permutations of the word *Mississippi*. So, the probability that the arrangement spells the word *Mississippi* is

$$P(\text{Mississippi}) = \frac{1}{34{,}650} \approx 0.000029.$$

Try It Yourself 8

A word consists of one L, two E's, two T's, and one R. If the letters are randomly arranged in order, what is the probability that the arrangement spells the word *Letter*?

a. *Find* the number of favorable outcomes and the number of distinguishable permutations.
b. *Divide* the number of favorable outcomes by the number of distinguishable permutations.

Answer: Page A34

EXAMPLE 9

Finding Probabilities

Find the probability of being dealt five diamonds from a standard deck of playing cards. (In poker, this is a diamond flush.)

SOLUTION The possible number of ways of choosing 5 diamonds out of 13 is ${}_{13}C_5$. The number of possible 5-card hands is ${}_{52}C_5$. So, the probability of being dealt 5 diamonds is

$$P(\text{diamond flush}) = \frac{{}_{13}C_5}{{}_{52}C_5} = \frac{1287}{2{,}598{,}960} \approx 0.0005.$$

Try It Yourself 9

A jury consists of five men and seven women. Three are selected at random for an interview. Find the probability that all three are men.

a. *Find* the product of the number of ways to choose three men from five and the number of ways to choose zero women from seven.
b. *Find* the number of ways to choose 3 jury members from 12.
c. *Divide* the result of Part a by the result of Part b.

Answer: Page A34

3.4 Exercises

Basic Skills and Concepts

1. When you use the Fundamental Counting Principle, what are you counting?
2. What is the difference between a permutation and a combination?

True or False In Exercises 3 and 4, determine whether the statement is true or false. If it is false, rewrite it so it is a true statement.

3. A combination is an ordered arrangement of objects.
4. The number of different ordered arrangements of n distinct objects is $n!$.

In Exercises 5 and 6, decide if the situation involves permutations, combinations, or neither. Explain your reasoning.

5. The number of ways 10 people can line up in a row for concert tickets
6. The number of ways a four-member committee can be chosen from 15 people

7. ***Space Shuttle Menu*** ◆ Space shuttle astronauts each consume an average of 3000 calories per day. One meal normally consists of a main dish, a vegetable dish, and two different desserts. The astronauts can choose from 10 main dishes, 8 vegetable dishes, and 13 desserts. How many different meals are possible? *(Source: NASA)*
8. ***Menu*** ◆ A menu has three choices for salad, five main dishes, and two desserts. How many different meals are available if you select a salad, a main dish, and a dessert?
9. ***Security System*** ◆ The access code for a car's security system consists of four digits. The first digit cannot be zero and the last digit must be odd. How many different codes are available?
10. ***True or False Quiz*** ◆ Assuming that no questions are left unanswered, in how many ways can a six-question true-false quiz be answered?
11. ***Horse Race*** ◆ A horse race has eight entries. Assuming that there are no ties, in how many different orders can the horses finish?
12. ***Security Code*** ◆ In how many ways can the letters A, B, C, D, E, and F be arranged for a six-letter security code?
13. ***Starting Lineup*** ◆ The starting lineup for a softball team consists of 10 players. How many different batting orders are possible using the starting lineup?
14. ***Assembly Process*** ◆ There are four processes involved in assembling a certain product. These processes can be performed in any order. If management wants to find which order is the least time consuming, how many different orders will have to be tested?
15. ***Horse Race*** ◆ A horse race has 10 entries. Assuming that there are no ties, in how many different ways can these horses finish first, second, and third?

16. ***Starting Lineup*** ◆ The starting lineup for a baseball team consists of nine players. Assuming that each member of a team with 25 players can play each position, in how many different ways can the starting lineup be filled?

17. ***Officers*** ◆ From a pool of 15 candidates, the offices of president, vice-president, secretary, and treasurer will be filled. In how many different ways can the offices be filled?

18. ***Team Photo*** ◆ In how many ways can a team of 13 players line up in a row to pose for a photo?

19. ***Tree Planting*** ◆ A landscaper wants to plant four oak trees, eight maple trees, and six poplar trees along the border of a lawn. If the trees are evenly spaced apart, in how many distinguishable ways can they be planted?

20. ***Letters*** ◆ In how many distinguishable ways can the letters in the word *Statistics* be written?

21. ***Jury Selection*** ◆ From a group of 40 people, a jury of 12 people is selected. In how many different ways can a jury of 12 people be selected?

22. ***Experimental Group*** ◆ In order to conduct an experiment, four subjects are randomly selected from a group of 20 subjects. How many different groups of four subjects are possible?

23. ***Pizza Toppings*** ◆ A pizza shop offers eight toppings. If no topping is used more than once, in how many different ways can a three-topping pizza be formed?

24. ***Lottery Number Selection*** ◆ A lottery has 52 numbers. In how many different ways can six of the numbers be selected? (Assume that order of selection is not important.)

25. ***Employee Selection*** ◆ Four sales representatives for a company are to be chosen to participate in a training program. The company has eight sales representatives, two in each of four regions. In how many ways can the four sales representatives be chosen if (a) there are no restrictions and (b) the selection must include a sales representative from each region? (c) What is the probability that the four sales representatives chosen to participate in the training program will be from only two of the four regions if they are chosen at random?

26. ***License Plates*** ◆ In a certain state, each automobile license plate number consists of two letters followed by a four-digit number. How many distinct license plate numbers can be formed if (a) there are no restrictions and (b) the letters "O" and "I" are not used? (c) What is the probability of selecting at random a license plate that ends in an even number?

27. ***Repairs*** ◆ In how many orders can three broken computers and two broken printers be repaired if (a) there are no restrictions, (b) the printers must be repaired first, and (c) the computers must be repaired first? (d) If the order of repairs has no restrictions and the order of repairs is done at random, what is the probability that a printer will be repaired first?

28. ***Defective Units*** ◆ A shipment of 10 microwave ovens contains two defective units. In how many ways can a restaurant buy three of these units and receive (a) no defective units, (b) one defective unit, and (c) at least two good units? (d) What is the probability of the restaurant buying at least two defective units?

Financial Shape In Exercises 29–32, use the following information. The graph shows the chances U.S. adults feel that their financial shape is good or not. *(Source: Pew Research Center)*

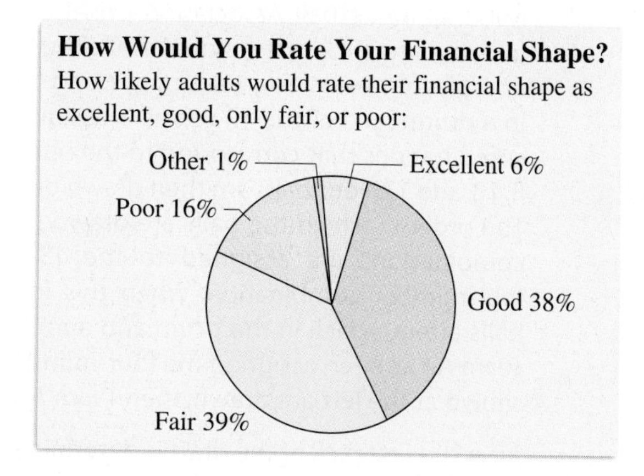

29. Suppose four people are chosen at random from a group of 1200. What is the probability that all four would rate their financial shape as excellent? (Make the assumption that the 1200 people are represented by the pie chart.)

30. Suppose 10 people are chosen at random from a group of 1200. What is the probability that all 10 would rate their financial shape as poor? (Make the assumption that the 1200 people are represented by the pie chart.)

31. Suppose 80 people are chosen at random from a group of 500. What is the probability that none of the 80 people would rate their financial shape as fair? (Make the assumption that the 500 people are represented by the pie chart.)

32. Suppose 55 people are chosen at random from a group of 500. What is the probability that none of the 55 people would rate their financial shape as good? (Make the assumption that the 500 people are represented by the pie chart.)

33. ***Probability*** ◆ In a state lottery, you must select 5 numbers (in any order) out of 40 correctly to win the top prize.

(a) How many ways can 5 numbers be chosen from 40 numbers?

(b) If you purchase one lottery ticket, what is the probability of winning the top prize?

34. ***Probability*** ◆ A company that has 200 employees chooses a committee of 15 to represent employee retirement issues. When the committee was formed, none of the 56 minority employees were selected.

(a) Use a technology tool to find the number of ways 15 employees can be chosen from 200.

(b) Use a technology tool to find the number of ways 15 employees can be chosen from 144 nonminorities.

(c) If the committee was chosen randomly (without bias), what is the probability that it contained no minorities?

(d) Does your answer to part (c) indicate that the committee selection was biased? Explain your reasoning.

Extending the Basics

NBA Draft Lottery In Exercises 35–40, use the following information. The National Basketball Association (NBA) uses a lottery to determine which team gets the first pick in its annual draft. The teams eligible for the lottery are the 13 non-playoff teams. Fourteen ping-pong balls numbered 1 through 14 are placed in a drum. Each of the 13 teams is assigned several of the possible four-number combinations that correspond to the numbers on the ping-pong balls, such as 3, 8, 10, and 12. Four balls are then drawn out to determine the first pick in the draft. The order in which the balls are drawn is not important. All of the four-number combinations are assigned to the 13 teams by computer except for one four-number combination. When this four-number combination is drawn, the balls are put back in the drum and another drawing takes place. For instance, if Team A has been assigned the four-number combination 3, 8, 10, 12 and the balls shown at the left are drawn, then Team A wins the first pick.

10 3 12 8

After the first pick of the draft is determined, the process continues to choose the teams that will select second and third. The remaining order of the draft is determined by the number of losses of each team.

35. In how many ways can four of the numbers 1 to 14 be selected if order is *not* important? How many sets of four numbers are assigned to the 13 teams?

36. In how many ways can four of the numbers be selected if order is important?

In the Pareto chart, the number of combinations assigned to each of the 13 teams is shown. The team with the most losses (the worst team) gets the most chances to win the lottery. So, the worst team receives the greatest frequency of four-number combinations, 250. The team with the best record of the 13 non-playoff teams has the fewest chances with 5 four-number combinations.

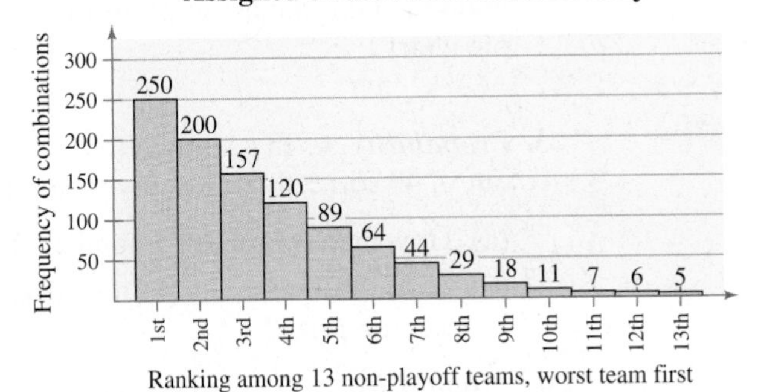

37. For each team, find the probability that the team will win the first pick.

38. What is the probability that the team with the worst record will win the second pick, given that the team with the best record, ranked 13th, wins the first pick?

39. What is the probability that the team with the worst record will win the third pick, given that the team with the best record, ranked 13th, wins the first pick and the team ranked 2nd wins the second pick?

40. What is the probability that neither the first- nor the second-worst teams will get the first pick?

TECHNOLOGY MINITAB EXCEL TI-83

Simulation: Composing Mozart Variations with Dice

Wolfgang Mozart (1756–1791) composed a wide variety of musical pieces. In his Musical Dice Game, he wrote a Wiener minuet with an almost endless number of variations. Each minuet has 16 bars. In the eighth and sixteenth bars, the player has a choice of two musical phrases. In each of the other 14 bars, the player has a choice of 11 phrases.

To create a minuet, Mozart suggested that the player toss 2 six-sided dice 16 times. For the eighth and sixteenth bars, choose Option 1 if the dice total is odd and Option 2 if it is even. For each of the other 14 bars, subtract 1 from the dice total. The following minuet is the result of the following sequence of numbers.

5 7 1 6 4 10 5 1
6 6 2 4 6 8 8 2

Exercises

1. How many phrases did Mozart write to create the Musical Dice Game minuet? Explain.

2. How many possible variations are there in Mozart's Musical Dice Game minuet? Explain.

3. Use technology to select randomly a number from 1 to 11.

(a) What is the theoretical probability of each number from 1 to 11 occurring?

(b) Use this procedure to select 100 integers between 1 and 11. Tally your results and compare them with the probabilities in part a.

4. What is the probability of randomly selecting options 6, 7, or 8 for the first bar? For all 14 bars? Find each probability using (a) theoretical probability and (b) the results of Exercise 3b.

5. Use technology to select randomly two numbers from 1, 2, 3, 4, 5, and 6. Find the sum and subtract 1 to obtain a total.

(a) What is the theoretical probability of each total from 1 to 11?

(b) Use this procedure to select 100 totals between 1 and 11. Tally your results and compare them with the probabilities in part a.

6. What is the probability of randomly selecting options 6, 7, or 8 for the first bar? For all 14 bars? Find each probability using (a) theoretical probability and (b) the results of Exercise 5b.

Extended solutions are given in the *Technology Supplement*.
Technical instruction is provided for *Minitab*, *Excel*, and the *TI-83*.

3 Chapter Summary

What did you learn?	Review Exercises
Section 3.1	
◆ How to identify the sample space of a probability experiment and to identify simple events	*1, 2*
◆ How to distinguish among classical probability, empirical probability, and subjective probability	*3–8*
◆ How to identify and use several properties of probability	*9, 10*
Section 3.2	
◆ How to find conditional probabilities	*11, 12*
◆ How to distinguish between independent and dependent events	*13, 14*
◆ How to use the Multiplication Rule to find the probability of two events occurring in sequence	*15, 16*
Section 3.3	
◆ How to determine if two events are mutually exclusive	*17, 18*
◆ How to use the Addition Rule to find the probability of two events	*19–22*
Section 3.4	
◆ How to use the Fundamental Counting Principle to find the number of ways two or more events can occur	*23, 24*
◆ How to find the number of ways a group of objects can be arranged in order and the number of ways to choose several objects from a group without regard to order	*25–29*
◆ How to use counting principles to find probabilities	*30, 31*

STATISTICS

Uses and Abuses

Uses

Probability affects decisions when the weather is forecast, when marketing strategies are determined, when medications are selected, and even when players are selected for professional sports teams. Although intuition is often used for determining probabilities, you will be better able to assess the likelihood that an event will occur by applying the rules of classical probability and empirical probability.

For instance, suppose you work for a real estate company and are asked to estimate the likelihood that a particular house will sell for a particular price within the next 90 days. You could use your intuition but you could better assess the probability by looking at sales records for similar houses.

Abuses

Assuming Probability has a "Memory" One common abuse of probability is thinking that probabilities have "memories." For example, if a coin is tossed eight times, the probability that it will land heads up all eight times is only 0.004. However, if the coin has already been tossed seven times and has landed heads up each time, the probability that it will land heads up on the eighth time is 0.5. Each toss is independent of all other tosses. The coin does not "remember" that it has already landed heads up seven times.

Adding Probabilities Incorrectly Another common abuse is adding probabilities incorrectly. For instance, suppose a company has 100 employees. Of these, 40 are women and 20 are minorities. If one employee is selected at random, the probability that the employee is a woman is 0.4 and the probability that the employee is a minority is 0.2. This does not mean, however, that the probability that a randomly selected employee is a woman or a minority is $0.4 + 0.2$ or 0.6. To determine this probability, you need to know how many employees belong to both groups, women *and* minorities.

Exercises

1. ***Assuming Probability has a "Memory"*** A "Daily Number" lottery has a 3-digit number from 000 to 999. You buy one ticket each day. Your number is 389.

    a. What is the probability of winning next Tuesday and Wednesday?

    b. You won on Tuesday. What's the probability of winning on Wednesday?

    c. You didn't win on Tuesday. What's the probability of winning on Wednesday?

2. ***Adding Probabilities Incorrectly*** A town has a population of 500 people. Suppose that the probability that a randomly chosen person owns a pick-up is 0.25 and the probability that a randomly chosen person owns an SUV is 0.30. What can you say about the probability that a randomly chosen person owns a pick-up or an SUV? Could this probability be 0.55? Could it be 0.60? Explain your reasoning.

3 Review Exercises

Section 3.1

In Exercises 1 and 2, identify the sample space of each probability experiment and list the outcomes of the event.

1. *Experiment:* Tossing four coins
Event: Getting three heads

2. *Experiment:* Rolling two six-sided dice
Event: Getting a sum of 4 or 5

In Exercises 3–8, classify each statement as an example of classical, empirical, or subjective probability.

3. Based on prior counts, a quality control officer says there is a 0.05 probability that a randomly chosen part is defective.

4. The probability of randomly selecting five cards of the same suit (a flush) from a standard deck is about 0.002.

5. The chance that Corporation A's stock price will fall today is 75%.

6. The probability of a person from the United States being left handed is 11%.

7. The probability of rolling two six-sided dice and getting a sum greater than nine is $\frac{1}{6}$.

8. The chance that a randomly selected person in the United States is between 15 and 24 years old is about 14%. *(Source: U.S. Census Bureau)*

In Exercises 9 and 10, the table shows the approximate U.S. age distribution from the 2000 Census. Use the table to determine the probability of the event. *(Source: U.S. Census Bureau)*

Age	19 and under	20–34	35–59	60–84	85 and over
Population	29%	21%	34%	15%	1%

9. What is the probability that a randomly selected person in the United States will be at least 20 years old?

10. What is the probability that a randomly selected person in the United States will be less than 60 years old?

Section 3.2

In Exercises 11 and 12 on the next page, the list shows the results of a study on the use of plus/minus grading at North Carolina State University. It shows the percents of graduate and undergraduate students who received grades with pluses and minuses (for example, C+, A–, etc.). *(Source: North Carolina State University)*

- Of all students who received one or more plus grades, 92% were undergraduates and 8% were graduates.
- Of all students who received one or more minus grades, 93% were undergraduates and 7% were graduates.

11. Find the probability that a student is an undergraduate student, given that the student received a plus grade.

12. Find the probability that a student is a graduate student, given that the student received a minus grade.

In Exercises 13 and 14, decide whether the events are independent or dependent.

13. Tossing a coin four times, getting four heads, and tossing it a fifth time and getting a head

14. Taking a driver's education course and passing the driver's license exam

In Exercises 15 and 16, find the probability of the sequence of events.

15. You are shopping, and your roommate has asked you to pick up toothpaste and dental rinse. However, your roommate did not tell you which brands to get. The store has six brands of toothpaste and four brands of dental rinse. What is the probability that you will purchase the correct brands of both products?

16. Your sock drawer has nine folded pairs of socks, with three pairs each of white, black, and blue. What is the probability, without looking in the drawer, that you will first select and remove a black pair, then select either a blue or white pair?

Section 3.3

In Exercises 17 and 18, decide if the events are mutually exclusive.

17. Event A: Randomly select a person who uses the Internet at least twice a week.
Event B: Randomly select a person who has not used the Internet in seven days.

18. Event A: Randomly select a person who loves cats.
Event B: Randomly select a person who owns a dog.

19. A random sample of 250 working adults found that 37% access the Internet at work, 44% access the Internet at home, and 21% access the Internet at both work and home. What is the probability that a person in this sample selected at random accesses the Internet at home or at work?

20. Another random sample of 225 working adults found that 21% access the Internet at work, 56% access the Internet at home, and 20% access the Internet both at work and home. What is the probability that a randomly selected person from this sample accesses the Internet at home or at work?

In Exercises 21 and 22, determine the probability.

21. A card is randomly selected from a standard deck. Find the probability that the card is between four and eight (inclusive) or is a club.

22. A twelve-sided die, numbered 1–12, is rolled. Find the probability that the roll results in an odd number or a number less than four.

Section 3.4

In Exercises 23 and 24, use the Fundamental Counting Principle.

23. Until recently, with the advent of cellular phones, modems, and pagers, the area codes in the United States and Canada followed a certain system. The first number could not be 0 or 1, the second could only be 0 or 1, and the third could not be 0. Under this system, how many area codes are possible?

24. Assuming that each character can be either a letter or digit, how many four-character license plates are possible?

In Exercises 25–29, use combinations and permutations.

25. A baseball card collector has six identical Mark McGwire cards and three identical Mike Schmidt cards. The collector's album has nine slots per page. In how many distinguishable ways can the collector arrange the nine cards on one page?

26. A florist has 12 different flowers from which floral arrangements can be made. If a centerpiece is to be made using five different flowers, how many different centerpieces could be made?

27. Fifteen cyclists enter a race. In how many ways can they finish first, second, and third?

28. Five players on a basketball team must choose a player on the opposing team to defend. In how many ways can they choose their defensive assignments?

29. A literary magazine editor must choose four short stories for this month's issue from 17 submissions. In how many ways can the editor choose this month's stories?

In Exercises 30 and 31, use counting principles to find the probabilities.

30. In poker, a full house consists of a three-of-a-kind and a two-of-a-kind. Find the probability of a full house consisting of three kings and two queens.

31. A batch of 200 calculators contains three defective calculators. What is the probability that a sample of three calculators will have (a) no defective calculators, (b) all defective calculators, (c) at least one defective calculator, and (d) at least one nondefective calculator?

PUTTING IT ALL TOGETHER

Real Statistics Real Decisions

You work for the company that runs the Powerball lottery. Powerball is a lottery game in which five white balls are chosen from a drum containing 49 balls and one red ball is chosen from a drum containing 42 balls. To win the jackpot, a player must match all five white balls and the red ball. Other winners and their prizes are also shown in the table.

Working in the public relations department, you handle many inquiries from the media and from lottery players. You receive the following e-mail.

> *You list the probability of matching only the red ball as* 1/74. *I know from my statistics class that the probability of winning is the ratio of the number of successful outcomes to the total number of outcomes. Could you please explain why the probability of matching only the red ball is* 1/74?

Your job is to answer this question, using the probability techniques you have learned in this chapter to justify your answer. In answering the question, assume only one ticket is purchased.

www.musl.com

Powerball Winners and Prizes

Match	Prize	Approximate Probability
5 white, 1 red	Jackpot	1/80,089,128
5 white	$100,000	1/1,953,393
4 white, 1 red	$5,000	1/364,042
4 white	$100	1/8879
3 white, 1 red	$100	1/8466
3 white	$7	1/207
2 white, 1 red	$7	1/605
1 white, 1 red	$4	1/118
1 red	$3	1/74

Source: Multi-State Lottery Association

Where Is Powerball played?
Powerball is played in 20 states and in Washington, D.C.

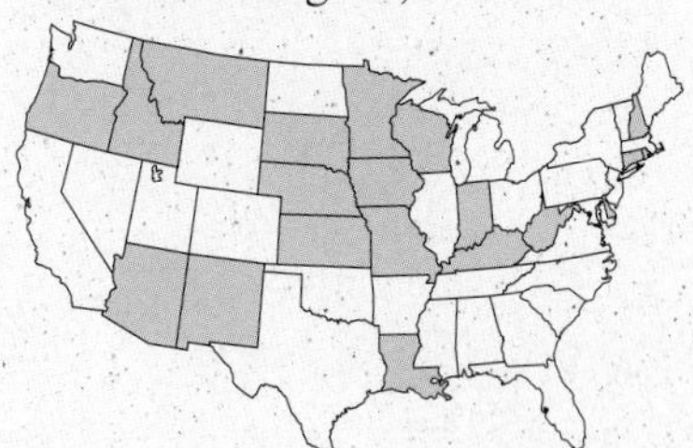

Source: Multi-State Lottery Association

Exercises

1. ***How Would You Do It?***

   (a) How would you investigate the question about the probability of matching only the red ball?

   (b) What statistical methods taught in this chapter would you use?

2. ***Answering the Question***

   Write an explanation that answers the question about the probability of matching only the red ball. Include in your explanation any probability formulas that justify your explanation.

3. ***Another Question***

   You receive another question asking how the overall probability of winning a prize in the Powerball lottery is determined. The overall probability of winning a prize in the Powerball lottery is 1/36. Write an explanation that answers the question and include any probability formulas that justify your explanation.

3 Chapter Quiz

Take this quiz as you would take a quiz in class. After you are done, check your work against the answers given in the back of the book.

1. The following table shows the estimated number (in thousands) of earned degrees conferred in the year 2000 by level and gender. *(Source: National Center for Education Statistics)*

		Gender		
		Male	**Female**	**Total**
Level of Degree	**Associate**	208	323	531
	Bachelor	502	659	1161
	Master	187	227	414
	Doctor	27	19	46
	Total	924	1228	2152

A person who earned a degree in the year 2000 is randomly selected. Find the probability of selecting someone who

(a) earned a bachelor's degree.

(b) earned a bachelor's degree given that the person is a female.

(c) earned a bachelor's degree given that the person is not a female.

(d) earned an associate's degree or a bachelor's degree.

(e) earned a doctorate given that the person is a male.

(f) earned a master's degree or is a female.

(g) earned an associate's degree and is a male.

(h) is a female given that the person earned a bachelor's degree.

2. Decide if the events are mutually exclusive. Then decide if the events are independent or dependent. Explain your reasoning.

Event A: A golfer scoring the best round in a four-round tournament
Event B: Losing the golf tournament

3. A shipment of 150 television sets contains three defective units. In how many ways can a vending company buy three of these units and receive

(a) no defective units?

(b) all defective units?

(c) at least one good unit?

4. In Exercise 3, find the probability of the vending company receiving

(a) no defective units.

(b) all defective units.

(c) at least one good unit.

5. The access code for a car's security system consists of four digits. The first digit cannot be zero and the last digit must be even. How many different codes are available?

6. From a pool of 25 candidates, the offices of president, vice-president, secretary, and treasurer will be filled. In how many different ways can the offices be filled?

3

Cumulative Test: Chapters 1–3

Take this test as you would take a test in class. After you are done, check your work against the answers given in the back of the book.

Refer to the following data set as you take this test. The data set is the number of dollars a sample of 30 people spent in one year on books.

156	150	109	98	136
170	178	199	110	191
187	119	104	160	132
117	111	120	103	123
153	91	162	93	118
127	90	181	110	116

1. Does the data set represent qualitative data or quantitative data? What is the data set's level of measurement?
2. Which method of data collection would you use to gather these data? Which sampling technique would you use? Explain your reasoning.
3. Make a frequency distribution of the data set using five classes. Include class limits, midpoints, frequencies, boundaries, relative frequencies, and cumulative frequencies.
4. Display the data using a frequency histogram and a frequency polygon on the same axes.
5. Display the data using a relative frequency histogram.
6. Display the data using a stem-and-leaf plot.
7. Describe the shape of the distribution as symmetric, uniform, or skewed.
8. Display the data using an ogive.
9. Find the mean, median, and mode of the data set. Are these values parameters or statistics? Explain.
10. Find the range, variance, and standard deviation of the data set. Interpret the results in the context of the real-life setting.
11. Find the quartiles and the interquartile range of the data set and construct a box-and-whisker plot.
12. Convert the minimum and maximum data entries to z-scores. Are either of the values unusual? Explain why or why not.
13. Find the probability of randomly selecting someone who spent less than \$120 on books. What is the probability of randomly selecting someone who spent more than \$120?
14. Find the probability of randomly selecting someone who spent less than \$120 or more than \$160 on books.
15. Five people are selected at random. Find the probability that at least one person spent more than \$175.
16. Assume that the 30 people in the data set were contacted through a telephone survey. After performing the survey, five of the respondents are randomly chosen for a more extensive survey. In how many ways can the five respondents be chosen?

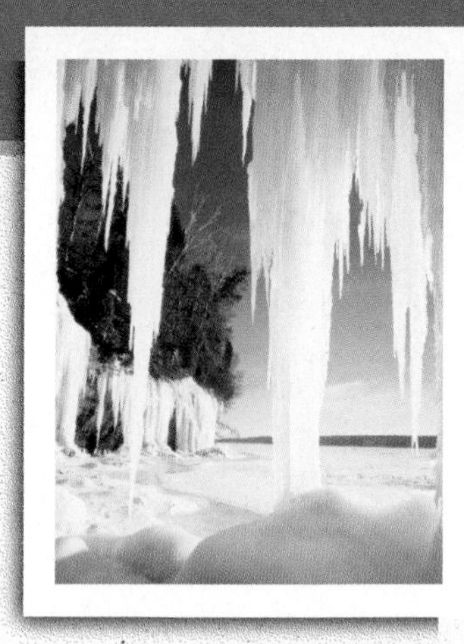

Where You've Been

In Chapters 1 through 3, you learned how to collect and describe data and how to find the probability of an event. These skills are used in many different types of careers. For example, data about climatic conditions are used to analyze and forecast the weather throughout the world. On a typical day, 5000 weather stations, 800 to 1100 upper-air balloon stations, 2000 ships, 600 aircraft, several polar-orbiting and geostationary satellites, and a variety of other data collection devices work together to provide meteorologists with data that are used to forecast the weather. Even with this much data, meteorologists cannot forecast the weather with certainty. Instead, they assign probabilities to certain weather conditions. For instance, a meteorologist might determine that there is a 40% chance of rain (based on the relative frequency of rain under similar weather conditions).

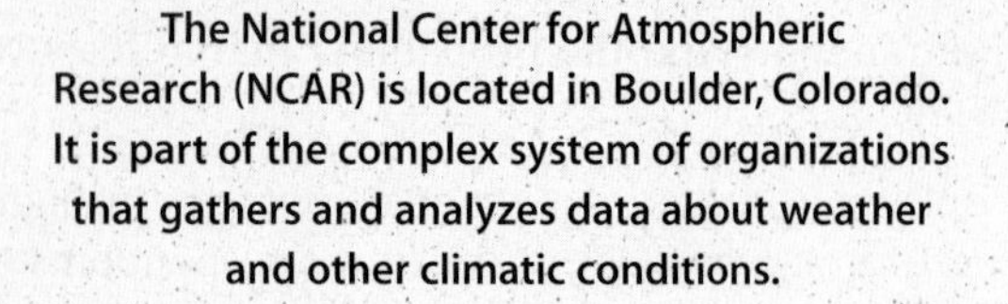

The National Center for Atmospheric Research (NCAR) is located in Boulder, Colorado. It is part of the complex system of organizations that gathers and analyzes data about weather and other climatic conditions.

Discrete Probability Distributions

CHAPTER
4

Where You're Going

In Chapter 4, you will learn how to create and use probability distributions. Suppose you are a meteorologist working on a three-day forecast. Assuming that having rain on one day is independent of having rain on another day, you have determined that there is a 40% probability of rain on each of the three days. What is the probability that it will rain on 0, 1, 2, or 3 of the days? To answer this, you can create a probability distribution for the possible outcomes.

Day 1	Day 2	Day 3	Probability	Days of Rain
☀	☀	☀	P(☀, ☀, ☀) = 0.216	0
☀	☀	💧	P(☀, ☀, 💧) = 0.144	1
☀	💧	☀	P(☀, 💧, ☀) = 0.144	1
☀	💧	💧	P(☀, 💧, 💧) = 0.096	2
💧	☀	☀	P(💧, ☀, ☀) = 0.144	1
💧	☀	💧	P(💧, ☀, 💧) = 0.096	2
💧	💧	☀	P(💧, 💧, ☀) = 0.096	2
💧	💧	💧	P(💧, 💧, 💧) = 0.064	3

Using the *Addition Rule* together with this tree diagram, you can determine the probability of having rain on various numbers of days.

Probability Distribution		
Days of Rain	**Tally**	**Probability**
0	1	0.216
1	3	0.432
2	3	0.288
3	1	0.064

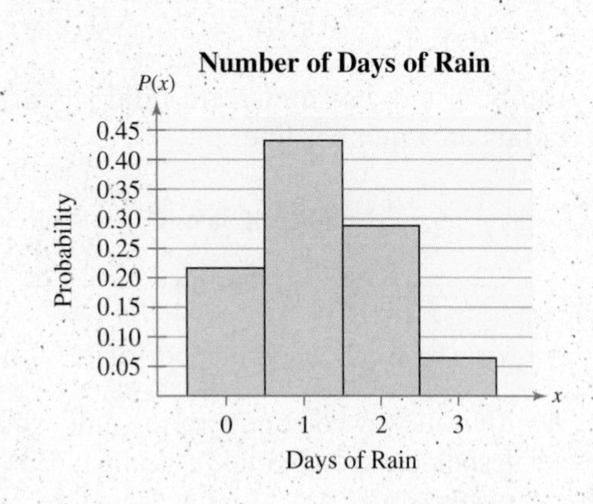

4.1 Probability Distributions

Random Variables • Discrete Probability Distributions • Mean, Variance, and Standard Deviation • Expected Value

What You Should Learn

- How to distinguish between discrete random variables and continuous random variables
- How to construct a discrete probability distribution and its graph
- How to determine if a distribution is a probability distribution
- How to find the mean, variance, and standard deviation of a discrete probability distribution
- How to find the expected value of a discrete probability distribution

Random Variables

The outcome of a probability experiment is often a count or a measure. When this occurs, the outcome is called a random variable.

> **DEFINITION**
>
> A **random variable** x represents a numerical value associated with each outcome of a probability experiment.

The word *random* indicates that x is determined by chance. There are two types of random variables: discrete and continuous.

> **DEFINITION**
>
> A random variable is **discrete** if it has a finite or countable number of possible outcomes that can be listed.
>
> A random variable is **continuous** if it has an uncountable number of possible outcomes, represented by an interval on the number line.

Study Tip

If a random variable is discrete, you can list the possible values it can assume. However, it is impossible to list all values for a continuous random variable.

Suppose you conduct a study of the number of calls a salesperson makes in one day. The possible values of the random variable x are 0, 1, 2, 3, 4, and so on. Because the set of possible outcomes

$$\{0, 1, 2, 3, \ldots\}$$

can be listed, x is a discrete random variable. You can represent its values as points on a number line.

Number of Sales Calls (Discrete)

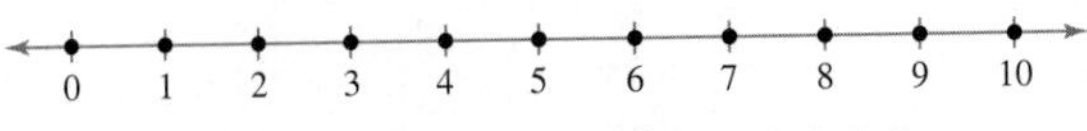

x can have only whole number values: 0, 1, 2, 3, ...

A different way to conduct the study would be to measure the time (in hours) a salesperson spends making calls in one day. Because the time spent making sales calls can be any number from 0 to 24 (including fractions and decimals), x is a continuous random variable. You can represent its values with an interval on a number line, but you cannot *list* all the possible values.

Hours Spent on Sales Calls (Continuous)

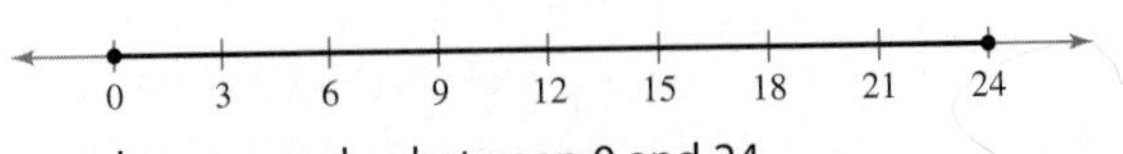

x can have any value between 0 and 24.

EXAMPLE 

Discrete Variables and Continuous Variables

Decide whether the random variable x is discrete or continuous. Explain your reasoning.

1. x represents the number of stocks in the Dow Jones Industrial Average that have share price increases on a given day.
2. x represents the volume of bottled water in a 32-ounce container.

Study Tip

In most practical applications, discrete random variables represent count data, while continuous random variables represent measured data.

SOLUTION

1. The number of stocks whose share price increases can be counted.

   $$\{0, 1, 2, 3, \ldots, 30\}$$

   So x is a *discrete* random variable.
2. The amount of water in the container can be any volume between 0 ounces and 32 ounces. So x is a *continuous* random variable.

Try It Yourself 1

Decide whether the random variable x is discrete or continuous.

1. x represents the length of time it takes to complete a test.
2. x represents the number of home runs hit during a baseball game.

a. Decide if x represents *counted* data or *measured* data.
b. Make a conclusion and *explain* your reasoning.

Answer: Page A34

It is important that you can distinguish between discrete and continuous random variables because different statistical techniques are used to analyze each. The remainder of this chapter focuses on discrete random variables and their probability distributions. You will study continuous distributions later.

Discrete Probability Distributions

Each value of a discrete random variable can be assigned a probability. By listing each value of the random variable with its corresponding probability, you are forming a probability distribution.

DEFINITION

A **discrete probability distribution** lists each possible value the random variable can assume, together with its probability. A probability distribution must satisfy the following conditions.

In Words	In Symbols
1. The probability of each value of the discrete random variable is between 0 and 1, inclusive.	$0 \leq P(x) \leq 1$
2. The sum of all the probabilities is 1.	$\Sigma P(x) = 1$

Because probabilities represent relative frequencies, a discrete probability distribution can be graphed with a relative frequency histogram.

GUIDELINES

Constructing a Discrete Probability Distribution

Let x be a discrete random variable with possible outcomes $x_1, x_2, \ldots, x_n$.

1. Make a frequency distribution for the possible outcomes.
2. Find the sum of the frequencies.
3. Find the probability of each possible outcome by dividing its frequency by the sum of the frequencies.
4. Check that each probability is between 0 and 1 and that the sum is 1.

EXAMPLE 2

Constructing and Graphing a Discrete Probability Distribution

Frequency Distribution

Score, x	Frequency, f
1	24
2	33
3	42
4	30
5	21

An industrial psychologist administered a personality inventory test for passive-aggressive traits to 150 employees. Individuals were rated on a score from 1 to 5, where 1 was extremely passive and 5 extremely aggressive. A score of 3 indicated neither trait. The results are shown at the left. Construct a probability distribution for the random variable x. Then graph the distribution.

SOLUTION Divide the frequency of each score by the total number of individuals in the study to find the probability for each value of the random variable.

$$P(1) = \frac{24}{150} = 0.16 \qquad P(2) = \frac{33}{150} = 0.22 \qquad P(3) = \frac{42}{150} = 0.28$$

$$P(4) = \frac{30}{150} = 0.2 \qquad P(5) = \frac{21}{150} = 0.14$$

The discrete probability distribution is shown in the following table. Note that each probability is between 0 and 1 and that the sum of the probabilities is 1.

x	1	2	3	4	5
$P(x)$	0.16	0.22	0.28	0.2	0.14

Passive-Aggressive Traits

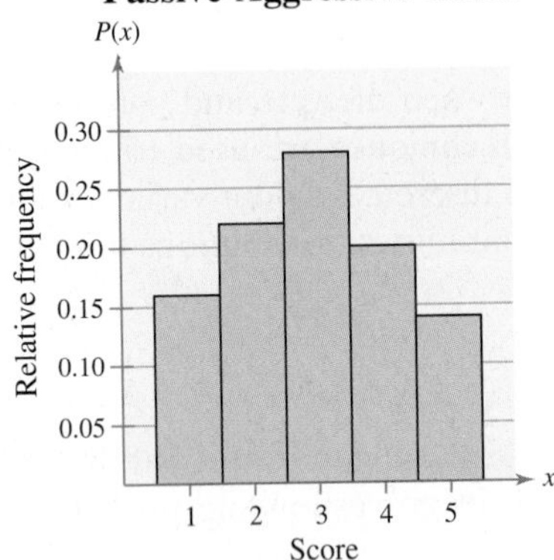

The relative frequency histogram is shown at the left. It is important to see that the area of each bar is equal to the probability of a particular outcome. Also, the probability of an event corresponds to the sum of the areas of the outcomes included in the event. For instance, the probability of the event "having a score of 2 or 3" is equal to the sum of the areas of the second and third bars,

$$(1)(0.22) + (1)(0.28) = 0.22 + 0.28 = 0.50.$$

Frequency Distribution

Sales per Day, x	Number of Days, f
0	16
1	19
2	15
3	21
4	9
5	10
6	8
7	2

Try It Yourself 2

A company tracks the number of sales new employees make each day during a 100-day probationary period. The results for one new employee are shown at the left. Construct and graph a probability distribution.

a. *Find* the probability of each outcome.
b. *Organize* the probabilities in a probability distribution.
c. *Graph* the probability distribution.

Answer: Page A34

Probability Distribution

Days of Rain, x	Probability, $P(x)$
0	0.216
1	0.432
2	0.288
3	0.064

EXAMPLE 

Verifying Probability Distributions

Verify that the distribution at the left (see page 161) is a probability distribution.

SOLUTION If the distribution is a probability distribution, then (1) each probability is between 0 and 1, inclusive, and (2) the sum of the probabilities equals 1.

1. Each probability is between 0 and 1.
2. $\Sigma P(x) = 0.216 + 0.432 + 0.288 + 0.064 = 1$

Because both conditions are met, the distribution is a probability distribution.

Try It Yourself 3

Verify that the distribution you constructed in Try It Yourself 2 is a probability distribution.

a. Verify that the *probability* of each outcome is between 0 and 1.
b. Verify that the *sum* of all the probabilities is 1.
c. Make a *conclusion*.

Answer: Page A34

Picturing the World

In a recent year in the United States, nearly 12 million accidents were reported to the police. A graph of the probability distribution of traffic accidents for various age groups from 16 to 64 is shown. *(Source: U.S. National Highway Traffic Safety Administration)*

U.S. Traffic Accidents by Age

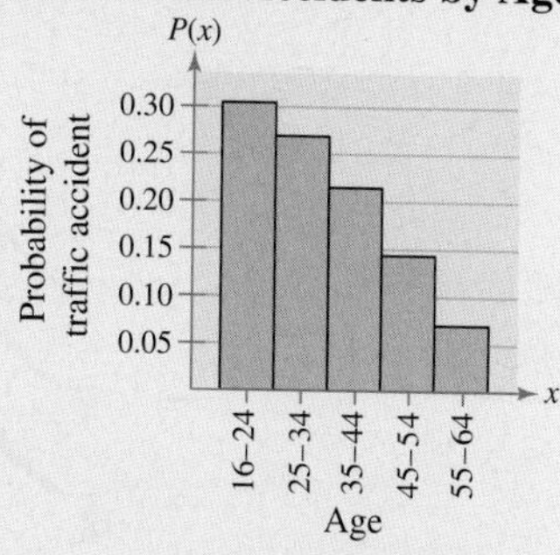

Estimate the probability that a randomly selected person involved in a traffic accident is in the 16 to 34 age group.

EXAMPLE 

Probability Distributions

Decide whether each distribution is a probability distribution.

1.

x	5	6	7	8
$P(x)$	0.28	0.21	0.43	0.15

2.

x	1	2	3	4
$P(x)$	$\frac{1}{2}$	$\frac{1}{4}$	$\frac{5}{4}$	-1

SOLUTION

1. Each probability is between 0 and 1, but the sum of the probabilities is 1.07, which is greater than 1. So, it is *not* a probability distribution.
2. The sum of the probabilities is equal to one, but $P(3)$ and $P(4)$ are not between 0 and 1. So, it is *not* a probability distribution. Probabilities can never be negative or greater than 1.

Try It Yourself 4

Decide whether the distribution is a probability distribution. Explain your reasoning.

1.

x	5	6	7	8
$P(x)$	$\frac{1}{16}$	$\frac{5}{8}$	$\frac{1}{4}$	$\frac{3}{16}$

2.

x	1	2	3	4
$P(x)$	0.09	0.36	0.49	0.06

a. Verify that the *probability* of each outcome is between 0 and 1.
b. Verify that the *sum* of all the probabilities is 1.
c. Make a *conclusion*.

Answer: Page A34

Mean, Variance, and Standard Deviation

You can measure the central tendency of a probability distribution with its mean and measure the variability with its variance and standard deviation. The mean of a discrete random variable is defined as follows.

Mean of a Discrete Random Variable

The **mean** of a discrete random variable is given by

$$\mu = \Sigma xP(x).$$

Each value of x is multiplied by its corresponding probability and the products are added.

The mean of the random variable represents the "theoretical average" of a probability experiment and sometimes is not a possible outcome. If the experiment were performed many thousands of times, the mean of all the outcomes would be close to the mean of the random variable.

EXAMPLE 

Finding the Mean of a Probability Distribution

x	$P(x)$
1	0.16
2	0.22
3	0.28
4	0.20
5	0.14

The probability distribution for the personality inventory test for passive-aggressive traits discussed in Example 2 is given at the left. Find the mean score. What can you conclude?

SOLUTION Use a table to organize your work, as shown below. From the table, you can see that the mean score is 2.94.

x	$P(x)$	$xP(x)$
1	0.16	$1(0.16) = 0.16$
2	0.22	$2(0.22) = 0.44$
3	0.28	$3(0.28) = 0.84$
4	0.20	$4(0.20) = 0.80$
5	0.14	$5(0.14) = 0.70$
	$\Sigma P(x) = 1$	$\Sigma xP(x) = 2.94$ ← Mean

A score of 3 represents an individual who exhibits neither passive nor aggressive traits. The mean is slightly under 3. So, you can conclude that the mean personality trait is neither extremely passive nor extremely aggressive, but is slightly closer to passive.

Try It Yourself 5

Find the mean of the probability distribution you constructed in Try It Yourself 2. What can you conclude?

a. *Find the product* of each outcome and its corresponding probability.
b. *Find the sum* of the products.
c. *What* can you conclude?

Answer: Page A34

While the mean of the random variable of a probability distribution describes a typical outcome, it gives no information about how the outcomes vary. To study the variation of the outcomes, you can use the variance and standard deviation of the random variable of a probability distribution.

Study Tip

A shortcut formula for the variance of a probability distribution is

$$\sigma^2 = [\Sigma x^2 P(x)] - \mu^2.$$

Standard Deviation of a Discrete Random Variable

The **variance** of a discrete random variable is

$$\sigma^2 = \Sigma(x - \mu)^2 P(x).$$

The **standard deviation** is

$$\sigma = \sqrt{\sigma^2}.$$

EXAMPLE 6

Finding the Variance and Standard Deviation

x	$P(x)$
1	0.16
2	0.22
3	0.28
4	0.20
5	0.14

The probability distribution for the personality inventory test for passive-aggressive traits discussed in Example 2 is given at the left. Find the variance and standard deviation of the probability distribution.

SOLUTION From Example 5, you know that the mean of the distribution is $\mu = 2.94$. Use a table to organize your work, as shown below.

x	$P(x)$	$x - \mu$	$(x - \mu)^2$	$P(x)(x - \mu)^2$
1	0.16	−1.94	3.764	0.602
2	0.22	−0.94	0.884	0.194
3	0.28	0.06	0.004	0.001
4	0.20	1.06	1.124	0.225
5	0.14	2.06	4.244	0.594
	$\Sigma P(x) = 1$			$\Sigma P(x)(x - \mu)^2 = 1.616$

Variance

So, the variance is

$$\sigma^2 = 1.616$$

and the standard deviation is

$$\sigma = \sqrt{\sigma^2} \approx \sqrt{1.62} \approx 1.27.$$

Try It Yourself 6

Find the variance and standard deviation for the probability distribution constructed in Try It Yourself 2.

a. *For each value of x, find the square* of the deviation from the mean and multiply that by the corresponding probability of x.
b. *Find the sum* of the products found in part (a) for the variance.
c. *Take the square root* of the variance for the standard deviation.

Answer: Page A34

Expected Value

DEFINITION

The **expected value** of a discrete random variable is equal to the mean of the random variable.

$$\text{Expected Value} = E(x) = \mu = \Sigma xP(x)$$

EXAMPLE 7

Finding an Expected Value

At a raffle, 1500 tickets are sold at \$2 each for four prizes of \$500, \$250, \$150, and \$75. You buy one ticket. What is the expected value of your gain?

Insight

In most applications, an expected value of 0 has a practical interpretation. For instance, in gambling games, an expected value of 0 implies that a game is a fair game (an unlikely occurrence!). In a profit and loss analysis, an expected value of 0 represents the break-even point.

SOLUTION To find the gain for each prize, subtract the price of the ticket from the prize. For instance, your gain for the \$500 prize is

$$\$500 - \$2 = \$498$$

and your gain for the \$250 prize is

$$\$250 - \$2 = \$248.$$

Then write a probability distribution for the possible gains (or outcomes).

Gain, x	\$498	\$248	\$148	\$73	−\$2
Probability, $P(x)$	$\frac{1}{1500}$	$\frac{1}{1500}$	$\frac{1}{1500}$	$\frac{1}{1500}$	$\frac{1496}{1500}$

Then, using the probability distribution, you can find the expected value.

$$\begin{aligned} E(x) &= \Sigma xP(x) \\ &= \$498 \cdot \frac{1}{1500} + \$248 \cdot \frac{1}{1500} + \$148 \cdot \frac{1}{1500} + \$73 \cdot \frac{1}{1500} + (-\$2) \cdot \frac{1496}{1500} \\ &= -\$1.35 \end{aligned}$$

Because the expected value is negative, you can expect to lose an average of \$1.35 for each ticket you buy.

Try It Yourself 7

During a one-year selling period (225 days), a sales representative made between 0 and 9 sales per day, as shown in the table. If this pattern continues, what is the expected value for the number of sales per day for the sales representative?

Number of sales, x	0	1	2	3	4	5	6	7	8	9
Frequency, f (in days)	25	48	60	45	20	10	8	5	3	1

a. *Identify the possible outcomes* and *find the probability* of each.
b. *Find the product* of each outcome and its corresponding probability.
c. *Find the sum* of the products.
d. *Interpret* the results.

Answer: Page A34

4.1 Exercises

Help

LarsonTutor 4.1

Companion Web Site

Student Solutions Manual 4.1

Videos 4.1

Technology Manuals

Basic Skills and Concepts

1. What is a random variable? Give an example of a discrete random variable and a continuous random variable. Justify your answer.

2. What is a discrete probability distribution? What are the two conditions that determine a probability distribution?

True or False In Exercises 3–6, determine whether the statement is true or false. If it is false, rewrite it so that it is a true statement.

3. In most applications, continuous random variables represent counted data, while discrete random variables represent measured data.

4. For a random variable x the word *random* indicates that the value of x is determined by chance.

5. The mean of a random variable represents the "theoretical average" of a probability experiment and sometimes is not a possible outcome.

6. The expected value of a discrete random variable is equal to the mean of the random variable.

Graphical Analysis In Exercises 7–10, decide whether the graph represents a discrete random variable or a continuous random variable. Explain your reasoning.

7. The home attendance for football games at a university

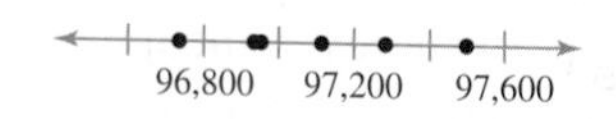

8. The length of time students use a computer each week

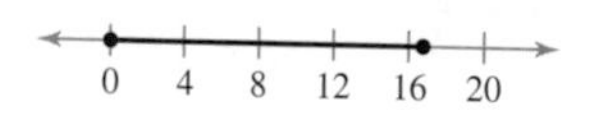

9. The annual vehicle-miles driven in the U.S. *(Source: U.S. Federal Highway Administration)*

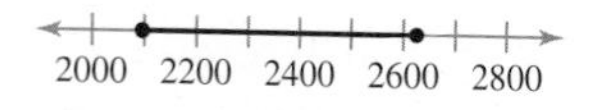

10. The annual traffic fatalities in the U.S. *(Source: National Highway Traffic Safety Administration)*

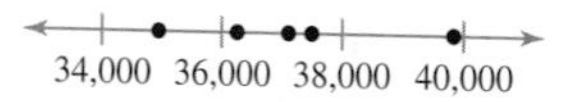

Distinguishing Between Discrete and Continuous Random Variables In Exercises 11–18, decide whether the random variable x is discrete or continuous. Explain your reasoning.

11. x represents the number of highway fatalities in one year in Texas.

12. x represents the length of time it takes to get to work.

13. x represents the volume of blood drawn for a blood test.

14. x represents the number of rainy days in the month of July in Florida.

15. x represents the number of books sold per quarter at a book store.

16. x represents the weight of a chemical compound.

17. x represents the amount of snow (in inches) that fell in Alaska last winter.

18. x represents the number of blackouts in California last summer.

19. *Employees Testing* ◆ A company gave psychological tests to prospective employees. The random variable x represents the possible test scores. Use the relative frequency histogram to find the probability that a person selected at random from the survey's sample had a test score of (a) more than two and (b) less than four.

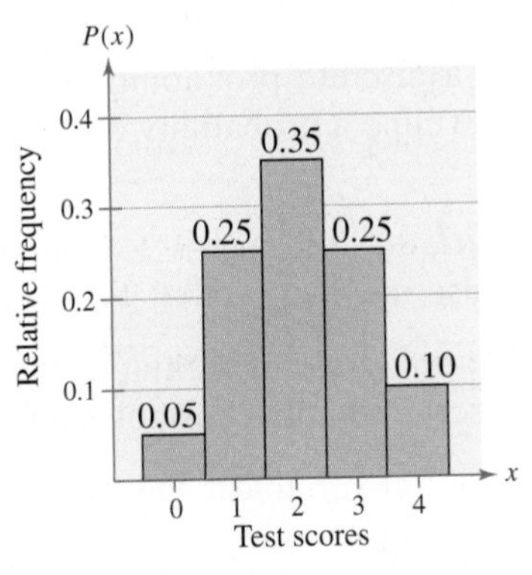

Figure for Exercise 19

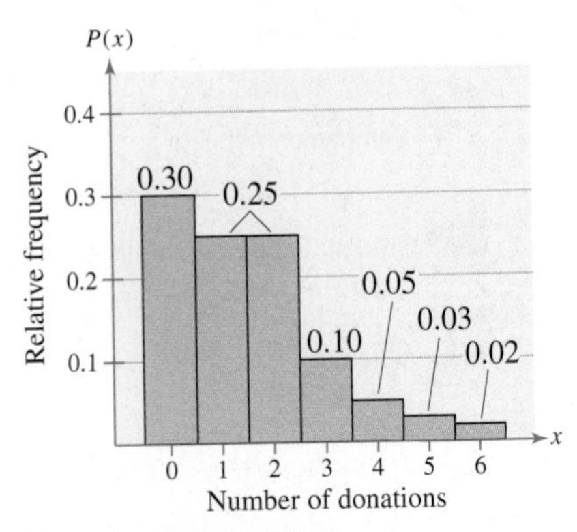

Figure for Exercise 20

20. *Blood Donations* ◆ A survey asked a sample of people how many times they donate blood each year. The random variable x represents the number of donations for one year. Use the relative frequency histogram to find the probability that a person selected at random from the survey's sample donated blood (a) more than once in a year and (b) less than three times in a year.

Determining a Missing Probability In Exercises 21 and 22, determine the probability distribution's missing probability value.

21. A sociologist surveyed the households in a small town. The random variable x represents the number of dependent children in the households.

x	0	1	2	3	4
$P(x)$	0.07	0.20	0.38	?	0.13

22. The sociologist in Exercise 21 surveyed the households in a neighboring town. The random variable x represents the number of dependent children in the households.

x	0	1	2	3	4	5	6
$P(x)$	0.05	?	0.23	0.21	0.17	0.11	0.08

Identifying Probability Distributions In Exercises 23–26, decide whether the distribution is a probability distribution. If it is not a probability distribution, identify the property (or properties) that are not satisfied.

23. *Psychological Tests* ◆ A company gave psychological tests to prospective employees. The random variable x represents the possible test scores.

x	0	1	2	3	4
$P(x)$	0.05	0.25	0.35	0.25	0.10

24. ***Vehicles Owned*** ◆ A survey asked a sample of people how many vehicles each owns. The random variable x represents the number of vehicles owned.

x	0	1	2	3
$P(x)$	0.005	0.435	0.555	0.206

25. ***Quality*** ◆ A quality inspector checked for imperfections in rolls of fabric. The random variable x represents the number of imperfections found.

x	0	1	2	3	4	5
$P(x)$	$\frac{3}{4}$	$\frac{1}{10}$	$\frac{1}{20}$	$\frac{1}{25}$	$\frac{1}{50}$	$-\frac{1}{100}$

26. ***Blood Donations*** ◆ A survey asked a sample of people how many times they donate blood each year. The random variable x represents the number of donations for one year.

x	0	1	2	3	4	5	6
$P(x)$	0.30	0.25	0.25	0.10	0.05	0.03	0.02

27. For the probability distribution in Exercise 23, find the (a) mean, (b) variance, and (c) standard deviation for the test score.

28. For the probability distribution in Exercise 26, find the (a) mean, (b) variance, and (c) standard deviation for the number of blood donations in a year.

Constructing Probability Distributions In Exercises 29–32, (a) use the frequency distribution to construct a probability distribution, find the (b) mean, (c) variance, and (d) standard deviation of the probability distribution, and (e) interpret the results in the context of the real-life situation.

29. ***Dogs*** ◆ The number of dogs per household in a small town. *(Adapted from American Veterinary Medical Association Center for Information Management)*

Dogs	0	1	2	3	4	5
Households	1491	425	168	48	29	14

30. ***Cats*** ◆ The number of cats per household in a small town. *(Adapted from American Veterinary Medical Association Center for Information Management)*

Cats	0	1	2	3	4	5
Households	1941	349	203	78	57	40

31. ***Computers*** ◆ The number of computers per household in a small town.

Computers	0	1	2	3
Households	300	280	95	20

32. ***Accidents*** ◆ The number of yearly motor vehicle accidents per student.

Accidents	0	1	2	3	4	5
Students	260	500	425	305	175	45

Constructing Probability Distributions In Exercises 33–36, construct a probability distribution for the given situation.

33. ***Rain*** ◆ A probability distribution for a three-day period where the random variable x represents the number of days with rain. Assume the chance of rain is 40%.

34. ***Snow*** ◆ A probability distribution for a three-day period where the random variable x represents the number of days with snow. Assume the chance of snow is 65%.

35. ***Girls*** ◆ A probability distribution for families with three children where the random variable x represents the number of girls.

36. ***Boys*** ◆ A probability distribution for families with four children where the random variable x represents the number of boys.

Finding Expected Value In Exercises 37–42, use the given probability distribution or relative frequency histogram to find the (a) mean, (b) variance, (c) standard deviation, and (d) expected value of the probability distribution.

37. ***Magazine Publishing*** ◆ A publisher introduces a new weekly magazine that sells for \$3.95. The company's marketers estimate that sales x (in thousands) will be approximated by the following probability distribution.

x	10	15	20	25	30	35
$P(x)$	0.200	0.300	0.250	0.150	0.075	0.025

38. ***Fire Insurance*** ◆ An insurance company offers fire protection policies with a face value of \$90,000. The company analyzes claims x of \$30,000, \$60,000, and \$90,000 and obtains the following probability distribution.

x	0	−30,000	−60,000	−90,000
$P(x)$	0.994	0.004	0.001	0.001

39. ***Hurricanes*** ◆ The relative frequency histogram shows the distribution of hurricanes that have hit the U.S. mainland by category, with 1 the weakest level and 5 the strongest. *(Source: USA TODAY Weather Almanac)*

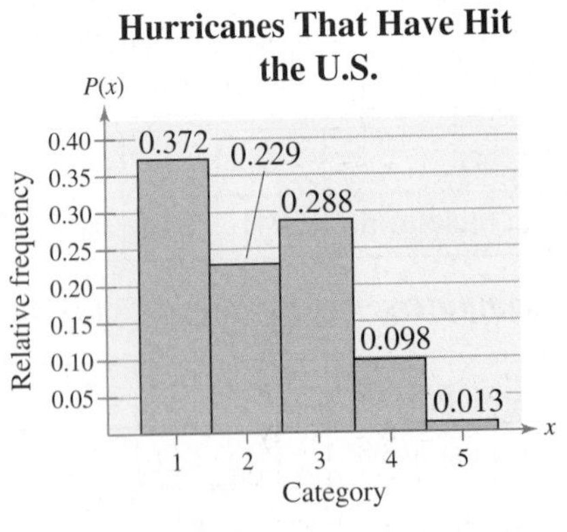

Figure for Exercise 39

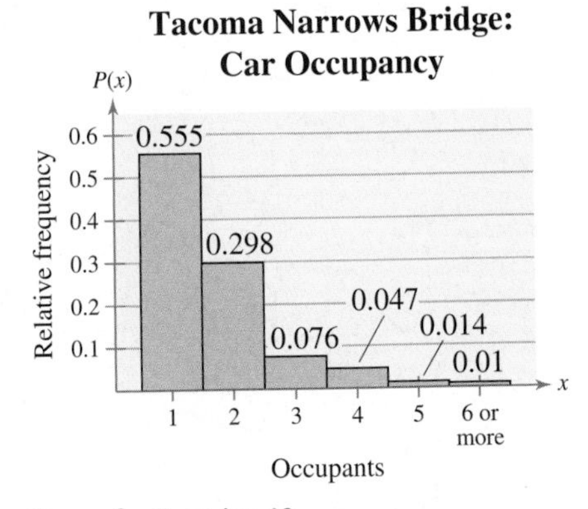

Figure for Exercise 40

40. ***Car Occupancy*** ◆ The relative frequency histogram shows the distribution of occupants in cars crossing the Tacoma Narrows Bridge in Washington. *(Adapted from Washington State Department of Transportation)*

41. ***Household Size*** ◆ The relative frequency histogram shows the distribution of household sizes in the U.S. for a recent year. *(Adapted from U.S. Census Bureau)*

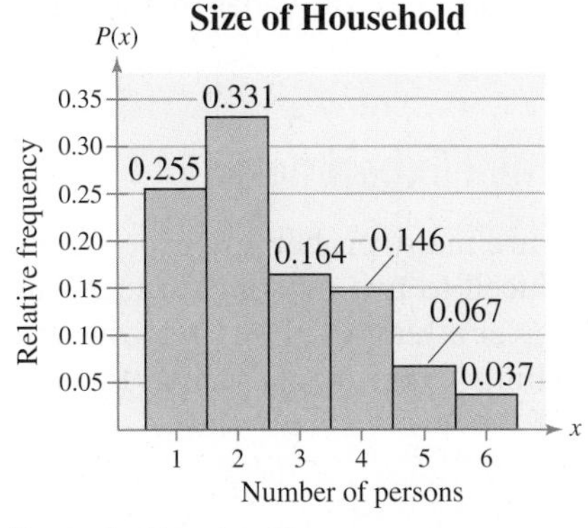

Figure for Exercise 41

Figure for Exercise 42

42. ***Car Occupancy*** ◆ The relative frequency histogram shows the distribution of carpooling by the number of cars owned by a household. *(Adapted from Federal Highway Administration)*

Extending the Basics

43. ***Finding Probabilities*** Use the probability distribution you made for Exercise 29 to find the probability of randomly selecting a household that has

(a) less than two dogs.

(b) at least two dogs.

(c) between two and four dogs, inclusive.

44. ***Finding Probabilities*** Use the probability distribution you made for Exercise 31 to find the probability of randomly selecting a household that has

(a) no computers.

(b) at least one computer.

(c) between one and three computers, inclusive.

Games of Chance In Exercises 45 and 46, find the expected net gain to the player for one play of the game. If x is the net gain to a player in a game of chance, then $E(x)$ is usually negative. This value gives the average amount per game the player can expect to lose.

45. In American roulette, the wheel has the 38 numbers

$$00, 0, 1, 2, \ldots, 34, 35, \text{ and } 36$$

marked on equally spaced slots. If a player bets \$1 on a number and wins, then the player keeps the dollar and receives an additional 35 dollars. Otherwise, the dollar is lost.

46. A charity organization is selling \$4 raffle tickets as part of a fund-raising program. The first prize is a boat valued at \$3150, and the second prize is a camping tent valued at \$450. The rest of the prizes are 15 \$25 gift certificates. The number of tickets sold is 5000.

4.2 Binomial Distributions

What You Should Learn

- How to determine if a probability experiment is a binomial experiment
- How to find binomial probabilities using the binomial probability formula, a binomial probability table, and technology
- How to construct a binomial distribution and its graph
- How to find the mean, variance, and standard deviation of a binomial probability distribution

Binomial Experiments • Binomial Probabilities • Graphing Binomial Distributions • Mean, Variance, and Standard Deviation

Binomial Experiments

There are many probability experiments for which the results of each trial can be reduced to two outcomes: success and failure. For instance, when a basketball player attempts a free throw, he or she either makes the basket or does not. Probability experiments such as these are called binomial experiments.

DEFINITION

A **binomial experiment** is a probability experiment that satisfies the following conditions.

1. The experiment is repeated for a fixed number of trials, where each trial is independent of the other trials.
2. There are only two possible outcomes of interest for each trial. The outcomes can be classified as a success (S) or as a failure (F).
3. The probability of a success $P(S)$ is the same for each trial.
4. The random variable x counts the number of successful trials.

The following notation is used for binomial experiments.

Notation for Binomial Experiments

Symbol	*Description*
n	The number of times a trial is repeated
$p = P(S)$	The probability of success in a single trial
$q = P(F)$	The probability of failure in a single trial $(q = 1 - p)$
x	The random variable represents a count of the number of successes in n trials: $x = 0, 1, 2, 3, \ldots, n$.

Here is a simple example of a binomial experiment. From a standard deck of cards, you pick a card, note whether it is a club or not, and replace the card. You repeat the experiment 5 times, so $n = 5$. The outcomes for each trial can be classified in two categories: $S =$ selecting a club and $F =$ selecting another suit. The probabilities of success and failure are

$$p = P(S) = \frac{1}{4} \quad \text{and} \quad q = P(F) = \frac{3}{4}.$$

The random variable x represents the number of clubs selected in the 5 trials. So, the possible values of the random variable are

0, 1, 2, 3, 4, and 5.

For instance, if $x = 2$, then exactly two of the five cards are clubs and the other three are not clubs. Note that x is a discrete random variable because its possible values can be listed.

EXAMPLE 1

Binomial Experiments

Decide whether the experiment is a binomial experiment. If it is, specify the values of n, p, and q, and list the possible values of the random variable x. If it is not, explain why.

1. A certain surgical procedure has an 85% chance of success. A doctor performs the procedure on eight patients. The random variable represents the number of successful surgeries.
2. A jar contains five red marbles, nine blue marbles, and six green marbles. You randomly select three marbles from the jar, *without replacement*. The random variable represents the number of red marbles.

SOLUTION

1. The experiment is a binomial experiment because it satisfies the four conditions of a binomial experiment. In the experiment, each surgery represents one trial. There are eight surgeries, and each surgery is independent of the others. There are only two possible outcomes for each surgery—either the surgery is a success or it is a failure. Also, the probability of success for each surgery is 0.85. Finally, the random variable x represents the number of successful surgeries.

$$n = 8$$

$$p = 0.85$$

$$q = 1 - 0.85 = 0.15$$

$$x = 0, 1, 2, 3, 4, 5, 6, 7, 8$$

2. The experiment is not a binomial experiment because it does not satisfy all four conditions of a binomial experiment. In the experiment, each marble selection represents one trial and selecting a red marble is a success. When selecting the first marble, the probability of success is 5/20. However, because the marble is not replaced, the probability of success is no longer 5/20. So, the trials are not independent, and the probability of a success is not the same for each trial.

Picturing the World

A recent survey of registered voters in the United States asked whether public school teachers should be required to take drug tests. The respondents' answers were either yes or no. *(Source: Family Research Council)*

Survey question: Should public school teachers be required to take drug tests?

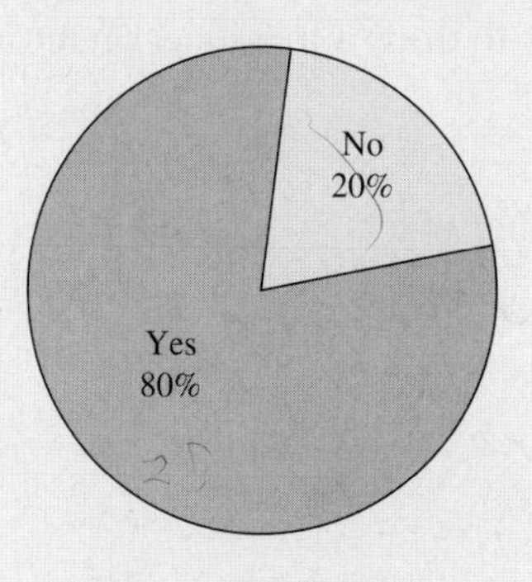

Why is this a binomial experiment? Identify the probability of success, p. Identify the probability of failure, q.

Try It Yourself 1

Decide whether the following is a binomial experiment. If it is, specify the values of n, p, and q and list the possible values of the random variable x. If it is not, explain why.

You take a multiple choice quiz that consists of 10 questions. Each question has four possible answers, only one of which is correct. To complete the quiz, you randomly guess the answer to each question. The random variable represents the number of correct answers.

a. Identify a *trial* of the experiment and what is a "success."
b. Decide if the experiment *satisfies the four conditions* of a binomial experiment.
c. *Make a conclusion and identify* n, p, q, and the possible values of x.

Answer: Page A35

Binomial Probabilities

There are several ways to find the probability of x successes in n trials of a binomial experiment. One way is to use the binomial probability formula.

> **Insight**
>
> In the binomial probability formula, ${}_nC_x$ determines the number of ways of getting x successes in n trials, regardless of order.
>
> $${}_nC_x = \frac{n!}{(n-x)!\,x!}$$

Binomial Probability Formula

In a binomial experiment, the probability of exactly x successes in n trials is

$$P(x) = {}_nC_x\, p^x q^{n-x} = \frac{n!}{(n-x)!\,x!}\, p^x q^{n-x}.$$

EXAMPLE 

Finding Binomial Probabilities

A six-sided die is rolled 3 times. Find the probability of rolling exactly one 6.

SOLUTION One way to answer the question is to draw a tree diagram and use the Multiplication Rule.

Roll 1, Roll 2, Roll 3 (tree diagram: Roll a six / Roll a number that is not a six)

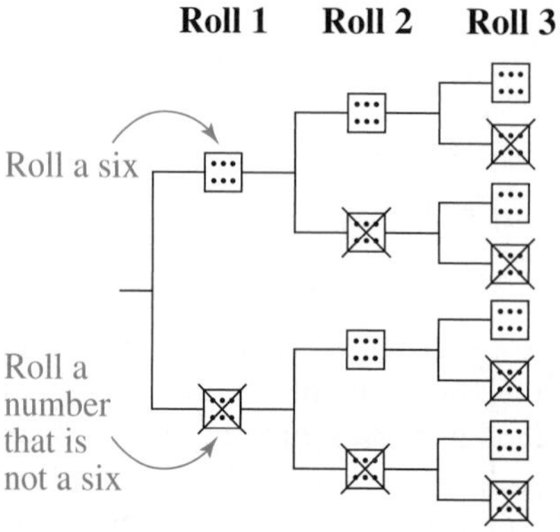

Number of 6's	Probability
3	$\frac{1}{6}\cdot\frac{1}{6}\cdot\frac{1}{6} = \frac{1}{216}$
2	$\frac{1}{6}\cdot\frac{1}{6}\cdot\frac{5}{6} = \frac{5}{216}$
2	$\frac{1}{6}\cdot\frac{5}{6}\cdot\frac{1}{6} = \frac{5}{216}$
1	$\frac{1}{6}\cdot\frac{5}{6}\cdot\frac{5}{6} = \frac{25}{216}$
2	$\frac{5}{6}\cdot\frac{1}{6}\cdot\frac{1}{6} = \frac{5}{216}$
1	$\frac{5}{6}\cdot\frac{1}{6}\cdot\frac{5}{6} = \frac{25}{216}$
1	$\frac{5}{6}\cdot\frac{5}{6}\cdot\frac{1}{6} = \frac{25}{216}$
0	$\frac{5}{6}\cdot\frac{5}{6}\cdot\frac{5}{6} = \frac{125}{216}$

There are three outcomes that have exactly one 6, and each has a probability of 25/216. So, the probability of rolling exactly one 6 is $3(25/216) \approx 0.347$. Another way to answer the question is to use the binomial probability formula. In this binomial experiment, rolling a 6 is a success, while rolling any other number is a failure. The values for n, p, q, and x are $n = 3$, $p = \frac{1}{6}$, $q = \frac{5}{6}$, and $x = 1$. The probability of rolling exactly one 6 is

$$P(1) = \frac{3!}{(3-1)!1!}\left(\frac{1}{6}\right)^1\left(\frac{5}{6}\right)^2 = 3\left(\frac{1}{6}\right)\left(\frac{25}{36}\right) = 3\left(\frac{25}{216}\right) = \frac{25}{72} \approx 0.347.$$

> **Study Tip**
>
> Recall that $n!$ is read "n factorial" and represents the product of all integers from n to 1. For instance,
>
> $$5! = 5\cdot4\cdot3\cdot2\cdot1 = 120.$$

Try It Yourself 2

A card is selected from a deck and replaced. If this experiment is repeated a total of five times, find the probability of selecting exactly three clubs.

a. *Identify* a trial, a success, and a failure.
b. *Identify* n, p, q, and x.
c. Use the *binomial probability formula.*

Answer: Page A35

By listing the possible values of x with the corresponding probability of each, you can construct a **binomial probability distribution.**

EXAMPLE 3

Constructing a Binomial Distribution

In a survey, workers in the United States were asked to name their expected sources of retirement income. The results are shown in the graph. Seven workers who participated in the survey are randomly selected and asked whether they expect to rely on social security for retirement income. Create a binomial probability distribution for the number of workers who respond yes.

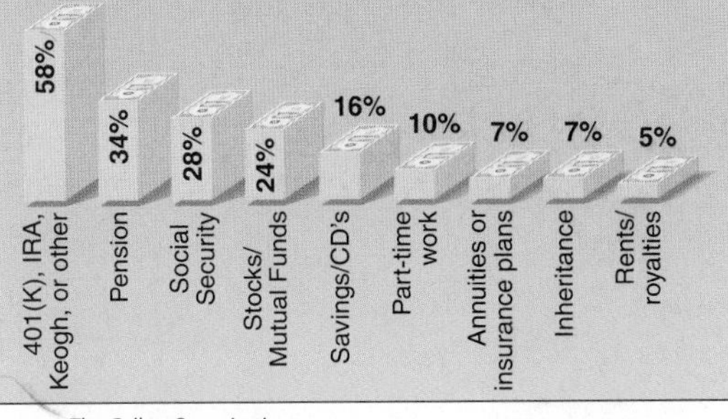

Source: The Gallup Organization

SOLUTION From the graph, you can see that 28% of working Americans expect to rely on social security for retirement income. So, $p = 0.28$ and $q = 0.72$. Because $n = 7$, the possible values of x are 0, 1, 2, 3, 4, 5, 6, and 7.

$$P(0) = {}_7C_0(0.28)^0(0.72)^7 = 1(0.28)^0(0.72)^7 \approx 0.1003$$
$$P(1) = {}_7C_1(0.28)^1(0.72)^6 = 7(0.28)^1(0.72)^6 \approx 0.2731$$
$$P(2) = {}_7C_2(0.28)^2(0.72)^5 = 21(0.28)^2(0.72)^5 \approx 0.3186$$
$$P(3) = {}_7C_3(0.28)^3(0.72)^4 = 35(0.28)^3(0.72)^4 \approx 0.2065$$
$$P(4) = {}_7C_4(0.28)^4(0.72)^3 = 35(0.28)^4(0.72)^3 \approx 0.0803$$
$$P(5) = {}_7C_5(0.28)^5(0.72)^2 = 21(0.28)^5(0.72)^2 \approx 0.0187$$
$$P(6) = {}_7C_6(0.28)^6(0.72)^1 = 7(0.28)^6(0.72)^1 \approx 0.0024$$
$$P(7) = {}_7C_7(0.28)^7(0.72)^0 = 1(0.28)^7(0.72)^0 \approx 0.0001$$

x	$P(x)$
0	0.1003
1	0.2731
2	0.3186
3	0.2065
4	0.0803
5	0.0187
6	0.0024
7	0.0001
	$\Sigma P(x) = 1$

Notice in the table at the left that all the probabilities are between 0 and 1 and that the sum of the probabilities is 1.

Study Tip

When probabilities are rounded to a fixed number of decimal places, the sum of the probabilities may differ slightly from 1.

Try It Yourself 3

Seven workers who participated in the survey are randomly selected and asked whether they expect to rely on a pension for retirement income. Create a binomial distribution for the number of retirees who respond yes.

a. *Identify* a trial, a success, and a failure.
b. *Identify* n, p, q, and possible values for x.
c. Use the *binomial probability formula* for each value of x.
d. *Organize your results in a table.*

Answer: Page A35

Finding binomial probabilities with the binomial probability formula can be a tedious and mistake-prone process. To make this process easier, you can use a binomial probability table. Table 2 in Appendix B lists the binomial probability for selected values of n and p.

EXAMPLE 4

Finding a Binomial Probability Using a Table

Fifty percent of working adults spend less than 20 minutes each way commuting to their jobs. If you randomly select six working adults, what is the probability that exactly three of them spend less than 20 minutes each way commuting to work? Use a table to find the probability. *(Source: Maritz AmeriPoll)*

SOLUTION A portion of Table 2 is shown here. Using the distribution for $n = 6$ and $p = 0.5$, you can find the probability that $x = 3$, as shown by the highlighted areas in the table.

		p												
n	*x*	**.01**	**.05**	**.10**	**.15**	**.20**	**.25**	**.30**	**.35**	**.40**	**.45**	**.50**	**.55**	**.60**
2	0	.980	.902	.810	.723	.640	.563	.490	.423	.360	.303	.250	.203	.160
	1	.020	.095	.180	.255	.320	.375	.420	.455	.480	.495	.500	.495	.480
	2	.000	.002	.010	.023	.040	.063	.090	.123	.160	.203	.250	.303	.360
3	0	.970	.857	.729	.614	.512	.422	.343	.275	.216	.166	.125	.091	.064
	1	.029	.135	.243	.325	.384	.422	.441	.444	.432	.408	.375	.334	.288
	2	.000	.007	.027	.057	.096	.141	.189	.239	.288	.334	.375	.408	.432
	3	.000	.000	.001	.003	.008	.016	.027	.043	.064	.091	.125	.166	.216
6	0	.941	.735	.531	.377	.262	.178	.118	.075	.047	.028	.016	.008	.004
	1	.057	.232	.354	.399	.393	.356	.303	.244	.187	.136	.094	.061	.037
	2	.001	.031	.098	.176	.246	.297	.324	.328	.311	.278	.234	.186	.138
	3	.000	.002	.015	.042	.082	.132	.185	.236	.276	.303	.312	.303	.276
	4	.000	.000	.001	.006	.015	.033	.060	.095	.138	.186	.234	.278	.311
	5	.000	.000	.000	.000	.002	.004	.010	.020	.037	.061	.094	.136	.187
	6	.000	.000	.000	.000	.000	.000	.001	.002	.004	.008	.016	.028	.047

So, the probability that exactly three of the six workers spend less than 20 minutes each way commuting to work is 0.312.

Try It Yourself 4

Twenty-five percent of all small businesses in the United States have a Web site. If you randomly select 10 small businesses, what is the probability that exactly four of them have a Web site? Use a table to find the probability. *(Source: Yankelovich Partners for IBM, U.S. Chamber of Commerce)*

a. *Identify* a trial, a success, and a failure.
b. *Identify* n, p, and x.
c. *Use Table 2* to find the binomial probability.

Answer: Page A35

An even more efficient way to find binomial probabilities is to use a calculator or a computer. For instance, you can find binomial probabilities using *Minitab*, *Excel*, and the *TI-83*.

EXAMPLE 

Using Technology to Find a Binomial Probability

The results of a recent survey indicate that 58% of households in the United States own a gas grill. If you randomly select 100 households, what is the probability that exactly 65 households will own a gas grill? Use a calculator or a computer to find the probability. *(Source: Leo Shapiro Research for Weber GrillWatch)*

SOLUTION *Minitab, Excel,* and the *TI-83* each have features that allow you to find binomial probabilities automatically.

MINITAB

Probability Density Function

Binomial with n = 100 and p = 0.580000

x	P(X=x)
65.00	0.0299

TI-83

```
binompdf(100,.58,65)
         .0299216472
```

EXCEL

	A	B	C	D
1	BINOMDIST(65,100,0.58,FALSE)			
2				0.029922

From the displays, you can see that the probability that exactly 65 households will own a gas grill is about 0.03.

Try It Yourself 5

The results of a recent survey indicate that 71% of people in the United States use more than one topping on their hotdogs. If you randomly select 250 people in the United States, what is the probability that exactly 178 of them will use more than one topping? Use a calculator or a computer to find the probability. *(Source: ICR Survey Research-Group for Hebrew International)*

a. *Identify* $n, p,$ and x.
b. *Calculate* the binomial probability.
c. *Interpret* the results.

Answer: Page A35

Study Tip

Detailed instructions for using *Minitab, Excel,* and the *TI-83* are shown in the Technology Guide that accompanies this text. For instance, here are instructions for finding a binomial probability on a *TI-83*.

[2nd] DISTR

0: binompdf(

Enter the values of *n, p,* and *x* separated by commas.

ENTER

EXAMPLE 

Finding Binomial Probabilities

A survey indicates that 41% of women in the United States consider reading as their favorite leisure-time activity. You randomly select four women and ask them if reading is their favorite leisure-time activity. Find the probability that (1) exactly two of them respond yes, (2) at least two of them respond yes, and (3) fewer than two of them respond yes. *(Source: Louis Harris & Associates)*

SOLUTION

1. Using $n = 4$, $p = 0.41$, $q = 0.59$, and $x = 2$, the probability that exactly two women will respond yes is

$$P(2) = {}_4C_2(0.41)^2(0.59)^2 = 6(0.41)^2(0.59)^2 \approx 0.351.$$

2. To find the probability that at least two women will respond yes, you can find the sum of $P(2)$, $P(3)$, and $P(4)$.

$$P(2) = {}_4C_2(0.41)^2(0.59)^2 = 6(0.41)^2(0.59)^2 \approx 0.351094$$
$$P(3) = {}_4C_3(0.41)^3(0.59)^1 = 4(0.41)^3(0.59)^1 \approx 0.162654$$
$$P(4) = {}_4C_4(0.41)^4(0.59)^0 = 1(0.41)^4(0.59)^0 \approx 0.028258$$

So, the probability that at least two will respond yes is

$$P(x \geq 2) = P(2) + P(3) + P(4)$$
$$\approx 0.351094 + 0.162654 + 0.028258$$
$$\approx 0.542.$$

3. To find the probability that fewer than two women will respond yes, you must find the sum of $P(0)$ and $P(1)$.

$$P(0) = {}_4C_0(0.41)^0(0.59)^4 = 1(0.41)^0(0.59)^4 \approx 0.121174$$
$$P(1) = {}_4C_1(0.41)^1(0.59)^3 = 4(0.41)^1(0.59)^3 \approx 0.336822$$

So, the probability that fewer than two will respond yes is

$$P(x < 2) = P(0) + P(1)$$
$$\approx 0.121174 + 0.336822$$
$$\approx 0.458.$$

Study Tip

The complement of "x is at least 2" is "x is less than 2." So, another way to find the probability in part (3) is

$$P(x < 2) = 1 - P(x \geq 2)$$
$$\approx 1 - 0.542$$
$$= 0.458.$$

Try It Yourself 6

A survey indicates that 21% of men in the United States consider fishing as their favorite leisure-time activity. You randomly select five U.S. men and ask them if fishing is their favorite leisure-time activity. Find the probability that (1) exactly two of them respond yes, (2) at least two of them respond yes, and (3) fewer than two of them respond yes. *(Source: Louis Harris & Associates)*

a. Determine the appropriate *values of x* for each situation.
b. Find the *binomial probability* for each value of x. Then find the sum, if necessary.
c. *Interpret* the results.

Answer: Page A35

Graphing Binomial Distributions

In Section 4.1, you learned how to construct and graph discrete probability distributions. Because a binomial distribution is a discrete probability distribution, you can use the same process.

EXAMPLE 7

Constructing and Graphing a Binomial Distribution

Sixty-five percent of households in the United States subscribe to cable TV. You randomly select six households and ask each if they subscribe to cable TV. Construct a probability distribution for the random variable x. Then graph the distribution. *(Source: Polk)*

SOLUTION To construct the binomial distribution, find the probability for each value of x. Using $n = 6$, $p = 0.65$, and $q = 0.35$, you can obtain the following.

$$P(0) = {}_6C_0(0.65)^0(0.35)^6 = 1(0.65)^0(0.35)^6 \approx 0.002$$

$$P(1) = {}_6C_1(0.65)^1(0.35)^5 = 6(0.65)^1(0.35)^5 \approx 0.020$$

$$P(2) = {}_6C_2(0.65)^2(0.35)^4 = 15(0.65)^2(0.35)^4 \approx 0.095$$

$$P(3) = {}_6C_3(0.65)^3(0.35)^3 = 20(0.65)^3(0.35)^3 \approx 0.235$$

$$P(4) = {}_6C_4(0.65)^4(0.35)^2 = 15(0.65)^4(0.35)^2 \approx 0.328$$

$$P(5) = {}_6C_5(0.65)^5(0.35)^1 = 6(0.65)^5(0.35)^1 \approx 0.244$$

$$P(6) = {}_6C_6(0.65)^6(0.35)^0 = 1(0.65)^6(0.35)^0 \approx 0.075$$

x	0	1	2	3	4	5	6
$P(x)$	0.002	0.020	0.095	0.235	0.328	0.244	0.075

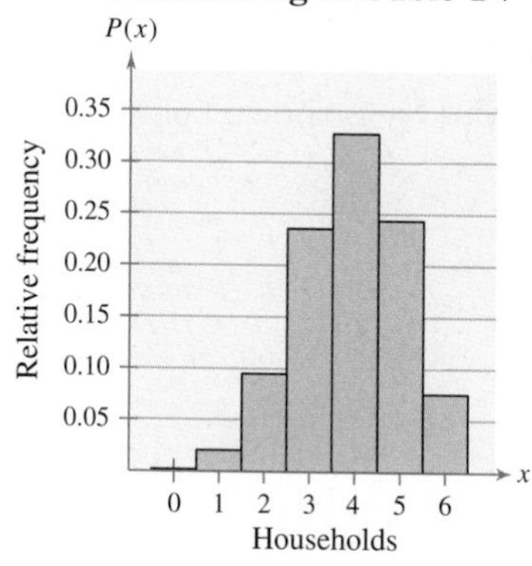

Because each probability is a relative frequency, you can graph the probability distribution using a relative frequency histogram as shown at the left.

Try It Yourself 7

Forty-one percent of households in the United States own a computer. You randomly select six households and ask each if they own a computer. Construct a probability distribution for the random variable x. Then graph the distribution. *(Source: Electronic Industries Association)*

a. *Find* the binomial probability for each value of the random variable x.
b. *Organize* the values of x and their corresponding probability in a binomial distribution.
c. Use a relative frequency histogram to *graph* the binomial distribution.

Answer: Page A35

Notice in Example 7 that the histogram is skewed left. The graph of a binomial distribution with $p > 0.5$ is skewed left, while the graph of a binomial distribution with $p < 0.5$ is skewed right. The graph of a binomial distribution with $p = 0.5$ is symmetric.

Mean, Variance, and Standard Deviation

Although you can use the formulas learned in Section 4.1 for mean, variance, and standard deviation of a probability distribution, the properties of a binomial distribution enable you to use much simpler formulas.

Population Parameters of a Binomial Distribution

Mean: $\mu = np$

Variance: $\sigma^2 = npq$

Standard deviation: $\sigma = \sqrt{npq}$

EXAMPLE 8

Finding Mean, Variance, and Standard Deviation

In Pittsburgh, Pennsylvania, about 57% of the days in a year are cloudy. Find the mean, variance, and standard deviation for the number of cloudy days during the month of June. What can you conclude? *(Source: National Climatic Data Center)*

SOLUTION There are 30 days in June. Using

$$n = 30,\ p = 0.57,\ \text{and}\ q = 0.43$$

you can find the mean, variance, and standard deviation as shown below.

$$\begin{aligned}\mu &= np \\ &= 30 \cdot 0.57 \\ &= 17.1\end{aligned}$$

$$\begin{aligned}\sigma^2 &= npq \\ &= 30 \cdot 0.57 \cdot 0.43 \\ &= 7.353\end{aligned}$$

$$\begin{aligned}\sigma &= \sqrt{npq} \\ &= \sqrt{7.353} \\ &\approx 2.71\end{aligned}$$

So, you can conclude that, on the average, there are 17.1 cloudy days during the month of June. The standard deviation is about 2.71 days.

Try It Yourself 8

In San Diego, California, 38% of the days in a year are clear. Find the mean, variance, and standard deviation for the number of clear days during the month of May. What can you conclude? *(Source: National Climatic Data Center)*

a. *Identify* a success and the value of n, p, and q.
b. *Find the product* of n and p to calculate the mean.
c. *Find the product* of n, p, and q *for the variance.*
d. *Find the square root* of the variance *for the standard deviation.*
e. What can you conclude?

Answer: Page A35

4.2 Exercises

Help

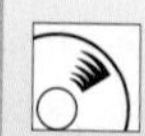
LarsonTutor 4.2

Companion Web Site

Student Solutions Manual 4.2

Videos 4.2

Technology Manuals

Basic Skills and Concepts

Graphical Analysis In Exercises 1 and 2, use the following information. The histograms each represent binomial distributions. Each distribution has the same number of trials n, but different probabilities of success p. Match the given probabilities with the correct graph.

1. $p = 0.20,\ p = 0.50,\ p = 0.80$

(a)

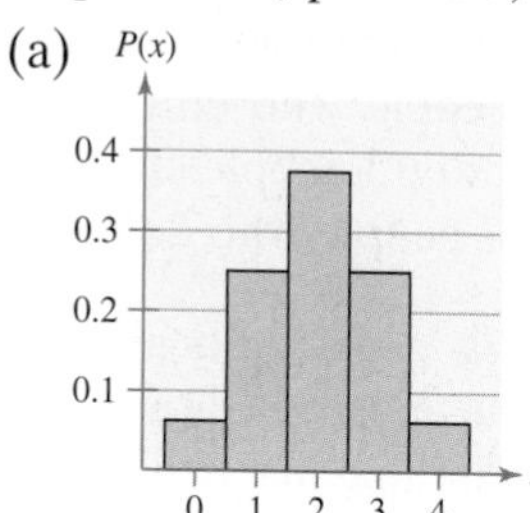

(b)

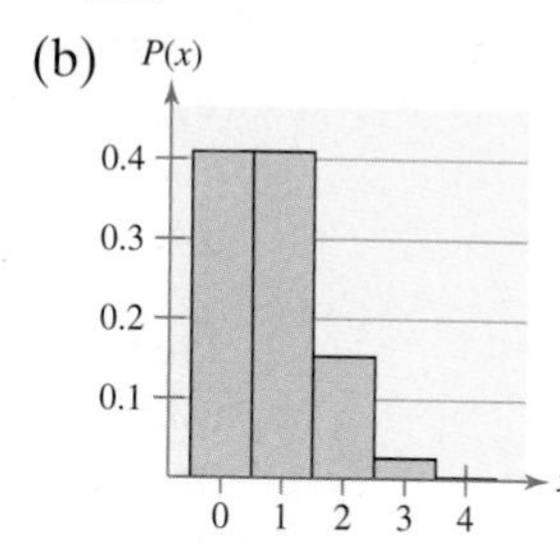

(c)

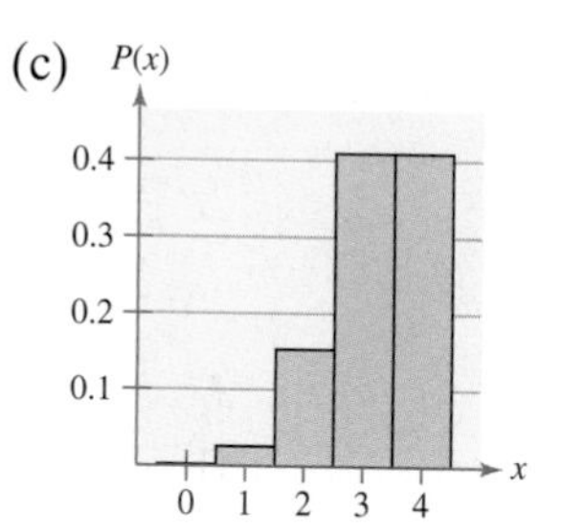

2. $p = 0.25,\ p = 0.50,\ p = 0.75$

(a)

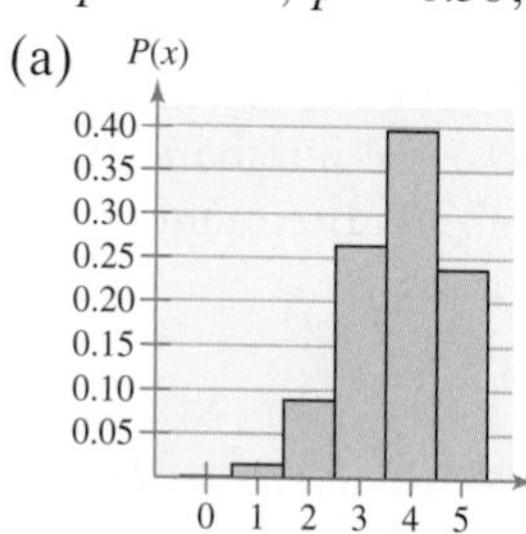

(b)

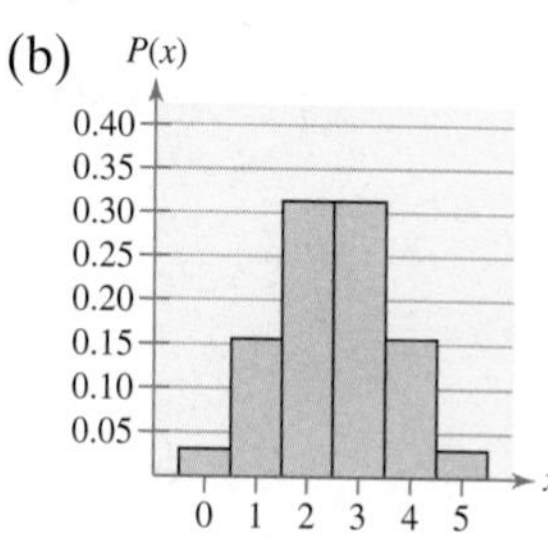

(c)

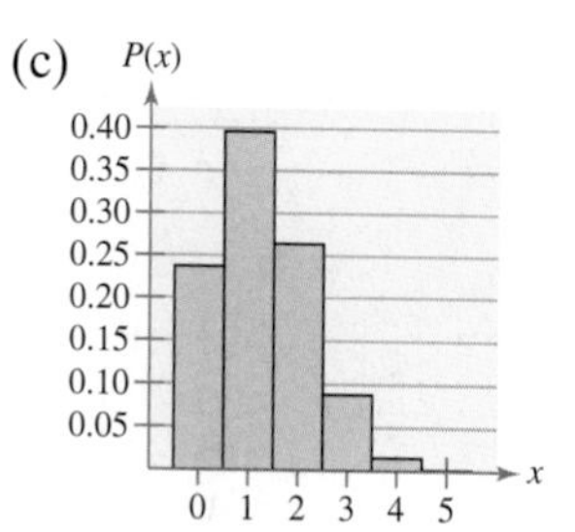

Graphical Analysis In Exercises 3 and 4, use the following information. Each histogram shown represents part of a binomial distribution. Each distribution has the same probability of success p, but different numbers of trials n. Match the given values of n with the correct graph. What happens as the value of n increases and the probability of success remains the same?

3. $n = 4,\ n = 8,\ n = 12$

(a)

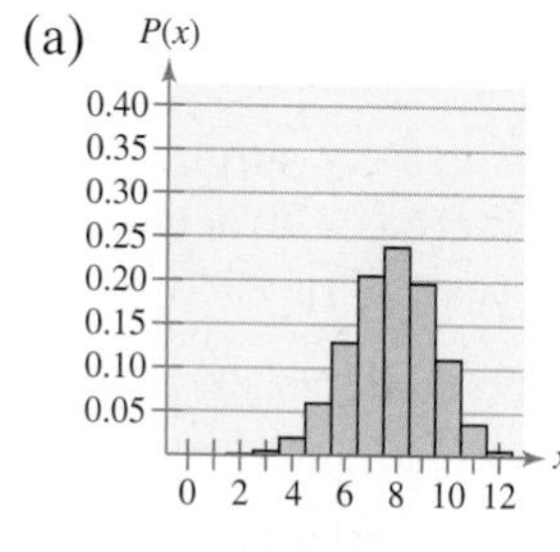

(b)

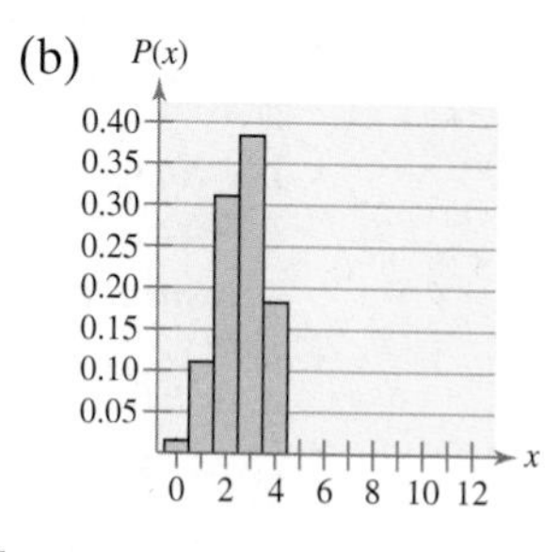

(c)

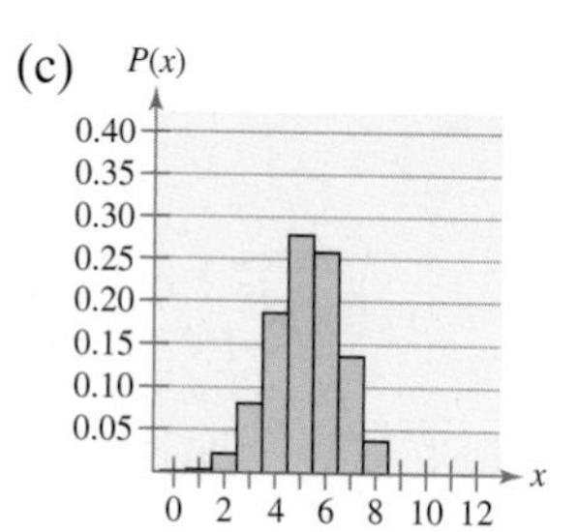

4. $n = 5,\ n = 10,\ n = 15$

(a)

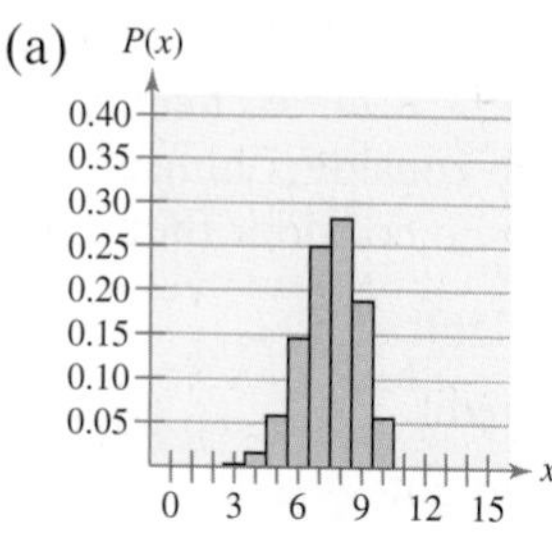

(b)

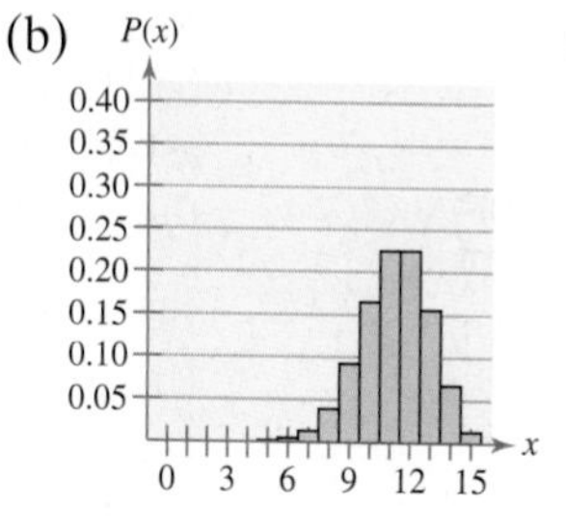

(c)

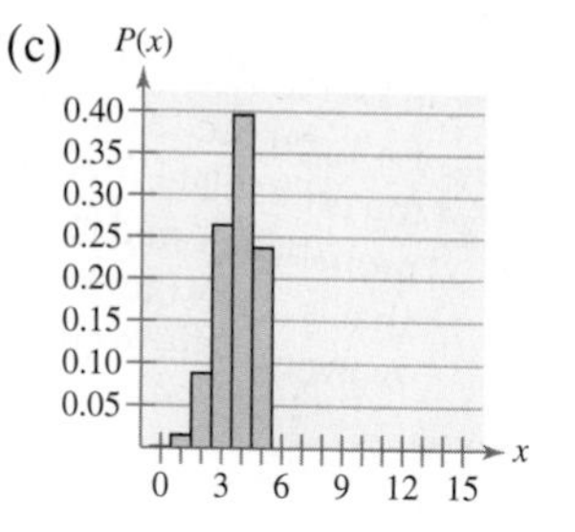

Identifying and Understanding Binomial Experiments In Exercises 5–8, decide whether the experiment is a binomial experiment. If it is, identify a success, specify the values of n, p, and q, and list the possible values of the random variable x. If it is not a binomial experiment, explain why.

5. ***Cyanosis*** ◆ Cyanosis is the condition of having bluish skin due to insufficient oxygen in the blood. About 80% of babies born with cyanosis recover fully. A hospital is caring for five babies born with cyanosis. The random variable represents the number of babies that fully recover. *(Source: The World Book Encyclopedia)*

6. ***Clothing Store Purchases*** ◆ From past records, a clothing store finds that 30% of the people who enter the store will make a purchase. During a one-hour period, 15 people enter the store. The random variable represents the number of people who do *not* make a purchase.

7. ***Political Polls*** ◆ A political polling organization calls 1012 people and asks, "Do you approve, disapprove, or have no opinion of the way the president is handling his job?" The random variable represents the number of people who approve of the way the president is handling his job. *(Adapted from The Gallup Organization)*

8. ***Lottery*** ◆ A state lottery randomly chooses six balls numbered from 1 to 40. You choose six numbers and purchase a lottery ticket. The random variable represents the number of matches on your ticket to the numbers drawn in the lottery.

Finding Binomial Probabilities In Exercises 9–14, find the indicated probabilities.

9. ***Answer Guessing*** ◆ You are taking a multiple-choice quiz that consists of five questions. Each question has four possible answers, only one of which is correct. To complete the quiz, you randomly guess the answer to each question. Find the probability of guessing

(a) exactly three answers correctly.

(b) at least three answers correctly.

(c) less than three answers correctly.

10. ***Surgery Success*** ◆ A surgical technique is performed on seven patients. You are told there is a 70% chance of success. Find the probability that the surgery is successful for

(a) exactly five patients.

(b) at least five patients.

(c) less than five patients.

Study Tip

In Exercises 11, 13, 15, 18, and 19, if you use Table 2 to find the probability, use the probability listed in the table that is closest to the value of p given in the exercise.

11. ***Basketball Fans*** ◆ Fifty-four percent of men consider themselves basketball fans. *(Source: Bruskin-Goldring Research)* You randomly select 10 men and ask each if he considers himself a basketball fan. Find the probability that the number who consider themselves basketball fans is

(a) exactly eight.

(b) at least eight.

(c) less than eight.

12. ***Favorite Cookie*** ◆ Ten percent of adults say oatmeal raisin is their favorite cookie. *(Source: WEAREVER)* You randomly select 12 adults and ask each to name his or her favorite cookie. Find the probability that the number who say oatmeal raisin is their favorite cookie is
    (a) exactly four.
    (b) at least four.
    (c) less than four.

13. ***Vacation Purpose*** ◆ Twenty-one percent of vacationers say the primary purpose of their vacation is outdoor recreation. *(Source: Travel Industry Association)* You randomly select 10 vacationers and ask each to name the primary purpose of his or her vacation. Find the probability that the number who say outdoor recreation is the primary purpose of their vacation is
    (a) exactly two.
    (b) at least two.
    (c) less than two.

14. ***Honeymoon Financing*** ◆ Seventy percent of married couples paid for their honeymoon themselves. *(Source: Bride's Magazine)* You randomly select 20 married couples and ask each if they paid for their honeymoon themselves. Find the probability that the number of couples who say they paid for their honeymoon themselves is
    (a) exactly seven.
    (b) at least seven.
    (c) less than seven.

Constructing Binomial Distributions In Exercises 15–20, (a) construct a binomial distribution, (b) graph the binomial distribution using a relative frequency histogram, find the (c) mean, (d) variance, and (e) standard deviation of the binomial distribution, and (f) interpret the results in the context of the real-life situation. What values of the random variable x would you consider unusual? Explain your reasoning.

15. ***Women Basketball Fans*** ◆ Thirty-six percent of women consider themselves basketball fans. You randomly select six women and ask each if she considers herself a basketball fan. *(Source: Bruskin-Goldring Research)*

16. ***No Trouble Sleeping*** ◆ One in four adults says he or she has no trouble sleeping at night. You randomly select five adults and ask each if he or she has no trouble sleeping at night. *(Source: Marist Institute for Public Opinion)*

17. ***Blood Donors*** ◆ Five percent of people in the United States eligible to donate blood actually do. You randomly select four eligible blood donors and ask if they donate blood. *(Adapted from American Association of Blood Banks)*

18. ***Blood Types*** ◆ Thirty-eight percent of people in the United States have type O^+ blood. You randomly select five Americans and ask them if their blood type is O^+. *(Source: American Association of Blood Banks)*

19. ***College Students' Income*** ◆ About 28% of college students earn at least $400 per month. You select at random five college students and ask each if he or she earns at least $400 per month. *(Source: Campus Concepts)*

20. ***Prescription Eyewear*** ◆ About 60% of people in the United States use prescription eyewear. You select at random four Americans and ask each if he or she uses prescrtiption eyewear. *(Source: Consumer Reports)*

No Vacation from Information In Exercises 21 and 22, use the following information. The graph shows the results of a survey of adults who were asked what types of information they would want to receive when on vacation. You randomly select six people who participated in the survey and ask each one of them what types of information they would want to receive when on vacation. Let x represent the number who would want general news information.

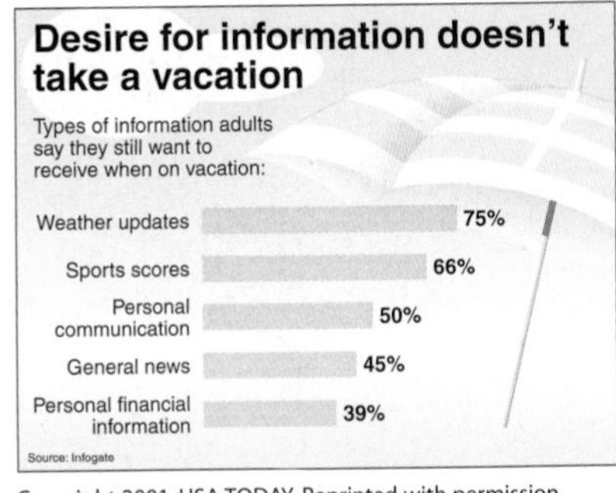

21. Construct a binomial distribution.

22. Find the probability that exactly two people will respond "general news."

Extending the Basics

Multinomial Experiments In Exercises 23 and 24, use the following information.

A multinomial experiment is a probability experiment that satisfies the following conditions.

1. The experiment is repeated a fixed number of times n where each trial is independent of the other trials.
2. Each trial has k possible mutually exclusive outcomes: $E_1, E_2, E_3, \ldots, E_k$.
3. Each outcome has a fixed probability. Therefore, $P(E_1) = p_1$, $P(E_2) = p_2$, $P(E_3) = p_3, \ldots, P(E_k) = p_k$. The sum of the probabilities for all outcomes is

$$p_1 + p_2 + p_3 + \cdots + p_k = 1.$$

4. x_1 is the number of times E_1 will occur, x_2 is the number of times E_2 will occur, x_3 is the number of times E_3 will occur, and so on.
5. The discrete random variable x counts the number of times $x_1, x_2, x_3, \ldots, x_k$ occurs in n independent trials where $x_1 + x_2 + x_3 + \cdots + x_k = n$. The probability that x will occur is

$$P(x) = \frac{n!}{x_1!x_2!x_3! \cdots x_k!} p_1^{x_1} p_2^{x_2} p_3^{x_3} \ldots p_k^{x_k}.$$

23. According to a theory in genetics, if tall and colorful plants are crossed with short and colorless plants, four types of plants will result: tall and colorful, tall and colorless, short and colorful, and short and colorless, with corresponding probabilities of $\frac{9}{16}$, $\frac{3}{16}$, $\frac{3}{16}$, and $\frac{1}{16}$. If 10 plants are selected, find the probability that five will be tall and colorful, two will be tall and colorless, two will be short and colorful, and one will be short and colorless.

24. Another proposed theory in genetics gives the corresponding probabilities for the four types of plants described as $\frac{5}{16}$, $\frac{4}{16}$, $\frac{1}{16}$, and $\frac{6}{16}$. If 10 plants are selected, find the probability that five will be tall and colorful, two will be tall and colorless, two will be short and colorful, and one will be short and colorless.

Case Study

WWW.AIR-TRANSPORT.ORG

AIR TRANSPORT ASSOCIATION

Binomial Distribution of Airplane Accidents

The Air Transport Association of America (ATA) is a support organization for the principal U.S. airlines. Some of the ATA's activities include promoting the air transport industry and conducting industry-wide studies.

The ATA also keeps statistics about commercial airline flights, including those that involve accidents. From 1970 through 1999 for aircraft with 10 or more seats, there were 108 fatal commercial airplane accidents involving U.S. airlines. The distribution of these accidents is shown in the histogram at the right.

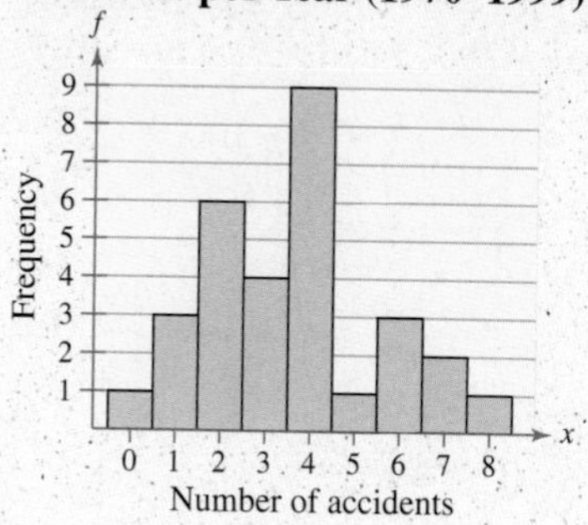

Year	1970	1971	1972	1973	1974	1975	1976	1977	1978	1979	1980	1981	1982	1983	1984
Accidents	2	6	7	6	7	2	2	3	5	4	0	4	4	4	1

Year	1985	1986	1987	1988	1989	1990	1991	1992	1993	1994	1995	1996	1997	1998	1999
Accidents	4	2	4	3	8	6	4	4	1	4	2	3	3	1	2

Exercises

1. In 1999, there were about 8 million commercial flights in the United States. If one is selected at random, what is the probability that it involved a fatal accident?

2. Suppose that the probability of a fatal accident in a given year is 0.0000004. A binomial probability distribution for $n = 8{,}000{,}000$ and $p = 0.0000004$ with $x = 0$ to 10 is shown.

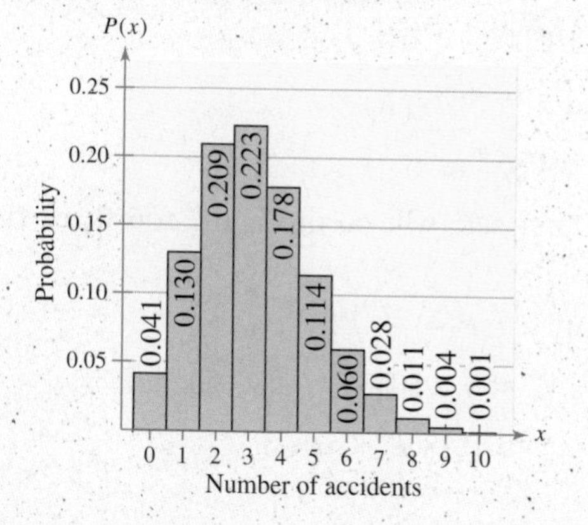

What is the probability that there will be (a) 4 fatal accidents in a year? (b) 10? (c) between 1 and 5, inclusive?

3. Construct a binomial distribution for $n = 8{,}000{,}000$ and $p = 0.0000008$ with $x = 0$ to 10. Compare your results to the distribution in Exercise 2.

4. Is a binomial distribution a good model for determining the probability of various numbers of fatal accidents during a year? Explain your reasoning and include a discussion of the four criteria for a binomial experiment.

5. The Federal Aviation Administration says that air flight is so safe that "a person could fly around the clock for over 438 years before being involved in a fatal accident." How can such a statement be justified?

4.3 More Discrete Probability Distributions

What You Should Learn

- How to find probabilities using the geometric distribution
- How to find probabilities using the Poisson distribution

The Geometric Distribution • The Poisson Distribution • Summary of Discrete Probability Distributions

The Geometric Distribution

In this section, you will study two more discrete probability distributions—the geometric distribution and the Poisson distribution.

Many actions in life are repeated until a success occurs. For instance, a CPA candidate might take the CPA exam several times before receiving a passing score, or you might have to dial your Internet connection several times before successfully logging on. Situations such as these can be represented by a geometric distribution.

DEFINITION

A **geometric distribution** is a discrete probability distribution of a random variable x that satisfies the following conditions.

1. A trial is repeated until a success occurs.
2. The repeated trials are independent of each other.
3. The probability of success p is constant for each trial.

The **probability that the first success will occur on trial number x** is

$P(x) = p(q)^{x-1}$, where $q = 1 - p$.

EXAMPLE 1

Finding Probabilities Using the Geometric Distribution

From experience, you know that the probability that you will make a sale on any given telephone call is 0.23. Find the probability that your first sale on any given day will occur on your fourth or fifth sales call.

SOLUTION To find the probability that your first sale will occur on the fourth or fifth call, first find the probability that the sale will occur on the fourth call and the probability that the sale will occur on the fifth call. Then find the sum of the resulting probabilities. Using $p = 0.23$, $q = 0.77$, and $x = 4$ you have

$$P(4) = 0.23 \cdot (0.77)^3 \approx 0.105003$$

Using $p = 0.23$, $q = 0.77$, and $x = 5$ you have

$$P(5) = 0.23 \cdot (0.77)^4 \approx 0.080852.$$

So, the probability that your first sale will occur on the fourth or fifth sales call is

$$\begin{aligned} P(\text{sale on fourth or fifth call}) &= P(4) + P(5) \\ &\approx 0.105003 + 0.080852 \\ &\approx 0.186. \end{aligned}$$

Study Tip

If the first success occurs on the fourth trial, the outcome is *FFFS* and the probability is

$$P = \overset{F}{(0.77)}\overset{F}{(0.77)}\overset{F}{(0.77)}\overset{S}{(0.23)}.$$

If the first success occurs on the fifth trial, the outcome is *FFFFS* and the probability is

$$P = \overset{F}{(0.77)}\overset{F}{(0.77)}\overset{F}{(0.77)}\overset{F}{(0.77)}\overset{S}{(0.23)}.$$

Try It Yourself 1

Find the probability that your first sale will occur before your fourth sales call.

a. *Use the geometric distribution* to find $P(1)$, $P(2)$, and $P(3)$.
b. *Find the sum* of $P(1)$, $P(2)$, and $P(3)$.
c. *Interpret* the results.

Answer: Page A35

The Poisson Distribution

In a binomial experiment you are interested in finding the probability of a specific number of successes in a given number of trials. Suppose instead that you want to know the probability that a specific number of occurrences takes place within a given unit of time or space. For instance, to determine the probability that an employee will take 15 sick days within a year, you can use the Poisson distribution.

DEFINITION

The **Poisson distribution** is a discrete probability distribution of a random variable x that satisfies the following conditions.

1. The experiment consists of counting the number of times, x, an event occurs in a given interval. The interval can be an interval of time, area, or volume.
2. The probability of the event occurring is the same for each interval.
3. The number of occurrences in one interval is independent of the number of occurrences in other intervals.

The probability of exactly x occurrences in an interval is

$$P(x) = \frac{\mu^x e^{-\mu}}{x!}$$

where e is an irrational number approximately equal to 2.71828 and μ is the mean number of occurrences per interval unit.

EXAMPLE 2

Using the Poisson Distribution

The mean number of accidents per month at a certain intersection is 3. What is the probability that in any given month 4 accidents will occur at this intersection?

SOLUTION Using $x = 4$ and $\mu = 3$, the probability that 4 accidents will occur in any given month at the intersection is

$$P(4) = \frac{3^4(2.71828)^{-3}}{4!}$$

$$\approx 0.168.$$

Try It Yourself 2

What is the probability that more than four accidents will occur in any given month at the intersection?

a. *Use the Poisson distribution* to find $P(0)$, $P(1)$, $P(2)$, $P(3)$, and $P(4)$.
b. *Find the sum* of $P(0)$, $P(1)$, $P(2)$, $P(3)$, and $P(4)$.
c. *Subtract* the sum from 1.
d. *Interpret* the results.

Answer: Page A35

In Example 2 you used a formula to determine a Poisson probability. You can also use a table to find Poisson probabilities. Table 3 in Appendix B lists the Poisson probability for selected values of x and μ. You can use technology tools, such as *Minitab*, *Excel*, and the *TI-83*, to find Poisson probabilities as well.

Picturing the World

The first successful suspension bridge built in the United States, the Tacoma Narrows Bridge, spans the Tacoma Narrows in Washington state. The average occupancy of vehicles that travel across the bridge is 1.6. The following probability distribution represents the vehicle occupancy on the bridge during a five-day period. *(Source: Washington State Department of Transportation)*

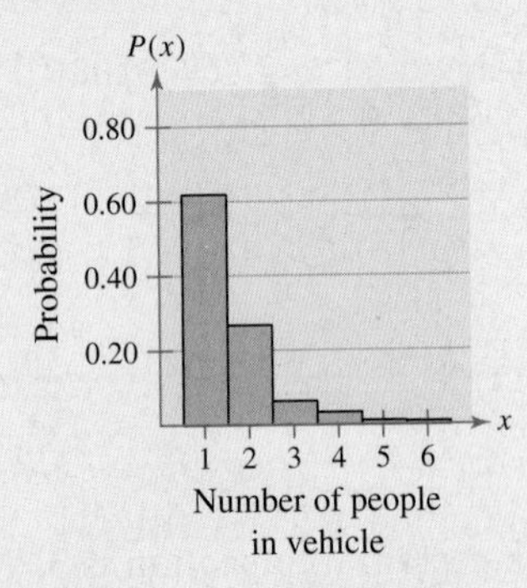

What is the probability that a randomly selected vehicle has two occupants or fewer?

EXAMPLE

Finding Poisson Probabilities Using a Table

A population count shows that there is an average of 3.6 rabbits per acre living in a field. Use a table to find the probability that two rabbits are found on any given acre of the field.

SOLUTION A portion of Table 3 in Appendix B is shown here. Using $\mu = 3.6$ and $x = 2$, you can find the Poisson probability as shown by the highlighted areas in the table.

	μ						
x	**3.1**	**3.2**	**3.3**	**3.4**	**3.5**	**3.6**	**3.7**
0	.0450	.0408	.0369	.0334	.0302	.0273	.0247
1	.1397	.1304	.1217	.1135	.1057	.0984	.0915
2	.2165	.2087	.2008	.1929	.1850	.1771	.1692
3	.2237	.2226	.2209	.2186	.2158	.2125	.2087
4	.1734	.1781	.1823	.1858	.1888	.1912	.1931
5	.1075	.1140	.1203	.1264	.1322	.1377	.1429
6	.0555	.0608	.0662	.0716	.0771	.0826	.0881
7	.0246	.0278	.0312	.0348	.0385	.0425	.0466
8	.0095	.0111	.0129	.0148	.0169	.0191	.0215
9	.0033	.0040	.0047	.0056	.0066	.0076	.0089
10	.0010	.0013	.0016	.0019	.0023	.0028	.0033

So, the probability that two rabbits are found on any given acre is 0.1771.

Try It Yourself 3

Two thousand brown trout are introduced into a small lake. The lake has a volume of 20,000 cubic meters. Use a table to find the probability that three brown trout are found in any given cubic meter of the lake.

a. *Find the average* number of brown trout per cubic meter.
b. *Identify* μ and x.
c. *Use Table 3* to find the Poisson probability.

Answer: Page A35

Summary of Discrete Probability Distributions

The following table summarizes the discrete probability distributions discussed in this chapter.

<table>
<tr><th>Distribution</th><th>Summary</th><th>Formulas</th></tr>
<tr><td>Binomial Distribution</td><td>A binomial experiment satisfies the following conditions.
1. The experiment is repeated for a fixed number (n) of independent trials.
2. There are only two possible outcomes for each trial. Each outcome can be classified as a success or as a failure.
3. The probability of a success must remain constant for each trial.
4. The random variable x counts the number of successful trials out of the n trials.

The parameters of a binomial distribution are n and p.</td><td>x = the number of successes in n trials
p = probability of success on a single trial
q = probability of failure on a single trial
$q = 1 - p$

The probability of exactly x successes in n trials is
$P(x) = {}_nC_x p^x q^{n-x}$
$= \frac{n!}{(n-x)!x!} p^x q^{n-x}$.</td></tr>
<tr><td>Geometric Distribution</td><td>A geometric distribution is a discrete probability distribution of the random variable x that satisfies the following conditions.
1. A trial is repeated until a success occurs.
2. The repeated trials are independent of each other.
3. The probability of success p is constant for each trial.
4. The random variable x represents the number of the trial in which the first success occurs.

The parameter of a geometric distribution is p.</td><td>x = the number of the trial in which the first success occurs
p = probability of success on a single trial
q = probability of failure on a single trial
$q = 1 - p$

The probability that the first success occurs on the xth trial is
$P(x) = p(q)^{x-1}$.</td></tr>
<tr><td>Poisson Distribution</td><td>The Poisson distribution is a discrete probability distribution that gives the probability of x occurrences of an event over a specified interval of time, area, or volume.

The parameter of a Poisson distribution is μ.</td><td>x = the number of occurrences in the given interval
μ = the average number of occurrences in a given time or space unit

The probability of x occurrences in an interval is
$P(x) = \frac{\mu^x e^{-\mu}}{x!}$.</td></tr>
</table>

4.3 Exercises

Help

LarsonTutor 4.3

Companion Web Site

Student Solutions Manual 4.3

Videos 4.3

Technology Manuals

Basic Skills and Concepts

Deciding on a Distribution In Exercises 1–6, decide which probability distribution—binomial, geometric, or Poisson—applies to the question. You do not need to answer the question. Instead, justify your choice.

1. ***Pilot's Test*** ◆ *Given:* The probability that a student passes the written test for a private pilot's license is 0.75. *Question:* What is the probability that a student will fail the test on the first attempt and pass it on the second attempt?

2. ***Wet Days*** ◆ *Given:* In Rapid City, South Dakota, the mean number of days with 0.01 inch or more precipitation for May is 12. *Question:* What is the probability that Rapid City has 18 days with 0.01 inch or more precipitation next May? *(Source: National Climatic Data Center)*

3. ***Oil Tankers*** ◆ *Given:* The mean number of oil tankers at a port city is 8 per day. The port has facilities to handle up to 12 oil tankers in a day. *Question:* What is the probability that too many tankers will arrive on a given day?

4. ***Exercise*** ◆ *Given:* Forty percent of adults in the U.S. exercise at least 30 minutes a week. In a survey of 120 randomly chosen adults, people were asked, "Do you exercise at least 30 minutes a week?" *Question:* What is the probability that exactly 50 of the people answer yes?

5. ***Cheaters*** ◆ *Given:* Of students ages 16 to 18 with A or B averages who plan to attend college after graduation, 80% cheated to get higher grades. Ten randomly chosen students with A or B averages who plan to attend college after graduation were asked, "Did you cheat to get higher grades?" *Question:* What is the probability that exactly two students answered no? *(Source: The Educational Testing Service)*

6. ***No Meat?*** ◆ *Given:* About 21% of Americans say they could not go one week without eating meat. You select at random 20 Americans. *Question:* What is the probability that the first person who says he or she cannot go one week without eating meat is the fifth person selected? *(Source: Reuters/Zogby)*

Using a Geometric Distribution to Find Probabilities In Exercises 7–10, find the indicated probabilities using the geometric distribution. If convenient, use technology.

7. ***Telephone Sales*** ◆ Assume the probability that you will make a sale on any given telephone call is 0.19. Find the probability that you (a) make your first sale on the fifth call, (b) make your first sale on the first, second, or third call, and (c) do not make a sale on the first three calls.

8. ***Free Throws*** ◆ Basketball player Shaquille O'Neal makes a free throw shot about 53.1% of the time. *(Source: CBS SportsLine.com)* Find the probability that (a) the first shot O'Neal makes is the second shot, (b) the first shot O'Neal makes is the first or second shot, and (c) O'Neal does not make two shots.

9. ***Auto Parts*** ◆ An auto parts seller finds that one in every 100 parts sold is defective. Find the probability that (a) the first defective part is the tenth part sold, (b) the first defective part is the first, second, or third part sold, and (c) none of the first 10 parts sold are defective.

10. ***Winning a Prize*** ◆ A cereal maker places a game piece in its cereal boxes. The probability of winning a prize in the game is one in four. Find the probability that you

(a) win your first prize with your fourth purchase.

(b) win your first prize with your first, second, or third purchase.

(c) do not win a prize with your first four purchases.

Using a Poisson Distribution to Find Probabilities In Exercises 11–14, find the indicated probabilities using the Poisson distribution. If convenient, use a Poisson probability table or technology tool to find the probability.

11. ***Business Failures*** ◆ The mean number of business failures per hour in the United States in a recent year was about 8. *(Source: The Wall Street Journal Almanac)*

Find the probability that

(a) exactly 4 businesses will fail in any given hour.

(b) at least 4 businesses will fail in any given hour.

(c) more than 4 businesses will fail in any given hour.

12. ***Typographical Errors*** ◆ A newspaper finds that the mean number of typographical errors per page is four. Find the probability that

(a) exactly three typographical errors will be found on a page.

(b) at most three typographical errors will be found on a page.

(c) more than three typographical errors will be found on a page.

13. ***Major Hurricanes*** ◆ A major hurricane is a hurricane with wind speeds of 111 miles per hour or greater. From 1900 to 1999, the mean number of major hurricanes to strike the U.S. mainland per year was about 0.6. *(Source: National Hurricane Center)*

Find the probability that in a given year

(a) exactly one major hurricane will strike the U.S. mainland.

(b) at most one major hurricane will strike the U.S. mainland.

(c) more than one major hurricane will strike the U.S. mainland.

14. ***Precipitation*** ◆ The mean number of days with 0.01 inch or more precipitation per month for Lewistown, Idaho, is about 8.7. Find the probability that in a given month (a) there are exactly 9 days with 0.01 inch or more precipitation, (b) there are at most 9 days with 0.01 inch or more precipitation, and (c) there are more than 9 days with 0.01 inch or more precipitation. *(Source: National Climatic Data Center)*

Extending the Basics

15. ***Approximating the Binomial Distribution*** ◆ A glass manufacturer finds that 1 in every 1000 glass items produced is warped.

(a) Use a binomial distribution to find the probability of finding five defective glass items in a random sample of 6500 glass items.

(b) The Poisson distribution can be used to approximate the binomial distribution for large values of n and small values of p. Repeat (a) using a Poisson distribution and compare the results.

16. ***Hypergeometric Distribution*** Binomial experiments require that any sampling be done with replacement because each trial must be independent of the others. The hypergeometric distribution also has two outcomes—success and failure. However, the sampling is done without replacement. Given a population of N items having k successes and $N - k$ failures, the probability of selecting a sample of size n that has x successes and $n - x$ failures is given by

$$P(x) = \frac{({}_kC_x)({}_{N-k}C_{n-x})}{{}_NC_n}.$$

In a shipment of 15 microchips, two are defective and 13 are not defective. A sample of three microchips is chosen at random. Find the probability that

(a) all three microchips are not defective.

(b) one microchip is defective and two are not defective.

(c) two microchips are defective and one is not defective.

Geometric Distribution: Mean and Variance In Exercises 17 and 18, use the fact that the mean of a geometric distribution is $\mu = 1/p$ and the variance is $\sigma^2 = q/p^2$.

17. ***Daily Lottery*** ◆ A daily number lottery chooses three balls numbered 0 to 9. The probability of winning the lottery is 1/1000. Let x be the number of times you play the lottery before winning the first time.

(a) Find the mean, variance, and standard deviation. Interpret the results.

(b) How many times would you expect to have to play the lottery before winning? Assume that it costs \$1 to play and winners are paid \$500. Would you expect to make or lose money playing this lottery? Explain.

18. ***Paycheck Errors*** ◆ A company assumes that 0.5% of the paychecks for a year were calculated incorrectly. The company has 200 employees and examines the payroll records from one month.

(a) Find the mean, variance, and standard deviation. Interpret the results.

(b) How many employee payroll records would you expect to examine before finding one with an error?

Poisson Distribution: Variance In Exercises 19 and 20, use the fact that the variance of a Poisson distribution is $\sigma^2 = \mu$.

19. ***Tiger Woods*** ◆ At one point in a recent year, the mean number of strokes per hole for golfer Tiger Woods was about 3.9. *(Source: PGATour.com)*

(a) Find the variance and standard deviation. Interpret the results.

(b) How likely is Woods to play an 18-hole round and have more than 72 strokes?

20. ***Snowfall*** ◆ The mean snowfall in January for Evansville, Indiana, is 4.0 inches. *(Source: National Climatic Data Center)*

(a) Find the variance and standard deviation. Interpret the results.

(b) Find the probability that the snowfall in January for Evansville, Indiana, will exceed seven inches.

TECHNOLOGY MINITAB EXCEL TI-83

Using Poisson Distributions as Queuing Models

Queuing means waiting in line to be served. There are many examples of queuing in everyday life: waiting at a traffic light, waiting in line at a grocery check-out counter, waiting for an elevator, being put on hold for a telephone call, and so on.

Poisson distributions are used to model and predict the number of people (calls, computer programs, vehicles) arriving at the line. In the following exercises, you are asked to use Poisson distributions to analyze the queues at a grocery store check-out counter.

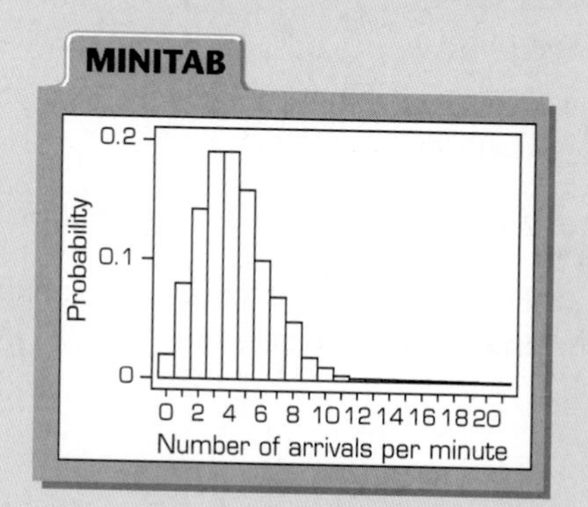

Exercises

In Exercises 1–6, consider a grocery store that can process a total of four customers at its check-out counters each minute.

1. Suppose that the mean number of customers who arrive at the check-out counters each minute is 4. Create a Poisson distribution with $\mu = 4$ for $x = 0$ to 20. Compare your results with the histogram shown at the upper right.

2. We used Minitab to generate 20 random numbers with a Poisson distribution for $\mu = 4$. Let the random number represent the number of arrivals at the check-out counter each minute for 20 minutes.

3 3 3 3 5 5 6 7 3 6

3 5 6 3 4 6 2 2 4 1

During each of the first four minutes, only three customers arrived. These customers could all be processed, so there were no customers waiting after four minutes.

(a) How many customers were waiting after 5 minutes? 6 minutes? 7 minutes? 8 minutes?

(b) Create a table that shows the number of customers waiting at the end of 1 through 20 minutes.

3. Generate a list of 20 random numbers with a Poisson distribution for $\mu = 4$. Create a table that shows the number of customers waiting at the end of 1 through 20 minutes.

4. Suppose that the mean increases to five arrivals per minute. If you can still only process four per minute, how many would you expect to be waiting in line after 20 minutes?

5. Simulate the setting in Exercise 4. Do this by generating a list of 20 random numbers with a Poisson distribution with $\mu = 5$. Then create a table that shows the number of customers waiting at the end of 20 minutes.

6. Suppose that the mean number of arrivals per minute is 5. What is the probability that 10 customers will arrive during the first minute?

7. Suppose that the mean number of arrivals per minute is 4.

(a) What is the probability that three, four, or five customers will arrive during the third minute?

(b) What is the probability that more than four customers will arrive during the first minute?

(c) What is the probability that more than four customers will arrive during each of the first four minutes?

Extended solutions are given in the *Technology Supplement.*
Technical instruction is provided for *Minitab*, *Excel*, and the *TI-83*.

4

Chapter Summary

What did you learn?

Review Exercises

Section 4.1

	Review Exercises
◆ How to distinguish between discrete random variables and continuous random variables	*1–4*
◆ How to determine if a distribution is a probability distribution	*5–10*
◆ How to construct a discrete probability distribution and its graph, and find the mean, variance, and standard deviation of a discrete probability distribution	*11–14*

$$\mu = \Sigma x P(x)$$
$$\sigma^2 = \Sigma(x - \mu)^2 P(x)$$
$$\sigma = \sqrt{\sigma^2}$$

◆ How to find the expected value of a discrete probability distribution — *15, 16*

Section 4.2

◆ How to determine if a probability experiment is a binomial experiment — *17, 18*

◆ How to find binomial probabilities using the binomial probability formula, a binomial probability table, and technology — *19–22*

$$P(x) = {}_nC_x p^x q^{n-x} = \frac{n!}{(n-x)!x!} p^x q^{n-x}$$

◆ How to construct a binomial distribution and its graph, and find the mean, variance, and standard deviation of a binomial probability distribution — *23–26*

$$\mu = np$$
$$\sigma^2 = npq$$
$$\sigma = \sqrt{npq}$$

Section 4.3

◆ How to find probabilities using the geometric distribution — *27, 28*

$$P(x) = pq^{x-1}$$

◆ How to find probabilities using the Poisson distribution — *29, 30*

$$P(x) = \frac{\mu^x e^{-\mu}}{x!}$$

STATISTICS

Uses and Abuses

Uses

There are countless occurrences of binomial probability distributions in business, science, engineering and many other fields.

For example, suppose you work for a marketing agency and are in charge of creating a television ad for Brand A toothpaste. The toothpaste manufacturer claims that 40% of toothpaste buyers prefer its brand. To check whether the manufacturer's claim is reasonable, your agency conducts a survey. Of 100 toothpaste buyers selected at random, you find that only 35 (or 35%) prefer Brand A. Could the manufacturer's claim still be true? What if your random sample of 100 found only 25 people (or 25%) who express a preference for Brand A? Would you still be justified in running the advertisement?

Knowing the characteristics of binomial probability distributions will help you answer this type of question. By the time you have completed this course, you will be able make educated decisions about the reasonableness of the manufacturer's claim.

Abuses

Interpreting the "Most Likely" Outcome A common misuse of binomial probability distributions is to think that the "most likely" outcome is the outcome that will occur most of the time. For instance, suppose you randomly choose a committee of four from a large population that is 50% women and 50% men. The most likely composition of the committee is that it will contain 2 men and 2 women. Although this is the most likely outcome, the probability that it will occur is only 0.375. There is a 0.5 chance that the committee will contain 1 man and 3 women or 3 men and 1 woman. So, if either of these outcomes occur, you should not assume that the selection was unusual or biased.

Exercises

In Exercises 1 and 2, suppose that the manufacturer's claim is true—40% of toothpaste buyers prefer Brand A toothpaste. Use the graph and technology to answer the questions.

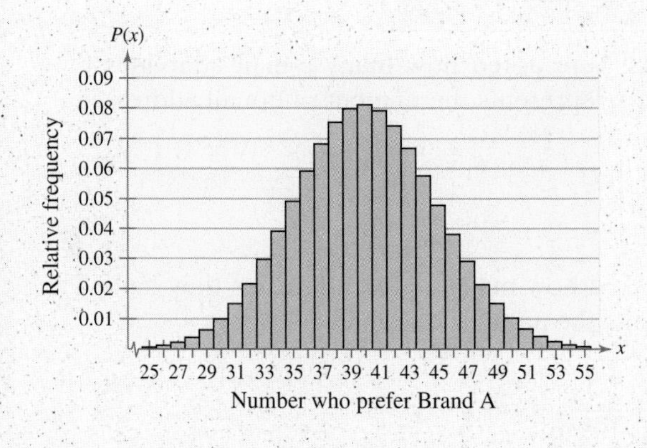

1. ***Interpreting the "Most Likely" Outcome*** In a random sample of 100, what is the most likely outcome? How likely is it?

2. ***Interpreting the "Most Likely" Outcome*** In a random sample of 100, what is the probability that between 35 and 45 people, inclusive, prefer Brand A? Explain your reasoning.

3. Suppose in a random sample of 100, you found 36 who prefer Brand A. Would the manufacturer's claim be believable? Explain your reasoning.

4. Suppose in a random sample of 100, you found 25 who prefer Brand A. Would the manufacturer's claim be believable? Explain your reasoning.

4 Review Exercises

Section 4.1

In Exercises 1–4, decide whether the random variable x is discrete or continuous.

1. x represents the number of pumps in use at a gas station.

2. x represents the weight of a truck at a weigh station.

3. x represents the amount of gas pumped at a gas station.

4. x represents the number of defects on a microchip.

In Exercises 5–10, decide whether the distribution is a probability distribution. If it is not, identify the property that is not satisfied.

5. The daily limit for catching bass at a certain lake is four. The random variable x represents the number of fish caught in a day.

x	0	1	2	3	4
$P(x)$	0.36	0.23	0.08	0.14	0.29

6. The vending machine at a workplace holds 48 bottles of iced tea. The random variable x represents the number of days until the iced tea is gone.

x	1	2	3	4	5	6
$P(x)$	0.05	0.15	0.40	0.25	0.10	0.05

7. A greeting card shop keeps records of customers' buying habits. The random variable x represents the number of cards sold to an individual customer in a shopping visit.

x	1	2	3	4	5	6	7
$P(x)$	0.68	0.14	0.08	0.05	0.02	0.02	0.01

8. The random variable x represents the number of classes in which a student is enrolled in a given semester at a university.

x	1	2	3	4	5	6	7	8
$P(x)$	$\frac{1}{80}$	$\frac{2}{75}$	$\frac{1}{10}$	$\frac{12}{25}$	$\frac{27}{20}$	$\frac{1}{5}$	$\frac{2}{25}$	$\frac{1}{120}$

9. In a survey, Internet users were asked how many e-mail addresses they have. The random variable x represents the number of e-mail addresses.

x	1	2	3
$P(x)$	0.26	0.31	0.43

10. In a survey, adults were asked how many e-mail addresses they have. The random variable x represents the number of e-mail addresses.

x	0	1	2	3
$P(x)$	0.11	0.20	0.28	0.39

In Exercises 11–14,

(a) use the frequency distribution table to construct a probability distribution.

(b) graph the probability distribution using a relative frequency histogram.

(c) find the mean, variance, and standard deviation of the probability distribution.

11. The number of pages in a section from a sample of statistics texts

Pages	Sections
2	3
3	12
4	72
5	115
6	169
7	120
8	83
9	48
10	22
11	6

12. The number of goals scored by a soccer team during a 32-game season

Goals	Games
0	7
1	8
2	10
3	3
4	3
5	1

13. A survey asked 200 households how many televisions they owned.

Televisions	Households
0	3
1	38
2	83
3	52
4	18
5	5
6	1

14. A television station sells advertising in 15-, 30-, 60-, 90-, and 120-second blocks. The distribution for one 24-hour day is given.

Length (in seconds)	Number
15	76
30	445
60	30
90	3
120	12

In Exercises 15 and 16, find the expected value of the random variable.

15. A person has shares of eight different stocks. The random variable x represents the number of stocks showing a loss on a selected day.

x	0	1	2	3	4	5	6	7	8
$P(x)$	0.02	0.11	0.18	0.32	0.15	0.09	0.05	0.05	0.03

16. A local pub has a chicken wing special on Tuesdays. The pub owners purchase wings in cases of 300. The random variable x represents the number of cases used during the special.

x	1	2	3	4
$P(x)$	$\frac{1}{9}$	$\frac{1}{3}$	$\frac{1}{2}$	$\frac{1}{18}$

Section 4.2

In Exercises 17 and 18, determine whether the experiment is a binomial experiment. If it is not, identify the property that is not satisfied. If it is, list the values of n, p, and q, and the values that x can assume.

17. Bags of plain M&M's contain 30% brown candies. One candy is selected from each of 12 bags. The random variable represents the number of brown candies selected. *(Source: Mars, Inc.)*

18. A fair coin is tossed repeatedly until 15 heads are obtained. The random variable x counts the number of tosses.

In Exercises 19–22, find the indicated probabilities.

19. One in four adults is currently on a diet. In a random sample of eight adults, what is the probability that the number currently on a diet is (a) exactly three, (b) at least three, and (c) more than three? *(Source: Wirthlin Worldwide)*

20. Three in five adults in the U.S. own an answering machine. In a random sample of 12 adults, what is the probability that the number owning answering machines is (a) exactly nine, (b) at least nine, and (c) more than nine? *(Source: Wirthlin Worldwide)*

21. Forty-three percent of adults in the U.S. receive fewer than five phone calls a day. In a random sample of seven adults, what is the probability that the number receiving fewer than five calls a day is (a) exactly three, (b) at least three, and (c) more than three? *(Source: Wirthlin Worldwide)*

22. In a typical day, 22% of people in the United States with Internet access go online to get news. In a random sample of five people in the United States with Internet access, what is the probability that the number going online to get news is (a) exactly two, (b) at least two, and (c) more than two? *(Source: Pew Internet & American Life Project)*

In Exercises 23–26,

(a) construct a binomial distribution.

(b) graph the binomial distribution using a relative frequency histogram.

(c) find the mean, variance, and standard deviation of the probability distribution.

23. Sixty-three percent of adults in the United States rent videotapes at least once a month. Consider a random sample of five Americans who are asked if they rent at least one videotape a month. *(Source: TELENATION/Market Facts, Inc.)*

24. Fifty-seven percent of families say that their children have an influence on their vacation destinations. Consider a random sample of six families who are asked if their children have an influence on their vacation destinations. *(Source: YP&B/Yankelovich 1997 Travel Monitor)*

25. Over a period of three months, 67% of San Franciscans went out to the movies. Consider a random sample of four San Franciscans who are asked if they went to the movies in those three months. *(Source: Scarborough Research)*

26. In a typical day, 17% of people in the United States with Internet access go online to check the weather. Consider a random sample of five people in the United States with Internet access who are asked if they check the weather when they go online. *(Source: Pew Internet & American Life Project)*

Section 4.3

In Exercises 27 and 28, find the indicated probabilities using the geometric distribution. If convenient, use technology to find the probabilities.

27. During a promotional contest, a soft drink company places winning caps on one of every six bottles. If you purchase one bottle a day, find the probability that you find your first winning cap (a) on the fourth day, (b) within four days, and (c) sometime after three days.

28. In a recent year, Mark McGwire hit 70 home runs in the 155 games he played. Assume that his home run production stayed at that level the following season. What is the probability that he would hit his first home run (a) on the first game of the season, (b) on the second game of the season, (c) on the first or second game of the season, and (d) within the first three games of the season? *(Source: Home Run Record)*

In Exercises 29 and 30, find the indicated probabilities using the Poisson distribution. If convenient, use a Poisson probability table or technology to find the probabilities.

29. During a 36-year period, lightning killed 3239 people in the United States. Assume that this rate holds true today and is constant throughout the year. Find the probability that tomorrow

(a) no one in the United States will be struck and killed by lightning.

(b) one person will be struck and killed.

(c) more than one person will be struck and killed. *(Source: National Oceanic and Atmospheric Administration)*

30. It is estimated that sharks kill 10 people each year worldwide. Find the probability that at least three people are killed by sharks this year

(a) assuming that this rate is true.

(b) if the rate is actually five people a year.

(c) if the rate is actually fifteen people a year. *(Source: International Shark Attack File)*

PUTTING IT ALL TOGETHER

Real Statistics Real Decisions

The Centers for Disease Control and Prevention (CDC) is required by law to publish a report on assisted reproductive technologies (ART). ART includes all fertility treatments in which both the egg and the sperm are used. These procedures generally involve the removal of eggs from a woman's ovaries, combining them with sperm in the laboratory, and returning them to the woman's body or giving them to another woman.

You are helping to prepare the CDC report and select at random 10 ART cycles for a special review. None of the cycles resulted in a clinical pregnancy. Your manager feels it is impossible to select at random 10 ART cycles that did not result in a clinical pregnancy. Use the information provided at the right and your knowledge of statistics to determine if your manager is correct.

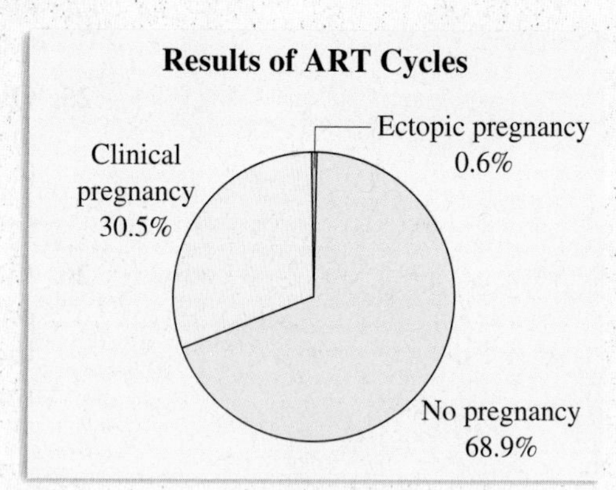

Source: Centers for Disease Control and Prevention

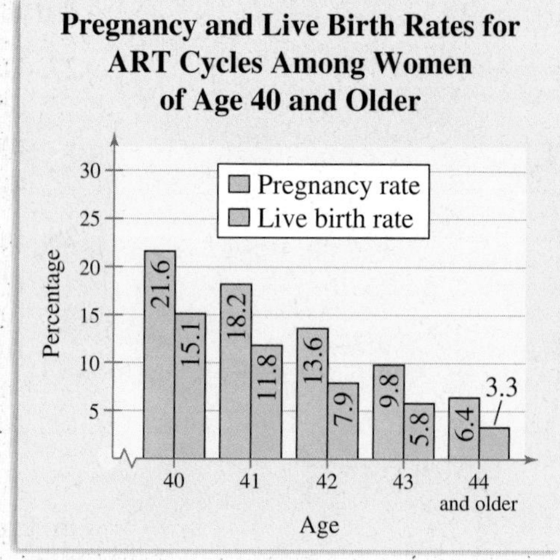

Source: Centers for Disease Control and Prevention

Exercises

1. ***How Would You Do It?***

   (a) How would you determine if your manager's view is correct, that it is impossible to select at random 10 ART cycles that did not result in a clinical pregnancy?

   (b) What probability distribution do you think best describes the situation? Do you think the distribution of the number of clinical pregnancies is discrete or continuous? Why?

2. ***Answering the Question***

   Write an explanation that answers the question, "Is it possible to select at random 10 ART cycles that did not result in a clinical pregnancy?" Include in your explanation the appropriate probability distribution and your calculation of the probability of no clinical pregnancies in 10 ART cycles.

3. ***Suspicious Samples?***

   Which of the following samples would you consider suspicious if someone told you that the sample was selected at random? Would you believe that the samples were selected at random? Why or why not?

   (a) Selecting at random 10 ART cycles among women of age 40, eight of which resulted in clinical pregnancies.

   (b) Selecting at random 10 ART cycles among women of age 41, none of which resulted in clinical pregnancies.

4 Chapter Quiz

Take this quiz as you would take a quiz in class. After you are done, check your work against the answers given in the back of the book.

1. Decide if the random variable, x, is discrete or continuous. Explain your reasoning.

(a) x represents the number of times Yellowstone Park's Old Faithful geyser erupts in one day.

(b) x represents the amount of sugar (in pounds) eaten each day in the United States.

2. The following table lists the number of U.S. mainland hurricane strikes (from 1900 to 1996) for various intensities according to the Saffir-Simpson Hurricane Scale. *(Source: National Hurricane Center)*

Intensity	Number of Hurricanes
1	57
2	37
3	47
4	15
5	2

(a) Construct a probability distribution of the data.

(b) Graph the discrete probability distribution using a relative frequency histogram.

(c) Find the mean, variance, and standard deviation of the probability distribution and interpret the results.

(d) Find the probability that a hurricane selected at random for further study has an intensity of at least 4.

3. A surgical technique is performed on eight patients. You are told there is an 80% chance of success.

(a) Construct a binomial distribution.

(b) Graph the binomial distribution using a relative frequency histogram.

(c) Find the mean, variance, and standard deviation of the probability distribution and interpret the results.

(d) Find the probability that the surgery is successful for exactly six of the patients.

(e) Find the probability that the surgery is successful for fewer than six patients.

4. A newspaper finds that the mean number of typographical errors per page is five. Find the probability that

(a) exactly five typographical errors will be found on a page.

(b) fewer than five typographical errors will be found on a page.

(c) no typographical errors will be found on a page.

Where You've Been

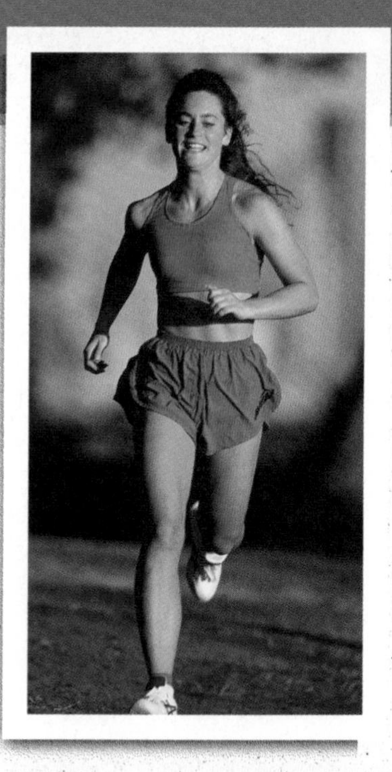

In Chapters 1 through 4, you learned how to collect and describe data, find the probability of an event, and analyze discrete probability distributions. You also learned that if a sample is used to make inferences about a population, then it is critical that the sample not be biased. Suppose, for instance, that you wanted to measure the serum cholesterol levels of adults in the United States. How would you organize the study? When the National Center for Health Statistics performed this study, it used random sampling and then classified the results according to the gender, ethnic background, and age of the participants. One conclusion from the study was that women's cholesterol levels tended to increase throughout their lives, whereas men's increased to age 65, and then decreased.

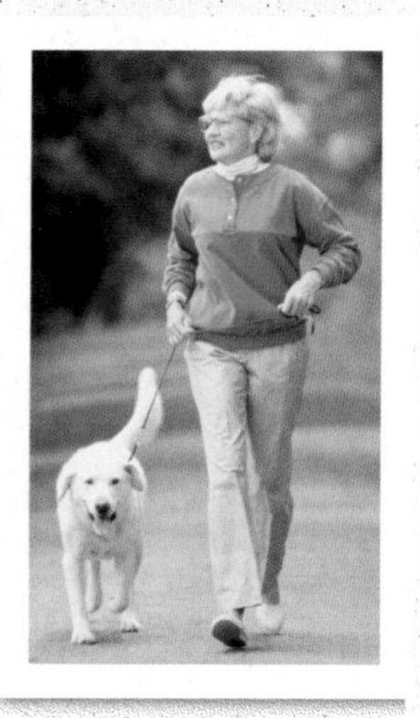

In 2000, the National Center for Health Statistics, located in Hyattsville, Maryland, began a 10-year program called *Healthy People 2010* to promote health through changes in people's lifestyles. While it is too early to analyze the results of this program, the results of a similar program that started in 1990, *Healthy People 2000*, are available. During the course of the program, some of the goals were met. For instance, heart disease and stroke death rates were down. Other goals were not met. For instance, although more adults were exercising, a quarter of all adults were still engaged in no physical activity.

Normal Probability Distributions

CHAPTER 5

Where You're Going

In Chapter 5, you will learn how to recognize normal (bell-shaped) distributions and how to use their properties in real-life applications. Suppose that you worked for the U.S. National Center for Health Statistics and were collecting data about various physical traits of people in the U.S. Which of the following would you expect to have bell-shaped, symmetric distributions: height, weight, cholesterol level, age, blood pressure, shoe size, reaction times, lung capacity? Of these, all except weight and age have distributions that are approximately normal. For instance, the four graphs below show the height and weight distributions for men and women in the United States aged 20 to 29. Notice that the height distributions are bell shaped, but the weight distributions are skewed right.

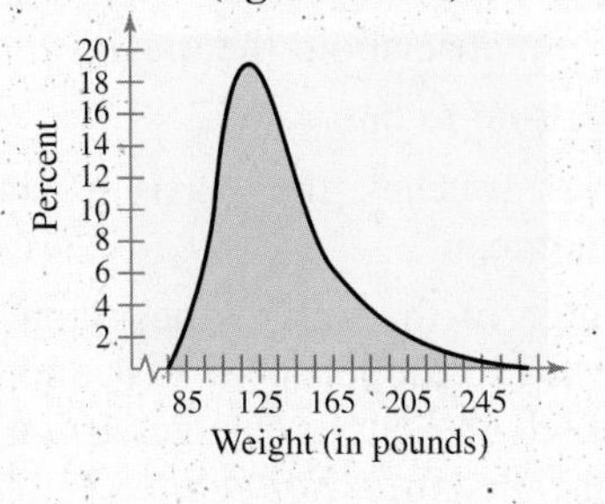

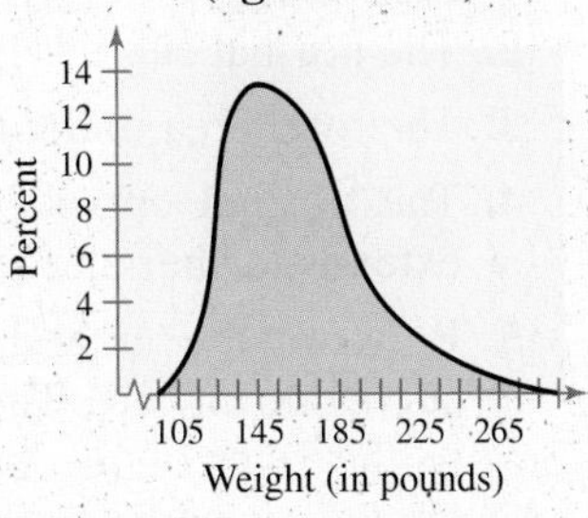

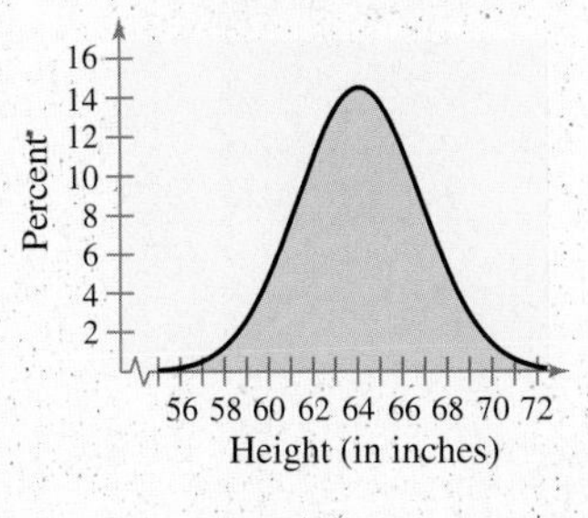

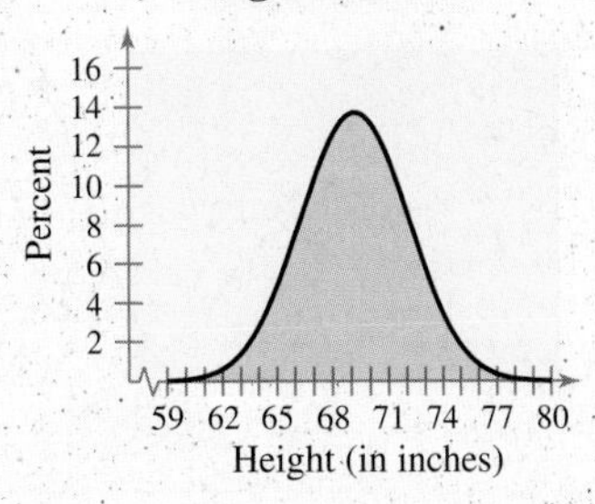

5.1 Introduction to Normal Distributions

Properties of a Normal Distribution • Normal Curves and Probability

What You Should Learn

- How to interpret graphs of normal probability distributions
- How to estimate areas under a normal curve and use them to estimate probabilities for random variables with normal distributions

Properties of a Normal Distribution

In Section 4.1, you learned that a **continuous random variable** has an infinite number of possible values that can be represented by an interval on the number line. Its probability distribution is called a **continuous probability distribution.** In this chapter, you will study the most important continuous probability distribution in statistics—the normal distribution. Normal distributions can be used to model many sets of measurements in nature, industry, and business. For instance, the systolic blood pressure of humans, the lifetime of television sets, and even housing costs are all normally distributed random variables.

GUIDELINES

Properties of a Normal Distribution

A **normal distribution** is a continuous probability distribution for a random variable x. The graph of a normal distribution is called the **normal curve.** A normal distribution has the following properties.

1. The mean, median, and mode are equal.
2. The normal curve is bell shaped and is symmetric about the mean.
3. The total area under the normal curve is equal to one.
4. The normal curve approaches, but never touches, the x-axis as it extends farther and farther away from the mean.
5. Between $\mu - \sigma$ and $\mu + \sigma$ (in the center of the curve) the graph curves downward. The graph curves upward to the left of $\mu - \sigma$ and to the right of $\mu + \sigma$. The points at which the curve changes from curving upward to curving downward are called *inflection points.*

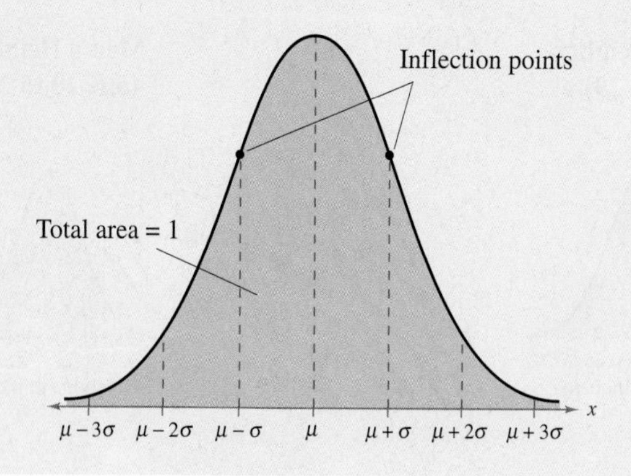

Insight

Because e and π are constants, a normal curve depends completely on μ and σ.

If x is a continuous random variable having a normal distribution with mean μ and standard deviation σ, you can graph a normal curve using the equation

$$y = \frac{1}{\sigma\sqrt{2\pi}} e^{-(x-\mu)^2/2\sigma^2}. \qquad e \approx 2.718 \text{ and } \pi \approx 3.14$$

A normal distribution can have any mean and any positive standard deviation. These two parameters, μ and σ, completely determine the shape of the normal curve. The mean gives the location of the line of symmetry and the standard deviation describes how much the data are spread out.

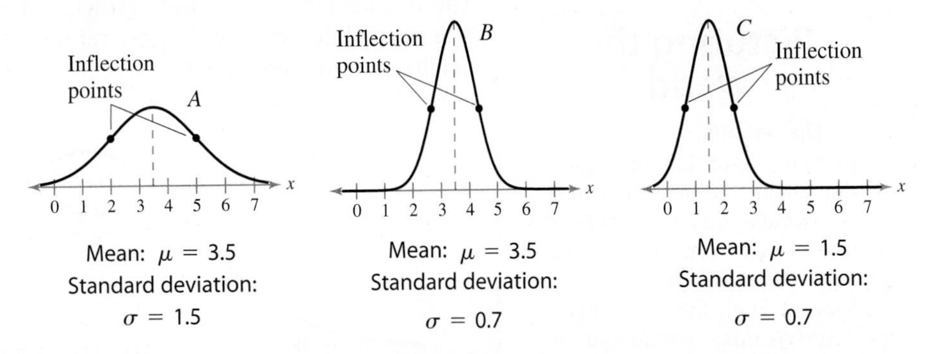

Notice that curve A and curve B above have the same mean, and curve B and curve C have the same standard deviation.

EXAMPLE  1

Understanding Mean and Standard Deviation

1. Which normal curve has a greater mean?
2. Which normal curve has a greater standard deviation?

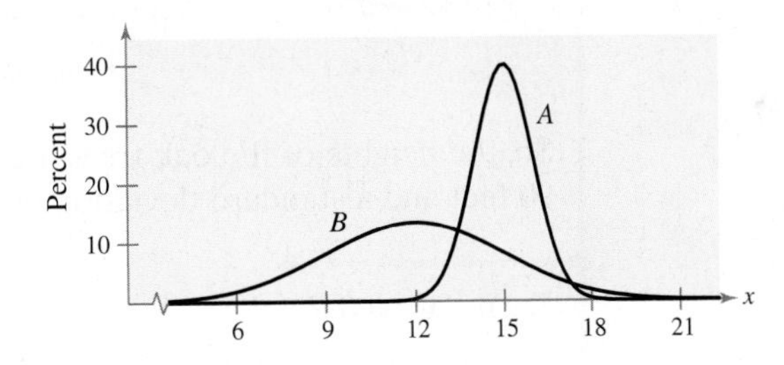

SOLUTION

1. The line of symmetry of curve A occurs at $x = 15$. The line of symmetry of curve B occurs at $x = 12$. So, curve A has a greater mean.
2. Curve B is more spread out than curve A, so curve B has a greater standard deviation.

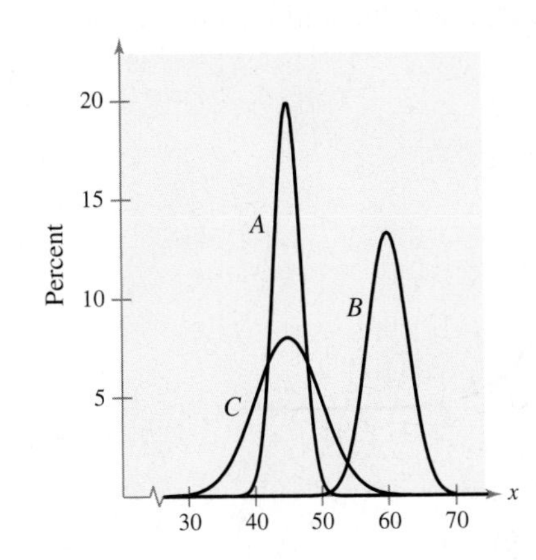

Try It Yourself 1

Consider the normal curves shown at the left. Which normal curve has the greatest mean? Which normal curve has the greatest standard deviation? Justify your answers.

a. Find the location of the *line of symmetry* of each curve. Make a conclusion about which mean is greatest.

b. Determine which normal curve is *more spread out.* Make a conclusion about which standard deviation is greatest.

Answer: Page A35

Picturing the World

The amount of dissolved oxygen is important in judging the quality of stream water. Acceptable dissolved oxygen levels range from 5 mg/L to 12 mg/L. Students from Strong Vincent High School in Erie, Pennsylvania, conducted a study of dissolved oxygen at Cascade Creek. The normal curve shows the students' results.

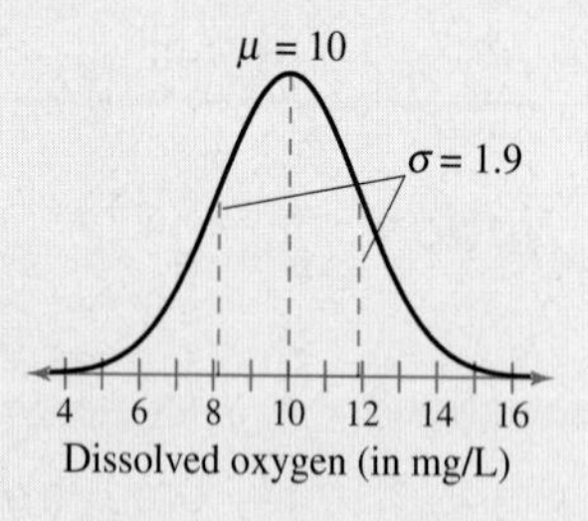

Are these dissolved oxygen levels acceptable? If not, are they too high or too low?

EXAMPLE 

Interpreting Graphs of Normal Distributions

The heights (in feet) of fully grown white oak trees are normally distributed. The normal curve shown below represents this distribution. What is the mean height of a fully grown white oak tree? Estimate the standard deviation of this normal distribution.

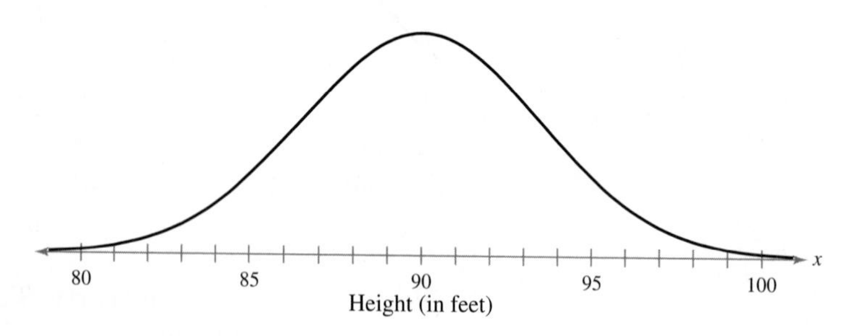

SOLUTION

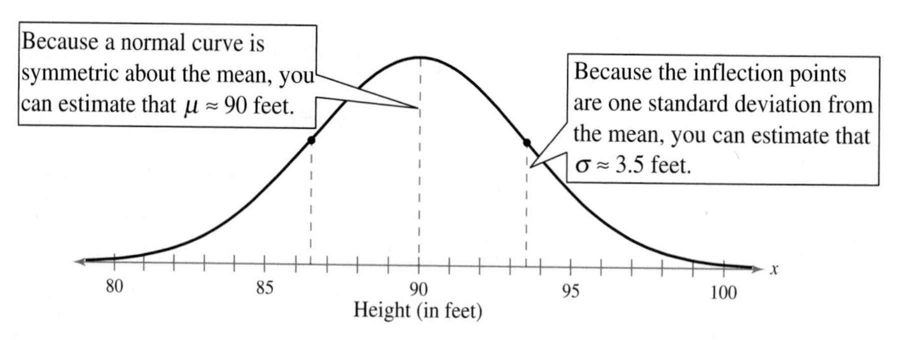

So, the heights of the oak trees are normally distributed with a mean of about 90 feet and a standard deviation of about 3.5 feet.

Try It Yourself 2

The diameters of fully grown white oak trees are normally distributed. The normal curve shown below represents this distribution. What is the mean diameter of a fully grown white oak tree? Estimate the standard deviation of this normal distribution.

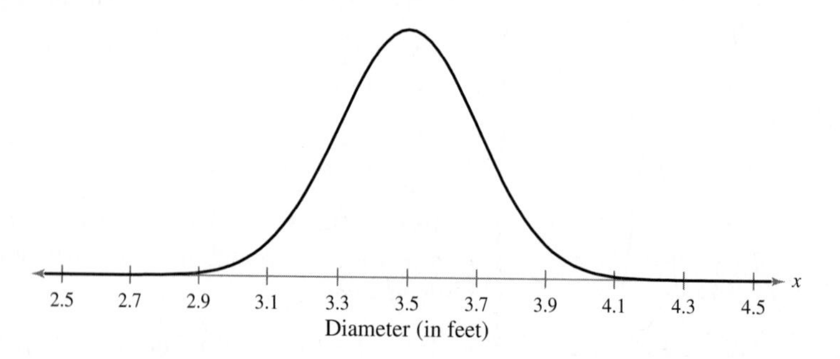

a. Find the *line of symmetry* and identify the mean.
b. Estimate the *inflection points* and identify the standard deviation.

Answer: Page A35

Normal Curves and Probability

The total area under a probability curve is equal to 1. The *area of a region under a probability curve* is equal to the probability that the random variable will have a value in the corresponding interval. In this chapter, you will learn several ways to find areas under normal curves. You have already studied one of these ways in Section 2.4—the *Empirical Rule.*

Insight

Even though a normal curve extends infinitely to the left and to the right, the area under every normal curve is equal to one.

Empirical Rule

In a normal distribution with mean μ and standard deviation σ, you can approximate areas under the normal curve as follows.

1. About 68% of the area lies between $\mu - \sigma$ and $\mu + \sigma$.
2. About 95% of the area lies between $\mu - 2\sigma$ and $\mu + 2\sigma$.
3. About 99.7% of the area lies between $\mu - 3\sigma$ and $\mu + 3\sigma$.

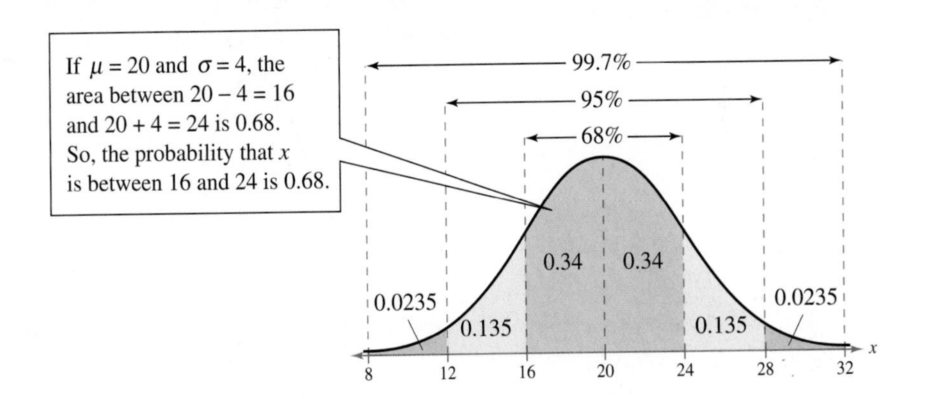

EXAMPLE 3

Estimating a Probability for a Normal Curve

Adult IQ scores are normally distributed with $\mu = 100$ and $\sigma = 15$. Estimate the probability that a randomly chosen adult has an IQ between 70 and 115.

SOLUTION A score of 70 is two standard deviations below the mean, and a score of 115 is one standard deviation above the mean. Using the Empirical Rule, the area under the normal curve between these two values is

$$\text{Area} = 0.135 + 0.34 + 0.34$$
$$= 0.815.$$

So, the probability the adult has an IQ between 70 and 115 is about 0.815.

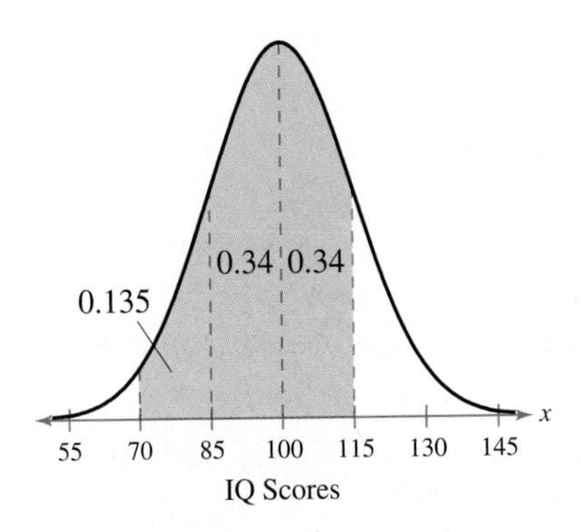

The probability that a randomly chosen adult has an IQ between 70 and 115 is

$$P(70 < x < 115) = 0.815.$$

Try It Yourself 3

Estimate the probability that a randomly chosen adult has an IQ between 85 and 145.

a. *How many standard deviations* are the scores from the mean?
b. Use the *Empirical Rule* to find the area.

Answer: Page A35

5.1 Exercises

LarsonTutor 5.1

Companion Web Site

Student Solutions Manual 5.1

Videos 5.1

Technology Manuals

Basic Skills and Concepts

1. Find three real-life examples of a continuous variable. Which do you think may be normally distributed? Why?

2. Determine whether the following statement is true or false. If it is false, explain why.

The total area under the normal curve is 0.997.

3. Draw two normal curves that have the same mean but different standard deviations. Describe the similarities and differences.

4. Draw two normal curves that have different means but the same standard deviations. Describe the similarities and differences.

Graphical Analysis In Exercises 5–10, decide whether the graph could represent a variable with a normal distribution. Explain your reasoning.

5.

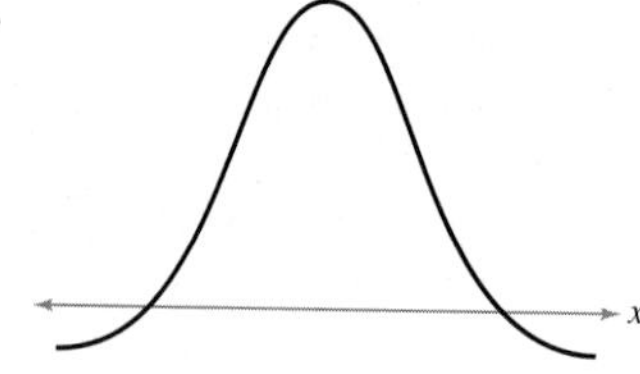

6.

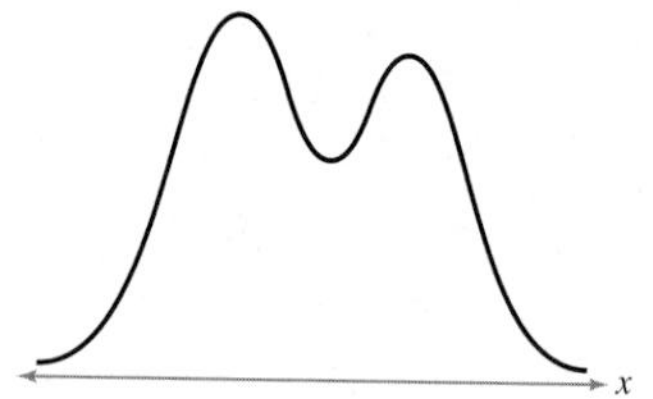

7.

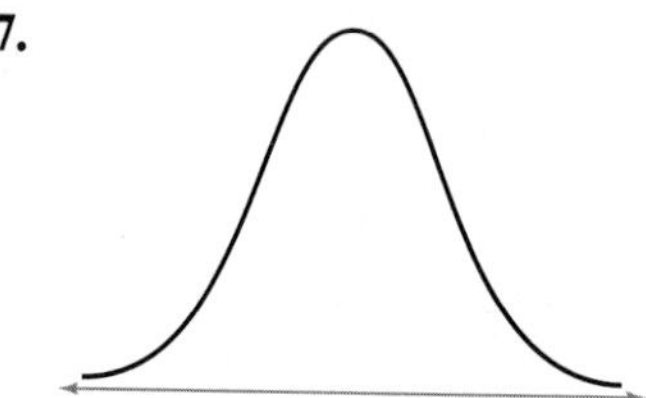

8.

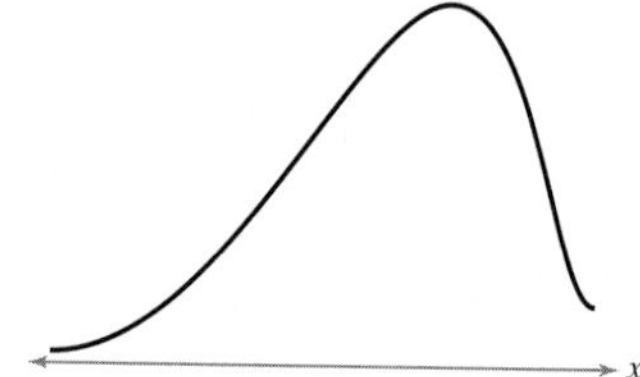

9.

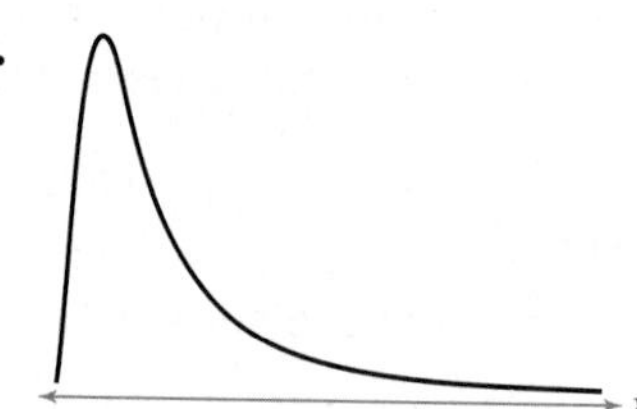

10. x

11. Name the two parameters that are necessary to determine probabilities for a particular normal distribution curve.

12. If a random variable is normally distributed, approximately 95% of the population values will lie between what two values?

In Exercises 13 and 14, you are given the mean and standard deviation of a continuous random variable that is normally distributed. Find an interval that contains about 95% of the distribution.

13. $\mu = 15, \sigma = 3$

14. $\mu = 19, \sigma = 2.5$

In Exercises 15 and 16, x is a random variable with a normal distribution. Estimate the probability that x falls in the indicated interval.

15. $\mu = 7, \sigma = 1.75$, estimate $P(5.25 < x < 8.75)$

16. $\mu = 20, \sigma = 5.4$, estimate $P(9.2 < x < 30.8)$

17. ***Graphical Analysis*** ◆ A company manufactures engine parts. The diameters of the engine parts are normally distributed with a mean of 3 inches and a standard deviation of 0.02 inch. Which of the following normal curves represents this distribution? Justify your conclusion.

(a)

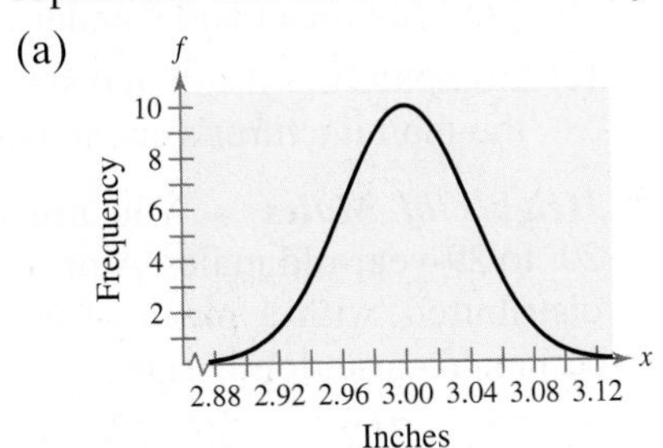

(b)

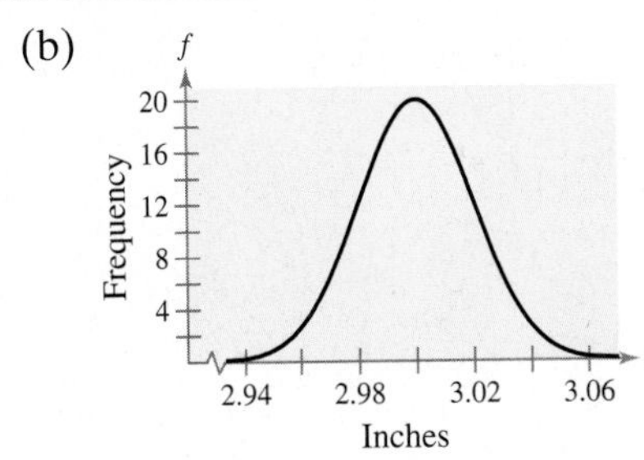

18. ***Graphical Analysis*** ◆ An instruction manual claims that the mean assembly time for a product is 4.2 hours and the standard deviation is 0.25 hour. Which normal curve represents this distribution? Justify your conclusion.

(a)

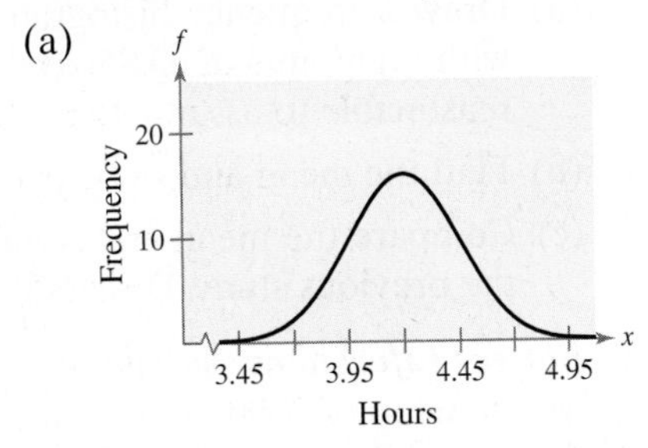

(b)

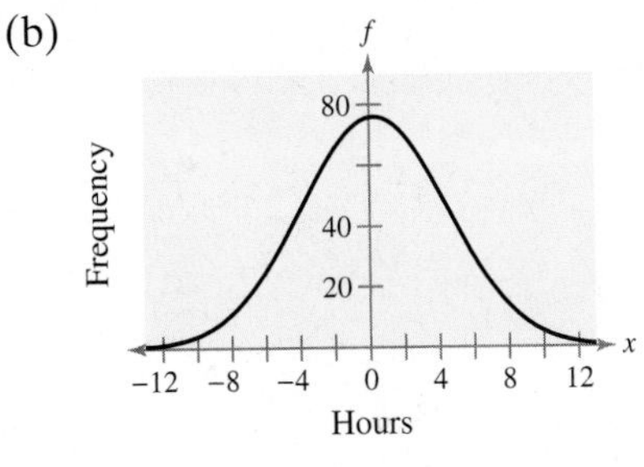

19. ***Engine Parts*** ◆ Refer to the correct normal distribution in Exercise 17. ($\mu = 3$ inches, $\sigma = 0.02$ inch)

(a) Estimate the probability that an engine part diameter is between 2.94 inches and 3.06 inches.

(b) Determine an interval in which about 95% of the engine part diameters will fall.

20. ***Assembly Times*** ◆ Refer to the correct normal distribution in Exercise 18. ($\mu = 4.2$ hours, $\sigma = 0.25$ hour)

(a) Estimate the probability that an assembly time is between 3.95 hours and 4.45 hours.

(b) Determine an interval in which about 95% of the assembly times will fall.

21. ***Cereal Boxes*** ◆ The contents of a cereal box are normally distributed, with a mean weight of 20 ounces and a standard deviation of 0.07 ounce. Determine an interval of values into which (a) about 95% of the cereal box weights will fall and (b) about 99.7% of the cereal box weights will fall.

22. ***Bags of Cookies*** ◆ The weights of bags of cookies are normally distributed, with a mean of 15 ounces and a standard deviation of 0.085 ounce. Determine an interval of values into which (a) about 95% of the bags of cookies will fall and (b) about 68% of the bags of cookies will fall.

DATA **23.** ***Manufacturer Claims*** ◆ You work for a consumer watchdog publication and are testing the advertising claims of a light bulb manufacturer. The manufacturer claims that the life span of the bulb is normally distributed, with a mean of 2000 hours and a standard deviation of 250 hours. You test 20 light bulbs and get the following life spans.

2210, 2406, 2267, 1930, 2005, 2502, 1106, 2140, 1949, 1921,
2217, 2121, 2004, 1397, 1659, 1577, 2840, 1728, 1209, 1639

(a) Draw a frequency histogram to display these data. Use five classes. Is it reasonable to assume that the life span is normally distributed? Why?

(b) Find the mean and standard deviation of your sample.

(c) Compare the mean and standard deviation of your sample to those in the manufacturer's claim. Discuss the differences.

DATA **24.** ***Heights of Males*** ◆ You are performing a study about the height of 20- to 29-year-old males. A previous study found the height to be normally distributed, with a mean of 69.2 inches and a standard deviation of 2.9 inches. You randomly sample 30 males and find their heights to be

72.1, 71.2, 67.9, 67.3, 69.5, 68.6, 68.8, 69.4, 73.5, 67.1,
69.2, 75.7, 71.1, 69.6, 70.7, 66.9, 71.4, 62.9, 69.2, 64.9,
68.2, 65.2, 69.7, 72.2, 67.5, 66.6, 66.5, 64.2, 65.4, 70.0.
(Source: National Center for Health Statistics)

(a) Draw a frequency histogram to display these data. Use seven classes with midpoints of 63.85, 65.85, 67.85, 69.85, 71.85, 73.85, and 75.85. Is it reasonable to assume that the heights are normally distributed? Why?

(b) Find the mean and standard deviation of your sample.

(c) Compare the mean and standard deviation of your sample to those in the previous study. Discuss the differences.

25. ***Battery Life Spans*** ◆ The life span of a battery is normally distributed, with a mean of 2000 hours and a standard deviation of 30 hours. Estimate the probability that a battery's life span is between 1970 and 2030 hours.

26. ***Nuts*** ◆ Assume the mean annual consumption of peanuts is normally distributed, with $\mu = 5.9$ pounds per person and $\sigma = 1.8$ pounds. Estimate the probability that a person consumes between 2.3 pounds and 9.5 pounds in a year.

27. ***Tire Life Spans*** ◆ The life span of a tire is normally distributed, with a mean of 30,000 miles and a standard deviation of 2000 miles. Estimate the probability that a tire's life span is between 30,000 and 34,000 miles.

28. ***Cheese*** ◆ Assume the mean annual consumption of cheese is normally distributed, with $\mu = 28.4$ pounds per person and $\sigma = 9.4$ pounds. Estimate the probability that a person consumes between 19.0 pounds and 28.4 pounds in a year.

29. ***Computer Lab Schedule*** ◆ The time per week a student uses a lab computer is normally distributed, with a mean of 6.2 hours and a standard deviation of 0.9 hour. You are planning the schedule for the computer lab. Of 2000 students, estimate the number of students who will use a lab computer for the given number of hours.

(a) Less than 5.3 hours

(b) Between 5.3 hours and 7.1 hours

(c) More than 7.1 hours

30. ***Health Club Schedule*** ◆ The time per workout an exerciser uses a stair climber is normally distributed, with a mean of 20 minutes and a standard deviation of 5 minutes. You are planning the schedule for a health club. Of 500 members who use a stair climber, estimate the number of people who will use a stair climber for the given number of minutes.

(a) Less than 10 minutes

(b) Between 10 and 15 minutes

(c) Between 15 and 20 minutes

Extending the Basics

31. ***Writing*** ◆ Draw a normal curve with a mean of 60 and a standard deviation of 12. Describe how you constructed the curve and discuss its features.

32. ***Writing*** ◆ Draw a normal curve with a mean of 450 and a standard deviation of 50. Describe how you constructed the curve and discuss its features.

33. ***Uniform Distribution*** ◆ Another continuous distribution is the **uniform distribution.** An example is $f(x) = 1$ for $0 \le x \le 1$. The mean of this distribution for this example is 0.5 and the standard deviation is approximately 0.29. The graph of this distribution for this example is a square with the height and width both equal to 1 unit. In general, the density function for a uniform distribution on the interval from $x = a$ to $x = b$ is given by

$$f(x) = \frac{1}{b - a}.$$

The mean is

$$\frac{a + b}{2}$$

and the variance is

$$\frac{(b - a)^2}{12}.$$

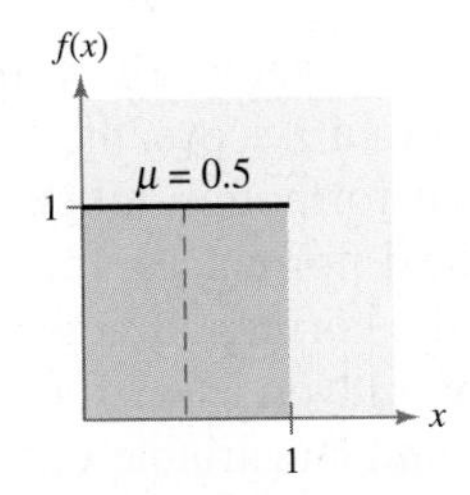

(a) Verify that the area under the curve is 1.

(b) Find the probability that x falls between 0.25 and 0.5.

(c) Find the probability that x falls between 0.3 and 0.7.

34. ***Uniform Distribution*** ◆ Consider the uniform density function $f(x) = 0.1$ for $10 \le x \le 20$. The mean of this distribution is 15 and the standard deviation is about 2.89.

(a) Draw a graph of the distribution and show that the area under the curve is 1.

(b) Find the probability that x falls between 12 and 15.

(c) Find the probability that x falls between 13 and 18.

5.2 The Standard Normal Distribution

What You Should Learn

- How to find areas under the standard normal curve

The Standard Normal Distribution

Insight

Because every normal distribution can be transformed to the standard normal distribution, you can use z-scores and the standard normal curve to find areas (and therefore probability) under any normal curve.

Study Tip

It is important that you know the difference between x and z. The random variable x is sometimes called a raw score and represents values in a *nonstandard* normal distribution, while z represents values in the *standard* normal distribution.

The Standard Normal Distribution

There are infinitely many normal distributions, each with its own mean and standard deviation. The normal distribution with a mean of 0 and a standard deviation of 1 is called **the standard normal distribution.** The horizontal scale of the graph of the standard normal distribution corresponds to z-scores. In Section 2.5, you learned that a z-score is a measure of position that indicates the number of standard deviations a value lies from the mean. Recall that you can transform an x-value to a z-score using the formula

$$z = \frac{\text{value} - \text{mean}}{\text{standard deviation}} = \frac{x - \mu}{\sigma}.$$

DEFINITION

The **standard normal distribution** is a normal distribution with a mean of 0 and a standard deviation of 1.

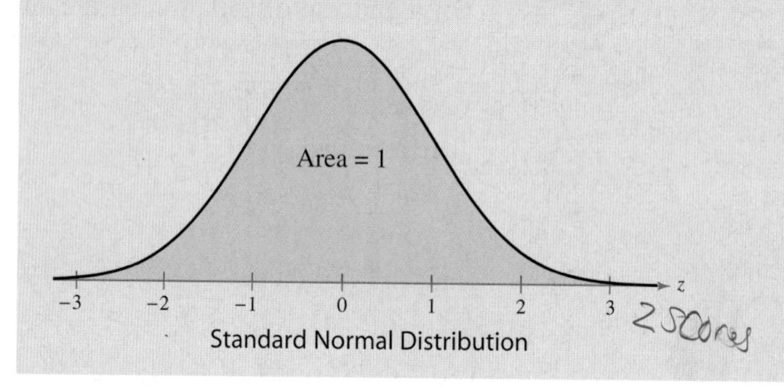

Standard Normal Distribution

If each data value of a normally distributed random variable x is transformed into a z-score, the result will be the standard normal distribution. When this transformation takes place, the area that falls in the interval under the nonstandard normal curve is the *same* as that under the standard normal curve within the corresponding z-boundaries.

In Section 5.1, you learned to approximate areas under a normal curve when values of the random variable x corresponded to $-3, -2, -1, 0, 1, 2,$ or 3 standard deviations from the mean. In this section, you will learn to calculate areas corresponding to other x-values. After you use the formula given above to transform an x-value to a z-score, you can use the Standard Normal Table in Appendix B. The table lists the cumulative area under the standard normal curve to the left of z for z-scores from -3.49 to 3.49. As you examine the table, notice the following.

Properties of the Standard Normal Distribution

1. The cumulative area is close to 0 for z-scores close to -3.49.
2. The cumulative area increases as the z-scores increase.
3. The cumulative area for $z = 0$ is 0.5000.
4. The cumulative area is close to 1 for z-scores close to $z = 3.49$.

EXAMPLE 1

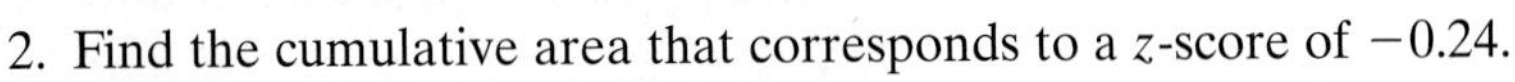

Using the Standard Normal Table

1. Find the cumulative area that corresponds to a z-score of 1.15.
2. Find the cumulative area that corresponds to a z-score of -0.24.

SOLUTION

Area = 0.8749

1. Find the area that corresponds to $z = 1.15$ by finding 1.1 in the left column and then moving across the row to the column under 0.05. The number in that row and column is 0.8749. So, the area to the left of $z = 1.15$ is 0.8749.

z	.00	.01	.02	.03	.04	.05	.06
0.0	.5000	.5040	.5080	.5120	.5160	.5199	.5239
0.1	.5398	.5438	.5478	.5517	.5557	.5596	.5636
0.2	.5793	.5832	.5871	.5910	.5948	.5987	.6026
0.9	.8159	.8186	.8212	.8238	.8264	.8289	.8315
1.0	.8413	.8438	.8461	.8485	.8508	.8531	.8554
1.1	.8643	.8665	.8686	.8708	.8729	.8749	.8770
1.2	.8849	.8869	.8888	.8907	.8925	.8944	.8962
1.3	.9032	.9049	.9066	.9082	.9099	.9115	.9131
1.4	.9192	.9207	.9222	.9236	.9251	.9265	.9279

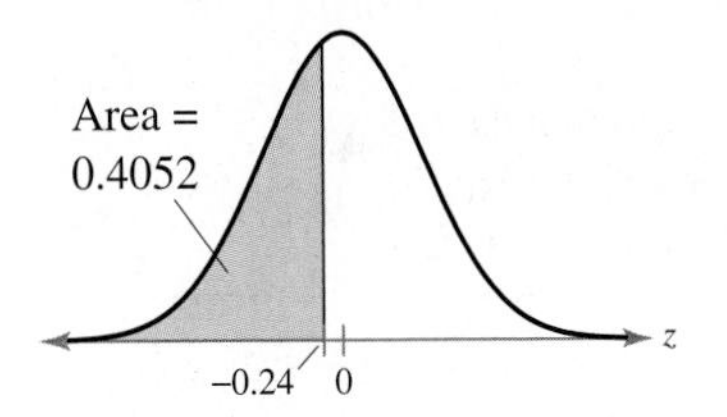

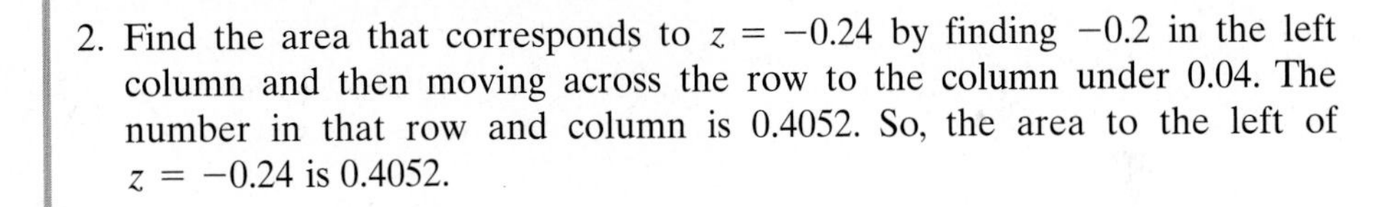

2. Find the area that corresponds to $z = -0.24$ by finding -0.2 in the left column and then moving across the row to the column under 0.04. The number in that row and column is 0.4052. So, the area to the left of $z = -0.24$ is 0.4052.

z	.09	.08	.07	.06	.05	.04	.03
−3.4	.0002	.0003	.0003	.0003	.0003	.0003	.0003
−3.3	.0003	.0004	.0004	.0004	.0004	.0004	.0004
−3.2	.0005	.0005	.0005	.0006	.0006	.0006	.0006
−0.5	.2776	.2810	.2843	.2877	.2912	.2946	.2981
−0.4	.3121	.3156	.3192	.3228	.3264	.3300	.3336
−0.3	.3483	.3520	.3557	.3594	.3632	.3669	.3707
−0.2	.3859	.3897	.3936	.3974	.4013	.4052	.4090
−0.1	.4247	.4286	.4325	.4364	.4404	.4443	.4483
−0.0	.4641	.4681	.4721	.4761	.4801	.4840	.4880

Study Tip

You can use a computer or calculator to find the cumulative area that corresponds to a z-score. For instance, here are instructions for finding the area that corresponds to $z = -0.24$ on a TI-83.

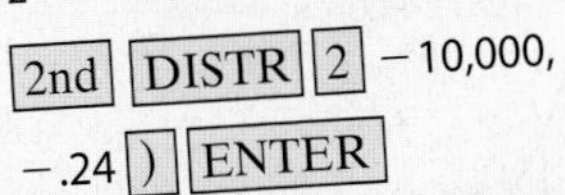

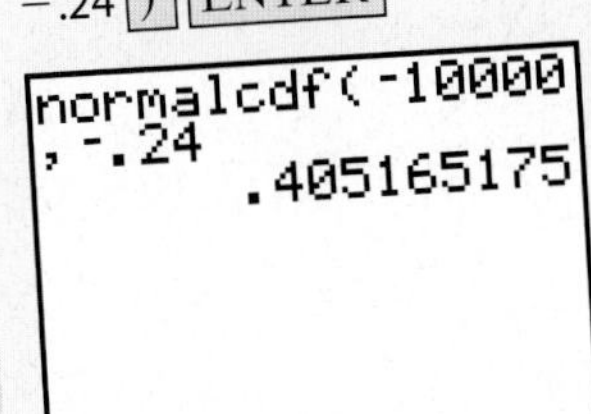

Try It Yourself 1

1. Find the area under the curve to the left of a z-score of -2.19.
2. Find the area under the curve to the left of a z-score of 2.17.

a. Locate the given z-score and *find the area* that corresponds to it in the Standard Normal Table.

Answer: Page A36

When the z-score is not in the table, use the entry closest to it. If the given z-score is exactly midway between two z-scores, then use the area midway between the corresponding areas.

You can use the following guidelines to find various types of areas under the standard normal curve.

GUIDELINES

Finding Areas Under the Standard Normal Curve

1. Sketch the standard normal curve and shade the appropriate area under the curve.
2. Find the area by following the directions for each case shown.
   a. To find the area to the *left* of z, find the area that corresponds to z in the Standard Normal Table.

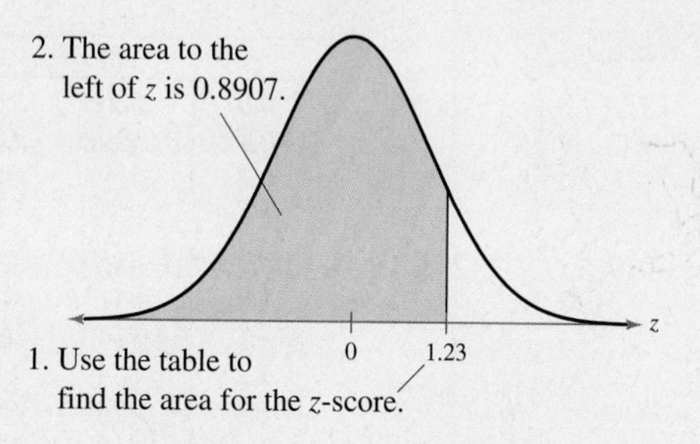

   b. To find the area to the *right* of z, use the Standard Normal Table to find the area that corresponds to z. Then subtract the area from 1.

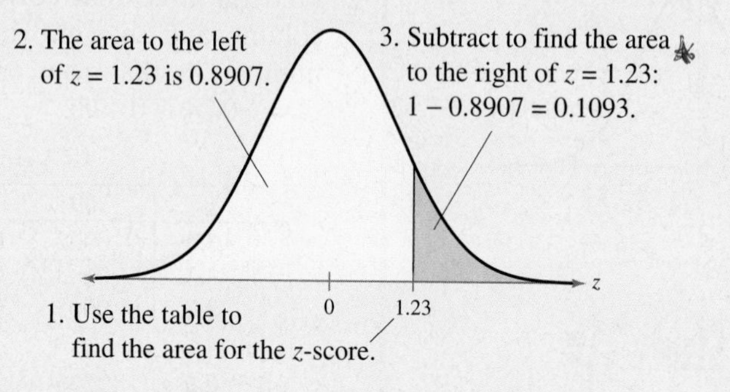

   c. To find the area *between* two z-scores, find the area corresponding to each z-score in the Standard Normal Table. Then subtract the smaller area from the larger area.

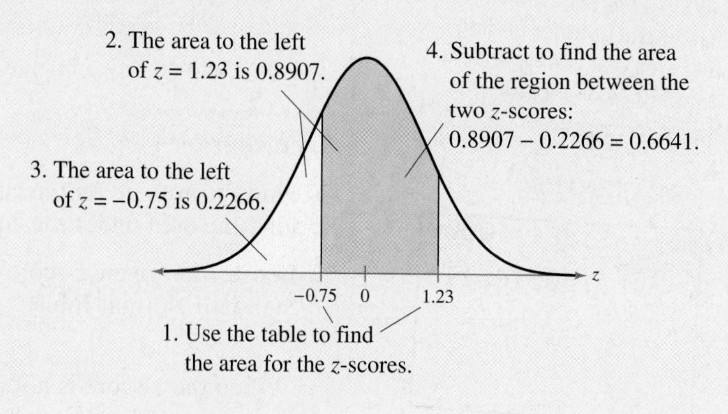

EXAMPLE 

Finding Area Under the Standard Normal Curve

Find the area under the standard normal curve to the left of $z = -0.99$.

SOLUTION The area under the standard normal curve to the left of $z = -0.99$ is shown.

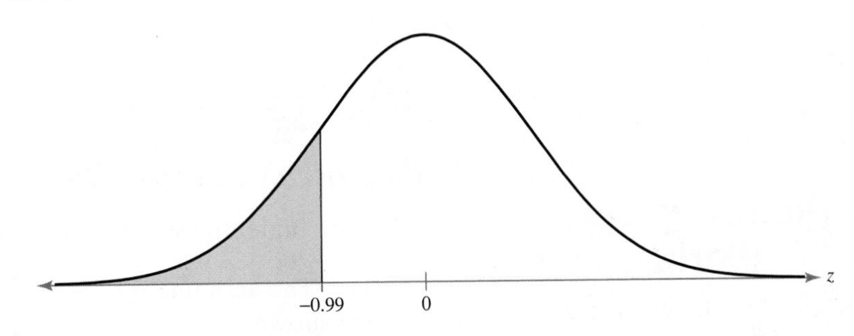

From the Standard Normal Table, this area is equal to 0.1611.

Insight

Because the normal distribution is a continuous probability distribution, the area under the standard normal curve to the left of a z-score gives the probability that z is less than that z-score. For instance, in Example 2, the area to the left of $z = -0.99$ is 0.1611. So, $P(z < -0.99) = 0.1611$, which is read as "the probability that z is less than -0.99 is 0.1611."

Try It Yourself 2

Find the area under the standard normal curve to the left of $z = 2.13$.

a. *Draw* the standard normal curve and shade the area under the curve and to the left of $z = 2.13$.

b. Use the Standard Normal Table to *find the area* that corresponds to $z = 2.13$.

Answer: Page A36

EXAMPLE 

Finding Area Under the Standard Normal Curve

Find the area under the standard normal curve to the right of $z = 1.06$.

SOLUTION The area under the standard normal curve to the right of $z = 1.06$ is shown.

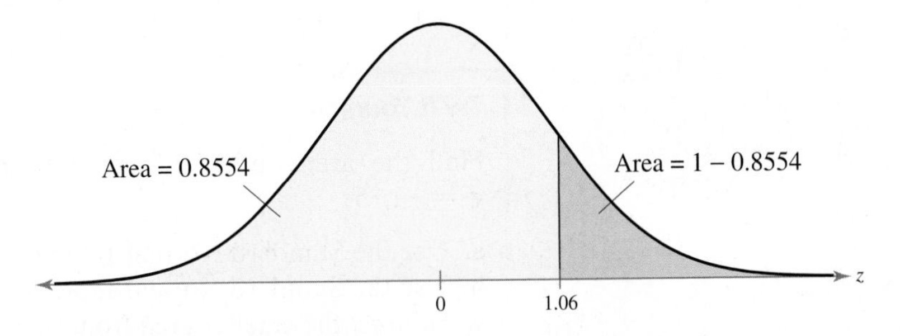

From the Standard Normal Table, the area to the left of $z = 1.06$ is 0.8554. Because the total area under the curve is 1, the area to the right of $z = 1.06$ is

$$\text{Area} = 1 - 0.8554$$
$$= 0.1446.$$

Try It Yourself 3

Find the area under the standard normal curve to the right of $z = -2.16$.

a. *Draw* the standard normal curve and shade the area below the curve and to the right of $z = -2.16$.
b. Use the Standard Normal Table to *find the area* to the left of $z = -2.16$.
c. *Subtract* the area from 1.

Answer: Page A36

EXAMPLE 4

Finding Area Under the Standard Normal Curve

Find the area under the standard normal curve between $z = -1.5$ and $z = 1.25$.

SOLUTION The area under the standard normal curve between $z = -1.5$ and $z = 1.25$ is shown.

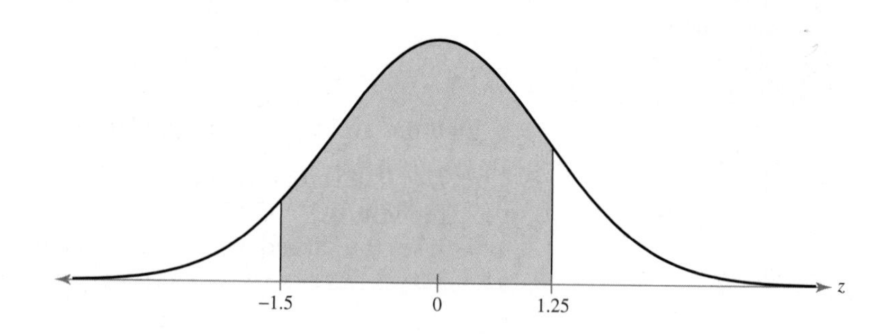

From the Standard Normal Table, the area to the left of $z = 1.25$ is 0.8944 and the area to the left of $z = -1.5$ is 0.0668. So, the area between $z = -1.5$ and $z = 1.25$ is

$$\text{Area} = 0.8944 - 0.0668$$
$$= 0.8276.$$

In other words, 82.76% of the area under the curve falls between $z = -1.5$ and $z = 1.25$.

Try It Yourself 4

Find the area under the standard normal curve between $z = -2.16$ and $z = -1.35$.

a. Use the Standard Normal Table to *find the area* to the left of $z = -1.35$.
b. Use the Standard Normal Table to *find the area* to the left of $z = -2.16$.
c. *Subtract* the smaller area from the larger area.

Answer: Page A36

Recall in Section 2.5 you learned, using the Empirical Rule, that values lying more than two standard deviations from the mean are considered unusual. Values lying more than three standard deviations from the mean are considered *very* unusual. So if a z-score is greater than 2 or less than -2, it is unusual. If a z-score is greater than 3 or less than -3, it is *very* unusual.

Picturing the World

Each year the Centers for Disease Control and Prevention and the National Center for Health Statistics jointly publish a report summarizing the vital statistics from the previous year. According to one publication, the number of births in a recent year was 3,899,589. The weights of the newborns can be approximated by a normal distribution, as shown by the following graph.

Weights of Newborns

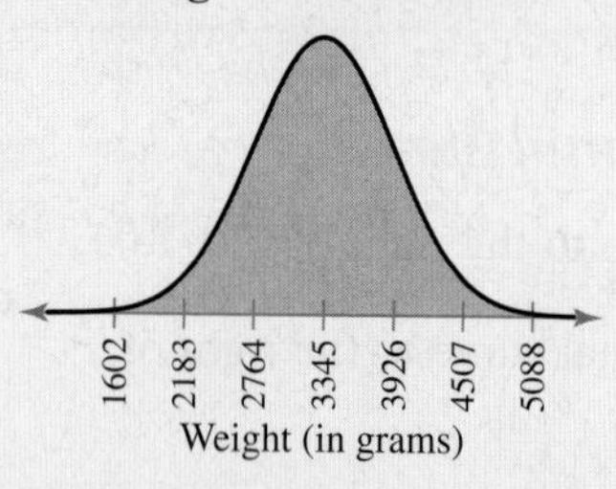

The weights of three newborns are 2000 grams, 3000 grams, and 4000 grams. Find the z-score that corresponds to each weight. Are any of these unusually heavy or light?

5.2 Exercises

Help

LarsonTutor 5.2

Companion Web Site

Student Solutions Manual 5.2

Videos 5.2

Technology Manuals

Basic Skills and Concepts

1. What is the mean of the standard normal distribution? What is the standard deviation of the standard normal distribution?

2. Describe how you can transform a nonstandard normal distribution to a standard normal distribution.

3. ***Getting at the Concept*** ◆ Why is it correct to say "a" normal distribution and "the" standard normal distribution?

4. ***Getting at the Concept*** ◆ If a z-score is zero, which of the following must be true? Explain your reasoning.

   (a) The mean is zero.

   (b) The corresponding x-value is zero.

   (c) The corresponding x-value is equal to the mean.

Graphical Analysis In Exercises 5–8, find the area of the indicated region under the standard normal curve.

5.

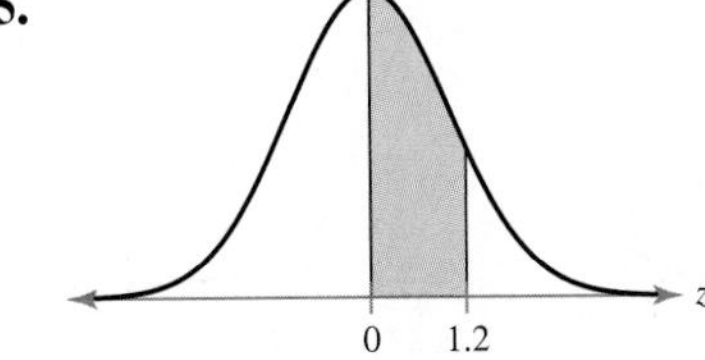

6.

7.

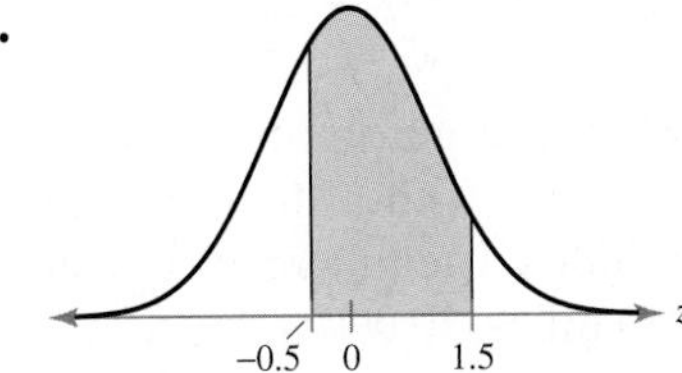

8.

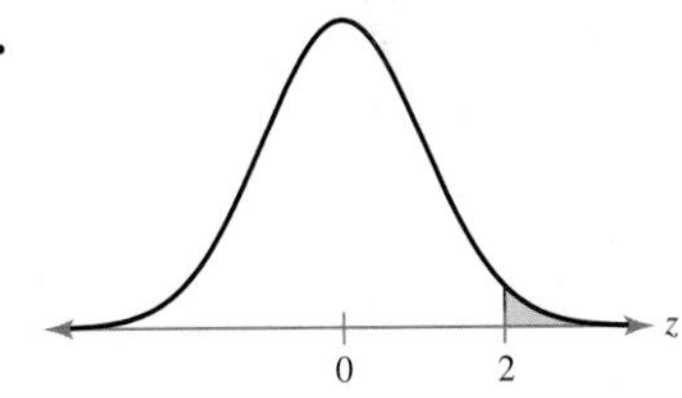

Finding Area In Exercises 9–32, find the indicated area under the standard normal curve.

9. To the left of $z = 1.54$

10. To the left of $z = 0.25$

11. To the left of $z = 1.96$

12. To the left of $z = 1.28$

13. To the right of $z = -0.95$

14. To the right of $z = -1.75$

15. To the right of $z = 1.28$

16. To the right of $z = 2.33$

17. To the left of $z = -2.575$

18. To the left of $z = -3.08$

19. To the right of $z = 1.645$

20. To the right of $z = 2.51$

21. Between $z = 0$ and $z = 1.96$

22. Between $z = 0$ and $z = 3.09$

23. Between $z = -1.53$ and $z = 0$

24. Between $z = -0.51$ and $z = 0$

25. Between $z = -1.96$ and $z = 1.96$

26. Between $z = -2.33$ and $z = 2.33$

27. Between $z = -0.44$ and $z = 1.18$

28. Between $z = -2.88$ and $z = 0.97$

29. To the left of $z = -1.28$ or to the right of $z = 1.28$

30. To the left of $z = -1.96$ or to the right of $z = 1.96$

31. To the left of $z = -2.97$ or to the right of $z = 1.66$

32. To the left of $z = -0.84$ or to the right of $z = 2.81$

Computing and Interpreting z-Scores of Normal Distributions

In Exercises 33–36, you will be given a normal distribution, the distribution's mean and standard deviation, four values from that distribution, and a graph of the Standard Normal Distribution. Use the information to answer the following.

(a) Without converting to *z*-scores, match each diameter with the letters A, B, C, and D on the given graph of the Standard Normal Distribution.

(b) Find the *z*-score that corresponds to each value and check your answers to part (a).

(c) Determine whether any of the values are unusual.

33. ***Ball Bearings*** ◆ Your company manufactures ball bearings. The diameters of the ball bearings are normally distributed, with a mean of 3 inches and a standard deviation of 0.02 inch. The diameters of four ball bearings selected at random are 3.01, 2.97, 2.98, and 3.05.

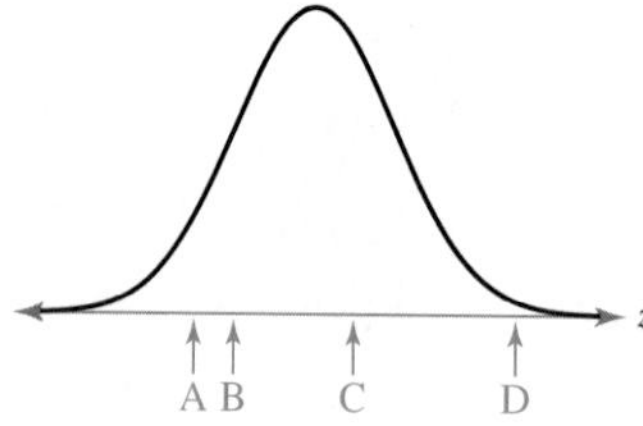

Figure for 33

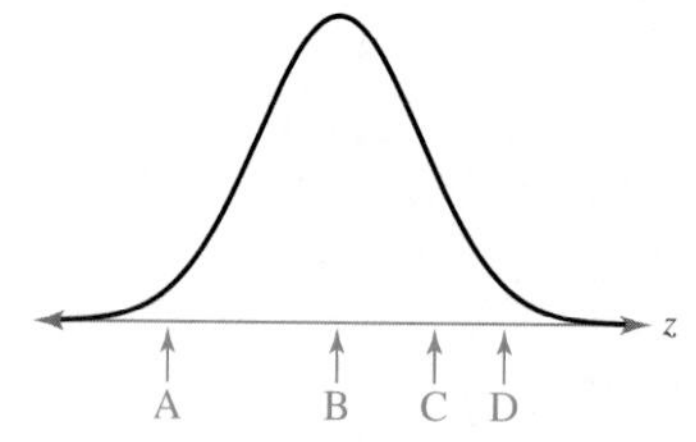

Figure for 34

34. ***Tires*** ◆ An automobile tire brand has a life expectancy that is normally distributed, with a mean life of 30,000 miles and a standard deviation of 2500 miles. The life spans of four tires selected at random are 35,150 miles, 24,750 miles, 30,000 miles, and 33,000 miles.

35. ***SAT I Scores*** ◆ The Scholastic Assessment Test (SAT) is an exam used by colleges and universities to evaluate undergraduate applicants. The test scores are normally distributed. In a recent year, the mean test score was 1019 and the standard deviation was 207. *(Source: College Board Online)* The test scores of four students selected at random are 950, 1250, 1467, and 801.

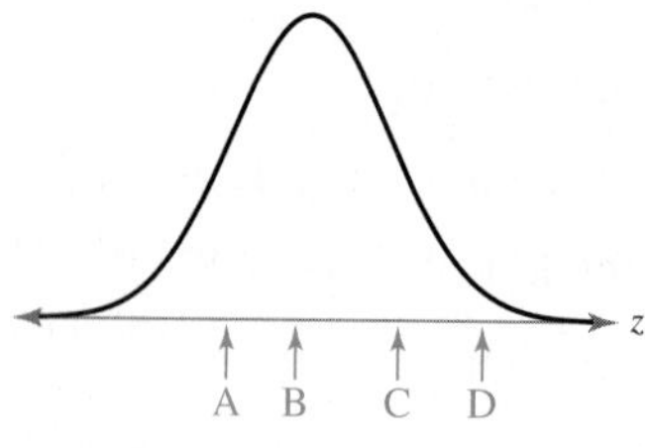

Figure for 35

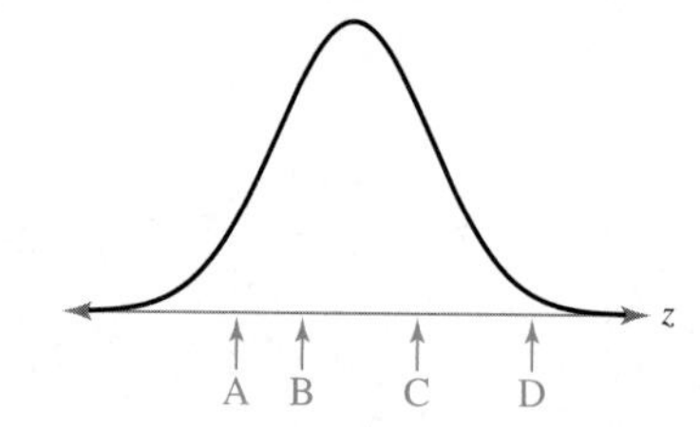

Figure for 36

36. ***ACT Scores*** ◆ The ACT is an exam used by colleges and universities to evaluate undergraduate applicants. The test scores are normally distributed. In a recent year, the mean test score was 21 and the standard deviation was 4.7. *(Adapted from ACT, Inc.)* The test scores of four students selected at random are 18, 32, 14, and 25.

Graphical Analysis In Exercises 37–40, find the probability of z occurring in the indicated region.

37.

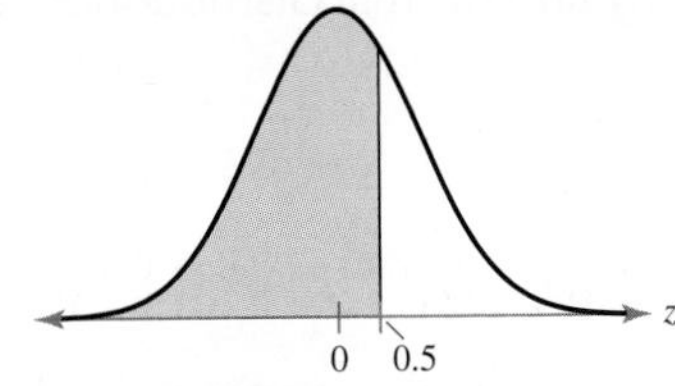

38.

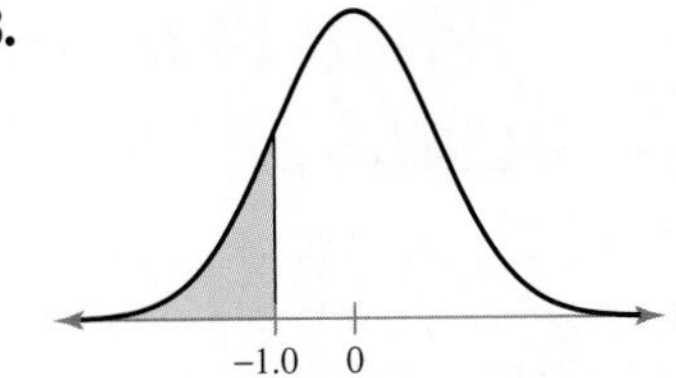

39.

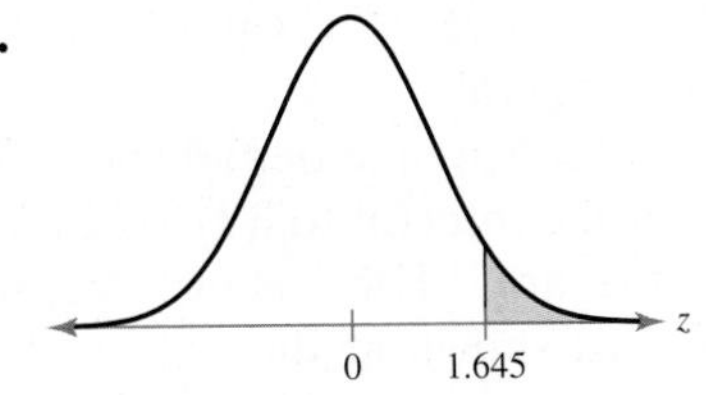

40.

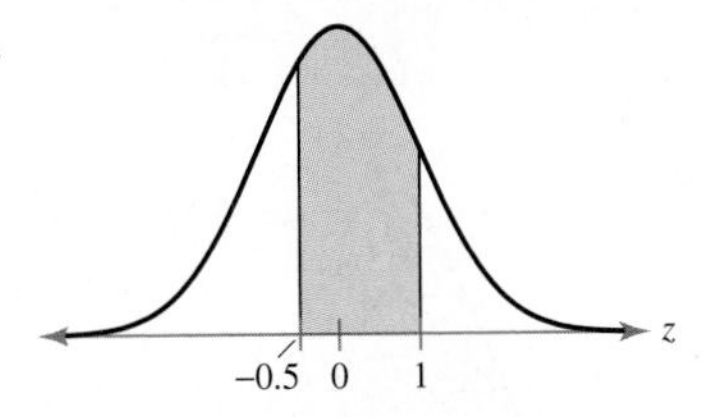

Finding Probabilities In Exercises 41–60, find the indicated probability using the standard normal distribution.

41. $P(z < 1.45)$
42. $P(z < 0.45)$
43. $P(z > -1.95)$
44. $P(z > -0.25)$
45. $P(z < -0.55)$
46. $P(z < -2.95)$
47. $P(z > 1.05)$
48. $P(z > 2.55)$
49. $P(0 < z < 2.05)$
50. $P(0 < z < 1.64)$
51. $P(-0.89 < z < 0)$
52. $P(-2.08 < z < 0)$
53. $P(-1.65 < z < 1.65)$
54. $P(-1.96 < z < 1.96)$
55. $P(-0.95 < z < 1.44)$
56. $P(-2.95 < z < 0.76)$
57. $P(z < -2.58 \text{ or } z > 2.58)$
58. $P(z < -1.96 \text{ or } z > 1.96)$
59. $P(z < -1.65 \text{ or } z > 1.65)$
60. $P(z < -2.05 \text{ or } z > 2.05)$

Extending the Basics

61. ***Chebychev's Theorem vs. the Normal Distribution*** ◆ Recall from Chebychev's Theorem that for an arbitrary probability distribution, at least $1 - (1/2^2)$, or 75%, of the distribution must fall within two standard deviations of the mean. For normal distributions, approximately how much of the total region under a normal curve lies within two standard deviations of the mean? Does this contradict Chebychev's Theorem? Explain.

62. ***Making a Table*** ◆ The Standard Normal Table in Appendix B lists the area under the standard normal curve to the left of z for z-scores between -3.49 and 3.49. Describe how to construct a standard normal table for the area under the standard normal curve to the left of z for z-scores between 0 and z, as shown in the figure.

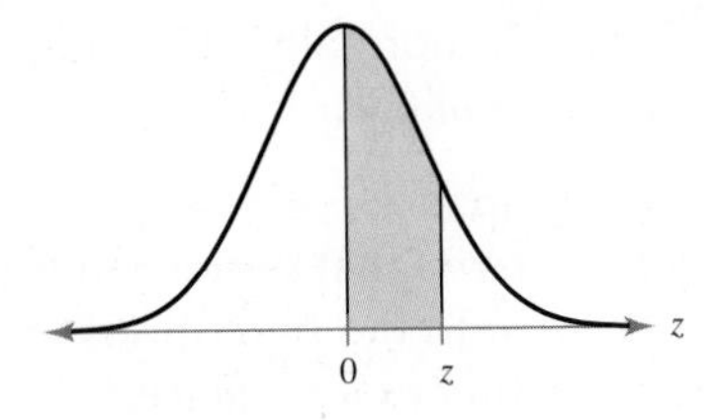

5.3 Normal Distributions: Finding Probabilities

What You Should Learn

- How to find probabilities for normally distributed variables using a table and using technology

Probability and Normal Distributions

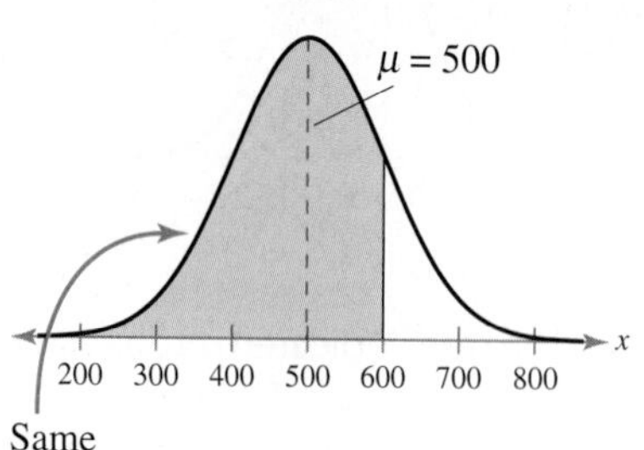

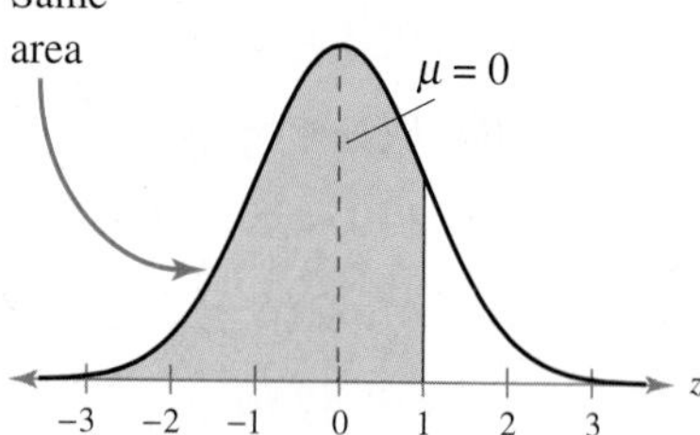

Probability and Normal Distributions

If a random variable x is normally distributed, you can find the probability that x will fall in a given interval by calculating the area under the normal curve for the given interval.

To find the area under any normal curve, first convert the upper and lower bounds of the interval to a z-scores. Then use the standard normal distribution to find the area. For instance, consider a normal curve with $\mu = 500$ and $\sigma = 100$, as shown at the upper left. The value of x one standard deviation above the mean is $\mu + \sigma = 500 + 100 = 600$. Now consider the standard normal curve shown at the lower left. The value of z one standard deviation above the mean is $\mu + \sigma = 0 + 1 = 1$. Because a z-score of 1 corresponds to an x-value of 600, and areas are not changed with a transformation to a standard normal curve, the shaded areas in the graphs are equal.

EXAMPLE 1

Finding Probabilities for Normal Distributions

A survey indicates that people use their computers an average of 2.4 years before upgrading to a new machine. The standard deviation is 0.5 year. If a computer owner is selected at random, find the probability that he or she will use it for less than 2 years before upgrading. Assume that the variable x is normally distributed.

SOLUTION The graph shows a normal curve with $\mu = 2.4$ and $\sigma = 0.5$ and a shaded area for x less than 2. The z-score that corresponds to 2 years is

$$z = \frac{x - \mu}{\sigma} = \frac{2 - 2.4}{0.5} = -0.8.$$

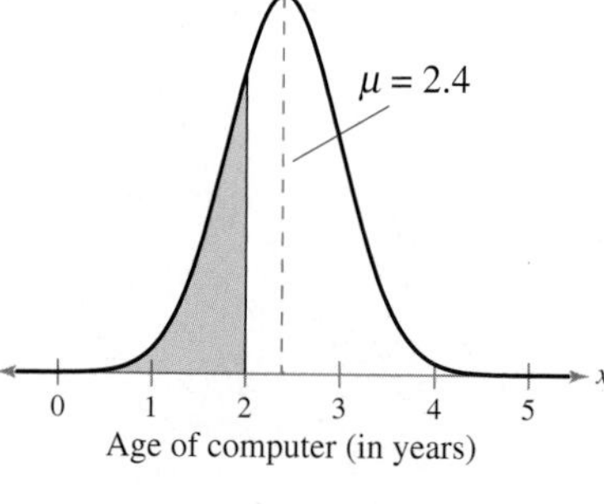

Using the Standard Normal Table, $P(z < -0.8) = 0.2119$. The probability that the computer will be upgraded in less than 2 years is 0.2119. So 21.19% of new owners will upgrade in less than two years.

Try It Yourself 1

A Ford Focus manual transmission gets an average of 28 miles per gallon (mpg) in city driving with a standard deviation of 1.6 mpg. If a Focus is selected at random, what is the probability that it will get more than 31 mpg? Assume that gas mileage is normally distributed. *(Source: U.S. Department of Energy, US Environmental Protection Agency)*

a. *Sketch* a graph.
b. *Find the z-score* that corresponds to 31 miles per gallon.
c. *Find the area* to the right of that z-score.
d. *Write the result as a sentence.*

Answer: Page A36

Study Tip

Another way to write the answer to Example 1 is $P(x < 2) = 0.2119$.

EXAMPLE 

Finding Probabilities for Normal Distributions

A survey indicates that for each trip to the supermarket, a shopper spends an average of $\mu = 45$ minutes with a standard deviation of $\sigma = 12$ minutes. The length of time spent in the store is normally distributed and is represented by the variable x. A shopper enters the store. (a) Find the probability that the shopper will be in the store for each interval of time listed below. (b) If 200 shoppers enter the store, how many shoppers would you expect to be in the store for each interval of time listed below?

1. Between 24 and 54 minutes
2. More than 39 minutes

SOLUTION

$\mu = 45$

x

10 20 30 40 50 60 70 80

Time (in minutes)

1(a). The graph at the left shows a normal curve with $\mu = 45$ minutes and $\sigma = 12$ minutes. The area for x between 24 and 54 minutes is shaded. The z-scores that correspond to 24 minutes and to 54 minutes are

$$z_1 = \frac{24 - 45}{12} = -1.75 \quad \text{and} \quad z_2 = \frac{54 - 45}{12} = 0.75.$$

So, the probability that a shopper will be in the store between 24 and 54 minutes is

$$\begin{aligned} P(24 < x < 54) &= P(-1.75 < z < 0.75) \\ &= P(z < 0.75) - P(z < -1.75) \\ &= 0.7734 - 0.0401 \\ &= 0.7333. \end{aligned}$$

(b) Another way of interpreting this probability is to say that 73.33% of the shoppers will be in the store between 24 and 54 minutes. If 200 shoppers enter the store, then you would expect $200(0.7333) = 146.66$ (or about 147) shoppers to be in the store between 24 and 54 minutes.

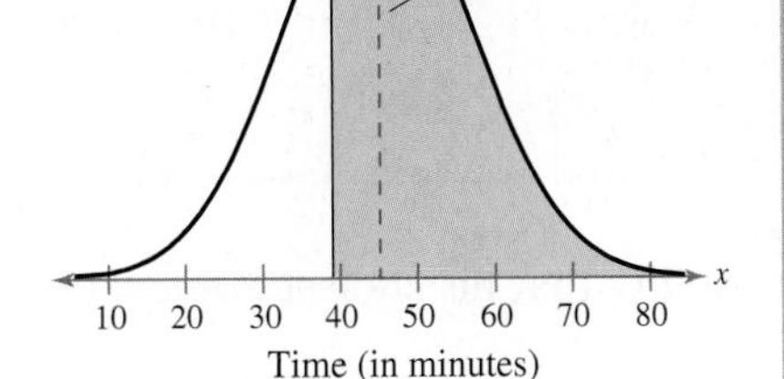

2(a). The graph at the left shows a normal curve with $\mu = 45$ minutes and $\sigma = 12$ minutes. The area for x greater than 39 minutes is shaded. The z-score that corresponds to 39 minutes is

$$z = \frac{39 - 45}{12} = -0.5.$$

So, the probability that a shopper will be in the store more than 39 minutes is

$$P(x > 39) = P(z > -0.5) = 1 - P(z < -0.5) = 1 - 0.3085 = 0.6915.$$

(b) If 200 shoppers enter the store, then you would expect $200(0.6915) = 138.3$ (or about 138) shoppers to be in the store more than 39 minutes.

Try It Yourself 2

What is the probability that the shopper will be in the supermarket between 33 and 60 minutes?

a. *Sketch* a graph.
b. *Find z-scores* that correspond to 60 minutes and 33 minutes.
c. *Find the cumulative area* for each z-score.
d. *Subtract the smaller area from the larger.*

Answer: Page A36

Picturing the World

In baseball, a batting average is the number of hits divided by the number of at-bats. The batting averages of the more than 750 Major League Baseball players in a recent year can be approximated by a normal distribution, as shown in the following graph. The mean of the batting averages is 0.266 and the standard deviation is 0.012.

Major League Baseball

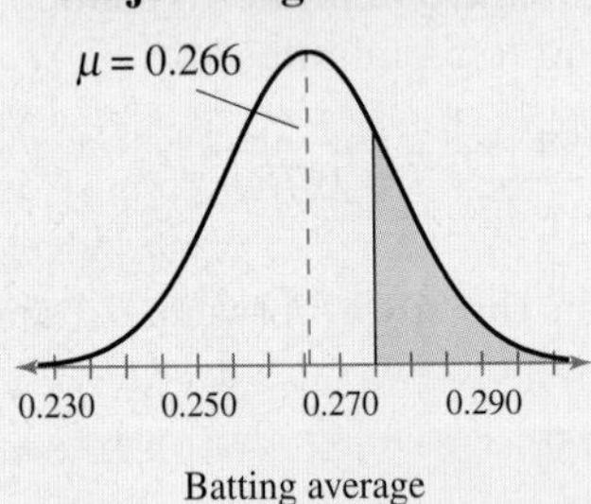

What percent of the players have a batting average of 0.275 or greater?

Another way to find normal probabilities is to use a calculator or a computer. You can find normal probabilities using *Minitab*, *Excel*, and the *TI-83*.

EXAMPLE  3

Using Technology to Find Normal Probabilities

Assume that cholesterol levels of men in the U.S. are normally distributed, with a mean of 215 milligrams per deciliter and a standard deviation of 25 milligrams per deciliter. If you randomly select a man from the U.S., what is the probability that his cholesterol level is less than 175? Use a calculator or a computer to find the probability.

SOLUTION *Minitab*, *Excel*, and the *TI-83* each have features that allow you to find normal probabilities without first converting to standard z-scores. For each, you must specify the mean and standard deviation of the population, as well as the x-value(s) that determine the interval.

MINITAB

Cumulative Distribution Function

Normal with mean = 215.000 and standard deviation = 25.0000

x	P(X <= x)
175.0000	0.0548

EXCEL

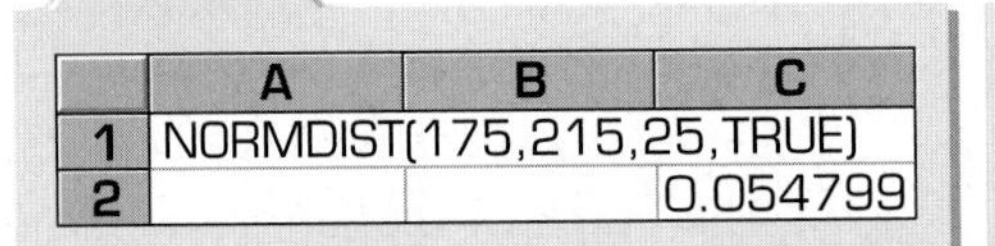

	A	B	C
1	NORMDIST(175,215,25,TRUE)		
2			0.054799

TI-83

normalcdf(0,175,215,25)
.0547992894

From the displays, you can see that the probability that his cholesterol level is less than 175 is about 0.055, or 5.5%.

Try It Yourself 3

If a man from the U.S. is selected at random, what is the probability that his cholesterol is between 190 and 225? Use a calculator or a computer.

a. *Read the user's guide* for the technology tool you are using.
b. *Enter the appropriate data* to obtain the probability.
c. *Write* the result as a sentence.

Answer: Page A36

Example 3 shows only one of several ways to find normal probabilities using *Minitab*, *Excel*, and the *TI-83*.

5.3 Exercises

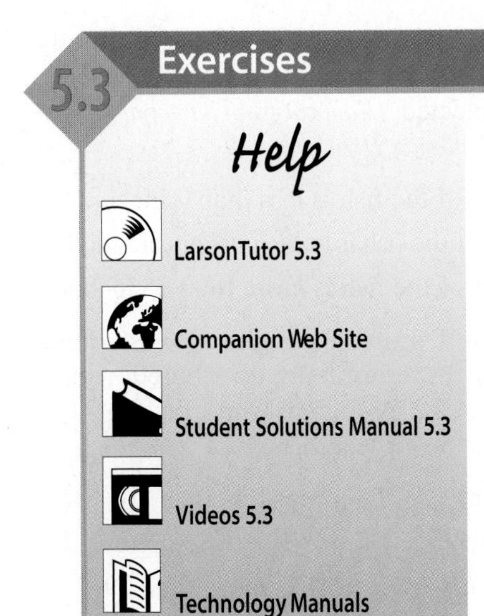

Basic Skills and Concepts

Graphical Analysis In Exercises 1–6, assume a member is selected at random from the population represented by the graph. Find the probability that the member selected at random is from the shaded area of the graph. Assume the variable x is normally distributed.

1. SAT Verbal Scores

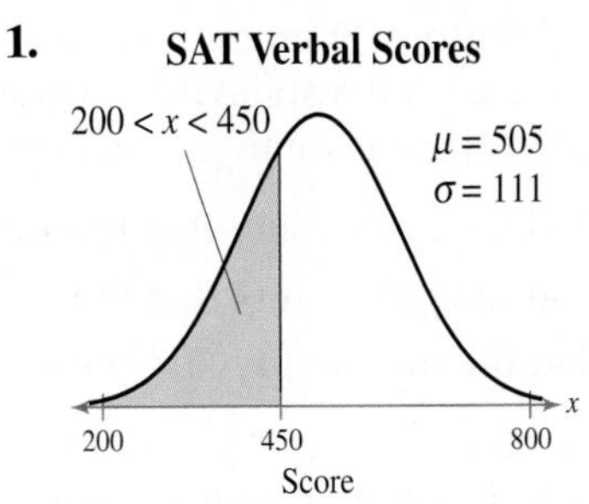

(Source: College Board Online)

2. SAT Math Scores

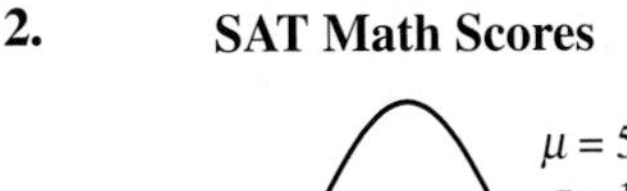
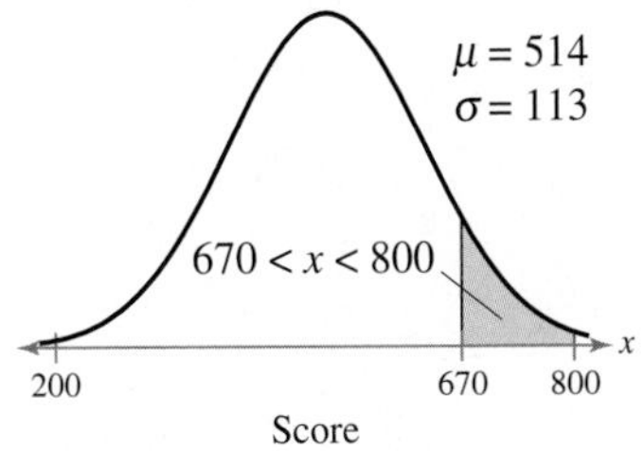

(Source: College Board Online)

3. U.S. Women Ages 20–29: Total Cholesterol

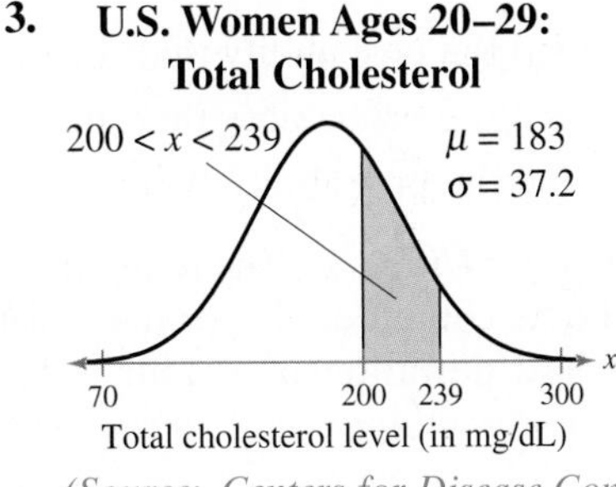

(Source: Centers for Disease Control and Prevention)

4. U.S. Women Ages 50–59: Total Cholesterol

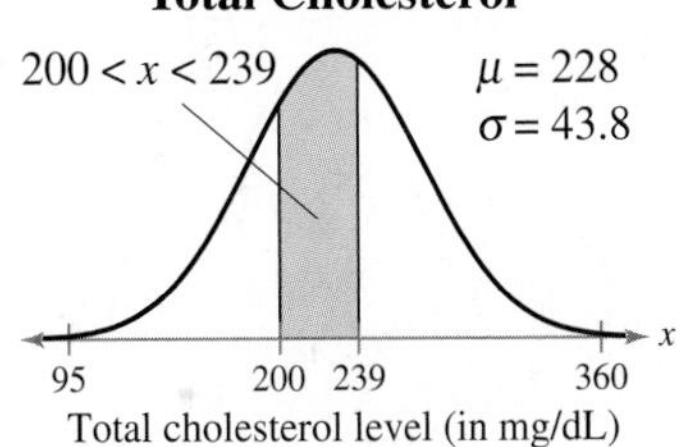

(Source: Centers for Disease Control and Prevention)

5. Chevrolet Blazer: Braking Distance on a Dry Surface

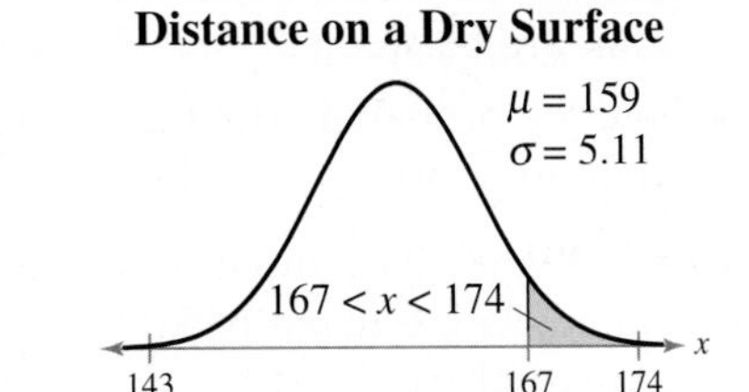

(Source: National Highway Traffic Safety Administration)

6. Chevrolet Blazer: Braking Distance on a Wet Surface

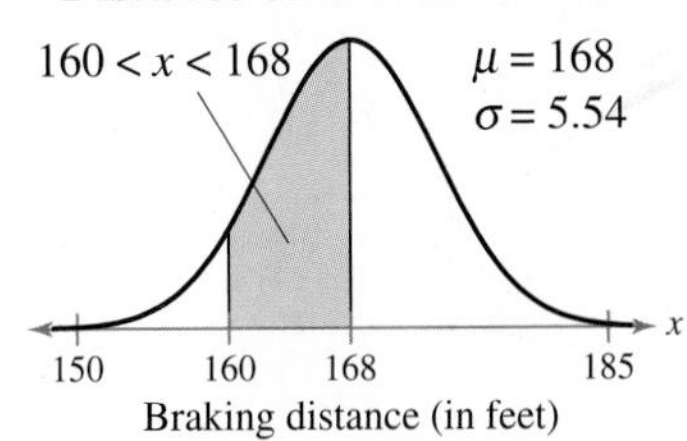

(Source: National Highway Traffic Safety Administration)

Finding Probabilities In Exercises 7–12, find the indicated probabilities. If convenient, use technology to find the probabilities.

7. ***Male Heights*** ◆ A survey was conducted to measure the height of U.S. males. In the survey, respondents were grouped by age. In the 20–29 age group, the heights were normally distributed, with a mean of 69.2 inches and a standard deviation of 2.9 inches. A study participant is randomly selected. *(Adapted from U.S. National Center for Health Statistics)*

(a) Find the probability that his height is less than 66 inches.

(b) Find the probability that his height is between 66 and 72 inches.

(c) Find the probability that his height is more than 72 inches.

8. *Fish Lengths* ◆ The lengths of Atlantic croaker fish are normally distributed, with a mean of 10 inches and a standard deviation of 2 inches. An Atlantic croaker fish is randomly selected. *(Adapted from National Marine Fisheries Service, Fisheries Statistics and Economics Division)*

(a) Find the probability that the length of the fish is less than 7 inches.

(b) Find the probability that the length of the fish is between 7 and 15 inches.

(c) Find the probability that the length of the fish is more than 15 inches.

9. *ACT Scores* ◆ In a recent year, the ACT scores for high school students with a 3.50 to 4.00 grade point average were normally distributed, with a mean of 24.2 and a standard deviation of 4.2. A student who took the ACT during this time is randomly selected. *(Adapted from ACT, Inc.)*

(a) Find the probability that the student's ACT score is less than 20.

(b) Find the probability that the student's ACT score is between 20 and 29.

(c) Find the probability that the student's ACT score is more than 29.

10. *Rhesus Monkeys* ◆ The weights of adult male rhesus monkeys are normally distributed, with a mean of 15 pounds and a standard deviation of 3 pounds. A rhesus monkey is randomly selected.

(a) Find the probability that the monkey's weight is less than 13 pounds.

(b) Find the probability that the weight is between 13 and 17 pounds.

(c) Find the probability that the monkey's weight is more than 17 pounds.

11. *Computer Usage* ◆ The number of hours per week adults in the U.S. spend on home computers is normally distributed, with a mean of 5 hours and a standard deviation of 1 hour. An adult in the U.S. is randomly selected. *(Adapted from American Demographics)*

(a) Find the probability that the hours spent on the home computer by the adult are less than 2.5 hours per week.

(b) Find the probability that the hours spent on the home computer by the adult are between 2.5 and 7.5 hours per week.

(c) Find the probability that the hours spent on the home computer by the adult are more than 7.5 hours per week.

12. *Utility Bills* ◆ The monthly utility bills in a certain city are normally distributed, with a mean of $100 and a standard deviation of $12. A utility bill is randomly selected.

(a) Find the probability that the utility bill is less than $80.

(b) Find the probability that the utility bill is between $80 and $115.

(c) Find the probability that the utility bill is more than $115.

Using Normal Distributions In Exercises 13–20, answer the questions about the specified normal distribution.

13. *Using SAT Verbal Scores* ◆ Use the normal distribution of SAT verbal scores in Exercise 1 for which the mean is 505 and the standard deviation is 111.

(a) What percent of the SAT verbal scores are less than 600?

(b) If 1000 SAT verbal scores are randomly selected, about how many would you expect to be greater than 550?

14. ***SAT Math Scores*** ◆ Use the normal distribution of SAT math scores in Exercise 2 for which the mean is 514 and the standard deviation is 113.

(a) What percent of the SAT math scores are less than 500?

(b) If 1500 SAT math scores are randomly selected, about how many would you expect to be greater than 600?

15. ***Cholesterol*** ◆ Use the normal distribution of women's total cholesterol levels in Exercise 3 for which the mean is 183 milligrams per deciliter and the standard deviation is 37.2 milligrams per deciliter.

(a) What percent of the women have a total cholesterol level less than 200 milligrams per deciliter of blood?

(b) If 250 U.S. women in the 20–29 age group are randomly selected, about how many would you expect to have a total cholesterol level greater than 240 milligrams per deciliter of blood?

16. ***Cholesterol*** ◆ Use the normal distribution of women's total cholesterol levels in Exercise 4 for which the mean is 228 milligrams per deciliter and the standard deviation is 43.8 milligrams per deciliter.

(a) What percent of the women have a total cholesterol level less than 239 milligrams per deciliter of blood?

(b) If 200 U.S. women in the 50–59 age group are randomly selected, about how many would you expect to have a total cholesterol level greater than 200 milligrams per deciliter of blood?

17. ***Length of Fish*** ◆ Use the normal distribution of fish lengths in Exercise 8 for which the mean is 10 inches and the standard deviation is 2 inches.

(a) What percent of the fish are longer than 11 inches?

(b) If 200 Atlantic croakers are randomly selected, about how many would you expect to be shorter than 8 inches?

18. ***Weight of Monkeys*** ◆ Use the normal distribution of monkey weights in Exercise 10 for which the mean is 15 pounds and the standard deviation is 3 pounds.

(a) What percent of the monkeys have a weight that is greater than 20 pounds?

(b) If 50 rhesus monkeys are randomly selected, about how many would you expect to weigh less than 12 pounds?

19. ***Home Computers*** ◆ Use the normal distribution of computer usage in Exercise 11 for which the mean is 5 hours and the standard deviation is 1 hour.

(a) What percent of the adults spend more than 2 hours per week on a home computer?

(b) If 35 adults in the U.S. are randomly selected, about how many would you expect to say they spend less than 3 hours per week on a home computer?

20. ***Monthly Bills*** ◆ Use the normal distribution of utility bills in Exercise 12 for which the mean is \$100 and the standard deviation is \$12.

(a) What percent of the utility bills are more than \$125?

(b) If 300 utility bills are randomly selected, about how many would you expect to be less than \$90?

Extending the Basics

Control Charts **Statistical process control (SPC)** is the use of statistics to monitor and improve the quality of a process, such as manufacturing an engine part. In SPC, information about a process is gathered and used to determine if a process is meeting all of the specified requirements. One tool used in SPC is a **control chart.** When individual measurements of a variable x are normally distributed, a control chart can be used to detect processes that are possibly out of statistical control. Three warning signals that a control chart uses to detect a process that may be out of control are as follows.

(1) A point lies beyond three standard deviations of the mean.

(2) There are nine consecutive points that fall on one side of the mean.

(3) At least two of three consecutive points lie more than two standard deviations from the mean.

In Exercises 21–24, a control chart is shown. Each chart has horizontal lines drawn at the mean μ, at $\mu \pm 2\sigma$, and at $\mu \pm 3\sigma$. Determine if the process shown is in control or out of control. Explain.

21. A gear has been designed to have a diameter of 3 inches. The standard deviation of the process is 0.2 inch.

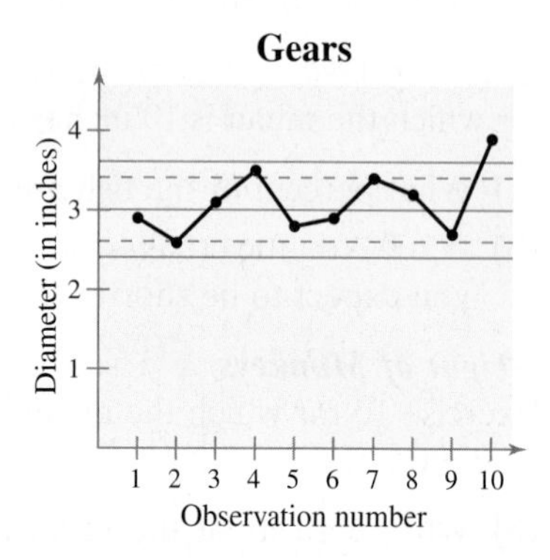

22. A nail has been designed to have a length of 4 inches. The standard deviation of the process is 0.12 inch.

23. A liquid-dispensing machine has been designed to fill bottles with 1 liter of liquid. The standard deviation of the process is 0.1 liter.

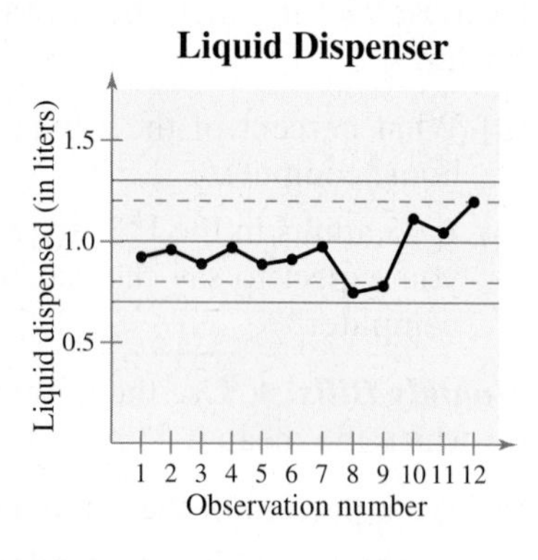

24. An engine part has been designed to have a diameter of 55 millimeters. The standard deviation of the process is 0.001 millimeter.

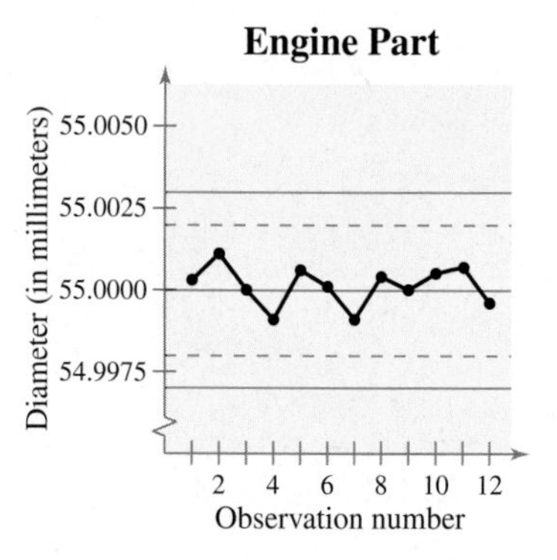

5.4 Normal Distributions: Finding Values

What You Should Learn

- How to find a *z*-score given the area under the normal curve
- How to transform a *z*-score to an *x*-value
- How to find a specific data value of a normal distribution given the probability

Finding *z*-Scores • Transforming a *z*-Score to an *x*-Value • Finding a Specific Data Value for a Given Probability

Finding *z*-Scores

In Section 5.3, you were given a normally distributed random variable x and you found the probability that x would fall in a given interval by calculating the area under the normal curve for the given interval.

But what if you are given a probability and want to find a value? For instance, a university might want to know what is the lowest test score a student can have on an entrance exam and still be in the top 10%, or a medical researcher might want to know the cutoff values to select the middle 90% of patients by age. In this section, you will learn how to find a value given an area under a normal curve (or a probability), as shown in the following example.

Study Tip

You can use a computer or calculator to find the *z*-score that corresponds to a cumulative area. For instance, here are instructions for finding the *z*-score that corresponds to an area of 0.3632 on a *TI-83*.

[2nd] [DISTR] [3] .3632

The calculator will display −.3499183227.

EXAMPLE 1

Finding a z-Score Given an Area

1. Find the z-score that corresponds to a cumulative area of 0.3632.
2. Find the z-score that corresponds to a cumulative area of 0.8925.

SOLUTION

1. Find the z-score that corresponds to an area of 0.3632 by locating 0.3632 in the Standard Normal Table. The values at the beginning of the corresponding row and at the top of the corresponding column give the z-score. For this area, the row value is −0.3 and the column value is 0.05. So, the z-score is −0.35.

z	.09	.08	.07	.06	.05	.04	.03
−3.4	.0002	.0003	.0003	.0003	.0003	.0003	.0003
−0.5	.2776	.2810	.2843	.2877	.2912	.2946	.2981
−0.4	.3121	.3156	.3192	.3228	.3264	.3300	.3336
−0.3	.3483	.3520	.3557	.3594	.3632	.3669	.3707
−0.2	.3859	.3897	.3936	.3974	.4013	.4052	.4090

2. Find the z-score that corresponds to an area of 0.8925 by locating 0.8925 in the Standard Normal Table. The values at the beginning of the corresponding row and at the top of the corresponding column give the z-score. For this area, the row value is 1.2 and the column value is 0.04. So, the z-score is 1.24.

z	.00	.01	.02	.03	.04	.05	.06
0.0	.5000	.5040	.5080	.5120	.5160	.5199	.5239
1.0	.8413	.8438	.8461	.8485	.8508	.8531	.8554
1.1	.8643	.8665	.8686	.8708	.8729	.8749	.8770
1.2	.8849	.8869	.8888	.8907	.8925	.8944	.8962
1.3	.9032	.9049	.9066	.9082	.9099	.9115	.9131

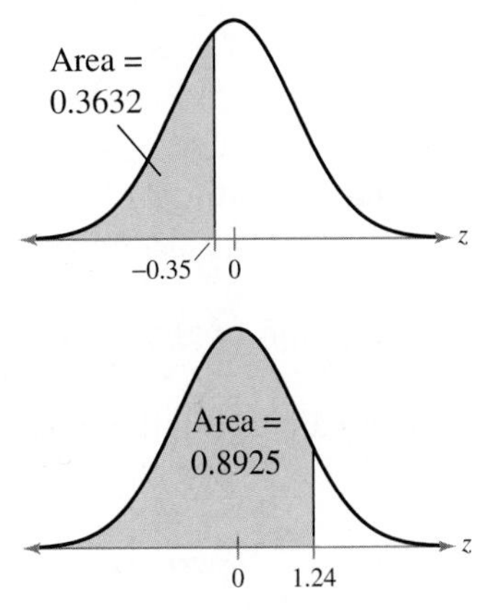

Try It Yourself 1

1. Find the z-score that corresponds to a cumulative area of 0.0384.
2. Find the z-score that corresponds to a cumulative area of 0.9901.

a. *Locate* the given area in the Standard Normal Table.
b. *Find the z-score* that corresponds to the area.

Answer: Page A36

Study Tip

In most cases, the given area will not be found in the table, so use the entry closest to it. If the given area is halfway between two area entries, use the z-score halfway between the corresponding z-scores. For instance, in part 1 of Example 2, the z-score between -1.64 and -1.65 is -1.645.

In Section 2.5, you learned that percentiles divide a data set into one hundred equal parts. To find a z-score that corresponds to a percentile, you can use the Standard Normal Table. Recall that if a value x represents the 83rd percentile, then 83% of the data values are below x and 17% of the data values are above x.

EXAMPLE 2

Finding a z-Score Given a Percentile

Find the z-score that corresponds to each of the following percentiles.

1. P_5 2. P_{50} 3. P_{90}

SOLUTION

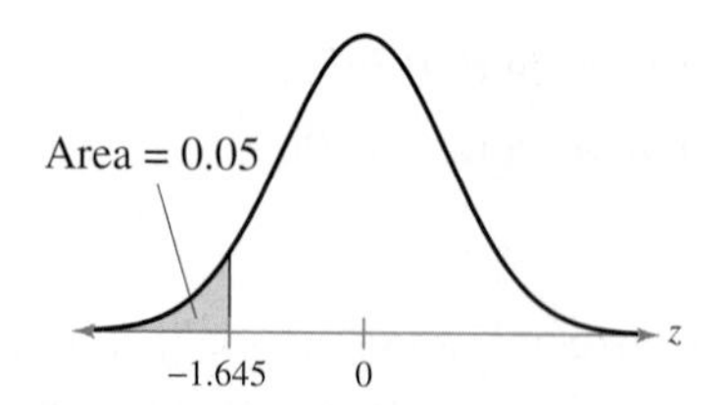

1. To find the z-score that corresponds to P_5, find the z-score that corresponds to an area of 0.05 (see figure) by locating 0.05 in the Standard Normal Table. The areas closest to 0.05 in the table are 0.0495 ($z = -1.65$) and 0.0505 ($z = -1.64$). Because 0.05 is halfway between the two areas in the table, use the z-score that is halfway between -1.64 and -1.65. So, the z-score that corresponds to an area of 0.05 is -1.645.

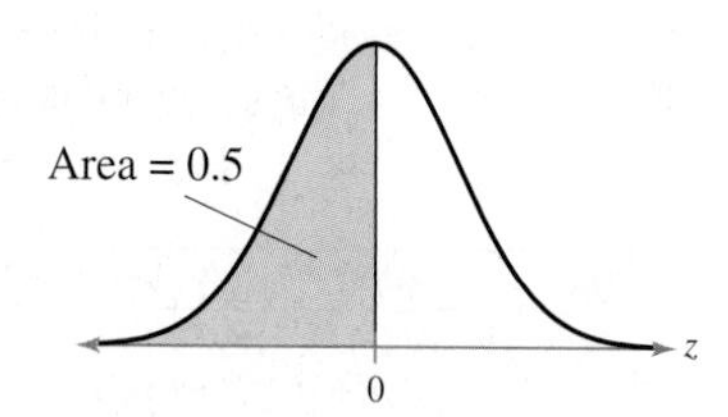

2. To find the z-score that corresponds to P_{50}, find the z-score that corresponds to an area of 0.5 by locating 0.5 in the Standard Normal Table. The area closest to 0.5 in the table is 0.5000, so the z-score that corresponds to an area of 0.5 is 0.00.

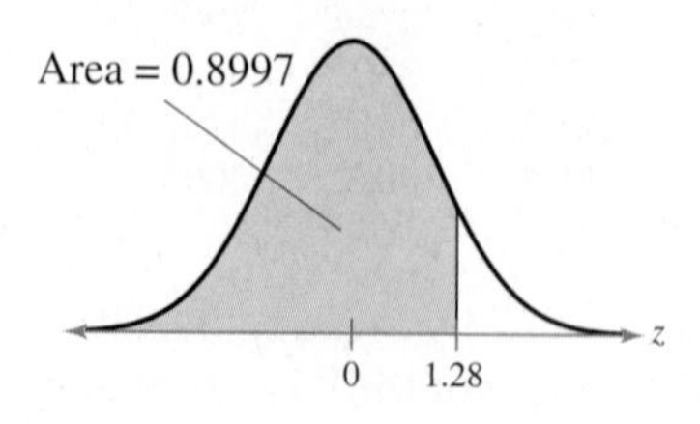

3. To find the z-score that corresponds to P_{90}, find the z-score that corresponds to an area of 0.9 by locating 0.9 in the Standard Normal Table. The area closest to 0.9 in the table is 0.8997, so the z-score that corresponds to an area of 0.9 is 1.28.

Try It Yourself 2

Find the z-score that corresponds to each of the following percentiles.

1. P_{10} 2. P_{20} 3. P_{99}

a. *Write* the percentile as an area. If necessary, draw a graph of the area to visualize the problem.
b. *Locate* the area in the Standard Normal Table. If the area is not in the table, use the closest area. (See Study Tip above.)
c. *Identify* the z-score that corresponds to the area.

Answer: Page A36

Transforming a z-Score to an x-Value

Recall that to transform an x-value to a z-score, you can use the formula

$$z = \frac{x - \mu}{\sigma}.$$

This formula gives z in terms of x. If you solve this formula for x, you get a new formula that gives x in terms of z.

$$z = \frac{x - \mu}{\sigma} \quad \text{Formula for } z \text{ in terms of } x$$

$$z\sigma = x - \mu \quad \text{Multiply each side by } \sigma.$$

$$\mu + z\sigma = x \quad \text{Add } \mu \text{ to each side.}$$

$$x = \mu + z\sigma \quad \text{Interchange sides.}$$

Transforming a z-Score to an x-Value

To transform a standard z-score to a data value x in a given population, use the formula

$$x = \mu + z\sigma.$$

EXAMPLE 3

Finding an x-Value

The speeds of vehicles along a stretch of highway have a mean of 56 mph and a standard deviation of 4 mph. Find the speeds x corresponding to z-scores of 1.96, −2.33, and 0. Interpret your results.

SOLUTION The x-value that corresponds to each standard score is calculated as follows.

$z = 1.96$: $\quad x = 56 + 1.96(4) = 63.84$ mph

$z = -2.33$: $\quad x = 56 + (-2.33)(4) = 46.68$ mph

$z = 0$: $\quad x = 56 + 0(4) = 56$ mph

You can see that 63.84 mph is above the mean, 46.68 is below the mean, and 56 is equal to the mean.

Try It Yourself 3

The monthly utility bills in a city have a mean of \$70 and a standard deviation of \$8. Find the *x-values* that correspond to z-scores of −0.75, 4.29, −1.82. What can you conclude?

a. *Identify* μ and σ of the nonstandard normal distribution.
b. *Transform* each z-score to an x-value.
c. *Interpret* the results.

Answer: Page A36

Finding a Specific Data Value for a Given Probability

You can also use the normal distribution to find a specific data value (x-value) for a given probability, as shown in Example 4.

EXAMPLE 

Finding a Specific Data Value

Scores for a civil service exam are normally distributed, with a mean of 75 and a standard deviation of 6.5. To be eligible for civil service employment, you must score in the top 5%. What is the lowest score you can earn and still be eligible for employment?

SOLUTION Exam scores in the top 5% correspond to the shaded region shown.

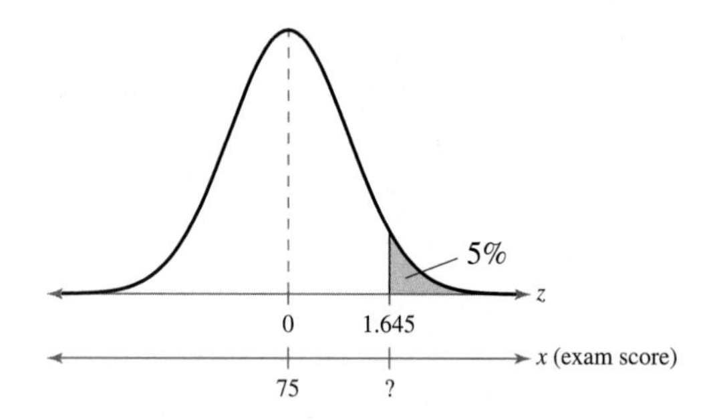

An exam score in the top 5% is any score above the 95th percentile. To find the score that represents the 95th percentile, you must first find the z-score that corresponds to a cumulative area of 0.95. From the Standard Normal Table, you can find that the areas closest to 0.95 are 0.9495 $(z = 1.64)$ and 0.9505 $(z = 1.65)$. Because 0.95 is halfway between the two areas in the table, use the z-score that is halfway between 1.64 and 1.65. That is, $z = 1.645$. Using the equation $x = \mu + z\sigma$, you have

$$\begin{aligned} x &= \mu + z\sigma \\ &= 75 + 1.645(6.5) \\ &\approx 85.69. \end{aligned}$$

So, the lowest score you can earn and still be eligible for employment is 86.

Try It Yourself 4

The braking distances of a sample of Ford F-150s are normally distributed. On a dry surface, the mean braking distance was 158 feet and the standard deviation was 6.51 feet. What is the longest braking distance on a dry surface one of these Ford F-150s could have and still be in the best 1%? *(Adapted from National Highway Traffic Safety Administration)*

a. *Sketch* a graph.
b. *Find the z-score* that corresponds to the given area.
c. *Find* x using the equation $x = \mu + z\sigma$.
d. *Write* the result as a sentence.

Answer: Page A36

Picturing the World

According to the American Medical Association, the mean number of hours all physicians spend in patient care each week is about 53.2 hours. The hours spent in patient care each week by physicians can be approximated by a normal distribution. Assume the standard deviation is 3 hours.

Hours Physicians Spend in Patient Care

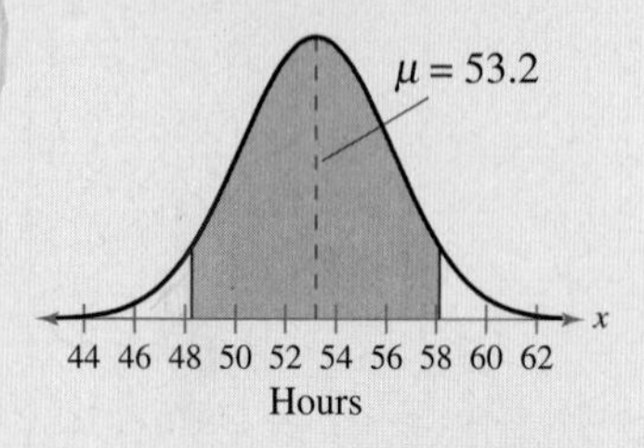

Between what two values does the middle 90% of the data lie?

EXAMPLE 5

Finding a Specific Data Value

In a randomly selected sample of 1169 men ages 40–49, the mean total cholesterol level was 211 milligrams per deciliter with a standard deviation of 39.2 milligrams per deciliter. Assume the total cholesterol levels are normally distributed. Find the highest total cholesterol level a man in this 40–49 age group can have and be in the lowest 1%. *(Adapted from Centers for Disease Control and Prevention)*

SOLUTION

Total cholesterol levels in the lowest 1% correspond to the shaded region shown.

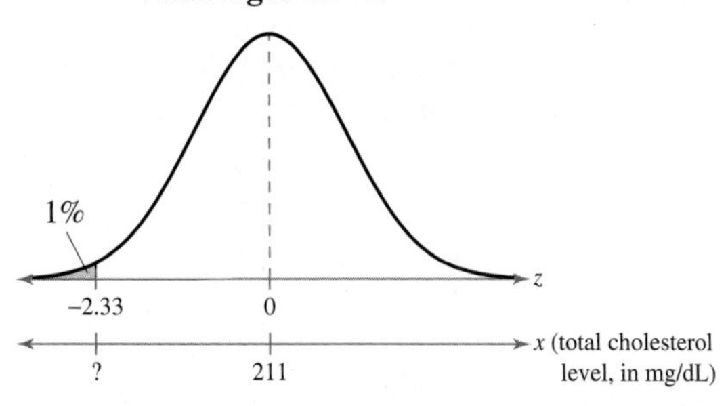

A total cholesterol level in the lowest 1% is any level below the 1st percentile. To find the level that represents the 1st percentile, you must first find the z-score that corresponds to a cumulative area of 0.01. From the Standard Normal Table, you can find that the area closest to 0.01 is 0.0099. So, the z-score that corresponds to an area of 0.01 is $z = -2.33$. Using the equation $x = \mu + z\sigma$, you have

$$\begin{aligned} x &= \mu + z\sigma \\ &= 211 + (-2.33)(39.2) \\ &= 119.66. \end{aligned}$$

So, the value that separates the lowest 1% of total cholesterol levels for men in the 40–49 age group from the highest 99% is about 120.

Try It Yourself 5

The length of time employees have worked at a corporation is normally distributed, with a mean of 11.2 years and a standard deviation of 2.1 years. In a company cutback, the lowest 10% in seniority are laid off. What is the maximum length of time an employee could have worked and still be laid off?

a. *Sketch* a graph.
b. *Find the z-score* that corresponds to the given area.
c. *Find* x using the equation $x = \mu + z\sigma$.
d. *Write* the result as a sentence.

Answer: Page A36

5.4 Exercises

Help

LarsonTutor 5.4

Companion Web Site

Student Solutions Manual 5.4

Videos 5.4

Technology Manuals

Basic Skills and Concepts

In Exercises 1–24, use the Standard Normal Table to find the z-score that corresponds to the given cumulative area or percentile. If the area is not in the table, use the entry closest to the area.

1. 0.0202 **2.** 0.2090 **3.** 0.8023 **4.** 0.6443

5. 0.4364 **6.** 0.0080 **7.** 0.9916 **8.** 0.7673

9. 0.05 **10.** 0.85 **11.** 0.8 **12.** 0.01

13. P_1 **14.** P_{15} **15.** P_{40} **16.** P_{55}

17. P_{88} **18.** P_{96} **19.** P_{25} **20.** P_{50}

21. P_{75} **22.** P_{10} **23.** P_{35} **24.** P_{65}

Graphical Analysis In Exercises 25–30, find the indicated z-score(s) shown in the graph.

25.

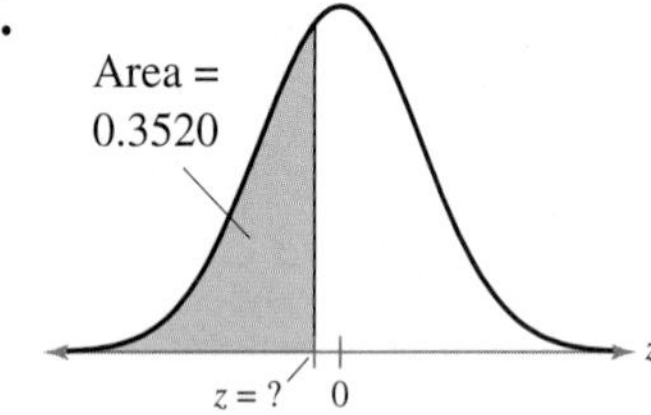

26.

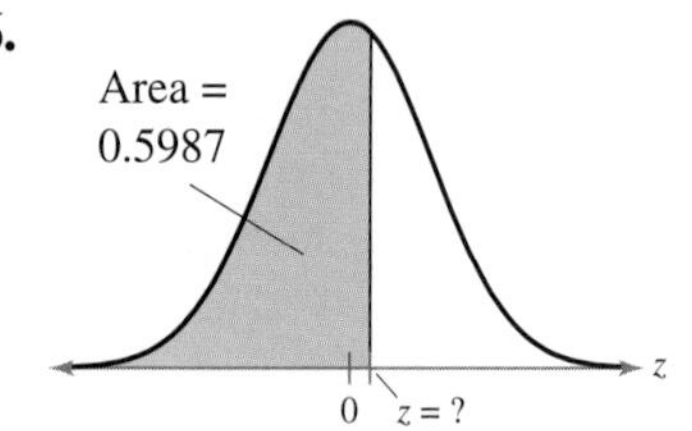

27.

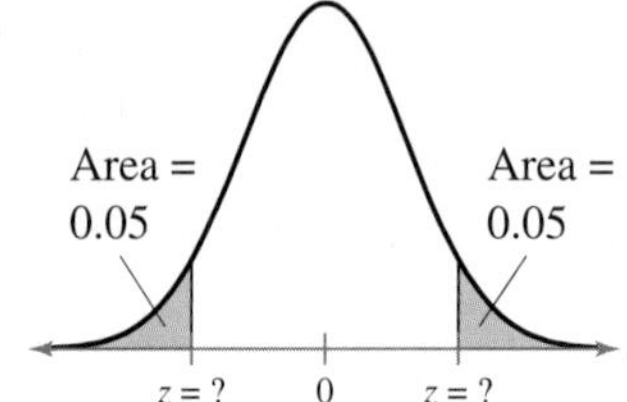

28.

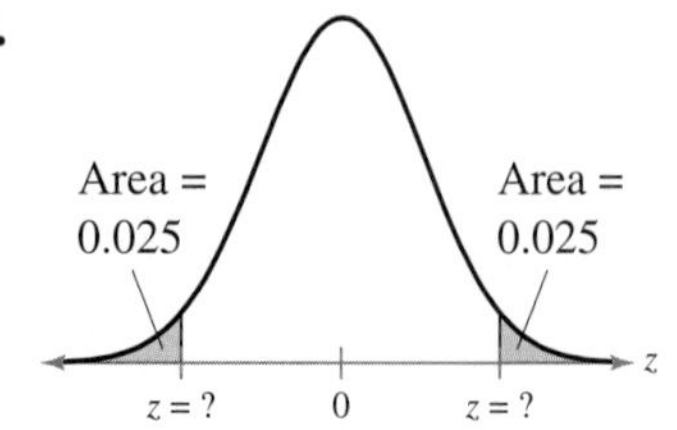

29.

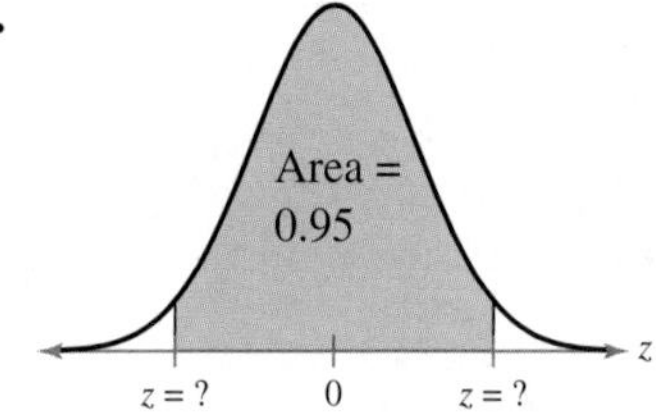

30.

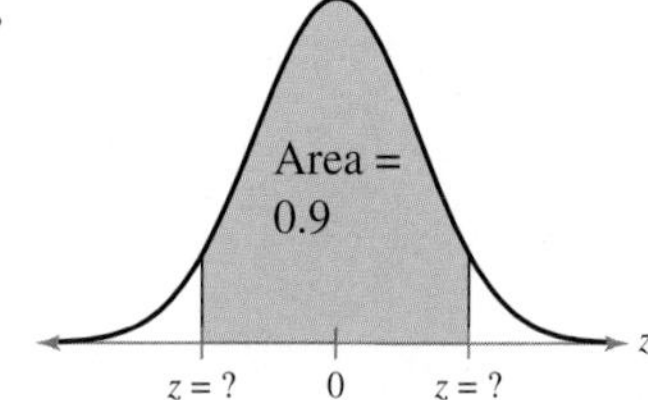

In Exercises 31–34, find the indicated z-score.

31. Find the z-score that has 62.8% of the distribution's area to its left.

32. Find the z-score that has 62.8% of the distribution's area to its right.

33. Find the z-score for which 80% of the distribution's area lies between $-z$ and z.

34. Find the z-score for which 99% of the distribution's area lies between $-z$ and z.

Using Normal Distributions In Exercises 35–40, answer the questions about the specified normal distribution.

35. ***Heights of Women*** ◆ In a survey of women in the U.S. (ages 20–29), the mean height was 64 inches with a standard deviation of 2.75 inches. *(Source: National Center for Health Statistics)*

(a) What height represents the 95th percentile?

(b) What height represents the first quartile?

36. ***Heights of Men*** ◆ In a survey of men in the U.S. (ages 20–29), the mean height was 69.2 inches with a standard deviation of 2.9 inches. *(Source: National Center for Health Statistics)*

(a) What height represents the 90th percentile?

(b) What height represents the first quartile?

37. ***Apples*** ◆ The annual per capita use of apples (in pounds) in the United States can be approximated by a normal distribution, as shown in the graph. *(Source: U.S. Department of Agriculture)*

(a) What annual per capita consumption of apples represents the 10th percentile?

(b) What annual per capita consumption of apples represents the third quartile?

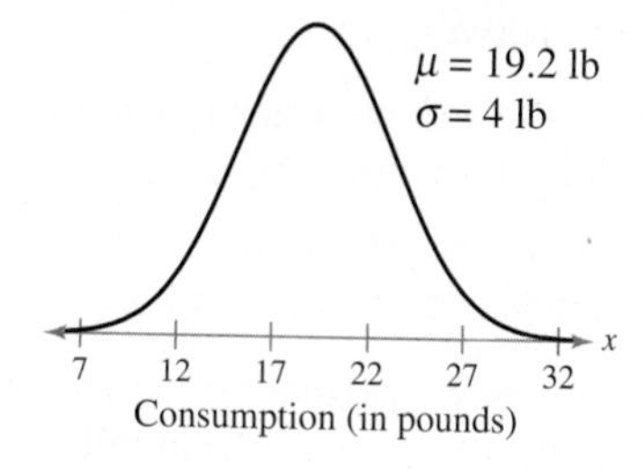

Figure for 37

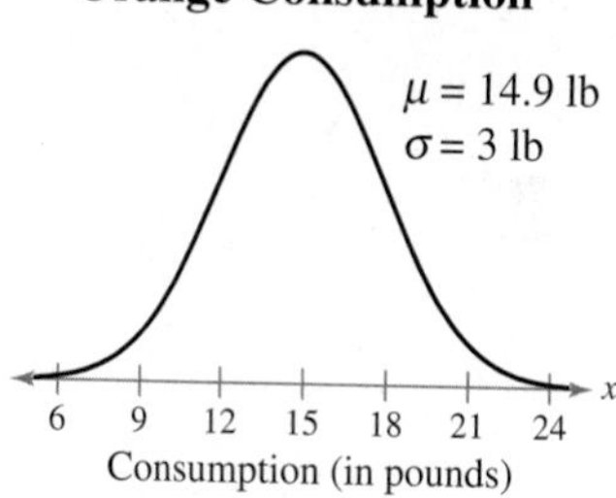

Figure for 38

38. ***Oranges*** ◆ The annual per capita use of oranges (in pounds) in the United States can be approximated by a normal distribution, as shown in the graph. *(Source: U.S. Department of Agriculture)*

(a) What annual per capita consumption of oranges represents the 5th percentile?

(b) What annual per capita consumption of oranges represents the third quartile?

39. ***Heart Transplant Waiting Times*** ◆ The time spent (in days) waiting for a heart transplant in Ohio and Michigan for patients with type A^+ blood can be approximated by a normal distribution, as shown in the graph. *(Source: Organ Procurement and Transplant Network)*

(a) What is the shortest time spent waiting for a heart that would still place a patient in the top 30% of waiting times?

(b) What is the longest time spent waiting for a heart that would still place a patient in the bottom 10% of waiting times?

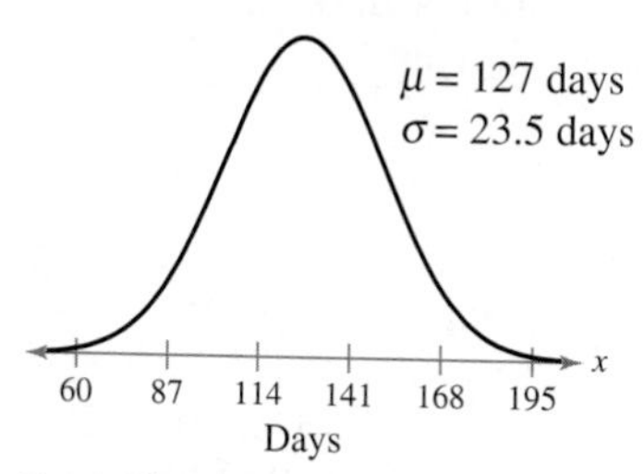

Figure for 39

Annual U.S. Per Capita Breakfast Cereal Consumption

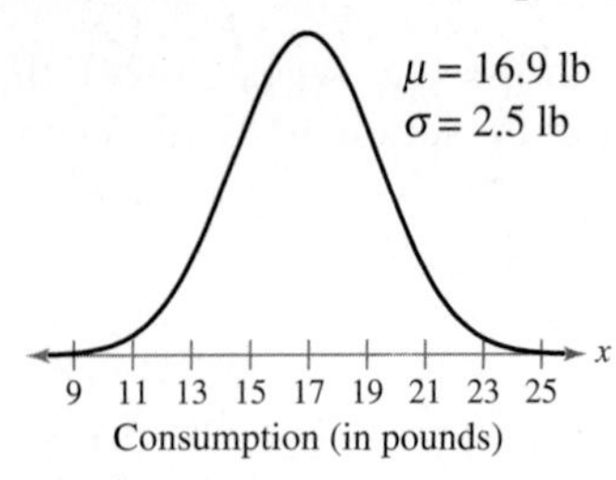

Figure for Exercise 40

40. ***Breakfast Cereal*** ◆ The annual per capita use of breakfast cereal (in pounds) in the United States can be approximated by a normal distribution, as shown in the graph. *(Adapted from U.S. Department of Agriculture)*

(a) What is the smallest annual per capita consumption of breakfast cereal that can be in the top 25% of consumptions?

(b) What is the largest annual per capita consumption of breakfast cereal that can be in the bottom 15% of consumptions?

Extending the Basics

41. ***Writing a Guarantee*** ◆ A brand of automobile tire has a life expectancy that is normally distributed, with a mean life of 30,000 miles and a standard deviation of 2500 miles. You sell this brand of tire and you want to give a guarantee for free replacement of tires that don't wear well. How should you word your guarantee if you are willing to replace approximately 10% of the tires you sell?

42. ***Planning a Study*** ◆ The life spans of a species of fruit fly are normally distributed, with a mean of 36 days and a standard deviation of 4 days. You are planning a study to obtain data for the life span of 200 fruit flies. For how many days should you plan to conduct the study? (Assume you want to include all data that lie within three standard deviations of the mean.)

43. ***Guarantee Period*** ◆ The life span of a machine is normally distributed, with a mean of 10.2 years and a standard deviation of 1.7 years. The manufacturer will replace a machine if it breaks before the guarantee period is over.

(a) If the manufacturer is willing to replace no more than 10%, find the length of time the manufacturer should set for the guarantee.

(b) If the manufacturer is willing to replace no more than 7%, find the length of time the manufacturer should set for the guarantee.

44. ***Test Score Requirement*** ◆ A college requires applicants to have an ACT score in the top 12% of all test scores. The ACT scores are normally distributed, with a mean of 21 and a standard deviation of 4.7.

(a) Find the lowest test score that a student could get and still meet the college's requirement.

(b) If 1500 students are randomly selected, how many would be expected to have a test score that would meet the college's requirement?

(c) How does the answer to part (a) change if the college decides to accept the top 18% of all test scores?

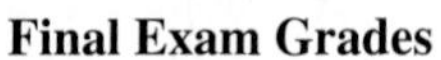

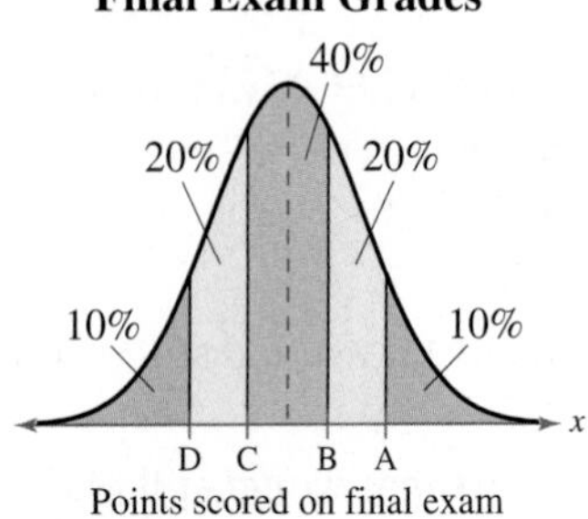

Figure for Exercise 45

45. In a large section of a statistics class, the points for the final exam are normally distributed with a mean of 72 and a standard deviation of 9. Grades are to be assigned according to the following rule.

- The top 10% receive As
- The next 20% receive Bs
- The middle 40% receive Cs
- The next 20% receive Ds
- The bottom 10% receive Fs

Find the lowest score on the final exam that would qualify a student for an A, a B, a C, and a D.

5 Case Study

WWW.CDC.GOV/NCHS

Birth Weights in America

The National Center for Health Statistics keeps records of many health-related aspects of people, including the birth weights of all babies born in the United States.

The birth weight of a baby is related to its gestation period (the time between conception and birth). For a given gestation period, the birth weights can be approximated by a normal distribution. The means and standard deviations of the birth weights for various gestation periods are shown at the right.

One of the many goals of NCHS is to reduce the percent of babies born with low birth weights. As you can see from the graph at the upper right, the problem of low birth weights increased from 1985 to 1998.

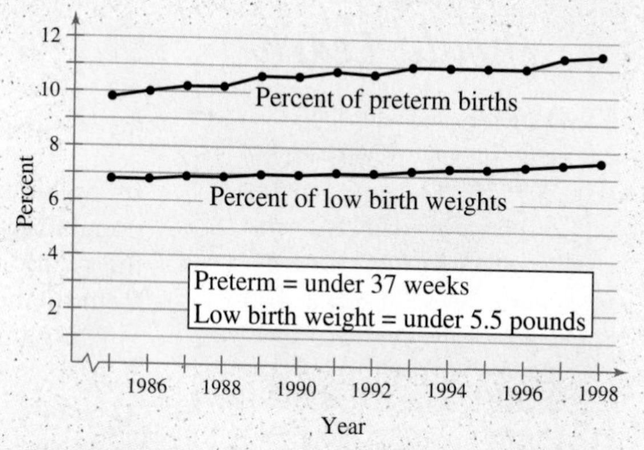

Gestation Period	Mean Birth Weight	Standard Deviation
Under 28 Weeks	2.01 lb	1.32 lb
28 to 31 Weeks	4.29 lb	1.92 lb
32 to 35 Weeks	5.82 lb	1.51 lb
36 Weeks	6.45 lb	1.22 lb
37 to 39 Weeks	7.31 lb	1.11 lb
40 Weeks	7.74 lb	1.07 lb
41 Weeks	7.89 lb	1.09 lb
42 Weeks and over	7.75 lb	1.14 lb

Exercises

1. The distributions of birth weights for three gestation periods are shown. Match the curves with the gestation periods. Explain your reasoning.

(a)

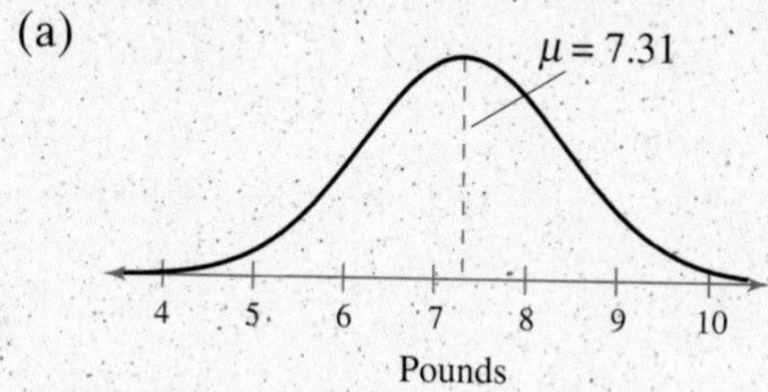

(b)

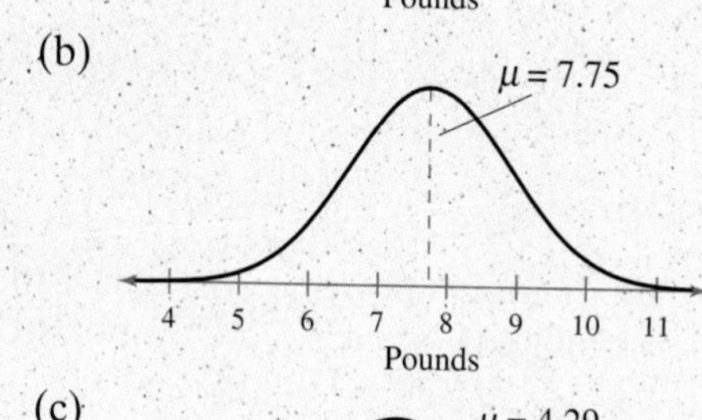

(c)

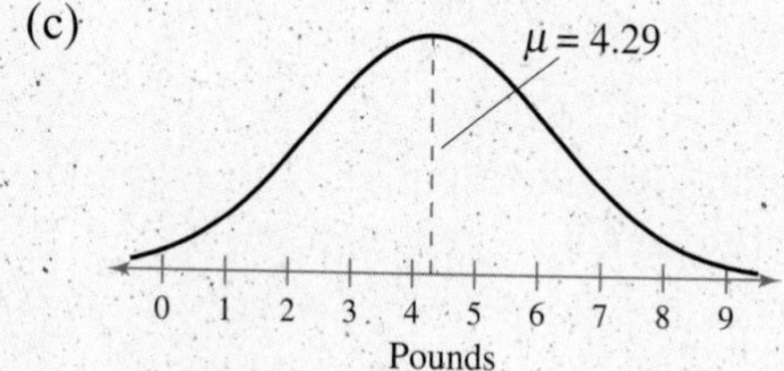

2. What percent of the babies born with each gestation period have a low birth weight (under 5.5 pounds)? Explain your reasoning.

(a) Under 28 weeks (b) 32 to 35 weeks
(c) 37 to 39 weeks (d) 42 weeks and over

3. Describe the weights of the top 10% of the babies born with each gestation period. Explain your reasoning.

(a) 37 to 39 weeks (b) 42 weeks and over

4. For each gestation period, what is the probability that a baby will weigh between 6 and 9 pounds at birth?

(a) 32 to 35 weeks (b) 37 to 39 weeks
(c) 42 weeks and over

5. A birth weight of less than 3.3 pounds is classified by NCHS as a "very low birth weight." What is the probability that a baby has a very low birth weight for each gestation period?

(a) Under 28 weeks (b) 32 to 35 weeks
(c) 37 to 39 weeks

5.5 The Central Limit Theorem

What You Should Learn

- How to find sampling distributions and verify their properties
- How to interpret the Central Limit Theorem
- How to apply the Central Limit Theorem to find the probability of a sample mean

Sampling Distributions • The Central Limit Theorem • Probability and the Central Limit Theorem

Sampling Distributions

In previous sections, you studied the relationship between the mean of a population and values of a random variable. In this section, you will study the relationship between a population mean and the means of samples taken from the population.

DEFINITION

A **sampling distribution** is the probability distribution of a sample statistic that is formed when samples of size n are repeatedly taken from a population. If the sample statistic is the sample mean, then the distribution is the **sampling distribution of sample means.**

For instance, consider the following Venn diagram. The rectangle represents a large population, and each circle represents a sample of size n. Because the sample entries can differ, the sample means can also differ. The mean of Sample 1 is $\bar{x}_1$, the mean of Sample 2 is $\bar{x}_2$, and so on. The sampling distribution of the sample means for samples of size n for this population consists of $\bar{x}_1$, $\bar{x}_2$, $\bar{x}_3$, and so on. If the samples are drawn with replacement, an infinite number of samples can be drawn from the population.

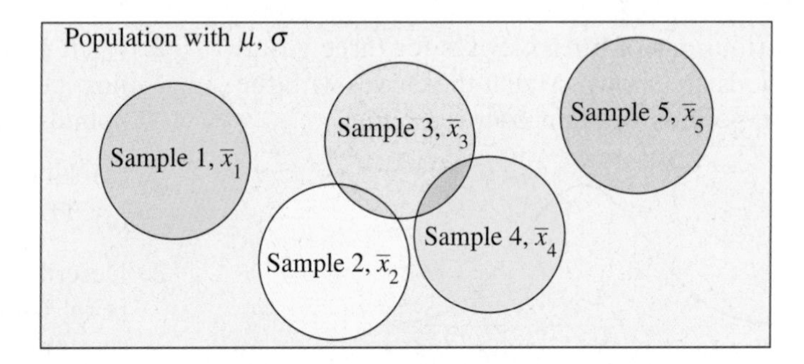

Properties of Sampling Distributions of Sample Means

1. The mean of the sample means $\mu_{\bar{x}}$ is equal to the population mean μ.

$$\mu_{\bar{x}} = \mu$$

2. The standard deviation of the sample means $\sigma_{\bar{x}}$ is equal to the population standard deviation σ divided by the square root of n.

$$\sigma_{\bar{x}} = \frac{\sigma}{\sqrt{n}}$$

The standard deviation of the sampling distribution of the sample means is called the **standard error of the mean.**

EXAMPLE 

A Sampling Distribution of Sample Means

You write the population values $\{1, 3, 5, 7\}$ on slips of paper and put them in a box. Then you randomly choose two slips of paper, with replacement. List all possible samples of size $n = 2$ and calculate the mean of each. These means form the sampling distribution of the sample means. Find the mean, variance, and standard deviation of the sample means. Compare your results with the mean $\mu = 4$, variance $\sigma^2 = 5$, and standard deviation $\sigma = \sqrt{5} \approx 2.236$ of the population.

SOLUTION List all 16 samples of size 2 from the population and the mean of each sample.

Sample	Mean
1, 1	1
1, 3	2
1, 5	3
1, 7	4
3, 1	2
3, 3	3
3, 5	4
3, 7	5

Sample	Mean
5, 1	3
5, 3	4
5, 5	5
5, 7	6
7, 1	4
7, 3	5
7, 5	6
7, 7	7

Relative Frequency Histogram of Population of x

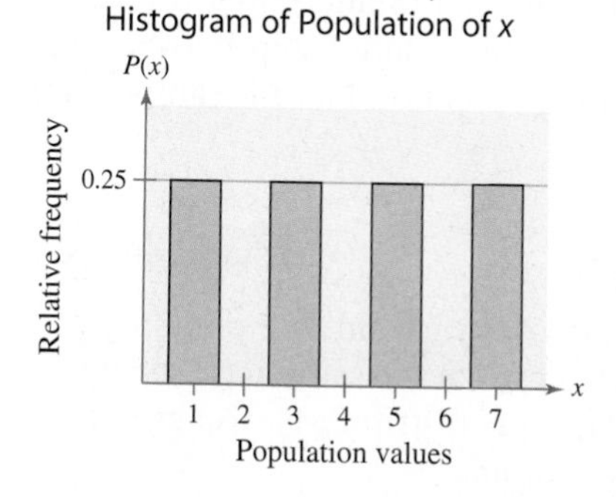

Relative Frequency Distribution of Sample Means

$\bar{x}$	f	Relative frequency
1	1	0.0625
2	2	0.1250
3	3	0.1875
4	4	0.2500
5	3	0.1875
6	2	0.1250
7	1	0.0625

Relative Frequency Histogram of Sampling Distribution of $\bar{x}$

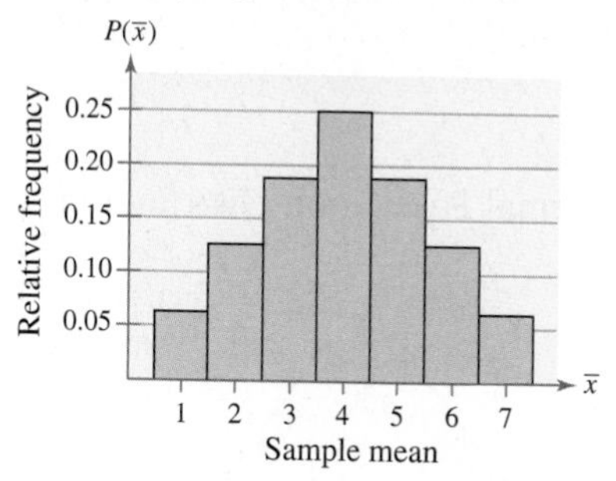

After constructing a relative frequency distribution of the sample means, you can graph the sampling distribution using a relative frequency histogram as shown at the left. Notice that the shape of the histogram is bell shaped and symmetric, similar to a normal curve. The mean, variance, and standard deviation of the 16 sample means are

$$\mu_{\bar{x}} = 4$$

$$(\sigma_{\bar{x}})^2 = \frac{5}{2} = 2.5 \quad \text{and} \quad \sigma_{\bar{x}} = \sqrt{\frac{5}{2}} = \sqrt{2.5} \approx 1.581.$$

These results satisfy the properties of sampling distributions because

$$\mu_{\bar{x}} = \mu = 4 \quad \text{and} \quad \sigma_{\bar{x}} = \frac{\sigma}{\sqrt{n}} = \frac{\sqrt{5}}{\sqrt{2}} \approx \frac{2.236}{\sqrt{2}} \approx 1.581.$$

Try It Yourself 1

List all possible samples of $n = 2$, with replacement, from the population $\{1, 2, 3, 5, 6, 7\}$. Calculate the mean, variance, and standard deviation of the sample means. Compare these values to the corresponding population parameters.

a. *Form* all possible samples of size 2 and find the mean of each.
b. *Find* the mean, variance, and standard deviation of the frequency distribution of the sample means.
c. *Compare* the mean, variance, and standard deviation of the sample means with those for the population.

Answer: Page A36

Study Tip

Review Sections 2.3 and 2.4 to find the mean and standard deviation of a frequency distribution.

The Central Limit Theorem

The Central Limit Theorem is one of the most important and useful theorems in statistics. This theorem forms the foundation for the inferential branch of statistics. The Central Limit Theorem describes the relationship between the sampling distribution of sample means and the population that the samples are taken from.

The Central Limit Theorem

1. If samples of size n, where $n \geq 30$, are drawn from any population with a mean μ and a standard deviation σ, then the sampling distribution of sample means approximates a normal distribution. The greater the sample size, the better the approximation.
2. If the population itself is normally distributed, the sampling distribution of sample means is normally distributed for *any* sample size n.

In either case, the sampling distribution of sample means has a mean equal to the population mean.

$$\mu_{\bar{x}} = \mu \qquad \text{Mean}$$

And the sampling distribution of sample means has a variance equal to $1/n$ times the variance of the population and a standard deviation equal to the population standard deviation divided by the square root of n.

$$\sigma_{\bar{x}}^2 = \frac{\sigma^2}{n} \qquad \text{Variance}$$

$$\sigma_{\bar{x}} = \frac{\sigma}{\sqrt{n}} \qquad \text{Standard deviation}$$

The standard deviation of the sampling distribution of the sample means, $\sigma_{\bar{x}}$, is also called the **standard error of the mean.**

Insight

The distribution of sample means has the same mean as the population. But its standard deviation is less than the standard deviation of the population. This tells you that the distribution of sample means has the same center as the population, but it is not as spread out.

Moreover, the distribution of sample means becomes less and less spread out (tighter concentration about the mean) as the sample size n increases.

1. Any Population Distribution

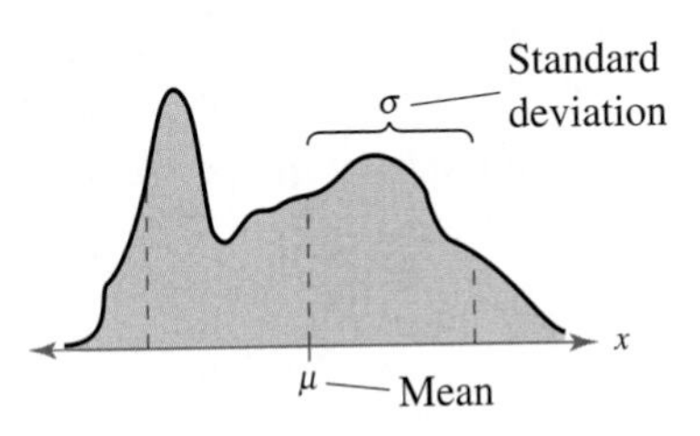

Distribution of Sample Means, $n \geq 30$

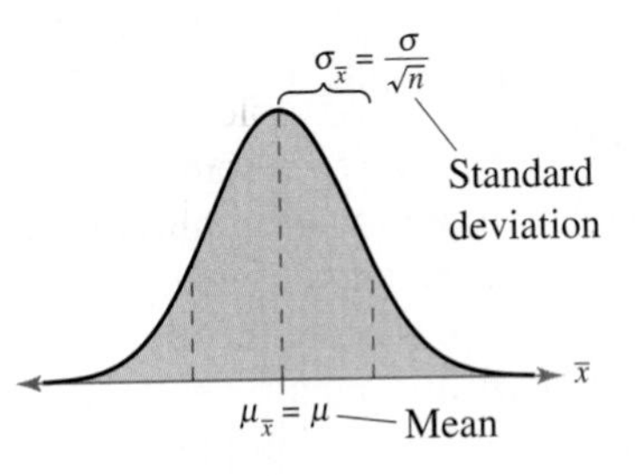

2. Normal Population Distribution

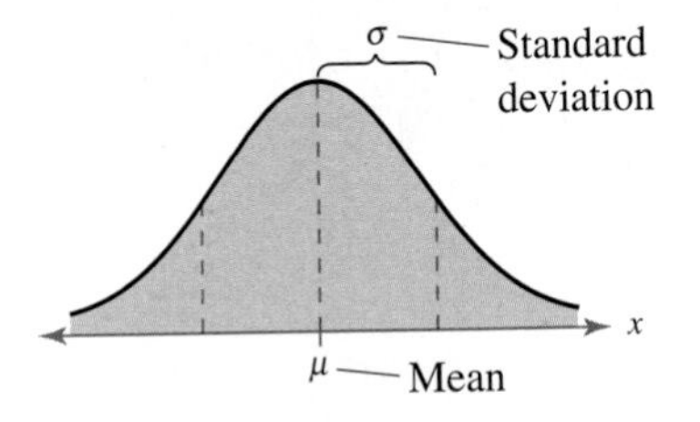

Distribution of Sample Means (any n)

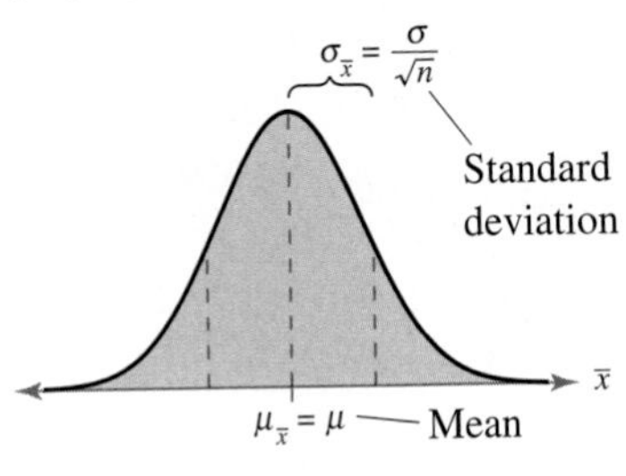

EXAMPLE 2

Interpreting the Central Limit Theorem

Phone bills for residents of Cincinnati have a mean of \$64 and a standard deviation of \$9, as shown in the following graph. Random samples of 36 phone bills are drawn from this population and the mean of each sample is determined. Find the mean and standard error of the mean of the sampling distribution. Then sketch a graph of the sampling distribution.

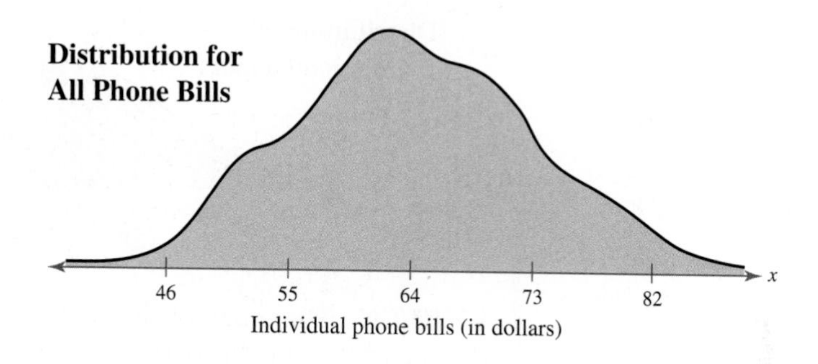

SOLUTION The mean of the sampling distribution is equal to the population mean, and the standard error of the mean is equal to the population standard deviation divided by $\sqrt{n}$. So,

$$\mu_{\bar{x}} = \mu = 64$$

and

$$\sigma_{\bar{x}} = \frac{\sigma}{\sqrt{n}} = \frac{9}{\sqrt{36}} = 1.5.$$

From the Central Limit Theorem, because the sample size is greater than 30, the sampling distribution can be approximated by a normal distribution with $\mu = \$64$ and $\sigma = \$1.50$.

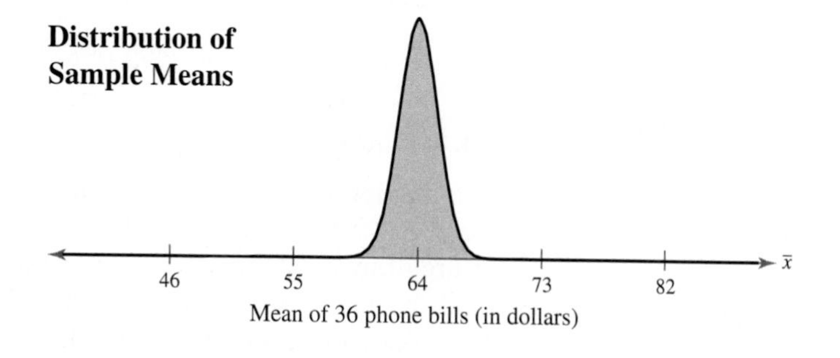

Try It Yourself 2

Suppose random samples of size 100 are drawn from the population in Example 2. Find the mean and standard error of the mean of the sampling distribution. Sketch a graph of the sampling distribution and compare it to the sampling distribution in Example 2.

a. *Find* $\mu_{\bar{x}}$ and $\sigma_{\bar{x}}$.

b. *Identify* the sample size. If $n \geq 30$, *sketch* a normal curve with mean $\mu_{\bar{x}}$ and standard deviation $\sigma_{\bar{x}}$.

Answer: Page A36

Picturing the World

In a recent year, there were more than 4 million parents in the United States who received child support payments. The following histogram shows the distribution of children per custodial parent. The mean number of children was 1.7 and the standard deviation was 0.9. *(Adapted from U.S. Bureau of the Census)*

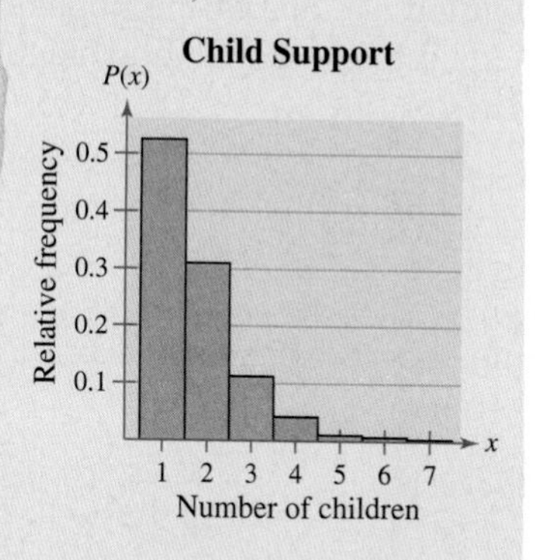

You randomly select 35 parents who receive child support and ask how many children in their custody are receiving child support payments. What is the probability that the mean of the sample is between 1.5 and 1.9 children?

EXAMPLE 3

Interpreting the Central Limit Theorem

The heights of fully grown white oak trees are normally distributed, with a mean of 90 feet and standard deviation of 3.5 feet. Random samples of size 4 are drawn from this population, and the mean of each sample is determined. Find the mean and standard error of the mean of the sampling distribution. Then sketch a graph of the sampling distribution.

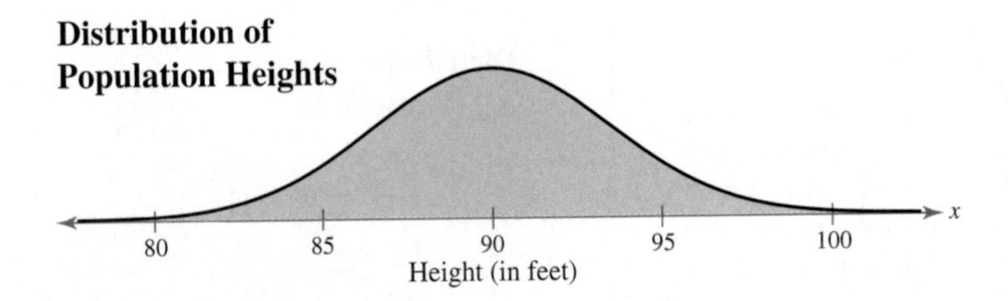

SOLUTION The mean of the sampling distribution is equal to the population mean and the standard error of the mean is equal to the population standard deviation divided by $\sqrt{n}$. So,

$$\mu_{\bar{x}} = \mu = 90 \text{ feet and } \sigma_{\bar{x}} = \frac{\sigma}{\sqrt{n}} = \frac{3.5}{\sqrt{4}} = 1.75 \text{ feet.}$$

From the Central Limit Theorem, because the population is normally distributed, the sampling distribution of the sample means is also normally distributed.

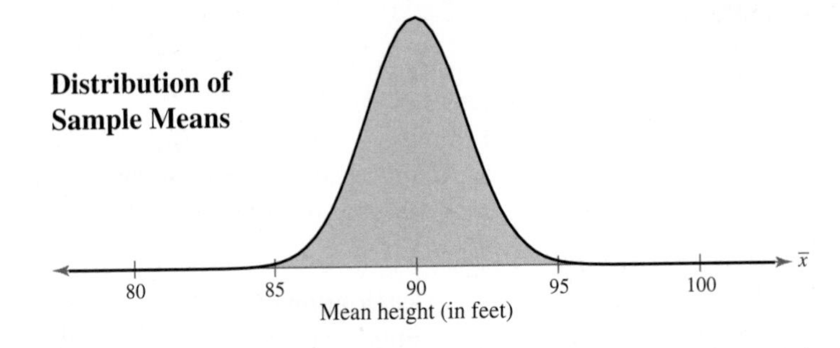

Try It Yourself 3

The diameters of fully grown white oak trees are normally distributed, with a mean of 3.5 feet and a standard deviation of 0.2 foot. Random samples of size 16 are drawn from this population, and the mean of each sample is determined. Find the mean and standard error of the mean of the sampling distribution. Then sketch a graph of the sampling distribution.

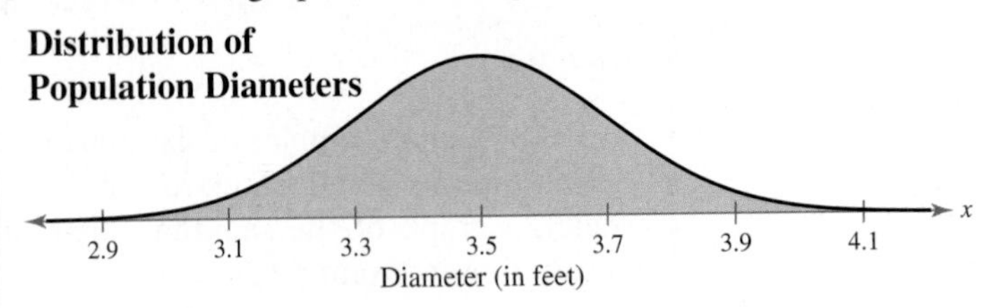

a. *Find* $\mu_{\bar{x}}$ and $\sigma_{\bar{x}}$.
b. *Sketch* a normal curve with mean $\mu_{\bar{x}}$ and standard deviation $\sigma_{\bar{x}}$.

Answer: Page A37

Probability and the Central Limit Theorem

In Sections 5.2 and 5.3, you learned how to find the probability that a random variable x will fall in a given interval of population values. In a similar manner, you can find the probability that a sample mean $\bar{x}$ will fall in a given interval of the $\bar{x}$ sampling distribution. To transform $\bar{x}$ to a z-score, you can use the formula

$$z = \frac{\text{value} - \text{mean}}{\text{standard error}} = \frac{\bar{x} - \mu_{\bar{x}}}{\sigma_{\bar{x}}} = \frac{\bar{x} - \mu}{\sigma/\sqrt{n}}.$$

EXAMPLE 4

Finding Probabilities for Sampling Distributions

The graph at the right lists the length of time adults spend reading newspapers. You randomly select 50 adults ages 18 to 24. What is the probability that the mean time they spend reading the newspaper is between 8.7 and 9.5 minutes? Assume that $\sigma = 1.5$ minutes.

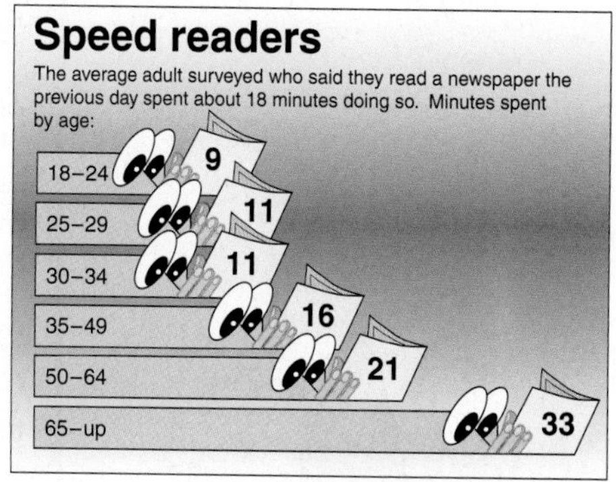

SOLUTION Because the sample size is greater than 30, you can use the Central Limit Theorem to conclude that the distribution of sample means is approximately normal with a mean and a standard deviation of

$$\mu_{\bar{x}} = \mu = 9 \text{ minutes}$$

$$\sigma_{\bar{x}} = \frac{\sigma}{\sqrt{n}} = \frac{1.5}{\sqrt{50}} \approx 0.21213 \text{ minute}.$$

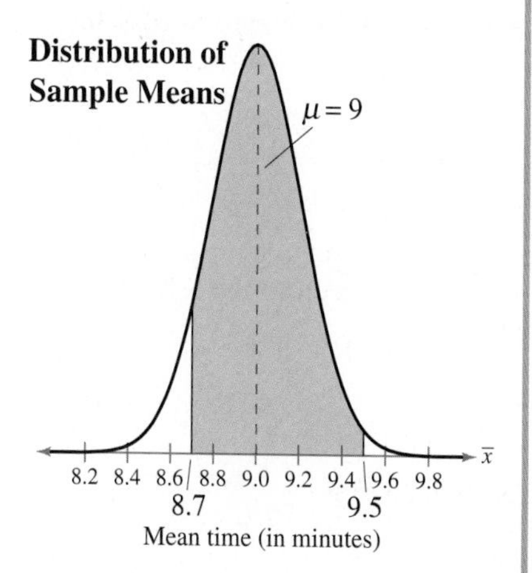

The graph of this distribution is shown at the left with a shaded area between 8.7 and 9.5 minutes.

The z-scores that correspond to sample means of 8.7 and 9.5 minutes are

$$z_1 = \frac{8.7 - 9}{1.5/\sqrt{50}} \approx \frac{8.7 - 9}{0.21213} \approx -1.41 \quad \text{and}$$

$$z_2 = \frac{9.5 - 9}{1.5/\sqrt{50}} \approx \frac{9.5 - 9}{0.21213} \approx 2.36.$$

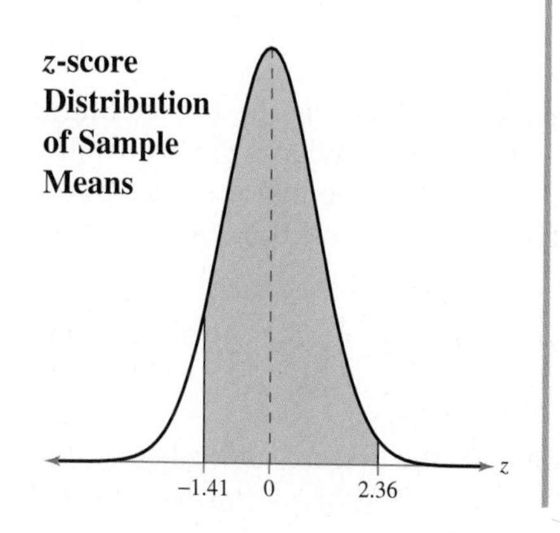

So, the probability that the mean time the 50 adults spend reading the newspaper is between 8.7 and 9.5 minutes is

$$\begin{aligned} P(8.7 < \bar{x} < 9.5) &= P(-1.41 < z < 2.36) \\ &= P(z < 2.36) - P(z < -1.41) \\ &= 0.9909 - 0.0793 \\ &= 0.9116. \end{aligned}$$

So, 91.16% of the samples of 50 adults ages 18 to 24 will have a mean of between 8.7 and 9.5 minutes. This implies that, assuming the value of $\mu = 9$ is correct, only 8.84% of such sample means will lie outside the given interval.

Try It Yourself 4

You randomly select 100 adults aged 18 to 24. What is the probability that the mean time they spend reading the newspaper is between 8.7 and 9.5 minutes? Use $\mu = 9$ and $\sigma = 1.5$ minutes.

a. Use the Central Limit Theorem to *find* $\mu_{\bar{x}}$ and $\sigma_{\bar{x}}$ and *sketch* the sampling distribution of the sample means.
b. *Find the z-scores* that correspond to $\bar{x} = 8.7$ minutes and $\bar{x} = 9.5$ minutes.
c. *Find the cumulative area* that corresponds to each z-score and calculate the probability.

Answer: Page A37

Study Tip

Before you find probabilities for intervals of the sample mean $\bar{x}$, use the Central Limit Theorem to determine the mean and the standard deviation of the sampling distribution of the sample means. That is, calculate $\mu_{\bar{x}}$ and $\sigma_{\bar{x}}$.

EXAMPLE 5

Finding Probabilities for Sampling Distributions

The mean rent of an apartment in a professionally managed apartment building is \$780. You randomly select nine professionally managed apartment buildings. What is the probability that the mean rent is less than \$825? Assume that the rents are normally distributed, with a standard deviation of \$150. *(Source: M/PH Research)*

SOLUTION Because the population is normally distributed, you can use the Central Limit Theorem to conclude that the distribution of sample means is normally distributed, with a mean of \$780 and a standard deviation of \$50.

$$\mu_{\bar{x}} = \mu = 780$$

and

$$\sigma_{\bar{x}} = \frac{\sigma}{\sqrt{n}} = \frac{150}{\sqrt{9}} = 50$$

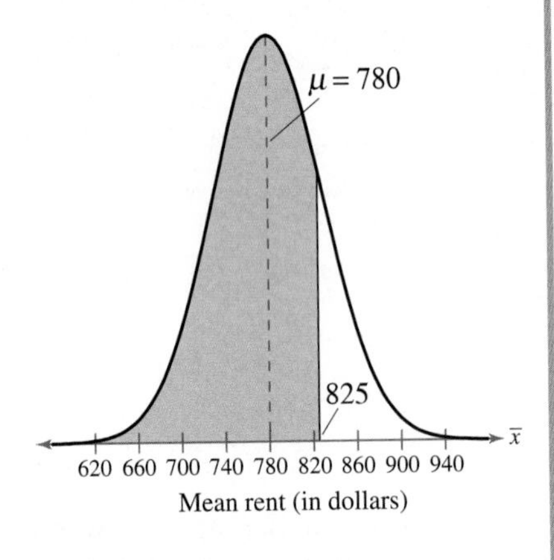

The graph of this distribution is shown at the left. The area to the left of \$825 is shaded. The z-score that corresponds to \$825 is

$$z = \frac{825 - 780}{150/\sqrt{9}} = \frac{825 - 780}{50} = 0.9.$$

So, the probability that the mean rent is less than \$825 is

$$P(\bar{x} < 825) = P(z < 0.9) = 0.8159.$$

This means that 81.59% of such samples with $n = 9$ will have a mean less than \$825 and 18.41% of these sample means will lie outside this interval.

Try It Yourself 5

The average sales price of an existing single-family house in the United States is \$125,700. You randomly select 12 single-family houses. What is the probability that the mean sales price is more than \$100,000? Assume that the sales prices are normally distributed with a standard deviation of \$26,000. *(Adapted from National Association of Realtors)*

a. Use the Central Limit Theorem to *find* $\mu_{\bar{x}}$ and $\sigma_{\bar{x}}$ and *sketch* the sampling distribution of the sample means.
b. *Find the z-score* that corresponds to $\bar{x} = \$100{,}000$.
c. *Find the cumulative area* that corresponds to the z-score and calculate the probability.

Answer: Page A37

EXAMPLE 

Finding Probabilities for x and $\overline{x}$

Credit card balances are normally distributed, with a mean of \$2870 and a standard deviation of \$900.

1. What is the probability that a randomly selected credit card holder has a credit card balance less than \$2500?
2. You randomly select 25 credit card holders. What is the probability that their mean credit card balance is less than \$2500?
3. Compare the probabilities from (1) and (2).

SOLUTION

1. In this case, you are asked to find the probability associated with a certain value of the random variable x. The z-score that corresponds to $x = \$2500$ is

$$z = \frac{x - \mu}{\sigma} = \frac{2500 - 2870}{900} \approx -0.41.$$

So, the probability that the card holder has a balance less than \$2500 is

$$P(x < 2500) = P(z < -0.41) = 0.3409.$$

2. Here, you are asked to find the probability associated with a sample mean $\overline{x}$. The z-score that corresponds to $\overline{x} = \$2500$ is

$$z = \frac{\overline{x} - \mu_{\overline{x}}}{\sigma_{\overline{x}}} = \frac{\overline{x} - \mu}{\sigma/\sqrt{n}}$$
$$= \frac{2500 - 2870}{900/\sqrt{25}}$$
$$= \frac{2500 - 2870}{180}$$
$$\approx -2.06.$$

So, the probability that the mean credit card balance of the 25 card holders is less than \$2500 is

$$P(\overline{x} < 2500) = P(z < -2.06) = 0.0197$$

3. Where there is a 34% chance that an individual will have a balance less than \$2500, there is only a 2% chance that the mean of a sample of 25 will have a balance less than \$2500.

Study Tip

To find probabilities for individual members of a population with a normally distributed random variable x, use the formula

$$z = \frac{x - \mu}{\sigma}.$$

To find probabilities for the mean $\overline{x}$ of a sample size n, use the formula

$$z = \frac{\overline{x} - \mu_{\overline{x}}}{\sigma_{\overline{x}}}.$$

Try It Yourself 6

Prices for sound-system receivers are normally distributed, with a mean of \$625 and a standard deviation of \$150. (1) What is the probability that a randomly selected receiver costs less than \$700? (2) You randomly select 10 receivers. What is the probability that their mean cost is less than \$700? (3) Compare these two probabilities.

a. *Find* the z-scores that correspond to x and $\overline{x}$.
b. Use the Standard Normal Table to *find the probability* associated with each z-score.
c. *Compare* the probabilities.

Answer: Page A37

5.5 Exercises

Basic Skills and Concepts

Verifying Properties of Sampling Distributions In Exercises 1 and 2, find the mean and standard deviation of the population. List all samples (with replacement) of the given size from that population. Find the mean and standard deviation of the sampling distribution and compare them to the mean and standard deviation of the population.

1. The number of movies that all five people in a family have seen in the past month is 4, 2, 8, 0, and 6. Use a sample size of 3.

2. Four people in a carpool paid the following amounts for textbooks this semester: \$120, \$140, \$180, and \$220. Use a sample size of 2.

Graphical Analysis In Exercises 3 and 4, the graph of a population distribution is shown at the left with its mean and standard deviation. Assume that a sample size of 100 is drawn from each population. Decide which of the graphs labeled (a)–(c) would most closely resemble the sampling distribution of the sample means for each graph. Explain your reasoning.

3. The waiting time (in seconds) at a traffic signal during a red light

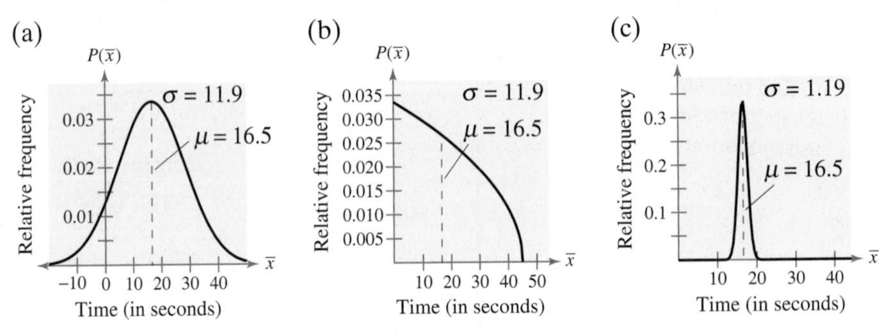

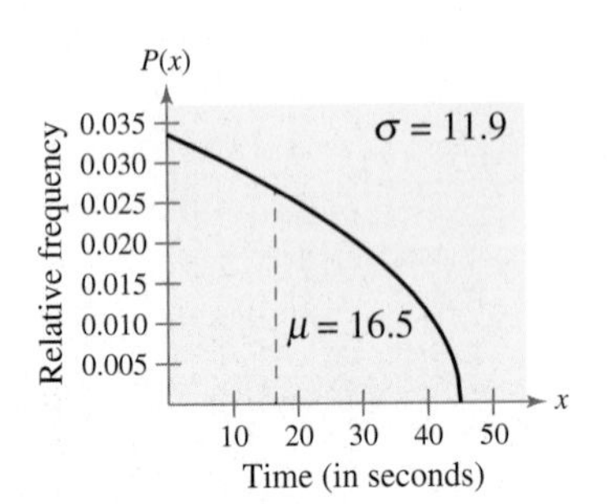

Figure for Exercise 3

4. The annual snowfall (in feet) for a central New York state county

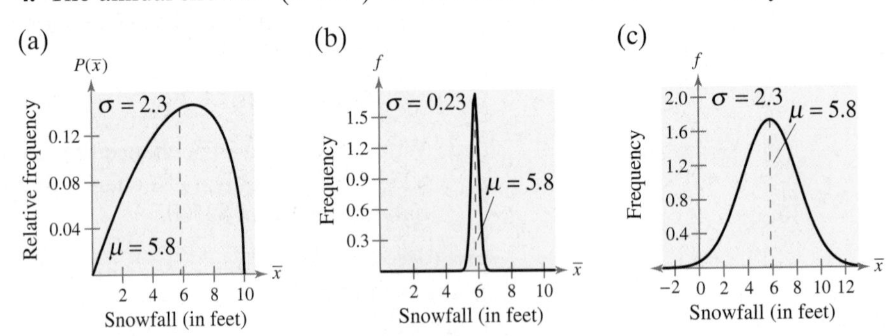

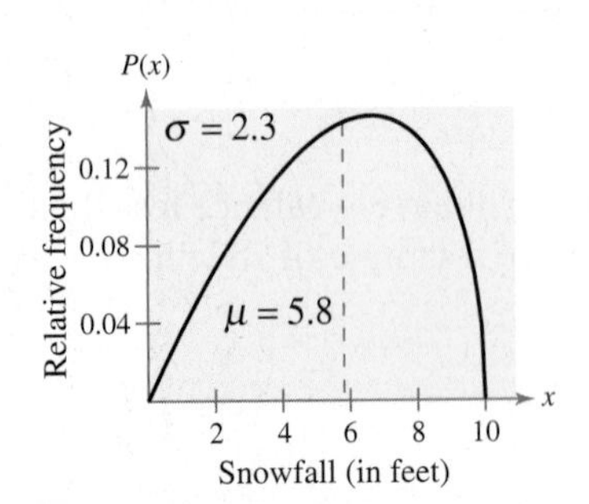

Figure for Exercise 4

Using the Central Limit Theorem In Exercises 5–8, use the Central Limit Theorem to find the mean and standard error of the mean of the indicated sampling distribution. Then sketch a graph of the sampling distribution.

5. ***Heights of Trees*** ◆ The heights of fully grown sugar maple trees are normally distributed, with a mean of 87.5 feet and a standard deviation of 6.25 feet. Random samples of size 12 are drawn from the population and the mean of each sample is determined.

6. ***Fly Eggs*** ◆ The number of eggs a female house fly lays during her lifetime is normally distributed, with a mean of 800 eggs and a standard deviation of 100 eggs. Random samples of size 15 are drawn from this population and the mean of each sample is determined.

7. ***Red Meat Consumed*** ◆ The per capita consumption of red meat by people in the U.S. in a recent year was normally distributed, with a mean of 115.6 pounds and a standard deviation of 38.5 pounds. Random samples of size 20 are drawn from this population and the mean of each sample is determined. *(Adapted from U.S. Department of Agriculture)*

8. ***Soft Drinks*** ◆ The per capita consumption of soft drinks by people in the U.S. in a recent year was normally distributed, with a mean of 53.0 gallons and a standard deviation of 17.1 gallons. Random samples of size 25 are drawn from this population and the mean of each sample is determined. *(Adapted from U.S. Department of Agriculture)*

9. Repeat Exercise 5 for samples of size 24 and 36. What happens to the mean and standard deviation of the distribution of sample means as the size of the sample increases?

10. Repeat Exercise 6 for samples of size 30 and 45. What happens to the mean and to the standard deviation of the distribution of sample means as the size of the sample increases?

Finding Probabilities In Exercises 11–16, find the probabilities.

11. ***Plumber Salaries*** ◆ The population mean annual salary for plumbers is $\mu = \$32{,}500$. A random sample of 42 plumbers is drawn from this population. What is the probability that the mean salary of the sample, $\overline{x}$, is less than $30,000? Assume $\sigma = \$5600$. *(Adapted from Jobs Rated Almanac)*

12. ***Nurse Salaries*** ◆ The population mean annual salary for registered nurses is $\mu = \$33{,}000$. A sample of 35 registered nurses is randomly selected. What is the probability that the mean annual salary of the sample, $\overline{x}$, is less than $29,500? Assume $\sigma = \$1700$. *(Adapted from Jobs Rated Almanac)*

13. ***Gas Prices: New England*** ◆ During a certain week the mean price of gasoline in the New England region was $\mu = \$1.535$ per gallon. What is the probability that the mean price $\overline{x}$ for a sample of 32 randomly selected gas stations in that area was between $1.530 and $1.545 that week? Assume $\sigma = \$0.045$. *(Adapted from Energy Information Administration)*

14. ***Gas Prices: California*** ◆ During a certain week the mean price of gasoline in California was $\mu = \$1.722$ per gallon. A random sample of 38 gas stations is drawn from this population. What is the probability that $\overline{x}$, the mean price for the sample, was between $1.727 and $1.737? Assume $\sigma = \$0.049$. *(Adapted from Energy Information Administration)*

15. ***Height of Women*** ◆ The mean height of women in the U.S. (ages 20–29) is $\mu = 64$ inches. If a random sample of 60 women (ages 20–29) is selected, what is the probability that $\overline{x}$, the mean height for the sample, is greater than 66 inches? Assume $\sigma = 2.75$ inches. *(Source: National Center for Health Statistics)*

16. ***Height of Men*** ◆ The mean height of men in the U.S. (ages 20–29) is $\mu = 69.2$ inches. If a random sample of 60 men in this age group is selected, what is the probability that $\overline{x}$, the mean height for the sample, is greater than 70 inches? Assume $\sigma = 2.9$ inches. *(Source: National Center for Health Statistics)*

17. ***Which Is More Likely?*** ◆ Assume that the heights given in Exercise 15 are normally distributed. Are you more likely to randomly select one woman with a height less than 70 inches or are you more likely to select a sample of 20 women with a mean height less than 70 inches? Explain.

18. ***Which Is More Likely?*** ◆ Assume that the heights given in Exercise 16 are normally distributed. Are you more likely to randomly select one man with a height less than 65 inches or are you more likely to select a sample of 15 men with a mean height less than 65 inches? Explain.

19. ***Make a Decision*** ◆ A machine used to fill gallon-sized paint cans is regulated so that the amount of paint dispensed has a mean of 128 ounces and a standard deviation of 0.20 ounce. You randomly select 40 cans and carefully measure the contents. The sample mean of the cans is 127.9 ounces. Does the machine need to be reset? Explain your reasoning.

20. ***Make a Decision*** ◆ A machine used to fill pint-sized milk containers is regulated so that the amount of milk dispensed has a mean of 64 ounces and a standard deviation of 0.11 ounce. You randomly select 40 containers and carefully measure the contents. The sample mean of the containers is 64.05 ounces. Does the machine need to be reset? Explain your reasoning.

21. ***Lumber Cutter*** ◆ Your lumber company has bought a machine that automatically cuts lumber. The seller of the machine claims that the machine cuts lumber to a mean length of 8 feet (96 inches) with a standard deviation of 0.5 inch. Assume the lengths are normally distributed. You randomly select 40 boards and find that the mean length is 96.25 inches.
    (a) Assuming the seller's claim is correct, what is the probability the mean of the sample is 96.25 inches or more?
    (b) Using your answer from part (a), what do you think of the seller's claim?
    (c) Would it be unusual to have an individual board with a length of 96.25 inches? Why or why not?

22. ***Ice Cream Carton Weights*** ◆ A manufacturer claims that the mean weight of its ice cream cartons is 10 ounces with a standard deviation of 0.5 ounces. Assume the weights are normally distributed. You test 25 cartons and find their mean weight is 10.15 ounces.
    (a) Assuming the manufacturer's claim is correct, what is the probability the mean of the sample is 10.15 ounces or more?
    (b) Using your answer from part (a), what do you think of the manufacturer's claim?
    (c) Would it be unusual to have an individual carton with a weight of 10.15 ounces? Why or why not?

23. ***Life of Tires*** ◆ A manufacturer claims that the life span of its tires is 50,000 miles. You work for a consumer protection agency and you are testing this manufacturer's tires. Assume the life spans of the tires are normally distributed. You select 100 tires at random and test them. The mean life span is 49,750. Assume $\sigma = 800$ miles.
    (a) Assuming the manufacturer's claim is correct, what is the probability the mean of the sample is 49,750 miles or less?
    (b) Using your answer from part (a), what do you think of the manufacturer's claim?
    (c) Would it be unusual to have an individual tire with a life span of 49,750 miles? Why or why not?

24. ***Brake Pads*** ◆ A brake pad manufacturer claims its brake pads will last for 38,000 miles. You work for a consumer protection agency and you are testing this manufacturer's brake pads. Assume the life spans of the brake pads are normally distributed. You randomly select 50 brake pads. In your tests, the mean life of the brake pads is 37,650 miles. Assume $\sigma = 1000$ miles.

(a) Assuming the manufacturer's claim is correct, what is the probability the mean of the sample is 37,650 miles or less?

(b) Using your answer from part (a), what do you think of the manufacturer's claim?

(c) Would it be unusual to have an individual brake pad last for 37,650 miles? Why or why not?

Extending the Basics

Finite Correction Factor The formula for the standard error of the mean

$$\sigma_{\bar{x}} = \frac{\sigma}{\sqrt{n}}$$

given in the Central Limit Theorem is based on an assumption that the population has infinitely many members. This is the case whenever sampling is done with replacement (each member is put back after it is selected) because the sampling process could be continued indefinitely. The formula is also valid if the sample size is small in comparison to the population. However, when sampling is done without replacement and the sample size n is more than 5% of the finite population of size N, there is a finite number of possible samples. A **finite correction factor,**

$$\sqrt{\frac{N-n}{N-1}}$$

should be used to adjust the standard error. The sampling distribution of the sample means will be normal with a mean equal to the population mean, and the standard error of the mean will be

$$\sigma_{\bar{x}} = \frac{\sigma}{\sqrt{n}}\sqrt{\frac{N-n}{N-1}}.$$

In Exercises 25 and 26, determine if the finite correction factor should be used. If so, use it in your calculations when you find the probability.

25. In a sample of 800 stations, the mean cash price for regular gasoline at the pump was \$1.384 per gallon and the standard deviation was \$0.009 per gallon. A random sample of size 55 is drawn from this population. What is the probability that the mean price per gallon is less than \$1.379? *(Adapted from U.S. Department of Energy)*

26. In a sample of 500 eruptions of the Old Faithful geyser at Yellowstone National Park, the mean duration of the eruptions was 3.32 minutes and the standard deviation was 1.09 minutes. A random sample of size 30 is drawn from this population. What is the probability that the mean duration of eruptions is between 2.5 minutes and 4 minutes? *(Adapted from Yellowstone National Park)*

TECHNOLOGY

MINITAB EXCEL TI-83

U.S. Census Bureau

www.census.gov

Age Distribution in the United States

One of the jobs of the U.S. Census Bureau is to keep track of the age distribution in the country. The age distribution in 2000 is shown below.

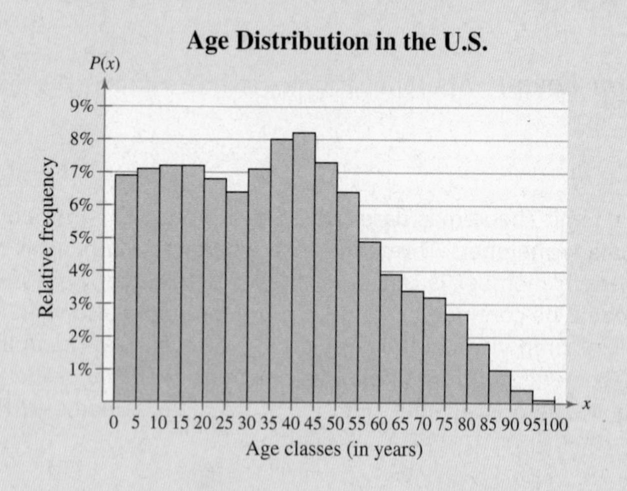

Class Boundaries	Class Midpoint	Relative Frequency
0–5	2.5	6.9%
5–10	7.5	7.1%
10–15	12.5	7.2%
15–20	17.5	7.2%
20–25	22.5	6.8%
25–30	27.5	6.4%
30–35	32.5	7.1%
35–40	37.5	8.0%
40–45	42.5	8.2%
45–50	47.5	7.3%
50–55	52.5	6.4%
55–60	57.5	4.9%
60–65	62.5	3.9%
65–70	67.5	3.4%
70–75	72.5	3.2%
75–80	77.5	2.7%
80–85	82.5	1.8%
85–90	87.5	1.0%
90–95	92.5	0.4%
95–100	97.5	0.1%

Exercises

We used a technology tool to select random samples with $n = 40$ from the age distribution of the United States. The means of the thirty-six samples were as follows.

DATA 28.14, 31.56, 36.86, 32.37, 36.12, 39.53, 36.19, 39.02, 35.62, 36.30, 34.38, 32.98, 36.41, 30.24, 34.19, 44.72, 38.84, 42.87, 38.90, 34.71, 34.13, 38.25, 38.04, 34.07, 39.74, 40.91, 42.63, 35.29, 35.91, 34.36, 36.51, 36.47, 32.88, 37.33, 31.27, 35.80

1. Enter the age distribution of the United States into a technology tool. Use the tool to find the mean age in the United States.
2. Enter the set of sample means into a technology tool. Find the mean of the set of sample means. How does it compare to the mean age in the United States? Does this agree with the result predicted by the Central Limit Theorem?
3. Are the ages of people in the U.S. normally distributed? Explain your reasoning.
4. Sketch a relative frequency histogram for the 36 sample means. Use nine classes. Is the histogram approximately bell shaped and symmetrical? Does this agree with the result predicted by the Central Limit Theorem?
5. Use a technology tool to find the standard deviation of the ages of people in the U.S.
6. Use a technology tool to find the standard deviation of the set of 36 sample means. How does it compare to the standard deviation of the ages? Does this agree with the result predicted by the Central Limit Theorem?

Extended solutions are given in the *Technology Supplement*.
Technical instruction is provided for *Minitab*, *Excel*, and the *TI-83*.

5.6 Normal Approximations to Binomial Distributions

What You Should Learn

- How to decide when the normal distribution can approximate the binomial distribution
- How to find the correction for continuity
- How to use the normal distribution to approximate binomial probabilities

Approximating a Binomial Distribution • Correction for Continuity • Approximating Binomial Probabilities

Approximating a Binomial Distribution

In Section 4.2, you learned how to find binomial probabilities. For instance, if a surgical procedure has an 85% chance of success and a doctor performs the procedure on 10 patients, it is easy to find the probability of exactly two successful surgeries.

But what if the doctor performs the surgical procedure on 150 patients and you want to find the probability of *fewer than 100* successful surgeries? To do this using the techniques described in Section 4.2, you would have to use the binomial formula 100 times and find the sum of the resulting probabilities. This approach is not practical, of course. A better approach is to use a normal distribution to approximate the binomial distribution.

Normal Approximation to a Binomial Distribution

If $np \geq 5$ and $nq \geq 5$, then the binomial random variable x is approximately normally distributed, with mean

$$\mu = np$$

and standard deviation

$$\sigma = \sqrt{npq}.$$

Study Tip

Properties of a binomial experiment

- n independent trials
- Two possible outcomes: success or failure
- Probability of success is p; probability of failure is $1 - p = 1$
- p is constant for each trial

To see why this result is valid, look at the following binomial distributions for $p = 0.25$ and $n = 4$, $n = 10$, $n = 25$, and $n = 50$. Notice that as n increases, the histogram approaches a normal curve.

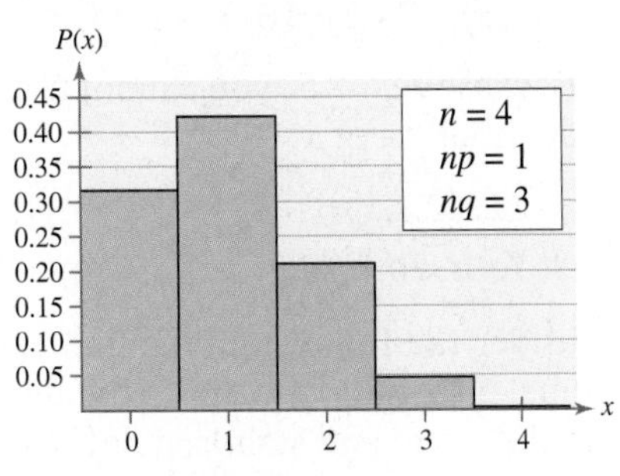

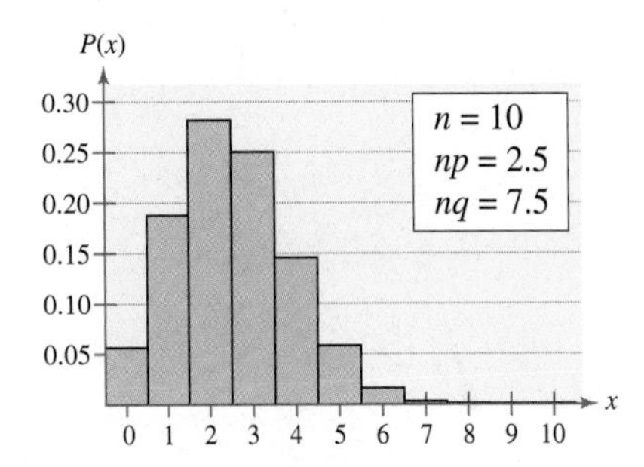

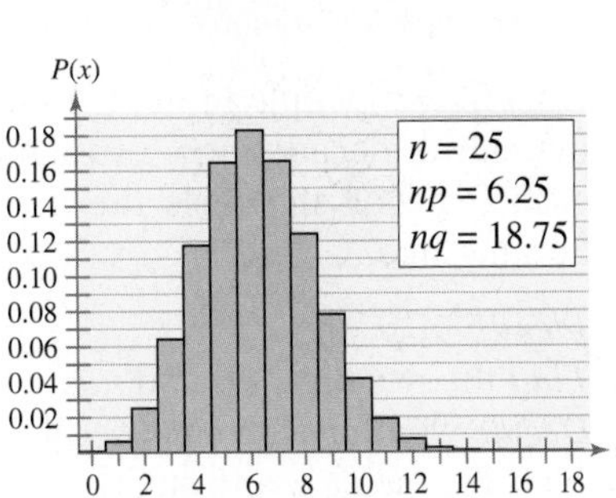

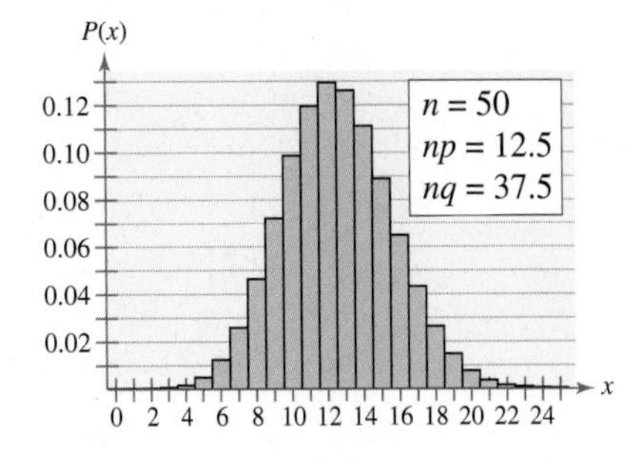

EXAMPLE 1

Approximating the Binomial Distribution

Two binomial experiments are listed. Decide whether you can use the normal distribution to approximate x, the number of people who reply yes. If so, find the mean and standard deviation. If not, explain why. *(Source: Marist College Institute of Public Opinion)*

1. Thirty-seven percent of people in the U.S. say that they always fly an American flag on the Fourth of July. You randomly select 15 people in the U.S. and ask each if he or she always flies an American flag on the Fourth of July.
2. Ninety-three percent of people in the U.S. want the national anthem to remain the same. You randomly select 65 people in the U.S. and ask each if he or she wants the national anthem to remain the same.

SOLUTION

1. In this binomial experiment, $n = 15$, $p = 0.37$, and $q = 0.63$. So,

   $$np = (15)(0.37) = 5.55$$

   and

   $$nq = (15)(0.63) = 9.45.$$

   Because np and nq are greater than 5, you can use the normal distribution with

   $$\mu = 5.55$$

   and

   $$\sigma = \sqrt{npq} = \sqrt{15 \cdot 0.37 \cdot 0.63} \approx 1.87$$

   to approximate the distribution of x.
2. In this binomial experiment, $n = 65$, $p = 0.93$, and $q = 0.07$. So,

   $$np = (65)(0.93) = 60.45$$

   and

   $$nq = (65)(0.07) = 4.55.$$

   Because $nq < 5$, you cannot use the normal distribution to approximate the distribution of x.

Try It Yourself 1

Consider the following binomial experiment. Decide whether you can use the normal distribution to approximate x, the number of people who reply yes. If so, find the mean and standard deviation. If not, explain why. *(Source: Marist College Institute of Public Opinion)*

Only 8% of people in the U.S. feel that the nation is more patriotic today than it was decades ago. You randomly select 70 people in the U.S. and ask each if he or she feels the nation is more patriotic today than it was decades ago.

a. *Identify* n, p, and q.
b. *Find* the products np and nq.
c. *Decide* whether you can use the normal distribution to approximate x.
d. *Find* the mean μ and standard deviation σ, if appropriate.

Answer: Page A37

Correction for Continuity

The binomial distribution is discrete and can be represented by a probability histogram. To calculate *exact* binomial probabilities, you can use the binomial formula for each value of x and add the results. Geometrically, this corresponds to adding the areas of bars in the probability histogram. When you do this, remember that each bar has a width of one unit and x is the midpoint of the interval.

When you use a *continuous* normal distribution to approximate a binomial probability, you need to move 0.5 units to the left and right of the midpoint to include all possible x-values in the interval. When you do this, you are making a **correction for continuity.**

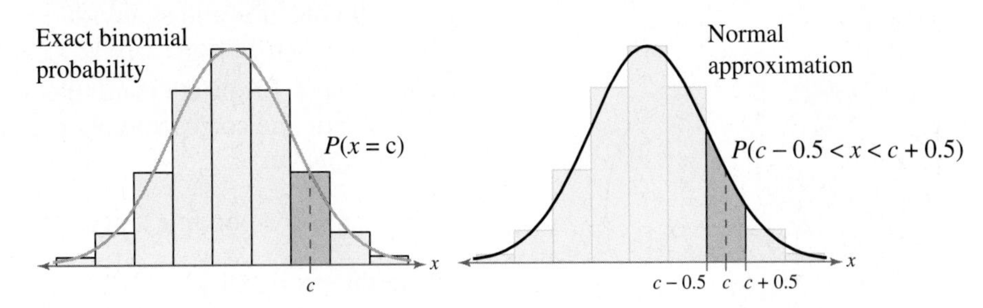

EXAMPLE 2

Using a Correction for Continuity

Use a correction for continuity to convert each of the following binomial intervals to a normal distribution interval.

1. The probability of getting between 270 and 310 successes, inclusive
2. The probability of at least 158 successes
3. The probability of getting less than 63 successes

Study Tip

To use a correction for continuity, simply subtract 0.5 from the lowest value and add 0.5 to the highest.

SOLUTION

1. The discrete midpoint values are 270, 271, . . . , 310. The interval for the continuous normal distribution are

   $269.5 < x < 310.5.$

2. The discrete midpoint values are 158, 159, 160, ... The interval for the continuous normal distribution are

   $x > 157.5.$

3. The discrete midpoint values are . . . , 60, 61, 62. The interval for the continuous normal distribution is

   $x < 62.5.$

Try It Yourself 2

Use a correction for continuity to convert each of the following binomial intervals to a normal distribution interval.

1. The probability of getting between 57 and 83 successes, inclusive
2. The probability of getting at most 54 successes

a. List the *midpoint values* for the binomial probability.
b. Use a *correction for continuity* to write the normal distribution interval.

Answer: Page A37

Picturing the World

In a survey of U.S. adults, people were asked if the law should allow doctors to aid dying patients who want to end their lives. The results of the survey are shown in the following pie chart. *(Source: Harris Poll)*

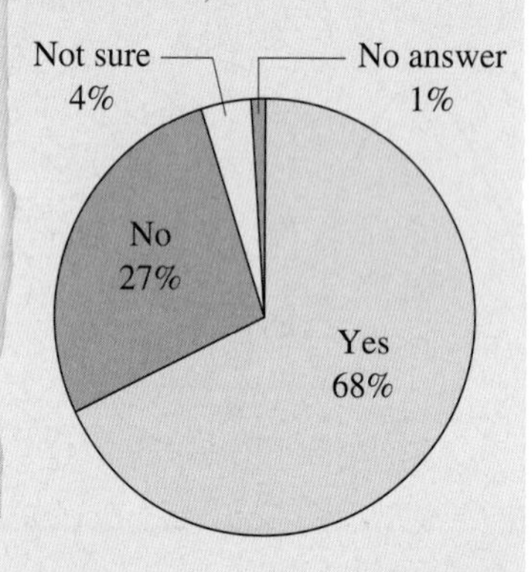

Assume that this Harris Poll is a true indication of the proportion of the population who believe in assisted death for terminally ill patients. If you sampled 50 adults at random, what is the probability that between 32 and 36, inclusive, would believe in assisted death?

Approximating Binomial Probabilities

GUIDELINES

Using the Normal Distribution to Approximate Binomial Probabilities

In Words	*In Symbols*
1. Verify that the binomial distribution applies.	Specify n, p, and q.
2. Determine if you can use the normal distribution to approximate x, the binomial variable.	Is $np \geq 5$? Is $nq \geq 5$?
3. Find the mean μ and standard deviation σ for the distribution.	$\mu = np$ $\sigma = \sqrt{npq}$
4. Apply the appropriate continuity correction. Shade the corresponding area under the normal curve.	Add or subtract 0.5 from endpoints.
5. Find the corresponding z-score(s).	$z = \dfrac{x - \mu}{\sigma}$
6. Find the probability.	Use the Standard Normal Table.

EXAMPLE 3

Approximating a Binomial Probability

Thirty-seven percent of people in the U.S. say that they always fly an American flag on the Fourth of July. You randomly select 15 people in the U.S. and ask each if he or she flies an American flag on the Fourth of July. What is the probability that fewer than eight of them respond yes? *(Source: Marist College Institute of Public Opinion)*

SOLUTION From Example 1, you know that you can use a normal distribution with $\mu = 5.55$ and $\sigma \approx 1.87$ to approximate the binomial distribution. Remember to apply the continuity correction for the value of x. In the binomial distribution, the possible midpoint values for "fewer than 8" are

$$\ldots 5, 6, 7.$$

To use the normal distribution, add 0.5 to the right hand boundary 7 to get $x = 7.5$. The graph at the left shows a normal curve with $\mu = 5.55$ and $\sigma \approx 1.87$ and a shaded area to the left of 7.5. The z-score that corresponds to $x = 7.5$ is

$$z = \frac{7.5 - 5.55}{1.87}$$

$$\approx 1.04.$$

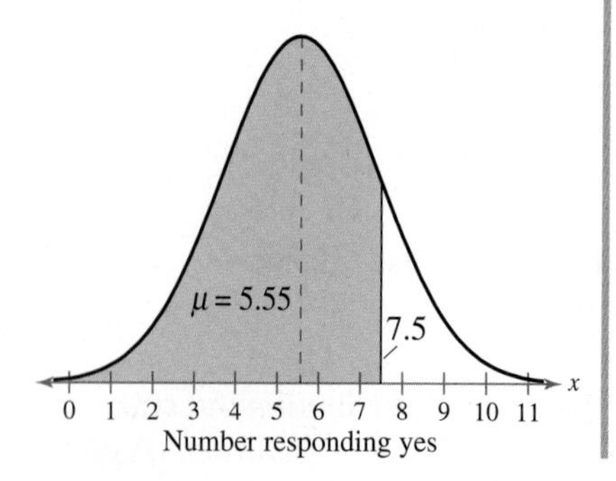

Using the Standard Normal Table,

$$P(z < 1.04) = 0.8508.$$

So, the probability that fewer than eight people respond yes is approximately 0.8508, or about 85%.

Try It Yourself 3

Only 8% of people in the U.S. feel that the nation is more patriotic today than it was decades ago. You randomly select 70 people in the U.S. and ask each if he or she feels the nation is more patriotic today than it was decades ago. What is the probability that more than 10 respond yes? (See Try It Yourself 1.)

a. *Determine* whether you can use the normal distribution to approximate the binomial variable (see part c of Try It Yourself 1).
b. *Find* the mean μ and the standard deviation σ for the distribution (see part d of Try It Yourself 1).
c. *Apply* the appropriate continuity correction and sketch a graph.
d. *Find* the corresponding z-score.
e. *Use* the Standard Normal Table to find the area to the left of z and calculate the probability.

Answer: Page A37

EXAMPLE 4

Approximating a Binomial Probability

Twenty-nine percent of people in the U.S. say they are confident that passenger trips to the moon will occur in their lifetime. You randomly select 200 people in the U.S. and ask each if he or she thinks passenger trips to the moon will occur in his or her lifetime. What is the probability that at least 50 will say yes? *(Source: Harper's Index, July 1998)*

SOLUTION Because $np = 200 \cdot 0.29 = 58$ and $nq = 200 \cdot 0.71 = 142$, the binomial variable x is approximately normally distributed with

$$\mu = np = 58 \quad \text{and} \quad \sigma = \sqrt{200 \cdot 0.29 \cdot 0.71} \approx 6.42.$$

Using the correction for continuity, you can rewrite the discrete probability $P(x \geq 50)$ as the continuous probability $P(x \geq 49.5)$. The graph shows a normal curve with $\mu = 58$ and $\sigma = 6.42$ and a shaded area to the right of 49.5. The z-score that corresponds to 49.5 is $z = (49.5 - 58)/6.42 \approx -1.32$. So, the probability that at least 50 will say yes is

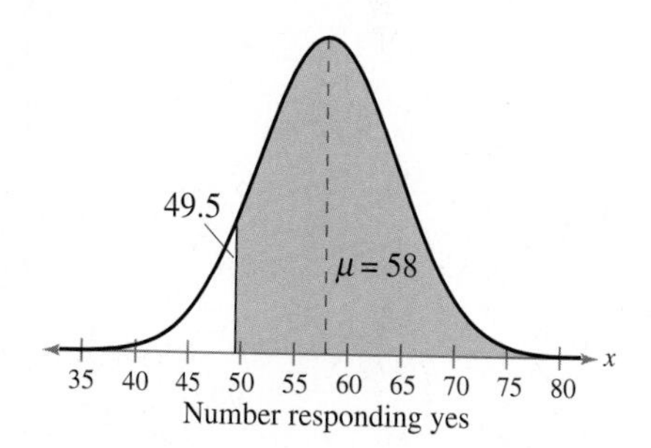

$$P(x \geq 49.5) = P(z \geq -1.32)$$
$$= 1 - P(z \leq -1.32) = 1 - 0.0934 = 0.9066.$$

Study Tip

In a discrete distribution, there is a difference between $P(x \geq c)$ and $P(x > c)$. This is true because the probability that x is exactly c is not zero. In a continuous distribution, however, there is no difference between $P(x \geq c)$ and $P(x > c)$ because the probability that x is exactly c is zero.

Try It Yourself 4

What is the probability that at most 65 people will say yes?

a. *Determine* whether you can use the normal distribution to approximate the binomial variable (see Example 4).
b. *Find* the mean μ and the standard deviation σ for the distribution.
c. *Apply* a continuity correction to rewrite $P(x \leq 65)$ and sketch a graph.
d. *Find* the corresponding z-score.
e. *Use* the Standard Normal Table to *find the area* to the left of z and calculate the probability.

Answer: Page A37

EXAMPLE 

Approximating a Binomial Probability

A survey reports that 39% of Internet users use Microsoft Internet Explorer as their browser. You randomly select 200 Internet users and ask each whether he or she uses Microsoft Internet Explorer as his or her browser. What is the probability that exactly 82 will say yes? *(Source: The Gallup Organization)*

SOLUTION Because $np = 200 \cdot 0.39 = 78$ and $nq = 200 \cdot 0.61 = 122$, the binomial variable x is approximately normally distributed with

$$\mu = np = 78 \quad \text{and} \quad \sigma = \sqrt{200 \cdot 0.39 \cdot 0.61} \approx 6.90.$$

Using the correction for continuity, you can rewrite the discrete probability $P(x = 82)$ as the continuous probability $P(81.5 < x < 82.5)$. The following graph shows a normal curve with $\mu = 78$ and $\sigma = 6.90$ and a shaded area between 81.5 and 82.5.

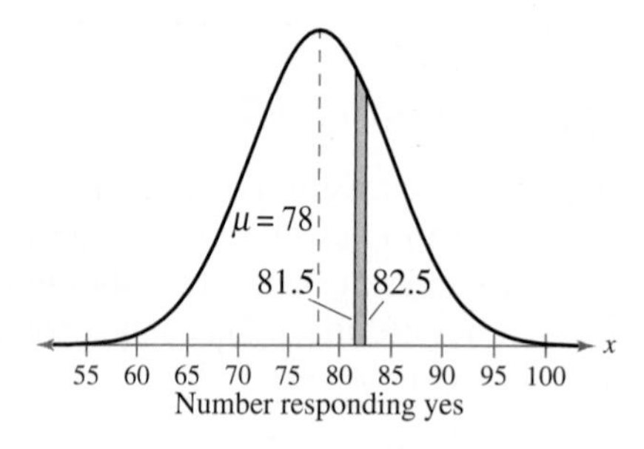

The z-scores that correspond to 81.5 and 82.5 are

$$z_1 = \frac{81.5 - 78}{6.90} \approx 0.51 \quad \text{and} \quad z_2 = \frac{82.5 - 78}{6.90} \approx 0.65.$$

So, the probability that exactly 82 Internet users will say they use Microsoft Internet Explorer is

$$\begin{aligned} P(81.5 < x < 82.5) &= P(0.51 < z < 0.65) \\ &= P(z < 0.65) - P(z < 0.51) \\ &= 0.7422 - 0.6950 \\ &= 0.0472. \end{aligned}$$

So, there is a probability of about 0.05 that exactly 82 of the Internet users will say they use Microsoft Internet Explorer.

Try It Yourself 5

What is the probability that exactly 79 people will say yes?

a. *Determine* whether you can use the normal distribution to approximate the binomial variable (see Example 5).
b. *Find* the mean μ and the standard deviation σ for the distribution (see Example 5).
c. *Apply* a continuity correction to rewrite $P(x = 79)$ and sketch a graph.
d. *Find* the corresponding z-scores.
e. *Use* the Standard Normal Table to *find the area* to the left of each z-score and calculate the probability.

Answer: Page A37

5.6 Exercises

Help

LarsonTutor 5.6

Companion Web Site

Student Solutions Manual 5.6

Videos 5.6

Technology Manuals

Basic Skills and Concepts

In Exercises 1–4, the sample size n, probability of success p, and probability of failure q are given for a certain binomial experiment. Decide whether you can use the normal distribution to approximate the random variable x.

1. $n = 20,\ p = 0.80,\ q = 0.20$

2. $n = 12,\ p = 0.60,\ q = 0.40$

3. $n = 15,\ p = 0.65,\ q = 0.35$

4. $n = 18,\ p = 0.85,\ q = 0.15$

Approximating a Binomial Distribution In Exercises 5–8, a binomial experiment is given. Decide whether you can use the normal distribution to approximate the binomial distribution. If so, find the mean and standard deviation. If not, explain why.

5. ***Metal Detectors*** ◆ A survey of U.S. adults found that 23% think putting metal detectors in schools would be the most effective way to stop violence in schools. You ask 10 adults selected at random if he or she thinks putting metal detectors in schools would be the most effective way to stop violence in schools. *(Source: Associated Press)*

6. ***Privacy Laws*** ◆ A survey of e-mail users found that 66% favored more government laws to ensure online privacy. You randomly select 20 e-mail users and ask each if he or she is in favor of more government laws to ensure online privacy. *(Source: The Gallup Organization)*

7. ***Prostate Cancer*** ◆ In a recent year, the American Cancer Society said that the five-year survival rate for all men diagnosed with prostate cancer was 93%. You randomly select 10 men who were diagnosed with prostate cancer and calculate their five-year survival rate. *(Source: American Cancer Society)*

8. ***Work Weeks*** ◆ A survey of workers in the U.S. found that 8.6% work fewer than 40 hours per week. You randomly select 30 workers in the U.S. and ask each if he or she works fewer than 40 hours per week.

In Exercises 9–12, match the binomial probability with the correct statement.

Probability	Statement
9. $P(x \geq 45)$	(a) P(there are fewer than 45 successes)
10. $P(x \leq 45)$	(b) P(there are at most 45 successes)
11. $P(x < 45)$	(c) P(there are more than 45 successes)
12. $P(x > 45)$	(d) P(there are at least 45 successes)

In Exercises 13–16, use the correction for continuity and match the binomial probability statement with the corresponding normal distribution statement.

Binomial Probability	Normal Probability
13. $P(x > 89)$	(a) $P(x > 89.5)$
14. $P(x \geq 89)$	(b) $P(x < 88.5)$
15. $P(x \leq 89)$	(c) $P(x \leq 89.5)$
16. $P(x < 89)$	(d) $P(x \geq 88.5)$

Graphical Analysis In Exercises 17 and 18, write the binomial probability and the normal probability for the shaded region of the graph. Find the value of each probability and compare the results.

17.

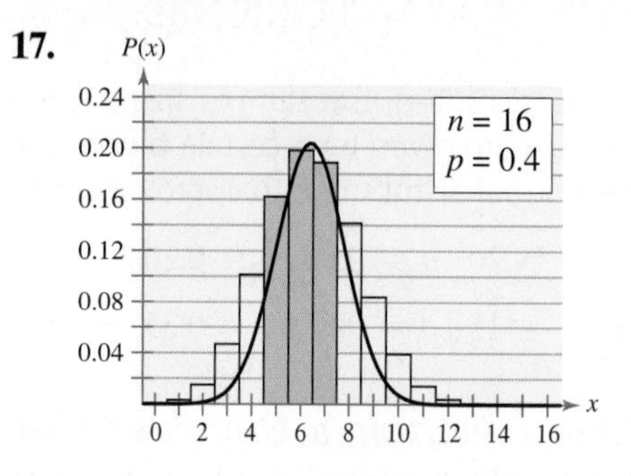

18.

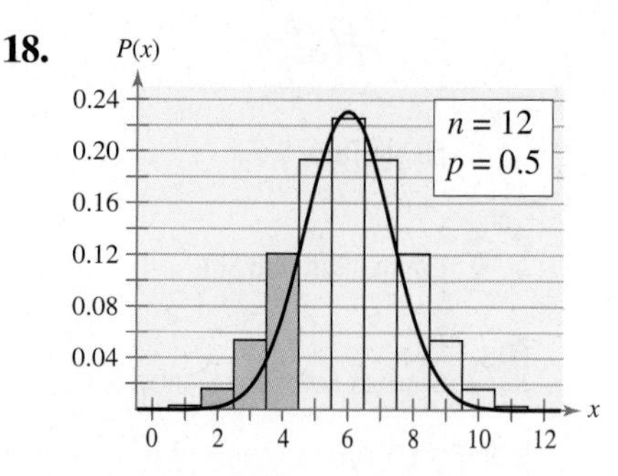

Approximating Binomial Probabilities In Exercises 19–22, decide whether you can use the normal distribution to approximate the binomial distribution. If so, use the normal distribution to approximate the indicated probabilities and sketch their graphs. If not, explain why and use the binomial distribution to find the indicated probabilities.

19. *Blood Type O^-* ◆ Seven percent of people in the U.S. have type O^- blood. You randomly select 30 people in the U.S. and ask them if their blood type is O^-. *(Source: American Association of Blood Banks)*

(a) Find the probability that exactly 10 people say they have O^- blood.

(b) Find the probability that at least 10 people say they have O^- blood.

(c) Find the probability that fewer than 10 people say they have O^- blood.

(d) A blood drive would like to get at least five donors with O^- blood. If there are 100 donors, what is the probability that there will not be enough O^- blood donors?

20. *Blood Type A^+* ◆ Thirty-four percent of people in the U.S. have type A^+ blood. You randomly select 32 people in the U.S. and ask them if their blood type is A^+. *(Source: American Association of Blood Banks)*

(a) Find the probability that exactly 12 people say they have A^+ blood.

(b) Find the probability that at least 12 people say they have A^+ blood.

(c) Find the probability that fewer than 12 people say they have A^+ blood.

(d) A blood drive would like to get at least 60 donors with A^+ blood. If there are 150 donors, what is the probability that there will not be enough A^+ blood donors?

21. *Favorite Cookie* ◆ Fifty-two percent of adults say chocolate chip is their favorite cookie. You randomly select 40 adults and ask each if chocolate chip is his or her favorite cookie. *(Source: WEAREVER)*

(a) Find the probability that at most 15 people say chocolate chip is their favorite cookie.

(b) Find the probability that at least 15 people say chocolate chip is their favorite cookie.

(c) Find the probability that more than 15 people say chocolate chip is their favorite cookie.

(d) A community bake sale has prepared 350 chocolate chip cookies. If the bake sale attracts 650 customers and they each buy one cookie, what is the probability there will not be enough chocolate chip cookies?

22. ***Long Work Weeks*** ◆ A survey of workers in the U.S. found that 2.9% work more than 70 hours per week. You randomly select 10 workers in the U.S. and ask each if he or she works more than 70 hours per week.

(a) Find the probability that at most three people say they work more than 70 hours per week.

(b) Find the probability that at least three people say they work more than 70 hours per week.

(c) Find the probability that more than three people say they work more than 70 hours per week.

(d) A large company is concerned about overworked employees who work more than 70 hours per week. If the company randomly selects 50 employees, what is the probability there will be no employee working more than 70 hours?

Extending the Basics

Getting Physical In Exercises 23 and 24, use the following information. The graph shows the results of a survey of people in the U.S. ages 33 to 51 who were asked if they participated in a sport.

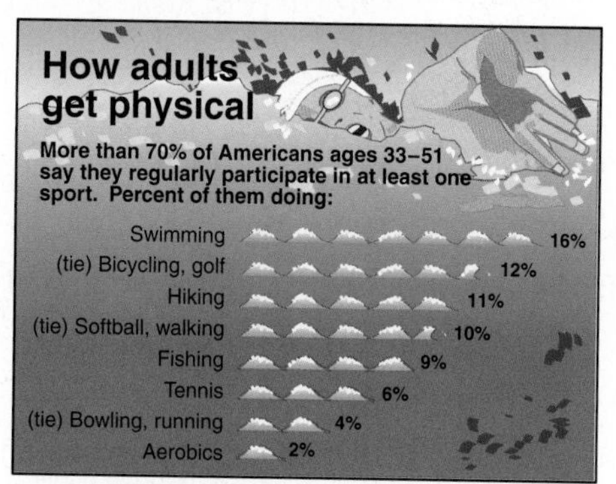

23. You randomly select 250 people in the U.S. ages 33 to 51 and ask each if he or she regularly participates in at least one sport. You find that 60% say no. How likely is this result? Do you think the sample is a good one? Explain your reasoning.

24. You randomly select 300 people in the U.S. ages 33 to 51 and ask each if he or she regularly participates in at least one sport. Of the 200 who say yes, 9% say they participate in hiking. How likely is this result? Is the sample a good one? Explain your reasoning.

Testing a Drug In Exercises 25 and 26, use the following information. A drug manufacturer claims that a certain drug cures a rare skin disease 75% of the time. To check the claim, the drug is tested on 100 patients. If at least 70 patients are cured, the claim will be accepted.

25. Find the probability that the claim will be rejected assuming that the manufacturer's claim is true.

26. Find the probability that the claim will be accepted assuming that the actual probability that the drug cures the skin disease is 65%.

5 Chapter Summary

What did you learn?	Review Exercises
Section 5.1	
◆ How to interpret graphs of normal probability distributions	1, 2
◆ How to estimate areas under a normal curve and use them to estimate probabilities for random variables with normal distributions	3–6
Section 5.2	
◆ How to find and interpret z-scores	7, 8
◆ How to find areas under the standard normal curve	9–20
Section 5.3	
◆ How to find probabilities for normally distributed variables	21–28
Section 5.4	
◆ How to find a z-score given the area under the normal curve	29–34
◆ How to transform a z-score to an x-value	35, 36
◆ How to find a specific data value of a normal distribution given the probability	37–40
Section 5.5	
◆ How to find sampling distributions and verify their properties	41, 42
◆ How to interpret the Central Limit Theorem	43, 44
◆ How to apply the Central Limit Theorem to find the probability of a sample mean	45–50
Section 5.6	
◆ How to decide when the normal distribution can approximate the binomial distribution	51, 52
◆ How to find the correction for continuity	53, 54
◆ How to use the normal distribution to approximate binomial probabilities	55, 56

STATISTICS

Uses and Abuses

Uses

Normal Distributions Normal distributions can be used to describe many real-life situations and are widely used in the fields of science, business, and psychology. They are the most important probability distributions in statistics and can be used to approximate other distributions, such as discrete binomial distributions.

The most incredible application of the normal distribution lies in the Central Limit Theorem. This theorem states that no matter what type of distribution a population may have, as long as the sample size is at least 30, the distribution of sample means will be normal. If the population is itself normal then the distribution of sample means will be normal no matter how small the sample is.

The normal distribution is essential to sampling theory. Sampling theory forms the basis of statistical inference, which you will begin to study in the next chapter.

Abuses

Confusing Likelihood with Certainty A common abuse of normal probability distributions is to confuse the concept of likelihood with the concept of certainty. For instance, if you randomly select a member from a population that is normally distributed, you know the probability is approximately 95% that you will obtain a value that lies within two standard deviations of the mean. This *does not* imply, however, that you cannot get an unusual result. In fact, 5% of the time you should expect to get a value that is more than two standard deviations from the mean.

Suppose a population is normally distributed with a mean of 100 and standard deviation of 15. It would not be unusual for an individual value taken from this population to be 112 or more. It *would* be, however, highly unusual to obtain a sample mean of 112 or more from a sample with 100 members.

Exercises

1. ***Confusing Likelihood with Certainty*** You are randomly selecting 100 people from a population that is normally distributed. Are you certain to get exactly 95 people that lie within two standard deviations of the mean? Explain your reasoning.

2. ***Confusing Likelihood with Certainty*** You are randomly selecting 10 people from a large population that is normally distributed. Which of the following is more likely? Explain your reasoning.

   a. All 10 lie within 2 standard deviations of the mean.

   b. At least one person does not lie within 2 standard deviations of the mean.

5 Review Exercises

Section 5.1

In Exercises 1 and 2, use the graph to estimate μ and σ.

1.

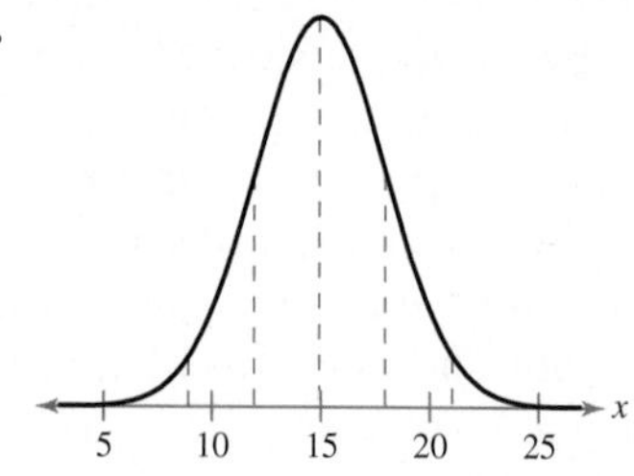

2.

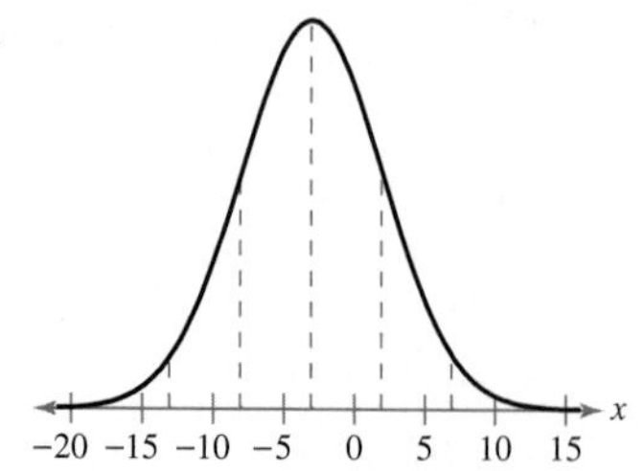

In Exercises 3–6, use the following information and the Empirical Rule to answer the questions. A certain light bulb's life span is normally distributed, with a mean of 670 hours and a standard deviation of 65 hours.

3. Between what two life spans will about 95% of these bulbs fall?

4. Between what two life spans will about 99.7% of these bulbs fall?

5. Estimate the probability that a randomly selected bulb will last between 605 hours and 735 hours.

6. Estimate the probability that a randomly selected bulb will last more than 865 hours.

Section 5.2

In Exercises 7 and 8, use the following information and standard scores to investigate observations about a normal population. A batch of 2500 resistors is normally distributed, with a mean resistance of 1.5 ohms and a standard deviation of 0.08 ohm. Four resistors are randomly selected and tested. Their resistances were measured at 1.32, 1.54, 1.66, and 1.78 ohms.

7. How many standard deviations from the mean are these observations?

8. Do any of these observations seem more or less likely than the others?

In Exercises 9–20, use the Standard Normal Table to find areas under the standard normal curve.

9. Find the area to the left of $z = -0.84$.

10. Find the area to the left of $z = 2.55$.

11. Find the area to the left of $z = -0.27$.

12. Find the area to the left of $z = 1.26$.

13. Find the area to the right of $z = 1.68$.

14. Find the area to the right of $z = -0.83$.

15. Find the area between $z = -1.64$ and the mean.

16. Find the area between $z = -1.22$ and $z = -0.43$.

17. Find the area between $z = 0.15$ and $z = 1.35$.

18. Find the area between $z = -1.96$ and $z = 1.96$.

19. Find the area to the left of $z = -1.5$ and to the right of $z = 1.5$.

20. Find the area to the left of $z = 0.12$ and to the right of $z = 1.72$.

Section 5.3

In Exercises 21–26, find the indicated probabilities.

21. $P(z < 1.28)$

22. $P(z > -0.74)$

23. $P(-2.15 < z < 1.55)$

24. $P(0.42 < z < 3.15)$

25. $P(z < -2.50 \text{ or } z > 2.50)$

26. $P(z < 0 \text{ or } z > 1.68)$

In Exercises 27 and 28, find the indicated probabilities.

27. The green turtle migrates across the Southern Atlantic in the winter, swimming great distances. A study found that the mean migration distance was 2200 kilometers and the standard deviation was 625 kilometers. Assuming that the distances are normally distributed, find the probability that a randomly selected green turtle migrates a distance of (a) less than 1900 kilometers, (b) between 2000 kilometers and 2500 kilometers, and (c) greater than 2450 kilometers. *(Adapted from DK Visual Encyclopedia)*

28. The world's smallest mammal is the Kitti's hog-nosed bat, with a mean weight of 1.5 grams and a standard deviation of 0.25 gram. Assuming that the weights are normally distributed, find the probability of randomly selecting a bat that weighs (a) between 1.0 gram and 2.0 grams, (b) between 1.6 grams and 2.2 grams, and (c) more than 2.2 grams. *(Adapted from DK Visual Encyclopedia)*

Section 5.4

In Exercises 29–34, use the Standard Normal Table to find the *z*-score that corresponds to the given cumulative area or percentile. If the area is not in the table, use the entry closest to the area.

29. 0.4721

30. 0.1

31. 0.8708

32. P_2

33. P_{85}

34. P_{20}

In Exercises 35–40, use the following information. On a dry surface, the braking distance (in meters) of a Pontiac Grand AM SE can be approximated by a normal distribution, as shown in the graph. *(Source: National Highway Traffic Safety Administration)*

Braking Distance of a Pontiac Grand Am SE

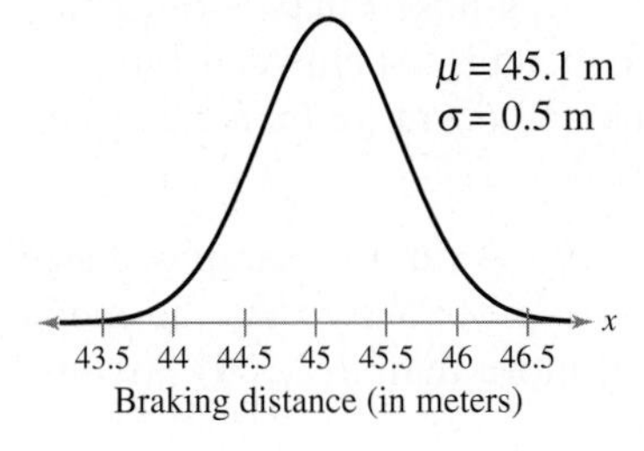

35. Find the braking distance of a Pontiac Grand AM SE that corresponds to $z = -2.4$.

36. Find the braking distance of a Pontiac Grand AM SE that corresponds to $z = 1.2$.

37. What braking distance of a Pontiac Grand AM SE represents the 95th percentile?

38. What braking distance of a Pontiac Grand AM SE represents the third quartile?

39. What is the shortest braking distance of a Pontiac Grand AM SE that can be in the top 10% of braking distances?

40. What is the longest braking distance of a Pontiac Grand AM SE that can be in the bottom 5% of braking distances?

Section 5.5

In Exercises 41 and 42, use the given population to find the sampling distribution of the sample means for the indicated sample sizes. Find the mean and standard deviation of the population and the mean and standard deviation of the sampling distribution. Compare the values.

41. A corporation has five executives. The number of minutes each exercises a week is reported as 40, 200, 80, 0, and 600. Draw three executives' names from this population, with replacement, and form a sampling distribution of the sample mean of the minutes they exercise.

42. There are four residents sharing a house. The number of times each washes his or her car each month is 1, 2, 0, and 3. Draw two names from this population, with replacement, and form a sampling distribution for the sample mean of the number of times their cars are washed each month.

In Exercises 43 and 44, use the Central Limit Theorem to find the mean and standard error of the mean of the indicated sampling distribution. Then sketch a graph of the sampling distribution.

43. The consumption of processed fruits by people in the U.S. in a recent year was normally distributed, with a mean of 154.8 pounds and a standard deviation of 51.6 pounds. Random samples of size 35 are drawn from this population. *(Adapted from U.S. Department of Agriculture)*

44. Repeat Exercise 43, assuming that the standard deviation was 25.8 pounds.

In Exercises 45–50, find the probabilities for the sampling distributions.

45. Refer to Exercise 27. If 12 green turtles are randomly selected, find the probability that the sample mean of the distance migrated is (a) less than 1900 kilometers, (b) between 2000 kilometers and 2500 kilometers, and (c) greater than 2450 kilometers. Compare your answers to those in Exercise 27.

46. Refer to Exercise 28. If a sample of 7 Kitti's hog-nosed bats is randomly selected, find the probability that the sample mean is (a) between 1.0 gram and 2.0 grams, (b) between 1.6 and 2.2 grams, and (c) more than 2.2 grams. Compare your answers to those in Exercise 28.

47. The mean annual salary for chauffeurs is \$21,000. A random sample of size 45 is drawn from this population. What is the probability that the mean annual salary is (a) less than \$20,000 and (b) more than \$22,500? Assume $\sigma = \$1500$. *(Adapted from Jobs Rated Almanac)*

48. Suppose the mean value of land and buildings per acre for farms is \$1300. A random sample of size 36 is drawn. What is the probability that the mean value of land and buildings per acre is (a) less than \$1400 and (b) more than \$1150? Assume $\sigma = \$250$.

49. The mean price of houses in a city is \$1.5 million with a standard deviation of \$500,000. The house prices are normally distributed. If you randomly select 15 houses in this city, what is the probability that the mean price will be less than \$1.125 million?

50. Mean rent in a city is \$5000 per month with a standard deviation of \$300. The rents are normally distributed. If you randomly select 15 apartments in this city, what is the probability that the mean price will be more than \$5250?

Section 5.6

In Exercises 51 and 52, a binomial experiment is given. Decide whether you can use the normal distribution to approximate the binomial distribution. If so, find the mean and standard deviation. If not, explain why.

51. In a recent year, the American Cancer Society predicted that the five-year survival rate for new cases of kidney cancer would be 59%. You randomly select 12 men who were new kidney cancer cases this year and calculate their five-year survival rate. *(Source: American Cancer Society)*

52. A survey indicates that 46% of women who take at least one vacation a year pack too much. You randomly select 12 women who take at least one vacation a year and ask them if they pack too much. *(Source: Opinion Research for DuPont)*

In Exercises 53 and 54, write the binomial probability as a normal probability using the continuity correction.

	Binomial Probability	Normal Probability
53.	$P(x \geq 25)$	$P(x > ?)$
54.	$P(x = 45)$	$P(? < x < ?)$

In Exercises 55 and 56, decide whether you can use the normal distribution to approximate the binomial distribution. If so, use the normal distribution to approximate the indicated probabilities and sketch their graphs. If not, explain why and use the binomial distribution to find the indicated probabilities.

55. Sixty-five percent of children aged 12 to 17 keep at least part of their savings in a savings account. You randomly select 45 children and ask each if he or she keeps at least part of his or her savings in a savings account. Find the probability that at most 20 children will say yes. *(Source: International Communications Research for Merrill Lynch)*

56. Thirty-three percent of adults graded public schools as excellent or good at preparing students for college. You randomly select 12 adults and ask them if they think public schools are excellent or good at preparing students for college. Find the probability that more than five adults will say yes. *(Source: Marist Institute for Public Opinion)*

PUTTING IT ALL TOGETHER

Real Statistics Real Decisions

Suppose you work for a manufacturing company as a statistical process analyst. Your job is to analyze processes and make sure they are in statistical control. In one process, a machine cuts wood boards to a thickness of 25 millimeters with an acceptable margin of error of ± 0.6 millimeter. (Assume this process can be approximated by a normal distribution.) So, the acceptable range of thicknesses for the boards is 24.4 millimeters to 25.6 millimeters, inclusive.

Due to machine vibrations and other factors, the setting of the wood cutting machine "shifts" from 25 millimeters. To check that the machine is cutting the boards to the correct thickness, you select at random three samples of four boards and find the mean thickness (in millimeters) of each sample. A coworker asks you why you take three samples of size four and find the mean instead of randomly choosing and measuring 12 boards individually to check the machine's settings. (*Note:* Both samples are chosen without replacement.)

Exercises

1. ***Sampling Individuals***

   You select one board and measure its thickness. Assume the machine shifts and is cutting boards with a mean thickness of 25.4 millimeters and a standard deviation of 0.2 millimeter.

   (a) What is the probability that you select a board that is *not* outside the acceptable range (in other words, you do not detect that the machine has shifted)? (See figure.)

   (b) If you randomly select 12 boards, what is the probability that you select at least one board that is *not* outside the acceptable range?

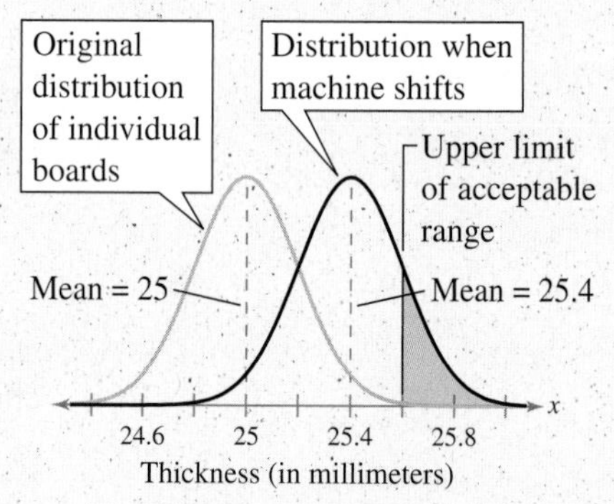

Figure for Exercise 1

2. ***Sampling Groups of Four***

   You select four boards and find their mean thickness. Assume the machine shifts and is cutting boards with a mean thickness of 25.4 millimeters and a standard deviation of 0.2 millimeter.

   (a) What is the probability that you select a sample of four boards that has a mean that is *not* outside the acceptable range? (See figure.)

   (b) If you randomly select three samples of four boards, what is the probability that you select at least one sample of four boards that has a mean that is *not* outside the acceptable range?

   (c) What is more sensitive to change—an individual measure or the mean?

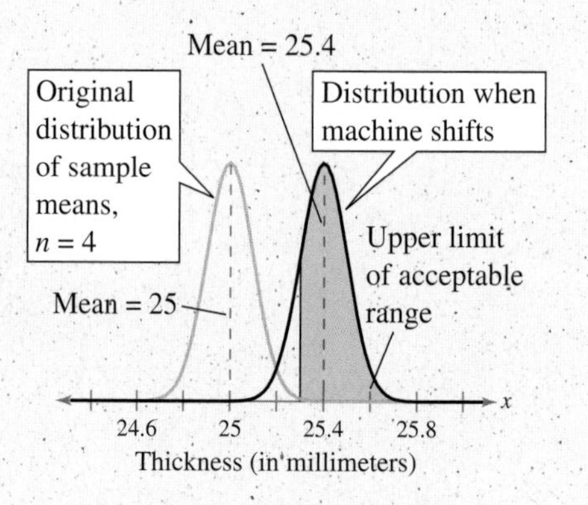

Figure for Exercise 2

3. ***Writing an Explanation***

   Write a paragraph to your coworker explaining why you take three samples of size four and find the mean of each sample instead of randomly choosing and measuring 12 boards individually to check the machine's settings.

5 Chapter Quiz

Take this quiz as you would take a quiz in class. After you are done, check your work against the answers given in the back of the book.

1. Find the following standard normal probabilities.

(a) $P(z > -2.10)$

(b) $P(z < 3.22)$

(c) $P(-2.33 < z < 2.33)$

(d) $P(z < -1.75 \text{ or } z > -0.75)$

2. Find the following normal probabilities for the given parameters.

(a) $\mu = 5.5, \sigma = 0.08, P(5.36 < x < 5.64)$

(b) $\mu = -8.2, \sigma = 7.84, P(-5.00 < x < 0)$

(c) $\mu = 18.5, \sigma = 9.25, P(x < 0 \text{ or } x > 37)$

In Exercises 3–10, use the following information. In a recent year, grade 8 Washington State public school students taking a mathematics assessment test had a mean score of 276.1 with a standard deviation of 34.4. Possible test scores could range from 0 to 500. Assume that the scores are normally distributed. *(Source: National Center for Educational Statistics)*

3. Find the probability that a student had a score higher than 315.

4. Find the probability that a student had a score between 250 and 305.

5. What percent of the students had a test score that is greater than 250?

6. If 2000 students are randomly selected, how many will have a test score that is less than 300?

7. What is the lowest score that would still place a student in the top 5% of the scores?

8. What is the highest score that would still place a student in the bottom 25% of the scores?

9. A random sample of 60 students is drawn from this population. What is the probability that the mean test score is greater than 300?

10. Are you more likely to randomly select one student with a test score greater than 300 or are you more likely to select a sample of 15 students with a mean test score greater than 300? Explain.

In Exercises 11 and 12, use the following information. In a survey of adults, 68% thought that DNA tests for identifying an individual were very reliable. You randomly select 24 adults and ask each if he or she thinks DNA tests for identifying an individual are very reliable. *(Source: CBS News)*

11. Decide whether you can use the normal distribution to approximate the binomial distribution. If so, find the mean and standard deviation. If not, explain why.

12. Find the probability that at most 15 people say DNA tests for identifying an individual are very reliable.

Where You've Been

In Chapters 1 through 5, you studied descriptive statistics (how to collect and describe data) and probability (how to find probabilities and analyze discrete and continuous probability distributions). For instance, the Wheat Quality Council uses descriptive statistics to analyze the data collected during its annual crop tour.

In a recent crop tour, 444 wheat fields were sampled. Of the 288 fields of spring wheat, the mean yield was 32.5 bushels per acre with a standard deviation of 10.9 bushels per acre. Of the 156 fields of durum wheat, the mean yield was 26.8 bushels per acre with a standard deviation of 8.0 bushels per acre.

Donn Pikop, a buyer for a milling company, checks a wheat spike in a Minnesota field.

The Wheat Quality Council has its headquarters in Pierre, South Dakota. Its primary function is to encourage the development and production of new and better varieties of wheat. The Council also assesses the yield and quality of wheat crops. For instance, in a recent crop survey, the wheat averaged about 61 pounds per bushel with a 14% protein content.

Confidence Intervals

CHAPTER

6

Where You're Going

In this chapter, you will begin your study of inferential statistics—the second major branch of statistics. For instance, from the mean of the sample in a recent crop survey, the Wheat Quality Council can estimate the mean crop yield to be 32.5 bushels per acre for *all* spring wheat that year. Because this estimate consists of a single number represented by a point on a number line, it is called a point estimate. The problem with using a point estimate is that it is rarely equal to the exact parameter (mean, standard deviation, proportion) of the population.

In this chapter, you will learn how to make a more meaningful estimate by specifying an interval of values on a number line, together with a statement of how confident you are that your interval contains the population parameter. Suppose the Wheat Council wanted to be 90% confident of its estimate for the mean yield of all spring wheat. Here is an overview of how it could construct an interval estimate.

Find the mean of a random sample. $\bar{x} = 32.5$ → Find the margin of error. $E = 1.3$ → Find the interval endpoints. Left: $32.5 - 1.3 = 31.2$ Right: $32.5 + 1.3 = 33.8$ → Form the interval estimate. $31.2 < \mu < 33.8$

31.2 | 32.5 | 33.8
30 | 31 | 32 | 33 | 34 | 35 | x
1.3 | 1.3

So, the Wheat Quality Council can be 90% confident that the mean yield for all spring wheat is between 31.2 and 33.8 bushels per acre.

6.1 Confidence Intervals for the Mean (Large Samples)

What You Should Learn

- How to find a point estimate and a maximum error of estimate
- How to construct and interpret confidence intervals for the population mean
- How to determine the required minimum sample size when estimating μ

Estimating Population Parameters • Confidence Intervals for the Population Mean • Sample Size

Estimating Population Parameters

In this chapter, you will learn an important technique of statistical inference—to use sample statistics to estimate the value of an unknown population parameter. In this section, you will learn how to use sample statistics to make an estimate of the population parameter μ when the sample size is at least 30 or when the population is normally distributed and the standard deviation σ is known. To make such an inference, begin by finding a point estimate.

DEFINITION

A **point estimate** is a single value estimate for a population parameter. The most unbiased point estimate of the population mean μ is the sample mean $\bar{x}$.

EXAMPLE 1

Finding a Point Estimate

Market researchers use the number of sentences per advertisement as a measure of readability for magazine advertisements. The following represents a random sample of the number of sentences found in 54 advertisements. Find a point estimate of the population mean μ. *(Source: Journal of Advertising Research)*

9	20	18	16	9	16	16	9	11	13	22	16	5	18	6	6	5	12
25	17	23	7	10	9	10	10	5	11	18	18	9	9	17	13	11	7
14	6	11	12	11	15	6	12	14	11	4	9	18	12	12	17	11	20

SOLUTION The sample mean of the data is

$$\bar{x} = \frac{\Sigma x}{n} = \frac{671}{54} \approx 12.4.$$

So, your point estimate for the mean length of all magazine advertisements is 12.4 sentences.

Try It Yourself 1

Another random sample of the number of sentences found in 30 magazine advertisements is listed at the left. Use this sample to find another point estimate for μ.

a. Find the sample mean.
b. Estimate the mean sentence length of the population.

Answer: Page A37

Sample Data

Number of Sentences					
16	9	14	11	17	12
99	18	13	12	5	9
17	6	11	17	18	20
6	14	7	11	12	12
5	11	18	6	4	13

In Example 1, the probability that the population mean is exactly 12.4 is virtually zero. So, instead of estimating μ to be exactly 12.4 using a *point* estimate, you can estimate that μ lies in an *interval*. This is called *making an interval estimate.*

DEFINITION

An **interval estimate** is an interval, or range of values, used to estimate a population parameter.

Although you can assume that the point estimate in Example 1 is not equal to the actual population mean, it is probably close to it. To form an interval estimate, use the point estimate as the center of the interval, then add and subtract a margin of error. For instance, if the margin of error is 2.1, then an interval estimate would be given by 12.4 ± 2.1 or $10.3 < \mu < 14.5$. The point estimate and interval estimate are as follows.

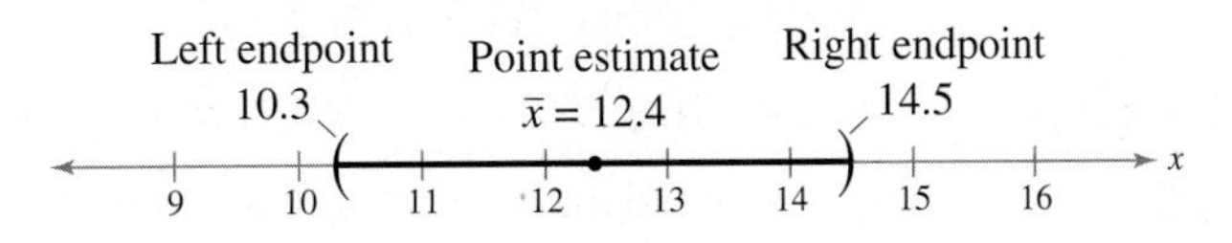

Interval estimate

Study Tip

In this course, you will usually use 90%, 95%, and 99% levels of confidence. The following z-scores correspond to these levels of confidence.

Level of Confidence	z_c
90%	1.645
95%	1.96
99%	2.575

Before finding an interval estimate, you should first determine how confident you need to be that your interval estimate contains the population mean μ.

DEFINITION

The **level of confidence** c is the probability that the interval estimate contains the population parameter.

You know from the Central Limit Theorem that when $n \geq 30$, the sampling distribution of sample means is a normal distribution. The level of confidence c is the area under the standard normal curve between the **critical values,** $-z_c$ and z_c. You can see from the graph that c is the percent of the area under the normal curve between $-z_c$ and z_c. The area remaining is $1 - c$, so the area in each tail is $\frac{1}{2}(1 - c)$. For instance, if $c = 90\%$, then 5% of the area lies to the left of $-z_c = -1.645$ and 5% lies to the right of $z_c = 1.645$.

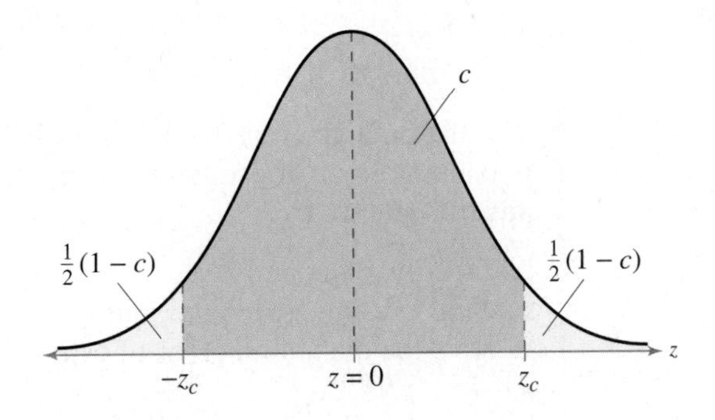

The distance between the point estimate and the actual parameter value is called the **error of estimate.** When estimating μ, the error of estimate is the distance $|\bar{x} - \mu|$. In most cases, of course, μ is unknown and $\bar{x}$ varies from sample to sample. However, you can calculate a maximum value for the error if you know the level of confidence and the sampling distribution.

DEFINITION

Given a level of confidence c, the **maximum error of estimate** (sometimes also called the margin of error or error tolerance) $\boldsymbol{E}$ is the greatest possible distance between the point estimate and the value of the parameter it is estimating.

$$E = z_c \sigma_{\bar{x}} = z_c \frac{\sigma}{\sqrt{n}}$$

When $n \geq 30$, the sample standard deviation s can be used in place of σ.

EXAMPLE 

Finding the Maximum Error of Estimate

Use the data in Example 1 and a 95% confidence level to find the maximum error of estimate for the mean number of sentences in all magazine advertisements.

SOLUTION The z-score that corresponds to a 95% confidence level is 1.96. This implies that 95% of the area under the standard normal curve falls within 1.96 standard deviations of the mean. (You can approximate the distribution of the sample means with a normal curve by the Central Limit Theorem, because $n = 54 \geq 30$.) You don't know the population standard deviation σ. But because $n \geq 30$, you can use s in place of σ.

$$s = \sqrt{\frac{\Sigma(x - \bar{x})^2}{n - 1}} \approx \sqrt{\frac{1333.2}{53}} \approx 5.0$$

Using the values $z_c = 1.96$, $\sigma \approx s \approx 5.0$, and $n = 54$,

$$E = z_c \frac{\sigma}{\sqrt{n}}$$

$$\approx 1.96 \cdot \frac{5.0}{\sqrt{54}} \approx 1.3.$$

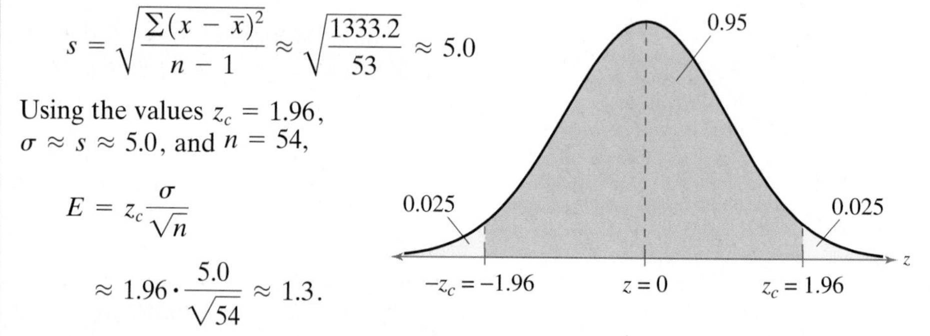

So, you are 95% confident that the maximum error of estimate for the population mean is about 1.3 sentences.

Try It Yourself 2

Use the data given in Try It Yourself 1 and a 95% confidence level to find the maximum error of estimate for the mean number of sentences in a magazine advertisement.

a. *Identify* z_c, n, and s.
b. *Find* E using z_c, $\sigma \approx s$, and n.
c. *State* the maximum error of estimate.

Answer: Page A37

Picturing the World

During most of the past 10 years, stocks and mutual funds produced some outstanding returns for investors. To estimate the mean annual rate of return for mutual funds, a random sample was taken of 35 mutual funds. The mean annual rate of return for the sample was 25.20%, with a standard deviation of 9.64%. *(Source: Fidelity Investments)*

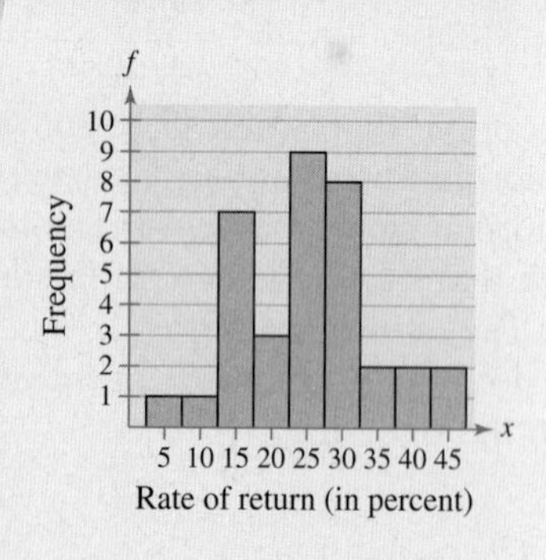

For a 95% confidence interval, what would be the maximum error of estimate for the population mean rate of return?

Confidence Intervals for the Population Mean

Using a point estimate and a maximum error of estimate, you can construct an interval estimate of a population parameter such as μ. This interval estimate is called a confidence interval.

DEFINITION

A c-confidence interval for the population mean μ is

$$\bar{x} - E < \mu < \bar{x} + E.$$

The probability that the confidence interval contains μ is c.

GUIDELINES

Finding a Confidence Interval for a Population Mean ($n \geq 30$ or σ known with a normally distributed population)

In Words	*In Symbols*
1. Find the sample statistics n and $\bar{x}$.	$\bar{x} = \dfrac{\Sigma x}{n}$
2. Specify σ, if known. Otherwise, if $n \geq 30$, find the sample standard deviation s and use it as an estimate for σ.	$s = \sqrt{\dfrac{\Sigma(x - \bar{x})^2}{n - 1}}$
3. Find the critical value z_c that corresponds to the given level of confidence.	Use the Standard Normal Table.
4. Find the maximum error of estimate E.	$E = z_c \dfrac{\sigma}{\sqrt{n}}$
5. Find the left and right endpoints and form the confidence interval.	Left endpoint: $\bar{x} - E$ Right endpoint: $\bar{x} + E$ Interval: $\bar{x} - E < \mu < \bar{x} + E$

Study Tip

The left and right endpoints of a confidence interval are $\bar{x} - E$ and $\bar{x} + E$, respectively. Other ways to represent a confidence interval are $(\bar{x} - E, \bar{x} + E)$ and $\bar{x} \pm E$. For instance, in Example 3, you could write the confidence interval as $(11.1, 13.7)$ or 12.4 ± 1.3.

EXAMPLE 3

Constructing a Confidence Interval

See *Minitab* steps on page 308.

Construct a 95% confidence interval for the mean number of sentences in all magazine advertisements.

SOLUTION In Examples 1 and 2, you found that $\bar{x} = 12.4$ and $E = 1.3$. The confidence interval is as follows.

Left Endpoint	Right Endpoint
$\bar{x} - E = 12.4 - 1.3 = 11.1$	$\bar{x} + E = 12.4 + 1.3 = 13.7$

$$11.1 < \mu < 13.7$$

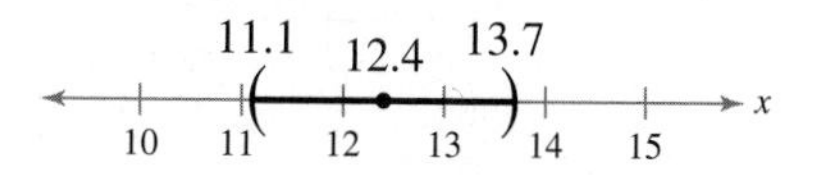

So, with 95% confidence, you can say that the population mean number of sentences is between 11.1 and 13.7.

Try It Yourself 3

Use the data given in Try It Yourself 1 to construct a 95% confidence interval for the mean number of sentences in all magazine advertisements. Compare your result to the interval found in Example 3.

a. *Find* $\overline{x}$ and E.
b. *Find* the left and right endpoints of the confidence interval.
c. *State* the 95% confidence interval for the mean number of sentences in all magazine advertisements and compare it to Example 3.

Answer: Page A37

Insight

The width of a confidence interval is $2E$. Examine the formula for E to see why a larger sample size tends to give you a narrower confidence interval for the same level of confidence.

EXAMPLE 4

Constructing a Confidence Interval Using Technology

Use a computer or graphing calculator to construct a 99% confidence interval for the mean number of sentences in all magazine advertisements, using the sample in Example 1.

SOLUTION To use a technology tool to solve the problem, enter the data and find that the sample standard deviation is $s \approx 5.0$. Then, use the confidence interval command to calculate the confidence interval (ZInterval for *TI-83*, 1-Sample Z for *Minitab*). The displays should look like the ones shown here.

MINITAB

Z Confidence Intervals

The assumed sigma = 5

Variable	N	Mean	StDev	SE Mean	99.0 % CI
C1	54	12.426	5.015	0.680	(10.673, 14.179)

TI-83

ZInterval
(10.673, 14.179)
$\bar{x}$=12.42592593
Sx=5.015454801
n=54

So, a 99% confidence interval for μ is $(10.7, 14.2)$. With 99% confidence, you can say that the population mean number of sentences is between 10.7 and 14.2.

Try It Yourself 4

Use the sample data in Example 1 and a computer or graphing calculator to construct 75% and 85% confidence intervals for the mean number of sentences in all magazine advertisements. How does the width of the confidence interval change as the level of confidence increases?

a. *Enter* the data.
b. *Use* the appropriate command to construct each confidence interval.
c. *Compare* the widths of the confidence intervals for $c = 0.75, 0.85$, and 0.99.

Answer: Page A37

In Example 4 and Try It Yourself 4, the same sample data was used to construct confidence intervals with different levels of confidence. Notice that as the level of confidence increases, the width of the confidence interval also increases. In other words, using the same sample data, *the greater the level of confidence, the wider the interval.*

If the population is normally distributed and the population standard deviation σ is known, you may use the normal sampling distribution for any sample size, as shown in Example 5.

EXAMPLE 

Constructing a Confidence Interval, σ Known

See *TI-83* steps on page 309.

A college admissions director wishes to estimate the mean age of all students currently enrolled. In a random sample of 20 students, the mean age is found to be 22.9 years. From past studies, the standard deviation is known to be 1.5 years and the population is normally distributed. Construct a 90% confidence interval of the population mean age.

SOLUTION Using $n = 20$, $\bar{x} = 22.9$, $\sigma = 1.5$, and $z_c = 1.645$, the maximum error of estimate at the 90% confidence interval is

$$E = z_c \frac{\sigma}{\sqrt{n}} = 1.645 \cdot \frac{1.5}{\sqrt{20}} \approx 0.55.$$

The 90% confidence interval can be written as $\bar{x} \pm E = 22.9 \pm 0.55$ or as follows.

Left Endpoint	Right Endpoint
$\bar{x} - E = 22.9 - 0.55 = 22.35$	$\bar{x} + E = 22.9 + 0.55 = 23.45$

$$22.35 < \mu < 23.45$$

So, with 90% confidence, you can say that the mean age of all the students is between 22.35 and 23.45 years.

Point estimate
$\bar{x} = 22.9$
22.35
23.45
22.5
23.0
23.5
x

Try It Yourself 5

Construct an 80% confidence interval of the population mean age for the college students in Example 5.

a. *Identify* n, $\bar{x}$, σ, and z_c.
b. *Find* E.
c. *Find* the left and right endpoints of the confidence interval.
d. *Specify* the 80% confidence interval.

Answer: Page A37

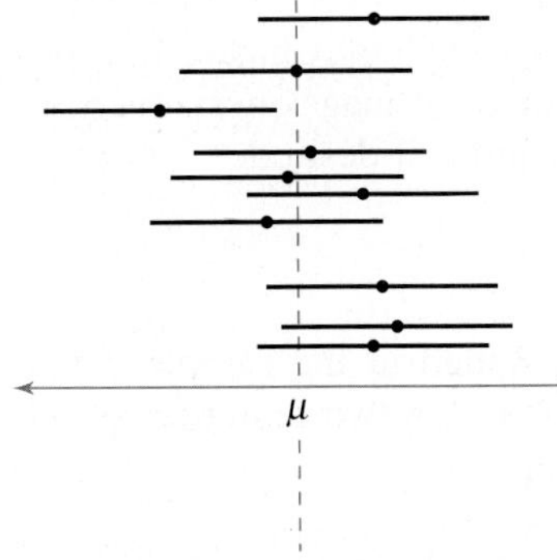

The horizontal segments represent 90% confidence intervals for different samples of the same size. In the long run, nine of every ten such intervals will contain μ.

After constructing a confidence interval, it is important that you interpret the results correctly. Consider the 90% confidence interval constructed in Example 5. Because μ already exists, it is either in the interval or not. It is *not* correct to say "There is a 90% probability that the actual mean is in the interval (22.35, 23.45)." The correct way to interpret your confidence interval is "There is a 90% probability that the confidence interval you described contains μ." This also means, of course, that there is a 10% probability that your confidence interval will not contain μ.

Sample Size

For the same sample statistics, as the level of confidence increases, the confidence interval widens. As the confidence interval widens, the precision of the estimate decreases. One way to improve the precision of an estimate without decreasing the level of confidence is to increase the sample size. But how large a sample size is needed to guarantee a certain level of confidence for a given maximum error of estimate?

Insight

Using the formula for the maximum error of estimate E, you can derive n as shown below.

$$E = z_c \frac{\sigma}{\sqrt{n}}$$

$$E\sqrt{n} = z_c \sigma$$

$$\sqrt{n} = \frac{z_c \sigma}{E}$$

$$n = \left(\frac{z_c \sigma}{E}\right)^2$$

Find a Minimum Sample Size to Estimate μ

Given a c-confidence level and a maximum error of estimate E, the minimum sample size n needed to estimate μ the population mean is

$$n = \left(\frac{z_c \sigma}{E}\right)^2.$$

If σ is unknown, you can estimate it using s, provided you have a preliminary sample with at least 30 members.

EXAMPLE 

Determining a Minimum Sample Size

You want to estimate the mean number of sentences in a magazine advertisement. How many magazine advertisements must be included in the sample if you want to be 95% confident that the sample mean is within one sentence of the population mean?

SOLUTION Using $c = 0.95$, $z_c = 1.96$, $\sigma \approx s \approx 5.0$ (from Example 2), and $E = 1$, you can solve for the minimum sample size n.

$$\begin{aligned} n &= \left(\frac{z_c \sigma}{E}\right)^2 \\ &\approx \left(\frac{1.96 \cdot 5.0}{1}\right)^2 \\ &= 96.04 \end{aligned}$$

When necessary, round up to obtain a whole number. So, you should include at least 97 magazine advertisements in your sample. (You already have 54, so you need 43 more.) Note that 97 is the *minimum* number of magazines advertisements to include in the sample. You could include more, if desired.

Try It Yourself 6

How many magazine advertisements must be included in the sample if you want to be 95% confident that the sample mean is within two sentences of the population mean? Compare your answer to Example 6.

a. *Identify* z_c, E, and s.
b. *Use* z_c, E, and $\sigma \approx s$ to find the minimum sample size n.
c. *State* how many magazine advertisements must be included in the sample and compare your answer to Example 6.

Answer: Page A37

6.1 Exercises

 LarsonTutor 6.1

 Companion Web Site

 Student Solutions Manual 6.1

 Videos 6.1

 Technology Manuals

Basic Skills and Concepts

1. When estimating a population mean, are you more likely to be correct if you use a point estimate or an interval estimate? Explain your reasoning.

2. Which statistic is the best unbiased estimator for μ?

(a) s (b) $\bar{x}$ (c) the median (d) the mode

3. Given the same sample statistics, which level of confidence would produce the widest confidence interval? Explain your reasoning.

(a) 90% (b) 95% (c) 98% (d) 99%

4. What is the effect on the width of the confidence interval when the sample size is increased? Explain your reasoning.

(a) The width increases (b) The width decreases (c) No effect

In Exercises 5–8, find the critical value z_c necessary to form a confidence interval at the given level of confidence.

5. $c = 0.80$

6. $c = 0.85$

7. $c = 0.75$

8. $c = 0.97$

Graphical Analysis In Exercises 9–14, use the values on the number line to find the error of estimate.

9. $\bar{x} = 3.8$ $\mu = 4.27$
Number line: 3.4 3.6 3.8 4.0 4.2 4.4 4.6 x

10. $\mu = 8.76$ $\bar{x} = 9.5$
Number line: 8.6 8.8 9.0 9.2 9.4 9.6 9.8 x

11. $\mu = 24.67$ $\bar{x} = 26.43$
Number line: 24 25 26 27 x

12. $\bar{x} = 46.56$ $\mu = 48.12$
Number line: 46 47 48 49 x

13. $\bar{x} = 0.7$ $\mu = 1.3$
Number line: 0.0 0.5 1.0 1.5 2.0 x

14. $\mu = 80.9$ $\bar{x} = 86.4$
Number line: 80 82 84 86 88 x

In Exercises 15–18, find the maximum error of estimate for the given values of c, s, and n.

15. $c = 0.90, s = 2.5, n = 36$

16. $c = 0.95, s = 3.0, n = 60$

17. $c = 0.65, s = 1.5, n = 50$

18. $c = 0.975, s = 3.4, n = 100$

In Exercises 19–22, construct the indicated confidence interval for the population mean μ.

19. $c = 0.90, \bar{x} = 15.2, s = 2.0, n = 60$

20. $c = 0.95, \bar{x} = 31.39, s = 0.8, n = 82$

21. $c = 0.95, \bar{x} = 4.27, s = 0.3, n = 42$

22. $c = 0.99, \bar{x} = 13.5, s = 1.5, n = 100$

Constructing Confidence Intervals In Exercises 23–26, you are given the sample mean and the sample standard deviation. Use this information to construct the 90% and 95% confidence intervals for the population mean. Which interval is wider?

23. ***Gas Grills*** ◆ A random sample of 32 gas grills has a mean price of $630.90 and a standard deviation of $56.70.

24. ***Stock Prices*** ◆ From a random sample of 36 days in a recent year, the closing stock prices for Hasbro had a mean of $12.53 and a standard deviation of $2.10. *(Adapted from Financial World)*

25. ***Wheat Yields*** ◆ A random sample of 156 fields of durum wheat has a mean yield of 26.8 bushels per acre and standard deviation of 8.0 bushels per acre. (See page 268.)

26. ***Sodium Chloride Concentration*** ◆ In 36 randomly selected seawater samples, the mean sodium chloride concentration was 23 cc/cubic meter and the standard deviation was 6.7 cc/cubic meter. *(Adapted from Dorling Kindersley Visual Encyclopedia)*

27. ***Washing Machine Repair Costs*** ◆ You work for a consumer advocate agency and want to find the mean repair cost of a washing machine. As part of your study, you randomly select 40 repair costs and find the mean to be $100.00. The sample standard deviation is $17.50. Construct a 95% confidence interval for the population mean. *(Adapted from Consumer Reports)*

28. ***VCR Repair Costs*** ◆ In a random sample of 50 VCRs, the mean repair cost was $\overline{x} = \$75$ and the standard deviation was $s = \$12.50$. Construct a 99% confidence interval for the population mean μ. *(Adapted from Consumer Reports)*

29. Repeat Exercise 27, changing the sample size to $n = 80$. Which confidence interval is wider? Explain.

30. Repeat Exercise 28, changing the sample size to $n = 90$. Which confidence interval is wider? Explain.

31. ***Heights of Beech Trees*** ◆ A random sample of 56 American beech trees has a mean height of 10.452 meters and a standard deviation of 2.130 meters. Construct a 99% confidence interval for the population mean height.

32. ***Diameters of Beech Trees*** ◆ In a random sample of 56 American beech trees, the mean diameter was 0.20 meter and the standard deviation was 0.06 meter. Construct a 95% confidence interval for the population mean.

33. Repeat Exercise 31, using a standard deviation of $s = 5.130$ meters. Which confidence interval is wider? Explain.

34. Repeat Exercise 32, using a standard deviation of $s = 0.1$ meter. Which confidence interval is wider? Explain.

35. How does the indicated change affect the confidence interval?

(a) Increase the level of confidence

(b) Increase the sample size

(c) Increase the standard deviation

36. Describe how you would form a 90% confidence interval to estimate the population mean age for students at your school.

Constructing Confidence Intervals In Exercises 37 and 38, you are given the sample mean and the population standard deviation. Use this information to construct the 90% and 99% confidence intervals for the population mean. Which interval is wider?

37. ***Newspaper Reading Times*** ◆ A publisher wants to estimate the mean length of time all adults spend reading newspapers. To do this, the publisher takes a random sample of 15 people and gets the following results.

$$11, 9, 8, 10, 10, 9, 7, 11, 11, 7, 6, 9, 10, 8, 10$$

From past studies, the publisher assumes σ is 1.5 minutes and that the population of times is normally distributed.

38. ***Computer Usage*** ◆ A computer company wants to estimate the mean number of hours per week all adults use computers at home. In a random sample of 21 adults, the mean length of time a computer was used at home was 5.3 hours. From past studies, the company assumes σ is 0.9 hour and that the population of times is normally distributed. *(Adapted from American Demographics)*

39. Determine the minimum required sample size if you want to be 95% confident that the sample mean is within one unit of the population mean given $\sigma = 4.8$. Assume the population is normally distributed.

40. Determine the minimum required sample size if you want to be 99% confident that the sample mean is within two units of the population mean given $\sigma = 1.4$. Assume the population is normally distributed.

41. ***Cholesterol Contents of Cheese*** ◆ A cheese processing company wants to estimate the mean cholesterol content of all one-ounce servings of cheese. The estimate must be within 0.5 milligram of the population mean.

(a) Determine the minimum required sample size to construct a 95% confidence interval for the population mean. Assume the population standard deviation is 2.8 milligrams.

(b) Repeat part (a) using a 99% confidence interval.

(c) Which level of confidence requires a larger sample size? Explain.

42. ***Ages of College Students*** ◆ An admissions director wants to estimate the mean age of all students enrolled at a college. The estimate must be within 1 year of the population mean. Assume the population of ages is normally distributed.

(a) Determine the minimum required sample size to construct a 90% confidence interval for the population mean. Assume the population standard deviation is 1.2 years.

(b) Repeat part (a) using a 99% confidence interval.

(c) Which level of confidence requires a larger sample size? Explain.

43. ***Paint Can Volumes*** ◆ A paint manufacturer uses a machine to fill gallon cans with paint. (a) The manufacturer wants to estimate the mean volume of paint the machine is putting in the cans within 0.25 ounce. Determine the minimum sample size required to construct a 90% confidence interval for the population mean. Assume the population standard deviation is 0.85 ounce. (b) Repeat part (a) using an error tolerance of 0.15 ounce. Which error tolerance requires a larger sample size? Explain.

Error tolerance = 0.25 oz

Figure for Exercise 43

Error tolerance = 1 mL

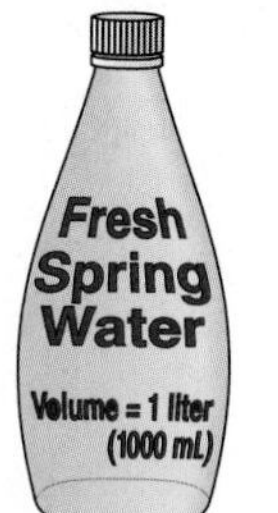

Figure for Exercise 44

44. ***Water Dispensing Machine*** ◆ A beverage company uses a machine to fill 1-liter bottles with water. Assume that the population of volumes is normally distributed. (a) The company wants to estimate the mean volume of water the machine is putting in the bottles within 1 milliliter. Determine the minimum sample size required to construct a 95% confidence interval for the population mean. Assume the population standard deviation is 3 milliliters. (b) Repeat part (a) using an error tolerance of 2 milliliters. Which error tolerance requires a larger sample size? Explain.

45. ***Plastic Sheet Cutting*** ◆ A machine cuts plastic into sheets that are 50 feet (600 inches) long. Assume that the population of lengths is normally distributed. (a) The company wants to estimate the mean length the machine is cutting the plastic within 0.125 inch. Determine the minimum sample size required to construct a 95% confidence interval for the population mean. Assume the population standard deviation is 0.25 inch. (b) Repeat part (a) using an error tolerance of 0.0625 inch. Which error tolerance requires a larger sample size? Explain.

46. ***Paint Sprayer*** ◆ A company uses an automated sprayer to apply paint to metal. The company sets the sprayer to apply the paint 1-inch thick. (a) The company wants to estimate the mean thickness the sprayer is applying the paint within 0.0625 inch. Determine the minimum sample size required to construct a 90% confidence interval for the population mean. Assume the population standard deviation is 0.25. (b) Repeat part (a) using an error tolerance of 0.03125. Which error tolerance requires a larger sample size? Explain.

47. ***Soccer Balls*** ◆ A soccer ball manufacturer wants to estimate the mean circumference of soccer balls within 0.1 inch. (a) Determine the minimum required sample size to construct a 99% confidence interval for the population mean. Assume the population standard deviation is 0.25 inch. (b) Repeat part (a) using a standard deviation of 0.3 inch. Which standard deviation requires a larger sample size? Explain.

48. ***Mini-Soccer Balls*** ◆ A soccer ball manufacturer wants to estimate the mean circumference of mini-soccer balls within 0.15 inch. Assume that the population of circumferences is normally distributed. (a) Determine the minimum required sample size to construct a 99% confidence interval for the population mean. Assume the population standard deviation is 0.20 inch. (b) Repeat part (a) using a standard deviation of 0.10 inch. Which standard deviation requires a larger sample size? Explain.

49. How does the indicated change affect the minimum sample size requirement?

(a) Increase the level of confidence
(b) Increase the error tolerance
(c) Increase the standard deviation

50. When estimating the population mean, why not construct a 99% confidence interval every time?

Using Technology In Exercises 51–54, you are given a data sample. Use a computer or graphing calculator to construct a 95% confidence interval for the population mean.

DATA **51.** ***Airfare*** ◆ A random sample of airfare prices (in dollars) for a one-way ticket from New York to Houston *(Adapted from Newsweek)*

Key: 29|2 = 292

Stem	Leaves
28	8
29	0 2 3 3 5 6 8
30	0 2 3 4 5 5 5 5 6 6 7 7
31	3 4 4 6
32	0 0 1 1 2 2 6 7

DATA **52.** ***Airfare*** ◆ A random sample of airfare prices (in dollars) for a one-way ticket from Atlanta to Chicago *(Adapted from Newsweek)*

Key: 8|7 = 87

Stem	Leaves
8	7 7
9	0 4 4
9	5 6 8 8 8 9
10	0 1 1 1 1 2 3 3 3 4 4 4
10	5 5 5 5 6 7 7 8 9
11	1 4
11	7

DATA **53.** ***Annual Precipitation*** ◆ A random sample of the annual precipitation (in inches) for Anchorage, Alaska *(Source: Alaska Climate Research Center)*

14.62	14.66	12.76	16.10	13.09
12.87	14.93	19.48	12.49	15.44
13.67	11.49	15.89	15.17	14.63
13.42	11.81	15.46	12.25	14.97
14.20	16.73	17.64	14.75	14.51
16.89	17.06	13.06	8.61	14.54

DATA **54. *Annual Precipitation*** ◆ A random sample of the annual precipitation (in inches) for Nome, Alaska *(Source: Alaska Climate Research Center)*

16.69	11.17	17.47	18.31	13.41	19.04	29.49	19.27	25.61
20.66	19.91	15.24	19.78	22.15	24.25	15.54	21.66	20.80
14.30	13.56	18.74	7.42	14.41	20.09	12.59	14.97	18.06
22.06	16.96	19.25	19.18	20.22	16.27			

Extending the Basics

Finite Population Correction Factor In Exercises 55 and 56, use the following information. In this section you studied the formation of a confidence interval to estimate a population mean when the population is large or infinite. When a population is finite, the formula that determines the standard error of the mean $\sigma_{\bar{x}}$ needs to be adjusted. If N is the size of the population and n is the size of the sample (where $n \geq 0.05\,N$), the standard error of the mean is

$$\sigma_{\bar{x}} = \frac{\sigma}{\sqrt{n}}\sqrt{\frac{N-n}{N-1}}.$$

The expression $\sqrt{(N-n)/(N-1)}$ is called the **finite population correction factor.** The maximum error of estimate is

$$E = z_c \frac{\sigma}{\sqrt{n}}\sqrt{\frac{N-n}{N-1}}.$$

55. Determine the finite population correction factor for each of the following.

(a) $N = 1000$ and $n = 500$

(b) $N = 1000$ and $n = 100$

(c) $N = 1000$ and $n = 75$

(d) $N = 1000$ and $n = 50$

(e) What happens to the finite population correction factor as the sample size n decreases but the population size N remains the same?

56. Determine the finite population correction factor for each of the following.

(a) $N = 100$ and $n = 50$

(b) $N = 400$ and $n = 50$

(c) $N = 700$ and $n = 50$

(d) $N = 2000$ and $n = 50$

(e) What happens to the finite population correction factor as the population size N increases but the sample size n remains the same?

57. *Sample Size* ◆ The equation for determining the sample size

$$n = \left(\frac{z_c \sigma}{E}\right)^2$$

can be obtained by solving the equation for the maximum error of estimate

$$E = \frac{z_c \sigma}{\sqrt{n}}$$

for n. Show that this is true and justify each step.

Case Study

WWW.NMFS.NOAA.GOV

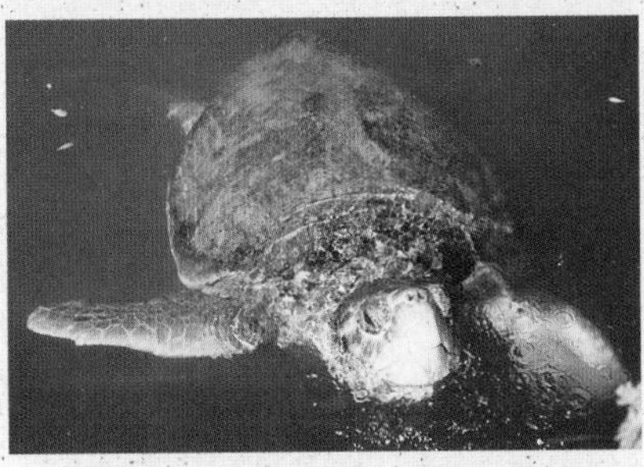

Shell Lengths of Loggerhead Sea Turtles

The National Marine Fisheries Services is part of the National Oceanic and Atmospheric Administration. NMFS's programs support the conservation and management of living marine resources.

There are six species of sea turtles in the United States and all are protected as endangered or threatened species. Rarely does a hatchling sea turtle live to maturity. In fact, it is believed that only 1 in 10,000 hatchlings lives long enough to reproduce.

In a study by Hays and Marsh reported in the *Canadian Journal of Zoology* (75: 40–46, 1997), 71 loggerhead sea turtles were captured and measured off the coast of Britain. The shell lengths of the turtles are shown in the stem-and-leaf plot at the right.

Part of the purpose of the study was to estimate the growth rate of juvenile turtles. The turtles were hatched off the coast of Florida and their drifting time in the Atlantic Ocean was estimated to be between 1.8 and 3.75 years. From this and the fact that a typical hatchling has a shell length of 4.5 centimeters, Hays and Marsh estimated that juvenile loggerhead sea turtles grow at a rate of between 4.3 and 8.9 centimeters per year.

Stem	Leaves
1	5 5 6 6 6 7 7 7 7 8 8 8 8 8 8 8 9 9 9 9 9
2	0 0 0 0 0 0 0 0 0 1 1 1 2 2 2 2 2 2 3 4 4
2	5 5 5 6 6 7
3	0 0 0 3 4
3	8
4	0
4	5 9
5	1 4
5	
6	0 1 1 4
6	
7	
7	5 8
8	
8	8 8
9	0 0
9	6
10	4

Key: $1|5 = 15$ cm

← Shell length →

Exercises

1. A loggerhead sea turtle is classified as a juvenile if its shell length is less than 40 centimeters. How many of the turtles in the sample were juveniles?
2. Use the sample to make a point estimate of the mean shell length of all juvenile loggerhead sea turtles that drift from their hatching site to the coast of Britain.
3. Find the standard deviation of the sample of juveniles.
4. Use the sample to make an interval estimate of the mean shell length of juvenile loggerhead sea turtles that drift from their hatching site to the coast of Britain.
   (a) Use a 90% confidence level.
   (b) Use a 95% confidence level.
   (c) Use a 99% confidence level.
5. How would your results have differed if you had used all the turtles in the sample instead of just the juvenile turtles? Explain your reasoning.
6. Complete the following table.

Juvenile Turtles	Length at Hatching	Length at Capture	Shell Growth
Minimum	4.5 cm	15 cm	?? cm
Maximum	4.5 cm	40 cm	?? cm

Use the table to estimate the rate of growth for juvenile loggerhead sea turtles under the following assumptions.

(a) Drift time = 1.8 years, minimum shell growth
(b) Drift time = 3.75 years, maximum shell growth

6.2 Confidence Intervals for the Mean (Small Samples)

What You Should Learn

- How to interpret the t-distribution and use a t-distribution table
- How to construct confidence intervals when $n < 30$ the population is normally distributed, and σ is unknown

The t-Distribution • Confidence Intervals and t-Distributions

The t-Distribution

In many real-life situations, the population standard deviation is unknown. Moreover, because of various constraints such as time and cost, it is often not practical to collect samples of size 30 or more. So, how can you construct a confidence interval for a population mean given such circumstances? If the random variable is normally distributed (or approximately normally distributed), the sampling distribution for $\overline{x}$ is a t-distribution.

Historical Reference

William S. Gosset (1876–1937) developed the t-distribution while employed by the Guinness Brewing Company in Dublin, Ireland. Gosset published his findings using the pseudonym Student. The t-distribution is sometimes referred to as Student's t-distribution. (See page 292 for others who were important in the history of statistics.)

DEFINITION

If the distribution of a random variable x is approximately normal, then the sampling distribution of $\overline{x}$ is a ***t*-distribution**, where

$$t = \frac{\overline{x} - \mu}{\frac{s}{\sqrt{n}}}.$$

Critical values of t are denoted by t_c. Several properties of the t-distribution are as follows.

1. The t-distribution is bell shaped and symmetric about the mean.
2. The t-distribution is a family of curves, each determined by a parameter called the degrees of freedom. The **degrees of freedom** are the number of free choices left after a sample statistic such as $\overline{x}$ is calculated. When you use a t-distribution to estimate a population mean, the degrees of freedom are equal to one less than the sample size.

   $$\text{d.f.} = n - 1 \quad \text{Degrees of Freedom}$$

3. The total area under a t-curve is 1 or 100%.
4. The mean, median, and mode of the t-distribution are equal to zero.
5. As the degrees of freedom increase, the t-distribution approaches the normal distribution. After 30 d.f. the t-distribution is very close to the standard normal z-distribution.

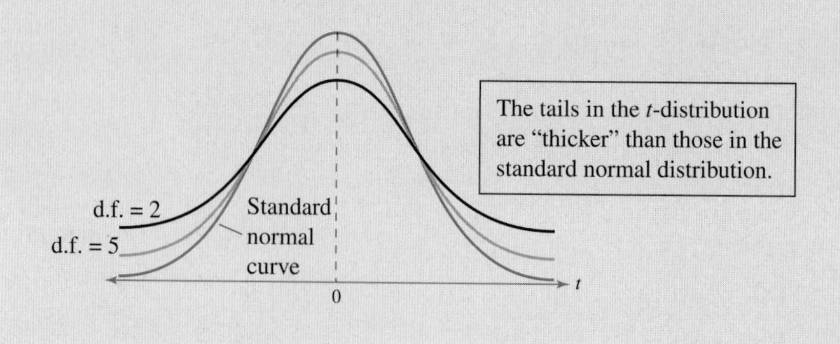

Table 5 of Appendix B lists critical values of t for selected confidence intervals and degrees of freedom.

EXAMPLE 

Finding Critical Values of t

Find the critical value t_c for a 95% confidence when the sample size is 15.

SOLUTION Because $n = 15$, the degrees of freedom are

$$\text{d.f.} = n - 1 = 15 - 1 = 14.$$

A portion of Table 5 is shown. Using d.f. = 14 and $c = 0.95$, you can find the critical value t_c as shown by the highlighted areas in the table.

	Level of confidence, c	0.50	0.80	0.90	0.95	0.98
	One tail, α	0.25	0.10	0.05	0.025	0.01
d.f.	Two tails, α	0.50	0.20	0.10	0.05	0.02
1		1.000	3.078	6.314	12.706	31.821
2		.816	1.886	2.920	4.303	6.965
3		.765	1.638	2.353	3.182	4.541
4		.741	1.533	2.132	2.776	3.747
5		.727	1.476	2.015	2.571	3.365
11		.697	1.363	1.796	2.201	2.718
12		.695	1.356	1.782	2.179	2.681
13		.694	1.350	1.771	2.160	2.650
14		.692	1.345	1.761	2.145	2.624
15		.691	1.341	1.753	2.131	2.602
16		.690	1.337	1.746	2.120	2.583

From the table, you can see that $t_c = 2.145$. The graph shows the t-distribution for 14 degrees of freedom, $c = 0.95$, and $t_c = 2.145$.

So, 95% of the area under the t-distribution curve with 14 degrees of freedom lies between $t = \pm 2.145$.

Study Tip

Unlike the z-table, critical values for a specific confidence interval can be found in the column headed by c in the appropriate d.f. row. (The symbol α will be explained in Chapter 7.)

Insight

For 30 or more degrees of freedom, the critical values for the t-distribution are close to the corresponding critical values for the normal distribution. Moreover, the values in the last row of the table marked ∞ d.f. correspond *exactly* to the normal distribution values.

Try It Yourself 1

Find the critical value t_c for a 90% confidence when the sample size is 22.

a. Identify the degrees of freedom.
b. Identify the level of confidence c.
c. Use Table 5 of Appendix B to find t_c.

Answer: Page A37

Confidence Intervals and *t*-Distributions

Constructing a confidence interval using the t-distribution is similar to constructing a confidence interval using the normal distribution—both use a point estimate $\bar{x}$ and a maximum error of estimate E.

Study Tip

Before using the t-distribution to construct a confidence interval, you should check that $n < 30$, σ is unknown, and the population is approximately normal.

GUIDELINES

Constructing a Confidence Interval for the Mean: *t*-Distribution

In Words	*In Symbols*
1. Identify the sample statistics n, $\bar{x}$, and s.	$\bar{x} = \dfrac{\Sigma x}{n}$, $s = \sqrt{\dfrac{\Sigma(x - \bar{x})^2}{n - 1}}$
2. Identify the degrees of freedom, the level of confidence c, and the critical value t_c.	d.f. $= n - 1$
3. Find the maximum error of estimate E.	$E = t_c \dfrac{s}{\sqrt{n}}$
4. Find the left and right endpoints and form the confidence interval.	Left endpoint: $\bar{x} - E$ Right endpoint: $\bar{x} + E$ Interval: $\bar{x} - E < \mu < \bar{x} + E$

EXAMPLE 

Constructing a Confidence Interval

See *Minitab* steps on page 308.

You randomly select 16 restaurants and measure the temperature of the coffee sold at each. The sample mean temperature is 162°F with a sample standard deviation of 10°F. Find the 95% confidence interval for the mean temperature. Assume the temperatures are approximately normally distributed.

SOLUTION Because the sample size is less than 30, σ is unknown, and the temperatures are approximately normally distributed, you can use the t-distribution. Using $n = 16$, $\bar{x} = 162$, $s = 10$, $c = 0.95$, and d.f. $= 15$, you can use Table 5 to find that $t_c = 2.131$. The maximum error of estimate at the 95% confidence interval is

$$E = t_c \frac{s}{\sqrt{n}} = 2.131 \cdot \frac{10}{\sqrt{16}} = 5.3275.$$

The confidence interval is as follows.

Left Endpoint	Right Endpoint
$\bar{x} - E = 162 - 5.3275 = 156.6725$	$\bar{x} + E = 162 + 5.3275 = 167.3275$

$$156.6725 < \mu < 167.3275$$

So, with 95% confidence, you can say that the mean temperature of coffee sold is between 156.7°F and 167.3°F.

$\bar{x} - E = 156.6725$ $\quad$ $\bar{x} = 162$ $\quad$ $\bar{x} + E = 167.3275$

156 158 160 162 164 166 168 x

Try It Yourself 2

Find the 90% and 99% confidence intervals for the mean temperature.

a. *Find* t_c and E for each level of confidence.
b. Use $\bar{x}$ and E to find the *left and right endpoints.*
c. *State* the 90% and 99% confidence intervals for the mean temperature.

Answer: Page A37

EXAMPLE 

See *TI-83* steps on page 309.

Constructing a Confidence Interval

You randomly select 20 mortgage institutions and determine the current mortgage interest rate at each. The sample mean rate is 6.93% with a sample standard deviation of 0.42%. Find the 99% confidence interval for the population mean mortgage interest rate. Assume the interest rates are approximately normally distributed.

SOLUTION Because the sample size is less than 30, σ is unknown, and the interest rates are approximately normally distributed, you can use the t-distribution. Using $n = 20$, $\bar{x} = 6.93$, $s = 0.42$, $c = 0.99$, and d.f. $= 19$, you can use Table 5 to find that $t_c = 2.861$. The maximum error of estimate at the 99% confidence interval is

$$E = t_c \frac{s}{\sqrt{n}}$$

$$= 2.861 \cdot \frac{0.42}{\sqrt{20}}$$

$$\approx 0.269.$$

The confidence interval is as follows.

Left Endpoint	Right Endpoint
$\bar{x} - E = 6.93 - 0.269 = 6.661$	$\bar{x} + E = 6.93 + 0.269 = 7.199$

$$6.661 < \mu < 7.199$$

So, with 99% confidence, you can say that the population mean mortgage interest rate is between 6.66% and 7.20%.

$\bar{x} - E = 6.661$ $\bar{x} = 6.93$ $\bar{x} + E = 7.199$

6.4 6.6 6.8 7.0 7.2 7.4 x

Try It Yourself 3

Find the 90% and 95% confidence intervals for the population mean mortgage interest rate. Compare the widths of the intervals.

a. Find t_c and E for each level of confidence.
b. Use $\bar{x}$ and E to find the left and right endpoints.
c. State the 90% and 95% confidence intervals for the population mean mortgage interest rate and compare their widths.

Answer: Page A38

Picturing the World

Two footballs, one filled with air and the other filled with helium, were kicked on a windless day at Ohio State University. The footballs were alternated with each kick. After 10 practice kicks, each football was kicked 29 more times. The distances (in yards) are listed. *(Source: OSC Scientists Get a Kick Out of Sports Controversy, "The Columbus Dispatch," November 21, 1993.)*

Air Filled

1	9
2	0 0 2 2 2
2	5 5 5 5 6 6
2	7 7 7 8 8 8 8 8 9 9 9
3	1 1 1 2
3	3 4 Key: 1|9 = 19

Helium Filled

1	1 2
1	4
1	
2	2
2	3 4 6 6 6
2	7 8 8 8 9 9 9 9
3	0 0 0 0 1 1 2 2
3	3 4 5
3	9

Assume that the distances are normally distributed for each football. Apply the flowchart at the right to each sample. Find a 95% confidence interval for the mean distance each football traveled. Do the confidence intervals overlap? What does this tell you?

The flowchart describes when to use the normal distribution to construct a confidence interval and when to use a t-distribution.

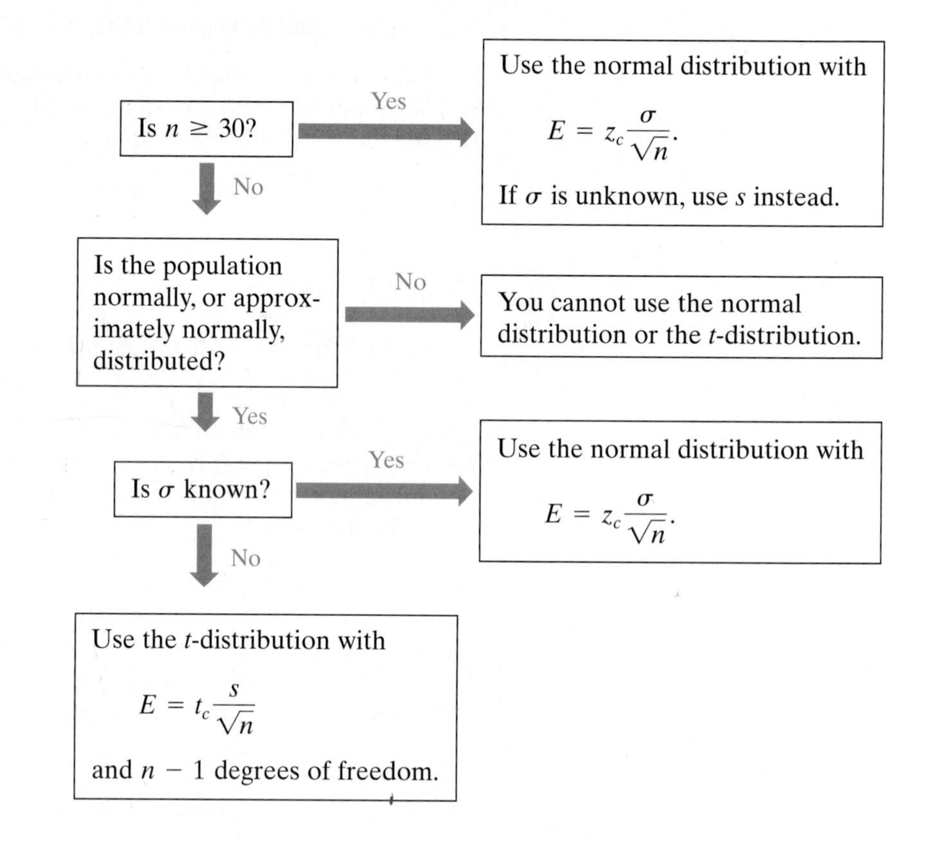

Choosing the Normal or t-Distribution

You randomly select 25 newly constructed houses. The sample mean construction cost is \$181,000 and the population standard deviation is \$28,000. Assuming construction costs are normally distributed, should you use the normal distribution, the t-distribution, or neither to construct a 95% confidence interval for the population mean construction cost? Explain your reasoning.

SOLUTION Because the population is normally distributed and the population standard deviation is known, you should use the normal distribution.

Try It Yourself 4

You randomly select 18 adult male athletes and measure the resting heart rate of each. The sample mean heart rate is 64 beats per minute with a sample standard deviation of 2.5 beats per minute. Assuming the heart rates are normally distributed, should you use the normal distribution, the t-distribution, or neither to construct a 90% confidence interval for the mean heart rate? Explain your reasoning.

a. Use the flowchart to determine which distribution you should use to construct the 90% confidence interval for the mean heart rate.

Answer: Page A38

6.2 Exercises

Help

LarsonTutor 6.2

Companion Web Site

Student Solutions Manual 6.2

Videos 6.2

Technology Manuals

Basic Skills and Concepts

In Exercises 1–4, find the critical value t_c for the given confidence level c and sample size n.

1. $c = 0.90, \ n = 10$

2. $c = 0.95, \ n = 12$

3. $c = 0.99, \ n = 16$

4. $c = 0.98, \ n = 20$

In Exercises 5 and 6, suppose you incorrectly used the normal distribution to find the maximum error of estimate for the given values of c, s, and n.

(a) Find the value of E using the normal distribution.

(b) Find the correct value using a t-distribution. Compare the results.

5. $c = 0.95, \ s = 5, \ n = 16$

6. $c = 0.99, \ s = 3, \ n = 6$

In Exercises 7–10, construct the indicated confidence interval for the population mean μ using (a) a t-distribution. (b) If you had incorrectly used a normal distribution, which interval would be wider?

7. $c = 0.90, \ \overline{x} = 12.5, \ s = 2.0, \ n = 6$

8. $c = 0.95, \ \overline{x} = 13.4, \ s = 0.85, \ n = 8$

9. $c = 0.98, \ \overline{x} = 4.3, \ s = 0.34, \ n = 14$

10. $c = 0.99, \ \overline{x} = 14, \ s = 2.0, \ n = 10$

Constructing Confidence Intervals In Exercises 11 and 12, you are given the sample mean and the sample standard deviation. Assume the variable is normally distributed and use a t-distribution to construct a 95% confidence interval for the population mean μ. What is the maximum error of estimate of μ?

11. ***Repair Costs: Microwaves*** ◆ In a random sample of five microwave ovens, the mean repair cost was \$75.00 and the standard deviation was \$12.50. *(Adapted from Consumer Reports)*

12. ***Repair Costs: Computers*** ◆ In a random sample of seven computers, the mean repair cost was \$100.00 and the standard deviation was \$42.50. *(Adapted from Consumer Reports)*

13. Suppose you did some research on repair costs of microwave ovens and found that the standard deviation is $\sigma = \$15$. Repeat Exercise 11, using a normal distribution with the appropriate calculations for a standard deviation that is known. Compare the results.

14. Suppose you did some research on repair costs of computers and found that the standard deviation is $\sigma = \$50$. Repeat Exercise 12, using a normal distribution with the appropriate calculations for a standard deviation that is known. Compare the results.

Constructing Confidence Intervals In Exercises 15 and 16, you are given the sample mean and the sample standard deviation. Assume the variable is normally distributed and use a normal distribution or a t-distribution to construct a 90% confidence interval for the population mean μ.

15. ***Waste Generated*** ◆ (a) In a random sample of 10 adults from the U.S., the mean waste generated per person per day was 4.3 pounds and the standard deviation was 1.2 pounds. (b) Repeat part (a), assuming the same statistics came from a sample size of 500. Compare the results. *(Adapted from U.S. Environmental Protection Agency)*

16. ***Waste Recycled*** ◆ (a) In a random sample of 12 adults from the U.S., the mean waste recycled per person per day was 1.2 pounds and the standard deviation was 0.3 pound. (b) Repeat part (a), assuming the same statistics came from a sample size of 600. Compare the results. *(Adapted from U.S. Environmental Protection Agency)*

Constructing Confidence Intervals In Exercises 17–20, a data set is given. For each data set, find (a) the sample mean, (b) the sample standard deviation, and (c) construct a 99% confidence interval for the population mean μ. Assume the population of each data set is normally distributed.

17. ***Biology*** ◆ The monthly incomes for 10 randomly selected people, each with a bachelor's degree in biology *(Adapted from U.S. Bureau of the Census)*

2148.51 1978.27 2093.63 2091.95 2282.18 2223.64 2276.50
2207.41 2285.69 2159.72

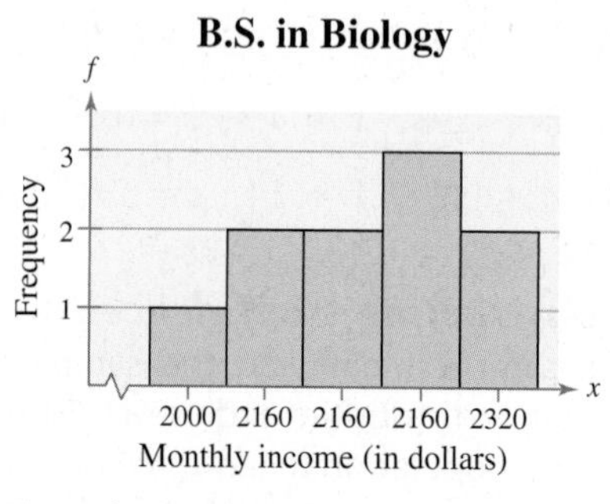

Figure for Exercise 17

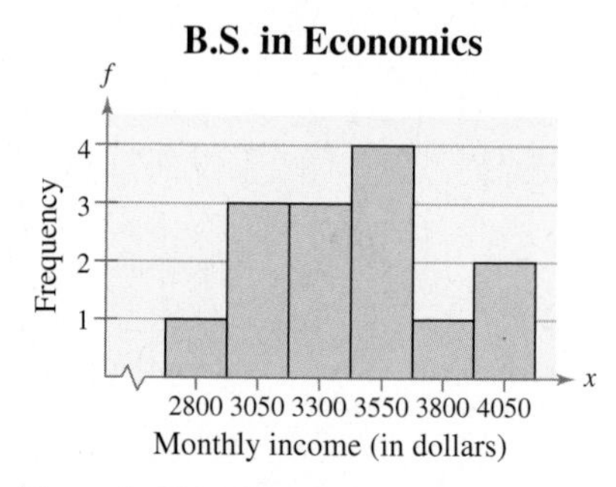

Figure for Exercise 18

18. ***Economics*** ◆ The monthly incomes for 14 randomly selected people, each with a bachelor's degree in economics *(Adapted from U.S. Bureau of the Census)*

3450.66 3596.73 3366.66 3455.40 3151.70 2727.08 3283.76
3527.64 3407.34 4036.64 4083.73 2946.47 3023.41 3806.22

19. ***SAT*** ◆ The SAT scores for 12 randomly selected senior high school students

1424 1223 987 692 947 723
837 721 747 540 623 1445

20. ***GPA*** ◆ The grade point averages for 15 randomly selected college students

2.3 3.3 2.6 1.8 0.2 3.1 4.0 0.7
2.3 2.0 3.1 1.4 1.3 1.6 1.6

Choosing a Distribution In Exercises 21–26, use a normal distribution or a t-distribution to construct a 95% confidence interval for the population mean. Justify your decision. If neither distribution can be used, explain why not.

21. ***Lengths of Bolts*** ◆ In a random sample of 70 bolts, the mean length was 1.25 inches and the standard deviation was 0.01 inch.

22. ***Prices of Toasters*** ◆ You took a random sample of 12 two-slice toasters and found the mean price was $61.12 and the standard deviation was $24.62. Assume the prices are normally distributed. *(Source: www.cooking.com)*

DATA **23.** ***Sports Cars: Miles per Gallon*** ◆ You take a random survey of 25 sports cars and record the miles per gallon for each. The data are listed below. Assume the miles per gallon are normally distributed. *(Adapted from Consumer Reports)*

24 24 27 20 26 23 18 29 24 22 22 27 26
20 28 30 23 24 19 22 24 26 23 24 25

DATA **24.** ***ACT*** ◆ In a recent year, the standard deviation of ACT scores for all students was 4.7. You take a random survey of 20 students and determine the ACT score of each. The scores are listed below. Assume the test scores are normally distributed. *(Source: ACT, Inc.)*

26 22 23 12 19 25 23 21 25 10
17 26 23 24 20 14 21 23 20 22

25. ***Hospital Waiting Times*** ◆ In a random sample of 19 patients at a hospital's minor emergency department, the mean waiting time (in minutes) before seeing a medical professional was 23 minutes and the standard deviation was 11 minutes. Assume the waiting times are not normally distributed.

26. ***Grocery Store Spending*** ◆ In a random sample of 17 shoppers at a grocery store, the mean amount spent was $28.13 and the standard deviation was $12.05. Assume the amounts spent are normally distributed.

Extending the Basics

27. ***Tennis Ball Manufacturing*** ◆ A company manufactures tennis balls. When its tennis balls are dropped onto a concrete surface from a height of 100 inches, the company wants the mean height the balls bounce upward to be 55.5 inches. To maintain this average, random samples of 25 tennis balls are periodically tested. If the t-value falls between $-t_{0.99}$ and $t_{0.99}$, the company will be satisfied that it is manufacturing acceptable tennis balls. A sample of 25 balls is randomly selected and tested. The mean bounce height of the sample is 56.0 inches and the standard deviation is 0.25 inch. Is the company making acceptable tennis balls? Explain your reasoning.

28. ***Light Bulb Manufacturing*** ◆ A company manufactures light bulbs. The company wants the bulbs to have a mean life span of 1000 hours. To maintain this average, random samples of 16 light bulbs are periodically tested. If the t-value falls between $-t_{0.99}$ and $t_{0.99}$, the company will be satisfied that it is manufacturing acceptable light bulbs. A sample of 16 light bulbs is randomly selected and tested. The mean life span of the sample is 1015 hours and the standard deviation is 25 hours. Is the company making acceptable light bulbs? Explain your reasoning.

HISTORY OF STATISTICS - TIMELINE

CONTRIBUTOR	TIME	CONTRIBUTION
John Graunt (1620–1674)	17th century	Studied records of deaths in London in the early 1600s. The first to make extensive statistical observations from massive amounts of data (Chapter 2), his work laid the foundation for modern statistics.
Blase Pascal (1632–1692) *Pierre Fermat* (1601–1695)	17th century	Pascal and Fermat corresponded about basic probability problems (Chapter 3)—especially those dealing with gaming and gambling.
Pierre Laplace (1749–1827)	18th century	Studied probability (Chapter 3) and is credited with putting probability on a sure mathematical footing.
Carl Gauss (1777–1855)	18th century	Studied regression and the method of least squares (Chapter 9) through astronomy. In his honor, the normal distribution is sometimes called the Gaussian distribution.
Lambert Quetelet (1796–1874)	19th century	Used descriptive statistics (Chapter 2) to analyze crime and mortality data and studied census techniques. Described normal distributions (Chapter 5) in connection with human traits such as height.
Francis Galton (1822–1911)	19th century	Used regression and correlation (Chapter 9) to study genetic variation in humans. He is credited with discovery of the Central Limit Theorem (Chapter 5).
Karl Pearson (1857–1936)	20th century	Studied natural selection using correlation (Chapter 9). Formed first academic department of statistics, and helped develop chi-square analysis (Chapter 6).
William Gosset (1876–1937)	20th century	Studied process of brewing and developed *t*-test to correct problems connected with small sample sizes (Chapter 6).
Ronald Fisher (1890–1962)	20th century	Studied biology and natural selection and developed ANOVA (Chapter 10), stressed the importance of experimental design (Chapter 1), and was the first to identify the null and alternative hypotheses (Chapter 7).
Charles Spearman (1863–1945)	20th century (later)	British psychologist who was one of the first to develop intelligence testing using factor analysis (Chapter 10).
Frank Wilcoxon (1892–1965)	20th century (later)	Biochemist who used statistics to study plant pathology. He introduced two-sample tests (Chapter 8), which led the way to the development of nonparametric statistics.
John Tukey (1915–2000)	20th century (later)	Worked at Princeton during World War II. Introduced exploratory data analysis techniques such as stem-and-leaf plots (Chapter 2). Also, worked at Bell Laboratories and is best known for his work in inferential statistics (Chapters 6–11).
David Kendall (1918–)	20th century (later)	Worked at Princeton and Cambridge. Is a leading authority on applied probability and data analysis (Chapters 2 and 3).

6.3 Confidence Intervals for Population Proportions

What You Should Learn

- How to find a point estimate for the population proportion
- How to construct a confidence interval for a population proportion
- How to determine the required minimum sample size when estimating a population proportion

Point Estimate for the Population Proportion p • Confidence Intervals for a Population Proportion p • Increasing Sample Size to Increase Precision

Point Estimate for the Population Proportion p

Recall from Section 4.2 that the probability of success in a single trial of a binomial experiment is p. This probability is a population **proportion.** In this section, you will learn how to estimate a population proportion p using a confidence interval. As with confidence intervals for μ, you will start with a point estimate.

DEFINITION

The point estimate for p, the population proportion of successes, is given by the proportion of successes in a sample and is denoted by

$$\hat{p} = \frac{x}{n}$$

where x is the number of successes in the sample and n is the number in the sample. The point estimate for the number of failures is $\hat{q} = 1 - \hat{p}$. The symbols $\hat{p}$ and $\hat{q}$ are read as "p hat" and "q hat."

EXAMPLE 

Finding a Point Estimate for p

In a survey of 1024 U.S. adults, 287 said that their favorite sport to watch is football. Find a point estimate for the population proportion of U.S. adults who say their favorite sport to watch is football. *(Source: The Gallup Organization)*

SOLUTION Using $n = 1024$ and $x = 287$,

$$\begin{aligned} \hat{p} &= \frac{x}{n} \\ &= \frac{287}{1024} \\ &\approx 0.28 \\ &= 28\%. \end{aligned}$$

Insight

In the first two sections, estimates were made for quantitative data. In this section, sample proportions are used to make estimates for qualitative data.

Try It Yourself 1

In a survey of 1470 adults from the U.S., 221 said that of all the presidents in our nation's history, they most admire Abraham Lincoln. Find a point estimate for the population proportion of adults who admire Lincoln over all other U.S. presidents. *(Adapted from Marist Institute for Public Opinion)*

a. *Identify* x and n.
b. Use x and n to *find* $\hat{p}$.

Answer: Page A38

Picturing the World

In 2001, a Gallup poll surveyed 1016 households in the U.S. about their pets. Of those surveyed, 599 said they had at least one dog or cat as a pet.

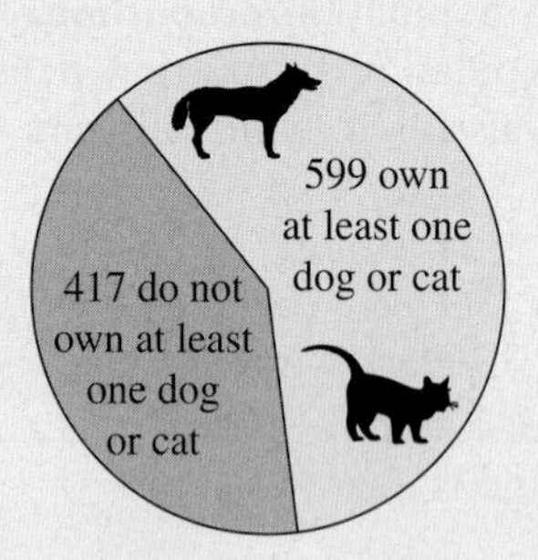

Find a 90% confidence interval for the proportion of households in the U.S. that own at least one cat or dog.

Confidence Intervals for a Population Proportion p

Constructing a confidence interval for a population proportion p is similar to constructing a confidence interval for a population mean. You start with a point estimate and calculate a maximum error of estimate.

DEFINITION

A ***c*-confidence interval** for the population proportion p is

$$\hat{p} - E < p < \hat{p} + E$$

where

$$E = z_c\sqrt{\frac{\hat{p}\hat{q}}{n}}.$$

The probability that the confidence interval contains p is c.

In Section 5.6, you learned that a binomial distribution can be approximated by the normal distribution if $np \geq 5$ and $nq \geq 5$. When $n\hat{p} \geq 5$ and $n\hat{q} \geq 5$, the sampling distribution for $\hat{p}$ is approximately normal with a mean of

$$\mu_{\hat{p}} = p$$

and a standard error of

$$\sigma_{\hat{p}} = \sqrt{\frac{pq}{n}}.$$

GUIDELINES

Constructing a Confidence Interval for a Population Proportion

In Words	*In Symbols*
1. Identify the sample statistics n and x.	
2. Find the point estimate $\hat{p}$.	$\hat{p} = \frac{x}{n}$
3. Verify that the sampling distribution of $\hat{p}$ can be approximated by the normal distribution.	$n\hat{p} \geq 5$, $n\hat{q} \geq 5$
4. Find the critical value z_c that corresponds to the given level of confidence c.	Use the Standard Normal Table.
5. Find the maximum error of estimate E.	$E = z_c\sqrt{\frac{\hat{p}\hat{q}}{n}}$
6. Find the left and right endpoints and form the confidence interval.	Left endpoint: $\hat{p} - E$ Right endpoint: $\hat{p} + E$ Interval: $\hat{p} - E < p < \hat{p} + E$

Minitab and *TI-83* steps are shown on pages 308 and 309.

EXAMPLE 

Constructing a Confidence Interval for p

Construct a 95% confidence interval for the proportion of adults in the U.S. who say that their favorite sport to watch is football.

SOLUTION From Example 1, $\hat{p} \approx 0.28$. So,

$$\hat{q} = 1 - 0.28 = 0.72.$$

Using $n = 1024$, you can verify that the sampling distribution of $\hat{p}$ can be approximated by the normal distribution.

$$n\hat{p} \approx 1024 \cdot 0.28 \approx 287 > 5$$

and

$$n\hat{q} \approx 1024 \cdot 0.72 \approx 737 > 5$$

Using $z_c = 1.96$, the maximum error of estimate is

$$E = z_c\sqrt{\frac{\hat{p}\hat{q}}{n}} \approx 1.96\sqrt{\frac{(0.28)(0.72)}{1024}} \approx 0.028.$$

The 95% confidence interval is as follows.

Left Endpoint	Right Endpoint
$\hat{p} - E = 0.28 - 0.028 = 0.252$	$\hat{p} + E = 0.28 + 0.028 = 0.308$

$$0.252 < p < 0.308$$

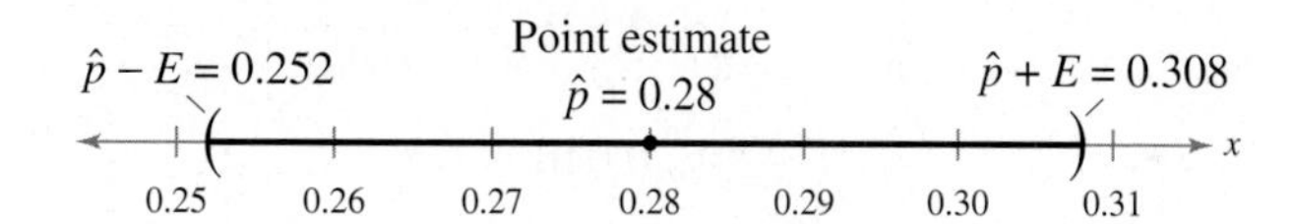

So, with 95% confidence, you can say that the proportion of adults who say football is their favorite sport is between 25.2% and 30.8%.

Try It Yourself 2

Construct a 90% confidence interval for the proportion of adults who say that of all the presidents in our nation's history, they most admire Abraham Lincoln.

a. *Find* $\hat{p}$ and $\hat{q}$.
b. *Verify* that the sampling distribution of $\hat{p}$ can be approximated by the normal distribution.
c. *Find* z_c and E.
d. *Use* $\hat{p}$ and E to find the left and right endpoints.
e. *Specify* the 90% confidence interval for the proportion of adults who admire Abraham Lincoln over all other U.S. presidents.

Answer: Page A38

The confidence level of 95% used in Example 2 is typical of opinion polls. The result, however, is usually not stated as a confidence interval. Instead, the result of Example 2 would usually be stated as "28% with a margin of error of ±2.8%."

EXAMPLE 

Constructing a Confidence Interval for p

The graph shown at the right is from a survey of 1001 U.S. adults. Construct a 99% confidence interval for the proportion of adults who think that airplanes are the safer mode of transportation.

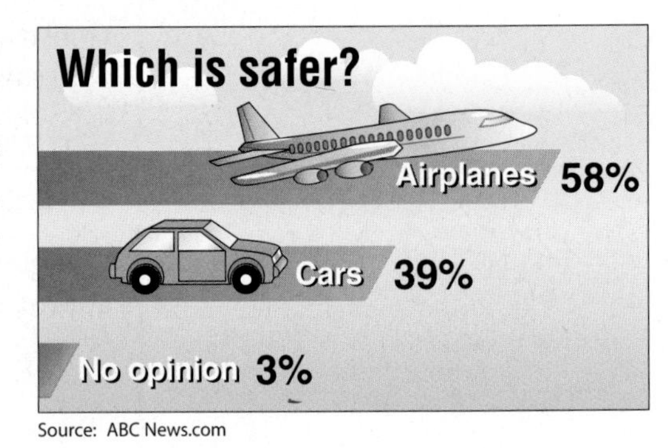

Source: ABC News.com

Insight

In Example 3, note that $np \geq 5$ and $nq \geq 5$. So, the sampling distribution of $\hat{p}$ is approximately normal.

SOLUTION From the graph, $\hat{p} = 0.58$. So,

$$\hat{q} = 1 - 0.58$$
$$= 0.42.$$

Using these values and the values $n = 1001$ and $z_c = 2.575$, the maximum error of estimate is

$$E = z_c\sqrt{\frac{\hat{p}\hat{q}}{n}}$$

$$\approx 2.575\sqrt{\frac{(0.58)(0.42)}{1001}}$$ Use Table 4 of Appendix B to estimate that z_c is halfway between 2.57 and 2.58.

$$\approx 0.040.$$

The 99% confidence interval is as follows.

Left Endpoint	Right Endpoint
$\hat{p} - E = 0.58 - 0.040 = 0.54$	$\hat{p} + E = 0.58 + 0.040 = 0.62$

$$0.54 < p < 0.62$$

With 99% confidence, you can say that the proportion of adults who think that airplanes are the safer mode of transportation is between 54% and 62%.

Try It Yourself 3

Use the survey information in Example 3 to construct a 99% confidence interval for the proportion of adults who think that cars are the safer mode of transportation.

a. *Identify* n and $\hat{p}$.
b. *Use* $\hat{p}$ to find $\hat{q}$.
c. *Verify* that the sampling distribution of $\hat{p}$ is approximately normal.
d. *Identify* the critical value z_c that corresponds to the given level of confidence.
e. *Find* the left and right endpoints of the confidence interval.
f. *Specify* the 99% confidence interval for the proportion of adults who think that cars are the safer mode of transportation.

Answer: Page A38

Increasing Sample Size to Increase Precision

One way to increase the precision of the confidence interval without decreasing the level of confidence is to increase the sample size.

Finding a Minimum Sample Size to Estimate p

Given a c-confidence level and a maximum error of estimate E, the minimum sample size n needed to estimate p is

$$n = \hat{p}\hat{q}\left(\frac{z_c}{E}\right)^2.$$

This formula assumes that you have a preliminary estimate for $\hat{p}$ and $\hat{q}$. If not, use $\hat{p} = 0.5$ and $\hat{q} = 0.5$.

The reason for using 0.5 as values for $\hat{p}$ and $\hat{q}$ when no preliminary estimate is available is that these values yield a maximum value for the product $\hat{p}\hat{q} = \hat{p}(1 - \hat{p})$. In other words, if you don't estimate the values of $\hat{p}$ and $\hat{q}$, you must pay the penalty of using a larger sample.

EXAMPLE 

Determining a Minimum Sample Size

You are running a political campaign and wish to estimate, with 95% confidence, the proportion of registered voters who will vote for your candidate. What is the minimum sample size needed if you are to be accurate within 3% of the population proportion?

SOLUTION Because you do not have a preliminary estimate for $\hat{p}$, use $\hat{p} = 0.5$ and $\hat{q} = 0.5$. Using $z_c = 1.96$, and $E = 0.03$, you can solve for n.

$$\begin{aligned} n &= \hat{p}\hat{q}\left(\frac{z_c}{E}\right)^2 \\ &= (0.5)(0.5)\left(\frac{1.96}{0.03}\right)^2 \\ &\approx 1067.11 \end{aligned}$$

Because n is a decimal, round up to the nearest whole number. So, at least 1068 registered voters should be included in the sample.

Try It Yourself 4

You wish to estimate, with 90% confidence, the proportion of adults age 18 to 29 who have high blood pressure. In a previous survey, 4% of adults in this age group had high blood pressure. What is the minimum sample size needed if you are to be accurate within 2% of the population proportion? *(Source: Centers for Disease Control and Prevention)*

a. *Identify* $\hat{p}$ and $\hat{q}$. If $\hat{p}$ is unknown, use 0.5.
b. *Identify* z_c and E.
c. Use $\hat{p}$, $\hat{q}$, z_c, and E to *find* the minimum sample size n.
d. *How many* adults should be included in the sample?

Answer: Page A38

6.3 Exercises

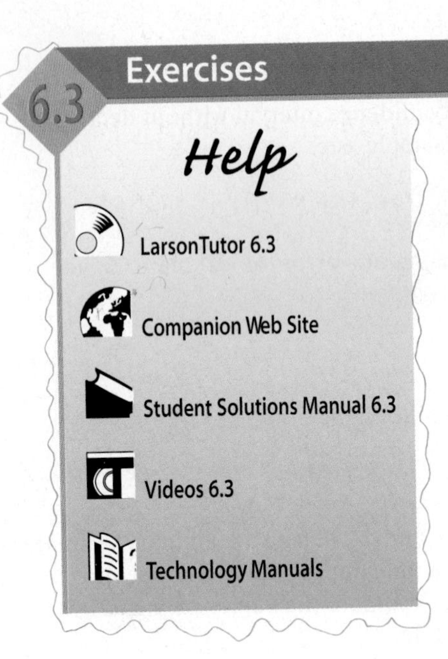

Basic Skills and Concepts

Finding $\hat{p}$ and $\hat{q}$ In Exercises 1–8, let p be the population proportion for the given condition. Find point estimates for p and q.

1. ***U.S. Food Safe?*** ◆ In a survey of 1040 adults from the U.S., 83 said they were not confident that the food they eat in the United States is safe. *(Source: Wirthlin Worldwide)*

2. ***Unions*** ◆ In a survey of 1001 U.S. adults, 501 believed the role of unions in protecting workers' rights is as important today as it was in years past. *(Source: Wirthlin Worldwide)*

3. ***Low Iodine Levels?*** ◆ A study of 34,000 U.S. adults found 4080 had low levels of iodine. *(Adapted from The Centers for Disease Control and Prevention)*

4. ***Accidents and Alcohol*** ◆ A study of 1907 traffic fatalities found 725 of the fatalities were alcohol related. *(Adapted from The National Highway Traffic Safety Administration)*

5. ***H.S. Baseball Injuries*** ◆ Of 1418 high school baseball players, 93 suffered an injury while playing the sport. *(Source: The Pennsylvania Athletic Trainers' Society, Inc.)*

6. ***H.S. Softball Injuries*** ◆ Of 1012 high school softball players, 83 suffered an injury while playing the sport. *(Source: The Pennsylvania Athletic Trainers' Society, Inc.)*

7. ***Addicted to Cigarettes?*** ◆ 189 smokers, in a survey of 262 smokers, considered themselves addicted to cigarettes. *(Source: The Gallup Organization)*

8. ***Girls Preferred?*** ◆ 277 U.S. adults, in a survey of 1026 U.S. adults, would prefer to have a girl if they could only have one child. *(Source: The Gallup Organization)*

Constructing a Confidence Interval In Exercises 9–16, construct the 95% and 99% confidence intervals for the population proportion p using the indicated sample statistics. Which interval is wider?

9. Use the statistics in Exercise 1.
10. Use the statistics in Exercise 2.
11. Use the statistics in Exercise 3.
12. Use the statistics in Exercise 4.
13. Use the statistics in Exercise 5.
14. Use the statistics in Exercise 6.
15. Use the statistics in Exercise 7.
16. Use the statistics in Exercise 8.

17. ***Travel Plans*** ◆ You are a travel agent and wish to estimate, with 95% confidence, the proportion of vacationers who plan to travel outside the United States in the next 12 months. Your estimate must be accurate within 3% of the true proportion.
    (a) Find the minimum sample size needed if no preliminary estimate is available.
    (b) Find the minimum sample size needed, using a prior study that found that 26% of the respondents said they planned to travel outside the United States in the next 12 months.
    (c) Compare the results from (a) and (b). *(Source: Wirthlin Worldwide)*

18. ***Online Service Usage*** ◆ You are a travel agent and wish to estimate, with 98% confidence, the proportion of vacationers who use an online service or the Internet to make reservations for lodging. Your estimate must be accurate within 4% of the population proportion.

(a) If no preliminary estimate is available, find the minimum sample size needed.

(b) Find the minimum sample size needed, using a prior study that found that 10% of the respondents said they used an online service or the Internet to make reservations for lodging.

(c) Compare the results from (a) and (b). *(Source: Wirthlin Worldwide)*

19. ***Camcorder Repairs*** ◆ You wish to estimate, with 96% confidence, the proportion of camcorders that need repairs or have problems by the time the product is five years old. Your estimate must be accurate within 2.5% of the true proportion.

(a) If no preliminary estimate is available, find the minimum sample size needed.

(b) Find the minimum sample size needed, using a prior study that found that 25% of camcorders needed repairs or had problems by the time the product was five years old.

(c) Compare the results from (a) and (b). *(Source: Consumer Reports)*

20. ***Computer Repairs*** ◆ You wish to estimate, with 97% confidence, the proportion of computers that need repairs or have problems by the time the product is three years old. Your estimate must be accurate within 3.5% of the true proportion.

(a) If no preliminary estimate is available, find the minimum sample size needed.

(b) Find the minimum sample size needed, using a prior study that found that 19% of computers needed repairs or had problems by the time the product was three years old.

(c) Compare the results from (a) and (b). *(Source: Consumer Reports)*

Constructing Confidence Intervals In Exercises 21 and 22, use the following information. The graph shows the results of 500 men, 500 women, 350 people who use microwaves often, and 350 people who rarely use microwaves who were asked if they favored irradiation of red meat to kill disease microbes. *(Source: Peter D. Hart Research Associates for Grocery Manufacturers of America)*

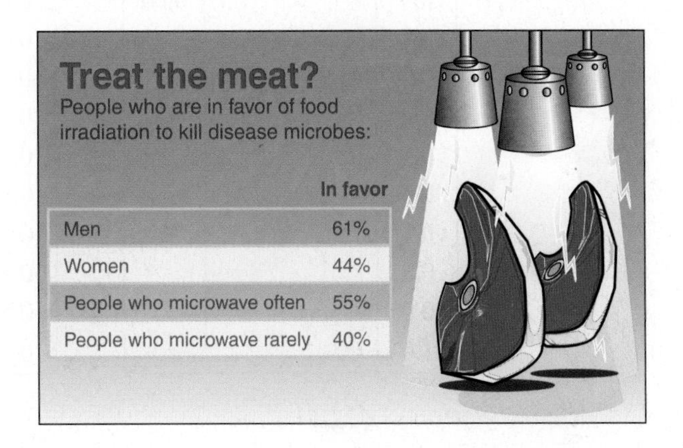

21. ***Men and Women*** ◆ Construct a 99% confidence interval for
    (a) the proportion of men who favor irradiation of red meat.
    (b) the proportion of women who favor irradiation of red meat. Is it possible that these two proportions are equal? Explain your reasoning.

22. ***Frequent Microwavers and Infrequent Microwavers*** ◆ Construct a 99% confidence interval for
    (a) the proportion of frequent microwave users who favor irradiation of red meat.
    (b) the proportion of infrequent microwave users who favor irradiation of red meat. Is it possible that these two proportions are equal? Explain your reasoning.

Extending the Basics

Newspaper Surveys In Exercises 23 and 24, translate the newspaper excerpt into a confidence interval for p.

23. In a survey of 8451 U.S. adults, 31.4% said they were taking Vitamin E as a supplement. The survey's margin of error is plus or minus 1%. *(Source: Decision Analyst, Inc.)*

24. In a survey of 1001 U.S. adults, 27% said they had smoked a cigarette in the past week. The survey's margin of error is plus or minus 3%. *(Source: The Gallup Organization)*

25. ***Why Check It?*** ◆ Why is it necessary to check that $n\hat{p} \geq 5$ and $n\hat{q} \geq 5$?

26. ***Sample Size*** ◆ The equation for determining the sample size $n = \hat{p}\hat{q}[(z_c)/E]^2$ can be obtained by solving the equation for the maximum error of estimate $E = z_c \sqrt{(\hat{p}\hat{q})/n}$ for n. Show that this is true and justify each step.

27. ***Maximum Value of $\hat{p}\hat{q}$*** ◆ Complete the tables for different values of $\hat{p}$ and $\hat{q} = 1 - \hat{p}$. From the table, which value of $\hat{p}$ appears to give the maximum value of the product $\hat{p}\hat{q}$?

$\hat{p}$	$\hat{q} = 1 - \hat{p}$	$\hat{p}\hat{q}$
0.0	1.0	0.00
0.1	0.9	0.09
0.2	0.8	
0.3		
0.4		
0.5		
0.6		
0.7		
0.8		
0.9		
1.0		

$\hat{p}$	$\hat{q} = 1 - \hat{p}$	$\hat{p}\hat{q}$
0.45		
0.46		
0.47		
0.48		
0.49		
0.50		
0.51		
0.52		
0.53		
0.54		
0.55		

TECHNOLOGY | MINITAB | EXCEL | TI-83

WWW.GALLUP.COM

THE GALLUP ORGANIZATION

Most Admired Polls

From 1948 to 2000 the Gallup Organization has conducted a "most admired" poll. The methodology for the 2000 poll is described at the right.

Survey Question
What man* that you have heard or read about, living today in any part of the world, do you admire most? And who is your second choice?

*Survey respondents are asked an identical question about most admired woman.

"The results are based on telephone interviews with a randomly selected national sample of 1011 adults, 18 and over, conducted December 15–17, 2000. For results based on samples of this size, one can say with 95 percent confidence that the error attributable to sampling and other random effects could be plus or minus 3 percentage points. In addition to sampling error, question wording and practical difficulties in conducting surveys can introduce error or bias into the findings of public opinion polls."

Exercises

1. In 2000, 61 people named Pope John Paul II as their most admired man. Use a technology tool to find a 95% confidence interval for the proportion that would have chosen the pope.

2. Does the confidence interval you obtained in Exercise 1 agree with the statement issued by the Gallup Organization that the proportion is 6% plus or minus 3%? Explain.

3. The most named woman was Hillary Clinton. The second named woman was Oprah Winfrey, who was named by 4% of the people in the sample. Use a technology tool to find a 95% confidence interval for the proportion of the population that would have chosen Oprah Winfrey.

4. Use a technology tool to simulate a most admired poll. Assume that the actual population proportion who most admire Oprah Winfrey is 7%. Run the simulation several times using $n = 1011$.

(a) What was the least value you obtained for $\hat{p}$?

(b) What was the greatest value you obtained for $\hat{p}$?

MINITAB

Generate 200 rows of data
Store in column(s): C1
Number of trials: 1011
Probability of success: .07

5. The 2000 results were a rarity because in the poll's 52-year history, it was unusual that no one (other than the first lady) received more than 6% of the vote. Is it possible, however, that the actual proportion of the population that most admired Oprah Winfrey was 7% or greater? Explain your reasoning.

Extended solutions are given in the *Technology Supplement*.
Technical instruction is provided for *Minitab*, *Excel*, and the *TI-83*.

6.4 Confidence Intervals for Variance and Standard Deviation

What You Should Learn

- How to interpret the chi-square distribution and use a chi-square distribution table
- How to use the chi-square distribution to construct a confidence interval for the variance and standard deviation

The Chi-Square Distribution • Confidence Intervals for σ^2 and σ

The Chi-Square Distribution

In manufacturing, it is necessary to control the amount that a process varies. For instance, an automobile part manufacturer must produce thousands of parts to be used in the manufacturing process. It is important that the parts vary little or not at all. How can you measure, and consequently control, the amount of variation in the parts? You can start with a point estimate.

DEFINITION

The **point estimate for σ^2** is s^2 and the **point estimate for σ** is s. s^2 is the most unbiased estimate for σ^2.

Study Tip

The Greek letter χ is pronounced "kī," which rhymes with the more familiar Greek letter π.

You can use a *chi-square distribution* to construct a confidence interval for the variance and standard deviation.

DEFINITION

If the random variable x has a normal distribution, then the distribution of

$$\chi^2 = \frac{(n-1)s^2}{\sigma^2}$$

forms a **chi-square distribution** for samples of any size $n > 1$. Several properties of the chi-square distribution are as follows.

1. All chi-square values χ^2 are greater than or equal to zero.
2. The chi-square distribution is a family of curves, each determined by the degrees of freedom. To form a confidence interval for σ^2, use the χ^2-distribution with degrees of freedom equal to one less than the sample size.

   $$\text{d.f.} = n - 1 \qquad \text{Degrees of Freedom}$$

3. The area under each curve of the chi-square distribution equals one.
4. Chi-square distributions are positively skewed.

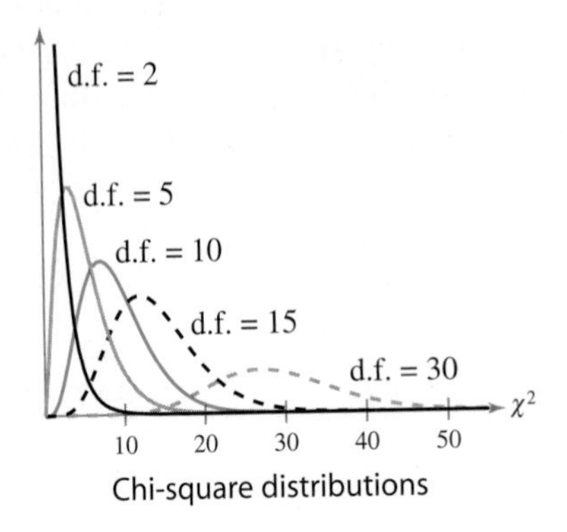

Chi-square distributions

Study Tip

For chi-square critical values with a c-confidence level, the following values are what you look up in Table 6 in Appendix B.

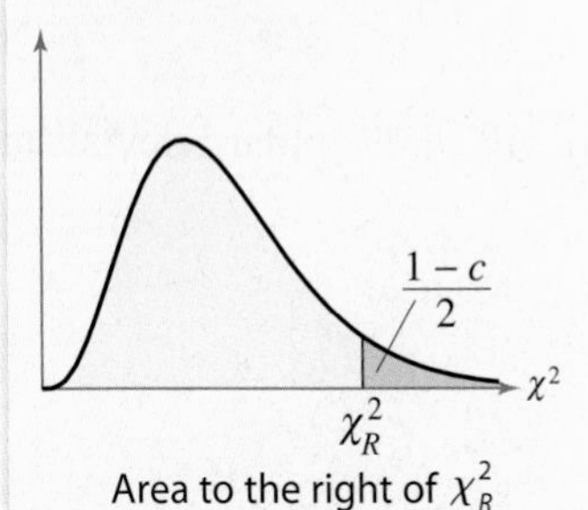

Area to the right of χ^2_R

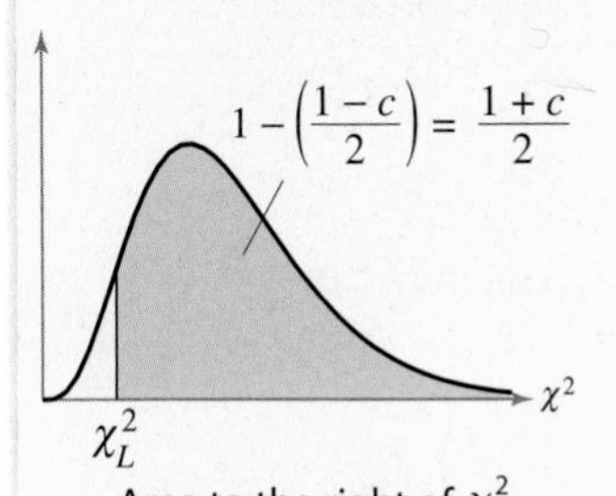

Area to the right of χ^2_L

The result is that you can conclude that the area between the left and right critical values is c.

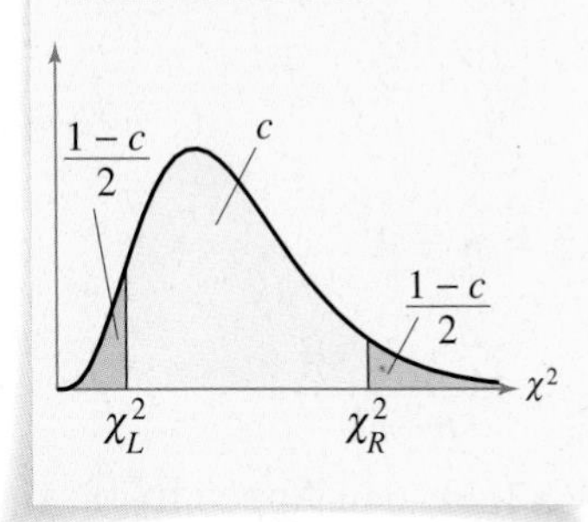

There are two critical values for each level of confidence. The value χ^2_R represents the right-tail critical value and χ^2_L represents the left-tail critical value. Table 6 in Appendix B lists critical values of χ^2 for various degrees of freedom and areas. Each area in the table represents the region under the chi-square curve to the *right* of the critical value.

EXAMPLE 1

Finding Critical Values for χ^2

Find the critical values χ^2_R and χ^2_L for a 90% confidence interval when the sample size is 20.

SOLUTION Because the sample size is 20, there are

$$\text{d.f.} = n - 1 = 20 - 1 = 19 \text{ degrees of freedom.}$$

The areas to the right of χ^2_R and χ^2_L are

$$\text{Area to right of } \chi^2_R = \frac{1-c}{2} = \frac{1-0.90}{2} = 0.05$$

and

$$\text{Area to right of } \chi^2_L = \frac{1+c}{2} = \frac{1+0.90}{2} = 0.95.$$

Part of Table 6 is shown. Using d.f. = 19 and the areas 0.95 and 0.05, you can find the critical values, as shown by the highlighted areas in the table.

Degrees of freedom	α						
	0.995	0.99	0.975	0.95	0.90	0.10	0.05
1	—	—	0.001	0.004	0.016	2.706	3.841
2	0.010	0.020	0.051	0.103	0.211	4.605	5.991
3	0.072	0.115	0.216	0.352	0.584	6.251	7.815
15	4.601	5.229	6.262	7.261	8.547	22.307	24.996
16	5.142	5.812	6.908	7.962	9.312	23.542	26.296
17	5.697	6.408	7.564	8.672	10.085	24.769	27.587
18	6.265	7.015	8.231	9.390	10.865	25.989	28.869
19	6.844	7.633	8.907	10.117	11.651	27.204	30.144
20	7.434	8.260	9.591	10.851	12.443	28.412	31.410

χ^2_L χ^2_R

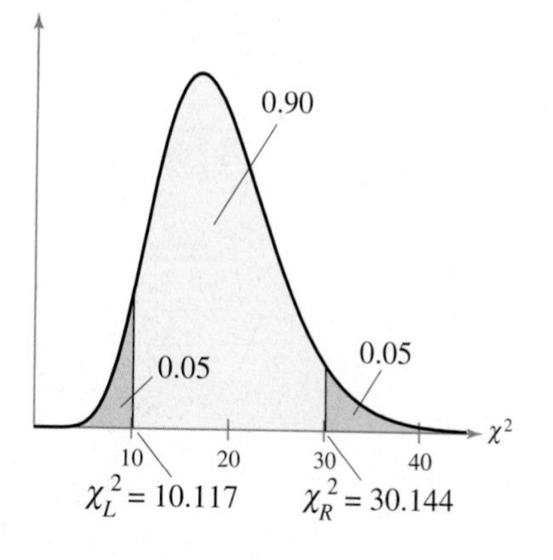

From the table, you can see that $\chi^2_R = 30.144$ and $\chi^2_L = 10.117$. So, 90% of the area under the curve lies between 10.117 and 30.144.

Try It Yourself 1

Find the critical values χ^2_R and χ^2_L for a 95% confidence interval when the sample size is 25.

a. *Identify* the degrees of freedom and the level of confidence.
b. *Find* the area to the right of χ^2_R and χ^2_L.
c. *Use* Table 6 of Appendix B to find χ^2_R and χ^2_L.

Answer: Page A38

Picturing the World

The gray whale has the longest annual migration distance of any mammal. Gray whales leave from Baja, California, and western Mexico in the spring, migrating to the Bering and Chukchi seas for the summer months. Tracking a sample of 50 whales for a year provided a mean migration distance of 11,064 miles with a standard deviation of 860 miles. *(Source: National Marine Fisheries Services)*

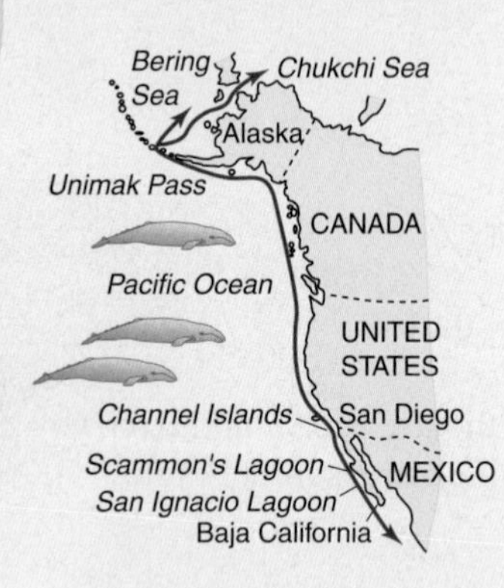

Construct a 90% confidence interval for the standard deviation for the migration distance of gray whales. Assume that the population of migration distances has a normal distribution.

Confidence Intervals for σ^2 and σ

You can use the critical values χ_R^2 and χ_L^2 to construct confidence intervals for a population variance and standard deviation. As you would expect, the best point estimate for the variance is s^2 and the best point estimate for the standard deviation is s.

DEFINITION

A c-confidence interval for a population variance and standard deviation is as follows.

Confidence Interval for σ^2:

$$\frac{(n-1)s^2}{\chi_R^2} < \sigma^2 < \frac{(n-1)s^2}{\chi_L^2}$$

Confidence Interval for σ:

$$\sqrt{\frac{(n-1)s^2}{\chi_R^2}} < \sigma < \sqrt{\frac{(n-1)s^2}{\chi_L^2}}$$

The probability that the confidence intervals contain σ^2 or σ is c.

GUIDELINES

Constructing a Confidence Interval for a Variance and Standard Deviation

In Words	*In Symbols*
1. Verify that the population has a normal distribution.	
2. Identify the sample statistic n and the degrees of freedom.	d.f. $= n - 1$
3. Find the point estimate s^2.	$s^2 = \dfrac{\Sigma(x-\bar{x})^2}{n-1}$
4. Find the critical values χ_R^2 and χ_L^2 that correspond to the given level of confidence c.	Use Table 6 in Appendix B.
5. Find the left and right endpointsand form the confidence interval for the population variance.	Left Endpoint, Right Endpoint: $\dfrac{(n-1)s^2}{\chi_R^2} < \sigma^2 < \dfrac{(n-1)s^2}{\chi_L^2}$
6. Find the confidence interval forthe population standard deviation by taking the square root of each endpoint.	$\sqrt{\dfrac{(n-1)s^2}{\chi_R^2}} < \sigma < \sqrt{\dfrac{(n-1)s^2}{\chi_L^2}}$

EXAMPLE 

Constructing a Confidence Interval

You randomly select and weigh 30 samples of an allergy medicine. The sample standard deviation is 1.2 milligrams. Assuming the weights are normally distributed, construct 99% confidence intervals for the population variance and standard deviation.

SOLUTION The areas to the right of χ_R^2 and χ_L^2 are

$$\text{Area to right of } \chi_R^2 = \frac{1-c}{2} = \frac{1-0.99}{2} = 0.005$$

and

$$\text{Area to right of } \chi_L^2 = \frac{1+c}{2} = \frac{1+0.99}{2} = 0.995.$$

Using the values $n = 30$, d.f. $= 29$, and $c = 0.99$, the critical values χ_R^2 and χ_L^2 are

$$\chi_R^2 = 52.366 \quad \text{and} \quad \chi_L^2 = 13.121.$$

Using these critical values and $s = 1.2$, the confidence interval for σ^2 is as follows.

Left Endpoint

$$\frac{(n-1)s^2}{\chi_R^2} = \frac{(30-1)(1.2)^2}{52.336} \approx 0.798$$

Right Endpoint

$$\frac{(n-1)s^2}{\chi_L^2} = \frac{(30-1)(1.2)^2}{13.121} \approx 3.183$$

$$0.798 < \sigma^2 < 3.183$$

The confidence interval for σ is

$$\sqrt{0.798} < \sigma < \sqrt{3.183}$$

$$0.89 < \sigma < 1.78.$$

So, with 99% confidence, you can say that the population variance is between 0.798 and 3.183. The population standard deviation is between 0.89 and 1.78 milligrams.

Try It Yourself 2

Find the 90% and 95% confidence intervals for the population variance and standard deviation of the medicine weights.

a. *Find* the critical values χ_R^2 and χ_L^2 for each confidence interval.
b. *Use* n, s, χ_R^2, and χ_L^2 to find the left and right endpoints for each confidence interval for the variance.
c. *Find* the square roots of the endpoints of each confidence interval.
d. *Specify* the 90% and 95% confidence intervals for the population variance and standard deviation.

Answer: Page A38

6.4 Exercises

Help

LarsonTutor 6.4

Companion Web Site

Student Solutions Manual 6.4

Videos 6.4

Technology Manuals

Basic Skills and Concepts

In Exercises 1–6, find the critical values χ_R^2 and χ_L^2 for the given confidence c and sample size n.

1. $c = 0.90,\ n = 10$

2. $c = 0.99,\ n = 13$

3. $c = 0.95,\ n = 22$

4. $c = 0.98,\ n = 26$

5. $c = 0.99,\ n = 30$

6. $c = 0.80,\ n = 29$

Constructing Confidence Intervals In Exercises 7–16, assume each sample is taken from a normally distributed population and construct the indicated confidence intervals for

(a) the population variance σ^2.

(b) the population standard deviation σ.

7. ***Vitamins*** ◆ To analyze the variation of vitamin supplement tablets, you randomly select and weigh 14 tablets. The results (in milligrams) are shown. Use a 90% level of confidence.

500.000 499.995 500.010 499.997 500.015 499.988 500.000
499.996 500.020 500.002 499.998 499.996 500.003 500.000

8. ***Cough Syrup*** ◆ You randomly select and measure the contents of 15 bottles of cough syrup. The results (in fluid ounces) are shown. Use a 90% level of confidence.

4.211 4.246 4.269 4.241 4.260 4.293 4.189 4.248
4.220 4.239 4.253 4.209 4.300 4.256 4.290

9. ***Car Batteries*** ◆ The number of hours of reserve capacity of 18 randomly selected automotive batteries is shown. Use a 99% level of confidence. *(Adapted from Consumer Reports)*

1.70 1.60 1.94 1.58 1.74 1.60 1.86 1.72 1.38
1.46 1.64 1.49 1.55 1.70 1.75 0.88 1.77 2.07

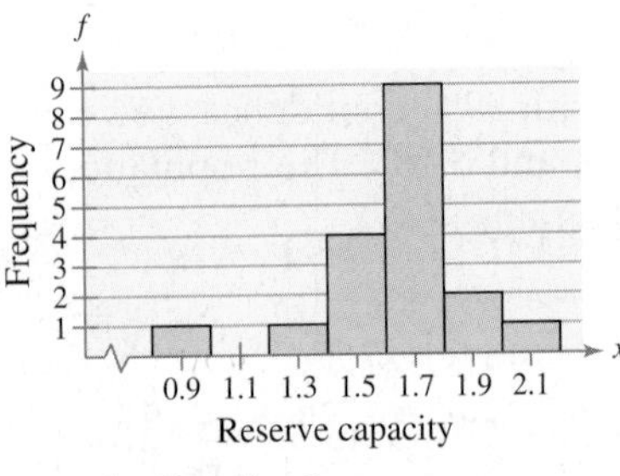

Figure for Exercise 9

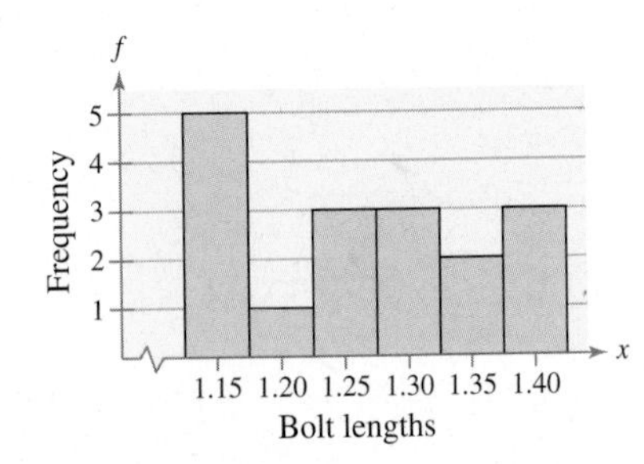

Figure for Exercise 10

10. ***Bolts*** ◆ You randomly select and measure 17 bolts. The results (in inches) are shown. Use a 95% level of confidence.

1.286 1.138 1.240 1.132 1.381 1.137 1.300 1.167 1.240
1.401 1.241 1.171 1.217 1.360 1.302 1.331 1.383

11. ***Lawn Mower*** ◆ A lawn mower manufacturer is trying to determine the standard deviation of the life of one of its lawn mower models. To do this, it randomly selects 12 lawn mowers that were sold several years ago and finds that the sample standard deviation is 3.25 years. Use a 99% level of confidence. *(Adapted from Consumer Reports)*

12. ***CD Players*** ◆ A magazine includes a report on the prices of compact disc players. The article states that 26 randomly selected CD players had a standard deviation of \$150. Use a 95% level of confidence. *(Adapted from Consumer Reports)*

13. ***Hotels*** ◆ As part of your vacation planning, you randomly contact 10 hotels in your destination area and record the room rate of each. The results are shown in the stem-and-leaf plot. Use $c = 0.90$. *(Adapted from Smith Travel Research)*

Stem	Leaves
6	0 3
7	
8	3
9	0
10	2 8
11	3 8
12	2
13	
14	1

Key: $8|3 = 83$

Data for Exercise 13

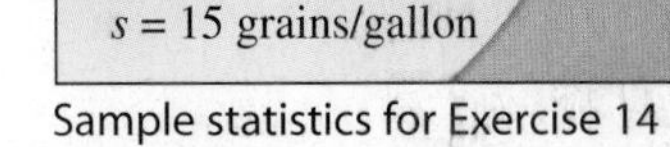

Sample statistics for Exercise 14

14. ***Water Quality*** ◆ As part of a water quality survey, you test the water hardness in several randomly selected streams. The results are shown above. Use $c = 0.95$.

15. ***Monthly Income*** ◆ The monthly incomes of 20 randomly selected individuals who have recently graduated with a bachelor's degree in social science have a sample standard deviation of \$107. Use a 95% level of confidence. *(Adapted from U.S. Bureau of the Census)*

16. ***Sodium Chloride Concentration*** ◆ The sodium chloride concentrations of 13 randomly selected seawater samples have a standard deviation of 6.7 cc/cubic meter. Use a 98% level of confidence. *(Adapted from Dorling Kindersley Visual Encyclopedia)*

Extending the Basics

17. ***Vitamin Tablet Weights*** ◆ You are analyzing the sample of vitamin supplement tablets in Exercise 7. The population standard deviation of the tablet's weights should be less than 0.015 milligram. Does the confidence interval you constructed for σ suggest that the variation in the tablet's weights is at an acceptable level? Explain your reasoning.

18. ***Cough Syrup Bottle Contents*** ◆ You are analyzing the sample of cough syrup bottles in Exercise 8. The population standard deviation of the bottle's contents should be less than 0.025 fluid ounce. Does the confidence interval you constructed for σ suggest that the variation in the bottle's contents is at an acceptable level? Explain your reasoning.

Using Technology to Construct Confidence Intervals

Here are some *Minitab* and *TI-83* printouts for some examples in this chapter. To duplicate the *Minitab* results, you need the original data. For the *TI-83*, you can simply enter the descriptive statistics. Answers may be slightly different due to rounding.

(See Example 3, page 273)

9	20	18	16	9	16	16	9	11	13	22	16	5	18	6	6	5	12
25	17	23	7	10	9	10	10	5	11	18	18	9	9	17	13	11	7
14	6	11	12	11	15	6	12	14	11	4	9	18	12	12	17	11	20

Display Descriptive Statistics...
Store Descriptive Statistics...
1-Sample Z...
1-Sample t...
2-Sample t...
Paired t...
1 Proportion...
2 Proportions...

MINITAB

Z Confidence Intervals

The assumed sigma = 5

Variable	N	Mean	StDev	SE Mean	95.0 % CI
C1	54	12.426	5.015	0.680	(11.092, 13.760)

(See Example 2, page 286)

159°F	173°F	162°F	151°F	173°F	162°F	148°F	172°F
167°F	170°F	151°F	153°F	172°F	143°F	166°F	170°F

Display Descriptive Statistics...
Store Descriptive Statistics...
1-Sample Z...
1-Sample t...
2-Sample t...
Paired t...
1 Proportion...
2 Proportions...

MINITAB

T Confidence Intervals

Variable	N	Mean	StDev	SE Mean	95.0 % CI
C3	16	162.00	10.00	2.50	(156.67, 167.33)

(See Example 2, page 295)

Display Descriptive Statistics...
Store Descriptive Statistics...
1-Sample Z...
1-Sample t...
2-Sample t...
Paired t...
1 Proportion...
2 Proportions...

MINITAB

Test and Confidence Interval for One Proportion

Test of p = 0.28 vs p not = 0.28

Sample	X	N	Sample p	95.0 % CI	Z–Value	P–Value
1	287	1024	0.280273	(0.252765, 0.307782)	0.02	0.984

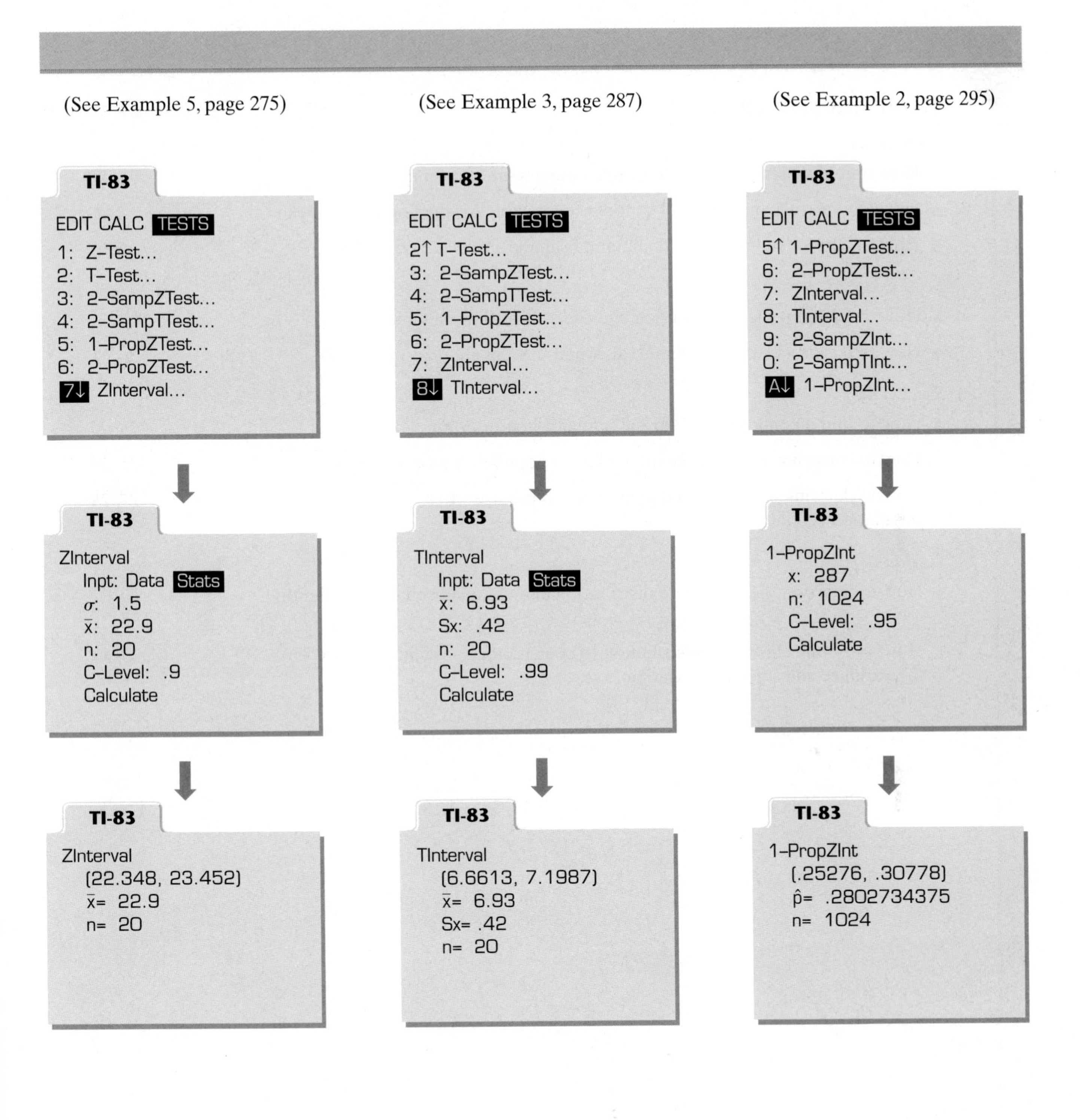
(See Example 5, page 275)
TI-83
EDIT CALC TESTS
1: Z-Test...
2: T-Test...
3: 2-SampZTest...
4: 2-SampTTest...
5: 1-PropZTest...
6: 2-PropZTest...
7↓ ZInterval...
TI-83
ZInterval
Inpt: Data Stats
σ: 1.5
x̄: 22.9
n: 20
C-Level: .9
Calculate
TI-83
ZInterval
(22.348, 23.452)
x̄= 22.9
n= 20
(See Example 3, page 287)
TI-83
EDIT CALC TESTS
2↑ T-Test...
3: 2-SampZTest...
4: 2-SampTTest...
5: 1-PropZTest...
6: 2-PropZTest...
7: ZInterval...
8↓ TInterval...
TI-83
TInterval
Inpt: Data Stats
x̄: 6.93
Sx: .42
n: 20
C-Level: .99
Calculate
TI-83
TInterval
(6.6613, 7.1987)
x̄= 6.93
Sx= .42
n= 20
(See Example 2, page 295)
TI-83
EDIT CALC TESTS
5↑ 1-PropZTest...
6: 2-PropZTest...
7: ZInterval...
8: TInterval...
9: 2-SampZInt...
0: 2-SampTInt...
A↓ 1-PropZInt...
TI-83
1-PropZInt
x: 287
n: 1024
C-Level: .95
Calculate
TI-83
1-PropZInt
(.25276, .30778)
p̂= .2802734375
n= 1024

6

Chapter Summary

What did you learn?	*Review Exercises*
Section 6.1	
◆ How to find a point estimate and a maximum error of estimate	*1, 2*
◆ How to construct and interpret confidence intervals for the population mean	*3, 4*
◆ How to determine the required minimum sample size when estimating μ	*5, 6*
Section 6.2	
◆ How to interpret the t-distribution and use a t-distribution table	*7, 8*
◆ How to construct confidence intervals when $n < 30$ and σ is unknown	*9–14*
Section 6.3	
◆ How to find a point estimate for the population proportion	*15–20*
◆ How to construct a confidence interval for a population proportion	*21–26*
◆ How to determine the required minimum sample size when estimating a population proportion	*27, 28*
Section 6.4	
◆ How to interpret the chi-square distribution and use a chi-square distribution table	*29–32*
◆ How to use the chi-square distribution to construct a confidence interval for the variance and standard deviation	*33, 34*

STATISTICS

Uses and Abuses

Uses

By now, you know that complete information about population parameters is often not available. The techniques of this chapter can be used to make interval estimates of these parameters so that you can make informed decisions.

From what you learned in this chapter, you know that point estimates of population parameters are rarely exact. Remembering this can help you make good decisions in your career and in everyday life. For instance, suppose the results of a survey tell you that 52% of the population plans to vote for a certain item on a ballot. You know that this is only a point estimate of the actual proportion that will vote for the item. If the interval estimate is $0.49 < p < 0.55$, then you know this means it is possible that the item will not receive a majority vote.

Abuses

Unrepresentative Samples There are many ways that surveys can result in incorrect predictions. When you read the results of a survey, remember to question the sample size, the sampling technique, and the questions asked. For example, suppose you want to know how a group of people will vote in an election. From the diagram at the right, you can see that even if your sample is large enough, it may not consist of actual voters.

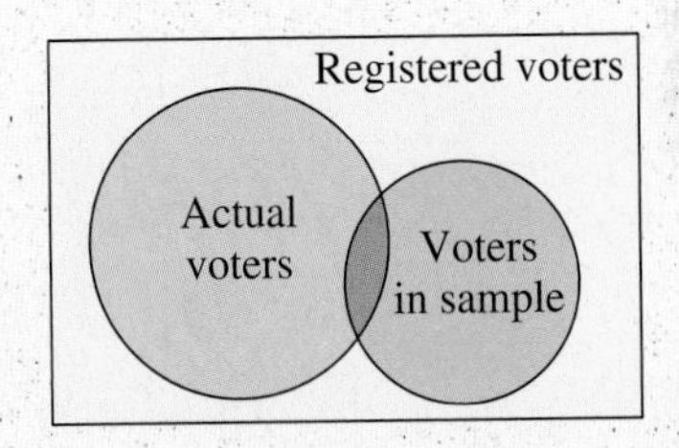

Biased Survey Questions In surveys, it is also important to analyze the wording of the questions. For example, a question about rezoning might be presented as: "Knowing that rezoning will result in more businesses contributing to school taxes, would you support the rezoning ?"

Exercises

1. ***Unrepresentative Samples*** Find an example of a survey that is reported in a newspaper or magazine. Describe different ways that the sample could have been unrepresentative of the population.
2. ***Biased Survey Questions*** Find an example of a survey that is reported in a newspaper or magazine. Describe different ways that the survey questions could have been biased.

6 Review Exercises

Section 6.1

In Exercises 1 and 2, find (a) the point estimate of the population mean and (b) the maximum error of estimate for a 90% confidence interval.

1. Waking times of 40 people who start work at 8:00 A.M. (in minutes past 5:00 A.M.)

135	145	95	140	135	95	110	50
90	165	110	125	80	125	130	110
25	75	65	100	60	125	115	135
95	90	140	40	75	50	130	85
100	160	135	45	135	115	75	130

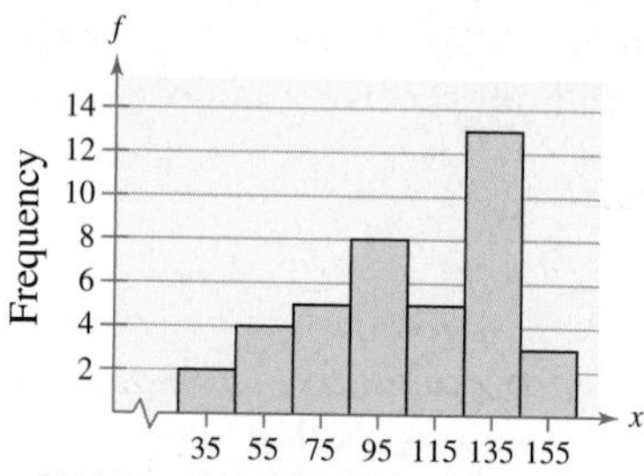

Figure for Exercise 1

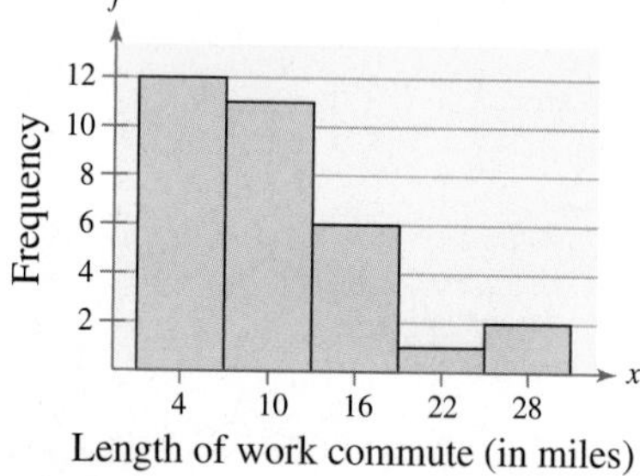

Figure for Exercise 2

2. Length of work commute of 32 people (in miles)

12	9	7	2	8	7	3	27
21	10	13	3	7	2	30	7
6	13	6	14	4	1	10	3
13	6	2	9	2	12	16	18

In Exercises 3 and 4, construct the indicated confidence interval for the population mean μ.

3. $c = 0.95, \bar{x} = 10.3, s = 0.277,$
$n = 100$

4. $c = 0.90, \bar{x} = 0.0925,$
$s = 0.0013, n = 45$

In Exercises 5 and 6, determine the minimum sufficient sample size.

5. Use the point estimate for μ from Exercise 1. Determine the minimum survey size that is necessary to be 95% confident that the sample mean waking time is within 10 minutes of the actual mean waking time.

6. Now suppose you want 99% confidence with a maximum error of 2 minutes. How many people would it be necessary to survey?

Section 6.2

In Exercises 7 and 8, find the critical value t_c for the given confidence level c and sample size n.

7. $c = 0.80, n = 8$

8. $c = 0.80, n = 22$

In Exercises 9 and 10, find the maximum error of estimate for μ.

9. $c = 0.90,\ s = 23.4,\ n = 16,\ \overline{x} = 52.8$

10. $c = 0.95,\ s = 0.05,\ n = 25,\ \overline{x} = 3.5$

11. Construct the confidence interval for μ using the statistics in Exercise 9.

12. Construct the confidence interval for μ using the statistics in Exercise 10.

13. In a random sample of 15 CD players brought in for repair, the average repair cost was \$80 and the standard deviation was \$14. Construct a 90% confidence interval for μ. Assume the repair costs are normally distributed. *(Adapted from Consumer Reports)*

14. Repeat Exercise 13 using a 99% confidence interval.

Section 6.3

In Exercises 15–20, let p be the proportion of the population who respond yes. Use the given information to find $\hat{p}$ and $\hat{q}$.

15. In a survey of 850 college students, 357 balanced their checkbooks monthly. *(Adapted from Bruskin Goldring)*

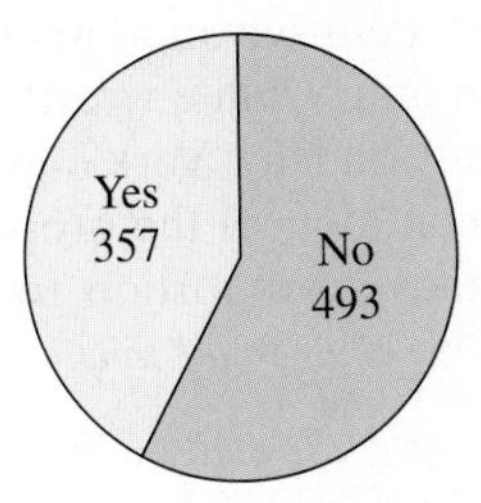

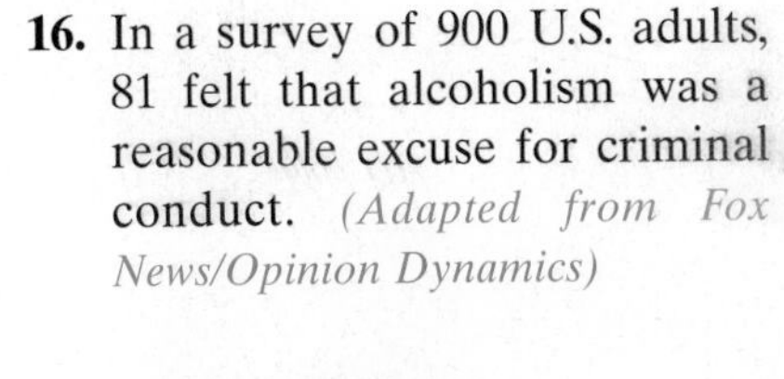

16. In a survey of 900 U.S. adults, 81 felt that alcoholism was a reasonable excuse for criminal conduct. *(Adapted from Fox News/Opinion Dynamics)*

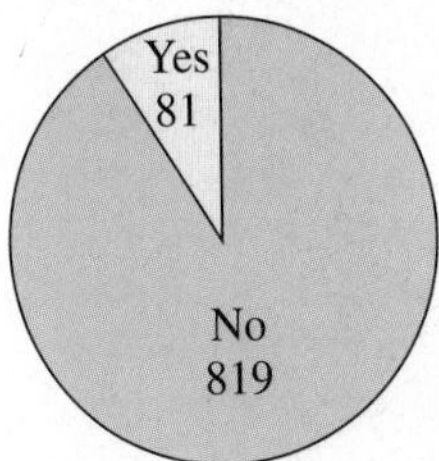

17. In a survey of 209 Montana residents, 61 felt their financial status was worse than a year ago. *(Source: StatLib/Bureau of Business and Economic Research, U. of Montana)*

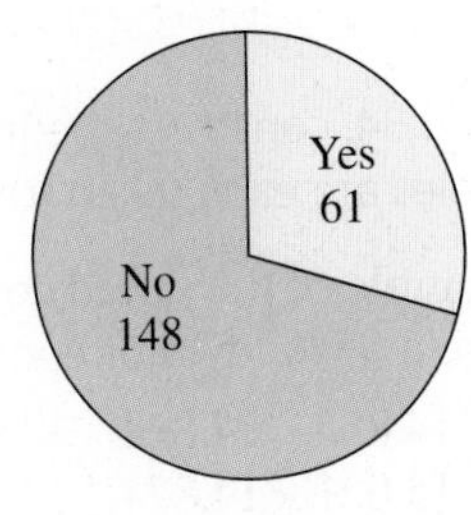

18. In a survey of 1400 U.S. adults, 546 had seen information comparing the quality of health plans, doctors, or hospitals. *(Adapted from Deloitte & Touche)*

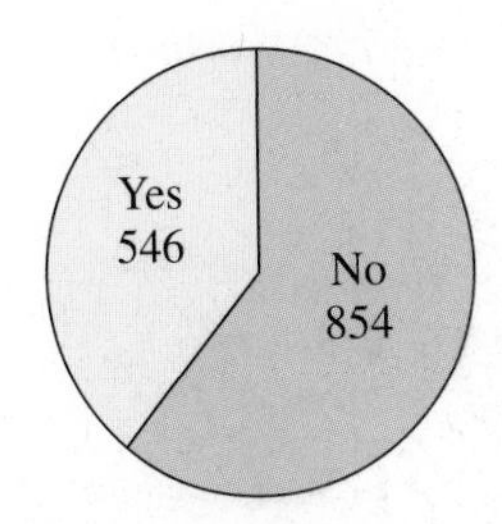

19. In a survey of 1741 U.S. adults, 700 said they think genetically modified foods are safe. *(Adapted from ABC News.com)*

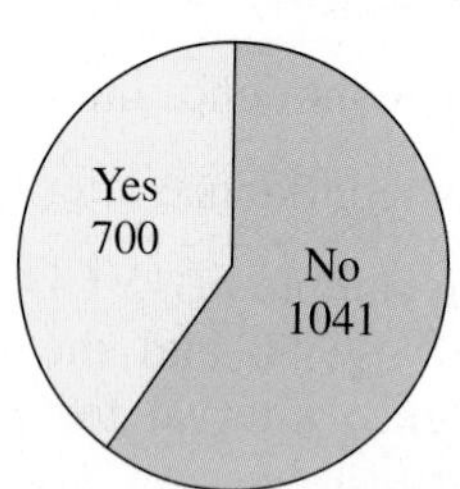

20. In a survey of 957 U.S. adults, 668 said they approve of labor unions. *(Adapted from The Gallup Organization)*

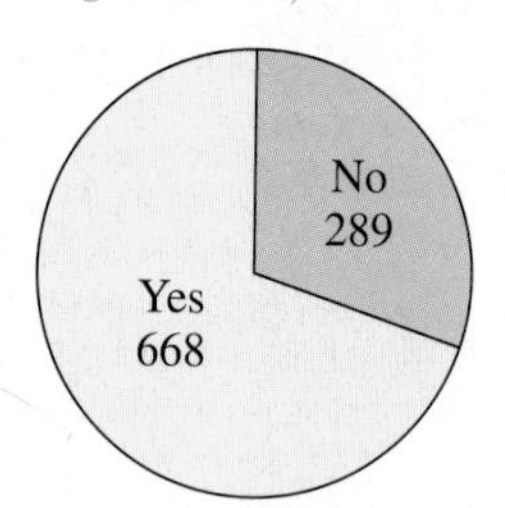

In Exercises 21–26, construct the indicated confidence interval for the population proportion p.

21. Use the sample in Exercise 15 with $c = 0.95$.

22. Use the sample in Exercise 16 with $c = 0.99$.

23. Use the sample in Exercise 17 with $c = 0.90$.

24. Use the sample in Exercise 18 with $c = 0.98$.

25. Use the sample in Exercise 19 with $c = 0.99$.

26. Use the sample in Exercise 20 with $c = 0.90$.

27. A study of Pennsylvania high school wrestlers found that 23% had sustained an injury while wrestling that year. Suppose you want to conduct a similar study on New York high school wrestlers, and you wish to estimate (with 95% confidence) the proportion of injured wrestlers within 5%. Use the prior study's proportion to find the minimum required sample size. *(Source: The Pennsylvania Athletic Trainers' Society, Inc.)*

28. Repeat Exercise 27, using a 99% confidence level and a maximum error of estimate of 2.5%. How does this sample size compare to your answer from Exercise 27?

Section 6.4

In Exercises 29–32, find the critical values χ^2_R and χ^2_L needed to estimate σ^2 for the given confidence level c and sample size n.

29. $c = 0.95, n = 13$

30. $c = 0.98, n = 25$

31. $c = 0.90, n = 8$

32. $c = 0.99, n = 10$

In Exercises 33 and 34, construct the indicated confidence intervals for σ^2 and σ. Assume the samples are each taken from a normally distributed population.

33. A random sample of the liquid content (in fluid ounces) of 16 beverage cans is shown. Use a 95% level of confidence.

14.816	14.863	14.814	14.998	14.965	14.824	14.884	14.838
14.916	15.021	14.874	14.856	14.860	14.772	14.980	14.919

34. Repeat Exercise 33 using a 99% level of confidence.

PUTTING IT ALL TOGETHER

Real Statistics ⟷ Real Decisions

As part of the U.S. Environmental Protection Agency's (EPA) efforts to "protect human health and safeguard the natural environment," the EPA conducts the Urban Air Toxics Monitoring Program (UATMP). The program has gathered thousands of air samples and analyzed them for concentrations of over 50 different organic compounds, such as acetylene (used for cutting and welding metals, and in the production of neoprene rubber, plastics, and resins). The results from UATMP are used to gain insight into the effects of air pollution and if efforts to clean up the air are working.

For instance, using air samples from a major city, the EPA can analyze the results and estimate the mean concentration of acetylene in the air using a 95% confidence interval. They can then compare the interval to previous years' results to see if there are any trends and if there has been a significant change in the amount of acetylene in the air.

You work for the EPA and are asked to interpret the results shown in the graph at the right. The graph shows the point estimate for the population mean concentration and the 95% confidence interval for μ for acetylene over a three-year period. The data is based on air samples taken at one major city.

EPA United States Environmental Protection Agency
www.epa.gov

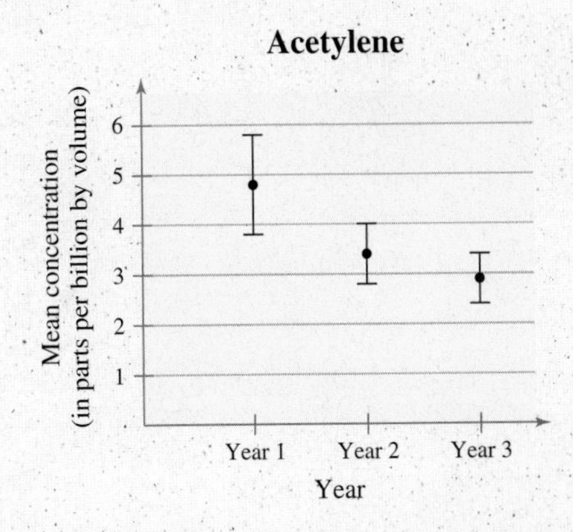

Exercises

1. ***Interpreting the Results***

Consider the graph of the mean concentration levels of acetylene. For the following years, decide if there has been a change in the mean concentration level of acetylene. Explain your reasoning.

(a) From Year 1 to Year 2

(b) From Year 2 to Year 3

(c) From Year 1 to Year 3

2. ***What Can You Conclude?***

Using the results of Exercise 1, what can you conclude about the efforts to reduce the concentration of acetylene in the air?

3. ***How Do You Think They Did It?***

How do you think the EPA constructed the 95% confidence interval for the population mean concentration of the organic compounds in the air? Do the following to answer the question (you do not need to make any calculations).

(a) What sampling distribution do you think they used? Why?

(b) Do you think they used the population standard deviation in calculating the maximum error of estimate? Why or why not? If not, what could they have used?

6 Chapter Quiz

Take this quiz as you would take a quiz in class. After you are done, check your work against the answers given in the back of the book.

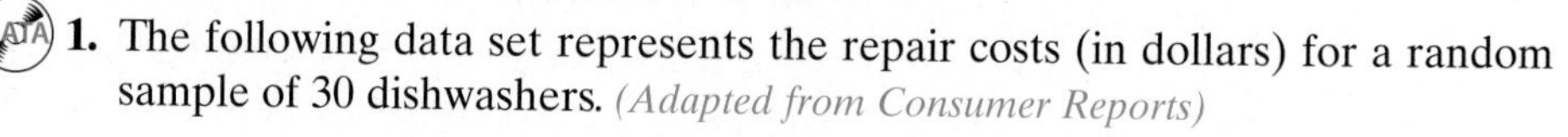

1. The following data set represents the repair costs (in dollars) for a random sample of 30 dishwashers. *(Adapted from Consumer Reports)*

41.82	52.81	57.80	68.16	73.48	78.88	88.13	88.79
90.07	90.35	91.68	91.72	93.01	95.21	95.34	96.50
100.05	101.32	103.59	104.19	105.62	111.32	117.14	118.42
118.77	119.01	120.70	140.52	141.84	147.06		

(a) Find the point estimate of the population mean.

(b) Find the maximum error of estimate for a 95% level of confidence.

(c) Construct a 95% confidence interval for the population mean and interpret the results.

2. You want to estimate the mean repair cost for dishwashers. The estimate must be within $10 of the population mean. Determine the required sample size to construct a 99% confidence interval for the population mean. Assume the population standard deviation is $22.50. *(Adapted from Consumer Reports)*

3. The following data set represents the time (in minutes) for a random sample of phone calls made by employees at a company.

7.5 2.0 12.1 8.8 9.4 7.3 1.9 2.8 7.0 7.3

(a) Find the sample mean.

(b) Find the sample standard deviation.

(c) Use the t-distribution to construct a 90% confidence interval for the population mean and interpret the results. Assume the population of the data set is normally distributed.

(d) Repeat part (c), assuming $\sigma = 3.5$ minutes. Compare the results.

4. In a random sample of eight people with advanced degrees in biology, the mean monthly income was $3705 and the standard deviation was $566. Assume the monthly incomes are normally distributed and use a t-distribution to construct a 95% confidence interval for the population mean monthly income for people with advanced degrees in biology. *(Adapted from U.S. Bureau of the Census)*

5. In a survey of 2000 adults from the U.S. age 65 and over, 1320 received a flu shot. *(Adapted from The Centers for Disease Control and Prevention)*

(a) Find a point estimate for the population proportion p of those receiving flu shots.

(b) Construct a 90% confidence interval for the population proportion.

(c) Find the minimum sample size needed to estimate the population proportion at the 99% confidence level in order to ensure that the estimate is accurate within 4% of the population proportion.

6. Refer to the data set in Exercise 1. Assume the population of dishwasher repair costs is normally distributed.

(a) Construct a 95% confidence interval for the population variance.

(b) Construct a 95% confidence interval for the population standard deviation.

6

Cumulative Test: Chapters 4–6

Take this test as you would take a test in class. After you are done, check your work against the answers given in the back of the book.

Refer to the following information as you take this test. In a survey of 474 U.S. women, 365 said that the media have a negative effect on women's health because they set unattainable standards for appearance. *(Source: Shape Up America!)*

1. Find a point estimate for p. Construct a 95% confidence interval for the population proportion p.

2. Find the minimum sample size needed to estimate the population proportion p with 99% confidence. The estimate must be accurate within 2% of p.

3. You randomly select 12 women. Use $\hat{p}$ from Exercise 1 to find the probability that at least 10 women will agree that the media have a negative effect on women's health because they set unattainable standards for appearance.

4. Use $\hat{p}$ from Exercise 1 to estimate the population mean, variance, and standard deviation of the binomial distribution and interpret the results.

5. Decide whether you can use the normal distribution to approximate this binomial distribution. If so, estimate the population mean and standard deviation and compare the results to Exercise 4. If not, explain why.

6. A mathematical way to measure a person's body fat is to use the body mass index (BMI). A person's BMI can be found by dividing weight (in kilograms) by the square of the height (in meters).

$$\text{BMI} = \frac{\text{weight (in kg)}}{(\text{height (in m)})^2}$$

   Assume the 474 women in the survey had a mean BMI of 25.6 and a standard deviation of 3.5. Construct a 90% confidence interval for the mean BMI for all U.S. women.

7. Which distribution did you use to construct the confidence interval in Exercise 6? Why?

8. Assume that women's BMIs are normally distributed. Are you more likely to randomly select one woman with a BMI less than 20 or are you more likely to select a sample of 15 women with a mean BMI less than 20? Explain.

9. You randomly select and record the BMIs of 30 women. The sample standard deviation is 3.2. Using a 95% level of confidence, construct the confidence intervals for (a) the population variance σ^2 and (b) the population standard deviation σ. Assume the sample is taken from a normally distributed population.

10. Determine which of the following values, if any, are unusual. Refer to the sample statistics in Exercise 6. ($\bar{x} = 25.6, s = 3.5$)

    (a) 29.1 (b) 18.2
    (c) 32.0 (d) 33.3

Where You've Been

In Chapter 6, you began your study of inferential statistics. There, you learned how to form a confidence interval estimate about a population parameter, such as the proportion of people in the United States who agree with a certain statement. For instance, in a poll taken for *USA TODAY*, people in the U.S. aged 18 and older were asked several questions about extraterrestrial life. Here are some of the results.

Survey Question	Number Surveyed	Number Who Said Yes
Do you believe UFOs really exist?	614	229
Have you ever seen a UFO?	229	28
Do extraterrestrial beings exist?	614	237
Have you ever seen one?	180	9

Fire in the Sky is Travis Walton's account of alien-UFO abduction that he claims to have happened on November 5, 1975.

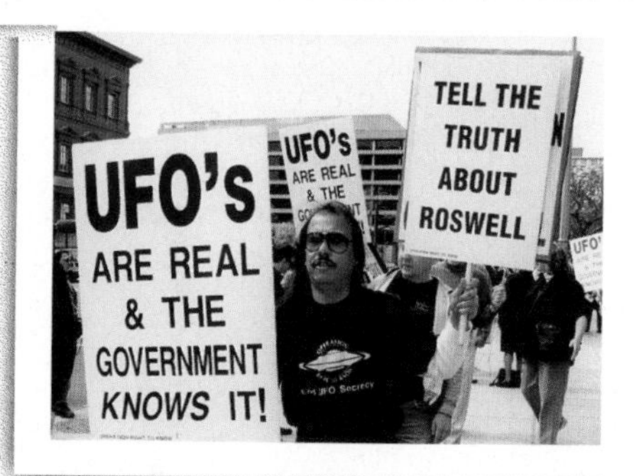

UFO sightings have been reported throughout the world, but some of the more famous ones were reported in the southwestern United States. Travis Walton reported his encounter to have taken place in northeastern Arizona, and the famous Roswell encounter was reported to have occurred in Roswell, New Mexico.

Hypothesis Testing with One Sample

CHAPTER 7

Where You're Going

In this chapter, you will continue your study of inferential statistics. But now, instead of making an estimate about a population parameter, you will learn how to test a claim about a parameter.

For instance, suppose that you work for *USA TODAY* and are asked to test a claim that the proportion of U.S. adults who believe UFOs really exist is $p = 0.30$. To test the claim, you take a random sample of $n = 614$ U.S. adults and find that 229 of them believe that UFOs really exist. Your sample statistic is $\hat{p} = \frac{229}{614} \approx 0.373$.

Is your sample statistic different enough from the claim $(p = 0.30)$ to decide that the claim is false? The answer lies in the sampling distribution of sample proportions taken from a population in which $p = 0.30$. The graph below shows that your sample statistic is almost 4 standard errors from the claim value. If the claim is true, the probability of the sample statistic being 4 standard errors (or more) from the claim value is extremely small. Something is wrong! If your sample was truly random, then you can conclude that the actual proportion of the population is not 0.30. In other words, you tested the original claim (hypothesis) and you decided to reject it.

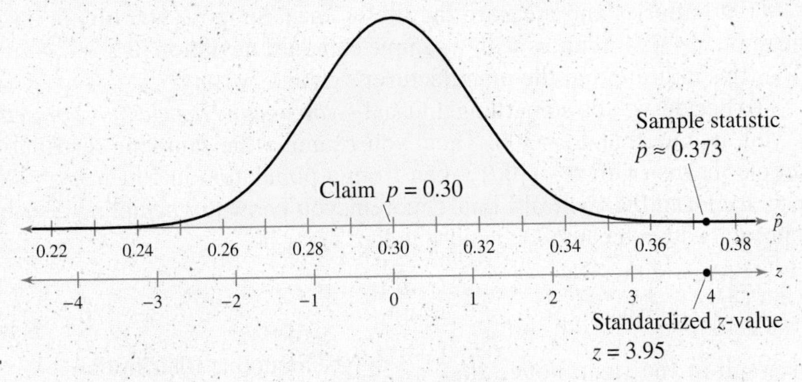

Sampling Distribution

This scene was added to *Close Encounters of the Third Kind* after the original Spielberg film was released.

7.1 Introduction to Hypothesis Testing

What You Should Learn

- A practical introduction to hypothesis tests
- How to state a null hypothesis and an alternative hypothesis
- How to identify type I and type II errors and interpret the level of significance
- How to know whether to use a one-tailed or two-tailed statistical test and finding a *P*-value
- How to make and interpret a decision based on the results of a statistical test
- How to write a claim for a hypothesis test

Hypothesis Tests • Stating a Hypothesis • Types of Errors and Level of Significance • Statistical Tests and *P*-values • Making a Decision and Interpreting the Decision • Strategies for Hypothesis Testing

Hypothesis Tests

Throughout the remainder of this course, you will study an important technique in inferential statistics called hypothesis testing. A **hypothesis test** is a process that uses sample statistics to test a claim about the value of a population parameter. Researchers in fields ranging from medicine to politics rely on hypothesis testing to make informed decisions about new medicines and the outcome of elections.

For instance, suppose a battery manufacturer claims that the mean life of its AA batteries is 300 minutes. If you suspect that this claim is not valid, how could you show that the claim is wrong?

Obviously, you can't test all the batteries. But you can still make a reasonable decision about the validity of the claim by taking a random sample from the population. If the sample mean differs enough from the claim, you can decide that the claim is wrong.

The average life of our new Ultra AA battery is 300 minutes.

For instance, to test the battery manufacturer's claim that the mean life of all batteries of this type is $\mu = 300$ minutes, you could take a random sample of $n = 100$ batteries and measure the life of each. Suppose you obtain a sample mean of $\bar{x} = 294$ minutes with a sample standard deviation of $s = 20$ minutes. Does this indicate that the manufacturer's claim is wrong?

To decide, you do something unusual—*you assume the claim is correct!* That is, you assume that $\mu = 300$. Then, you examine the sampling distribution of sample means (with $n = 100$) taken from a population in which $\mu = 300$ and $\sigma = 20$. From the Central Limit Theorem, you know this sampling distribution is normal with a mean of 300 and standard error of

$$\frac{20}{\sqrt{100}} = 2.$$

In graph at the right, notice that your sample mean of $\bar{x} = 294$ minutes is highly unlikely—it is 3 standard errors from the claimed mean! Using the techniques you studied in Chapter 5, you can determine that if the claim is true, the probability of obtaining a sample mean of 294 or less is 0.0013. Your assumption that the manufacturer's claim is correct has led you to an improbable result. So, either you had a very unusual sample, or the claim is false. The logical conclusion is that the claim is probably false.

Sampling Distribution

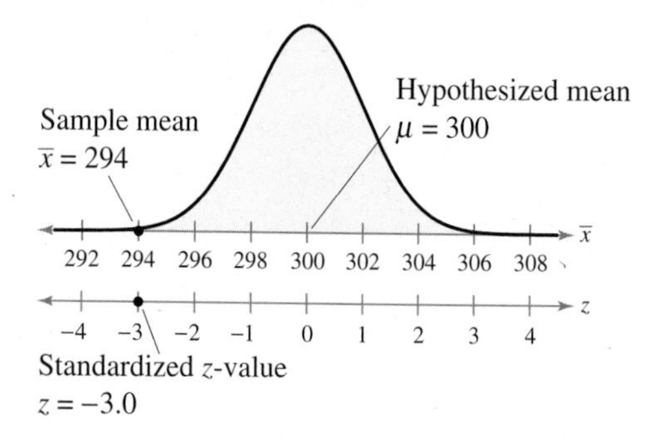

Insight

As you study this chapter, don't get confused regarding concepts of certainty and importance. For instance, even if you were very certain that the mean life of a type of AA battery is not 300 minutes, the actual mean life might be very close to this value and the difference might not be important.

Stating a Hypothesis

A claim about a population parameter is called a **statistical hypothesis.** To test a statistical hypothesis, you should carefully state a pair of hypotheses—one that represents the claim and the other, its complement. When one of these hypotheses is false, the other must be true. Of these two hypotheses, the one that contains a statement of *equality* is the *null hypothesis.* The complement of the null hypothesis is the *alternative hypothesis.* Either hypothesis—the null or the alternative—may represent the original claim.

Insight

The term *null hypothesis* was introduced by Ronald Fisher (see page 292). If the statement in the null hypothesis is not true, then the alternative hypothesis must be true.

DEFINITION

1. A **null hypothesis H_0** is a statistical hypothesis that contains a statement of equality, such as $\leq$, $=$, or $\geq$.
2. The **alternative hypothesis H_a** is the complement of the null hypothesis. It is a statement that must be true if H_0 is false and it contains a statement of inequality, such as $>$, $\neq$, or $<$.

H_0 is read as "H subzero" or "H naught" and H_a is read as "H sub-a."

Picturing the World

In a study performed at the Cleveland Clinic, a sample of 50 randomly chosen adults who developed symptoms of the common cold were given five zinc gluconate throat lozenges per day. The sample mean duration of nasal congestion was four days. So, the researchers claimed that the mean duration of nasal congestion of all adults who take zinc lozenges after developing cold symptoms is four days. *(Adapted from Annals of Internal Medicine)*

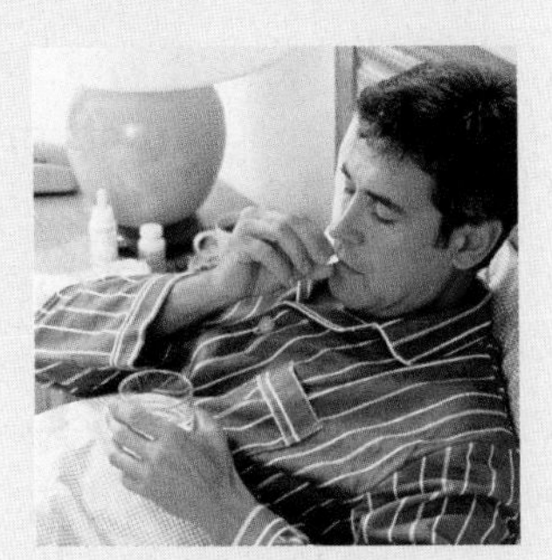

Determine a null hypothesis and alternative hypothesis for this claim.

To write the null and alternative hypotheses, translate the claim made about the population parameter from a verbal statement to a mathematical statement. Then, write its complement. For instance, if the claim value is k and the population parameter is μ, then some possible pairs of null and alternative hypotheses are

$$\begin{cases} H_0\colon \mu \leq k \\ H_a\colon \mu > k \end{cases} \qquad \begin{cases} H_0\colon \mu \geq k \\ H_a\colon \mu < k \end{cases} \qquad \begin{cases} H_0\colon \mu = k \\ H_a\colon \mu \neq k \end{cases}.$$

The following table shows the relationship between possible verbal statements about the parameter μ and the corresponding null and alternative hypotheses. Similar statements can be made to test other population parameters, such as p, σ, or σ^2.

Verbal Statement H_0	Mathematical Statements	Verbal Statement H_a
The mean is... greater than or equal to k at least k not less than k	$\begin{cases} H_0\colon \mu \geq k \\ H_a\colon \mu < k \end{cases}$	The mean is... less than k below k fewer than k
The mean is... less than or equal to k at most k not more than k	$\begin{cases} H_0\colon \mu \leq k \\ H_a\colon \mu > k \end{cases}$	The mean is... greater than k above k more than k
The mean is... equal to k k exactly k	$\begin{cases} H_0\colon \mu = k \\ H_a\colon \mu \neq k \end{cases}$	The mean is... not equal to k different from k not k

EXAMPLE 

Stating the Null and Alternative Hypotheses

Write the claim as a mathematical sentence. State the null and alternative hypotheses, and identify which represents the claim.

1. A university claims that the proportion of its students who graduate in four years is 82%.
2. A water faucet manufacturer claims that the mean flow rate of a certain type of faucet is less than 2.5 gallons per minute.
3. A cereal company claims that the mean weight of the contents of its 20-ounce size cereal boxes is more than 20 ounces.

SOLUTION

1. The claim "the proportion . . . is 82%" can be written as $p = 0.82$. Its complement is $p \neq 0.82$. Because $p = 0.82$ contains the statement of equality, it becomes the null hypothesis. In this case, the null hypothesis represents the claim.

   H_0: $p = 0.82$ (Claim)

   H_a: $p \neq 0.82$

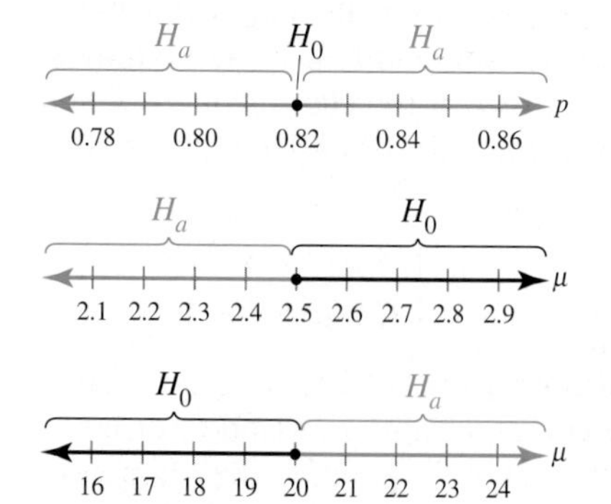

In each of these graphs, notice that each point on the number line is in H_0 or H_a, but no point is in both.

2. The claim "the mean . . . is less than 2.5 gpm" can be written as $\mu < 2.5$. Its complement is $\mu \geq 2.5$. Because $\mu \geq 2.5$ contains the statement of equality, it becomes the null hypothesis. In this case, the alternative hypothesis represents the claim.

   H_0: $\mu \geq 2.5$ gpm

   H_a: $\mu < 2.5$ gpm (Claim)

3. The claim "the mean . . . is more than 20 ounces" can be written as $\mu > 20$. Its complement is $\mu \leq 20$. Because $\mu \leq 20$ contains the statement of equality, it becomes the null hypothesis. In this case, the alternative hypothesis represents the claim.

   H_0: $\mu \leq 20$ ounces

   H_a: $\mu > 20$ ounces (Claim)

Try It Yourself 1

Write the claim as a mathematical sentence. State the null and alternative hypotheses, and identify which represents the claim.

1. An automobile battery manufacturer claims that the mean life of a certain type of battery is 74 months.
2. A television manufacturer claims that the variance of the life of a certain type of television is less than or equal to 3.5.
3. A radio station claims that its proportion of the local listening audience is greater than 39%.

a. *Identify* the verbal claim and *write* it as a mathematical statement.
b. *Write* the complement of the claim.
c. *Identify* the null and alternative hypotheses and *determine* which one represents the claim.

Answer: Page A38

Types of Errors and Level of Significance

No matter which hypothesis represents the claim, you always begin a hypothesis test by assuming that the equality condition in the null hypothesis is true. So, when you perform a hypothesis test, you make one of two decisions:

1. reject the null hypothesis or
2. fail to reject the null hypothesis.

Because your decision is based on incomplete information (a sample rather than the entire population), there is always the possibility you will make the wrong decision.

For instance, suppose your friend claims that a certain coin is fair. To test your friend's claim, you flip the coin 100 times and get 49 heads and 51 tails. You would probably agree that you do not have enough evidence to reject the claim. Even so, it is possible that the coin is actually not fair and you had an unusual sample.

But what if you flip the coin 100 times and get 21 heads and 79 tails? It is very unlikely that you would get only 21 heads out of 100 tosses with a fair coin. So, you have obtained sufficient evidence to reject your friend's claim that the coin is fair. However, you can't be 100% sure that the claim is false. It is possible that the coin is fair and you had an unusual sample.

If p represents the proportion of heads, the claim that "the coin is fair" can be written as the mathematical statement $p = 0.5$. Its complement is written as $p \neq 0.5$. So, your null hypothesis and alternative hypothesis are

H_0: $p = 0.5$ (Claim)

and

H_a: $p \neq 0.5$.

Remember, the only way to be certain of whether H_0 is true or false is to test the entire population. Because your decision (to reject H_0 or fail to reject H_0) is based on a sample, you must accept the fact that your decision might be incorrect. You might have rejected the null hypothesis when it is actually true. Or, you might have failed to reject the null hypothesis when it is actually false.

DEFINITION

A **type I error** occurs if the null hypothesis is rejected when it is actually true.

A **type II error** occurs if the null hypothesis is not rejected when it is actually false.

The following table shows the four possible outcomes of a hypothesis test.

	Actual Truth of H_0	
Decision	*H_0 is true.*	*H_0 is false.*
Do not reject H_0	Correct decision	Type II error
Reject H_0	Type I error	Correct decision

	Defendant is innocent.	*Defendant is guilty.*
Not guilty verdict	Justice	Type II error
Guilty verdict	Type I error	Justice

Hypothesis testing is sometimes compared to the legal system used in the United States. Under this system, the following steps are used.

1. A carefully worded accusation is written.
2. The defendant is assumed innocent (H_0) until proven guilty. The burden of proof lies with the prosecution. If the evidence is not strong enough, there is no conviction. A "not guilty" verdict does not prove that a defendant is innocent.
3. The evidence needs to be conclusive beyond a reasonable doubt. The system assumes that more harm is done by convicting the innocent (type I error) than by not convicting the guilty (type II error).

EXAMPLE 2

Identifying Type I and Type II Errors

The USDA limit for salmonella contamination for chicken is 20%. A meat-packing company claims that its chicken falls within the limit. You perform a hypothesis test to determine whether the company's claim is true. When will a type I or type II error occur? Which is more serious? *(Source: United States Department of Agriculture)*

SOLUTION Let p represent the proportion of the chicken that is contaminated. The company's claim is "less than or equal to 20% is contaminated." You can write the null and alternative hypotheses as follows.

H_0: $p \leq 0.2$ (Claim) The proportion is less than or equal to 20%.
H_a: $p > 0.2$ The proportion is greater than 20%.

Chicken meets USDA limits. Chicken exceeds USDA limits.

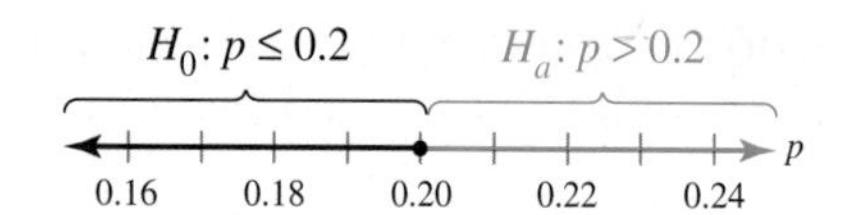

A type I error will occur if the actual proportion of contaminated chickens is less than or equal to 0.2, but you decide to reject H_0. A type II error will occur if the actual proportion of contaminated chickens is greater than 0.2, but you do not reject H_0. With a type I error, you might create a health scare and hurt the sales of chicken producers who were actually meeting the USDA limits. With a type II error, you could be allowing chicken that exceeded the USDA contamination limit to be sold to consumers. A type II error could result in sickness or even death.

Try It Yourself 2

A company specializing in parachute assembly claims that its main parachute failure rate is not more than 1%. You perform a hypothesis test to determine whether the company's claim is true. When will a type I or type II error occur? Which is more serious?

a. *State* the null and alternative hypotheses.
b. *Write* the possible type I and type II errors.
c. *Determine* which error is more serious.

Answer: Page A38

Because there is variation from sample to sample, there is always a possibility that you will reject a null hypothesis when it is actually true. You can decrease the probability of doing so by lowering the *level of significance*.

Insight

When you decrease α (the maximum allowable probability of making a type I error), you are likely to be increasing β.

DEFINITION

In a hypothesis test, the **level of significance** is your maximum allowable probability of making a type I error. It is denoted by α, the lowercase Greek letter alpha.

The probability of a type II error is denoted by β, the lowercase Greek letter beta.

By setting the level of significance at a small value, you are saying that you want the probability of rejecting a true null hypothesis to be small. Three commonly used levels of significance are $\alpha = 0.10$, $\alpha = 0.05$, and $\alpha = 0.01$.

Statistical Tests and *P*-values

After stating the null and alternative hypotheses and specifying the level of significance, the next step in a hypothesis test is to obtain a random sample from the population and calculate sample statistics such as the mean and the standard deviation. The statistic that is compared to the parameter in the null hypothesis is called the **test statistic.** The type of test used and the sampling distribution is based on the test statistic.

In this chapter, you will learn about several one-sample statistical tests. The following table shows the relationships between population parameters and their corresponding test statistics, sampling distributions, and standardized test statistics.

Population parameter	Test statistic	Sampling distribution	Standardized test statistic
μ	$\bar{x}$	Normal ($n \geq 30$) Student t ($n < 30$)	z (Section 7.2) t (Section 7.3)
p	$\hat{p}$	Normal	z (Section 7.4)
σ^2	s^2	Chi-square	χ^2 (Section 7.5)

One way to decide whether to reject the null hypothesis is to determine whether the probability of obtaining the standardized test statistic (or one that is more extreme) is less than the level of significance.

DEFINITION

Assuming the null hypothesis is true, a ***P*-value** (or **probability value**) of a hypothesis test is the probability of obtaining a sample statistic with a value as extreme or more extreme than the one determined from the sample data.

The nature of a hypothesis test depends on whether the hypothesis test is a left-, right-, or two-tailed test. The type of test depends on the region of the sampling distribution that favors a rejection of H_0. This region is indicated by the alternative hypothesis.

DEFINITION

1. If the alternative hypothesis H_a contains the less-than inequality symbol $(<)$, the hypothesis test is a **left-tailed test.**

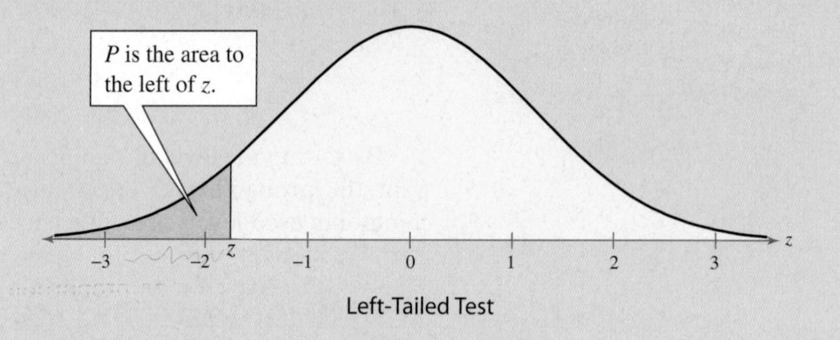

2. If the alternative hypothesis H_a contains the greater-than inequality symbol $(>)$, the hypothesis test is a **right-tailed test.**

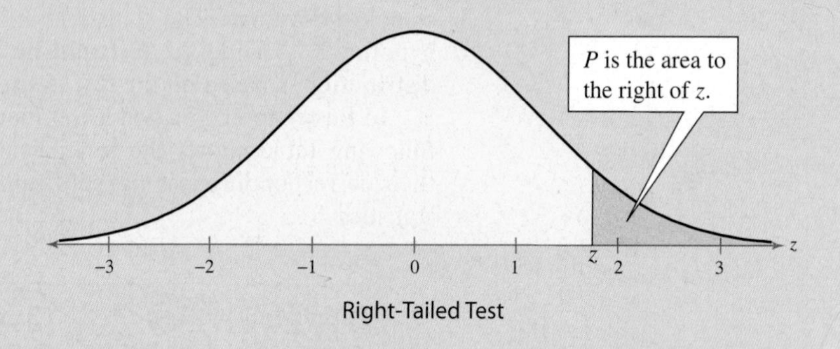

3. If the alternative hypothesis H_a contains the not-equal-to symbol $(\neq)$, the hypothesis test is a **two-tailed test.** In a two-tailed test, each tail has an area of $\frac{1}{2}P$.

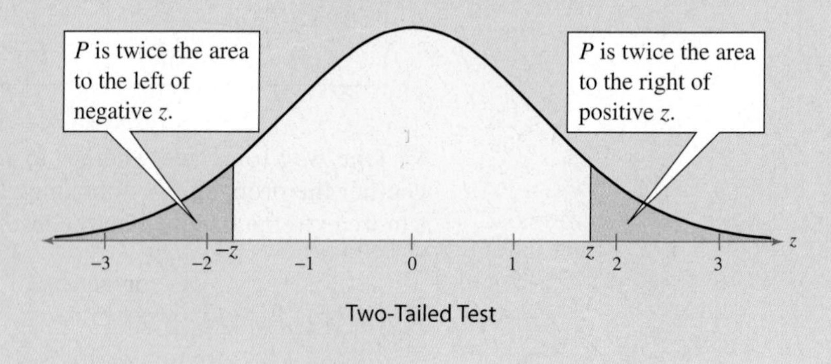

The smaller the P-value of the test, the more evidence there is to reject the null hypothesis. Remember, however, that even a very low P-value does not constitute proof that the null hypothesis is false.

EXAMPLE 3

Identifying the Nature of a Hypothesis Test

For each claim, state H_0 and H_a in words and in symbols. Then determine whether the hypothesis test is a left-tailed test, right-tailed test, or two-tailed test. Sketch a normal sampling distribution and shade the area for the P-value.

1. A university claims that the proportion of its students who graduate in 4 years is 82%.
2. A water faucet manufacturer claims that the mean flow rate of a certain type of faucet is less than 2.5 gallons per minute.
3. A cereal company claims that the mean weight of the contents of its 20-ounce size cereal boxes is more than 20 ounces.

SOLUTION

In Symbols	*In Words*
1. H_0: $p = 0.82$	The proportion of students who graduate in 4 years is 82%.
H_a: $p \neq 0.82$	The proportion of students who graduate in 4 years is not 82%.

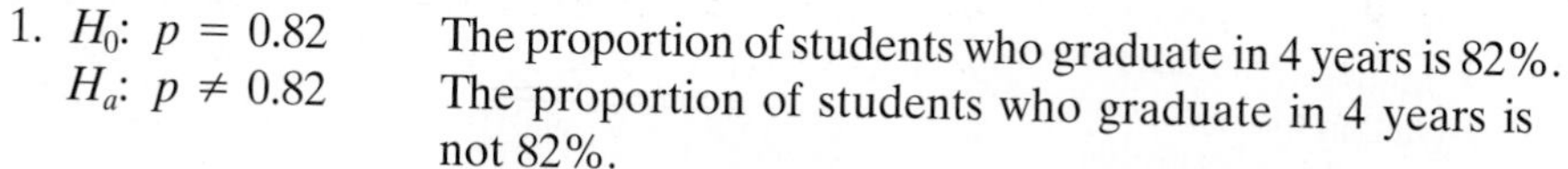

$\frac{1}{2}$ P-value area, $\frac{1}{2}$ P-value area; $-z$, 0, z

Because H_a contains the $\neq$ symbol, the test is a two-tailed hypothesis test. The graph of the normal sampling distribution at the left shows the shaded area for the P-value.

2. H_0: $\mu \geq 2.5$ gpm	The mean flow rate of a certain type of faucet is greater than or equal to 2.5 gallons per minute.
H_a: $\mu < 2.5$ gpm	The mean flow rate of a certain type of faucet is less than 2.5 gallons per minute.

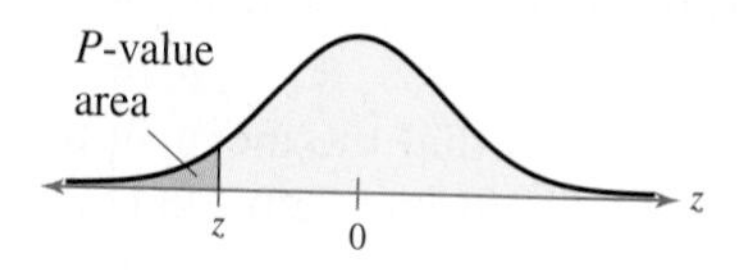

Because H_a contains the $<$ symbol, the test is a left-tailed hypothesis test. The graph of the normal sampling distribution at the left shows the shaded area for the P-value.

3. H_0: $\mu \leq 20$ oz	The mean weight of the contents of the cereal boxes is less than or equal to 20 ounces.
H_a: $\mu > 20$ oz	The mean weight of the contents of the cereal boxes is greater than 20 ounces.

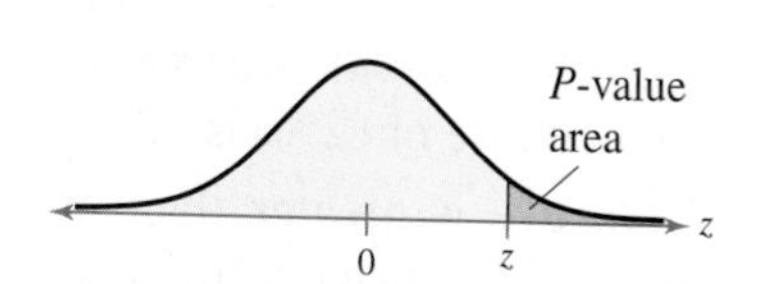

Because H_a contains the $>$ symbol, the test is a right-tailed hypothesis test. The graph of the normal sampling distribution at the left shows the shaded area for the P-value.

Try It Yourself 3

For each claim, determine whether the hypothesis test is a left-, right-, or two-tailed test. Sketch a normal sampling distribution and shade the area for the P-value.

1. An automobile battery manufacturer claims that the mean life of a certain type of battery is 74 months.
2. A radio station claims that its proportion of the local listening audience is greater than 39%.

a. *Write* H_0 and H_a.
b. *Determine* whether the test is left-tailed, right-tailed, or two-tailed.
c. *Sketch* the sampling distribution and shade the area for the P-value.

Answer: Page A38

Making a Decision and Interpreting the Decision

To conclude a hypothesis test, you make a decision and interpret that decision. There are only two possible outcomes to a hypothesis test: (1) reject the null hypothesis, and (2) fail to reject the null hypothesis.

Insight

In this chapter, you will learn that there are two basic types of **decision rules** for deciding to reject H_0 or fail to reject H_0. The decision rule described on this page is based on *P*-values. The second basic type of decision rule is based on rejection regions. When the standardized test statistic falls in the rejection region, the observed probability (*P*-value) of a type I error is less than α. You will learn more about rejection regions in the next section.

Decision Rule Based on *P*-value

To use a *P*-value to make a conclusion in a hypothesis test, compare the *P*-value to α.

1. If $P \leq \alpha$, then reject H_0.
2. If $P > \alpha$, then fail to reject H_0.

If you fail to reject the null hypothesis, it does not mean that you have accepted the null hypothesis as true. It simply means that there is not enough evidence to reject the null hypothesis.

EXAMPLE 4

Interpreting a Decision

You perform a hypothesis test for each of the following claims. How should you interpret your decision if you reject H_0? If you fail to reject H_0?

1. H_0 (Claim): A university claims that the proportion of its students who graduate in four years is 82%.
2. H_a (Claim): A government safety administration claims that the mean stopping distance (on a dry surface) for a Chevrolet Malibu is less than 148 feet.

SOLUTION

1. If you reject H_0, then you should conclude "there is sufficient evidence to indicate that the university's 4-year graduation rate is not 82%." If you fail to reject H_0, then you should conclude "there is insufficient evidence to indicate that the university's claim (of a 4-year graduation rate of 82%) is false."
2. The null hypothesis is "the mean stopping distance ... is greater than or equal to 148 feet." If you reject H_0, then you should conclude "there is enough evidence to support the government safety administration's claim (the stopping distance for a Chevrolet Malibu is less than 148 feet)." If you fail to reject H_0, then you should conclude "there is not enough evidence to support the government safety administration's claim that the stopping distance for a Chevrolet Malibu is less than 148 feet."

Try It Yourself 4

You perform a hypothesis test for the following claim. How should you interpret your decision if you reject H_0? How should you interpret your decision if you fail to reject H_0?

H_a (Claim): A radio station claims that its proportion of the local listening audience is greater than 39%.

a. Interpret your decision if you reject the null hypothesis.
b. Interpret your decision if you do not reject the null hypothesis.

Answer: Page A38

The general steps for a hypothesis test are summarized below.

1. State the claim mathematically and verbally. Identify the null and alternative hypotheses.

H_0: ? H_a: ?

⬇

2. Specify the level of significance.

$\alpha =$?

⬇

3. Determine the standardized sampling distribution and sketch its graph.

⬇

4. Calculate the test statistic and its standardized value. Add it to your sketch.

⬇

5. Find the P-value.

⬇

6. Use the following decision rule.

Is the P-value less than or equal to the level of significance?

Yes → Reject H_0.

No → Fail to reject H_0.

⬇

7. Write a statement to interpret the decision in the context of the original claim.

In the flowchart above, the graphs show a right-tailed test. However, the same basic steps also apply to left-tailed and two-tailed tests.

Strategies for Hypothesis Testing

When you are making a claim or testing someone else's claim, you need to consider the context of the claim and decide whether it is "one tailed" or "two tailed."

EXAMPLE 5

Writing a Claim as One Tailed or Two Tailed

Consider the claim described at the beginning of this section by a battery manufacturer that the mean life of its AA battery is 300 minutes. You work for a consumer advocate magazine and are asked to test the battery manufacturer's claim. From your point of view, which of the following best represents the manufacturer's claim?

$\mu = 300$

$\mu \leq 300$

$\mu \geq 300$

SOLUTION To answer this question, you need to think about the context of the manufacturer's claim. As a consumer advocate, your concern is that consumers will get at least 300 minutes of use out of a battery. It would be fine with you if the mean life of the battery turned out to be greater than 300 minutes. So, from your point of view, the manufacturer's claim is best stated as $\mu \geq 300$.

Try It Yourself 5

1. You represent a chemical company that is being sued for paint damage to automobiles. You believe the mean cost of repair per automobile is about \$650. From your point of view, which of the following would best express your claim?

   $\mu = 650$, $\mu \leq 650$, $\mu \geq 650$

2. You are on a research team that is investigating the mean temperature of adult humans (see page 349). The commonly accepted claim is that the mean temperature is 98.6°F. From your point of view, which of the following best represents this claim?

   $\mu = 98.6$, $\mu \leq 98.6$, $\mu \geq 98.6$

a. *Think* about the context of the claim.
b. *Choose* the statement that best represents your point of view.

Answer: Page A38

In a courtroom, the strategy used by an attorney depends on whether the attorney is representing the defense or the prosecution. In a similar way, the strategy that you will use in hypothesis testing should depend on whether you are trying to support or reject a claim. Remember that you cannot use a hypothesis test to support your claim if your claim is the null hypothesis. So, as a researcher, if you want a conclusion that supports your claim, word your claim so it is the alternative hypothesis.

7.1 Exercises

Help

LarsonTutor 7.1

Companion Web Site

Student Solutions Manual 7.1

Videos 7.1

Technology Manuals

Basic Skills and Concepts

Stating Hypotheses In Exercises 1–6, use the given statement to represent a claim. Write its complement and state which is H_0 and which is H_a.

1. $\mu \leq 645$

2. $\mu < 128$

3. $\sigma \neq 5$

4. $\sigma^2 \geq 1.2$

5. $p < 0.45$

6. $p = 0.21$

Graphical Analysis In Exercises 7–10, match the null hypothesis with its graph. Then state the alternative hypothesis and sketch its graph.

7. H_0: $\mu \geq 3$

(a) 1 2 3 4 μ

8. H_0: $\mu \leq 3$

(b)
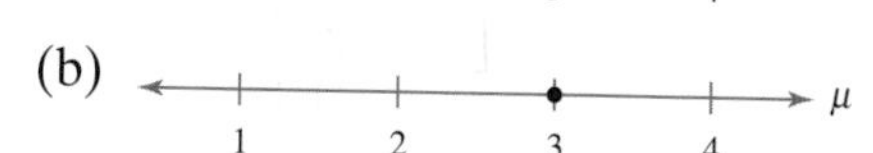

9. H_0: $\mu = 3$

(c) 1 2 3 4 μ

10. H_0: $\mu \geq 2$

(d)
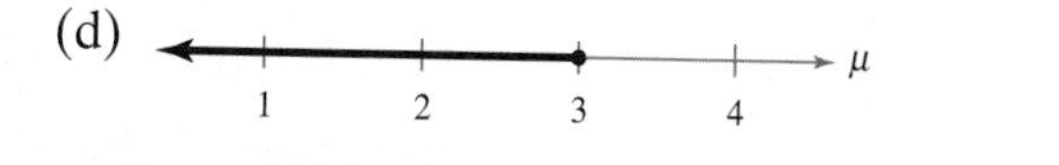

Stating the Hypotheses In Exercises 11–16, state the claim mathematically. Write the null and alternative hypotheses. Identify which is the claim.

11. ***Light Bulbs*** ◆ A light bulb manufacturer claims that the mean life of a certain type of light bulb is more than 750 hours.

12. ***Shipping Errors*** ◆ As stated by a company's shipping department, the number of shipping errors per million shipments has a standard deviation that is less than 3.

13. ***Base Price of a Car*** ◆ The standard deviation of the base price of a certain type of car is no more than \$1220. *(Adapted from Consumer Reports)*

14. ***What Shoppers Buy*** ◆ A research organization reports that 9% of all adult grocery shoppers in the U.S. never buy the store brand. *(Source: Wirthlin Worldwide)*

15. ***Misdiagnosed or Overlooked Cancer Cases*** ◆ The results of a recent study examining the proportion of clinically misdiagnosed or missed cancer cases indicate that 44% of all cancer cases are missed or misdiagnosed. *(Source: The Journal of the American Medical Association)*

16. ***Survival Times*** ◆ A study claims that the mean survival time for certain cancer patients treated immediately with chemotherapy and radiation is 24 months.

Identifying Errors In Exercises 17–22, write sentences describing type I and type II errors for a hypothesis test of the indicated claim.

17. ***Repeat Buyers*** ◆ A car dealer claims that at least 24% of its new customers will return to buy their next car.

18. ***Literary Skills*** ◆ A study claims that the proportion of adults in the U.S. with rudimentary literary skills is 21%. *(Source: U.S. Department of Education)*

19. ***Chess*** ◆ A local chess club claims that the length of time to play a game has a standard deviation of more than 23 minutes.

20. ***ER Visits*** ◆ A hospital spokesperson states that 2% of emergency room visits by college undergraduates are for alcohol-related health problems.

21. ***Cell Phone Calls*** ◆ According to a recent consumer magazine report, approximately 60% of all personal cell phone calls are made during evenings and weekends. *(Source: Consumer Reports)*

22. ***Wristwatch Batteries*** ◆ A battery manufacturer guarantees that the standard deviation of the life of its wristwatch batteries is less than 5 months.

Identifying Tests In Exercises 23–28, determine whether the hypothesis test for each claim is left tailed, right tailed, or two tailed. Explain your reasoning.

23. ***Security Alarms*** ◆ At least 14% of all homeowners have a home security alarm.

24. ***Clocks*** ◆ A manufacturer of grandfather clocks claims that the mean time its clocks lose is no more than 0.02 second per day.

25. ***Lung Cancer*** ◆ A government report claims the proportion of lung cancer cases that are due to smoking is 90%.

26. ***Tires*** ◆ The mean life of a certain tire is no less than 50,000 miles. *(Adapted from Goodyear)*

27. ***Return Rate*** ◆ A financial analyst claims that the return rate of a 15-year U.S. bond has a standard deviation of 5.3%.

28. ***Dreams*** ◆ A research institute claims the mean length of most dreams is greater than 10 minutes. *(Adapted from The Lucidity Institute)*

Interpreting a Decision In Exercises 29–34, consider each claim. If a hypothesis test is performed, how should you interpret a decision that

(a) rejects the null hypothesis?

(b) fails to reject the null hypothesis?

29. ***Pictures Developed*** ◆ The mean number of pictures developed for a standard roll of film with 24 exposures is at least 22.

30. ***Shipment Weights*** ◆ The standard deviation of the mean weight of all U.S. Postal Service shipments is 0.40 pound.

31. ***Hourly Wages*** ◆ The U.S. Department of Labor claims the proportion of hourly workers earning over $10.00 per hour is greater than 42%. *(Adapted from U.S. Department of Labor)*

32. ***Gas Mileage*** ◆ An automotive manufacturer claims the standard deviation for the gas mileage of its models is 3.9 miles per gallon.

33. ***Car Prices*** ◆ The mean price of a model year car is $20,440. *(Source: Dodge)*

34. ***Calories*** ◆ A soft-drink maker claims the mean calorie content of its beverages is 26 calories per serving.

35. ***Writing a Claim: Medical Research*** ◆ Your medical research team is investigating the proper dose of a certain heart medication. The medicine manufacturer thinks that the mean dose should be 10 milligrams. From your point of view, which of the following best represents your claim?

$$\mu = 10, \quad \mu \leq 10, \quad \mu \geq 10$$

36. ***Writing a Claim: Taxicab Company*** ◆ A taxicab company claims that the mean travel time between two destinations is about 21 minutes. From the taxicab company's point of view, which of the following best represents this claim?

$$\mu = 21, \quad \mu < 21, \quad \mu > 21$$

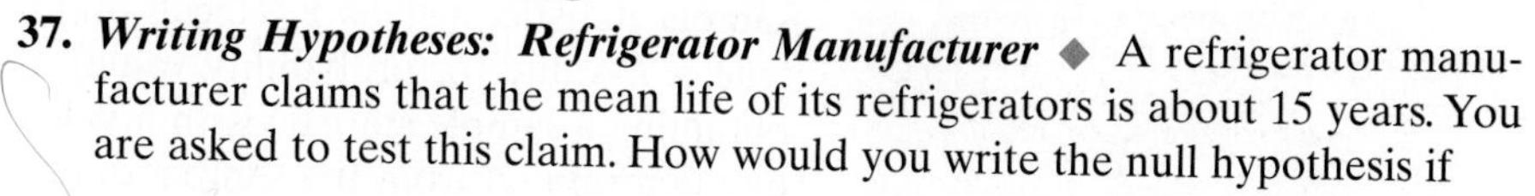

37. ***Writing Hypotheses: Refrigerator Manufacturer*** ◆ A refrigerator manufacturer claims that the mean life of its refrigerators is about 15 years. You are asked to test this claim. How would you write the null hypothesis if

(a) you represent the manufacturer and want to support the claim?

(b) you represent a consumer group and want to reject the claim?

38. ***Writing Hypotheses: Internet Provider*** ◆ An Internet provider is trying to gain advertising deals and claims that the mean time a customer spends on line per day is about 28 minutes. You are asked to test this claim. How would you write the null hypothesis if

(a) you represent the internet provider and want to support the claim?

(b) you represent an advertiser and want to reject the claim?

Extending the Basics

39. ***Getting at the Concept*** ◆ Why can decreasing the probability of a type I error cause an increase in the probability of a type II error?

40. ***Getting at the Concept*** ◆ Why not use a level of significance of $\alpha = 0$?

Graphical Analysis In Exercises 41 and 42, you are given a null hypothesis and three confidence intervals that represent three samplings. Decide whether each confidence interval indicates that you should reject H_0. Explain your reasoning.

41. H_0: $\mu \geq 70$

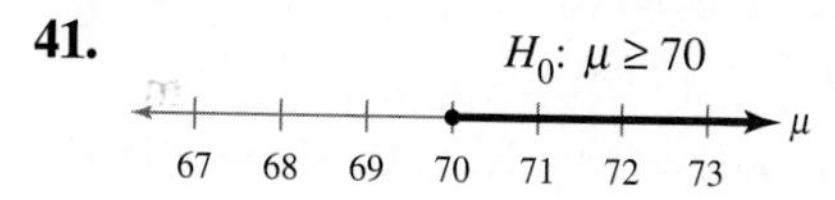

(a) $\mu = 68 \pm 1$

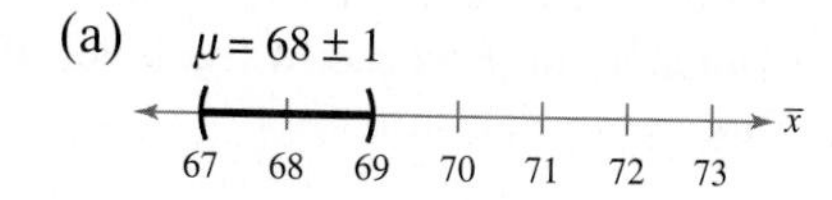

(b) $\mu = 69 \pm 2$

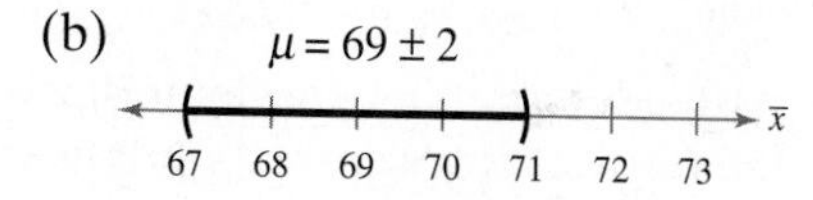

(c) $\mu = 71 \pm 1.5$

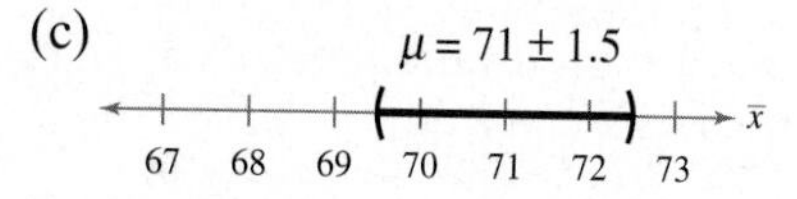

42. H_0: $\mu \leq 54$

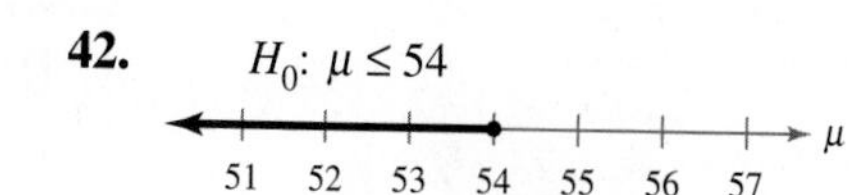

(a) $\mu = 55 \pm 0.5$

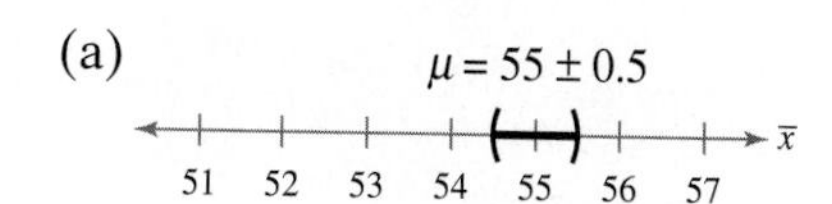

(b) $\mu = 53 \pm 1.5$

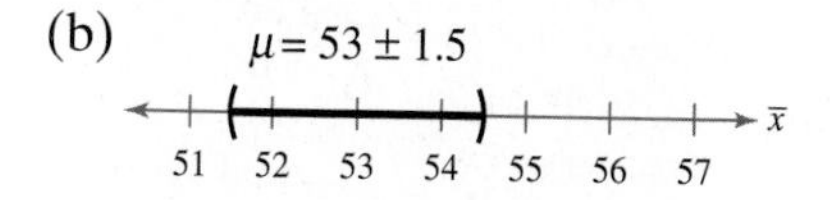

(c) $\mu = 55 \pm 1.5$

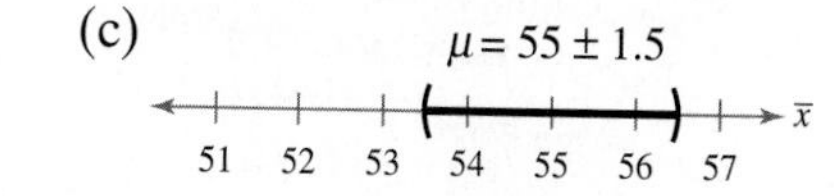

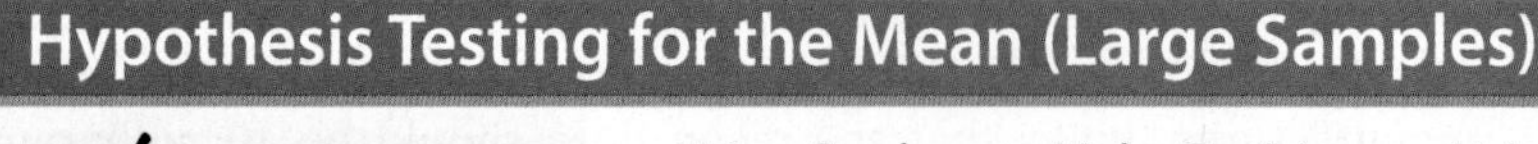

7.2 Hypothesis Testing for the Mean (Large Samples)

What You Should Learn

- How to find P-values and use them to test a mean μ
- How to use P-values for a z-test
- How to find critical values and rejection regions in a normal distribution
- How to use rejection regions for a z-test

Using P-values to Make Decisions • Using P-values for a z-Test • Rejection Regions and Critical Values • Using Rejection Regions for a z-Test

Using P-values to Make Decisions

In Chapter 5, you learned that when the sample size is at least 30, the sampling distribution for $\overline{x}$ (the sample mean) is normal. In Section 7.1, you learned that a way to reach a conclusion in a hypothesis test is to use a P-value for the sample statistic, such as $\overline{x}$. Recall that when you assume the null hypothesis is true, a P-value (or probability value) of a hypothesis test is the probability of obtaining a sample statistic with a value as extreme or more extreme than the one determined from the sample data. The decision rule for a hypothesis test based on a P-value is as follows.

Decision Rule Based on P-value

To use a P-value to make a conclusion in a hypothesis test, compare the P-value to α.

1. If $P \leq \alpha$, then reject H_0.
2. If $P > \alpha$, then fail to reject H_0.

EXAMPLE 1

Interpreting a P-value

The P-value for a hypothesis test is $P = 0.0237$. What is your decision if the level of significance is (1) $\alpha = 0.05$ and (2) $\alpha = 0.01$?

SOLUTION

1. Because $0.0237 < 0.05$, you should reject the null hypothesis.
2. Because $0.0237 > 0.01$, you should fail to reject the null hypothesis.

Try It Yourself 1

The P-value for a hypothesis test is $P = 0.0347$. What is your decision if the level of significance is (1) $\alpha = 0.01$ and (2) $\alpha = 0.05$?

a. *Compare* the P-value to the level of significance.
b. *Make* your decision.

Answer: Page A38

Insight

The lower the P-value, the more evidence there is in favor of rejecting H_0. The P-value gives you the lowest level of significance for which the sample statistic allows you to reject the null hypothesis. In Example 1, you would reject H_0 at any level of significance greater than or equal to 0.0237.

Finding the P-value for a Hypothesis Test

After determining the hypothesis test's standardized test statistic and the test statistic's corresponding area, do one of the following to find the P-value.

a. For a left-tailed test, P = (Area in left tail).
b. For a right-tailed test, P = (Area in right tail).
c. For a two-tailed test, P = 2(Area in tail of test statistic).

EXAMPLE 2

Finding a P-value for a Left-Tailed Test

Find the P-value for a left-tailed hypothesis test with a test statistic of $z = -2.23$. Decide whether to reject H_0 if the level of significance is $\alpha = 0.01$.

The area to the left of $z = -2.23$ is $P = 0.0129$.

$z = -2.23$

Left-Tailed Test

SOLUTION

The graph shows a standard normal curve with a shaded area to the left of $z = -2.23$. For a left-tailed test,

$$P = (\text{Area in left tail}).$$

From Table 4 in Appendix B, the area corresponding to $z = -2.23$ is 0.0129, which is the area in the left tail. So, the P-value for a left-tailed hypothesis test with a test statistic of $z = -2.23$ is $P = 0.0129$. Because $0.0129 > 0.01$, you should fail to reject H_0.

Try It Yourself 2

Find the P-value for a left-tailed hypothesis test with a test statistic of $z = -1.62$. Decide whether to reject H_0 if the level of significance is $\alpha = 0.05$.

a. *Use* Table 4 to locate the area that corresponds to $z = -1.62$.
b. *Calculate* the P-value for a left-tailed test, the area in the left tail.
c. *Decide* whether to reject H_0.

Answer: Page A38

EXAMPLE 3

Finding a P-value for a Two-Tailed Test

Find the P-value for a two-tailed hypothesis test with a test statistic of $z = 2.14$. Decide whether to reject H_0 if the level of significance is $\alpha = 0.05$.

The area to the right of $z = 2.14$ is 0.0162, so $P = 2(0.0162) = 0.0324$.

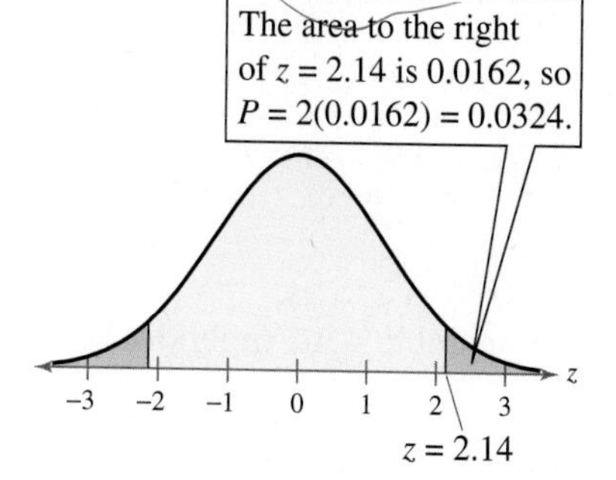

Two-Tailed Test

SOLUTION

The graph shows a standard normal curve with shaded areas to the left of $z = -2.14$ and to the right of $z = 2.14$. For a two-tailed test,

$$P = 2(\text{Area in tail of test statistic}).$$

From Table 4 in Appendix B, the area corresponding to $z = 2.14$ is 0.9838. The area in the right tail is $1 - 0.9838 = 0.0162$. So, the P-value for a two-tailed hypothesis test with a test statistic of $z = 2.14$ is $P = 2(0.0162) = 0.0324$. Because $0.0324 < 0.05$, you should reject H_0.

Try It Yourself 3

Find the P-value for a two-tailed hypothesis test with a test statistic of $z = 2.31$. Decide whether to reject H_0 if the level of significance is $\alpha = 0.01$.

a. *Use* Table 4 to locate the area that corresponds to $z = 2.31$.
b. *Calculate* the P-value for a two-tailed test, twice the area in the tail of the test statistic.
c. *Decide* whether to reject H_0.

Answer: Page A38

Using *P*-values for a *z*-Test

The z-test for the mean is used in populations for which the sampling distribution of sample means is normal. To use the z-test, you need to find the standardized value for your test statistic $\overline{x}$.

$$z = \frac{(\text{Sample mean}) - (\text{Hypothesized mean})}{\text{Standard error}}$$

Study Tip

With all hypothesis tests, it is helpful to sketch the sampling distribution. Your sketch should include the standardized test statistic.

z-Test for a Mean

The **z-test** is a statistical test for a population mean. The z-test can be used when the population is normal and σ is known, or for any population when the sample size n is at least 30. The **test statistic** is the sample mean $\overline{x}$ and the **standardized test statistic** is z.

$$z = \frac{\overline{x} - \mu}{\sigma/\sqrt{n}}$$

Recall that $\frac{\sigma}{\sqrt{n}}$ = standard error = $\sigma_{\overline{x}}$.

When $n \geq 30$, you can use the sample standard deviation s in place of σ.

GUIDELINES

Using *P*-values for a *z*-Test for Mean μ

In Words	*In Symbols*
1. State the claim mathematically and verbally. Identify the null and alternative hypotheses.	State H_0 and H_a.
2. Specify the level of significance.	Identify α.
3. Determine the standardized test statistic.	$z = \frac{\overline{x} - \mu}{\sigma/\sqrt{n}}$ or if $n \geq 30$, use $\sigma \approx s$.
4. Find the area that corresponds to z.	Use Table 4 in Appendix B.

5. Find the *P*-value.

a. For a left-tailed test, P = (Area in left tail).

b. For a right-tailed test, P = (Area in right tail).

c. For a two-tailed test, P = 2(Area in tail of test statistic).

6. Make a decision to reject or fail to reject the null hypothesis.	Reject H_0 if *P*-value is less than or equal to α. Otherwise, fail to reject H_0.
7. Interpret the decision in the context of the original claim.	

EXAMPLE 4

Hypothesis Testing Using P-values

In an advertisement, a pizza shop claims that its mean delivery time is less than 30 minutes. A random selection of 36 delivery times has a sample mean of 28.5 minutes and a standard deviation of 3.5 minutes. Is there enough evidence to support the claim at $\alpha = 0.01$? Use a P-value.

SOLUTION The claim is "the mean delivery time is less than 30 minutes." So, the null and alternative hypotheses are

H_0: $\mu \geq 30$ minutes and H_a: $\mu < 30$ minutes. (Claim)

Using the z-test, the standardized test statistic is

$$z = \frac{\bar{x} - \mu}{\sigma/\sqrt{n}}$$

$$= \frac{28.5 - 30}{3.5/\sqrt{36}} \qquad \text{Because } n \geq 30, \text{ use } \sigma \approx s = 3.5.$$

$$\approx -2.57.$$

Using Table 4, the area corresponding to $z = -2.57$ is 0.0051. Because this test is a left-tailed test, the P-value is equal to the area to the left of $z = -2.57$. So, $P = 0.0051$. Because the P-value is less than $\alpha = 0.01$, you should decide to reject the null hypothesis. So, at the 1% level of significance, you have sufficient evidence to conclude that the mean delivery time is less than 30 minutes.

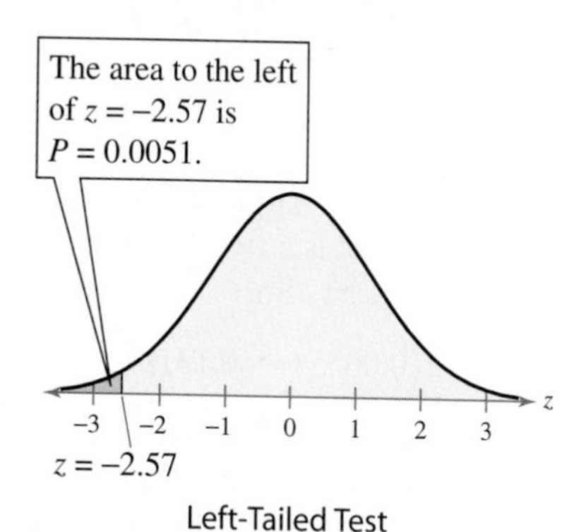

Left-Tailed Test

Try It Yourself 4

Home owners claim that the mean speed of automobiles traveling on their street is greater than the speed limit of 35 miles per hour. A random sample of 100 automobiles has a mean speed of 36 miles per hour and a standard deviation of 4 miles per hour. Is there enough evidence to support the claim at $\alpha = 0.05$? Use a P-value.

a. *Identify* the claim. Then *state* the null and alternative hypotheses.
b. *Identify* the level of significance.
c. *Find* the standardized test statistic z.
d. *Find* the P-value.
e. *Decide* whether to reject the null hypothesis.
f. *Interpret* the decision in the context of the original claim.

Answer: Page A38

EXAMPLE 

See *Minitab* steps on page 378.

Hypothesis Testing Using P-values

You think that the average franchise investment information given in the graph is incorrect, so you randomly select 30 franchises and determine the necessary investment for each. The sample mean investment is \$135,000 with a standard deviation of \$30,000. Is there enough evidence to support your claim at $\alpha = 0.05$? Use a P-value.

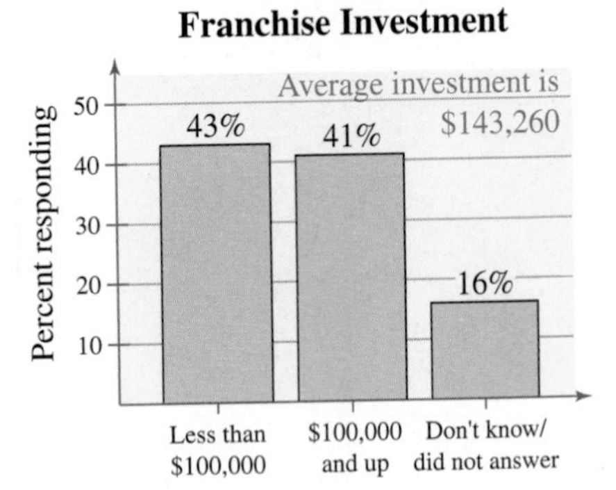

SOLUTION The claim is "the mean is different from \$143,260." So, the null and alternative hypotheses are

$$H_0\colon \mu = \$143{,}260 \quad \text{and} \quad H_a\colon \mu \neq \$143{,}260. \text{ (Claim)}$$

The level of significance is $\alpha = 0.05$. Using the z-test, the standardized test statistic is

$$z = \frac{\bar{x} - \mu}{\sigma/\sqrt{n}}$$

$$= \frac{135{,}000 - 143{,}260}{30{,}000/\sqrt{30}} \qquad \text{Because } n \geq 30, \text{ use } \sigma \approx s = 30{,}000.$$

$$\approx -1.51.$$

The area to the left of $z = -1.51$ is 0.0655, so $P = 2(0.0655) = 0.1310$.

$-3 \quad -2 \quad -1 \quad 0 \quad 1 \quad 2 \quad 3 \quad z$

$z = -1.51$

Two-Tailed Test

Using Table 4 in Appendix B, the area corresponding to $z = -1.51$ is 0.0655. Because the test is a two-tailed test, the P-value is equal to twice the area to the left of $z = -1.51$. So,

$$P = 2(0.0655) = 0.1310.$$

Because the P-value is greater than α, you should fail to reject the null hypothesis. So, there is not enough evidence at the 5% level of significance to conclude that the mean franchise investment is not \$143,260.

Try It Yourself 5

One of your distributors reports an average of 150 sales per day for the distributorship. You suspect that this average is not accurate, so you randomly select 35 days and determine the number of sales each day. The sample mean is 143 daily sales with a standard deviation of 15 sales. At $\alpha = 0.01$, is there enough evidence to doubt the distributor's reported average? Use a P-value.

a. *Identify* the claim. Then *state* a null and alternative hypotheses.
b. *Identify* the level of significance.
c. *Find* the standardized test statistic z.
d. *Find* the P-value.
e. *Decide* whether to reject the null hypothesis.
f. *Interpret* the decision in the context of the original claim.

Answer: Page A38

EXAMPLE 6

Using a Technology Tool to Find a P-value

What decision should you make for the following Minitab printout, using a level of significance of $\alpha = 0.05$?

MINITAB

Test of mu = 6.200 vs mu not = 6.2000
The assumed sigma = 0.470

Variable	N	Mean	StDev	SE Mean	Z	P
Sample	53	6.0666	0.4146	0.0646	-2.07	0.039

SOLUTION The *P*-value for this test is given as 0.039. Because the *P*-value is less than 0.05, reject the null hypothesis.

Study Tip

For this test, a sample of 53 values was entered into a column called "Sample." The hypotheses are

H_0: $\mu = 6.2$

and

H_a: $\mu \neq 6.2$.

Try It Yourself 6

For the *Minitab* hypothesis test shown in Example 6, make a decision at the $\alpha = 0.01$ level of significance.

a. *Compare* the *P*-value to the level of significance.
b. *Make* your decision.

Answer: Page A39

Rejection Regions and Critical Values

Another method to decide whether to reject the null hypothesis is to determine whether the standardized test statistic falls within a range of values called the rejection region of the sampling distribution.

DEFINITION

A **rejection region** (or **critical region**) of the sampling distribution is the range of values for which the null hypothesis is not probable. If a test statistic falls in this region, the null hypothesis is rejected. A **critical value** z_0 separates the rejection region from the nonrejection region.

GUIDELINES

Finding Critical Values in a Normal Distribution

1. Specify the level of significance α.
2. Decide whether the test is left tailed, right tailed, or two tailed.
3. Find the critical value(s) z_0. If the hypothesis test is
   a. *left tailed*, find the z-score that corresponds to an area of α.
   b. *right tailed*, find the z-score that corresponds to an area of $1 - \alpha$.
   c. *two tailed*, find the z-scores that correspond to $\frac{1}{2}\alpha$ and $1 - \frac{1}{2}\alpha$.
4. Sketch the standard normal distribution. Draw a vertical line at each critical value and shade the rejection region(s).

EXAMPLE 7

Finding a Critical Value for a Left-Tailed Test

Find the critical value and rejection region for a left-tailed test with $\alpha = 0.01$.

SOLUTION The graph at the right shows a standard normal curve with a shaded area of 0.01 in the left tail. Using Table 4 in Appendix B, the z-score that corresponds to an area of 0.01 is -2.33. So, the critical value is $z_0 = -2.33$. The rejection region is to the left of this critical value.

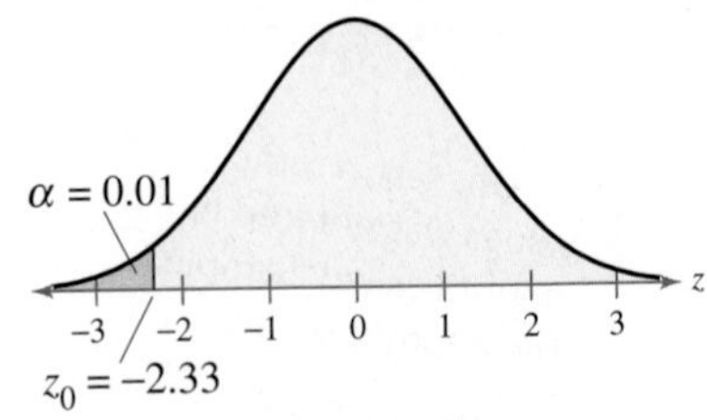

1% Level of Significance

Try It Yourself 7

Find the critical value and rejection region for a left-tailed test with $\alpha = 0.10$.

a. *Draw* a standard normal curve with an area of α in the left tail.
b. *Use* Table 4 in Appendix B to locate the area that is closest to α.
c. *Find* the z-score that corresponds to this area.
d. *Identify* the rejection region.

Answer: Page A39

If you cannot find the exact area in Table 4, use the area that is closest. For instance, in Example 7, the area closest to 0.01 is 0.0099.

EXAMPLE 8

Finding Critical Values for a Two-Tailed Test

Find the critical values and rejection regions for a two-tailed test with $\alpha = 0.05$.

SOLUTION The graph at the right shows a standard normal curve with shaded areas of $\frac{1}{2}\alpha = 0.025$ in each tail. The area to the left of $-z_0$ is $\frac{1}{2}\alpha = 0.025$, and the area to the left of z_0 is $1 - \frac{1}{2}\alpha = 0.975$. Using Table 4, the z-scores that correspond to the areas 0.025 and 0.975 are -1.96 and 1.96, respectively. So, the critical values are $-z_0 = -1.96$ and $z_0 = 1.96$. The rejection regions are to the left of $-z_0$ and to the right of z_0.

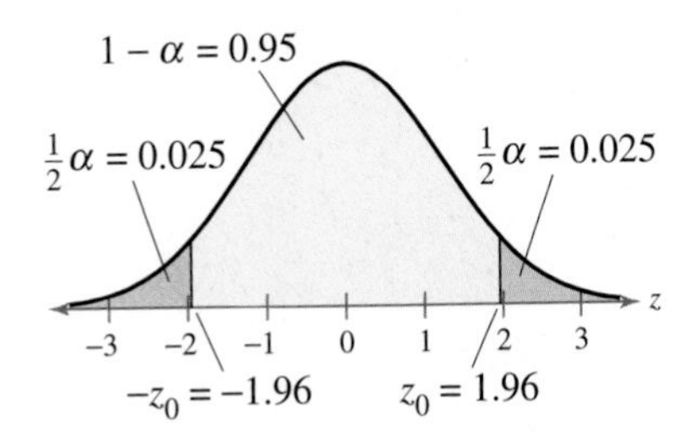

5% Level of Significance

Study Tip

Notice in Example 8 that the critical values are opposites. This is always true for two-tailed z-tests.

The table lists the critical values for commonly used levels of significance.

Alpha	Tail	z
0.10	Left	−1.28
	Right	1.28
	Two	±1.645
0.05	Left	−1.645
	Right	1.645
	Two	±1.96
0.01	Left	−2.33
	Right	2.33
	Two	±2.575

Try It Yourself 8

Find the critical values and rejection regions for a two-tailed test with $\alpha = 0.08$.

a. *Draw* a graph of the standard normal curve with an area of $\frac{1}{2}\alpha$ in each tail.
b. *Use* Table 4 to locate the areas that are closest to $\frac{1}{2}\alpha$ and $1 - \frac{1}{2}\alpha$.
c. *Find* the z-scores that correspond to these areas.
d. *Identify* the rejection regions.

Answer: Page A39

Using Rejection Regions for a *z*-Test

To conclude a hypothesis test using rejection region(s), you make a decision and interpret the decision as follows.

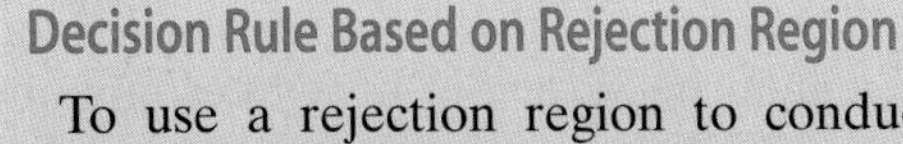

Decision Rule Based on Rejection Region

To use a rejection region to conduct a hypothesis test, calculate the standardized test statistic z. If the standardized test statistic

1. is in the rejection region, then reject H_0.
2. is *not* in the rejection region, then fail to reject H_0.

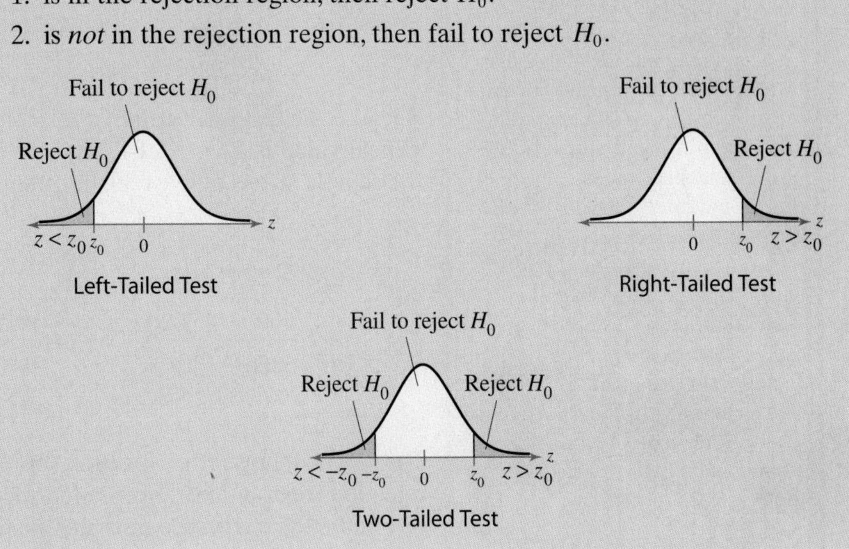

If you fail to reject the null hypothesis, it does not mean that you have accepted the null hypothesis as true. It simply means that there is not enough evidence to reject the null hypothesis.

GUIDELINES

Using Rejection Regions for a *z*-Test for a Mean μ

In Words	*In Symbols*
1. State the claim mathematically and verbally. Identify the null and alternative hypotheses.	State H_0 and H_a.
2. Specify the level of significance.	Identify α.
3. Sketch the sampling distribution.	
4. Determine the critical value(s).	Use Table 4 in Appendix B.
5. Determine the rejection region(s).	
6. Find the standardized test statistic.	$z = \dfrac{\bar{x} - \mu}{\sigma/\sqrt{n}}$ or if $n \geq 30$ use $\sigma \approx s$.
7. Make a decision to reject or fail to reject the null hypothesis.	If z is in the rejection region, reject H_0. Otherwise, do not reject H_0.
8. Interpret the decision in the context of the original claim.	

EXAMPLE 

Testing μ with a Large Sample

See *TI-83* steps on page 379.

Employees in a large accounting firm claim that the mean salary of the firm's accountants is less than that of its competitor's, which is \$45,000. A random sample of 30 of the firm's accountants has a mean salary of \$43,500 with a standard deviation of \$5200. At $\alpha = 0.05$, test the employees' claim.

SOLUTION The claim is "the mean salary is less than \$45,000." So, the null and alternative hypotheses can be written as

$$H_0: \mu \geq \$45{,}000 \quad \text{and} \quad H_a: \mu < \$45{,}000. \text{ (Claim)}$$

Because the test is a left-tailed test and the level of significance is $\alpha = 0.05$, the critical value is $z_0 = -1.645$ and the rejection region is $z < -1.645$. Because the sample size is at least 30, the standardized test statistic for the z-test is

$$z = \frac{\bar{x} - \mu}{\sigma/\sqrt{n}}$$

$$= \frac{43{,}500 - 45{,}000}{5200/\sqrt{30}} \qquad \text{Because } n \geq 30, \text{ use } \sigma \approx s = 5200.$$

$$\approx -1.58.$$

The graph shows the location of the rejection region and the standardized test statistic z. Because z is not in the rejection region, you fail to reject the null hypothesis. In other words, there is not enough evidence at the 5% level of significance to support the employees' claim that the mean salary is less than \$45,000.

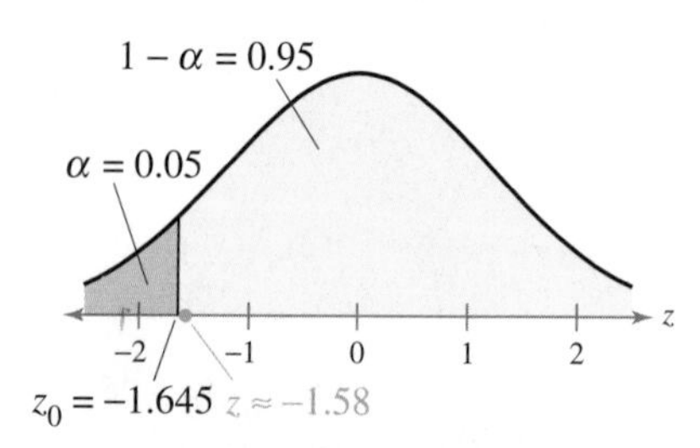

5% Level of Significance

Be sure you understand the decision made in Example 9. Even though your sample has a mean of \$43,500, you cannot (at a 5% level of significance) support the claim that the mean of all the accountants' salaries is less than \$45,000. The difference between your test statistic and the hypothesized mean is probably due to sampling error.

Try It Yourself 9

The CEO of the firm claims that the mean work day of the firm's accountants is less than 8.5 hours. A random sample of 35 of the firm's accountants has a mean work day of 8.2 hours with a standard deviation of 0.5 hours. At $\alpha = 0.01$, test the CEO's claim.

a. *Identify* the claim and state H_0 and H_a.
b. *Identify* the level of significance α.
c. *Find* the critical value z_0 and identify the rejection region.
d. *Find* the standardized test statistic z.
e. Sketch a graph. *Decide* whether to reject the null hypothesis.
f. Is there enough evidence to support the claim that the mean work day is less than 8.5 hours?

Answer: Page A39

Picturing the World

Each year, the Environmental Protection Agency (EPA) publishes reports of gas mileage for all makes and models of passenger vehicles. In a recent year, compact cars with automatic transmissions that posted the best mileage were the Chevrolet Prizm and Toyota Corolla. Each had a mean mileage of 28 mpg (city) and 36 mpg (highway). Suppose that Chevrolet believes a Prizm exceeds 36 mpg on the highway. To support its claim, it tested 34 cars on highway driving and got a sample mean of 37.2 mpg with a standard deviation of 2.3 mpg. *(Source: EPA)*

Is the evidence strong enough to prove that Prizm's highway mpg exceeds the EPA estimate? Use a z-test with $\alpha = 0.01$.

EXAMPLE 10

Testing μ with a Large Sample

The U.S. Department of Agriculture reports that the mean cost of raising a child from birth to age 2 in a rural area is \$8390. You believe this value is incorrect, so you select a random sample of 900 children (age 2) and find that the mean cost is \$8275 with a standard deviation of \$1540. At $\alpha = 0.05$, is there enough evidence to conclude that the mean cost is different from \$8390? *(Adapted from U.S. Department of Agriculture Center for Nutrition Policy and Promotion)*

SOLUTION You want to support the claim that "the mean cost is different from \$8390." So, the null and alternative hypotheses are

$$H_0\text{: } \mu = \$8390$$

and

$$H_a\text{: } \mu \neq \$8390. \text{ (Claim)}$$

Because the test is a two-tailed test and the level of significance is $\alpha = 0.05$, the critical values are $-z_0 = -1.96$ and $z_0 = 1.96$. The rejection regions are $z < -1.96$ and $z > 1.96$. Because $n \geq 30$, the standardized test statistic for the z-test is

$$z = \frac{\overline{x} - \mu}{\sigma/\sqrt{n}}$$

$$= \frac{8275 - 8390}{1540/\sqrt{900}} \qquad \text{Because } n \geq 30, \text{ use } \sigma \approx s = 1540.$$

$$\approx -2.24.$$

The graph shows the location of the rejection regions and the standardized test statistic z. Because z is in the rejection region, you should decide to reject the null hypothesis. In other words, you have enough evidence to conclude that the mean cost of raising a child from birth to age 2 in a rural area is significantly different from \$8390 at the 5% level of significance.

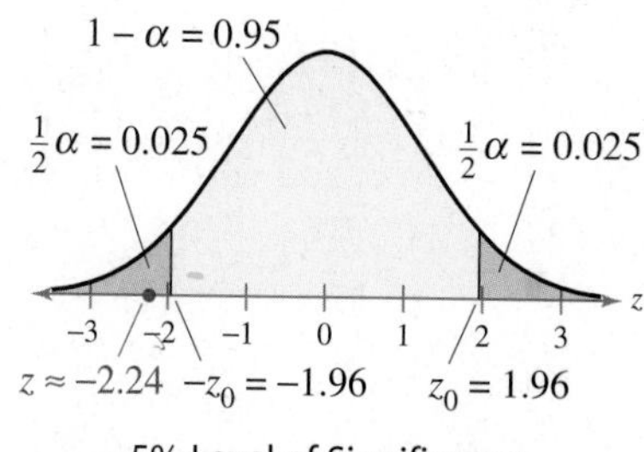

5% Level of Significance

Try It Yourself 10

Using the information and results of Example 10, is there enough evidence to support the claim that the mean cost of raising a child from birth to age 2 in a rural area is different from \$8390 at $\alpha = 0.01$?

a. *Identify* the level of significance α.
b. *Find* the critical values $\pm z_0$ and identify the rejection regions.
c. Sketch a graph. *Decide* whether to reject the null hypothesis.
d. Is there enough evidence to support the claim that the mean cost is significantly different from \$8390 at the 1% level of significance?

Answer: Page A39

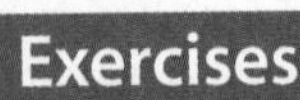

7.2 Exercises

Help

LarsonTutor 7.2

Companion Web Site

Student Solutions Manual 7.2

Videos 7.2

Technology Manuals

Basic Skills and Concepts

In Exercises 1–6, find the P-value for the indicated hypothesis test with the given standardized test statistic z. Decide whether to reject H_0 for the given level of significance α.

1. Left-tailed test, $z = -1.20$, $\alpha = 0.10$

2. Left-tailed test, $z = -1.69$, $\alpha = 0.05$

3. Right-tailed test, $z = 2.34$, $\alpha = 0.01$

4. Right-tailed test, $z = 1.23$, $\alpha = 0.10$

5. Two-tailed test, $z = -1.56$, $\alpha = 0.05$

6. Two-tailed test, $z = 2.30$, $\alpha = 0.01$

Graphical Analysis In Exercises 7–10, match each P-value with the graph that displays its area. The graphs are labeled (a)–(d).

7. $P = 0.0089$

8. $P = 0.0132$

9. $P = 0.0233$

10. $P = 0.0287$

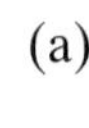

(a)

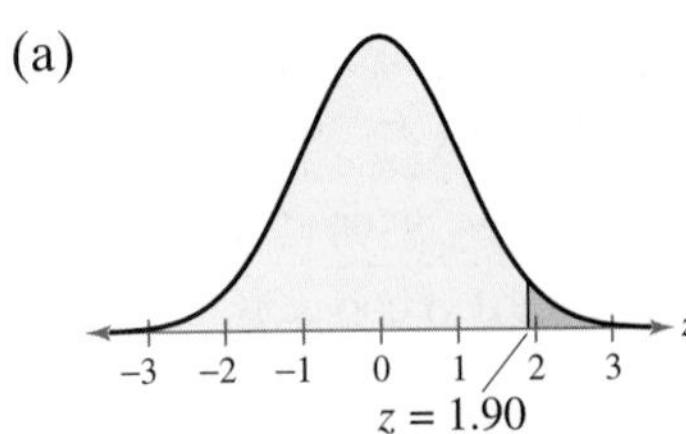

(b)

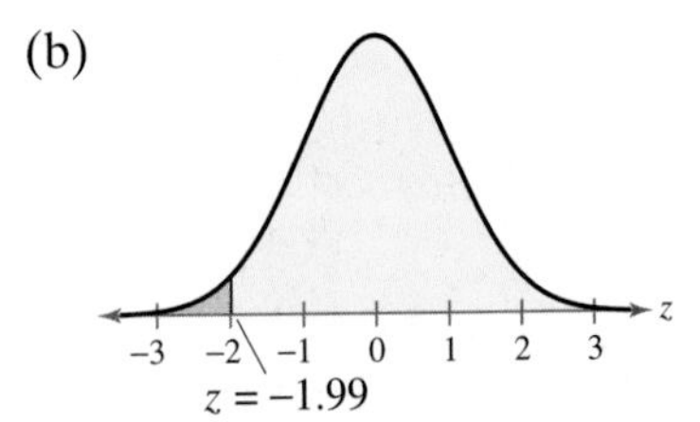

(c)

(d)

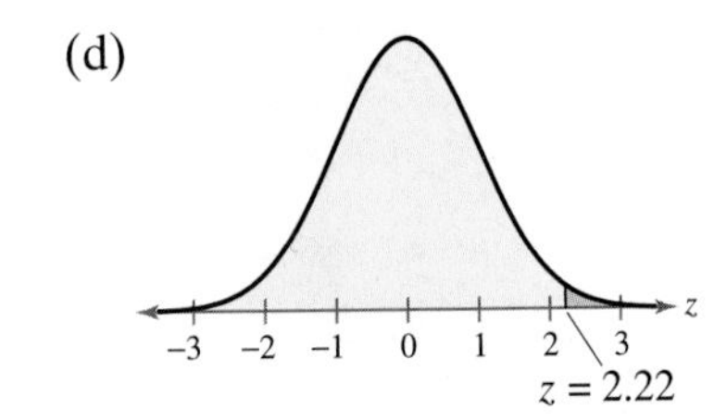

11. Given H_0: $\mu = 100$, H_a: $\mu \neq 100$, and $P = 0.0461$.

(a) Do you reject or fail to reject H_0 at the 0.01 level of significance?

(b) Do you reject or fail to reject H_0 at the 0.05 level of significance?

12. Given H_0: $\mu \geq 8.5$, H_a: $\mu < 8.5$, and $P = 0.0691$.

(a) Do you reject or fail to reject H_0 at the 0.01 level of significance?

(b) Do you reject or fail to reject H_0 at the 0.05 level of significance?

Finding Critical Values In Exercises 13–18, find the critical value(s) for the indicated type of test and level of significance α.

13. Right-tailed test, $\alpha = 0.05$

14. Right-tailed test, $\alpha = 0.08$

15. Left-tailed test, $\alpha = 0.03$

16. Left-tailed test, $\alpha = 0.09$

17. Two-tailed test, $\alpha = 0.02$

18. Two-tailed test, $\alpha = 0.10$

Graphical Analysis In Exercises 19–22,

(a) state whether the graph shows a left-tailed, right-tailed, or two-tailed test.

(b) state whether $\alpha = 0.01$, 0.05, or 0.10.

19.

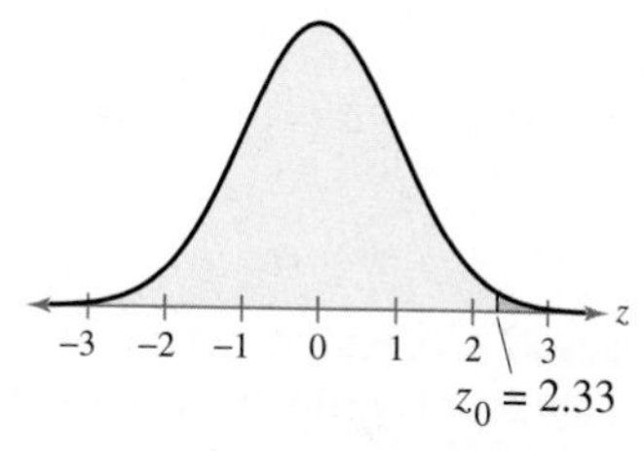

20.

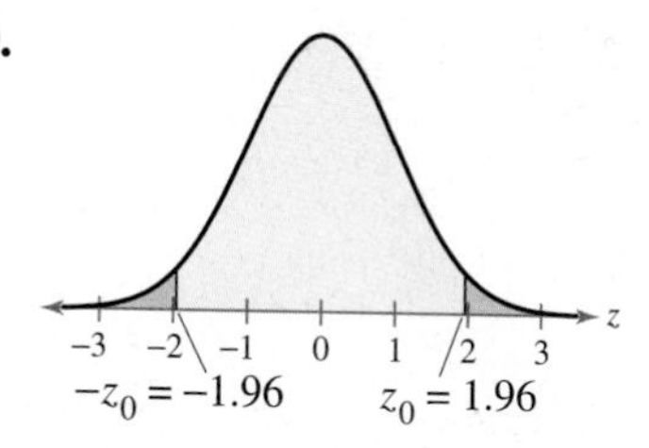

21.

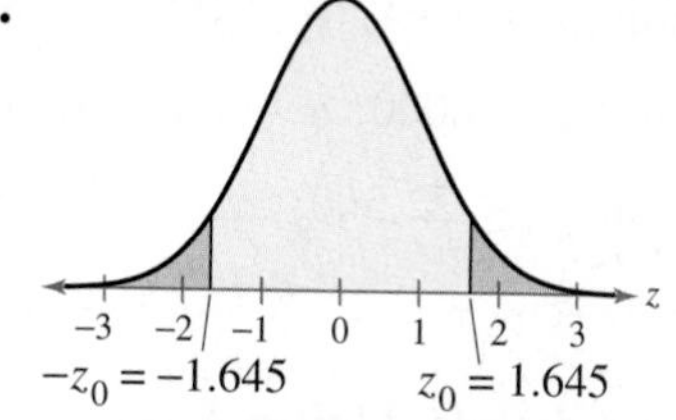

22.

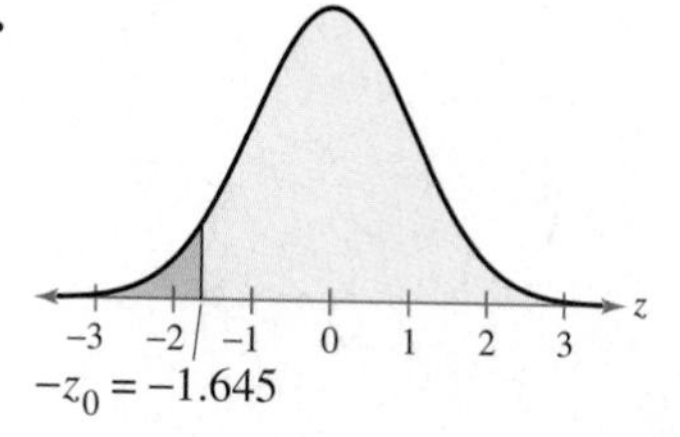

Graphical Analysis In Exercises 23–26, state whether each standardized test statistic z allows you to reject the null hypothesis. Explain your reasoning.

23. (a) $z = 1.631$
(b) $z = 1.723$
(c) $z = -1.464$
(d) $z = -1.655$

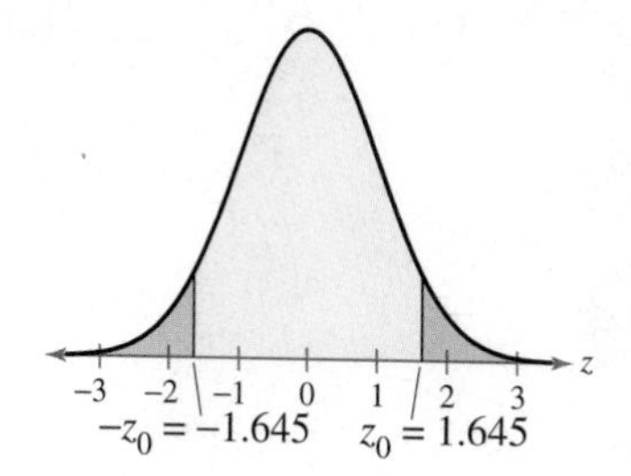

24. (a) $z = 1.98$
(b) $z = -1.89$
(c) $z = 1.65$
(d) $z = -1.99$

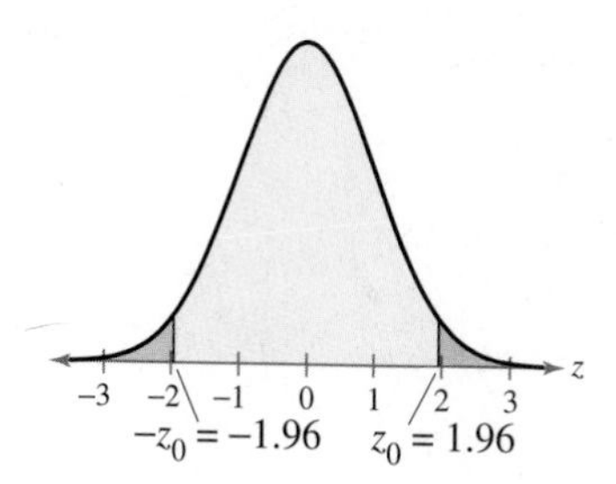

25. (a) $z = -1.301$
(b) $z = 1.203$
(c) $z = 1.280$
(d) $z = 1.286$

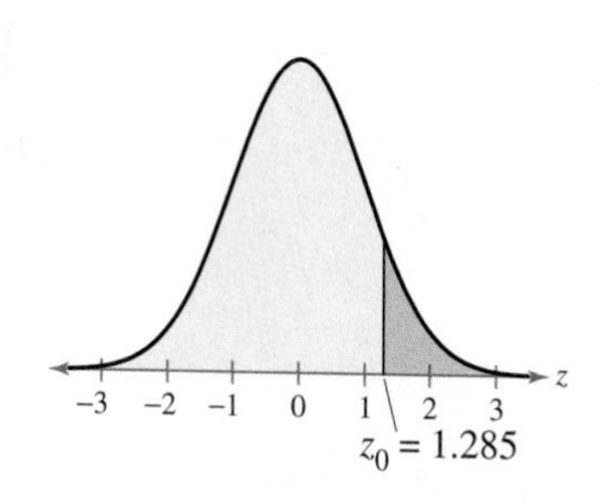

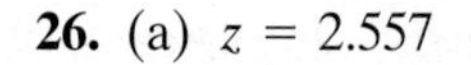

26. (a) $z = 2.557$
(b) $z = -2.755$
(c) $z = 2.585$
(d) $z = -2.475$

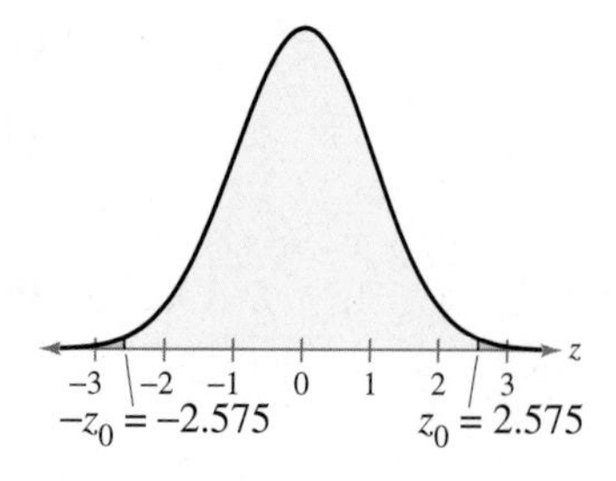

In Exercises 27–30, test the claim about the population mean μ at the given level of significance using the given sample statistics.

27. Claim: $\mu = 40$; $\alpha = 0.05$. Sample statistics: $\bar{x} = 39.2$, $s = 3.23$, $n = 75$

28. Claim: $\mu > 1030$; $\alpha = 0.05$. Sample statistics: $\bar{x} = 1035$, $s = 23$, $n = 50$

29. Claim: $\mu \neq 6000$; $\alpha = 0.01$. Sample statistics: $\bar{x} = 5800$, $s = 350$, $n = 35$

30. Claim: $\mu \leq 22{,}500$; $\alpha = 0.01$. Sample statistics: $\bar{x} = 23{,}250$, $s = 1200$, $n = 45$

Testing Claims Using P-values In Exercises 31–36,

(a) write the claim mathematically and identify H_0 and H_a.
(b) find the standardized test statistic z and its corresponding area.
(c) find the P-value.
(d) decide whether to reject or fail to reject the null hypothesis. Then interpret the decision in the context of the original claim.

31. ***Mathematical Assessment Tests*** ◆ In Illinois, a random sample of 85 eighth-grade students has a mean score of 265 with a standard deviation of 55 on a national mathematical assessment test. This prompts a state school administrator to declare that the mean score for the state's eighth-graders on the examination is more than 260. At $\alpha = 0.04$, is there enough evidence to support the administrator's claim? *(Adapted from National Center for Education Statistics)*

32. ***Automotive Battery Reserves*** ◆ An automotive battery manufacturer guarantees that the mean reserve capacity of a certain battery is greater than 1.5 hours. To test this claim, you randomly select a sample of 50 batteries and find the mean reserve capacity to be 1.55 hours with a standard deviation of 0.32 hour. At $\alpha = 0.10$, do you have enough evidence to support the manufacturer's claim?

33. ***Tea Drinkers*** ◆ A tea drinker's society estimates that the mean consumption of tea by a person in the U.S. is more than 7 gallons per year. In a sample of 100 people, you find that the mean consumption of tea is 7.8 gallons per year with a standard deviation of 2.67 gallons. At $\alpha = 0.07$, can you support the society's claim? *(Adapted from U.S. Department of Agriculture)*

34. ***Tuna Consumption*** ◆ A sample of 60 people shows that the mean tuna consumption by a person in the U.S. is 3.2 pounds per year with a standard deviation of 1.13 pounds. A nutritionist considers this information and claims that the mean tuna consumption by a person in the U.S. is 3.4 pounds per year. At $\alpha = 0.08$, can you reject the nutritionist's claim? *(Adapted from U.S. Department of Agriculture)*

DATA **35. *Quitting Smoking*** ◆ The number of years it took a random sample of 32 former smokers to quit permanently is listed. At $\alpha = 0.05$, test the claim that the mean time it takes smokers to quit smoking permanently is 15 years. *(Adapted from The Gallup Organization)*

15.7 13.2 22.6 13.0 10.7 18.1 14.7 7.0 17.3 7.5 21.8
12.3 19.8 13.8 16.0 15.5 13.1 20.7 15.5 9.8 11.9 16.9
7.0 19.3 13.2 14.6 20.9 15.4 13.3 11.6 10.9 21.6

DATA **36. *Salaries*** ◆ An Alabama politician claims that the mean annual salary for engineering managers in Alabama is at least the national mean, \$69,000. The annual salary (in dollars) for a random sample of 34 engineering managers in Alabama is listed. At $\alpha = 0.03$, test the politician's claim. *(Adapted from America's Career InfoNet)*

65,612 67,610 60,739 76,997 82,977 65,692 81,732 83,302 71,772
82,978 79,608 66,402 67,331 83,160 74,074 55,055 79,496 47,938
65,828 76,414 82,449 71,593 53,018 67,836 46,160 74,877 60,823
71,044 64,214 75,162 64,075 63,056 74,005 57,142

Testing Claims In Exercises 37–42, (a) write the claim mathematically and identify H_0 and H_a, (b) find the critical values and identify the rejection regions, (c) find the standardized test statistic, and (d) decide whether to reject or fail to reject the null hypothesis. Then interpret the decision in the context of the original claim.

37. *Caffeine Content in Colas* ◆ A company that makes cola drinks states that the mean caffeine content per one 12-ounce bottle of cola is 40 milligrams. Suppose you work as a quality control manager and are asked to verify this claim. During your tests, you find that a random sample of thirty 12-ounce bottles of cola has a mean caffeine content of 39.2 milligrams with a standard deviation of 7.5 milligrams. At $\alpha = 0.01$, can you reject the company's claim? *(Adapted from Reader's Digest Eating for Good Health)*

38. *Caffeine Content* ◆ A coffee shop claims that its fresh-brewed drinks have a mean caffeine content of 80 milligrams per five ounces. You work for a city health agency and are asked to test this claim. You find that a random sample of 42 five-ounce servings has a mean caffeine content of 83 milligrams and a standard deviation of 35 milligrams. At $\alpha = 0.05$, do you have enough evidence to reject the shop's claim? *(Adapted from Reader's Digest Eating for Good Health)*

39. *Light Bulbs* ◆ A light bulb manufacturer guarantees that the mean life of a certain type of light bulb is at least 750 hours. If a random sample of 36 light bulbs has a mean life of 745 hours with a standard deviation of 60 hours, do you have enough evidence to reject the manufacturer's claim? Use $\alpha = 0.02$.

40. *Sodium Content in Cereal* ◆ In your work for a national health organization, you are asked to monitor the amount of sodium in a certain brand of cereal. You find that a random sample of 52 cereal servings has a mean sodium content of 232 milligrams with a standard deviation of 10 milligrams. At $\alpha = 0.04$, can you conclude that the mean sodium content per serving of cereal is no more than 230 milligrams?

41. *Nitrogen Dioxide Levels* ◆ A scientist estimates that the mean nitrogen dioxide level in West London is greater than 28 parts per billion. You want to test this estimate. To do so, you determine the nitrogen dioxide levels for 36 randomly selected days. The results (in parts per billion) are listed below. At $\alpha = 0.06$, can you support the scientist's estimate? *(Adapted from National Environmental Technology Centre)*

27 29 53 31 16 47 22 17 13 46 99 15
20 17 28 10 14 9 35 29 32 67 24 31
43 29 12 39 65 94 12 27 13 16 40 62

42. *Weight Loss* ◆ A weight loss program claims that program participants have a mean weight loss of at least 10 pounds after one month. You work for a medical association and are asked to test this claim. A random sample of 30 program participants and their weight losses (in pounds) after one month is listed below. At $\alpha = 0.03$, do you have enough evidence to reject the program's claim?

Weight Loss (in pounds) after One Month

Key: 5|7 = 5.7

Stem	Leaves
5	7 7
6	6 7
7	0 1 9
8	2 2 7 9
9	0 3 5 6 8
10	2 5 6 6
11	1 2 5 7 8
12	0 7 8
13	8
14	
15	0

43. *Electric Usage* ◆ You believe the mean annual kilowatt usage of U.S. residential customers is less than 10,000. You do some research and find that a random sample of 30 residential customers has a mean kilowatt usage of 9900 with a standard deviation of 280. You conduct a statistical experiment where H_0: $\mu \geq 10{,}000$ and H_a: $\mu < 10{,}000$. At $\alpha = 0.01$, explain why you cannot reject H_0. *(Adapted from Edison Electric Institute)*

44. *Using Different Values of α and n* ◆ In Exercise 43, you believe that H_0 is not valid. Which of the following allows you to reject H_0?

(a) Use the same values but increase α from 0.01 to 0.02.

(b) Use the same values but increase α from 0.01 to 0.03.

(c) Use the same values but increase n from 30 to 50.

(d) Use the same values but increase n from 30 to 100.

Extending the Basics

45. *Writing* ◆ Explain the difference between the classical z-test for μ and the z-test for μ using a P-value.

7 **Case Study** WWW.AMSTAT.ORG/PUBLICATIONS/JSE

Human Body Temperature: What's Normal?

In an article in the *Journal of Statistics Education* (vol. 4, no. 2, 1996), Allen Shoemaker describes a study that was reported in the *Journal of the American Medical Association*.* It is generally accepted that the mean body temperature of adult humans is 98.6°F. In his article, Shoemaker uses the data from the JAMA article to test this hypothesis. Here is a summary of his test.

Claim: The body temperature of adults is 98.6°F.

H_0: $\mu = 98.6$°F (Claim) H_a: $\mu \neq 98.6$°F

Sample Size: $n = 130$

Population: Adult human temperatures (Fahrenheit)

Distribution: Approximately normal

Test Statistics: $\bar{x} = 98.25$, $s = 0.73$

* Data for the JAMA article was collected from healthy men and women, aged 18 to 40, at the University of Maryland Center for Vaccine Development, Baltimore.

Men's Temperatures

Stem	Leaves
96	3
96	7 9
97	0 1 1 1 2 3 4 4 4 4
97	5 5 6 6 6 7 8 8 8 8 9 9
98	0 0 0 0 0 0 1 1 2 2 2 2 3 3 4 4 4 4
98	5 5 6 6 6 6 6 6 7 7 8 8 8 9
99	0 0 0 1 2 3 4
99	5
100	
100	

Key: 96|3 = 96.3

Women's Temperatures

Stem	Leaves
96	4
96	7 8
97	2 2 4
97	6 7 7 8 8 8 9 9 9
98	0 0 0 0 0 1 2 2 2 2 2 2 3 3 3 4 4 4 4 4
98	5 6 6 6 6 7 7 7 7 7 7 8 8 8 8 8 8 8 9
99	0 0 1 1 2 2 3 4
99	9
100	0
100	8

Key: 96|4 = 96.4

Exercises

1. Complete the hypothesis test for all adults (men and women) by performing the following steps. Use a level of significance of $\alpha = 0.05$.
   (a) Sketch the sampling distribution.
   (b) Determine the critical values and add them to your sketch.
   (c) Determine the rejection regions and shade them in your sketch.
   (d) Find the standardized test statistic. Add it to your sketch.
   (e) Make a decision to reject or fail to reject the null hypothesis.
   (f) Interpret the decision in the context of the original claim.
2. If you lower the level of significance to $\alpha = 0.01$, does your decision change? Explain your reasoning.
3. Test the hypothesis that the mean temperature of men is 98.6°F. What can you conclude at a level of significance of $\alpha = 0.01$?
4. Test the hypothesis that the mean temperature of women is 98.6°F. What can you conclude at a level of significance of $\alpha = 0.01$?
5. Use the sample of 130 temperatures to form a 99% confidence interval for the mean body temperature of adult humans.
6. The conventional "normal" body temperature was established by Carl Wunderlich over 100 years ago. What, in Wunderlich's sampling procedure, do you think might have led him to an incorrect conclusion?

7.3 Hypothesis Testing for the Mean (Small Samples)

What You Should Learn

- How to find critical values in a *t*-distribution
- How to use the *t*-test to test a mean μ
- How to use technology to find *P*-values and use them with a *t*-test to test a mean μ

Critical Values in a *t*-Distribution • The *t*-Test for a Mean μ ($n < 30$, σ unknown) • Using *P*-values with *t*-Tests

Critical Values in a *t*-Distribution

In real life, it is often not practical to collect samples of size 30 or more. However, if the population has a normal, or nearly normal, distribution, you can still test the population mean μ. To do so, you can use the *t*-sampling distribution with $n - 1$ degrees of freedom.

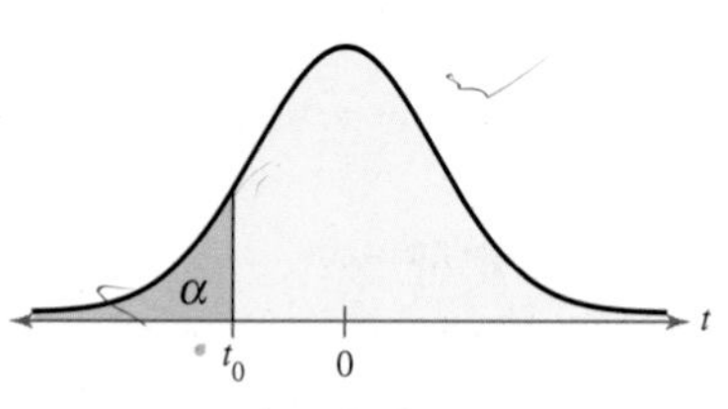

Left-Tailed Test

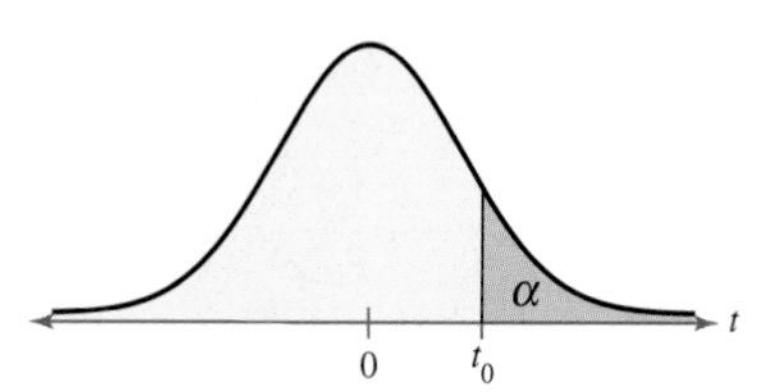

Right-Tailed Test

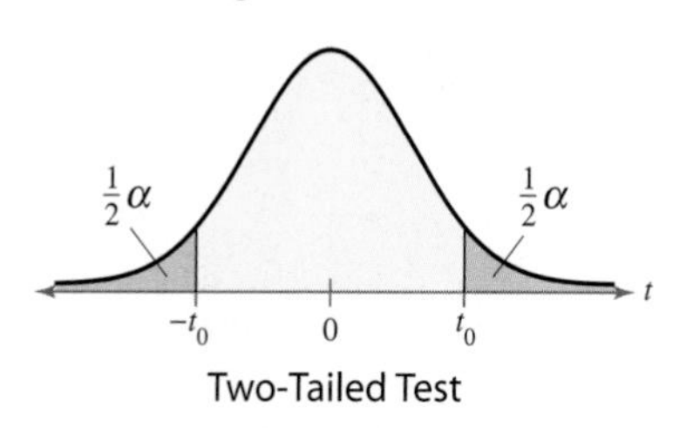

Two-Tailed Test

GUIDELINES

Finding Critical Values in a *t*-Distribution

1. Identify the level of significance α.
2. Identify the degrees of freedom d.f. $= n - 1$.
3. Find the critical value(s) using Table 5 in Appendix B in the row with $n - 1$ degrees of freedom. If the hypothesis test is
   a. *left tailed,* use "One Tail, α" column with a negative sign.
   b. *right tailed,* use "One Tail, α" column with a positive sign.
   c. *two tailed,* use "Two Tails, α" column with a negative and a positive sign.

EXAMPLE 1

Finding Critical Values for t

Find the critical value t_0 for a left-tailed test given $\alpha = 0.05$ and $n = 21$.

SOLUTION The degrees of freedom are

$$\text{d.f.} = n - 1 = 21 - 1 = 20.$$

To find the critical value, use Table 5 with d.f. $= 20$ and 0.05 in the "One Tail, α" column. Because the test is a left-tailed test, the critical value is negative. So,

$$t_0 = -1.725.$$

$\alpha = 0.05$

$t_0 = -1.725$

Try It Yourself 1

Find the critical value t_0 for a left-tailed test with $\alpha = 0.01$ and $n = 14$.

a. *Find* the *t*-value in Table 5. Use d.f. $= 13$ and $\alpha = 0.01$ in the "One Tail, α" column.
b. *Use* a negative sign.

Answer: Page A39

EXAMPLE 2

Finding Critical Values for t

Find the critical value t_0 for a right-tailed test with $\alpha = 0.01$ and $n = 17$.

SOLUTION The degrees of freedom are

$$\begin{aligned} \text{d.f.} &= n - 1 \\ &= 17 - 1 \\ &= 16. \end{aligned}$$

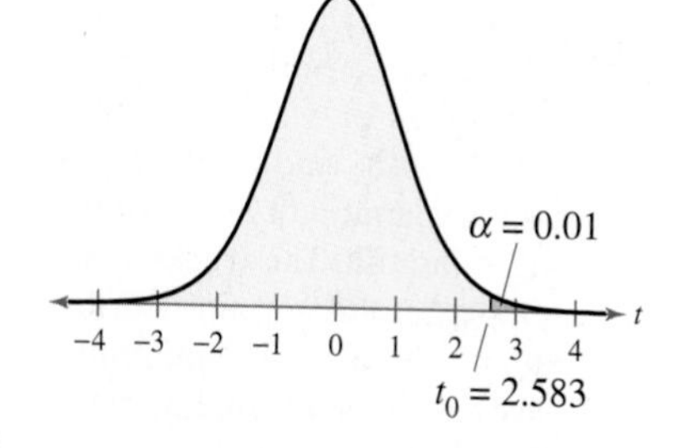

To find the critical value, use Table 5 with d.f. = 16 and $\alpha = 0.01$ in the "One tail, α" column. Because the test is right tailed, the critical value is positive. So,

$$t_0 = 2.583.$$

Try It Yourself 2

Find the critical value t_0 for a right-tailed test with $\alpha = 0.05$ and $n = 9$.

a. *Find* the t-value in Table 5 using d.f. = 8 and $\alpha = 0.05$ in the "One tail, α" column.

b. *Use* a positive sign.

Answer: Page A39

EXAMPLE 3

Finding Critical Values for t

Find the critical values t_0 and $-t_0$ for a two-tailed test with $\alpha = 0.05$ and $n = 26$.

SOLUTION The degrees of freedom are

$$\begin{aligned} \text{d.f.} &= n - 1 \\ &= 26 - 1 \\ &= 25. \end{aligned}$$

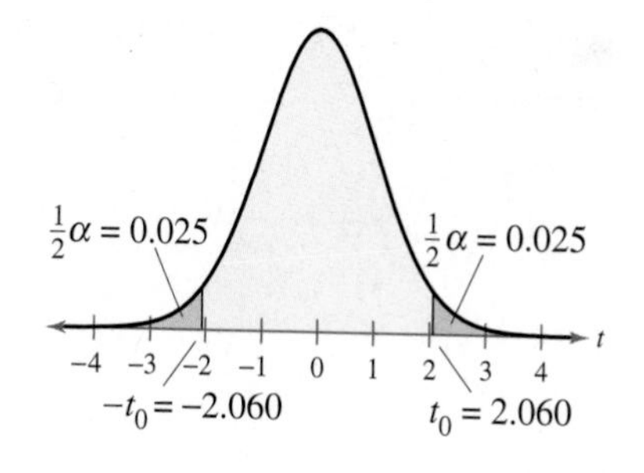

To find the critical value, use Table 5 with d.f. = 25 and $\alpha = 0.05$ in the "Two tail, α" column. Because the test is two tailed, one critical value is negative and one is positive. So,

$$-t_0 = -2.060 \text{ and } t_0 = 2.060.$$

Try It Yourself 3

Find the critical values $\pm t_0$ for a two-tailed test with $\alpha = 0.01$ and $n = 16$.

a. *Find* the t-value in Table 5 using d.f. = 15 and 0.01 in the "Two tail, α" column.

b. *Use* a negative and a positive sign.

Answer: Page A39

The t-Test for a Mean μ ($n < 30$, σ unknown)

To test a claim about a mean μ using a small sample ($n < 30$) from a normal, or nearly normal, distribution and σ is unknown, you can use a t-sampling distribution.

$$t = \frac{(\text{Sample mean}) - (\text{Hypothesized mean})}{\text{Standard error}}$$

t-Test for a Mean μ

The ***t*-test for the mean** is a statistical test for a population mean. The t-test can be used when

- the population is normal or nearly normal
- σ is unknown and

$n < 30$.

The **test statistic** is $\bar{x}$ and the **standardized test statistic** is t.

$$t = \frac{\bar{x} - \mu}{s/\sqrt{n}}$$

The degrees of freedom are d.f. $= n - 1$.

Picturing the World

Using a t-test, a decision was made whether to send truckloads of waste contaminated with cadmium to a sanitary landfill or a hazardous waste landfill. The trucks were sampled to determine if the mean level of cadmium exceeds the allowable amount of 1 milligram per liter for a sanitary landfill. In the study, the null hypothesis was $\mu \leq 1$. *(Source: Pacific Northwest National Laboratory)*

	H_0 True	H_0 False
Fail to reject H_0		
Reject H_0		

Describe the possible Type I and Type II errors of this situation.

GUIDELINES

Using the t-Test for a Mean μ (Small Sample)

In Words	*In Symbols*
1. State the claim mathematically and verbally. Identify the null and alternative hypotheses.	State H_0 and H_a.
2. Specify the level of significance.	Identify α.
3. Identify the degrees of freedom and sketch the sampling distribution.	d.f. $= n - 1$
4. Determine any critical values.	Use Table 5 in Appendix B.
5. Determine any rejection regions.	
6. Find the standardized test statistic.	$t = \frac{\bar{x} - \mu}{s/\sqrt{n}}$
7. Make a decision to reject or fail to reject the null hypothesis.	If t is in the rejection region, reject H_0. Otherwise, fail to reject H_0.
8. Interpret the decision in the context of the original claim.	

Remember that when you make a decision, the possibility of a type I or a type II error exists.

Note that if you prefer using P-values, turn to page 355 to learn how to use P-values for a t-test for a mean μ (small sample).

EXAMPLE 

Testing μ with a Small Sample

See *Minitab* steps on page 378.

A used car dealer says that the mean price of a 1999 Ford F-150 Super Cab is at least \$16,500. You suspect this claim is incorrect and find that a random sample of 14 similar vehicles has a mean price of \$15,700 and a standard deviation of \$1250. Is there enough evidence to reject the dealer's claim at $\alpha = 0.05$? Assume the population is normally distributed. *(Adapted from Kelly Blue Book)*

SOLUTION The claim is "the mean price is at least \$16,500." So, the null and alternative hypotheses are

H_0: $\mu \geq \$16{,}500$ (Claim)

and

H_a: $\mu < \$16{,}500$.

The test is a left-tailed test, the level of significance is $\alpha = 0.05$, and there are d.f. = 14 − 1 = 13 degrees of freedom. So, the critical value is $t_0 = -1.771$. The rejection region is $t < -1.771$. Using the t-test, the standardized test statistic is

$$t = \frac{\bar{x} - \mu}{s/\sqrt{n}}$$

$$= \frac{15{,}700 - 16{,}500}{1250/\sqrt{14}}$$

$$\approx -2.39.$$

The graph shows the location of the rejection region and the standardized test statistic t. Because t is in the rejection region, you should decide to reject the null hypothesis. There is enough evidence at the 5% level of significance to reject the claim that the mean price of a 1999 Ford F-150 Super Cab is at least \$16,500.

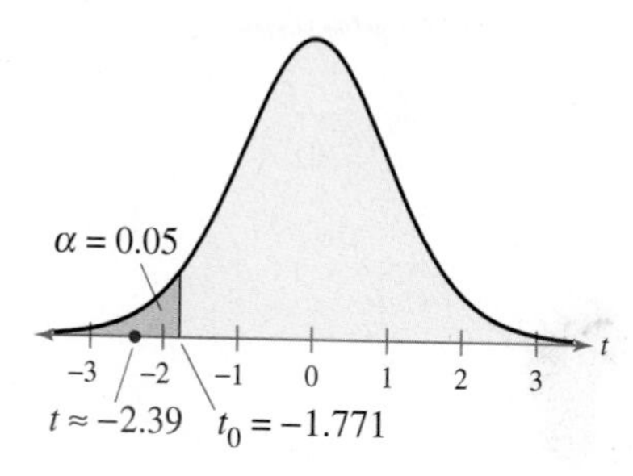

Try It Yourself 4

An insurance agent says that the mean cost of insuring a 1999 Ford F-150 Super Cab is at least \$875. A random sample of nine similar insurance quotes has a mean cost of \$825 and a standard deviation of \$62. Is there enough evidence to reject the agent's claim at $\alpha = 0.01$? Assume the population is normally distributed.

a. *Identify* the claim and state H_0 and H_a.
b. *Identify* the level of significance α and the degrees of freedom d.f.
c. *Find* the critical value t_0 and identify the rejection region.
d. *Use* the t-test to find the standardized test statistic t.
e. Sketch a graph. *Decide* whether to reject the null hypothesis.
f. Is there enough evidence to reject the claim that the mean cost of insuring a 1999 Ford F-150 Super Cab is at least \$875?

Answer: Page A39

EXAMPLE 5

See *TI-83* steps on page 379.

Testing μ with a Small Sample

An industrial company claims that the mean pH level of the water in a nearby river is 6.8. You randomly select 19 water samples and measure the pH of each. The sample mean and standard deviation are 6.7 and 0.24, respectively. Is there enough evidence to reject the company's claim at $\alpha = 0.05$? Assume the population is normally distributed.

SOLUTION The claim is "the mean pH level is 6.8." So, the null and alternative hypotheses are

$$H_0\text{: } \mu = 6.8 \text{ (Claim)}$$

and

$$H_a\text{: } \mu \neq 6.8.$$

The test is a two-tailed test, the level of significance is $\alpha = 0.05$, and there are d.f. $= 19 - 1 = 18$ degrees of freedom. So, the critical values are $-t_0 = -2.101$ and $t_0 = 2.101$. The rejection regions are $t < -2.101$ and $t > 2.101$. Using the t-test, the standardized test statistic is

$$\begin{aligned} t &= \frac{\overline{x} - \mu}{s/\sqrt{n}} \\ &= \frac{6.7 - 6.8}{0.24/\sqrt{19}} \\ &\approx -1.82. \end{aligned}$$

The graph shows the location of the rejection region and the standardized test statistic t. Because t is not in the rejection region, you should decide not to reject the null hypothesis. There is not enough evidence at the 5% level of significance to reject the claim that the mean pH is 6.8.

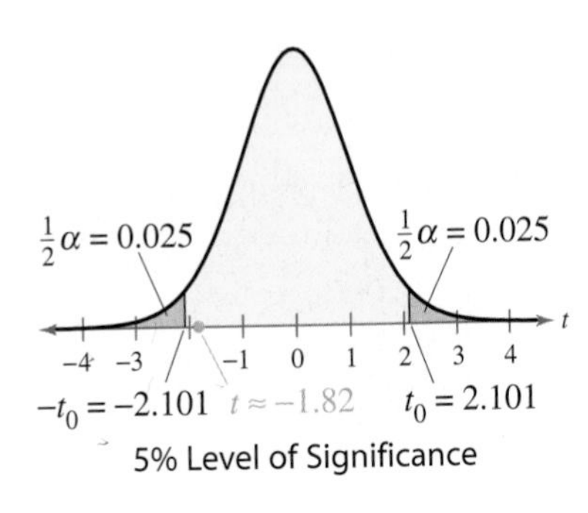

5% Level of Significance

Try It Yourself 5

The company also claims that the mean conductivity of the river is 1890 mg/L. The conductivity of a water sample is a measure of the total dissolved solids in the sample. You randomly select 19 water samples and measure the conductivity of each. The sample mean and standard deviation are 2500 mg/L and 700 mg/L, respectively. Is there enough evidence to reject the company's claim at $\alpha = 0.01$? Assume the population is normally distributed.

a. *Identify* the claim and state H_0 and H_a.
b. *Identify* the level of significance α and the degrees of freedom d.f.
c. *Find* the critical values $\pm t_0$ and identify the rejection regions.
d. *Use* the t-test to find the standardized test statistic t.
e. Sketch a graph. *Decide* whether to reject the null hypothesis.
f. Is there enough evidence to reject the company's claim?

Answer: Page A39

Using P-values with t-Tests

Suppose you wanted to find a P-value given $t = 1.98$, 15 degrees of freedom, and a right-tailed test. Using Table 5 in Appendix B, you can determine that P falls between $\alpha = 0.025$ and $\alpha = 0.05$, but you cannot determine an exact value for P. In such cases, you can use technology to perform a hypothesis test and find exact P-values.

EXAMPLE 

Using P-values with a t-Test

The American Automobile Association claims that the mean daily meal cost for a family of four traveling on vacation in California is \$132. A random sample of 11 such families has a mean daily meal cost of \$141 with a standard deviation of \$20. Is there enough evidence to reject the claim at $\alpha = 0.10$? Assume the population is normally distributed.

SOLUTION The *TI-83* display at the far left shows how to set up the hypothesis test. The two displays on the right show the possible results, depending on whether you select "CALCULATE" or "DRAW."

TI-83

T–Test
Inpt:Data **Stats**
μ_0:132
$\bar{x}$:141
Sx:20
n:11
μ: **$\neq\mu_0$** $<\mu_0$ $>\mu_0$
Calculate Draw

TI-83

T–Test
$\mu \neq 132$
t=1.492481156
p=.1664335286
$\bar{x}$=141
Sx=20
n=11

TI-83

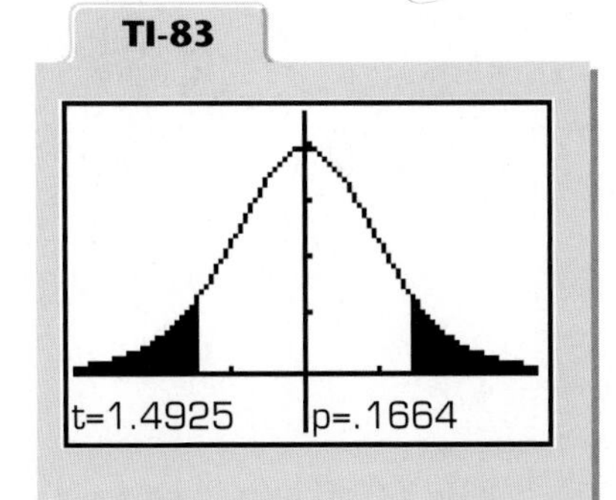

From the displays you can see that $P \approx 0.1664$. Because $P > 0.10$, there is not enough evidence to reject the claim at the 10% level of significance. In other words, you should not reject the null hypothesis.

Try It Yourself 6

The American Automobile Association claims that the mean nightly lodging rate for a family of four traveling on vacation in California is at least \$136. A random sample of six such families has a mean nightly lodging rate of \$126 with a standard deviation of \$12. Is there enough evidence to reject the claim at $\alpha = 0.05$? Assume the population is normally distributed.

a. *Use* a *TI-83* to find the P-value.
b. *Compare* the P-value to the level of significance α.
c. *Make* a decision.
d. Is there enough evidence to reject the claim?

Answer: Page A39

Study Tip

Note that the *TI-83* display on the far right in Example 6 also displays the standardized test statistic $t \approx 1.4925$.

To use *Minitab* to perform a one-sample t-test, you must have the raw data.

7.3 Exercises

LarsonTutor 7.3

Companion Web Site

Student Solutions Manual 7.3

Videos 7.3

Technology Manuals

Basic Skills and Concepts

1. Explain how to find critical values for a t sampling distribution.

2. Explain how to use a t-test to test a hypothesized mean μ given a small sample $(n < 30)$. What assumption about the population is necessary?

In Exercises 3–14, find the critical value(s) for the indicated t-test, level of significance α, and sample size n.

3. Right-tailed test, $\alpha = 0.05$, $n = 23$
4. Right-tailed test, $\alpha = 0.01$, $n = 11$
5. Left-tailed test, $\alpha = 0.025$, $n = 19$
6. Left-tailed test, $\alpha = 0.05$, $n = 14$
7. Two-tailed test, $\alpha = 0.01$, $n = 27$
8. Two-tailed test, $\alpha = 0.05$, $n = 10$
9. Right-tailed test, $\alpha = 0.10$, $n = 20$
10. Right-tailed test, $\alpha = 0.05$, $n = 15$
11. Left-tailed test, $\alpha = 0.01$, $n = 28$
12. Left-tailed test, $\alpha = 0.005$, $n = 12$
13. Two-tailed test, $\alpha = 0.02$, $n = 5$
14. Two-tailed test, $\alpha = 0.10$, $n = 22$

Graphical Analysis In Exercises 15–18, state whether the standardized test statistic t indicates that you should reject the null hypothesis. Explain.

15. (a) $t = 2.091$
(b) $t = 0$
(c) $t = -1.08$
(d) $t = -2.096$

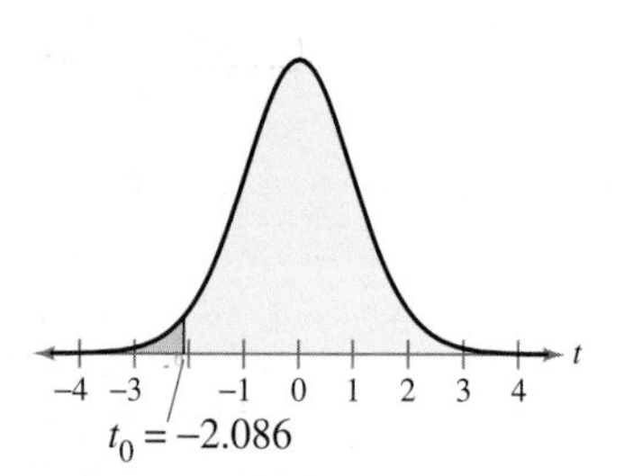

16. (a) $t = 1.308$
(b) $t = -1.389$
(c) $t = 1.650$
(d) $t = -0.998$

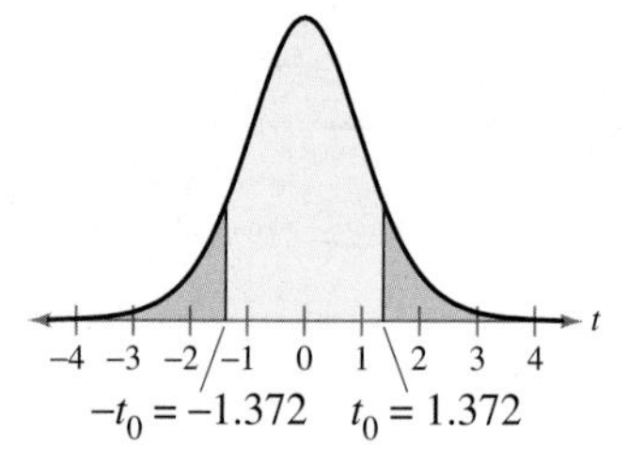

17. (a) $t = -2.502$
(b) $t = 2.203$
(c) $t = 2.680$
(d) $t = -2.703$

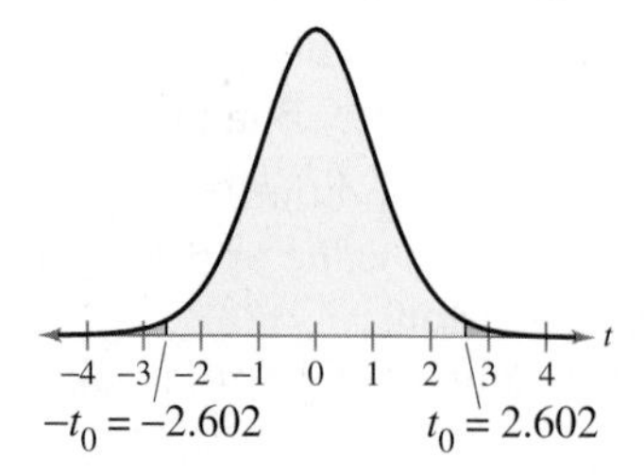

18. (a) $t = 1.705$
(b) $t = -1.755$
(c) $t = -1.585$
(d) $t = 1.745$

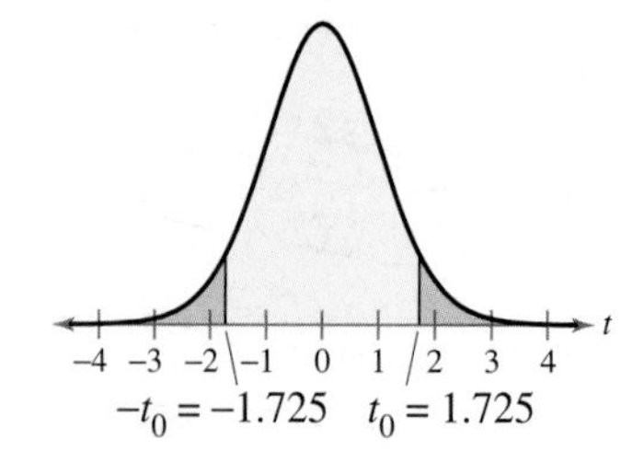

In Exercises 19–22, use a t-test to test the claim about the population mean μ for the given level of significance α and sample statistics. For each claim, assume the distribution is approximately normal.

19. Claim: $\mu = 15$; $\alpha = 0.01$. Sample statistics: $\bar{x} = 13.9$, $s = 3.23$, $n = 6$

20. Claim: $\mu > 25$; $\alpha = 0.05$. Sample statistics: $\bar{x} = 26.2$, $s = 2.32$, $n = 17$

21. Claim: $\mu \geq 8000$; $\alpha = 0.01$. Sample statistics: $\bar{x} = 7700$, $s = 450$, $n = 25$

22. Claim: $\mu \neq 52{,}200$; $\alpha = 0.05$. Sample statistics: $\bar{x} = 53{,}220$, $s = 1200$, $n = 4$

Testing Claims In Exercises 23–28,

(a) write the claim mathematically and identify H_0 and H_a.

(b) find the critical values and identify the rejection regions.

(c) find the standardized test statistic.

(d) decide whether to reject or fail to reject the null hypothesis. Then interpret the decision in the context of the original claim.

For each claim, assume the distribution of the population is approximately normal.

23. ***Microwave Repair Costs*** ◆ A microwave oven repairer says that the mean repair cost for damaged microwave ovens is less than \$100. You work for the repairer and want to test this claim. You find that a random sample of five microwave ovens has a mean repair cost of \$75 and a standard deviation of \$12.50. At $\alpha = 0.01$, do you have enough evidence to support the repairer's claim? *(Adapted from Consumer Reports)*

24. ***Computer Repair Costs*** ◆ A computer repairer believes that the mean repair cost for damaged computers is more than \$95. To test this claim, you determine the repair costs for seven randomly selected computers and find that the mean repair cost is \$100 per computer with a standard deviation of \$42.50. At $\alpha = 0.01$, do you have enough evidence to support the repairer's claim? *(Adapted from Consumer Reports)*

25. ***Waste Recycled*** ◆ An environmentalist estimates that the mean waste recycled by adults in the U.S. is more than 1 pound per person per day. You want to test this claim. You find that the mean waste recycled per person per day for a random sample of 12 adults in the U.S. is 1.2 pounds and the standard deviation is 0.3 pound. At $\alpha = 0.05$, can you support the claim? *(Adapted from U.S. Environmental Protection Agency)*

26. ***Waste Generated*** ◆ As part of your work for an environmental awareness group, you want to test a claim that the mean waste generated by adults in the U.S. is more than 4 pounds per day. In a random sample of 10 adults in the U.S., you find that the mean waste generated per person per day is 4.3 pounds with a standard deviation of 1.2 pounds. At $\alpha = 0.05$, can you support the claim? *(Adapted from U.S. Environmental Protection Agency)*

27. ***Annual Pay*** ◆ An employment information service claims that the mean annual pay for full-time male workers over age 25 and without high school diplomas is \$24,600. The annual pay for a random sample of 10 full-time male workers without high school diplomas is listed. At $\alpha = 0.05$, test the claim that the mean salary is \$24,600. *(Adapted from U.S. Bureau of the Census)*

22,954 23,438 24,655 23,695 25,275
19,212 21,456 25,493 26,480 28,585

28. *Annual Pay* ◆ An employment information service says the mean annual pay for full-time female workers over age 25 and without high school diplomas is $17,100. The annual pay for a random sample of 12 full-time female workers without high school diplomas is listed. At $\alpha = 0.05$, test the claim that the mean salary is $17,100. *(Adapted from U.S. Bureau of the Census)*

16,009 16,790 17,328 18,161 16,631 21,028
16,114 17,176 17,503 19,764 15,316 18,801

Testing Claims Using P-values In Exercises 29–34, (a) write the claim mathematically and identify H_0 and H_a, (b) use technology to find the *P*-value, and (c) decide whether to reject or fail to reject the null hypothesis. Then interpret the decision in the context of the original claim. Assume the population is normally distributed.

DATA **29. *Soda Consumption*** ◆ As part of your study on the food consumption habits of teenage males, you randomly select 20 teenage males and ask each how many 12-ounce servings of soda they drink per day. The results are listed below. At $\alpha = 0.05$, is there enough evidence to claim that teenage males drink less than 3.0 twelve-ounce servings of soda per day? *(Adapted from Center for Science in the Public Interest)*

3.3 2.1 2.5 2.1 3.4 3.3 4.4 3.4 2.5 3.2
2.5 3.8 2.0 2.9 1.9 1.3 1.9 2.8 4.2 2.2

DATA **30. *Water*** ◆ A bottled water association says that the mean number of 8-ounce glasses of water U.S. adults drink each day is less than 5.0. The number of 8-ounce glasses of water a random sample of 24 U.S. adults drank in one day is listed. At $\alpha = 0.05$, is there enough evidence to support the association's claim? *(Adapted from USA TODAY)*

3.6 4.5 5.9 5.3 3.1 4.1 3.9 4.3 4.5 3.6 2.5 5.2
4.7 5.6 3.0 2.6 4.0 2.7 3.9 6.5 5.4 3.1 6.0 2.9

31. *Class Size* ◆ You receive a brochure from a large university. The brochure indicates that the mean class size for full-time faculty is less than 32. You want to test this claim. You randomly select 18 classes taught by full-time faculty and determine the class size of each. The results are listed below. At $\alpha = 0.01$, can you support the university's claim? *(Adapted from National Center for Education Statistics)*

35 28 29 33 32 40 26 25 29 28 30 36 33 29 27 30 28 25

32. *Faculty Classroom Hours* ◆ The dean of a university estimates that the mean number of classroom hours per week for full-time faculty is 11.0. As a member of the student council, you want to test this claim. A random sample of the number of classroom hours for full-time faculty for one week is listed below. At $\alpha = 0.01$, can you reject the dean's claim? *(Adapted from National Center for Education Statistics)*

11.8 8.6 12.6 7.9 6.4 10.4 13.6 9.1

33. *Eating Out* ◆ A restaurant association says the typical household in the U.S. spends a mean of $2116 per year on food away from home. You are a consumer reporter for a national publication and want to test this claim. You randomly select 12 U.S. households and find out how much each spent on food away from home per year. Can you reject the restaurant association's claim at $\alpha = 0.02$? *(Adapted from the National Restaurant Association)*

3474 1262 853 1988 2778 2283 2797 1916 3415 2645 2726 2948

34. ***Lodging Costs*** ◆ A travel association says the daily lodging costs for a family in the U.S. is \$113. You work for a tourist publication and want to test this claim. You randomly select 10 U.S. families and find out how much each spent on lodging for one overnight trip. At $\alpha = 0.02$, can you reject the travel association's claim? *(Adapted from the American Automobile Association)*

145 118 123 136 100 85 55 185 129 162

35. ***Credit Card Balances*** ◆ To test the claim that the mean credit card balance among families which have a balance is greater than \$1500, you do some research and find that a random sample of six cardholders has a mean credit card balance of \$1700 with a standard deviation of \$325. You conduct a statistical experiment where H_0: $\mu \leq \$1500$ and H_a: $\mu > \$1500$. At $\alpha = 0.01$, explain why you cannot reject H_0. Assume the population is normally distributed. *(Adapted from Board of Governors of the Federal Reserve System)*

36. ***Using Different Values of*** α ***and n*** ◆ In Exercise 35, you believe that H_0 is not valid. Which of the following allows you to reject H_0? (a) use the same values but increase α from 0.01 to 0.05, (b) use the same values but increase α from 0.01 to 0.10, (c) use the same values but increase n from 6 to 12, and (d) use the same values but increase n from 6 to 24.

Extending the Basics

Deciding on a Distribution In Exercises 37 and 38, decide whether you should use a normal sampling distribution or a *t*-sampling distribution to perform the hypothesis test. Justify your decision. Then use the distribution to test the claim. Write a short paragraph about the results of the test and what you can conclude about the claim.

37. ***Gas Mileage*** ◆ A car company says that the mean gas mileage for its luxury sedan is at least 21 miles per gallon (mpg). You believe the claim is incorrect and find that a random sample of 5 cars has a mean gas mileage of 19 mpg and a standard deviation of 4 mpg. Assume the gas mileage of all of the company's luxury sedans is normally distributed. At $\alpha = 0.05$, test the company's claim. *(Adapted from Consumer Reports)*

38. ***Faculty Time*** ◆ An administrator of a state university system says that the mean time full-time faculty spend with students outside of class is 337 hours per week. (The hours were counted for each student, so that if a professor spent 5 hours with a group of 4 students it would count as 20 hours.) The mean time a random sample of 50 full-time faculty members spent with students outside of class was 332 hours per week and the standard deviation was 10 hours per week. At $\alpha = 0.01$, test the claim that the mean time is 337 hours per week. *(Adapted from National Center for Education Statistics)*

39. Repeat Exercise 37, but assume the population standard deviation is $\sigma = 5$ miles per gallon. Compare the results.

40. Repeat Exercise 38, but assume the sample size is 5 and the amount of time is normally distributed. Compare the results.

41. ***Is a Test Necessary?*** ◆ In Exercise 23, suppose the repairer's claim was that the mean repair cost was at least \$50. Assuming the same sample statistics, is it necessary to use a hypothesis test to test the repairer's claim? Why or why not?

7.4 Hypothesis Testing for Proportions

Hypothesis Test for Proportions

What You Should Learn

- How to use the z-test to test a population proportion p

Hypothesis Test for Proportions

Hypothesis tests for proportions occur when a politician wants to know the proportion of his or her constituents that favor a certain bill or when a quality assurance engineer tests the proportion of parts that are defective.

In this section, you will learn how to test a population proportion p. If $np \geq 5$ and $nq \geq 5$ for a binomial distribution, then the sampling distribution for $\hat{p}$ is normal with

$$\mu_{\hat{p}} = p$$

and

$$\sigma_{\hat{p}} = \sqrt{pq/n}.$$

z-Test for a Proportion p

Given a binomial distribution such that $np \geq 5$ and $nq \geq 5$, you can use a z-test to test a population proportion p. The **test statistic** is the sample proportion $\hat{p}$ and the **standardized test statistic** is z.

$$z = \frac{\hat{p} - \mu_{\hat{p}}}{\sigma_{\hat{p}}} = \frac{\hat{p} - p}{\sqrt{pq/n}}$$

Insight

A hypothesis test for a proportion p can also be performed using P-values. Use the same guidelines on page 336 for using P-values for a z-test for a mean μ, but in Step 3 find the standardized test statistic by using the formula

$$z = \frac{\hat{p} - p}{\sqrt{pq/n}}.$$

The other steps in the test are the same.

GUIDELINES

Using a z-Test for a Proportion p

Verify that $np \geq 5$ and $nq \geq 5$.

In Words	*In Symbols*
1. State the claim mathematically and verbally. Identify the null and alternative hypotheses.	State H_0 and H_a.
2. Specify the level of significance.	Identify α.
3. Sketch the sampling distribution.	
4. Determine any critical values.	Use Table 4 in Appendix B.
5. Determine any rejection regions.	
6. Find the standardized test statistic.	$z = \frac{\hat{p} - p}{\sqrt{pq/n}}$
7. Make a decision to reject or fail to reject the null hypothesis.	If z is in the rejection region, reject H_0. Otherwise, fail to reject H_0.
8. Interpret the decision in the context of the original claim.	

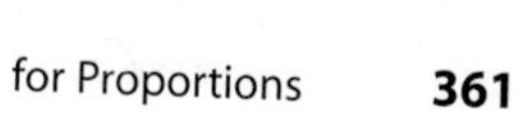

EXAMPLE 1

Hypothesis Test for a Proportion

See *TI-83* steps on page 379.

A medical researcher claims that less than 20% of adults in the U.S. are allergic to a medication. In a random sample of 100 adults, 15% say they have such an allergy. Test the researcher's claim at $\alpha = 0.01$. *(Adapted from Marist College Institute for Public Opinion)*

SOLUTION The products $np = 100(0.20) = 20$ and $nq = 100(0.80) = 80$ are both greater than 5. So, you can use a z-test. The claim is "less than 20% are allergic to a medication." So, the null and alternative hypotheses are

$$H_0:\ p \geq 0.2 \quad \text{and} \quad H_a:\ p < 0.2.\ \text{(Claim)}$$

Because the test is a left-tailed test and the level of significance is $\alpha = 0.01$, the critical value is $z_0 = -2.33$ and the rejection region is $z < -2.33$. Using the z-test, the standardized test statistic is

$$z = \frac{\hat{p} - p}{\sqrt{pq/n}} = \frac{0.15 - 0.2}{\sqrt{(0.2)(0.8)/100}} = -1.25.$$

The graph shows the location of the rejection region and the standardized test statistic z. Because z is not in the rejection region, you should decide not to reject the null hypothesis. In other words, there is not enough evidence to support the claim that less than 20% of adults in the U.S. are allergic to the medication.

Study Tip

Remember that when you fail to reject H_0, a type II error is possible. For instance, in Example 1 the null hypothesis, $p \geq 0.20$, may be false.

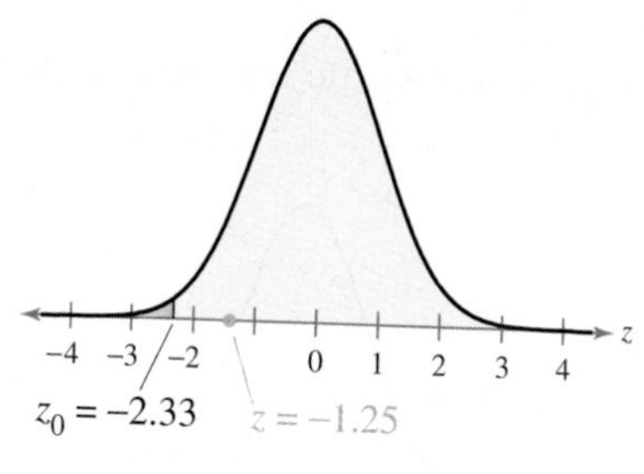

Try It Yourself 1

A researcher claims that less than 30% of adults in the U.S. are allergic to trees, weeds, flowers, and grasses. In a random sample of 86 adults, 20% say they have such an allergy. At $\alpha = 0.05$, is there enough evidence to support the researcher's claim? *(Adapted from Marist College Institute for Public Opinion)*

a. *Verify* that $np \geq 5$ and $nq \geq 5$.
b. *Identify* the claim and state H_0 and H_a.
c. *Identify* the level of significance α.
d. *Find* the critical value z_0 and *identify* the rejection region.
e. *Use* the z-test to find the standardized test statistic z.
f. *Decide* whether to reject the null hypothesis. Use a graph if necessary.
g. Is there enough evidence to support the claim?

Answer: Page A39

To use a P-value to perform the hypothesis test in Example 1, use Table 4 to find the area corresponding to $z = -1.25$. The area is 0.1056. Because this is a left-tailed test, the P-value is equal to the area to the left of $z = -1.25$. So, $P = 0.1056$. Because the P-value is greater than $\alpha = 0.01$, you should fail to reject the null hypothesis. Note this is the same result obtained in Example 1.

EXAMPLE 2

See *Minitab* steps on page 378.

Hypothesis Test for a Proportion

Harper's Index claims that 23% of people in the U.S. are in favor of outlawing cigarettes. You decide to test this claim and ask a random sample of 200 people in the U.S. whether they are in favor of outlawing cigarettes. Of the 200 people, 27% are in favor. At $\alpha = 0.05$, is there enough evidence to reject the claim?

SOLUTION The products $np = 200(0.23) = 46$ and $nq = 200(0.77) = 154$ are both greater than 5. So, you can use a z-test. The claim is "23% of people in the U.S. are in favor of outlawing cigarettes." So, the null and alternative hypotheses are

H_0: $p = 0.23$ (Claim) and H_a: $p \neq 0.23$.

Because the test is a two-tailed test and the level of significance is $\alpha = 0.05$, the critical values are $-z_0 = -1.96$ and $z_0 = 1.96$. The rejection regions are $z < -1.96$ and $z > 1.96$. Using the z-test, the standardized test statistic is

$$z = \frac{\hat{p} - p}{\sqrt{pq/n}} = \frac{0.27 - 0.23}{\sqrt{(0.23)(0.77)/200}} \approx 1.34.$$

The graph shows the location of the rejection regions and the standardized test statistic z.

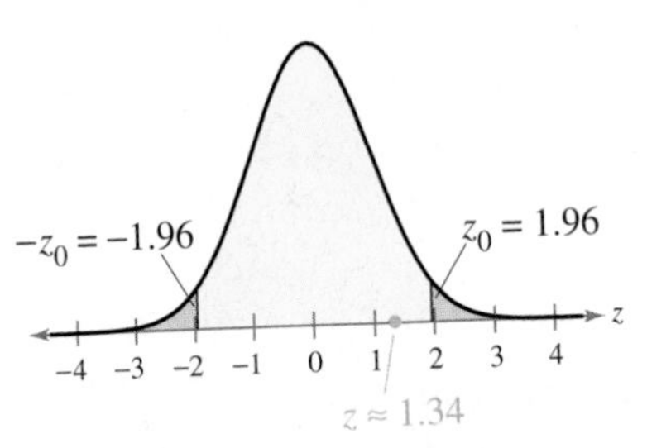

Because z is not in the rejection region, you should fail to reject the null hypothesis. At the 5% level of significance, there is not enough evidence to reject the claim that 23% of people in the U.S. are in favor of outlawing cigarettes.

Picturing the World

A recent study claimed that 53.7% of 3-year-olds in the U.S. were given over-the-counter drugs (such as Tylenol) in a period of 30 days. To test this claim you conduct a random telephone survey of 300 parents of 3-year-olds. In the survey, you find that 166 of the 3-year-olds were given an over-the-counter drug during the past 30 days. *(Source: JAMA)*

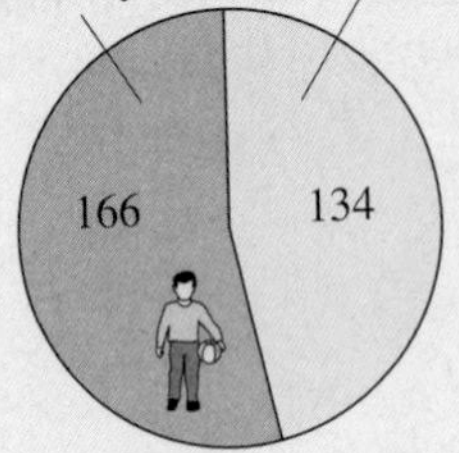

At $\alpha = 0.05$, is there enough evidence to support the study's findings?

Try It Yourself 2

USA TODAY reports that 5% of U.S. adults have seen an extraterrestrial being. You decide to test this claim and ask a random sample of 250 U.S. adults whether they have ever seen an extraterrestrial being. Of those surveyed, 8% reply yes. At $\alpha = 0.01$, is there enough evidence to reject the claim?

a. *Verify* that $np \geq 5$ and $nq \geq 5$.
b. *Identify* the claim and state H_0 and H_a.
c. *Identify* the level of significance α.
d. *Find* the critical values $-z_0$ and z_0 and *identify* the rejection regions.
e. *Use* the z-test to find the standardized test statistic z.
f. *Decide* whether to reject the null hypothesis. Use a graph if necessary.
g. Is there enough evidence to reject the claim?

Answer: Page A39

EXAMPLE 3

Using a Hypothesis Test to Test a Proportion

The Pew Research Center claims that more than 55% of U.S. adults regularly watch a network news broadcast. You decide to test this claim and ask a random sample of 425 adults in the U.S. whether they regularly watch a network news broadcast. Of the 425 adults, 255 respond yes. At $\alpha = 0.05$, is there enough evidence to support the claim? *(Source: Pew Research Center for the People and the Press)*

SOLUTION The products $np = 425(0.55) \approx 234$ and $nq = 425(0.45) \approx 191$ are both greater than 5. So, you can use a z-test. The claim is "more than 55% of U.S. adults watch a network news broadcast." So, the null and alternative hypotheses are

H_0: $p \leq 0.55$ and H_a: $p > 0.55$. (Claim)

Because the test is a right-tailed test and the level of significance is $\alpha = 0.05$, the critical value is $z_0 = 1.645$ and the rejection region is $z > 1.645$. Using the z-test, the standardized test statistic is

$$z = \frac{\hat{p} - p}{\sqrt{pq/n}} = \frac{(x/n) - p}{\sqrt{pq/n}} = \frac{(255/425) - 0.55}{\sqrt{(0.55)(0.45)/425}} \approx 2.07.$$

The graph shows the location of the rejection region and the standardized test statistic z. Because z is in the rejection region, you should decide to reject the null hypothesis. There is enough evidence at the 5% level of significance to support the claim that more than 55% of U.S. adults regularly watch a network news broadcast.

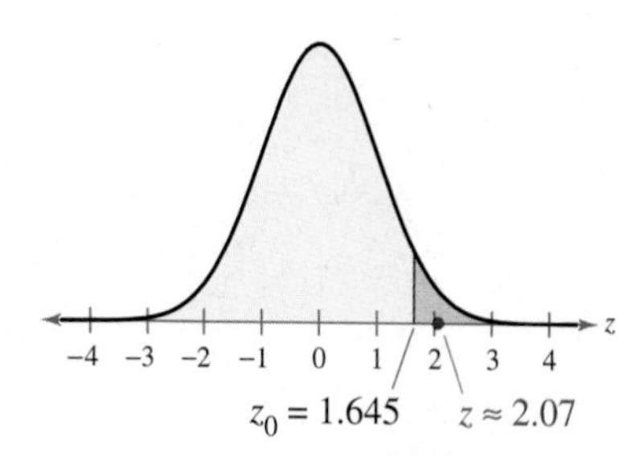

Try It Yourself 3

The Pew Research Center claims that more than 38% of U.S. adults regularly watch a cable news broadcast. You decide to test this claim and ask a random sample of 75 adults whether they regularly watch a cable news broadcast. Of the 75 adults, 33 respond yes. At $\alpha = 0.01$, is there enough evidence to support the claim? *(Source: Pew Research Center for the People and the Press)*

a. *Verify* that $np \geq 5$ and $nq \geq 5$.
b. *Identify* the claim and state H_0 and H_a.
c. *Identify* the level of significance α.
d. *Find* the critical value z_0 and *identify* the rejection region.
e. *Use* the z-test to find the standardized test statistic z.
f. *Decide* whether to reject the null hypothesis. Use a graph if necessary.
g. Is there enough evidence to support the claim?

Answer: Page A39

Exercises

Help

LarsonTutor 7.4

Companion Web Site

Student Solutions Manual 7.4

Videos 7.4

Technology Manuals

Basic Skills and Concepts

1. Explain how to test a population proportion p.

2. Explain how to decide when a normal distribution can be used to approximate a binomial distribution.

In Exercises 3–8, decide whether the normal sampling distribution can be used. If it can be used, test the claim about the population proportion p for the given values and level of significance α.

3. Claim: $p \neq 0.25$; $\alpha = 0.05$. Sample statistics: $\hat{p} = 0.239$, $n = 105$

4. Claim: $p \leq 0.30$; $\alpha = 0.05$. Sample statistics: $\hat{p} = 0.35$, $n = 500$

5. Claim: $p < 0.60$; $\alpha = 0.01$. Sample statistics: $\hat{p} = 0.58$, $n = 35$

6. Claim: $p > 0.125$; $\alpha = 0.01$. Sample statistics: $\hat{p} = 0.2325$, $n = 45$

7. Claim: $p \geq 0.48$; $\alpha = 0.10$. Sample statistics: $\hat{p} = 0.40$, $n = 70$

8. Claim: $p = 0.80$; $\alpha = 0.10$. Sample statistics: $\hat{p} = 0.875$, $n = 400$

Testing Claims In Exercises 9–14, (a) write the claim mathematically and identify H_0 and H_a, (b) find the critical values and identify the rejection regions, (c) find the standardized test statistic, and (d) decide whether to reject or fail to reject the null hypothesis. Then interpret the decision in the context of the original claim.

9. ***Smokers*** ◆ A medical researcher says that at least 25% of U.S. adults are smokers. In a random sample of 200 U.S. adults, 24.5% say that they are smokers. At $\alpha = 0.01$, do you have enough evidence to reject the researcher's claim? *(Adapted from U.S. National Center for Health Statistics)*

10. ***Do You Eat Breakfast?*** ◆ A medical researcher estimates that no more than 55% of U.S. adults eat breakfast every day. In a random sample of 250 U.S. adults, 56.4% say that they eat breakfast every day. At $\alpha = 0.01$, is there enough evidence to reject the researcher's claim? *(Adapted from U.S. National Center for Health Statistics)*

11. ***Environmentally Conscious Consumers*** ◆ You are employed by an environmental conservation agency that recently claimed more than 30% of U.S. consumers have stopped buying a certain product because the manufacturing of the product pollutes the environment. You want to test this claim. To do so, you randomly select 1050 U.S. consumers and find that 32% have stopped buying this product because of pollution concerns. At $\alpha = 0.03$, can you support the claim? *(Adapted from Wirthlin Worldwide)*

12. ***Just Say No to GMO*** ◆ An environmentalist claims that more than 50% of British consumers want supermarkets to stop selling genetically modified foods. You want to test this claim. You find that in a random sample of 100 British consumers, 53% say that they want supermarkets to stop selling genetically modified foods. At $\alpha = 0.10$, can you support the environmentalist's claim? *(Adapted from Friends of the Earth)*

13. ***Government Regulations Harmful?*** ◆ In your work for a business regulatory agency, you find that in a sample of 1762 people in the U.S., 1004 of them believe that government regulation of business does more harm than good. At $\alpha = 0.02$, can you reject the claim that 60% of people in the U.S. have this view? *(Adapted from Pew Research Center for the People and the Press)*

14. ***Government Inefficient and Wasteful?*** ◆ A government watchdog association claims that 70% of people in the U.S. agree that the government is inefficient and wasteful. You work for a government agency and are asked to test this claim. You find that in a random sample of 1165 people in the U.S., 746 agreed with this view. At $\alpha = 0.05$, do you have enough evidence to reject the association's claim? *(Adapted from Pew Research Center for the People and the Press)*

Trade Show Giveaways In Exercises 15 and 16, use the graph which shows what attendees think about the effectiveness of giveaways at trade shows.

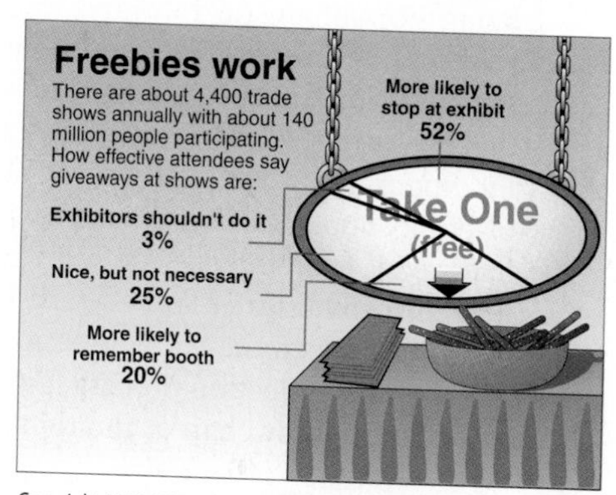

15. ***Do Giveaways Bring More Visitors?*** ◆ At a trade show, you interview a random sample of 50 attendees. The results of the survey show that 48% of the attendees said they were more likely to visit an exhibit when there is a giveaway. At $\alpha = 0.05$, test the claim that at least 52% of the attendees at trade shows are more likely to visit an exhibit when there is a giveaway.

16. ***Should Giveaways be Used?*** ◆ Use your conclusion in Exercise 15 to write a paragraph on the use of giveaways at trade shows. Do you think your company should continue to use giveaways to get people to visit the company's exhibits? Explain.

Extending the Basics

Alternate Formula In Exercises 17 and 18, use the following information. When you know the number of successes x, the sample size n, and probability p, it can be easier to use the formula

$$z = \frac{x - np}{\sqrt{npq}}$$

to find the standardized test statistic when using a z-test for a proportion p.

17. Rework Exercise 13 using the alternate formula and compare the results.

18. The alternate formula is derived from the formula

$$z = \frac{\hat{p} - p}{\sqrt{pq/n}} = \frac{(x/n) - p}{\sqrt{pq/n}}.$$

Use this formula to derive the alternate formula. Justify each step.

TECHNOLOGY MINITAB EXCEL TI-83

The Case of the Vanishing Women

53% → 29% → 9% → 0%

From 1966 to 1968, Dr. Benjamin Spock and others were tried for conspiracy to violate the Selective Service Act by encouraging resistance to the Vietnam War. By a series of three selections, there ended up being no women on the jury. In 1969, Hans Zeisel wrote an article in *The University of Chicago Law Review* using statistics and hypothesis testing to argue that the jury selection was biased against Dr. Spock. Dr. Spock was a well-known pediatrician and author of books about raising children. Millions of mothers had read his books and followed his advice. By keeping women off the jury, Zeisel argued that the court prejudiced the verdict.

The jury selection process for Dr. Spock's trial is shown in the flowchart at the right.

Stage 1. The clerk of the Federal District Court selected 350 people "at random" from the Boston City Directory. The directory contained several hundred names, 53% of whom were women. However, only 102 of the 350 people selected were women.

Stage 2. The trial judge, Judge Ford, selected 100 people "at random" from the 350 people. This group was called a venire and it contained only nine women.

Stage 3. The court clerk assigned numbers to the members of the venire and one by one, they were interrogated by the attorneys for the prosecution and defense until 12 members of the jury were chosen. At this stage, only one potential female juror was questioned and she was eliminated by the prosecutor under his quota of peremptory challenges (for which he did not have to give a reason).

Exercises

1. The *Minitab* display below shows a hypothesis test for a claim that the proportion of women in the city directory is $p = 0.53$. In the test, $n = 350$ and $\hat{p} = 0.2914$. Should you reject the claim? What is the level of significance? Explain.
2. In Exercise 1, you rejected the claim that $p = 0.53$. But this claim was true. What type of error is this?
3. If you reject a true claim with a level of significance that is virtually zero, what does this tell you about the randomness of your sampling process?
4. Describe a hypothesis test for Judge Ford's "random" selection of the venire. Use a claim of

$$p = \frac{102}{350} \approx 0.2914.$$

   (a) Write the null and alternative hypotheses.
   (b) Use a technology tool to perform the test.
   (c) Make a decision.
   (d) Interpret the decision in the context of the original claim. Could Judge Ford's selection of 100 venire members have been random?

MINITAB

Test and Confidence Interval for One Proportion

Test of p = 0.53 vs p not = 0.53

Sample	X	N	Sample p	99.0 % CI	Z-Value	P-Value
1	102	350	0.291429	(0.228862, 0.353995)	-8.94	0.000

Extended solutions are given in the *Technology Supplement*.
Technical instruction is provided for *Minitab*, *Excel*, and the *TI-83*.

7.5 Hypothesis Testing for Variance and Standard Deviation

What You Should Learn

- How to find critical values for a χ^2-test
- How to use the χ^2-test to test a variance or a standard deviation

Critical Values for a χ^2-Test • The Chi-Square Test

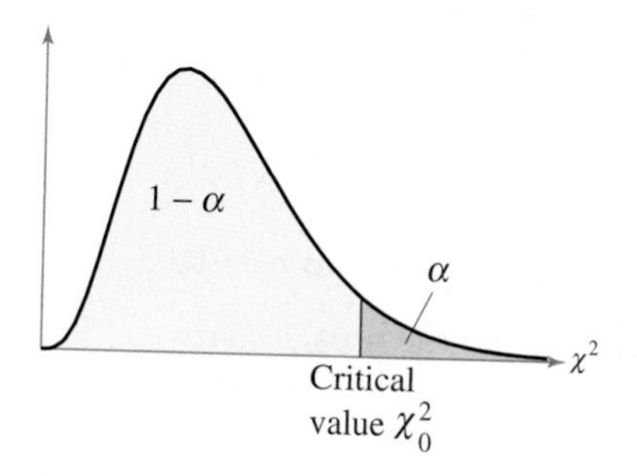

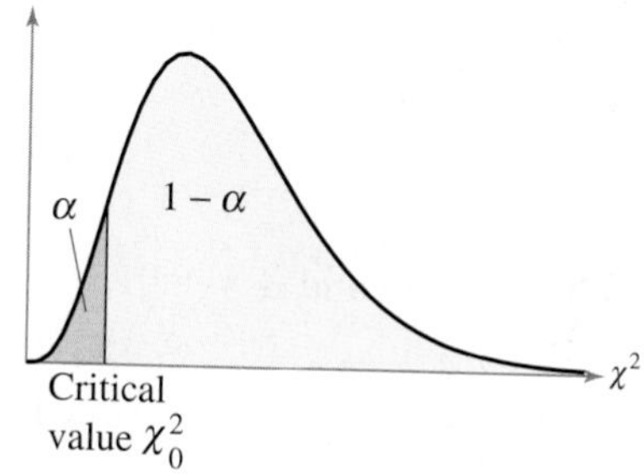

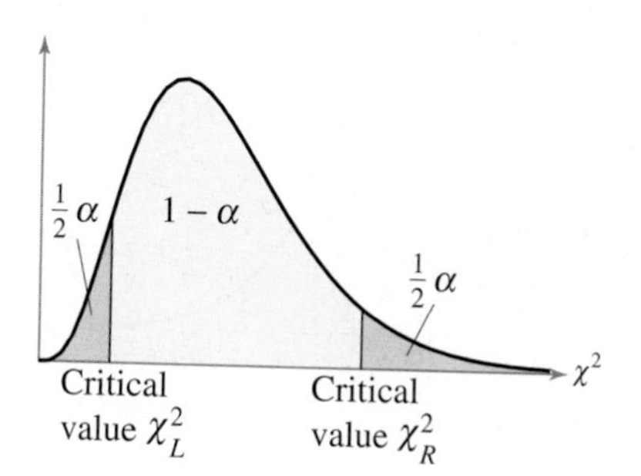

Critical Values for a χ^2-Test

In real life, it is often important to produce consistent predictable results. For instance, consider a company that manufactures golf balls. The manufacturer must produce millions of golf balls each having the same size and the same weight. There is a very low tolerance for variation. If the population is normal, you can test the variance and standard deviation of the process using the chi-square distribution with $n - 1$ degrees of freedom.

GUIDELINES

Finding Critical Values for the χ^2-Test

1. Specify the level of significance α.
2. Determine the degrees of freedom d.f. $= n - 1$.
3. The critical values for the χ^2-distribution are found in Table 6 of Appendix B. To find the critical value(s) for a
   a. *right-tailed test,* use the value that corresponds to d.f. and α.
   b. *left-tailed test,* use the value that corresponds to d.f. and $1 - \alpha$.
   c. *two-tailed test,* use the values that correspond to d.f. and $\frac{1}{2}\alpha$ and d.f. and $1 - \frac{1}{2}\alpha$.

EXAMPLE 1

Finding Critical Values for χ^2

Find the critical χ^2-value for a right-tailed test when $n = 26$ and $\alpha = 0.10$.

SOLUTION The degrees of freedom are

$$\text{d.f.} = n - 1 = 26 - 1 = 25.$$

The graph at the right shows a χ^2-distribution with 25 degrees of freedom and a shaded area of $\alpha = 0.10$ in the right tail. Using Table 6 with d.f. $= 25$ and $\alpha = 0.10$, the critical value is

$$\chi_0^2 = 34.382.$$

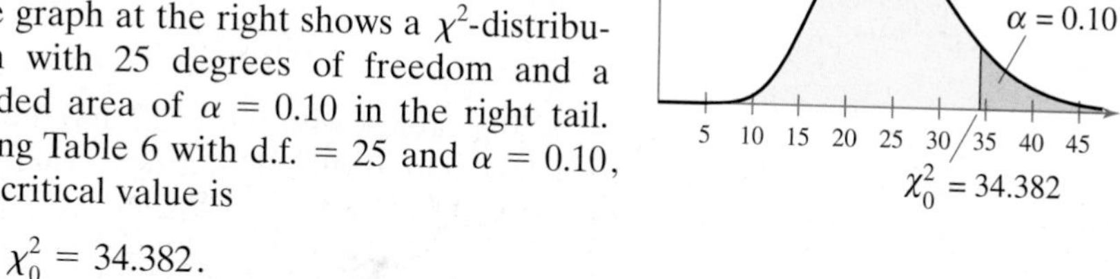

Try It Yourself 1

Find the critical χ^2-value for a right-tailed test when $n = 18$ and $\alpha = 0.01$.

a. Find the value using Table 6 with d.f. $= n - 1$ and the area α.

Answer: Page A39

EXAMPLE 2

Finding Critical Values for χ^2

Find the critical χ^2-value for a left-tailed test when $n = 11$ and $\alpha = 0.01$.

SOLUTION The degrees of freedom are d.f. $= n - 1 = 11 - 1 = 10$. The graph at the right shows a χ^2-distribution with 10 degrees of freedom and a shaded area of $\alpha = 0.01$ in the left tail. The area to the right of the critical value is

$$1 - \alpha = 1 - 0.01 = 0.99.$$

Using Table 6 with d.f. $= 10$ and the area $1 - \alpha = 0.99$, the critical value is $\chi_0^2 = 2.558$.

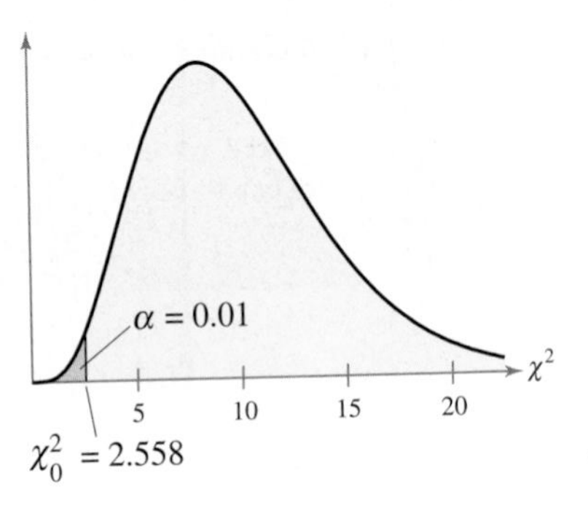

Try It Yourself 2

Find the critical χ^2-value for a left-tailed test when $n = 30$ and $\alpha = 0.05$.

a. *Find* the value using Table 6 with d.f. $= n - 1$ and the area $1 - \alpha$.

Answer: Page A39

EXAMPLE 3

Finding Critical Values for χ^2

Find the critical χ^2-values for a two-tailed test when $n = 13$ and $\alpha = 0.01$.

SOLUTION The degrees of freedom are d.f. $= n - 1 = 13 - 1 = 12$. The graph at the right shows a χ^2-distribution with 12 degrees of freedom and a shaded area of $\frac{1}{2}\alpha = 0.005$ in each tail. The areas to the right of the critical values are

$$\tfrac{1}{2}\alpha = 0.005$$

and

$$1 - \tfrac{1}{2}\alpha = 0.995.$$

Using Table 6 with d.f. $= 12$ and the areas 0.005 and 0.995, the critical values are $\chi_L^2 = 3.074$ and $\chi_R^2 = 28.299$.

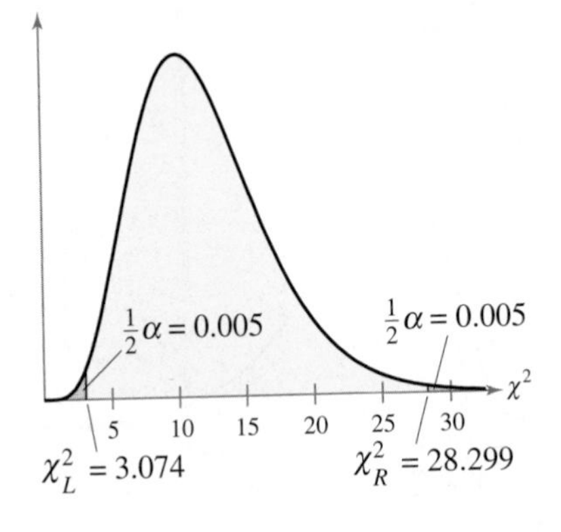

Try It Yourself 3

Find the critical χ^2-values for a two-tailed test when $n = 19$ and $\alpha = 0.05$.

a. *Find* the first critical value χ_R^2 using Table 6 with d.f. $= n - 1$ and the area $\frac{1}{2}\alpha$.

b. *Find* the second critical value χ_L^2 using Table 6 with d.f. $= n - 1$ and the area $1 - \frac{1}{2}\alpha$.

Answer: Page A39

Note that because chi-square distributions are not symmetric (like normal or t-distributions), in a two-tailed test the two critical values are not opposites. Each critical value must be calculated separately.

The Chi-Square Test

To test a variance σ^2 or a standard deviation σ of a population that is normally distributed, you can use the χ^2-test. The χ^2-test for a variance or standard deviation is not as robust as the tests for the population mean μ or the population proportion p. So it is essential that when performing a χ^2-test for a variance or standard deviation that the population is normally distributed. The results can be misleading if the population is not normal.

χ^2-Test for a Variance or Standard Deviation

The **χ^2-test** is a statistical test for a population variance or standard deviation. The χ^2-test can be used when the population is normal. The **test statistic** is s^2 and the **standardized test statistic** is

$$\chi^2 = \frac{(n-1)s^2}{\sigma^2}.$$

The sampling distribution for s^2 and s is a chi-square distribution with degrees of freedom

$$\text{d.f.} = n - 1.$$

GUIDELINES

Using the χ^2-Test for a Variance or Standard Deviation

In Words	*In Symbols*
1. State the claim mathematically and verbally. Identify the null and alternative hypotheses.	State H_0 and H_a.
2. Specify the level of significance.	Identify α.
3. Determine the degrees of freedom and sketch the sampling distribution.	$\text{d.f.} = n - 1$
4. Determine any critical values.	Use Table 6 in Appendix B.
5. Determine any rejection regions.	
6. Find the standardized test statistic.	$\chi^2 = \dfrac{(n-1)s^2}{\sigma^2}$
7. Make a decision to reject or fail to reject the null hypothesis.	If χ^2 is in the rejection region, reject H_0. Otherwise, fail to reject H_0.
8. Interpret the decision in the context of the original claim.	

Picturing the World

A community center claims that the chlorine level in its pool has a standard deviation of 0.46 parts per million (ppm). A sampling of the pool's chlorine levels at 25 random times during a month yields a standard deviation of 0.61 ppm. *(Adapted from American Pool Supply)*

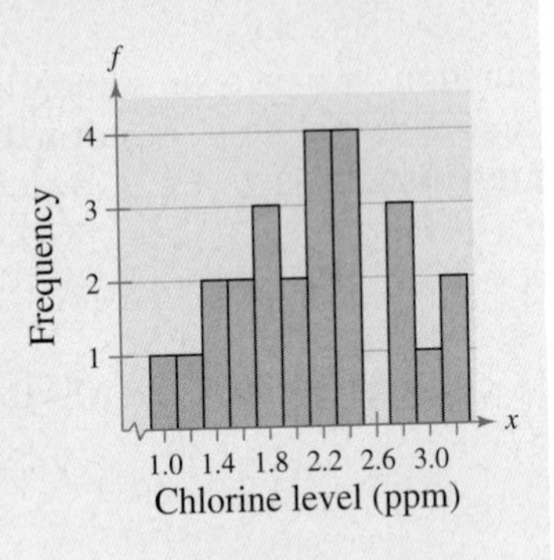

At $\alpha = 0.05$ is there enough evidence to reject the claim?

EXAMPLE 4

Using a Hypothesis Test for the Population Variance

A dairy processing company claims that the variance of the amount of fat in the whole milk processed by the company is no more than 0.25. You suspect this is wrong and find that a random sample of 41 milk containers has a variance of 0.27. At $\alpha = 0.05$, is there enough evidence to reject the company's claim? Assume the population is normally distributed.

SOLUTION The claim is "the variance is no more than 0.25." So, the null and alternative hypotheses are

H_0: $\sigma^2 \le 0.25$ (Claim) and H_a: $\sigma^2 > 0.25$.

The test is a right-tailed test, the level of significance is $\alpha = 0.05$, and there are d.f. $= 41 - 1 = 40$ degrees of freedom. So, the critical value is

$$\chi_0^2 = 55.758.$$

The rejection region is

$$\chi^2 > 55.758.$$

Using the χ^2-test, the standardized test statistic is

$$\chi^2 = \frac{(n-1)s^2}{\sigma^2} = \frac{(41-1)(0.27)}{0.25} = 43.2.$$

The graph shows the location of the rejection region and the standardized test statistic χ^2. Because χ^2 is not in the rejection region, you should decide not to reject the null hypothesis. You don't have enough evidence to reject the company's claim at the 5% level of significance.

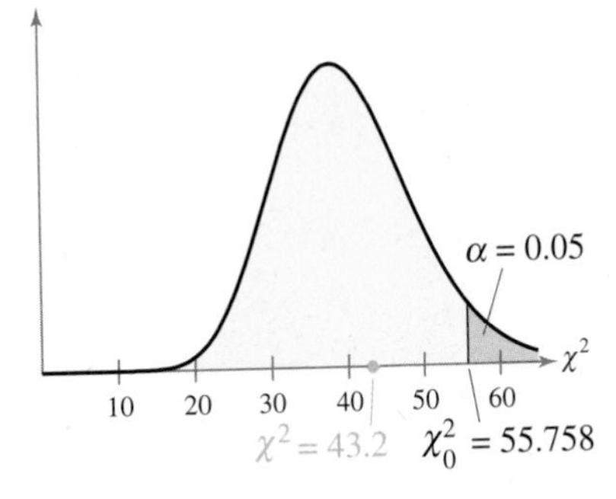

Try It Yourself 4

A bottling company claims that the variance of the amount of sports drink in a 12-ounce bottle is no more than 0.40. A random sample of 31 bottles has a variance of 0.75. At $\alpha = 0.01$, is there enough evidence to reject the company's claim? Assume the population is normally distributed.

a. *Identify* the claim and state H_0 and H_a.
b. *Identify* the level of significance α and the degrees of freedom d.f.
c. *Find* the critical value and *identify* the rejection region.
d. *Use* the χ^2-test to find the standardized test statistic χ^2.
e. *Decide* whether to reject the null hypothesis. Use a graph if necessary.
f. Is there enough evidence to reject the claim?

Answer: Page A40

EXAMPLE 5

Using a Hypothesis Test for the Standard Deviation

A restaurant claims that the standard deviation in the length of serving times is less than 2.9 minutes. A random sample of 23 serving times has a standard deviation of 2.1 minutes. At $\alpha = 0.10$, is there enough evidence to support the restaurant's claim? Assume the population is normally distributed.

Study Tip

Although you are testing a standard deviation in Example 5, the χ^2-statistic requires variances. Don't forget to square the given standard deviations to calculate these variances.

SOLUTION The claim is "the standard deviation is less than 2.9 minutes." So, the null and alternative hypotheses are

H_0: $\sigma \geq 2.9$ minutes and H_a: $\sigma < 2.9$ minutes. (Claim)

The test is a left-tailed test, the level of significance is $\alpha = 0.10$, and there are

$$\text{d.f.} = 23 - 1 = 22$$

degrees of freedom. So, the critical value is

$$\chi_0^2 = 14.042.$$

The rejection region is $\chi^2 < 14.042$. Using the χ^2-test, the standardized test statistic is

$$\chi^2 = \frac{(n-1)s^2}{\sigma^2} = \frac{(23-1)(2.1)^2}{2.9^2} \approx 11.54.$$

The graph shows the location of the rejection region and the standardized test statistic χ^2. Because χ^2 is in the rejection region, you should decide to reject the null hypothesis. So, there is enough evidence at the 10% level of significance to support the claim that the standard deviation for the length of serving times is less than 2.9 minutes.

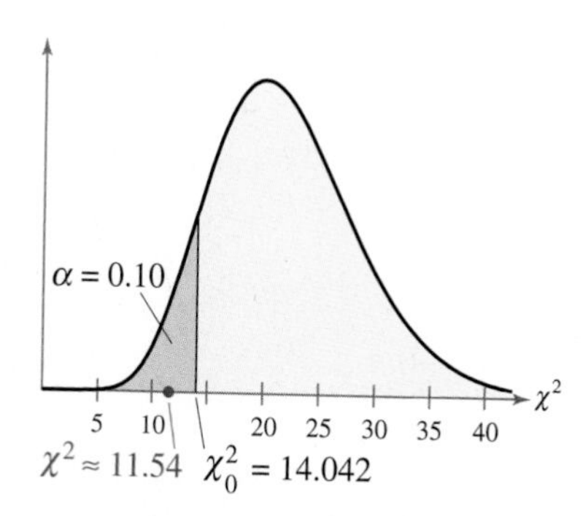

Try It Yourself 5

A police chief claims that the standard deviation in the length of response times is less than 3.7 minutes. A random sample of nine response times has a standard deviation of 3.0 minutes. At $\alpha = 0.05$, is there enough evidence to support the police chief's claim? Assume the population is normally distributed.

a. *Identify* the claim and state H_0 and H_a.
b. *Identify* the level of significance α and the degrees of freedom d.f.
c. *Find* the critical value and *identify* the rejection region.
d. *Use* the χ^2-test to find the standardized test statistic χ^2.
e. *Decide* whether to reject the null hypothesis. Use a graph if necessary.
f. Is there enough evidence to support the claim?

Answer: Page A40

EXAMPLE 6

Using a Hypothesis Test for the Population Variance

A sporting goods manufacturer claims that the variance of the strength in a certain fishing line is 15.9. A random sample of 15 fishing line spools has a variance of 21.8. At $\alpha = 0.05$, is there enough evidence to reject the manufacturer's claim? Assume the population is normally distributed.

SOLUTION The claim is "the variance is 15.9." So, the null and alternative hypotheses are

H_0: $\sigma^2 = 15.9$ (Claim) and H_a: $\sigma^2 \neq 15.9$.

The test is a two-tailed test, the level of significance is $\alpha = 0.05$, and there are d.f. $= 15 - 1 = 14$ degrees of freedom. So, the critical values are $\chi^2_L = 5.629$ and $\chi^2_R = 26.119$. The rejection regions are $\chi^2 < 5.629$ and $\chi^2 > 26.119$. Using the χ^2-test, the standardized test statistic is

$$\chi^2 = \frac{(n-1)s^2}{\sigma^2} = \frac{(15-1)(21.8)}{15.9} \approx 19.19.$$

The graph shows the location of the rejection regions and the standardized test statistic χ^2. Because χ^2 is not in the rejection regions, you should decide not to reject the null hypothesis. At the 5% level of significance, there is not enough evidence to reject the claim that the variance is 15.9.

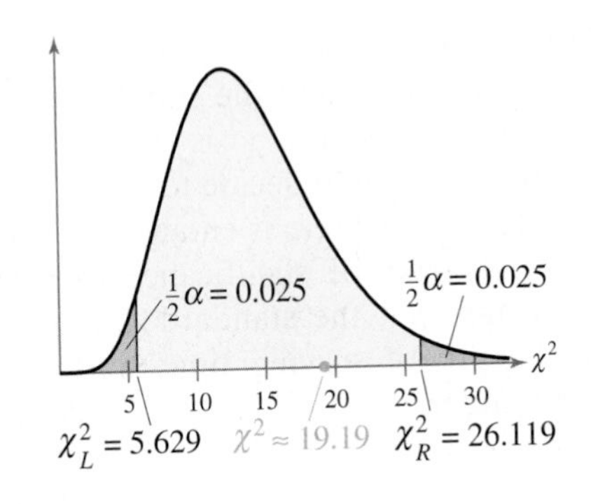

Try It Yourself 6

A tire manufacturer claims that the variance of the diameters in a certain tire model is 8.6. A random sample of 10 tires has a variance of 4.3. At $\alpha = 0.01$, is there enough evidence to reject the manufacturer's claim? Assume the population is normally distributed.

a. *Identify* the claim and state H_0 and H_a.
b. *Identify* the level of significance α and the degrees of freedom d.f.
c. *Find* the critical values and *identify* the rejection regions.
d. *Use* the χ^2-test to find the standardized test statistic χ^2.
e. *Decide* whether to reject the null hypothesis. Use a graph if necessary.
f. Is there enough evidence to reject the claim?

Answer: Page A40

7.5 Exercises

Help

LarsonTutor 7.5

Companion Web Site

Student Solutions Manual 7.5

Videos 7.5

Technology Manuals

Basic Skills and Concepts

1. Explain how to find critical values in a χ^2 sampling distribution.

2. Explain how to test a population variance or a population standard deviation.

In Exercises 3–8, find the critical values for the indicated test for a population variance, sample size n, and level of significance α.

3. Right-tailed test, $n = 27, \alpha = 0.05$

4. Right-tailed test, $n = 10, \alpha = 0.10$

5. Left-tailed test, $n = 7, \alpha = 0.01$

6. Left-tailed test, $n = 24, \alpha = 0.05$

7. Two-tailed test, $n = 16, \alpha = 0.10$

8. Two-tailed test, $n = 29, \alpha = 0.01$

Graphical Analysis In Exercises 9–12, state whether the standardized test statistic χ^2 allows you to reject the null hypothesis.

9. (a) $\chi^2 = 2.091$
(b) $\chi^2 = 0$
(c) $\chi^2 = 1.086$
(d) $\chi^2 = 6.3471$

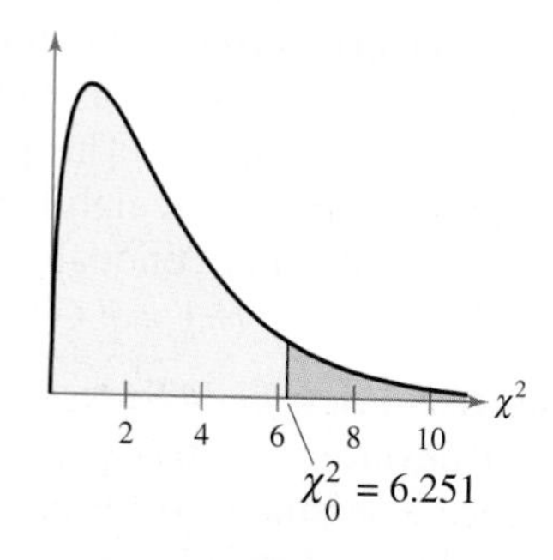

10. (a) $\chi^2 = 0.771$
(b) $\chi^2 = 9.486$
(c) $\chi^2 = 0.701$
(d) $\chi^2 = 9.508$

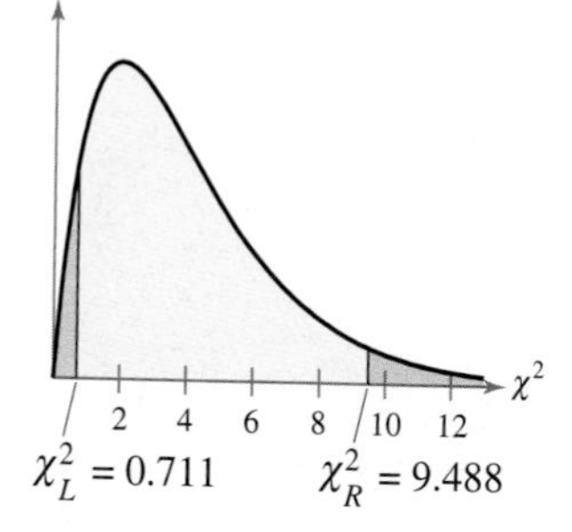

11. (a) $\chi^2 = 22.302$
(b) $\chi^2 = 23.309$
(c) $\chi^2 = 8.457$
(d) $\chi^2 = 8.577$

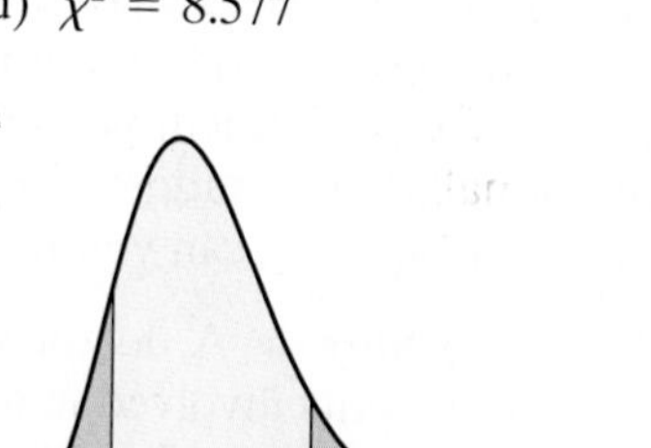

12. (a) $\chi^2 = 10.065$
(b) $\chi^2 = 10.075$
(c) $\chi^2 = 10.585$
(d) $\chi^2 = 10.745$

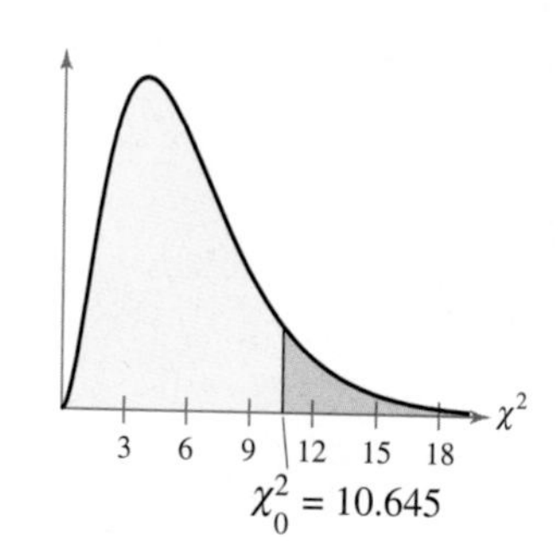

In Exercises 13–16, use a χ^2-test to test the claim about the population variance σ^2 or standard deviation σ for the given values and level of significance α. Assume the population is normal.

13. Claim: $\sigma^2 = 0.52$; $\alpha = 0.05$. Sample statistics: $s^2 = 0.508$, $n = 18$

14. Claim: $\sigma^2 \geq 3.5$; $\alpha = 0.05$. Sample statistics: $s^2 = 3.42$, $n = 21$

15. Claim: $\sigma < 40$; $\alpha = 0.01$. Sample statistics: $s = 40.8$, $n = 12$

16. Claim: $\sigma < 0.12$; $\alpha = 0.10$. Sample statistics: $s = 0.103$, $n = 19$

Testing Claims In Exercises 17–26, (a) write the claim mathematically and identify H_0 and H_a, (b) find the critical value(s) and identify the rejection region(s), (c) use the χ^2-test to find the standardized test statistic, and (d) decide whether to reject or fail to reject the null hypothesis. Then interpret the decision in the context of the original claim. Assume the populations are normally distributed.

17. ***Life of Appliances*** ◆ A large appliance company estimates that the variance of the life of its appliances is 3. You work for a consumer advocacy group and are asked to test this claim. You find that a random sample of the lives of 27 of the company's appliances has a variance of 2.8. At $\alpha = 0.05$, do you have enough evidence to reject the company's claim? *(Adapted from Consumer Reports)*

18. ***Luxury Sedan Gas Mileage*** ◆ An automotive manufacturer believes that the variance of the gas mileage for its luxury sedans is 5. You work for an energy conservation agency and want to test this claim. You find that a random sample of the miles per gallon of 23 of the manufacturer's sedans has a variance of 4.5. At $\alpha = 0.05$, do you have enough evidence to reject the manufacturer's claim? *(Adapted from Consumer Reports)*

19. ***Physical Science Assessment Tests*** ◆ On a physical science assessment test, the scores of a random sample of 22 eighth-grade students have a standard deviation of 27.7. This prompts a test administrator to claim that the standard deviation for eighth graders on the examination is less than 29. At $\alpha = 0.10$, is there enough evidence to support the administrator's claim? *(Adapted from National Center for Educational Statistics)*

20. ***Life Science Assessment Tests*** ◆ A state school administrator says that the standard deviation of test scores for eighth-grade students who took a life science assessment test is less than 30. You work for the administrator and are asked to test this claim. To do so, you randomly select 10 tests and find that the tests have a standard deviation of 28.8. At $\alpha = 0.01$, is there enough evidence to support the administrator's claim? *(Adapted from National Center for Educational Statistics)*

21. ***Hospital Waiting Times*** ◆ A hospital spokesperson claims that the standard deviation of the waiting times experienced by patients in its minor emergency department is no more than 0.5 minute. You doubt the validity of this claim. If a random sample of 25 waiting times has a standard deviation of 0.7 minute, can you reject the spokesperson's claim? Use $\alpha = 0.10$.

22. ***Length of Stay*** ◆ A doctor says the standard deviation of the lengths of stay for patients involved in a crash where the vehicle struck a tree is 6.14 days. A random sample of 20 lengths of stay for patients involved in this type of crash has a standard deviation of 6.5 days. At $\alpha = 0.05$, can you reject the doctor's claim? *(Adapted from National Highway Traffic Safety Administration)*

23. ***Total Charge*** ◆ An insurance agent says the standard deviation of the total charge for patients involved in a crash where the vehicle struck a wall is less than \$18,000. A random sample of 28 total charges for patients involved in this type of crash has a standard deviation of \$18,500. At $\alpha = 0.10$, can you support the agent's claim? *(Adapted from National Highway Traffic Safety Administration)*

24. ***Hotel Room Rates*** ◆ A travel agency estimates that the standard deviation of the room rates of hotels in a certain city is no more than \$25. You work for a consumer advocacy group and are asked to test this claim. You find that a random sample of 13 hotels has a standard deviation of \$27.50. At $\alpha = 0.01$, do you have enough evidence to reject the agency's claim? *(Adapted from Smith Travel Research)*

25. ***Salaries*** ◆ The annual salaries of 16 randomly chosen actuaries are listed below. At $\alpha = 0.05$, can you conclude that the standard deviation of the annual salaries is greater than \$20,000? *(Adapted from America's Career InfoNet)*

45,018	34,952	85,517	45,553	41,900	76,384	48,862	37,615
61,104	96,710	97,875	68,245	39,945	53,582	65,252	72,522

26. ***Salaries*** ◆ An employment information service says that the standard deviation of the annual salaries for public relations managers is at least \$14,500. The annual salaries for 18 randomly chosen public relations managers are listed. At $\alpha = 0.10$, can you reject the claim? *(Adapted from America's Career InfoNet)*

37,517	50,217	29,177	51,744	69,422	60,770
50,549	50,263	62,939	62,372	65,014	49,164
34,811	55,413	51,310	80,433	34,185	31,805

Extending the Basics

***P*-values** You can calculate the *P*-value for a χ^2-test using technology. After calculating the χ^2-test value, you can use the cumulative density function (CDF) to calculate the area under the curve. From Example 4 on page 370, $\chi^2 = 43.2$. Using a *TI-83* (choose 7 from the DISTR menu), enter 0 for the lower bound, 43.2 for the upper bound, and 40 for the degrees of freedom as shown.

The *P*-value is approximately $1 - 0.664 = 0.336$. Because $P > \alpha = 0.05$, the conclusion is to fail to reject H_0.

27. Use the *P*-value method to perform the hypothesis test for Exercise 23.

28. Use the *P*-value method to perform the hypothesis test for Exercise 24.

7 A Summary of Hypothesis Testing

Insight

Large sample sizes will usually increase the cost and effort of testing a hypothesis, but they also tend to make your decision more reliable.

With hypothesis testing, perhaps more than any other area of statistics, it can be difficult to see the forest for all the trees. To help you see the forest—the overall picture—we provide a summary of what you studied in this chapter.

Writing the Hypotheses

- You are given a claim about a population parameter μ, p, σ^2, or σ.
- Rewrite the claim and its complement using $\underbrace{\leq, \geq, =}_{H_0}$ and $\underbrace{>, <, \neq}_{H_a}$.
- Identify the claim. Is it H_0 or H_a?

Specifying a Level of Significance

- Specify α, the maximum acceptable probability of rejecting a valid H_0 (a type I error).

Specifying the Sample Size

- Specify your sample size n.

Choosing the Test ■ Any population ■ Normally distributed population

- **Mean:** H_0 describes a hypothesized population mean μ.
  - ■ Use a ***z*-test** for *any* population if $n \geq 30$.
  - ■ Use a ***z*-test** if the population is normal and σ is known for any n.
  - ■ Use a ***t*-test** if the population is normal and $n < 30$, but σ is unknown.
- **Proportion:** H_0 describes a hypothesized population proportion p.
  - ■ Use a ***z*-test** for *any* population if $np \geq 5$ and $nq \geq 5$.
- **Variance or Standard Deviation:** H_0 describes a hypothesized population variance σ^2 or standard deviation σ.
  - ■ Use a **χ^2-test** if the population is normal.

Sketching the Sampling Distribution

- Use H_a to decide if the test is left tailed, right tailed, or two tailed.

Finding the Standardized Test Statistic

- Take a random sample of size n from the population.
- Compute the test statistic $\bar{x}$, p, or s^2.
- Find the standardized test statistic z, t, or χ^2.

Making a Decision

Option 1. Decision based on rejection region

- Use α to find the critical value(s) z_0, t_0, or χ_0^2 and rejection region(s).
- **Decision Rule:**
  Reject H_0 if the standardized test statistic is in the rejection region.
  Fail to reject H_0 if the standardized test statistic is not in the rejection region.

Option 2. Decision based on P-value

- Use the standardized test statistic or a technology tool to find the P-value.
- **Decision Rule:**
  Reject H_0 if $P \leq \alpha$.
  Fail to reject H_0 if $P > \alpha$.

Study Tip

If your standardized test statistic is z or t, remember that these values measure standard deviations from the mean. Values that are outside of ± 3 indicate that H_0 is very unlikely. Values that are outside of ± 5 indicate that H_0 is almost impossible.

z-Test for a Hypothesized Mean μ *(Section 7.2)*

Test Statistic: $\bar{x}$

Critical value: z_0 (Use Table 4.)

If $n \geq 30$, s can be used in place of σ.
Sampling distribution of sample means is a normal distribution.

Standardized Test Statistic: z

$$z = \frac{\bar{x} - \mu}{\sigma/\sqrt{n}}$$

Sample mean: $\bar{x}$; Hypothesized mean: μ; Population standard deviation: σ; Sample size: n

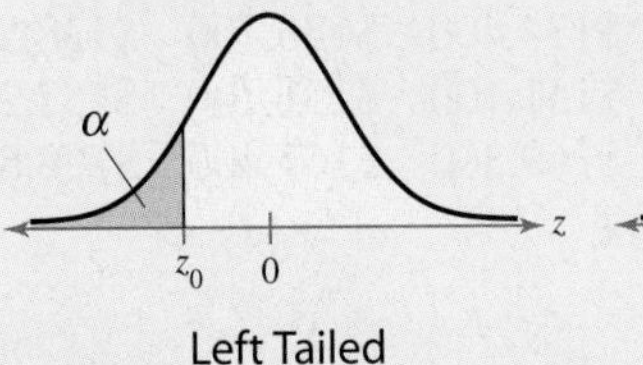

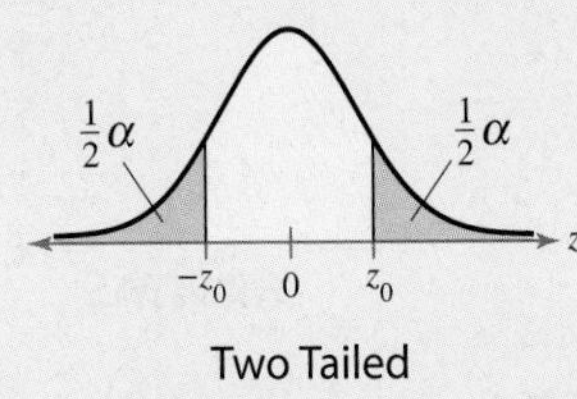

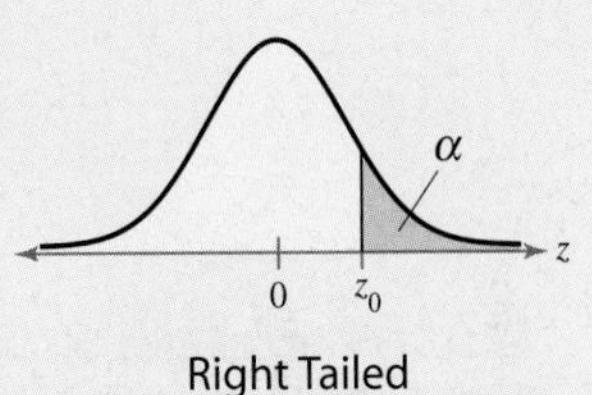

z-Test for a Hypothesized Proportion p *(Section 7.4)*

Test Statistic: $\hat{p}$

Critical value: z_0 (Use Table 4.)

Sampling distribution of sample proportions is a normal distribution.

Standardized Test Statistic: z

$$z = \frac{\hat{p} - p}{\sqrt{pq/n}}$$

Sample proportion: $\hat{p}$; Hypothesized proportion: p; $q = 1 - p$; Sample size: n

t-Test for a Hypothesized Mean μ *(Section 7.3)*

Test Statistic: $\bar{x}$

Critical value: t_0 (Use Table 5.)

Sampling distribution of sample means is a t-distribution with d.f. $= n - 1$.

Standardized Test Statistic: t

$$t = \frac{\bar{x} - \mu}{s/\sqrt{n}}$$

Sample mean: $\bar{x}$; Hypothesized mean: μ; Sample standard deviation: s; Sample size: n

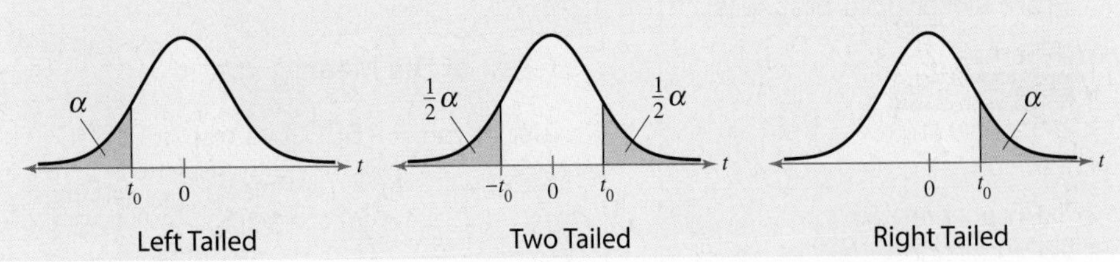

χ^2-Test for a Hypothesized Variance σ^2 or Standard Deviation σ *(Section 7.5)*

Test Statistic: s^2

Critical value: χ^2_0 (Use Table 6.)

Sampling distribution is a chi-square distribution with d.f. $= n - 1$.

Standardized Test Statistic: χ^2

$$\chi^2 = \frac{(n-1)s^2}{\sigma^2}$$

Sample size: n; Sample variance: s^2; Hypothesized variance: σ^2

α
χ^2_0
χ^2
Left Tailed

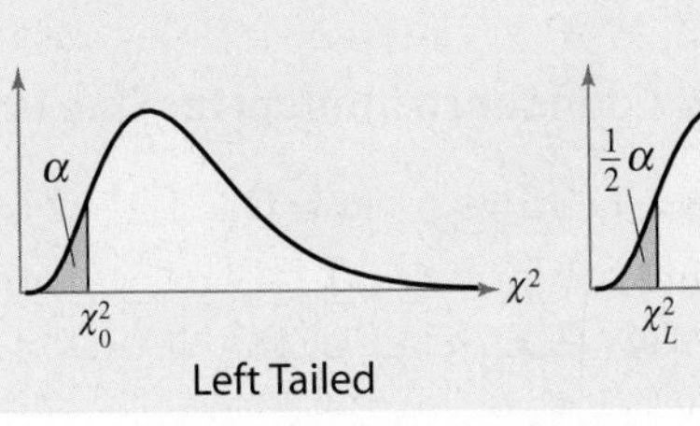

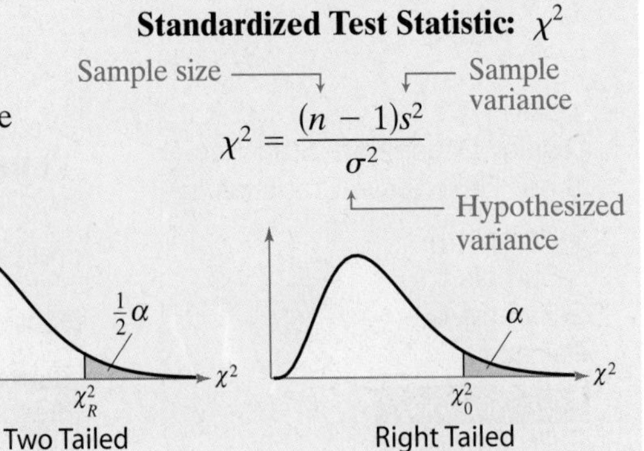

Using Technology to Perform Hypothesis Tests

Here are some *Minitab* and *TI-83* printouts for some of the examples in this chapter. To duplicate the *Minitab* results, you need the original data. For the *TI-83,* you can simply enter the descriptive statistics.

(See Example 5, page 338)

Data for Investments for 30 Franchises

\$70,700	\$69,400	\$90,600	\$85,500	\$97,100	\$100,800
\$114,700	\$119,600	\$123,600	\$127,200	\$131,200	\$132,000
\$134,400	\$136,700	\$138,500	\$140,900	\$143,300	\$143,900
\$151,100	\$151,700	\$157,200	\$159,800	\$163,000	\$163,500
\$169,300	\$167,400	\$168,800	\$168,800	\$159,100	\$170,200

Display Descriptive Statistics...
Store Descriptive Statistics...
1-Sample Z...
1-Sample t...
2-Sample t...
Paired t...
1 Proportion...
2 Proportions...

MINITAB

Z-Test of the Mean

Test of mu = 143260 vs mu not = 143260
The assumed sigma = 30000

Variable	N	Mean	StDev	SE Mean	Z	P
C1	30	135000	30000	5477	−1.51	0.13

(See Example 4, page 353)

Data for prices of 14 Ford F-150 Super Cab Pickups

\$14,500	\$13,790	\$13,800	\$14,030	\$15,400
\$15,560	\$15,855	\$15,935	\$16,150	\$16,300
\$17,100	\$17,100	\$17,070	\$17,210	

Display Descriptive Statistics...
Store Descriptive Statistics...
1-Sample Z...
1-Sample t...
2-Sample t...
Paired t...
1 Proportion...
2 Proportions...

MINITAB

T-Test of the Mean

Test of mu = 16500 vs mu < 16500

Variable	N	Mean	StDev	SE Mean	T	P
C1	14	15700	1250	334	−2.39	0.016

Display Descriptive Statistics...
Store Descriptive Statistics...
1-Sample Z...
1-Sample t...
2-Sample t...
Paired t...
1 Proportion...
2 Proportions...

MINITAB (See Example 2, page 362)

Test and Confidence Interval for One Proportion

Test of p = 0.23 vs p not = 0.23

Sample	X	N	Sample p	95.0 % CI	Z-Value	P-Value
1	54	200	0.270000	(0.208471, 0.331529)	1.34	0.179

(See Example 9, page 342)

TI-83

EDIT CALC TESTS
1: Z-Test...
2: T-Test...
3: 2-SampZTest...
4: 2-SampTTest...
5: 1-PropZTest...
6: 2-PropZTest...
7↓ZInterval...

TI-83

Z-Test
Inpt:Data Stats
μ_0: 45000
σ: 5200
$\bar{x}$: 43500
n: 30
μ: $\neq \mu_0$ $< \mu_0$ $> \mu_0$
Calculate Draw

TI-83

Z-Test
$\mu < 45000$
z= −1.579968916
p= .0570569964
$\bar{x}$= 43500
n= 30

TI-83

z=-1.58 p=.0571

(See Example 5, page 354)

TI-83

EDIT CALC TESTS
1: Z-Test...
2: T-Test...
3: 2-SampZTest...
4: 2-SampTTest...
5: 1-PropZTest...
6: 2-PropZTest...
7↓ZInterval...

TI-83

T-Test
Inpt:Data Stats
μ_0: 6.8
$\bar{x}$: 6.7
Sx: .24
n: 19
μ: $\neq \mu_0$ $< \mu_0$ $> \mu_0$
Calculate Draw

TI-83

T-Test
$\mu \neq 6.8$
t= −1.816207893
p= .0860316039
$\bar{x}$= 6.7
Sx= .24
n= 19

TI-83

t=-1.8162 p=.086

(See Example 1, page 361)

TI-83

EDIT CALC TESTS
1: Z-Test...
2: T-Test...
3: 2-SampZTest...
4: 2-SampTTest...
5: 1-PropZTest...
6: 2-PropZTest...
7↓ZInterval...

TI-83

1-PropZTest
p_0: .2
x: 15
n: 100
prop$\neq p_0$ $< p_0$ $> p_0$
Calculate Draw

TI-83

1-PropZTest
prop< .2
z= −1.25
p= .105649839
$\hat{p}$= .15
n= 100

TI-83

z=-1.25 p=.1056

7

Chapter Summary

What did you learn?	*Review Exercises*
Section 7.1	
◆ How to state a null hypothesis and an alternative hypothesis	*1–6*
◆ How to interpret the level of significance of a hypothesis test and identify type I and type II errors, how to know whether to use a one-tailed or a two-tailed statistical test, and how to interpret a decision based on the results of a statistical test	*7–10*
Section 7.2	
◆ How to find critical values for a z-test	*11–14*
◆ How to use the z-test to test a mean μ	*15–18, 21, 22*
◆ How to find P-values and use them to test a mean μ	*19, 20*
Section 7.3	
◆ How to find critical values for a t-test	*23–26*
◆ How to use the t-test to test a mean μ	*27–34*
◆ How to use technology to find P-values and use them with a t-test to test a mean μ	*35, 36*
Section 7.4	
◆ How to use the z-test to test a population proportion p	*37–46*
Section 7.5	
◆ How to find critical values for a χ^2-test	*47–50*
◆ How to use the χ^2-test to test a variance or a standard deviation	*51–56*

STATISTICS

Uses and Abuses

Uses

Hypothesis Testing Hypothesis testing is important in many different fields because it gives a scientific procedure for assessing the validity of a claim about a population. Some of the concepts in hypothesis testing are intuitive, but some are not. For instance, in this chapter you learned that a test statistic does not have to differ much from a hypothesized population parameter to conclude that the null hypothesis should be rejected.

Abuses

Not Using a Random Sample The entire theory of hypothesis testing is based on the fact that the sample is randomly selected. If the sample is not random, then you cannot use it to imply anything about a population parameter.

Attempting to Prove the Null Hypothesis Another abuse is to assume that you have proven the null hypothesis when the P-value is greater than the level of confidence (or the test statistic does not fall in the rejection region). Remember that hypothesis testing can never prove that the null hypothesis is true—only that there is not enough evidence to reject it. This is similar to a defendant who is found not guilty in a court of law. The defendant is not declared to be innocent—only that there was not enough evidence to convict him or her.

Making Type I or Type II Errors Remember that a Type I error is rejecting a null hypothesis that is true and a Type II error is failing to reject a null hypothesis that is false. You can decrease the chance of making both types of errors by increasing the sample size.

Exercises

In Exercises 1–4, assume that you work in a market research department. You are asked to write a report about the claim that 47% of all new car buyers prefer a 4-door car.

1. ***Not Using a Random Sample*** How could you choose a random sample to test this hypothesis?
2. ***Attempting to Prove the Null Hypothesis*** What is the null hypothesis in this situation? Describe how your report could be incorrect by trying to prove the null hypothesis.
3. ***Making a Type I Error*** Describe how your report could make a Type I error.
4. ***Making a Type II Error*** Describe how your report could make a Type II error.

7 Review Exercises

Section 7.1

In Exercises 1–6, use the given claim to state a null and an alternative hypothesis. Identify which hypothesis represents the claim.

1. Claim: $\mu \leq 1593$

2. Claim: $\mu = 35$

3. Claim: $p < 0.205$

4. Claim: $\mu \neq 150{,}020$

5. Claim: $\sigma > 4.5$

6. Claim: $p \geq 0.70$

In Exercises 7–10, do the following.

(a) State the null and alternative hypotheses.

(b) Determine when a type I or type II error occurs for a hypothesis test of the claim.

(c) Determine whether the hypothesis test is a left-tailed test, a right-tailed test, or a two-tailed test. Explain your reasoning.

(d) How should you interpret a decision that rejects the null hypothesis?

(e) How should you interpret a decision that fails to reject the null hypothesis?

7. An organization believes that the proportion of U.S. adults who use nonprescription pain relievers is 85%. *(Source: American Pharmaceutical Association)*

8. A tire manufacturer guarantees that the mean life of a certain type of tire is at least 30,000 miles.

9. A soup maker says that the standard deviation of the sodium content in one serving of a certain soup is no more than 50 milligrams. *(Adapted from Consumer Reports)*

10. A cereal maker claims that the mean number of fat calories in one serving of its cereal is less than 20.

Section 7.2

In Exercises 11–14, find the critical value(s) for the indicated *z*-test with level of significance α.

11. Left-tailed test, $\alpha = 0.02$

12. Two-tailed test, $\alpha = 0.005$

13. Right-tailed test, $\alpha = 0.025$

14. Two-tailed test, $\alpha = 0.08$

In Exercises 15–18, test the claim about the population mean μ with a *z*-test using the given sample statistics and level of significance α.

15. Claim: $\mu \leq 45$; $\alpha = 0.05$. Statistics: $\bar{x} = 47.2$, $s = 6.7$, $n = 42$

16. Claim: $\mu \neq 0$; $\alpha = 0.05$. Statistics: $\bar{x} = -0.69$, $s = 2.62$, $n = 60$

17. Claim: $\mu < 5.500$; $\alpha = 0.01$. Statistics: $\bar{x} = 5.497$, $s = 0.011$, $n = 36$

18. Claim: $\mu = 7450$; $\alpha = 0.05$. Statistics: $\bar{x} = 7512$, $s = 243$, $n = 57$

In Exercises 19 and 20, use a *P*-value to test the claim about the population mean μ using the given sample statistics. State your decision for $\alpha = 0.10$, $\alpha = 0.05$, and $\alpha = 0.01$ levels of significance.

19. Claim: $\mu \leq 0.05$; Statistics: $\bar{x} = 0.057$, $s = 0.018$, $n = 32$

20. Claim: $\mu \neq 230$; Statistics: $\bar{x} = 216.5$, $s = 17.3$, $n = 48$

In Exercises 21 and 22, test the claim about the population mean μ using rejection region(s) or a *P*-value.

21. A tourist agency in Florida claims the mean daily cost of meals and lodging for a family of four traveling in Florida is \$252. You work for a consumer protection advocate and want to test this claim. In a random sample of 50 families of four traveling in Florida, the mean daily cost of meals and lodging is \$260 and the standard deviation is \$25. Do you have enough evidence to reject the agency's claim? Use $\alpha = 0.05$. *(Adapted from American Automobile Association)*

22. A tourist agency in California claims the mean daily cost of meals and lodging for a family of four traveling in California is at most \$268. In an annual survey performed by the consumer protection advocacy you work for, a random sample of 45 families of four traveling in California had a mean daily cost of meals and lodging of \$277 with a standard deviation of \$40. At $\alpha = 0.05$, do you have enough evidence to reject the agency's claim? *(Adapted from American Automobile Association)*

Section 7.3

In Exercises 23–26, find the critical value(s) for the *t*-test with the indicated sample size *n* and level of significance α.

23. Two-tailed test, $n = 20$, $\alpha = 0.05$

24. Right-tailed test, $n = 8$, $\alpha = 0.01$

25. Left-tailed test, $n = 15$, $\alpha = 0.10$

26. Two-tailed test, $n = 12$, $\alpha = 0.05$

In Exercises 27–32, test the claim about the population mean μ using the given sample statistics and level of significance α. Assume the population is normally distributed.

27. Claim: $\mu \neq 95$; $\alpha = 0.05$. Statistics: $\bar{x} = 94.1$, $s = 1.53$, $n = 12$

28. Claim: $\mu > 12{,}700$; $\alpha = 0.05$. Statistics: $\bar{x} = 12{,}804$, $s = 248$, $n = 21$

29. Claim: $\mu \geq 0$; $\alpha = 0.10$. Statistics: $\bar{x} = -0.45$, $s = 1.38$, $n = 16$

30. Claim: $\mu = 4.20$; $\alpha = 0.02$. Statistics: $\bar{x} = 4.41$, $s = 0.26$, $n = 9$

31. Claim: $\mu \leq 48$; $\alpha = 0.01$. Statistics: $\bar{x} = 52$, $s = 2.5$, $n = 7$

32. Claim: $\mu < 850$; $\alpha = 0.025$. Statistics: $\bar{x} = 875$, $s = 25$, $n = 14$

In Exercises 33 and 34, use a *t*-test to investigate the claim. Assume each population is normally distributed.

33. A fitness magazine advertises that the average monthly cost of joining a health club is \$25. You work for a consumer advocacy group and are asked to test this claim. You find that a random sample of 18 clubs has a mean monthly cost of \$26.25 and a standard deviation of \$3.23. At $\alpha = 0.10$, do you have enough evidence to reject the advertisement's claim?

34. A certain restaurant claims that its hamburgers have no more than 10 grams of fat. You work for a nutritional health agency and are asked to test this claim. You find that a random sample of nine hamburgers has a mean fat content of 13.5 grams and a standard deviation of 5.8 grams. At $\alpha = 0.10$, do you have enough evidence to reject the restaurant's claim?

In Exercises 35 and 36, use a *t*-statistic and its *P*-value to test the claim about the population mean μ using the given sample statistics. Assume the population is normally distributed.

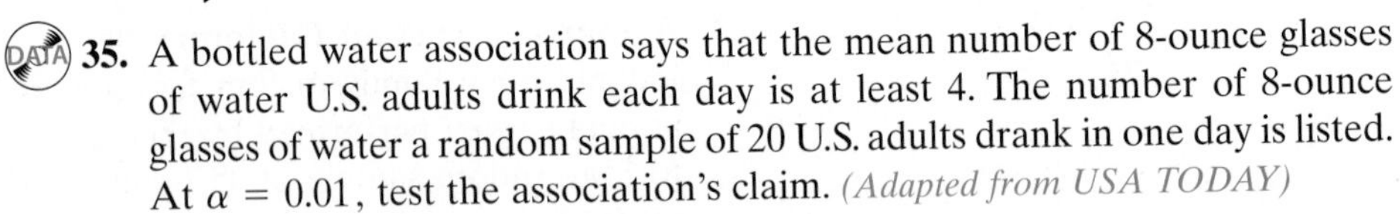

35. A bottled water association says that the mean number of 8-ounce glasses of water U.S. adults drink each day is at least 4. The number of 8-ounce glasses of water a random sample of 20 U.S. adults drank in one day is listed. At $\alpha = 0.01$, test the association's claim. *(Adapted from USA TODAY)*

4.7 5.4 3.2 3.9 4.3 4.5 2.5 5.2 4.5 5.3
3.2 2.2 4.1 2.3 3.7 5.1 4.1 3.6 3.1 2.8

36. A large university says the mean number of classroom hours per week for full-time faculty is more than 9. A random sample of the number of classroom hours for full-time faculty for one week is listed. At $\alpha = 0.05$, test the university's claim. *(Adapted from National Center for Education Statistics)*

10.7 9.8 11.6 9.7 7.6 11.3 14.1 8.1 11.5 8.5 6.9

Section 7.4

In Exercises 37–44, decide whether the normal distribution can be used to approximate the binomial distribution. If it can, use the *z*-test to test the claim about the population proportion *p* for the given sample statistics and level of significance α.

37. Claim: $p = 0.15$; $\alpha = 0.05$. Statistics: $\hat{p} = 0.09$, $n = 40$

38. Claim: $p < 0.70$; $\alpha = 0.01$. Statistics: $\hat{p} = 0.50$, $n = 68$

39. Claim: $p < 0.08$; $\alpha = 0.05$. Statistics: $\hat{p} = 0.03$, $n = 45$

40. Claim: $p = 0.50$; $\alpha = 0.10$. Statistics: $\hat{p} = 0.71$, $n = 129$

41. Claim: $p \geq 0.04$; $\alpha = 0.10$. Statistics: $\hat{p} = 0.03$, $n = 30$

42. Claim: $p \neq 0.34$; $\alpha = 0.01$. Statistics: $\hat{p} = 0.29$, $n = 60$

43. Claim: $p \neq 0.20$; $\alpha = 0.01$. Statistics: $\hat{p} = 0.23$, $n = 56$

44. Claim: $p \leq 0.80$; $\alpha = 0.10$. Statistics: $\hat{p} = 0.85$, $n = 43$

In Exercises 45 and 46, test the claim about the population proportion p.

45. A communications industry spokesperson claims that over 40% of people in the U.S. either own a cellular phone or have a family member that does. In a random survey of 1036 people in the U.S., 456 said that they or a family member owned a cellular phone. Test the spokesperson's claim at the $\alpha = 0.10$ level. What can you conclude? *(Adapted from Wirthlin Worldwide)*

46. The Western blot assay is a blood test for the presence of HIV. It has been found that this test sometimes gives false positive results for HIV; specifically, when it does not find a certain antibody called p-31 antibody. A medical researcher claims that the rate of false positives in this case is 50%. A recent study of 39 randomly selected U.S. blood donors who tested positive for HIV but did not have p-31 antibodies in their blood found that 20 were actually HIV negative. Test the researcher's claim of a 50% false positive rate at the $\alpha = 0.05$ level. What can you conclude? *(Source: The Journal of the American Medical Association)*

Section 7.5

In Exercises 47–50, find the critical value(s) for the χ^2-test with the indicated sample size n and level of significance α.

47. Right-tailed test, $n = 20$, $\alpha = 0.05$

48. Two-tailed test, $n = 14$, $\alpha = 0.01$

49. Right-tailed test, $n = 25$, $\alpha = 0.10$

50. Left-tailed test, $n = 6$, $\alpha = 0.05$

In Exercises 51–54, test the claim about the indicated population parameter (σ or σ^2) with a χ^2-test using the given sample statistics and level of significance α. Assume the population is normally distributed.

51. Claim: $\sigma^2 > 2$; $\alpha = 0.10$. Statistics: $s^2 = 2.38$, $n = 18$

52. Claim: $\sigma^2 \leq 60$; $\alpha = 0.025$. Statistics: $s^2 = 72.7$, $n = 15$

53. Claim: $\sigma = 1.25$; $\alpha = 0.05$. Statistics: $s = 1.03$, $n = 6$

54. Claim: $\sigma \neq 0.035$; $\alpha = 0.01$. Statistics: $s = 0.026$, $n = 16$

In Exercises 55 and 56, test the claim about the population variance or standard deviation. Assume each population is normally distributed.

55. A bolt manufacturer makes a type of bolt to be used in airtight containers. The manufacturer needs to be sure that all of its bolts are very similar in width, so it sets an upper tolerance limit for the variance of bolt width at 0.01. A random sample of 28 bolts yields a variance of 0.064 for bolt width. Test the manufacturer's claim that the variance is at most 0.01 at the $\alpha = 0.005$ level. What can you conclude?

56. A bottler needs to be sure that its liquid dispensers are set properly. The standard deviation of liquid dispensed must be no more than 0.0025 liter. A random sample of 14 bottles has a standard deviation of 0.0031 liter. Test the bottler's claim that the standard deviation is no more than 0.0025 liter at the $\alpha = 0.01$ level. What can you conclude?

PUTTING IT ALL TOGETHER

Real Statistics Real Decisions

You work for the public relations department of the Social Security Administration. In an effort to design better advertising campaigns, your department decides to conduct a survey to find out the opinions people in the U.S. have about the Social Security system. One of the questions asked and the results of each response and the respondent's age are given in the table.

Your department believes that less than 40% of people in the U.S. think the Social Security system will have the money available to provide the benefits they expect for their retirements. Also, your department believes that the mean age of people in the U.S. who would say yes to this question is 60 years or older. As the department's research analyst, it is your job to work with the data and determine if these claims can be supported or rejected.

www.ssa.gov

Do you think the Social Security system will have the money available to provide the benefits you expect for your retirement?

Age	Response
83	Yes
37	No
49	No
27	Yes
27	No
59	Yes
44	Yes
46	No
25	No
30	No
72	Yes
66	No
47	No
47	No
50	Yes
76	Yes
51	Yes
58	No
79	Yes
69	Yes
66	No
35	Yes
47	No
29	Yes
76	No
31	Yes
40	No
61	Yes
30	Yes
71	No
64	No
18	No

Adapted from Newsweek

Exercises

1. *How Would You Do It?*
   (a) What sampling technique would you use to select the sample for the study? Why? What sampling technique would you use if you wanted to select samples from four age groups: 18–29, 30–49, 50–64, and 65 and over?
   (b) Which technique in part (a) will give you a sample that is representative of the population?
   (c) Identify possible flaws or biases in your study.
2. *Testing a Proportion*
   Test the claim that less than 40% of people in the U.S. think the Social Security system will have the money available to provide the benefits they expect for their retirements. Use $\alpha = 0.10$. Write a paragraph that interprets the test's decision. Does the decision support your department's claim?
3. *Testing a Mean*
   Test the claim that the mean age of people in the U.S. who would say yes to the survey question given in the table is 60 years or older. Use $\alpha = 0.10$ and assume that the population is normally distributed. Write a paragraph that interprets the test's decision. Is there enough evidence to reject your department's claim?
4. *Your Conclusions*
   Based on your analysis of the responses to this survey question, what would you tell your department?

7 Chapter Quiz

Take this quiz as you would take a quiz in class. After you are done, check your work against the answers given in the back of the book.

For this quiz, do the following.

(a) State the claim mathematically. Identify H_0 and H_a.

(b) Determine when a type I or type II error occurs.

(c) Determine whether the hypothesis test is a one-tailed or a two-tailed test and whether to use a z-test, a t-test, or a χ^2-test. Explain your reasoning.

(d) If necessary, find the critical value(s) and identify the rejection region(s).

(e) Find the appropriate test statistic. If necessary, find the P-value.

(f) Decide whether to reject or fail to reject the null hypothesis. Then interpret the decision in the context of the original claim.

1. A citrus grower's association believes that the mean consumption of fresh citrus fruits by people in the U.S. is at least 94 pounds per year. A random sample of 103 people in the U.S. has a mean consumption of fresh citrus fruits of 93.5 pounds per year and a standard deviation of 30 pounds. At $\alpha = 0.02$, can you reject the association's claim that the mean consumption of fresh citrus fruits by people in the U.S. is at least 94 pounds per year? *(Adapted from U.S. Department of Agriculture)*

2. An auto maker estimates that the mean gas mileage of its luxury sedan is at least 25 miles per gallon. A random sample of eight such cars had a mean of 23 miles per gallon and a standard deviation of 5 miles per gallon. At $\alpha = 0.05$, can you reject the auto maker's claim that the mean gas mileage of its luxury sedan is at least 25 miles per gallon? Assume the population is normally distributed. *(Adapted from Consumer Reports)*

3. A maker of microwave ovens advertises that no more than 10% of its microwaves need repair during the first five years of use. In a random sample of 57 microwaves five years old, 13% needed repairs. At $\alpha = 0.04$, can you reject the maker's claim that no more than 10% of its microwaves need repair during the first five years of use? *(Adapted from Consumer Reports)*

4. A state school administrator says that the standard deviation of SAT verbal test scores is 105. A random sample of 14 SAT verbal test scores has a standard deviation of 113. At $\alpha = 0.01$, test the administrator's claim. What can you conclude? Assume the population is normally distributed. *(Adapted from National Center for Educational Statistics)*

5. An employment information service reports that the mean annual salary for full-time male workers over the age of 25 with a bachelor's degree is \$53,102. You doubt the validity of this claim. If a random sample of 12 full-time male workers with bachelor's degrees has a mean annual salary of \$52,201 with a standard deviation of \$6500, can you reject the service's claim? Use a P-value and $\alpha = 0.05$. Assume the population is normally distributed. *(Adapted from U.S. Bureau of the Census)*

6. A tourist agency in Massachusetts claims the mean daily cost of meals and lodging for a family of four traveling in the state is \$276. You work for a consumer protection advocate and want to test this claim. In a random sample of 35 families of four traveling in Massachusetts, the mean daily cost of meals and lodging is \$285 and the standard deviation is \$30. Do you have enough evidence to reject the agency's claim? Use a P-value and $\alpha = 0.05$. *(Adapted from American Automobile Association)*

Where You've Been

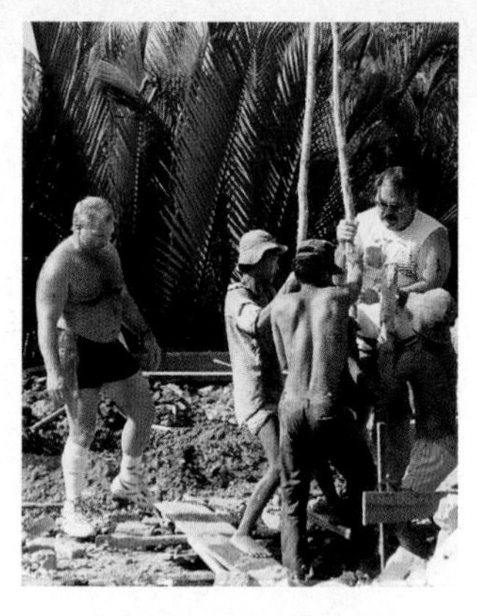

Vets With a Mission doing volunteer work on a classroom in Vietnam.

In Chapter 6, you were introduced to inferential statistics and you learned how to form confidence intervals to estimate a parameter. Then, in Chapter 7, you learned how to test a claim about a population parameter, basing your decision on sample statistics and their distributions.

A study reported in the *Journal of the American Medical Association (JAMA)* describes a random sample of 7924 enlisted men who entered the Army from 1965 to 1971 and served in Vietnam. The following proportions were found.

Army Vets Who Served in Vietnam
($n = 7924$)

Condition	Frequency	Proportion
Depression	357	$\hat{p}_1 = 0.045$
Anxiety	388	$\hat{p}_1 = 0.049$
Alcohol abuse or dependence	1086	$\hat{p}_1 = 0.137$

The Vietnam War was the longest and most unpopular war the United States ever fought. More than 2 million Americans fought in the war—fifty-eight thousand of them were killed. Since the war, there have been many studies about the effects on Vietnam veterans returning to a country that was so divided in its support of the war.

Universal Pictures presents an A. Kitman Ho & Ixtlan production, an Oliver Stone picture. Tom Cruise, *Born on the Fourth of July*. Kyra Sedgwick, Raymond J. Barry, Jerry Levine, Frank Whaley, and Willem Dafoe. Music by John Williams. Production designer, Bruno Rubeo. Director of Photography, Robert Richardson. Based on the novel by Ron Kovic. Screenplay by Oliver Stone, David Freeman, & Ron Kovic.

Hypothesis Testing with Two Samples

CHAPTER

8

Where You're Going

In this chapter, you will continue your study of inferential statistics and hypothesis testing. Now, however, instead of testing a hypothesis about a single population, you will learn how to test a hypothesis that compares two populations.

For instance, in the *JAMA* study a second random sample of 7364 enlisted men who entered the Army from 1965 to 1971 was taken. These men were similar to the Vietnam veteran sample in terms of level of education, employment, income, marital status, and satisfaction with personal relationships. The second group, however, had *not* served in Vietnam. Here are the study's findings for the second group.

Army Vets Who Did Not Serve in Vietnam
(n = 7364)

Condition	**Frequency**	**Proportion**
Depression	169	$\hat{p}_2 = 0.023$
Anxiety	236	$\hat{p}_2 = 0.032$
Alcohol abuse or dependence	677	$\hat{p}_2 = 0.092$

From these two samples, can you conclude that there was a significantly greater proportion of depression, anxiety, or alcohol abuse or dependence among Army vets who served in Vietnam than among those who did not serve in Vietnam? Or, might the differences in the proportions be due to chance?

In this chapter, you will learn that you can answer these questions by testing the hypothesis that the two proportions are equal. For the proportions of depression, the P-value for the hypothesis that $p_1 = p_2$ is about 0.0000. So, it is almost impossible that the two groups experienced the same proportion of depression.

8.1 Testing the Difference Between Means (Large Independent Samples)

What You Should Learn

- An introduction to two-sample hypothesis testing for the difference between two population parameters
- How to perform a two-sample z-test for the difference between two means μ_1 and μ_2 using large independent samples

An Overview of Two-Sample Hypothesis Testing • Two-Sample z-Test for the Difference Between Means

An Overview of Two-Sample Hypothesis Testing

In Chapter 7, you studied methods for testing a claim about the value of a population parameter. In this chapter, you will learn how to test a claim comparing parameters from two populations.

For instance, suppose you are developing a marketing plan for an Internet service provider and want to determine whether there is a difference in the amount of time male and female college students spend online each day. The only way you could conclude with certainty that there is a difference is to take a census of all college students, calculate the mean daily times male students and female students spend online, and find the difference. Of course, it is not practical to take such a census. However, you can still determine with some degree of certainty whether such a difference exists.

You can begin by assuming that there is no difference in the mean times of the two populations. That is, $\mu_1 - \mu_2 = 0$. Then by taking a random sample from each population, and using the resulting two-sample test statistic $\bar{x}_1 - \bar{x}_2$, you can perform a two-sample hypothesis test. Suppose you obtain the following results.

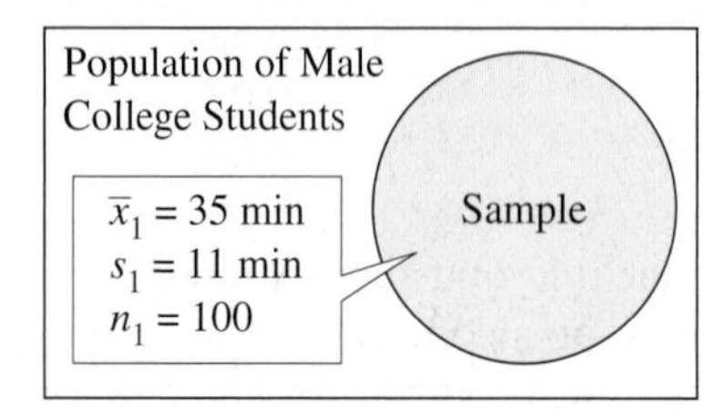

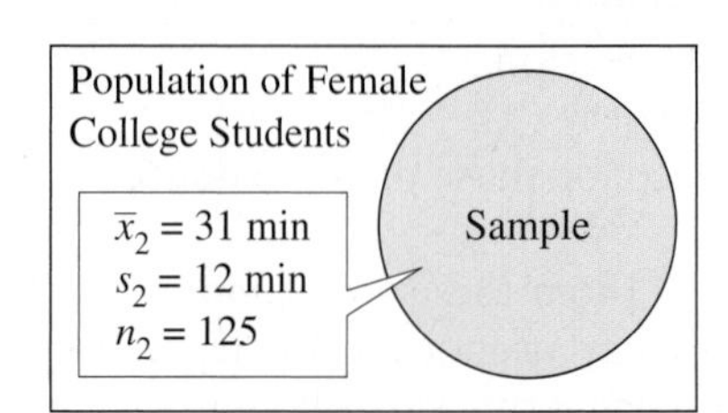

The graph below shows the sampling distribution of $\bar{x}_1 - \bar{x}_2$ for many similar samples taken from each population, under the assumption that $\mu_1 - \mu_2 = 0$. From the graph, you can see that it would be quite unlikely to obtain sample means that differ by 4 minutes if the actual difference were 0. The difference of the sample means would be more than 2.5 standard errors from the hypothesized difference of 0! So, you can conclude that there is a significant difference in the amount of time male college students and female college students spend online each day.

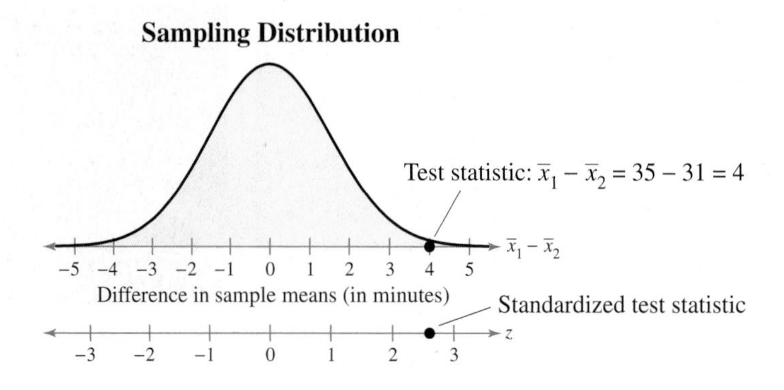

It is important to remember that when you perform a two-sample hypothesis test, you are testing a claim concerning the difference between the parameters in two populations, not the values of the parameters themselves.

DEFINITION

For a two-sample hypothesis test,

1. the **null hypothesis** H_0 is a statistical hypothesis that usually states there is no difference between the parameters of two populations. The null hypothesis always contains the symbol $\leq$, $=$, or $\geq$.
2. the **alternative hypothesis** H_a is a statistical hypothesis that is true when H_0 is false. The alternative hypothesis contains the symbol $>$, $\neq$, or $<$.

Study Tip

You can also write the null and alternative hypotheses as follows.

$$\begin{cases} H_0\colon \mu_1 - \mu_2 = 0 \\ H_a\colon \mu_1 - \mu_2 \neq 0 \end{cases}$$

$$\begin{cases} H_0\colon \mu_1 - \mu_2 \leq 0 \\ H_a\colon \mu_1 - \mu_2 > 0 \end{cases}$$

$$\begin{cases} H_0\colon \mu_1 - \mu_2 \geq 0 \\ H_a\colon \mu_1 - \mu_2 < 0 \end{cases}$$

To write a null and an alternative hypothesis for a two-sample hypothesis test, translate the claim made about the population parameters from a verbal statement to a mathematical statement. Then, write its complementary statement. For instance, if the claim is about two population parameters μ_1 and μ_2, then some possible pairs of null and alternative hypotheses are

$$\begin{cases} H_0\colon \mu_1 = \mu_2 \\ H_a\colon \mu_1 \neq \mu_2 \end{cases}, \quad \begin{cases} H_0\colon \mu_1 \leq \mu_2 \\ H_a\colon \mu_1 > \mu_2 \end{cases}, \quad \text{and} \quad \begin{cases} H_0\colon \mu_1 \geq \mu_2 \\ H_a\colon \mu_1 < \mu_2 \end{cases}.$$

Two-Sample *z*-Test for the Difference Between Means

In the remainder of this section, you will learn how to perform a z-test for the difference between two population means μ_1 and μ_2. To perform such a test, three conditions are necessary.

1. The samples must be randomly selected.
2. The samples must be independent. Two samples are **independent** if the sample selected from one population is not related to the sample selected from the second population. (You will study the distinction between independent samples and dependent samples in Section 8.3.)
3. Each sample size must be at least 30 or, if not, each population must have a normal distribution with a known standard deviation.

Sampling Distribution for $\bar{x}_1 - \bar{x}_2$

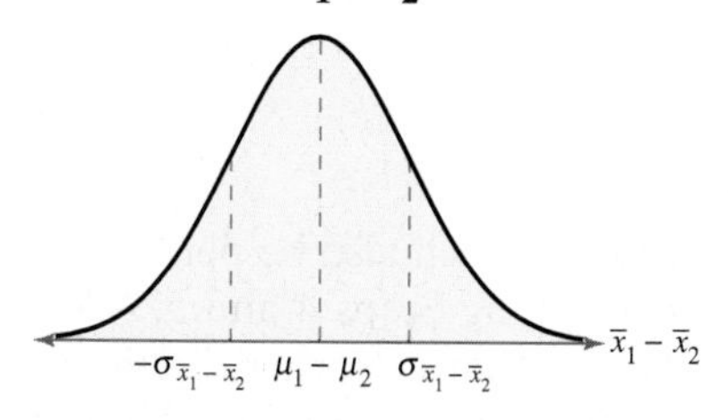

If these requirements are met, then the sampling distribution for $\bar{x}_1 - \bar{x}_2$ (the difference of the sample means) is a normal distribution with mean and standard error of

$$\mu_{\bar{x}_1-\bar{x}_2} = \mu_{\bar{x}_1} - \mu_{\bar{x}_2} = \mu_1 - \mu_2$$

and

$$\sigma_{\bar{x}_1-\bar{x}_2} = \sqrt{\sigma_{\bar{x}_1}^2 + \sigma_{\bar{x}_2}^2} = \sqrt{\frac{\sigma_1^2}{n_1} + \frac{\sigma_2^2}{n_2}}.$$

Notice that the variance of the sampling distribution $\sigma_{\bar{x}_1-\bar{x}_2}^2$ is the sum of the variances of the individual sampling distributions for $\bar{x}_1$ and $\bar{x}_2$.

Because the sampling distribution for $\bar{x}_1 - \bar{x}_2$ is a normal distribution, you can use the z-test to test the difference between two population means μ_1 and μ_2. Notice that the standardized test statistic takes the form of

$$z = \frac{(\text{Observed difference}) - (\text{Hypothesized difference})}{(\text{Standard error})}.$$

Two-Sample z-Test for the Difference Between Means

A **two-sample z-test** can be used to test the difference between two population means μ_1 and μ_2 when a large sample (at least 30) is randomly selected from each population and the samples are *independent*. The **test statistic** is $\bar{x}_1 - \bar{x}_2$, and the **standardized test statistic** is

$$z = \frac{(\bar{x}_1 - \bar{x}_2) - (\mu_1 - \mu_2)}{\sigma_{\bar{x}_1 - \bar{x}_2}} \quad \text{where} \quad \sigma_{\bar{x}_1 - \bar{x}_2} = \sqrt{\frac{\sigma_1^2}{n_1} + \frac{\sigma_2^2}{n_2}}.$$

When the samples are large, you can use s_1 and s_2 in place of σ_1 and σ_2. If the samples are not large, you can still use a two-sample z-test, provided the populations are normally distributed and the population standard deviations are known.

If the null hypothesis states $\mu_1 = \mu_2$, $\mu_1 \leq \mu_2$, or $\mu_1 \geq \mu_2$, then $\mu_1 = \mu_2$ is assumed and the expression $\mu_1 - \mu_2$ is equal to 0 in the preceding test.

GUIDELINES

Using a Two-Sample z-Test for the Difference Between Means (Large Independent Samples)

In Words	*In Symbols*
1. Identify the claim. State the null and the alternative hypotheses.	State H_0 and H_a.
2. Specify the level of significance.	Identify α.
3. Sketch the sampling distribution.	
4. Find the critical value(s).	Use Table 4 in Appendix B.
5. Determine the rejection region(s).	
6. Find the standardized test statistic.	$z = \frac{(\bar{x}_1 - \bar{x}_2) - (\mu_1 - \mu_2)}{\sigma_{\bar{x}_1 - \bar{x}_2}}$
7. Make a decision to reject or fail to reject the null hypothesis.	If z is in the rejection region, reject H_0. Otherwise, do not reject H_0.
8. Interpret the decision in the context of the original claim.	

A hypothesis test for the difference between means can also be performed using P-values. Use the same guidelines above, skipping Steps 4 and 5. After finding the standardized test statistic, use the Standard Normal Table to calculate the P-value. Then make a decision to reject or fail to reject the null hypothesis. If P is less than or equal to α, reject H_0. Otherwise, fail to reject H_0.

Picturing the World

There are about 21,500 civilian federal employees working in Michigan and about 25,500 in Massachusetts. In a survey, 200 federal employees in each state were asked to report their salary. The results were as follows. *(Adapted from United States Office of Personal Management)*

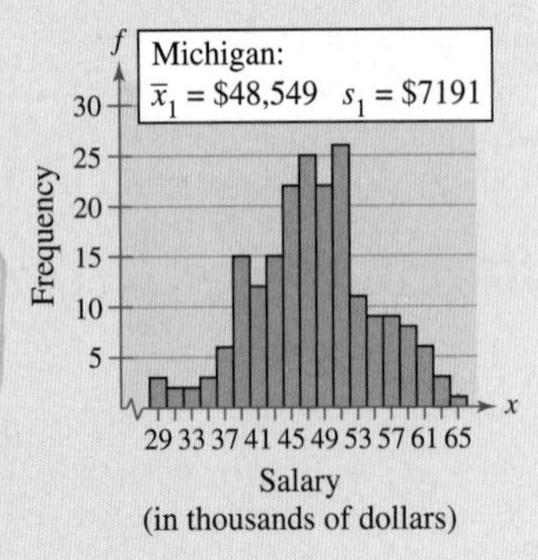

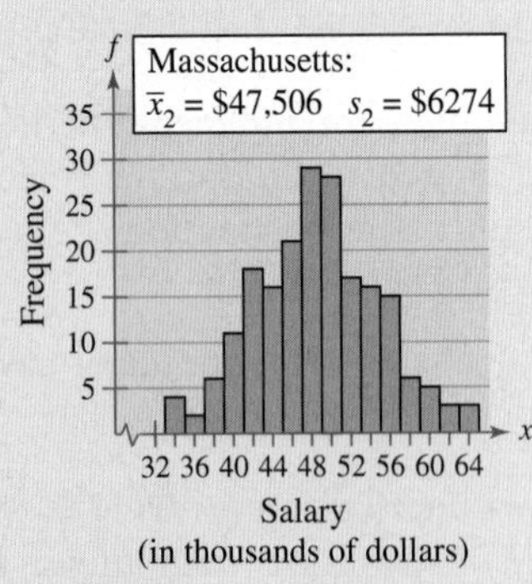

Is there enough evidence to conclude that there is a difference in the mean incomes of civilian federal employees in Michigan and Massachusetts using $\alpha = 0.05$?

See *TI-83* steps on page 431.

EXAMPLE 

A Two-Sample z-Test for the Difference Between Means

An advertising executive claims that there is a difference in the mean household income for credit card holders of Visa Gold and of MasterCard Gold. The results of a random survey of 100 customers from each group are shown at the left. The two samples are independent. Do the results support the executive's claim? Use $\alpha = 0.05$. *(Adapted from Claritas Inc.)*

Sample Statistics for Household Incomes

Visa Gold	MasterCard Gold
$\bar{x}_1 = \$60{,}900$ $s_1 = \$12{,}000$ $n_1 = 100$	$\bar{x}_2 = \$64{,}300$ $s_2 = \$15{,}000$ $n_2 = 100$

SOLUTION You want to test the claim that there is a difference in the mean household incomes for Visa Gold and MasterCard Gold credit card holders. So, the null and alternative hypotheses are

$$H_0: \mu_1 = \mu_2 \quad \text{and} \quad H_a: \mu_1 \neq \mu_2. \text{ (Claim)}$$

Because the test is a two-tailed test and the level of significance is $\alpha = 0.05$, the critical values are -1.96 and 1.96. The rejection regions are $z < -1.96$ and $z > 1.96$. Because both samples are large, s_1 and s_2 are used to calculate the standard error.

$$\sigma_{\bar{x}_1 - \bar{x}_2} = \sqrt{\frac{s_1^2}{n_1} + \frac{s_2^2}{n_2}} = \sqrt{\frac{12{,}000^2}{100} + \frac{15{,}000^2}{100}} \approx 1921$$

Using the z-test, the standardized test statistic is

$$z = \frac{(\bar{x}_1 - \bar{x}_2) - (\mu_1 - \mu_2)}{\sigma_{\bar{x}_1 - \bar{x}_2}}$$

$$\approx \frac{(60{,}900 - 64{,}300) - 0}{1921} \qquad \mu_1 - \mu_2 = 0 \text{ because } H_0 \text{ states } \mu_1 = \mu_2$$

$$\approx -1.770.$$

Study Tip

In Example 1, the P-value is 0.0768. Because this is greater than 0.05, you should fail to reject H_0.

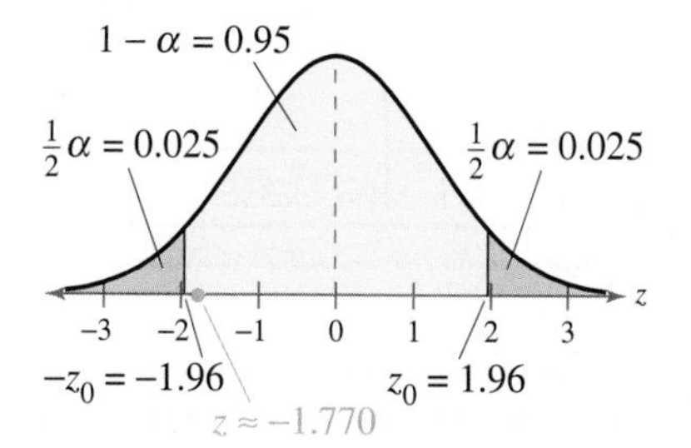

The graph at the left shows the location of the rejection regions and the standardized test statistic z. Because z is not in the rejection region, you should fail to reject the null hypothesis. At the 5% level, there is not enough evidence to conclude that there is a significant difference in the mean household incomes of Visa Gold and MasterCard Gold credit card holders.

Try It Yourself 1

A survey indicates that the mean per capita credit card charge for residents of New Hampshire and New York is \$3900 and \$3500 per year, respectively. The survey included a randomly selected sample of size 50 from each state and sample standard deviations are \$900 (NH) and \$500 (NY). The two samples are independent. At $\alpha = 0.01$, is there enough evidence to conclude that there is a difference in the mean credit-card charges? *(Adapted from Card Management Information Services and the U.S. Census Bureau)*

a. *Identify* the claim and state H_0 and H_a.
b. *Specify* the level of significance α.
c. *Find* the critical values and *identify* the rejection regions.
d. *Use* the z-test to find the standardized test statistic z.
e. *Decide* whether to reject the null hypothesis. Use a graph if necessary.
f. *Interpret* your decision.

Answer: Page A40

EXAMPLE 

Using Technology to Perform a Two-Sample z-Test

Sample Statistics for Daily Cost of Meals and Lodging for a Family of Four

Texas	Washington
$\bar{x}_1 = \$208$	$\bar{x}_2 = \$218$
$s_1 = \$15$	$s_2 = \$28$
$n_1 = 50$	$n_2 = 35$

The American Automobile Association claims that the average daily cost for meals and lodging when vacationing in Texas is less than the same average costs when vacationing in Washington state. The table at the left shows the results of a random survey of vacationers in each state. The two samples are independent. At $\alpha = 0.01$, is there enough evidence to support the claim? (H_0: $\mu_1 \geq \mu_2$ and H_a: $\mu_1 < \mu_2$ (claim))

SOLUTION The top two displays show how to set up the hypothesis test using a *TI-83*. The remaining displays show the possible results, depending on whether you select "Calculate" or "Draw."

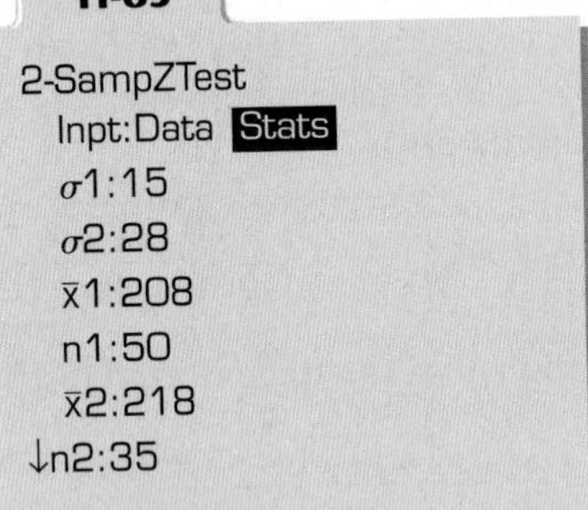

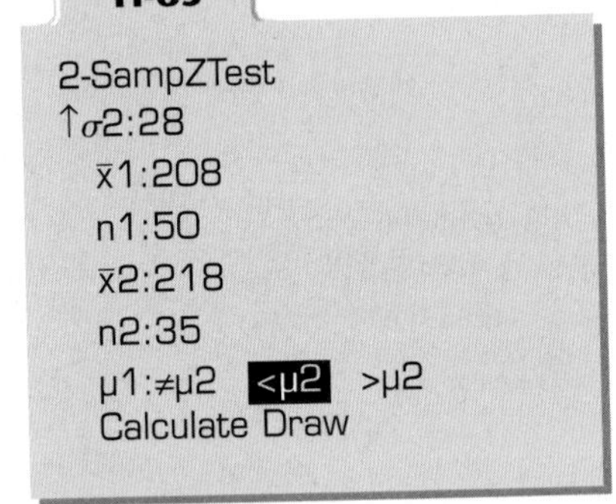

TI-83

2-SampZTest
μ1<μ2
z=−1.928074718
p=.0269228477
x̄1:208
x̄2:218
↓n1:50

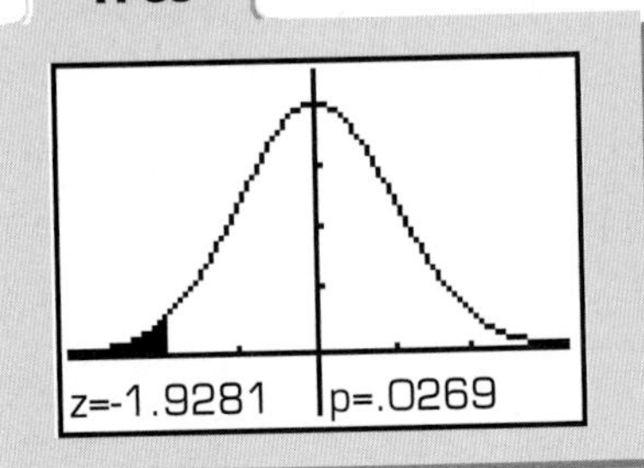

Study Tip

Note that the *TI-83* displays $P \approx 0.0269$. Because $P > \alpha$, you should fail to reject the null hypothesis.

To use *Minitab* to perform a two-sample z-test, you must use raw data.

Because the test is a left-tailed test and $\alpha = 0.01$, the rejection region is $z < -2.33$. The standardized test statistic $z \approx -1.93$ is not in the rejection region, so you should fail to reject the null hypothesis. At the 1% level, there is not enough evidence to support the American Automobile Association's claim.

Try It Yourself 2

The American Automobile Association claims that the average daily meal and lodging costs while vacationing in Florida are greater than the same average costs while vacationing in Delaware. The table at the left shows the results of a random survey of vacationers in each state. The two samples are independent. At $\alpha = 0.05$, is there enough evidence to support the claim?

Sample Statistics for Daily Cost of Meals and Lodging for a Family of Four

Florida	Delaware
$\bar{x}_1 = \$252$	$\bar{x}_2 = \$242$
$s_1 = 22$	$s_2 = 18$
$n_1 = 150$	$n_2 = 200$

a. *Use* a *TI-83* to find the test statistic or the *P*-value.
b. *Determine* whether the test statistic is in the rejection region or *compare* the *P*-value to the level of significance α.
c. *Make* a decision.

Answer: Page A40

8.1 Exercises

Help

LarsonTutor 8.1

Companion Web Site

Student Solutions Manual 8.1

Videos 8.1

Technology Manuals

Basic Skills and Concepts

1. Explain how to perform a two-sample z-test for the difference between the means of two populations using large independent samples.

2. What two conditions are necessary in order to use the z-test to test the difference between two population means?

In Exercises 3–6, (a) find the test statistic, (b) find the standardized test statistic, (c) decide whether the standardized test statistic is in the rejection region, and (d) decide whether you should reject or fail to reject the claim. The samples are random and independent.

3. Claim: $\mu_1 = \mu_2$, $\alpha = 0.05$
Sample statistics: $\overline{x}_1 = 16$, $s_1 = 1.1$, $n_1 = 50$, and $\overline{x}_2 = 14$, $s_2 = 1.5$, $n_2 = 50$

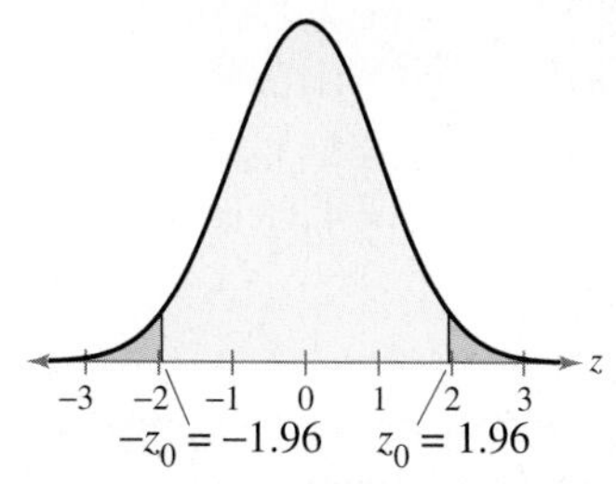

Figure for Exercise 3

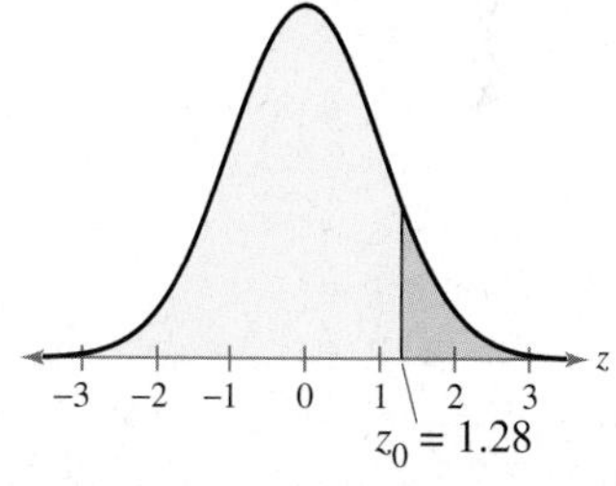

Figure for Exercise 4

4. Claim: $\mu_1 > \mu_2$, $\alpha = 0.10$
Sample statistics: $\overline{x}_1 = 500$, $s_1 = 30$, $n_1 = 100$, and $\overline{x}_2 = 510$, $s_2 = 15$, $n_2 = 75$

5. Claim: $\mu_1 < \mu_2$, $\alpha = 0.01$
Sample statistics: $\overline{x}_1 = 1225$, $s_1 = 75$, $n_1 = 35$, and $\overline{x}_2 = 1195$, $s_2 = 105$, $n_2 = 105$

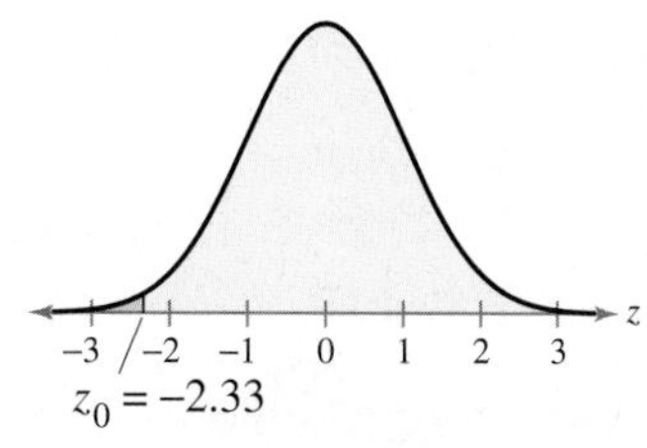

Figure for Exercise 5

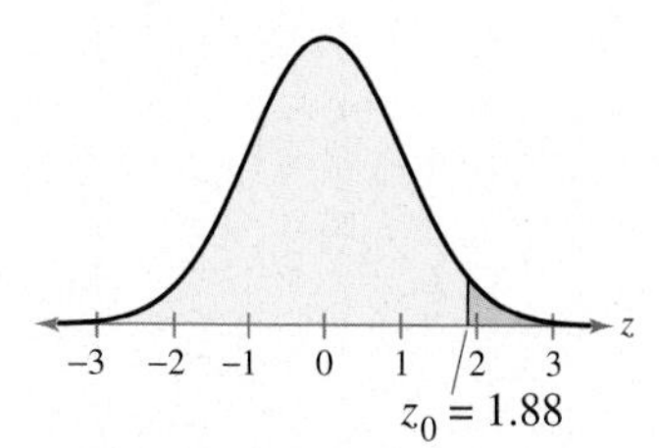

Figure for Exercise 6

6. Claim: $\mu_1 \leq \mu_2$, $\alpha = 0.03$
Sample statistics: $\overline{x}_1 = 5004$, $s_1 = 136$, $n_1 = 144$, and $\overline{x}_2 = 4895$, $s_2 = 215$, $n_2 = 156$

In Exercises 7 and 8, use the given sample statistics to test the claim about the difference between two population means μ_1 and μ_2 at the given level of significance α.

7. Claim: $\mu_1 \leq \mu_2$; $\alpha = 0.01$. Statistics: $\overline{x}_1 = 5.2$, $s_1 = 0.2$, $n_1 = 45$, and $\overline{x}_2 = 5.5$, $s_2 = 0.3$, $n_2 = 37$

8. Claim: $\mu_1 = \mu_2$; $\alpha = 0.05$. Statistics: $\overline{x}_1 = 52$, $s_1 = 2.5$ $n_1 = 70$, and $\overline{x}_2 = 45$, $s_2 = 5.5$, $n_2 = 60$

Testing the Difference Between Two Means In Exercises 9–22, (a) identify the claim and state H_0 and H_a, (b) use Table 4 in Appendix B to find the critical value(s) and identify the rejection region(s), (c) find the standardized test statistic z, and (d) decide whether to reject or fail to reject the null hypothesis. Then interpret the decision in the context of the original claim. If convenient, use technology to solve the problem. In each exercise, assume the samples are randomly selected and that the samples are independent.

9. ***Braking Distances*** ◆ A safety engineer records the braking distances of two types of tires. Each randomly selected sample has 35 tires. The results of the tests are shown in the figure. At $\alpha = 0.10$, can the engineer support the claim that the mean braking distance is the same for both types of tires? *(Adapted from Consumer Reports)*

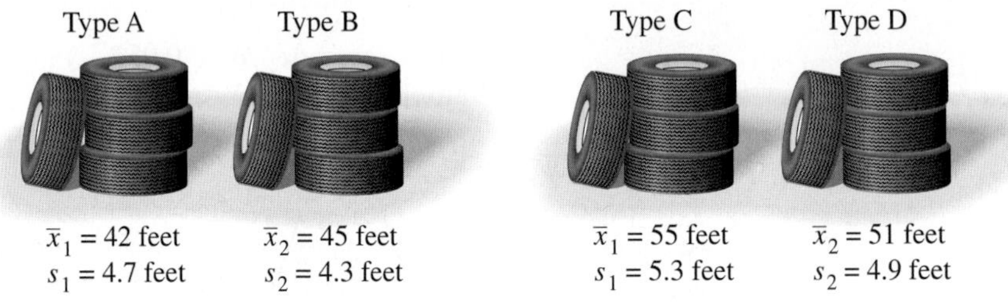

Figure for Exercise 9

Figure for Exercise 10

10. ***Braking Distances*** ◆ To compare the braking distances for two types of tires, a safety engineer conducts 50 braking tests for each of two types of tires. The results of the tests are shown in the figure. At $\alpha = 0.10$, can the engineer support the claim that the mean braking distance for Type C is greater than the mean braking distance for Type D? *(Adapted from Consumer Reports)*

11. ***Repair Costs: Microwave Ovens*** ◆ You want to buy a microwave oven and will choose Model A if its repair costs are lower than Model B's. You research the repair costs of 47 Model A ovens and 55 Model B ovens. The results of your research are shown in the figure. At $\alpha = 0.01$, would you buy Model A? *(Adapted from Consumer Reports)*

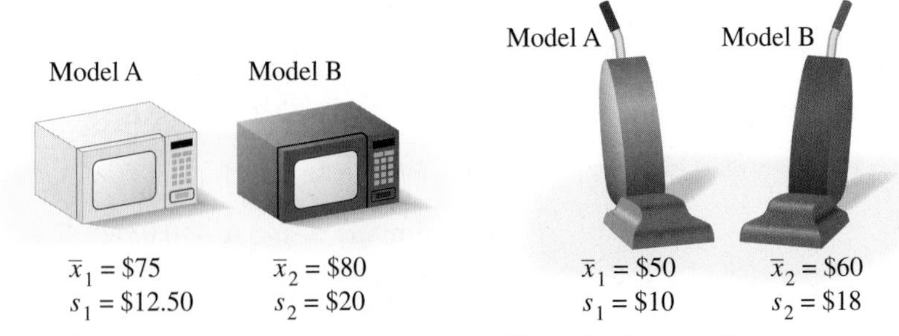

Figure for Exercise 11

Figure for Exercise 12

12. ***Repair Costs: Vacuum Cleaners*** ◆ You want to buy a vacuum cleaner, and a salesperson tells you the repair costs for Model A and Model B are equal. You research the repair costs of 34 Model A vacuum cleaners and 46 Model B vacuum cleaners. The results of your research are shown in the figure. At $\alpha = 0.01$, can you reject the salesperson's claim? *(Adapted from Consumer Reports)*

13. ***ACT Scores*** ◆ The mean ACT score for 43 male high school students is 21.2 and the standard deviation is 4.9. The mean ACT score for 56 female high school students is 20.9 and the standard deviation is 4.6. At $\alpha = 0.01$, do male and female high school students have equal ACT scores? *(Adapted from ACT Inc.)*

14. ***ACT Scores*** ◆ A guidance counselor claims high school students in a college prep program have higher ACT scores than those in a general program. The mean ACT score for 49 high school students who are in a college prep program is 22.1 and the standard deviation is 4.8. The mean ACT score for 44 high school students who are in a general program is 19.6 and the standard deviation is 5.4. At $\alpha = 0.10$, can you support the guidance counselor's claim? *(Adapted from ACT Inc.)*

15. ***Lodging Costs*** ◆ A travel association says the daily lodging cost for a family traveling in California is the same as Florida. The daily lodging cost for 35 families traveling in California is \$136 and the standard deviation is \$25. The daily lodging cost for 35 families traveling in Florida is \$140 and the standard deviation is \$30. At $\alpha = 0.10$, is there enough evidence to reject the travel association's claim? *(Adapted from the American Automobile Association)*

16. ***Money Spent Eating Out*** ◆ A restaurant association says that households in the U.S. headed by people under the age of 25 spend less on food away from home than households headed by people ages 55–64. The mean amount spent by 30 households headed by people under the age of 25 is \$1526 and the standard deviation is \$225. The mean amount spent by 30 households headed by people ages 55–64 is \$2136 and the standard deviation is \$350. Can you support the restaurant association's claim at $\alpha = 0.05$? *(Adapted from the National Restaurant Association)*

17. ***Lodging Costs*** ◆ Refer to Exercise 15. Two more samples are taken, one from California and one from Florida. For 50 families traveling in California, $\bar{x}_1 = \$146$ and $s_1 = \$30$. For 50 families traveling in Florida, $\bar{x}_2 = \$142$ and $s_2 = \$32$. Use $\alpha = 0.10$. Do the new samples lead to a different conclusion?

18. ***Money Spent Eating Out*** ◆ Refer to Exercise 16. Two more samples are taken, one from each age group. For 40 households headed by people under the age of 25, $\bar{x}_1 = \$1600$ and $s_1 = \$230$. For 40 households headed by people ages 55–64, $\bar{x}_2 = \$2040$ and $s_2 = \$380$. Use $\alpha = 0.05$. Do the new samples lead to a different conclusion?

DATA **19.** ***Watching More TV?*** ◆ A sociologist claims that children ages 3–12 spent more time watching television in 1981 than children ages 3–12 do today. A study was conducted in 1981 to find the time that children ages 3–12 watched television on weekdays. The results (in hours per weekday) are given below.

2.0 2.5 2.1 2.3 2.1 1.6 2.6 2.1 2.1 2.4 2.1 2.1 1.5
1.7 2.1 2.3 2.5 3.3 2.2 2.9 1.5 1.9 2.4 2.2 1.2 3.0
1.0 2.1 1.9 2.2

Recently, a similar study was conducted. The results are given below.

1.9 1.8 0.9 1.6 2.0 1.7 1.1 1.1 1.6 2.0 1.4 1.5 1.7
1.6 1.6 1.7 1.2 2.0 2.2 1.6 1.5 2.0 1.6 1.8 1.7 1.3
1.1 1.4 1.2 2.0

At $\alpha = 0.025$, can you support the sociologist's claim? *(Adapted from University of Michigan's Institute for Social Research)*

20. ***Spending More Time Studying?*** ◆ A sociologist thinks middle school boys spent less time studying in 1981 than middle school boys do today. A study was conducted in 1981 to find the time that middle school boys spent studying on weekdays. The results (in minutes per weekday) are given below.

13.9	17.3	10.0	21.2	12.8	13.7	15.0	11.9	17.3	12.1	13.5
17.9	19.1	18.2	14.8	19.1	6.2	10.6	10.7	23.1	15.7	14.1
10.7	17.2	18.6	16.3	17.8	16.2	15.8	7.5	9.7	3.7	5.3
18.8	8.9									

Recently, a similar study was conducted. The results are given below.

17.5	27.9	14.1	19.9	20.9	19.6	15.9	21.8	23.0	20.6	20.5
31.0	24.0	14.4	22.5	24.6	12.0	18.6	22.8	27.6	20.1	18.5
25.5	22.5	20.0	27.8	13.2	18.4	21.6	23.0	24.5	28.0	17.4
15.2	25.0									

At $\alpha = 0.03$, can you support the sociologist's claim? *(Adapted from University of Michigan's Institute for Social Research)*

21. ***Washer Diameters*** ◆ A production engineer claims that there is no difference in the mean washer diameter manufactured by two different methods. The first method produces washers with the following diameters (in inches).

0.861	0.864	0.882	0.887	0.858	0.879	0.887	0.876	0.870
0.894	0.884	0.882	0.869	0.859	0.887	0.875	0.863	0.887
0.882	0.862	0.906	0.880	0.877	0.864	0.873	0.860	0.866
0.869	0.877	0.863	0.875	0.883	0.872	0.879	0.861	

The second method produces washers with these diameters (in inches).

0.705	0.703	0.715	0.711	0.690	0.720	0.702	0.686	0.704
0.712	0.718	0.695	0.708	0.695	0.699	0.715	0.691	0.696
0.680	0.703	0.697	0.694	0.714	0.694	0.672	0.688	0.700
0.715	0.709	0.698	0.696	0.700	0.706	0.695	0.715	

At $\alpha = 0.01$, can you reject the production engineer's claim?

22. ***Nut Diameters*** ◆ A production engineer claims that there is no difference in the mean nut diameter manufactured by two different methods. The first method produces nuts with the following diameters (in centimeters).

3.330	3.337	3.329	3.354	3.325	3.343	3.333	3.347	3.332	3.358
3.353	3.335	3.341	3.331	3.327	3.326	3.337	3.336	3.323	3.347
3.329	3.345	3.329	3.338	3.353	3.339	3.338	3.338	3.350	3.320
3.364	3.340	3.348	3.339	3.336	3.321	3.316	3.352	3.320	3.336

The second method produces nuts with these diameters (in centimeters).

3.513	3.490	3.498	3.504	3.483	3.512	3.494	3.514	3.495	3.489
3.493	3.499	3.497	3.495	3.496	3.485	3.506	3.517	3.484	3.498
3.522	3.505	3.501	3.491	3.500	3.499	3.475	3.486	3.501	3.496
3.504	3.513	3.511	3.501	3.487	3.508	3.515	3.505	3.496	3.505

At $\alpha = 0.04$, can you reject the production engineer's claim?

23. ***Getting at the Concept*** ◆ Explain why the null hypothesis H_0: $\mu_1 = \mu_2$ is equivalent to the null hypothesis H_0: $\mu_1 - \mu_2 = 0$.

24. ***Getting at the Concept*** ◆ Explain why the null hypothesis H_0: $\mu_1 \geq \mu_2$ is equivalent to the null hypothesis H_0: $\mu_1 - \mu_2 \geq 0$.

Extending the Basics

Testing a Difference Other than Zero Sometimes a researcher is interested in testing a difference in means other than zero. For instance, you may want to determine if children today spend an average of 9 hours a week more in day care (or preschool) than children did 20 years ago. In Exercises 25–28, you will test the difference between two means using a null hypothesis of $H_0: \mu_1 - \mu_2 = k$, $H_0: \mu_1 - \mu_2 \geq k$, or $H_0: \mu_1 - \mu_2 \leq k$. The formula for the z-test is still

$$z = \frac{(\bar{x}_1 - \bar{x}_2) - (\mu_1 - \mu_2)}{\sigma_{\bar{x}_1 - \bar{x}_2}} \quad \text{where} \quad \sigma_{\bar{x}_1 - \bar{x}_2} = \sqrt{\frac{\sigma_1^2}{n_1} + \frac{\sigma_2^2}{n_2}}.$$

25. ***Time in Day Care or Preschool*** ◆ In 1981, a study of 70 randomly selected children (under 3 years old) found that the mean length of time spent in day care or preschool per week was 11.5 hours with a standard deviation of 3.8 hours. A recent study of 65 randomly selected children (under 3 years old) found that the mean length of time spent in day care or preschool per week was 20 hours and the standard deviation was 6.7 hours. At $\alpha = 0.01$, test the claim that children spend 9 hours a week more in day care or preschool today than they did in 1981. *(Adapted from University of Michigan's Institute for Social Research)*

26. ***Time Watching TV*** ◆ A recent study of 48 randomly selected children (ages 6–8) found that the mean length of time spent watching television each week was 12.63 hours and the standard deviation was 4.21 hours. The mean time 56 randomly selected children (ages 9–12) watched television each week was 13.60 hours and the standard deviation was 4.53 hours. At $\alpha = 0.05$, test the claim that the mean time per week children ages 6–8 watch television is one hour less than that of children ages 9–12. *(Adapted from University of Michigan's Institute for Social Research)*

27. ***Statistician Salaries*** ◆ Is the difference between the mean annual salaries of statisticians in California and Pennsylvania more than $6000? To decide, you select a random sample of statisticians from each state. The results of each survey are shown in the figure. At $\alpha = 0.10$, what should you conclude? *(Adapted from America's Career InfoNet)*

Statisticians in California

$\bar{x}_1 = \$52,300$
$s_1 = \$8875$
$n_1 = 45$

Statisticians in Pennsylvania

$\bar{x}_2 = \$50,900$
$s_2 = \$9175$
$n_2 = 37$

28. ***Photographer and Computer Programmer Salaries*** ◆ At $\alpha = 0.05$, test the claim that the difference between the mean salary for computer programmers and the mean salary for photographers in North Carolina is less than $30,000. The results of a survey of randomly selected computer programmers and photographers in North Carolina are shown in the figure. *(Adapted from America's Career InfoNet)*

Computer Programmers in North Carolina

$\bar{x}_1 = \$50,100$
$s_1 = \$8250$
$n_1 = 31$

Photographers in North Carolina

$\bar{x}_2 = \$21,200$
$s_2 = \$3200$
$n_2 = 33$

Constructing Confidence Intervals for $\mu_1 - \mu_2$ You can construct a confidence interval for the difference in two population means $\mu_1 - \mu_2$ by using the following if $n_1 \geq 30$ and $n_2 \geq 30$, or both populations are normally distributed. Also, the samples must be randomly selected and independent.

$$(\bar{x}_1 - \bar{x}_2) - z_c\sqrt{\frac{\sigma_1^2}{n_1} + \frac{\sigma_2^2}{n_2}} < \mu_1 - \mu_2 < (\bar{x}_1 - \bar{x}_2) + z_c\sqrt{\frac{\sigma_1^2}{n_1} + \frac{\sigma_2^2}{n_2}}$$

In Exercises 29 and 30, construct the indicated confidence interval for $\mu_1 - \mu_2$.

29. ***Herbal Supplement and Weight Loss*** ◆ A study was conducted to see if an herbal supplement helped to reduce weight. After 12 weeks, 42 subjects using the herbal supplement and a high-fiber, low-calorie diet had a mean weight loss of 3.2 kilograms and a standard deviation of 3.3 kilograms. During the same time period, 42 subjects using a placebo and a high-fiber, low-calorie diet had a mean weight loss of 4.1 kilograms and a standard deviation of 3.9 kilograms. Construct a 95% confidence interval for $\mu_1 - \mu_2$, where μ_1 is the mean weight loss for the group using the herbal supplement and μ_2 is the mean weight loss for the group using the placebo. *(Source: The Journal of the American Medical Association)*

30. ***Comparing Cancer Drugs*** ◆ Two groups of patients with colorectal cancer are treated with a different drug. Group A's 140 patients are treated using the drug Irinotecan and Group B's 127 patients are treated using the drug Fluorouracil. The mean number of months that Group A reported no cancer-related pain was 10.3 and the standard deviation was 1.2. The mean number of months that Group B reported no cancer-related pain was 8.5 and the standard deviation was 1.5. Construct a 95% confidence interval for $\mu_1 - \mu_2$, where μ_1 is the mean number of months that Group A reported no cancer-related pain and μ_2 is the mean number of months that Group B reported no cancer-related pain. *(Adapted from The Lancet)*

31. ***Make a Decision*** ◆ Refer to the study in Exercise 29. At $\alpha = 0.05$, test the claim that the mean weight loss for the group using the herbal supplement is greater than the mean weight loss for the group using the placebo. Would you recommend using the herbal supplement with a high-fiber, low-calorie diet to lose weight? Explain your reasoning.

32. ***Make a Decision*** ◆ Refer to the study in Exercise 30. At $\alpha = 0.05$, test the claim that the mean number of months of cancer-related pain relief with Irinotecan is greater than the mean number of months of cancer-related pain relief obtained with Fluorouracil. Would you recommend using Irinotecan over Fluorouracil to relieve cancer-related pain? Explain your reasoning.

33. ***Getting at the Concept*** ◆ Compare the confidence interval you constructed in Exercise 29 to the hypothesis test result in Exercise 31. Explain why you would fail to reject the null hypothesis if the confidence interval contains 0.

34. ***Getting at the Concept*** ◆ Compare the confidence interval you constructed in Exercise 30 to the hypothesis test result in Exercise 32. Explain why you would reject the null hypothesis if the confidence interval contains only positive numbers.

8 **Case Study** WWW.AMERICANHEART.ORG

Oat Bran and Cholesterol Level

In an article in the *Journal of Family Practice* (vol. 33, no. 6, pp. 600–608), a study about cholesterol level is described. In the study, men and women (ages 20 to 70) were randomly divided into three groups. Each group was asked to follow the American Heart Association Step I Diet. In addition, the first group was asked to eat two 1-ounce servings of oat bran each day and the second group was asked to eat two 1-ounce servings of wheat cereal each day.

Before the study, each group had mean total cholesterol levels of 238 mg/dL and LDL cholesterol levels of 164 mg/dL. The cholesterol levels of each group after the six-week study are shown in the table at the right.

Step I Diet

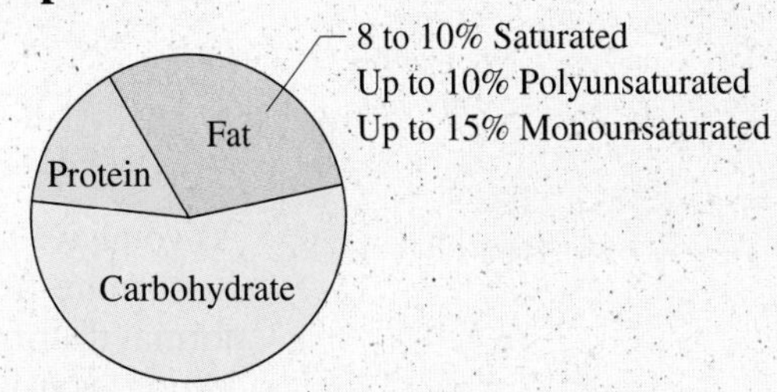

Total Fat	30% or less
Carbohydrate	55% or more
Protein	About 15%
Cholesterol	Less than 300 mg per day
Total Calories	To achieve or maintain desired weight

	Oat Bran and Diet $n_1 = 48$	Wheat Cereal and Diet $n_2 = 49$	Diet Alone $n_3 = 48$
Total Cholesterol	$\bar{x}_1 = 224$ $s_1 = 26.7$	$\bar{x}_2 = 231$ $s_2 = 30.5$	$\bar{x}_3 = 236$ $s_3 = 30.2$
LDL Cholesterol	$\bar{x}_1 = 148$ $s_1 = 26.7$	$\bar{x}_2 = 158$ $s_2 = 23.6$	$\bar{x}_3 = 162$ $s_3 = 27.5$

Cholesterol is measured in milligrams per deciliter.

Exercises

In Exercises 1–4, perform a two-sample z-test to determine whether the mean cholesterol levels of the two indicated groups are different. For each exercise, write your conclusion as a sentence. Use $\alpha = 0.05$.

1. Test the total cholesterol levels of people who ate oat bran while on the Step I Diet against those who ate wheat cereal while on the Step I Diet.
2. Test the total cholesterol levels of people who ate oat bran while on the Step I Diet against those who were on the Step I Diet alone.
3. Test the LDL cholesterol levels of people who ate oat bran while on the Step I Diet against those who ate wheat cereal while on the Step I Diet.
4. Test the LDL cholesterol levels of people who ate oat bran while on the Step I Diet against those who were on the Step I Diet alone.
5. In a different study, reported in the *Annals of the Internal Medicine*, 91 people were given a cholesterol-reducing drug called Pravastatin. The mean level of total cholesterol before the study was 214 mg/dL and the mean level of LDL cholesterol was 140 mg/dL. After the study, the mean levels were 167 mg/dL and 95 mg/dL.
   (a) Test the mean total cholesterol level of this group with the oat bran group described previously. Use $\alpha = 0.01$ and assume $s = 20$ mg/dL for the Pravastatin group.
   (b) Test the mean LDL cholesterol level of this group with the oat bran group described previously. Use $\alpha = 0.01$ and assume $s = 20$ mg/dL for the Pravastatin group.
6. In which comparisons did you find a difference in cholesterol levels? Write a summary of your findings.

8.2 Testing the Difference Between Means (Small Independent Samples)

What You Should Learn

- How to perform a t-test for the difference between two population means μ_1 and μ_2 using small independent samples

The Two-Sample t-Test for the Difference Between Means

The Two-Sample t-Test for the Difference Between Means

As you have learned, in real life, it is often not practical to collect samples of size 30 or more from each of two populations. However, if both populations have a normal distribution, you can still test the difference between their means. In this section, you will learn how to use a t-test to test the difference between two population means μ_1 and μ_2 using a sample from each population. To use a t-test for small independent samples, the following conditions are necessary.

1. The samples must be randomly selected.
2. The samples must be independent. Recall that two samples are independent if the sample selected from one population is not related to the sample selected from the second population.
3. Each population must have a normal distribution.

When these conditions are met, the sampling distribution for $\bar{x}_1 - \bar{x}_2$, the difference between the sample means, is a t-distribution with mean $\mu_1 - \mu_2$. The standard error and the degrees of freedom of the sampling distribution depend on whether the population variances σ_1^2 and σ_2^2 are equal.

If the population variances are equal, information from both samples is combined to calculate a **pooled estimate of the standard deviation.**

$$\hat{\sigma} = \sqrt{\frac{(n_1 - 1)s_1^2 + (n_2 - 1)s_2^2}{n_1 + n_2 - 2}}$$ Pooled estimate of σ

The standard error for the sampling distribution of $\bar{x}_1 - \bar{x}_2$ is

$$\sigma_{\bar{x}_1 - \bar{x}_2} = \hat{\sigma}\sqrt{\frac{1}{n_1} + \frac{1}{n_2}}$$ Variances are equal.

and d.f. $= n_1 + n_2 - 2$.

If the variances are not equal, the standard error is

$$\sigma_{\bar{x}_1 - \bar{x}_2} = \sqrt{\frac{s_1^2}{n_1} + \frac{s_2^2}{n_2}}$$ Variances are not equal.

and d.f. $=$ smaller of $n_1 - 1$ or $n_2 - 1$.

The requirements for the z-test described in Section 8.1 and the t-test described in this section are compared below.

	z-Test	t-Test
Samples	Must be independent	Must be independent
Distribution and sample size	Both samples must have at least 30 members *or* the populations must be normal with known standard deviations.	The population must be normal. (One or both of the samples can have less than 30 members.)

Study Tip

You will learn to test for differences in variances in two populations in Chapter 10. In this chapter, each example and exercise will state whether the variances are equal.

If the sampling distribution for $\bar{x}_1 - \bar{x}_2$ is a t-distribution, you can use a two-sample t-test to test the difference between two population means μ_1 and μ_2.

Picturing the World

A study published in the journal *Nature* reported that students showed an increase in scores on IQ tests equivalent to almost 9 points after listening to a Mozart piano sonata for 10 minutes. The study generated interest among the general public and researchers. Suppose you tried to duplicate the results as follows. A test with a possible 100 points was administered to 55 students. The 25 students who listened to classical music 10 minutes before taking the test had an average of 83.12 with a standard deviation of 5.7. The 30 students acting as the control group had an average of 79.9 with a standard deviation of 6.2.

At $\alpha = 0.05$ is there enough evidence to support the claim that listening to music increases IQ scores? Assume the populations are normally distributed and the population variances are equal.

Two-Sample t-Test for the Difference Between Means

A **two-sample t-test** is used to test the difference between two population means μ_1 and μ_2 when a sample is randomly selected from each population. To perform this test, each population must be normally distributed, the samples should be independent, and the size of at least one of the samples must be less than 30. The standardized test statistic is

$$t = \frac{(\bar{x}_1 - \bar{x}_2) - (\mu_1 - \mu_2)}{\sigma_{\bar{x}_1 - \bar{x}_2}}.$$

If the population variances are equal, then

$$\text{d.f.} = n_1 + n_2 - 2$$

and

$$\sigma_{\bar{x}_1 - \bar{x}_2} = \sqrt{\frac{(n_1 - 1)s_1^2 + (n_2 - 1)s_2^2}{n_1 + n_2 - 2}} \cdot \sqrt{\frac{1}{n_1} + \frac{1}{n_2}}.$$

If the population variances are not equal, then d.f. is the smaller of $n_1 - 1$ or $n_2 - 1$ and

$$\sigma_{\bar{x}_1 - \bar{x}_2} = \sqrt{\frac{s_1^2}{n_1} + \frac{s_2^2}{n_2}}.$$

GUIDELINES

Using a Two-Sample t-Test for the Difference Between Means (Small Independent Samples)

In Words	*In Symbols*
1. Identify the claim. State the null and alternative hypotheses.	State H_0 and H_a.
2. Specify the level of significance.	Identify α.
3. Determine the degrees of freedom.	d.f. $= n_1 + n_2 - 2$ or d.f. $=$ smaller of $n_1 - 1$ or $n_2 - 1$
4. Find the critical value(s).	Use Table 5 in Appendix B.
5. Identify the rejection region(s).	
6. Find the standardized test statistic.	$t = \frac{(\bar{x}_1 - \bar{x}_2) - (\mu_1 - \mu_2)}{\sigma_{\bar{x}_1 - \bar{x}_2}}$
7. Make a decision to reject or fail to reject the null hypothesis.	If t is in the rejection region, reject H_0. Otherwise, do not reject H_0.
8. Interpret the decision in the context of the original claim.	

See *Minitab* steps on page 430.

Sample Statistics for Stopping Distances Under Winter Conditions

Winterfire	Alpin
$\bar{x}_1 = 51$	$\bar{x}_2 = 55$
$s_1 = 8$	$s_2 = 3$
$n_1 = 10$	$n_2 = 12$

EXAMPLE 

A Two-Sample t-Test for the Difference Between Means

Consumer Reports tested several types of snow tires to determine how well each performed under winter conditions. When traveling on ice at 15 mph, 10 Firestone Winterfire tires had a mean stopping distance of 51 feet with a standard deviation of 8 feet with an anti-lock brake system. The mean stopping distance for 12 Michelin XM+S Alpin tires was 55 feet with a standard deviation of 3 feet with an anti-lock brake system. Can you conclude that there is a difference in the stopping distances of the two types of tires? Use $\alpha = 0.01$. Assume the populations are normally distributed and the population variances are not equal.

SOLUTION You want to test whether the mean stopping distances are different. So, the null and alternative hypotheses are

$$H_0\colon \mu_1 = \mu_2 \quad \text{and} \quad H_a\colon \mu_1 \neq \mu_2. \text{ (Claim)}$$

Because the variances are not equal and the smaller sample size is 10, use d.f. $= 10 - 1 = 9$. Because the test is a two-tailed test with d.f. $= 9$, and $\alpha = 0.01$, the critical values are -3.250 and 3.250. The rejection regions are $t < -3.250$ and $t > 3.250$. The standard error is

$$\sigma_{\bar{x}_1 - \bar{x}_2} = \sqrt{\frac{s_1^2}{n_1} + \frac{s_2^2}{n_2}} = \sqrt{\frac{8^2}{10} + \frac{3^2}{12}} \approx 2.674.$$

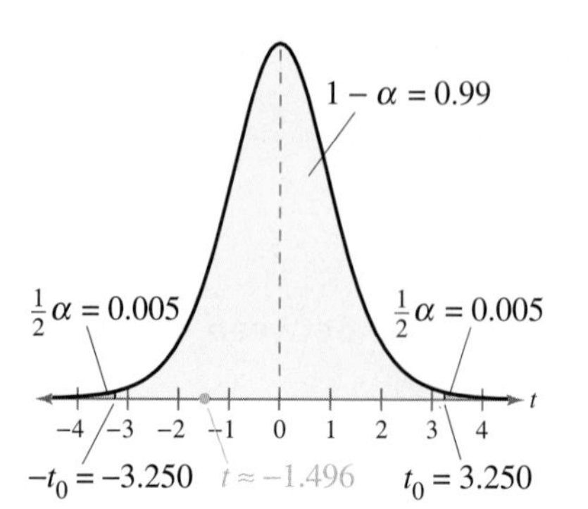

Using the t-test, the standardized test statistic is

$$t = \frac{(\bar{x}_1 - \bar{x}_2) - (\mu_1 - \mu_2)}{\sigma_{\bar{x}_1 - \bar{x}_2}} \approx \frac{(51 - 55) - 0}{2.674} \approx -1.496.$$

The graph at the left shows the location of the rejection regions and the standardized test statistic t. Because t is not in a rejection region, you should fail to reject the null hypothesis. At the 1% level, there is not enough evidence to conclude that the mean stopping distances of the tires are different.

Try It Yourself 1

Sample Statistics for Stopping Distances on Wet Pavement

Winterfire	Alpin
$\bar{x}_1 = 102$	$\bar{x}_2 = 94$
$s_1 = 10$	$s_2 = 4$
$n_1 = 10$	$n_2 = 12$

When traveling on wet pavement at 40 mph, Winterfire had a mean stopping distance of 102 feet with a standard deviation of 10 feet without an anti-lock brake system. The mean stopping distance for the Alpin was 94 feet with a standard deviation of 4 feet without an anti-lock brake system. If 10 Winterfire tires and 12 Alpin tires were used in the test, can you conclude that the mean stopping distances are different? Use $\alpha = 0.05$. (Assume the populations are normally distributed and the population variances are not equal.) *(Adapted from Consumer Reports)*

a. *Identify* the claim and state H_0 and H_a.
b. *Specify* the level of significance α.
c. *Determine* the degrees of freedom.
d. *Find* the critical values and *identify* the rejection regions.
e. *Use* the t-test to find the standardized test statistic t.
f. *Decide* whether to reject the null hypothesis. Use a graph if necessary.
g. Is there enough evidence to conclude that the mean stopping distances are different?

Answer: Page A40

See *TI-83* steps on page 431.

EXAMPLE 

A Two-Sample t-Test for the Difference Between Means

A manufacturer claims that the calling range (in miles) of its 900-MHz cordless telephone is greater than that of its leading competitor. You perform a study using 14 randomly selected phones from the manufacturer and 16 randomly selected similar phones from its competitor. The results are shown at the left. At $\alpha = 0.05$, is there enough evidence to support the manufacturer's claim? Assume the populations are normally distributed and the population variances are equal.

Sample Statistics for Calling Range

Manufacturer	Competition
$\bar{x}_1 = 1275$	$\bar{x}_2 = 1250$
$s_1 = 45$	$s_2 = 30$
$n_1 = 14$	$n_2 = 16$

SOLUTION The claim is "the mean range of our cordless phone is greater than the mean range of yours." So, the null and alternative hypotheses are

$$H_0\colon \mu_1 \leq \mu_2 \quad \text{and} \quad H_a\colon \mu_1 > \mu_2. \text{ (Claim)}$$

Because the variances are equal, d.f. $= n_1 + n_2 - 2 = 14 + 16 - 2 = 28$. Because the test is a right-tailed test, d.f. $= 28$, and $\alpha = 0.05$, the critical value is $t_0 = 1.701$. The rejection region is $t > 1.701$. The standard error is

$$\sigma_{\bar{x}_1-\bar{x}_2} = \sqrt{\frac{(n_1-1)s_1^2 + (n_2-1)s_2^2}{n_1+n_2-2}} \cdot \sqrt{\frac{1}{n_1} + \frac{1}{n_2}}$$

$$= \sqrt{\frac{(13)(45^2) + (15)(30^2)}{14+16-2}} \cdot \sqrt{\frac{1}{14} + \frac{1}{16}} \approx 13.802.$$

Using the t-test, the standardized test statistic is

$$t = \frac{(\bar{x}_1 - \bar{x}_2) - (\mu_1 - \mu_2)}{\sigma_{\bar{x}_1-\bar{x}_2}}$$

$$\approx \frac{(1275 - 1250) - 0}{13.802}$$

$$\approx 1.811.$$

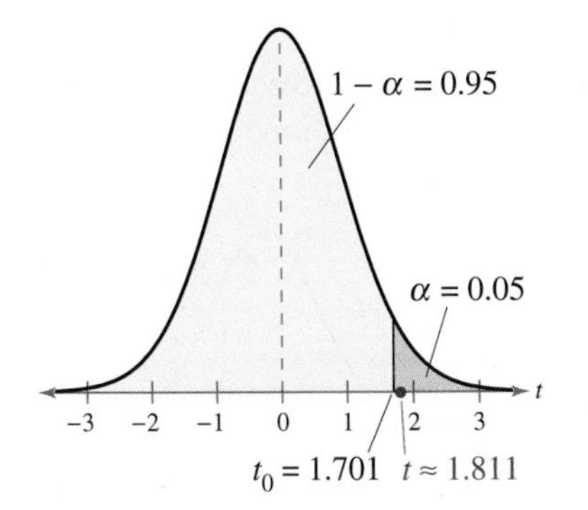

The graph at the left shows the location of the rejection region and the standardized test statistic t. Because t is in the rejection region, you should decide to reject the null hypothesis. At the 5% level, there is enough evidence to support the manufacturer's claim that its phone has a greater calling range than its competitor's.

Try It Yourself 2

A manufacturer claims that the watt usage of its 17-inch monitors is less than that of its leading competitor. You perform a study and obtain the results shown at the left. At $\alpha = 0.10$, is there enough evidence to support the manufacturer's claim? Assume the populations are normally distributed and the population variances are equal.

Sample Statistics for Watt Usage

Manufacturer	Competition
$\bar{x}_1 = 73$	$\bar{x}_2 = 74$
$s_1 = 2.4$	$s_2 = 3.2$
$n_1 = 12$	$n_2 = 15$

a. *Identify* the claim and state H_0 and H_a.
b. *Specify* the level of significance α.
c. *Determine* the degrees of freedom.
d. *Find* the critical value and *identify* the rejection region.
e. *Use* the t-test to find the standardized test statistic t.
f. *Decide* whether to reject the null hypothesis. Use a graph if necessary.
g. Is there enough evidence to support the manufacturer's claim?

Answer: Page A40

8.2 Exercises

LarsonTutor 8.2

Companion Web Site

Student Solutions Manual 8.2

Videos 8.2

Technology Manuals

Basic Skills and Concepts

1. Explain how to perform a two-sample t-test for the difference between the means of two populations.

2. What conditions are necessary in order to use a t-test to test the difference between two population means?

In Exercises 3–10, use Table 5 to find the critical value(s) for the indicated test, level of significance α, and sample sizes n_1 and n_2. Assume that the samples are independent, normal, and random, and that the population variances are (a) equal and (b) not equal.

3. Two-tailed, $\alpha = 0.10$, $n_1 = 10$, $n_2 = 12$

4. Right-tailed, $\alpha = 0.01$, $n_1 = 12$, $n_2 = 15$

5. Left-tailed, $\alpha = 0.025$, $n_1 = 15$, $n_2 = 9$

6. Two-tailed, $\alpha = 0.05$, $n_1 = 19$, $n_2 = 22$

7. Right-tailed, $\alpha = 0.05$, $n_1 = 13$, $n_2 = 8$

8. Left-tailed, $\alpha = 0.10$, $n_1 = 7$, $n_2 = 2$

9. Two-tailed, $\alpha = 0.01$, $n_1 = 12$, $n_2 = 17$

10. Right-tailed, $\alpha = 0.005$, $n_1 = 5$, $n_2 = 11$

In Exercises 11–14, (a) find the test statistic, (b) find the standardized test statistic, (c) decide whether the standardized test statistic is in the rejection region, and (d) decide whether you should reject or fail to reject the claim.

11. Claim: $\mu_1 = \mu_2$, $\alpha = 0.01$
Sample statistics: $\bar{x}_1 = 33.7$, $s_1 = 3.5$,
$n_1 = 10$, and $\bar{x}_2 = 35.5$, $s_2 = 2.2$, $n_2 = 7$
Assume $\sigma_1^2 = \sigma_2^2$

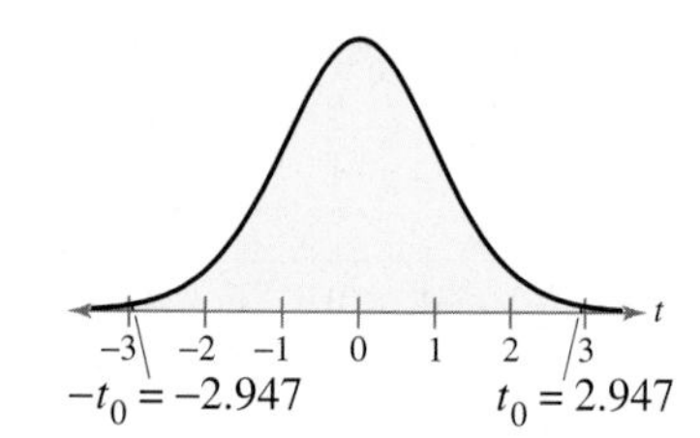

12. Claim: $\mu_1 \geq \mu_2$, $\alpha = 0.10$
Sample statistics: $\bar{x}_1 = 0.515$, $s_1 = 0.305$,
$n_1 = 11$, and $\bar{x}_2 = 0.475$, $s_2 = 0.215$, $n_2 = 9$
Assume $\sigma_1^2 = \sigma_2^2$

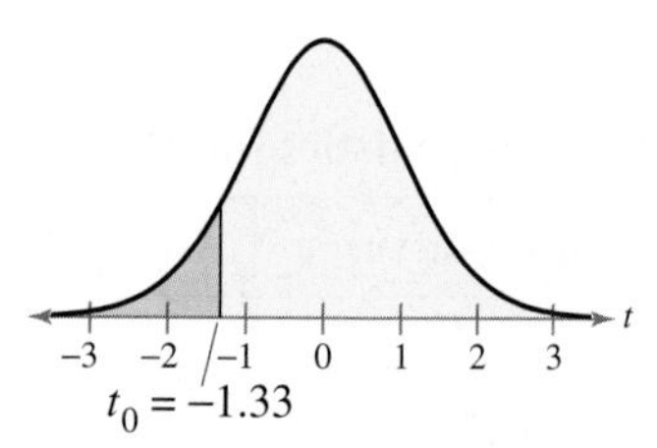

13. Claim: $\mu_1 \leq \mu_2$, $\alpha = 0.05$
Sample statistics: $\bar{x}_1 = 2250$, $s_1 = 175$,
$n_1 = 13$, and $\bar{x}_2 = 2305$, $s_2 = 52$, $n_2 = 10$
Assume $\sigma_1^2 \neq \sigma_2^2$

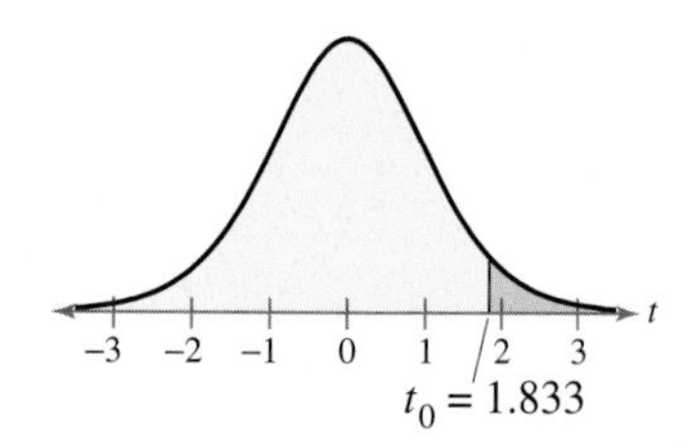

14. Claim: $\mu_1 \leq \mu_2$, $\alpha = 0.01$
Sample statistics: $\bar{x}_1 = 45$, $s_1 = 4.8$, $n_1 = 16$, and $\bar{x}_2 = 50$, $s_2 = 1.2$, $n_2 = 14$
Assume $\sigma_1^2 \neq \sigma_2^2$

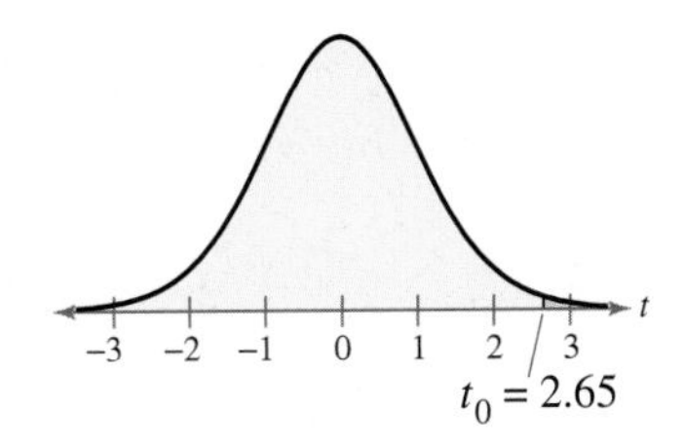

Testing the Difference Between Two Means

In Exercises 15–24,

(a) write the claim mathematically and identify H_0 and H_a.

(b) find the critical value(s) and identify the rejection region(s).

(c) find the standardized test statistic.

(d) decide whether to reject or fail to reject the null hypothesis and interpret the decision in the context of the original claim.

Assume the populations have approximately normal distributions.

15. ***Footwell Intrusion*** ◆ An insurance actuary claims that the mean footwell intrusion for small and midsize cars is equal. Crash tests at 40 miles per hour were performed on 14 randomly selected small cars and 23 randomly selected midsize cars. The amount that the footwell intruded on the driver's legs was measured. The mean footwell intrusion for the small cars was 23.1 centimeters and the standard deviation was 8.69 centimeters. The mean footwell intrusion for the midsize cars was 25.3 centimeters and the standard deviation was 7.21 centimeters. At $\alpha = 0.10$, can you reject the insurance actuary's claim? Assume the population variances are equal. *(Adapted from Insurance Institute for Highway Safety)*

Footwell Intrusion

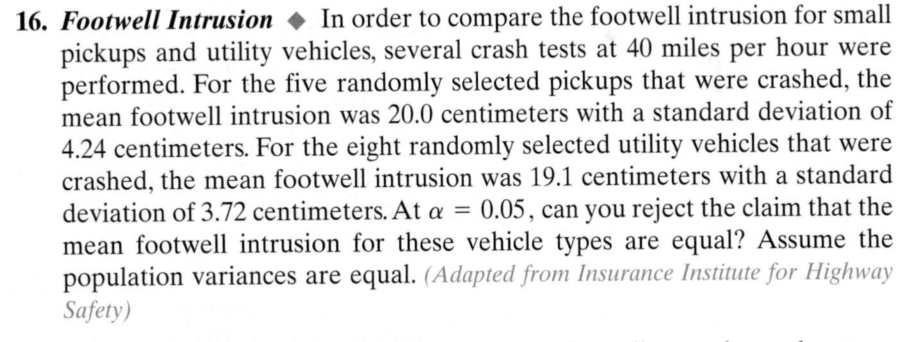

16. ***Footwell Intrusion*** ◆ In order to compare the footwell intrusion for small pickups and utility vehicles, several crash tests at 40 miles per hour were performed. For the five randomly selected pickups that were crashed, the mean footwell intrusion was 20.0 centimeters with a standard deviation of 4.24 centimeters. For the eight randomly selected utility vehicles that were crashed, the mean footwell intrusion was 19.1 centimeters with a standard deviation of 3.72 centimeters. At $\alpha = 0.05$, can you reject the claim that the mean footwell intrusion for these vehicle types are equal? Assume the population variances are equal. *(Adapted from Insurance Institute for Highway Safety)*

17. ***Repair Costs: Bumpers*** ◆ In crash tests at five miles per hour, the mean bumper repair cost for 14 randomly selected small cars is \$574 with a standard deviation of \$185. In similar tests of 23 randomly selected midsize cars, the mean bumper repair cost is \$734 with a standard deviation of \$268. At $\alpha = 0.10$, can you conclude that the mean bumper repair cost is less for small cars than it is for midsize cars? Assume that the population variances are equal. *(Adapted from Insurance Institute for Highway Safety)*

18. ***Repair Costs: Bumpers*** ◆ Crash tests at five miles per hour were performed on 5 randomly selected small pickups and 8 randomly selected small utility vehicles. For the small pickups, the mean bumper repair cost was \$1520 and the standard deviation was \$403. For the small utility vehicles, the mean bumper repair cost was \$937 and the standard deviation was \$382. At $\alpha = 0.10$, is there enough evidence to conclude that the mean bumper repair cost is greater for small pickups than for small utility vehicles? Assume the population variances are equal. *(Adapted from Insurance Institute for Highway Safety)*

19. *Annual Wages* ◆ A personnel director claims that the mean annual wage is greater in Allegheny County than it is in Erie County. In Allegheny County, a random sample of 19 residents has a mean annual income of $30,800 and a standard deviation of $8600. In Erie County, a random sample of 15 residents has a mean annual income of $25,700 and a standard deviation of $5500. At $\alpha = 0.10$, can you support the personnel director's claim? Assume the population variances are not equal. *(Adapted from U.S. Bureau of Labor Statistics, Pennsylvania Department of Labor and Industry)*

20. *Annual Wages* ◆ Test the claim at $\alpha = 0.01$ that the mean annual incomes in Delaware and Greene Counties are the same. A random sample of 17 residents of Delaware County has a mean annual income of $31,200 and a standard deviation of $7800. In Greene County, a random sample of 18 residents has a mean annual income of $29,500 and a standard deviation of $7375. Assume the population variances are equal. Hint: For $n > 29$, use the last row (∞) in the t-distribution table. *(Adapted from U.S. Bureau of Labor Statistics, Pennsylvania Department of Labor and Industry)*

DATA **21. *Tensile Strength*** ◆ The tensile strength of a metal is a measure of its ability to resist tearing when it is pulled lengthwise. Using a new experimental type of treatment, steel bars were produced with the following tensile strengths (in newtons per square millimeter).

Experimental Method:

363 355 305 350 340
373 311 348 338 320

With the old method, steel bars were produced with the following tensile strengths (in newtons per square millimeter).

Old Method:

362 382 368 398 381 391 400
410 396 411 385 385 395

At $\alpha = 0.01$, does the new treatment make a difference in the tensile strength of steel bars? Assume the population variances are equal.

DATA **22. *Tensile Strength*** ◆ An engineer wants to compare the tensile strengths of steel bars that are produced using a conventional method and an experimental method. (The tensile strength of a metal is a measure of its ability to resist tearing when pulled lengthwise.) To do so, the engineer randomly selects steel bars that are manufactured using each method and records the following tensile strengths in newtons per square millimeter.

Experimental method:

395 389 421 394 407 411
389 402 422 416 402 408
400 386 411 405 389

Conventional method:

362 352 380 382 413
384 400 378 419 379
384 388 372 383

At $\alpha = 0.10$, can the engineer claim that the experimental method produces steel with greater mean tensile strength? Should the engineer recommend using the experimental method? Assume the population variances are not equal.

23. ***Teaching Methods*** ◆ A new method of teaching reading is being tested on third-grade students. A group of randomly selected students is taught using the experimental curriculum. A control group of randomly selected students is taught using the old curriculum. The reading test scores of the two groups are given in the stem-and-leaf plot.

Old Curriculum		New Curriculum
9	3	
9 9	4	3
9 8 8 4 3 3 2 1	5	2 4
7 6 4 2 2 1 0 0	6	0 1 1 4 7 7 7 7 7 8 9 9
	7	0 1 1 2 3 3 4 9
	8	2 4

Key:
9|4 = 49 (old curriculum)
4|3 = 43 (new curriculum)

At $\alpha = 0.10$, is there enough evidence to conclude that the new method of teaching reading produces higher reading test scores than the old method? Would you recommend changing to the new method? Assume the population variances are equal.

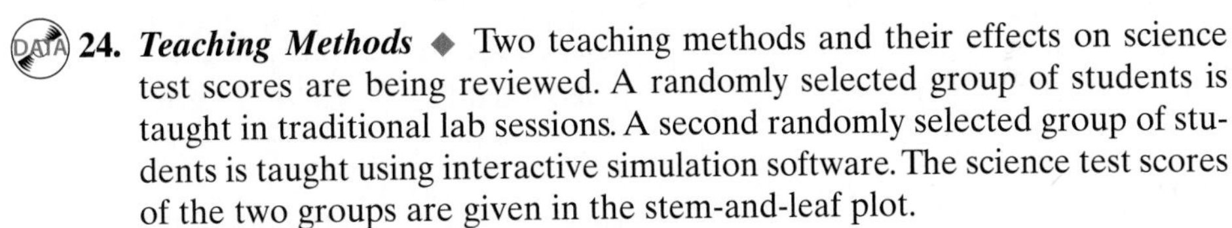

24. ***Teaching Methods*** ◆ Two teaching methods and their effects on science test scores are being reviewed. A randomly selected group of students is taught in traditional lab sessions. A second randomly selected group of students is taught using interactive simulation software. The science test scores of the two groups are given in the stem-and-leaf plot.

Traditional Lab		Interactive Simulation Software
4	6	
9 9 8 8 7 6 6 3 2 1 0	7	0 4 5 5 7 7 8
9 8 5 1 1 1 0 0	8	0 0 3 4 7 8 8 9 9
2 0	9	1 3 9

Key:
0|9 = 90 (traditional)
9|1 = 91 (interactive)

At $\alpha = 0.05$, can you support the claim that the mean science test score is lower for students taught using the lab method than it is for students taught using the interactive simulation software? Assume the population variances are equal.

Extending the Basics

Constructing Confidence Intervals for $\mu_1 - \mu_2$ If the sampling distribution for $\bar{x}_1 - \bar{x}_2$ is a *t*-distribution and the populations have equal variances, you can construct a confidence interval for $\mu_1 - \mu_2$ using the following.

$$(\bar{x}_1 - \bar{x}_2) - t_c\hat{\sigma}\sqrt{\frac{1}{n_1} + \frac{1}{n_2}} < \mu_1 - \mu_2 < (\bar{x}_1 - \bar{x}_2) + t_c\hat{\sigma}\sqrt{\frac{1}{n_1} + \frac{1}{n_2}}$$

where $\hat{\sigma} = \sqrt{\dfrac{(n_1 - 1)s_1^2 + (n_2 - 1)s_2^2}{n_1 + n_2 - 2}}$

and

$$\text{d.f.} = n_1 + n_2 - 2$$

In Exercises 25 and 26, construct a confidence interval for $\mu_1 - \mu_2$. Assume the populations are approximately normal with equal variances.

25. ***Calories*** ◆ In a study of various fast foods, you find that the mean calorie content of 15 grilled chicken sandwiches from Arby's is $\bar{x}_1 = 430$ calories with a standard deviation of $s_1 = 6.2$ calories. You also find that the mean calorie content of 12 similar chicken sandwiches from McDonald's is $\bar{x}_2 = 440$ calories with a standard deviation of $s_2 = 8.1$ calories. Construct a 95% confidence interval for the difference in mean calorie content of grilled chicken sandwiches. *(Adapted from Fast Food Facts, Minnesota Attorney General's Office)*

26. ***Protein*** ◆ A nutritionist wants to compare the mean protein content of grilled chicken sandwiches from Arby's and McDonald's. To do so, you randomly select several grilled chicken sandwiches from each restaurant and measure the protein content of each. The results are listed below. Construct a 95% confidence interval for the difference in mean protein content of grilled chicken sandwiches. *(Adapted from Fast Food Facts, Minnesota Attorney General's Office)*

Restaurant	Mean protein content	Standard deviation	Sample size
Arby's	$\bar{x}_1 = 23$ grams	$s_1 = 2.1$ grams	$n_1 = 15$
McDonald's	$\bar{x}_2 = 27$ grams	$s_2 = 1.8$ grams	$n_2 = 12$

Constructing Confidence Intervals for $\mu_1 - \mu_2$ If the sampling distribution for $\bar{x}_1 - \bar{x}_2$ is a *t*-distribution and the population variances are not equal, you can construct a confidence interval for $\mu_1 - \mu_2$, using the following.

$$(\bar{x}_1 - \bar{x}_2) - t_c\sqrt{\frac{s_1^2}{n_1} + \frac{s_2^2}{n_2}} < \mu_1 - u_2 < (\bar{x}_1 - \bar{x}_2) + t_c\sqrt{\frac{s_1^2}{n_1} + \frac{s_2^2}{n_2}}$$

and d.f. is the smaller of $n_1 - 1$ or $n_2 - 1$.

In Exercises 27 and 28, construct the indicated confidence interval for $\mu_1 - \mu_2$. Assume the populations are approximately normal with unequal variances.

27. ***Cholesterol*** ◆ To compare the mean cholesterol content of grilled chicken sandwiches from Arby's and McDonald's, you randomly select several sandwiches from each restaurant and measure their cholesterol contents. The results are listed below. Construct an 80% confidence interval for the difference in mean cholesterol content of grilled chicken sandwiches. *(Adapted from Fast Food Facts, Minnesota Attorney General's Office)*

Restaurant	Mean cholesterol content	Standard deviation	Sample size
Arby's	$\bar{x}_1 = 61$ mg	$s_1 = 3.59$ mg	$n_1 = 15$
McDonald's	$\bar{x}_2 = 60$ mg	$s_2 = 2.41$ mg	$n_2 = 12$

28. ***Carbohydrates*** ◆ A study of fast food finds that the mean carbohydrate content of 20 grilled chicken sandwiches from Arby's is $\bar{x}_1 = 41$ grams with a standard deviation of $s_1 = 2.42$ grams. The study also finds that the mean carbohydrate content of 15 grilled chicken sandwiches from McDonald's is $\bar{x}_2 = 38$ grams with a standard deviation of $s_2 = 1.65$ grams. Construct an 80% confidence interval for the difference in mean carbohydrate content. *(Adapted from Fast Food Facts, Minnesota Attorney General's Office)*

8.3 Testing the Difference Between Means (Dependent Samples)

What You Should Learn

- How to decide whether two samples are independent or dependent
- How to perform a *t*-test to test the mean of the differences for a population of paired data

Independent and Dependent Samples • The *t*-Test for the Difference Between Means

Independent and Dependent Samples

In Sections 8.1 and 8.2, you studied two-sample hypothesis tests in which the samples were independent. In this section, you will learn how to perform two-sample hypothesis tests using dependent samples.

DEFINITION

Two samples are **independent** if the sample selected from one population is not related to the sample selected from the second population. The two samples are **dependent** if each member of one sample corresponds to a member of the other sample. Dependent samples are also called **paired samples** or **matched samples.**

EXAMPLE 

Independent and Dependent Samples

Classify each pair of samples as independent or dependent.

1. Sample 1: Resting heart rates of 35 individuals before drinking coffee
   Sample 2: Resting heart rates of the same individuals after drinking two cups of coffee
2. Sample 1: Test scores for 35 statistics students
   Sample 2: Test scores for 42 biology students who do not study statistics

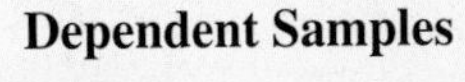

Dependent Samples

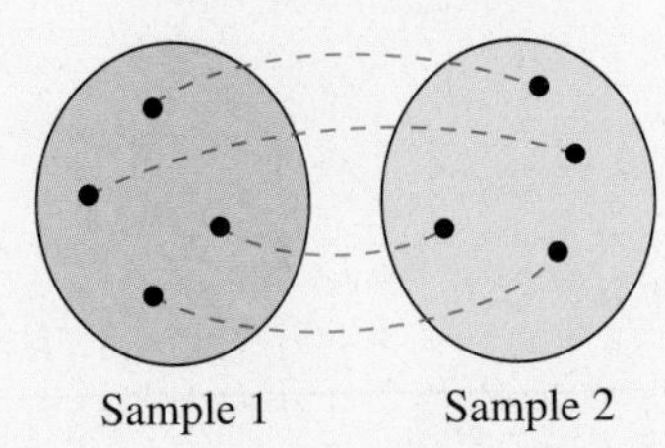

SOLUTION

1. These samples are dependent. Because the resting heart rates of the same individuals are taken, the samples are related. The samples can be paired with respect to each individual.
2. These samples are independent. It is not possible to form a pairing between the members of samples—the sample sizes are different and the data represent test scores for different individuals.

Try It Yourself 1

Classify each pair of samples as independent or dependent.

1. Sample 1: Heights of 27 adult females
   Sample 2: Heights of 27 adult males
2. Sample 1: Midterm exam scores of 14 chemistry students
   Sample 2: Final exam scores of the same 14 chemistry students

a. Determine whether the samples are related.

Answer: Page A40

Picturing the World

The manufacturer of an appetite suppressant claims that when its product is taken with a low-fat diet and regular exercise for four months, the average weight loss is 20 pounds. *(Adapted from American Family Physician)* To test this claim, you studied 12 dieters for 4 months. The dieters followed a low-fat diet with regular exercise all 4 months. The results are shown in the following table.

Weights (in pounds) of 12 dieters

	Original weight	4th month
1	185	168
2	194	177
3	213	196
4	198	180
5	244	229
6	162	144
7	211	197
8	273	252
9	178	161
10	192	178
11	181	161
12	209	193

Does your study provide sufficient evidence to reject the manufacturer's claim at a level of significance of $\alpha = 0.10$? Assume the weights are normally distributed.

The *t*-Test for the Difference Between Means

In Sections 8.1 and 8.2, you performed two-sample hypothesis tests with independent samples using the test statistic $\bar{x}_1 - \bar{x}_2$ (the difference in the means of the two samples). To perform a two-sample hypothesis test with dependent samples, you will use a different technique. You will first find the difference for each data pair,

$$d = x_1 - x_2. \quad \text{Difference between entries for a data pair}$$

The test statistic is the mean of these differences,

$$\bar{d} = \frac{\Sigma d}{n}. \quad \text{Mean of the differences between paired data entries in the dependent samples}$$

To conduct the test, the following conditions are required.

1. The samples must be randomly selected.
2. The samples must be dependent (paired).
3. Both populations must be normally distributed.

If these requirements are met, then the sampling distribution for $\bar{d}$, the mean of the differences of the paired data entries in the dependent samples, has a *t*-distribution with $n - 1$ degrees of freedom, where n is the number of data pairs.

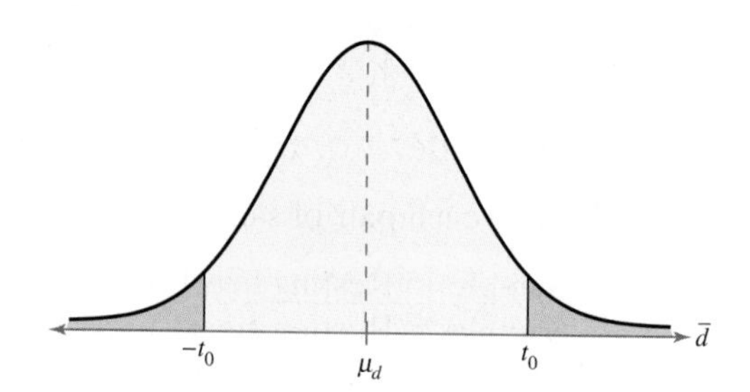

The following symbols are used for the *t*-test for μ_d.

Symbol	Description
n	The number of pairs of data
d	The difference between entries for a data pair, $d = x_1 - x_2$
μ_d	The hypothesized mean of the differences of paired data in the population
$\bar{d}$	The mean of the differences between the paired data entries in the dependent samples $\bar{d} = \frac{\Sigma d}{n}$
s_d	The standard deviation of the differences between the paired data entries in the dependent samples $s_d = \sqrt{\frac{n(\Sigma d^2) - (\Sigma d)^2}{n(n-1)}}$

Although formulas are given for the mean and standard deviation of differences, we suggest that you use a technology tool to calculate these statistics.

Because the sampling distribution for $\overline{d}$ is a t-distribution, you can use a t-test to test a claim about the mean of the differences for a population of paired data.

t-Test for the Difference Between Means

A t-test can be used to test the difference of two population means when a sample is randomly selected from each population. To perform the test, each population must be normal and each member of the first sample is paired with a member of the second sample. The **test statistic** is

$$\overline{d} = \frac{\Sigma d}{n}$$

and the **standardized test statistic** is

$$t = \frac{\overline{d} - \mu_d}{s_d / \sqrt{n}}.$$

The degrees of freedom are

$$\text{d.f.} = n - 1.$$

GUIDELINES

Using the t-Test for the Difference Between Means (Dependent Samples)

In Words	*In Symbols*
1. Identify the claim. State the null and the alternative hypotheses.	State H_0 and H_a.
2. Specify the level of significance.	Identify α.
3. Identify the degrees of freedom.	$\text{d.f.} = n - 1$
4. Find the critical value(s).	Use Table 5 in Appendix B.
5. Identify the rejection region(s).	
6. Calculate $\overline{d}$ and s_d. Use a table.	$\overline{d} = \frac{\Sigma d}{n}$ $s_d = \sqrt{\frac{n(\Sigma d^2) - (\Sigma d)^2}{n(n-1)}}$
7. Calculate the standardized test statistic.	$t = \frac{\overline{d} - \mu_d}{s_d / \sqrt{n}}$
8. Make a decision to reject or fail to reject the null hypothesis.	If t is in the rejection region, reject H_0. Otherwise, do not reject H_0.
9. Interpret the decision in the context of the original claim.	

Study Tip

If $n > 29$, use the last row (∞) in the t-distribution table.

See *Minitab* steps on page 430.

EXAMPLE 

The t-Test for the Difference Between Means

A golf club manufacturer claims that golfers can lower their scores by using the manufacturer's newly designed golf clubs. Eight golfers are randomly selected and each is asked to give his or her most recent score. After using the new clubs for one month, the golfers are again asked to give their most recent score. The scores for each golfer are given in the table below. Assuming the golf scores are normally distributed, is there enough evidence to support the manufacturer's claim at $\alpha = 0.10$?

Golfer	1	2	3	4	5	6	7	8
Score (old design)	89	84	96	82	74	92	85	91
Score (new design)	83	83	92	84	76	91	80	91

Study Tip

You can also use technology and a *P*-value to perform a hypothesis test for the difference between means. For instance, in Example 2, you can enter the data in *Minitab* (as shown on page 430) and find $P \approx 0.089$. Because $P < \alpha$, you should decide to reject the null hypothesis.

SOLUTION The claim is that "golfers can lower their scores." In other words, the manufacturer claims that the score using the old clubs will be greater than the score using the new clubs. Each difference is given by

$$d = (\text{old score}) - (\text{new score}).$$

The null and alternative hypotheses are

H_0: $\mu_d \leq 0$ and H_a: $\mu_d > 0$. (Claim)

Old	New	d	d^2
89	83	6	36
84	83	1	1
96	92	4	16
82	84	−2	4
74	76	−2	4
92	91	1	1
85	80	5	25
91	91	0	0
		$\Sigma = 13$	$\Sigma = 87$

Because the test is a right-tailed test, $\alpha = 0.10$, and d.f. $= 8 - 1 = 7$, the critical value is $t_0 = 1.415$. The rejection region is $t > 1.415$. Using the table at the left, you can calculate $\bar{d}$ and s_d as follows.

$$\bar{d} = \frac{\Sigma d}{n} = \frac{13}{8} = 1.625$$

$$s_d = \sqrt{\frac{n(\Sigma d^2) - (\Sigma d)^2}{n(n-1)}} = \sqrt{\frac{8(87) - (13)^2}{8(8-1)}} \approx 3.07$$

Using the *t*-test, the standardized test statistic is

$$t = \frac{\bar{d} - \mu_d}{s_d/\sqrt{n}} \approx \frac{1.625 - 0}{3.07/\sqrt{8}} \approx 1.50.$$

The graph below shows the location of the rejection region and the standardized test statistic *t*. Because *t* is in the rejection region, you should decide to reject the null hypothesis. There is enough evidence to support the golf club manufacturer's claim at the 10% level. The results of this test indicate that after using the new clubs, golf scores were significantly lower.

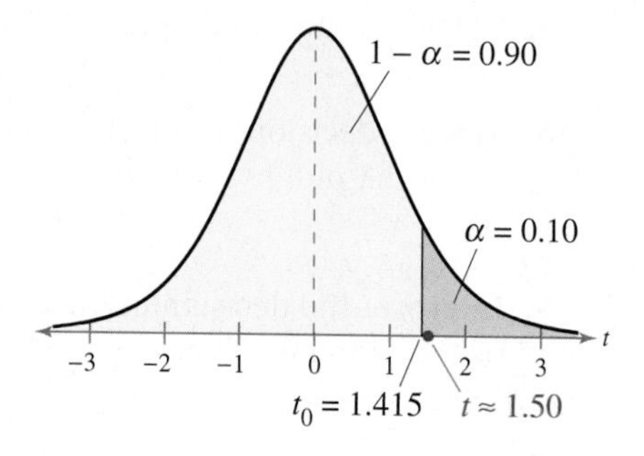

Before	After
72	73
81	80
76	79
74	76
75	76
80	80
68	74
75	77
78	75
76	74
74	76
77	78

Try It Yourself 2

A physician claims that an experimental medication increases an individual's heart rate. Twelve test subjects are randomly selected and the heart rate of each is measured. The subjects are then injected with the medication and, after one hour, the heart rate of each is measured again. The results are listed at the left. Assuming the heart rates are normally distributed, is there enough evidence to support the physician's claim at $\alpha = 0.05$?

a. *Identify* the claim and state H_0 and H_a.
b. *Specify* the level of significance α and the degrees of freedom d.f.
c. *Find* the critical value t_0 and *identify* the rejection region.
d. Calculate $\bar{d}$ and s_d.
e. *Use* the t-test to find the standardized test statistic t.
f. *Decide* whether to reject the null hypothesis. Use a graph if necessary.
g. Is there enough evidence to support the physician's claim?

Answer: Page A40

EXAMPLE 

The t-Test for the Difference Between Means

A state legislator wants to determine whether her voter's performance rating (0–100) has changed from last year to this year. The following table shows the legislator's performance rating for the same 16 randomly selected voters for last year and this year. At $\alpha = 0.01$, is there enough evidence to conclude that the legislator's performance rating has changed? Assume the performance ratings are normally distributed.

Voter	1	2	3	4	5	6	7	8
Rating (last year)	60	54	78	84	91	25	50	65
Rating (this year)	56	48	70	60	85	40	40	55

Voter	9	10	11	12	13	14	15	16
Rating (last year)	68	81	75	45	62	79	58	63
Rating (this year)	80	75	78	50	50	85	53	60

Study Tip

If you prefer to use a technology tool for this type of test, enter the data in two columns and form a third column in which you calculate the difference for each pair. You can now perform a one-sample t-test on the difference column as shown in Chapter 7.

SOLUTION If there is a change in the legislator's rating, there will be a difference between "this year's" ratings and "last year's" ratings. Because the legislator wants to see if there is a difference, the null and alternative hypotheses are

$$H_0\colon \mu_d = 0$$

and

$$H_a\colon \mu_d \neq 0. \text{ (Claim)}$$

Because the test is a two-tailed test, $\alpha = 0.01$, and d.f. $= 16 - 1 = 15$, the critical values are -2.947 and 2.947. The rejection regions are $t < -2.947$ and $t > 2.947$.

Before	After	d	d^2
60	56	4	16
54	48	6	36
78	70	8	64
84	60	24	576
91	85	6	36
25	40	−15	225
50	40	10	100
65	55	10	100
68	80	−12	144
81	75	6	36
75	78	−3	9
45	50	−5	25
62	50	12	144
79	85	−6	36
58	53	5	25
63	60	3	9
		$\Sigma = 53$	$\Sigma = 1581$

Using the table at the left, you can calculate $\bar{d}$ and s_d as shown below.

$$\bar{d} = \frac{\Sigma d}{n} = \frac{53}{16} = 3.3125$$

$$s_d = \sqrt{\frac{n(\Sigma d^2) - (\Sigma d)^2}{n(n-1)}} = \sqrt{\frac{16(1581) - (53)^2}{16(16-1)}} \approx 9.68$$

Using the t-test, the standardized test statistic is

$$t = \frac{\bar{d} - \mu_d}{s_d/\sqrt{n}} \approx \frac{3.3125 - 0}{9.68/\sqrt{16}} \approx 1.369.$$

The graph below shows the location of the rejection region and the standardized test statistic t. Because t is not in the rejection region, you should fail to reject the null hypothesis at the 1% level. There is not enough evidence to conclude that the legislator's performance rating has changed.

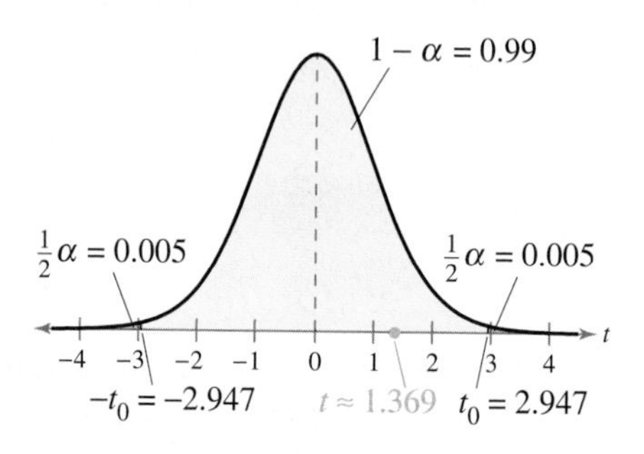

Try It Yourself 3

A medical researcher wants to determine whether a drug changes the body's temperature. Seven test subjects are randomly selected and the body temperature of each is measured. The subjects are then given the drug and, after 20 minutes, the body temperature of each is measured again. The results are listed below. At $\alpha = 0.05$, is there enough evidence to conclude that the drug changes the body's temperature? Assume the body temperatures are normally distributed.

Subject	1	2	3	4	5	6	7
Initial temperature	101.8	98.5	98.1	99.4	98.9	100.2	97.9
Second temperature	99.2	98.4	98.2	99	98.6	99.7	97.8

a. *Identify* the claim and state H_0 and H_a.
b. *Specify* the level of significance α and the degrees of freedom d.f.
c. *Find* the critical values and the rejection regions.
d. Calculate $\bar{d}$ and s_d.
e. *Use* the t-test to find the standardized test statistic t.
f. *Decide* whether to reject the null hypothesis. Use a graph if necessary.
g. Is there enough evidence to conclude that the drug changes the body's temperature at the specified level of significance?

Answer: Page A40

8.3 Exercises

LarsonTutor 8.3

Companion Web Site

Student Solutions Manual 8.3

Videos 8.3

Technology Manuals

Basic Skills and Concepts

1. What is the difference between two samples that are dependent and two samples that are independent? Give an example of two dependent samples and two independent samples.

2. What conditions are necessary in order to use the dependent samples t-test for the mean of the difference of two populations?

In Exercises 3–10, classify the two given samples as independent or dependent. Explain your reasoning.

3. Sample 1: The SAT scores for 35 high school students who did not take an SAT preparation course

Sample 2: The SAT scores for 40 high school students who did take an SAT preparation course

4. Sample 1: The SAT scores for 44 high school students

Sample 2: The SAT scores for the same 44 high school students after taking an SAT preparation course

5. Sample 1: The weights of 51 adults

Sample 2: The weights of the same 51 adults after participating in a diet and exercise program for one month

6. Sample 1: The weights of 40 females

Sample 2: The weights of 40 males

7. Sample 1: The average speed of 23 powerboats using an old hull design

Sample 2: The average speed of 14 powerboats using a new hull design

8. Sample 1: The fuel mileage of 10 cars

Sample 2: The fuel mileage of the same 10 cars using a fuel additive

9. The table shows the braking distances (in feet) for each of four different sets of tires with the car's anti-lock braking system (ABS) on and with ABS off. The tests were done on ice with cars traveling at 15 miles per hour. *(Source: Consumer Reports)*

Tire Set	1	2	3	4
Braking distance with ABS	42	55	43	61
Braking distance without ABS	58	67	59	75

10. The table shows the heart rates (in beats per minute) of five people before exercising and after.

Person	1	2	3	4	5
Heart rate before exercising	65	72	85	78	93
Heart rate after exercising	127	135	140	136	150

In Exercises 11–16, test the claim about the mean of the difference of two populations. Use a t-test for dependent, random samples at the given level of significance with the given statistics. Is the test right tailed, left tailed, or two tailed? Assume the populations are normally distributed.

11. Claim: $\mu_d < 0$, $\alpha = 0.05$. Statistics: $\bar{d} = 10$, $s_d = 1.5$, $n = 10$

12. Claim: $\mu_d = 0$, $\alpha = 0.01$. Statistics: $\bar{d} = 3.2$, $s_d = 0.25$, $n = 8$

13. Claim: $\mu_d \leq 0$, $\alpha = 0.10$. Statistics: $\bar{d} = 6.1$, $s_d = 0.36$, $n = 16$

14. Claim: $\mu_d > 0$, $\alpha = 0.05$. Statistics: $\bar{d} = 5.3$, $s_d = 0.41$, $n = 35$

15. Claim: $\mu_d \geq 0$, $\alpha = 0.01$. Statistics: $\bar{d} = -2.3$, $s_d = 1.2$, $n = 15$

16. Claim: $\mu_d \neq 0$, $\alpha = 0.10$. Statistics: $\bar{d} = 4.2$, $s_d = 0.8$, $n = 20$

Testing the Difference Between Two Means In Exercises 17–24, (a) identify the claim and state H_0 and H_a, (b) find the critical value(s) and identify the rejection region(s), (c) calculate $\bar{d}$ and s_d, (d) use the t-test to find the standardized test statistic t, and (e) decide whether to reject or fail to reject the null hypothesis and interpret the decision in the context of the original claim. For each randomly selected sample, assume the distribution of the population is normal.

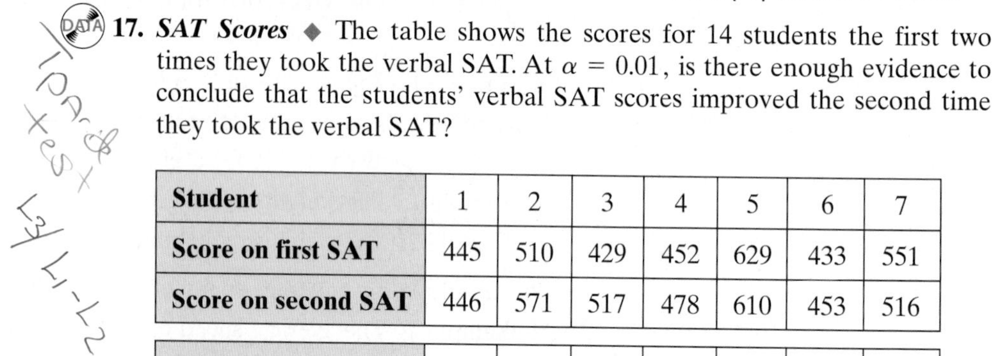

DATA **17.** ***SAT Scores*** ◆ The table shows the scores for 14 students the first two times they took the verbal SAT. At $\alpha = 0.01$, is there enough evidence to conclude that the students' verbal SAT scores improved the second time they took the verbal SAT?

Student	1	2	3	4	5	6	7
Score on first SAT	445	510	429	452	629	433	551
Score on second SAT	446	571	517	478	610	453	516

Student	8	9	10	11	12	13	14
Score on first SAT	358	477	325	513	636	571	442
Score on second SAT	478	532	399	531	648	603	461

18. ***SAT Scores*** ◆ An SAT prep course claims to improve the test scores of students. The table shows the scores for ten students the first two times they took the verbal SAT. Before taking the SAT for the second time, each student took a course to try to improve his or her verbal SAT scores. Test the claim at $\alpha = 0.01$.

Student	1	2	3	4	5	6	7
Score on first SAT	308	456	352	433	306	471	422
Score on second SAT	400	524	409	491	348	583	451

Student	8	9	10
Score on first SAT	370	320	418
Score on second SAT	408	391	450

19. ***Gas Mileage*** ◆ The table shows the gas mileage (in miles per gallon) of eight cars with and without using a fuel additive. At $\alpha = 0.10$, is there enough evidence to conclude that the fuel additive improved gas mileage?

Car	1	2	3	4	5
Gas mileage without additive	35.8	37.7	39.4	36.8	36.6
Gas mileage with fuel additive	36.2	39.8	40.1	39.3	36.9

Car	6	7	8
Gas mileage without additive	33.7	38.4	37.3
Gas mileage with fuel additive	34.5	38.8	39.1

20. ***Gas Mileage*** ◆ To test whether a fuel additive improves gas mileage, the gas mileage of nine cars was measured with and without the fuel additive. The results are given below. At $\alpha = 0.10$, can you conclude that the fuel additive improved gas mileage?

Car	1	2	3	4	5
Gas mileage without additive	34.5	36.7	34.4	39.8	33.6
Gas mileage with fuel additive	36.4	38.8	36.1	40.1	34.7

Car	6	7	8	9
Gas mileage without additive	35.4	38.4	35.3	37.9
Gas mileage with fuel additive	38.3	40.2	37.2	38.7

DATA **21.** ***Blood Pressure*** ◆ A pharmaceutical company guarantees that its new drug reduces systolic blood pressure. The table shows the systolic blood pressure (in millimeters of mercury) for 15 patients before taking the new drug and 3 hours after taking the drug. At $\alpha = 0.05$, can you conclude that the new drug reduces systolic blood pressure?

Patient	1	2	3	4	5
Systolic blood pressure (before)	201	171	186	162	165
Systolic blood pressure (after)	192	165	167	155	148

Patient	6	7	8	9	10
Systolic blood pressure (before)	167	175	148	172	204
Systolic blood pressure (after)	144	152	134	151	178

Patient	11	12	13	14	15
Systolic blood pressure (before)	188	145	178	144	192
Systolic blood pressure (after)	175	121	158	127	167

22. *Blood Pressure* ◆ A pharmaceutical company claims that its new drug reduces diastolic blood pressure. The diastolic blood pressure (in millimeters of mercury) for 10 patients before taking the new drug and 2 hours after taking the drug is shown in the table below. At $\alpha = 0.05$, is there enough evidence to support the company's claim?

Patient	1	2	3	4	5
Diastolic blood pressure (before)	103	122	106	112	125
Diastolic blood pressure (after)	98	121	107	105	108

Patient	6	7	8	9	10
Diastolic blood pressure (before)	97	107	118	112	104
Diastolic blood pressure (after)	89	102	114	101	108

23. *Headaches* ◆ A physical therapist suggests that soft tissue therapy and spinal manipulation help to reduce the length of time patients suffer from headaches. The table shows the number of hours per day a sample of 11 patients suffered from headaches, before and after seven weeks of receiving soft tissue therapy and spinal manipulation. At $\alpha = 0.01$, is there enough evidence to support the therapist's claim? *(Adapted from The Journal of the American Medical Association)*

Patient	1	2	3	4	5	6
Daily headache hours (before)	2.8	2.4	2.8	2.6	2.7	2.9
Daily headache hours (after)	1.6	1.3	1.6	1.4	1.5	1.6

Patient	7	8	9	10	11
Daily headache hours (before)	3.2	2.9	4.1	1.6	2.5
Daily headache hours (after)	1.7	1.6	1.8	1.2	1.4

24. *Headaches* ◆ The table shows the number of hours per day a sample of 13 patients suffered from headaches before and after seven weeks of soft tissue therapy and laser treatment. At $\alpha = 0.01$, test the claim that the new drug helps to reduce the number of daily headache hours. *(Adapted from The Journal of the American Medical Association)*

Patient	1	2	3	4	5	6	7
Daily headache hours (before)	4.1	3.9	3.8	4.5	2.4	3.6	3.4
Daily headache hours (after)	2.2	2.8	2.5	2.6	1.9	1.8	2.0

Patient	8	9	10	11	12	13
Daily headache hours (before)	3.4	3.5	2.7	3.7	4.4	3.2
Daily headache hours (after)	1.6	1.5	2.1	1.8	3.0	1.7

Extending the Basics

Constructing Confidence Intervals for μ_d To construct a confidence interval for μ_d, use the following inequality.

$$\bar{d} - t_c \frac{s_d}{\sqrt{n}} < \mu_d < \bar{d} + t_c \frac{s_d}{\sqrt{n}}$$

In Exercises 25 and 26, construct the indicated confidence interval for μ_d. Assume the populations are normally distrubuted.

25. ***Drug Testing*** ◆ A sleep disorder specialist wants to test the effectiveness of a new drug that is reported to increase the number of hours of sleep patients get during the night. To do so, the specialist randomly selects 16 patients and records the number of hours of sleep each gets with and without the new drug. The results of the two-night study are listed below. Construct a 90% confidence interval for μ_d.

Patient	1	2	3	4	5	6	7	8	9	10	11	12	13	14	15	16
Hours of sleep without the drug	1.8	2.0	3.4	3.5	3.7	3.8	3.9	3.9	4.0	4.9	5.1	5.2	5.0	4.5	4.2	4.7
Hours of sleep using the new drug	3.0	3.6	4.0	4.4	4.5	5.2	5.5	5.7	6.2	6.3	6.6	7.8	7.2	6.5	5.6	5.9

26. ***Herbal Medicine Testing*** ◆ An herbal medicine is tested on 14 randomly selected patients with sleeping disorders. The table shows the hours of sleep patients got during one night without using the herbal medicine and the hours of sleep the patients got on another night after the herbal medicine was administered. Construct a 95% confidence interval for μ_d.

Patient	1	2	3	4	5	6	7	8	9	10	11	12	13	14
Hours of sleep without medicine	1.0	1.4	3.4	3.7	5.1	5.1	5.2	5.3	5.5	5.8	4.2	4.8	2.9	4.5
Hours of sleep using the herbal medicine	2.9	3.3	3.5	4.4	5.0	5.0	5.2	5.3	6.0	6.5	4.4	4.7	3.1	4.7

8.4 Testing the Difference Between Proportions

What You Should Learn

- How to perform a z-test for the difference between two population proportions p_1 and p_2

Two-Sample z-Test for the Difference Between Proportions

Two-Sample z-Test for the Difference Between Proportions

In this section, you will learn how to use a z-test to test the difference between two population proportions p_1 and p_2 using a sample proportion from each population. For instance, suppose you want to determine whether the proportion of female college students who earn a bachelor's degree in four years is different from the proportion of male college students who earn a bachelor's degree in four years. To use a z-test to test such a difference, the following conditions are necessary.

1. The samples must be randomly selected.
2. The samples must be independent.
3. The samples must be large enough to use a normal sampling distribution. That is,

$$n_1 p_1 \geq 5$$
$$n_1 q_1 \geq 5$$
$$n_2 p_2 \geq 5 \quad \text{and}$$
$$n_2 q_2 \geq 5.$$

If these conditions are met, then the sampling distribution for $\hat{p}_1 - \hat{p}_2$, the difference between the sample proportions, is a normal distribution with mean

$$\mu_{\hat{p}_1 - \hat{p}_2} = p_1 - p_2$$

and standard error

$$\sigma_{\hat{p}_1 - \hat{p}_2} = \sqrt{\frac{p_1 q_1}{n_1} + \frac{p_2 q_2}{n_2}}.$$

Notice that you need to know the population proportions to calculate the standard error. Because a hypothesis test for $p_1 - p_2$ is based on the condition of equality that $p_1 = p_2$, you can calculate a weighted estimate of p_1 and p_2 using

$$\bar{p} = \frac{x_1 + x_2}{n_1 + n_2} \qquad \text{where } x_1 = n_1 \hat{p}_1 \text{ and } x_2 = n_2 \hat{p}_2.$$

Using the weighted estimate $\bar{p}$, the standard error of the sampling distribution for $\hat{p}_1 - \hat{p}_2$ is

$$\sigma_{\hat{p}_1 - \hat{p}_2} = \sqrt{\bar{p}\,\bar{q}\left(\frac{1}{n_1} + \frac{1}{n_2}\right)} \qquad \text{where } \bar{q} = 1 - \bar{p}.$$

Also observe that you need to know the population proportions in verifying that the samples are large enough to be approximated by the normal distribution. But when determining whether the z-test can be used for the difference between proportions for a binomial experiment, you should use $\bar{p}$ in place of p_1 and p_2 and use $\bar{q}$ in place of q_1 and q_2.

Study Tip

The following symbols are used in the z-test for $p_1 - p_2$. See Sections 4.2 and 5.6 to review the binomial distribution.

Symbol	Description
p_1, p_2	Population proportions
x_1, x_2	Number of successes in each sample
n_1, n_2	Size of each sample
$\hat{p}_1, \hat{p}_2$	Sample proportions of successes
$\bar{p}$	Weighted estimate for p_1 and p_2

Picturing the World

A study found a difference in the proportion of adult men and women who use alternative medicines, such as herbal medicines, massage, spiritual healing, and acupuncture. In the study, 400 randomly selected women and 450 randomly selected men were asked if they had used at least one alternative medicine therapy during the past year. The results are shown below. *(Adapted from Journal of the American Medical Association)*

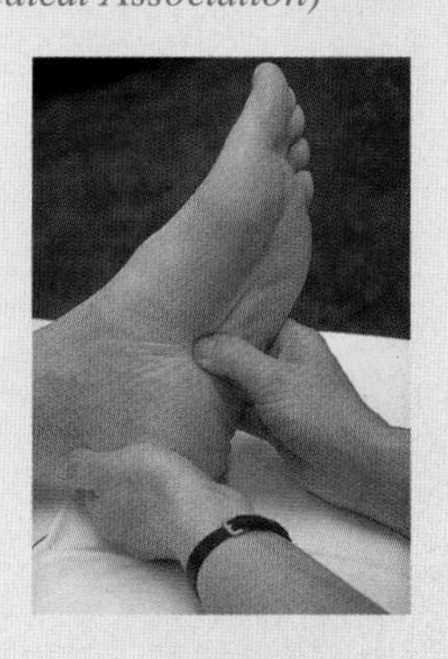

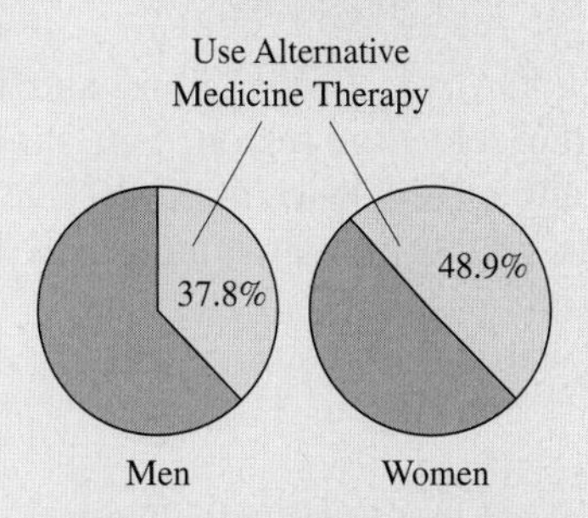

At $\alpha = 0.05$, is there enough evidence to support the claim that there is a significant difference in use of alternative medicines between men and women?

If the sampling distribution for $\hat{p}_1 - \hat{p}_2$ is normal, you can use a two-sample z-test to test the difference between two population proportions p_1 and p_2.

Two-Sample z-Test for the Difference Between Proportions

A two-sample z-test is used to test the difference between two population proportions p_1 and p_2 when a sample is randomly selected from each population. The **test statistic** is

$$\hat{p}_1 - \hat{p}_2$$

and the **standardized test statistic** is

$$z = \frac{(\hat{p}_1 - \hat{p}_2) - (p_1 - p_2)}{\sqrt{\bar{p}\,\bar{q}\left(\frac{1}{n_1} + \frac{1}{n_2}\right)}}$$

where

$$\bar{p} = \frac{x_1 + x_2}{n_1 + n_2} \quad \text{and} \quad \bar{q} = 1 - \bar{p}.$$

Note: $n_1\bar{p}$, $n_1\bar{q}$, $n_2\bar{p}$, and $n_2\bar{q}$ must be at least 5.

If the null hypothesis states $p_1 = p_2$, $p_1 \le p_2$, or $p_1 \ge p_2$, then $p_1 = p_2$ is assumed and the expression $p_1 - p_2$ is equal to 0 in the preceding test.

GUIDELINES

Using a Two-Sample z-Test for the Difference Between Proportions

In Words	*In Symbols*
1. Identify the claim. State the null and alternative hypotheses.	State H_0 and H_a.
2. Specify the level of significance.	Identify α.
3. Find the critical value(s).	Use Table 4 in Appendix B.
4. Identify the rejection region(s).	
5. Find the weighted estimate of p_1 and p_2.	$\bar{p} = \frac{x_1 + x_2}{n_1 + n_2}$
6. Find the standardized test statistic.	$z = \frac{(\hat{p}_1 - \hat{p}_2) - (p_1 - p_2)}{\sqrt{\bar{p}\,\bar{q}\left(\frac{1}{n_1} + \frac{1}{n_2}\right)}}$
7. Make a decision to reject or fail to reject the null hypothesis.	If z is in the rejection region, reject H_0. Otherwise, do not reject H_0.
8. Interpret the decision in the context of the claim.	

To use P-values in a test for the difference in proportions, use the same guidelines as above, skipping steps 3 and 4. After finding the standardized test statistic, use the Standard Normal Table to calculate the P-value. Then make a decision to reject or fail to reject the null hypothesis. If P is less than or equal to α, reject H_0. Otherwise fail to reject H_0.

See *TI-83* steps on page 431.

EXAMPLE 1

A Two-Sample z-Test for the Difference Between Proportions

In a study of 200 randomly selected adult female and 250 randomly selected adult male Internet users, 30% of the females and 38% of the males said that they plan to shop online at least once during the next month. At $\alpha = 0.10$, test the claim that there is a difference in the proportion of female Internet users who plan to shop online and the proportion of male Internet users who plan to shop online.

Sample Statistics for Internet Users

Female	Male
$n_1 = 200$	$n_2 = 250$
$\hat{p}_1 = 0.30$	$\hat{p}_2 = 0.38$
$n_1\hat{p}_1 = 60$	$n_2\hat{p}_2 = 95$

SOLUTION You want to determine whether there is a difference in the proportions. So, the null and alternative hypotheses are

$$H_0\colon p_1 = p_2 \quad \text{and} \quad H_a\colon p_1 \neq p_2. \text{ (Claim)}$$

Because the test is two tailed and the level of significance is $\alpha = 0.10$, the critical values are $z_0 = -1.645$ and $z_0 = 1.645$. The rejection regions are $z < -1.645$ and $z > 1.645$. The weighted estimate of the population proportions is

$$\overline{p} = \frac{x_1 + x_2}{n_1 + n_2} = \frac{60 + 95}{200 + 250} = \frac{155}{450} \approx 0.344$$

and

$$\overline{q} = 1 - \overline{p} = 1 - 0.344 = 0.656.$$

Because $200(0.344)$, $200(0.656)$, $250(0.344)$, and $250(0.656)$ are at least 5, you can use a two-sample z-test. The standardized test statistic is

$$z = \frac{(\hat{p}_1 - \hat{p}_2) - (p_1 - p_2)}{\sqrt{\overline{p}\,\overline{q}\left(\frac{1}{n_1} + \frac{1}{n_2}\right)}} \approx \frac{(0.30 - 0.38) - 0}{\sqrt{(0.344)(0.656)\left(\frac{1}{200} + \frac{1}{250}\right)}} \approx -1.775.$$

Study Tip

To find x_1 and x_2, use $x_1 = n_1\hat{p}_1$ and $x_2 = n_1\hat{p}_2$.

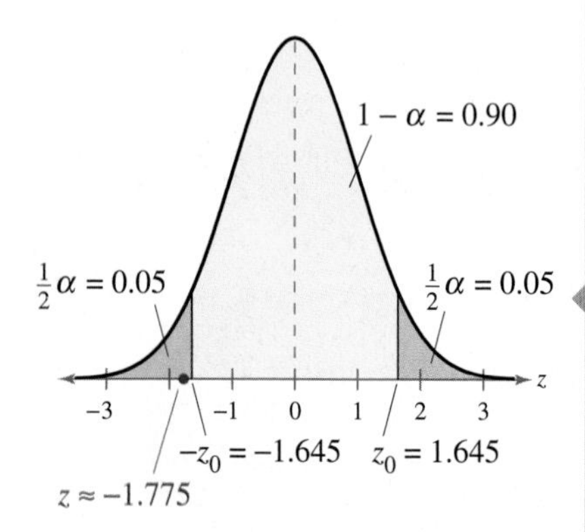

The graph at the left shows the location of the rejection regions and the standardized test statistic. Because z is in the rejection region, you should decide to reject the null hypothesis. You have enough evidence at the 10% level to conclude that there is a difference in the proportion of female and the proportion of male Internet users who plan to shop online.

Try It Yourself 1

Consider the results of the *JAMA* study discussed in the Chapter Opener. At $\alpha = 0.05$, is there a difference between the proportion of Vietnam Army veterans who suffer from anxiety and the proportion of non-Vietnam Army veterans who suffer from anxiety?

a. *Identify* the claim and state H_0 and H_a.
b. *Specify* the level of significance α.
c. *Find* the critical values and *identify* the rejection regions.
d. *Find* $\overline{p}$ and $\overline{q}$.
e. *Verify* that $n_1\overline{p}$, $n_1\overline{q}$, $n_2\overline{p}$, and $n_2\overline{q}$ are at least 5.
f. *Use* the two-sample z-test to find the standardized test statistic z.
g. *Decide* whether to reject the null hypothesis.
h. Is there enough evidence to conclude that there is a difference between the two groups?

Answer: Page A40

Sample Statistics for Cholesterol-Reducing Medication

Received medication	Received placebo
$n_1 = 4700$	$n_2 = 4300$
$x_1 = 301$	$x_2 = 357$
$\hat{p}_1 = 0.064$	$\hat{p}_2 = 0.083$

Study Tip

To find $\hat{p}_1$ and $\hat{p}_2$, use

$$\hat{p}_1 = \frac{x_1}{n_1} \text{ and}$$

$$\hat{p}_2 = \frac{x_2}{n_2}.$$

EXAMPLE 

Two-Sample z-Test for the Difference Between Proportions

A medical research team conducted a study to test the effect of a cholesterol-reducing medication. At the end of the study, the researchers found that of the 4700 subjects who took the medication, 301 died of heart disease. Of the 4300 subjects who took a placebo, 357 died of heart disease. At $\alpha = 0.01$, can you conclude that the death rate is lower for those who took the medication than for those who took the placebo? *(Source: New England Journal of Medicine)*

SOLUTION You want to determine whether the death rate is lower for those who took the medication than for those who took the placebo. So, the null and alternative hypotheses are

$$H_0: p_1 \geq p_2$$

and

$$H_a: p_1 < p_2. \text{ (Claim)}$$

Because the test is left tailed and the level of significance is $\alpha = 0.01$, the critical value is $z_0 = -2.33$. The rejection region is $z < -2.33$. The weighted estimate of p_1 and p_2 is

$$\overline{p} = \frac{x_1 + x_2}{n_1 + n_2} = \frac{301 + 357}{4700 + 4300} = \frac{658}{9000} \approx 0.073$$

and $\overline{q} = 1 - \overline{p} = 1 - 0.073 = 0.927$. Because $4700(0.073)$, $4700(0.927)$, $4300(0.073)$, and $4300(0.927)$ are at least 5, you can use a two-sample z-test.

$$z = \frac{(\hat{p}_1 - \hat{p}_2) - (p_1 - p_2)}{\sqrt{\overline{p}\,\overline{q}\left(\frac{1}{n_1} + \frac{1}{n_2}\right)}} \approx \frac{(0.064 - 0.083) - 0}{\sqrt{(0.073)(0.927)\left(\frac{1}{4700} + \frac{1}{4300}\right)}} \approx -3.461$$

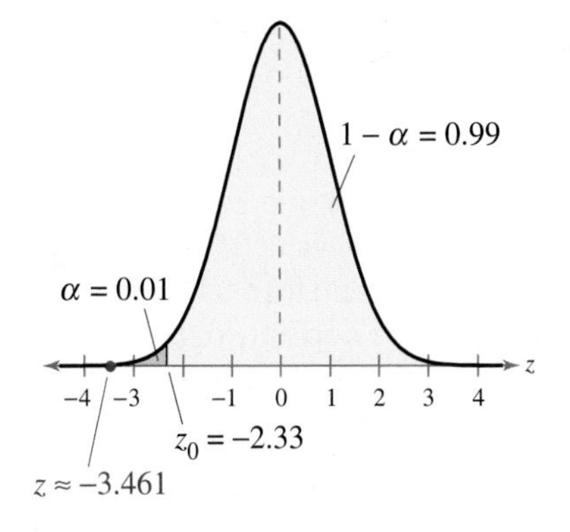

The graph at the left shows the location of the rejection region and the standardized test statistic. Because z is in the rejection region, you should decide to reject the null hypothesis. At the 1% level, there is enough evidence to conclude that the death rate is lower for those who took the medication than for those who took the placebo.

Try It Yourself 2

Consider the results of the *JAMA* study discussed in the Chapter Opener. At $\alpha = 0.05$, can you conclude that the proportion of Vietnam Army veterans who suffer from alcohol abuse or dependence is greater than the proportion of non-Vietnam Army veterans who suffer from alcohol abuse or dependence?

a. *Identify* the claim and state H_0 and H_a.
b. *Specify* the level of significance α.
c. *Find* the critical value and *identify* the rejection region.
d. *Find* $\overline{p}$ and $\overline{q}$.
e. *Verify* that $n_1\overline{p}$, $n_1\overline{q}$, $n_2\overline{p}$, and $n_2\overline{q}$ are at least 5.
f. *Use* the two-sample z-test to find the standardized test statistic z.
g. *Decide* whether to reject the null hypothesis.
h. For the given level of significance, is there enough evidence to conclude that alcohol abuse or dependence is greater for Vietnam Army veterans?

Answer: Page A40

8.4 Exercises

Help

LarsonTutor 8.4

Companion Web Site

Student Solutions Manual 8.4

Videos 8.4

Technology Manuals

Basic Skills and Concepts

1. Explain how to perform a two-sample z-test for the difference between two population proportions.

2. What conditions are necessary in order to use the z-test to test the difference between two population proportions?

In Exercises 3–8, test the claim about the difference between two population proportions p_1 and p_2 for the given level of significance α and the given sample statistics. Is the test right tailed, left tailed, or two tailed? Assume the sample statistics are from independent random samples.

3. Claim: $p_1 \neq p_2$, $\alpha = 0.01$. Sample statistics: $x_1 = 35$, $n_1 = 70$, and $x_2 = 36$, $n_2 = 60$

4. Claim: $p_1 < p_2$, $\alpha = 0.05$. Sample statistics: $x_1 = 471$, $n_1 = 785$, and $x_2 = 372$, $n_2 = 465$

5. Claim: $p_1 \leq p_2$, $\alpha = 0.10$. Sample statistics: $x_1 = 344$, $n_1 = 860$, and $x_2 = 304$, $n_2 = 800$

6. Claim: $p_1 \leq p_2$, $\alpha = 0.04$. Sample statistics: $x_1 = 27$, $n_1 = 50$, and $x_2 = 45$, $n_2 = 85$

7. Claim: $p_1 = p_2$, $\alpha = 0.05$. Sample statistics: $x_1 = 29$, $n_1 = 45$, and $x_2 = 25$, $n_2 = 30$

8. Claim: $p_1 \neq p_2$, $\alpha = 0.01$. Sample statistics: $x_1 = 520$, $n_1 = 1055$, and $x_2 = 450$, $n_2 = 1150$

Testing the Difference Between Two Proportions In Exercises 9–14, (a) identify the claim and state H_0 and H_a, (b) find the critical value(s) and identify the rejection region(s), (c) find the standardized test statistic, and (d) decide whether to reject or fail to reject the null hypothesis and interpret the decision in the context of the original claim. Assume the random samples are independent. If convenient, use technology to solve the problem.

9. ***Alternative Medicine Use*** ◆ In a 1991 study of 1539 adults, 520 said they used alternative medicines (for example, folk remedies and homeopathy) in the previous year. In a more recent study of 2055 adults, 865 said they used alternative medicines in the previous year. At $\alpha = 0.05$, can you reject the claim that the proportion of adults using alternative medicines has not changed since 1991? *(Source: The Journal of the American Medical Association)*

Have You Used Alternative Medicines in the Past Year?

1991 Study
(1539 Adults)

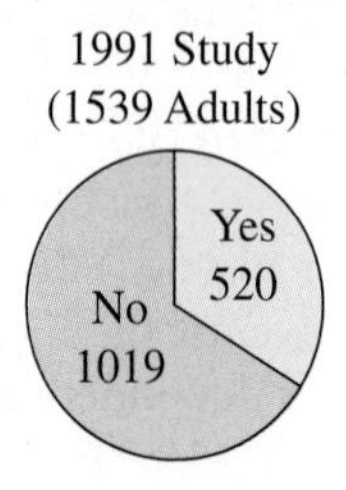

Recent Study
(2055 Adults)

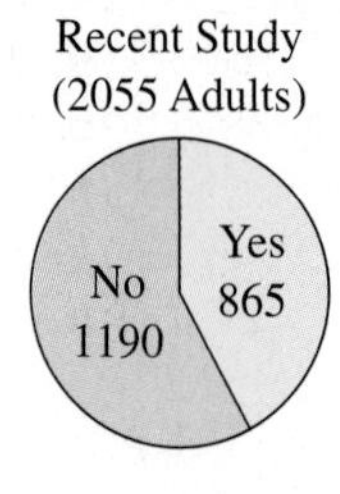

10. ***Antidepressants*** ◆ A medical research team studied patients using an antidepressant drug to recover from chronic depression. Of 77 patients who used the drug for two years, five suffered a new bout of depression. Of 84 patients who used the drug for seven months, 19 suffered a new bout of depression. At $\alpha = 0.10$, can you reject the claim that the proportions of patients suffering new bouts of depression are the same for both groups? *(Source: The Journal of the American Medical Association)*

11. ***Smokers in Alabama and Missouri*** ◆ A state-by-state survey found that the proportion of adults who are smokers in Alabama and Missouri was 24.7% and 28.7%, respectively. (Suppose the number of respondents from each state was 2000.) At $\alpha = 0.01$, can you support the claim that the proportion of adults who are smokers is lower in Alabama than in Missouri? *(Adapted from The Centers for Disease Control and Prevention)*

12. ***Smokers in California and Oregon*** ◆ In a random survey of 1500 adults in California and 1000 adults in Oregon, you find that the percent who are smokers are 18.4% and 20.7%, respectively. At $\alpha = 0.05$, is there enough evidence to conclude that the proportion of adults who are smokers is lower in California than in Oregon? *(Adapted from The Centers for Disease Control and Prevention)*

13. ***Smokers in Public and Private Schools*** ◆ In a survey of 3420 college students attending private schools, 917 said they had smoked in the last 30 days. In a survey of 5131 college students attending public schools, 1503 said they had smoked in the last 30 days. At $\alpha = 0.01$, can you support the claim that the proportion of college students who said they had smoked in the last 30 days in the private schools is less than the proportion in the public schools? *(Adapted from The Journal of the American Medical Association)*

Have You Smoked in the Last 30 Days?

3420 Private College Students

Yes 917

No 2503

5131 Public College Students

Yes 1503

No 3628

14. ***Smokers: Then and Now*** ◆ In a survey of 10,572 college students, 2358 said they had smoked in the last 30 days. In another survey of 8551 college students taken four years later, 2437 said they had smoked in the last 30 days. At $\alpha = 0.10$, can you reject the claim that the proportion of college students who said they had smoked in the last 30 days has not changed? *(Source: The Journal of the American Medical Association)*

Have You Smoked in the Last 30 Days?

Original Survey (10,572 College Students)

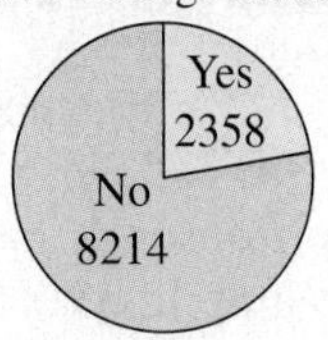

Survey Four Years Later (8551 College Students)

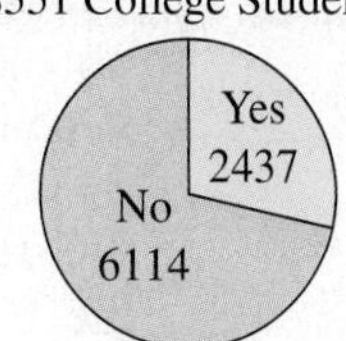

Utility Satisfaction In Exercises 15–18, refer to the figure. Assume the samples are random and independent.

15. ***Telephone Service: Midwest and Northeast*** ◆ At $\alpha = 0.01$, can you support the claim that the proportion of customers completely satisfied with their local telephone service is greater in the Midwest than in the Northeast? Assume the survey included 1020 midwesterners and 900 northeasterners.

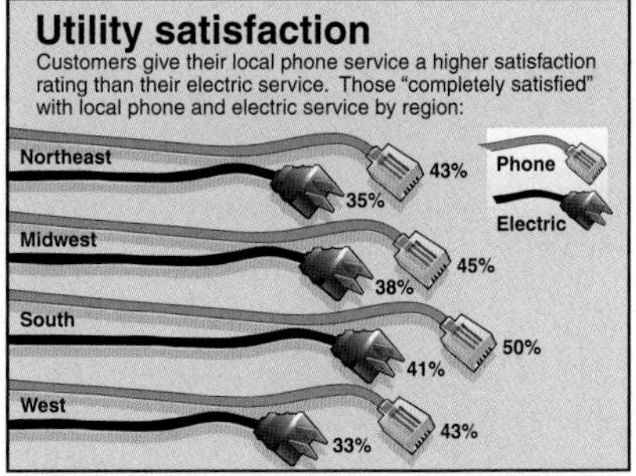

16. ***Telephone Service: South and West*** ◆ A telephone service representative believes that the proportion of customers completely satisfied with their local telephone service is greater in the South than in the West. If the representative's belief is based on the results of the survey, is there enough evidence to support the representative's belief? Assume the survey included 978 southern residents and 1100 western residents. Use $\alpha = 0.01$.

17. ***Electric Service: South and West*** ◆ Based on the survey, a consumer advocate reports that the proportion of customers completely satisfied with their electric service is the same in the South and in the West. If the survey included 1000 southern residents and 1100 western residents, can you reject the advocate's claim? Use $\alpha = 0.05$.

18. ***Electric Service: Midwest and Northeast*** ◆ At $\alpha = 0.05$, can you reject the claim that the proportion of customers completely satisfied with their electric service is the same in the Midwest and in the Northeast? Assume the survey included 1000 midwesterners and 1020 northeasterners.

Extending the Basics

Constructing Confidence Intervals for $p_1 - p_2$ You can construct a confidence interval for the difference of two population means $p_1 - p_2$ by using the following.

$$(\hat{p}_1 - \hat{p}_2) - z_c\sqrt{\frac{\hat{p}_1\hat{q}_1}{n_1} + \frac{\hat{p}_2\hat{q}_2}{n_2}} < p_1 - p_2 < (\hat{p}_1 - \hat{p}_2) + z_c\sqrt{\frac{\hat{p}_1\hat{q}_1}{n_1} + \frac{\hat{p}_2\hat{q}_2}{n_2}}$$

In Exercises 19 and 20, construct the indicated confidence interval for $p_1 - p_2$. Assume the samples are random and independent.

19. ***Students Planning to Study Engineering*** ◆ Several years ago, a survey of 977,000 students taking the SAT revealed that 11.7% of the students were planning to study engineering in college. In a recent survey of 1,085,000 students taking the SAT, 8.5% of the students were planning to study engineering. Construct a 95% confidence interval for the difference in proportions $p_1 - p_2$. *(Source: College Entrance Examination Board)*

20. ***Students Planning to Study Social Science*** ◆ In a certain year, the percent of students taking the SAT who said they intended to study social science in college was 7.5%. Eleven years later, 11.3% said they intended to study social science in college. Construct a 95% confidence interval for the difference in proportions $p_1 - p_2$. Assume 997,000 students were surveyed the first year and 1,085,000 students were surveyed eleven years later. *(Source: College Entrance Examination Board)*

TECHNOLOGY

MINITAB | EXCEL | TI-83

Tails Over Heads

In the article "Tails Over Heads" in the *Washington Post* (Oct. 13, 1996), journalist William Casey describes one of his hobbies—keeping track of every coin he finds on the street! From January 1, 1985 until the article was written, Casey found 11,902 coins.

As each coin is found, Casey records the time, date, location, value, mint location, and whether the coin is lying heads up or tails up. In the article, Casey notes that 6130 coins were found tails up and 5772 were found heads up. Of the 11,902 coins found, 43 were minted in San Francisco, 7133 were minted in Philadelphia, and 4726 were minted in Denver.

A simulation of Casey's experiment can be done in *Minitab* as shown below. A frequency histogram of one simulation's results is shown at the right.

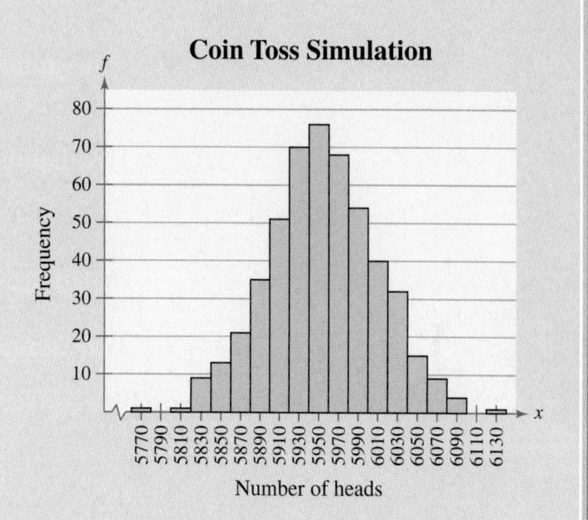

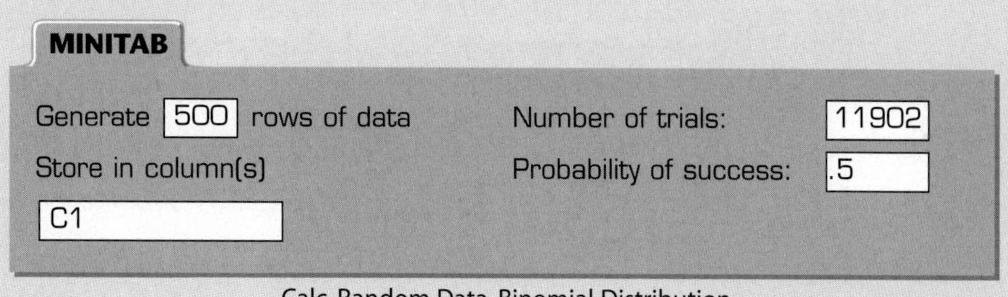

Calc, Random Data, Binomial Distribution

Exercises

1. Use a technology tool to perform a one-sample z-test to test the hypothesis that the probability of a "found coin" lying heads up is 0.5. Use $\alpha = 0.01$. Use Casey's data as your sample and write your conclusion as a sentence.

2. Does Casey's data differ significantly from chance? If so, what might be the reason?

3. In the simulation shown above, what percent of the trials had heads less than or equal to the number of heads in Casey's data? Use a technology tool to repeat the simulation. Are your results comparable?

In Exercises 4 and 5, use a technology tool to perform a two-sample z-test to decide whether there is a difference in the mint dates and in the values of coins found on a street from 1985 through 1996. Write your conclusion as a sentence. Use $\alpha = 0.05$.

4. Mint dates of coins (years)

Philadelphia:	Denver:
$\bar{x}_1 = 1984.8$	$\bar{x}_2 = 1983.4$
$s_1 = 8.6$	$s_2 = 8.4$

5. Value of coins (dollars)

Philadelphia:	Denver:
$\bar{x}_1 = \$0.034$	$\bar{x}_2 = \$0.033$
$s_1 = \$0.054$	$s_2 = \$0.052$

Extended solutions are given in the *Technology Supplement*.
Technical instruction is provided for *Minitab*, *Excel*, and the *TI-83*.

Using Technology to Perform Two-Sample Hypothesis Tests

Here are some *Minitab* and *TI-83* printouts for several examples in this chapter. To duplicate the *Minitab* results, you need the original data. For the *TI-83*, you can simply enter the descriptive statistics.

(See Example 1, page 404)

Stopping Distances for 10 Firestone Winterfire Tires, in feet

63 48 57 42 54 58 45 55 37 51

Stopping Distances for 12 Michelin XM+S Alpin Tires, in feet

49 52 52.5 53 54.5 55 55.5 56 57 57.5 58.5 59.5

Display Descriptive Statistics...
Store Descriptive Statistics...
1-Sample Z...
1-Sample t...
2-Sample t...
Paired t...
1 Proportion...
2 Proportions...

MINITAB

Two-Sample T-Test and Confidence Interval

```
Two-sample T for C1 vs C2

        N     Mean    StDev   SE Mean
C1     10    51.00     8.00       2.5
C2     12    55.00     3.01      0.87

99% CI for difference: (-12.31, 4.31)
T-Test of difference = 0 (vs not = ): T-Value = -1.50
P-Value = 0.163  DF = 11
Difference = mu C1 - mu C2
Estimate for difference: -4.00
```

(See Example 2, page 414)

Golf Scores, Old and New Club Design

Golfer	1	2	3	4	5	6	7	8
Score (old design)	89	84	96	82	74	92	85	91
Score (new design)	83	83	92	84	76	91	80	91

Display Descriptive Statistics...
Store Descriptive Statistics...
1-Sample Z...
1-Sample t...
2-Sample t...
Paired t...
1 Proportion...
2 Proportions...

MINITAB

Paired T-Test

```
Paired T for C1 - C2

              N     Mean    StDev   SE Mean
C1            8    86.63     6.89      2.43
C2            8    85.00     5.81      2.05
Difference    8     1.63     3.07      1.08

T-Test of mean difference = 0 (vs>0):  T-Value = 1.50  P-Value = 0.089
```

(See Example 1, page 393)

TI-83

EDIT CALC TESTS
1: Z-Test...
2: T-Test...
3: 2-SampZTest...
4: 2-SampTTest...
5: 1-PropZTest...
6: 2-PropZTest...
7↓ZInterval...

TI-83

2-SampZTest
Inpt:Data Stats
σ1: 12000
σ2: 15000
x̄1: 60900
n1: 100
x̄2: 64300
↓n2: 100

TI-83

2-SampZTest
↑σ2: 15000
x̄1: 60900
n1: 100
x̄2: 64300
n2: 100
μ1: ≠μ2 <μ2 >μ2
Calculate Draw

TI-83

2-SampZTest
μ1 ≠ μ2
z= −1.769969301
p= .0767321594
x̄1= 60900
x̄2= 64300
↓n1= 100

(See Example 2, page 405)

TI-83

EDIT CALC TESTS
1: Z-Test...
2: T-Test...
3: 2-SampZTest...
4: 2-SampTTest...
5: 1-PropZTest...
6: 2-PropZTest...
7↓ZInterval...

TI-83

2-SampTTest
Inpt:Data Stats
x̄1: 1275
Sx1: 45
n1: 14
x̄2: 1250
Sx2: 30
↓n2: 16

TI-83

2-SampTTest
↑n1: 14
x̄2: 1250
Sx2: 30
n2: 16
μ1: ≠μ2 <μ2 >μ2
Pooled: No Yes
Calculate Draw

TI-83

2-SampTTest
μ1 > μ2
t= 1.811358919
p= .0404131295
df= 28
x̄1= 1275
↓x̄2= 1250

(See Example 1, page 424)

TI-83

EDIT CALC TESTS
1: Z-Test...
2: T-Test...
3: 2-SampZTest...
4: 2-SampTTest...
5: 1-PropZTest...
6: 2-PropZTest...
7↓ZInterval...

TI-83

2-PropZTest
x1: 60
n1: 200
x2: 95
n2: 250
p1: ≠p2 <p2 >p2
Calculate Draw

TI-83

2-PropZTest
p1 ≠ p2
z= −1.774615984
p= .0759612188
p̂1= .3
p̂2= .38
↓p̂= .3444444444

8

Chapter Summary

What did you learn?

Review Exercises

Section 8.1

- How to perform a two-sample z-test for the difference between two means μ_1 and μ_2 using large independent samples — *1–6*

Section 8.2

- How to perform a t-test for the difference between two population means μ_1 and μ_2 using small independent samples — *7–14*

Section 8.3

- How to decide whether two samples are independent or dependent — *15, 16*
- How to perform a t-test to test the mean of the differences for a population of paired data — *17–22*

Section 8.4

- How to perform a z-test for the difference between two population proportions p_1 and p_2 — *23–28*

Two Sample Hypothesis Testing for Population Means

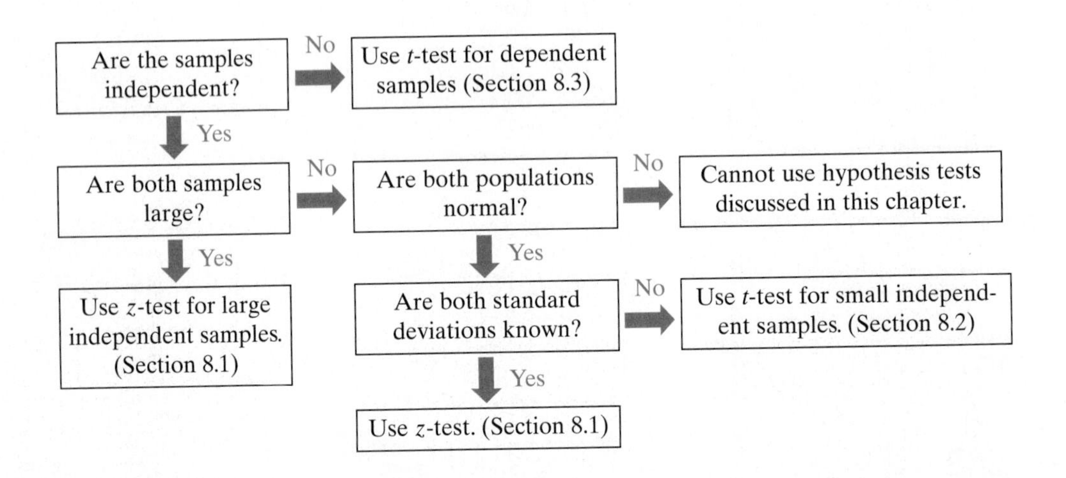

STATISTICS

Uses and Abuses

Uses

Hypothesis Testing with Two Samples Hypothesis testing enables you to decide whether differences in samples indicate actual differences in populations or are merely due to sampling error. Here are some of the kinds of questions that two-sample hypothesis testing can help you analyze. Is one medication more effective than another? Are boys more likely to have attention deficit disorder than girls? Does a new marketing method produce better results than the previous marketing method?

Abuses

Using Nonrepresentative Samples When comparing data collected from two different samples, care should be taken to ensure that the samples are similar in all characteristics except the variable being tested. For instance, suppose you are examining a claim that a new arthritis medication lessens joint pain.

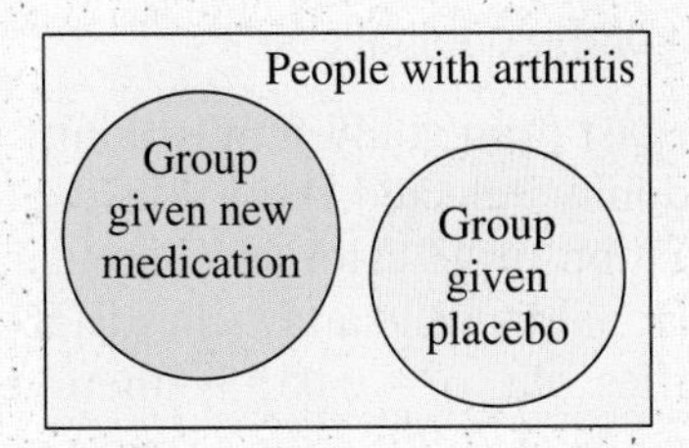

If the group that is given a medication is over sixty years old and the group given the placebo is under forty, other variables might affect the outcome of the study. When you look for other abuses in a study, consider how the claim in the study was determined. What were the sample sizes? Were the samples random? Were they independent? Was the sampling conducted by an unbiased researcher?

Exercises

1. ***Using Nonrepresentative Samples*** Assume that you work for the Food and Drug Administration (*www.fda.gov*). A pharmaceutical company has applied for approval to market a new arthritis mediation. The research involved a test group that was given the medication and another test group that was given a placebo. Describe some ways that the test groups might not have been representative of the entire population of people with arthritis.

2. Medical research often involves blind and double-blind testing. Explain what these two terms mean. If necessary, use your library or another research resource to help you answer this question.

8 Review Exercises

Section 8.1

In Exercises 1–4, use the given sample statistics (independent samples) to test the claim about the difference between two population means μ_1 and μ_2 at the given level of significance α.

1. Claim: $\mu_1 \geq \mu_2$, $\alpha = 0.05$. Sample statistics: $\overline{x}_1 = 1.28$, $s_1 = 0.28$, $n_1 = 76$, and $\overline{x}_2 = 1.36$, $s_2 = 0.23$, $n_2 = 65$

2. Claim: $\mu_1 = \mu_2$, $\alpha = 0.01$. Sample statistics: $\overline{x}_1 = 5595$, $s_1 = 52$, $n_1 = 156$, and $\overline{x}_2 = 5575$, $s_2 = 68$, $n_2 = 216$

3. Claim: $\mu_1 < \mu_2$, $\alpha = 0.10$. Sample statistics: $\overline{x}_1 = 0.28$, $s_1 = 0.11$, $n_1 = 41$, and $\overline{x}_2 = 0.33$, $s_2 = 0.10$, $n_2 = 34$

4. Claim: $\mu_1 \neq \mu_2$, $\alpha = 0.05$. Sample statistics: $\overline{x}_1 = 82$, $s_1 = 11$, $n_1 = 410$, and $\overline{x}_2 = 90$, $s_2 = 10$, $n_2 = 340$

In Exercises 5 and 6, (a) identify the claim and state H_0 and H_a, (b) identify the rejection region(s), (c) find the standardized test statistic, (d) decide whether to reject, or fail to reject, the null hypothesis, and (e) interpret this decision in the context of the original claim.

5. In a fast food study, a nutritionist finds that the mean calorie content of 36 randomly selected Arby's fish sandwiches is 529 calories with a standard deviation of 43 calories. The mean calorie content of 41 randomly selected McDonald's fish sandwiches is 560 calories with a standard deviation of 57 calories. At $\alpha = 0.05$, is there enough evidence for the nutritionist to conclude that the Arby's sandwich has fewer calories than the McDonald's sandwich? *(Adapted from Fast Food Facts, Minnesota Attorney General's Office)*

6. A study of fast food nutrition compared the caloric content of french fries. Thirty-eight randomly selected servings of Burger King medium french fries had a mean of 370 calories and a standard deviation of 50 calories, while 35 randomly selected servings of Hardee's medium french fries had a mean of 350 calories and a standard deviation of 45 calories. At $\alpha = 0.10$, can you support the claim that the caloric content of the two types of french fries is different? *(Adapted from Fast Food Facts, Minnesota Attorney General's Office)*

Section 8.2

In Exercises 7–12, use the given sample statistics to test the claim about the differences between two population means μ_1 and μ_2 at the given level of significance α. Assume the samples are random and independent and that the populations are approximately normally distributed.

7. Claim: $\mu_1 = \mu_2$, $\alpha = 0.05$. Sample statistics: $\overline{x}_1 = 250$, $s_1 = 26$, $n_1 = 21$, and $\overline{x}_2 = 240$, $s_2 = 22$, $n_2 = 12$. Assume equal variances.

8. Claim: $\mu_1 = \mu_2$, $\alpha = 0.10$. Sample statistics: $\overline{x}_1 = 0.015$, $s_1 = 0.011$, $n_1 = 8$, and $\overline{x}_2 = 0.019$, $s_2 = 0.004$, $n_2 = 6$. Assume variances are not equal.

9. Claim: $\mu_1 \leq \mu_2$, $\alpha = 0.05$. Sample statistics: $\overline{x}_1 = 183.5$, $s_1 = 1.3$, $n_1 = 25$, and $\overline{x}_2 = 184.7$, $s_2 = 3.9$, $n_2 = 25$. Assume variances are not equal.

10. Claim: $\mu_1 \geq \mu_2$, $\alpha = 0.01$. Sample statistics: $\bar{x}_1 = 25.6$, $s_1 = 8.25$, $n_1 = 15$, and $\bar{x}_2 = 22.4$, $s_2 = 7.85$, $n_2 = 34$. Assume equal variances.

11. Claim: $\mu_1 \neq \mu_2$, $\alpha = 0.01$. Sample statistics: $\bar{x}_1 = 61$, $s_1 = 3.3$, $n_1 = 5$, and $\bar{x}_2 = 55$, $s_2 = 1.2$, $n_2 = 7$. Assume equal variances.

12. Claim: $\mu_1 \geq \mu_2$, $\alpha = 0.10$. Sample statistics: $\bar{x}_1 = 520$, $s_1 = 25$, $n_1 = 7$, and $\bar{x}_2 = 500$, $s_2 = 55$, $n_2 = 6$. Assume variances are not equal.

In Exercises 13 and 14, (a) identify the claim and state H_0 and H_a, (b) identify the rejection region(s), (c) find the standardized test statistic, (d) decide whether to reject, or fail to reject, the null hypothesis, and (e) interpret this decision in the context of the original claim.

13. A study of methods for teaching reading in the third grade was conducted. One classroom of 21 students participated in directed reading activities for eight weeks. Another classroom of 23 students followed the same curriculum without the activities. Students in both classrooms then took the same reading test.

The following are reading scores for the first (treated) classroom.

24 43 58 71 43 49 61 44 67 49 53
56 59 52 62 54 57 33 46 43 57

The following are reading scores for the second (control) classroom.

42 43 55 26 62 37 33 41 19 54 20 85
46 10 17 60 53 42 37 42 55 28 48

Diagnostics suggest that the data are sampled from a normal population and the population variances are equal. Test the claim that third graders taught with the directed reading activities score higher than those taught without the activities. Use $\alpha = 0.05$. *(Source: StatLib/Schmitt, Maribeth C., The Effects of an Elaborated Directed Reading Activity on the Metacomprehension Skills of Third Graders)*

14. A real estate agent claims that there is no difference between the mean household incomes of two neighborhoods. The mean income of 12 randomly selected households from the first neighborhood was \$18,250 with a standard deviation of \$1200. In the second neighborhood, 10 randomly selected households had a mean income of \$17,500 with a standard deviation of \$950. Assume normal distributions and equal population variances. Test the claim at $\alpha = 0.05$.

Section 8.3

In Exercises 15 and 16, classify the samples as independent or dependent. Explain your reasoning.

15. Sample 1: Maze completion times for 14 standard laboratory mice

Sample 2: Maze completion times for 14 laboratory mice bred for higher metabolic rate

16. Sample 1: Maze completion times for 43 mice

Sample 2: Maze completion times for those 43 mice after two weeks of maze practice

In Exercises 17–20, using a test for paired data, test the claim about the mean of the difference of the two populations at the given level of significance using the given statistics. Is the test right tailed, left tailed, or two tailed? Assume the sample statistics are from populations that are normally distributed.

17. Claim: $\mu_d = 0, \alpha = 0.05$. Statistics: $\bar{d} = 10, s_d = 12.4, n = 100$

18. Claim: $\mu_d < 0, \alpha = 0.01$. Statistics: $\bar{d} = 3.2, s_d = 1.38, n = 25$

19. Claim: $\mu_d \leq 6, \alpha = 0.10$. Statistics: $\bar{d} = 10.3, s_d = 1.24, n = 33$

20. Claim: $\mu_d \neq 15, \alpha = 0.05$. Statistics: $\bar{d} = 17.5, s_d = 4.05, n = 37$

In Exercises 21 and 22, (a) identify the claim and state H_0 and H_a, (b) find the critical value(s) and identify the rejection region(s), (c) calculate $\bar{d}$ and s_d, (d) use the t-test to find the standardized test statistic, (e) decide whether to reject, or fail to reject, the null hypothesis, and (f) interpret the decision in the context of the original claim. For each sample, assume the distribution of the population is normal.

21. A medical researcher wants to test the effects of calcium supplements on men's blood pressure. As part of the study, 10 randomly selected men are given a calcium supplement for 12 weeks. The researcher measures the men's diastolic blood pressure at the beginning and at the end of the 12-week study and records the results shown below. At $\alpha = 0.10$, can the researcher claim that the men's diastolic blood pressure decreased? *(Source: The Journal of American Medicine)*

Patient	1	2	3	4	5	6	7	8	9	10
Before	107	110	123	129	112	111	107	112	136	102
After	100	114	105	112	115	116	106	102	125	104

22. In a study testing the effects of an herbal supplement on blood pressure in men, 11 randomly selected men were given an herbal supplement for 15 weeks. The following measurements are for each subject's diastolic blood pressure taken before and after the 15-week treatment period.

Patient	1	2	3	4	5	6	7	8	9	10	11
Before	123	109	112	102	98	114	119	112	110	117	130
After	124	97	113	105	95	119	114	114	121	118	133

At $\alpha = 0.10$, can you support the claim that systolic blood pressure was lowered?

Section 8.4

In Exercises 23–26, test the claim about the difference between two population proportions p_1 and p_2 at the given level of significance α and for the given sample statistics. Is the test right tailed, left tailed, or two tailed? Assume the sample statistics are from independent samples.

23. Claim: $p_1 = p_2$, $\alpha = 0.05$. Sample statistics: $x_1 = 375$, $n_1 = 720$, and $x_2 = 365$, $n_2 = 660$

24. Claim: $p_1 \leq p_2$, $\alpha = 0.01$. Sample statistics: $x_1 = 15$, $n_1 = 100$, and $x_2 = 42$, $n_2 = 200$

25. Claim: $p_1 > p_2$, $\alpha = 0.10$. Sample statistics: $x_1 = 227$, $n_1 = 556$, and $x_2 = 198$, $n_2 = 420$

26. Claim: $p_1 < p_2$, $\alpha = 0.05$. Sample statistics: $x_1 = 86$, $n_1 = 900$, and $x_2 = 107$, $n_2 = 1200$

In Exercises 27 and 28, (a) identify the claim and state H_0 and H_a, (b) use Table 4 to find the critical value(s) and identify the rejection region(s), (c) find the standardized test statistic, (d) decide whether to reject, or fail to reject, the null hypothesis, and (e) interpret the decision in the context of the original claim. Assume the samples are independent.

27. In a random sample of 200 Canadians ten years ago, 22 had college degrees. Five years ago, in a random sample of 300 Canadians, 40 had college degrees. At $\alpha = 0.10$, can you reject the claim that the proportion of Canadians with college degrees was the same for both years? *(Adapted from Statistics Canada)*

Do You Have a College Degree?

Study Done Ten Years Ago
(200 Canadian Adults)

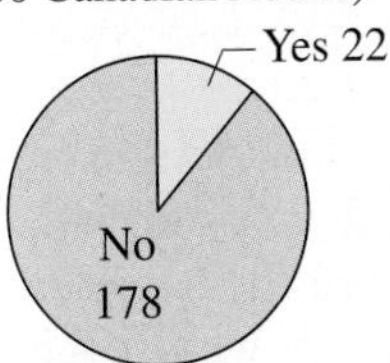

Study Done Five Years Ago
(300 Canadian Adults)

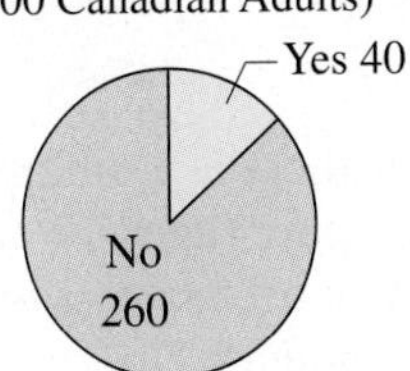

28. In a recent year, of 92,880 students taking the SAT I who were planning to major in education, 71,518 were female. That year, of the 57,962 students planning to major in biological sciences, 37,096 were female. At $\alpha = 0.05$, test the claim that the proportion of females with intended education majors was higher than the proportion of females with intended biological sciences majors. *(Source: College Board)*

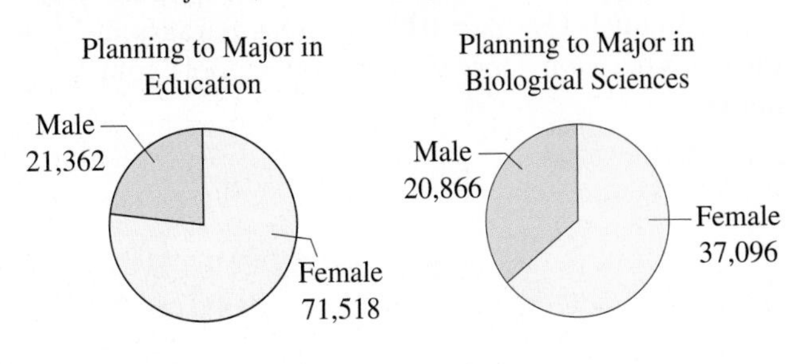

PUTTING IT ALL TOGETHER

Real Statistics Real Decisions

You work for the statistics department of the National Center for Health Statistics. In an effort to determine if the mean length of hospital stays for patients diagnosed with pneumonia are different today compared to 10 years ago, your department decides to analyze data from a random selection of hospital records. The results for several patients ages 17 and under are shown in the histograms for 10 years ago and this year.

Your department believes that there is a difference in the mean length of hospital stays for patients diagnosed with pneumonia. As the department's research analyst, it is your job to work with the data and determine if this claim can be supported.

Exercises

1. ***How Would You Do It?***

   (a) What sampling technique would you use to select the sample for the study? Why? What sampling technique would you use if you divided hospital records according to four geographic regions (northeast, south, midwest, and west), randomly selected one region, and obtained records from each hospital in that region?

   (b) Which technique in part (a) would be easiest to implement? Why?

   (c) Identify possible flaws or biases in your study.

2. ***Choosing a Test***

   To test the claim that there is a difference in the mean length of hospital stays for patients (ages 17 and under) diagnosed with pneumonia, should you use a z-test or a t-test? Are the samples independent or dependent? Do you need to know what each population's distribution is? Do you need to know anything about the population variances?

3. ***Testing a Mean***

   Test the claim that there is a difference in the mean length of hospital stays for patients (ages 17 and under) diagnosed with pneumonia. Assume the populations are normal and the population variances are not equal. Use $\alpha = 0.05$. Write a paragraph that interprets the test's decision. Does the decision support your department's claim?

Patients 17 and Under With Pneumonia (10 Years Ago)

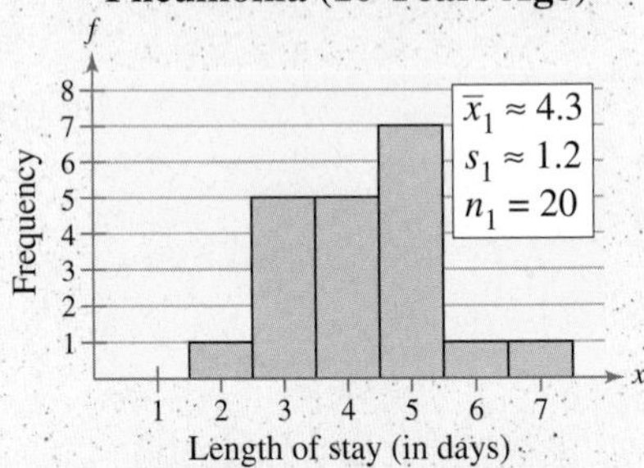

Patients 17 and Under With Pneumonia (This Year)

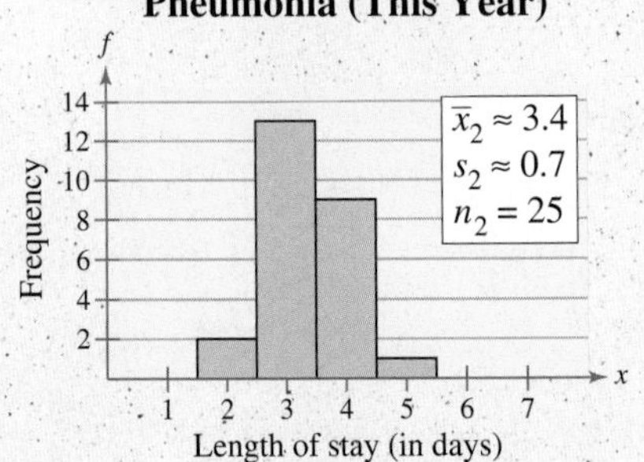

8

Chapter Quiz

Take this quiz as you would take a quiz in class. After you are done, check your work against the answers given in the back of the book.

For this quiz, do the following.

(a) Write the claim mathematically and identify H_0 and H_a.

(b) Determine whether the hypothesis test is a one-tailed test or a two-tailed test and whether to use a *z*-test or a *t*-test. Explain your reasoning.

(c) Find the critical value(s) and identify the rejection region(s).

(d) Use the appropriate test to find the appropriate standardized test statistic.

(e) Decide whether to reject or fail to reject the null hypothesis. Then interpret the decision in the context of the original claim.

1. The mean score on a science assessment for 49 randomly selected male high school students was 299.5 and the standard deviation was 2.0. The mean score on the same test for 50 randomly selected female high school students was 288.9 and the standard deviation was 1.7. At $\alpha = 0.05$, can you support the claim that the mean score on the science assessment for male high school students was higher than for the female high school students? *(Adapted from National Center for Educational Statistics)*

2. A science teacher suggests that the mean scores on a science assessment test for nine-year-old boys and girls are equal. If the mean score for 13 randomly selected boys is 232.2 with a standard deviation of 1.3 and the mean score for 15 randomly selected girls is 230 with a standard deviation of 1.4, can you reject the teacher's suggestion? Assume the populations are normally distributed and the variances are equal. Use $\alpha = 0.01$. *(Adapted from National Center for Educational Statistics)*

3. Of 1,296,000 accidents involving drivers aged 21 to 24, 5% involved alcohol. Of 856,000 accidents involving drivers aged 65 years and over, 1% involved alcohol. At $\alpha = 0.10$, can you support the claim that the proportion of accidents involving alcohol is higher for drivers in the 21 to 24 age group than for drivers aged 65 and over? *(Source: U.S. Highway Safety Administration)*

DATA **4.** The table shows the scores for 12 randomly selected students the first and second times they took the mathematics SAT. At $\alpha = 0.05$, is there enough evidence to conclude that the students' SAT scores improved on the second test? Assume the populations are normally distributed.

Student	1	2	3	4	5	6
Score on first SAT	457	419	343	539	394	413
Score on second SAT	532	523	427	607	444	490

Student	7	8	9	10	11	12
Score on first SAT	392	421	439	340	493	339
Score on second SAT	428	524	532	397	550	357

Where You've Been

In Chapters 1–8, you studied descriptive statistics, probability, and inferential statistics. One of the techniques you learned in descriptive statistics was graphing paired data with a scatter plot (Section 2.2). For instance, the winning times for the men's and women's 100-meter run in the summer Olympics from 1928 through 2000 are given in tabular form at the right and in graphical form below.

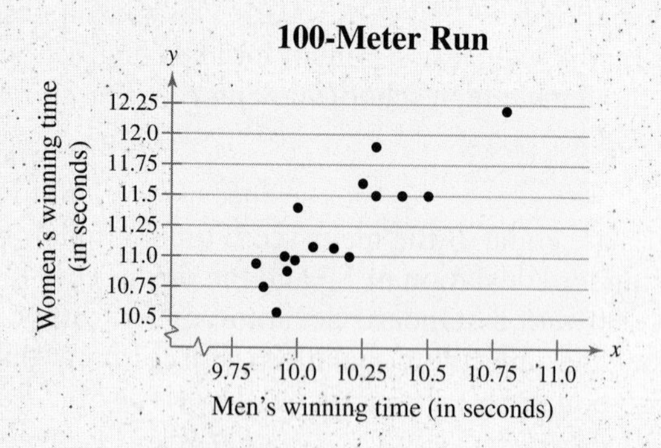

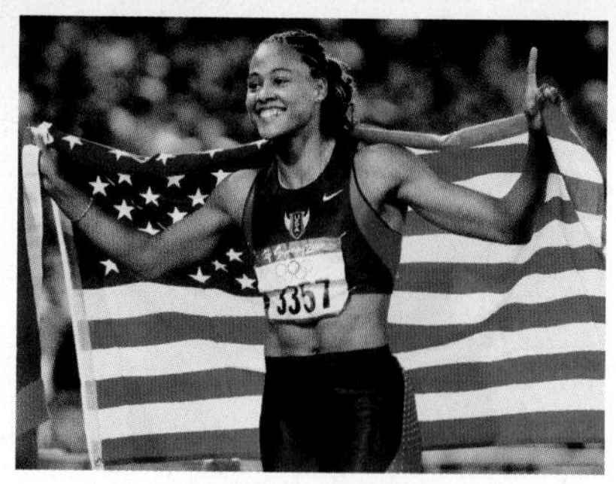

Marion Jones (U.S.)
2000 100-Meter Run

Winning Times for 100-Meter Run Summer Olympics

Year	Men (in seconds), x	Women (in seconds), y
1928	10.80	12.20
1932	10.30	11.90
1936	10.30	11.50
1948	10.30	11.90
1952	10.40	11.50
1956	10.50	11.50
1960	10.20	11.00
1964	10.00	11.40
1968	9.95	11.00
1972	10.14	11.07
1976	10.06	11.08
1980	10.25	11.60
1984	9.99	10.97
1988	9.92	10.54
1992	9.96	10.82
1996	9.84	10.94
2000	9.87	10.75

Lenny Krayzelburg (U.S.)
2000 200-Meter Backstroke

The modern summer Olympic games began in Athens, Greece, in 1896. The winter Olympic games began in Chamonix, France, in 1924. The games have been held in the United States eight times: St. Louis (S 1904), Lake Placid (W 1932), Los Angeles (S 1932), Squaw Valley (W 1960), Lake Placid (W 1980), Los Angeles (S 1984), Atlanta (S 1996), and Salt Lake City (W 2002).

Correlation and Regression

CHAPTER 9

Where You're Going

In this chapter, you will study how to describe and test the significance of relationships between two variables when data are presented as ordered pairs. For instance, in the scatter plot for the 100-meter run, it appears that women's fast times tend to correspond to men's fast times and women's slow times tend to correspond to men's slow times. This relationship is described by saying the women's times are positively correlated to the men's times. Graphically, the relationship can be described by drawing a line, called a regression line, that fits the points as closely as possible. The second scatter plot below shows a similar result—the women's running high jump heights are positively correlated to the men's running high jump heights.

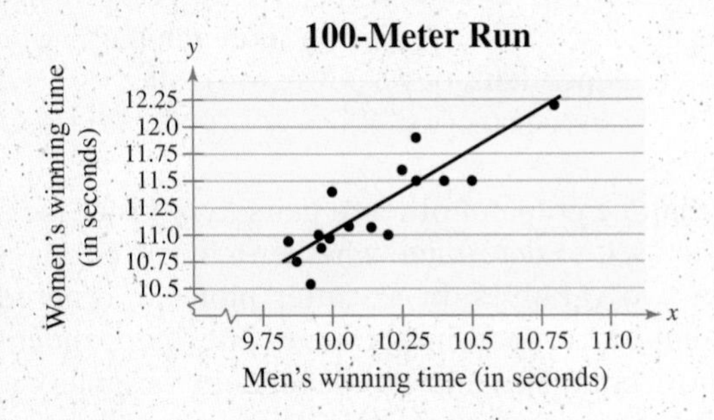

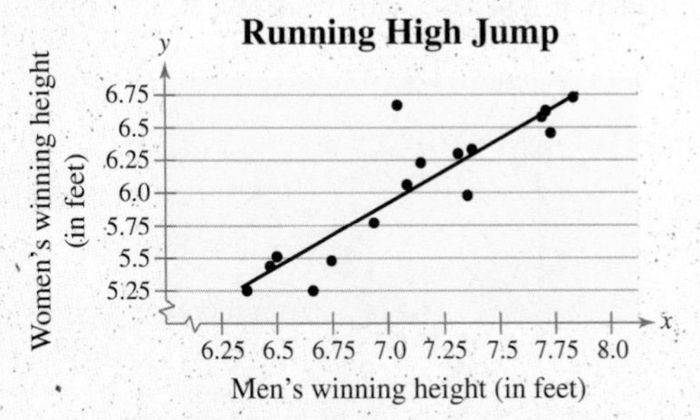

2000 Men's Rowing

Stacy Dragila (U.S.)
2000 Pole Vault

9.1 Correlation

What You Should Learn

- An introduction to linear correlation, independent and dependent variables, and the types of correlation
- How to find a correlation coefficient
- How to perform a hypothesis test for a population correlation coefficient ρ

An Overview of Correlation • Correlation Coefficient • Using a Table to Test a Population Correlation Coefficient • Hypothesis Testing for a Population Correlation Coefficient • Correlation and Causation

An Overview of Correlation

Suppose a safety inspector wants to determine whether a relationship exists between the number of hours of training for an employee and the number of accidents involving that employee. Or suppose a psychologist wants to know whether a relationship exists between the number of hours a person sleeps each night and that person's reaction time. How would you help them determine if any relationship exists?

In this section, you will study how to describe what type of relationship, or correlation, exists between two variables, and how to determine whether the correlation is significant.

DEFINITION

A **correlation** is a relationship between two variables. The data can be represented by the ordered pairs (x, y) where x is the **independent,** or **explanatory, variable** and y is the **dependent,** or **response, variable.**

In Section 2.2, you learned that the graph of ordered pairs (x, y) is called a **scatter plot.** A scatter plot can be used to determine whether a linear (straight line) correlation exists between two variables. In a scatter plot, the ordered pairs (x, y) are graphed as points in a coordinate plane. The independent variable x is measured by the horizontal axis and the dependent variable y is measured by the vertical axis. The following scatter plots show several types of correlation.

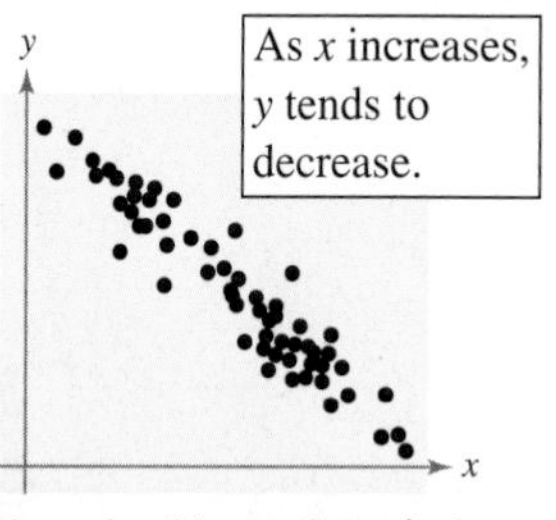

Negative Linear Correlation

As x increases, y tends to increase.

Positive Linear Correlation

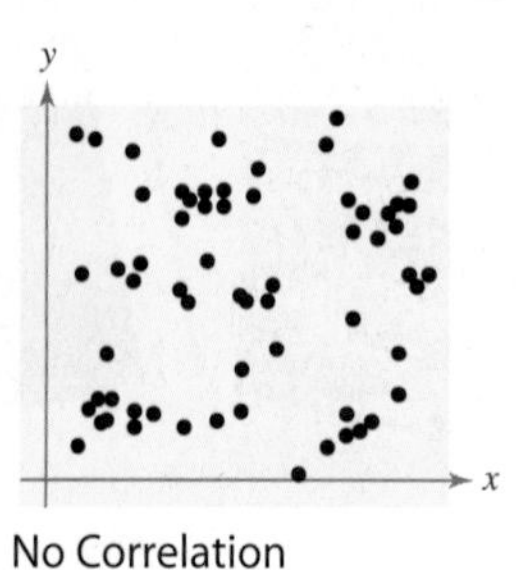

No Correlation

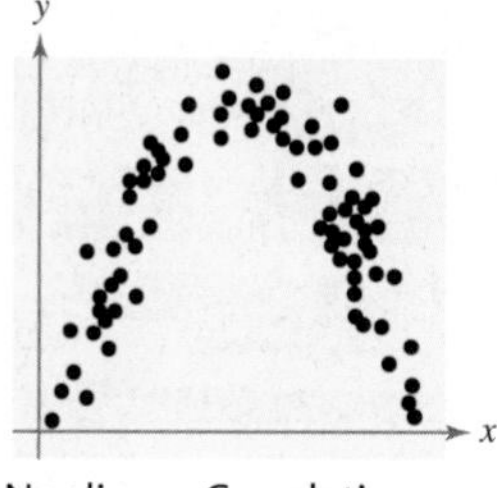

Nonlinear Correlation

EXAMPLE 1

Constructing a Scatter Plot

Advertising expenses (1000s of $), x	Company sales (1000s of $), y
2.4	225
1.6	184
2.0	220
2.6	240
1.4	180
1.6	184
2.0	186
2.2	215

A marketing manager conducted a study to determine whether there is a linear relationship between money spent on advertising and company sales. The data are listed in the table at the left. Display the data in a scatter plot and determine whether there appears to be a positive or negative linear correlation or no linear correlation.

SOLUTION The scatter plot is shown at the right. From the scatter plot, it appears that there is a positive linear correlation between the variables. Reading from left to right, as the advertising expenses increase, the sales tend to increase.

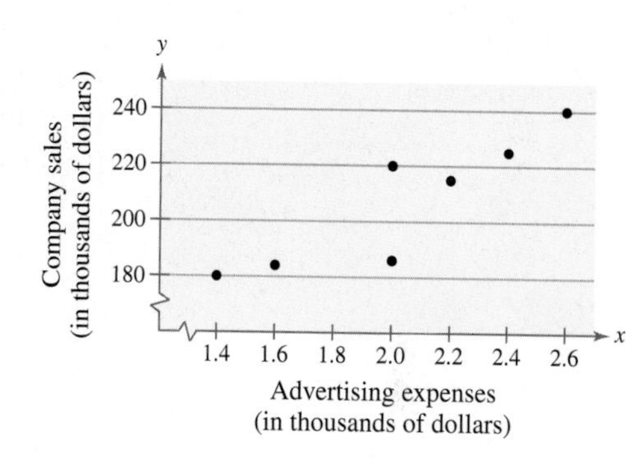

Try It Yourself 1

Income level (in 1000s of $), x	Donating percent, y
42	9
48	10
50	8
59	5
65	6
72	3

A sociologist conducted a study to determine whether there is a linear relationship between family income level (in thousands of dollars) and percent of income donated to charities. The data are listed in the table at the left. Display the data in a scatter plot and determine the type of correlation.

a. *Draw* and *label* the x- and y-axes.
b. *Plot* each ordered pair on the graph.
c. Does there appear to be a correlation?

Answer: Page A40

EXAMPLE 2

Constructing a Scatter Plot

A student nurse conducts a study to determine whether there is a linear relationship between an individual's weight (in pounds) and daily water consumption (in ounces). The data are listed in the following table. Organize the data in a scatter plot and describe the type of correlation.

Weight, x	102	119	124	141	142	154	201	220
Water, y	50	32	82	64	54	21	86	39

SOLUTION The scatter plot is shown at the right. From the scatter plot it appears that there is no linear correlation between the variables. A person's weight does not appear to be related to the amount of water that person consumes.

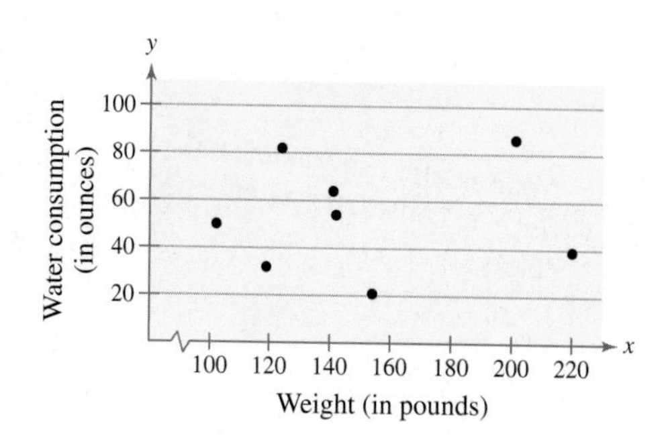

Study Tip

Save any data put into a technology tool as these data will be used throughout the chapter.

Try It Yourself 2

A marketing manager conducts a study to determine whether there is a linear relationship between a person's age and the number of magazines to which that person subscribes. The data are listed in the following table. Display the data in a scatter plot and determine the type of correlation.

Age, x	21	26	33	35	48	50	55	64
Subscriptions, y	4	0	3	1	3	0	2	6

a. *Draw* and *label* the x- and y-axes.
b. *Plot* each ordered pair on the graph.
c. Does there appear to be a correlation?

Answer: Page A40

EXAMPLE 3

Constructing a Scatter Plot Using Technology

Duration, x	Time, y	Duration, x	Time, y
1.80	56	3.78	79
1.82	58	3.83	85
1.88	60	3.87	81
1.90	62	3.88	80
1.92	60	4.10	89
1.93	56	4.27	90
1.98	57	4.30	84
2.03	60	4.30	89
2.05	57	4.43	84
2.13	60	4.43	89
2.30	57	4.47	86
2.35	57	4.47	80
2.37	61	4.53	89
2.82	73	4.55	86
3.13	76	4.60	88
3.27	77	4.60	92
3.65	77	4.63	91
3.70	82		

Old Faithful, located in Yellowstone National Park, is the world's most famous geyser. The duration (in minutes) of several of Old Faithful's eruptions and the times (in minutes) until the next eruption are listed in the table at the left. Using a *TI-83*, display the data in a scatter plot. Determine the type of correlation.

SOLUTION Begin by entering the x-values into List 1 and the y-values into List 2. Use Stat Plot to construct the scatter plot. The plot should look similar to the one shown below. From the scatter plot it appears that the variables have a positive linear correlation. In other words, you can conclude that the longer the duration of the eruption, the longer the time before the next eruption begins.

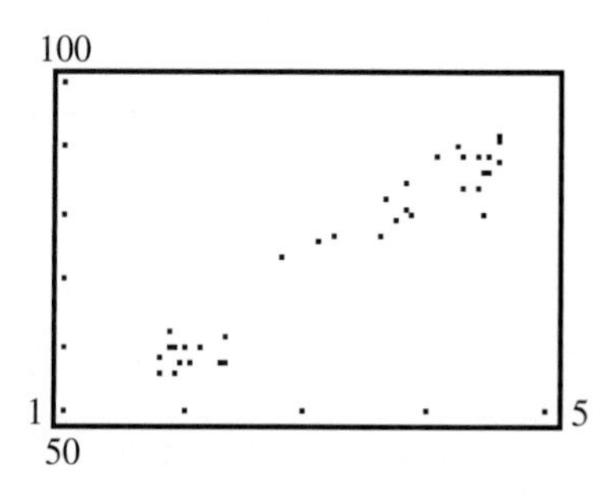

Try It Yourself 3

Consider the data from the Chapter Opener on women's and men's winning times in the 100-meter run. Use a technology tool to display the data in a scatter plot. Determine the type of correlation.

a. *Enter* the data into List 1 and List 2.
b. *Construct* the scatter plot.
c. Does there appear to be a correlation?

Answer: Page A41

Study Tip

You can also use *Minitab* and *Excel* to construct scatter plots.

Correlation Coefficient

Interpreting correlation using a scatter plot can be subjective. A more precise way to measure the type and strength of a linear correlation between two variables is to calculate the correlation coefficient.

> **Insight**
>
> The formal name for r is the **Pearson product moment correlation coefficient.** It is named after the English statistician Karl Pearson (1857–1936). (See page 292.)

DEFINITION

The **correlation coefficient** is a measure of the strength and the direction of a linear relationship between two variables. The symbol r represents the sample correlation coefficient. The formula for r is

$$r = \frac{n\sum xy - (\sum x)(\sum y)}{\sqrt{n\sum x^2 - (\sum x)^2}\sqrt{n\sum y^2 - (\sum y)^2}}$$

where n is the number of pairs of data.

The population correlation coefficient is represented by ρ (the lowercase Greek letter rho, pronounced "row").

The range of the correlation coefficient is -1 to 1. If x and y have a strong positive linear correlation, r is close to 1. If x and y have a strong negative linear correlation, r is close to -1. If there is no linear correlation or a weak linear correlation, r is close to 0. Several examples are shown below.

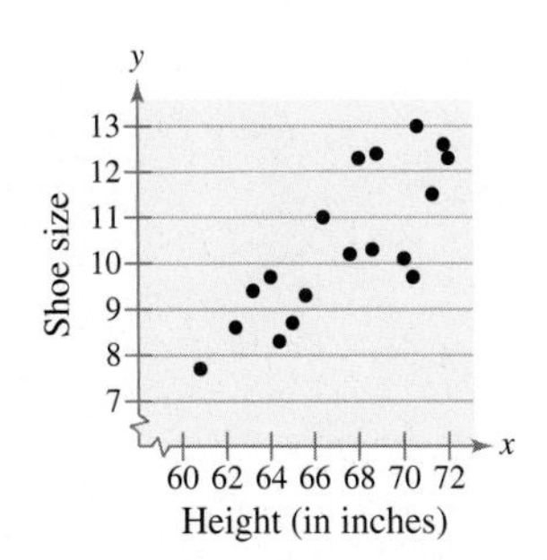

Strong positive correlation
$r = 0.81$

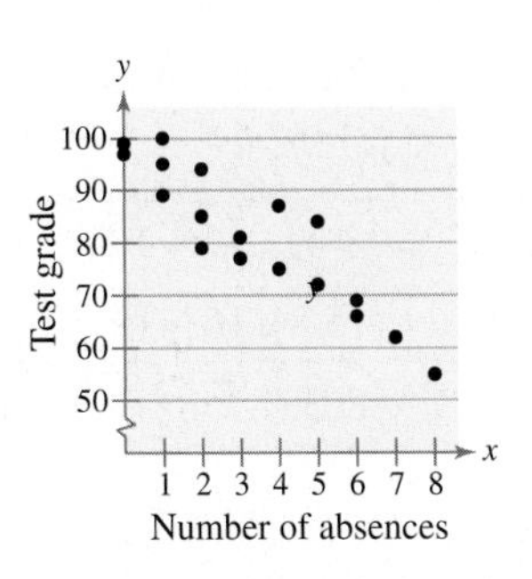

Strong negative correlation
$r = -0.92$

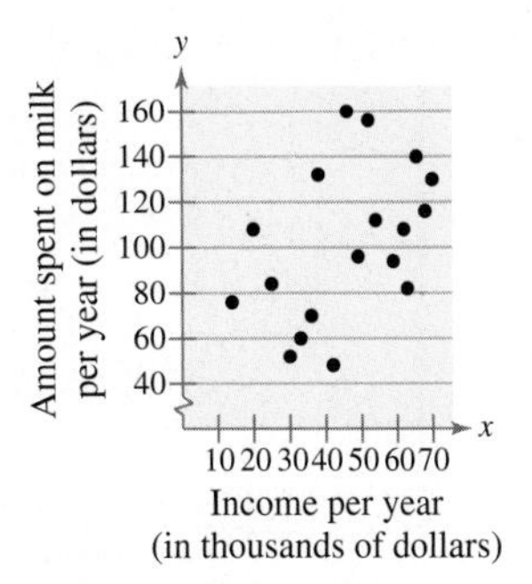

Weak positive correlation
$r = 0.45$

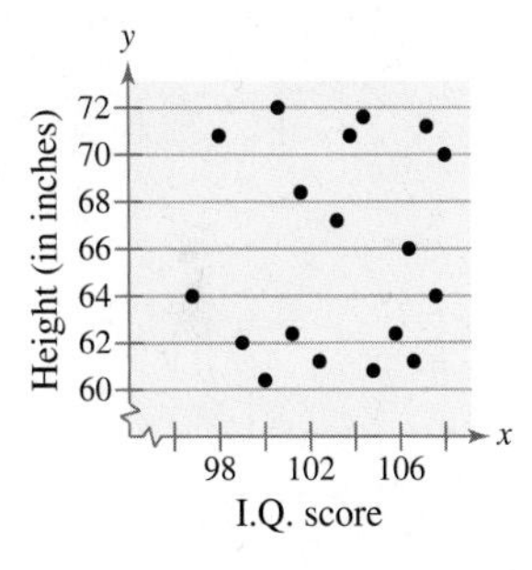

No correlation
$r = 0.04$

GUIDELINES

Calculating a Correlation Coefficient

In Words	*In Symbols*
1. Find the sum of the x-values.	$\sum x$
2. Find the sum of the y-values.	$\sum y$
3. Multiply each x-value by its corresponding y-value and find the sum.	$\sum xy$
4. Square each x-value and find the sum.	$\sum x^2$
5. Square each y-value and find the sum.	$\sum y^2$
6. Use these five sums to calculate the correlation coefficient.	$r = \frac{n\sum xy - (\sum x)(\sum y)}{\sqrt{n\sum x^2 - (\sum x)^2}\sqrt{n\sum y^2 - (\sum y)^2}}$

> **Study Tip**
>
> $\sum x^2$ means square each value and add the squares. $(\sum x)^2$ means add the values and square the sum.

EXAMPLE 4

Finding the Correlation Coefficient

Calculate the correlation coefficient for the advertising expenditures and company sales data given in Example 1. What can you conclude?

SOLUTION Use a table to help calculate the correlation coefficient.

Advertising expenses (1000s of $), x	Company sales (1000s of $), y	xy	x^2	y^2
2.4	225	540	5.76	50,625
1.6	184	294.4	2.56	33,856
2.0	220	440	4	48,400
2.6	240	624	6.76	57,600
1.4	180	252	1.96	32,400
1.6	184	294.4	2.56	33,856
2.0	186	372	4	34,596
2.2	215	473	4.84	46,225
$\Sigma x = 15.8$	$\Sigma y = 1634$	$\Sigma xy = 3289.8$	$\Sigma x^2 = 32.44$	$\Sigma y^2 = 337{,}558$

Using these sums and $n = 8$, the correlation coefficient is

$$r = \frac{n\Sigma xy - (\Sigma x)(\Sigma y)}{\sqrt{n\Sigma x^2 - (\Sigma x)^2}\sqrt{n\Sigma y^2 - (\Sigma y)^2}}$$

$$= \frac{8(3289.8) - (15.8)(1634)}{\sqrt{8(32.44) - 15.8^2}\sqrt{8(337{,}558) - 1634^2}}$$

$$= \frac{501.2}{\sqrt{9.88}\sqrt{30{,}508}}$$

$$\approx 0.913.$$

Because r is close to 1, there is a strong positive linear correlation. As the amount spent on advertising increases, the company sales also increase.

Try It Yourself 4

Calculate the correlation coefficient for the income level and donating percent data given in Try It Yourself 1. What can you conclude?

Income level (in 1000s of $), x	Donating percent, y
42	9
48	10
50	8
59	5
65	6
72	3

a. *Identify n.*
b. *Use a table* to calculate Σx, Σy, Σxy, Σx^2, and Σy^2.
c. *Use the resulting sums and n* to calculate r.
d. What can you conclude?

Answer: Page A41

Using Technology to Find a Correlation Coefficient

Use a technology tool to calculate the correlation coefficient for the Old Faithful data given in Example 3. What can you conclude?

SOLUTION *Minitab*, *Excel*, and the *TI-83* each have features that allow you to calculate a correlation coefficient for paired data sets. Try using this technology to find r. You should obtain results similar to the following.

MINITAB

Correlations: C1, C2

Pearson correlation of C1 and C2 = 0.970

EXCEL

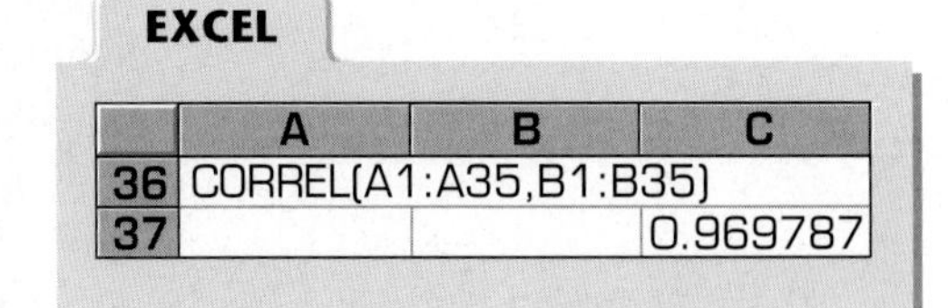

	A	B	C
36	CORREL(A1:A35,B1:B35)		
37			0.969787

Before using the *TI-83* to calculate r, you must enter the Diagnostic On command. To do so, enter the following keystrokes:

2nd 0

cursor to *DiagnosticOn*

ENTER ENTER.

The following screens describe how to find r using a *TI-83* with the data stored in List 1 and List 2.

TI-83

EDIT CALC TESTS
1: 1–Var Stats
2: 2–Var Stats
3: Med–Med
4: LinReg(ax+b)
5: QuadReg
6: CubicReg
7↓Quart Reg

TI-83

LinReg(ax+b) L1,
L2

TI-83

LinReg
y=ax+b
a=11.8244078
b=35.30117105
r2=.9404868083
r=.9697869912

Correlation coefficient

The result, $r \approx 0.970$, suggests a strong positive linear correlation.

Try It Yourself 5

Calculate the correlation coefficient for the data given in the Chapter Opener on page 440. What can you conclude?

a. *Enter* the data.
b. Use the appropriate feature to *calculate r.*
c. What can you conclude?

Answer: Page A41

Using a Table to Test a Population Correlation Coefficient

Once you have calculated r, the sample correlation coefficient, you will want to determine whether there is enough evidence to decide that the population correlation coefficient ρ is significant at a specified level of significance α. Remember that you are using sample data to make a decision about population data. It is always possible that your inference may be wrong. Using a level of significance of $\alpha = 0.05$ means that 5% of the time you will say the population correlation coefficient is significant when it is really not. For $\alpha = 0.01$, you will make this type of error 1% of the time.

One way to determine whether the population correlation coefficient ρ is significant is to use the critical values given in Table 11 in Appendix B. As you examine Table 11, notice that the first column represents the number n of pairs of data in the sample, and the second and third columns represent the critical values for a level of significance $\alpha = 0.05$ and $\alpha = 0.01$, respectively.

If $|r|$ is greater than the critical value, there is enough evidence to decide that the correlation is significant. Otherwise, there is *not* enough evidence to say that the correlation is significant. For instance, to determine whether ρ is significant for five pairs of data ($n = 5$) at a level of significance of $\alpha = 0.01$, you need to compare $|r|$ to a critical value of 0.959 as shown in the table.

n	$\alpha = 0.05$	$\alpha = 0.01$
4	0.950	0.990
5	0.878	0.959
6	0.811	0.917

If $|r| > 0.959$, the correlation is significant. Otherwise, there is *not* enough evidence to support that the correlation is significant. The guidelines for this process are as follows.

Study Tip

If you determine that the linear correlation is significant, then you will be able to proceed to write the equation for the line that best describes the data. This line, called the line of regression, can be used to predict the value of y when given a value of x. You will learn how to write this equation in the next section.

GUIDELINES

Using Table 11 for the Correlation Coefficient ρ

In Words	*In Symbols*
1. Determine the number of pairs of data in the sample.	Determine n.
2. Specify the level of significance.	Specify α.
3. Find the critical value.	Use Table 11 in Appendix B.
4. Decide if the correlation is significant.	If $|r| >$ critical value, the correlation is significant. Otherwise, there is *not* enough evidence to support that the correlation is significant.
5. Interpret the decision in the context of the original claim.	

EXAMPLE 6

Using Table 11 for a Correlation Coefficient

In Example 5, you used 35 pairs of data to find $r \approx 0.970$. Test the significance of this correlation coefficient. Use $\alpha = 0.05$.

SOLUTION

The number of pairs of data is 35, so $n = 35$. The level of significance is $\alpha = 0.05$. Using Table 11, find the critical value in the $\alpha = 0.05$ column that corresponds to the row with $n = 35$. The number in that column and row is 0.334.

n	$\alpha = 0.05$	$\alpha = 0.01$
4	0.950	0.990
5	0.878	0.959
6	0.811	0.917
7	0.754	0.875
8	0.707	0.834
9	0.666	0.798
10	0.632	0.765
11	0.602	0.735
12	0.576	0.708
13	0.553	0.684
14	0.532	0.661
23	0.413	0.526
24	0.404	0.515
25	0.396	0.505
26	0.388	0.496
27	0.381	0.487
28	0.374	0.479
29	0.367	0.471
30	0.361	0.463
35	0.334	0.430
40	0.312	0.403
45	0.294	0.380

Because $|r| \approx 0.970 > 0.334$, you can decide that the population correlation is significant. So, there is enough evidence at the 5% level of significance to conclude that there is a significant linear correlation between the duration of Old Faithful's eruptions and the time between eruptions.

Try It Yourself 6

In Try It Yourself 4, you calculated the correlation of the income level and donating percent data to be $r \approx -0.916$. Test the significance of this correlation coefficient. Use $\alpha = 0.01$.

a. *Determine* the number of pairs of data in the sample.
b. *Specify* the level of significance.
c. *Find* the critical value. Use Table 11.
d. *Compare* $|r|$ to the critical value and *decide* if the correlation is significant.
e. Is there enough evidence to conclude that there is a significant linear correlation between the income level and the donating percent?

Answer: Page A41

Hypothesis Testing for a Population Correlation Coefficient

A hypothesis test for ρ can be one tailed or two tailed. The null and alternative hypotheses for these tests are as follows.

$$\begin{cases} H_0: \rho \geq 0 \text{ (no significant negative correlation)} \\ H_a: \rho < 0 \text{ (significant negative correlation)} \end{cases}$$ **Left-tailed test**

$$\begin{cases} H_0: \rho \leq 0 \text{ (no significant positive correlation)} \\ H_a: \rho > 0 \text{ (significant positive correlation)} \end{cases}$$ **Right-tailed test**

$$\begin{cases} H_0: \rho = 0 \text{ (no significant correlation)} \\ H_a: \rho \neq 0 \text{ (significant correlation)} \end{cases}$$ **Two-tailed test**

In this text, you will only consider hypothesis tests for ρ that are two tailed.

The *t*-Test for the Correlation Coefficient

A ***t*-test** can be used to test whether the correlation between two variables is significant. The **test statistic** is r and the **standardized test statistic** is

$$t = \frac{r}{\sigma_r} = \frac{r}{\sqrt{\dfrac{1 - r^2}{n - 2}}}.$$

The sampling distribution for r is a t-distribution with $n - 2$ degrees of freedom.

GUIDELINES

Using the *t*-Test for the Correlation Coefficient ρ

In Words	*In Symbols*
1. State the null and the alternative hypotheses.	State H_0 and H_a.
2. Specify the level of significance.	Specify α.
3. Determine the degrees of freedom.	d.f. $= n - 2$
4. Find the critical value(s) and identify the rejection region(s).	Use Table 5 in Appendix B.
5. Find the standardized test statistic.	$t = \frac{r}{\sqrt{\frac{1 - r^2}{n - 2}}}$
6. Make a decision to reject or fail to reject the null hypothesis.	If t is in the rejection region, reject H_0. Otherwise, do not reject H_0.
7. Interpret the decision in the context of the original claim.	

EXAMPLE 

The t-Test for a Correlation Coefficient

In Example 4, you used eight pairs of data to find $r \approx 0.913$. Test the significance of this correlation coefficient. Use $\alpha = 0.05$.

SOLUTION The null and alternative hypotheses are

$$H_0: \rho = 0 \text{ (no correlation)} \quad \text{and} \quad H_a: \rho \neq 0 \text{ (significant correlation).}$$

Because there are eight pairs of data in the sample, there are $8 - 2 = 6$ degrees of freedom. Because the test is a two-tailed test, $\alpha = 0.05$, and d.f. $= 6$, the critical values are -2.447 and 2.447. The rejection regions are $t < -2.447$ and $t > 2.447$. Using the t-test, the standardized test statistic is

$$t = \frac{r}{\sqrt{\frac{1 - r^2}{n - 2}}}$$

$$= \frac{0.913}{\sqrt{\frac{1 - (0.913)^2}{8 - 2}}} \approx 5.482.$$

The following graph shows the location of the rejection regions and the standardized test statistic.

Because t is in the rejection region, you should decide to reject the null hypothesis. At the 5% level, there is enough evidence to conclude that there is a significant linear correlation between advertising expenditures and company sales.

Insight

In Example 7, you can use Table 11 in Appendix B to test the population correlation coefficient ρ. Given $n = 8$ and $\alpha = 0.05$, the critical value from Table 11 is 0.707. Because

$$|r| \approx 0.913 > 0.707,$$

the correlation is significant. Note that this is the same result you obtained using a t-test for the population correlation coefficient ρ.

Study Tip

Be sure you see in Example 7 that when you reject the null hypothesis, it means that there is sufficient evidence that the correlation is significant.

Try It Yourself 7

In Try It Yourself 5, you calculated the correlation coefficient of the Olympic 100-meter winning times data to be $r \approx 0.846$. Test the significance of this correlation coefficient. Use $\alpha = 0.01$.

a. *State* the null and alternative hypotheses.
b. *Specify* the level of significance.
c. *Determine* the degrees of freedom.
d. *Find* the critical values and identify the rejection regions.
e. *Find* the standardized test statistic.
f. *Make a decision* to reject or fail to reject the null hypothesis.
g. Is there enough evidence to conclude that there is a significant linear correlation between the men's winning 100-meter run times and the women's winning 100-meter run times?

Answer: Page A41

Correlation and Causation

The fact that two variables are strongly correlated does not in itself imply a cause-and-effect relationship between the variables. More in-depth study is usually needed to determine whether there is a causal relationship between the variables.

If there is a significant correlation between two variables, a researcher should consider the following possibilities.

1. **Is there a direct cause-and-effect relationship between the variables?**

   That is, does x cause y? For instance, consider the relationship between advertising expenditures and company sales that has been discussed throughout this section. It is reasonable to conclude that spending more money on advertising will result in more sales.

2. **Is there a reverse cause-and-effect relationship between the variables?**

   That is, does y cause x? For instance, consider the Old Faithful data that have been discussed throughout this section. These variables have a positive linear correlation, and it is possible to conclude that the duration of an eruption affects the time before the next eruption. However, it is also possible that the time between eruptions affects the duration of the next eruption.

3. **Is it possible that the relationship between the variables can be caused by a third variable or perhaps a combination of several other variables?**

   For instance, consider the men's and women's winning times listed in the Chapter Opener. While these variables have a positive linear correlation, it is doubtful that just because the men's winning time decreases the women's winning time will also decrease. The relationship is probably due to several other variables, such as training techniques, improvements in athletic equipment, or climate conditions at the Olympic location.

4. **Is it possible that the relationship between two variables may be a coincidence?**

   For instance, while it may be possible to find a significant correlation between the number of animal species living in certain regions and the number of people who own more than two cars in those regions, it is highly unlikely that the variables are directly related. The relationship is probably due to coincidence.

Determining which of the above cases is valid for a data set can be difficult. For instance, consider the following example. Suppose a person breaks out in a rash each time he eats shrimp at a certain restaurant. The natural conclusion is that the person is allergic to shrimp. However, upon further study by an allergist, it is found that the person in not allergic to shrimp, but to a type of seasoning the chef is putting into the shrimp.

Picturing the World

The following scatter plot shows the results of a survey conducted as a group project by students in a high school statistics class in the San Francisco area. In the survey, 125 high school students were asked their GPA (grade point average) and the number of caffeine drinks they consumed each day.

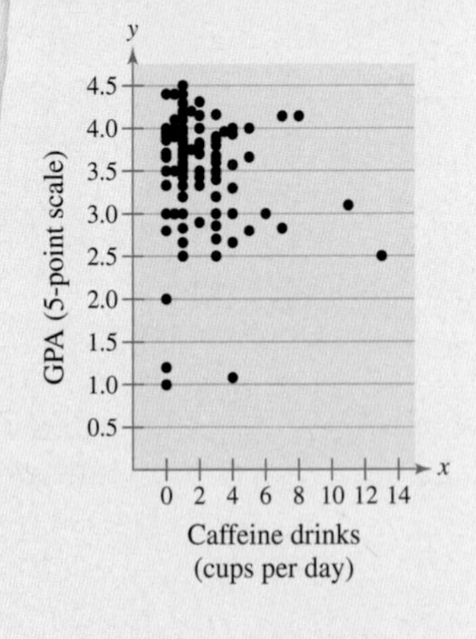

What type of correlation, if any, does the scatter plot show between caffeine consumption and GPA?

9.1 Exercises

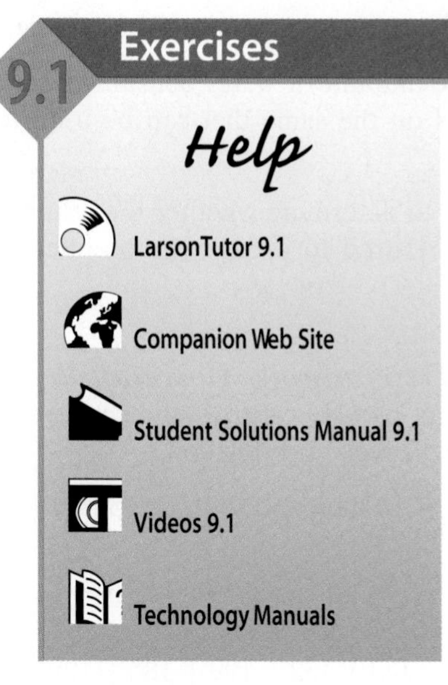

Basic Skills and Concepts

Graphical Analysis In Exercises 1–4, the scatter plots of paired data sets are given. Determine whether there is a positive linear correlation, negative linear correlation, or no linear correlation between the variables.

1.

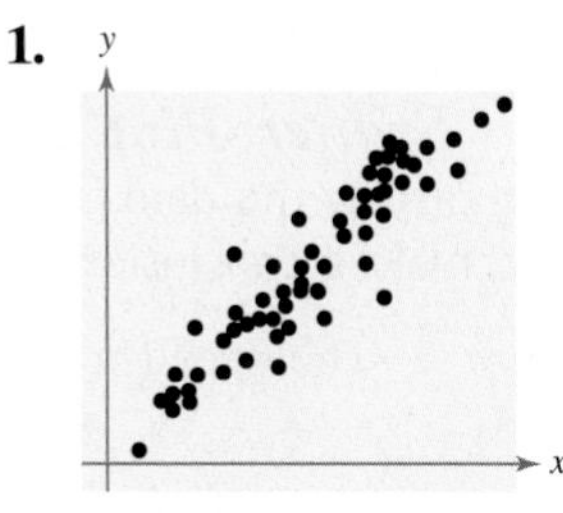

2.

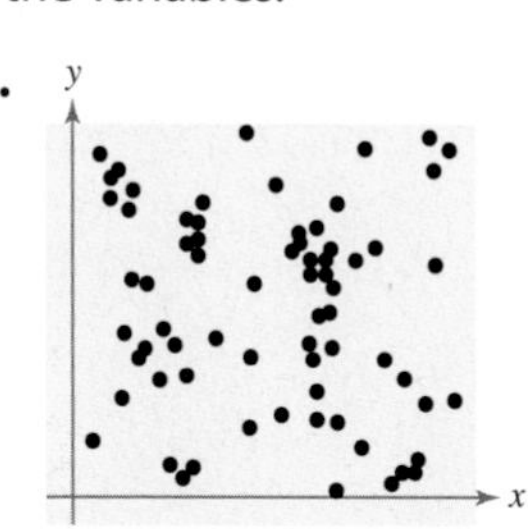

3.

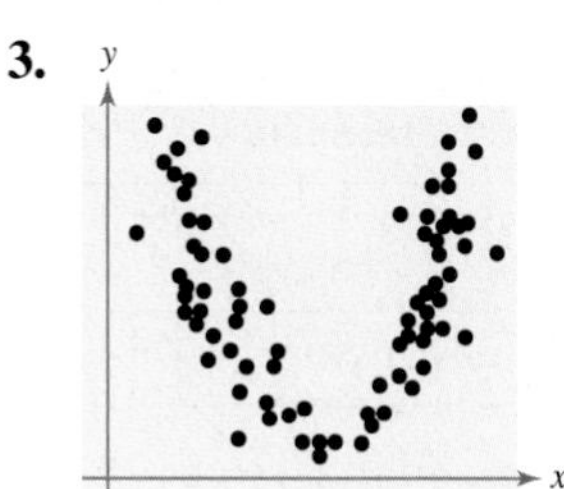

4.

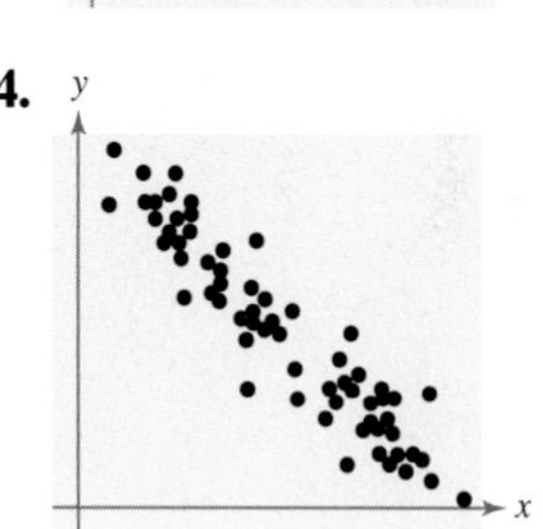

Graphical Analysis In Exercises 5–8, the scatter plots show the results of a survey of 20 randomly selected males ages 24–35. Using age as the explanatory variable, match each scatter plot to the appropriate description. Explain your reasoning.

(a) Age and body temperature

(b) Age and balance on student loans

(c) Age and income

(d) Age and height

5.

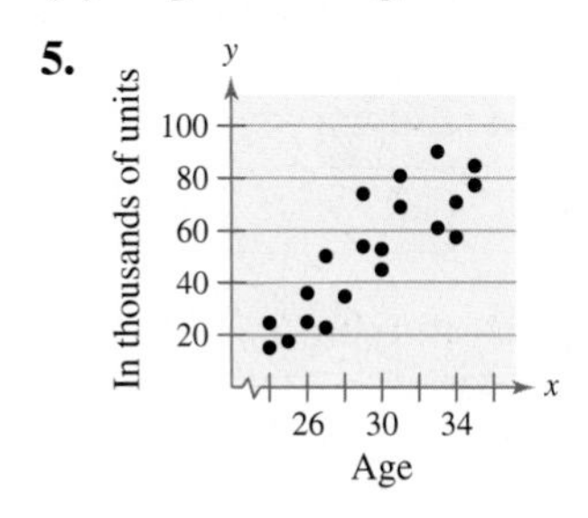

6.

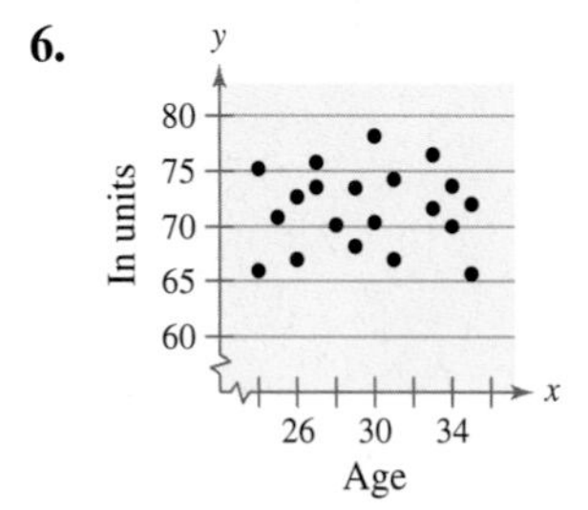

7.

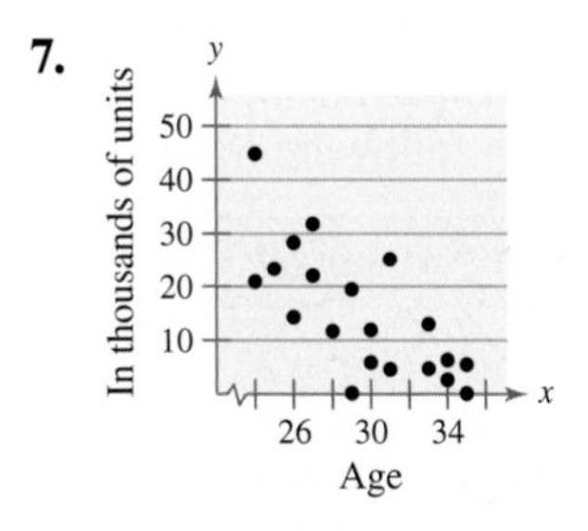

8.

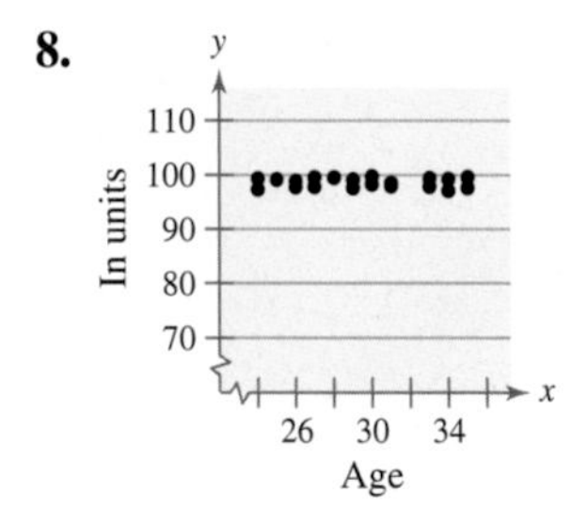

In Exercises 9 and 10, identify the explanatory variable and the response variable.

9. A nutritionist wants to determine if the amount of water consumed each day by persons of the same weight and on the same diet can be used to predict individual weight loss.

10. An insurance company hires an actuary to determine whether the number of hours of safety driving classes can be used to predict the number of driving accidents for each driver.

Constructing a Scatter Plot and Determining Correlation In Exercises 11–18, (a) display the data in a scatter plot, (b) calculate the correlation coefficient r, and (c) make a conclusion about the type of correlation.

11. ***Blood Pressure*** ◆ The ages (in years) of 10 men and their systolic blood pressures

Age, x	16	25	39	45	49	64	70
Systolic blood pressure, y	109	122	143	132	199	185	199

Age, x	29	57	22
Systolic blood pressure, y	130	175	118

12. ***Vocabulary Size*** ◆ The ages (in years) of 11 children and the number of words in their vocabulary

Age, x	1	2	3	4	5	6	3
Vocabulary size, y	3	440	1200	1500	2100	2600	1100

Age, x	5	2	4	6
Vocabulary size, y	2000	500	1525	2500

DATA **13.** ***Test Scores and Study Times*** ◆ The number of hours 13 students spent studying for a test and their scores on that test

Hours spent studying, x	0	1	2	4	4	5	5
Test score, y	40	41	51	48	64	69	73

Hours spent studying, x	5	6	6	7	7	8
Test score, y	75	68	93	84	90	95

DATA **14.** ***Test Scores and Hours Watching TV*** ◆ The number of hours 12 students watched television during the weekend and the scores of each student who took a test the following Monday

Hours spent watching TV, x	0	1	2	3	3	5
Test score, y	96	85	82	74	95	68

Hours spent watching TV, x	5	5	6	7	7	10
Test score, y	76	84	58	65	75	50

15. ***Shoe Size and Height*** ◆ The shoe sizes and heights (in inches) for 14 men

Shoe size, x	8.5	9.0	9.0	9.5	10.0	10.0	10.5
Height, y	66.0	68.5	67.5	70.0	70.0	72.0	71.5

Shoe size, x	10.5	11.0	11.0	11.0	12.0	12.0	12.5
Height, y	69.5	71.5	72.0	73.0	73.5	74.0	74.0

16. ***Coffee Sales and Temperature*** ◆ The high temperature (in °F) and coffee sales (in hundreds of dollars) for a coffee shop for eight randomly selected days

Temperature, x	32	39	51	60	65	72	78	81
Coffee sales, y	26.2	24.8	19.7	20.0	13.3	13.9	11.4	11.2

17. ***Age and Sleep*** ◆ The age (in years) and the number of hours slept in a day by six infants

Age, x	0.1	0.2	0.5	0.7	0.8	0.9
Hours slept, y	14.9	14.5	13.4	14.1	13.4	13.7

18. ***Earnings and Dividends*** ◆ The earnings per share and dividends per share for 10 electric utility companies in a recent year *(Source: The Value Line Investment Survey)*

Earnings per share, x	1.67	1.73	1.77	1.79	1.84
Dividend per share, y	1.73	1.46	1.48	1.42	1.63

Earnings per share, x	1.90	1.92	1.97	1.99	2.00
Dividend per share, y	1.60	1.83	1.46	1.67	1.72

Testing Claims In Exercises 19–24, use Table 11 in Appendix B as shown in Example 6 or perform a hypothesis test using Table 5 in Appendix B as shown in Example 7 to make a conclusion about the indicated correlation coefficient. Also, if convenient, use technology to solve the problem.

19. ***Braking Distances: Dry Surface*** ◆ The weights (in pounds) of eight vehicles and the variability of their braking distances (in feet) when stopping on a dry surface are shown in the table. Can you conclude that there is a significant linear correlation between vehicle weight and variability in braking distance on a dry surface? Use $\alpha = 0.01$. *(Adapted from National Highway Traffic Safety Administration)*

Weight, x	5940	5340	6500	5100	5850	4800	5600	5890
Variability in braking distance, y	1.78	1.93	1.91	1.59	1.66	1.50	1.61	1.70

20. ***Braking Distances: Wet Surface*** ◆ The weights (in pounds) of eight vehicles and the variability of their braking distances (if feet) when stopping on a wet surface are shown in the table. At $\alpha = 0.05$, can you conclude that there is a significant linear correlation between vehicle weight and variability in braking distance on a wet surface? *(Adapted from National Highway Traffic Safety Administration)*

Weight, x	5890	5340	6500	4800	5940	5600	5100	5850
Variability in braking distance, y	2.92	2.40	4.09	1.72	2.88	2.53	2.32	2.78

DATA **21.** ***Studying and Test Scores*** ◆ The number of hours 13 students spent studying for a test and their scores on that test are shown in the table. Is there enough evidence to conclude that there is a significant linear correlation between the data? Use $\alpha = 0.01$. (Use the value of r found in Exercise 13.)

Hours spent studying, x	0	1	2	4	4	5	5
Test score, y	40	41	51	48	64	69	73

Hours spent studying, x	5	6	6	7	7	8
Test score, y	75	68	93	84	90	95

DATA **22.** ***Watching TV and Test Scores*** ◆ An instructor wants to show students that there is a linear correlation between the number of hours they watch television during a certain weekend and their scores on a test taken the following Monday. The number of television viewing hours and the test scores for 12 randomly selected students are listed in the table. At $\alpha = 0.05$, is there enough evidence for the instructor to conclude that there is a significant linear correlation between the data? (Use the value of r found in Exercise 14.)

Hours spent watching TV, x	0	1	2	3	3	5
Test score, y	96	85	82	74	95	68

Hours spent watching TV, x	5	5	6	7	7	10
Test score, y	76	84	58	65	75	50

DATA **23.** ***Shoe Size and Height*** ◆ You randomly select 14 men and measure the height (in inches) and shoe size of each. The data are listed below. At $\alpha = 0.05$, can you conclude that there is a significant linear correlation between the men's shoe sizes and heights? (Use the value of r found in Exercise 15.)

Shoe size, x	8.5	9.0	9.0	9.5	10.0	10.0	10.5
Height, y	66.0	68.5	67.5	70.0	70.0	72.0	71.5

Shoe size, x	10.5	11.0	11.0	11.0	12.0	12.0	12.5
Height, y	69.5	71.5	72.0	73.0	73.5	74.0	74.0

24. *Earnings and Dividends* ◆ The following table lists the earnings per share and dividends per share for 10 electric utility companies in a recent year. At $\alpha = 0.01$, can you conclude that there is a significant linear correlation between earnings per share and dividends per share? (Use the value of r found in Exercise 18.) *(Source: The Value Line Investment Survey)*

Earnings per share, x	1.67	1.73	1.77	1.79	1.84
Dividend per share, y	1.73	1.46	1.48	1.42	1.63

Earnings per share, x	1.90	1.92	1.97	1.99	2.00
Dividend per share, y	1.60	1.83	1.46	1.67	1.72

25. Which value of r indicates a stronger correlation: $r = 0.73$ or $r = -0.84$? Explain your reasoning.

26. Which of the following values could not represent a correlation coefficient? Explain why.

(a) $r = 0.92$ (b) $r = 1.05$ (c) $r = -0.73$ (d) $r = -0.05$

27. Explain how to perform a hypothesis test for a population correlation coefficient ρ.

28. Discuss the difference between r and ρ.

Extending the Basics

Interchanging x and y In Exercises 29 and 30, calculate the correlation coefficient r, letting Row 1 represent the x-values and Row 2 the y-values. Then calculate the correlation coefficient r, letting Row 2 represent the x-values and Row 1 the y-values. What effect does switching the explanatory and response variables have on the correlation coefficient?

29.

Row 1	16	25	39	45	49	64	70
Row 2	109	122	143	132	199	185	199

30.

Row 1	0	1	2	3	3	5	5	5	6	7
Row 2	96	85	82	74	95	68	76	84	58	65

31. *Writing* ◆ Use your school's library or some other reference source to find a real-life data set with the indicated cause-and-effect relationship. Write a paragraph describing each variable and explain why you think the variables have the indicated cause-and-effect relationship.

(a) *Direct Cause-and-Effect:* Changes in one variable cause changes in the other variable.

(b) *Other Factors:* The relationship between the variables is caused by a third variable.

(c) *Coincidence:* The relationship between the variables is a coincidence.

Linear Regression

What You Should Learn

- How to find the equation of a regression line
- How to predict y-values using a regression equation

Regression Lines • Applications of Regression Lines

Regression Lines

After verifying that the correlation between two variables is significant, the next step is to determine the equation of the line that best models the data. This line is called a regression line and its equation can be used to predict the value of y for a given value of x. While many lines can be drawn through a set of points, a regression line is determined by specific criteria.

Consider the scatter plot and the line shown below. For each data point, d represents the difference between the observed y-value and the predicted y-value on the line. These differences are called **residuals** and can be positive, negative, or zero. When the point is above the line, $d > 0$. When the point is below the line, $d < 0$. If the observed y-value equals the predicted y-value, $d = 0$. Of all possible lines that can be drawn through a set of points, the regression line is the line for which the sum of the squares of all the residuals

$$\Sigma d^2$$

is a minimum.

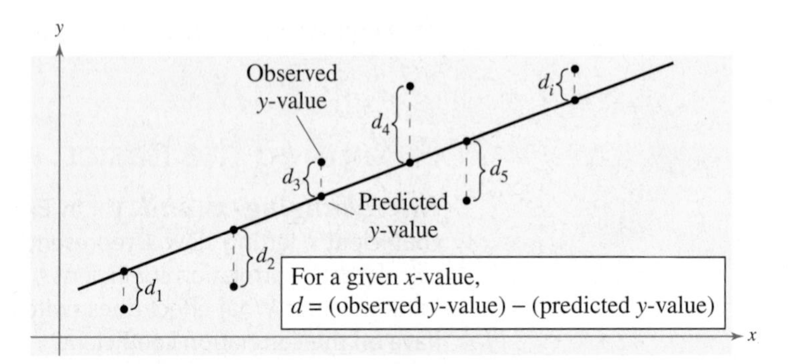

DEFINITION

A **regression line,** also called a **line of best fit,** is the line for which the sum of the squares of the residuals is a minimum.

In algebra, you learned that you can write an equation of a line by finding its slope m and y-intercept b. The equation has the form

$$y = mx + b.$$

Recall that the slope of a line is the ratio of its rise over its run and the y-intercept is the y-value of the point at which the line crosses the y-axis. It is the y-value when $x = 0$.

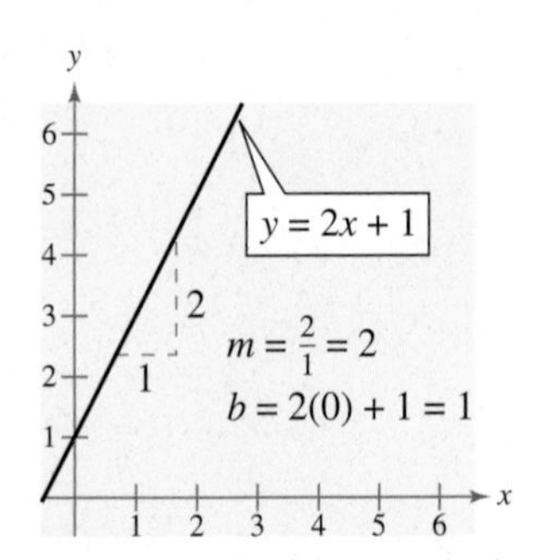

In algebra, you used two points to determine the equation of a line. In statistics, you will use every point in the data set to determine the equation of the line of regression.

A regression line allows you to use the explanatory variable x to make predictions for the response variable y.

Study Tip

Notice that the sums in the formula for m are the same as those used to calculate r.

The Equation of a Regression Line

The equation of a regression line for an independent variable x and a dependent variable y is

$$\hat{y} = mx + b$$

where $\hat{y}$ is the predicted y-value for a given x-value. The slope m and y-intercept b are given by

$$m = \frac{n\sum xy - (\sum x)(\sum y)}{n\sum x^2 - (\sum x)^2} \quad \text{and} \quad b = \bar{y} - m\bar{x} = \frac{\sum y}{n} - m\frac{\sum x}{n}$$

where $\bar{y}$ is the mean of the y-values in the data set and $\bar{x}$ is the mean of the x-values. The regression line always passes through the point $(\bar{x}, \bar{y})$.

EXAMPLE 1

Finding the Equation of a Regression Line

Find the equation of the regression line for the advertising expenditures and company sales data used in Section 9.1.

Advertising expenses (1000s of \$), x	Company sales (1000s of \$), y
2.4	225
1.6	184
2.0	220
2.6	240
1.4	180
1.6	184
2.0	186
2.2	215

SOLUTION In Example 4 of Section 9.1, you found that $n = 8$, $\sum x = 15.8$, $\sum y = 1634$, $\sum xy = 3289.8$, and $\sum x^2 = 32.44$. You can use these values to calculate the slope and y-intercept of the regression line as shown.

$$m = \frac{n\sum xy - (\sum x)(\sum y)}{n\sum x^2 - (\sum x)^2} = \frac{8(3289.8) - (15.8)(1634)}{8(32.44) - 15.8^2} = \frac{501.2}{9.88} \approx 50.7287$$

$$b = \bar{y} - m\bar{x} = \frac{1634}{8} - (50.7287)\frac{15.8}{8} = 204.25 - (50.7287)(1.975) \approx 104.0608$$

So, the equation of the regression line is

$$\hat{y} = 50.729x + 104.061.$$

The regression line and scatter plot of the data are shown at the right. If you plot the point $(\bar{x}, \bar{y}) = (1.975, 204.25)$, you will notice that the line passes through this point.

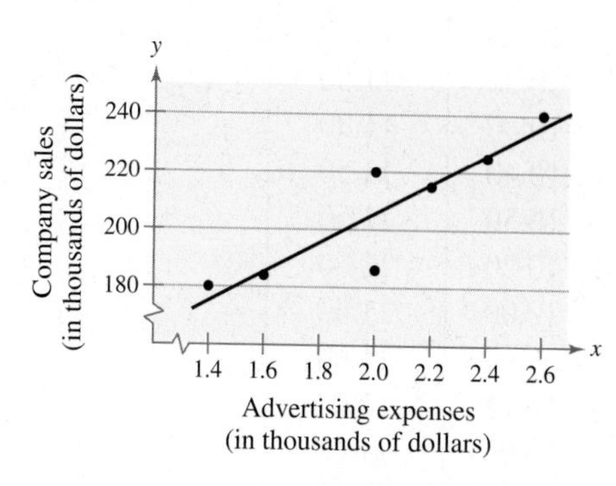

Try It Yourself 1

Find the equation of the regression line for the income level and donating percent data used in Section 9.1.

Income level (in 1000s of \$), x	Donating percent, y
42	9
48	10
50	8
59	5
65	6
72	3

a. *Identify* n, $\sum x$, $\sum y$, $\sum xy$, and $\sum x^2$ from Try It Yourself 4 of Section 9.1.
b. *Calculate* the slope m.
c. *Calculate* the y-intercept b.
d. *Write* the regression line.

Answer: Page A41

EXAMPLE 

Using Technology to Find a Regression Equation

Duration, x	Time, y	Duration, x	Time, y
1.80	56	3.78	79
1.82	58	3.83	85
1.88	60	3.87	81
1.90	62	3.88	80
1.92	60	4.10	89
1.93	56	4.27	90
1.98	57	4.30	84
2.03	60	4.30	89
2.05	57	4.43	84
2.13	60	4.43	89
2.30	57	4.47	86
2.35	57	4.47	80
2.37	61	4.53	89
2.82	73	4.55	86
3.13	76	4.60	88
3.27	77	4.60	92
3.65	77	4.63	91
3.70	82		

Use a technology tool to find the equation of the regression line for the Old Faithful data used in Section 9.1.

SOLUTION *Minitab*, *Excel*, and the *TI-83* each have features that automatically calculate a regression equation. Try using this technology to find the regression equation. You should obtain results similar to the following.

MINITAB

Regression Analysis: C2 versus C1

The regression equation is
C2 = 35.3 + 11.8 C1

Predictor	Coef	SE Coef	T	P
Constant	35.301	1.804	19.57	0.000
C1	11.8244	0.5178	22.84	0.000

S = 3.277 R-Sq = 94.0% R-Sq(adj) = 93.9%

EXCEL

	A	B	C	D
1	Slope:			
2	INDEX(LINEST(known_y's,known_x's),1)			
3				11.82441
4				
5	Y-intercept:			
6	INDEX(LINEST(known_y's,known_x's),2)			
7				35.30117

TI-83

LinReg
y=ax+b
a=11.8244078
b=35.30117105
r²=.9404868083
r=.9697869912

From the displays, you can see that the regression equation is

$$\hat{y} = 11.824x + 35.301.$$

The *TI-83* display at the right shows the regression line and a scatter plot of the data.

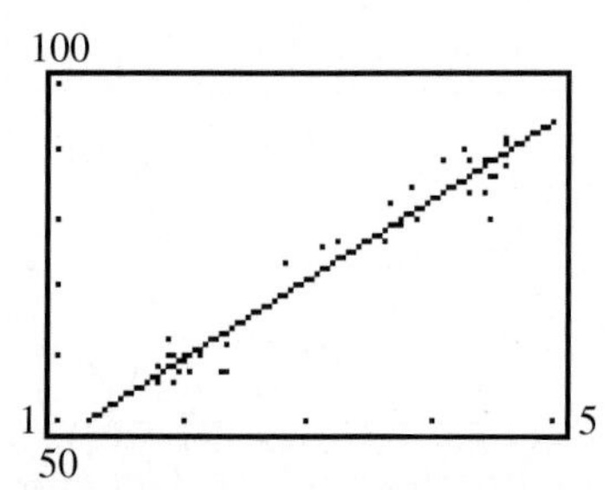

Men's times, x	Women's times, y
10.80	12.20
10.30	11.90
10.30	11.50
10.30	11.90
10.40	11.50
10.50	11.50
10.20	11.00
10.00	11.40
9.95	11.00
10.14	11.07
10.06	11.08
10.25	11.60
9.99	10.97
9.92	10.54
9.96	10.82
9.84	10.94
9.87	10.75

Try It Yourself 2

Use a technology tool to find the equation of the regression line for the men's and women's winning times given in the Chapter Opener.

a. *Enter* the data.
b. *Perform the necessary steps* to calculate the slope and y-intercept.
c. *Specify* the regression line.

Answer: Page A41

Applications of Regression Lines

After finding the equation of a regression line, you can use the equation to predict y-values over the range of the data *if the correlation between x and y is significant*. For example, an advertising executive could forecast company sales based on advertising expenditures. To predict y-values, substitute the given x-value into the regression equation, then calculate $\hat{y}$, the predicted y-value.

Picturing the World

The following scatter plot shows the relationship between the number of motor vehicles in a state and the number of motor vehicle deaths. *(Source: Federal Highway Admin. & National Safety Council)*

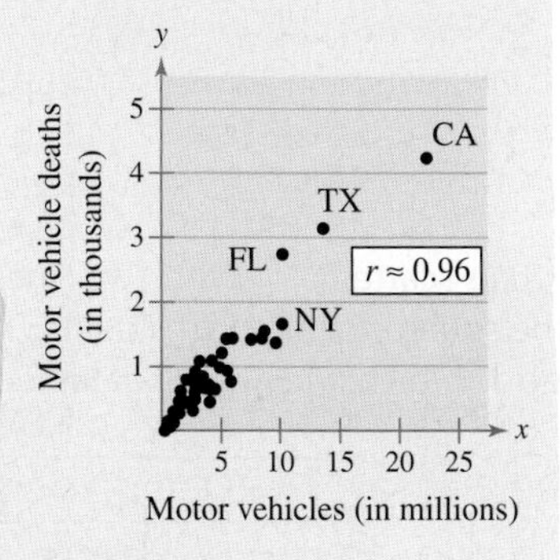

Describe the correlation between these two variables in words. Use the scatter plot to predict the number of motor vehicle deaths in a state that has 16 million motor vehicles. The regression line for this scatter plot is $\hat{y} = 0.053 + 0.192x$. Use this equation to make your prediction. (Assume x and y have a significant linear correlation.) How does your algebraic prediction compare with your graphical one?

EXAMPLE 

Predicting y-Values Using Regression Equations

The regression equation for the advertising expenditures (in 1000s of dollars) and company sales (in 1000s of dollars) data is

$$\hat{y} = 50.729x + 104.061.$$

Use this equation to predict the *expected* company sales for the following advertising expenditures. (Recall from Section 9.1, Example 7 that x and y have a significant linear correlation.)

1. 1.5 thousand dollars 2. 1.8 thousand dollars 3. 2.5 thousand dollars

SOLUTION To predict the expected company sales, substitute each advertising expenditure for x in the regression equation. Then calculate $\hat{y}$.

1. $\hat{y} = 50.729x + 104.061$
$= 50.729(1.5) + 104.061$
≈ 180.155
When the advertising expenditures are \$1500, the company sales are about \$180,155.

2. $\hat{y} = 50.729x + 104.061$
$= 50.729(1.8) + 104.061$
≈ 195.373
When the advertising expenditures are \$1800, the company sales are about \$195,373.

3. $\hat{y} = 50.729x + 104.061$
$= 50.729(2.5) + 104.061$
≈ 230.884
When the advertising expenditures are \$2500, the company sales are about \$230,884.

Prediction values are meaningful only for x-values in (or close to) the range of the data. The x-values in the original data set range from 1.4 to 2.6. So, it would not be appropriate to use the regression line $\hat{y} = 50.729x + 104.061$ to predict company sales for advertising expenditures such as 0.5 (\$500) or 5.0 (\$5000).

Try It Yourself 3

The regression equation for the Old Faithful data is

$$\hat{y} = 11.824x + 35.301.$$

Use this to predict the time until the next eruption for each of the following eruption durations. (Recall from Section 9.1, Example 6 that x and y have a significant linear correlation.)

1. 2 minutes
2. 3.32 minutes

a. *Substitute* each value for x in the regression equation.
b. *Calculate* $\hat{y}$.
c. *Specify* the predicted time until the next eruption for each eruption duration.

Answer: Page A41

9.2 Exercises

Help

LarsonTutor 9.2

Companion Web Site

Student Solutions Manual 9.2

Videos 9.2

Technology Manuals

Basic Skills and Concepts

In Exercises 1–8, match the description in the left column with a description in the right column.

1. Regression line
2. Residual
3. The y-value of a data point corresponding to x_i
4. The y-value for a point on the regression line corresponding to x_i
5. Slope
6. y-intercept
7. The mean of the y-values
8. The point a regression line always passes through

a. The difference between the observed y-value of a data point and the predicted y-value on the line for the same data point
b. $\hat{y}_i$
c. The line of best fit
d. y_i
e. b
f. $(\bar{x}, \bar{y})$
g. m
h. $\bar{y}$

Graphical Analysis In Exercises 9 and 10, a scatter plot with two lines is given. For each line, calculate $\sum d^2$, the sum of the squares of all the residuals. Then compare the values of $\sum d^2$ for each line and decide which line fits the data better.

9. ***Trees: Heights and Diameters*** ◆ The heights (in feet) and diameters (in inches) of eight trees

Height, x	70	72	75	76
Diameter, y	8.3	10.5	11.0	11.4

Height, x	85	78	77	80
Diameter, y	12.9	14.0	16.3	18.0

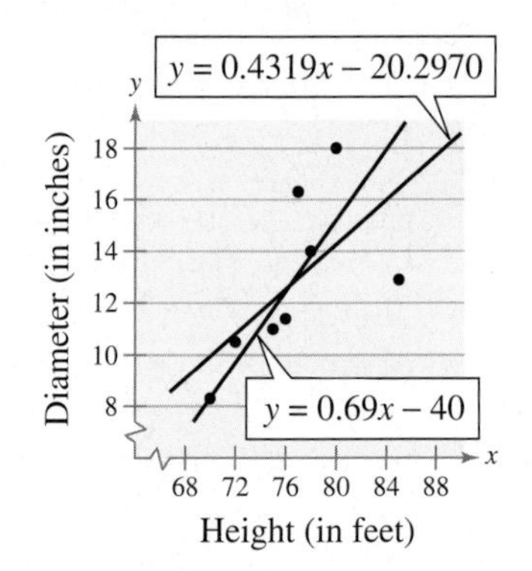

10. ***Hot Dogs: Caloric and Sodium Content*** ◆ The caloric content and the sodium content (in milligrams) for 10 beef hot dogs *(Source: Consumer Reports)*

Calories, x	186	181	176	149	184
Sodium, y	495	477	425	322	482

Calories, x	190	158	139	175	148
Sodium, y	587	370	322	479	375

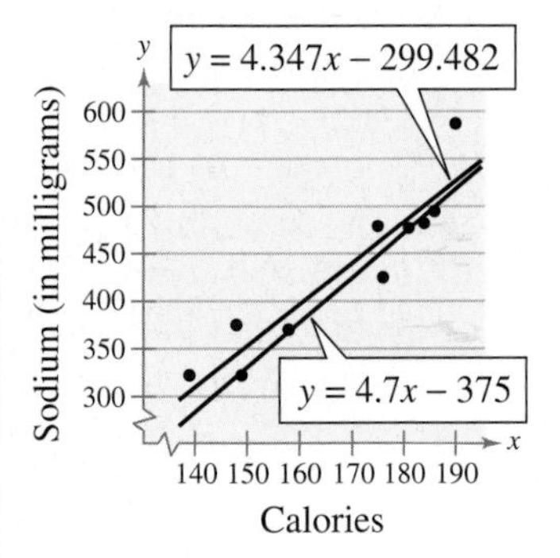

In Exercises 11–14, match the regression equation with the appropriate graph. (Note that the x- and y-axes are broken.)

11. $\hat{y} = -1.04x + 50.3$

12. $\hat{y} = -0.902x + 90.5$

13. $\hat{y} = 0.000114x + 2.53$

14. $\hat{y} = -0.667x + 52.6$

a.

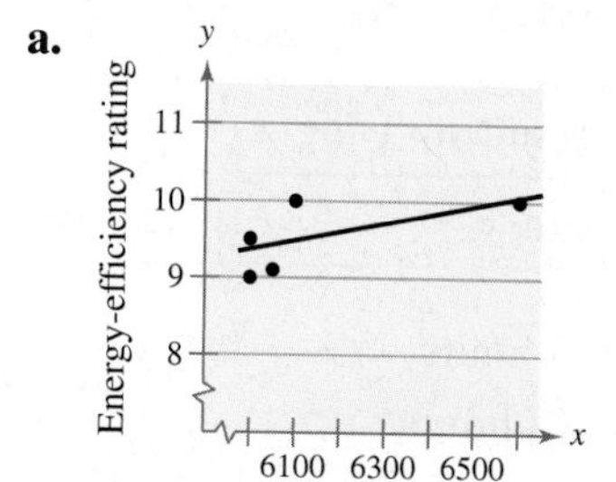

b.

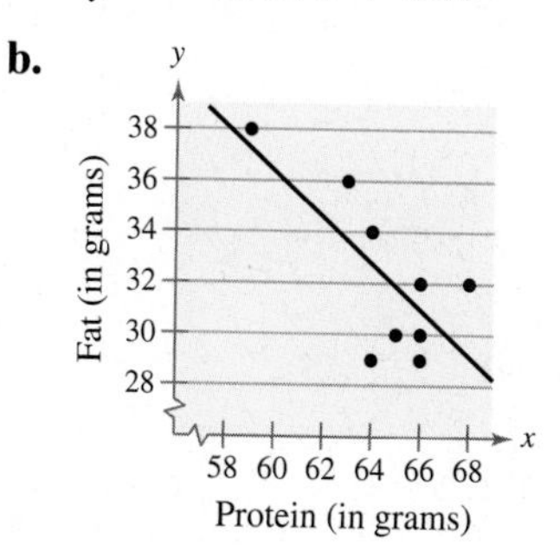

c.

d.

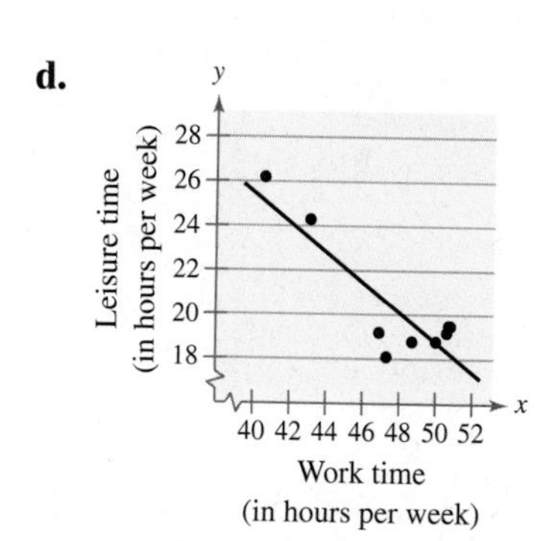

Finding the Equation of a Regression Line In Exercises 15–22, find the equation of the regression line for the given data and graph a scatter plot of the data and the regression line. (Each pair of variables has a significant correlation.) Then use the regression equation to predict the value of y for each of the given x-values, if meaningful. If the x-value is not meaningful to predict the value of y, explain why not.

15. ***Age and Blood Pressure*** ◆ The ages (in years) of seven men and their systolic blood pressures

Age, x	16	25	39	45	49	64	70
Systolic blood pressure, y	109	122	143	132	199	185	199

(a) $x = 18$ years
(b) $x = 71$ years
(c) $x = 29$ years
(d) $x = 55$ years

16. ***Age and Vocabulary*** ◆ The ages (in years) of eight children and the number of words in their vocabulary

Age, x	1	2	3	4	5	6	3	5
Vocabulary size, y	3	440	1200	1500	2100	2600	1100	2000

(a) $x = 2$ years
(b) $x = 3$ years
(c) $x = 6$ years
(d) $x = 12$ years

17. ***Hours Studying and Test Scores*** ◆ The number of hours 13 students spent studying for a test and their scores on that test

Hours spent studying, x	0	1	2	4	4	5	5
Test score, y	40	41	51	48	64	69	73

Hours spent studying, x	5	6	6	7	7	8
Test score, y	75	68	93	84	90	95

(a) $x = 3$ hours
(b) $x = 6.5$ hours
(c) $x = 13$ hours
(d) $x = 4.5$ hours

18. ***TV and Test Scores*** ◆ The number of hours 12 students spent watching television during a weekend and their scores on a test on Monday

Hours spent watching TV, x	0	1	2	3	3	5
Test score, y	96	85	82	74	95	68

Hours spent watching TV, x	5	5	6	7	7	10
Test score, y	76	84	58	65	75	50

(a) $x = 4$ hours
(b) $x = 8$ hours
(c) $x = 9$ hours
(d) $x = 15$ hours

19. ***Shoe Size and Height*** ◆ The shoe sizes and heights (in inches) for 14 men

Shoe size, x	8.5	9.0	9.0	9.5	10.0	10.0	10.5
Height, y	66.0	68.5	67.5	70.0	70.0	72.0	71.5

Shoe size, x	10.5	11.0	11.0	11.0	12.0	12.0	12.5
Height, y	69.5	71.5	72.0	73.0	73.5	74.0	74.0

(a) $x =$ size 11.5
(b) $x =$ size 8.0
(c) $x =$ size 15.5
(d) $x =$ size 10.0

20. ***Age and Hours Slept*** ◆ The age (in years) and the number of hours slept in a day by 10 infants

Age, x	0.1	0.2	0.5	0.7	0.8	0.9
Hours slept, y	14.9	14.5	13.4	14.1	13.4	13.7

Age, x	0.1	0.6	0.5	0.7
Hours slept, y	14.1	13.9	14.0	14.2

(a) $x = 0.3$ year
(b) $x = 3.9$ years
(c) $x = 0.6$ year
(d) $x = 0.4$ year

21. ***Braking Distances: Dry Surface*** ◆ The weights (in pounds) of eight vehicles and the variability of their braking distances (in feet) when stopping on a dry surface. *(Adapted from National Highway Traffic Safety Administration)*

Weight, x	5720	4050	6130	5000	5010	4270	5500	5550
Variability in braking distance, y	2.19	1.36	2.58	1.74	1.78	1.69	1.80	1.87

(a) $x = 4500$ pounds
(b) $x = 6000$ pounds
(c) $x = 7500$ pounds
(d) $x = 5750$ pounds

22. ***Braking Distances: Wet Surface*** ◆ The weights (in pounds) of eight vehicles and the variability of their braking distances (in feet) when stopping on a wet surface. *(Adapted from National Highway Traffic Safety Administration)*

Weight, x	5720	4050	6130	5000	5010	4270	5500	5550
Variability in braking distance, y	3.78	2.43	4.63	2.88	3.25	2.76	3.42	3.51

(a) $x = 4800$ pounds
(b) $x = 5850$ pounds
(c) $x = 4075$ pounds
(d) $x = 3000$ pounds

23. ***Writing*** ◆ Explain how to predict y-values using the equation of a regression line.

24. ***Writing*** ◆ Given a set of data and a corresponding regression line, describe all values of x that provide meaningful predictions for y.

Extending the Basics

Interchanging x and y In Exercise 25 and 26, do the following.

(a) Find the equation of the regression line for the given data, letting Row 1 represent the x-values and Row 2 the y-values. Graph a scatter plot of the data and draw the regression line with it.

(b) Find the equation of the regression line for the given data, letting Row 2 represent the x-values and Row 1 the y-values. Graph a scatter plot of the data and draw the regression line with it.

(c) What effect does switching the explanatory and response variables have on the regression line?

25.

Row 1	16	25	39	45	49	64	70
Row 2	109	122	143	132	199	185	199

26.

Row 1	0	1	2	3	3	5	5	5	6	7
Row 2	96	85	82	74	95	68	76	84	58	65

Cellular Phones In Exercises 27–31, use the following information. You work for a cellular phone industry analyst and gather the data shown in the table. The table shows the number of cellular phone subscribers and the average monthly cellular phone bill in the United States for 13 years. *(Source: Cellular Telecommunications Industry Association)*

Number of subscribers (in millions), x	Average monthly bill (in dollars), y
1.2	96.83
2.1	98.02
3.5	89.30
5.3	80.90
7.6	72.74
11.0	68.68
16.0	61.48
24.1	56.21
33.8	51.00
44.0	47.70
55.3	42.78
69.2	39.43
86.0	41.24

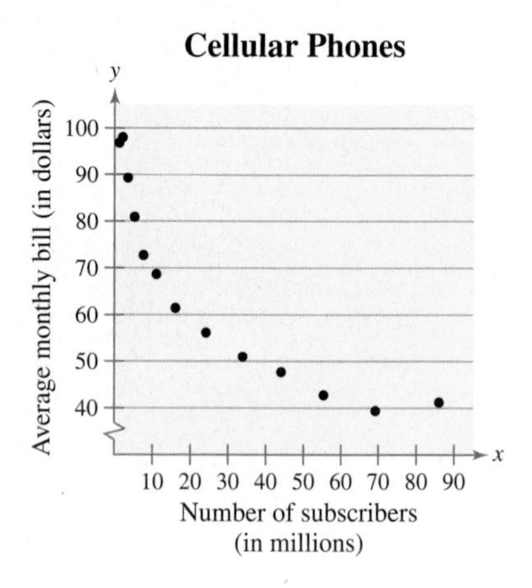

27. *Correlation* ◆ Using the scatter plot of the cellular phone data given above, what type of correlation, if any, do you think the data have? Explain your reasoning.

DATA **28. *Regression Line*** ◆ Find an equation of the regression line for the data. Graph a scatter plot of the data and the regression line.

29. *Using the Regression Line* ◆ The analyst uses the regression line you found in Exercise 28 to predict the average monthly bill for $x = 140$ million subscribers. Is this a valid prediction? Explain your reasoning.

30. *Significant Correlation?* ◆ The analyst claims that the data have a significant correlation for $\alpha = 0.01$. Verify this claim.

31. *Cause and Effect* ◆ Write a paragraph describing the cause-and-effect relationship between the number of cellular phone subscribers and the average monthly cellular phone bill.

32. *Writing* ◆ Use your school's library or some other reference source to find a real-life data set and do the following.

- Find the equation of the regression line for the data.
- Graph a scatter plot of the data with the regression line.
- Write a paragraph describing each variable and what kind of cause-and-effect relationship you think the variables have.
- Pick any two points in the data set and find the equation of the line that passes through both. Then confirm the fact that Σd^2 for the regression line is less than Σd^2 for the line constructed through the two data points.

9 Case Study

Correlation of Body Measurements

In a study published in *Medicine and Science in Sports and Exercise,* 17(2), 189, the measurements of 252 men (ages 22–81) are given. Of the 14 measurements taken of each man, some have significant correlations and others don't. For instance, the scatter plot at the right shows that the hip and abdomen circumferences of the men have a strong linear correlation ($r = 0.85$). The partial table shown here lists only the first 9 rows of the data.

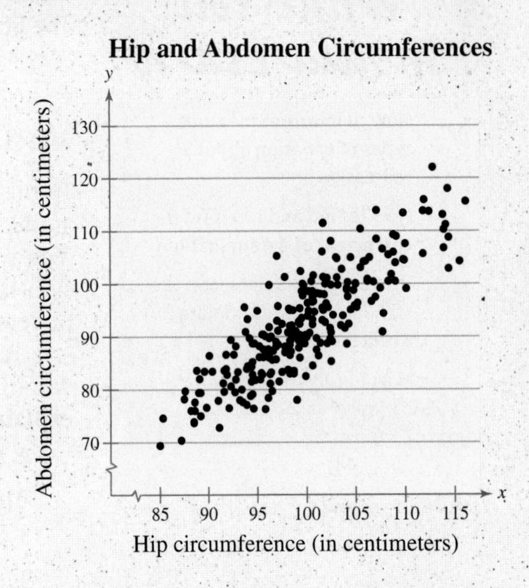

Age (yr)	Weight (lb)	Height (in.)	Neck (cm)	Chest (cm)	Abdom. (cm)	Hip (cm)	Thigh (cm)	Knee (cm)	Ankle (cm)	Bicep (cm)	Forearm (cm)	Wrist (cm)	Body fat %
22	173.25	72.25	38.5	93.6	83.0	98.7	58.7	37.3	23.4	30.5	28.9	18.2	6.1
22	154.00	66.25	34.0	95.8	87.9	99.2	59.6	38.9	24.0	28.8	25.2	16.6	25.3
23	154.25	67.75	36.2	93.1	85.2	94.5	59.0	37.3	21.9	32.0	27.4	17.1	12.3
23	198.25	73.50	42.1	99.6	88.6	104.1	63.1	41.7	25.0	35.6	30.0	19.2	11.7
23	159.75	72.25	35.5	92.1	77.1	93.9	56.1	36.1	22.7	30.5	27.2	18.2	9.4
23	188.15	77.50	38.0	96.6	85.3	102.5	59.1	37.6	23.2	31.8	29.7	18.3	10.3
24	184.25	71.25	34.4	97.3	100.0	101.9	63.2	42.2	24.0	32.2	27.7	17.7	28.7
24	210.25	74.75	39.0	104.5	94.4	107.8	66.0	42.0	25.6	35.7	30.6	18.8	20.9
24	156.00	70.75	35.7	92.7	81.9	95.3	56.4	36.5	22.0	33.5	28.3	17.3	14.2

Exercises

1. Using your intuition, classify the following (x, y) pairs as having a weak correlation ($0 < r < 0.5$), a moderate correlation ($0.5 < r < 0.8$), or a strong correlation ($0.8 < r < 1.0$).

(a) (weight, neck)
(b) (weight, height)
(c) (age, body fat)
(d) (chest, hip)
(e) (age, wrist)
(f) (ankle, wrist)
(g) (forearm, height)
(h) (bicep, forearm)
(i) (weight, body fat)
(j) (knee, thigh)
(k) (hip, abdomen)
(l) (abdomen, hip)

2. Now, use a technology tool to find the correlation coefficient for each pair in Exercise 1. Compare your results to those obtained by intuition.

3. Use a technology tool to find the regression line for each pair in Exercise 1 that has a strong correlation.

4. Use the results of Exercise 3 to predict the following.

(a) The neck circumference of a man whose weight is 180 pounds.

(b) The abdomen circumference of a man whose hip circumference is 100 centimeters.

5. Are there pairs of measurements that have stronger correlation coefficients than 0.85? Use a technology tool and intuition to reach a conclusion.

9.3 Measures of Regression and Prediction Intervals

What You Should Learn

- How to interpret the three types of variation about a regression line
- How to find and interpret the coefficient of determination
- How to find and interpret the standard error of estimate for a regression line
- How to construct and interpret a prediction interval for y

Variation about a Regression Line • The Coefficient of Determination • The Standard Error of Estimate • Prediction Intervals

Variation about a Regression Line

In this section, you will study two measures used in correlation and regression studies—the coefficient of determination and the standard error of estimate. You will also learn how to construct a prediction interval for y using a regression line and a given value of x. Before studying these concepts, you need to understand the three types of variation about a regression line.

To find the total variation, the explained variation, and the unexplained variation about a regression line, you must first calculate the **total deviation,** the **explained deviation,** and the **unexplained deviation** for each ordered pair (x_i, y_i) in a data set. These deviations are shown in the graph.

$$\text{Total deviation} = y_i - \bar{y}$$
$$\text{Explained deviation} = \hat{y}_i - \bar{y}$$
$$\text{Unexplained deviation} = y_i - \hat{y}_i$$

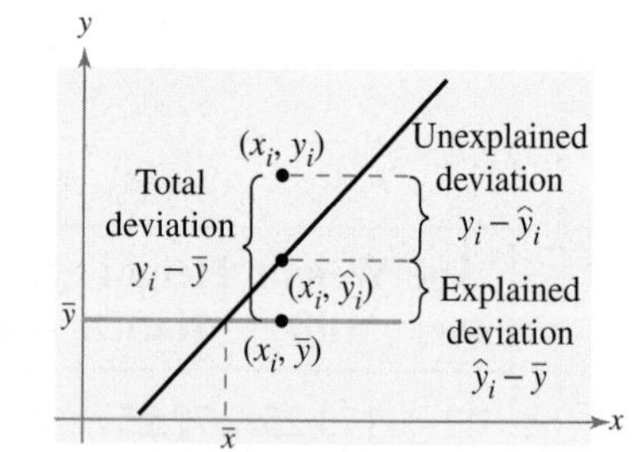

After calculating the deviations for each data point (x_i, y_i), you can find the total variation, the explained variation, and the unexplained variation.

Study Tip

Consider the advertising and sales data used throughout this chapter with a regression line of

$\hat{y} = 50,729x + 104.061.$

Using the data point (2.0, 220), you can find the total, explained, and unexplained deviations as follows.

Total deviation:

$y_i - \bar{y} = 220 - 204.25$
$= 15.75$

Explained deviation:

$\hat{y}_i - \bar{y} = 205.519 - 204.25$
$= 1.269$

Unexplained deviation:

$y_i - \hat{y}_i = 220 - 205.519$
$= 14.481$

DEFINITION

The **total variation** about a regression line is the sum of the squares of the differences between the y-value of each ordered pair and the mean of y.

$$\text{Total variation} = \Sigma(y_i - \bar{y})^2$$

The **explained variation** is the sum of the squares of the differences between each predicted y-value and the mean of y.

$$\text{Explained variation} = \Sigma(\hat{y}_i - \bar{y})^2$$

The **unexplained variation** is the sum of the squares of the differences between the y-value of each ordered pair and each corresponding predicted y-value.

$$\text{Unexplained variation} = \Sigma(y_i - \hat{y}_i)^2$$

The sum of the explained and unexplained variations is equal to the total variation.

$$\text{Total variation} = \text{Explained variation} + \text{Unexplained variation}$$

As its name implies, the *explained variation* can be explained by the relationship between x and y. The *unexplained variation* cannot be explained by the relationship between x and y and is due to chance or other variables.

The Coefficient of Determination

You already know how to calculate the correlation coefficient r. The square of this coefficient is called the coefficient of determination. It can be shown that the coefficient of determination is equal to the ratio of the explained variation to the total variation.

DEFINITION

The **coefficient of determination r^2** is the ratio of the explained variation to the total variation. That is,

$$r^2 = \frac{\text{Explained variation}}{\text{Total variation}}.$$

It is important that you interpret the coefficient of determination correctly. For instance, if the correlation coefficient is $r = 0.90$, then the coefficient of determination is

$$r^2 = 0.90^2$$
$$= 0.81.$$

This means that 81% of the variation of y can be explained by the relationship between x and y. The remaining 19% of the variation is unexplained and is due to other factors or to sampling error.

Picturing the World

Janette Benson (Psychology Department, University of Denver) performed a study relating the age that infants crawl (in weeks after birth) with the average monthly temperature six months after birth. Her results are based on a sample of 414 infants. Janette Benson believes that the reason for the correlation in temperature to crawling age is that parents tend to bundle infants in more restrictive clothing and blankets during cold months. This bundling doesn't allow the infant as much opportunity to move and experiment with crawling.

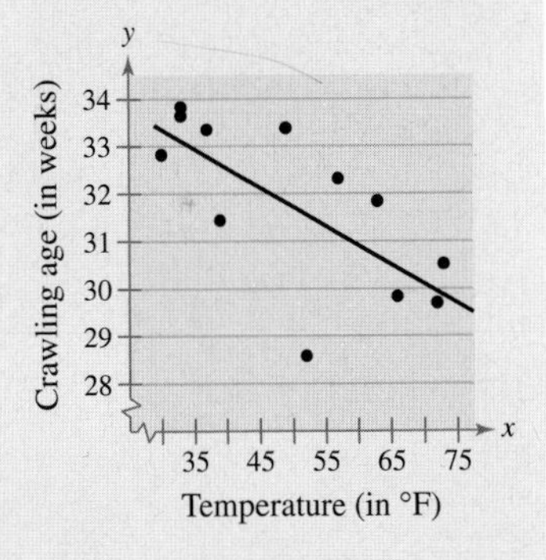

The correlation coefficient is $r = 0.70$. What percent of the variation in the data can be explained? What percent is due to chance, sampling error, or other factors?

EXAMPLE

Finding the Coefficient of Determination

The correlation coefficient for the advertising expenditures and company sales data given in Section 9.1 is $r \approx 0.913$, as shown in Example 4. Find the coefficient of determination. What does this tell you about the explained variation of the data about the regression line? The unexplained variation?

SOLUTION The coefficient of determination is

$$r^2 = (0.913)^2$$
$$\approx 0.834.$$

This means that about 83.4% of the variation in the company sales can be explained by the variation in the advertising expenditures. About 16.6% of the variation is unexplained and is due to chance or other variables.

Try It Yourself 1

The correlation coefficient for the Old Faithful data given in Section 9.1 is $r \approx 0.970$, as shown in Example 5. Find the coefficient of determination. What does this tell you about the explained variation of the data about the regression line? The unexplained variation?

a. *Identify* the correlation coefficient r.
b. *Square* r.
c. What percent of the variation in the times is explained? What percent is unexplained?

Answer: Page A41

The Standard Error of Estimate

When a $\hat{y}$-value is predicted from an x-value, the prediction is a point estimate. You can construct an interval estimate for $\hat{y}$, but first you need to calculate the standard error of estimate.

DEFINITION

The **standard error of estimate s_e** is the standard deviation of the observed y_i-values about the predicted $\hat{y}$-value for a given x_i-value. It is given by

$$s_e = \sqrt{\frac{\Sigma(y_i - \hat{y}_i)^2}{n - 2}}$$

where n is the number of ordered pairs in the data set.

From this formula, you can see that the standard error of estimate is the square root of the unexplained variation divided by $n - 2$. So, the closer the observed y-values are to the predicted y-values, the smaller the standard error of estimate will be.

GUIDELINES

Finding the Standard Error of Estimate s_e

In Words	*In Symbols*
1. Make a table that includes the column headings shown at the right.	$x_i, y_i, \hat{y}_i, (y_i - \hat{y}_i), (y_i - \hat{y}_i)^2$
2. Use the regression equation to calculate the predicted y-values.	$\hat{y}_i = mx_i + b$
3. Calculate the sum of the squares of the differences between each observed y-value and the corresponding predicted y-value.	$\Sigma(y_i - \hat{y}_i)^2$
4. Find the standard error of estimate.	$s_e = \sqrt{\frac{\Sigma(y_i - \hat{y}_i)^2}{n - 2}}$

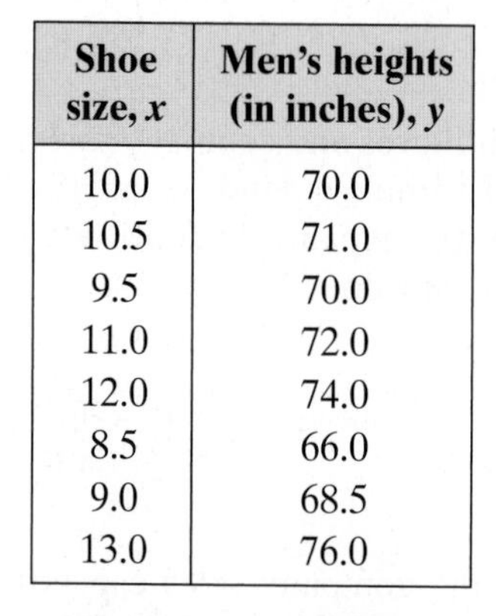

Shoe size, x	Men's heights (in inches), y
10.0	70.0
10.5	71.0
9.5	70.0
11.0	72.0
12.0	74.0
8.5	66.0
9.0	68.5
13.0	76.0

Shoe size, x	Men's heights (in inches), y
10.5	71.5
10.5	70.5
10.0	72.0
9.5	70.0
10.0	71.0
10.5	69.5
11.0	71.5
12.0	73.5

You can also find the standard error of estimate using the following formula.

$$s_e = \sqrt{\frac{\Sigma y^2 - b\Sigma y - m\Sigma xy}{n - 2}}$$

This formula is easy to use if you have already calculated the slope m, the y-intercept b, and several of the sums. For instance, the regression line for the data set given at the left is $\hat{y} = 1.84247x + 51.77413$ and the values of the sums are $\Sigma y^2 = 80{,}877.5$, $\Sigma y = 1137$, and $\Sigma xy = 11{,}940.25$. Using the alternate formula, the standard error of estimate is

$$s_e = \sqrt{\frac{\Sigma y^2 - b\Sigma y - m\Sigma xy}{n - 2}}$$

$$= \sqrt{\frac{80{,}877.5 - 51.77413(1137) - 1.84247(11{,}940.25)}{16 - 2}} \approx 0.877.$$

EXAMPLE 

Finding the Standard Error of Estimate

The regression equation for the advertising expenditures and company sales data as calculated in Example 1 of Section 9.2 is

$$\hat{y} = 50.729x + 104.061.$$

Find the standard error of estimate.

SOLUTION Use a table to calculate the sum of the squared differences of each observed y-value and the corresponding predicted y-value.

x_i	y_i	$\hat{y}_i$	$(y_i - \hat{y}_i)^2$
2.4	225	225.81	0.6561
1.6	184	185.23	1.5129
2.0	220	205.52	209.6704
2.6	240	235.96	16.3216
1.4	180	175.08	24.2064
1.6	184	185.23	1.5129
2.0	186	205.52	381.0304
2.2	215	215.66	0.4356
			$\Sigma = 635.3463$

Using $n = 8$ and $\Sigma(y_i - \hat{y}_i)^2 = 635.3463$, the standard error of estimate is

$$s_e = \sqrt{\frac{\Sigma(y_i - \hat{y}_i)^2}{n - 2}}$$

$$= \sqrt{\frac{635.3463}{8 - 2}}$$

$$\approx 10.290.$$

So, the standard error of estimate is about 10.290. That means the standard deviation of the company sales for a specific advertising expenditure is about \$10,290.

Try It Yourself 2

A researcher collects the data shown at the left and concludes that there is a significant relationship between the amount of radio advertising time (in minutes per week) and the weekly sales of a product (in hundreds of dollars). Find the standard error of estimate. Use the regression equation

$$\hat{y} = 1.41x + 7.31.$$

Radio ad time	Weekly sales
15	26
20	32
20	38
30	56
40	54
45	78
50	80
60	88

a. *Use a table* to calculate the sum of the squared differences of each observed y-value and the corresponding predicted y-value.
b. *Identify* the number of ordered pairs in the data set n.
c. *Calculate* s_e.
d. *Interpret* the results.

Answer: Page A41

Prediction Intervals

Two variables have a **bivariate normal distribution** if for any fixed values of x, the corresponding values of y are normally distributed and for any fixed values of y, the corresponding values of x are normally distributed. Because regression equations are determined using sample data and because x and y are assumed to have a bivariate normal distribution, you can construct a prediction interval for the true value of y. To construct the prediction interval, use a t-distribution with $n - 2$ degrees of freedom.

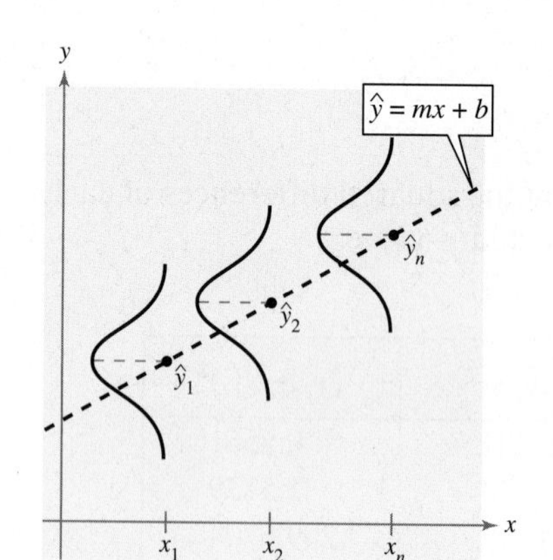

Bivariate Normal Distribution

DEFINITION

Given a linear regression equation $\hat{y} = mx + b$ and x_0, a specific value of x, a **c-prediction interval** for y is

$$\hat{y} - E < y < \hat{y} + E$$

where

$$E = t_c s_e \sqrt{1 + \frac{1}{n} + \frac{n(x_0 - \overline{x})^2}{n\Sigma x^2 - (\Sigma x)^2}}.$$

The point estimate is $\hat{y}$ and the maximum error of estimate is E. The probability that the prediction interval contains y is c.

GUIDELINES

Constructing a Prediction Interval for y for a Specific Value of x

In Words	*In Symbols*
1. Identify the number of ordered pairs in the data set n and the degrees of freedom.	d.f. $= n - 2$
2. Use the regression equation and the given x-value to find the point estimate $\hat{y}$.	$\hat{y}_i = mx_i + b$
3. Find the critical value t_c that corresponds to the given level of confidence c.	Use Table 5 in Appendix B.
4. Find the standard error of estimate s_e.	$s_e = \sqrt{\frac{\Sigma(y_i - \hat{y}_i)^2}{n - 2}}$
5. Find the maximum error of estimate E.	$E = t_c s_e \sqrt{1 + \frac{1}{n} + \frac{n(x_0 - \overline{x})^2}{n\Sigma x^2 - (\Sigma x)^2}}$
6. Find the left and right endpoints and form the prediction interval.	Left endpoint: $\hat{y} - E$ Right endpoint: $\hat{y} + E$ Interval: $\hat{y} - E < y < \hat{y} + E$

Study Tip

The formulas for s_e and E use the quantities $\Sigma(y_i - \hat{y}_i)^2$, $(\Sigma x)^2$, and Σx^2. Use a table to calculate these quantities.

EXAMPLE 

Constructing a Prediction Interval

Construct a 95% prediction interval for the company sales when the advertising expenditures are \$2100 (see Example 2). What can you conclude?

SOLUTION Because $n = 8$, there are

$$8 - 2 = 6$$

degrees of freedom. Using the regression equation

$$\hat{y} = 50.729x + 104.061$$

and

$$x = 2.1$$

the point estimate is

$$\begin{aligned}\hat{y} &= 50.729x + 104.061\\ &= 50.729(2.1) + 104.061\\ &\approx 210.592.\end{aligned}$$

From Table 5, the critical value is

$$t_c = 2.447$$

and from Example 2,

$$s_e = 10.290.$$

Using these values, the maximum error of estimate is

$$\begin{aligned}E &= t_c s_e\sqrt{1 + \frac{1}{n} + \frac{n(x_0 - \bar{x})^2}{n(\Sigma x^2) - (\Sigma x)^2}}\\ &= (2.447)(10.290)\sqrt{1 + \frac{1}{8} + \frac{8(2.1 - 1.975)^2}{8(32.44) - (15.8)^2}}\\ &\approx 26.857.\end{aligned}$$

Using $\hat{y} = 210.592$ and $E = 26.857$, the confidence interval is

Left Endpoint	Right Endpoint
$210.592 - 26.857 = 183.735$	$210.592 + 26.857 = 237.449$

$$183.735 < y < 237.449.$$

So, you can be 95% confident that when advertising expenditures are \$2100, the company sales will be between \$183,735 and \$237,449.

Insight

The greater the difference between x and $\bar{x}$, the wider the prediction interval is. For instance, in Example 3 the 95% prediction intervals for $1.4 < x < 2.6$ are shown below.

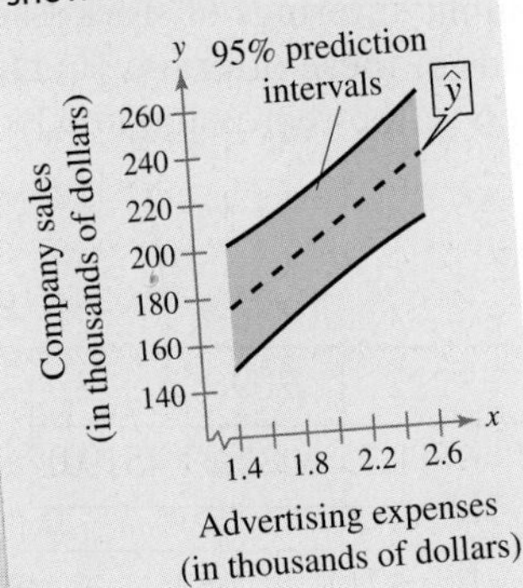

Try It Yourself 3

Construct a 95% prediction interval for the company sales when the advertising expenses are \$2500. What can you conclude?

a. *Specify* n, d.f., t_c, s_e.
b. *Calculate* $\hat{y}$ when $x = 2.5$.
c. *Calculate* the maximum error of estimate E.
d. *Construct* the prediction interval.
e. What can you conclude?

Answer: Page A41

9.3 Exercises

Help

LarsonTutor 9.3

Companion Web Site

Student Solutions Manual 9.3

Videos 9.3

Technology Manuals

Basic Skills and Concepts

Graphical Analysis In Exercises 1–3, use the graph below to answer the question.

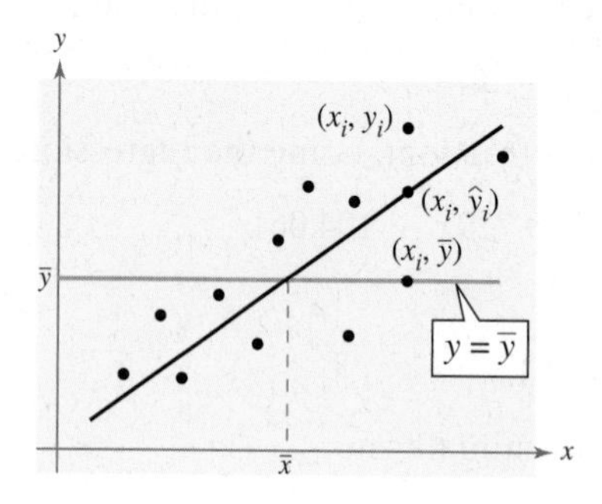

1. Use the graph to describe the total variation about a regression line in words and in symbols.

2. Use the graph to describe the explained variation about a regression line in words and in symbols.

3. Use the graph to describe the unexplained variation about a regression line in words and in symbols.

4. The coefficient of determination is the ratio of which two types of variations? What does the coefficient of determination measure?

In Exercises 5–8, use the value of the linear correlation coefficient to calculate the coefficient of determination. What does this tell you about the explained variation of the data about the regression line? The unexplained variation?

5. $r = 0.250$

6. $r = -0.375$

7. $r = -0.891$

8. $r = 0.964$

Finding Types of Variations and the Coefficient of Determination
In Exercises 9–16, find the (a) coefficient of determination and interpret the results, and (b) standard error of estimate s_e and interpret the results.

9. ***Stock Offerings*** ◆ The number of initial public offerings of stock issued in a recent 12-year period and the total proceeds of these offerings (in millions of U.S. dollars) are listed below. The equation of the regression line is

$$\hat{y} = 55.884x - 7189.033.$$

(Source: Securities Data Co.)

No. of issues, x	332	694	518	222	209	172
Proceeds, y	6284.8	17,738.8	16,745.7	6111.7	6082.0	4519.0

No. of issues, x	366	512	667	571	575	865
Proceeds, y	16,283.2	23,379.8	34,461.1	22,771.9	29,270.8	48,789.8

10. *Cigarettes* ◆ The following table represents the number of cigarettes consumed (in billions) in the United States and the number of cigarettes exported (in billions) from the United States for seven years. The equation of the regression line is

$$\hat{y} = -1.535x + 968.999.$$

(Source: U.S. Department of Agriculture)

Cigarettes consumed in U.S., x	525	510	500	485	486	487	487
Cigarettes exported by U.S., y	164	179	206	196	220	231	244

11. *Retail Space and Sales* ◆ The following table represents the total square footage (in billions) of retailing space at shopping centers and their sales (in billions of U.S. dollars) for 11 years. The equation of the regression line is

$$\hat{y} = 230.8x - 289.8.$$

(Source: International Council of Shopping Centers)

Total square footage, x	1.6	2.3	3.0	3.4	3.9	4.6
Sales, y	123.2	211.5	385.5	475.1	641.1	716.9

Total square footage, x	4.7	4.8	4.9	5.0	5.1
Sales, y	768.2	806.6	851.3	893.8	933.9

12. *Work and Leisure Time* ◆ The median number of work hours per week and the median number of leisure hours per week for people in the U.S. for 10 recent years are shown below. The equation of the regression line is

$$\hat{y} = -0.646x + 50.734.$$

(Source: Louis Harris & Associates)

Median no. of work hrs. per week, x	40.6	43.1	46.9	47.3	46.8
Median no. of leisure hrs. per week, y	26.2	24.3	19.2	18.1	16.6

Median no. of work hrs. per week, x	48.7	50.0	50.7	50.6	50.8
Median no. of leisure hrs. per week, y	18.8	18.8	19.5	19.2	19.5

13. *Earnings of Men and Women* ◆ The following table represents median weekly earnings (in U.S. dollars) of full-time male and female workers for five years. The equation of the regression line is

$$\hat{y} = 0.898x - 82.291.$$

(Source: U.S. Bureau of Labor Statistics)

Median weekly earnings of male workers, x	312	419	485	538	557
Median weekly earnings of female workers, y	201	290	348	406	418

14. ***Voter Turnout*** ◆ The U.S. voting age population (in millions) and the turnout of the voting age population (in millions) for federal elections for eight years are listed in the table. The data can be modeled by the regression equation $\hat{y} = 0.373x + 26.377$. *(Source: Federal Election Commission)*

Voting age population, x	120.3	140.8	152.3	164.6
Turnout in federal elections, y	73.2	77.7	81.6	86.5

Voting age population, x	174.5	182.8	189.5	196.5
Turnout in federal elections, y	92.7	91.6	104.4	96.5

15. ***Campaign Money*** ◆ The money raised and spent (both in millions of U.S. dollars) by all congressional campaigns for eight recent years are shown in the table. The data can be modeled by the regression equation $\hat{y} = 1.020x - 25.854$. *(Source: Federal Election Commission)*

Money raised, x	354.7	397.2	472.0	477.6
Money spent, y	342.4	374.1	450.9	459.0

Money raised, x	471.7	659.3	740.5	790.5
Money spent, y	446.3	680.2	725.2	765.3

16. ***Fund Assets*** ◆ The following table represents the total assets (in billions of U.S. dollars) of equity funds and bond and income funds for nine years. The equation of the regression line is $\hat{y} = 0.689x + 68.861$. *(Source: Investment Company Institute)*

Equity funds, x	35.9	41.2	77.0	116.9	180.7
Bond and income funds, y	13.1	14	36.6	134.8	273.1

Equity funds, x	249.0	411.6	749.0	1269.0
Bond and income funds, y	304.8	441.4	761.1	798.3

Constructing and Interpreting Prediction Intervals In Exercises 17–24, construct the indicated prediction interval and interpret the results.

17. ***Proceeds*** ◆ Construct a 95% prediction interval for the proceeds from initial public offerings in Exercise 9 when the number of issues is 712.

18. ***Cigarettes Consumed*** ◆ Construct a 95% prediction interval for the cigarettes exported by the United States in Exercise 10 when the number of cigarettes consumed in the United States is 490 billion.

19. ***Sales*** ◆ Using the results of Exercise 11, construct a 90% prediction interval for shopping center sales when the total square footage of shopping centers is 4.5 billion.

20. ***Leisure Hours*** ◆ Using the results of Exercise 12, construct a 90% prediction interval for the median number of leisure hours per week when the median number of work hours per week is 45.1.

21. ***Earnings of Female Workers*** ◆ When the median weekly earnings of male workers is $500, find a 99% prediction interval for the median weekly earnings of female workers. Use the results of Exercise 13.

22. ***Predicting Voter Turnout*** ◆ When the voting age population is 190 million, construct a 99% prediction interval for the voter turnout in federal elections. Use the results of Exercise 14.

23. ***Campaign Spending*** ◆ A total of $775.8 million is raised in one year for congressional campaigns. Construct a prediction interval for the money spent by the campaigns. Use the results of Exercise 15 and $c = 0.95$.

24. ***Total Assets*** ◆ The total assets in equity funds is $900 billion. Construct a prediction interval for the total assets in bond and income funds. Use the results of Exercise 16 and $c = 0.90$.

Extending the Basics

Old Vehicles In Exercises 25–31, use the information given at the right.

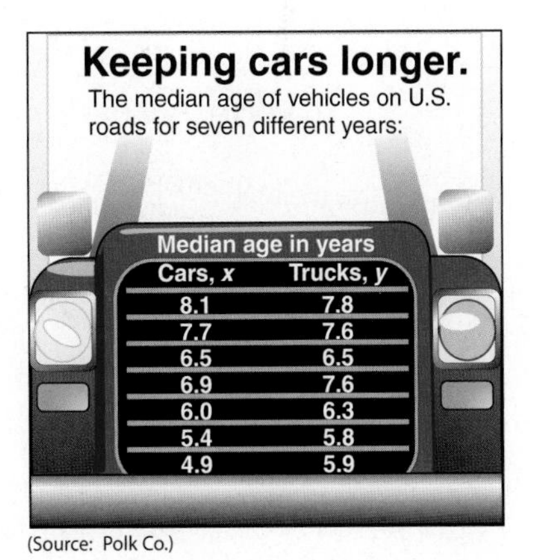

Cars, x	Trucks, y
8.1	7.8
7.7	7.6
6.5	6.5
6.9	7.6
6.0	6.3
5.4	5.8
4.9	5.9

(Source: Polk Co.)

25. ***Scatter Plot*** ◆ Construct a scatter plot of the data. Show $\overline{y}$ and $\overline{x}$ on the graph.

26. ***Regression Line*** ◆ Find and graph the regression line.

27. ***Deviation*** ◆ Calculate the explained deviation, the unexplained deviation, and the total deviation for each data point.

28. ***Variation*** ◆ Find the (a) explained variation, (b) unexplained variation, and (c) total variation.

29. ***Coefficient of Determination*** ◆ Find the coefficient of determination. What can you conclude?

30. ***Error of Estimate*** ◆ Find the standard error of estimate s_e and interpret the results.

31. ***Prediction Interval*** ◆ Construct a 95% prediction interval for the median age of trucks in use when the median age of cars in use is 7.3.

32. ***Correlation Coefficient and Slope*** ◆ Recall that the formula for the correlation coefficient r is

$$r = \frac{n\sum xy - (\sum x)(\sum y)}{\sqrt{n\sum x^2 - (\sum x)^2}\sqrt{n\sum y^2 - (\sum y)^2}}$$

and the formula for the slope m of a regression line is

$$m = \frac{n\sum xy - (\sum x)(\sum y)}{n\sum x^2 - (\sum x)^2}.$$

Given a set of data, why must the slope m of the data's regression line always have the same sign as the data's correlation coefficient r?

9.4 Multiple Regression

What You Should Learn

- How to use technology to find a multiple regression equation, the standard error of estimate and the coefficient of determination
- How to use a multiple regression equation to predict y-values

Finding a Multiple Regression Equation • Predicting y-Values

Finding a Multiple Regression Equation

In many instances, a better prediction model can be found for a dependent (response) variable by using more than one independent (explanatory) variable. For example, a more accurate prediction for the company sales discussed in previous sections might be made by considering the number of employees on the sales staff as well as the advertising expenditures. Models that contain more than one independent variable are multiple regression models.

DEFINITION

A multiple regression equation has the form

$$\hat{y} = b + m_1x_1 + m_2x_2 + m_3x_3 + \cdots + m_kx_k$$

where $x_1, x_2, x_3, \ldots, x_k$ are the independent variables, b is the y-intercept, and y is the dependent variable.

Insight

Because the mathematics associated with multiple regression is complicated, this section focuses on how to use technology to find a multiple regression equation and how to interpret the results.

The y-intercept b is the value of y when all x_i are 0. Each coefficient m_i is the amount of change in y when the independent variable x_i is changed by one unit and all other independent variables are held constant.

EXAMPLE 

Finding a Multiple Regression Equation

A researcher wants to determine how employee salaries at a certain company are related to the length of employment, previous experience, and education. The researcher selects eight employees from the company and obtains the following data.

Employee	Salary, y	Employment (in years), x_1	Experience (in years), x_2	Education (in years), x_3
A	37,310	10	2	16
B	37,380	5	6	16
C	34,135	3	1	12
D	36,985	6	5	14
E	38,715	8	8	16
F	40,620	20	0	12
G	39,200	8	4	18
H	40,320	14	6	17

Use *Minitab* to find a multiple regression equation that models the data.

SOLUTION Enter the y-values in C1 and the x_1-, x_2-, and x_3-values in C2, C3, and C4, respectively. Select "Regression▶Regression . . . " from the Stat menu. Using the salaries as the response variable and the remaining data as the predictors, you should obtain results similar to the following.

MINITAB

Regression Analysis: Salary, y versus x1, x2, x3

```
The regression equation is
Salary, y = 29764 + 364 x1 + 228 x2 + 267 x3

Predictor      Coef         SE Coef        T        P
Constant       29764  (b)     1981      15.02    0.000
x1            364.41  (m1)   48.32       7.54    0.002
x2             227.6  (m2)   123.8       1.84    0.140
x3             266.9  (m3)   147.4       1.81    0.144

S = 659.5      R-Sq = 94.4%      R-Sq(adj) = 90.2%
```

The regression equation is $\hat{y} = 29{,}764 + 364x_1 + 228x_2 + 267x_3$.

Study Tip

In Example 1, it is important that you interpret the coefficients m_1, m_2, and m_3 correctly. For instance, if x_2 and x_3 are held constant and x_1 increases by 1, then y increases by \$364. Similarly, if x_1 and x_3 are held constant and x_2 increases by 1, then y increases by \$228. If x_1 and x_2 are held constant and x_3 increases by 1, then y increases by \$267.

Try It Yourself 1

A statistics professor wants to determine how students' final grades are related to the midterm exam grades and number of classes missed. The professor selects 10 students from her class and obtains the following data.

Student	Final grade, y	Midterm exam, x_1	Classes missed, x_2
1	81	75	1
2	90	80	0
3	86	91	2
4	76	80	3
5	51	62	6
6	75	90	4
7	44	60	7
8	81	82	2
9	94	88	0
10	93	96	1

Use technology to find a multiple regression equation that models the data.

a. *Enter* the data.

b. *Calculate* the regression line.

Answer: Page A41

Minitab displays much more than the regression equation and the coefficients of the independent variables. For example, it also displays the standard error of estimate, denoted by S, and the coefficient of determination, denoted by R-Sq. In Example 1, S = 659.5 and R-Sq = 94.4%. So, the standard error of estimate is \$659.50. The coefficient of determination tells you that 94.4% of the variation in y can be explained by the multiple regression model. The remaining 5.6% is unexplained and is due to other factors or chance.

Picturing the World

In a lake in Finland, 159 fish of seven species were caught and measured for weight G, length L, height H, and width W (H and W are percents of L). The regression equation for G and L is

$$G = -488 + 28.5\,L,$$

$$r = 0.925.$$

Using all four variables, the regression equation is

$$G = -711 + 28.2L + 1.46H + 13.4W,$$

$$r = 0.931.$$

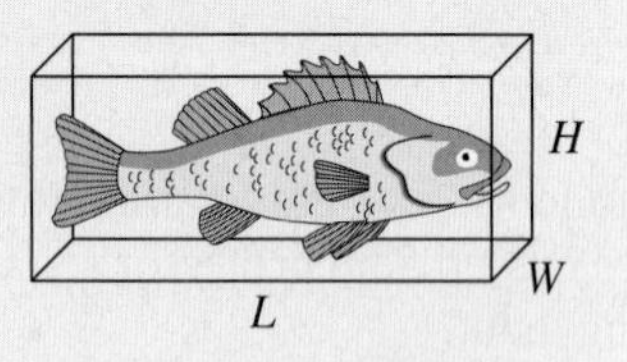

Predict the weight of a fish with the following measurements: $L = 40$, $H = 17$, and $W = 11$. How do your predictions vary when you use a single variable versus many variables? Which do you think is more accurate?

Predicting y-Values

After finding the equation of the multiple regression line, you can use the equation to predict y-values over the range of the data. To predict y-values, substitute the given value for each independent variable into the equation, then calculate $\hat{y}$.

EXAMPLE 

Predicting y-Values Using Multiple Regression Equations

Use the regression equation found in Example 1 to predict an employee's salary given the following conditions.

1. 12 years of current employment, 5 years of experience, and 16 years of education
2. 4 years of current employment, 2 years of experience, and 12 years of education
3. 8 years of current employment, 7 years of experience, and 17 years of education

SOLUTION To predict each employee's salary, substitute the values for x_1, x_2, and x_3 into the regression equation. Then calculate $\hat{y}$.

1. $\hat{y} = 29{,}764 + 364x_1 + 228x_2 + 267x_3$
   $= 29{,}764 + 364(12) + 228(5) + 267(16)$
   $= 39{,}544$

   The employee's predicted salary is \$39,544.

2. $\hat{y} = 29{,}764 + 364x_1 + 228x_2 + 267x_3$
   $= 29{,}764 + 364(4) + 228(2) + 267(12)$
   $= 34{,}880$

   The employee's predicted salary is \$34,880.

3. $\hat{y} = 29{,}764 + 364x_1 + 228x_2 + 267x_3$
   $= 29{,}764 + 364(8) + 228(7) + 267(17)$
   $= 38{,}811$

   The employee's predicted salary is \$38,811.

Try It Yourself 2

Use the regression equation found in Try It Yourself 1 to predict a student's final grade given the following conditions.

1. A student has a midterm exam score of 89 and misses 1 class.
2. A student has a midterm exam score of 78 and misses 3 classes.
3. A student has a midterm exam score of 83 and misses 2 classes.

a. *Substitute* the midterm score for x_1 into the regression equation.
b. *Substitute* the corresponding number of missed classes for x_2 into the regression equation.
c. *Calculate* $\hat{y}$.
d. What is each student's final grade?

Answer: Page A41

Exercises

Help

LarsonTutor 9.4

Companion Web Site

Student Solutions Manual 9.4

Videos 9.4

Technology Manuals

Basic Skills and Concepts

Predicting y-Values In Exercises 1–4, use the multiple regression equation to predict the y-values for the given values of the independent variables.

1. ***Peanut Yield*** ◆ The equation used to predict peanut yield (in pounds) is

$$\hat{y} = 6503 - 14.8x_1 + 12.2x_2$$

where x_1 is the number of acres planted (in thousands) and x_2 is the number of acres harvested (in thousands). *(Source: U.S. National Agricultural Statistics Service)*

(a) $x_1 = 1458,\ x_2 = 1450$ (b) $x_1 = 1500,\ x_2 = 1475$
(c) $x_1 = 1400,\ x_2 = 1385$ (d) $x_1 = 1525,\ x_2 = 1500$

2. ***Rice Yield*** ◆ To predict the annual rice yield (in pounds), use the equation

$$\hat{y} = 859 + 5.76x_1 + 3.82x_2$$

where x_1 is the number of acres planted (in thousands) and x_2 is the number of acres harvested (in thousands). *(Source: U.S. National Agricultural Statistics Service)*

(a) $x_1 = 2532,\ x_2 = 2255$ (b) $x_1 = 3581,\ x_2 = 3021$
(c) $x_1 = 3213,\ x_2 = 3065$ (d) $x_1 = 2758,\ x_2 = 2714$

3. ***Black Cherry Tree Volume*** ◆ The volume (in cubic feet) of black cherry trees can be modeled by the equation

$$\hat{y} = -52.2 + 0.3x_1 + 4.5x_2$$

where x_1 is the tree's height (in feet) and x_2 is the tree's diameter (in inches). *(Source: Journal of the Royal Statistical Society)*

(a) $x_1 = 70,\ x_2 = 8.6$ (b) $x_1 = 65,\ x_2 = 11.0$
(c) $x_1 = 83,\ x_2 = 17.6$ (d) $x_1 = 87,\ x_2 = 19.6$

4. ***Earnings Per Share*** ◆ The earnings per share (in dollars) for McDonald's Corporation are given by the equation

$$\hat{y} = -0.396 + 0.186x_1 + 0.071x_2$$

where x_1 represents total revenue (in billions of dollars) and x_2 represents total net worth (in billions of dollars). *(Source: McDonald's Corporation)*

(a) $x_1 = 11.4,\ x_2 = 8.6$ (b) $x_1 = 8.1,\ x_2 = 6.2$
(c) $x_1 = 10.7,\ x_2 = 8.5$ (d) $x_1 = 7.3,\ x_2 = 5.1$

Finding a Multiple Regression Equation In Exercises 5 and 6, use technology to find the multiple regression equation for the data given in the table. Then answer the following.

(a) What is the standard error of estimate?

(b) What is the coefficient of determination?

(c) Interpret the results of (a) and (b).

DATA **5.** ***Sales*** ◆ The total square footage (in billions) of retailing space at shopping centers, the number (in thousands) of shopping centers, and the sales (in billions of U.S. dollars) for shopping centers for a recent 11-year period are listed in the table. *(Source: International Council of Shopping Centers)*

Sales, y	Total square footage, x_1	Number of shopping centers, x_2
123.2	1.6	13.2
211.5	2.3	17.5
385.5	3.0	22.1
475.1	3.4	25.5
641.1	3.9	32.6
716.9	4.6	38.0
768.2	4.7	39.0
806.6	4.8	39.6
851.3	4.9	40.4
893.8	5.0	41.2
933.9	5.1	42.1

DATA **6.** ***Shareholder's Equity*** ◆ The following table represents the net sales (in billions of dollars), total assets (in billions of dollars), and shareholder's equity (in billions of dollars) for Wal-Mart for a recent six-year period. *(Source: Wal-Mart Annual Reports)*

Shareholder's equity, y	Net sales, x_1	Total assets, x_2
5.4	32.6	11.4
7.0	43.9	15.4
8.8	55.5	20.6
10.8	67.3	26.4
12.7	82.5	32.8
14.8	93.6	37.5

Extending the Basics

Adjusted r^2 The calculation of r^2, the coefficient of determination, depends on the number of data pairs and the number of independent variables. An adjusted value of r^2 can be calculated, based on the number of degrees of freedom, as follows.

$$r^2_{adj} = 1 - \left[\frac{(1 - r^2)(n - 1)}{n - k - 1}\right]$$

where n is the number of data pairs and k is the number of independent variables.

In Exercises 7 and 8, after calculating r^2_{adj}, determine the percentage of the variation in y that can be explained by the relationships between variables according to r^2_{adj}. Compare this result to the one obtained with r^2.

7. Calculate r^2_{adj} for the data in Exercise 5.

8. Calculate r^2_{adj} for the data in Exercise 6.

TECHNOLOGY

MINITAB | EXCEL | TI-83

Federal Trade Commission

Tar, Nicotine, and Carbon Monoxide

Each year, the Federal Trade Commission tests the tar, nicotine, and carbon monoxide content of the various brands of cigarettes manufactured in the United States. The testing is conducted by the Tobacco Institute Testing Laboratory (TITL) using methods approved by the FTC. The complete report of the more than 1200 brands tested each year is available free from the FTC.

In the table at the right, we chose 22 different brands and styles of cigarettes. We included some with high levels of tar, nicotine, and carbon monoxide and some with low levels. The numbers in the table are as follows.

T = tar in mg
N = nicotine in mg
W = weight in g
C = carbon monoxide in mg

www.ftc.gov

Brand	T	N	W	C
Alpine	16	1.0	0.99	15
Benson & Hedges	15	1.1	1.09	15
Camel	9	0.7	0.93	11
Carlton	5	0.5	0.95	3
Chesterfield	23	1.3	0.89	15
Kent	12	0.9	0.92	12
Kool	16	1.2	0.94	15
L&M	13	0.9	0.89	13
Lark	11	0.9	0.96	11
Marlboro	15	1.0	0.93	14
Merit	9	0.7	0.97	10
Newport Lights	9	0.8	0.85	10
Now	1	0.1	0.79	2
Old Gold	16	1.2	0.92	17
Pall Mall	10	0.9	1.04	12
Raleigh	15	1.0	0.96	15
Salem Ultra	4	0.4	0.91	6
Tareyton	14	1.0	1.01	14
True	5	0.5	0.98	6
Viceroy	10	0.8	0.97	10
Virginia Slims	15	1.1	0.95	13
Winston	11	0.8	1.12	13

Exercises

1. Use a technology tool to draw a scatter plot of the following (x, y) pairs in the data set.
   (a) (tar, nicotine)
   (b) (tar, weight)
   (c) (tar, carbon monoxide)
   (d) (nicotine, weight)
   (e) (nicotine, carbon monoxide)
   (f) (weight, carbon monoxide)

2. From the scatter plots in Exercise 1, which pairs of variables appear to have a strong correlation?

3. Use a technology tool to find the correlation coefficient for each pair of variables in Exercise 1. Which has the strongest correlation?

4. Use a technology tool to find the regression line for the following variables.
   (a) (tar, nicotine)
   (b) (tar, carbon monoxide)

5. Use the results of Exercise 4 to predict the following.
   (a) The nicotine content of a cigarette that has a tar content of 13 mg
   (b) The carbon monoxide content of a cigarette that has a tar content of 13 mg

6. Use a technology tool to find the multiple regression equations of the following forms.
   (a) $T = b + m_1N + m_2W + m_3C$
   (b) $T = b + m_1N + m_2C$

7. Use the results of Exercise 6 to predict the tar content of a cigarette that has 1.0 mg of nicotine and 10 mg of carbon monoxide.

Extended solutions are given in the *Technology Supplement*.
Technical instruction is provided for *Minitab*, *Excel*, and the *TI-83*.

9

Chapter Summary

What did you learn? — Review Exercises

Section 9.1

- How to find a correlation coefficient — 1, 2

$$r = \frac{n\Sigma xy - (\Sigma x)(\Sigma y)}{\sqrt{n\Sigma x^2 - (\Sigma x)^2}\sqrt{n\Sigma y^2 - (\Sigma y)^2}}$$

- How to perform a hypothesis test for a population correlation coefficient ρ — 3–6

$$t = \frac{r}{\sqrt{\frac{1 - r^2}{n - 2}}}$$

Section 9.2

- How to find the equation of a regression line, $\hat{y} = mx + b$ — 7, 8

$$m = \frac{n\Sigma xy - (\Sigma x)(\Sigma y)}{n\Sigma x^2 - (\Sigma x)^2}$$

$$b = \bar{y} - m\bar{x}$$

$$= \frac{\Sigma y}{n} - m\frac{\Sigma x}{n}$$

- How to predict y-values using a regression equation — 9, 10

Section 9.3

- How to find and interpret the coefficient of determination r^2 — 11–16
- How to find and interpret the standard error of estimate for a regression line — 15, 16

$$s_e = \sqrt{\frac{\Sigma(y_i - \hat{y}_i)^2}{n - 2}} = \sqrt{\frac{\Sigma y^2 - b\Sigma y - m\Sigma xy}{n - 2}}$$

- How to construct and interpret a prediction interval for y, $\hat{y} - E < y < \hat{y} + E$ — 17–20

$$E = t_c s_e \sqrt{1 + \frac{1}{n} + \frac{n(x_0 - \bar{x})^2}{n\Sigma x^2 - (\Sigma x)^2}}$$

Section 9.4

- How to use technology to find a multiple regression equation, the standard error of estimate, and the coefficient of determination — 21, 22
- How to use a multiple regression equation to predict y-values — 23, 24

$$\hat{y} = b + m_1x_1 + m_2x_2 + m_3x_3 + \cdots + m_kx_k$$

STATISTICS

Uses and Abuses

Uses

Correlation and Regression Correlation and regression analysis can be used to determine whether there is a significant relationship between two variables. If there is, you can use one of the variables to predict the value of the other variable. For example, educators have used correlation and regression analysis to determine that there is a significant correlation between a student's SAT score and the grade point average from a student's freshman year at college. Consequently, many colleges and universities use SAT scores of high school applicants as a predictor of the applicant's initial success at college.

Abuses

Confusing Correlation and Causation The most common abuse of correlation in studies is to confuse the concepts of correlation with those of causation. Good SAT scores do not cause good college grades. Rather, there are other variables, such as good study habits and motivation, that contribute to both. When a strong correlation is found between two variables, look for other variables that correlate with both.

Considering Only Linear Correlation The correlation studied in this chapter is linear correlation. When the correlation coefficient is close to 1 or close to -1, the data points can be modeled by a straight line. It is possible that a correlation coefficient is close to zero but there is still a strong correlation of a different type. Consider the data listed in the following table. The value of the correlation coefficient is 0; however, the data are perfectly correlated with the equation $x^2 + y^2 = 1$.

x	y
1	0
0	1
-1	0
0	-1

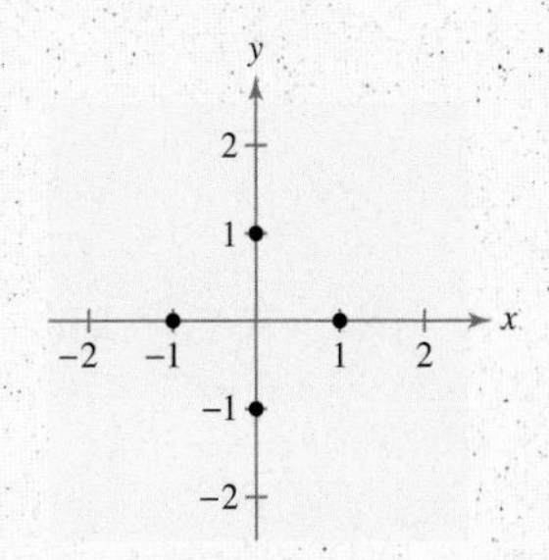

Exercises

1. ***Confusing Correlation and Causation*** Find an example of an article in a newspaper or magazine that confuses correlation and causation. Discuss other variables that could contribute to the relationship between the variables.

2. ***Considering Only Linear Correlation*** Find an example of two real-life variables that have a nonlinear correlation.

9 Review Exercises

Section 9.1

In Exercises 1 and 2, organize the data in a scatter plot. Then find the sample correlation coefficient *r*. Determine whether there is a positive linear correlation, negative linear correlation, or no linear correlation between the variables. What can you conclude?

1. The ages of eight cows (in years) and their milk production (in gallons) per week

Age, *x*	4	4	6	7	7	8	10	11
Milk production, *y*	37.0	35.4	33.3	33.1	32.3	33.7	30.2	29.6

2. The annual per capita sugar consumption (in kilograms) and the average number of cavities of 11- and 12-year-old children in seven countries

Sugar, *x*	2.1	5.0	6.3	6.5	7.7	8.7	11.6
Cavities, *y*	0.59	1.51	1.55	1.70	2.18	2.10	2.43

In Exercises 3 and 4, use the given sample statistics to test the claim about the population correlation coefficient ρ at the indicated level of significance α for the given sample statistics.

3. Claim: $\rho = 0, \alpha = 0.01$. Sample statistics: $r = 0.24, n = 26$

4. Claim: $\rho \neq 0, \alpha = 0.05$. Sample statistics: $r = -0.55, n = 22$

In Exercises 5 and 6, test the claim about the population correlation coefficient ρ at the indicated level of significance α. Then interpret the decision in the context of the original claim.

5. Refer to the data in Exercise 1. At $\alpha = 0.05$, test the claim that there is a linear correlation between a cow's age and milk production.

6. Refer to the data in Exercise 2. Is there enough evidence to conclude that there is a linear correlation between sugar consumption and tooth decay? Use $\alpha = 0.01$.

Section 9.2

In Exercises 7 and 8, use the data to find the equation of the regression line. Then construct a scatter plot of the data and draw the regression line. Can you make a guess about the sign and magnitude of *r*? Calculate *r* and check your guess.

7. The heights in inches of adult brothers and sisters from 11 families

Sister, *x*	65	62	63	68	63	67	64	60	67	65	68
Brother, *y*	70	69	66	74	72	74	70	67	68	72	74

8. The engine displacement (in cubic inches) and the fuel economy (in miles per gallon) of seven automobiles

Displacement, x	170	134	220	305	109	256	322
Fuel efficiency, y	29.5	34.5	23.0	17.0	33.5	23.0	15.5

In Exercises 9 and 10, use the regression equations found in Exercises 7 and 8 to predict the value of y for each value of x, if meaningful. If not, explain why not. (Correlation between x and y in Exercises 7 and 8 is significant for $\alpha = 0.05$.)

9. Refer to Exercise 7. What height would you predict for a male whose sister is (a) 61 in.? (b) 66 in.? (c) 71 in.? (d) 50 in.?

10. Refer to Exercise 8. What fuel efficiency rating would you predict for a car with an engine displacement of (a) 86 in.^{3}? (b) 198 in.^{3}? (c) 289 in.^{3}? (d) 407 in.^{3}?

Section 9.3

In Exercises 11–14, use the value of the linear correlation coefficient r to find the coefficient of determination. Interpret the result.

11. $r = -0.553$

12. $r = -0.962$

13. $r = 0.181$

14. $r = 0.740$

In Exercises 15 and 16, use the data to (a) find the coefficient of determination r^2 and interpret the result with regard to the regression line, and (b) find the standard error of estimate s_e and interpret the result.

15. The following table shows the area of eight living spaces (in square feet) and the cooling capacity (in Btu per hour) of the air conditioners used in those spaces. The regression equation is $\hat{y} = 3003.0 + 9.468x$. *(Adapted from Consumer Reports)*

Living area, x	730	485	205	420	550	590	385	630
Cooling capacity, y	10,200	7000	5300	6800	7250	9000	6900	9400

16. The following table shows the prices of 16 gas grills (in U.S. dollars) and their usable cooking area (in square inches). The regression equation is $\hat{y} = -209.5 + 1.3052x$. *(Source: Consumer Reports)*

Area, x	430	338	426	446	465	372	305	403
Price, y	480	360	570	450	350	250	175	200

Area, x	389	424	306	424	309	386	328	261
Price, y	270	270	200	200	180	200	190	150

In Exercises 17–20, construct the indicated prediction intervals.

17. Construct a 90% prediction interval for the height of a brother in Exercise 7 whose sister is 64 inches tall.

18. Construct a 90% prediction interval for the fuel efficiency of an automobile in Exercise 8 that has an engine displacement of 265 cubic inches.

19. Construct a 95% prediction interval for the cooling capacity of an air conditioner in Exercise 15 that is used in a living area of 720 square feet.

20. Construct a 95% prediction interval for the price of a gas grill in Exercise 16 with a usable cooking area of 400 square inches.

Section 9.4

In Exercises 21 and 22, refer to the following information. The table shows the tar, nicotine, weight, and carbon monoxide content, all in milligrams, of 13 brands of U.S. cigarettes. *(Source: Federal Trade Commission)*

Carbon monoxide, y	Tar, x_1	Nicotine, x_2	Weight, x_3
13.6	14.1	0.86	985.3
16.6	16.0	1.06	1093.8
10.2	8.0	0.67	928.0
5.4	4.1	0.40	946.2
15.0	15.0	1.04	888.5
9.0	8.8	0.76	1026.7
12.3	12.4	0.95	922.5
16.3	16.6	1.12	937.2
15.4	14.9	1.02	885.8
13.0	13.7	1.01	964.3
14.4	15.1	0.90	931.6
10.0	7.8	0.57	970.5
10.2	11.4	0.78	1124.0

21. Use technology to find the multiple regression equation from the table data.

22. Find the standard error of estimate s_e and the coefficient of determination r^2. What percentage of the variation of y can be explained by the regression equation?

In Exercises 23 and 24, use the multiple regression equation to predict the value of y for the given values of the independent variables.

23. An equation that can be used to predict fuel economy (in miles per gallon) for automobiles is $\hat{y} = 41.3 - 0.004x_1 - 0.0049x_2$, where x_1 is the engine displacement (in cubic inches) and x_2 is the vehicle weight (in pounds).

(a) $x_1 = 305, x_2 = 3750$ (b) $x_1 = 225, x_2 = 3100$

(c) $x_1 = 105, x_2 = 2200$ (d) $x_1 = 185, x_2 = 3000$

24. Use the regression equation found in Exercise 21.

(a) $x_1 = 16.5, x_2 = 0.69, x_3 = 946.4$ (b) $x_1 = 7.2, x_2 = 0.55, x_3 = 950.5$

(c) $x_1 = 16.1, x_2 = 0.88, x_3 = 961.2$ (d) $x_1 = 9.7, x_2 = 0.99, x_3 = 1118.6$

PUTTING IT ALL TOGETHER

Real Statistics Real Decisions

www.troyerfarms.com

You work for snack company Troyer Farms. The company's main product is potato chips. The company's profitability depends, in part, on producing the chips at the lowest possible cost. This means getting the most chips possible from each potato.

You are analyzing the potatoes and trying to find a way to predict the pounds of chips produced using the characteristics of the potatoes. You have discovered that two potatoes of the same weight yield a different amount (weight) of chips because more of the weight in one potato may be due to water. You decide to measure the **specific gravity** of each potato. The specific gravity of a potato gives the mass of the potato compared with the mass of a similar volume of water. You want to determine if there is a significant correlation between the specific gravity of a potato and the weight of chips yielded by the potato.

Specific gravity, x	Pounds of potato chips, y
1.069	6226
1.098	6895
1.073	5981
1.070	5893
1.089	7124
1.096	6982
1.069	6057
1.088	7012
1.074	6099
1.073	5538
1.070	6100
1.098	7002
1.094	6991
1.066	5910
1.072	6138
1.075	6024
1.066	6003
1.096	6884
1.097	7056
1.071	6084
1.095	7019
1.090	6895
1.099	7129
1.096	6998

Adapted from Troyer Farms

Exercises

1. *Analyzing the Data*

(a) Construct a scatter plot of the data in the table. The data give the specific gravity for 24 batches of potatoes and the pounds of chips produced. Use the graph to determine whether there is a positive linear correlation, negative linear correlation, or no linear correlation between specific gravity of the potatoes and the pounds of chips.

(b) Calculate the correlation coefficient r and verify your conclusion in part (a).

(c) Test the significance of the correlation coefficient found in part (b). Use $\sigma = 0.05$.

(d) Find the equation of the regression line for the specific gravity of the potatoes and the pounds of chips. Add the graph of the regression line to your scatter plot in part (a). Does the regression line appear to be a good fit?

(e) Can you use the equation of the regression line to predict the weight of chips produced given the specific gravity? Why or why not?

(f) Find the coefficient of determination and the standard error of estimate s_e and interpret the results.

2. *What Do You Think?*

According to your analysis, would you agree or disagree with the following? Why or why not?

Because there is a significant linear correlation between specific gravity of the potatoes and the pounds of chips, Troyer Farms should consider the specific gravity of potatoes when growing or purchasing them.

9 Chapter Quiz

Take this quiz as you would take a quiz in class. After you are done, check your work against the answers given in the back of the book.

For Exercises 1–8, refer to the data in the following table. The table lists the personal income and outlays (both in trillions of dollars) for Americans for 11 recent years. *(Source: U.S. Commerce Department, Bureau of Economic Analysis)*

Personal income, x	Personal outlays, y
4.5	3.7
4.9	4.0
5.0	4.1
5.3	4.3
5.6	4.6
5.9	4.8
6.2	5.1
6.5	5.4
7.0	5.7
7.4	6.1
7.8	6.5

1. Construct a scatter plot for the data. Do the data appear to have a positive linear correlation, a negative linear correlation, or no linear correlation? Explain.

2. Calculate the correlation coefficient r. What can you conclude?

3. Test the level of significance of the correlation coefficient r. Use $\alpha = 0.05$.

4. Find the equation of the regression line for the data. Include the regression line in the scatter plot.

5. Use the regression line to predict the personal outlays when the personal income is 5.3 trillion dollars.

6. Find the coefficient of determination and interpret the results.

7. Find the standard error of estimate s_e and interpret the results.

8. Construct a 95% prediction interval for personal outlays when personal income is 6.4 trillion dollars. Interpret the results.

9. The equation used to predict sunflower yield (in pounds) is

$$\hat{y} = 1257 - 1.34x_1 + 1.41x_2$$

where x_1 is the number of acres planted (in thousands) and x_2 is the number of acres harvested (in thousands). Use the regression equation to predict the y-values for the given values of the independent variables listed below. Then determine which variable has a greater influence on the value of y. *(Source: U.S. National Agricultural Statistics Service)*

(a) $x_1 = 2103, x_2 = 2037$

(b) $x_1 = 3387, x_2 = 3009$

(c) $x_1 = 2185, x_2 = 1980$

(d) $x_1 = 3485, x_2 = 3404$

9 Cumulative Test: Chapters 7–9

Take this test as you would take a test in class. After you are done, check your work against the answers given in the back of the book.

For Exercises 1–4, refer to the following information.

- A rice growers' association claims that the mean per capita consumption of rice by people in the United States is at least 20.1 pounds per year.
- A sample of 73 people has a mean per capita consumption of rice of 18.9 pounds per year and a standard deviation of 6.7 pounds.

(Adapted from U.S. Department of Agriculture)

1. At $\alpha = 0.01$, test the claim made by the rice growers' association. What can you conclude?

2. Describe the conditions for which a type I or type II error occurs for the hypothesis test in Exercise 1.

3. A corn growers' association studied 102 people and found that their mean consumption of corn products was 22.7 pounds and the standard deviation was 7.6 pounds. At $\alpha = 0.05$, test the claim that the mean consumption of corn products is the same as the mean consumption of rice. What can you conclude? *(Adapted from U.S. Department of Agriculture)*

4. Which distribution did you use to perform the hypothesis test in

(a) Exercise 1? Why?

(b) Exercise 3? Why?

For Exercises 5–10, use the following table. The table lists the number of acres (in thousands) of rice planted and harvested in the United States for eight years. *(Source: U.S. National Agriculture Statistics Service)*

Acres planted, x	2897	2884	3176	2920	3353	3121	2819	3056
Acres harvested, y	2823	2781	3132	2833	3316	3093	2799	3034

5. Calculate the correlation coefficient r and determine whether there is a positive linear correlation, negative linear correlation, or no linear correlation between the variables. What can you conclude?

6. Find the equation of the regression line for the given data. Graph a scatter plot of the data and the regression line.

7. Find the coefficient of determination and interpret the results.

8. Find the standard error of estimate s_e and interpret the results.

9. At $\alpha = 0.01$, test the claim that there is no linear correlation between the number of acres of rice planted and the number of acres of rice harvested.

10. Construct a 95% prediction interval for the number of acres of rice harvested when the number of acres of rice planted is 3100 thousand acres. Interpret the results.

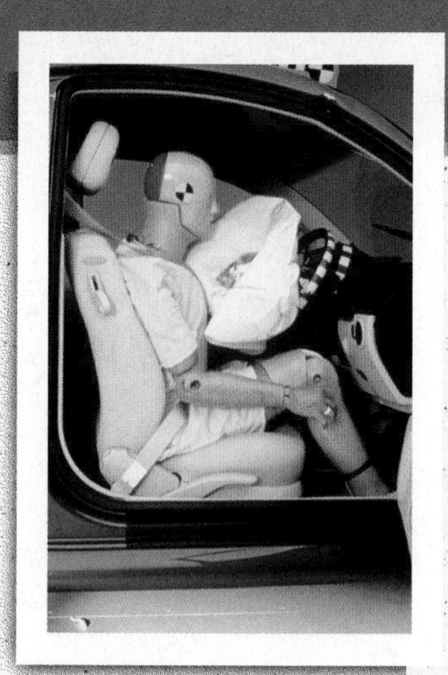

Where You've Been

As part of the New Car Assessment Program, the government buys new cars each year and crashes them into a wall at 35 miles per hour to compare how different vehicles protect front-seat passengers in a head-on collision. To measure the forces and impacts that occur during a crash test, dummies are equipped with special instruments and placed in the car. The crash test results include data on head, chest, and leg injuries. For a low crash test number, the injury potential in a 35 miles per hour frontal crash is low. If the crash test number is high, then the injury potential is high. Using the techniques of Chapter 8, you can determine if the mean chest injury potential is the same for pickups and vans. The sample statistics are as follows. *(Source: National Highway Traffic Safety Administration)*

Vehicle	Number	Mean chest injury	Standard deviation
Vans	$n_1 = 30$	$\bar{x}_1 = 54.2$	$s_1 = 12.1$
Pickups	$n_2 = 36$	$\bar{x}_2 = 50.6$	$s_2 = 8.94$

For the means of chest injury, the *P*-value for the hypothesis that $\mu_1 = \mu_2$ is about 0.177. At $\alpha = 0.05$, you fail to reject the null hypothesis. So, you do not have enough evidence to conclude that there is a significant difference in the means of the chest injury potential in a 35 miles per hour frontal crash for vans and pickups.

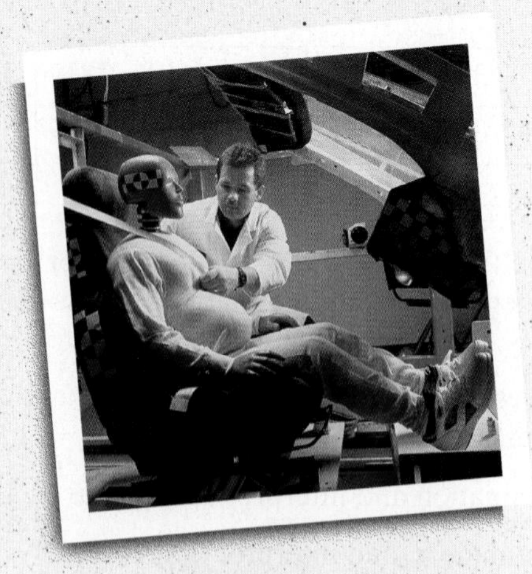

Federal law requires all passenger cars to pass a 30-mph frontal crash test. These tests are conducted through the New Car Assessment Program, in which the federal government buys brand new vehicles directly off a lot and crashes them. Results of these tests are classified using a one- to five-star rating, with one star indicating the least protection and five stars the most protection.

Chi-Square Tests and the *F*-Distribution

CHAPTER 10

Where You're Going

In Chapter 8, you learned how to test a hypothesis that compares two populations by basing your decisions on sample statistics and their distributions. In this chapter, you will learn how to test a hypothesis that compares three or more populations.

For instance, in addition to the crash tests for vans and pickups, a third group of vehicles was also tested. The results for all three types of vehicles are as follows.

Vehicle	Number	Mean chest injury	Standard deviation
Vans	$n_1 = 30$	$\bar{x}_1 = 54.2$	$s_1 = 12.1$
Pickups	$n_2 = 36$	$\bar{x}_2 = 50.6$	$s_2 = 8.94$
Light	$n_3 = 35$	$\bar{x}_3 = 46.4$	$s_3 = 6.90$

From these three samples, is there evidence of a difference in chest injury potential among vans, pickups, and light vehicles in a 35-mph frontal crash?

In this chapter, you will learn that you can answer this question by testing the hypothesis that the three means are equal. For the means of chest injury, the *P*-value for the hypothesis that $\mu_1 = \mu_2 = \mu_3$ is about 0.005. At $\alpha = 0.05$, you can reject the null hypothesis. So, you can conclude that for the three types of vehicles tested, at least one of the means of the chest injury potential in a 35 miles per hour frontal crash is different from the others.

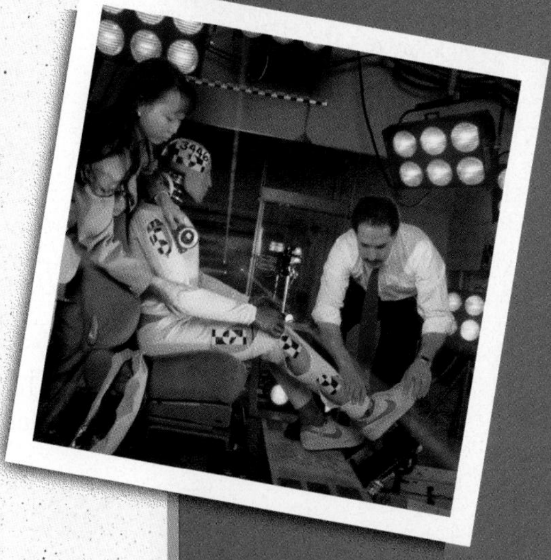

10.1 Goodness of Fit

What You Should Learn

- How to use the chi-square distribution to test whether a frequency distribution fits a claimed distribution

The Chi-Square Goodness-of-Fit Test

The Chi-Square Goodness-of-Fit Test

Suppose a marketing executive is planning a new advertising campaign and wants to determine the proportions of radio music listeners in a specific broadcast region who prefer each of six types of music. To determine these proportions, the executive can perform a multinomial experiment. A **multinomial experiment** is a probability experiment consisting of a fixed number of trials in which there are more than two possible outcomes for each independent trial. The probability for each outcome is fixed and each outcome is classified into **categories.** (Remember from Section 4.2 that a **binomial** experiment has only two possible outcomes.)

Now, suppose the marketing executive wants to determine whether a radio station's claim concerning the distribution of proportions of music preferences is correct. To do so, the executive could compare the distribution of proportions obtained in the multinomial experiment to the radio station's specified distribution. How can the executive compare the distributions? The answer is, perform a chi-square goodness-of-fit test.

DEFINITION

A **chi-square goodness-of-fit test** is used to test whether a frequency distribution fits an expected distribution.

Insight

The hypothesis tests described in Sections 10.1 and 10.2 can be used for qualitative data.

To begin a goodness-of-fit test, you must first state a null and an alternative hypothesis. Generally, the null hypothesis states that the frequency distribution fits the specified distribution and the alternative hypothesis states that the frequency distribution does not fit the specified distribution.

For example, suppose the radio station claims that the distribution of music preferences for listeners in the broadcast region is as shown below.

Distribution of music preferences	
Classical	4%
Country	36%
Gospel	11%
Oldies	2%
Pop	18%
Rock	29%

To test the radio station's claim, the executive can perform a chi-square goodness-of-fit test using the following null and alternative hypotheses.

H_0: The distribution of music preferences in the broadcast region is 4% classical, 36% country, 11% gospel, 2% oldies, 18% pop, and 29% rock. (Claim)

H_a: The distribution of music preferences differs from the claimed or expected distribution.

Picturing the World

How often do motorists see a driver run a red traffic light? According to the distribution in the following display, 26% of all motorists see at least one driver run a red traffic light daily.

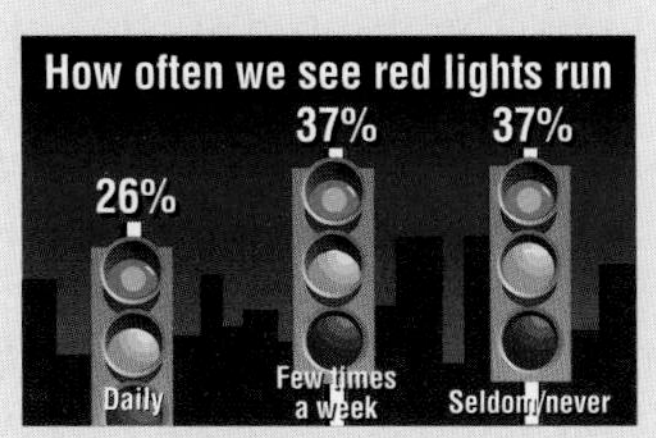

A sample of 268 motorists is asked how often they see drivers run red traffic lights: daily, a few times per week, or seldom/never. What is the expected frequency for each response?

To calculate the test statistic for the chi-square goodness-of-fit test, you can use observed frequencies and expected frequencies. To calculate the expected frequencies, you must assume the null hypothesis is true.

DEFINITION

The **observed frequency O** of a category is the frequency for the category observed in the sample data.

The **expected frequency E** of a category is the *calculated* frequency for the category. Expected frequencies are obtained assuming the specified (or hypothesized) distribution. The expected frequency for the ith category is

$$E_i = np_i$$

where n is the number of trials (the sample size) and p_i is the assumed probability of the ith category.

EXAMPLE

Finding Observed Frequencies and Expected Frequencies

A marketing executive randomly selects 500 music listeners from the broadcast region and asks each whether he or she prefers classical, country, gospel, oldies, pop, or rock music. The results are listed at the right. Find the observed frequencies and the expected frequencies for each type of music.

Survey results ($n = 500$)	
Classical	8
Country	210
Gospel	72
Oldies	10
Pop	75
Rock	125

SOLUTION The observed frequency for each type of music is the number of music listeners naming a particular type of music. The expected frequency for each type of music is the product of the number of listeners in the survey and the probability that a listener will name a particular type of music. The observed frequencies and expected frequencies are listed in the following table.

Type of music	% of listeners	Observed frequency	Expected frequency
Classical	4%	8	500(0.04) = 20
Country	36%	210	500(0.36) = 180
Gospel	11%	72	500(0.11) = 55
Oldies	2%	10	500(0.02) = 10
Pop	18%	75	500(0.18) = 90
Rock	29%	125	500(0.29) = 145

Insight

The sum of the expected frequencies always equals the sum of the observed frequencies. For instance, in Example 1 the sum of the observed frequencies and the sum of the expected frequencies are both 500.

Try It Yourself 1

Suppose the executive randomly selects 300 music listeners in the listening region. Find the expected frequencies for each type of music.

a. Multiply 300 by the probability that a listener will name each particular type of music.

Answer: Page A41

To use the chi-square goodness-of-fit test, the following must be true.

1. The observed frequencies must be obtained using a random sample.
2. Each expected frequency must be greater than or equal to 5.

If the expected frequency of a category is less than 5, it may be possible to combine it with another category to meet the requirements.

The Chi-Square Goodness-of-Fit Test

If the conditions listed above are satisfied, then the sampling distribution for the goodness-of-fit test is a chi-square distribution with $k - 1$ degrees of freedom, where k is the number of categories. The test statistic for the chi-square goodness-of-fit test is

$$\chi^2 = \sum \frac{(O - E)^2}{E}$$

where O represents the observed frequency of each category and E represents the expected frequency of each category.

When the observed frequencies closely match the expected frequencies, the differences between O and E will be small and the chi-square test statistic will be close to 0. As such, the null hypothesis is unlikely to be rejected. However, when there are large discrepancies between the observed frequencies and the expected frequencies, the differences between O and E will be large, resulting in a large chi-square test statistic. A large chi-square test statistic is evidence for rejecting the null hypothesis. So, the chi-square goodness-of-fit test is always a right-tailed test.

GUIDELINES

Performing a Chi-Square Goodness-of-Fit Test

In Words	*In Symbols*
1. Identify the claim. State the null and alternative hypothesis.	State H_0 and H_a.
2. Specify the level of significance.	Identify α.
3. Determine the degrees of freedom.	d.f. $= k - 1$
4. Find the critical value.	Use Table 6 in Appendix B.
5. Identify the rejection region.	
6. Calculate the test statistic.	$\chi^2 = \sum \frac{(O - E)^2}{E}$
7. Make a decision to reject or fail to reject the null hypothesis.	If χ^2 is in the rejection region, reject H_0. Otherwise, do not reject.
8. Interpret the decision in the context of the original claim.	

EXAMPLE 

Performing a Chi-Square Goodness-of-Fit Test

A radio station claims that the music preferences of the listeners in the station's broadcast region are distributed as shown in the table at the left below. You randomly select 500 radio music listeners from the broadcast region and ask each whether he or she prefers classical, country, gospel, oldies, pop, or rock music. The survey results are listed in the table at the right below. Using $\alpha = 0.01$, perform a chi-square goodness-of-fit test to test the claimed distribution. What can you conclude?

Distribution of music preferences	
Classical	4%
Country	36%
Gospel	11%
Oldies	2%
Pop	18%
Rock	29%

Survey results ($n = 500$)	
Classical	8
Country	210
Gospel	72
Oldies	10
Pop	75
Rock	125

Type of music	Observed frequency	Expected frequency
Classical	8	20
Country	210	180
Gospel	72	55
Oldies	10	10
Pop	75	90
Rock	125	145

SOLUTION The observed and expected frequencies are shown in the table at the left. The expected frequencies were calculated in Example 1. Because the observed frequencies were obtained using a random sample and each expected frequency is at least 5, you can use the chi-square goodness-of-fit test to test the proposed distribution. For this test, the null and alternative hypotheses are as follows.

H_0: The distribution of music preferences in the broadcast region is 4% classical, 36% country, 11% gospel, 2% oldies, 18% pop, and 29% rock. (Claim)

H_a: The distribution of music preferences differs from the claimed or expected distribution.

Because there are six categories, the chi-square distribution has $k - 1 = 6 - 1 = 5$ degrees of freedom. Using d.f. $= 5$ and $\alpha = 0.01$, the critical value is 15.086. Using the observed and expected frequencies, the chi-square test statistic is

$$\chi^2 = \Sigma\frac{(O - E)^2}{E}$$

$$= \frac{(8 - 20)^2}{20} + \frac{(210 - 180)^2}{180} + \frac{(72 - 55)^2}{55} + \frac{(10 - 10)^2}{10} + \frac{(75 - 90)^2}{90} + \frac{(125 - 145)^2}{145}$$

$$\approx 22.713.$$

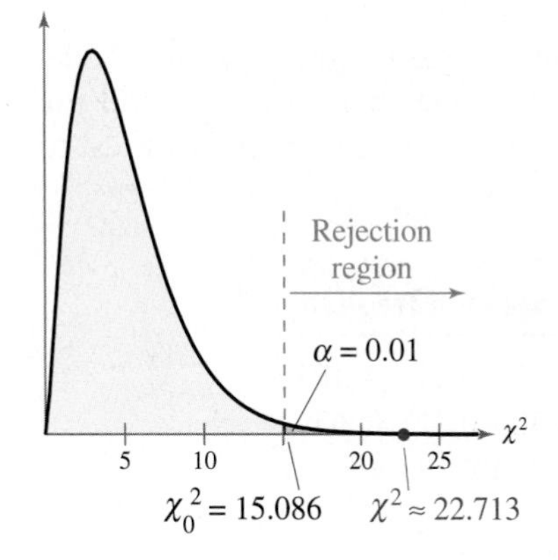

The graph shows the location of the rejection region and the chi-square test statistic. Because χ^2 is in the rejection region, you should decide to reject the null hypothesis. In other words, at the 1% level, there is enough evidence to conclude that the distribution of music preferences differs from the radio station's claimed or expected distribution.

Ages	Claimed distribution	Survey results
0–9	16%	76
10–19	20%	84
20–29	8%	30
30–39	14%	60
40–49	15%	54
50–59	12%	40
60–69	10%	42
70+	5%	14

Try It Yourself 2

A sociologist claims that the age distribution for the residents of a certain city is the same as it was 10 years ago. The distribution of ages 10 years ago is shown in the table at the left. You randomly select 400 residents and record the age of each. The survey results are listed in the table. At $\alpha = 0.05$, perform a chi-square goodness-of-fit test to determine whether the distribution has changed.

a. *Verify* that the expected frequency is at least 5 for each category.
b. *Identify* the claimed distribution and state H_0 and H_a.
c. *Specify* the level of significance α.
d. *Determine* the degrees of freedom.
e. *Find* the critical value and *identify* the rejection region.
f. *Find* the chi-square test statistic.
g. *Decide* whether to reject the null hypothesis. Use a graph if necessary.
h. Is there enough evidence to conclude that the distribution of ages has changed?

Answer: Page A42

EXAMPLE 3

Performing a Chi-Square Goodness-of-Fit Test

Men's survey results ($n = 300$)	
For irradiation	168
Against irradiation	71
No opinion	61

Irradiation is a controversial method for preserving meat. The display at the right shows two distributions describing attitudes toward the irradiation of red meat. You work for a meat-packing plant and want to test the distribution describing men's attitudes. To test the distribution, you randomly select 300 men and ask each whether he is in favor of irradiation, against irradiation, or has no opinion. The results are listed in the table at the left. At $\alpha = 0.05$, test the claimed or expected distribution.

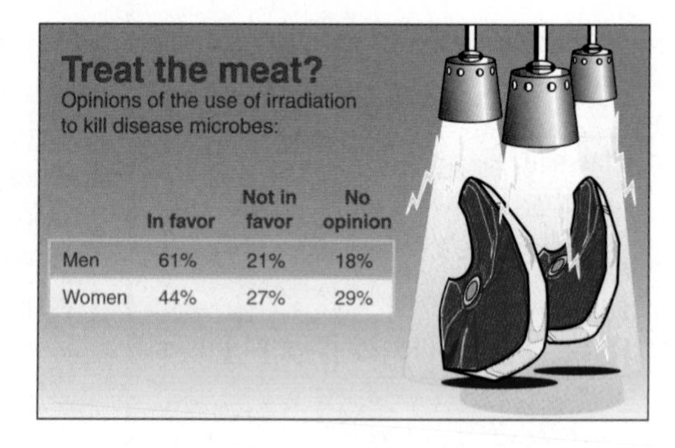

SOLUTION The observed frequencies and the expected frequencies are shown in the following table. Because each expected frequency is at least 5 and the men were randomly selected, you can use the chi-square goodness-of-fit test to test the claimed distribution.

Response	Observed frequency	Expected frequency
In favor of	168	$300(0.61) = 183$
Against	71	$300(0.21) = 63$
No opinion	61	$300(0.18) = 54$

The null and alternative hypotheses are as follows.

H_0: The distribution of men's attitudes toward the irradiation of red meat is 61% in favor, 21% against, and 18% no opinion. (Claim)

H_a: The distribution of men's attitudes toward the irradiation of red meat differs from the claimed or expected distribution.

Because there are three categories, the chi-square distribution has $k - 1 = 3 - 1 = 2$ degrees of freedom. Using d.f. $= 2$ and $\alpha = 0.05$, the critical value is 5.991. Using the observed and expected frequencies, the chi-square test statistic is as shown in the following table.

Study Tip

Another way to calculate the chi-square test statistic is to organize the calculations in a table as shown in Example 3.

O	E	$O - E$	$(O - E)^2$	$\frac{(O - E)^2}{E}$
168	183	−15	225	1.229508197
71	63	8	64	1.015873016
61	54	7	49	0.9074074074
				$\chi^2 = \Sigma \frac{(O - E)^2}{E} \approx 3.1528$

The graph shows the location of the rejection region and the chi-square test statistic. Because χ^2 is not in the rejection region, you should decide not to reject the null hypothesis. In other words, at the 5% level, there is not enough evidence to dispute the claimed or expected distribution of men's opinions.

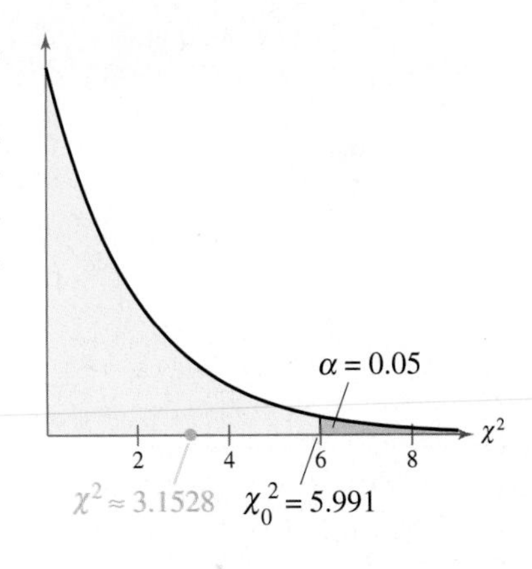

Try It Yourself 3

You also want to test the distribution describing women's attitudes. To test the distribution, you randomly select 200 women and ask each whether she is in favor of irradiation, against irradiation, or has no opinion. The results are listed in the table at the left. At $\alpha = 0.01$, test the claimed or expected distribution. What can you conclude?

Women's survey results ($n = 200$)	
For irradiation	100
Against irradiation	48
No opinion	52

a. *Verify* that the expected frequency is at least 5 for each category.
b. *Identify* the claimed distribution and state H_0 and H_a.
c. *Specify* the level of significance α.
d. *Determine* the degrees of freedom.
e. *Find* the critical value and *identify* the rejection region.
f. Use the observed and expected frequencies to *find the chi-square test statistic.*
g. *Decide* whether to reject the null hypothesis. Use a graph if necessary.
h. Is there enough evidence to dispute the claimed distribution?

Answer: Page A42

The chi-square goodness-of-fit test is often used to determine whether a distribution is uniform. For such tests, the expected frequencies of the categories are equal. When testing a uniform distribution, you can find the expected frequency of each category by dividing the sample size by the number of categories. For example, suppose a company believes that the number of sales made by its sales force is uniform throughout the five-day work week. If the sample consists of 1000 sales, then the expected value of the sales for each day would be $1000/5 = 200$.

10.1 Exercises

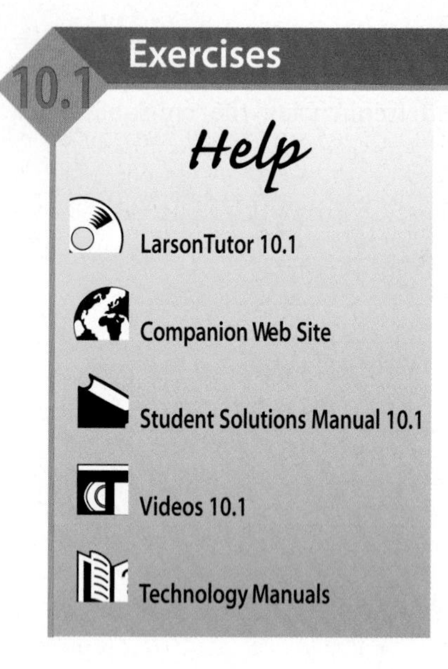

Basic Skills and Concepts

Performing a Chi-Square Goodness-of-Fit Test In Exercises 1–10, (a) identify the claim and state H_0 and H_a, (b) find the critical value and identify the rejection region, (c) find the test statistic χ^2, and (d) decide whether to reject or fail to reject the null hypothesis. Then interpret the decision in the context of the original claim.

1. ***Coffee*** ◆ Results from a survey five years ago asking where coffee drinkers typically drink their first cup of coffee are shown in the graph. To determine whether this distribution has changed, you randomly select 581 coffee drinkers and ask each where they typically drink their first cup of coffee. The results are listed in the table. Can you conclude that there has been a change in the claimed or expected distribution? Use $\alpha = 0.05$. *(Adapted from USA TODAY)*

Survey results	
Response	**Frequency, f**
At home	389
At workplace	110
While commuting	55
Restaurant/coffee bar/ other	27

2. ***Reasons Workers Leave*** ◆ A personnel director believes that the distribution of the reasons workers leave their jobs is different from the one shown in the graph. The director randomly selects 200 workers who recently left their jobs and asks each his or her reason for doing so. The results are shown in the table. At $\alpha = 0.05$, are the distributions different? *(Adapted from USA TODAY)*

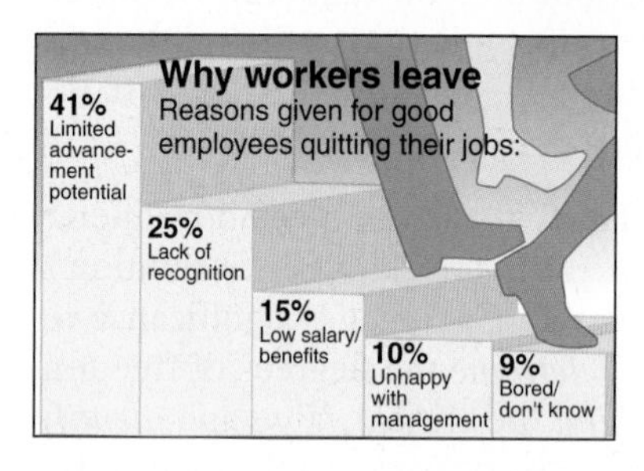

Survey results	
Response	**Frequency, f**
Limited advancement potential	78
Lack of recognition	52
Low salary/benefits	30
Unhappy with mgmt.	25
Bored/don't know	15

3. ***Bicycle Accidents: Day of Week*** ◆ A bicycle safety organization claims that fatal bicycle accidents are uniformly distributed throughout the week. The following table lists the day of the week for which 911 randomly selected fatal bicycle accidents occurred. At $\alpha = 0.10$, is the distribution uniform? *(Adapted from Insurance Institute for Highway Safety)*

Day	Frequency, f	Day	Frequency, f
Sunday	118	Thursday	129
Monday	119	Friday	146
Tuesday	127	Saturday	135
Wednesday	137		

4. *Bicycle Accidents: Month of Year* ◆ A bicycle safety organization conducted a study of 996 randomly selected fatal bicycle accidents. The month each accident occurred is listed in the following table. At $\alpha = 0.10$, can you conclude that fatal bicycle accidents are not uniformly distributed by month? *(Adapted from Insurance Institute for Highway Safety)*

Month	Frequency, f	Month	Frequency, f
January	50	July	129
February	48	August	122
March	81	September	89
April	72	October	87
May	90	November	73
June	101	December	54

5. *Crash Deaths: Object Struck* ◆ The pie chart shows the distribution of roadside hazard crash deaths with respect to the object hit. After highway warning signs were erected, a study was conducted to see if there was a change in the distribution. The results are listed in the table. Can you conclude that there is a change in the distribution? Use $\alpha = 0.01$. *(Adapted from Insurance Institute for Highway Safety)*

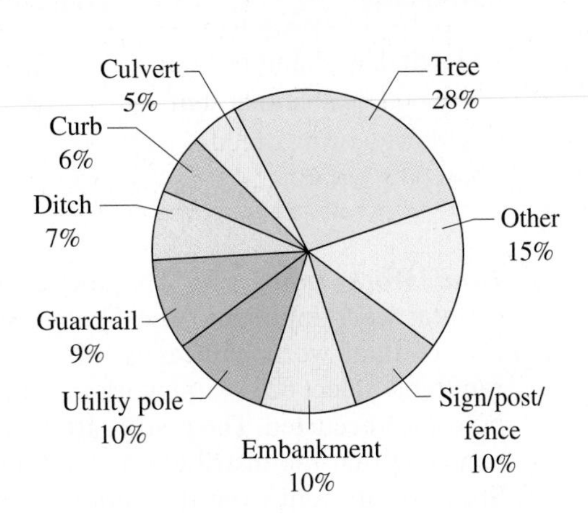

Study results	
Object struck	**Frequency, f**
Tree	179
Embankment	100
Utility pole	107
Guardrail	57
Ditch	36
Curb	43
Culvert	28
Sign/post/fence	68
Other	73

6. *Crash Deaths: Time of Day* ◆ The pie chart shows the distribution of the time of day of roadside hazard crash deaths for a previous year. The results of a recent study of 627 randomly selected hazard crash deaths are listed in the table. At $\alpha = 0.01$, has the distribution changed? *(Adapted from Insurance Institute for Highway Safety)*

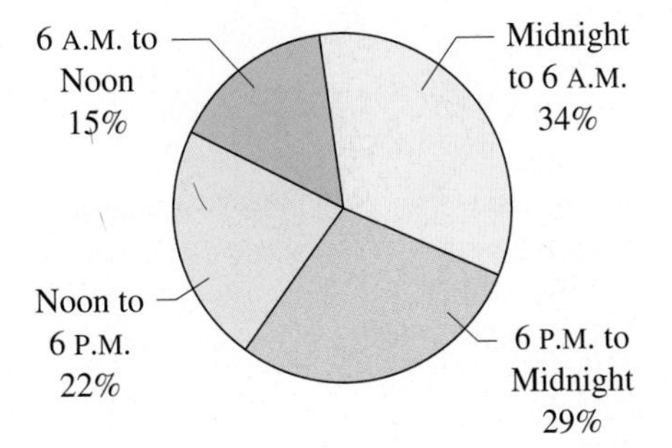

Study results	
Time of day	**Frequency, f**
Midnight to 6 A.M.	224
6 A.M. to Noon	128
Noon to 6 P.M.	115
6 P.M. to Midnight	160

7. *Educational Attainment* ◆ A social service organization reports that the level of educational attainment of mothers receiving food stamps is uniformly distributed. To test this claim, you randomly select 99 mothers who currently receive food stamps and record the educational attainment of each. The results are listed in the following table. At $\alpha = 0.025$, can you reject the claim that the distribution is uniform? *(Adapted from U.S. Census Bureau)*

Response	Frequency, f
Not a high school graduate	37
High school graduate	40
College (1 year or more)	22

8. *Marital Status* ◆ A social service worker believes that the marital status of mothers receiving food stamps is uniformly distributed. To test this claim, you randomly select 101 mothers who currently receive food stamps and record the marital status of each. The results are listed below. At $\alpha = 0.025$, can you reject the claim that the distribution is uniform? *(Adapted from U.S. Census Bureau)*

Response	Frequency, f
Married, husband present	20
Married, husband absent	19
Widowed or divorced	23
Never married	39

9. *Fatal Work Injuries* ◆ The pie chart shows the national distribution of fatal work injuries in the United States. You believe that the distribution of fatal work injuries is different in the western United States and randomly select 6231 fatal work injuries occurring in that region and record how each occurred. The results are listed in the table. At $\alpha = 0.05$, can you conclude that the distribution of fatal work injuries in the western United States is different from the national distribution? *(Adapted from U.S. Bureau of Labor Statistics)*

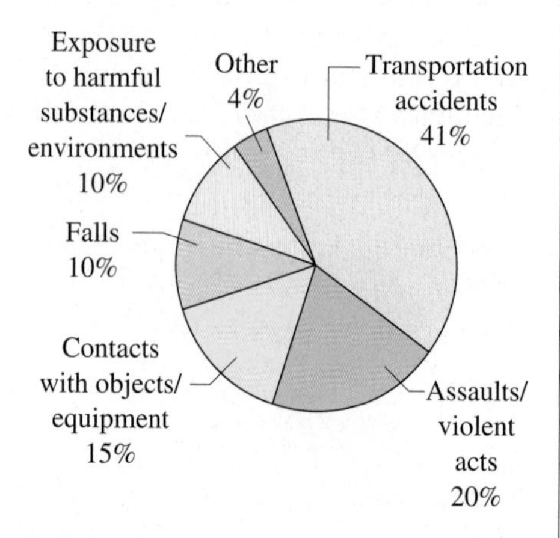

Study results: Western U.S.	
Cause	**Frequency, f**
Transportation accidents	2500
Assaults/violent acts	1300
Contacts with objects and equipment	985
Falls	620
Exposure to harmful substances or environments	602
Other	224

10. ***Marriage*** ◆ A marriage counselor says that 50% of all marriages are the first marriage for both the bride and the groom, 12% are the first for the bride and a remarriage for the groom, 14% are a remarriage for the bride and a first for the groom, and 24% are a remarriage for both. You randomly select 103 married couples and gather the results shown in the table. At $\alpha = 0.01$, can you reject the counselor's claim?

Response	Frequency, f
First marriage of bride and groom	55
First marriage of bride, remarriage of groom	12
Remarriage of bride, first marriage of groom	12
Remarriage of bride and groom	24

Extending the Basics

Testing for Normality Using a chi-square goodness-of-fit test, you can decide, with some degree of certainty, whether a variable is normally distributed. In all chi-square tests for normality, the null and alternative hypotheses are as follows.

H_0: The variable has a normal distribution.

H_a: The variable does not have a normal distribution.

To determine the expected frequencies when performing a chi-square test for normality, first find the mean and standard deviation of the frequency distribution. Then use the mean and standard deviation to compute the *z*-score for each class boundary. Then use the *z*-scores to calculate the area under the standard normal curve for each class. Multiplying the resulting class areas by the sample size yields the expected frequency for each class.

In Exercises 11 and 12, (a) find the expected frequencies, (b) find the critical value and identify the rejection region, (c) calculate the test statistic χ^2, and (d) decide whether to reject or fail to reject the null hypothesis. Then interpret the decision in the context of the original claim.

11. ***Test Scores*** ◆ The frequency distribution shows the results of 200 test scores. Are the test scores normally distributed? Use $\alpha = 0.01$.

Class boundaries	49.5–58.5	58.5–67.5	67.5–76.5
Frequency, f	19	61	82

Class boundaries	76.5–85.5	85.5–94.5
Frequency, f	34	4

12. ***Test Scores*** ◆ At $\alpha = 0.05$, test the claim that the 400 test scores shown in the frequency distribution are normally distributed.

Class boundaries	50.5–60.5	60.5–70.5	70.5–80.5
Frequency, f	28	106	151

Class boundaries	80.5–90.5	90.5–100.5
Frequency, f	97	18

10.2 Independence

What You Should Learn

- How to use a contingency table to find expected frequencies
- How to use a chi-square distribution to test whether two variables are independent

Contingency Tables • The Chi-Square Test for Independence

Contingency Tables

In Section 3.2, you learned that two variables are **independent** if the occurrence of one variable does not affect the probability of the occurrence of the other variable. For instance, the outcomes of a roll of a die and a toss of a coin are independent. But, suppose a medical researcher wants to determine if there is a relationship between caffeine consumption and heart attack risk. Are these variables independent or are they dependent? In this section, you will learn how to use the chi-square test for independence to answer such a question. To perform a chi-square test for independence, you will use sample data that are organized in a contingency table.

DEFINITION

An $r \times c$ **contingency table** shows the observed frequencies for two variables. The observed frequencies are arranged in r rows and c columns. The intersection of a row and a column is called a **cell.**

For example, the following table is a 2×5 contingency table. It has two rows and five columns and shows the results of a random sample of 550 company CEOs classified by age and size of company. From the table, you can see that 108 of the CEOs between the ages of 50 and 59 direct small or midsize companies, while 85 of the CEOs in this age group direct large companies.

Study Tip

In a contingency table, the notation $E_{r,c}$ represents the expected frequency for the cell in row r, column c. For instance, in the table at the right, $E_{1,4}$ represents the expected frequency for the cell in row 1, column 4.

Company size	**Age**				
	39 and under	**40–49**	**50–59**	**60–69**	**70 and over**
Small/midsize	42	69	108	60	21
Large	5	18	85	120	22

(Adapted from Grant Thornton LLP, The Segal Company)

Assuming the two variables of study in a contingency table are independent, you can use the contingency table to find the expected frequency for each cell. The formula for calculating the expected frequency for each cell is given below.

Finding the Expected Frequency for Contingency Table Cells

The expected frequency for a cell $E_{r,c}$ in a contingency table is

$$\text{Expected frequency } E_{r,c} = \frac{(\text{Sum of row } r) \times (\text{Sum of column } c)}{\text{Sample size}}.$$

Insight

In Example 1, once the expected frequency for $E_{1,1}$ is calculated to be 25.64, you can determine the expected frequency for $E_{2,1}$ to be $47 - 25.64 = 21.36$. That is, the expected frequency for the last cell in each row or column can be found by subtracting from the total.

EXAMPLE 

Finding Expected Frequencies

Find the expected frequency for each cell in the contingency table. Assume that the variables, age and company size, are independent.

Company size	Age					
	39 and under	**40–49**	**50–59**	**60–69**	**70 and over**	**Total**
Small/midsize	42	69	108	60	21	300
Large	5	18	85	120	22	250
Total	47	87	193	180	43	550

SOLUTION Using the formula

$$\text{Expected frequency } E_{r,c} = \frac{(\text{Sum of row } r) \times (\text{Sum of column } c)}{\text{Sample size}}$$

you can find each expected frequency as shown.

$$E_{1,1} = \frac{300 \cdot 47}{550} \approx 25.64 \quad E_{1,2} = \frac{300 \cdot 87}{550} \approx 47.45 \quad E_{1,3} = \frac{300 \cdot 193}{550} \approx 105.27$$

$$E_{1,4} = \frac{300 \cdot 180}{550} \approx 98.18 \quad E_{1,5} = \frac{300 \cdot 43}{550} \approx 23.45 \quad E_{2,1} = \frac{250 \cdot 47}{550} \approx 21.36$$

$$E_{2,2} = \frac{250 \cdot 87}{550} \approx 39.55 \quad E_{2,3} = \frac{250 \cdot 193}{550} \approx 87.73 \quad E_{2,4} = \frac{250 \cdot 180}{550} \approx 81.82$$

$$E_{2,5} = \frac{250 \cdot 43}{550} \approx 19.55$$

Try It Yourself 1

The marketing consultant for a travel agency wants to determine whether certain travel concerns are related to travel purpose. A random sample of 300 travelers is selected and the results are classified as shown in the following contingency table. Assuming that the variables travel concern and travel purpose are independent, find the expected frequency for each cell. *(Adapted from NPD Group for Embassy Suites)*

	Travel concern			
Travel purpose	**Hotel room**	**Leg room on plane**	**Rental car size**	**Other**
Business	36	108	14	22
Leisure	38	54	14	14

a. *Calculate* the sum of each row.
b. *Calculate* the sum of each column.
c. *Determine* the sample size.
d. *Use the formula* to find the expected frequency for each cell.

Answer: Page A42

The Chi-Square Test for Independence

After finding the expected frequencies, you can test whether the variables are independent using a chi-square independence test.

DEFINITION

A **chi-square independence test** is used to test the independence of two variables. Using a chi-square test, you can determine whether the occurrence of one variable affects the probability of the occurrence of the other variable.

To use the chi-square independence test, the following conditions must be true.

1. The observed frequencies must be obtained using a random sample.
2. Each expected frequency must be greater than or equal to 5.

The Chi-Square Independence Test

If the conditions listed above are satisfied, then the sampling distribution for the chi-square independence test is a chi-square distribution with

$$(r - 1)(c - 1)$$

degrees of freedom, where r and c are the number of rows and columns, respectively, of a contingency table. The test statistic for the chi-square independence test is

$$\chi^2 = \Sigma\frac{(O - E)^2}{E}$$

where O represents the observed frequencies and E represents the expected frequencies.

To begin the independence test, you must first state a null and an alternative hypothesis. For a chi-square independence test, the null and alternative hypotheses are always some variation of the following statements.

H_0: The variables are independent.

H_a: The variables are dependent.

The expected frequencies are calculated assuming that the two variables are independent. If the variables are independent, then you can expect little difference between the observed frequencies and the expected frequencies. When the observed frequencies closely match the expected frequencies, the differences between O and E will be small and the chi-square test statistic will be close to 0. As such, the null hypothesis is unlikely to be rejected.

However, if the variables are dependent, there will be large discrepancies between the observed frequencies and the expected frequencies. When the differences between O and E are large, the chi-square test statistic is also large. A large chi-square test statistic is evidence for rejecting the null hypothesis. So, the chi-square independence test is always a right-tailed test.

Picturing the World

A researcher wishes to determine if a relationship exists between job status (full or part-time) and two different work arrangements. The results of a random sample of 772 workers are listed in the following contingency table. *(Adapted from the Bureau of Labor Statistics)*

	Work arrangement	
Status	**Contract**	**Traditional**
Full-time	5	630
Part-time	1	136

Can the researcher use this sample to test for independence using a chi-square independence test? Why or why not?

GUIDELINES

Performing a Chi-Square Test for Independence

In Words	*In Symbols*
1. Identify the claim. State the null and alternative hypotheses.	State H_0 and H_a.
2. Specify the level of significance.	Specify α.
3. Determine the degrees of freedom.	d.f. $= (r - 1)(c - 1)$
4. Find the critical value.	Use Table 6 in Appendix B.
5. Identify the rejection region.	
6. Calculate the test statistic.	$\chi^2 = \Sigma\frac{(O - E)^2}{E}$
7. Make a decision to reject or fail to reject the null hypothesis.	If χ^2 is in the rejection region, reject H_0. Otherwise, do not reject H_0.
8. Interpret the decision in the context of the original claim.	

Study Tip

A contingency table with three rows and four columns will have

$(3 - 1)(4 - 1) = (2)(3)$
$= 6$ d.f.

EXAMPLE 2

Performing a Chi-Square Independence Test

The following contingency table shows the results of a random sample of 550 company CEOs classified by age and size of company. The expected frequencies are displayed in parentheses. At $\alpha = 0.01$, can you conclude that the CEOs' ages are related to company size?

Company size	**Age of CEOs**					
	39 and under	**40–49**	**50–59**	**60–69**	**70 and over**	**Total**
Small/midsize	42 (25.64)	69 (47.45)	108 (105.27)	60 (98.18)	21 (23.45)	300
Large	5 (21.36)	18 (39.55)	85 (87.73)	120 (81.82)	22 (19.55)	250
Total	47	87	193	180	43	550

SOLUTION The expected frequencies were calculated in Example 1. Because each expected frequency is at least 5 and the CEOs were randomly selected, you can use the chi-square independence test to test whether the variables are independent. The null and alternative hypotheses are as follows.

H_0: The CEOs' ages are independent of the company size.

H_a: The CEOs' ages are dependent on the company size.

Because the contingency table has two rows and five columns, the chi-square distribution has $(r - 1)(c - 1) = (2 - 1)(5 - 1) = 4$ degrees of freedom. Because d.f. = 4 and $\alpha = 0.01$, the critical value is 13.277. Using the observed and expected frequencies, the chi-square test statistic is as shown in the following table.

O	E	$O - E$	$(O - E)^2$	$\frac{(O - E)^2}{E}$
42	25.64	16.36	267.6496	10.4388
69	47.45	21.55	464.4025	9.7872
108	105.27	2.73	7.4529	0.0708
60	98.18	−38.18	1457.7124	14.8473
21	23.45	−2.45	6.0025	0.2560
5	21.36	−16.36	267.6496	12.5304
18	39.55	−21.55	464.4025	11.7422
85	87.73	−2.73	7.4529	0.0850
120	81.82	38.18	1457.7124	17.8161
22	19.55	2.45	6.0025	0.3070
		$\chi^2 = \Sigma \frac{(O - E)^2}{E} \approx 77.9$		

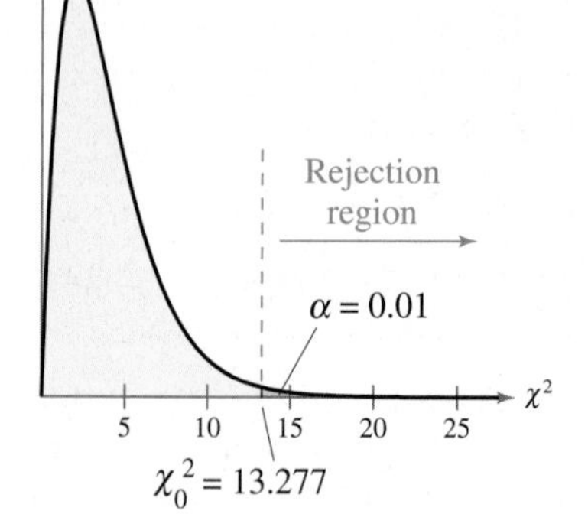

The graph shows the location of the rejection region. Because $\chi^2 \approx 77.9$ is in the rejection region, you should decide to reject the null hypothesis. In other words, there is enough evidence at the 1% level of significance to conclude that the CEOs' ages and the company size are dependent.

Try It Yourself 2

The marketing consultant for a travel agency wants to determine whether travel concerns are related to travel purpose. A random sample of 300 travelers is selected and asked his or her primary travel concern. The results are classified as shown in the following contingency table. At $\alpha = 0.01$, can the consultant conclude that the travel concerns depend on the purpose of travel? (The expected frequencies are displayed in parentheses.) *(Adapted from NPD Group for Embassy Suites)*

	Travel concern				
Travel purpose	**Hotel room**	**Leg room on plane**	**Rental car size**	**Other**	**Total**
Business	36 (44.4)	108 (97.2)	14 (16.8)	22 (21.6)	180
Leisure	38 (29.6)	54 (64.8)	14 (11.2)	14 (14.4)	120
Total	74	162	28	36	300

a. *Identify* the claim and *state* H_0 and H_a.
b. *Specify* the level of significance α.
c. *Determine* the degrees of freedom.
d. *Find* the critical value and *identify* the rejection region.
e. Use the observed and expected frequencies to *find the chi-square test statistic.*
f. *Decide* whether to reject the null hypothesis. Use a graph if necessary.
g. Is there enough evidence to conclude that the travel concerns depend on the purpose of travel?

Answer: Page A42

EXAMPLE 

Using Technology for a Chi-Square Independence Test

Setup

```
X²-Test
 Observed: [A]
 Expected: [B]
 Calculate Draw
```

Calculate

```
X²-Test
 x²=3.493357223
 p=.321624691
 df=3
```

Draw

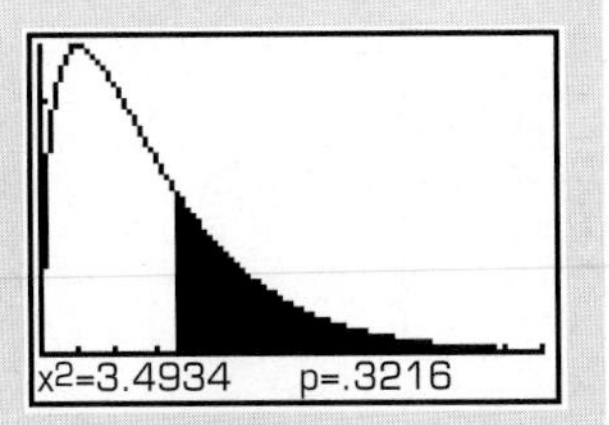

A health club manager wants to determine whether the number of days per week that college students spend exercising is related to gender. A random sample of 275 college students is selected and the results are classified as shown in the following table. At $\alpha = 0.05$, is there enough evidence to conclude that the number of days spent exercising per week is related to gender?

	Days per week spent exercising				
Gender	**0–1**	**2–3**	**4–5**	**6–7**	**Total**
Male	40	53	26	6	125
Female	34	68	37	11	150
Total	74	121	63	17	275

SOLUTION The null and alternative hypotheses can be stated as follows.

H_0: The number of days spent exercising per week is independent of gender.

H_a: The number of days spent exercising per week depends on gender.

Enter the observed frequencies into matrix A. Then set up the chi-square test using a *TI-83* as shown at the left. Display the expected value matrix to verify that the expected value for each cell is at least 5.

The displays at the left show the results of selecting "Calculate" or "Draw." Because d.f. $= 3$ and $\alpha = 0.05$, the rejection region is $\chi^2 > 7.815$. The test statistic $\chi^2 \approx 3.49$ is not in the rejection region, so you should fail to reject the null hypothesis. There is not enough evidence to conclude that the number of days spent exercising per week is related to gender.

Study Tip

You can also use *P*-values to perform a chi-square test for independence. For instance, in Example 3, note that the TI-83 displays $P \approx 0.322$. Because $P > \alpha$, you should fail to reject the null hypothesis.

Try It Yourself 3

A researcher wants to determine whether the number of minutes adults spend online per day is related to gender. A random sample of 450 adults is selected and the results are classified as shown in the following table. At $\alpha = 0.05$, is there enough evidence to conclude that the number of minutes spent online per day is related to gender?

	Minutes spent online per day					
Gender	**0–15**	**15–30**	**30–45**	**45–60**	**60 and over**	**Total**
Male	19	36	75	90	55	275
Female	21	72	45	19	18	175
Total	40	108	120	109	73	450

a. *Find* the critical value and *identify* the rejection region.
b. *Enter* the observed frequencies.
c. *Use* a technology tool to find the χ^2 test statistic
d. *Make* a decision. Is there enough evidence to conclude that the number of minutes spent online per day is related to gender?

Answer: Page A42

10.2 Exercises

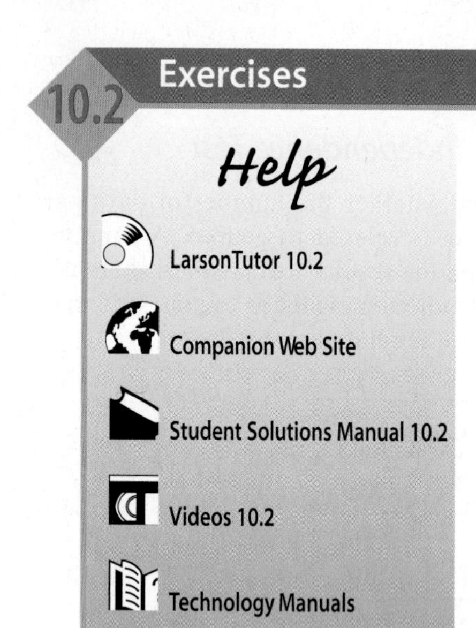

Basic Skills and Concepts

Performing a Chi-Square Test for Independence In Exercises 1–11, perform the indicated chi-square test for independence by doing the following.

(a) Identify the claim and state the null and alternative hypotheses.

(b) Determine the degrees of freedom, find the critical value, and identify the rejection region.

(c) Calculate the test statistic. (If possible, use a technology tool.)

(d) Decide to reject or fail to reject the null hypothesis. Then interpret the decision in the context of the original claim.

1. ***Achievement and School Location*** ◆ Is achieving a basic skill level in a subject related to the location of the school? A random sample of students by the location of school and the number achieving basic skill levels in three subjects is shown in the following table. At $\alpha = 0.01$, test the hypothesis that the variables are independent. *(Adapted from USA TODAY)*

	Subject		
Location of school	**Reading**	**Math**	**Science**
Urban	43	42	38
Suburban	63	66	65

2. ***Attitudes About Safety*** ◆ The results of a random sample of students by type of school and their attitudes on safety steps taken by the school staff are shown in the following table. At $\alpha = 0.01$, can you conclude that attitudes about the safety steps taken by the school staff are related to the type of school? *(Adapted from USA TODAY)*

	School staff has	
Type of school	**Taken all steps necessary for student safety**	**Taken some steps toward student safety**
Public	40	51
Private	64	34

3. ***Rating Public Schools*** ◆ The following contingency table shows how a random sample of adults rated their local public schools and how they rated America's public schools. At $\alpha = 0.05$, can you conclude that the adults' ratings are related to the type of school? *(Adapted from USA TODAY)*

	Rating			
Type of school	**Excellent**	**Good**	**Fair**	**Poor**
Local	120	405	263	151
National	41	238	481	179

4. ***Grades for Our Leaders*** ◆ The contingency table shows how a random sample of college freshmen graded the leaders of three types of institutions. At $\alpha = 0.05$, can you conclude that the grades are related to the institution? *(Adapted from Louis Harris for Northwestern Mutual Life Insurance)*

	Grade				
Institution	**A**	**B**	**C**	**D**	**F**
Military	25	46	19	5	3
Religious	18	44	24	7	5
Media/press	5	23	37	21	12

5. ***Obsessive-Compulsive Disorder*** ◆ The results of a random sample of patients with obsessive-compulsive disorder treated with a drug or with a placebo are shown in the contingency table. At $\alpha = 0.10$, can you conclude that the treatment is related to the result? Based on these results, would you recommend using the drug as part of a treatment for obsessive-compulsive disorder? *(Adapted from The Journal of the American Medical Association)*

	Treatment	
Result	**Drug**	**Placebo**
Improvement	39	25
No change	54	70

6. ***Chronic Fatigue Syndrome*** ◆ The contingency table shows a random sample of patients with chronic fatigue syndrome treated with a drug or with a placebo. At $\alpha = 0.10$, can you conclude that the variables treatment and result are dependent? Based on these results, would you recommend using the drug as part of a treatment for chronic fatigue syndrome? *(Adapted from The Journal of the American Medical Association)*

	Treatment	
Result	**Drug**	**Placebo**
Improvement	20	19
No change	10	16

7. ***Continuing Education*** ◆ You work for a college's continuing education department and want to determine whether the reasons given by workers for continuing their education is related to job type. In your study, you randomly collect the data shown in the contingency table. At $\alpha = 0.01$, can you conclude that the variables reason and type of worker are dependent? How could you use this information in your marketing efforts? *(Adapted from USA TODAY)*

	Reason		
Type of worker	**Professional**	**Personal**	**Professional and personal**
Technical	30	36	41
Other	47	25	30

8. ***Blood Alcohol Concentration*** ◆ You are investigating the relationship between the ages and the blood alcohol concentration of fatally injured pedestrians. During your investigation, you randomly collect the data shown in the contingency table. At $\alpha = 0.01$, is there enough evidence to conclude that blood alcohol concentration is related to age in nighttime pedestrian deaths? *(Adapted from Insurance Institute for Highway Safety)*

	Blood Alcohol Concentration		
Age	**0.00**	**0.01–0.09**	**0.10 and greater**
16–34 years	439	85	696
35 years and over	513	98	622

9. ***Vehicles and Crashes*** ◆ You work for an insurance company and are studying the relationship between types of crashes and the vehicles involved. As part of your study, you randomly select 3207 vehicle crashes and organize the resulting data as shown in the contingency table. At $\alpha = 0.05$, can you conclude that the type of crash depends on the type of vehicle? *(Adapted from Insurance Institute for Highway Safety)*

	Vehicle		
Type of crash	**Car**	**Pickup truck/ utility vehicle**	**Cargo/large passenger van**
Single-vehicle	895	493	45
Multiple-vehicle	1400	336	38

10. ***Alcohol-Related Accidents*** ◆ The following contingency table shows a random sample of fatally injured passenger vehicle drivers (with blood alcohol concentrations greater than or equal to 0.10) by age and gender. At $\alpha = 0.05$, can you conclude that age is related to gender in such alcohol-related accidents? *(Adapted from Insurance Institute for Highway Safety)*

	Age					
Gender	**16–20**	**21–30**	**31–40**	**41–50**	**51–60**	**61 and over**
Male	32	51	52	43	28	10
Female	13	22	33	21	10	6

11. ***Coauthored Books*** ◆ The following contingency table shows a random sample of engineering, psychology, and biology books and if and how they were coauthored. At $\alpha = 0.10$, can you conclude that the subject matter and coauthorship are related? *(Adapted from CHI Research, Inc.)*

	Subject matter		
Coauthorship	**Engineering**	**Psychology**	**Biology**
Coauthored	47	44	50
Internationally coauthored	17	9	16
Not coauthored	36	47	34

Extending the Basics

Homogeneity of Proportions Test Another chi-square test that involves a contingency table is the **homogeneity of proportions test.** The test is used to determine if several proportions are equal when samples are taken from different populations. Before sampling the populations and making the contingency table, the sample sizes are determined. After randomly sampling different populations, you can test whether the proportion of elements in a category is the same for each population using the same guidelines in performing a chi-square independence test. The null and alternative hypotheses are always some variation of the following statements.

H_0: The proportions are equal.

H_a: At least one of the proportions is different from the others.

To perform a homogeneity of proportions test, the observed frequencies must be obtained using a random sample, and each expected frequency must be greater than or equal to 5.

12. ***Motor Vehicle Deaths*** ◆ The table shows a random sample of motor vehicle deaths by age and gender. At $\alpha = 0.05$, perform a homogeneity of proportions test on the claim that the proportions of motor vehicle deaths involving males or females are the same for each age group. *(Adapted from Insurance Institute for Highway Safety)*

	Age			
Gender	**16–24**	**25–34**	**35–44**	**45–54**
Male	80	54	41	39
Female	35	21	18	17

	Age			
Gender	**55–64**	**65–74**	**75–84**	**85 and over**
Male	39	45	73	55
Female	20	27	40	19

13. ***Testing a Drug*** ◆ The contingency table shows the results of a random sample of patients with obsessive-compulsive disorder after being treated with a drug or with a placebo. At $\alpha = 0.10$, perform a homogeneity of proportions test on the claim that the proportions of the results for drug and placebo treatment are the same. *(Adapted from The Journal of the American Medical Association)*

	Treatment	
Result	**Drug**	**Placebo**
Improvement	39	25
No change	54	70

10 Case Study

WWW.NHTSA.DOT.GOV

Traffic Safety Facts

Each year, the National Highway Traffic Safety Administration (NHTSA) together with the National Center for Statistics and Analysis (NCSA) publishes *Traffic Safety Facts,* which summarizes the motor vehicle traffic crash experience for the United States. *Traffic Safety Facts 1999* includes trend data, crash data, vehicle data, and people data. Also, the NHTSA and NCSA publish a report summarizing the motor vehicle crash data of the 17 states in the NHTSA's State Data System.

In 1999, there were 41,611 fatalities in the United States as a result of motor vehicle crashes. The pie chart at the right shows the national distribution of traffic fatalities with respect to age group. For example, 24% of all motor vehicle fatalities were young adults aged 16–24. Using the data from the 17 states in the NHTSA's State Data System as a sample, the contingency table lists the number of motor vehicle fatalities according to age and geographic location within the United States.

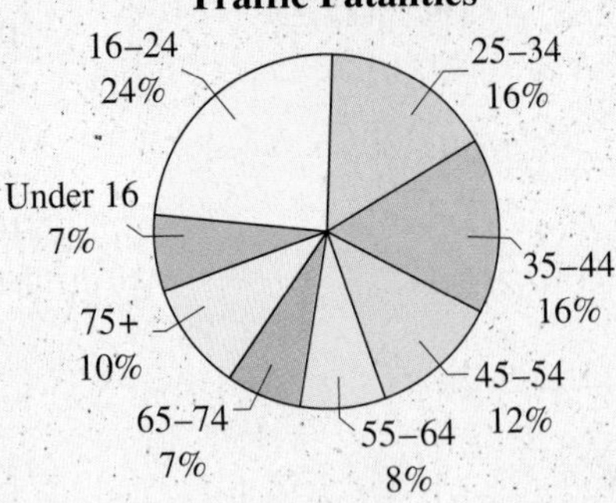

Motor Vehicle Fatalities

	Region		
Age	**Eastern U.S.**	**Central U.S.**	**Western U.S.**
Under 16	646	732	518
16–24	1907	2269	1262
25–34	1524	1583	1125
35–44	1242	1274	792
45–54	846	802	555
55–64	590	568	371
65–74	737	574	408
75+	931	739	481

Exercises

1. In 1999, how many people in the United States aged 16–24 died as a result of a motor vehicle crash?
2. Assuming the variables region and age are independent, in which region did the number of motor vehicle fatalities for the 16–24 age group exceed the expected number of fatalities?
3. Assuming the variables region and age are independent, in which region did the number of motor vehicle fatalities for the 25–34 age group exceed the expected number of fatalities?
4. At $\alpha = 0.05$, perform a chi-square test to determine whether the variables region and age are independent. What can you conclude?

In Exercises 5–7, perform a chi-square goodness-of-fit test to compare the national distribution of motor vehicle fatalities with the distribution of each region of the United States. Use the national distribution as the claimed distribution. Use $\alpha = 0.05$.

5. Compare the distribution of the sample of motor vehicle fatalities from the eastern United States with the national distribution. What can you conclude?
6. Compare the distribution of the sample of motor vehicle fatalities from the central United States with the national distribution. What can you conclude?
7. Compare the distribution of the sample of motor vehicle fatalities from the western United States with the national distribution. What can you conclude?
8. In addition to the variables used in this case study, what other variables do you think are important considerations when studying the distribution of motor vehicle fatalities?

10.3 Comparing Two Variances

What You Should Learn

- How to interpret the *F*-distribution and use an *F*-table to find critical values
- How to perform a two-sample *F*-test to compare two variances

The *F*-Distribution • The Two-Sample *F*-Test for Variances

The *F*-Distribution

In Chapter 8, you learned how to perform hypothesis tests to compare population means and population proportions. Recall from Section 8.2 that the *t*-test for the difference between two population means depends on whether the population variances are equal. To determine whether the population variances are equal, you can perform a two-sample *F*-test.

In this section, you will learn about the *F*-distribution and how to use the *F*-distribution to compare two variances.

DEFINITION

Let s_1^2 and s_2^2 represent the sample variances of two different populations. If both populations are normal and the population variances σ_1^2 and σ_2^2 are equal, then the sampling distribution of

$$F = \frac{s_1^2}{s_2^2}$$

is called an ***F*-distribution.** Several properties of the F-distribution are as follows.

1. The F-distribution is a family of curves each of which is determined by two types of degrees of freedom: the degrees of freedom corresponding to the variance in the numerator, denoted by $\mathbf{d.f._N}$, and the degrees of freedom corresponding to the variance in the denominator, denoted by $\mathbf{d.f._D}$.
2. F-distributions are positively skewed.
3. The total area under each curve of an F-distribution is equal to 1.
4. F-values are always greater than or equal to zero.
5. For all F-distributions, the mean value of F is approximately equal to 1.

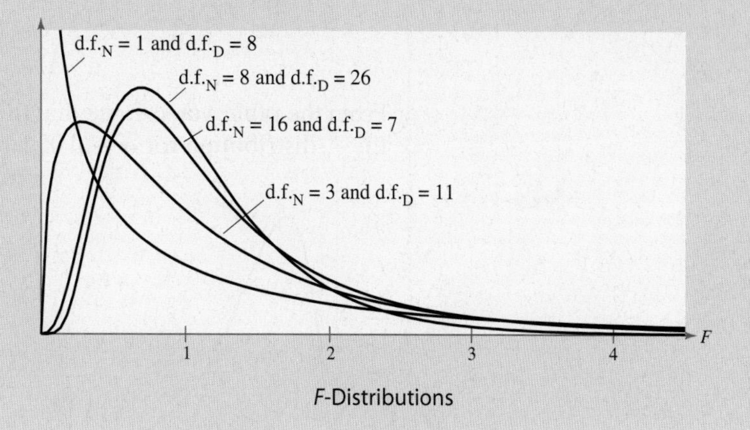

F-Distributions

Table 7 in Appendix B lists the critical values for the *F*-distribution for selected levels of significance α and degrees of freedom, $d.f._N$ and $d.f._D$.

GUIDELINES

Finding Critical Values for the *F*-Distribution

1. Specify the level of significance α.
2. Determine the degrees of freedom for the numerator, $d.f._N$.
3. Determine the degrees of freedom for the denominator, $d.f._D$.
4. Use Table 7 in Appendix B to find the critical value. If the hypothesis test is
   a. one tailed, use the α *F*-table.
   b. two tailed, use the $\frac{1}{2}\alpha$ *F*-table.

Study Tip

In the sampling distribution of $F = \frac{s_1^2}{s_2^2}$, the larger variance is always in the numerator. So, *F* is always greater than or equal to 1. As such, all one-tailed tests are right-tailed tests, and for all two-tailed tests, you need only to find the right-tailed critical value.

EXAMPLE 

Finding Critical F-Values for a Right-Tailed Test

Find the critical *F*-value for a right-tailed test when $\alpha = 0.05$, $d.f._N = 6$, and $d.f._D = 29$.

SOLUTION A portion of Table 7 is shown below. Using the $\alpha = 0.05$ *F*-table with $d.f._N = 6$ and $d.f._D = 29$, you can find the critical value as shown by the highlighted areas in the table.

$d.f._D$: Degrees of freedom, denominator	$\alpha = 0.05$							
	$d.f._N$: Degrees of freedom, numerator							
	1	**2**	**3**	**4**	**5**	**6**	**7**	**8**
1	161.4	199.5	215.7	224.6	230.2	234.0	236.8	238.9
2	18.51	19.00	19.16	19.25	19.30	19.33	19.35	19.37
26	4.23	3.37	2.98	2.74	2.59	2.47	2.39	2.32
27	4.21	3.35	2.96	2.73	2.57	2.46	2.37	2.31
28	4.20	3.34	2.95	2.71	2.56	2.45	2.36	2.29
29	4.18	3.33	2.93	2.70	2.55	2.43	2.35	2.28
30	4.17	3.32	2.92	2.69	2.53	2.42	2.33	2.27

From the table, you can see that the critical value is $F_0 = 2.43$. The graph shows the *F*-distribution for $\alpha = 0.05$, $d.f._N = 6$, $d.f._D = 29$, and $F_0 = 2.43$.

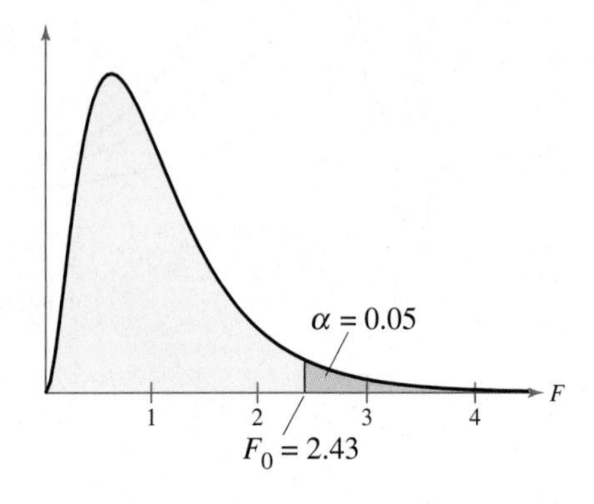

Try It Yourself 1

Find the critical F-value for a right-tailed test when $\alpha = 0.01$, $\text{d.f.}_N = 3$, and $\text{d.f.}_D = 15$.

a. *Specify* the level of significance α.
b. *Use* Table 7 in Appendix B to find the critical value.

Answer: Page A42

When performing a two-tailed hypothesis test using the F-distribution, you need only to find the right-tailed critical value. You must, however, remember to use the $\frac{1}{2}\alpha$ F-table.

EXAMPLE 2

Finding Critical F-Values for a Two-Tailed Test

Find the critical F-value for a two-tailed test when $\alpha = 0.05$, $\text{d.f.}_N = 4$, and $\text{d.f.}_D = 8$.

SOLUTION A portion of Table 7 is shown below. Using the

$$\frac{1}{2}\alpha = \frac{1}{2}(0.05) = 0.025$$

F-table with $\text{d.f.}_N = 4$, and $\text{d.f.}_D = 8$, you can find the critical value as shown by the highlighted areas in the table.

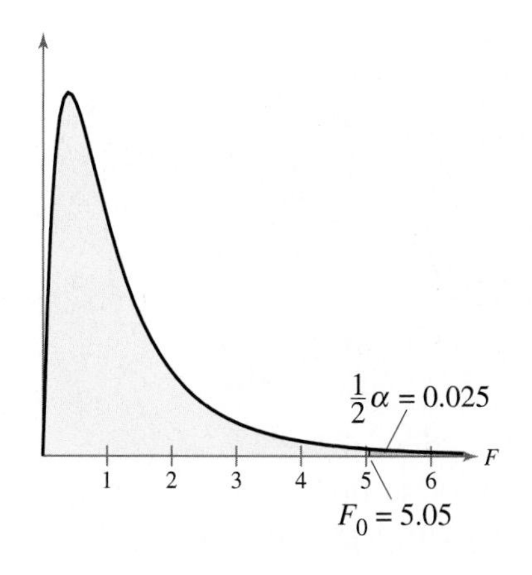

d.f._D: Degrees of freedom, denominator	$\alpha = 0.025$							
	d.f._N: Degrees of freedom, numerator							
	1	**2**	**3**	**4**	**5**	**6**	**7**	**8**
1	647.8	799.5	864.2	899.6	921.8	937.1	948.2	956.7
2	38.51	39.00	39.17	39.25	39.30	39.33	39.36	39.37
3	17.44	16.04	15.44	15.10	14.88	14.73	14.62	14.54
4	12.22	10.65	9.98	9.60	9.36	9.20	9.07	8.98
5	10.01	8.43	7.76	7.39	7.15	6.98	6.85	6.76
6	8.81	7.26	6.60	6.23	5.99	5.82	5.70	5.60
7	8.07	6.54	5.89	5.52	5.29	5.12	4.99	4.90
8	7.57	6.06	5.42	5.05	4.82	4.65	4.53	4.43
9	7.21	5.71	5.08	4.72	4.48	4.32	4.20	4.10

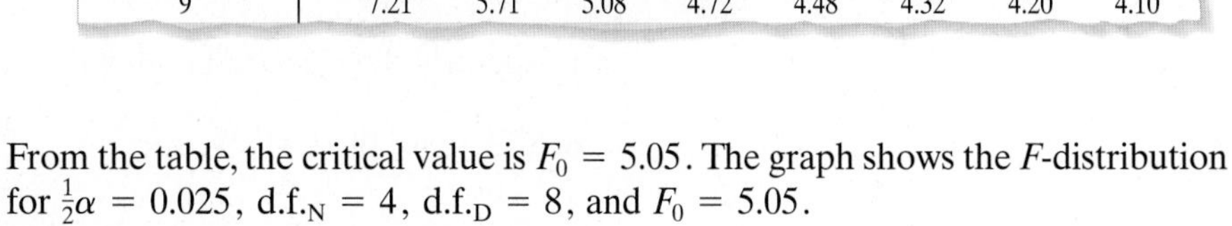

From the table, the critical value is $F_0 = 5.05$. The graph shows the F-distribution for $\frac{1}{2}\alpha = 0.025$, $\text{d.f.}_N = 4$, $\text{d.f.}_D = 8$, and $F_0 = 5.05$.

Try It Yourself 2

Find the critical F-value for a right-tailed test when $\alpha = 0.01$, $\text{d.f.}_N = 2$, and $\text{d.f.}_D = 5$.

a. *Specify* the level of significance α.
b. *Use* Table 7 in Appendix B with $\frac{1}{2}\alpha$ to find the critical value.

Answer: Page A42

The Two-Sample *F*-Test for Variances

In the remainder of this section, you will learn how to perform a two-sample *F*-test for comparing two population variances using a sample from each population. To perform such a test, three conditions must be met.

1. The samples must be randomly selected.
2. The samples must be independent.
3. Each population must have a normal distribution.

If these requirements are met, you can use the *F*-test to compare the population variances σ_1^2 and σ_2^2.

Two-Sample *F*-Test for Variances

A two-sample *F*-test is used to compare two population variances σ_1^2 and σ_2^2 when a sample is randomly selected from each population. The populations must be independent and normally distributed. The test statistic is

$$F = \frac{s_1^2}{s_2^2}$$

where s_1^2 and s_2^2 represent the sample variances with $s_1^2 \geq s_2^2$. The degrees of freedom for the numerator is $\text{d.f.}_N = n_1 - 1$ and the degrees of freedom for the denominator is $\text{d.f.}_D = n_2 - 1$, where n_1 is the size of the sample having variance s_1^2 and n_2 is the size of the sample having variance s_2^2.

GUIDELINES

Using a Two-Sample *F*-Test to Compare σ_1^2 and σ_2^2

In Words	*In Symbols*
1. Identify the claim. State the null and the alternative hypotheses.	State H_0 and H_a.
2. Specify the level of significance.	Specify α.
3. Determine the degrees of freedom.	$\text{d.f.}_N = n_1 - 1$ $\text{d.f.}_D = n_2 - 1$
4. Find the critical value.	Use Table 7 in Appendix B.
5. Identify the rejection region.	
6. Calculate the test statistic.	$F = \frac{s_1^2}{s_2^2}$
7. Make a decision to reject or fail to reject the null hypothesis.	If F is in the rejection region, reject H_0. Otherwise, do not reject H_0.
8. Interpret the decision in the context of the original claim.	

Picturing the World

Does location have an effect on the variance of real estate selling prices? A random sample of selling prices (in thousands of dollars) of houses sold in Albuquerque, New Mexico, is given in the following table. The first column represents the selling prices of houses in northeastern Albuquerque, and the second column lists the selling prices of houses in the remaining Albuquerque regions. *(Source: Albuquerque Board of Realtors)*

Northeast section	Other sections
205.0	129.5
208.0	97.5
215.0	93.9
215.0	82.0
199.9	78.0
190.0	77.0
180.0	70.0
156.0	62.0
145.0	54.0
144.9	107.0

Assuming the population of selling prices is normally distributed, is it possible to use a two-sample F-test to compare the population variances?

EXAMPLE 3

Performing a Two-Sample F-Test

A bank manager is designing a system that is intended to decrease the variance of the time customers wait in line. Under the old system, a random sample of 10 customers had a variance of 144. Under the new system, a random sample of 21 customers had a variance of 100. At $\alpha = 0.10$, is there enough evidence to convince the manager to switch to the new system? Assume both populations are normally distributed.

SOLUTION Because $144 > 100$, $s_1^2 = 144$ and $s_2^2 = 100$. Therefore, s_1^2 and σ_1^2 represent the sample and population variances for the old system, respectively. Using the claim "the variance of the waiting times under the new system is less than the variance of the waiting times under the old system," the null and alternative hypotheses are

$$H_0: \sigma_1^2 \le \sigma_2^2 \quad \text{and} \quad H_a: \sigma_1^2 > \sigma_2^2. \text{ (Claim)}$$

Because the test is a right-tailed test with $\alpha = 0.10$, $\text{d.f.}_N = n_1 - 1 = 9$, and $\text{d.f.}_D = n_2 - 1 = 20$, the critical value is 1.96. Using the F-test, the test statistic is

$$F = \frac{s_1^2}{s_2^2} = \frac{144}{100} = 1.44.$$

The graph shows the location of the rejection region and the test statistic. Because F is not in the rejection region, you should fail to reject the null hypothesis. In other words, there is not enough evidence to convince the manager to switch to the new system.

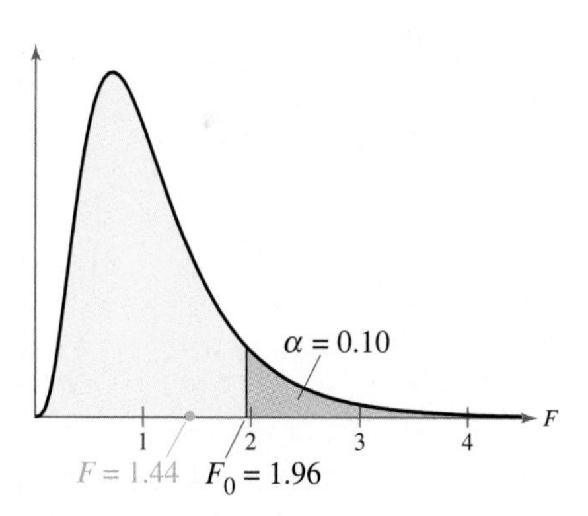

Try It Yourself 3

A medical researcher claims that a specially treated intravenous solution decreases the variance of the time required for nutrients to enter the bloodstream. Independent samples from each type of solution are randomly selected, and the results are shown in the table at the left. At $\alpha = 0.01$, is there enough evidence to support the researcher's claim? Assume the populations are normally distributed.

Normal solution	Treated solution
$n = 25$	$n = 20$
$s^2 = 180$	$s^2 = 56$

a. *Identify* the claim and *state* H_0 and H_a.
b. *Specify* the level of significance α.
c. *Determine* the degrees of freedom for the numerator and for the denominator.
d. *Find* the critical value and *identify* the rejection region.
e. *Use* the F-test to find the test statistic F.
f. *Decide* whether to reject the null hypothesis. Use a graph if necessary.
g. Is there enough evidence to support the claim?

Answer: Page A42

EXAMPLE 4

Performing a Two-Sample F-Test

Stock A	Stock B
$n = 30$ $s = 3.5$	$n = 31$ $s = 5.7$

You want to purchase stock in a company and are deciding between two different stocks. Because a stock's risk can be associated with the standard deviation of its daily closing prices, you randomly select samples of the daily closing prices for each stock to obtain the results shown at the left. At $\alpha = 0.05$, can you conclude that one of the two stocks is a riskier investment? Assume the stock closing prices are normally distributed.

Study Tip

When sample standard deviations are given, be sure to square them to obtain the sample variances.

SOLUTION Because $5.7^2 > 3.5^2$, $s_1^2 = 5.7^2$ and $s_2^2 = 3.5^2$. Therefore, s_1^2 and σ_1^2 represent the sample and population variances for Stock B, respectively. Using the claim "one of the stocks is a riskier investment," the null and alternative hypotheses are

H_0: $\sigma_1^2 = \sigma_2^2$ and H_a: $\sigma_1^2 \neq \sigma_2^2$. (Claim)

Because the test is a two-tailed test with $\frac{1}{2}\alpha = \frac{1}{2}(0.05) = 0.025$, d.f.$_N = n_1 - 1 = 30$, and d.f.$_D = n_2 - 1 = 29$, the critical value is 2.09. Using the F-test, the test statistic is

$$F = \frac{s_1^2}{s_2^2} = \frac{5.7^2}{3.5^2} \approx 2.65.$$

The graph shows the location of the rejection region and the test statistic. Because F is in the rejection region, you should reject the null hypothesis. In other words, one of the two stocks is a riskier investment.

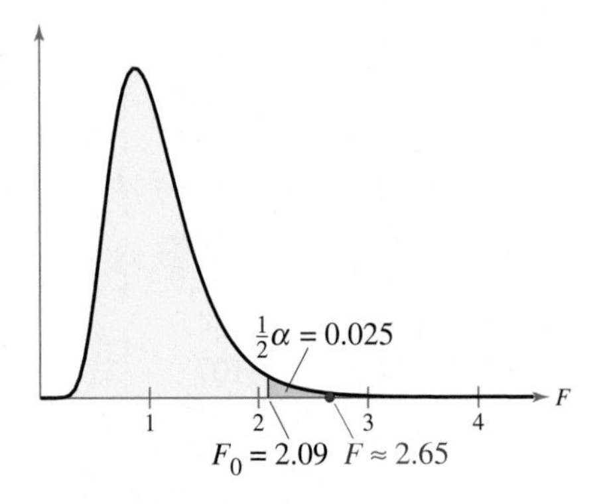

Try It Yourself 4

Location A	Location B
$n = 16$ $s = 0.95$	$n = 22$ $s = 0.78$

A biologist claims that the pH levels of the soil in two geographic locations have equal standard deviations. Independent samples from each location are randomly selected, and the results are shown at the left. At $\alpha = 0.01$, is there enough evidence to reject the biologist's claim? Assume the pH levels are normally distributed.

a. *Identify* the claim and *state* H_0 and H_a.
b. *Specify* the level of significance α.
c. *Determine* the degrees of freedom for the numerator and for the denominator.
d. *Find* the critical value and *identify* the rejection region.
e. *Use* the F-test to find the test statistic F.
f. *Decide* whether to reject the null hypothesis. Use a graph if necessary.
g. Is there enough evidence to reject the claim?

Answer: Page A42

10.3 Exercises

LarsonTutor 10.3

Companion Web Site

Student Solutions Manual 10.3

Videos 10.3

Technology Manuals

Basic Skills and Concepts

1. Explain how to find the critical value for an F-test.

2. List five properties of the F-distribution.

In Exercises 3–8, find the critical F-value for a right-tailed test using the indicated level of significance α and degrees of freedom d.f._N and d.f._D.

3. $\alpha = 0.05, \text{d.f.}_N = 4, \text{d.f.}_D = 18$

4. $\alpha = 0.01, \text{d.f.}_N = 3, \text{d.f.}_D = 24$

5. $\alpha = 0.01, \text{d.f.}_N = 5, \text{d.f.}_D = 11$

6. $\alpha = 0.05, \text{d.f.}_N = 6, \text{d.f.}_D = 19$

7. $\alpha = 0.10, \text{d.f.}_N = 10, \text{d.f.}_D = 15$

8. $\alpha = 0.025, \text{d.f.}_N = 7, \text{d.f.}_D = 3$

In Exercises 9–14, test the claim about the differences between two population variances σ_1^2 and σ_2^2 at the given level of significance α using the given sample statistics. Assume the sample statistics are from independent samples that are randomly selected and each population has a normal distribution.

9. Claim: $\sigma_1^2 > \sigma_2^2, \alpha = 0.10$.
Sample statistics: $s_1^2 = 773$, $n_1 = 5$; $s_2^2 = 765, n_2 = 6$

10. Claim: $\sigma_1^2 = \sigma_2^2, \alpha = 0.05$.
Sample statistics: $s_1^2 = 310$, $n_1 = 7$; $s_2^2 = 297, n_2 = 8$

11. Claim: $\sigma_1^2 \leq \sigma_2^2, \alpha = 0.01$.
Sample statistics: $s_1^2 = 842$, $n_1 = 11$; $s_2^2 = 836, n_2 = 10$

12. Claim: $\sigma_1^2 \neq \sigma_2^2, \alpha = 0.05$.
Sample statistics: $s_1^2 = 141$, $n_1 = 15$; $s_2^2 = 117, n_2 = 14$

13. Claim: $\sigma_1^2 = \sigma_2^2, \alpha = 0.05$.
Sample statistics: $s_1^2 = 5.2$, $n_1 = 7$; $s_2^2 = 4.8, n_2 = 9$

14. Claim: $\sigma_1^2 > \sigma_2^2, \alpha = 0.05$.
Sample statistics: $s_1^2 = 44.6$, $n_1 = 16$; $s_2^2 = 39.3, n_2 = 12$

Comparing Two Variances In Exercises 15–22, (a) identify the claim and state H_0 and H_a, (b) find the critical value and identify the rejection region, (c) find the test statistic, and (d) decide whether to reject or fail to reject the null hypothesis. Then interpret the decision in the context of the original claim. Assume the samples are independent and each population has a normal distribution.

15. ***Life of Appliances*** ◆ Company A claims that the variance of the life of its appliances is less than the variance of the life of Company B's appliances. A random sample of the lives of 20 of Company A's appliances has a variance of 2.6. A random sample of the lives of 23 of Company B's appliances has a variance of 2.8. At $\alpha = 0.05$, can you support Company A's claim? *(Adapted from Consumer Reports)*

16. ***Fuel Consumption*** ◆ An automobile manufacturer claims that the variance of the fuel consumption for its luxury sedans is less than the variance of the fuel consumption for the luxury sedans of a top competitor. A random sample of the fuel consumption of 19 of the manufacturer's sedans has a variance of 4.2. A random sample of the fuel consumption of 22 of its competitor's sedans has a variance of 4.5. At $\alpha = 0.05$, can you support the manufacturer's claim? *(Adapted from Consumer Reports)*

17. ***Physical Science Assessment*** ◆ In a recent interview, a state school administrator stated that the standard deviations of physical science assessment test scores for eighth-grade students are the same in Districts 1 and 2. If a random sample of 12 test scores from District 1 has a standard deviation of 27.7 points and a random sample of 14 test scores from District 2 has a standard deviation of 26.1 points, can you reject the administrator's claim? Use $\alpha = 0.10$. *(Adapted from National Center for Educational Statistics)*

18. ***Comparison of Test Scores*** ◆ A school administrator reports that the standard deviation of the test scores for eighth-grade students are the same in District 1 and 2. As proof, the administrator gives the results of a study of test scores in each district. The study shows that a random sample of 10 test scores from District 1 has a standard deviation of 28.8 points and a random sample of 13 test scores from District 2 has a standard deviation of 26.8 points. At $\alpha = 0.01$, can you reject the administrator's claim? *(Adapted from National Center for Educational Statistics)*

19. ***Waiting Times*** ◆ A random sample of 25 waiting times (in minutes) before patients saw a medical professional in a hospital's minor emergency department had a standard deviation of 0.7 minute. After implementing a new admissions procedure, a random sample of 21 waiting times had a standard deviation of 0.5 minute. At $\alpha = 0.10$, can you support the hospital's claim that the standard deviation of the waiting times has decreased?

20. ***Room Rates*** ◆ A travel agency's marketing brochure indicates that the standard deviations of hotel room rates for two cities are the same. If a random sample of 13 hotel room rates in one city has a standard deviation of \$27.50 and a random sample of 15 hotel room rates in the other city has a standard deviation of \$29.75, can you reject the agency's claim? Use $\alpha = 0.01$. *(Adapted from Smith Travel Research)*

21. ***Annual Salaries*** ◆ The annual salaries for a random sample of 16 actuaries working in California have a standard deviation of \$13,900. The annual salaries for a random sample of 17 actuaries working in New York have a standard deviation of \$8800. Using this information, can you conclude that the standard deviation of the annual salaries for actuaries is greater in California than in New York? Use $\alpha = 0.05$. *(Adapted from America's Career InfoNet)*

Actuaries in California
$s_1 = \$13{,}900$
$n_1 = 16$

Actuaries in New York
$s_2 = \$8800$
$n_2 = 17$

Figure for 21

Actuaries in Connecticut
$s_1 = \$9900$
$n_1 = 22$

Actuaries in Colorado
$s_2 = \$7800$
$n_2 = 24$

Figure for 22

22. ***Annual Salaries*** ◆ An employment information service claims the standard deviation of the annual salaries for public relations managers is greater in Connecticut than in Colorado. The annual salaries for a random sample of 22 public relations managers in Connecticut have a standard deviation of \$9900. The annual salaries for a random sample of 24 public relations managers in Colorado have a standard deviation of \$7800. At $\alpha = 0.05$, can you support the service's claim? *(Adapted from America's Career InfoNet)*

Extending the Basics

Finding Left-Tailed Critical F-values In this section you learned that if s_1^2 is larger than s_2^2, then you only need to calculate the right-tailed critical F-value for a two-tailed test. For other applications of the F-distribution, you will need to calculate the left-tailed critical F-value. To calculate the left-tailed critical F-value, do the following.

(1) Interchange the values for d.f._N and d.f._D.

(2) Find the corresponding F-value in Table 7.

(3) Calculate the reciprocal of the F-value to obtain the left-tailed critical F-value.

In Exercises 23 and 24, find the right- and left-tailed critical F-values for a two-tailed test with the given values of α, d.f._N and d.f._D.

23. $\alpha = 0.05, \text{d.f.}_N = 6, \text{d.f.}_D = 3$

24. $\alpha = 0.10, \text{d.f.}_N = 20, \text{d.f.}_D = 17$

Confidence Interval for σ_1^2/σ_2^2 When s_1^2 and s_2^2 are the variances of randomly selected, independent samples from normally distributed populations, then a confidence interval for σ_1^2/σ_2^2 is

$$\frac{s_1^2}{s_2^2}F_L < \frac{\sigma_1^2}{\sigma_1^2} < \frac{s_1^2}{s_2^2}F_R$$

where F_L is the left-tailed critical F-value and F_R is the right-tailed critical F-value.

In Exercises 25 and 26, construct the indicated confidence interval for σ_1^2/σ_2^2. Assume the samples are independent and each population has a normal distribution.

25. ***Cholesterol Content*** ◆ In a recent study of the cholesterol content in grilled chicken sandwiches served in fast food restaurants, a nutritionist found that a random sample of sandwiches from Arby's and from McDonald's had the sample statistics shown in the table. Construct a 95% confidence interval for σ_1^2/σ_2^2, where σ_1^2 and σ_2^2 are the variances of the cholesterol content of grilled chicken sandwiches from Arby's and McDonald's, respectively. *(Adapted from Fast Food Facts, Minnesota Attorney General's Office)*

Cholesterol content for grilled chicken sandwiches		
Restaurant	Arby's	McDonald's
Sample variance	$s_1^2 = 9.61$	$s_2^2 = 8.41$
Sample size	$n_1 = 15$	$n_2 = 12$

Table for Exercise 25

Carbohydrate content for grilled chicken sandwiches		
Restaurant	Arby's	McDonald's
Sample variance	$s_1^2 = 4.84$	$s_2^2 = 3.24$
Sample size	$n_1 = 15$	$n_2 = 12$

Table for Exercise 26

26. ***Carbohydrate Content*** ◆ A fast food study found that the carbohydrate content of 15 randomly selected grilled chicken sandwiches from Arby's had a variance of 4.84. The study also found that the carbohydrate content of 12 randomly selected grilled chicken sandwiches from McDonald's had a variance of 3.24. Construct a 95% confidence interval for σ_1^2/σ_2^2, where σ_1^2 and σ_2^2 are the variances of the carbohydrate content of grilled chicken sandwiches from Arby's and McDonald's, respectively. *(Adapted from Fast Food Facts, Minnesota Attorney General's Office)*

10.4 Analysis of Variance

What You Should Learn

- How to use one-way analysis of variance to test claims involving three or more means
- An introduction to two-way analysis of variance

One-Way ANOVA • Two-Way ANOVA

One-Way ANOVA

Suppose a medical researcher is analyzing the effectiveness of three types of pain relievers and wants to determine whether there is a difference in the mean length of the time it takes each medication to provide relief. To determine whether such a difference exists, the researcher can use the F-distribution together with a technique called *analysis of variance.* Because one independent variable is being studied, the process is called *one-way analysis of variance.*

DEFINITION

One-way analysis of variance is a hypothesis-testing technique that is used to compare means from three or more populations. Analysis of variance is usually abbreviated as **ANOVA.**

To begin a one-way analysis of variance test, you should first state a null and an alternative hypothesis. For a one-way ANOVA test, the null and alternative hypotheses are always similar to the following statements.

H_0: $\mu_1 = \mu_2 = \mu_3 = \cdots = \mu_k$ (All population means are equal.)

H_a: At least one of the means is different from the others.

When you reject the null hypothesis in an ANOVA test, you can conclude that one of the means is different from the others. Without performing more statistical tests, however, you cannot determine which of the means is different.

To use a one-way ANOVA test, the following conditions must be true.

1. Each sample must be randomly selected from a normal, or approximately normal, population.
2. The samples must be independent of each other.
3. Each population must have the same variance.

The test statistic for a one-way ANOVA test is the ratio of two variances: the variance between samples and the variance within samples.

$$\text{Test statistic} = \frac{\text{variance between samples}}{\text{variance within samples}}$$

1. The variance between samples MS_B measures the differences related to the treatment given to each sample and is sometimes called the **mean square between.**
2. The variance within samples MS_W measures the differences related to entries within the same sample. This variance, sometimes called the **mean square within,** is usually due to sampling error.

One-Way Analysis of Variance Test

If the conditions for a one-way analysis of variance test are satisfied, then the sampling distribution for the test is the F-distribution. The test statistic is

$$F = \frac{MS_B}{MS_W}.$$

The degrees of freedom for the F-test are

$$\text{d.f.}_N = k - 1$$

and

$$\text{d.f.}_D = N - k$$

where k is the number of samples and N is the sum of the sample sizes.

If there is little or no difference between the means, then MS_B will be approximately equal to MS_W and the test statistic will be approximately 1. Values of F close to 1 suggest that you should fail to reject the null hypothesis. However, if one of the means differs significantly from the others, MS_B will be greater than MS_W and the test statistic will be greater than 1. Values of F significantly greater than 1 suggest that you reject the null hypothesis. As such, all one-way ANOVA tests are right-tailed tests. That is, if the test statistic is greater than the critical value, H_0 will be rejected.

Study Tip

The notations n_i, $\bar{x}_i$, and s_i^2 represent the sample size, mean, and variance of the ith sample, respectively. $\bar{\bar{x}}$ is sometimes called the grand mean.

GUIDELINES

Finding the Test Statistic for a One-Way ANOVA Test

In Words	*In Symbols*
1. Find the mean and variance of each sample.	$\bar{x} = \frac{\Sigma x}{n} \quad s^2 = \frac{\Sigma(x - \bar{x})^2}{n - 1}$
2. Find the mean of all entries in all samples (the grand mean).	$\bar{\bar{x}} = \frac{\Sigma x}{N}$
3. Find the sum of squares between the samples.	$SS_B = \Sigma n_i(\bar{x}_i - \bar{\bar{x}})^2$
4. Find the sum of squares within the samples.	$SS_W = \Sigma(n_i - 1)s_i^2$
5. Find the variance between the samples.	$MS_B = \frac{SS_B}{k - 1} = \frac{SS_B}{\text{d.f.}_N}$
6. Find the variance within the samples.	$MS_W = \frac{SS_W}{N - k} = \frac{SS_W}{\text{d.f.}_D}$
7. Find the test statistic.	$F = \frac{MS_B}{MS_W}$

The notation SS_B represents the sum of squares between groups.

$$SS_B = n_1(\overline{x}_1 - \overline{\overline{x}})^2 + n_2(\overline{x}_2 - \overline{\overline{x}})^2 + \cdots + n_k(\overline{x}_k - \overline{\overline{x}})^2$$

$$= \Sigma n_i(\overline{x}_i - \overline{\overline{x}})^2$$

The notation SS_W represents the sum of squares within groups.

$$SS_W = (n_1 - 1)s_1^2 + (n_2 - 1)s_2^2 + \cdots + (n_k - 1)s_k^2$$

$$= \Sigma(n_i - 1)s_i^2$$

GUIDELINES

Performing a One-Way Analysis of Variance Test

In Words	*In Symbols*
1. Identify the claim. State the null and alternative hypotheses.	State H_0 and H_a.
2. Specify the level of significance.	Specify α.
3. Determine the degrees of freedom.	$\text{d.f.}_N = k - 1$ $\text{d.f.}_D = N - k$
4. Find the critical value.	Use Table 7 in Appendix B.
5. Identify the rejection region.	
6. Calculate the test statistic.	$F = \dfrac{MS_B}{MS_W}$
7. Make a decision to reject or fail to reject the null hypothesis.	If F is in the rejection region, reject H_0. Otherwise, do not reject H_0.
8. Interpret the decision in the context of the original claim.	

Tables are a convenient way to summarize the results of a one-way analysis of variance test. ANOVA summary tables are set up as shown below.

ANOVA Summary Table

Variation	Sum of squares	Degrees of freedom	Mean squares	F
Between	SS_B	d.f._N	$MS_B = \dfrac{SS_B}{\text{d.f.}_N}$	$MS_B \div MS_W$
Within	SS_W	d.f._D	$MS_W = \dfrac{SS_W}{\text{d.f.}_D}$	

EXAMPLE 

Performing a One-Way ANOVA Test

A medical researcher wants to determine whether there is a difference in the mean length of time it takes three types of pain relievers to provide relief from headache pain. Several headache sufferers are randomly selected and given one of the three medications. Each headache sufferer records the time (in minutes) it takes the medication to begin working. The results are listed in the following table. At $\alpha = 0.01$, can you conclude that the mean times are different? Assume that each population of relief times is normally distributed and that the population variances are equal.

Medication 1	Medication 2	Medication 3
12	16	14
15	14	17
17	21	20
12	15	15
	19	
$\bar{x}_1 = \frac{56}{4} = 14$	$\bar{x}_2 = \frac{85}{5} = 17$	$\bar{x}_3 = \frac{66}{4} = 16.5$
$s_1^2 = 6$	$s_2^2 = 8.5$	$s_3^2 = 7$

SOLUTION The null and alternative hypotheses are as follows.

H_0: $\mu_1 = \mu_2 = \mu_3$

H_a: At least one mean is different from the others.

Because there are $k = 3$ samples, $\text{d.f.}_\text{N} = k - 1 = 3 - 1 = 2$. The sum of the sample sizes is $N = n_1 + n_2 + n_3 = 4 + 5 + 4 = 13$. So, $\text{d.f.}_\text{D} = N - k = 13 - 3 = 10$. Using $\text{d.f.}_\text{N} = 2$, $\text{d.f.}_\text{D} = 10$, and $\alpha = 0.01$, the critical value is 7.56. To find the test statistic, first calculate $\bar{\bar{x}}$, MS_B, and MS_W.

$$\bar{\bar{x}} = \frac{\Sigma x}{N} = \frac{56 + 85 + 66}{13} \approx 15.92$$

$$\begin{aligned} MS_B = \frac{SS_B}{\text{d.f.}_\text{N}} &= \frac{\Sigma n_i(\bar{x}_i - \bar{\bar{x}})^2}{k - 1} \\ &\approx \frac{4(14 - 15.92)^2 + 5(17 - 15.92)^2 + 4(16.5 - 15.92)^2}{3 - 1} \\ &\approx \frac{21.92}{2} = 10.96 \end{aligned}$$

$$\begin{aligned} MS_W = \frac{SS_W}{\text{d.f.}_\text{D}} &= \frac{\Sigma(n_i - 1)s_i^2}{N - k} \\ &= \frac{(4 - 1)(6) + (5 - 1)(8.5) + (4 - 1)(7)}{13 - 3} = \frac{73}{10} = 7.3 \end{aligned}$$

Using $MS_B = 10.96$ and $MS_W = 7.3$, the test statistic is

$$F = \frac{MS_B}{MS_W} = \frac{10.96}{7.3} \approx 1.50.$$

The graph shows the location of the rejection region. Because $F = 1.50$ is not in the rejection region, you should decide not to reject the null hypothesis. In other words, there is not enough evidence at the 1% level of significance to conclude that there is a difference in the mean length of time it takes the three pain relievers to provide relief from headache pain.

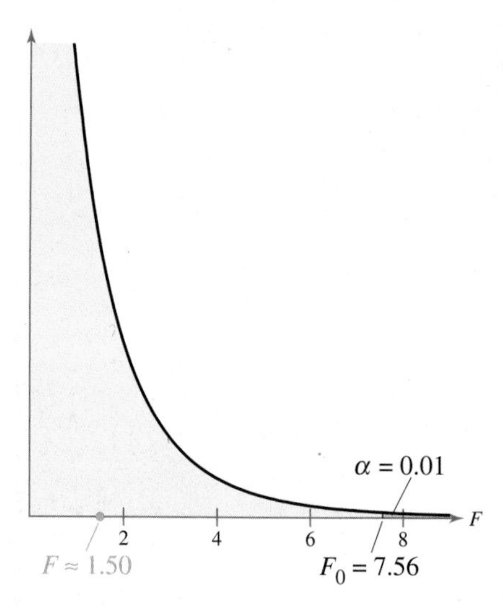

The ANOVA summary table for Example 1 is shown below.

Variation	Sum of squares	Degrees of freedom	Mean squares	F
Between	21.92	2	10.96	1.50
Within	73	10	7.3	

Picturing the World

Adult common cuckoos do not care for their own eggs. Cuckoos lay their eggs in the nests of other birds, such as sparrows, robins, and wrens. These birds adopt the cuckoo eggs, hatching and caring for the eggs along with their own. A biologist is studying the lengths of cuckoo eggs found in sparrow, robin, and wren nests. The randomly selected egg lengths (in millimeters) are listed in the following table.

Host nest		
Sparrow	**Robin**	**Wren**
24.08	22.66	20.89
23.40	22.51	20.97
22.95	21.44	22.31
22.82	22.70	21.54
23.98	22.15	21.19
24.59	22.75	20.19
22.95	23.02	21.38
25.16	21.72	20.50
23.39	21.49	20.99
23.74	22.15	21.16
21.15	22.15	20.83
23.81	22.28	20.74
22.40	22.32	21.40
24.17	22.98	21.90

At $\alpha = 0.05$, can you conclude that the mean length of the cuckoo eggs found in one type of nest is different from the others? Assume that each population of cuckoo egg lengths is normally distributed and that the population variances are equal.

Try It Yourself 1

A sales analyst wants to determine whether there is a difference in the mean monthly sales of a company's four sales regions. Several salespersons from each region are randomly selected and each provides his or her sales amounts (in thousands of dollars) for the previous month. The results are listed in the following table. At $\alpha = 0.05$, can the analyst conclude that there is a difference in the mean monthly sales among the sales regions? Assume that each population of sales is normally distributed and that the population variances are equal.

North	East	South	West
34	47	40	21
28	36	30	30
18	30	41	24
24	38	29	37
	44		23
$\bar{x}_1 = 26$	$\bar{x}_2 = 39$	$\bar{x}_3 = 35$	$\bar{x}_4 = 27$
$s_1^2 \approx 45.33$	$s_2^2 = 45$	$s_3^2 \approx 40.67$	$s_4^2 = 42.5$

a. *State* H_0 and H_a.
b. *Specify* the level of significance α.
c. *Determine* the degrees of freedom for the numerator and for the denominator.
d. *Find* the critical value and *identify* the rejection region.
e. *Find* the test statistic F.
f. *Decide* whether to reject the null hypothesis. Use a graph if necessary.
g. Is there enough evidence to conclude that there is a difference in the mean monthly sales among the sales regions?

Answer: Page A42

Using technology greatly simplifies the one-way ANOVA process. When using a technology tool such as *Excel*, *Minitab*, or the *TI-83* to perform a one-way analysis of variance test, you can use *P*-values to decide whether to reject the null hypothesis. If the *P*-value is less than α, you should reject H_0.

EXAMPLE 

Using Technology to Perform ANOVA Tests

Three airline companies offer flights between Corydon and Lincolnville. Several randomly selected flight times (in minutes) between the towns for each airline are shown at the left. Assume that the populations of flight times are normally distributed, the samples are independent, and that the population variances are equal. At $\alpha = 0.01$, can you conclude that there is a difference in the means of the flight times? Use a technology tool.

Airline 1	Airline 2	Airline 3
122	119	120
135	133	158
126	143	155
131	149	126
125	114	147
116	124	164
120	126	134
108	131	151
142	140	131
113	136	141

SOLUTION The results obtained using the *TI-83* are shown below. From the results, you can see that $P \approx 0.006$. Because $P < \alpha$, you should reject the null hypothesis. In other words, you can conclude that there is a difference in the means of the flight times.

TI-83

```
One-way ANOVA
 F=6.13023478
 p=.0063828069
 Factor
  df=2
  SS=1806.46667
↓ MS=903.233333
```

TI-83

```
One-way ANOVA
↑ MS=903.233333
 Error
  df=27
  SS=3978.2
  MS=147.340741
 Sxp=12.1383994
```

Try It Yourself 2

The data listed in the following table represent the GPAs of randomly selected freshmen, sophomores, juniors, and seniors. At $\alpha = 0.05$, can you conclude that there is a difference in the means of the GPAs? Assume that the populations of GPAs are normally distributed and that the population variances are equal. Use a technology tool.

Freshmen	2.34	2.38	3.31	2.39	3.40	2.70	2.34			
Sophomores	3.26	2.22	3.26	3.29	2.95	3.01	3.13	3.59	2.84	3.00
Juniors	2.80	2.60	2.49	2.83	2.34	3.23	3.49	3.03	2.87	
Seniors	3.31	2.35	3.27	2.86	2.78	2.75	3.05	3.31		

a. *Enter* the data.
b. *Perform* the ANOVA test.
c. *Compare* the resulting *P*-value to the given level of significance α.
d. Can you conclude that there is a difference in the means of the GPAs?

Answer: Page A42

Two-Way ANOVA

Insight

The conditions for a two-way ANOVA test are the same as those for a one-way ANOVA test with the additional condition that all samples must be of equal size.

When you want to test the effect of *two* independent variables, or factors, on one dependent variable, you can use a **two-way analysis of variance test.** For example, suppose a medical researcher wants to test the effect of gender *and* type of medication on the mean length of time it takes pain relievers to provide relief. To perform such an experiment, the researcher can use the following two-way ANOVA block design.

Type of medication	Gender: M	Gender: F
I	Males taking type I	Females taking type I
II	Males taking type II	Females taking type II
III	Males taking type III	Females taking type III

Insight

If gender and type of medication have no effect on the length of time it takes a pain reliever to provide relief, then there will be no significant difference in the means of the relief times.

A two-way ANOVA test has three null hypotheses—two for each main effect and one for the interaction effect. A **main effect** is the effect of one independent variable on the dependent variable and the **interaction effect** is the effect of both independent variables on the dependent variable. For example, the hypotheses for the pain reliever experiment are as follows.

Hypotheses for main effects:

H_0: Gender has no effect on the mean length of time it takes a pain reliever to provide relief.

H_a: Gender has an effect on the mean length of time it takes a pain reliever to provide relief.

H_0: The type of medication has no effect on the mean length of time it takes a pain reliever to provide relief.

H_a: The type of medication has an effect on the mean length of time it takes a pain reliever to provide relief.

Hypotheses for interaction effect:

H_0: There is no interaction effect between gender and type of medication on the mean length of time it takes a pain reliever to provide relief.

H_a: There is an interaction effect between gender and type of medication on the mean length of time it takes a pain reliever to provide relief.

To test these hypotheses, you can perform a two-way ANOVA test. Using the F-distribution, a two-way ANOVA test calculates an F-test statistic for each hypothesis. As a result, it is possible to reject none, one, two, or all of the null hypotheses. The statistics involved with a two-way ANOVA test is beyond the scope of this course. You can, however, use a technology tool such as *Minitab* to perform a two-way ANOVA test.

10.4 Exercises

Help

LarsonTutor 10.4

Companion Web Site

Student Solutions Manual 10.4

Videos 10.4

Technology Manuals

Basic Skills and Concepts

Performing a One-Way ANOVA Test In Exercises 1–13, perform the indicated one-way ANOVA test by

(a) identifying the claim and stating H_0 and H_a.

(b) determining the degrees of freedom for the numerator and for the denominator, identifying the critical value, and identifying the rejection region.

(c) calculating the test statistic.

(d) deciding to reject or fail to reject the null hypothesis and interpreting the decision in the context of the original claim.

Assume each sample is drawn from a normal, or approximately normal, population, that the samples are independent of each other, and that the populations have the same variances. If convenient, use technology to solve the problem.

1. ***Toothpaste*** ◆ The following table shows the cost per month (in dollars) for a random sample of toothpastes exhibiting moderate abrasiveness, low abrasiveness, or very low abrasiveness. At $\alpha = 0.05$, can you conclude that the mean costs per month are different? *(Source: Consumer Reports)*

Moderate abrasiveness	Low abrasiveness	Very low abrasiveness
0.75	0.74	0.44
0.46	0.96	1.40
3.16	0.65	4.00
0.97	0.70	0.91
0.72	0.72	
2.74	1.71	
0.62	1.46	
0.72	1.57	
1.00	0.77	
0.78	0.98	
1.01	0.61	
1.11	1.04	
1.45	0.61	

2. ***Automobile Batteries*** ◆ The prices (in dollars) for 16 randomly selected automobile batteries are listed in the table. The prices are classified according to battery type. At $\alpha = 0.05$, is there enough evidence to conclude that at least one of the mean battery prices is different from the others? *(Source: Consumer Reports)*

Group size 24	50	65	75	64		
Group size 58	83	63	65	63	85	65
Group size 34/78	89	84	75	90	70	60

3. ***Exterior Deck Treatments*** ◆ The following table shows the price per gallon (in dollars) for a random sample of exterior deck treatments. At $\alpha = 0.10$, can you reject the claim that the mean prices are the same for the three types of treatments? *(Source: Consumer Reports)*

Semitransparent treatments	Lightly tinted treatments	Clear treatments
24	51	13
23	14	13
22	21	10
17	16	12
21		22
17		

DATA 4. ***Reading Expenditures*** ◆ The following table shows the annual amount spent on reading (in dollars) for a random sample of United States consumers from four regions. At $\alpha = 0.10$, can you reject the claim that the mean annual amounts are the same in all regions? *(Adapted from U.S. Bureau of Labor Statistics)*

Northeast	Midwest	South	West
308	246	103	223
58	169	143	184
141	246	164	221
109	158	119	269
220	167	99	199
144	76	214	171
316		108	204

DATA 5. ***Days Spent at a Hospital*** ◆ In a recent study, a health insurance company investigated the number of days patients spend at a hospital. As part of the study, the company randomly selected patients from various parts of the United States and recorded the number of days each patient spent at a hospital. The results of the study are shown below. At $\alpha = 0.01$, can the company reject the claim that the mean number of days patients spend in the hospital is the same for all four regions? *(Adapted from U.S. National Center for Health Statistics)*

Northeast	Midwest	South	West
8	7	3	5
6	7	5	4
9	8	6	6
4	5	6	4
5	5	3	6
6	5	6	6
8	5	4	5
10	5	6	4
11	5		6
7	7		

6. *Building Space* ◆ The following table shows the square footage (in thousands) for a random sample of American buildings from four regions. At $\alpha = 0.10$, can you conclude that the mean square footage for one of the regions is different? *(Adapted from U.S. Energy Information Administration)*

Northeast	Midwest	South	West
13.9	10.4	14.1	13.0
18.0	11.6	16.4	15.0
15.7	5.3	15.6	11.4
11.7	10.8	3.1	10.4
24.6	12.3	16.6	11.5
12.2	14.5	17.0	16.0
18.1	3.8	13.2	5.5
10.0	15.2	8.6	20.0
8.7	18.8	25.5	7.0
17.1	13.6	9.6	2.8
17.4	11.4	12.8	15.3
16.3	7.9	15.4	14.6
19.4	15.2	16.2	11.0

7. *Housing Prices* ◆ A realtor is comparing the prices of one-family houses in four cities. After randomly selecting one-family houses in the four cities and determining the price for each, the realtor organizes the prices (in thousands of dollars) in a table as shown below. At $\alpha = 0.10$, can the realtor reject the claim that the mean price is the same for all four cities? *(Adapted from National Association of Realtors)*

City A	City B	City C	City D
131.2	137.9	164.4	83.9
122.6	88.0	49.4	98.3
103.7	95.1	144.7	168.9
148.8	79.3	108.6	93.7
144.5	81.5	137.2	204.3
206.9	101.9	83.6	85.1

City A	City B	City C	City D
164.6	128.8	100.5	100.4
45.0	147.2	99.4	211.7
162.8	38.7	67.8	140.1
132.7	57.9	147.9	131.3
178.2	67.3	39.2	167.0
131.2		149.9	84.3

8. *Personal Income* ◆ The following table lists the salaries of randomly selected individuals from four large metropolitan areas. At $\alpha = 0.05$, can you conclude that the mean salary is different in at least one of the areas? *(Adapted from U.S. Bureau of Economic Analysis)*

Pittsburgh	Dallas	Chicago	Minneapolis
27,800	30,000	32,000	30,000
28,000	33,900	35,800	40,000
25,500	29,750	28,000	35,000
29,150	25,000	38,900	33,000
30,295	34,055	27,245	29,805

9. *Mobile Home Prices* ◆ The following table shows the prices (in dollars) for a random sample of new mobile homes from four manufacturers. At $\alpha = 0.10$, can you conclude that at least one of the mean prices is different? *(Adapted from U.S. Bureau of the Census)*

Manufacturer A	Manufacturer B	Manufacturer C	Manufacturer D
34,918	37,223	39,029	43,176
34,904	38,079	38,711	52,844
33,971	35,646	21,774	42,936
39,404	39,420	28,836	41,685
58,000	34,732	35,535	45,320
32,710	51,223	25,185	54,841
45,049	40,630	35,791	
48,373		32,607	
		27,728	

10. *Energy Consumption* ◆ The following table shows the energy consumed (in millions of Btu's) in one year for a random sample of households from four regions. At $\alpha = 0.01$, can you conclude that, for at least one region, the mean energy consumption is different? *(Adapted from U.S. Energy Information Administration)*

Northeast	Midwest	South	West
115.2	53.0	43.6	24.7
123.2	231.1	55.7	43.4
167.3	107.9	82.7	124.9
139.5	125.6	83.5	58.6
142.7	207.2	43.3	115.6
174.0	246.2	38.2	67.0
138.3	144.9	50.1	15.7
98.9	180.4	12.5	125.0
196.6	137.8	179.7	105.8
57.3	201.8		55.5
	84.2		

11. *Amount Spent on Energy* ◆ The following table shows the amount spent (in dollars) on energy in one year for a random sample of households from four regions. At $\alpha = 0.01$, can you reject the claim that the mean amounts spent are equal for all regions? *(Adapted from U.S. Energy Information Administration)*

Northeast	Midwest	South	West
2623	1310	1252	947
573	940	1997	1801
2142	1342	1678	1706
1146	1294	1105	1244
853	1190	1593	1345
1073	1514	1667	768
2193	689		1109

12. ***Sports Team Involvement*** ◆ The following table shows the number of students who played on a sports team for a random sample of female students in grades 9 through 12. At $\alpha = 0.01$, can you reject the claim that the mean numbers of female students who played on a sports team are equal for all grades? *(Adapted from U.S. Centers for Disease Control and Prevention)*

Grade 9	82	91	53	133	64	212
Grade 10	125	51	58	206	87	77
Grade 11	187	81	46	56	115	
Grade 12	42	77	165	65	49	102

13. ***Credit Card Balances*** ◆ The following table shows the credit card balance for a random sample of families whose head is in one of four age groups. At $\alpha = 0.01$, can you conclude that for at least one age group the mean credit card balance is different? *(Adapted from Board of Governors of the Federal Reserve System)*

Under 35	1655	1498	1523	1575	1430	1582
35–44	2013	2072	2072	2241	1965	
45–54	2050	1822	1822	2085	1993	2002
55–64	2047	2212	2212	2278	2192	

Extending the Basics

The Scheffé Test If the null hypothesis is rejected in a one-way ANOVA test of three or more means, a Scheffé Test can be performed to find which means have a significant difference. In a Scheffé Test, the means are compared two at a time. For example, with three means you would have the following comparisons: $\bar{x}_1$ versus $\bar{x}_2$, $\bar{x}_1$ versus $\bar{x}_3$, and $\bar{x}_2$ versus $\bar{x}_3$. For each comparison, calculate

$$\frac{(\bar{x}_a - \bar{x}_b)^2}{\frac{SS_w}{\Sigma(n_i - 1)}[(1/n_a) + (1/n_b)]}$$

where $\bar{x}_a$ and $\bar{x}_b$ are the means being compared and n_a and n_b are the corresponding sample sizes. Then compare the value to the critical value. Calculate the critical value using the same steps as in a one-way ANOVA test and multiply the result by $k - 1$. Use this information to solve Exercises 14–17.

14. Refer to the data in Exercise 6. At $\alpha = 0.10$, perform a Scheffé Test to determine which means have a significant difference.

15. Refer to the data in Exercise 7. At $\alpha = 0.10$, perform a Scheffé Test to determine which means have a significant difference.

16. Refer to the data in Exercise 10. At $\alpha = 0.01$, perform a Scheffé Test to determine which means have a significant difference.

17. Refer to the data in Exercise 9. At $\alpha = 0.01$, perform a Scheffé Test to determine which means have a significant difference.

TECHNOLOGY

MINITAB EXCEL TI-83

Crash Tests

At the beginning of this chapter, you learned that as part of the New Car Assessment Program, the government buys new cars each year and crashes them into a wall at 35 miles per hour to compare how well different vehicles protect front-seat passengers in a head-on collision. You also learned that the dummies used in a crash test are equipped with instruments that measure the forces and impacts that occur during the crash test.

The table at the right displays the left leg injury data from the crash tests of a random sample of three vehicle types: medium (e.g., Ford Taurus), heavy (e.g., Lincoln Town Car), and multiple purpose (e.g., Toyota 4-Runner).

Vehicle Type		
Medium	**Heavy**	**Multiple Purpose**
926	541	1911
996	406	1539
332	1529	267
1353	1132	388
519	767	932
705	1224	401
611	314	1100
1657	1728	1595
571	764	430
961	260	1909
1580	1527	606
775	766	469
1132	862	430
1512	1138	1277
1500	1326	1001
1554	883	1283

Exercises

In Exercises 1–3, refer to the following samples. Use $\alpha = 0.05$.

(a) Medium and heavy vehicles
(b) Medium and multiple-purpose vehicles
(c) Heavy and multiple-purpose vehicles

1. Are the samples independent of each other? Explain.

2. Use a technology tool to determine whether each sample is from a normal, or approximately normal, population.

3. Use a technology tool to determine whether the samples were selected from populations having equal variances.

4. Using the results of Exercises 1–3, discuss whether the three conditions for a one-way ANOVA test are satisfied. If so, use a technology tool to test the claim that medium, heavy, and multiple-purpose vehicles have the same mean left leg injury potential in a 35 miles per hour frontal crash. Use $\alpha = 0.05$.

5. Repeat Exercises 1–4 using the data in the following table. The table displays the right leg injury data from the crash tests of a random sample of three vehicle types: medium, heavy, and multiple purpose.

Vehicle Type		
Medium	**Heavy**	**Multiple Purpose**
708	1629	257
642	493	985
611	613	428
1063	667	501
892	979	322
582	1264	540
1409	1247	1015
751	855	2856
757	1113	821
144	495	515
454	626	718
438	575	420
716	863	141
913	776	692
1049	1090	302
1107	644	1252

Extended solutions are given in the *Technology Supplement*.
Technical instruction is provided for *Minitab*, *Excel*, and the *TI-83*.

Chapter Summary

What did you learn? | *Review Exercises*

Section 10.1

- How to use the chi-square distribution to test whether a frequency distribution fits a claimed distribution — *1, 2*

$$\chi^2 = \Sigma \frac{(O - E)^2}{E}$$

Section 10.2

- How to use a contingency table to find expected frequencies and how to use a chi-square distribution to test whether two variables are independent — *3, 4*

Section 10.3

- How to interpret the *F*-distribution and use an *F*-table to find critical values — *5–8*

$$F = \frac{s_1^2}{s_2^2}$$

- How to perform a two-sample *F*-test to compare two variances — *9–14*

Section 10.4

- How to use one-way analysis of variance to test claims involving three or more means — *15, 16*

STATISTICS

Uses and Abuses

Uses

One-Way Analysis of Variance (ANOVA) ANOVA can help you make important decisions about the allocation of resources. For instance, suppose you work for a large manufacturing company and part of your responsibility is to determine the distribution of the company's sales throughout the world and decide where to focus the company's efforts. Because wrong decisions will cost your company money, you want to make sure that you make the right decisions.

Abuses

Preconceived Notions There are several ways that the tests presented in this chapter can be abused. For example, it is easy to allow preconceived notions to affect the results of a chi-square goodness-of-fit test and a test for independence. When testing to see whether a distribution has changed, do not let the existing distribution "cloud" the study results. Similarly, when determining whether two variables are independent, do not let your intuition "get in the way." As with any hypothesis test, you must properly gather appropriate data and perform the corresponding test before you can reach a logical conclusion

Incorrect Interpretation of Rejection of Null Hypothesis It is important to remember that when you reject the null hypothesis of an ANOVA test, you are simply stating that you have enough evidence to determine that at least one of the population means is different from the others. You are not finding them all to be different. One way to further test which of the population means differs from the others is explained in Extending the Basics of Section 10.4 Exercises.

Exercises

1. ***Preconceived Notions*** ANOVA depends on having independent variables. Describe an abuse that might occur by having dependent variables. Then describe how the abuse could be avoided.

2. ***Incorrect Interpretation of Rejection of Null Hypothesis*** Find an example of the use of ANOVA. In that use, describe what would be meant by "rejection of the null hypothesis." How should rejection of the null hypothesis be correctly interpreted?

10 Review Exercises

Section 10.1

In Exercises 1 and 2, use a χ^2 goodness-of-fit test to test the claim about the population distribution. Interpret the decision in the context of the original claim.

1. A health care investigator wishes to test the following claim: Of all doctor's office visits in the United States, 25% are from new patients, 25% are from old patients with a new problem, and the remainder are old patients with a recurring problem. A random sample of various doctors finds that 97 patients were new, 142 were old with a new problem, and 457 were old patients with a recurring problem. Test the claim at $\alpha = 0.05$. *(Adapted from U.S. National Center for Health Statistics)*

2. A legal researcher is studying the age distribution of juries by comparing them to the overall age distribution of available jurors. The researcher claims that the jury distribution is different from the overall distribution; that is, there is a noticeable age bias in jury selection in this area. The following table shows the number of jurors at a county court in one year and the percent of persons residing in that county, by age. Use the population distribution to find the expected juror frequencies. Test the researcher's claim at $\alpha = 0.01$.

	21–29	30–39	40–49	50–59	60 and above
Jury	45	128	244	224	359
Population	20.5%	21.7%	18.1%	17.3%	22.4%

Section 10.2

In Exercises 3 and 4, use the given contingency table to (a) find the expected frequencies of each table element, (b) perform a χ^2 independence test, and (c) comment on the relationship between the two variables. Assume the variables are independent.

3. The following table shows the highest level of education of people in the United States by age category in a recent year. The numbers listed are in thousands of persons. Use $\alpha = 0.10$. *(Source: U.S. Census Bureau)*

	H.S.—did not complete	H.S. completed	College 1–3 years	College 4 or more years
25–44	10,102	26,980	23,014	23,123
45 and above	18,782	30,914	20,169	20,636

4. The contingency table shows the results of a random sample of 480 individuals classified by gender and type of vehicle owned. Use $\alpha = 0.05$.

	Type of vehicle owned			
	Car	**Truck**	**SUV**	**Van**
Males	85	96	45	6
Females	110	75	60	3

Section 10.3

In Exercises 5–8, find the critical *F*-value for a right-tailed test using the indicated level of significance α and degrees of freedom d.f._N and d.f._D.

5. $\alpha = 0.05, \text{d.f.}_N = 6, \text{d.f.}_D = 50$

6. $\alpha = 0.01, \text{d.f.}_N = 12, \text{d.f.}_D = 10$

7. $\alpha = 0.10, \text{d.f.}_N = 5, \text{d.f.}_D = 12$

8. $\alpha = 0.05, \text{d.f.}_N = 20, \text{d.f.}_D = 25$

In Exercises 9 and 10, use the given sample statistics to test the claim about two population variances σ_1^2 and σ_2^2 at the indicated level of significance α. Assume that both populations are normally distributed and the randomly selected samples are independent.

9. Claim: $\sigma_1^2 \leq \sigma_2^2$, $\alpha = 0.01$. Sample statistics: $s_1^2 = 653$, $n_1 = 16$, $s_2^2 = 270$, $n_2 = 21$

10. Claim: $\sigma_1^2 \neq \sigma_2^2$, $\alpha = 0.10$. Sample statistics: $s_1^2 = 112{,}676$, $n_1 = 6$, $s_2^2 = 49{,}572$, $n_2 = 11$

In Exercises 11–14, test the claim about two population variances at the indicated level of significance α. Interpret the results in the context of the claim.

11. An agricultural analyst is comparing the wheat production in Oklahoma counties. The analyst claims that the variation in wheat production is greater in Garfield County than in Kay County. A random sample of 21 Garfield County farms yields a standard deviation of 0.76 bushels/acre; 16 Kay County farms are found to have a standard deviation of 0.58 bushels/acre. Diagnostics suggest that wheat production is normally distributed in both counties. Test the analyst's claim at $\alpha = 0.10$. *(Adapted from Environmental Verification and Analysis Center—University of Oklahoma)*

12. A steel pipe fittings company claims that the yield strength of its nontempered couplings is more variable than that of its tempered couplings. A random sample of 9 tempered couplings has a standard deviation of 13.1 megapascals, while a similar sample of 9 nontempered couplings had a standard deviation of 25.4 megapascals. From past data, it is known that the company's production process results in normally distributed yield strengths. Test the company's claim at $\alpha = 0.05$.

DATA **13.** The following table shows the SAT verbal test scores for 9 randomly selected female students and 14 randomly selected male students. Assume that SAT verbal test scores are normally distributed. At $\alpha = 0.01$, test the claim that the test score variance for females is different than that for males.

Female	480	610	340	630	520	690	540
Male	560	680	360	530	380	460	630

Female	600	680					
Male	310	730	740	520	560	400	510

 **14.** A plastics company that produces automobile dashboard inserts has just received a new injection mold that is supposedly more consistent than its current mold. A quality technician wishes to test whether this new mold will produce inserts that are less variable in diameter than those produced with the company's current mold. The following table shows independent random samples (of size 12) of insert diameters (in centimeters) for both the current and new molds. Assume that the samples are from normally distributed populations. At $\alpha = 0.05$, test the claim that the new mold produces inserts that are less variable in diameter than the current mold produces.

New	9.611	9.618	9.594	9.580	9.611	9.597
Current	9.571	9.642	9.650	9.651	9.596	9.636

New	9.638	9.568	9.605	9.603	9.647	9.590
Current	9.570	9.537	9.641	9.625	9.626	9.579

Section 10.4

In Exercises 15 and 16, use the given sample data to perform a one-way ANOVA test using the indicated level of significance α. What can you conclude?

 **15.** The table at the right shows the residential energy expenditures (in dollars) in one year of a random sample of households in four U.S. regions. Diagnostics suggest that the sample is drawn from a normally distributed population. Use $\alpha = 0.10$ to test for differences among the means for the four regions. *(Adapted from U.S. Energy Information Administration)*

Northeast	Midwest	South	West
2088	759	888	1115
1259	1346	773	898
1762	975	1141	1605
1783	1523	1232	857
1623	1233	1310	465
2175	1903	953	359
1265	1236	1005	909
1258	1176	1475	1193

 **16.** The table at the right shows the annual income (in dollars) of a random sample of households in four U.S. regions. Diagnostics suggest that the sample is drawn from an approximately normally distributed population. Use $\alpha = 0.05$ to test for differences among the means for the four regions. *(Adapted from U.S. Census Bureau)*

Northeast	Midwest	South	West
50,533	27,175	37,348	43,970
41,798	50,788	61,090	36,378
50,661	48,847	25,393	32,587
26,789	29,602	35,098	30,665
47,620	27,152	23,261	
	42,364	47,798	
	9,078		

PUTTING IT ALL TOGETHER

Real Statistics Real Decisions

The National Fraud Information Center (NFIC) was established in 1992 by the National Consumers League (NCL) to combat the growing problem of telemarketing and Internet fraud by improving prevention and enforcement. NCL works to protect and promote the economic and social interests of consumers in the U.S., using education, research, science, and publications. The NCL was formed in 1899.

You work for the NFIC as a statistical analyst. You are studying data on telemarketing fraud. As part of your analysis, you are testing the goodness of fit, testing for independence, comparing variances, and performing ANOVA.

www.fraud.org

Exercises

1. ***Goodness of Fit*** A claimed distribution for the ages of telemarketing fraud victims is given in the table at the right. The results of a survey of 1000 randomly selected telemarketing fraud victims are also shown in the table. Using $\alpha = 0.01$, perform a chi-square goodness-of-fit test to test the claimed distribution. What can you conclude? Do you think the claimed distribution is valid? Why or why not?

Ages	Claimed distribution	Survey results
Under 20	1%	30
20–29	13%	200
30–39	16%	300
40–49	19%	270
50–59	16%	150
60–69	13%	40
70+	22%	10

Table for Exercise 1

2. ***Independence*** The following contingency table shows the results of a random sample of 2000 telemarketing fraud victims classified by age and type of fraud. The frauds were committed using bogus sweepstakes or credit card offers.

   (a) Calculate the expected frequency for each cell in the contingency table. Assume the variables age and type of fraud are independent.

   (b) Can you conclude that the age of the victims is related to the type of fraud? Use $\alpha = 0.01$.

Type of fraud	Age of victims								
	Under 20	20–29	30–39	40–49	50–59	60–69	70–79	80+	Total
Sweepstakes	10	60	70	130	90	160	280	200	1000
Credit cards	20	180	260	240	180	70	30	20	1000
Total	30	240	330	370	270	230	310	220	2000

10 Chapter Quiz

Take this quiz as you would take a quiz in class. After you are done, check your work against the answers given in the back of the book.

For each Exercise,

(a) state H_0 and H_a.

(b) specify the level of significance.

(c) find the critical value.

(d) identify the rejection region.

(e) calculate the test statistic.

(f) make a decision.

(g) interpret the results in the context of the problem.

For Exercises 1 and 2, use the following data. The data list the annual wages (in thousands of dollars) for randomly selected individuals from three metropolitan areas. Assume the wages are normally distributed and that the samples are independent. *(Adapted from U.S. Bureau of Labor Statistics)*

San Jose, CA: 48.6, 61.9, 19.8, 35.2, 25.5, 58.4, 47.4, 48.6, 17.6, 57.6, 81.1, 26.1, 76.8

Dallas, TX: 32.9, 46.4, 19.7, 24.0, 19.7, 18.4, 25.2, 10.3, 26.9, 20.5, 52.6, 40.7, 24.4, 36.5, 33.9, 28.0, 35.3, 25.5

Ann Arbor, MI: 22.7, 36.0, 20.7, 26.8, 24.0, 49.2, 29.5, 28.8, 38.8, 35.8, 23.1, 27.6, 30.8, 33.5, 23.3, 38.7

1. At $\alpha = 0.01$, is there enough evidence to conclude that the variances in annual wages for San Jose, CA and Dallas, TX are different?

2. Are the mean annual wages equal for all three cities? Use $\alpha = 0.10$. Assume that the population variances are equal.

For Exercises 3 and 4, use the following table. The table lists the distribution of educational achievement for people in the U.S. aged 25 and over. It also lists the results of a random survey for two additional age categories. *(Adapted from U.S. Census Bureau)*

	25 and over	35–44	65–74
Not a H.S. graduate	16.6%	37	124
High school graduate	33.3%	102	148
Some college, no degree	17.3%	59	61
Associate degree	7.5%	27	15
Bachelor's degree	17.0%	51	37
Advanced degree	8.3%	25	22

3. Does the distribution for people in the U.S. aged 25 and over describe people in the U.S. aged 35–44? Use $\alpha = 0.01$.

4. Does the distribution for people in the U.S. aged 25 and over describe people in the U.S. aged 65–74? Use $\alpha = 0.05$.

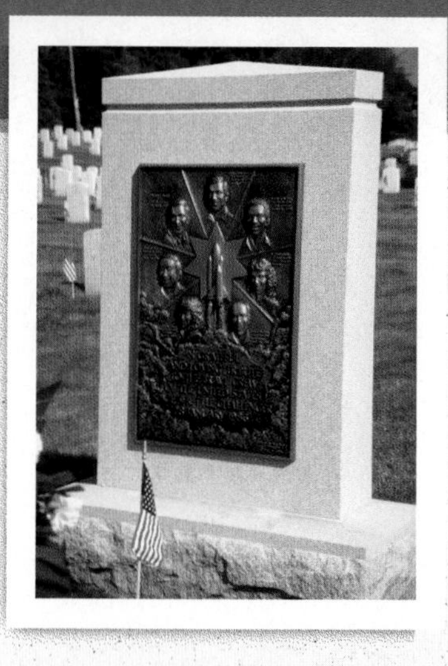

Where You've Been

Up to this point in the text, you have studied dozens of different statistical formulas and tests that can help you in a decision-making process. Knowing how to apply various statistical tests is important. However, it is even more important that you always use critical thinking when using a statistical test. Using critical thinking when applying statistical techniques can help prevent costly mistakes and tragic errors.

The explosion of the space shuttle Challenger on January 28, 1986 is a classic example of how tragic results might have been avoided by using correct statistical techniques. Seven astronauts died because two large rubber O-rings leaked during takeoff. The rings leaked because of the low temperature at the takeoff time.

The correlation between O-ring failure and air temperature was known before the time of takeoff. In fact, an engineering group had recommended that the flight be delayed. In their argument, however, the engineers failed to present the data clearly. How would you present the following data?

Temperature (°C)	12	14	14	17	19	19	19	19	20	21	21	21
Damage Index	11	4	4	2	0	0	0	0	0	4	0	4

Temperature (°C)	21	21	22	23	24	24	24	24	26	26	27
Damage Index	0	0	0	0	4	0	0	0	0	0	0

Nonparametric Tests

CHAPTER 11

Where You're Going

In this chapter, you will study additional statistical tests. Each of these has usefulness in real-life applications. Remember, however, that no statistical test, including those in this chapter, can be applied blindly. Moreover, the more important your decision, the more important it is that you look at the data from several perspectives.

With the O-ring data on the previous page, the air temperature T and the damage index D can be related by the regression line $D = 10.28 - 0.426T$. The correlation coefficient, however, is relatively weak $(r \approx -0.6)$. The P-value is 0.001, which would indicate that the correlation is significant, but the D-values do not pass the normality requirement.

So, while a simple correlation test might indicate a relationship between air temperature and O-ring damage, one might question the results because the data do not fit the requirements for the test. Similar tests you will study in this chapter, such as Spearman's rank correlation test, will give you additional information. If you had seen the following scatter plot before the Challenger's takeoff, what decision would you have made? Is the relationship between air temperature and O-ring damage evident enough for you to have postponed the flight? (Note: The air temperature at takeoff on January 28, 1986 was 0° Celsius, and the O-ring temperature was about 6° colder.) *(Source: Tufte, E.R., Visual Explanations)*

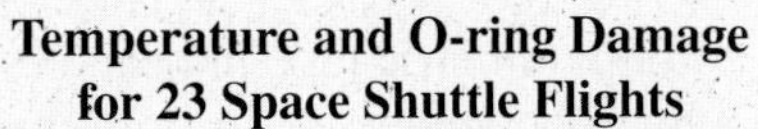

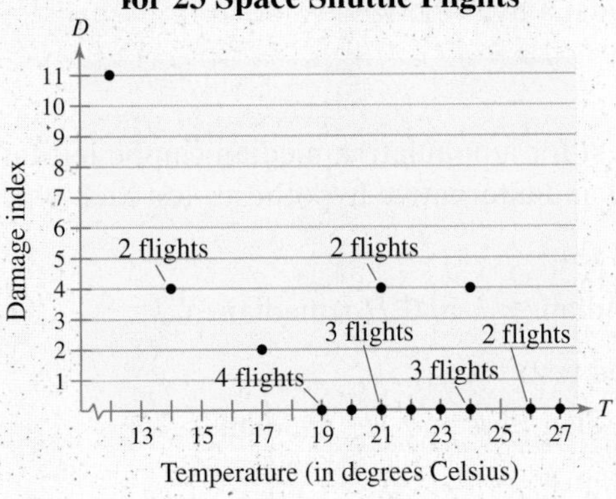

11.1 The Sign Test

What You Should Learn

- How to use the sign test to test a population median
- How to use the sign test to test the difference between two population medians (dependent samples)

The Sign Test for a Population Median • The Paired-Sample Sign Test

The Sign Test for a Population Median

Many of the hypothesis tests studied so far have imposed one or more requirements for a population distribution. For example, some tests require that a population must have a normal distribution, and other tests require that population variances be equal. What if, for a given test, such requirements cannot be met? For these cases, statisticians have developed hypothesis tests that are "distribution free." Such tests are called nonparametric tests.

DEFINITION

A **nonparametric test** is a hypothesis test that does not require any specific conditions concerning the shape of populations or the value of any population parameters.

Nonparametric tests are usually easier to perform than corresponding parametric tests. However, they are usually less efficient than parametric tests. Stronger evidence is required to reject a null hypothesis using the results of a nonparametric test. Consequently, whenever possible, you should use a parametric test. One of the easiest nonparametric tests to perform is the sign test.

Insight

For many nonparametric tests, statisticians test the median instead of the mean.

DEFINITION

The **sign test** is a nonparametric test that can be used to test a population median against a hypothesized value k.

The sign test for a population median can be left tailed, right tailed, or two tailed. The null and alternative hypotheses for each type of test are as follows.

Left-tailed test:

H_0: median $\geq k$ and H_a: median $< k$

Right-tailed test:

H_0: median $\leq k$ and H_a: median $> k$

Two-tailed test:

H_0: median $= k$ and H_a: median $\neq k$

To use the sign test, first compare each entry in the sample to the hypothesized median k. If the entry is below the median, assign it a $-$ sign; if the entry is above the median, assign it a $+$ sign; and if the entry is equal to the median, assign it a 0. Then compare the number of $+$ and $-$ signs. (The 0's are ignored.) If there is a large difference between the number of $+$ signs and the number of $-$ signs, it is likely that the median is different from the hypothesized value and the null hypothesis should be rejected.

Table 8 in Appendix B lists the critical values for the sign test for selected levels of significance and sample sizes. When using the sign test, the sample size n is the total number of $+$ and $-$ signs. If the sample size is greater than 25, you can use the standard normal distribution to find the critical values.

Insight

Because the 0's are ignored, there are two possible outcomes when comparing a data entry to a hypothesized median: a $+$ or a $-$ sign. If the median is k, then about half of the values will be above k and half will be below. As such, the probability for each sign is 0.5. Table 8 in Appendix B is constructed using the binomial distribution where $p = 0.5$.

When $n > 25$, you can use the normal approximation (with a correction for conti-nuity) for the binomial. In this case, use $\mu = np = 0.5n$ and $\sigma = \sqrt{npq} = \frac{\sqrt{n}}{2}$.

Test Statistic for the Sign Test

When $n \leq 25$, the test statistic x for the sign test is the smaller number of $+$ or $-$ signs.

When $n > 25$, the test statistic for the sign test is

$$z = \frac{(x + 0.5) - 0.5n}{\frac{\sqrt{n}}{2}}$$

where x is the smaller number of $+$ or $-$ signs and n is the sample size, i.e., the total number of $+$ and $-$ signs.

Because x is defined to be the smaller number of $+$ or $-$ signs, the rejection region is always in the left tail. Consequently, the sign test for a population median is always a left-tailed test or a two-tailed test. When the test is two tailed, use only the left-tailed critical value. (If x is defined to be the larger number of $+$ or $-$ signs, the rejection region is always in the right tail. Right-tailed sign tests are presented in the exercises.)

GUIDELINES

Performing a Sign Test for a Population Median

In Words	*In Symbols*
1. Identify the claim. State the null and alternative hypotheses.	State H_0 and H_a.
2. Specify the level of significance.	Identify α.
3. Determine the sample size n by assigning $+$ signs and $-$ signs to the sample data.	$n =$ total number of $+$ and $-$ signs
4. Find the critical value.	If $n \leq 25$, use Table 8 in App. B. If $n > 25$, use Table 4 in App. B.
5. Calculate the test statistic.	If $n \leq 25$, use x. If $n > 25$, use $z = \frac{(x + 0.5) - 0.5n}{\frac{\sqrt{n}}{2}}$.
6. Make a decision to reject or fail to reject the null hypothesis.	If x or z is less than or equal to the critical value, reject H_0. Otherwise, do not reject H_0.
7. Interpret the decision in the context of the original claim.	

EXAMPLE 

Using the Sign Test

A bank manager claims that the median number of customers per day is no more than 750. A teller doubts the accuracy of this claim. The number of bank customers per day for 16 randomly selected days are listed below. At $\alpha = 0.05$, can the teller reject the bank manager's claim?

775	765	801	742
754	753	739	751
745	750	777	769
756	760	782	789

SOLUTION The teller must disprove the bank manager's claim that "the median number of customers per day is no more than 750." So, the null and alternative hypotheses are

H_0: median ≤ 750 (claim) and H_a: median > 750.

The table below shows the results of comparing each data entry to the hypothesized median 750.

+	+	+	−
+	+	−	+
−	0	+	+
+	+	+	+

From the table, you can see that there are 3 $-$ signs and 12 $+$ signs. So, $n = 12 + 3 = 15$. Using Table 8 with $\alpha = 0.05$ (one tailed) and $n = 15$, the critical value is 3. Because $n \leq 25$, the test statistic x is the smaller number of $+$ or $-$ signs. So, $x = 3$. Because $x = 3$ is equal to the critical value, the teller should reject the null hypothesis. At the 5% level, the teller can reject the bank manager's claim and conclude that the median number of customers per day is more than 750.

Try It Yourself 1

A supermarket manager claims that the median number of customers per day is greater than 2500. A supplier wants to verify the accuracy of this claim. The number of customers per day for 24 randomly selected days are listed below.

Number of customers per day for 24 days							
2174	2491	2682	2510	2557	2418	2709	2562
2390	2467	2500	2205	2246	2054	2243	2627
1949	2500	2592	2567	2478	2348	2692	2580

At $\alpha = 0.025$, can the supplier support the manager's claim?

a. *Identify* the claim and *state* H_0 and H_a.
b. *Specify* the level of significance α.
c. *Determine* the sample size n.
d. *Find* the critical value.
e. *Determine* the test statistic x.
f. *Decide* whether to reject the null hypothesis.
g. Can the supplier support the manager's claim?

Answer: Page A43

Picturing the World

The National Confectioners Association reports that people in the U.S. spend a lot of money on candy. In 2000, people in the U.S. spent a total of \$13.5 billion on candy. And the U.S. Department of Commerce reported that in 2000, the average person in the U.S. ate about 23.5 pounds of candy.

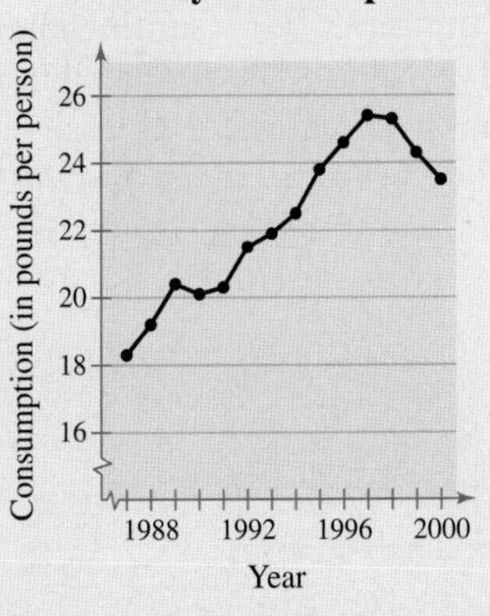

If you were to test the U.S. Department of Commerce's claim concerning per capita candy consumption, would you use a parametric test or a nonparametric test? What factors must you consider?

Study Tip

When performing a two-tailed sign test, remember to use only the left-tailed critical value.

EXAMPLE 

Using the Sign Test

A car dealership claims to give customers a median trade-in offer of at least \$6000. A random sample of 103 transactions revealed that the trade-in offer for 60 automobiles was less than \$6000 and the trade-in offer for 40 automobiles was more than \$6000. At $\alpha = 0.01$, can you reject the dealership's claim?

SOLUTION To reject the dealership's claim, you must disprove the claim that "the median trade-in offer is at least \$6000." The null and alternative hypotheses are

H_0: median ≥ 6000 (claim) and H_a: median < 6000.

Because $n > 25$, you should use Table 4, the Standard Normal Table, to find the critical value. Because the test is a left-tailed test with $\alpha = 0.01$, the critical value is -2.33. Of the 103 transactions, there are 60 $-$ signs and 40 $+$ signs. Ignoring the zeros, the sample size is

$$n = 60 + 40 = 100$$

and

$$x = 40.$$

Using these values, the test statistic is

$$z = \frac{(40 + 0.5) - 0.5(100)}{\sqrt{100}/2}$$

$$= \frac{-9.5}{5}$$

$$= -1.9.$$

The graph at the right shows the location of the rejection region and the test statistic z. Because z is greater than the critical value, it is not in the rejection region, and you should fail to reject the null hypothesis. At the 1% level of significance, you cannot reject the dealership's claim that the median trade-in offer is at least \$6000.

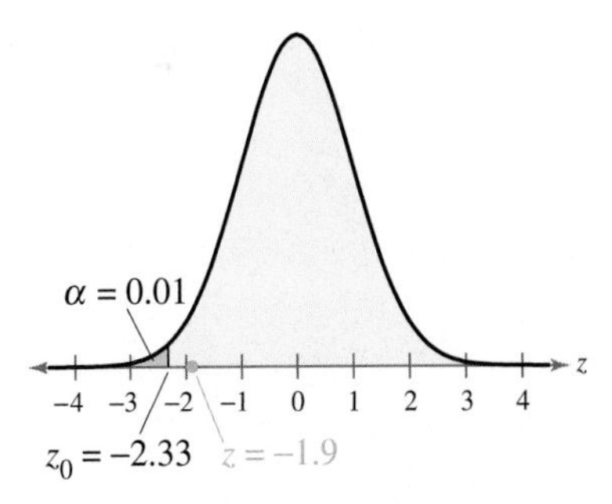

Try It Yourself 2

A realtor claims that the median sales price of houses sold in a certain region is \$134,500. A random sample of 85 house sales reveals that 30 houses were sold for less than \$134,500 and 51 houses were sold for more than \$134,500. At $\alpha = 0.10$, can you reject the realtor's claim?

a. *Identify* the claim and *state* H_0 and H_a.
b. *Specify* the level of significance α.
c. *Determine* the sample size n.
d. *Find* the critical value.
e. *Determine* the test statistic z.
f. *Decide* whether to reject the null hypothesis.
g. Can you reject the realtor's claim?

Answer: Page A43

The Paired-Sample Sign Test

In Section 8.3, you learned how to use a t-test for the difference between means of dependent samples. That test required both populations to be normally distributed. If the parametric condition of normality cannot be satisfied, you can use the sign test to test the difference between two population medians. To perform the paired-sample sign test for the difference between two population medians, the following conditions must be met.

1. A sample must be randomly selected from each population.
2. The samples must be dependent (paired).

The paired-sample sign test can be left tailed, right tailed, or two tailed. This test is similar to the sign test for a single population median. However, instead of comparing each data entry to a hypothesized median and recording a $+$, $-$, or 0, you find the difference between corresponding data entries and record the sign of the difference. Generally, to find the difference, subtract the entry representing the second variable from the entry representing the first variable. Then compare the number of $+$ and $-$ signs. (The 0's are ignored.) If the number of $+$ signs is approximately equal to the number of $-$ signs, the null hypothesis should not be rejected. If, however, there is a significant difference between the number of $+$ signs and the number of $-$ signs, the null hypothesis should be rejected.

GUIDELINES

Performing a Paired-Sample Sign Test

In Words	*In Symbols*
1. Identify the claim. State the null and alternative hypotheses.	State H_0 and H_a.
2. Specify the level of significance.	Identify α.
3. Determine the sample size n by finding the difference for each data pair. Assign a $+$ sign for a positive difference, a $-$ sign for a negative difference, and a 0 for no difference.	$n =$ total number of $+$ and $-$ signs
4. Find the critical value.	Use Table 8 in Appendix B.
5. Determine the test statistic.	$x =$ lesser number of $+$ and $-$ signs
6. Make a decision to reject or fail to reject the null hypothesis.	If the test statistic is less than or equal to the critical value, reject H_0. Otherwise, do not reject H_0.
7. Interpret the decision in the context of the original claim.	

EXAMPLE 

Using the Paired-Sample Sign Test

A psychologist claims that the number of repeat offenders will decrease if first-time offenders complete a special course. You randomly select 10 prisons and record the number of repeat offenders during a two-year period. Then after first-time offenders complete the course, you record the number of repeat offenders at each prison for another two-year period. The results are listed in the following table. At $\alpha = 0.025$, can you support the psychologist's claim?

Prison	1	2	3	4	5	6	7	8	9	10
Before	21	34	9	45	30	54	37	36	33	40
After	19	22	16	31	21	30	22	18	17	21

SOLUTION To support the psychologist's claim, you could use the following null and alternative hypotheses.

H_0: The number of repeat offenders will not decrease.

H_a: The number of repeat offenders will decrease. (claim)

The table below shows the sign of the differences between the "before" and "after" data.

Prison	1	2	3	4	5	6	7	8	9	10
Before	21	34	9	45	30	54	37	36	33	40
After	19	22	16	31	21	30	22	18	17	21
Sign	+	+	−	+	+	+	+	+	+	+

From the table, you can see that there is 1 − sign and 9 + signs. So, $n = 1 + 9 = 10$. Using Table 8 with $\alpha = 0.025$ (one tailed) and $n = 10$, the critical value is 1. The test statistic x is the smaller number of + or − signs. So, $x = 1$. Because $x = 1$ is less than or equal to the critical value, you should reject the null hypothesis. At the 2.5% level, you can support the psychologist's claim that the number of repeat offenders will decrease.

Try It Yourself 3

A medical researcher claims that a new vaccine will decrease the number of colds in adults. You randomly select 14 adults and record the number of colds each has in a one-year period. After giving the vaccine to each adult, you again record the number of colds each has in a one-year period. The results are listed in the table at the left. At $\alpha = 0.05$, can you support the researcher's claim?

Adult	Before vaccine	After vaccine
1	3	2
2	4	1
3	2	0
4	1	1
5	3	1
6	6	3
7	4	3
8	5	2
9	2	2
10	0	2
11	2	3
12	5	4
13	3	3
14	3	2

a. *Identify* the claim and *state* H_0 and H_a.
b. *Specify* the level of significance α.
c. *Determine* the sample size n.
d. *Find* the critical value.
e. *Determine* the test statistic x.
f. *Decide* whether to reject the null hypothesis.
g. Can you support the researcher's claim?

Answer: Page A43

Exercises

Help

LarsonTutor 11.1

Companion Web Site

Student Solutions Manual 11.1

Videos 11.1

Technology Manuals

Basic Skills and Concepts

1. What is a nonparametric test? How does a nonparametric test differ from a parametric test? What are the advantages and disadvantages of using a nonparametric test?

2. Explain how to use the sign test to test a population median.

Performing a Sign Test In Exercises 3–18,

(a) write the claim mathematically and identify H_0 and H_a.

(b) find the critical value.

(c) calculate the test statistic.

(d) decide whether to reject or fail to reject the null hypothesis. Then interpret the decision in the context of the original claim.

3. ***Credit Card Charges*** ◆ In order to estimate the median amount of new credit card charges for the previous month, a financial service accountant randomly selects 12 credit card accounts and records the amount of new charges for each account for the previous month. The amounts are listed below. At $\alpha = 0.01$, can the accountant conclude that the median amount of new credit card charges for the previous month was more than \$200? *(Adapted from Board of Governors of the Federal Reserve System)*

$246.71 $282.59 $155.03 $102.17 $209.80 $165.88
$199.41 $170.83 $196.54 $216.46 $145.92 $209.47

4. ***Temperatures*** ◆ A meteorologist estimates that the daily median temperature for the month of July in Pittsburgh is 72° Fahrenheit. The temperatures (in degrees Fahrenheit) for 15 randomly selected July days are listed below. At $\alpha = 0.01$, is there enough evidence to reject the meteorologist's hypothesis? *(Adapted from U.S. National Oceanic and Atmospheric Administration)*

64 72 72 75 76 70 66 73
67 65 74 74 67 77 66

5. ***Sales Prices of Homes*** ◆ A real estate agent believes that the median sales price of new privately owned one-family homes sold in the past year is \$140,000 or less. The sales prices of eight randomly selected homes are listed below. At $\alpha = 0.05$, is there enough evidence to reject the agent's claim? *(Adapted from U.S. Bureau of the Census and the U.S. Department of Housing and Urban Development)*

$170,000 $133,450 $89,500 $144,600
$75,800 $161,000 $149,000 $160,000

6. ***Temperature*** ◆ During a weather report, a meteorologist states that the daily median temperature for the month of January in San Diego is 57° Fahrenheit. The temperatures (in degrees Fahrenheit) for 18 randomly selected January days are listed below. At $\alpha = 0.01$, can you reject the meteorologist's claim? *(Adapted from U.S. National Oceanic and Atmospheric Administration)*

58 62 55 55 53 52 52 59 55
55 60 56 57 61 58 63 63 55

7. ***Credit Card Debt*** ◆ A financial services institution reports that the median amount of credit card debt for families holding such debts is at least \$1500. In a random sample of 104 families holding debt, you see that the debt for 68 families is less than \$1500 and the debt for 36 families is greater than \$1500. At $\alpha = 0.02$, can you reject the institution's claim? *(Adapted from Board of Governors of the Federal Reserve System, Federal Reserve Bulletin)*

8. ***Financial Debt*** ◆ A financial services accountant estimates that the median amount of financial debt for families holding such debts is less than \$22,500. In a random sample of 70 families holding debt, the debt for 28 families was less than \$22,500 and the debt for 42 families was greater than \$22,500. At $\alpha = 0.025$, can you support the accountant's estimate? *(Adapted from Board of Governors of the Federal Reserve System, Federal Reserve Bulletin)*

9. ***Social Science Doctorates*** ◆ A social sciences association conducted a study to determine the median age of recipients of social science doctorates. As part of the study, the association randomly selected 20 social service doctorates and found that 9 were conferred before age 36, 8 were conferred after age 36, and 3 were conferred at age 36. Test the association's claim that the median age of recipients of social science doctorates is greater than 36 years. Use $\alpha = 0.01$. *(Adapted from U.S. National Science Foundation)*

10. ***Physical Science Doctorates*** ◆ A science association claims that the median age of recipients of physical science doctorates is less than 31 years. In a random sample of 24 physical science doctorates, 8 were conferred at less than 31 years, 12 were conferred at greater than 31 years, and 4 were conferred at 31 years. At $\alpha = 0.05$, can you support the association's claim? *(Adapted from U.S. National Science Foundation)*

11. ***Unit Size*** ◆ A renters' organization claims that the median number of rooms in renter-occupied units is 4. You randomly select 50 renter-occupied units and obtain the results shown below. At $\alpha = 0.05$, can you reject the organization's claim? *(Adapted from U.S. Bureau of the Census)*

Unit size	Number of units
Less than 4 rooms	15
4 rooms	3
More than 4 rooms	32

Data for Exercise 11

Square footage	Number of units
Less than 1300	11
1300	2
More than 1300	9

Data for Exercise 12

12. ***Square Footage*** ◆ A renters' organization believes that the median square footage of renter-occupied units is 1300 square feet. To test this claim, you randomly select 22 renter-occupied units and obtain the results shown above. At $\alpha = 0.10$, can you reject the organization's claim? *(Adapted from U.S. Bureau of the Census)*

13. ***Hourly Earnings*** ◆ A labor organization estimates that the median hourly earnings of male workers paid hourly rates is \$9.81. In a random sample of 41 male workers paid hourly rates, 16 are paid less than \$9.81 per hour, 23 are paid more than \$9.81 per hour, and 2 are paid \$9.81 per hour. At $\alpha = 0.01$, can you reject the organization's claim? *(Adapted from U.S. Bureau of Labor Statistics)*

14. ***Hourly Earnings*** ◆ A labor organization claims that the median hourly earnings of female workers paid hourly rates is at most \$7.81. In a random sample of 23 female workers paid hourly rates, 9 are paid less than \$7.81 per hour, 11 are paid more than \$7.81 per hour, and 3 are paid \$7.81 per hour. At $\alpha = 0.05$, can you reject the organization's claim? *(Adapted from U.S. Bureau of Labor Statistics)*

15. ***Spinal Manipulation*** ◆ The table shows the daily headache hours suffered by eight patients before and after receiving soft tissue therapy and spinal manipulation for seven weeks. At $\alpha = 0.05$, is there enough evidence to conclude that daily headache hours were reduced after the soft tissue therapy and spinal manipulation? *(Adapted from The Journal of the American Medical Association)*

Patient	1	2	3	4	5	6	7	8
Headache hours (before)	0.8	2.4	2.8	2.6	2.7	0.9	1.2	2.2
Headache hours (after)	1.6	1.3	1.6	1.4	1.5	1.6	1.7	1.8

DATA **16.** ***Headaches*** ◆ The table shows the daily headache hours suffered by 12 patients before and after receiving soft tissue therapy and laser treatment for seven weeks. At $\alpha = 0.01$, is there enough evidence to conclude that daily headache hours were reduced after receiving the soft tissue therapy and laser treatment? *(Adapted from The Journal of the American Medical Association)*

Patient	1	2	3	4	5	6
Headache hours (before)	2.1	3.9	3.8	2.5	2.4	3.6
Headache hours (after)	2.2	2.8	2.5	2.6	1.9	1.8

Patient	7	8	9	10	11	12
Headache hours (before)	3.4	2.4	3.5	2.0	2.7	2.4
Headache hours (after)	2.0	1.6	1.5	2.1	1.8	3.0

17. ***Improving SAT Scores*** ◆ A tutoring agency believes that by completing a special course, students can improve their verbal SAT skills. As part of a study, 12 students take the verbal part of the SAT, complete the special course, then take the verbal part of the SAT again. The students' scores are listed below. At $\alpha = 0.05$, is there enough evidence to conclude that the students' verbal SAT scores improved?

Student	1	2	3	4	5	6
Score on first SAT	308	456	352	433	306	471
Score on second SAT	300	524	409	419	304	483

Student	7	8	9	10	11	12
Score on first SAT	538	207	205	351	360	251
Score on second SAT	708	253	399	350	480	303

18. *SAT Scores* ◆ Students at a certain school are required to take the SAT twice. The table shows both verbal SAT scores for 12 students. At $\alpha = 0.01$, can you conclude that the students' scores improved the second time they took the SAT?

Student	1	2	3	4	5	6
Score on first SAT	445	510	429	452	629	453
Score on second SAT	446	571	517	478	610	453

Student	7	8	9	10	11	12
Score on first SAT	358	477	325	513	636	571
Score on second SAT	378	532	299	501	648	603

19. *Office Parties* ◆ An alcohol awareness organization conducted a survey by randomly selecting companies and asking them if they give their employees a holiday party where alcohol is served. The organization asked companies that said yes if the companies provided transportation home or offered alternatives to driving home. The results are shown in the figure. *(Adapted from USA TODAY)*

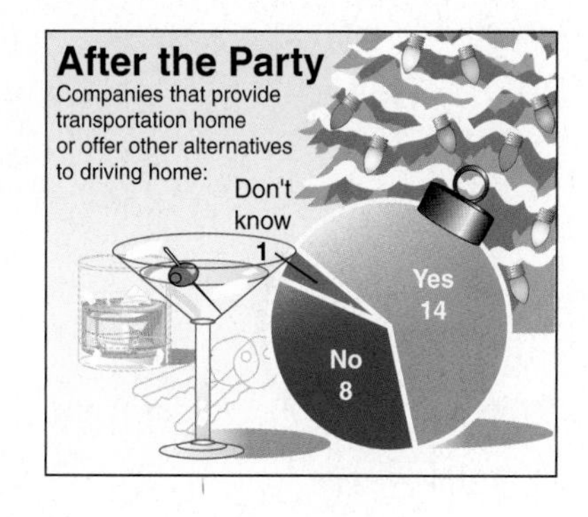

(a) Use a sign test to test the null hypothesis that the proportion of companies providing transportation home or offering alternatives to driving home is equal to the proportion of companies that do not. Assign a + sign to a company that does, assign a − sign to a company that does not, and assign a 0 to a company that does not know. Use $\alpha = 0.05$.

(b) What can you conclude?

20. *Credit Cards* ◆ A financial institution conducts a survey by randomly selecting credit cardholders and asking them if they almost always pay off their credit card balances. The results are shown in the figure. *(Adapted from Board of Governors of the Federal Reserve System)*

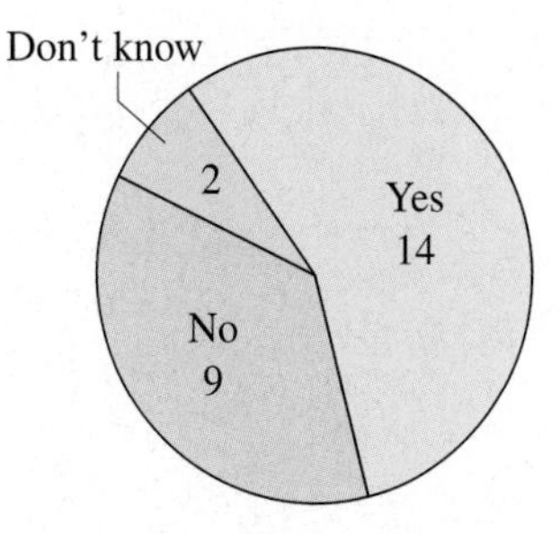

(a) Use a sign test to test the null hypothesis that the proportion of credit cardholders almost always paying off their credit card balances is equal to the proportion of credit cardholders that do not. Assign a + sign to a credit cardholder that does, assign a − sign to a credit cardholder that does not, and assign a 0 to a credit cardholder that does not know. Use $\alpha = 0.05$.

(b) What can you conclude?

Extending the Basics

More on Sign Tests When you are using a sign test for $n > 25$ and the test is left tailed, you know you can reject the null hypothesis if the test statistic

$$z = \frac{(x + 0.5) - 0.5n}{\frac{\sqrt{n}}{2}}$$

is less than or equal to the left-tailed critical value, where x is the smaller number of + and − signs. For a right-tailed test you can reject the null hypothesis if the test statistic

$$z = \frac{(x + 0.5) - 0.5n}{\frac{\sqrt{n}}{2}}$$

is greater than or equal to the *right-tailed* critical value, where x is the *larger* number of + and − signs.

In Exercises 21–24, (a) write the claim mathematically and identify H_0 and H_a, (b) find the critical value, (c) calculate the test statistic, and (d) decide whether to reject or fail to reject the null hypothesis. Then interpret the decision in the context of the original claim.

21. ***Weekly Earnings*** ◆ A labor organization claims that the median weekly earnings of female workers is less than or equal to \$418. To test this claim, you randomly select 50 female workers and ask each to provide her weekly earnings. The results are shown in the table. At $\alpha = 0.01$, can you reject the organization's claim? *(Adapted from U.S. Bureau of Labor Statistics)*

Weekly earnings	Number of workers
Less than \$418	18
\$418	3
More than \$418	29

Data for Exercise 21

Weekly earnings	Number of workers
Less than \$557	21
\$557	1
More than \$557	48

Data for Exercise 22

22. ***Weekly Earnings*** ◆ A labor organization states that the median weekly earnings of male workers is greater than \$557. To test this claim, you randomly select 70 male workers and ask each to provide his weekly earnings. The results are shown in the table. At $\alpha = 0.01$, can you support the organization's claim? *(Adapted from U.S. Bureau of Labor Statistics)*

23. ***Ages of Brides*** ◆ A marriage counselor estimates that the median age of brides at the time of their first marriage is greater than 24 years. In a random sample of 65 brides, 22 are less than 24 years, 38 are more than 24 years, and 5 are 24 years. At $\alpha = 0.05$, can you support the counselor's claim? *(Adapted from U.S. National Center for Health Statistics)*

24. ***Ages of Grooms*** ◆ A marriage counselor estimates that the median age of grooms at the time of their first marriage is less than or equal to 25.9 years. In a random sample of 56 grooms, 23 are less than 25.9 years, 33 are more than 25.9 years, and none are 25.9 years. At $\alpha = 0.05$, is there enough evidence to reject the counselor's claim? *(Adapted from U.S. National Center for Health Statistics)*

11.2 The Wilcoxon Tests

What You Should Learn

- How to use the Wilcoxon signed-rank test to determine if two dependent samples are selected from populations having the same distribution
- How to use the Wilcoxon rank sum test to determine if two independent samples are selected from populations having the same distribution

The Wilcoxon Signed-Rank Test • The Wilcoxon Rank Sum Test

The Wilcoxon Signed-Rank Test

In this section, you will study the Wilcoxon signed-rank test and the Wilcoxon rank sum test. Unlike the sign test, the strength of these two nonparametric tests is that each considers the magnitude, or size, of the data entries.

In Section 8.3, you used a t-test together with dependent samples to determine whether there was a difference between two populations. To use the t-test to test such a difference, you must assume (or know) that the dependent samples are randomly selected from populations having a normal distribution. But, what if this assumption cannot be made? Instead of using the two-sample t-test, you can use the Wilcoxon signed-rank test.

DEFINITION

The **Wilcoxon signed-rank test** is a nonparametric test that can be used to determine whether two *dependent* samples were selected from populations having the same distribution.

Study Tip

The absolute value of a number is its value, disregarding its sign. For example, $|3| = 3$ and $|-7| = 7$.

GUIDELINES

Performing a Wilcoxon Signed-Rank Test

In Words	*In Symbols*
1. Identify the claim. State the null and alternative hypotheses.	State H_0 and H_a.
2. Specify the level of significance.	Specify α.
3. Determine the sample size n.	
4. Find the critical value.	Use Table 9 in Appendix B.
5. Calculate the test statistic w_s. **a.** Complete a table using the headers listed at the right. **b.** Find the sum of the positive ranks and the sum of the negative ranks. **c.** Select the smaller of absolute values of the sums.	Headers: **Sample 1, Sample 2, Difference, Absolute value, Rank,** and **Signed rank.**
6. Make a decision to reject or fail to reject the null hypothesis.	If w_s is less than or equal to the critical value, reject H_0. Otherwise, do not reject H_0.
7. Interpret the decision in the context of the original claim.	

EXAMPLE 

Performing a Wilcoxon Signed-Rank Test

A sports psychologist believes that listening to music affects the length of athletes' workout sessions. The length of time (in minutes) of ten athletes' workout sessions, while listening to music and while not listening to music, are listed in the table. At $\alpha = 0.05$, can you support the sports psychologist's claim?

Length of workout session, with music	45	38	28	39	41	47	62	54	33	44
Length of workout session, without music	38	40	33	36	42	41	54	47	28	35

SOLUTION The claim is "Music affects the length of athletes' workout sessions." To test this claim, use the following null and alternative hypotheses.

H_0: There is no difference in the length of the athletes' workout sessions.

H_a: There is a difference in the length of the athletes' workout sessions.

This Wilcoxon signed-rank test is a two-tailed test with $\alpha = 0.05$ and $n = 10$. Using Table 9 in Appendix B, the critical value is 8. To find the test statistic w_s complete a table as shown below.

Study Tip

Do not assign a rank to any differences of zero. The sample size n should be the total number of nonzero differences. In the case of a tie between data entries, use the average of the corresponding ranks. For instance, if two data entries are tied for the fifth rank, use the average of 5 and 6, which is 5.5. If three entries are tied for the fifth rank, use the average of 5, 6, and 7, or 6.

Length of session, with music	Length of session, without music	Difference	Absolute value	Rank	Signed rank
45	38	7	7	7.5	7.5
38	40	−2	2	2	−2
28	33	−5	5	4.5	−4.5
39	36	3	3	3	3
41	42	−1	1	1	−1
47	41	6	6	6	6
62	54	8	8	9	9
54	47	7	7	7.5	7.5
33	28	5	5	4.5	4.5
44	35	9	9	10	10

The sum of the negative ranks is

$$-1 + (-2) + (-4.5) = -7.5.$$

The sum of the positive ranks is

$$(+3) + (+4.5) + (+6) + (+7.5) + (+7.5) + (+9) + (+10) = 47.5.$$

The test statistic is the smaller of the absolute value of these two sums. Because $|-7.5| < |47.5|$, the test statistic is $w_s = 7.5$.

Because the test statistic is less than the critical value, that is, $7.5 < 8$, you should decide to reject the null hypothesis. In other words, at the 5% level of significance, you have enough evidence to support the claim that music makes a difference in the length of athletes' workout sessions.

Picturing the World

In an effort to determine whether smoking is a risk factor for pneumonia in patients with chicken pox, doctors recorded the carbon monoxide transfer factor levels in seven patients (all smokers) upon their admission to the hospital and one week later. *(Source: British Medical Journal)*

Patient	Level at admission	Level after one week
1	62	63
2	60	64
3	58	85
4	56	80
5	66	60
6	40	73
7	50	52

At $\alpha = 0.05$, can you conclude that the carbon monoxide transfer level changed significantly in one week?

Try It Yourself 1

A quality control inspector wants to test the effectiveness of a spray-on water repellent. To test this claim, he selects 12 pieces of fabric, sprays water on each, and measures the amount of water repelled (in milliliters). He then applies the water repellent and repeats the experiment. The results are listed in the table. At $\alpha = 0.01$, can he conclude that the water repellent is effective?

No repellent	8	7	7	4	6	10
Repellent applied	15	12	11	6	6	8
No repellent	9	5	9	11	8	4
Repellent applied	8	6	12	8	14	8

a. *Identify* the claim and *state* H_0 and H_a.
b. *Specify* the level of significance α.
c. *Determine* the sample size n.
d. *Find* the critical value.
e. *Calculate* the test statistic w_s by making a table, finding the sum of the positive ranks and the sum of the negative ranks, and finding the absolute value of each.
f. *Decide* whether to reject the null hypothesis. Use a graph if necessary.
g. Is there enough evidence to support the claim?

Answer: Page A43

The Wilcoxon Rank Sum Test

In Sections 8.1 and 8.2, you used a z-test or a t-test together with independent samples to determine whether there was a difference between two populations. To use these tests to test such a difference, you had to make several assumptions concerning the distribution of each population. But, what if these assumptions cannot be made? You can still compare the populations using the Wilcoxon rank sum test.

DEFINITION

The **Wilcoxon rank sum test** is a nonparametric test that can be used to determine whether two *independent* samples were selected from populations having the same distribution.

Study Tip

Use the Wilcoxon signed-rank test for dependent samples and the Wilcoxon rank sum test for independent samples.

A requirement for the Wilcoxon rank sum test is that the sample size of both samples must be at least 10. When calculating the test statistic for the Wilcoxon rank sum test, let n_1 represent the sample size of the smaller sample and n_2 represent the sample size of the larger sample. If the two samples have the same size, it does not matter which one is n_1 or n_2.

When calculating the sum of the ranks R, use the ranks for the smaller of the two samples. If the two samples have the same size, you can use the ranks from either sample, but you must use the ranks from the sample you associate with n_1.

Test Statistic for the Wilcoxon Rank Sum Test

Given two independent samples, the test statistic z for the Wilcoxon rank sum test is

$$z = \frac{R - \mu_R}{\sigma_R}$$

where

R = sum of the ranks for the smaller sample

$$\mu_R = \frac{n_1(n_1 + n_2 + 1)}{2}$$

and

$$\sigma_R = \sqrt{\frac{n_1 n_2(n_1 + n_2 + 1)}{12}}.$$

GUIDELINES

Performing a Wilcoxon Rank Sum Test

In Words	*In Symbols*
1. Identify the claim. State the null and alternative hypotheses.	State H_0 and H_a.
2. Specify the level of significance.	Specify α.
3. Find the critical value.	Use Table 4 of Appendix B.
4. Determine the sample sizes.	$n_1 \leq n_2$
5. Find the sum of the ranks for the smaller sample. **a.** List the combined data in ascending order. **b.** Rank the combined data. **c.** Add the sum of the ranks for the smaller sample.	R
6. Calculate the test statistic.	$z = \frac{R - \mu_R}{\sigma_R}$
7. Make a decision to reject or fail to reject the null hypothesis.	If z is in the rejection region, reject H_0. Otherwise, do not reject H_0.
8. Interpret the decision in the context of the original claim.	

EXAMPLE 2

Performing a Wilcoxon Rank Sum Test

The table lists the earnings (in thousands of dollars) of a random sample of 10 male and 12 female salespersons. At $\alpha = 0.10$ can you conclude that there is a difference between the males' and females' earnings?

Male earnings	28	43	64	51	48	44	36	45	67	49		
Female earnings	36	27	51	43	35	48	41	37	34	47	50	40

SOLUTION The claim is "there is a difference between the males' and females' earnings." The null and alternative hypotheses for this test are as follows.

H_0: There is no difference between the males' and the females' earnings.

H_a: There is a difference between the males' and the females' earnings.

Because the test is a two-tailed test with $\alpha = 0.10$, the critical values are

$z_0 = -1.645$ and $z_0 = 1.645$.

The rejection regions are

$z \leq -1.645$ and $z \geq 1.645$.

Before calculating the test statistic, you must find the values of R, μ_R, and σ_R. The table shows the combined data listed in ascending order and the corresponding ranks.

Ordered data	Sample	Rank
27	F	1
28	M	2
34	F	3
35	F	4
36	M	5.5
36	F	5.5
37	F	7
40	F	8
41	F	9
43	M	10.5
43	F	10.5

Ordered data	Sample	Rank
44	M	12
45	M	13
47	F	14
48	M	15.5
48	F	15.5
49	M	17
50	F	18
51	M	19.5
51	F	19.5
64	M	21
67	M	22

Because the smaller sample is the sample of males, R is the sum of the male rankings.

$$R = 2 + 5.5 + 10.5 + 12 + 13 + 15.5 + 17 + 19.5 + 21 + 22$$
$$= 138$$

Using $n_1 = 10$ and $n_2 = 12$, you can find μ_R and σ_R as follows.

$$\mu_R = \frac{n_1(n_1 + n_2 + 1)}{2} = \frac{10(10 + 12 + 1)}{2} = \frac{230}{2} = 115$$

$$\sigma_R = \sqrt{\frac{n_1 n_2(n_1 + n_2 + 1)}{12}}$$
$$= \sqrt{\frac{(10)(12)(10 + 12 + 1)}{12}}$$
$$= \sqrt{\frac{2760}{12}}$$
$$= \sqrt{230}$$
$$\approx 15.2$$

Using $R = 138$, $\mu_R = 115$, and $\sigma_R = 15.2$, the test statistic is

$$z = \frac{R - \mu_R}{\sigma_R}$$
$$= \frac{138 - 115}{15.2}$$
$$\approx 1.51.$$

From the graph at the right, you can see that the test statistic z is not in the rejection region. At the 10% level, you should decide to fail to reject the null hypothesis. In other words, you cannot conclude that there is a difference between the males' and females' earnings.

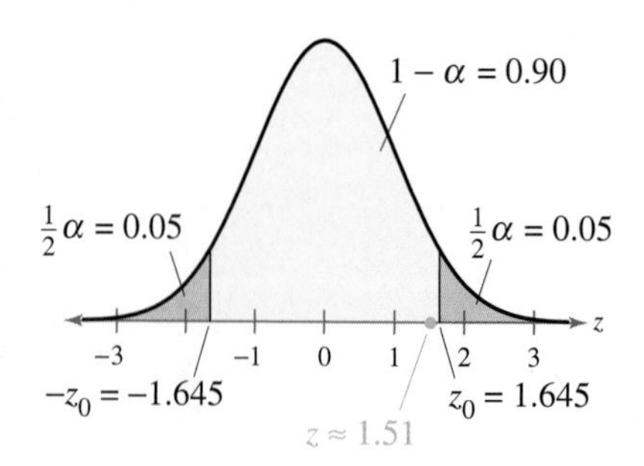

Try It Yourself 2

You are investigating the automobile insurance claims paid (in thousands of dollars) by two insurance companies. The table lists a random, independent sample of 12 claims paid by two insurance companies. At $\alpha = 0.05$, can you conclude that there is a difference in the claims paid by the companies?

Company A	6.2	10.6	2.5	4.5	6.5	7.4	9.9	3.0	5.8	3.9	6.0	6.3
Company B	7.3	5.6	3.4	1.8	2.2	4.7	10.8	4.1	1.7	3.0	4.4	5.3

a. *Identify* the claim and *state* H_0 and H_a.
b. *Specify* the level of significance α.
c. *Find* the critical value.
d. *Determine* the sample sizes n_1 and n_2.
e. *List* the combined data in ascending order, *rank* the data, and *find* the sum of the ranks of the smaller sample.
f. *Calculate* the test statistic.
g. *Decide* whether to reject the null hypothesis. Use a graph if necessary.
h. Is there enough evidence to conclude that there is a difference in the claims paid by the companies?

Answer: Page A43

Exercises

Help

LarsonTutor 11.2

Companion Web Site

Student Solutions Manual 11.2

Videos 11.2

Technology Manuals

Basic Skills and Concepts

Performing a Wilcoxon Test In Exercises 1–6,

(a) write the claim mathematically and identify H_0 and H_a.

(b) decide whether to use a Wilcoxon signed-rank test or a Wilcoxon rank sum test.

(c) find the critical value.

(d) calculate the test statistic.

(e) decide whether to reject or fail to reject the null hypothesis. Then interpret the decision in the context of the original claim.

1. ***Calcium Supplements and Blood Pressure*** ◆ In a study testing the effects of calcium supplements on blood pressure in men, eight men were randomly chosen and given a calcium supplement for 12 weeks. The measurements listed in the table are for each subject's diastolic blood pressure taken before and after the 12-week treatment period. At $\alpha = 0.01$, can you reject the claim that there was no reduction in diastolic blood pressure? *(Adapted from The Journal of American Medicine)*

Patient	1	2	3	4	5	6	7	8
Before treatment	108	109	120	129	112	111	117	135
After treatment	99	115	105	116	115	117	108	122

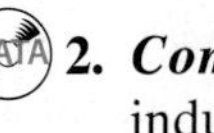

2. ***Comparing Salaries: Manufacturing and Construction*** ◆ A private industry analyst claims that there is no difference in the salaries earned by workers in the manufacturing and construction industries. A random sample of ten manufacturing and ten construction workers and their salaries is listed in the table. At $\alpha = 0.10$, can you reject the analyst's claim? *(Adapted from U.S. Bureau of Labor Statistics)*

Industry	**Salary (in thousands of dollars)**									
Manufacturing	31	38	33	33	35	47	33	29	38	45
Construction	31	30	27	32	28	34	30	33	26	35

3. ***Earnings by Degree*** ◆ A college administrator believes that there is a difference in the earnings of people with bachelor's degrees and those with associate's degrees. The table lists the earnings (in thousands of dollars) of a random sample of eleven people with bachelor's degrees and ten people with associate's degrees. At $\alpha = 0.05$, is there enough evidence to support the administrator's belief? *(Adapted from Bureau of the Census)*

Level of highest degree	**Salary (in thousands of dollars)**										
Bachelor's	37	41	55	34	56	63	40	28	43	27	34
Associate's	22	26	24	18	18	18	21	17	24	24	

4. *Mother's Age and Baby Care* ◆ A natal health care researcher conducts a study of a random selection of mothers under 20 years and 20–24 years. The number of weeks each mother breastfed her baby is listed in the table. At $\alpha = 0.01$, can you reject the researcher's claim that there is no difference in the number of months these mothers breast-feed their babies? *(Adapted from U.S. National Center for Health Statistics)*

Age of mother	Duration (in weeks)											
Under 20 years	21	24	24	30	11	17	10	28	21	12	15	
20–24 years	25	28	26	23	22	13	28	21	27	32	19	27

5. *Teacher Salaries* ◆ A teacher's union representative claims that there is a difference in the salaries earned by teachers in Ohio and Pennsylvania. A random sample of 12 Ohio and 12 Pennsylvania teachers and their salaries is listed in the table. At $\alpha = 0.05$, is there enough evidence to support the representative's claim? *(Adapted from National Education Association)*

State	Salary (in thousands of dollars)											
Ohio	36	39	44	41	38	45	36	45	39	31	53	30
Pennsylvania	41	60	38	46	43	48	63	59	50	45	38	28

6. *Headaches* ◆ A medical researcher wants to determine whether a new drug affects the number of headache hours experienced by headache sufferers. To do so, the researcher selects seven patients and asks each to give the number of headache hours (per day) each experiences before and after taking the drug. The results are listed in the table. At $\alpha = 0.05$, can the researcher conclude that the new drug affects the number of headache hours?

Patient	1	2	3	4	5	6	7
Headache hours (before)	0.8	2.4	2.8	2.6	2.7	0.9	1.2
Headache hours (after)	1.6	1.3	1.6	1.4	1.5	1.6	1.7

Extending the Basics

Wilcoxon Signed-Rank Test for $n > 30$ If you are performing a Wilcoxon signed-rank test and the sample size n is greater than 30, you can use the Standard Normal Table and the following formula to find the test statistic.

$$z = \frac{w_s - \dfrac{n(n+1)}{4}}{\sqrt{\dfrac{n(n+1)(2n+1)}{24}}}$$

In Exercises 7 and 8, perform the indicated Wilcoxon signed-rank test using the test statistic for $n > 30$.

7. *Fuel Additive* ◆ A petroleum engineer wants to know whether a certain fuel additive improves a car's gas mileage. To decide, the engineer records the gas mileage of 33 cars with and without the additive. The results are listed in the table. At $\alpha = 0.10$, can the engineer conclude that the gas mileage is improved?

Car	1	2	3	4	5	6	7	8	9	10	11
Without additive	36.4	36.4	36.6	36.6	36.8	36.9	37.0	37.1	37.2	37.2	36.7
With additive	36.7	36.9	37.0	37.5	38.0	38.1	38.4	38.7	38.8	38.9	36.3

Car	12	13	14	15	16	17	18	19	20	21	22
Without additive	37.5	37.6	37.8	37.9	37.9	38.1	38.4	40.2	40.5	40.9	35.0
With additive	38.9	39.0	39.1	39.4	39.4	39.5	39.8	40.0	40.0	40.1	36.3

Car	23	24	25	26	27	28	29	30	31	32	33
Without additive	32.7	33.6	34.2	35.1	35.2	35.3	35.5	35.9	36.0	36.1	37.2
With additive	32.8	34.2	34.7	34.9	34.9	35.3	35.9	36.4	36.6	36.6	38.3

8. *Fuel Additive* ◆ A petroleum engineer claims that a fuel additive improves gas mileage. The table lists the gas mileage (in miles per gallon) of 32 cars measured with and without the fuel additive. Test the petroleum engineer's claim at $\alpha = 0.05$.

Car	1	2	3	4	5	6	7	8
Without additive	34.0	34.2	34.4	34.4	34.6	34.8	35.6	35.7
With additive	36.6	36.7	37.2	37.2	37.3	37.4	37.6	37.7

Car	9	10	11	12	13	14	15	16
Without additive	30.2	31.6	32.3	33.0	33.1	33.7	33.7	33.8
With additive	34.2	34.9	34.9	34.9	35.7	36.0	36.2	36.5

Car	17	18	19	20	21	22	23	24
Without additive	35.7	36.1	36.1	36.6	36.6	36.8	37.1	37.1
With additive	37.8	38.1	38.2	38.3	38.3	38.7	38.8	38.9

Car	25	26	27	28	29	30	31	32
Without additive	37.2	37.9	37.9	38.0	38.0	38.4	38.8	42.1
With additive	39.1	39.1	39.2	39.4	39.8	40.3	40.8	43.2

Case Study 11

WWW.CDC.GOV

Health and Nutrition

The Recommended Dietary Allowances (RDA) are a set of nutrient standards that represent the average daily intakes of energy and nutrients considered adequate to meet the needs of most healthy Americans. For instance, for individuals ages 20–29 years, the RDA for calcium is between 800 and 1200 milligrams.

To determine how well people in the United States were meeting the RDA standards, the Centers for Disease Control and Prevention together with the National Center for Health Statistics conducted several three-year studies. The results of the studies are classified by age, gender, and race-ethnicity. A complete summary of the studies can be found at the Web site listed above.

The table at the right lists the calcium intake (in milligrams) for randomly selected members in various groups of individuals ages 20–29 years.

Calcium Intake (in milligrams)			
All males	**Non-Hispanic white males**	**All females**	**Non-Hispanic black females**
1362	819	416	887
763	1659	1187	347
1093	1264	507	378
877	944	481	1158
894	959	570	484
990	1039	710	680
796	858	1351	457
1289	1332	370	799
1687	881	1209	1042
664	1668	979	323

Exercises

1. Construct a box-and-whisker plot for each of the following groups on the same graph. Do any of the median calcium intakes appear to be the same? Different?

(a) All males

(b) Non-Hispanic white males

(c) All females

(d) Non-Hispanic black females

In Exercises 2–5, use the sign test to test the claim. What can you conclude? Use $\alpha = 0.05$.

2. Test the claim that the median calcium intake for all males ages 20–29 is less than or equal to 1000 milligrams.

3. Is the median calcium intake 1100 milligrams for non-Hispanic white males ages 20–29?

4. For all females ages 20–29, test the claim that the median calcium intake is greater than or equal to 700 milligrams.

5. Test the claim that the median calcium intake for non-Hispanic black females ages 20–29 is different from 800 milligrams.

In Exercises 6 and 7, use the Wilcoxon rank sum test to test the claim. What can you conclude? Use $\alpha = 0.01$.

6. Test the claim that there is no difference in the median calcium intake for all males ages 20–29 and the median calcium intake for non-Hispanic white males ages 20–29.

7. Is there a difference in the median calcium intake for all females ages 20–29 and the median calcium intake for non-Hispanic black females ages 20–29?

11.3 The Kruskal-Wallis Test

What You Should Learn

- How to use the Kruskal-Wallis test to determine whether three or more samples were selected from populations having the same distribution

The Kruskal-Wallis Test

The Kruskal-Wallis Test

In Section 10.4, you learned how to use one-way ANOVA techniques to compare the means of three or more populations. When using one-way ANOVA, you should verify that each independent sample is selected from a population that is normally, or approximately normally, distributed. If, however, you cannot verify that the populations are normal, you can still compare the distributions of three or more populations. To do so, you can use the Kruskal-Wallis test.

DEFINITION

The **Kruskal-Wallis** test is a nonparametric test that can be used to determine whether three or more independent samples were selected from populations having the same distribution.

The null and alternative hypotheses for the Kruskal-Wallis test are as follows.

H_0: There is no difference in the distribution of the populations.

H_a: There is a difference in the distribution of the populations.

Two conditions for using the Kruskal-Wallis test are that each sample must be randomly selected and the size of each sample must be at least 5. If these conditions are met, the sampling distribution for the Kruskal-Wallis test is a chi-square distribution with $k - 1$ degrees of freedom, where k is the number of samples. You can calculate the Kruskal-Wallis test statistic using the following formula.

Test Statistic for the Kruskal-Wallis Test

Given three or more independent samples, the test statistic H for the Kruskal-Wallis test is

$$H = \frac{12}{N(N+1)}\left(\frac{R_1^2}{n_1} + \frac{R_2^2}{n_2} + \cdots + \frac{R_k^2}{n_k}\right) - 3(N+1)$$

where

k represents the number of samples,

n_i is the size of the ith sample,

N is the sum of the sample sizes,

and

R_i is the sum of the ranks of the ith sample.

Performing a Kruskal-Wallis test consists of combining and ranking the sample data. The data are then separated according to sample and the sum of the ranks of each sample is calculated. These sums are then used to calculate the test statistic H, which is an approximation of the variance of the rank sums. If the samples are selected from populations having the same distribution, the sums of the ranks will be approximately equal, H will be small, and the null hypothesis should not be rejected. If, however, the sums of the ranks are quite different, H will be large, and the null hypothesis should be rejected.

Because the null hypothesis is rejected only when H is significantly large, the Kruskal-Wallis test is always a right-tailed test.

GUIDELINES

Performing a Kruskal-Wallis Test

In Words	*In Symbols*
1. Identify the claim. State the null and alternative hypotheses.	State H_0 and H_a.
2. Specify the level of significance.	Specify α.
3. Determine the degrees of freedom.	d.f. $= k - 1$
4. Find the critical value and identify the rejection region.	Use Table 6 in Appendix B.
5. Find the sums of the ranks for each sample. **a.** List the combined data in ascending order. **b.** Rank the combined data.	
6. Calculate the test statistic.	$H = \frac{12}{N(N+1)} \cdot \left(\frac{R_1^2}{n_1} + \frac{R_2^2}{n_2} + \cdots + \frac{R_k^2}{n_k}\right) - 3(N+1)$
7. Make a decision to reject or fail to reject the null hypothesis.	If H is in the rejection region, reject H_0. Otherwise, do not reject H_0.
8. Interpret the decision in the context of the original claim.	

EXAMPLE 

Performing a Kruskal-Wallis Test

You want to compare the hourly pay rates of accountants who work in Michigan, New York, and Virginia. To do so, you randomly select several accountants in each state and record their hourly pay rate. The hourly pay rates are listed in the table. At $\alpha = 0.01$, can you conclude that the distributions of accountants' hourly pay rates in these three states are different?

Sample hourly pay rates		
MI (Sample 1)	**NY (Sample 2)**	**VA (Sample 3)**
14.24	21.18	17.02
14.06	20.94	20.63
14.85	16.26	17.47
17.47	21.03	15.54
14.83	19.95	15.38
19.01	17.54	14.90
13.08	14.89	20.48
15.94	18.88	18.50
13.48	20.06	12.80
16.94		15.57

SOLUTION You want to test the claim that there is no difference in the hourly pay rates in Michigan, New York, and Virginia. The null and alternative hypotheses are as follows.

H_0: There is no difference in the hourly pay rates in the three states.

H_a: There is a difference in the hourly pay rates in the three states.

The test is a right-tailed test with $\alpha = 0.01$ and d.f. $= k - 1 = 3 - 1 = 2$. Using Table 6, the critical value is 9.210. Before calculating the test statistic, you must find the sum of the ranks for each sample. The table shows the combined data listed in ascending order and the corresponding ranks.

Ordered data	Sample	Rank
12.80	VA	1
13.08	MI	2
13.48	MI	3
14.06	MI	4
14.24	MI	5
14.83	MI	6
14.85	MI	7
14.89	NY	8
14.90	VA	9
15.38	VA	10

Ordered data	Sample	Rank
15.54	VA	11
15.57	VA	12
15.94	MI	13
16.26	NY	14
16.94	MI	15
17.02	VA	16
17.47	VA	17.5
17.47	MI	17.5
17.54	NY	19
18.50	VA	20

Ordered data	Sample	Rank
18.88	NY	21
19.01	MI	22
19.95	NY	23
20.06	NY	24
20.48	VA	25
20.63	VA	26
20.94	NY	27
21.03	NY	28
21.18	NY	29

Picturing the World

To compare the water temperature of cities bordering the Gulf of Mexico, the following data were randomly collected. *(Adapted from The USA Today Weather Almanac)*

Port Mansfield, TX	Eugene Island, LA	Dauphin Island, AL
62	51	63
69	55	51
77	57	54
52	63	60
60	74	75
75	82	80
83	85	70
65	60	78
79	64	82
83	76	84
82	83	
	86	

At $\alpha = 0.05$, can you conclude that the temperature distributions of the three cities are different?

The sum of the ranks for each sample is as follows.

$$R_1 = 2 + 3 + 4 + 5 + 6 + 7 + 13 + 15 + 17.5 + 22 = 94.5$$
$$R_2 = 8 + 14 + 19 + 21 + 23 + 24 + 27 + 28 + 29 = 193$$
$$R_3 = 1 + 9 + 10 + 11 + 12 + 16 + 17.5 + 20 + 25 + 26 = 147.5$$

Using these sums and the values $n_1 = 10$, $n_2 = 9$, $n_3 = 10$, and $N = 29$, the test statistic is

$$H = \frac{12}{29(29+1)}\left(\frac{94.5^2}{10} + \frac{193^2}{9} + \frac{147.5^2}{10}\right) - 3(29+1) \approx 9.41.$$

From the graph at the right, you can see that the test statistic H is in the rejection region. So, at the 1% level, you should decide to reject the null hypothesis. In other words, you can conclude that there is a difference in accountants' hourly pay rates in Michigan, New York, and Virginia.

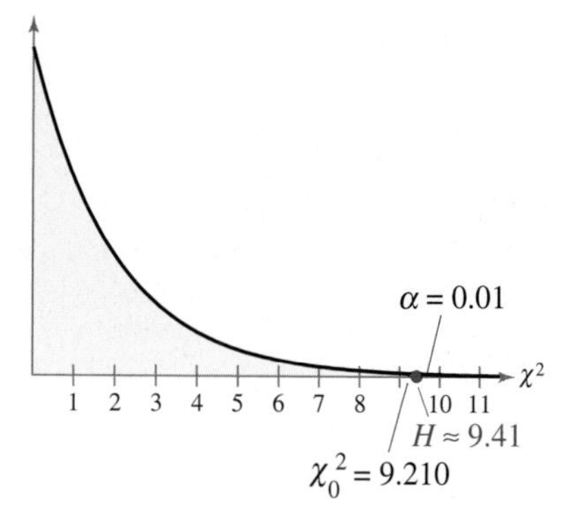

Try It Yourself 1

You want to compare the salaries of underwriters who work in California, Connecticut, and New Jersey. To compare the salaries, you randomly select ten underwriters in each state and record their salary. The salaries (in thousands of dollars) are listed in the table. At $\alpha = 0.10$, can you conclude that the distributions of the underwriters' salaries in these three states are different?

Sample salaries		
CA (Sample 1)	**CT (Sample 2)**	**NJ (Sample 3)**
26.42	25.57	50.16
48.46	29.86	46.72
33.68	51.03	29.73
37.18	57.07	44.57
36.55	45.04	36.35
41.33	37.39	37.24
38.36	30.00	43.26
40.31	33.68	34.29
46.55	45.29	33.91
47.17	61.46	43.89

a. *Identify* the claim and *state* H_0 and H_a.
b. *Specify* the level of significance α.
c. *Determine* the degrees of freedom.
d. *Find* the critical value and identify the rejection region.
e. *List* the combined data in ascending order, *rank* the data, and *find* the sum of the ranks of each sample.
f. *Calculate* the test statistic.
g. *Decide* whether to reject the null hypothesis. Use a graph if necessary.
h. Is there enough evidence to conclude that the salaries are different?

Answer: Page A43

11.3 Exercises

Help

LarsonTutor 11.3

Companion Web Site

Student Solutions Manual 11.3

Videos 11.3

Technology Manuals

Basic Skills and Concepts

Performing a Kruskal-Wallis Test In Exercises 1–4, (a) write the claim mathematically and identify H_0 and H_a, (b) find the critical value, (c) find the sums of the ranks for each sample and calculate the test statistic, and (d) decide whether to reject or fail to reject the null hypothesis. Then interpret the decision in the context of the original claim.

1. ***Home Insurance*** ◆ The following table lists the annual premium for a random sample of high-risk home insurance policies in California, Florida, and Illinois. At $\alpha = 0.05$, can you conclude that the distributions of the annual premiums in these three states are different? *(Adapted from Consumer Reports)*

State	Annual Premium (in dollars)						
California	1654	553	839	777	884	753	643
Florida	1682	2067	1392	1237	1609	1550	1441
Illinois	756	496	381	437	480	566	647

2. ***Home Insurance*** ◆ An independent insurance representative wants to determine whether there is a difference in the annual premiums for high-risk home insurance in three states: New Jersey, New York, and Pennsylvania. To do so, the representative randomly selects several homes with high-risk insurance in each state and determines the annual premium for each. At $\alpha = 0.05$, can the representative conclude that the distributions of the annual premiums in these states are different? *(Adapted from Consumer Reports)*

State	Annual Premium (in dollars)						
New Jersey	441	420	474	411	371	470	
New York	753	684	869	719	1036	613	663
Pennsylvania	653	405	380	484	383	382	387

3. ***Annual Salaries*** ◆ You are writing an article concerning the annual salaries of workers in four U.S. states: Alabama, Colorado, Georgia, and Iowa. The annual salaries of randomly selected workers from each state are listed in the table. Can you write in your report that the distributions of the annual salaries in these four states are different? Use $\alpha = 0.10$. *(Adapted from U.S. Bureau of Labor Statistics)*

State	Annual Salary (in thousands of dollars)							
Alabama	25.9	28.9	21.2	29.0	23.5	14.5	27.4	26.8
Colorado	25.0	19.8	10.8	28.8	25.5	26.9	23.4	35.5
Georgia	34.4	25.1	40.2	39.8	22.8	32.2	36.5	33.2
Iowa	23.4	20.3	23.3	34.1	37.5	22.5	26.1	20.9

 **4.** ***Annual Salaries*** ◆ The following table lists the annual salaries for a random sample of workers in Maryland, Missouri, New Mexico, and Ohio. At $\alpha = 0.10$, can you conclude that the distributions of the annual salaries in these four states are different? *(Adapted from U.S. Bureau of Labor Statistics)*

State	Annual Salary (in thousands of dollars)						
Maryland	12.4	28.5	29.6	28.8	39.0	38.9	32.0
Missouri	25.5	45.7	28.4	25.5	23.8	43.2	24.9
New Mexico	25.3	28.9	26.3	35.1	35.2	23.2	17.8
Ohio	32.7	41.6	16.4	33.1	18.3	34.2	29.6

Extending the Basics

Comparing Two Tests In Exercises 5 and 6, perform the indicated test using

(a) a Kruskal-Wallis test.

(b) a one-way ANOVA test, assuming that each population is normally distributed and the population variances are equal.

(c) Compare the results.

If convenient, use technology to solve the problem.

 **5.** ***Hospital Patient Stays*** ◆ An insurance underwriter reports that the mean number of days patients spend in a hospital differs according to the region of the United States in which the patient lives. The table lists the number of days randomly selected patients spent in a hospital. At $\alpha = 0.01$, can you support the underwriter's claim? *(Adapted from U.S. National Center for Health Statistics)*

Region	Number of Days									
Northeast	9	7	7	4	6	13	3	9	1	7
Midwest	6	5	4	10	1	5	7	4	5	8
South	5	8	1	5	8	7	5	1		
West	2	3	6	6	5	4	3	6	5	

 **6.** ***Energy Consumption*** ◆ The following table shows the energy consumed (in millions of Btu's) in one year for a random sample of households from four regions. At $\alpha = 0.01$, can you conclude that the mean energy consumptions are different? *(Adapted from U.S. Energy Information Administration)*

Region	Energy Consumed (in millions of Btu's)									
Northeast	72	106	151	138	104	108	95	134	100	174	
Midwest	84	183	194	165	120	212	148	129	113	62	97
South	91	40	72	91	147	74	70	67			
West	74	32	78	28	106	39	118	63	70	56	

11.4 Rank Correlation

What You Should Learn

- How to use the Spearman rank correlation coefficient to determine whether the correlation between two variables is significant

The Spearman Rank Correlation Coefficient

The Spearman Rank Correlation Coefficient

In Section 9.1, you learned how to measure the strength of the relationship between two variables using the Pearson correlation coefficient r. Two requirements for the Pearson correlation coefficient are that the variables are linearly related and that the population represented by each variable is normally distributed. If these requirements cannot be met, you can examine the relationship between two variables using the nonparametric equivalent to the Pearson correlation coefficient—the Spearman rank correlation coefficient.

The Spearman rank correlation coefficient has several advantages over the Pearson correlation coefficient. For instance, the Spearman rank correlation coefficient can be used to describe the relationship between linear or nonlinear data. The Spearman rank correlation coefficient can be used for data at the ordinal level. And, the Spearman rank correlation coefficient is easier to calculate by hand than the Pearson coefficient.

DEFINITION

The **Spearman rank correlation coefficient** r_s is a measure of the strength of the relationship between two variables. The Spearman rank correlation coefficient is calculated using the ranks of paired sample data entries. The formula for the Spearman rank correlation coefficient is

$$r_s = 1 - \frac{6\Sigma d^2}{n(n^2 - 1)}$$

where

n is the number of paired data entries

and

d is the difference between the ranks of a paired data entry.

The values of r_s range from -1 to $+1$, inclusive. If the ranks of corresponding data pairs are exactly identical, r_s is equal to $+1$. If the ranks are in "reverse" order, r_s is equal to -1. If the ranks of corresponding data pairs have no relationship, r_s is equal to 0.

After calculating the Spearman rank correlation coefficient, you can determine whether the correlation between the variables is significant. You can do this by performing a hypothesis test for the population correlation coefficient ρ_s. The null and alternative hypotheses for this test are as follows.

H_0: $\rho_s = 0$ (There is no correlation between the variables.)

H_a: $\rho_s \neq 0$ (There is a significant correlation between the variables.)

The critical values for the Spearman rank correlation coefficient are listed in Table 10 of Appendix B. Table 10 lists critical values for selected levels of significance and for sample sizes of 30 or less. The test statistic for the hypothesis test is the Spearman rank correlation coefficient r_s.

GUIDELINES

Testing the Significance of the Spearman Rank Correlation Coefficient

In Words	*In Symbols*
1. State the null and the alternative hypotheses.	State H_0 and H_a.
2. Specify the level of significance.	Specify α.
3. Find the critical value.	Use Table 10 in Appendix B.
4. Find the test statistic.	$r_s = 1 - \dfrac{6\Sigma d^2}{n(n^2 - 1)}$
5. Make a decision to reject or fail to reject the null hypothesis.	If $\lvert r_s \rvert$ is greater than the critical value, reject H_0. Otherwise, do not reject H_0.
6. Interpret the decision in the context of the original claim.	

EXAMPLE 

The Spearman Rank Correlation Coefficient

The table lists the prices (in dollars per 100 pounds) received by U.S. farmers for beef and turkey from 1990 to 1996. At $\alpha = 0.05$, can you conclude that there is a correlation between the beef and turkey prices? *(Adapted from National Agricultural Statistics Service)*

Year	Beef	Turkey
1990	74.6	39.4
1991	72.7	38.4
1992	71.3	37.7
1993	72.6	39.0
1994	66.7	40.4
1995	61.8	41.6
1996	58.7	43.3

SOLUTION The null and alternative hypotheses are as follows.

H_0: $\rho_s = 0$ (There is no correlation between the beef and turkey prices.)

H_a: $\rho_s \neq 0$ (There is a correlation between the beef and turkey prices.)

Picturing the World

The table lists the number of men and women (in thousands) who have graduated from a U.S. college with a bachelor's degree from 1990 to 2000. *(Source: U.S. National Center for Education Statistics)*

Year	Male	Female
1990	492	560
1991	504	590
1992	521	616
1993	533	632
1994	532	637
1995	526	634
1996	522	642
1997	521	652
1998	520	664
1999	517	661
2000	517	668

Does a correlation exist between the number of men and women who graduate with bachelor's degrees each year? Use $\alpha = 0.05$.

Each data set has seven entries. Using Table 10 with $\alpha = 0.05$ and $n = 7$, the critical value is 0.786. Before calculating the test statistic, you must find d^2, the sum of the squares of the differences of the ranks of the data sets. You can use a table to calculate d^2 as shown below.

Beef prices	Rank	Turkey prices	Rank	d	d^2
74.6	7	39.4	4	3	9
72.7	6	38.4	2	4	16
71.3	4	37.7	1	3	9
72.6	5	39.0	3	2	4
66.7	3	40.4	5	−2	4
61.8	2	41.6	6	−4	16
58.7	1	43.3	7	−6	36
					$\Sigma = 94$

Using $n = 7$ and $\Sigma d^2 = 94$, the test statistic is

$$r_s = 1 - \frac{6\Sigma d^2}{n(n^2 - 1)}$$

$$= 1 - \frac{6(94)}{7(7^2 - 1)}$$

$$\approx -0.679.$$

Because $|-0.679| < 0.786$, you should fail to reject the null hypothesis. At the 5% level, you cannot conclude that there is a significant correlation between beef and turkey prices between 1990 and 1996.

Try It Yourself 1

The table lists the prices (in cents per pound) received by U.S. farmers for oat and wheat from 1989 to 1996. At $\alpha = 0.05$, can you conclude that there is a correlation between the oat and wheat prices?

Year	1989	1990	1991	1992	1993	1994	1995	1996
Oat	1.49	1.14	1.21	1.32	1.36	1.22	1.67	1.90
Wheat	3.72	2.61	3.00	3.24	3.26	3.45	4.55	4.30

a. *State* the null and alternative hypotheses.
b. *Specify* the level of significance.
c. *Find* the critical value.
d. *Use a table* to calculate d^2.
e. *Find* the standardized test statistic.
f. *Make a decision* to reject or fail to reject the null hypothesis.
g. Is there enough evidence to conclude that there is a significant correlation between oat and wheat prices between 1989 and 1996?

Answer: Page A44

Exercises

Help

LarsonTutor 11.4

Companion Web Site

Student Solutions Manual 11.4

Videos 11.4

Technology Manuals

Basic Skills and Concepts

Testing a Claim In Exercises 1–4, (a) identify the claim and state H_0 and H_a, (b) find the critical value using Table 10, (c) find the standardized test statistic r_s, and (d) decide whether to reject the null hypothesis. Then interpret the decision in the context of the original claim.

1. ***Farming: Debt and Income*** ◆ In an agricultural report, a commodities analyst suggests that there is a correlation between debt and income in the farming business. The table lists the total debts and total incomes for farms in seven states for a recent year. At $\alpha = 0.01$, is there enough evidence to support the analyst's claim? *(Source: U.S. Department of Agriculture)*

State	Debt (in millions of dollars)	Income (in millions of dollars)
California	14,917	23,310
Illinois	8665	9050
Iowa	11,642	12,853
Minnesota	8104	8809
Nebraska	8111	9454
North Carolina	3468	7831
Texas	9960	13,053

2. ***Suitcases*** ◆ You work for a consumer product review organization and are asked to write a review on suitcases. As part of your review, you need to analyze the relationship between quality and price. The following table lists the overall scores and the prices for eight different suitcases. (The overall score represents the ease of use, features, construction, and durability of a suitcase.) At $\alpha = 0.01$, can you conclude that there is a correlation between the overall score and price? *(Adapted from Consumer Reports)*

Overall score	90	85	81	78	72	68	64	61
Price (in dollars)	495	230	190	160	350	230	260	200

3. ***Air Conditioners*** ◆ The following table lists the overall scores and the prices for 12 different models of air conditioners. The overall score represents the air conditioner's performance and quietness. At $\alpha = 0.10$, can you conclude that there is a correlation between overall score and price? *(Adapted from Consumer Reports)*

Overall score	83	82	79	78	78	74
Price (in dollars)	320	360	360	300	230	285

Overall score	71	67	64	63	61	33
Price (in dollars)	320	300	340	295	220	280

4. *Portable CD Players* ◆ Is the price of a portable CD player related to its quality? To answer this question, you randomly select 11 portable CD players and determine the overall score and price of each. (The overall score represents the error correction, locate speed, battery life, and headphone quality of a CD player.) The results of the study are listed in the table. At $\alpha = 0.01$, can you conclude that there is a correlation between the overall score and the price? *(Adapted from Consumer Reports)*

Overall score	Price (in dollars)
82	150
78	100
68	120
67	140
61	145
60	100
60	150
58	80
57	200
55	80
49	75

Test Scores and GNP In Exercises 5–8, use the following table. The table lists the average achievement score in eighth-grade science and mathematics along with the gross national product (GNP) of nine countries for a recent year. (The GNP is a measure of a nation's total economic activity.) *(Source: IEA Third International Mathematics and Science Study, Boston College; U.S. Bureau of the Census)*

Country	Science average	Mathematics average	GNP (in billions of dollars)
Australia	545	530	342
Canada	531	527	542
Czech Republic	574	564	101
France	498	538	1521
Japan	571	605	5153
Portugal	480	454	103
Spain	517	487	554
Switzerland	522	545	316
USA	534	500	7247

5. *Science and GNP* ◆ At $\alpha = 0.05$, can you conclude that there is a correlation between science achievement scores and GNP?

6. *Math and GNP* ◆ At $\alpha = 0.05$, can you conclude that there is a correlation between mathematics achievement scores and GNP?

7. *Science and Math* ◆ At $\alpha = 0.05$, can you conclude that there is a correlation between science and mathematics achievement scores?

8. *Writing a Summary* ◆ Use the results from Exercises 5–7 to write a summary about the correlation (or the lack of a correlation) between test scores and GNP.

Extending the Basics

Testing the Rank Correlation Coefficient for $n > 30$ If you are testing the significance of the Spearman rank correlation coefficient and the sample size n is greater than 30, you can use the Standard Normal Table and the following to find the critical value.

$$\frac{\pm z}{\sqrt{n-1}}$$

In Exercises 9 and 10, perform the indicated test.

9. ***Work Injuries in Industry*** ◆ The following table lists the average hours worked per week and the number of on-the-job injuries for a random sample of U.S. industries in a recent year. At $\alpha = 0.05$, can you conclude that there is a correlation between average hours worked and the number of on-the-job injuries? *(Adapted from U.S. Bureau of Labor Statistics; National Safety Council)*

Hours worked	47.6	44.1	45.6	45.5	44.5	47.3	44.6	45.9	45.5
Injuries	16	33	25	33	18	20	21	18	21

Hours worked	43.7	44.8	42.5	46.5	42.3	45.5	41.8	43.1	44.4
Injuries	28	15	26	34	32	26	28	22	19

Hours worked	44.5	43.7	44.9	47.8	46.6	45.5	43.5	42.8	44.8
Injuries	23	20	28	24	26	29	21	28	23

Hours worked	43.5	47.0	44.5	50.1	46.7	43.1
Injuries	26	24	20	28	26	25

10. ***Work Injuries in Construction*** ◆ The following table lists the average hours worked per week and the number of on-the-job injuries for a random sample of U.S. construction companies in a recent year. At $\alpha = 0.05$, can you conclude that there is a correlation between average hours worked and the number of on-the-job injuries? *(Adapted from U.S. Bureau of Labor Statistics; National Safety Council)*

Hours worked	40.5	38.3	37.8	38.2	38.6	41.2	39.0	41.0	40.6
Injuries	12	13	19	18	22	22	17	13	15

Hours worked	44.1	39.7	41.2	41.1	38.2	42.3	39.2	36.1	36.2
Injuries	10	18	19	13	24	12	12	13	15

Hours worked	38.7	36.0	37.3	36.5	37.9	38.0	36.7	40.1	35.5
Injuries	18	11	24	16	13	23	14	10	5

Hours worked	38.2	42.3	39.0	39.6	39.1	39.6	39.1
Injuries	14	13	18	15	23	15	23

TECHNOLOGY

MINITAB | EXCEL | TI-83

Selling Prices of Homes

The National Association of Realtors is the world's largest professional organization. Its members, who number over 720,000, include salespeople, brokers, appraisers, counselors, and property managers. One of the things the National Association of Realtors does is keep track of the selling prices of homes in the United States. These can be used to identify regional differences in the cost of homes.

The table at the right shows the selling prices (in thousands of dollars) of a random sample of metropolitan homes sold in a recent year in four U.S. regions: Northeast, Midwest, South, and West.

Selling prices of metropolitan homes (in thousands of dollars)

Northeast	Midwest	South	West
132.3	124.7	145.9	176.2
151.3	116.4	139.5	165.7
148.6	131.7	127.5	201.1
158.9	129.8	118.0	164.4
138.8	126.6	139.6	156.8
143.7	119.3	130.1	167.7
136.8	130.2	136.4	190.7
135.5	124.6	121.8	198.9
130.1	130.5	147.4	184.7
142.7	121.2	131.5	160.4
140.1	129.1	123.5	156.5
133.9	105.3	133.2	170.7

Exercises

In Exercises 1–5, refer to the selling prices of metropolitan homes in the table. Use $\alpha = 0.05$ for all tests.

1. Construct a box-and-whisker plot for each region. Do the median selling prices appear to differ between regions?
2. Use a technology tool to perform a sign test to test the claim that the median selling price in the South is at least $125,000.
3. Use a technology tool to perform a Wilcoxon rank sum test to test the claim that the median selling prices in the Northeast and Midwest are the same.
4. Use a technology tool to perform a Kruskal-Wallis test to test the claim that the median selling prices for all four regions are the same.
5. Use a technology tool to perform a one-way ANOVA to test the claim that the average selling prices for all four regions are the same. How do your results compare to those in Exercise 4?
6. Repeat Exercises 1, 3, 4, and 5 using the data in the following table. The table shows the selling prices (in thousands of dollars) of a random sample of existing apartment condominiums and co-ops sold in a recent year in four U.S. regions: Northeast, Midwest, South, and West.

Selling prices of condominiums and co-ops (in thousands of dollars)

Northeast	Midwest	South	West
95.6	129.4	102.8	102.9
125.5	147.8	88.7	142.4
126.4	142.6	93.2	160.6
103.1	131.6	101.2	148.0
129.8	116.6	101.5	129.8
123.0	124.4	95.5	148.2
116.8	122.7	90.6	153.5
134.0	112.0	89.4	135.7
117.8	145.5	80.8	152.0
115.0	157.0	96.1	128.7
109.4	140.5	103.3	136.5
121.2	134.7	94.0	129.8
130.6	129.5	99.8	130.2
118.6	118.3	113.5	123.5
123.7	158.7	90.4	151.4

Extended solutions are given in the *Technology Supplement*. Technical instruction is provided for *Minitab*, *Excel*, and the *TI-83*.

11

Chapter Summary

What did you learn? | *Review Exercises*

Section 11.1

- How to use the sign test to test a population median and to test the difference between two population medians (dependent samples) — *1–6*

Section 11.2

- How to use the Wilcoxon signed-rank test and Wilcoxon rank sum test to test the difference between two population distributions — *7–10*

Section 11.3

- How to use the Kruskal-Wallis test to test for differences among three or more population distributions — *11, 12*

Section 11.4

- How to use the Spearman rank correlation coefficient to determine whether the correlation between two variables is significant — *13, 14*

The table summarizes parametric and nonparametric tests. Always use the parametric test if the conditions for that test are satisfied.

Test application	Parametric test	Nonparametric test
One-sample tests	z-test for a population mean t-test for a population mean	Sign test for a population median
Two-sample tests		
Dependent samples	t-test for the difference between means	Paired-sample sign test Wilcoxon signed-rank test
Independent samples	z-test for the difference between means t-test for the difference between means	Wilcoxon rank sum test
Tests involving three or more samples	One-way ANOVA	Kruskal-Wallis test
Correlation	Pearson correlation coefficient	Spearman rank correlation coefficient

STATISTICS

Uses and Abuses

Uses

Nonparametric Tests Before you could perform many of the hypothesis tests you learned about in previous chapters, you had to ensure that certain conditions about the population were satisfied. For example, before you could run a t-test, you had to verify that the population is normally distributed. One advantage of the nonparametric tests shown in this chapter is that they are distribution free. That is they do not require any particular information about the population or populations being tested. Another advantage of nonparametric tests is that they are easier to perform than their parametric counterparts. This means that they are easier to understand and quicker to use. Nonparametric tests can often be used when data are at the nominal or ordinal level.

Abuses

Insufficient Evidence Stronger evidence is needed to reject a null hypothesis in a nonparametric test than it is to reject a null hypothesis in a corresponding parametric test. That is, when you are trying to support a claim represented by the alternative hypothesis, you might need a larger sample when performing a nonparametric test. If the outcome of a nonparametric test results in failing to reject the null hypothesis, you should investigate the sample size used. It may be that a larger sample will produce different results.

Using an Inappropriate Test In general, when information about the populations (such as the condition of normality) is known, it is more efficient to use a parametric test. However, if information about the population is not known, nonparametric tests can be helpful.

Exercises

1. ***Insufficient Evidence*** Give an example of a nonparametric test in which there is not enough evidence to reject the null hypothesis.
2. ***Using an Inappropriate Test*** Discuss the nonparametric tests described in this chapter and match each test with its parametric counterpart, which you studied in earlier chapters.

Review Exercises

Section 11.1

In Exercises 1–6, use a sign test to test the claim by doing the following.

(a) Write the claim mathematically and identify H_0 and H_a.

(b) Find the critical value.

(c) Calculate the test statistic.

(d) Decide whether to reject or fail to reject the null hypothesis. Then interpret the decision in the context of the original claim.

1. A financial services institution estimates that the median value of stock among families that own stock is $13,500. The stock values (in thousands of dollars) among 17 randomly selected families that own stock are listed below. At $\alpha = 0.01$, can you reject the institution's claim? *(Adapted from Board of Governors of the Federal Reserve System)*

4.09 18.09 8.52 19.30 14.47 9.82 12.46 0.95 20.21
9.21 9.29 9.94 10.84 9.16 14.07 17.33 13.74

2. A financial services institution claims that the median credit card debt among families that earn $10,000 to $24,999 is more than $1200. The credit card debts (in thousands of dollars) among 13 randomly selected families are listed below. At $\alpha = 0.01$, can you support the institution's claim? *(Adapted from Board of Governors of the Federal Reserve System)*

1.97 1.10 1.02 1.05 0.98 1.36 1.74
1.69 1.48 0.99 1.29 0.92 1.45

3. A mail-order company believes that the median turnover time between receipt of a telephone order and packing of that order is six hours or less. Over a five-day period, 78 orders are randomly selected and their turnover time is recorded in $\frac{1}{2}$-hour increments. Eight orders took six hours, 26 orders took under six hours, and 44 orders took over six hours. At $\alpha = 0.10$, can you reject the company's claim?

4. In a study testing the effects of calcium supplements on blood pressure in men, ten randomly selected men were given a calcium supplement for 12 weeks. The following measurements are for each subject's diastolic blood pressure taken before and after the 12-week treatment period. At $\alpha = 0.10$, can you reject the claim that there was no reduction in diastolic blood pressure? *(Adapted from The Journal of American Medicine)*

Patient	1	2	3	4	5	6	7
Before treatment	107	110	123	129	112	111	107
After treatment	100	114	105	112	115	116	106

Patient	8	9	10
Before treatment	112	136	102
After treatment	102	125	104

5. In a study testing the effects of an herbal supplement on blood pressure in men, eleven randomly selected men were given an herbal supplement for 15 weeks. The following measurements are for each subject's diastolic blood pressure taken before and after the 15-week treatment period. At $\alpha = 0.05$, can you reject the claim that there was no reduction in diastolic blood pressure? *(Adapted from The Journal of American Medicine)*

Patient	1	2	3	4	5	6	7
Before treatment	123	109	112	102	98	114	119
After treatment	124	97	113	105	95	119	114

Patient	8	9	10	11
Before treatment	112	110	117	130
After treatment	114	121	118	133

6. The career placement office at a large university claims that the median starting salary of graduates with a bachelor's degree in marketing is $27,900. Of last year's graduates in marketing, 57 were randomly surveyed about their salaries. Of the 54 graduates who were currently employed, 21 were paid less than $27,900 annually, and 33 were paid more than $27,900. At $\alpha = 0.05$, can you reject the office's claim? *(Adapted from National Association of Colleges and Employers)*

Section 11.2

In Exercises 7–10, use a Wilcoxon test to test the claim by doing the following.

(a) Decide whether the samples are dependent or independent; then choose the appropriate Wilcoxon test.

(b) Write the claim mathematically and identify H_0 and H_a.

(c) Find the critical value.

(d) Calculate the test statistic.

(e) Decide whether to reject or fail to reject the null hypothesis. Then interpret the decision in the context of the original claim.

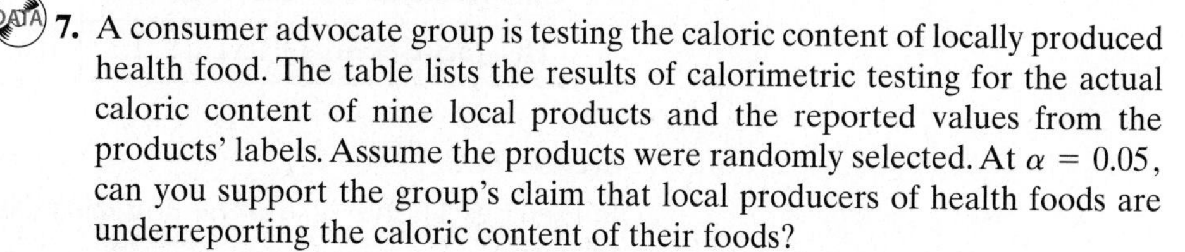

7. A consumer advocate group is testing the caloric content of locally produced health food. The table lists the results of calorimetric testing for the actual caloric content of nine local products and the reported values from the products' labels. Assume the products were randomly selected. At $\alpha = 0.05$, can you support the group's claim that local producers of health foods are underreporting the caloric content of their foods?

Product	1	2	3	4	5	6	7	8	9
Actual	88	82	107	171	150	140	211	172	199
Reported	60	75	75	120	135	140	170	180	185

8. A career placement advisor suggests that there is no difference in the starting salaries earned by female and male humanities graduates. A random sample of 11 female and 11 male humanities graduates and their starting salaries is listed in the table. At $\alpha = 0.05$, can you reject the advisor's claim? *(Adapted from U.S. Department of Education)*

Gender	Salary (in thousands of dollars)										
Female	20.2	20.7	21.8	20.5	20.7	20.7	20.3	20.3	20.6	20.1	20.6
Male	20.2	22.2	21.1	21.7	22.7	20.5	22.0	21.6	22.3	21.3	21.9

9. A career placement advisor estimates that there is a difference in the total time to earn a doctorate degree by female and male graduate students. A random sample of 12 female and 12 male graduate students and their total time to earn a doctorate degree is listed in the table. At $\alpha = 0.01$, can you support the advisor's claim? *(Adapted from U.S. Department of Education)*

Gender	Total time (in years)											
Female	15	14	15	15	15	11	13	16	9	9	11	13
Male	11	8	9	11	10	8	8	10	11	9	10	8

10. A medical researcher claims that a new drug affects the number of headache hours experienced by headache sufferers. The number of headache hours (per day) experienced by eight randomly selected patients before and after taking the drug are listed in the table. At $\alpha = 0.05$, can you support the researcher's claim?

Patient	1	2	3	4
Headache hours (before)	0.9	2.3	2.7	2.4
Headache hours (after)	1.4	1.5	1.4	1.8

Patient	5	6	7	8
Headache hours (before)	2.9	1.9	1.3	3.1
Headache hours (after)	1.3	0.6	0.7	1.9

Section 11.3

In Exercises 11 and 12, use the Kruskal-Wallis test to test the claim by doing the following.

(a) Write the claim mathematically and identify H_0 and H_a.

(b) Find the critical value.

(c) Find the sums of the ranks for each sample and calculate the test statistic.

(d) Decide whether to reject or fail to reject the null hypothesis. Then interpret the decision in the context of the original claim.

11. The following table lists the starting salaries for a random sample of college graduates in three fields of study. At $\alpha = 0.05$, can you conclude that the distributions of the starting salaries in these three fields of study are different? *(Adapted from U.S. Department of Education)*

Field of study	Starting salary (in thousands of dollars)									
Education	20.4	20.5	20.4	20.2	20.4	20.2	20.1	20.1	19.4	19.9
Humanities	21.1	21.3	21.1	20.8	20.9	21.3	20.7	21.5	20.6	21.0
Natural science	22.2	22.1	22.2	21.0	21.1	21.8	22.6	22.0	22.5	21.2

12. The following table lists the total time to earn a doctorate degree for a random sample of college graduates in three fields of study. At $\alpha = 0.05$, can you conclude that the distributions of the total times in these three fields of study are different? *(Adapted from U.S. Department of Education)*

Field of study	Total time (in years)										
Computer science	9	10	9	7	11	9	8	10	9	9	10
Natural science	8	8	7	8	7	7	8	8	8	8	7
Social science	12	13	11	11	10	9	14	10	10	11	12

Section 11.4

In Exercises 13 and 14, use the Spearman rank correlation coefficient to test the claim by doing the following.

(a) Write the claim mathematically and identify H_0 and H_a.

(b) Find the critical value using Table 10.

(c) Find the standardized test statistic r_s.

(d) Decide whether to reject the null hypothesis. Then interpret the decision in the context of the original claim.

13. The following table lists the overall scores and the prices for eight randomly selected television sets. The overall score represents the set's picture quality, sound quality, ease of use, and cable performance. At $\alpha = 0.01$, can you conclude that there is a correlation between overall score and price? *(Adapted from Consumer Reports)*

Overall score	80	79	70	65	63	62	61	58
Price (in dollars)	580	740	630	660	600	590	700	750

14. The following table lists the overall scores and the prices for nine randomly selected cordless phones. The overall score represents the phone's speech clarity and handset convenience. At $\alpha = 0.05$, can you conclude that there is a correlation between overall score and price? *(Adapted from Consumer Reports)*

Overall score	81	78	72	68	66	65	67	63	57
Price (in dollars)	130	200	70	80	105	130	100	150	60

PUTTING IT ALL TOGETHER

Real Statistics Real Decisions

In a recent year, according to the Bureau of Labor Statistics, the median number of years that wage and salary workers had been with their current employer (called employee tenure) was 3.5 years. Information on employee tenure has been gathered since the early 1950s using the Current Population Survey (CPS), a monthly survey of 50,000 households that provides information on employment, unemployment, earnings, demographics, and other characteristics of the U.S. population. With respect to employee tenure, the questions measure how long workers had been with their current employer, not how long they plan to stay with their employer.

stats.bls.gov

Exercises

1. *How Would You Do It?*

   (a) What sampling technique would you use to select the sample for the CPS?

   (b) Do you think the technique in part (a) will give you a sample that is representative of the U.S. population? Why or why not?

   (c) Identify possible flaws or biases in the survey based on the technique you chose in part (a).

2. *Is There a Difference?*

   A congressional representative claims that the median tenure for workers from the representative's district is less than the national median tenure of 3.5 years. The claim is based on the representative's data and is shown in the table at the right. (Assume that the employees were randomly selected.)

   (a) Is it possible that the claim is true? What questions should you ask about how the data was collected?

   (b) How would you test the representative's claim? Can you use a parametric test, or do you need to use a nonparametric test?

   (c) State the null hypothesis and the alternative hypothesis.

   (d) Test the claim using $\alpha = 0.05$. What can you conclude?

3. *Comparing Male and Female Employee Tenures*

   A congressional representative claims that the median tenure for male workers is greater than the median tenure for female workers. The claim is based on the representative's data and is shown in the table at the right. (Assume that the employees were randomly selected from the representative's district.)

   (a) How would you test the representative's claim? Can you use a parametric test, or do you need to use a nonparametric test?

   (b) State the null hypothesis and the alternative hypothesis.

   (c) Test the claim using $\alpha = 0.05$. What can you conclude?

Employee tenure of 20 workers		
4.1	2.1	2.8
2.3	1.0	1.4
3.5	4.5	3.4
4.6	3.2	4.9
3.1	3.4	5.7
1.2	4.1	2.6
3.9	3.1	

Table for Exercise 2

Employee tenure for a sample of male workers	Employee tenure for a sample of female workers
3.3	3.7
3.9	4.2
4.1	2.7
3.3	3.6
4.4	3.3
3.3	1.1
3.1	4.4
4.1	4.4
2.7	2.6
4.9	1.5
0.9	4.5
4.6	2.0
	4.0
	3.0

Table for Exercise 3

11 Chapter Quiz

Take this quiz as you would take a quiz in class. After you are done, check your work against the answers given in the back of the book.

For this quiz, do the following.

(a) Write the claim mathematically and identify H_0 and H_a.

(b) Decide which test to use.

(c) Find the critical value(s) and identify the critical region(s).

(d) Calculate the test statistic.

(e) Decide whether to reject or fail to reject the null hypothesis. Then interpret the decision in the context of the original claim.

1. A women's organization claims that there is a difference in the salaries earned by female and male employees of state and local governments. A random sample of nine female and nine male state and local government employees and their salaries is listed in the table. At $\alpha = 0.10$, can you support the organization's claim? *(Adapted from U.S. Equal Employment Opportunity Commission)*

Gender	Salary (in thousands of dollars)								
Female	29.2	28.8	28.4	32.0	23.8	34.3	24.8	25.6	22.8
Male	43.9	35.3	20.4	31.3	38.8	30.7	31.8	27.3	32.4

2. A government official believes that the median age in Puerto Rico is 28 years. In a random sample of 25 Puerto Ricans, 10 were less than 28 years, 13 were greater than 28 years, and 2 were 28 years. At $\alpha = 0.05$, can you reject the official's claim? *(Adapted from U.S. Bureau of the Census)*

3. The following table lists the average hours worked per week and the number of on-the-job injuries for a random sample of U.S. industries in a recent year. At $\alpha = 0.01$, can you conclude that there is a correlation between average hours worked and the number of on-the-job injuries? *(Adapted from U.S. Bureau of Labor Statistics; National Safety Council)*

Hours worked	47.3	46.7	48.7	39.6	42.3	43.6	47.5	47.9
Injuries	1	5	3	2	3	4	3	3

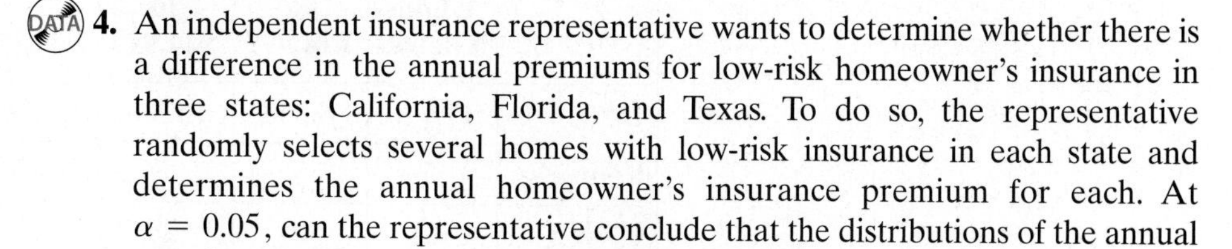

DATA **4.** An independent insurance representative wants to determine whether there is a difference in the annual premiums for low-risk homeowner's insurance in three states: California, Florida, and Texas. To do so, the representative randomly selects several homes with low-risk insurance in each state and determines the annual homeowner's insurance premium for each. At $\alpha = 0.05$, can the representative conclude that the distributions of the annual premiums in these states are different? *(Adapted from Consumer Reports)*

State	Annual Premium (in dollars)						
California	756	429	476	427	501	512	492
Florida	677	701	573	405	604	506	458
Texas	591	490	735	546	633	502	447

Cumulative Test: Chapters 10–11

Take this test as you would take a test in class. After you are done, check your work against the answers given in the back of the book.

For Exercises 1–4, refer to the following table. The table lists the annual incomes (in thousands of dollars) for a random sample of households in California and Florida. *(Adapted from U.S. Bureau of the Census)*

California	62.1	44.0	45.5	37.9	40.2
Florida	44.3	37.6	28.5	17.0	22.8

California	26.3	48.8	49.3	31.6	44.8
Florida	15.8	34.7	23.2	23.7	31.9

1. At $\alpha = 0.05$, test the claim that the median household income in California is less than $40,000.

2. At $\alpha = 0.05$, test the claim that the median household income in Florida is greater than $29,000.

3. At $\alpha = 0.01$, test the claim that the median household incomes in California and Florida are the same.

4. At $\alpha = 0.01$, test the claim that the variances of the median household incomes in California and Florida are the same. (Assume each population of incomes is normally distributed.)

In Exercises 5 and 6, use the following table. The table lists three distributions:

- the percent distribution of annual incomes of randomly selected households in the United States.
- two frequency distributions. The frequency distributions represent the results of a survey of 200 randomly selected households in Kentucky and Utah.

(Adapted from U.S. Bureau of the Census)

	Under $10,000	**$10,000 to $14,999**	**$15,000 to $24,999**	**$25,000 to $34,999**	**$35,000 to $49,999**	**$50,000 to $74,999**	**$75,000 and over**
U.S.	11.7%	8.6%	15.3%	13.7%	16.3%	18.0%	16.4%
Kentucky	5	7	52	73	59	4	0
Utah	13	16	28	35	36	61	11

5. At $\alpha = 0.10$, test the claim that the distribution of household incomes in Kentucky is the same as the U.S. distribution.

6. At $\alpha = 0.10$, test the claim that the distribution of household incomes in Utah is the same as the U.S. distribution.

Appendix A

In this appendix we use a 0-to-*z* table as an alternate development of the standard normal distribution. It is intended that this appendix be used after completing Section 5.1 in the text. If used, this appendix should replace the material in Section 5.2 except for the exercises.

Standard Normal Distribution (0-to-*z*)

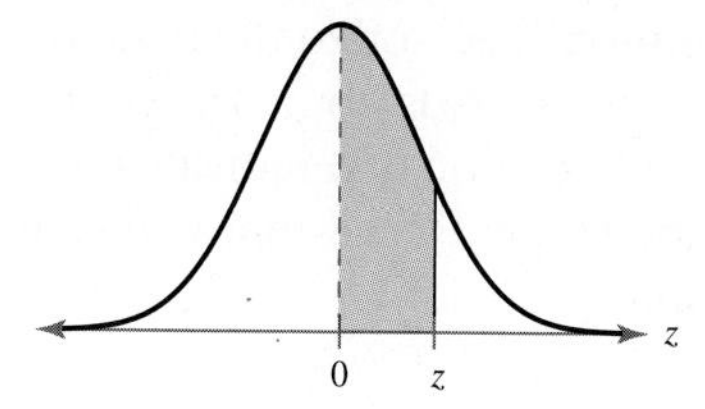

z	.00	.01	.02	.03	.04	.05	.06	.07	.08	.09
0.0	.0000	.0040	.0080	.0120	.0160	.0199	.0239	.0279	.0319	.0359
0.1	.0398	.0438	.0478	.0517	.0557	.0596	.0636	.0675	.0714	.0753
0.2	.0793	.0832	.0871	.0910	.0948	.0987	.1026	.1064	.1103	.1141
0.3	.1179	.1217	.1255	.1293	.1331	.1368	.1406	.1443	.1480	.1517
0.4	.1554	.1591	.1628	.1664	.1700	.1736	.1772	.1808	.1844	.1879
0.5	.1915	.1950	.1985	.2019	.2054	.2088	.2123	.2157	.2190	.2224
0.6	.2257	.2291	.2324	.2357	.2389	.2422	.2454	.2486	.2517	.2549
0.7	.2580	.2611	.2642	.2673	.2704	.2734	.2764	.2794	.2823	.2852
0.8	.2881	.2910	.2939	.2967	.2995	.3023	.3051	.3078	.3106	.3133
0.9	.3159	.3186	.3212	.3238	.3264	.3289	.3315	.3340	.3365	.3389
1.0	.3413	.3438	.3461	.3485	.3508	.3531	.3554	.3577	.3599	.3621
1.1	.3643	.3665	.3686	.3708	.3729	.3749	.3770	.3790	.3810	.3830
1.2	.3849	.3869	.3888	.3907	.3925	.3944	.3962	.3980	.3997	.4015
1.3	.4032	.4049	.4066	.4082	.4099	.4115	.4131	.4147	.4162	.4177
1.4	.4192	.4207	.4222	.4236	.4251	.4265	.4279	.4292	.4306	.4319
1.5	.4332	.4345	.4357	.4370	.4382	.4394	.4406	.4418	.4429	.4441
1.6	.4452	.4463	.4474	.4484	.4495	.4505	.4515	.4525	.4535	.4545
1.7	.4554	.4564	.4573	.4582	.4591	.4599	.4608	.4616	.4625	.4633
1.8	.4641	.4649	.4656	.4664	.4671	.4678	.4686	.4693	.4699	.4706
1.9	.4713	.4719	.4726	.4732	.4738	.4744	.4750	.4756	.4761	.4767
2.0	.4772	.4778	.4783	.4788	.4793	.4798	.4803	.4808	.4812	.4817
2.1	.4821	.4826	.4830	.4834	.4838	.4842	.4846	.4850	.4854	.4857
2.2	.4861	.4864	.4868	.4871	.4875	.4878	.4881	.4884	.4887	.4890
2.3	.4893	.4896	.4898	.4901	.4904	.4906	.4909	.4911	.4913	.4916
2.4	.4918	.4920	.4922	.4925	.4927	.4929	.4931	.4932	.4934	.4936
2.5	.4938	.4940	.4941	.4943	.4945	.4946	.4948	.4949	.4951	.4952
2.6	.4953	.4955	.4956	.4957	.4959	.4960	.4961	.4962	.4963	.4964
2.7	.4965	.4966	.4967	.4968	.4969	.4970	.4971	.4972	.4973	.4974
2.8	.4974	.4975	.4976	.4977	.4977	.4978	.4979	.4979	.4980	.4981
2.9	.4981	.4982	.4982	.4983	.4984	.4984	.4985	.4985	.4986	.4986
3.0	.4987	.4987	.4987	.4988	.4988	.4989	.4989	.4989	.4990	.4990
3.1	.4990	.499I	.4991	.4991	.4992	.4992	.4992	.4992	.4993	.4993
3.2	.4993	.4993	.4994	.4994	.4994	.4994	.4994	.4995	.4995	.4995
3.3	.4995	.4995	.4995	.4996	.4996	.4996	.4996	.4996	.4996	.4997
3.4	.4997	.4997	.4997	.4997	.4997	.4997	.4997	.4997	.4997	.4998

From Frederick C. Mosteller and Robert E. K. Rourke, *Sturdy Statistics,* 1973, Addison-Wesley Publishing Co., Reading, MA. Reprinted with permission of Frederick Mosteller.

Alternate Presentation of The Standard Normal Distribution

What You Should Learn

- How to find areas under the standard normal curve

The Standard Normal Distribution

The Standard Normal Distribution

There are infinitely many normal distributions, each with its own mean and standard deviation. The normal distribution with a mean of 0 and a standard deviation of 1 is called **the standard normal distribution.** The horizontal scale of the graph of the standard normal distribution corresponds to z-scores. In Section 2.5, you learned that a z-score is a measure of position that indicates the number of standard deviations a value lies from the mean. Recall that you can transform an x-value to a z-score using the formula

$$z = \frac{\text{value} - \text{mean}}{\text{standard deviation}} = \frac{x - \mu}{\sigma}.$$

Insight

Because every normal distribution can be transformed to the standard normal distribution, you can use z-scores and the standard normal curve to find areas (and therefore probability) under any normal curve.

DEFINITION

The **standard normal distribution** is a normal distribution with a mean of 0 and a standard deviation of 1.

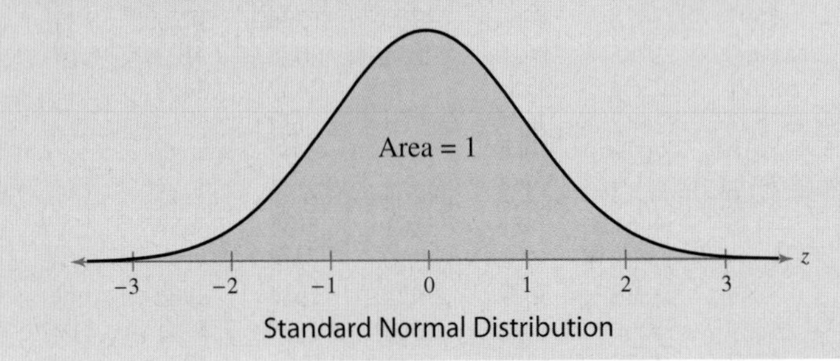

Standard Normal Distribution

If each data value of a normally distributed random variable x is transformed into a z-score, the result will be the standard normal distribution. When this transformation takes place, the area that falls in the interval under the nonstandard normal curve is the *same* as that under the standard normal curve within the corresponding z-boundaries.

In Section 5.1, you learned to approximate areas under a normal curve when values of the random variable x corresponded to −3, −2, −1, 0, 1, 2, or 3 standard deviations from the mean. In this section, you will learn to calculate areas corresponding to other x-values. After you transform an x-value to a z-score, you can use the Standard Normal Table (0-to-z) on page A1. The table lists the area under the standard normal curve between 0 and the given z-score. As you examine the table, notice the following.

Study Tip

It is important that you know the difference between x and z. The random variable x is sometimes called a raw score and represents values in a *nonstandard* normal distribution, while z represents values in the *standard* normal distribution.

Properties of the Standard Normal Distribution

1. The distribution is symmetric about the mean ($z = 0$).
2. The area under the standard normal curve to the left of $z = 0$ is 0.5 and the area to the right of $z = 0$ is 0.5.
3. The area under the standard normal curve increases as the distance between 0 and z increases.

At first glance, the table on page A1 appears to give areas for positive z-scores only. However, because of the symmetry of the standard normal curve, the table also gives areas for negative z-scores (see Example 1).

Study Tip

When the z-score is not in the table, use the entry closest to it. If the given z-score is exactly midway between two z-scores, then use the area midway between the corresponding areas.

EXAMPLE 

Using the Standard Normal Table (0-to-z)

1. Find the area under the standard normal curve between 0 and $z = 1.15$.
2. Find the z-scores that correspond to an area of 0.0948.

SOLUTION

1. Find the area that corresponds to $z = 1.15$ by finding 1.1 in the left column and then moving across the row to the column under 0.05. The number in that row and column is 0.3749. So, the area between $z = 0$ and $z = 1.15$ is 0.3749.

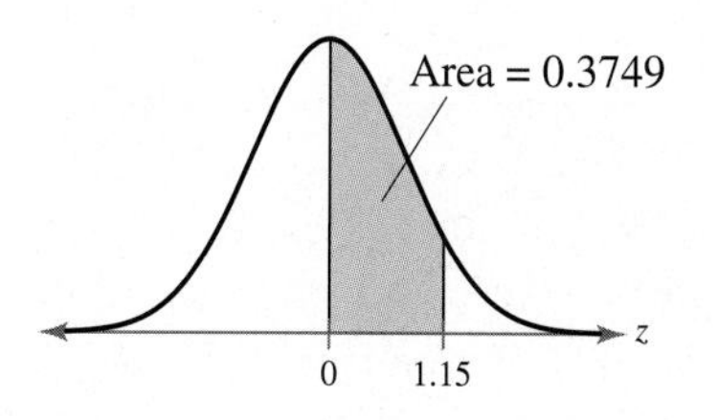

z	.00	.01	.02	.03	.04	.05	.06
0.0	.0000	.0040	.0080	.0120	.0160	.0199	.0239
0.1	.0398	.0438	.0478	.0517	.0557	.0596	.0636
0.2	.0793	.0832	.0871	.0910	.0948	.0987	.1026
0.3	.1179	.1217	.1255	.1293	.1331	.1368	.1406
0.9	.3159	.3186	.3212	.3238	.3264	.3289	.3315
1.0	.3413	.3438	.3461	.3485	.3508	.3531	.3554
1.1	.3643	.3665	.3686	.3708	.3729	.3749	.3770
1.2	.3849	.3869	.3888	.3907	.3925	.3944	.3962
1.3	.4032	.4049	.4066	.4082	.4099	.4115	.4131
1.4	.4192	.4207	.4222	.4236	.4251	.4265	.4279

2. Find the z-scores that correspond to an area of 0.0948 by location 0.0948 in the table. The values at the beginning of the corresponding row and at the top of the corresponding column give the z-score. For an area of 0.0948, the row value is 0.2 and the column value is 0.04. So, the z-scores are $z = -0.24$ and $z = 0.24$.

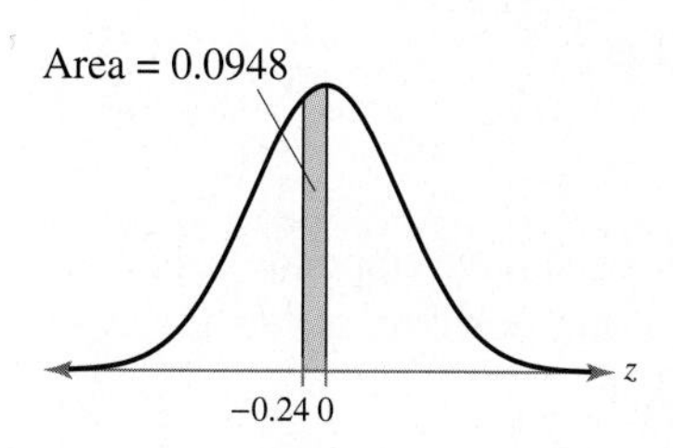

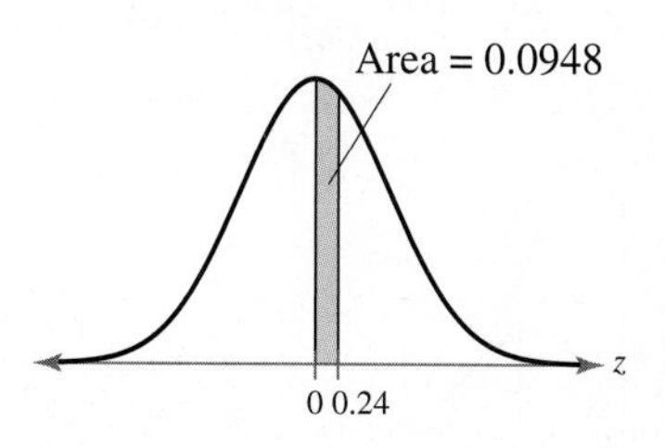

z	.00	.01	.02	.03	.04	.05	.06
0.0	.0000	.0040	.0080	.0120	.0160	.0199	.0239
0.1	.0398	.0438	.0478	.0517	.0557	.0596	.0636
0.2	.0793	.0832	.0871	.0910	.0948	.0987	.1026
0.3	.1179	.1217	.1255	.1293	.1331	.1368	.1406
0.4	.1554	.1591	.1628	.1664	.1700	.1736	.1772
0.5	.1915	.1950	.1985	.2019	.2054	.2088	.2123

Try It Yourself 1

(1) Find the area under the standard normal curve between $z = 0$ and $z = 2.19$.

a. Locate the given z-score and *find the corresponding area* in the Standard Normal Table on page A1.

(2) Find the z-scores that correspond to an area of 0.4850.

b. Locate the given area in the Standard Normal Table (0-to-z) and *find the corresponding z-score.*

Answer: Page A##

Use the following guidelines to find various types of areas under the standard normal curve.

GUIDELINES

Finding Areas Under the Standard Normal Curve

1. Sketch the standard normal curve and shade the appropriate area under the curve.
2. Use the Standard Normal Table (0-to-z) to find the area that corresponds to the given z-score(s).
3. Find the desired area by following the directions for each case shown.

a. Area to the left of z

i. When $z < 0$, *subtract* the area from 0.5.

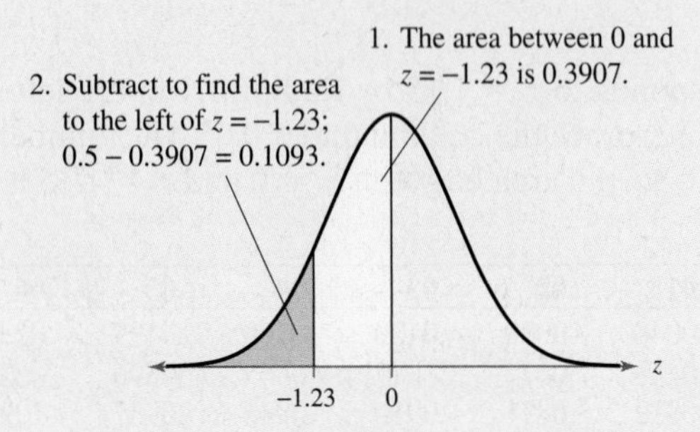

ii. When $z > 0$, *add* 0.5 to the area.

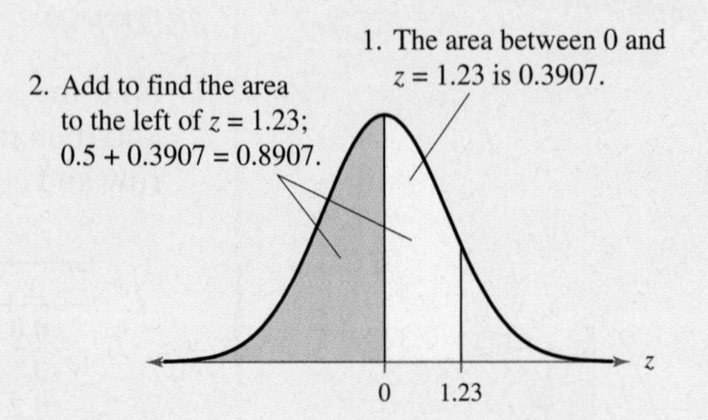

b. Area to the right of z

i. When $z < 0$, *add* 0.5 to the area.

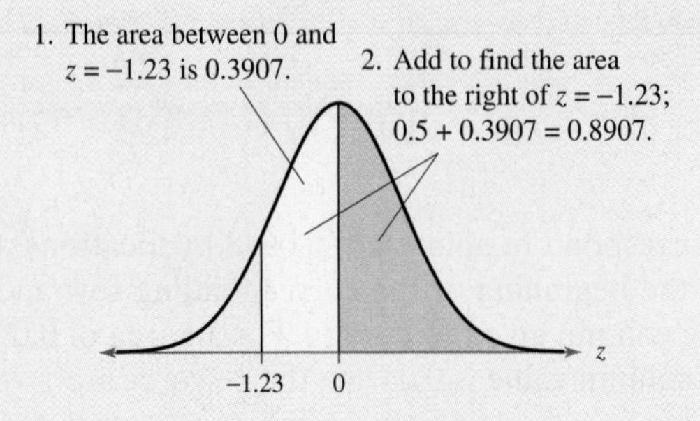

ii. When $z > 0$, *subtract* the area from 0.5.

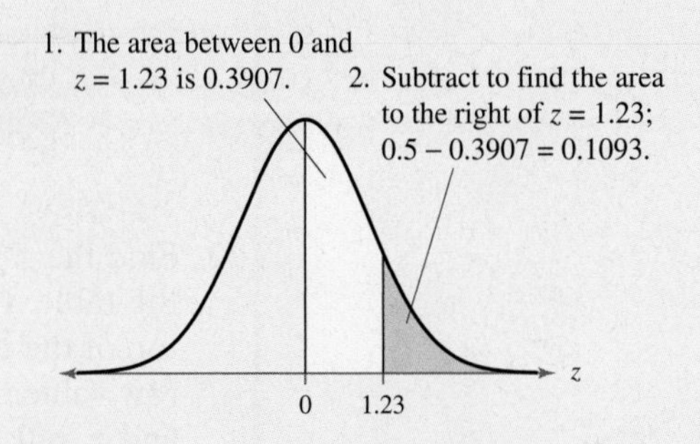

c. Area between two z-scores

i. When the two z-scores have the same sign (both positive or both negative), *subtract* the smaller area from the larger area.

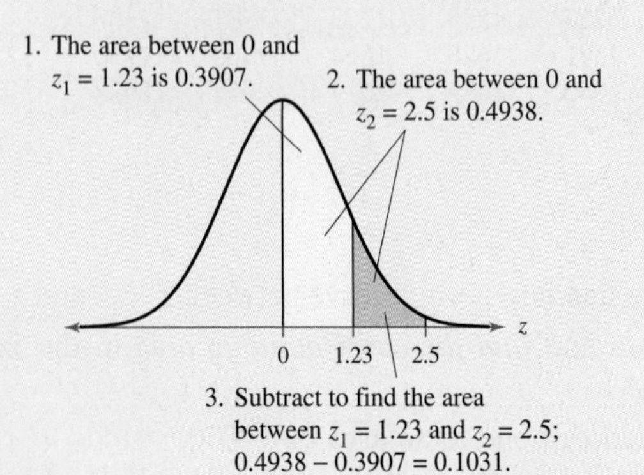

ii. When the two z-scores have opposite signs (one negative and one positive), *add* the areas.

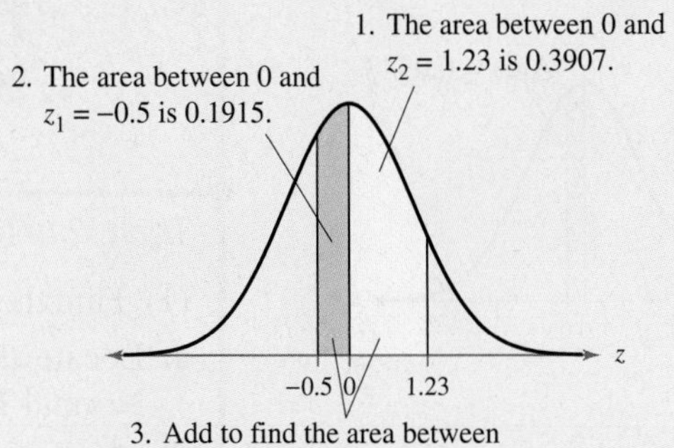

EXAMPLE 

Finding Area Under the Standard Normal Curve

Find the area under the standard normal curve to the left of $z = -0.99$.

SOLUTION The area under the standard normal curve to the left of $z = -0.99$ is shown.

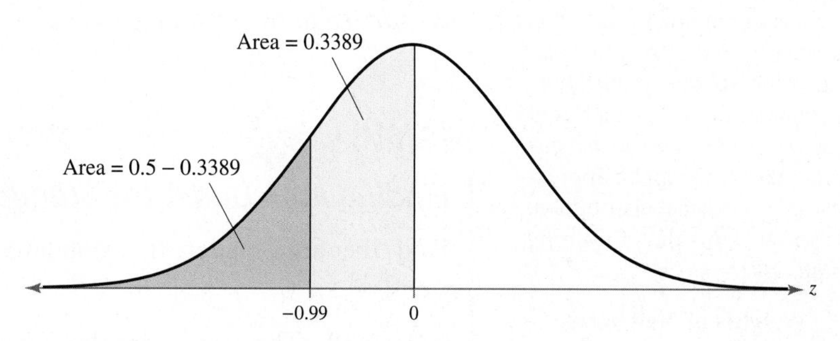

From the Standard Normal Table (0-to-z), the area corresponding to $z = -0.99$ is 0.3389. Because the area to the left of 0 is 0.5, the area to the left of $z = -0.99$ is $0.5 - 0.3389 = 0.1611$.

Insight

Because the normal distribution is a continuous probability distribution, the area under the standard normal curve to the left of a z-score gives the probability that z is less than that z-score. For instance, in Example 2, the area to the left of $z = -0.99$ is 0.1611. So, $P(z < -0.99) = 0.1611$, which is read as "the probability that z is less than -0.99 is 0.1611."

Try It Yourself 2

Find the area under the standard normal curve to the left of $z = 2.13$.

a. *Draw* the standard normal curve and shade the area under the curve and to the left of $z = 2.13$.
b. Use the Standard Normal Table (0-to-z) to *find the area* that corresponds to $z = 2.13$.
c. *Add* 0.5 to the resulting area.

Answer: Page A44

EXAMPLE 

Finding Area Under the Standard Normal Curve

Find the area under the standard normal curve to the right of $z = 1.06$.

SOLUTION The area under the standard normal curve to the right of $z = 1.06$ is shown.

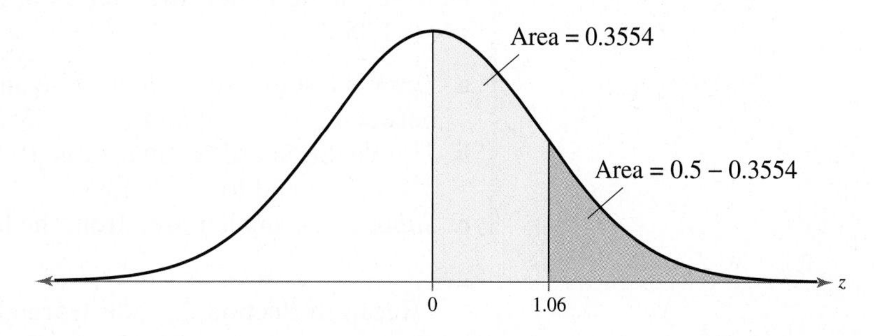

From the Standard Normal Table (0-to-z), the area corresponding to $z = 1.06$ is 0.3554. Because the area to the right of 0 is 0.5, the area to the right of $z = 1.06$ is $0.5 - 0.3554 = 0.1446$.

Picturing the World

Each year the Centers for Disease Control and Prevention and the National Center for Health Statistics jointly publish a report summarizing the vital statistics from the previous year. According to one publication, the number of births in a recent year was 3,899,589. The weights of the newborns can be approximated by a normal distribution, as shown by the following graph.

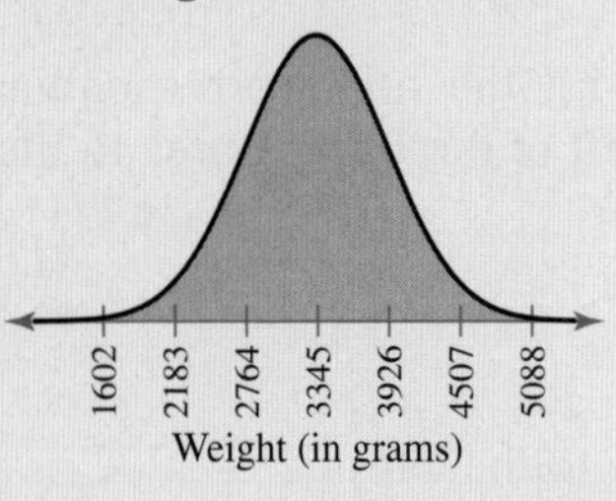

The weights of three newborns are 2000 grams, 3000 grams, and 4000 grams. Find the z-score that corresponds to each weight. Are any of these unusually heavy or light?

Try It Yourself 3

Find the area under the standard normal curve to the right of $z = -2.16$.

a. *Draw* the standard normal curve and shade the area below the curve and to the right of $z = -2.16$.
b. Use the Standard Normal Table (0-to-z) to *find the area* that corresponds to $z = -2.16$.
c. *Add* 0.5 to the resulting area. *Answer: Page A44*

EXAMPLE 

Finding Area Under the Standard Normal Curve

Find the area under the standard normal curve between $z = -1.5$ and $z = 1.25$.

SOLUTION The area under the standard normal curve between $z = -1.5$ and $z = 1.25$ is shown.

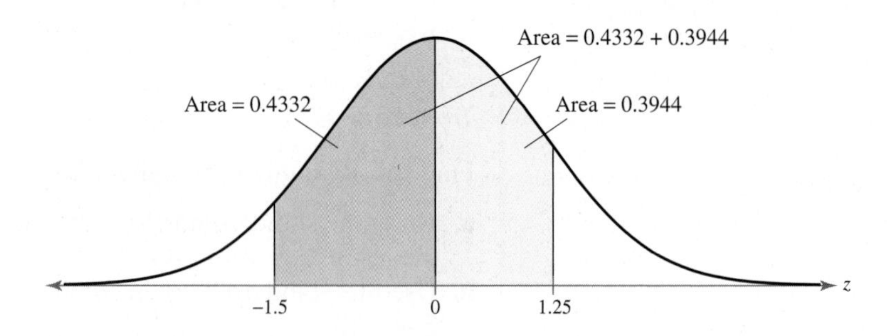

From the Standard Normal Table, the area corresponding to $z = -1.5$ is 0.4332 and the area corresponding to $z = 1.25$ is 0.3944. To find the area between these two z-scores, add the resulting areas.

$$\text{Area} = 0.4332 + 0.3944 = 0.8276$$

In other words, 82.76% of the area under the curve falls between $z = -1.5$ and $z = 1.25$.

Try It Yourself 4

Find the area under the standard normal curve between $z = -2.16$ and $z = -1.35$.

a. *Draw* the standard normal curve and shade the area below the curve that is between $z = -2.16$ and $z = -1.35$.
b. Use the Standard Normal Table (0-to-z) to *find the areas* that correspond to $z = -2.16$ and to $z = -1.35$.
c. *Subtract* the smaller area from the larger area. *Answer: Page A44*

Recall in Section 2.5 you learned, using the Empirical Rule, that values lying more than two standard deviations from the mean are considered unusual. Values lying more than three standard deviations from the mean are considered *very* unusual. So if a z-score is greater than 2 or less than -2, it is unusual. If a z-score is greater than 3 or less than -3, it is *very* unusual.

Appendix B

Table 1—Random Numbers

92630	78240	19267	95457	53497	23894	37708	79862	76471	66418
79445	78735	71549	44843	26104	67318	00701	34986	66751	99723
59654	71966	27386	50004	05358	94031	29281	18544	52429	06080
31524	49587	76612	39789	13537	48086	59483	60680	84675	53014
06348	76938	90379	51392	55887	71015	09209	79157	24440	30244
28703	51709	94456	48396	73780	06436	86641	69239	57662	80181
68108	89266	94730	95761	75023	48464	65544	96583	18911	16391
99938	90704	93621	66330	33393	95261	95349	51769	91616	33238
91543	73196	34449	63513	83834	99411	58826	40456	69268	48562
42103	02781	73920	56297	72678	12249	25270	36678	21313	75767
17138	27584	25296	28387	51350	61664	37893	05363	44143	42677
28297	14280	54524	21618	95320	38174	60579	08089	94999	78460
09331	56712	51333	06289	75345	08811	82711	57392	25252	30333
31295	04204	93712	51287	05754	79396	87399	51773	33075	97061
36146	15560	27592	42089	99281	59640	15221	96079	09961	05371
29553	18432	13630	05529	02791	81017	49027	79031	50912	09399
23501	22642	63081	08191	89420	67800	55137	54707	32945	64522
57888	85846	67967	07835	11314	01545	48535	17142	08552	67457
55336	71264	88472	04334	63919	36394	11196	92470	70543	29776
10087	10072	55980	64688	68239	20461	89381	93809	00796	95945
34101	81277	66090	88872	37818	72142	67140	50785	21380	16703
53362	44940	60430	22834	14130	96593	23298	56203	92671	15925
82975	66158	84731	19436	55790	69229	28661	13675	99318	76873
54827	84673	22898	08094	14326	87038	42892	21127	30712	48489
25464	59098	27436	89421	80754	89924	19097	67737	80368	08795
67609	60214	41475	84950	40133	02546	09570	45682	50165	15609
44921	70924	61295	51137	47596	86735	35561	76649	18217	63446
33170	30972	98130	95828	49786	13301	36081	80761	33985	68621
84687	85445	06208	17654	51333	02878	35010	67578	61574	20749
71886	56450	36567	09395	96951	35507	17555	35212	69106	01679
00475	02224	74722	14721	40215	21351	08596	45625	83981	63748
25993	38881	68361	59560	41274	69742	40703	37993	03435	18873
92882	53178	99195	93803	56985	53089	15305	50522	55900	43026
25138	26810	07093	15677	60688	04410	24505	37890	67186	62829
84631	71882	12991	83028	82484	90339	91950	74579	03539	90122
34003	92326	12793	61453	48121	74271	28363	66561	75220	35908
53775	45749	05734	86169	42762	70175	97310	73894	88606	19994
59316	97885	72807	54966	60859	11932	35265	71601	55577	67715
20479	66557	50705	26999	09854	52591	14063	30214	19890	19292
86180	84931	25455	26044	02227	52015	21820	50599	51671	65411
21451	68001	72710	40261	61281	13172	63819	48970	51732	54113
98062	68375	80089	24135	72355	95428	11808	29740	81644	86610
01788	64429	14430	94575	75153	94576	61393	96192	03227	32258
62465	04841	43272	68702	01274	05437	22953	18946	99053	41690
94324	31089	84159	92933	99989	89500	91586	02802	69471	68274
05797	43984	21575	09908	70221	19791	51578	36432	33494	79888
10395	14289	52185	09721	25789	38562	54794	04897	59012	89251
35177	56986	25549	59730	64718	52630	31100	62384	49483	11409
25633	89619	75882	98256	02126	72099	57183	55887	09320	73463
16464	48280	94254	45777	45150	68865	11382	11782	22695	41988

Table 2—Binomial Distribution

This table shows the probability of x successes in n independent trials, each with probability of success p.

		p																			
n	***x***	**.01**	**.05**	**.10**	**.15**	**.20**	**.25**	**.30**	**.35**	**.40**	**.45**	**.50**	**.55**	**.60**	**.65**	**.70**	**.75**	**.80**	**.85**	**.90**	**.95**
2	0	.980	.902	.810	.723	.640	.563	.490	.423	.360	.303	.250	.203	.160	.123	.090	.063	.040	.023	.010	.002
	1	.020	.095	.180	.255	.320	.375	.420	.455	.480	.495	.500	.495	.480	.455	.420	.375	.320	.255	.180	.095
	2	.000	.002	.010	.023	.040	.063	.090	.123	.160	.203	.250	.303	.360	.423	.490	.563	.640	.723	.810	.902
3	0	.970	.857	.729	.614	.512	.422	.343	.275	.216	.166	.125	.091	.064	.043	.027	.016	.008	.003	.001	.000
	1	.029	.135	.243	.325	.384	.422	.441	.444	.432	.408	.375	.334	.288	.239	.189	.141	.096	.057	.027	.007
	2	.000	.007	.027	.057	.096	.141	.189	.239	.288	.334	.375	.408	.432	.444	.441	.422	.384	.325	.243	.135
	3	.000	.000	.001	.003	.008	.016	.027	.043	.064	.091	.125	.166	.216	.275	.343	.422	.512	.614	.729	.857
4	0	.961	.815	.656	.522	.410	.316	.240	.179	.130	.092	.062	.041	.026	.015	.008	.004	.002	.001	.000	.000
	1	.039	.171	.292	.368	.410	.422	.412	.384	.346	.300	.250	.200	.154	.112	.076	.047	.026	.011	.004	.000
	2	.001	.014	.049	.098	.154	.211	.265	.311	.346	.368	.375	.368	.346	.311	.265	.211	.154	.098	.049	.014
	3	.000	.000	.004	.011	.026	.047	.076	.112	.154	.200	.250	.300	.346	.384	.412	.422	.410	.368	.292	.171
	4	.000	.000	.000	.001	.002	.004	.008	.015	.026	.041	.062	.092	.130	.179	.240	.316	.410	.522	.656	.815
5	0	.951	.774	.590	.444	.328	.237	.168	.116	.078	.050	.031	.019	.010	.005	.002	.001	.000	.000	.000	.000
	1	.048	.204	.328	.392	.410	.396	.360	.312	.259	.206	.156	.113	.077	.049	.028	.015	.006	.002	.000	.000
	2	.001	.021	.073	.138	.205	.264	.309	.336	.346	.337	.312	.276	.230	.181	.132	.088	.051	.024	.008	.001
	3	.000	.001	.008	.024	.051	.088	.132	.181	.230	.276	.312	.337	.346	.336	.309	.264	.205	.138	.073	.021
	4	.000	.000	.000	.002	.006	.015	.028	.049	.077	.113	.156	.206	.259	.312	.360	.396	.410	.392	.328	.204
	5	.000	.000	.000	.000	.000	.001	.002	.005	.010	.019	.031	.050	.078	.116	.168	.237	.328	.444	.590	.774
6	0	.941	.735	.531	.377	.262	.178	.118	.075	.047	.028	.016	.008	.004	.002	.001	.000	.000	.000	.000	.000
	1	.057	.232	.354	.399	.393	.356	.303	.244	.187	.136	.094	.061	.037	.020	.010	.004	.002	.000	.000	.000
	2	.001	.031	.098	.176	.246	.297	.324	.328	.311	.278	.234	.186	.138	.095	.060	.033	.015	.006	.001	.000
	3	.000	.002	.015	.042	.082	.132	.185	.236	.276	.303	.312	.303	.276	.236	.185	.132	.082	.042	.015	.002
	4	.000	.000	.001	.006	.015	.033	.060	.095	.138	.186	.234	.278	.311	.328	.324	.297	.246	.176	.098	.031
	5	.000	.000	.000	.000	.002	.004	.010	.020	.037	.061	.094	.136	.187	.244	.303	.356	.393	.399	.354	.232
	6	.000	.000	.000	.000	.000	.000	.001	.002	.004	.008	.016	.028	.047	.075	.118	.178	.262	.377	.531	.735
7	0	.932	.698	.478	.321	.210	.133	.082	.049	.028	.015	.008	.004	.002	.001	.000	.000	.000	.000	.000	.000
	1	.066	.257	.372	.396	.367	.311	.247	.185	.131	.087	.055	.032	.017	.008	.004	.001	.000	.000	.000	.000
	2	.002	.041	.124	.210	.275	.311	.318	.299	.261	.214	.164	.117	.077	.047	.025	.012	.004	.001	.000	.000
	3	.000	.004	.023	.062	.115	.173	.227	.268	.290	.292	.273	.239	.194	.144	.097	.058	.029	.011	.003	.000
	4	.000	.000	.003	.011	.029	.058	.097	.144	.194	.239	.273	.292	.290	.268	.227	.173	.115	.062	.023	.004
	5	.000	.000	.000	.001	.004	.012	.025	.047	.077	.117	.164	.214	.261	.299	.318	.311	.275	.210	.124	.041
	6	.000	.000	.000	.000	.000	.001	.004	.008	.017	.032	.055	.087	.131	.185	.247	.311	.367	.396	.372	.257
	7	.000	.000	.000	.000	.000	.000	.000	.001	.002	.004	.008	.015	.028	.049	.082	.133	.210	.321	.478	.698
8	0	.923	.663	.430	.272	.168	.100	.058	.032	.017	.008	.004	.002	.001	.000	.000	.000	.000	.000	.000	.000
	1	.075	.279	.383	.385	.336	.267	.198	.137	.090	.055	.031	.016	.008	.003	.001	.000	.000	.000	.000	.000
	2	.003	.051	.149	.238	.294	.311	.296	.259	.209	.157	.109	.070	.041	.022	.010	.004	.001	.000	.000	.000
	3	.000	.005	.033	.084	.147	.208	.254	.279	.279	.257	.219	.172	.124	.081	.047	.023	.009	.003	.000	.000
	4	.000	.000	.005	.018	.046	.087	.136	.188	.232	.263	.273	.263	.232	.188	.136	.087	.046	.018	.005	.000
	5	.000	.000	.000	.003	.009	.023	.047	.081	.124	.172	.219	.257	.279	.279	.254	.208	.147	.084	.033	.005
	6	.000	.000	.000	.000	.001	.004	.010	.022	.041	.070	.109	.157	.209	.259	.296	.311	.294	.238	.149	.051
	7	.000	.000	.000	.000	.000	.000	.001	.003	.008	.016	.031	.055	.090	.137	.198	.267	.336	.385	.383	.279
	8	.000	.000	.000	.000	.000	.000	.000	.000	.001	.002	.004	.008	.017	.032	.058	.100	.168	.272	.430	.663
9	0	.914	.630	.387	.232	.134	.075	.040	.021	.010	.005	.002	.001	.000	.000	.000	.000	.000	.000	.000	.000
	1	.083	.299	.387	.368	.302	.225	.156	.100	.060	.034	.018	.008	.004	.001	.000	.000	.000	.000	.000	.000
	2	.003	.063	.172	.260	.302	.300	.267	.216	.161	.111	.070	.041	.021	.010	.004	.001	.000	.000	.000	.000
	3	.000	.008	.045	.107	.176	.234	.267	.272	.251	.212	.164	.116	.074	.042	.021	.009	.003	.001	.000	.000
	4	.000	.001	.007	.028	.066	.117	.172	.219	.251	.260	.246	.213	.167	.118	.074	.039	.017	.005	.001	.000
	5	.000	.000	.001	.005	.017	.039	.074	.118	.167	.213	.246	.260	.251	.219	.172	.117	.066	.028	.007	.001
	6	.000	.000	.000	.001	.003	.009	.021	.042	.074	.116	.164	.212	.251	.272	.267	.234	.176	.107	.045	.008
	7	.000	.000	.000	.000	.000	.001	.004	.010	.021	.041	.070	.111	.161	.216	.267	.300	.302	.260	.172	.063
	8	.000	.000	.000	.000	.000	.000	.000	.001	.004	.008	.018	.034	.060	.100	.156	.225	.302	.368	.387	.299
	9	.000	.000	.000	.000	.000	.000	.000	.000	.000	.001	.002	.005	.010	.021	.040	.075	.134	.232	.387	.630

Table 2—Binomial Distribution *(continued)*

		p																			
n	*x*	.01	.05	.10	.15	.20	.25	.30	.35	.40	.45	.50	.55	.60	.65	.70	.75	.80	.85	.90	.95
10	0	.904	.599	.349	.197	.107	.056	.028	.014	.006	.003	.001	.000	.000	.000	.000	.000	.000	.000	.000	.000
	1	.091	.315	.387	.347	.268	.188	.121	.072	.040	.021	.010	.004	.002	.000	.000	.000	.000	.000	.000	.000
	2	.004	.075	.194	.276	.302	.282	.233	.176	.121	.076	.044	.023	.011	.004	.001	.000	.000	.000	.000	.000
	3	.000	.010	.057	.130	.201	.250	.267	.252	.215	.166	.117	.075	.042	.021	.009	.003	.001	.000	.000	.000
	4	.000	.001	.011	.040	.088	.146	.200	.238	.251	.238	.205	.160	.111	.069	.037	.016	.006	.001	.000	.000
	5	.000	.000	.001	.008	.026	.058	.103	.154	.201	.234	.246	.234	.201	.154	.103	.058	.026	.008	.001	.000
	6	.000	.000	.000	.001	.006	.016	.037	.069	.111	.160	.205	.238	.251	.238	.200	.146	.088	.040	.011	.001
	7	.000	.000	.000	.000	.001	.003	.009	.021	.042	.075	.117	.166	.215	.252	.267	.250	.201	.130	.057	.010
	8	.000	.000	.000	.000	.000	.000	.001	.004	.011	.023	.044	.076	.121	.176	.233	.282	.302	.276	.194	.075
	9	.000	.000	.000	.000	.000	.000	.000	.000	.002	.004	.010	.021	.040	.072	.121	.188	.268	.347	.387	.315
	10	.000	.000	.000	.000	.000	.000	.000	.000	.000	.000	.001	.003	.006	.014	.028	.056	.107	.197	.349	.599
11	0	.895	.569	.314	.167	.086	.042	.020	.009	.004	.001	.000	.000	.000	.000	.000	.000	.000	.000	.000	.000
	1	.099	.329	.384	.325	.236	.155	.093	.052	.027	.013	.005	.002	.001	.000	.000	.000	.000	.000	.000	.000
	2	.005	.087	.213	.287	.295	.258	.200	.140	.089	.051	.027	.013	.005	.002	.001	.000	.000	.000	.000	.000
	3	.000	.014	.071	.152	.221	.258	.257	.225	.177	.126	.081	.046	.023	.010	.004	.001	.000	.000	.000	.000
	4	.000	.001	.016	.054	.111	.172	.220	.243	.236	.206	.161	.113	.070	.038	.017	.006	.002	.000	.000	.000
	5	.000	.000	.002	.013	.039	.080	.132	.183	.221	.236	.226	.193	.147	.099	.057	.027	.010	.002	.000	.000
	6	.000	.000	.000	.002	.010	.027	.057	.099	.147	.193	.226	.236	.221	.183	.132	.080	.039	.013	.002	.000
	7	.000	.000	.000	.000	.002	.006	.017	.038	.070	.113	.161	.206	.236	.243	.220	.172	.111	.054	.016	.001
	8	.000	.000	.000	.000	.000	.001	.004	.010	.023	.046	.081	.126	.177	.225	.257	.258	.221	.152	.071	.014
	9	.000	.000	.000	.000	.000	.000	.001	.002	.005	.013	.027	.051	.089	.140	.200	.258	.295	.287	.213	.087
	10	.000	.000	.000	.000	.000	.000	.000	.000	.001	.002	.005	.013	.027	.052	.093	.155	.236	.325	.384	.329
	11	.000	.000	.000	.000	.000	.000	.000	.000	.000	.000	.000	.001	.004	.009	.020	.042	.086	.167	.314	.569
12	0	.886	.540	.282	.142	.069	.032	.014	.006	.002	.001	.000	.000	.000	.000	.000	.000	.000	.000	.000	.000
	1	.107	.341	.377	.301	.206	.127	.071	.037	.017	.008	.003	.001	.000	.000	.000	.000	.000	.000	.000	.000
	2	.006	.099	.230	.292	.283	.232	.168	.109	.064	.034	.016	.007	.002	.001	.000	.000	.000	.000	.000	.000
	3	.000	.017	.085	.172	.236	.258	.240	.195	.142	.092	.054	.028	.012	.005	.001	.000	.000	.000	.000	.000
	4	.000	.002	.021	.068	.133	.194	.231	.237	.213	.170	.121	.076	.042	.020	.008	.002	.001	.000	.000	.000
	5	.000	.000	.004	.019	.053	.103	.158	.204	.227	.223	.193	.149	.101	.059	.029	.011	.003	.001	.000	.000
	6	.000	.000	.000	.004	.016	.040	.079	.128	.177	.212	.226	.212	.177	.128	.079	.040	.016	.004	.000	.000
	7	.000	.000	.000	.001	.003	.011	.029	.059	.101	.149	.193	.223	.227	.204	.158	.103	.053	.019	.004	.000
	8	.000	.000	.000	.000	.001	.002	.008	.020	.042	.076	.121	.170	.213	.237	.231	.194	.133	.068	.021	.002
	9	.000	.000	.000	.000	.000	.000	.001	.005	.012	.028	.054	.092	.142	.195	.240	.258	.236	.172	.085	.017
	10	.000	.000	.000	.000	.000	.000	.000	.001	.002	.007	.016	.034	.064	.109	.168	.232	.283	.292	.230	.099
	11	.000	.000	.000	.000	.000	.000	.000	.000	.000	.001	.003	.008	.017	.037	.071	.127	.206	.301	.377	.341
	12	.000	.000	.000	.000	.000	.000	.000	.000	.000	.000	.000	.001	.002	.006	.014	.032	.069	.142	.282	.540
15	0	.860	.463	.206	.087	.035	.013	.005	.002	.000	.000	.000	.000	.000	.000	.000	.000	.000	.000	.000	.000
	1	.130	.366	.343	.231	.132	.067	.031	.013	.005	.002	.000	.000	.000	.000	.000	.000	.000	.000	.000	.000
	2	.009	.135	.267	.286	.231	.156	.092	.048	.022	.009	.003	.001	.000	.000	.000	.000	.000	.000	.000	.000
	3	.000	.031	.129	.218	.250	.225	.170	.111	.063	.032	.014	.005	.002	.000	.000	.000	.000	.000	.000	.000
	4	.000	.005	.043	.116	.188	.225	.219	.179	.127	.078	.042	.019	.007	.002	.001	.000	.000	.000	.000	.000
	5	.000	.001	.010	.045	.103	.165	.206	.212	.186	.140	.092	.051	.024	.010	.003	.001	.000	.000	.000	.000
	6	.000	.000	.002	.013	.043	.092	.147	.191	.207	.191	.153	.105	.061	.030	.012	.003	.001	.000	.000	.000
	7	.000	.000	.000	.003	.014	.039	.081	.132	.177	.201	.196	.165	.118	.071	.035	.013	.003	.001	.000	.000
	8	.000	.000	.000	.001	.003	.013	.035	.071	.118	.165	.196	.201	.177	.132	.081	.039	.014	.003	.000	.000
	9	.000	.000	.000	.000	.001	.003	.012	.030	.061	.105	.153	.191	.207	.191	.147	.092	.043	.013	.002	.000
	10	.000	.000	.000	.000	.000	.001	.003	.010	.024	.051	.092	.140	.186	.212	.206	.165	.103	.045	.010	.001
	11	.000	.000	.000	.000	.000	.000	.001	.002	.007	.019	.042	.078	.127	.179	.219	.225	.188	.116	.043	.005
	12	.000	.000	.000	.000	.000	.000	.000	.000	.002	.005	.014	.032	.063	.111	.170	.225	.250	.218	.129	.031
	13	.000	.000	.000	.000	.000	.000	.000	.000	.000	.001	.003	.009	.022	.048	.092	.156	.231	.286	.267	.135
	14	.000	.000	.000	.000	.000	.000	.000	.000	.000	.000	.000	.002	.005	.013	.031	.067	.132	.231	.343	.366
	15	.000	.000	.000	.000	.000	.000	.000	.000	.000	.000	.000	.000	.000	.002	.005	.013	.035	.087	.206	.463

Table 2—Binomial Distribution *(continued)*

		p																			
n	*x*	**.01**	**.05**	**.10**	**.15**	**.20**	**.25**	**.30**	**.35**	**.40**	**.45**	**.50**	**.55**	**.60**	**.65**	**.70**	**.75**	**.80**	**.85**	**.90**	**.95**
16	0	.851	.440	.185	.074	.028	.010	.003	.001	.000	.000	.000	.000	.000	.000	.000	.000	.000	.000	.000	.000
	1	.138	.371	.329	.210	.113	.053	.023	.009	.003	.001	.000	.000	.000	.000	.000	.000	.000	.000	.000	.000
	2	.010	.146	.275	.277	.211	.134	.073	.035	.015	.006	.002	.001	.000	.000	.000	.000	.000	.000	.000	.000
	3	.000	.036	.142	.229	.246	.208	.146	.089	.047	.022	.009	.003	.001	.000	.000	.000	.000	.000	.000	.000
	4	.000	.006	.051	.131	.200	.225	.204	.155	.101	.057	.028	.011	.004	.001	.000	.000	.000	.000	.000	.000
	5	.000	.001	.014	.056	.120	.180	.210	.201	.162	.112	.067	.034	.014	.005	.001	.000	.000	.000	.000	.000
	6	.000	.000	.003	.018	.055	.110	.165	.198	.198	.168	.122	.075	.039	.017	.006	.001	.000	.000	.000	.000
	7	.000	.000	.000	.005	.020	.052	.101	.152	.189	.197	.175	.132	.084	.044	.019	.006	.001	.000	.000	.000
	8	.000	.000	.000	.001	.006	.020	.049	.092	.142	.181	.196	.181	.142	.092	.049	.020	.006	.001	.000	.000
	9	.000	.000	.000	.000	.001	.006	.019	.044	.084	.132	.175	.197	.189	.152	.101	.052	.020	.005	.000	.000
	10	.000	.000	.000	.000	.000	.001	.006	.017	.039	.075	.122	.168	.198	.198	.165	.110	.055	.018	.003	.000
	11	.000	.000	.000	.000	.000	.000	.001	.005	.014	.034	.067	.112	.162	.201	.210	.180	.120	.056	.014	.001
	12	.000	.000	.000	.000	.000	.000	.000	.001	.004	.011	.028	.057	.101	.155	.204	.225	.200	.131	.051	.006
	13	.000	.000	.000	.000	.000	.000	.000	.000	.001	.003	.009	.022	.047	.089	.146	.208	.246	.229	.142	.036
	14	.000	.000	.000	.000	.000	.000	.000	.000	.000	.001	.002	.006	.015	.035	.073	.134	.211	.277	.275	.146
	15	.000	.000	.000	.000	.000	.000	.000	.000	.000	.000	.000	.001	.003	.009	.023	.053	.113	.210	.329	.371
	16	.000	.000	.000	.000	.000	.000	.000	.000	.000	.000	.000	.000	.000	.001	.003	.010	.028	.074	.185	.440
20	0	.818	.358	.122	.039	.012	.003	.001	.000	.000	.000	.000	.000	.000	.000	.000	.000	.000	.000	.000	.000
	1	.165	.377	.270	.137	.058	.021	.007	.002	.000	.000	.000	.000	.000	.000	.000	.000	.000	.000	.000	.000
	2	.016	.189	.285	.229	.137	.067	.028	.010	.003	.001	.000	.000	.000	.000	.000	.000	.000	.000	.000	.000
	3	.001	.060	.190	.243	.205	.134	.072	.032	.012	.004	.001	.000	.000	.000	.000	.000	.000	.000	.000	.000
	4	.000	.013	.090	.182	.218	.190	.130	.074	.035	.014	.005	.001	.000	.000	.000	.000	.000	.000	.000	.000
	5	.000	.002	.032	.103	.175	.202	.179	.127	.075	.036	.015	.005	.001	.000	.000	.000	.000	.000	.000	.000
	6	.000	.000	.009	.045	.109	.169	.192	.171	.124	.075	.036	.015	.005	.001	.000	.000	.000	.000	.000	.000
	7	.000	.000	.002	.016	.055	.112	.164	.184	.166	.122	.074	.037	.015	.005	.001	.000	.000	.000	.000	.000
	8	.000	.000	.000	.005	.022	.061	.114	.161	.180	.162	.120	.073	.035	.014	.004	.001	.000	.000	.000	.000
	9	.000	.000	.000	.001	.007	.027	.065	.116	.160	.177	.160	.119	.071	.034	.012	.003	.000	.000	.000	.000
	10	.000	.000	.000	.000	.002	.010	.031	.069	.117	.159	.176	.159	.117	.069	.031	.010	.002	.000	.000	.000
	11	.000	.000	.000	.000	.000	.003	.012	.034	.071	.119	.160	.177	.160	.116	.065	.027	.007	.001	.000	.000
	12	.000	.000	.000	.000	.000	.001	.004	.014	.035	.073	.120	.162	.180	.161	.114	.061	.022	.005	.000	.000
	13	.000	.000	.000	.000	.000	.000	.001	.005	.015	.037	.074	.122	.166	.184	.164	.112	.055	.016	.002	.000
	14	.000	.000	.000	.000	.000	.000	.000	.001	.005	.015	.037	.075	.124	.171	.192	.169	.109	.045	.009	.000
	15	.000	.000	.000	.000	.000	.000	.000	.000	.001	.005	.015	.036	.075	.127	.179	.202	.175	.103	.032	.002
	16	.000	.000	.000	.000	.000	.000	.000	.000	.000	.001	.005	.014	.035	.074	.130	.190	.218	.182	.090	.013
	17	.000	.000	.000	.000	.000	.000	.000	.000	.000	.000	.001	.004	.012	.032	.072	.134	.205	.243	.190	.060
	18	.000	.000	.000	.000	.000	.000	.000	.000	.000	.000	.000	.001	.003	.010	.028	.067	.137	.229	.285	.189
	19	.000	.000	.000	.000	.000	.000	.000	.000	.000	.000	.000	.000	.000	.002	.007	.021	.058	.137	.270	.377
	20	.000	.000	.000	.000	.000	.000	.000	.000	.000	.000	.000	.000	.000	.000	.001	.003	.012	.039	.122	.358

Table 3—Poisson Distribution

					μ					
x	**0.1**	**0.2**	**0.3**	**0.4**	**0.5**	**0.6**	**0.7**	**0.8**	**0.9**	**1.0**
0	.9048	.8187	.7408	.6703	.6065	.5488	.4966	.4493	.4066	.3679
1	.0905	.1637	.2222	.2681	.3033	.3293	.3476	.3595	.3659	.3679
2	.0045	.0164	.0333	.0536	.0758	.0988	.1217	.1438	.1647	.1839
3	.0002	.0011	.0033	.0072	.0126	.0198	.0284	.0383	.0494	.0613
4	.0000	.0001	.0003	.0007	.0016	.0030	.0050	.0077	.0111	.0153
5	.0000	.0000	.0000	.0001	.0002	.0004	.0007	.0012	.0020	.0031
6	.0000	.0000	.0000	.0000	.0000	.0000	.0001	.0002	.0003	.0005
7	.0000	.0000	.0000	.0000	.0000	.0000	.0000	.0000	.0000	.0001

					μ					
x	**1.1**	**1.2**	**1.3**	**1.4**	**1.5**	**1.6**	**1.7**	**1.8**	**1.9**	**2.0**
0	.3329	.3012	.2725	.2466	.2231	.2019	.1827	.1653	.1496	.1353
1	.3662	.3614	.3543	.3452	.3347	.3230	.3106	.2975	.2842	.2707
2	.2014	.2169	.2303	.2417	.2510	.2584	.2640	.2678	.2700	.2707
3	.0738	.0867	.0998	.1128	.1255	.1378	.1496	.1607	.1710	.1804
4	.0203	.0260	.0324	.0395	.0471	.0551	.0636	.0723	.0812	.0902
5	.0045	.0062	.0084	.0111	.0141	.0176	.0216	.0260	.0309	.0361
6	.0008	.0012	.0018	.0026	.0035	.0047	.0061	.0078	.0098	.0120
7	.0001	.0002	.0003	.0005	.0008	.0011	.0015	.0020	.0027	.0034
8	.0000	.0000	.0001	.0001	.0001	.0002	.0003	.0005	.0006	.0009
9	.0000	.0000	.0000	.0000	.0000	.0000	.0001	.0001	.0001	.0002

					μ					
x	**2.1**	**2.2**	**2.3**	**2.4**	**2.5**	**2.6**	**2.7**	**2.8**	**2.9**	**3.0**
0	.1225	.1108	.1003	.0907	.0821	.0743	.0672	.0608	.0550	.0498
1	.2572	.2438	.2306	.2177	.2052	.1931	.1815	.1703	.1596	.1494
2	.2700	.2681	.2652	.2613	.2565	.2510	.2450	.2384	.2314	.2240
3	.1890	.1966	.2033	.2090	.2138	.2176	.2205	.2225	.2237	.2240
4	.0992	.1082	.1169	.1254	.1336	.1414	.1488	.1557	.1622	.1680
5	.0417	.0476	.0538	.0602	.0668	.0735	.0804	.0872	.0940	.1008
6	.0146	.0174	.0206	.0241	.0278	.0319	.0362	.0407	.0455	.0504
7	.0044	.0055	.0068	.0083	.0099	.0118	.0139	.0163	.0188	.0216
8	.0011	.0015	.0019	.0025	.0031	.0038	.0047	.0057	.0068	.0081
9	.0003	.0004	.0005	.0007	.0009	.0011	.0014	.0018	.0022	.0027
10	.0001	.0001	.0001	.0002	.0002	.0003	.0004	.0005	.0006	.0008
11	.0000	.0000	.0000	.0000	.0000	.0001	.0001	.0001	.0002	.0002
12	.0000	.0000	.0000	.0000	.0000	.0000	.0000	.0000	.0000	.0001

					μ					
x	**3.1**	**3.2**	**3.3**	**3.4**	**3.5**	**3.6**	**3.7**	**3.8**	**3.9**	**4.0**
0	.0450	.0408	.0369	.0334	.0302	.0273	.0247	.0224	.0202	.0183
1	.1397	.1304	.1217	.1135	.1057	.0984	.0915	.0850	.0789	.0733
2	.2165	.2087	.2008	.1929	.1850	.1771	.1692	.1615	.1539	.1465
3	.2237	.2226	.2209	.2186	.2158	.2125	.2087	.2046	.2001	.1954
4	.1734	.1781	.1823	.1858	.1888	.1912	.1931	.1944	.1951	.1954
5	.1075	.1140	.1203	.1264	.1322	.1377	.1429	.1477	.1522	.1563
6	.0555	.0608	.0662	.0716	.0771	.0826	.0881	.0936	.0989	.1042
7	.0246	.0278	.0312	.0348	.0385	.0425	.0466	.0508	.0551	.0595
8	.0095	.0111	.0129	.0148	.0169	.0191	.0215	.0241	.0269	.0298
9	.0033	.0040	.0047	.0056	.0066	.0076	.0089	.0102	.0116	.0132
10	.0010	.0013	.0016	.0019	.0023	.0028	.0033	.0039	.0045	.0053
11	.0003	.0004	.0005	.0006	.0007	.0009	.0011	.0013	.0016	.0019
12	.0001	.0001	.0001	.0002	.0002	.0003	.0003	.0004	.0005	.0006
13	.0000	.0000	.0000	.0000	.0001	.0001	.0001	.0001	.0002	.0002
14	.0000	.0000	.0000	.0000	.0000	.0000	.0000	.0000	.0000	.0001

Table 3—Poisson Distribution *(continued)*

	μ									
x	4.1	4.2	4.3	4.4	4.5	4.6	4.7	4.8	4.9	5.0
0	.0166	.0150	.0136	.0123	.0111	.0101	.0091	.0082	.0074	.0067
1	.0679	.0630	.0583	.0540	.0500	.0462	.0427	.0395	.0365	.0337
2	.1393	.1323	.1254	.1188	.1125	.1063	.1005	.0948	.0894	.0842
3	.1904	.1852	.1798	.1743	.1687	.1631	.1574	.1517	.1460	.1404
4	.1951	.1944	.1933	.1917	.1898	.1875	.1849	.1820	.1789	.1755
5	.1600	.1633	.1662	.1687	.1708	.1725	.1738	.1747	.1753	.1755
6	.1093	.1143	.1191	.1237	.1281	.1323	.1362	.1398	.1432	.1462
7	.0640	.0686	.0732	.0778	.0824	.0869	.0914	.0959	.1002	.1044
8	.0328	.0360	.0393	.0428	.0463	.0500	.0537	.0575	.0614	.0653
9	.0150	.0168	.0188	.0209	.0232	.0255	.0280	.0307	.0334	.0363
10	.0061	.0071	.0081	.0092	.0104	.0118	.0132	.0147	.0164	.0181
11	.0023	.0027	.0032	.0037	.0043	.0049	.0056	.0064	.0073	.0082
12	.0008	.0009	.0011	.0014	.0016	.0019	.0022	.0026	.0030	.0034
13	.0002	.0003	.0004	.0005	.0006	.0007	.0008	.0009	.0011	.0013
14	.0001	.0001	.0001	.0001	.0002	.0002	.0003	.0003	.0004	.0005
15	.0000	.0000	.0000	.0000	.0001	.0001	.0001	.0001	.0001	.0002

	μ									
x	5.1	5.2	5.3	5.4	5.5	5.6	5.7	5.8	5.9	6.0
0	.0061	.0055	.0050	.0045	.0041	.0037	.0033	.0030	.0027	.0025
1	.0311	.0287	.0265	.0244	.0225	.0207	.0191	.0176	.0162	.0149
2	.0793	.0746	.0701	.0659	.0618	.0580	.0544	.0509	.0477	.0446
3	.1348	.1293	.1239	.1185	.1133	.1082	.1033	.0985	.0938	.0892
4	.1719	.1681	.1641	.1600	.1558	.1515	.1472	.1428	.1383	.1339
5	.1753	.1748	.1740	.1728	.1714	.1697	.1678	.1656	.1632	.1606
6	.1490	.1515	.1537	.1555	.1571	.1584	.1594	.1601	.1605	.1606
7	.1086	.1125	.1163	.1200	.1234	.1267	.1298	.1326	.1353	.1377
8	.0692	.0731	.0771	.0810	.0849	.0887	.0925	.0962	.0998	.1033
9	.0392	.0423	.0454	.0486	.0519	.0552	.0586	.0620	.0654	.0688
10	.0200	.0220	.0241	.0262	.0285	.0309	.0334	.0359	.0386	.0413
11	.0093	.0104	.0116	.0129	.0143	.0157	.0173	.0190	.0207	.0225
12	.0039	.0045	.0051	.0058	.0065	.0073	.0082	.0092	.0102	.0113
13	.0015	.0018	.0021	.0024	.0028	.0032	.0036	.0041	.0046	.0052
14	.0006	.0007	.0008	.0009	.0011	.0013	.0015	.0017	.0019	.0022
15	.0002	.0002	.0003	.0003	.0004	.0005	.0006	.0007	.0008	.0009
16	.0001	.0001	.0001	.0001	.0001	.0002	.0002	.0002	.0003	.0003
17	.0000	.0000	.0000	.0000	.0000	.0000	.0001	.0001	.0001	.0001

	μ									
x	6.1	6.2	6.3	6.4	6.5	6.6	6.7	6.8	6.9	7.0
0	.0022	.0020	.0018	.0017	.0015	.0014	.0012	.0011	.0010	.0009
1	.0137	.0126	.0116	.0106	.0098	.0090	.0082	.0076	.0070	.0064
2	.0417	.0390	.0364	.0340	.0318	.0296	.0276	.0258	.0240	.0223
3	.0848	.0806	.0765	.0726	.0688	.0652	.0617	.0584	.0552	.0521
4	.1294	.1249	.1205	.1162	.1118	.1076	.1034	.0992	.0952	.0912
5	.1579	.1549	.1519	.1487	.1454	.1420	.1385	.1349	.1314	.1277
6	.1605	.1601	.1595	.1586	.1575	.1562	.1546	.1529	.1511	.1490
7	.1399	.1418	.1435	.1450	.1462	.1472	.1480	.1486	.1489	.1490
8	.1066	.1099	.1130	.1160	.1188	.1215	.1240	.1263	.1284	.1304
9	.0723	.0757	.0791	.0825	.0858	.0891	.0923	.0954	.0985	.1014
10	.0441	.0469	.0498	.0528	.0558	.0588	.0618	.0649	.0679	.0710
11	.0245	.0265	.0285	.0307	.0330	.0353	.0377	.0401	.0426	.0452
12	.0124	.0137	.0150	.0164	.0179	.0194	.0210	.0227	.0245	.0264
13	.0058	.0065	.0073	.0081	.0089	.0098	.0108	.0119	.0130	.0142
14	.0025	.0029	.0033	.0037	.0041	.0046	.0052	.0058	.0064	.0071
15	.0010	.0012	.0014	.0016	.0018	.0020	.0023	.0026	.0029	.0033
16	.0004	.0005	.0005	.0006	.0007	.0008	.0010	.0011	.0013	.0014
17	.0001	.0002	.0002	.0002	.0003	.0003	.0004	.0004	.0005	.0006
18	.0000	.0001	.0001	.0001	.0001	.0001	.0001	.0002	.0002	.0002
19	.0000	.0000	.0000	.0000	.0000	.0000	.0000	.0001	.0001	.0001

Table 3—Poisson Distribution *(continued)*

	μ									
x	7.1	7.2	7.3	7.4	7.5	7.6	7.7	7.8	7.9	8.0
0	.0008	.0007	.0007	.0006	.0006	.0005	.0005	.0004	.0004	.0003
1	.0059	.0054	.0049	.0045	.0041	.0038	.0035	.0032	.0029	.0027
2	.0208	.0194	.0180	.0167	.0156	.0145	.0134	.0125	.0116	.0107
3	.0492	.0464	.0438	.0413	.0389	.0366	.0345	.0324	.0305	.0286
4	.0874	.0836	.0799	.0764	.0729	.0696	.0663	.0632	.0602	.0573
5	.1241	.1204	.1167	.1130	.1094	.1057	.1021	.0986	.0951	.0916
6	.1468	.1445	.1420	.1394	.1367	.1339	.1311	.1282	.1252	.1221
7	.1489	.1486	.1481	.1474	.1465	.1454	.1442	.1428	.1413	.1396
8	.1321	.1337	.1351	.1363	.1373	.1382	.1388	.1392	.1395	.1396
9	.1042	.1070	.1096	.1121	.1144	.1167	.1187	.1207	.1224	.1241
10	.0740	.0770	.0800	.0829	.0858	.0887	.0914	.0941	.0967	.0993
11	.0478	.0504	.0531	.0558	.0585	.0613	.0640	.0667	.0695	.0722
12	.0283	.0303	.0323	.0344	.0366	.0388	.0411	.0434	.0457	.0481
13	.0154	.0168	.0181	.0196	.0211	.0227	.0243	.0260	.0278	.0296
14	.0078	.0086	.0095	.0104	.0113	.0123	.0134	.0145	.0157	.0169
15	.0037	.0041	.0046	.0051	.0057	.0062	.0069	.0075	.0083	.0090
16	.0016	.0019	.0021	.0024	.0026	.0030	.0033	.0037	.0041	.0045
17	.0007	.0008	.0009	.0010	.0012	.0013	.0015	.0017	.0019	.0021
18	.0003	.0003	.0004	.0004	.0005	.0006	.0006	.0007	.0008	.0009
19	.0001	.0001	.0001	.0002	.0002	.0002	.0003	.0003	.0003	.0004
20	.0000	.0000	.0001	.0001	.0001	.0001	.0001	.0001	.0001	.0002
21	.0000	.0000	.0000	.0000	.0000	.0000	.0000	.0000	.0001	.0001

	μ									
x	8.1	8.2	8.3	8.4	8.5	8.6	8.7	8.8	8.9	9.0
0	.0003	.0003	.0002	.0002	.0002	.0002	.0002	.0002	.0001	.0001
1	.0025	.0023	.0021	.0019	.0017	.0016	.0014	.0013	.0012	.0011
2	.0100	.0092	.0086	.0079	.0074	.0068	.0063	.0058	.0054	.0050
3	.0269	.0252	.0237	.0222	.0208	.0195	.0183	.0171	.0160	.0150
4	.0544	.0517	.0491	.0466	.0443	.0420	.0398	.0377	.0357	.0337
5	.0882	.0849	.0816	.0784	.0752	.0722	.0692	.0663	.0635	.0607
6	.1191	.1160	.1128	.1097	.1066	.1034	.1003	.0972	.0941	.0911
7	.1378	.1358	.1338	.1317	.1294	.1271	.1247	.1222	.1197	.1171
8	.1395	.1392	.1388	.1382	.1375	.1366	.1356	.1344	.1332	.1318
9	.1256	.1269	.1280	.1290	.1299	.1306	.1311	.1315	.1317	.1318
10	.1017	.1040	.1063	.1084	.1104	.1123	.1140	.1157	.1172	.1186
11	.0749	.0776	.0802	.0828	.0853	.0878	.0902	.0925	.0948	.0970
12	.0505	.0530	.0555	.0579	.0604	.0629	.0654	.0679	.0703	.0728
13	.0315	.0334	.0354	.0374	.0395	.0416	.0438	.0459	.0481	.0504
14	.0182	.0196	.0210	.0225	.0240	.0256	.0272	.0289	.0306	.0324
15	.0098	.0107	.0116	.0126	.0136	.0147	.0158	.0169	.0182	.0194
16	.0050	.0055	.0060	.0066	.0072	.0079	.0086	.0093	.0101	.0109
17	.0024	.0026	.0029	.0033	.0036	.0040	.0044	.0048	.0053	.0058
18	.0011	.0012	.0014	.0015	.0017	.0019	.0021	.0024	.0026	.0029
19	.0005	.0005	.0006	.0007	.0008	.0009	.0010	.0011	.0012	.0014
20	.0002	.0002	.0002	.0003	.0003	.0004	.0004	.0005	.0005	.0006
21	.0001	.0001	.0001	.0001	.0001	.0002	.0002	.0002	.0002	.0003
22	.0000	.0000	.0000	.0000	.0001	.0001	.0001	.0001	.0001	.0001

Table 3—Poisson Distribution *(continued)*

	μ									
x	9.1	9.2	9.3	9.4	9.5	9.6	9.7	9.8	9.9	10.0
0	.0001	.0001	.0001	.0001	.0001	.0001	.0001	.0001	.0001	.0000
1	.0010	.0009	.0009	.0008	.0007	.0007	.0006	.0005	.0005	.0005
2	.0046	.0043	.0040	.0037	.0034	.0031	.0029	.0027	.0025	.0023
3	.0140	.0131	.0123	.0115	.0107	.0100	.0093	.0087	.0081	.0076
4	.0319	.0302	.0285	.0269	.0254	.0240	.0226	.0213	.0201	.0189
5	.0581	.0555	.0530	.0506	.0483	.0460	.0439	.0418	.0398	.0378
6	.0881	.0851	.0822	.0793	.0764	.0736	.0709	.0682	.0656	.0631
7	.1145	.1118	.1091	.1064	.1037	.1010	.0982	.0955	.0928	.0901
8	.1302	.1286	.1269	.1251	.1232	.1212	.1191	.1170	.1148	.1126
9	.1317	.1315	.1311	.1306	.1300	.1293	.1284	.1274	.1263	.1251
10	.1198	.1210	.1219	.1228	.1235	.1241	.1245	.1249	.1250	.1251
11	.0991	.1012	.1031	.1049	.1067	.1083	.1098	.1112	.1125	.1137
12	.0752	.0776	.0799	.0822	.0844	.0866	.0888	.0908	.0928	.0948
13	.0526	.0549	.0572	.0594	.0617	.0640	.0662	.0685	.0707	.0729
14	.0342	.0361	.0380	.0399	.0419	.0439	.0459	.0479	.0500	.0521
15	.0208	.0221	.0235	.0250	.0265	.0281	.0297	.0313	.0330	.0347
16	.0118	.0127	.0137	.0147	.0157	.0168	.0180	.0192	.0204	.0217
17	.0063	.0069	.0075	.0081	.0088	.0095	.0103	.0111	.0119	.0128
18	.0032	.0035	.0039	.0042	.0046	.0051	.0055	.0060	.0065	.0071
19	.0015	.0017	.0019	.0021	.0023	.0026	.0028	.0031	.0034	.0037
20	.0007	.0008	.0009	.0010	.0011	.0012	.0014	.0015	.0017	.0019
21	.0003	.0003	.0004	.0004	.0005	.0006	.0006	.0007	.0008	.0009
22	.0001	.0001	.0002	.0002	.0002	.0002	.0003	.0003	.0004	.0004
23	.0000	.0001	.0001	.0001	.0001	.0001	.0001	.0001	.0002	.0002
24	.0000	.0000	.0000	.0000	.0000	.0000	.0000	.0001	.0001	.0001

	μ									
x	11	12	13	14	15	16	17	18	19	20
0	.0000	.0000	.0000	.0000	.0000	.0000	.0000	.0000	.0000	.0000
1	.0002	.0001	.0000	.0000	.0000	.0000	.0000	.0000	.0000	.0000
2	.0010	.0004	.0002	.0001	.0000	.0000	.0000	.0000	.0000	.0000
3	.0037	.0018	.0008	.0004	.0002	.0001	.0000	.0000	.0000	.0000
4	.0102	.0053	.0027	.0013	.0006	.0003	.0001	.0001	.0000	.0000
5	.0224	.0127	.0070	.0037	.0019	.0010	.0005	.0002	.0001	.0001
6	.0411	.0255	.0152	.0087	.0048	.0026	.0014	.0007	.0004	.0002
7	.0646	.0437	.0281	.0174	.0104	.0060	.0034	.0018	.0010	.0005
8	.0888	.0655	.0457	.0304	.0194	.0120	.0072	.0042	.0024	.0013
9	.1085	.0874	.0661	.0473	.0324	.0213	.0135	.0083	.0050	.0029
10	.1194	.1048	.0859	.0663	.0486	.0341	.0230	.0150	.0095	.0058
11	.1194	.1144	.1015	.0844	.0663	.0496	.0355	.0245	.0164	.0106
12	.1094	.1144	.1099	.0984	.0829	.0661	.0504	.0368	.0259	.0176
13	.0926	.1056	.1099	.1060	.0956	.0814	.0658	.0509	.0378	.0271
14	.0728	.0905	.1021	.1060	.1024	.0930	.0800	.0655	.0514	.0387
15	.0534	.0724	.0885	.0989	.1024	.0992	.0906	.0786	.0650	.0516
16	.0367	.0543	.0719	.0866	.0960	.0992	.0963	.0884	.0772	.0646
17	.0237	.0383	.0550	.0713	.0847	.0934	.0963	.0936	.0863	.0760
18	.0145	.0256	.0397	.0554	.0706	.0830	.0909	.0936	.0911	.0844
19	.0084	.0161	.0272	.0409	.0557	.0699	.0814	.0887	.0911	.0888
20	.0046	.0097	.0177	.0286	.0418	.0559	.0692	.0798	.0866	.0888

Table 3—Poisson Distribution *(continued)*

	μ									
x	**11**	**12**	**13**	**14**	**15**	**16**	**17**	**18**	**19**	**20**
21	.0024	.0055	.0109	.0191	.0299	.0426	.0560	.0684	.0783	.0846
22	.0012	.0030	.0065	.0121	.0204	.0310	.0433	.0560	.0676	.0769
23	.0006	.0016	.0037	.0074	.0133	.0216	.0320	.0438	.0559	.0669
24	.0003	.0008	.0020	.0043	.0083	.0144	.0226	.0328	.0442	.0557
25	.0001	.0004	.0010	.0024	.0050	.0092	.0154	.0237	.0336	.0446
26	.0000	.0002	.0005	.0013	.0029	.0057	.0101	.0164	.0246	.0343
27	.0000	.0001	.0002	.0007	.0016	.0034	.0063	.0109	.0173	.0254
28	.0000	.0000	.0001	.0003	.0009	.0019	.0038	.0070	.0117	.0181
29	.0000	.0000	.0001	.0002	.0004	.0011	.0023	.0044	.0077	.0125
30	.0000	.0000	.0000	.0001	.0002	.0006	.0013	.0026	.0049	.0083
31	.0000	.0000	.0000	.0000	.0001	.0003	.0007	.0015	.0030	.0054
32	.0000	.0000	.0000	.0000	.0001	.0001	.0004	.0009	.0018	.0034
33	.0000	.0000	.0000	.0000	.0000	.0001	.0002	.0005	.0010	.0020
34	.0000	.0000	.0000	.0000	.0000	.0000	.0001	.0002	.0006	.0012
35	.0000	.0000	.0000	.0000	.0000	.0000	.0000	.0001	.0003	.0007
36	.0000	.0000	.0000	.0000	.0000	.0000	.0000	.0001	.0002	.0004
37	.0000	.0000	.0000	.0000	.0000	.0000	.0000	.0000	.0001	.0002
38	.0000	.0000	.0000	.0000	.0000	.0000	.0000	.0000	.0000	.0001
39	.0000	.0000	.0000	.0000	.0000	.0000	.0000	.0000	.0000	.0001

Table 4—Standard Normal Distribution

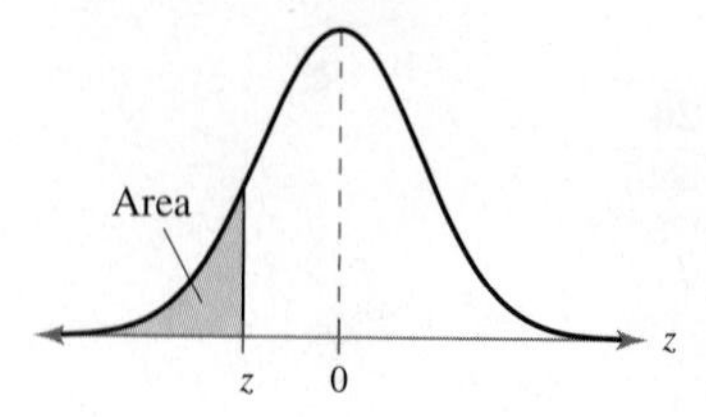

z	.09	.08	.07	.06	.05	.04	.03	.02	.01	.00
−3.4	.0002	.0003	.0003	.0003	.0003	.0003	.0003	.0003	.0003	.0003
−3.3	.0003	.0004	.0004	.0004	.0004	.0004	.0004	.0005	.0005	.0005
−3.2	.0005	.0005	.0005	.0006	.0006	.0006	.0006	.0006	.0007	.0007
−3.1	.0007	.0007	.0008	.0008	.0008	.0008	.0009	.0009	.0009	.0010
−3.0	.0010	.0010	.0011	.0011	.0011	.0012	.0012	.0013	.0013	.0013
−2.9	.0014	.0014	.0015	.0015	.0016	.0016	.0017	.0017	.0018	.0019
−2.8	.0019	.0020	.0021	.0021	.0022	.0023	.0023	.0024	.0025	.0026
−2.7	.0026	.0027	.0028	.0029	.0030	.0031	.0032	.0033	.0034	.0035
−2.6	.0036	.0037	.0038	.0039	.0040	.0041	.0043	.0044	.0045	.0047
−2.5	.0048	.0049	.0051	.0052	.0054	.0055	.0057	.0059	.0060	.0062
−2.4	.0064	.0066	.0068	.0069	.0071	.0073	.0075	.0078	.0080	.0082
−2.3	.0084	.0087	.0089	.0091	.0094	.0096	.0099	.0102	.0104	.0107
−2.2	.0110	.0113	.0116	.0119	.0122	.0125	.0129	.0132	.0136	.0139
−2.1	.0143	.0146	.0150	.0154	.0158	.0162	.0166	.0170	.0174	.0179
−2.0	.0183	.0188	.0192	.0197	.0202	.0207	.0212	.0217	.0222	.0228
−1.9	.0233	.0239	.0244	.0250	.0256	.0262	.0268	.0274	.0281	.0287
−1.8	.0294	.0301	.0307	.0314	.0322	.0329	.0336	.0344	.0352	.0359
−1.7	.0367	.0375	.0384	.0392	.0401	.0409	.0418	.0427	.0436	.0446
−1.6	.0455	.0465	.0475	.0485	.0495	.0505	.0516	.0526	.0537	.0548
−1.5	.0559	.0571	.0582	.0594	.0606	.0618	.0630	.0643	.0655	.0668
−1.4	.0681	.0694	.0708	.0722	.0735	.0749	.0764	.0778	.0793	.0808
−1.3	.0823	.0838	.0853	.0869	.0885	.0901	.0918	.0934	.0951	.0968
−1.2	.0985	.1003	.1020	.1038	.1056	.1075	.1093	.1112	.1131	.1151
−1.1	.1170	.1190	.1210	.1230	.1251	.1271	.1292	.1314	.1335	.1357
−1.0	.1379	.1401	.1423	.1446	.1469	.1492	.1515	.1539	.1562	.1587
−0.9	.1611	.1635	.1660	.1685	.1711	.1736	.1762	.1788	.1814	.1841
−0.8	.1867	.1894	.1922	.1949	.1977	.2005	.2033	.2061	.2090	.2119
−0.7	.2148	.2177	.2206	.2236	.2266	.2296	.2327	.2358	.2389	.2420
−0.6	.2451	.2483	.2514	.2546	.2578	.2611	.2643	.2676	.2709	.2743
−0.5	.2776	.2810	.2843	.2877	.2912	.2946	.2981	.3015	.3050	.3085
−0.4	.3121	.3156	.3192	.3228	.3264	.3300	.3336	.3372	.3409	.3446
−0.3	.3483	.3520	.3557	.3594	.3632	.3669	.3707	.3745	.3783	.3821
−0.2	.3859	.3897	.3936	.3974	.4013	.4052	.4090	.4129	.4168	.4207
−0.1	.4247	.4286	.4325	.4364	.4404	.4443	.4483	.4522	.4562	.4602
−0.0	.4641	.4681	.4721	.4761	.4801	.4840	.4880	.4920	.4960	.5000

Critical Values

Level of Confidence c	z_c
0.80	1.28
0.90	1.645
0.95	1.96
0.99	2.575

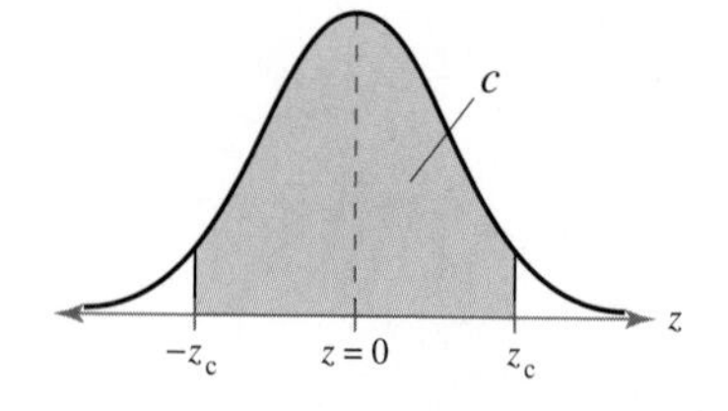

Table 4—Standard Normal Distribution *(continued)*

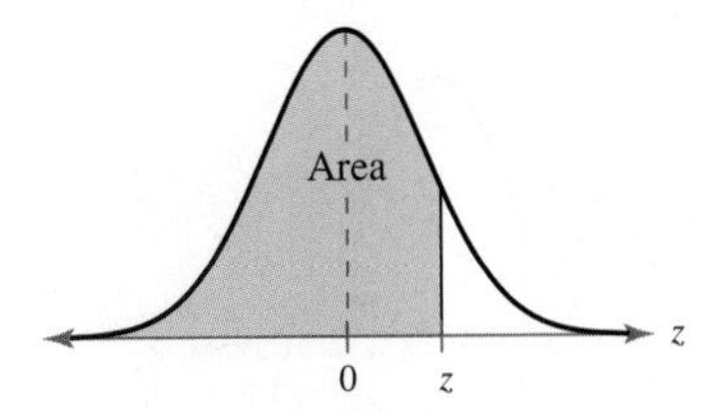

z	.00	.01	.02	.03	.04	.05	.06	.07	.08	.09
0.0	.5000	.5040	.5080	.5120	.5160	.5199	.5239	.5279	.5319	.5359
0.1	.5398	.5438	.5478	.5517	.5557	.5596	.5636	.5675	.5714	.5753
0.2	.5793	.5832	.5871	.5910	.5948	.5987	.6026	.6064	.6103	.6141
0.3	.6179	.6217	.6255	.6293	.6331	.6368	.6406	.6443	.6480	.6517
0.4	.6554	.6591	.6628	.6664	.6700	.6736	.6772	.6808	.6844	.6879
0.5	.6915	.6950	.6985	.7019	.7054	.7088	.7123	.7157	.7190	.7224
0.6	.7257	.7291	.7324	.7357	.7389	.7422	.7454	.7486	.7517	.7549
0.7	.7580	.7611	.7642	.7673	.7704	.7734	.7764	.7794	.7823	.7852
0.8	.7881	.7910	.7939	.7967	.7995	.8023	.8051	.8078	.8106	.8133
0.9	.8159	.8186	.8212	.8238	.8264	.8289	.8315	.8340	.8365	.8389
1.0	.8413	.8438	.8461	.8485	.8508	.8531	.8554	.8577	.8599	.8621
1.1	.8643	.8665	.8686	.8708	.8729	.8749	.8770	.8790	.8810	.8830
1.2	.8849	.8869	.8888	.8907	.8925	.8944	.8962	.8980	.8997	.9015
1.3	.9032	.9049	.9066	.9082	.9099	.9115	.9131	.9147	.9162	.9177
1.4	.9192	.9207	.9222	.9236	.9251	.9265	.9278	.9292	.9306	.9319
1.5	.9332	.9345	.9357	.9370	.9382	.9394	.9406	.9418	.9429	.9441
1.6	.9452	.9463	.9474	.9484	.9495	.9505	.9515	.9525	.9535	.9545
1.7	.9554	.9564	.9573	.9582	.9591	.9599	.9608	.9616	.9625	.9633
1.8	.9641	.9649	.9656	.9664	.9671	.9678	.9686	.9693	.9699	.9706
1.9	.9713	.9719	.9726	.9732	.9738	.9744	.9750	.9756	.9761	.9767
2.0	.9772	.9778	.9783	.9788	.9793	.9798	.9803	.9808	.9812	.9817
2.1	.9821	.9826	.9830	.9834	.9838	.9842	.9846	.9850	.9854	.9857
2.2	.9861	.9864	.9868	.9871	.9875	.9878	.9881	.9884	.9887	.9890
2.3	.9893	.9896	.9898	.9901	.9904	.9906	.9909	.9911	.9913	.9916
2.4	.9918	.9920	.9922	.9925	.9927	.9929	.9931	.9932	.9934	.9936
2.5	.9938	.9940	.9941	.9943	.9945	.9946	.9948	.9949	.9951	.9952
2.6	.9953	.9955	.9956	.9957	.9959	.9960	.9961	.9962	.9963	.9964
2.7	.9965	.9966	.9967	.9968	.9969	.9970	.9971	.9972	.9973	.9974
2.8	.9974	.9975	.9976	.9977	.9977	.9978	.9979	.9979	.9980	.9981
2.9	.9981	.9982	.9982	.9983	.9984	.9984	.9985	.9985	.9986	.9986
3.0	.9987	.9987	.9987	.9988	.9988	.9989	.9989	.9989	.9990	.9990
3.1	.9990	.9991	.9991	.9991	.9992	.9992	.9992	.9992	.9993	.9993
3.2	.9993	.9993	.9994	.9994	.9994	.9994	.9994	.9995	.9995	.9995
3.3	.9995	.9995	.9995	.9996	.9996	.9996	.9996	.9996	.9996	.9997
3.4	.9997	.9997	.9997	.9997	.9997	.9997	.9997	.9997	.9997	.9998

Table 5—*t*-Distribution

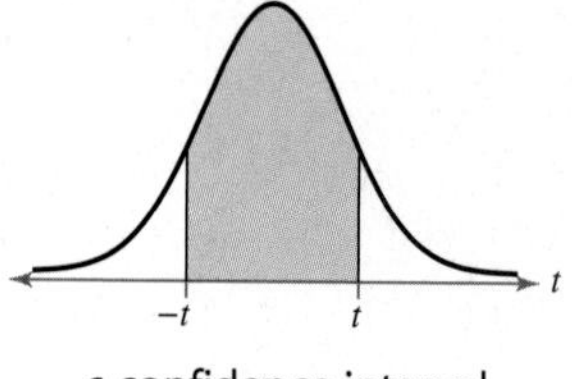

c-confidence interval

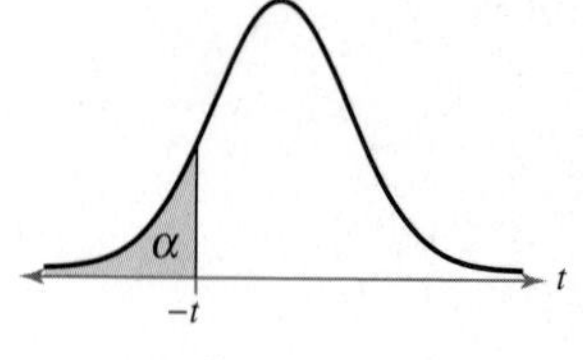

Left-tailed test

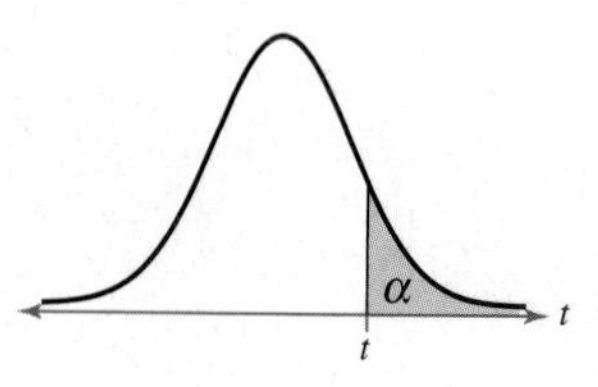

Right-tailed test

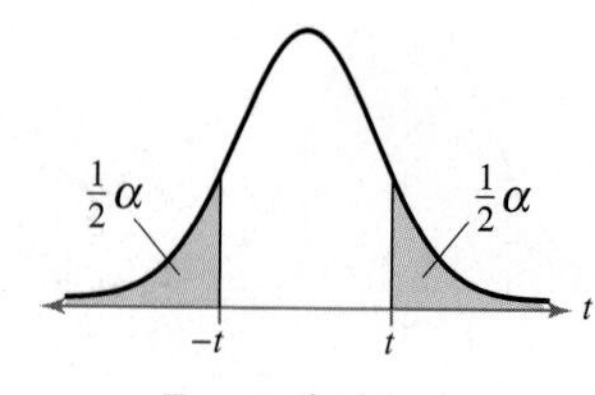

Two-tailed test

	Level of confidence, c	0.50	0.80	0.90	0.95	0.98	0.99
	One tail, α	0.25	0.10	0.05	0.025	0.01	0.005
d.f.	Two tails, α	0.50	0.20	0.10	0.05	0.02	0.01
1		1.000	3.078	6.314	12.706	31.821	63.657
2		.816	1.886	2.920	4.303	6.965	9.925
3		.765	1.638	2.353	3.182	4.541	5.841
4		.741	1.533	2.132	2.776	3.747	4.604
5		.727	1.476	2.015	2.571	3.365	4.032
6		.718	1.440	1.943	2.447	3.143	3.707
7		.711	1.415	1.895	2.365	2.998	3.499
8		.706	1.397	1.860	2.306	2.896	3.355
9		.703	1.383	1.833	2.262	2.821	3.250
10		.700	1.372	1.812	2.228	2.764	3.169
11		.697	1.363	1.796	2.201	2.718	3.106
12		.695	1.356	1.782	2.179	2.681	3.055
13		.694	1.350	1.771	2.160	2.650	3.012
14		.692	1.345	1.761	2.145	2.624	2.977
15		.691	1.341	1.753	2.131	2.602	2.947
16		.690	1.337	1.746	2.120	2.583	2.921
17		.689	1.333	1.740	2.110	2.567	2.898
18		.688	1.330	1.734	2.101	2.552	2.878
19		.688	1.328	1.729	2.093	2.539	2.861
20		.687	1.325	1.725	2.086	2.528	2.845
21		.686	1.323	1.721	2.080	2.518	2.831
22		.686	1.321	1.717	2.074	2.508	2.819
23		.685	1.319	1.714	2.069	2.500	2.807
24		.685	1.318	1.711	2.064	2.492	2.797
25		.684	1.316	1.708	2.060	2.485	2.787
26		.684	1.315	1.706	2.056	2.479	2.779
27		.684	1.314	1.703	2.052	2.473	2.771
28		.683	1.313	1.701	2.048	2.467	2.763
29		.683	1.311	1.699	2.045	2.462	2.756
∞		.674	1.282	1.645	1.960	2.326	2.576

Table 6—Chi-Square Distribution

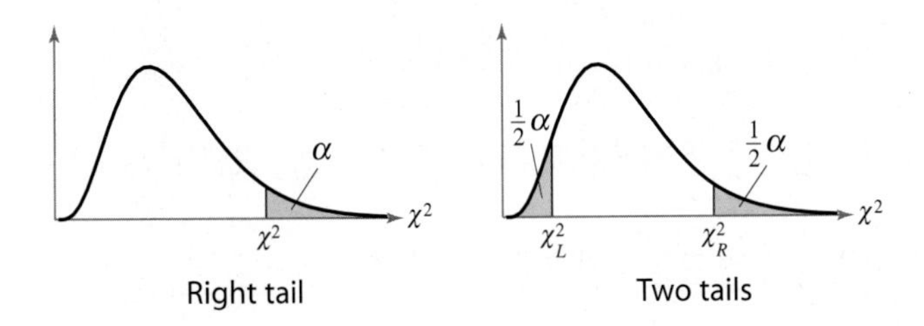

Degrees of freedom	α									
	0.995	**0.99**	**0.975**	**0.95**	**0.90**	**0.10**	**0.05**	**0.025**	**0.01**	**0.005**
1	—	—	0.001	0.004	0.016	2.706	3.841	5.024	6.635	7.879
2	0.010	0.020	0.051	0.103	0.211	4.605	5.991	7.378	9.210	10.597
3	0.072	0.115	0.216	0.352	0.584	6.251	7.815	9.348	11.345	12.838
4	0.207	0.297	0.484	0.711	1.064	7.779	9.488	11.143	13.277	14.860
5	0.412	0.554	0.831	1.145	1.610	9.236	11.071	12.833	15.086	16.750
6	0.676	0.872	1.237	1.635	2.204	10.645	12.592	14.449	16.812	18.548
7	0.989	1.239	1.690	2.167	2.833	12.017	14.067	16.013	18.475	20.278
8	1.344	1.646	2.180	2.733	3.490	13.362	15.507	17.535	20.090	21.955
9	1.735	2.088	2.700	3.325	4.168	14.684	16.919	19.023	21.666	23.589
10	2.156	2.558	3.247	3.940	4.865	15.987	18.307	20.483	23.209	25.188
11	2.603	3.053	3.816	4.575	5.578	17.275	19.675	21.920	24.725	26.757
12	3.074	3.571	4.404	5.226	6.304	18.549	21.026	23.337	26.217	28.299
13	3.565	4.107	5.009	5.892	7.042	19.812	22.362	24.736	27.688	29.819
14	4.075	4.660	5.629	6.571	7.790	21.064	23.685	26.119	29.141	31.319
15	4.601	5.229	6.262	7.261	8.547	22.307	24.996	27.488	30.578	32.801
16	5.142	5.812	6.908	7.962	9.312	23.542	26.296	28.845	32.000	34.267
17	5.697	6.408	7.564	8.672	10.085	24.769	27.587	30.191	33.409	35.718
18	6.265	7.015	8.231	9.390	10.865	25.989	28.869	31.526	34.805	37.156
19	6.844	7.633	8.907	10.117	11.651	27.204	30.144	32.852	36.191	38.582
20	7.434	8.260	9.591	10.851	12.443	28.412	31.410	34.170	37.566	39.997
21	8.034	8.897	10.283	11.591	13.240	29.615	32.671	35.479	38.932	41.401
22	8.643	9.542	10.982	12.338	14.042	30.813	33.924	36.781	40.289	42.796
23	9.262	10.196	11.689	13.091	14.848	32.007	35.172	38.076	41.638	44.181
24	9.886	10.856	12.401	13.848	15.659	33.196	36.415	39.364	42.980	45.559
25	10.520	11.524	13.120	14.611	16.473	34.382	37.652	40.646	44.314	46.928
26	11.160	12.198	13.844	15.379	17.292	35.563	38.885	41.923	45.642	48.290
27	11.808	12.879	14.573	16.151	18.114	36.741	40.113	43.194	46.963	49.645
28	12.461	13.565	15.308	16.928	18.939	37.916	41.337	44.461	48.278	50.993
29	13.121	14.257	16.047	17.708	19.768	39.087	42.557	45.722	49.588	52.336
30	13.787	14.954	16.791	18.493	20.599	40.256	43.773	46.979	50.892	53.672
40	20.707	22.164	24.433	26.509	29.051	51.805	55.758	59.342	63.691	66.766
50	27.991	29.707	32.357	34.764	37.689	63.167	67.505	71.420	76.154	79.490
60	35.534	37.485	40.482	43.188	46.459	74.397	79.082	83.298	88.379	91.952
70	43.275	45.442	48.758	51.739	55.329	85.527	90.531	95.023	100.425	104.215
80	51.172	53.540	57.153	60.391	64.278	96.578	101.879	106.629	112.329	116.321
90	59.196	61.754	65.647	69.126	73.291	107.565	113.145	118.136	124.116	128.299
100	67.328	70.065	74.222	77.929	82.358	118.498	124.342	129.561	135.807	140.169

Table 7—*F*-Distribution

d.f.$_D$: Degrees of freedom, denominator	$\alpha = 0.005$ d.f.$_N$: Degrees of freedom, numerator																		
	1	2	3	4	5	6	7	8	9	10	12	15	20	24	30	40	60	120	∞
1	16211	20000	21615	22500	23056	23437	23715	23925	24091	24224	24426	24630	24836	24940	25044	25148	25253	25359	25465
2	198.5	199.0	199.2	199.2	199.3	199.3	199.4	199.4	199.4	199.4	199.4	199.4	199.4	199.5	199.5	199.5	199.5	199.5	199.5
3	55.55	49.80	47.47	46.19	45.39	44.84	44.43	44.13	43.88	43.69	43.39	43.08	42.78	42.62	42.47	42.31	42.15	41.99	41.83
4	31.33	26.28	24.26	23.15	22.46	21.97	21.62	21.35	21.14	20.97	20.70	20.44	20.17	20.03	19.89	19.75	19.61	19.47	19.32
5	22.78	18.31	16.53	15.56	14.94	14.51	14.20	13.96	13.77	13.62	13.38	13.15	12.90	12.78	12.66	12.53	12.40	12.27	12.14
6	18.63	14.54	12.92	12.03	11.46	11.07	10.79	10.57	10.39	10.25	10.03	9.81	9.59	9.47	9.36	9.24	9.12	9.00	8.88
7	16.24	12.40	10.88	10.05	9.52	9.16	8.89	8.68	8.51	8.38	8.18	7.97	7.75	7.65	7.53	7.42	7.31	7.19	7.08
8	14.69	11.04	9.60	8.81	8.30	7.95	7.69	7.50	7.34	7.21	7.01	6.81	6.61	6.50	6.40	6.29	6.18	6.06	5.95
9	13.61	10.11	8.72	7.96	7.47	7.13	6.88	6.69	6.54	6.42	6.23	6.03	5.83	5.73	5.62	5.52	5.41	5.30	5.19
10	12.83	9.43	8.08	7.34	6.87	6.54	6.30	6.12	5.97	5.85	5.66	5.47	5.27	5.17	5.07	4.97	4.86	4.75	4.64
11	12.73	8.91	7.60	6.88	6.42	6.10	5.86	5.68	5.54	5.42	5.24	5.05	4.86	4.76	4.65	4.55	4.44	4.34	4.23
12	11.75	8.51	7.23	6.52	6.07	5.76	5.52	5.35	5.20	5.09	4.91	4.72	4.53	4.43	4.33	4.23	4.12	4.01	3.90
13	11.37	8.19	6.93	6.23	5.79	5.48	5.25	5.08	4.94	4.82	4.64	4.46	4.27	4.17	4.07	3.97	3.87	3.76	3.65
14	11.06	7.92	6.68	6.00	5.56	5.26	5.03	4.86	4.72	4.60	4.43	4.25	4.06	3.96	3.86	3.76	3.66	3.55	3.44
15	10.80	7.70	6.48	5.80	5.37	5.07	4.85	4.67	4.54	4.42	4.25	4.07	3.88	3.79	3.69	3.58	3.48	3.37	3.26
16	10.58	7.51	6.30	5.64	5.21	4.91	4.69	4.52	4.38	4.27	4.10	3.92	3.73	3.64	3.54	3.44	3.33	3.22	3.11
17	10.38	7.35	6.16	5.50	5.07	4.78	4.56	4.39	4.25	4.14	3.97	3.79	3.61	3.51	3.41	3.31	3.21	3.10	2.98
18	10.22	7.21	6.03	5.37	4.96	4.66	4.44	4.28	4.14	4.03	3.86	3.68	3.50	3.40	3.30	3.20	3.10	2.99	2.87
19	10.07	7.09	5.92	5.27	4.85	4.56	4.34	4.18	4.04	3.93	3.76	3.59	3.40	3.31	3.21	3.11	3.00	2.89	2.78
20	9.94	6.99	5.82	5.17	4.76	4.47	4.26	4.09	3.96	3.85	3.68	3.50	3.32	3.22	3.12	3.02	2.92	2.81	2.69
21	9.83	6.89	5.73	5.09	4.68	4.39	4.18	4.01	3.88	3.77	3.60	3.43	3.24	3.15	3.05	2.95	2.84	2.73	2.61
22	9.73	6.81	5.65	5.02	4.61	4.32	4.11	3.94	3.81	3.70	3.54	3.36	3.18	3.08	2.98	2.88	2.77	2.66	2.55
23	9.63	6.73	5.58	4.95	4.54	4.26	4.05	3.88	3.75	3.64	3.47	3.30	3.12	3.02	2.92	2.82	2.71	2.60	2.48
24	9.55	6.66	5.52	4.89	4.49	4.20	3.99	3.83	3.69	3.59	3.42	3.25	3.06	2.97	2.87	2.77	2.66	2.55	2.43
25	9.48	6.60	5.46	4.84	4.43	4.15	3.94	3.78	3.64	3.54	3.37	3.20	3.01	2.92	2.82	2.72	2.61	2.50	2.38
26	9.41	6.54	5.41	4.79	4.38	4.10	3.89	3.73	3.60	3.49	3.33	3.15	2.97	2.87	2.77	2.67	2.56	2.45	2.33
27	9.34	6.49	5.36	4.74	4.34	4.06	3.85	3.69	3.56	3.45	3.28	3.11	2.93	2.83	2.73	2.63	2.52	2.41	2.25
28	9.28	6.44	5.32	4.70	4.30	4.02	3.81	3.65	3.52	3.41	3.25	3.07	2.89	2.79	2.69	2.59	2.48	2.37	2.29
29	9.23	6.40	5.28	4.66	4.26	3.98	3.77	3.61	3.48	3.38	3.21	3.04	2.86	2.76	2.66	2.56	2.45	2.33	2.24
30	9.18	6.35	5.24	4.62	4.23	3.95	3.74	3.58	3.45	3.34	3.18	3.01	2.82	2.73	2.63	2.52	2.42	2.30	2.18
40	8.83	6.07	4.98	4.37	3.99	3.71	3.51	3.35	3.22	3.12	2.95	2.78	2.60	2.50	2.40	2.30	2.18	2.06	1.93
60	8.49	5.79	4.73	4.14	3.76	3.49	3.29	3.13	3.01	2.90	2.74	2.57	2.39	2.29	2.19	2.08	1.96	1.83	1.69
120	8.18	5.54	4.50	3.92	3.55	3.28	3.09	2.93	2.81	2.71	2.54	2.37	2.19	2.09	1.98	1.87	1.75	1.61	1.43
∞	7.88	5.30	4.28	3.72	3.35	3.09	2.90	2.74	2.62	2.52	2.36	2.19	2.00	1.90	1.79	1.67	1.53	1.36	1.00

Table 7—*F*-Distribution *(Continued)*

d.f.$_D$: Degrees of freedom, denominator	$\alpha = 0.01$ d.f.$_N$: Degrees of freedom, numerator																		
	1	2	3	4	5	6	7	8	9	10	12	15	20	24	30	40	60	120	∞
1	4052	4999.5	5403	5625	5764	5859	5928	5982	6022	6056	6106	6157	6209	6235	6261	6287	6313	6339	6366
2	98.50	99.00	99.17	99.25	99.30	99.33	99.36	99.37	99.39	99.40	99.42	99.43	99.45	99.46	99.47	99.47	99.48	99.49	99.50
3	34.12	30.82	29.46	28.71	28.24	27.91	27.67	27.49	27.35	27.23	27.05	26.87	26.69	26.60	26.50	26.41	26.32	26.22	26.13
4	21.20	18.00	16.69	15.98	15.52	15.21	14.98	14.80	14.66	14.55	4.37	14.20	14.02	13.93	13.84	13.75	13.65	13.56	13.46
5	16.26	13.27	12.06	11.39	10.97	10.67	10.46	10.29	10.16	10.05	9.89	9.72	9.55	9.47	9.38	9.29	9.20	9.11	9.02
6	13.75	10.92	9.78	9.15	8.75	8.47	8.26	8.10	7.98	7.87	7.72	7.56	7.40	7.31	7.23	7.14	7.06	6.97	6.88
7	12.25	9.55	8.45	7.85	7.46	7.19	6.99	6.84	6.72	6.62	6.47	6.31	6.16	6.07	5.99	5.91	5.82	5.74	5.65
8	11.26	8.65	7.59	7.01	6.63	6.37	6.18	6.03	5.91	5.81	5.67	5.52	5.36	5.28	5.20	5.12	5.03	4.95	4.86
9	10.56	8.02	6.99	6.42	6.06	5.80	5.61	5.47	5.35	5.26	5.11	4.96	4.81	4.73	4.65	4.57	4.48	4.40	4.31
10	10.04	7.56	6.55	5.99	5.64	5.39	5.20	5.06	4.94	4.85	4.71	4.56	4.41	4.33	4.25	4.17	4.08	4.00	3.91
11	9.65	7.21	6.22	5.67	5.32	5.07	4.89	4.74	4.63	4.54	4.40	4.25	4.10	4.02	3.94	3.86	3.78	3.69	3.60
12	9.33	6.93	5.95	5.41	5.06	4.82	4.64	4.50	4.39	4.30	4.16	4.01	3.86	3.78	3.70	3.62	3.54	3.45	3.36
13	9.07	6.70	5.74	5.21	4.86	4.62	4.44	4.30	4.19	4.10	3.96	3.82	3.66	3.59	3.51	3.43	3.34	3.25	3.17
14	8.86	6.51	5.56	5.04	4.69	4.46	4.28	4.14	4.03	3.94	3.80	3.66	3.51	3.43	3.35	3.27	3.18	3.09	3.00
15	8.68	6.36	5.42	4.89	4.56	4.32	4.14	4.00	3.89	3.80	3.67	3.52	3.37	3.29	3.21	3.13	3.05	2.96	2.87
16	8.53	6.23	5.29	4.77	4.44	4.20	4.03	3.89	3.78	3.69	3.55	3.41	3.26	3.18	3.10	3.02	2.93	2.84	2.75
17	8.40	6.11	5.18	4.67	4.34	4.10	3.93	3.79	3.68	3.59	3.46	3.31	3.16	3.08	3.00	2.92	2.83	2.75	2.65
18	8.29	6.01	5.09	4.58	4.25	4.01	3.84	3.71	3.60	3.51	3.37	3.23	3.08	3.00	2.92	2.84	2.75	2.66	2.57
19	8.18	5.93	5.01	4.50	4.17	3.94	3.77	3.63	3.52	3.43	3.30	3.15	3.00	2.92	2.84	2.76	2.67	2.58	2.49
20	8.10	5.85	4.94	4.43	4.10	3.87	3.70	3.56	3.46	3.37	3.23	3.09	2.94	2.86	2.78	2.69	2.61	2.52	2.42
21	8.02	5.78	4.87	4.37	4.04	3.81	3.64	3.51	3.40	3.31	3.17	3.03	2.88	2.80	2.72	2.64	2.55	2.46	2.36
22	7.95	5.72	4.82	4.31	3.99	3.76	3.59	3.45	3.35	3.26	3.12	2.98	2.83	2.75	2.67	2.58	2.50	2.40	2.31
23	7.88	5.66	4.76	4.26	3.94	3.71	3.54	3.41	3.30	3.21	3.07	2.93	2.78	2.70	2.62	2.54	2.45	2.35	2.26
24	7.82	5.61	4.72	4.22	3.90	3.67	3.50	3.36	3.26	3.17	3.03	2.89	2.74	2.66	2.58	2.49	2.40	2.31	2.21
25	7.77	5.57	4.68	4.18	3.85	3.63	3.46	3.32	3.22	3.13	2.99	2.85	2.70	2.62	2.54	2.45	2.36	2.27	2.17
26	7.72	5.53	4.64	4.14	3.82	3.59	3.42	3.29	3.18	3.09	2.96	2.81	2.66	2.58	2.50	2.42	2.33	2.23	2.13
27	7.68	5.49	4.60	4.11	3.78	3.56	3.39	3.26	3.15	3.06	2.93	2.78	2.63	2.55	2.47	2.38	2.29	2.20	2.10
28	7.64	5.45	4.57	4.07	3.75	3.53	3.36	3.23	3.12	3.03	2.90	2.75	2.60	2.52	2.44	2.35	2.26	2.17	2.06
29	7.60	5.42	4.54	4.04	3.73	3.50	3.33	3.20	3.09	3.00	2.87	2.73	2.57	2.49	2.41	2.33	2.23	2.14	2.03
30	7.56	5.39	4.51	4.02	3.70	3.47	3.30	3.17	3.07	2.98	2.84	2.70	2.55	2.47	2.39	2.30	2.21	2.11	2.01
40	7.31	5.18	4.31	3.83	3.51	3.29	3.12	2.99	2.89	2.80	2.66	2.52	2.37	2.29	2.20	2.11	2.02	1.92	1.80
60	7.08	4.98	4.13	3.65	3.34	3.12	2.95	2.82	2.72	2.63	2.50	2.35	2.20	2.12	2.03	1.94	1.84	1.73	1.60
120	6.85	4.79	3.95	3.48	3.17	2.96	2.79	2.66	2.56	2.47	2.34	2.19	2.03	1.95	1.86	1.76	1.66	1.53	1.38
∞	6.63	4.61	3.78	3.32	3.02	2.80	2.64	2.51	2.41	2.32	2.18	2.04	1.88	1.79	1.70	1.59	1.47	1.32	1.00

Table 7—*F*-Distribution *(Continued)*

d.f.$_D$: Degrees of freedom, denominator	$\alpha = 0.025$																		
	d.f.$_N$: Degrees of freedom, numerator																		
	1	**2**	**3**	**4**	**5**	**6**	**7**	**8**	**9**	**10**	**12**	**15**	**20**	**24**	**30**	**40**	**60**	**120**	∞
1	647.8	799.5	864.2	899.6	921.8	937.1	948.2	956.7	963.3	968.6	976.7	984.9	993.1	997.2	1001	1006	1010	1014	1018
2	38.51	39.00	39.17	39.25	39.30	39.33	39.36	39.37	39.39	39.40	39.41	39.43	39.45	39.46	39.46	39.47	39.48	39.49	39.50
3	17.44	16.04	15.44	15.10	14.88	14.73	14.62	14.54	14.47	14.42	14.34	14.25	14.17	14.12	14.08	14.04	13.99	13.95	13.90
4	12.22	10.65	9.98	9.60	9.36	9.20	9.07	8.98	8.90	8.84	8.75	8.66	8.56	8.51	8.46	8.41	8.36	8.31	8.26
5	10.01	8.43	7.76	7.39	7.15	6.98	6.85	6.76	6.68	6.62	6.52	6.43	6.33	6.28	6.23	6.18	6.12	6.07	6.02
6	8.81	7.26	6.60	6.23	5.99	5.82	5.70	5.60	5.52	5.46	5.37	5.27	5.17	5.12	5.07	5.01	4.96	4.90	4.85
7	8.07	6.54	5.89	5.52	5.29	5.12	4.99	4.90	4.82	4.76	4.67	4.57	4.47	4.42	4.36	4.31	4.25	4.20	4.14
8	7.57	6.06	5.42	5.05	4.82	4.65	4.53	4.43	4.36	4.30	4.20	4.10	4.00	3.95	3.89	3.84	3.78	3.73	3.67
9	7.21	5.71	5.08	4.72	4.48	4.32	4.20	4.10	4.03	3.96	3.87	3.77	3.67	3.61	3.56	3.51	3.45	3.39	3.33
10	6.94	5.46	4.83	4.47	4.24	4.07	3.95	3.85	3.78	3.72	3.62	3.52	3.42	3.37	3.31	3.26	3.20	3.14	3.08
11	6.72	5.26	4.63	4.28	4.04	3.88	3.76	3.66	3.59	3.53	3.43	3.33	3.23	3.17	3.12	3.06	3.00	2.94	2.88
12	6.55	5.10	4.47	4.12	3.89	3.73	3.61	3.51	3.44	3.37	3.28	3.18	3.07	3.02	2.96	2.91	2.85	2.79	2.72
13	6.41	4.97	4.35	4.00	3.77	3.60	3.48	3.39	3.31	3.25	3.15	3.05	2.95	2.89	2.84	2.78	2.72	2.66	2.60
14	6.30	4.86	4.24	3.89	3.66	3.50	3.38	3.29	3.21	3.15	3.05	2.95	2.84	2.79	2.73	2.67	2.61	2.55	2.49
15	6.20	4.77	4.15	3.80	3.58	3.41	3.29	3.20	3.12	3.06	2.98	2.86	2.76	2.70	2.64	2.59	2.52	2.46	2.40
16	6.12	4.69	4.08	3.73	3.50	3.34	3.22	3.12	3.05	2.99	2.89	2.79	2.68	2.63	2.57	2.51	2.45	2.38	2.32
17	6.04	4.62	4.01	3.66	3.44	3.28	3.16	3.06	2.98	2.92	2.82	2.72	2.62	2.56	2.50	2.44	2.38	2.32	2.25
18	5.98	4.56	3.95	3.61	3.38	3.22	3.10	3.01	2.93	2.87	2.77	2.67	2.56	2.50	2.44	2.38	2.32	2.26	2.19
19	5.92	4.51	3.90	3.56	3.33	3.17	3.05	2.96	2.88	2.82	2.72	2.62	2.51	2.45	2.39	2.33	2.27	2.20	2.13
20	5.87	4.46	3.86	3.51	3.29	3.13	3.01	2.91	2.84	2.77	2.68	2.57	2.46	2.41	2.35	2.29	2.22	2.16	2.09
21	5.83	4.42	3.82	3.48	3.25	3.09	2.97	2.87	2.80	2.73	2.64	2.53	2.42	2.37	2.31	2.25	2.18	2.11	2.04
22	5.79	4.38	3.78	3.44	3.22	3.05	2.93	2.84	2.76	2.70	2.60	2.50	2.39	2.33	2.27	2.21	2.14	2.08	2.00
23	5.75	4.35	3.75	3.41	3.18	3.02	2.90	2.81	2.73	2.67	2.57	2.47	2.36	2.30	2.24	2.18	2.11	2.04	1.97
24	5.72	4.32	3.72	3.38	3.15	2.99	2.87	2.78	2.70	2.64	2.54	2.44	2.33	2.27	2.21	2.15	2.08	2.01	1.94
25	5.69	4.29	3.69	3.35	3.13	2.97	2.85	2.75	2.68	2.61	2.51	2.41	2.30	2.24	2.18	2.12	2.05	1.98	1.91
26	5.66	4.27	3.67	3.33	3.10	2.94	2.82	2.73	2.65	2.59	2.49	2.39	2.25	2.22	2.16	2.09	2.03	1.95	1.88
27	5.63	4.24	3.65	3.31	3.08	2.92	2.80	2.71	2.63	2.57	2.47	2.36	2.25	2.19	2.13	2.07	2.00	1.93	1.85
28	5.61	4.22	3.63	3.29	3.06	2.90	2.78	2.69	2.61	2.55	2.45	2.34	2.23	2.17	2.11	2.05	1.98	1.91	1.83
29	5.59	4.20	3.61	3.27	3.04	2.88	2.76	2.67	2.59	2.53	2.43	2.32	2.21	2.15	2.09	2.03	1.96	1.89	1.81
30	5.57	4.18	3.59	3.25	3.03	2.87	2.75	2.65	2.57	2.51	2.41	2.31	2.20	2.14	2.07	2.01	1.94	1.87	1.79
40	5.42	4.05	3.46	3.13	2.90	2.74	2.62	2.53	2.45	2.39	2.29	2.18	2.07	2.01	1.94	1.88	1.80	1.72	1.64
60	5.29	3.93	3.34	3.01	2.79	2.63	2.51	2.41	2.33	2.27	2.17	2.06	1.94	1.88	1.82	1.74	1.67	1.58	1.48
120	5.15	3.80	3.23	2.89	2.67	2.52	2.39	2.30	2.22	2.16	2.05	1.94	1.82	1.76	1.69	1.61	1.53	1.43	1.31
∞	5.02	3.69	3.12	2.79	2.57	2.41	2.29	2.19	2.11	2.05	1.94	1.83	1.71	1.64	1.57	1.48	1.39	1.27	1.00

Table 7—*F*-Distribution *(Continued)*

$d.f._D$: Degrees of freedom, denominator	$\alpha = 0.05$																		
	$d.f._N$: Degrees of freedom, numerator																		
	1	2	3	4	5	6	7	8	9	10	12	15	20	24	30	40	60	120	∞
1	161.4	199.5	215.7	224.6	230.2	234.0	236.8	238.9	240.5	241.9	243.9	245.9	248.0	249.1	250.1	251.1	252.2	253.3	254.3
2	18.51	19.00	19.16	19.25	19.30	19.33	19.35	19.37	19.38	19.40	19.41	19.43	19.45	19.45	19.46	19.47	19.48	19.49	19.50
3	10.13	9.55	9.28	9.12	9.01	8.94	8.89	8.85	8.81	8.79	8.74	8.70	8.66	8.64	8.62	8.59	8.57	8.55	8.53
4	7.71	6.94	6.59	6.39	6.26	6.16	6.09	6.04	6.00	5.96	5.91	5.86	5.80	5.77	5.75	5.72	5.69	5.66	5.63
5	6.61	5.79	5.41	5.19	5.05	4.95	4.88	4.82	4.77	4.74	4.68	4.62	4.56	4.53	4.50	4.46	4.43	4.40	4.36
6	5.99	5.14	4.76	4.53	4.39	4.28	4.21	4.15	4.10	4.06	4.00	3.94	3.87	3.84	3.81	3.77	3.74	3.70	3.67
7	5.59	4.74	4.35	4.12	3.97	3.87	3.79	3.73	3.68	3.64	3.57	3.51	3.44	3.41	3.38	3.34	3.30	3.27	3.23
8	5.32	4.46	4.07	3.84	3.69	3.58	3.50	3.44	3.39	3.35	3.28	3.22	3.15	3.12	3.08	3.04	3.01	2.97	2.93
9	5.12	4.26	3.86	3.63	3.48	3.37	3.29	3.23	3.18	3.14	3.07	3.01	2.94	2.90	2.86	2.83	2.79	2.75	2.71
10	4.96	4.10	3.71	3.48	3.33	3.22	3.14	3.07	3.02	2.98	2.91	2.85	2.77	2.74	2.70	2.66	2.62	2.58	2.54
11	4.84	3.98	3.59	3.36	3.20	3.09	3.01	2.95	2.90	2.85	2.79	2.72	2.65	2.61	2.57	2.53	2.49	2.45	2.40
12	4.75	3.89	3.49	3.26	3.11	3.00	2.91	2.85	2.80	2.75	2.69	2.62	2.54	2.51	2.47	2.43	2.38	2.34	2.30
13	4.67	3.81	3.41	3.18	3.03	2.92	2.83	2.77	2.71	2.67	2.60	2.53	2.46	2.42	2.38	2.34	2.30	2.25	2.21
14	4.60	3.74	3.34	3.11	2.96	2.85	2.76	2.70	2.65	2.60	2.53	2.46	2.39	2.35	2.31	2.27	2.22	2.18	2.13
15	4.54	3.68	3.29	3.06	2.90	2.79	2.71	2.64	2.59	2.54	2.48	2.40	2.33	2.29	2.25	2.20	2.16	2.11	2.07
16	4.49	3.63	3.24	3.01	2.85	2.74	2.66	2.59	2.54	2.49	2.42	2.35	2.28	2.24	2.19	2.15	2.11	2.06	2.01
17	4.45	3.59	3.20	2.96	2.81	2.70	2.61	2.55	2.49	2.45	2.38	2.31	2.23	2.19	2.15	2.10	2.06	2.01	1.96
18	4.41	3.55	3.16	2.93	2.77	2.66	2.58	2.51	2.46	2.41	2.34	2.27	2.19	2.15	2.11	2.06	2.02	1.97	1.92
19	4.38	3.52	3.13	2.90	2.74	2.63	2.54	2.48	2.42	2.38	2.31	2.23	2.16	2.11	2.07	2.03	1.98	1.93	1.88
20	4.35	3.49	3.10	2.87	2.71	2.60	2.51	2.45	2.39	2.35	2.28	2.20	2.12	2.08	2.04	1.99	1.95	1.90	1.84
21	4.32	3.47	3.07	2.84	2.68	2.57	2.49	2.42	2.37	2.32	2.25	2.18	2.10	2.05	2.01	1.96	1.92	1.87	1.81
22	4.30	3.44	3.05	2.82	2.66	2.55	2.46	2.40	2.34	2.30	2.23	2.15	2.07	2.03	1.98	1.94	1.89	1.84	1.78
23	4.28	3.42	3.03	2.80	2.64	2.53	2.44	2.37	2.32	2.27	2.20	2.13	2.05	2.01	1.96	1.91	1.86	1.81	1.76
24	4.26	3.40	3.01	2.78	2.62	2.51	2.42	2.36	2.30	2.25	2.18	2.11	2.03	1.98	1.94	1.89	1.84	1.79	1.73
25	4.24	3.39	2.99	2.76	2.60	2.49	2.40	2.34	2.28	2.24	2.16	2.09	2.01	1.96	1.92	1.87	1.82	1.77	1.71
26	4.23	3.37	2.98	2.74	2.59	2.47	2.39	2.32	2.27	2.22	2.15	2.07	1.99	1.95	1.90	1.85	1.80	1.75	1.69
27	4.21	3.35	2.96	2.73	2.57	2.46	2.37	2.31	2.25	2.20	2.13	2.06	1.97	1.93	1.88	1.84	1.79	1.73	1.67
28	4.20	3.34	2.95	2.71	2.56	2.45	2.36	2.29	2.24	2.19	2.12	2.04	1.96	1.91	1.87	1.82	1.77	1.71	1.65
29	4.18	3.33	2.93	2.70	2.55	2.43	2.35	2.28	2.22	2.18	2.10	2.03	1.94	1.90	1.85	1.81	1.75	1.70	1.64
30	4.17	3.32	2.92	2.69	2.53	2.42	2.33	2.27	2.21	2.16	2.09	2.01	1.93	1.89	1.84	1.79	1.74	1.68	1.62
40	4.08	3.23	2.84	2.61	2.45	2.34	2.25	2.18	2.12	2.08	2.00	1.92	1.84	1.79	1.74	1.69	1.64	1.58	1.51
60	4.00	3.15	2.76	2.53	2.37	2.25	2.17	2.10	2.04	1.99	1.92	1.84	1.75	1.70	1.65	1.59	1.53	1.47	1.39
120	3.92	3.07	2.68	2.45	2.29	2.17	2.09	2.02	1.96	1.91	1.83	1.75	1.66	1.61	1.55	1.50	1.43	1.35	1.25
∞	3.84	3.00	2.60	2.37	2.21	2.10	2.01	1.94	1.88	1.83	1.75	1.67	1.57	1.52	1.46	1.39	1.32	1.22	1.00

Table 7—*F*-Distribution *(Continued)*

$d.f._D$: Degrees of freedom, denominator	$\alpha = 0.10$																		
	$d.f._N$: Degrees of freedom, numerator																		
	1	2	3	4	5	6	7	8	9	10	12	15	20	24	30	40	60	120	∞
1	39.86	49.50	53.59	55.83	57.24	58.20	58 91	59.44	59.86	60.19	60.71	61.22	61.74	62.00	62.26	62.53	62.79	63.06	63.33
2	8.53	9.00	9.16	9.24	9.29	9.33	9.35	9.37	9.38	9.39	9.41	9.42	9.44	9.45	9.46	9.47	9.47	9.48	9.49
3	5.54	5.46	5.39	5.34	5.31	5.28	5.27	5.25	5.24	5.23	5.22	5.20	5.18	5.18	5.17	5.16	5.15	5.14	5.13
4	4.54	4.32	4.19	4.11	4.05	4.01	3.98	3.95	3.94	3.92	3.90	3.87	3.84	3.83	3.82	3.80	3.79	3.78	3.76
5	4.06	3.78	3.62	3.52	3.45	3.40	3.37	3.34	3.32	3.30	3.27	3.24	3.21	3.19	3.17	3.16	3.14	3.12	3.10
6	3.78	3.46	3.29	3.18	3.11	3.05	3.01	2.98	2.96	2.94	2.90	2.87	2.84	2.82	2.80	2.78	2.76	2.74	2.72
7	3.59	3.26	3.07	2.96	2.88	2.83	2.78	2.75	2.72	2.70	2.67	2.63	2.59	2.58	2.56	2.54	2.51	2.49	2.47
8	3.46	3.11	2.92	2.81	2.73	2.67	2.62	2.59	2.56	2.54	2.50	2.46	2.42	2.40	2.38	2.36	2.34	2.32	2.29
9	3.36	3.01	2.81	2.69	2.61	2.55	2.51	2.47	2.44	2.42	2.38	2.34	2.30	2.28	2.25	2.23	2.21	2.18	2.16
10	3.29	2.92	2.73	2.61	2.52	2.46	2.41	2.38	2.35	2.32	2.28	2.24	2.20	2.18	2.16	2.13	2.11	2.08	2.06
11	3.23	2.86	2.66	2.54	2.45	2.39	2.34	2.30	2.27	2.25	2.21	2.17	2.12	2.10	2.08	2.05	2.03	2.00	1.97
12	3.18	2.81	2.61	2.48	2.39	2.33	2.28	2.24	2.21	2.19	2.15	2.10	2.06	2.04	2.01	1.99	1.96	1.93	1.90
13	3.14	2.76	2.56	2.43	2.35	2.28	2.23	2.20	2.16	2.14	2.10	2.05	2.01	1.98	1.96	1.93	1.90	1.88	1.85
14	3.10	2.73	2.52	2.39	2.31	2.24	2.19	2.15	2.12	2.10	2.05	2.01	1.96	1.94	1.91	1.89	1.86	1.83	1.80
15	3.07	2.70	2.49	2.36	2.27	2.21	2.16	2.12	2.09	2.06	2.02	1.97	1.92	1.90	1.87	1.85	1.82	1.79	1.76
16	3.05	2.67	2.46	2.33	2.24	2.18	2.13	2.09	2.06	2.03	1.99	1.94	1.89	1.87	1.84	1.81	1.78	1.75	1.72
17	3.03	2.64	2.44	2.31	2.22	2.15	2.10	2.06	2.03	2.00	1.96	1.91	1.86	1.84	1.81	1.78	1.75	1.72	1.69
18	3.01	2.62	2.42	2.29	2.20	2.13	2.08	2.04	2.00	1.98	1.93	1.89	1.84	1.81	1.78	1.75	1.72	1.69	1.66
19	2.99	2.61	2.40	2.27	2.18	2.11	2.06	2.02	1.98	1.96	1.91	1.86	1.81	1.79	1.76	1.73	1.70	1.67	1.63
20	2.97	2.59	2.38	2.25	2.16	2.09	2.04	2.00	1.96	1.94	1.89	1.84	1.79	1.77	1.74	1.71	1.68	1.64	1.61
21	2.96	2.57	2.36	2.23	2.14	2.08	2.02	1.98	1.95	1.92	1.87	1.83	1.78	1.75	1.72	1.69	1.66	1.62	1.59
22	2.95	2.56	2.35	2.22	2.13	2.06	2.01	1.97	1.93	1.90	1.86	1.81	1.76	1.73	1.70	1.67	1.64	1.60	1.57
23	2.94	2.55	2.34	2.21	2.11	2.05	1.99	1.95	1.92	1.89	1.84	1.80	1.74	1.72	1.69	1.66	1.62	1.59	1.55
24	2.93	2.54	2.33	2.19	2.10	2.04	1.98	1.94	1.91	1.88	1.83	1.78	1.73	1.70	1.67	1.64	1.61	1.57	1.53
25	2.92	2.53	2.32	2.18	2.09	2.02	1.97	1.93	1.89	1.87	1.82	1.77	1.72	1.69	1.66	1.63	1.59	1.56	1.52
26	2.91	2.52	2.31	2.17	2.08	2.01	1.96	1.92	1.88	1.86	1.81	1.76	1.71	1.68	1.65	1.61	1.58	1.54	1.50
27	2.90	2.51	2.30	2.17	2.07	2.00	1.95	1.91	1.87	1.85	1.80	1.75	1.70	1.67	1.64	1.60	1.57	1.53	1.49
28	2.89	2.50	2.29	2.16	2.06	2.00	1.94	1.90	1.87	1.84	1.79	1.74	1.69	1.66	1.63	1.59	1.56	1.52	1.48
29	2.89	2.50	2.28	2.15	2.06	1.99	1.93	1.89	1.86	1.83	1.78	1.73	1.68	1.65	1.62	1.58	1.55	1.51	1.47
30	2.88	2.49	2.28	2.14	2.05	1.98	1.93	1.88	1.85	1.82	1.77	1.72	1.67	1.64	1.61	1.57	1.54	1.50	1.46
40	2.84	2.44	2.23	2.09	2.00	1.93	1.87	1.83	1.79	1.76	1.71	1.66	1.61	1.57	1.54	1.51	1.47	1.42	1.38
60	2.79	2.39	2.18	2.04	1.95	1.87	1.82	1.77	1.74	1.71	1.66	1.60	1.54	1.51	1.48	1.44	1.40	1.35	1.29
120	2.75	2.35	2.13	1.99	1.90	1.82	1.77	1.72	1.68	1.65	1.60	1.55	1.48	1.45	1.41	1.37	1.32	1.26	1.19
∞	2.71	2.30	2.08	1.94	1.85	1.77	1.72	1.67	1.63	1.60	1.55	1.49	1.42	1.38	1.34	1.30	1.24	1.17	1.00

From M. Merrington and C. M. Thompson (1943). Table of Percentage Points of the Inverted Beta (F) Distribution. *Biometrika* 33. pp. 74–87. Reprinted with permission from *Biometrika.*

Table 8—Critical Values for the Sign Test

Reject the null hypothesis if the test statistic x is less than or equal to the value in the table.

	One-tailed, $\alpha = 0.005$	$\alpha = 0.01$	$\alpha = 0.025$	$\alpha = 0.05$
n	Two-tailed, $\alpha = 0.01$	$\alpha = 0.02$	$\alpha = 0.05$	$\alpha = 0.10$
8	0	0	0	1
9	0	0	1	1
10	0	0	1	1
11	0	1	1	2
12	1	1	2	2
13	1	1	2	3
14	1	2	3	3
15	2	2	3	3
16	2	2	3	4
17	2	3	4	4
18	3	3	4	5
19	3	4	4	5
20	3	4	5	5
21	4	4	5	6
22	4	5	5	6
23	4	5	6	7
24	5	5	6	7
25	5	6	6	7

Note: Table 8 is for one-tailed or two-tailed tests. The sample size n represents the total number of + and − signs. The test value x is the smaller number of + or − signs..

Source: From *Journal of American Statistical Association* Vol. 41 (1946) pp. 557–66. W. J. Dixon and A. M. Mood.

Table 9—Critical Values for the Wilcoxon Signed-Rank Test

Reject the null hypothesis if the test statistic w_s value is less than or equal to the value given in the table.

	One tailed, $\alpha = 0.05$	$\alpha = 0.025$	$\alpha = 0.01$	$\alpha = 0.005$
n	**Two tailed,** $\alpha = 0.10$	$\alpha = 0.05$	$\alpha = 0.02$	$\alpha = 0.01$
5	1			
6	2	1		
7	4	2	0	
8	6	4	2	0
9	8	6	3	2
10	11	8	5	3
11	14	11	7	5
12	17	14	10	7
13	21	17	13	10
14	26	21	16	13
15	30	25	20	16
16	36	30	24	19
17	41	35	28	23
18	47	40	33	28
19	54	46	38	32
20	60	52	43	37
21	68	59	49	43
22	75	66	56	49
23	83	73	62	55
24	92	81	69	61
25	101	90	77	68
26	110	98	85	76
27	120	107	93	84
28	130	117	102	92
29	141	127	111	100
30	152	137	120	109

Source: From *Some Rapid Approximate Statistical Procedures.*

Table 10—Critical Values for the Spearman Rank Correlation

Reject H_0: $\rho_s = 0$ if the absolute value of r_s is greater than the value given in the table.

n	$\alpha = 0.10$	$\alpha = 0.05$	$\alpha = 0.01$
5	0.900	—	—
6	0.829	0.886	—
7	0.714	0.786	0.929
8	0.643	0.738	0.881
9	0.600	0.700	0.833
10	0.564	0.648	0.794
11	0.536	0.618	0.818
12	0.497	0.591	0.780
13	0.475	0.566	0.745
14	0.457	0.545	0.716
15	0.441	0.525	0.689
16	0.425	0.507	0.666
17	0.412	0.490	0.645
18	0.399	0.476	0.625
19	0.388	0.462	0.608
20	0.377	0.450	0.591
21	0.368	0.438	0.576
22	0.359	0.428	0.562
23	0.351	0.418	0.549
24	0.343	0.409	0.537
25	0.336	0.400	0.526
26	0.329	0.392	0.515
27	0.323	0.385	0.505
28	0.317	0.377	0.496
29	0.311	0.370	0.487
30	0.305	0.364	0.478

Source: From N. L. Johnson and F. C. Leone, *Statistical and Experimental Design,* Vol. I (1964), p. 412. Reprinted with permission from the Institute of Mathematical Statistics.

Table 11—Critical Values for the Pearson Correlation Coefficient

Reject H_0: $\rho = 0$ if the absolute value of r is greater than the value given in the table.

n	$\alpha = 0.05$	$\alpha = 0.01$
4	0.950	0.990
5	0.878	0.959
6	0.811	0.917
7	0.754	0.875
8	0.707	0.834
9	0.666	0.798
10	0.632	0.765
11	0.602	0.735
12	0.576	0.708
13	0.553	0.684
14	0.532	0.661
15	0.514	0.641
16	0.497	0.623
17	0.482	0.606
18	0.468	0.590
19	0.456	0.575
20	0.444	0.561
21	0.433	0.549
22	0.423	0.537
23	0.413	0.526
24	0.404	0.515
25	0.396	0.505
26	0.388	0.496
27	0.381	0.487
28	0.374	0.479
29	0.367	0.471
30	0.361	0.463
35	0.334	0.430
40	0.312	0.403
45	0.294	0.380
50	0.279	0.361
55	0.266	0.345
60	0.254	0.330
65	0.244	0.317
70	0.235	0.306
75	0.227	0.296
80	0.220	0.286
85	0.213	0.278
90	0.207	0.270
95	0.202	0.263
100	0.197	0.256

The critical values in Table 11 were generated using Excel.

Try It Yourself Answers

CHAPTER 1

Section 1.1

1a. The population consists of the prices per gallon of regular gasoline at all gasoline stations in the United States.

b. The sample consists of the prices per gallon of regular gasoline at the 800 surveyed stations.

c. The data set consists of the 800 prices.

2a. Population **b.** Parameter

3a. Descriptive statistics involve the statement "76% of women and 60% of men had a physical examination within the previous year."

b. An inference drawn from the study is that a higher percentage of women had a physical examination within the previous year.

Section 1.2

1a. City names and city population

b. City name: Nonnumerical
City population: Numerical

c. City name: Qualitative
City population: Quantitative

2. (1a) The final standings represent a ranking of hockey teams.

(1b) ordinal

(2a) The collection of phone numbers represents labels. No mathematical computations can be made.

(2b) nominal

3. (1a) The collection of body temperatures represents data that can be ordered, but makes no sense written as a ratio.

(1b) Interval

(2a) The collection of heart rates represents data that can be ordered and written as a ratio that makes sense.

(2b) Ratio

Section 1.3

1. (1a) Focus: Effect of exercise on senior citizens.

(1b) Population: Collection of all senior citizens.

(1c) Experiment

(2a) Focus: Effect of radiation fallout on senior citizens.

(2b) Population: Collection of all senior citizens.

(2c) Sampling

2a. Example: start with the first digits 92630782 ...

b. 92|63|07|82|40|19|26

c. 63, 7, 40, 19, 26

3. (1a) The sample was selected by only using available students.

(1b) Convenience sampling

(2a) The sample was selected by numbering each student in the school, randomly choosing a starting number, and selecting students at regular intervals from the starting number.

(2b) Systematic sampling

CHAPTER 2

Section 2.1

1a. 6 classes **b.** Min = 0 Max = 63
Class width = 11

c.

Lower limit	Upper limit
0	10
11	21
22	32
33	43
44	54
55	65

d. See part (e).

e.

Class	Frequency, f
0–10	27
11–21	13
22–32	16
33–43	7
44–54	11
55–65	3

2a. See part (b).

b.

Class	Frequency, f	Midpoint	Relative frequency	Cumulative frequency
0–10	27	5	0.3506	27
11–21	13	16	0.1688	40
22–32	16	27	0.2078	56
33–43	7	38	0.0909	63
44–54	11	49	0.1429	74
55–65	3	60	0.0390	77
	$N = 77$		$\Sigma \frac{f}{N} = 1$	

c. Over 35% of the population is under 11 years old. Less than 4% of the population is over 54 years old.

3a.

Class Boundaries
−0.5–10.5
10.5–21.5
21.5–32.5
32.5–43.5
43.5–54.5
54.5–65.5

b. Use class midpoints for the horizontal scale and frequency for the vertical scale.

c.

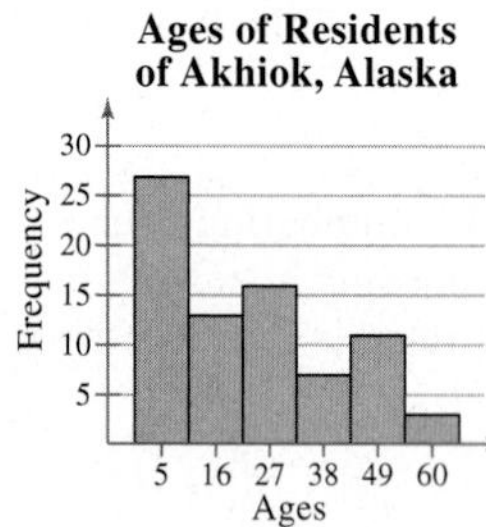

d. Most of the residents are less than 32 years old.

4a. Use class midpoints for the horizontal scale and frequency for the vertical scale.

b. See part (c).

c.

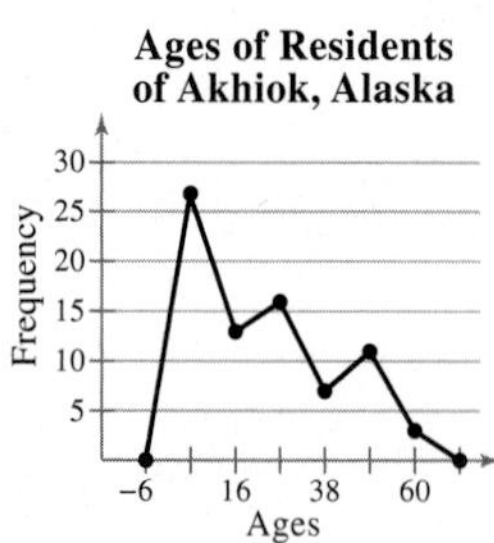

d. The population of Akhiok, Alaska is predominantly made up of young people.

5abc.

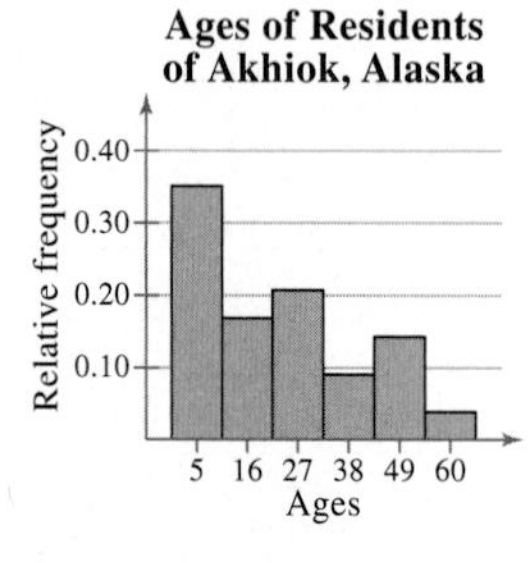

6a. Use upper class boundaries for the horizontal scale and cumulative frequency for the vertical scale.

b. See part (c).

c.

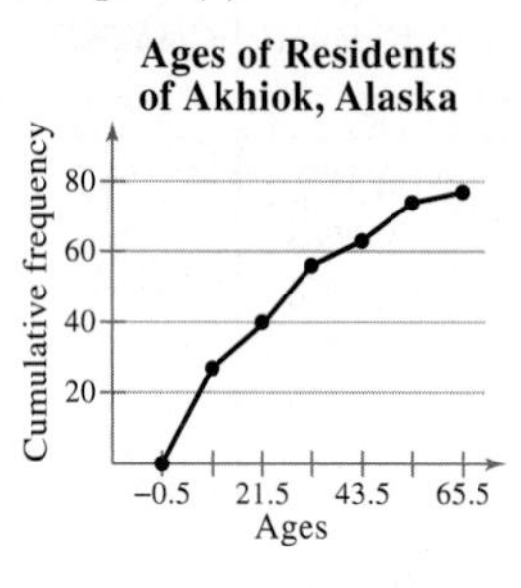

d. Approximately 63 residents are less than 45 years old.

7. See the solution to Try It Yourself 3.

Section 2.2

1a.

```
0 |
1 |
2 |
3 |
4 |
5 |
6 |
```

b. Key: 3|5 = 35

```
0 | 6 3 7 1 7 5 0 3 2 1 8 6 9 2 0 4 1 6 5 5 4 6 8 2 4
1 | 7 2 5 0 1 7 0 3 6 1 1 0 2 6
2 | 8 7 1 4 7 2 5 1 5 3 9 6 8
3 | 9 3 6 2 4 0 3 1 2 1
4 | 8 7 5 6 2 9 1
5 | 0 4 3 0 2 6 5 1
6 | 3
```

c. Key: 3|5 = 35

```
0 | 0 1 1 1 2 2 2 3 3 4 4 4 5 5 5 6 6 6 6 7 7 8 8 9
1 | 0 0 0 1 1 1 2 2 3 5 6 6 7 7
2 | 1 1 2 3 4 5 5 6 7 7 8 8 9
3 | 0 1 1 2 2 3 3 4 6 9
4 | 1 2 5 6 7 8 9
5 | 0 0 1 2 3 4 5 6
6 | 3
```

d. It appears that the residents of Akhiok are a young population with most of the ages being below 40 years old.

2ab. Key: 3|5 = 35

```
0 | 0 1 1 1 2 2 2 3 3 4 4 4
0 | 5 5 5 6 6 6 6 7 7 8 8 9
1 | 0 0 0 1 1 1 2 2 3
1 | 5 6 6 7 7
2 | 1 1 2 3 4
2 | 5 5 6 7 7 8 8 9
3 | 0 1 1 2 2 3 3 4
3 | 6 9
4 | 1 2
4 | 5 6 7 8 9
5 | 0 0 1 2 3 4
5 | 5 6
6 | 3
```

3a. Use ages for the horizontal axis.

b.

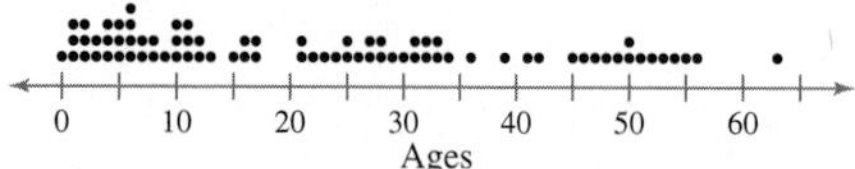

c. It appears that a large percentage of the population is younger than 40 years old.

4a.

Vehicle type	Killed (frequency)	Relative frequency	Central angle
Cars	25,063	0.6580	237°
Trucks	9409	0.2470	89°
Motorcycles	3141	0.0825	30°
Other	474	0.0124	4°
	$\Sigma f = 38{,}087$	$\Sigma \frac{f}{n} \approx 1$	

b.

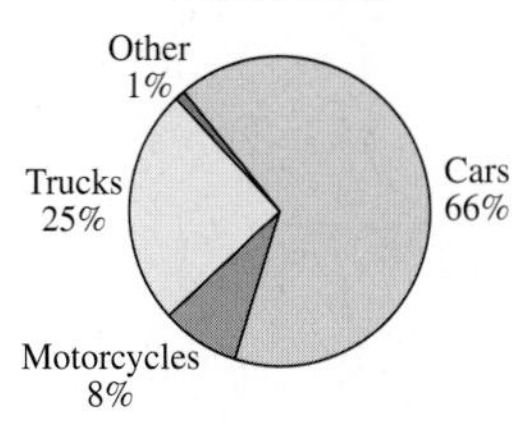

c. It appears that motorcycle deaths for both years is nearly the same but car deaths increased by 8% and truck deaths decreased by 9%.

5a.

Cause	Frequency, f
Auto Dealers	14668
Auto Repair	9728
Home Furnishing	7792
Computer Sales	5733
Dry Cleaning	4649

b.

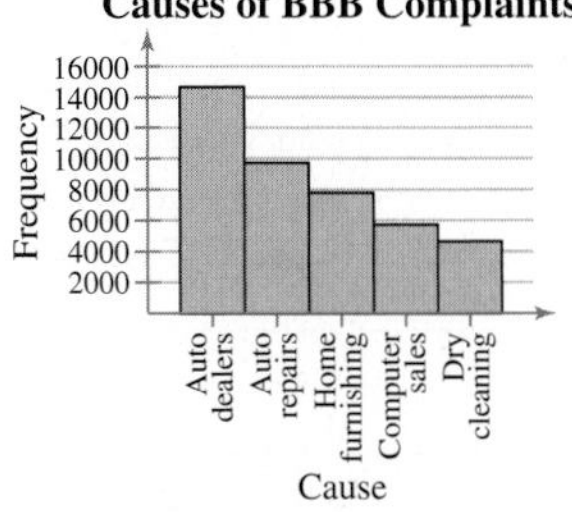

c. It appears that the auto industry (dealers and repair shops) account for the largest portion of complaints filed at the BBB.

6ab.

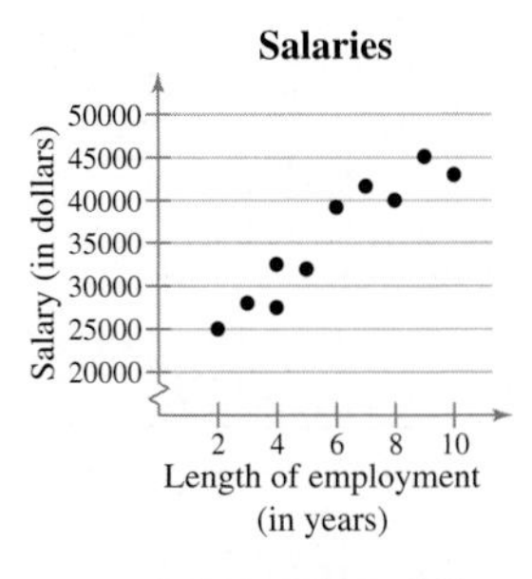

c. It appears that the longer an employee is with the company, the larger his/her salary will be.

7ab.

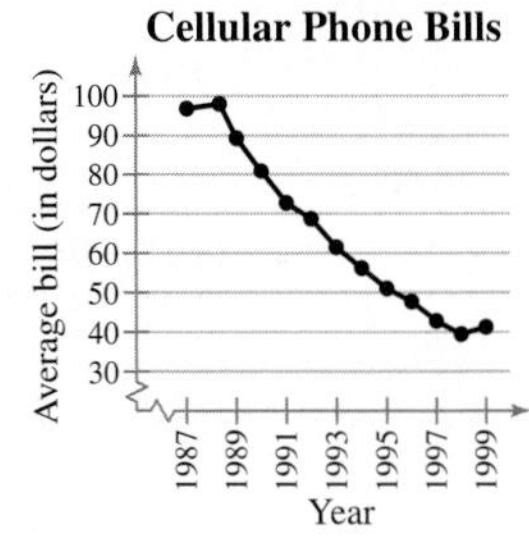

c. It appears that the average monthly bill for cellular telephone subscribers has decreased significantly from 1987 to 1999.

Section 2.3

1a. 1745 **b.** 22.66

c. The typical age of a resident of Akhiok is 22.66 years old.

2a. 0, 1, 1, 1, 2, 2, 2, 3, 3, 4, 4, 4, 5, 5, 5, 6, 6, 6, 6, 7, 7, 8, 8, 9, 10, 10, 10, 11, 11, 11, 12, 12, 13, 15, 16, 16, 17, 17, 21, 21, 22, 23, 24, 25, 25, 26, 27, 27, 28, 28, 29, 30, 31, 31, 32, 32, 33, 33, 34, 36, 39, 41, 42, 45, 46, 47, 48, 49, 50, 50, 51, 52, 53, 54, 55, 56, 63

b. 21

c. Half of the residents of Akhiok are younger than 21 and half are older than 21.

3a. 0, 1, 1, 1, 2, 2, 2, 3, 3, 4, 4, 4, 5, 5, 5, 6, 6, 6, 6, 7, 7, 8, 8, 9, 10, 10, 10, 11, 11, 12, 12, 13, 15, 16, 16, 17, 17, 21, 21, 22, 23, 24, 25, 25, 26, 27, 27, 28, 28, 29, 30, 31, 31, 32, 32, 33, 36, 39, 41, 42, 45, 46, 47, 48, 49, 50, 50, 51, 52, 53, 54, 55, 63

b. 19

4a.

Age	Frequency, f	Age	Frequency, f	Age	Frequency, f
0	1	17	2	39	1
1	3	21	2	41	1
2	3	22	1	42	1
3	2	23	1	45	1
4	3	24	1	46	1
5	3	25	2	47	1
6	4	26	1	48	1
7	2	27	2	49	1
8	2	28	2	50	2
9	1	29	1	51	1
10	3	30	1	52	1
11	3	31	2	53	1
12	2	32	2	54	1
13	1	33	2	55	1
15	1	34	1	56	1
16	2	36	1	63	1

b. 6

c. The mode of the ages of the residents of Akhiok is 6 years old.

5a. Yes **b.** The mode of the responses to the survey is "Yes."

6a. 21.58; 21; 20

b. The mean in example 6 ($\bar{x} = 23.75$) was heavily influenced by the age 65. Neither the median nor the mode was affected as much by the age 65.

7ab.

Source	Score x	Weight w	$x \cdot w$
Test Mean	86	0.50	43
Mid-Term	96	0.15	14.4
Final	98	0.20	19.6
Computer Lab	98	0.10	9.8
Homework	100	0.05	5
		$\Sigma w = 1.00$	$\Sigma(x \cdot w) = 91.8$

c. 91.8 **d.** The weighted mean for the course is 91.8.

8abc.

Class	Midpoint x	Frequency, f	$x \cdot f$
0–10	5	27	135
11–21	16	13	208
22–32	27	16	432
33–43	38	7	266
44–54	49	11	539
55–65	60	3	180
		$N = 77$	$\Sigma(x \cdot f) = 1760$

d. 22.86

Section 2.4

1a. Min = 23 or $23,000 Max = 58 or $58,000

b. 35, or $35,000

c. The range of the starting salaries for Corporation B is 35 or $35,000 (much larger than the range of Corporation A).

2a. 41.5, or $41,500

bc.

Salary, x (1000s of dollars)	Deviation, $x - \mu$ (1000s of dollars)
23	−18.5
29	−12.5
32	−9.5
40	−1.5
41	−0.5
41	−0.5
49	7.5
50	8.5
52	10.5
58	16.5
$\Sigma x = 415$	$\Sigma(x - \mu) = 0$

3ab. $\mu - 41.5$, or $41,500

Salary, x	$x - \mu$	$(x - \mu)^2$
23	−18.5	342.25
29	−12.5	156.25
32	−9.5	90.25
40	−1.5	2.25
41	−0.5	0.25
41	−0.5	0.25
49	7.5	56.25
50	8.5	72.25
52	10.5	110.25
58	16.5	272.25
$\Sigma x = 415$	$\Sigma(x - \mu) = 0$	$\Sigma(x - \mu)^2 = 1102.5$

c. 110.25 **d.** 10.5, or $10,500

e. The population variance is 110.25 and the population standard deviation is 10.5 or $10,500.

4a. See 3ab. **b.** 122.5 **c.** 11.07, or $11,070

5a. Enter data **b.** 37.89; 3.98

6a. 7, 7, 7, 7, 7, 13, 13, 13, 13, 13 **b.** 3

7a. 1 standard deviation **b.** 34%

c. The estimated percent of the heights that are between 61.25 and 64 inches is 34%.

8a. 0 **b.** 70.6

c. At least 75% of the data lie within 2 standard deviations of the mean.

d. At least 75% of the population of Alaska is between 0 and 70.6 years old.

9a.

x	f	xf
0	10	0
1	19	19
2	7	14
3	7	21
4	5	20
5	1	5
6	1	6
	$n = 50$	$\Sigma xf = 85$

b. 1.7

c.

$x - \bar{x}$	$(x - \bar{x})^2$	$(x - \bar{x})^2 \cdot f$
−1.70	2.8900	28.9
−0.70	0.4900	9.31
0.30	0.0900	0.63
1.30	1.6900	11.83
2.30	5.2900	26.45
3.30	10.9800	10.89
4.30	18.4900	18.49
		$\Sigma(x - \bar{x})^2 f = 106.5$

d. 1.47

10a.

Class	x	f	xf
0–99	49.5	380	18,810
100–199	149.5	230	34,385
200–299	249.5	210	52,395
300–399	349.5	50	17,475
400–499	449.5	60	26,970
500+	650	70	45,500
		$n = 1000$	$\Sigma xf = 195{,}535$

b. 195.54

c.

$x - \bar{x}$	$(x - \bar{x})^2$	$(x - \bar{x})^2 f$
−146.04	21,327.68	8,104,518.4
−46.04	2119.68	487,526.4
53.96	2911.68	611,452.8
153.96	23,703.68	1,185,184
253.96	64,495.68	3,869,740.8
454.46	206,533.89	14,457,372.3
		$\Sigma(x - \bar{x})^2 f = 28{,}715{,}794.7$

d. 169.54

Section 2.5

1a. 0, 1, 1, 1, 2, 2, 2, 3, 3, 4, 4, 4, 5, 5, 5, 6, 6, 6, 6, 7, 7, 8, 8, 9, 10, 10, 10, 11, 11, 11, 12, 12, 13, 15, 16, 16, 17, 17, 21, 21, 22, 23, 24, 25, 25, 26, 27, 27, 28, 28, 29, 30, 31, 31, 32, 32, 33, 33, 34, 36, 39, 41, 42, 45, 46, 47, 48, 49, 50, 50, 51, 52, 53, 54, 55, 56, 63

b. 21 **c.** 6.5, 33.5

2a. Enter data **b.** 17, 23, 28.5

c. One quarter of the tuition costs is $17,000 or less, one half is $23,000 or less, and three quarters is $28,500 or less.

3a. 6.5, 33.5 **b.** 27

c. The ages in the middle half of the data set vary by 27 years.

4a. 0, 6.5, 21, 33.5, 63

bc. Ages of Residents of Akhiok, Alaska

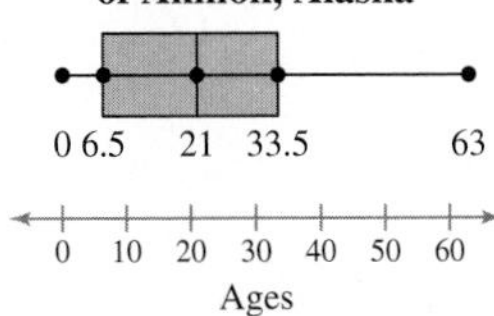

d. It appears that half of the ages are between 6.5 and 33.5 years.

5a. 85th percentile

b. 85% of the ages are 47 years or younger.

6a. (a) $\mu = 70$, $\sigma = 8$

(b) $z_1 = \dfrac{60 - 70}{8} = -1.25$

$z_2 = \dfrac{71 - 70}{8} = 0.125$

$z_3 = \dfrac{92 - 70}{8} = 2.75$

(c) From the z-score, $60 is 1.25 standard deviations below the mean, $71 is 0.125 standard deviation above the mean, and $92 is 2.75 standard deviations above the mean.

CHAPTER 3

Section 3.1

1ab. (1)

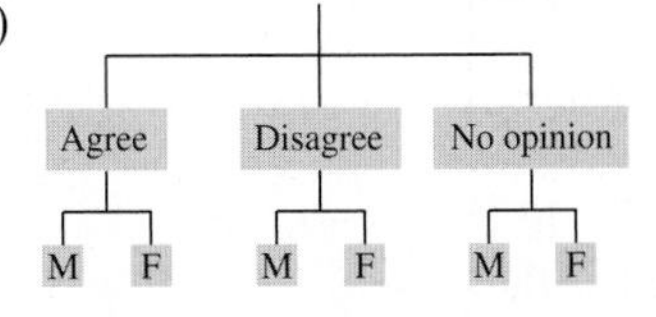

(2)

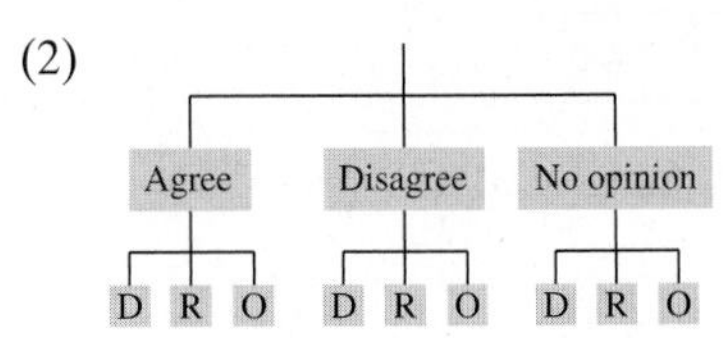

c. (1) 6 (2) 9

d. (1) Let A = Agree, D = Disagree, N = No Opinion, M = Male, F = Female

Sample Space = $\{AM, AF, DM, DF, NM, NF\}$

(2) Let A = Agree, D = Disagree, N = No Opinion, De = Democrat R = Republican, O = Other

Sample Space = $\{ADe, AR, AO, DDe, DR, DO, NDe, NR, NO\}$

2a. (1) 6 outcomes (2) 1 outcome

b. (1) Not a simple event (2) Simple event

3a. (1) 52 (2) 1 (3) 0.0192

b. (1) 52 (2) 13 (3) 0.25

c. (1) 52 (2) 52 (3) 1

4a. The event is "the next claim processed is fraudulent." 4

b. 100 **c.** 0.04

5a. 54 **b.** 1000 **c.** 0.054

6a. Event = salmon successfully passing through a dam on the Columbia River.

b. Estimated

c. Empirical probability

7a. 0.425 **b.** 0.575 **c.** $\frac{23}{40}$ or 0.575

Section 3.2

1a. (1) 30 and 102 (2) 11 and 50

b. (1) 0.294 (2) 0.22

2a. (1) No (2) Yes

b. (1) Independent (2) Dependent

c. (1) A salmon successfully swimming through a dam does not affect the probability of another successfully swimming through the same dam.

(2) It has been shown in studies that exercising frequently lowers the resting rate of the heart.

3a. (1) Independent (2) Dependent

b. (1) 0.7225 (2) 0.108

4a. (1) event (2) complement

b. (1) 0.729 (2) 0.999

Section 3.3

1a. (1) None are true. (2) None are true.

(3) A and B cannot occur at the same time.

b. (1) Not mutually exclusive

(2) Not mutually exclusive

(3) Mutually exclusive

2a. (1) Mutually exclusive (2) Not mutually exclusive

b. (1) $\frac{1}{6}, \frac{1}{2}$ (2) $\frac{12}{52}, \frac{13}{52}, \frac{3}{52}$

c. (1) 0.667 (2) 0.423

3a. $A = \{\text{sales between \$0 and \$24,999}\}$
$B = \{\text{sales between \$25,000 and \$49,000}\}$

b. A and B cannot occur at the same time. A and B are mutually exclusive.

c. $\frac{3}{36}$ and $\frac{5}{36}$ **d.** 0.222

4a. (1) Mutually exclusive (2) Not mutually exclusive

b. (1) 0.149 (2) 0.910

5a. 0.15 **b.** 0.85

Section 3.4

1a. Manufacturer: 4 **b.** 72
Size: 3
Color: 6

2a. (1) Each letter is an event (26 choices).
(2) Each letter is an event (26, 25, 24, 23, 22, and 21 choices).

b. (1) 308,915,776 (2) 165,765,600

3a. 6 **b.** 720

4a. 336

b. There are 336 possible ways that three horses can finish in first, second, and third place.

5a. $n = 12, r = 4$ **b.** 11,880

6a. $n = 20, n_1 = 6, n_2 = 9, n_3 = 5$ **b.** 77,597,520

7a. $n = 16, r = 3$ **b.** 560

c. There are 560 different possible 3-person committees that can be selected from 16 employees.

8a. 1 and 180 **b.** 0.0056

9a. 10 **b.** 220 **c.** 0.045

CHAPTER 4

Section 4.1

1a. (1) measured (2) counted

b. (1) Continuous because x can be any amount of time needed to complete a test.
(2) Discrete because x can be counted.

2ab.

x	f	$P(x)$
0	16	0.16
1	19	0.19
2	15	0.15
3	21	0.21
4	9	0.09
5	10	0.10
6	8	0.08
7	2	0.02
	100	1

c.

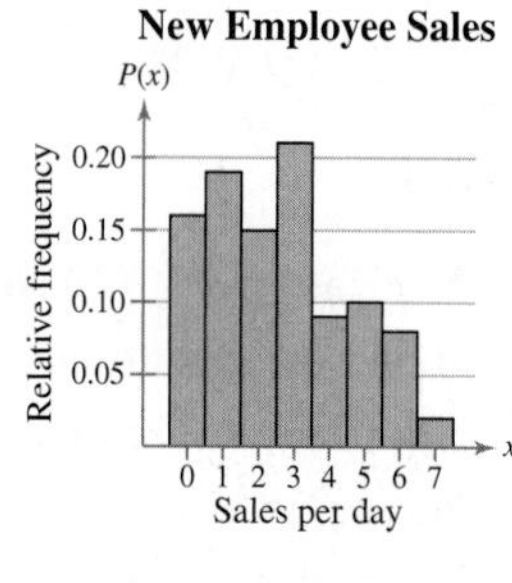

3a. Each $P(x)$ is between 0 and 1.

b. $\Sigma P(x) = 1$

c. Is a probability distribution.

4a. (1) Yes, each outcome is between 0 and 1.
(2) Yes, each outcome is between 0 and 1.

b. (1) No (2) Yes

c. (1) Not a probability distribution
(2) Is a probability distribution

5ab.

x	$P(x)$	$xP(x)$
0	0.16	0.00
1	0.19	0.19
2	0.15	0.30
3	0.21	0.63
4	0.09	0.36
5	0.10	0.50
6	0.08	0.48
7	0.02	0.14
	1	2.60

c. $\mu = 2.60$
On average, 2.60 sales are made per day.

6ab.

x	$P(x)$	$x - \mu$	$(x - \mu)^2$	$P(x)(x - \mu)^2$
0	0.16	−2.6	6.76	1.0816
1	0.19	−1.6	2.56	0.4864
2	0.15	−0.6	0.36	0.054
3	0.21	0.4	0.16	0.0336
4	0.09	1.4	1.96	0.1764
5	0.10	2.4	5.76	0.576
6	0.08	3.4	11.56	0.9248
7	0.02	4.4	19.36	0.3872
	$\Sigma P(x) = 1$			$\Sigma P(x)(x - \mu)^2 = 3.72$

c. 1.93

7ab.

x	f	$P(x)$	$xP(x)$
0	25	0.11	0.000
1	48	0.213	0.213
2	60	0.267	0.533
3	45	0.200	0.600
4	20	0.089	0.356
5	10	0.044	0.222
6	8	0.036	0.213
7	5	0.022	0.156
8	3	0.013	0.107
9	1	0.004	0.040
	225	1	2.440

c. 2.44

d. You can expect an average of 2.44 sales per day.

Section 4.2

1a. Trial: Individual questions (10 trials)

Success: Question answered correctly

b. Yes

c. $n = 10, p = 0.25, q = 0.75, x = 0, 1, 2, \ldots, 9, 10$

2a. Trial: 5 cards being drawn with replacement

Success: Card drawn is a club.

Failure: Card drawn is not a club.

b. $n = 5, p = 0.25, q = 0.75, x = 3$

c. $P(3) = \frac{5!}{2!\,3!}(0.25)^3(0.75)^2 \approx 0.088$

3a. Trial: 7 workers

Success: Selecting a worker who will rely on pension

Failure: Selecting a worker who will not rely on pension

b. $n = 7, p = 0.34, q = 0.66, x = 0, 1, 2, \ldots, 6, 7$

c. $P(0) = {}_7C_0\,(0.34)^0\,(0.66)^7 = 0.0546$

$P(1) = {}_7C_1\,(0.34)^1\,(0.66)^6 = 0.197$

$P(2) = {}_7C_2\,(0.34)^2\,(0.66)^5 = 0.304$

$P(3) = {}_7C_3\,(0.34)^3\,(0.66)^4 = 0.261$

$P(4) = {}_7C_4\,(0.34)^4\,(0.66)^3 = 0.134$

$P(5) = {}_7C_5\,(0.34)^5\,(0.66)^2 = 0.0416$

$P(6) = {}_7C_6\,(0.34)^6\,(0.66)^1 = 0.00714$

$P(7) = {}_7C_7\,(0.34)^7\,(0.66)^0 = 0.000525$

d.

x	$P(x)$
0	0.0546
1	0.197
2	0.304
3	0.261
4	0.134
5	0.0416
6	0.00714
7	0.000525

4a. Trial: 10 businesses

Success: Selecting a business with a Web site

Failure: Selecting a business with out a site

b. $n = 10, p = 0.25, x = 4$ **c.** 0.146

5a. $n = 250, p = 0.71, x = 178$ **b.** 0.056

c. The probability that exactly 178 people in the United States will use more than one topping on their hotdog is about 0.056.

6a. (1) $x = 2$ (2) $x = 2, 3, 4,$ or 5 (3) $x = 0$ or 1

b. (1) 0.217

(2) 0.217, 0.058, 0.008, 1.67×10^{-7}; 0.283

(3) 0.308, 0.409; 0.717

c. (1) The probability that exactly two men consider fishing their favorite leisure-time activity is about 0.217.

(2) The probability that at least two men consider fishing their favorite leisure-time activity is about 0.283.

(3) The probability that fewer than two men consider fishing their favorite leisure-time activity is about 0.717.

7a. 0.042, 0.176, 0.306, 0.283, 0.148, 0.041, 0.005

b.

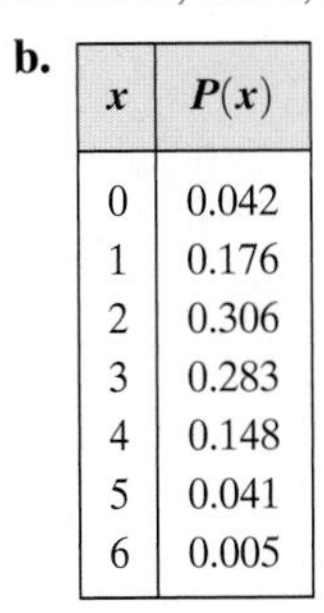

x	$P(x)$
0	0.042
1	0.176
2	0.306
3	0.283
4	0.148
5	0.041
6	0.005

c.

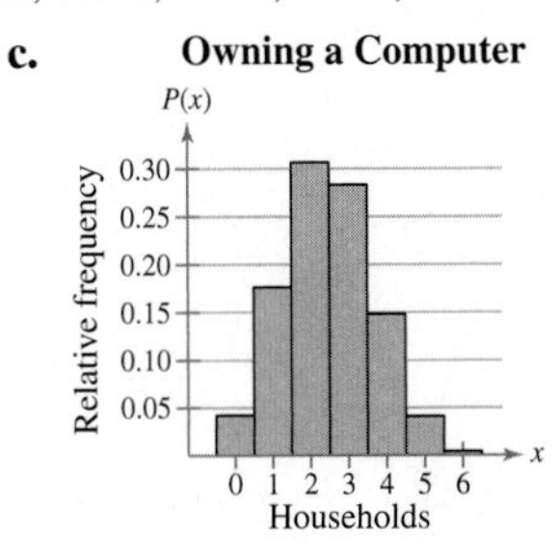

8a. Success: Selecting a clear day, $n = 31,\ p = 0.38,\ q = 0.62$

b. 11.78 **c.** 7.3036 **d.** 2.703

e. On average, there are about 12 clear days during the year. The standard deviation is about 3 days.

Section 4.3

1a. 0.23, 0.177, 0.136 **b.** 0.543

c. The probability that your first sale will occur before your fourth sales call is 0.543.

2a. $P(0) \approx 0.050$

$P(1) \approx 0.149$

$P(2) \approx 0.224$

$P(3) \approx 0.224$

$P(4) \approx 0.168$

b. 0.815 **c.** 0.185

d. The probability that more than four accidents will occur in any given month at the intersection is 0.185.

3a. 0.10 **b.** 0.10, 3 **c.** 0.0002

CHAPTER 5

Section 5.1

1a. A: 45, B: 60, C: 45 B has the greatest mean.

b. Curve C is more spread out, so curve C has the greatest standard deviation.

2a. 3.5 **b.** 3.3, 3.7; 0.2

3a. 85 is 1 standard deviation below the mean and 145 is 3 standard deviations above the mean.

b. 0.8385

Section 5.2

1a. (1) 0.0143 (2) 0.985

2a.

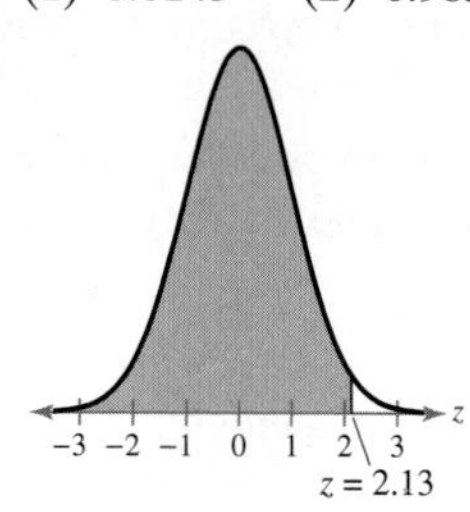

b. 0.9834

3a.

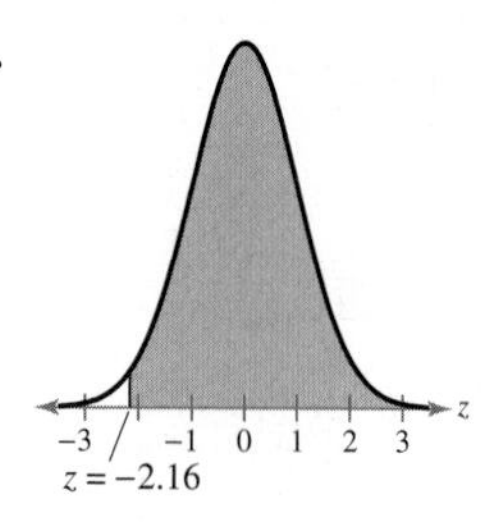

b. 0.0154 **c.** 0.9846

4a. 0.0885 **b.** 0.0154 **c.** 0.0731

Section 5.3

1a.

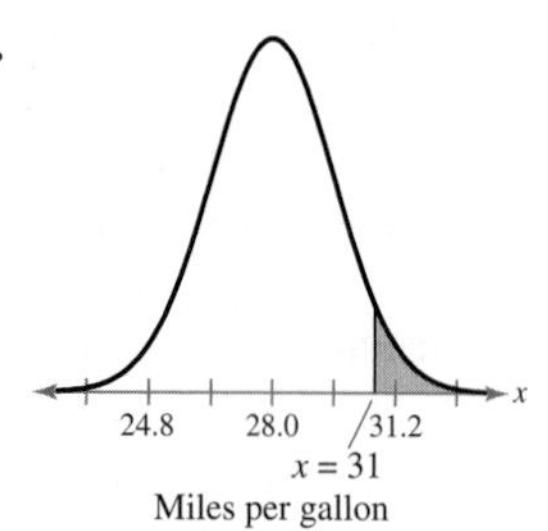

b. 1.875

c. 0.0304

d. The probability that a randomly selected manual transmission Focus will get more than 31 mpg in city driving is 0.0304.

2a.

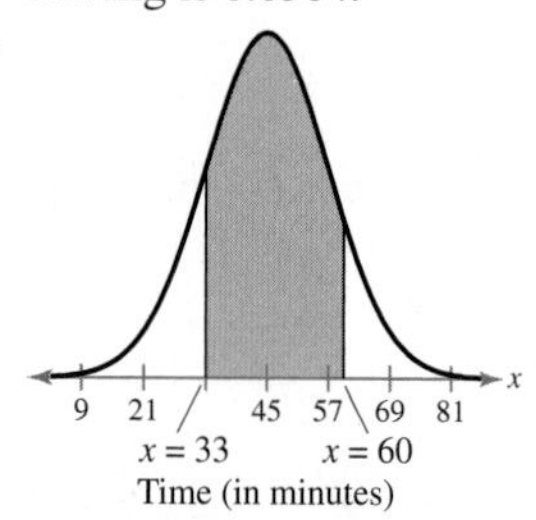

b. $-1, 1.25$

c. 0.1587; 0.8944

d. 0.7357

3a. Read user's guide for the technology tool.

b. Enter the data.

c. The probability that a randomly selected U.S. man's cholesterol is between 190 and 225 is about 0.4968.

Section 5.4

1a. -1.77 **b.** 2.33

2a. (1) Area = 0.10 (2) Area = 0.20

(3) Area = 0.99

bc. (1) -1.28 (2) -0.84 (3) 2.33

3a. $\mu = 70$ $\sigma = 8$

b. 64; 104.32; 55.44

c. 64 is below the mean, 104.32 is above the mean, and 55.44 is below the mean.

4ab.

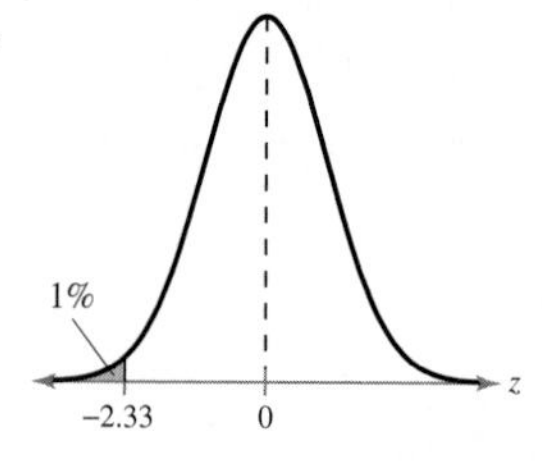

c. 142.832

d. So, the longest braking distance a Ford F-150 could have and still be in the best 1% is 143 ft.

5ab.

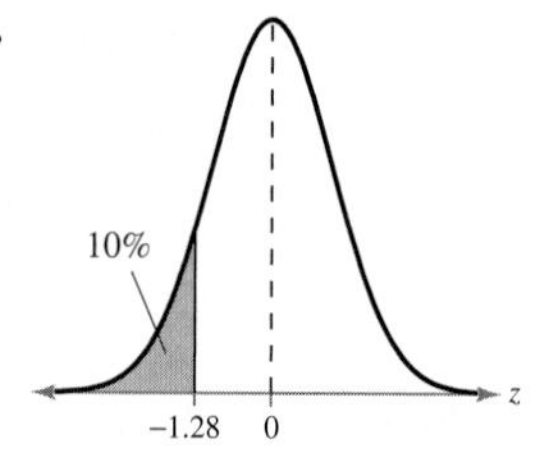

c. 8.512

d. So, the maximum length of time an employee could have worked and still be laid off is 8 years.

Section 5.5

1a.

Sample	Mean	Sample	Mean	Sample	Mean
1, 1	1	3, 1	2	6, 1	3.5
1, 2	1.5	3, 2	2.5	6, 2	4
1, 3	2	3, 3	3	6, 3	4.5
1, 5	3	3, 5	4	6, 5	5.5
1, 6	3.5	3, 6	4.5	6, 6	6
1, 7	4	3, 7	5	6, 7	6.5
2, 1	1.5	5, 1	3	7, 1	4
2, 2	2	5, 2	3.5	7, 2	4.5
2, 3	2.5	5, 3	4	7, 3	5
2, 5	3.5	5, 5	5	7, 5	6
2, 6	4	5, 6	5.5	7, 6	6.5
2, 7	4.5	5, 7	6	7, 7	7

b. 4, 2.33, 1.53 **c.** 4, 4.67, 2.16

2a. 64, 0.9

b. $n = 100$

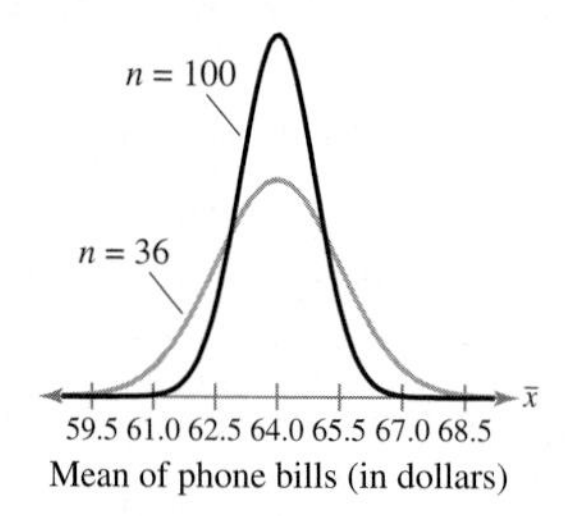

3a. 3.5, 0.05

b.

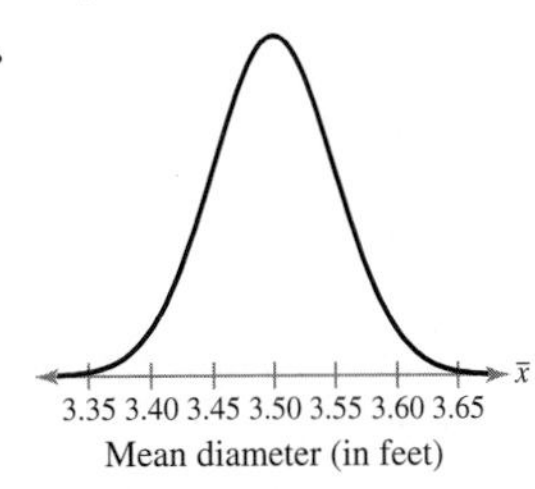

4a. 9, 0.15 **b.** −2, 3.33 **c.** 0.9768

5a. 125,700; 7505.55 **b.** −3.42 **c.** 0.9997

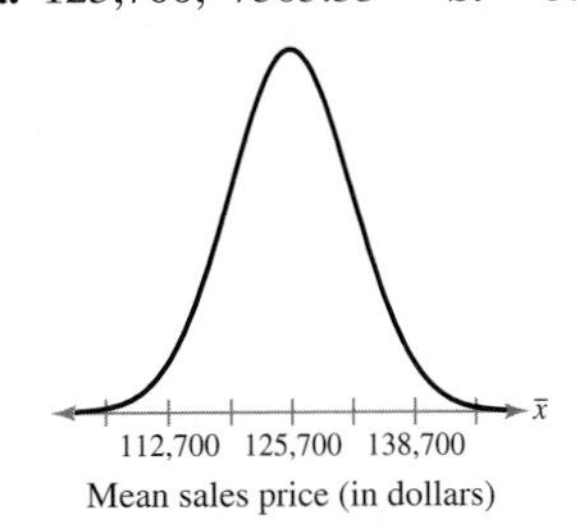

6a. 0.5, 1.58 **b.** 0.6915, 0.9429

c. There is a 69% chance an *individual receiver* will cost less than \$700. There is a 94% chance that the *mean of a sample of 10 receivers* is less than \$700.

Section 5.6

1a. 70, 0.08, 0.92 **b.** 5.6, 64.4

c. Normal distribution can be used. **d.** 5.6, 2.270

2a. (1) 57, 58, . . . , 83 (2) . . . , 52, 53, 54

b. (1) $56.5 < x < 83.5$ (2) $x < 54.5$

3a. yes **b.** 5.6, 2.27

c. $x > 10.5$

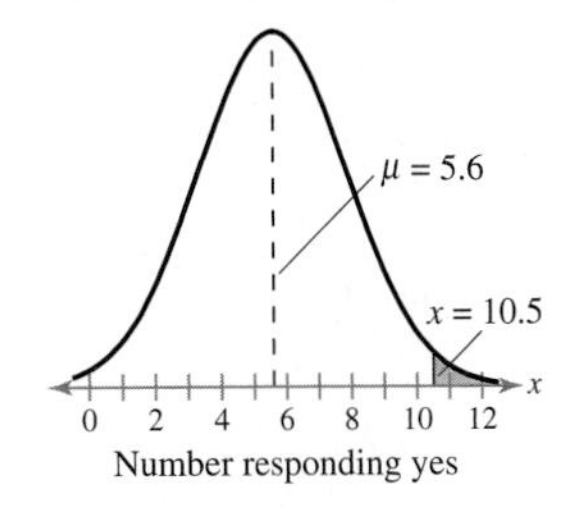

d. 2.16

e. 0.9846, 0.0154

4a. yes **b.** 58, 6.42

c. $x < 65.5$

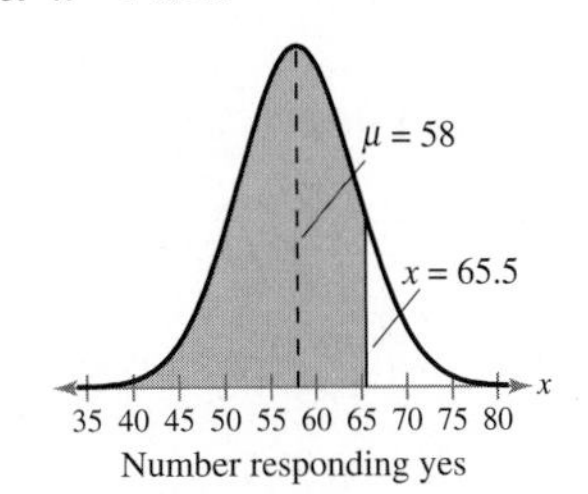

d. 1.17

e. 0.8790

5a. yes **b.** 78, 6.90

c. $78.5 < x < 79.5$

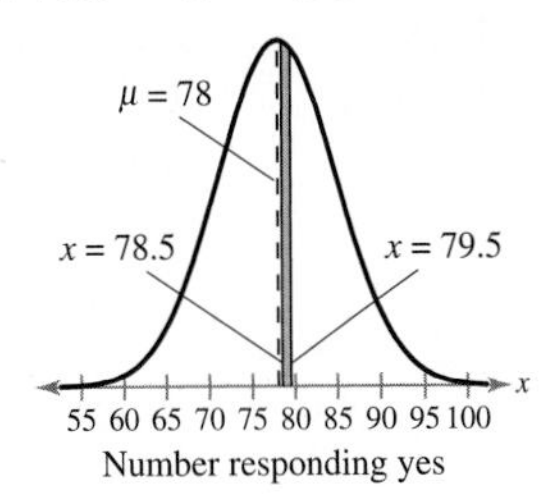

d. 0.07; 0.22

e. 0.5279; 0.5871; 0.0592

CHAPTER 6

Section 6.1

1a. 14.767

b. The mean number of sentences per magazine advertisement is 14.767.

2a. 1.96, 30, 16.536 **b.** 5.917

c. You are 95% confident that the maximum error of the estimate is about 5.917 sentences per magazine advertisement.

3a. $\bar{x} = 14.767,\ E = 5.917$ **b.** 8.850, 20.684

c. You are 95% confident that the mean number of sentences per magazine advertisements is between 8.850 and 20.684.

4b. (11.6, 13.2); (11.4, 13.4)

c. As the confidence level increases, so does the width of the interval.

5a. 20, 22.9, 1.5, 1.282 **b.** 0.430 **c.** 22.47; 23.33

d. You are 80% confident that the mean age of the student is between 22.47 and 23.33 years.

6a. 1.96, 2, $s \approx 5.0$ **b.** 25

c. You should have at least 25 magazine advertisements in your sample.

Section 6.2

1a. 21 **b.** 0.90 **c.** 1.721

2a. 1.753; 4.383; 2.947; 7.368

b. (157.617, 166.383); (154.632, 169.368)

c. You are 90% confident that the mean temperature of coffee sold is between 157.617° and 166.383°.

You are 99% confident that the mean temperature of coffee sold is between 154.632° and 169.368°.

3a. 1.729; 0.162; 2.093; 0.197

b. (6.768, 7.092); (6.733, 7.127)

c. You are 90% confident that the mean mortgage interest rate is contained between 6.768% and 7.092%.

You are 95% confident that the mean mortgage interest rate is contained between 6.733% and 7.127%.

4a. No; Yes; No; Use *t*-distribution

Section 6.3

1a. 221, 1470 **b.** 0.150

2a. 0.150, 0.850

b. $n\hat{p} = 220.5 > 5$ and $n\hat{q} = 1249.5 > 5$

c. 1.645; 0.015 **d.** (0.135, 0.165)

e. You are 90% confident that the proportion of adults that admired Abraham Lincoln the most is contained between 13.5% and 16.5%.

3a. 1001, 0.39 **b.** 0.61

c. $n\hat{p} = 1001 \cdot 0.39 \approx 390 > 5$

$n\hat{q} = 1001 \cdot 0.61 \approx 611 > 5$

Distribution of $\hat{p}$ is approximately normal.

d. 2.575 **e.** (0.350, 0.430)

f. You are 99% confident that the proportion of adults who think that cars are the safer mode of transportation is contained between 35.0% and 43.0%.

4a. 0.04, 0.96 **b.** 1.645, 0.02 **c.** 260

d. At least 260 adults should be included in the sample.

Section 6.4

1a. 24, 0.95 **b.** 0.025, 0.975 **c.** 39.364, 12.401

2a. 90%CI: 42.557, 17.708; 95%CI: 45.722, 16.047

b. (0.981, 2.358); (0.913, 2.602)

c. (0.990, 1.536); (0.956, 1.613)

d. You are 90% confident that the population variance is between 0.981 and 2.358, and that the population standard deviation is between 0.990 and 1.536. You are 95% confident that the population variance is between 0.913 and 2.602 and that the population standard deviation is between 0.956 and 1.613.

CHAPTER 7

Section 7.1

1a. (1) The mean . . . is 74 months.

$\mu = 74$

(2) The variance . . . is less than or equal to 3.5.

$\sigma^2 \leq 3.5$

(3) The proportion . . . is greater than 39%.

$p > 0.39$

b. (1) $\mu \neq 74$ (2) $\sigma^2 > 3.5$ (3) $p \leq 0.39$

c. (1) H_0: $\mu = 74$ (claim); H_a: $\mu \neq 74$

(2) H_0: $\sigma^2 \leq 3.5$ (claim); H_a: $\sigma^2 > 3.5$

(3) H_0: $p \leq 0.39$; H_a: $p > 0.39$ (claim)

2a. H_0: $p \leq 0.01$; H_a: $p > 0.01$

b. Type I error will occur if the actual proportion is less than or equal to 0.01, but you decided to reject H_0.

Type II error will occur if the actual proportion is greater than 0.01, but you do not reject H_0.

c. Type II error is more serious since you would be misleading the consumer, possibly causing serious injury or death.

3a. (1) H_0: $\mu = 74$; H_a: $\mu \neq 74$

(2) H_0: $p \leq 0.39$; H_a: $p > 0.39$

b. (1) two tailed (2) right tailed

c. (1) (2)

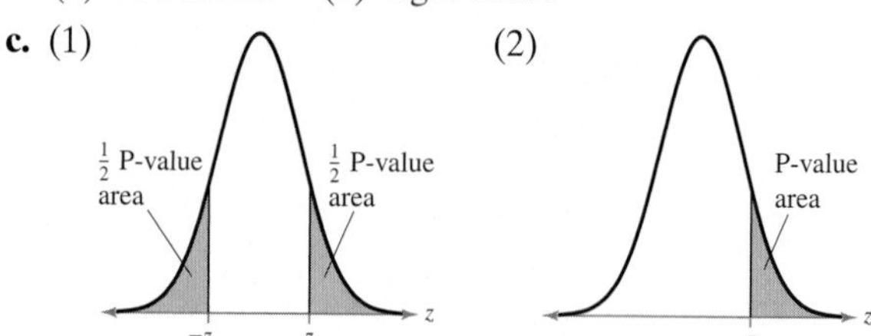

4a. There is enough evidence to support the radio station's claim.

b. There is not enough evidence to support the radio station's claim.

5a. (No answer required)

b. (1) $\mu \leq 650$ (2) $\mu = 98.6$

Section 7.2

1a. (1) $0.01 < 0.0347$ (2) $0.0347 < 0.05$

b. (1) because $0.0347 > 0.01$, fail to reject H_0

(2) because $0.0347 < 0.05$, reject H_0

2ab. 0.0526

c. because $0.0526 > 0.05$, fail to reject H_0

3a. 0.9896 **b.** 0.0208

c. because $0.0208 > 0.01$, fail to reject H_0

4a. The claim is "the mean speed is greater than 35 miles per hour."

H_0: $\mu \leq 35$; H_a: $\mu > 35$ (claim)

b. 0.05 **c.** 2.5 **d.** 0.0062

e. because $0.0062 < 0.05$, fail to reject H_0.

f. Because you fail to reject H_0, there is not enough evidence to claim the average speed limit is greater than 35 miles per hour.

5a. H_0: $\mu = 150$ (claim); H_a: $\mu \neq 150$

b. 0.01 **c.** −2.76 **d.** *P*-value = 0.0058

e. Reject H_0

f. There is enough evidence to state the claim is false.

6a. $0.039 > 0.01$ **b.** Fail to reject H_0

7a.

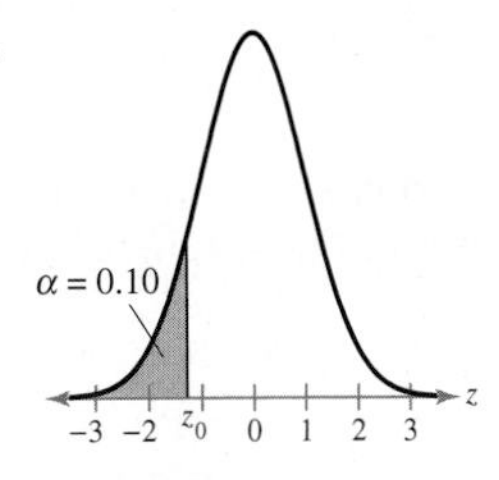

b. 0.1003 **c.** −1.28 **d.** $z < -1.28$

8a.

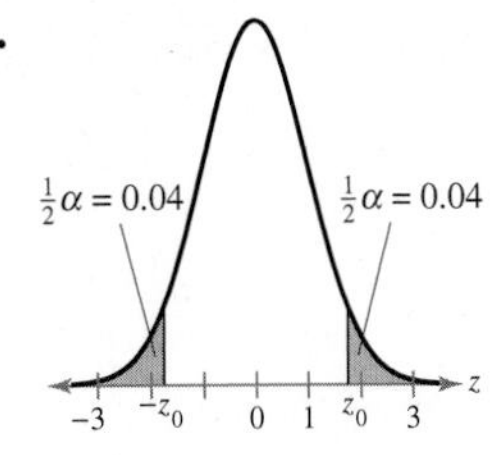

b. 0.0401 and 0.9599 **c.** −1.75 and 1.75

d. $z < -1.75$, $z > 1.75$

9a. $H_0: \mu \geq 8.5$; $H_a: \mu < 8.5$ (claim) **b.** $\alpha = 0.01$

c. $z_0 = -2.33$; Rejection region: $z < -2.33$

d. −3.550

e.

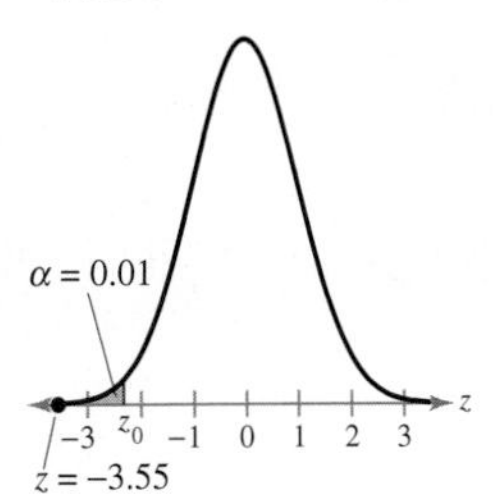

f. Reject H_0; There is enough evidence to support the claim.

10a. 0.01

b. ±2.575; Rejection regions: $z < -2.575$, $z > 2.575$

c. Fail to reject H_0

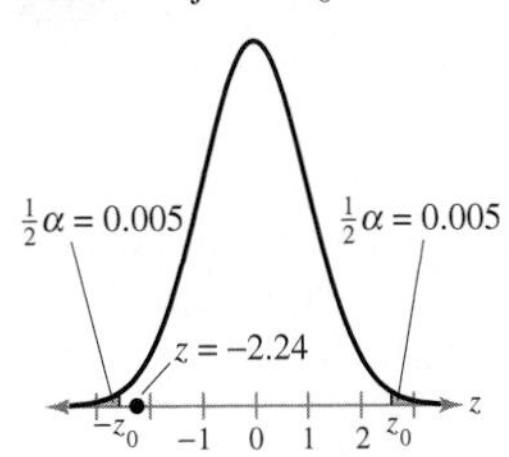

d. There is not enough evidence to support the claim that the mean cost is significantly different from $8390 at the 1% level of significance.

Section 7.3

1a. 2.650 **b.** −2.650

2a. 1.860 **b.** 1.860

3a. 2.947 **b.** ±2.947

4a. $H_0: \mu \geq \$875$ (claim); $H_a: \mu < \$875$

b. 0.01 and 8 **c.** −2.896 **d.** −2.419

e. Fail to reject H_0

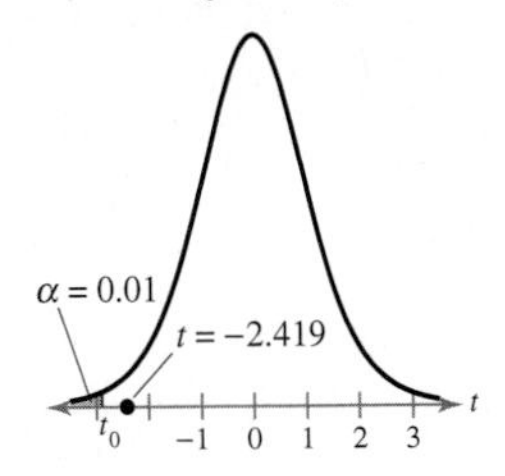

f. There is not enough evidence to reject the claim.

5a. $H_0: \mu = 1890$ (claim); $H_a: \mu \neq 1890$

b. 0.01 and 18 **c.** ±2.878 **d.** 3.798

e. Reject H_0

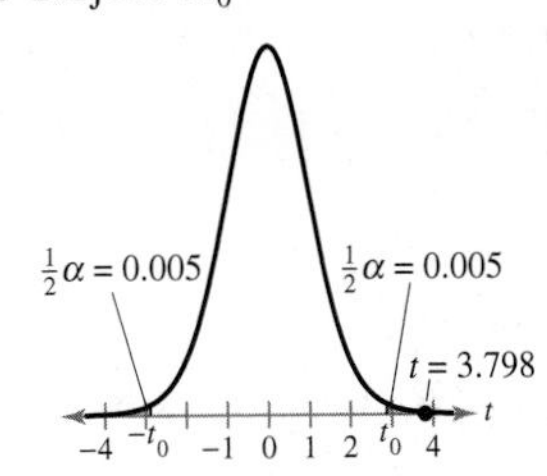

f. There is enough evidence to reject the company's claim.

6a. 0.0484 **b.** $0.0484 < 0.05$ **c.** Reject H_0

d. There is enough evidence to reject the claim.

Section 7.4

1a. $np = 25.8 > 5$, $nq = 60.2 > 5$

b. $H_0: p \geq 0.30$; $H_a: p < 0.30$ (claim)

c. 0.05 **d.** $z_0 = -1.645$; $z < -1.645$

e. −2.024 **f.** Reject H_0

g. There is enough evidence to support the claim.

2a. $np = 12.5 > 5$, $nq = 237.5 > 5$

b. $H_0: p = 0.05$ (claim); $H_a: p \neq 0.05$ **c.** 0.01

d. ±2.575; $z < -2.575$, $z > 2.575$

e. 2.176 **f.** Fail to reject H_0

g. There is not enough evidence to reject the claim.

3a. $np = 28.5 > 5$, $nq = 46.5 > 5$

b. $H_0: p \leq 0.38$; $H_a: p > 0.38$ (claim) **c.** 0.01

d. $z_0 = 2.33$; $z > 2.33$ **e.** 1.071 **f.** Fail to reject H_0

g. There is not enough evidence to support the claim.

Section 7.5

1a. 33.409

2a. 17.708

3a. 31.526 **b.** 8.231

4a. $H_0: \sigma^2 \leq 0.40$ (claim); $H_a: \sigma^2 > 0.40$
b. 0.01, 30 **c.** 50.892 **d.** 56.250 **e.** Reject H_0
f. There is enough evidence to reject the claim.
5a. $H_0: \sigma \geq 3.7$; $H_a: \sigma < 3.7$ (claim)
b. 0.05, 8
c. $\chi_0^2 = 2.733$; $\chi^2 < 2.733$
d. 5.259 **e.** Fail to reject H_0
f. There is not enough evidence to support the claim.
6a. $H_0: \sigma^2 = 8.6$ (claim); $H_a: \sigma^2 \neq 8.6$
b. 0.01, 9
c. $\chi_L^2 = 1.735$, $\chi_R^2 = 23.589$; $\chi^2 > 23.589$ or $\chi^2 < 1.735$
d. 4.50 **e.** Fail to reject H_0
f. There is not enough evidence to reject the claim.

CHAPTER 8

Section 8.1

1a. $H_0: \mu_1 = \mu_2$; $H_a: \mu_1 \neq \mu_2$ (claim)
b. 0.01 **c.** $z_0 = \pm 2.575$; $z < -2.575$ or $z > 2.575$
d. 2.747 **e.** Reject H_0
f. There is enough evidence to support the claim.
2a. $z \approx 4.542$; P-value ≈ 0.00000278
b. Rejection region is $z > 1.645$ or P-value < 0.05
c. Reject H_0

Section 8.2

1a. $H_0: \mu_1 = \mu_2$; $H_a: \mu_1 \neq \mu_2$ (claim)
b. 0.05 **c.** 9
d. $t_0 = \pm 2.262$; $t < -2.262$ or $t > 2.262$
e. 2.376 **f.** Reject H_0
g. There is enough evidence to support the claim.
2a. $H_0: \mu_1 \geq \mu_2$; $H_a: \mu_1 < \mu_2$ (claim)
b. 0.10 **c.** 25
d. $t_0 = -1.316$; $t < -1.316$
e. -0.898 **f.** Fail to reject H_0
g. There is not enough evidence to support the claim.

Section 8.3

1a. (1) Independent (2) Dependent
2a. $H_0: \mu_d \geq 0$; $H_a: \mu_d < 0$ (claim)
b. $\alpha = 0.05$, d.f. $= 11$ **c.** $t_0 \approx -1.796$; $t < -1.796$
d. $\bar{d} = -1$; $s_d \approx 2.374$ **e.** $t \approx -1.459$
f. Fail to reject H_0
g. There is not enough evidence to support the claim.
3a. $H_0: \mu_d = 0$; $H_a: \mu_d \neq 0$ (claim)
b. $\alpha = 0.05$, d.f. $= 6$
c. $t_0 = \pm 2.447$; $t < -2.447$ or $t > 2.447$
d. $\bar{d} \approx 0.557$; $s_d \approx 0.924$ **e.** $t \approx 1.595$
f. Fail to reject H_0
g. There is not enough evidence to conclude that the drug changes the body's temperature at the specified level of significance.

Section 8.4

1a. $H_0: p_1 = p_2$; $H_a: p_1 \neq p_2$ (claim)
b. $\alpha = 0.05$ **c.** $z_0 = \pm 1.96$; $z < -1.96$ or $z > 1.96$
d. $\bar{p} = 0.041$; $\bar{q} = 0.959$
e. $n_1\bar{p} \approx 323 > 5$, $n_1\bar{q} \approx 7599 > 5$, $n_2\bar{p} \approx 300 > 5$, and $n_2\bar{q} \approx 7062 > 5$.
f. $z \approx 5.297$ **g.** Reject H_0
h. There is enough evidence to support the claim.
2a. $H_0: p_1 \leq p_2$; $H_a: p_1 > p_2$ (claim)
b. $\alpha = 0.05$ **c.** $z_0 = 1.645$; $z > 1.645$
d. $\bar{p} = 0.115$; $\bar{q} = 0.885$
e. $n_1\bar{p} \approx 911 > 5$, $n_1\bar{q} \approx 7013 > 5$, $n_2\bar{p} \approx 847 > 5$, and $n_2\bar{q} \approx 6517 > 5$.
f. $z \approx 8.715$ **g.** Reject H_0
h. There is enough evidence to support the claim.

CHAPTER 9

Section 9.1

1ab.

y
Percent donated
10
8
6
4
2
x
40 45 50 55 60 65 70 75
Family income
(in thousands of dollars)

c. Yes, it appears that there is a negative linear correlation. As family income increases, the percent of income donated to charity decreases.

2ab.

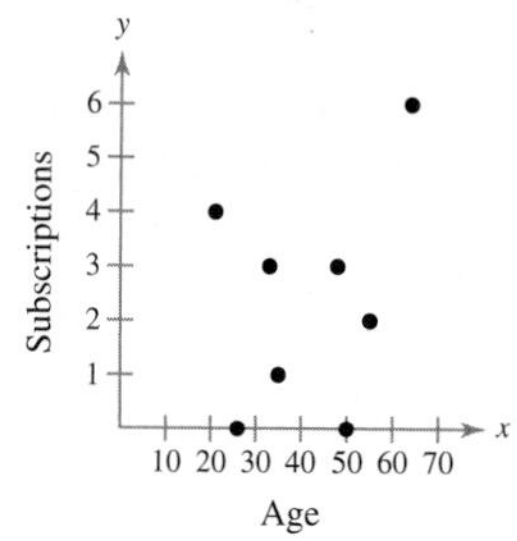

c. No, it appears that there is no correlation between age and subscriptions.

3ab.

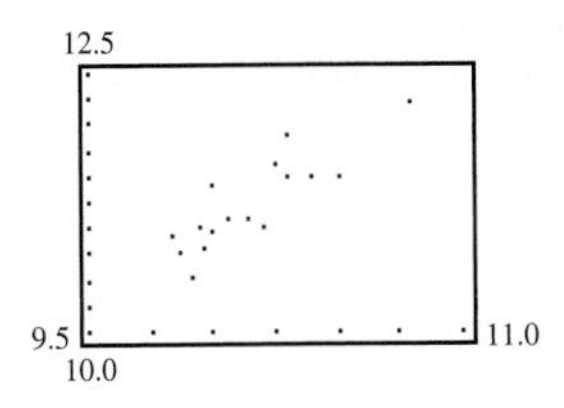

c. Yes, there appears to be a positive linear relationship between men's winning time and women's winning time.

4a. $n = 6$

b.

x	y	xy	x^2	y^2
42	9	378	1764	81
48	10	480	2304	100
50	8	400	2500	64
59	5	295	3481	25
65	6	390	4225	36
72	3	216	5184	9
$\Sigma x = 336$	$\Sigma y = 41$	$\Sigma xy = 2159$	$\Sigma x^2 = 19{,}458$	$\Sigma y^2 = 315$

c. -0.916

d. Since r is close to -1, there appears to be a strong negative linear correlation between income level and donating percent.

5a. Enter data **b.** 0.846

c. Since r is close to 1, there appears to be a strong positive linear correlation between men's winning time and women's winning time.

6a. $n = 6$ **b.** $\alpha = 0.01$ **c.** 0.917

d. $|r| \approx 0.916 < 0.917$; the correlation is not significant.

e. There is not enough evidence to conclude that there is a significant correlation between the income level and the donating percent.

7a. H_0: $\rho = 0$ and H_a: $\rho \neq 0$ **b.** 0.01 **c.** 15

d. ± 2.947; $t < -2.947$ or $t > 2.947$

e. 6.146 **f.** Reject H_0

g. There is enough evidence in the sample to conclude that a significant correlation exists.

Section 9.2

1a. $n = 6$

x	y	xy	x^2
42	9	378	1764
48	10	480	2304
50	8	400	2500
59	5	295	3481
65	6	390	4225
72	3	216	5184
$\Sigma x = 336$	$\Sigma y = 41$	$\Sigma xy = 2159$	$\Sigma x^2 = 19{,}458$

b. $m \approx -0.2134$ **c.** $b \approx 18.7837$

d. $\hat{y} = -0.213x + 18.784$

2a. Enter the data. **b.** $m \approx 1.520$; $b \approx -4.175$

c. $\hat{y} = 1.520x - 4.175$

3a. (1) $\hat{y} = 11.824(2) + 35.301$

(2) $\hat{y} = 11.824(3.32) + 35.301$

b. (1) 58.949 (2) 74.557

c. (1) 58.949 minutes (2) 74.557 minutes

Section 9.3

1a. 0.970 **b.** 0.941

c. 94.1% of the variation in the times is explained. 5.9% of the variation is unexplained.

2a.

x_i	y_i	$\hat{y}_i$	$(y_i - \hat{y}_i)^2$
15	26	28.392	5.722
20	32	35.419	11.690
20	38	35.419	6.662
30	56	49.473	42.602
40	54	63.527	90.764
45	78	70.554	55.443
50	80	77.581	5.852
60	88	91.635	13.213
			$\Sigma = 231.948$

b. 8 **c.** 6.218

d. The standard deviation of the weekly sales for a specific radio ad time is about $621.80.

3a. $n = 8$, d.f. $= 6$, $t_c = 2.447$, $s_e \approx 10.290$

b. 230.884 **c.** 29.236 **d.** $(201.648, 260.120)$

e. You can be 95% confident that the company sales will be between $201,648 and $260,120 when advertising expenditures are $2500.

Section 9.4

1a. Enter data.

b. $\hat{y} = 46.385 + 0.540x_1 - 4.897x_2$

2ab. (1) $\hat{y} = 46.385 + 0.540(89) - 4.897(1)$

(2) $\hat{y} = 46.385 + 0.540(78) - 4.897(3)$

(3) $\hat{y} = 46.385 + 0.540(83) - 4.897(2)$

c. (1) $\hat{y} = 89.548$ (2) $\hat{y} = 73.814$ (3) $\hat{y} = 81.411$

d. (1) 90 (2) 74 (3) 81

CHAPTER 10

Section 10.1

1a.

Music	% of listeners	Expected frequency
Classical	4%	12
Country	36%	108
Gospel	11%	33
Oldies	2%	6
Pop	18%	54
Rock	29%	87

2a. The expected frequencies are 64, 80, 32, 56, 60, 48, 40, and 20, all of which are at least 5.

b. Claimed distribution:

Ages	Distribution
0–9	16%
10–19	20%
20–29	8%
30–39	14%
40–49	15%
50–59	12%
60–69	10%
70+	5%

H_0: Distribution of ages is as shown in table above.

H_a: Distribution of ages differs from the claimed distribution.

c. 0.05 **d.** 7

e. $\chi_0^2 = 14.067$; $\chi^2 > 14.067$

f. $\chi^2 \approx 6.694$ **g.** Fail to reject H_0

h. There is not enough evidence to conclude that the distribution of ages differs from the claimed distribution.

3a. The expected frequencies are 88, 54, and 58, all of which are at least 5.

b. Claimed distribution:

Response	Distribution
In favor of	44%
Against	27%
No opinion	29%

H_0: Distribution of responses is as shown in table above.

H_a: Distribution of responses differs from the claimed distribution.

c. 0.01 **d.** 2

e. $\chi_0^2 = 9.210$; $\chi^2 > 9.210$

f. $\chi^2 \approx 2.924$ **g.** Fail to reject H_0

h. There is not enough evidence to conclude that the distribution of responses differs from the claimed distribution.

Section 10.2

1ab.

	Hotel	Leg room	Rental size	Other	Total
Business	36	108	14	22	180
Leisure	38	54	14	14	120
Total	74	162	28	36	300

c. 300

d.

	Hotel	Leg room	Rental size	Other
Business	44.4	97.2	16.8	21.6
Leisure	29.6	64.8	11.2	14.4

2a. H_0: Travel concern is independent of travel purpose.

H_a: Travel concern is dependent on travel purpose. (claim)

b. 0.01 **c.** 3

d. $\chi_0^2 = 11.345$; $\chi^2 > 11.345$ **e.** $\chi^2 \approx 8.158$

f. Fail to reject H_0

g. There is not enough evidence to conclude that travel concern is dependent on travel purpose.

3a. $\chi_0^2 = 9.488$; $\chi^2 > 9.488$

b. Enter the data.

c. $\chi^2 \approx 65.619$

d. Reject H_0; Yes

Section 10.3

1a. 0.01 **b.** 5.42

2a. 0.01 **b.** 13.27

3a. H_0: $\sigma_1^2 \le \sigma_2^2$; H_a: $\sigma_1^2 > \sigma_2^2$ (claim)

b. 0.01 **c.** 24, 19

d. $F_0 = 2.92$; $F > 2.92$

e. $F \approx 3.214$ **f.** Reject H_0

g. There is enough evidence to support the claim.

4a. H_0: $\sigma_1 = \sigma_2$ (claim); H_a: $\sigma_1 \ne \sigma_2$

b. 0.01 **c.** 15, 21 **d.** $F_0 = 3.43$; $F > 3.43$

e. $F \approx 1.483$ **f.** Fail to reject H_0

g. There is not enough evidence to reject the claim.

Section 10.4

1a. H_0: $\mu_1 = \mu_2 = \mu_3 = \mu_4$

H_a: At least one mean is different from the others.

b. 0.05 **c.** 3, 14

d. $F_0 = 3.34$; $F > 3.34$

e. $F \approx 4.22$ **f.** Reject H_0

g. There is enough evidence to conclude that at least one mean is different from the others.

2a. Enter the data.

b. $F = 1.34$; P-value $= 0.280$

c. Fail to reject H_0

d. There is not enough evidence to conclude that at least one mean is different from the others.

CHAPTER 11

Section 11.1

1a. H_0: median ≤ 2500; H_a: median > 2500 (claim)

b. $\alpha = 0.025$ **c.** $n = 22$

d. CV = 5 **e.** $x = 10$

f. Fail to reject H_0

g. There is not enough evidence to support the claim.

2a. H_0: median $= 134{,}500$ (claim); H_a: median $\neq 134{,}500$

b. $\alpha = 0.10$ **c.** $n = 81$

d. $z_0 = -1.645$

e. $x = 30$

$$z = \frac{(x + 0.5) - 0.5(n)}{\frac{\sqrt{n}}{2}}$$

$$= \frac{(30 + 0.5) - 0.5(81)}{\frac{\sqrt{81}}{2}} = -2.22$$

f. Reject H_0

g. There is enough evidence to reject the claim.

3a. H_0: The number of colds will not decrease.

H_a: The number of colds will decrease. (claim)

b. $\alpha = 0.05$ **c.** $n = 11$

d. CV = 2 **e.** $x = 2$

f. Reject H_0

g. There is enough evidence to support the claim.

Section 11.2

1a. H_0: The water repellent is not effective.

H_a: The water repellent is effective. (claim)

b. $\alpha = 0.01$ **c.** $n = 12$

d. CV = 10 **e.** $w_s = 10.5$

f. Fail to reject H_0

g. There is not enough evidence at the 1% level to support the claim.

2a. H_0: There is no difference in the claims paid by the companies.

H_a: There is a difference in the claims paid by the companies. (claim)

b. $\alpha = 0.05$ **c.** $z_0 = \pm 1.96$

d. $n_1 = 12$ and $n_2 = 12$

e. $R = 120.5$

f. $\mu_R = 150$; $\sigma_R = 17.321$; $z = -1.703$

g. Fail to reject H_0

h. There is not enough evidence to conclude that there is a difference in the claims paid by the companies.

Section 11.3

1a. H_0: There is no difference in the salaries in the three states.

H_a: There is a difference in the salaries in the three states. (claim)

b. $\alpha = 0.10$ **c.** d.f. $= k - 1 = 2$

d. $\chi_0^2 = 4.605 \rightarrow$ Reject H_0 if $\chi^2 > 4.605$

e.

State	Salary	Rank
CT	25.57	1
CA	26.42	2
NJ	29.73	3
CT	29.86	4
CT	30.00	5
CA	33.68	6.5
CT	33.68	6.5
NJ	33.91	8
NJ	34.29	9
NJ	36.35	10
CA	36.55	11
CA	37.18	12
NJ	37.24	13
CT	37.39	14
CA	38.36	15
CA	40.31	16
CA	41.33	17
NJ	43.26	18
NJ	43.89	19
NJ	44.57	20
CT	45.04	21
CT	45.29	22
CA	46.55	23
NJ	46.72	24
CA	47.17	25
CA	48.46	26
NJ	50.16	27
CT	51.03	28
CT	57.07	29
CT	61.46	30

$R_1 = 153.5$

$R_2 = 160.5$

$R_1 = 151$

f. 0.063

g. Fail to reject H_0

h. There is not enough evidence to support the claim.

Section 11.4

1a. H_0: $\rho_s = 0$; H_a: $\rho_s \neq 0$

b. $\alpha = 0.05$

c. CV = 0.738

d.

Oat	Rank	Wheat	Rank	d	d^2
1.49	6	3.72	6	0	0
1.14	1	2.61	1	0	0
1.21	2	3	2	0	0
1.32	4	3.24	3	1	1
1.36	5	3.26	4	1	1
1.22	3	3.45	5	−2	4
1.67	7	4.55	8	−1	1
1.9	8	4.3	7	1	1
					8

$\Sigma d^2 = 8$

e. $r_s = 1 - \dfrac{6\Sigma d^2}{n(n^2 - 1)} = 0.905$

f. Reject H_0

g. There is enough evidence to conclude that a significant correction exists.

APPENDIX A

1a. (1) 0.4857

b. (2) $z = \pm 2.17$

2a.

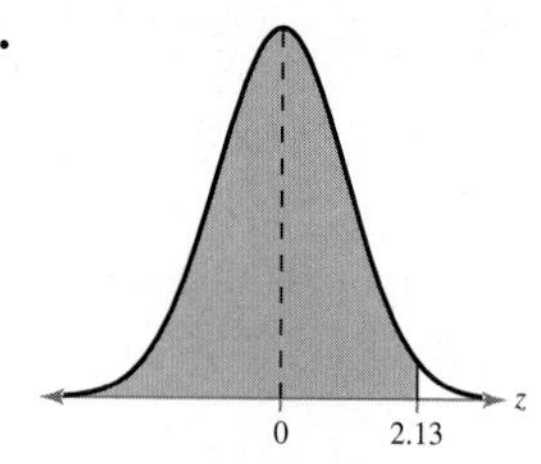

b. 0.4834 **c.** 0.9834

3a.

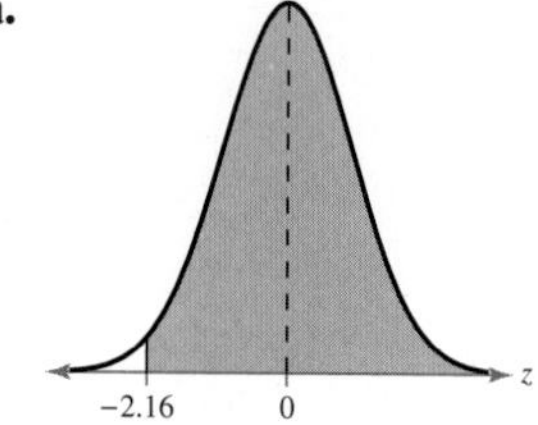

b. 0.4846 **c.** 0.9846

4a.

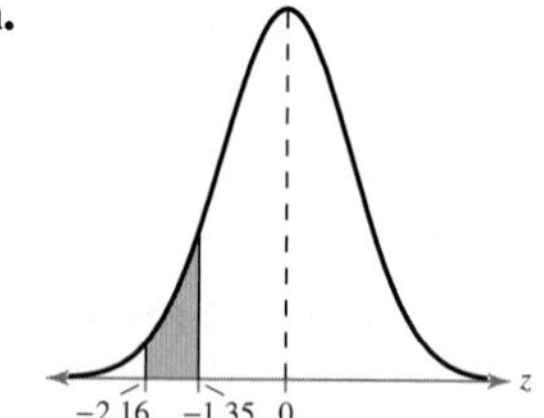

b. 0.4846; 0.4115 **c.** 0.0731

Odd Answers

CHAPTER 1

Section 1.1 *(page 6)*

1. A sample is a subset of a population.

3. False. A statistic is a measure that describes a sample characteristic.

5. True

7. False. A population is the collection of *all* outcomes, responses, measurements, or counts that are of interest.

9. Population, because it is a collection of the ages of all the state governors.

11. Sample, because the collection of the 500 students is a subset within the population.

13. Population: Party of registered voters in Bucks County.

Sample: Party of Bucks County voters responding to phone survey.

15. Population: Ages of adults in the U.S. who own computers.

Sample: Ages of adults in the U.S. who own Dell computers.

17. Population: Collection of all infants in Italy.

Sample: Collection of the 33,043 infants in the study.

19. Population: Collection of all women in the U.S.

Sample: Collection of the 546 U.S. women surveyed.

21. Statistic. The value $57,000 is a numerical description of a sample of annual salaries.

23. Statistic. 10% is a numerical description of a sample of computer users.

25. Statistic. 47% is a numerical description of a sample of people in the United States.

27. The statement "56% are the primary investor in their household" is an application of descriptive statistics.

An inference drawn from the sample is that an association exists between U.S. women and being the primary investor in their household.

29. Answers vary.

Section 1.2 *(page 12)*

1. Nominal and ordinal **3.** True

5. False. Data at the ordinal level are qualitative or quantitative.

7. Qualitative **9.** Quantitative

11. Ordinal. Data can be arranged in order, but the differences between data entries make no sense.

13. Ratio. A ratio of two data values can be formed so one data value can be expressed as a multiple of another.

15. Interval. The data can be ordered and you can calculate meaningful differences between data entries.

17. Ordinal **19.** Nominal

21. Interval data can be ordered and differences between entries can be calculated. Ratio data have all the properties of interval data with the addition that a ratio of two data values can be formed so one data value can be expressed as a multiple of another.

Answers will vary.

Section 1.3 *(page 20)*

1. False. Using stratified sampling guarantees that members of each group within a population will be sampled.

3. False. To select a systematic sample, a population is ordered in some way and then members of the population are selected at regular intervals.

5. Perform an experiment because you want to measure the effect of a treatment on the human digestive system.

7. Use a simulation since the situation is impractical.

9. Simple random sampling is used since each telephone number has an equal chance of being dialed and all samples of 1599 phone numbers have an equal chance of being selected. The sample may be biased because only homes with telephones will be sampled.

11. Convenience sampling is used since the students were chosen due to their convenience of location. Bias may enter into the sample because the students sampled may not be representative of the population of students.

13. Simple random sampling is used since each outpatient had an equal chance of being contacted and all samples of 1819 outpatients had an equal chance of being selected.

15. Stratified sampling is used since a sample is taken from each one-acre subplot.

17. Systematic sampling is used since every twentieth name on a list is being used.

19. Question is biased since it already suggests that drinking fruit juice is good for you. The question might be rewritten as "How does drinking fruit juice affect your health?"

21. The households sampled represent various locations, ethnic groups, and income brackets. Each of these variables is considered a stratum.

23. (a) Advantage: Allows respondent to express some depth and shades of meaning in the answer.

Disadvantage: Not easily quantified and difficult to compare surveys.

(b) Advantage: Easy to analyze results.

Disadvantage: May not provide appropriate alternatives and may influence the opinion of the respondent.

Uses and Abuses for Chapter 1 *(page 25)*

1. Answers will vary.
2. Answers will vary.

Review Answers for Chapter 1 *(page 26)*

1. Population: Collection of all U.S. adult VCR owners.

   Sample: Collection of the 898 U.S. adult VCR owners that were sampled.
3. Population: Collection of all U.S. ATMs.

   Sample: Collection of 860 ATMs that were sampled.
5. Parameter 7. Parameter
9. Quantitative because monthly salaries are numerical measurements.
11. Quantitative because ages are numerical measurements.
13. Interval. It makes no sense saying that 100 degrees is twice as hot as 50 degrees.
15. Nominal. The data is categorical and cannot be arranged in a meaningful order.
17. Take a census since judges keep accurate records of charitable donations.
19. Perform an experiment since you want to measure the effect of a treatment on chrysanthemums.
21. Simple random sampling is used because random telephone numbers were generated and called.
23. Cluster sampling is used because each community is considered a cluster.
25. Stratified sampling is used because 25 students are randomly selected from each grade level.
27. Telephone sampling only samples individuals who have telephones, are available, and are willing to respond.
29. The selected communities may not be representative of the entire area.

Real Statistics–Real Decisions for Chapter 1
(page 28)

1. (a) If the survey identifies the type of reader that is responding, then stratified sampling ensures that representatives from each group of readers (engineers, manufacturers, researchers, and developers) are included in the sample.

   (b) Yes (c) Use sampling and surveys

   (d) You may take too large of a percentage of your sample from a subgroup of the population that is relatively small.
2. (a) Fuel–qualitative; Percent responding–quantitative

   (b) Fuel–nominal; Percent responding–ratio
   Time–ordinal; Percent responding–ratio

   (c) Sample (d) Statistics
3. (a) Answers will vary.

   (b) Answers will vary

Chapter Quiz for Chapter 1 *(page 29)*

1. Population: Collection of all individuals with sleep disorders.

   Sample: Collection of 163 patients in study.
2. (a) Statistic (b) Parameter
3. (a) Qualitative (b) Quantitative
4. (a) Nominal because no mathematical computations can be made.

   (b) Ratio because one data value can be expressed as a multiple of another.

   (c) Interval because meaningful differences between entries can be calculated, but a zero entry is not an inherent zero.
5. (a) Perform an experiment because you want to measure the effect a treatment has on patients.

   (b) Use sampling because it would be impossible to question everyone in the population.
6. (a) Convenience sample because all of the people sampled are in one convenient location.

   (b) Systematic sample because every fifth part is sampled.

   (c) Stratified sample because the population is first stratified and then a sample is collected from each strata.
7. Convenience
8. (a) False. A statistic is a numerical measure that describes a sample characteristic.

   (b) False. Ratio data represent the highest level of measurement.

CHAPTER 2

Section 2.1 *(page 41)*

1. By organizing the data into a frequency distribution, patterns within the data may become more evident.
3. False. The midpoint of a class is the sum of the lower and upper limits of the class divided by two.
5.

Class	f	Midpoint	Relative frequency	Cumulative frequency
20–29	10	24.5	0.01	10
30–39	132	34.5	0.13	142
40–49	284	44.5	0.29	426
50–59	300	54.5	0.30	726
60–69	175	64.5	0.18	901
70–79	65	74.5	0.07	966
80–89	25	84.5	0.03	991
	$\Sigma f = 991$		$\Sigma \frac{f}{n} = 1$	

7. (a) Least frequency ≈ 10

   (b) Greatest frequency ≈ 300

   (c) Class width $= 10$

9. (a) 50 (b) 12.5–13.5 lb (c) 24 (d) 19.5 lb

11. (a) Class with greatest relative frequency: 8–9 in.
Class with least relative frequency: 17–18 in.
(b) Greatest relative frequency ≈ 0.195
Least relative frequency ≈ 0.005
(c) Approximately 0.015

13. Class with greatest frequency: 500–550
Class with least frequency: 250–300 or 700–750

15.

Class	Frequency, f	Midpoint	Relative frequency	Cumulative frequency
0–7	8	3.5	0.32	8
8–15	8	11.5	0.32	16
16–23	3	19.5	0.12	19
24–31	3	27.5	0.12	22
32–39	3	35.5	0.12	25
	$\Sigma f = 25$		$\Sigma \frac{f}{n} = 1$	

17.

Class	Frequency, f	Midpoint	Relative frequency	Cumulative frequency
1000–2019	12	1509.5	0.5455	12
2020–3039	3	2529.5	0.1364	15
3040–4059	2	3549.5	0.0909	17
4060–5079	3	4569.5	0.1364	20
5080–6099	1	5589.5	0.0455	21
6100–7119	1	6609.5	0.0455	22
	$\Sigma f = 22$		$\Sigma \frac{f}{N} \approx 1$	

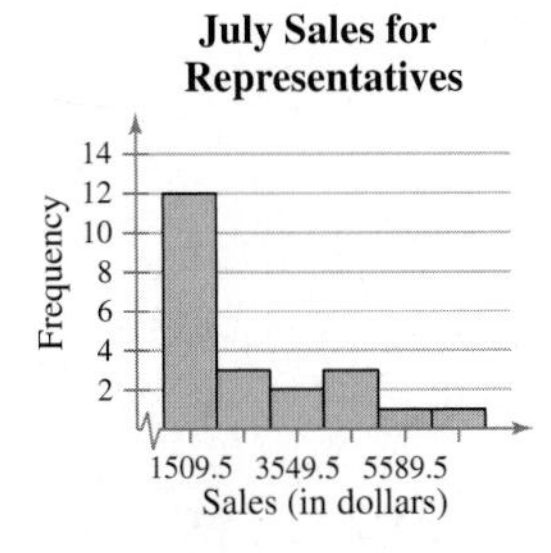

Class with greatest frequency: 1000–2019
Class with least frequency: 5080–6099; 6100–7119

19.

Class	Frequency, f	Midpoint	Relative frequency	Cumulative frequency
291–318	4	304.5	0.1818	4
319–346	3	332.5	0.1364	7
347–374	2	360.5	0.0909	9
375–402	4	388.5	0.1818	13
403–430	3	416.5	0.1364	16
431–458	3	444.5	0.1364	19
459–486	1	472.5	0.0455	20
487–514	2	500.5	0.0909	22
	$\Sigma f = 22$		$\Sigma \frac{f}{n} \approx 1$	

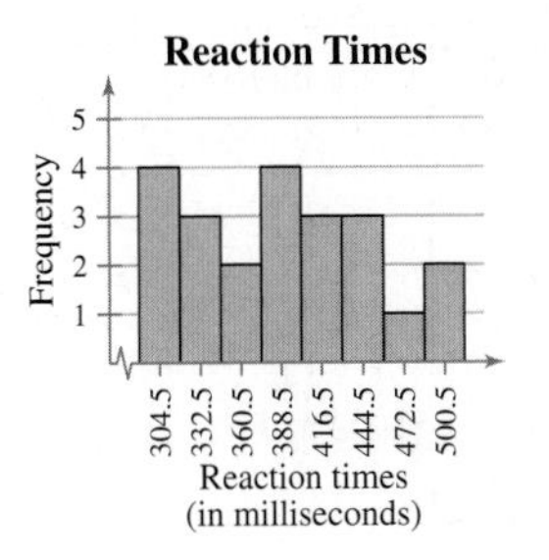

Class with greatest frequency: 291–318; 375–402
Class with least frequency: 459–486

21.

Class	Frequency, f	Midpoint	Relative frequency	Cumulative frequency
146–169	6	157.5	0.2308	6
170–193	9	181.5	0.3462	15
194–217	3	205.5	0.1154	18
218–241	6	229.5	0.2308	24
242–265	2	253.5	0.0769	26
	$\Sigma f = 26$		$\Sigma \frac{f}{n} \approx 1$	

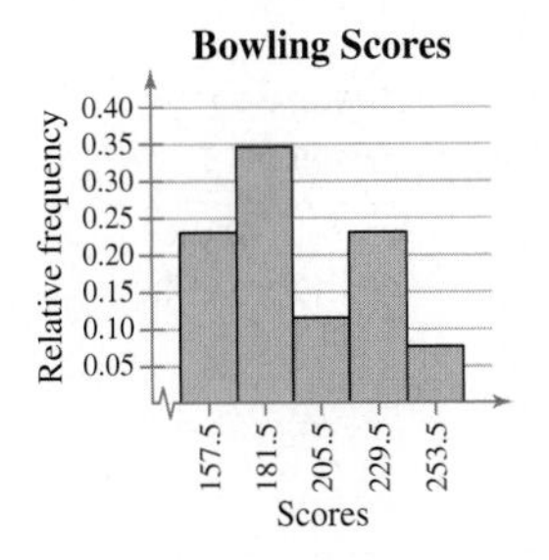

Class with greatest relative frequency: 170–193
Class with least relative frequency: 242–265

23.

Class	Frequency, f	Midpoint	Relative frequency	Cumulative frequency
33–36	8	34.5	0.3077	8
37–40	6	38.5	0.2308	14
41–44	5	42.5	0.1923	19
45–48	2	46.5	0.0769	21
49–52	5	50.5	0.1923	26
	$\Sigma f = 26$		$\Sigma \frac{f}{n} = 1$	

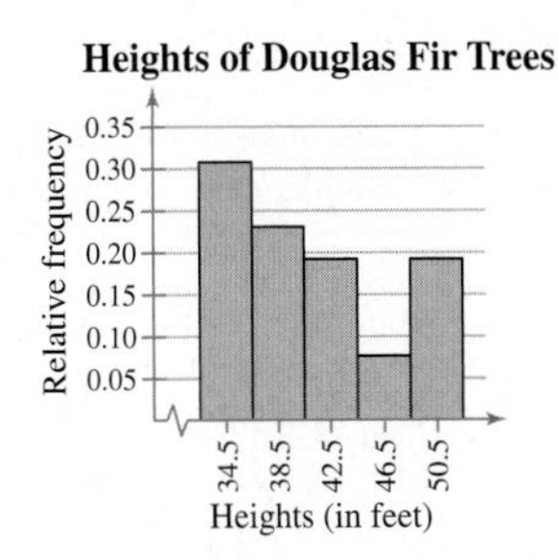

Class with greatest relative frequency: 33–36
Class with least relative frequency: 45–48

25.

Class	Frequency, f	Relative frequency	Cumulative frequency
50–53	1	0.0417	1
54–57	0	0.0000	1
58–61	4	0.1667	5
62–65	9	0.3750	14
66–69	7	0.2917	21
70–73	3	0.1250	24
	$\Sigma f = 24$	$\Sigma \frac{f}{n} = 1$	

Location of the greatest increase in frequency: 62–65

27.

Class	Frequency, f	Relative frequency	Cumulative frequency
2–4	9	0.3214	9
5–7	6	0.2143	15
8–10	7	0.2500	22
11–13	3	0.1071	25
14–16	2	0.0714	27
17–19	1	0.0357	28
	$\Sigma f = 28$	$\Sigma \frac{f}{n} \approx 1$	

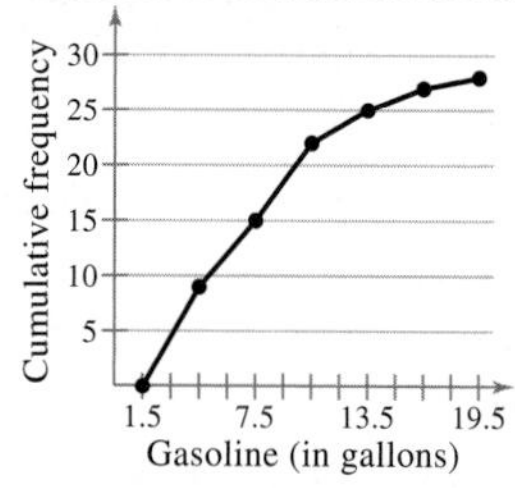

Location of the greatest increase in frequency: 2–4

29.

Class	Frequency, f	Midpoint	Relative frequency	Cumulative frequency
47–57	1	52	0.05	1
58–68	1	63	0.05	2
69–79	5	74	0.25	7
80–90	8	85	0.4	15
91–101	5	96	0.25	20
	$\Sigma f = 20$		$\Sigma \frac{f}{N} = 1$	

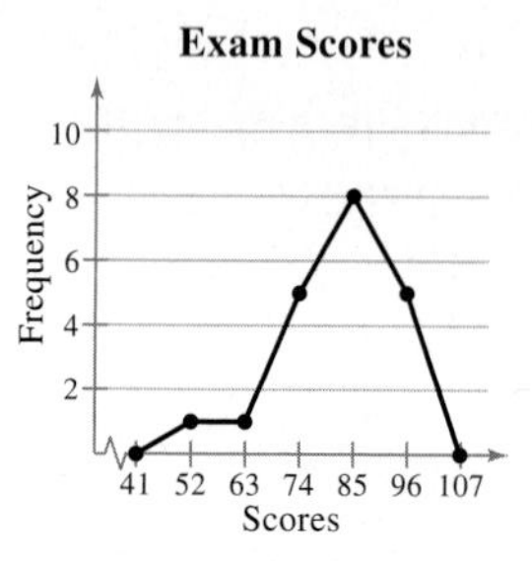

Class with greatest frequency: 80–90

Class with least frequency: 47–57 and 58–68

31. (a)

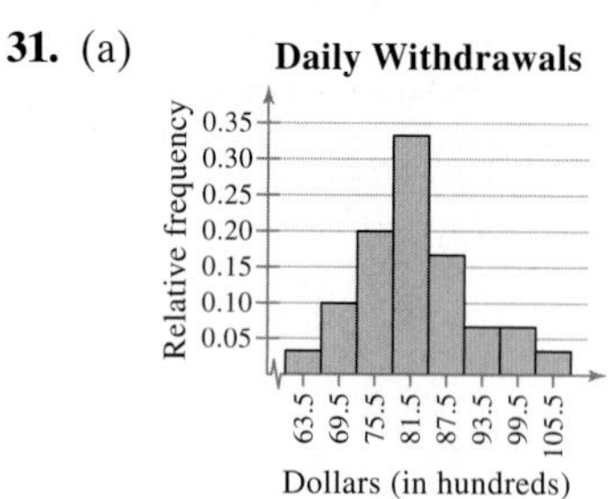

(b) \$9600, since the sum of the relative frequencies for the last two classes is 0.10.

(c) 16.7%, since the sum of the relative frequencies for the last three classes is 0.167.

33.

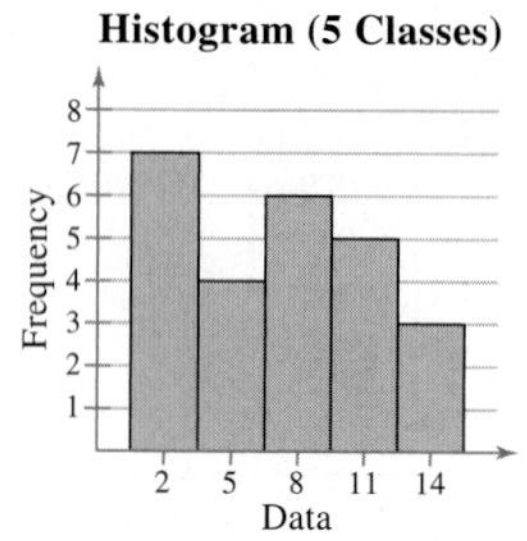

Histogram (10 Classes)

Frequency

1.5 5.5 9.5 13.5 17.5

Data

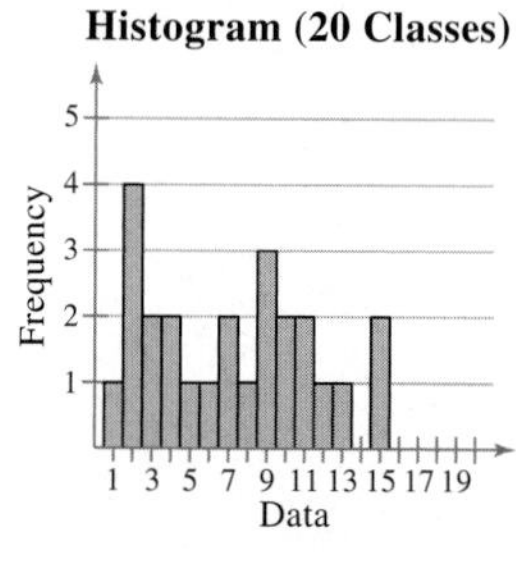

In general, a greater number of classes better-preserves the actual values of the data set, but is not as helpful for observing general trends and making conclusions. When choosing the number of classes, an important consideration is the size of the data set. For instance, you would not want to use 20 classes if your data set contained 20 entries. In this particular example, as the number of classes increases, the histogram shows more fluctuation. The histograms with 10 and 20 classes have classes with zero frequencies. Not much is gained by using more than five classes. Therefore, it appears that five classes would be best. (Answers will vary.)

Section 2.2 *(page 53)*

1. Quantitative: Stem-and-Leaf Plot, Dot Plot, Histogram, Scatter Plot, Time Series Chart

Qualitative: Pie Chart, Pareto Chart

3. a **5.** b

7. 27, 32, 41, 43, 43, 44, 47, 47, 48, 50, 51, 51, 52, 53, 53, 53, 54, 54, 54, 54, 55, 56, 56, 58, 59, 68, 68, 68, 73, 78, 78, 85

Max: 85 Min: 27

9. 13, 13, 14, 14, 14, 15, 15, 15, 15, 15, 16, 17, 17, 18, 19

Max: 19 Min: 13

11. Amheuser-Busch spends the most on advertising and Ford spends the least. (Answers will vary.)

13. Tailgaters irk drivers the most while too cautious drivers irk drivers the least. (Answers will vary.)

15. Key: 3|3 = 33

3	233459
4	01134556678
5	133
6	0069

Most elephants tend to drink less than 55 gallons of water per day. (Answers will vary.)

17. Key: 17|5 = 17.5

16	48
17	113455679
18	13446669
19	0023356
20	18

It appears that most farmers charge 17 to 19 cents per pound of apples. (Answers will vary.)

19.

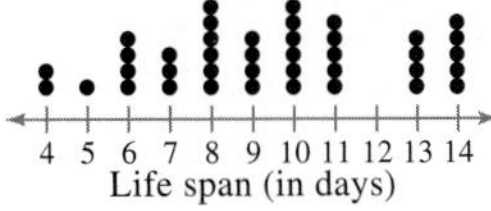

It appears that the life span of a fly tends to be between 4 and 14 days. (Answers will vary.)

21.

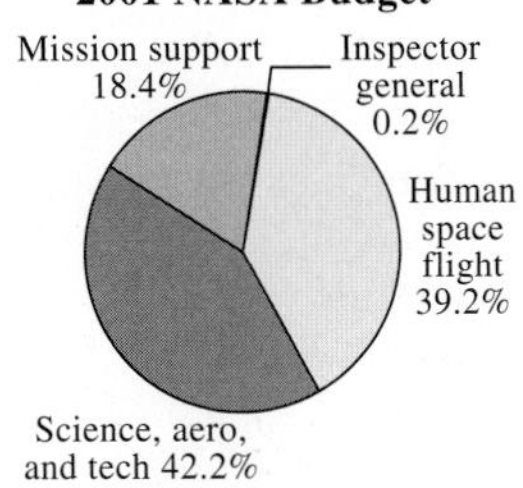

It appears that 42.2% of NASA's 2001 budget went to science aeronautics, and technology. (Answers will vary.)

23.

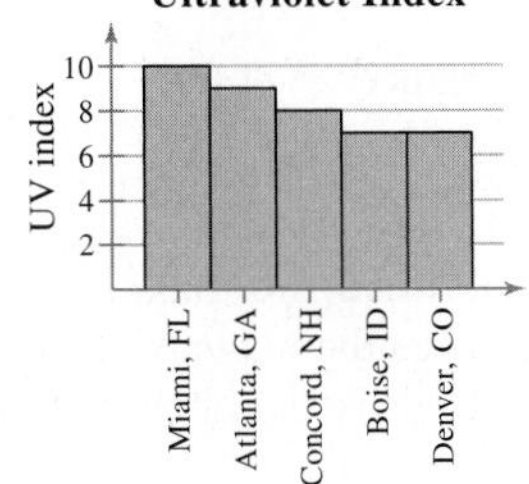

It appears that Boise, ID and Denver, CO have the same UV index. (Answers will vary.)

25.

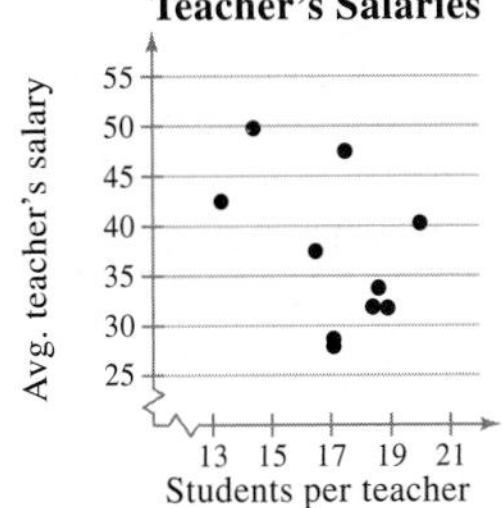

It appears that a teacher's average salary decreases as the number of students per teacher increases. (Answers will vary.)

27.

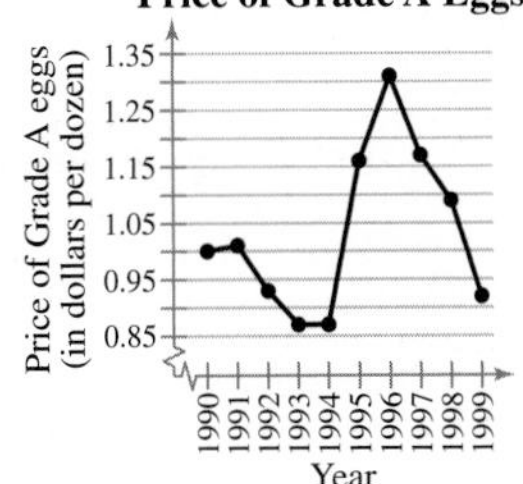

It appears that in 1996, the price for Grade A eggs was the highest. (Answers will vary.)

29. (a) When data are taken at regular intervals over a period of time, a time series chart should be used. (Answers will vary.)

(b)

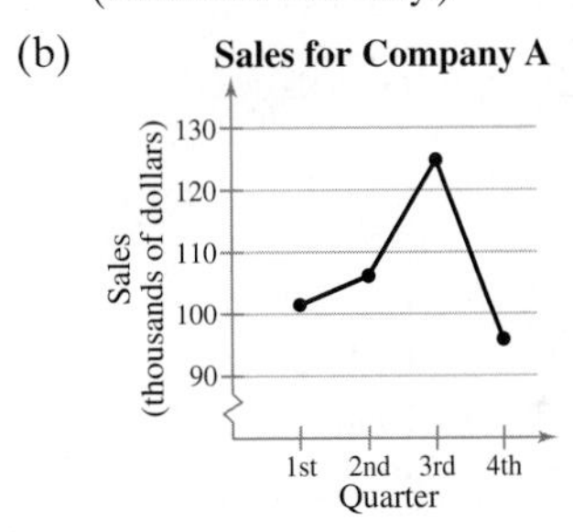

Section 2.3 *(page 64)*

1. False. The mean is the measure of central tendency most likely to be affected by an extreme value (or outlier).

3. False. All quantitative data sets have a median.

5. A data set with an outlier within it would be an example. (Answers will vary.)

7. The shape of the distribution is skewed right since the bars have a "tail" to the right.

9. The shape of the distribution is uniform since the bars are approximately the same height.

11. (9), since the distribution of values ranges from 1 to 12 and has (approximately) equal frequencies.

13. (10), since the distribution has a maximum value of 90 and is skewed left due to a few students scoring much lower than the majority of the students.

15. (a) $\bar{x} \approx 6.23$

median = 6

mode = 5

(b) Median, since the distribution is skewed.

17. (a) $\bar{x} \approx 4.57$
median = 4.8
mode = 4.8
(b) Median, since there are outliers present.

19. (a) $\bar{x} = 97$
median = 97.2
mode = 94.8, 95.4, 97.2, 103.1
(b) Median, since the distribution is skewed.

21. (a) $\bar{x}$ = not possible
median = not possible
mode = "Worse"
(b) Mode, since the data is at the nominal level of measurement.

23. (a) $\bar{x} \approx 170.63$
median = 169.3
mode = not possible
(b) Mean, since there are no outliers.

25. (a) $\bar{x} = 22.6$
median = 19
mode = 14
(b) Median, since the distribution is skewed.

27. (a) $\bar{x} \approx 14.11$
median = 14.25
mode = 2.5
(b) Mean, since there are no outliers.

29. A = mode, since it's the highest bar.
B = median, since the distribution is skewed right.
C = mean, since the distribution is skewed right.

31. 89.3 **33.** 2.8 **35.** 65.5 **37.** 35.01

39.

Class	Frequency, f	Midpoint
3–4	3	3.5
5–6	8	5.5
7–8	4	7.5
9–10	2	9.5
11–12	2	11.5
13–14	1	13.5
	$\Sigma f = 20$	

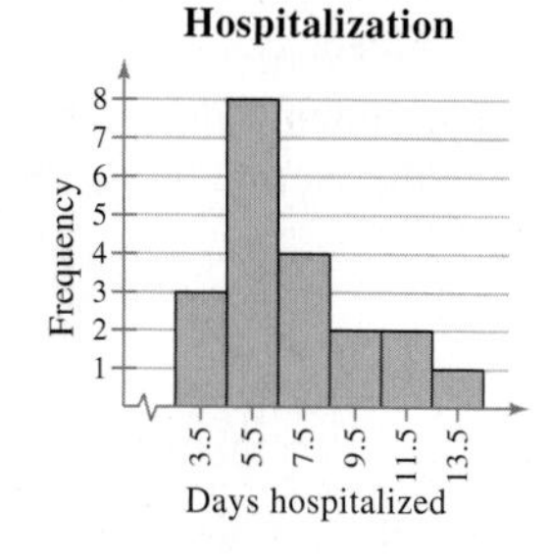

Positively skewed

41.

Class	Frequency, f	Midpoint
62–64	3	63
65–67	7	66
68–70	9	69
71–73	8	72
74–76	3	75
	$\Sigma f = 30$	

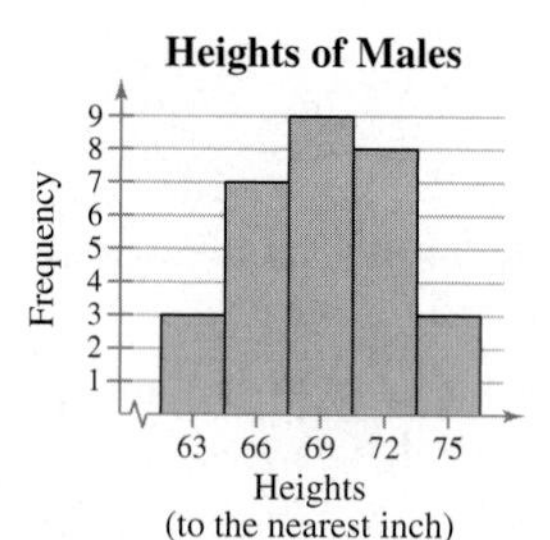

Symmetric

43. (a) $\bar{x} = 6.005$
median = 6.01
(b) $\bar{x} = 5.945$
median = 6.01
(c) Mean

45. (a) Mean, since Car A has the highest mean of the three.
(b) Median, since Car B has the highest median of the three.
(c) Mode, since Car C has the highest mode of the three.

47. (a) $\bar{x} = 49.2\overline{3}$
(b) median = 46.5
(c) Key: 3|6 = 36
(d) Positively skewed

```
1 | 1 3
2 | 2 8
3 | 6 6 6 7 7 7 8
4 | 1 3 4 6 7 ——— mean
5 | 1 1 1 3        median
6 | 1 2 3 4
7 | 2 2 4 6
8 | 5
9 | 0
```

49. Two different symbols are needed since they describe a measure of central tendency for two different sets of data (sample is a subset of the population).

Section 2.4 *(page 80)*

1. range = 7, mean = 8.1, variance = 5.69, standard deviation ≈ 2.39

3. range = 14, mean ≈ 11.11, variance ≈ 21.61, standard deviation ≈ 4.65

5. 73

7. The range is the difference between the maximum and minimum values of a data set. The advantage of the range is that it is easy to calculate. The disadvantage is that it uses only two entries from the data set.

9. The units of variance are squared. Its units are meaningless. (Example: dollars^2)

11. (a) Range = 25.1

(b) Range = 45.1

(c) Changing the maximum value of the data set greatly affects the range.

13. Graph (a) has a standard deviation of 2.4 and graph (b) has a standard deviation of 5. Graph (b) has more variability.

15. Company B

17. (a) 17.6, 37.35, 6.11

8.7, 8.71, 2.95

(b) It appears from the data that the annual salaries in LA are more variable than the salaries in Long Beach.

19. (a) M: range = 405

$s^2 \approx 16{,}225.27$

$s \approx 127.38$

F: range = 552

$s^2 \approx 34{,}575.14$

$s \approx 185.94$

(b) It appears from the data that the SAT scores for females are more variable than the SAT scores for males.

21. (a) Greatest sample standard deviation: (ii)

Data set (ii) has more entries that are farther away from the mean.

Least same standard deviation: (iii)

Data set (iii) has more entries that are close to the mean.

(b) The three data sets have the same mean but have different standard deviations.

23. Similarity: Both estimate proportions of the data contained within k standard deviations of the mean.

Difference: The Empirical Rule assumes the distribution is bell shaped; Chebychev's Theorem makes no such assumption.

25. 68% **27.** (a) 51 (b) 17 **29.** 24

31. $\bar{x} = 2.075$

$s \approx 1.328$

33. Class width $= \dfrac{\text{max} - \text{min}}{5} = \dfrac{13 - 1}{5} = 2.4 \rightarrow 3$

Class	f	Midpoint	xf	$x - \mu$	$(x - \mu)^2$	$(x - \mu)^2 f$
1–3	3	2	6	−6	36	108
4–6	7	5	35	−3	9	63
7–9	9	8	72	0	0	0
10–12	11	11	121	3	9	99
13–15	1	14	14	6	36	36
	31		248			306

$$\mu = \frac{\Sigma xf}{N} = \frac{248}{31} = 8$$

$$\sigma = \sqrt{\frac{\Sigma(x - \mu)^2 f}{N}} = \sqrt{\frac{306}{31}} \approx 3.14$$

35.

f	Mid-point	xf	$x - \bar{x}$	$(x - \bar{x})^2$	$(x - \bar{x})^2 f$
1	70.5	70.5	−44	1936	1936
12	92.5	1110	−22	484	5808
25	114.5	2862.5	0	0	0
10	136.5	1365	22	484	4840
2	158.5	317	44	1936	3872
$n = 50$		$\Sigma xf = 5725$			$\Sigma(x - \bar{x})^2 f = 16{,}456$

$$\bar{x} = \frac{\Sigma xf}{n} = \frac{5725}{50} = 114.5$$

$$s = \sqrt{\frac{(x - \bar{x})^2 f}{n - 1}} = \sqrt{\frac{16{,}456}{49}} \approx 18.33$$

37.

Class	Midpoint, x	f	xf	$x - \bar{x}$	$(x - \bar{x})^2$	$(x - \bar{x})^2 f$
0–4	2	19.3	38.6	−34.44	1186.1	22,891.7
5–13	9	35.3	317.7	−27.44	753	26,580.9
14–17	15.5	17.2	266.6	−20.94	438.5	7542.2
18–24	21	28.7	602.7	−15.44	238.4	6842.1
25–34	29.5	36.9	1088.55	−6.94	48.2	1778.6
35–44	39.5	42.3	1670.85	3.06	9.4	397.6
45–64	54.5	73.6	4011.2	18.06	326.2	24,008.3
65+	70	36.8	2576	33.56	1126.3	41,447.8
		290.1	10,572.2			131,489.2

$$\bar{x} = \frac{\Sigma xf}{n} = \frac{10{,}572.2}{290.1} \approx 36.44$$

$$s = \sqrt{\frac{\Sigma(x - \bar{x})f}{n - 1}} \approx \sqrt{\frac{131{,}489.2}{289.1}} \approx 21.33$$

39. $CV_{\text{heights}} = \dfrac{3.44}{72.75} \cdot 100 \approx 4.73$

$CV_{\text{weights}} = \dfrac{18.47}{187.83} \cdot 100 \approx 9.83$

It appears that weight is more variable than height.

41. (a) $\bar{x} = 550,\ s \approx 302.765$

(b) $\bar{x} = 5500,\ s \approx 3027.65$

(c) $\bar{x} = 55,\ s \approx 30.2765$

(d) By multiplying each entry by a constant k, the new sample mean is $k \cdot \bar{x}$, and the new sample standard deviation is $k \cdot s$.

43. (a) $P = \dfrac{3(17 - 19)}{2.3} \approx -2.61$; skewed left

(b) $P = \dfrac{3(32 - 25)}{5.1} \approx 4.12$; skewed right

Section 2.5 *(page 93)*

1. (a) $Q_1 = 4.5$, $Q_2 = 6$, $Q_3 = 7.5$

(b)

1 4.5 6 7.5 9

0 1 2 3 4 5 6 7 8 9

3. The basketball team scored more points per game than 75% of the teams in the league.

5. The student scored above 63% of the students who took the ACT placement test.

7. True

9. (a) Min = 10
(b) Max = 21
(c) $Q_1 = 13$
(d) $Q_2 = 15$
(e) $Q_3 = 17$
(f) IQR = 4

11. (a) Min = 900
(b) Max = 2100
(c) $Q_1 = 1250$
(d) $Q_2 = 1500$
(e) $Q_3 = 1950$
(f) IQR = 700

13. (a) Min = −1.9
(b) Max = 2.1
(c) $Q_1 = -0.5$
(d) $Q_2 = 0.1$
(e) $Q_3 = 0.7$
(f) IQR = 1.2

15. $Q_1 = \text{B}$, $Q_2 = \text{A}$, $Q_3 = \text{C}$, since about one quarter of the data falls on or below 17, 18.5 is the median of the entire data set, and about three quarters of the data falls on or below 20.

17. (a) 2, 4, 5
(b)

Watching Television

0 2 4 5 9

0 1 2 3 4 5 6 7 8 9

Hours

19. (a) 5
(b) 50%
(c) 25%

21. $z = 0$; B
$z = 2.14$; C
$z = -1.43$; A
A z-score of 2.14 would be unusual.

23. (a) Statistics: $z = \dfrac{73 - 63}{7} \approx 1.43$

Biology: $z = \dfrac{26 - 23}{3.9} \approx 0.77$

(b) The student did better on the statistics test.

25. (a) Statistics: $z = \dfrac{78 - 63}{7} \approx 2.14$

Biology: $z = \dfrac{29 - 23}{3.9} \approx 1.54$

(b) The student did better on the statistics test.

27. (a) $z_1 = \dfrac{34{,}000 - 35{,}000}{2250} \approx -0.44$

$z_2 = \dfrac{37{,}000 - 35{,}000}{2250} \approx 0.89$

$z_3 = \dfrac{31{,}000 - 35{,}000}{2250} \approx -1.78$

None of the selected tires have unusual life spans.

(b) For 30,500, 2.5th percentile
For 37,250, 84th percentile
For 35,000, 50th percentile

29. About 67 inches; 20% of the heights are below 67 inches.

31. $z_1 = \dfrac{74 - 69.2}{2.9} \approx 1.66$

$z_2 = \dfrac{62 - 69.2}{2.9} \approx -2.48$

$z_3 = \dfrac{80 - 69.2}{2.9} \approx 3.72$

The heights that are 62 and 80 inches are unusual.

33. (a) 42, 49, 56
(b)

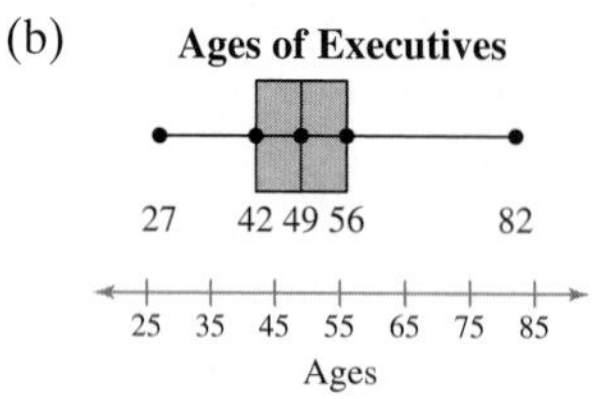

(c) Half of the ages are between 42 and 56 years.
(d) 49, since half of the executives are older and half are younger.

Uses and Abuses for Chapter 2 *(page 101)*

1. Answers will vary.

2. The salaries of employees at a business could contain an outlier.

The median is not affected by an outlier because the median does not take into account the outlier's numerical value.

Review Answers for Chapter 2 *(page 102)*

1.

Class	Midpoint	Boundaries	Frequency, f	Rel freq	Cum freq
20–23	21.5	19.5–23.5	1	0.05	1
24–27	25.5	23.5–27.5	2	0.10	3
28–31	29.5	27.5–31.5	6	0.30	9
32–35	33.5	31.5–35.5	7	0.35	16
36–39	37.5	35.5–39.5	4	0.20	20
			$\Sigma f = 20$	$\Sigma \frac{f}{n} = 1$	

3.

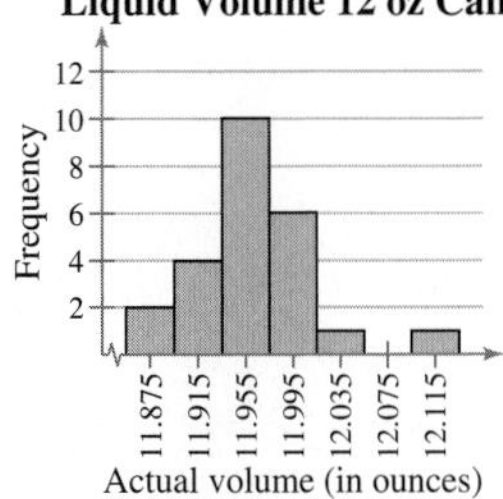

5.

Class	Midpoint	Frequency, f
79–93	86	9
94–108	101	12
109–123	116	5
124–138	131	3
139–153	146	2
154–168	161	1
		$\Sigma f = 32$

7.

1	3789
2	012333445557889
3	11234578
4	347
5	1

9.

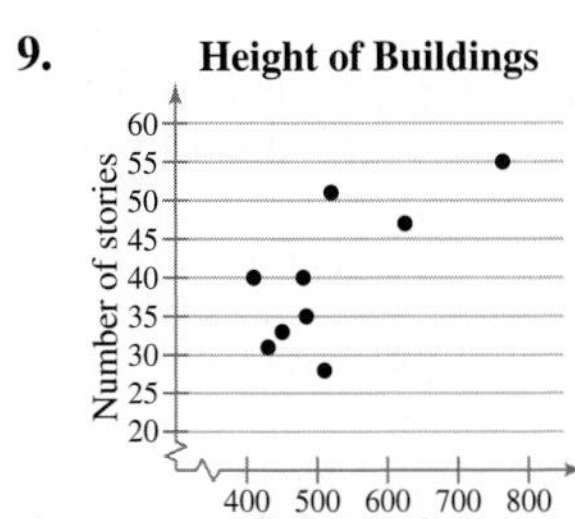

The number of stories appears to increase with height.

11.

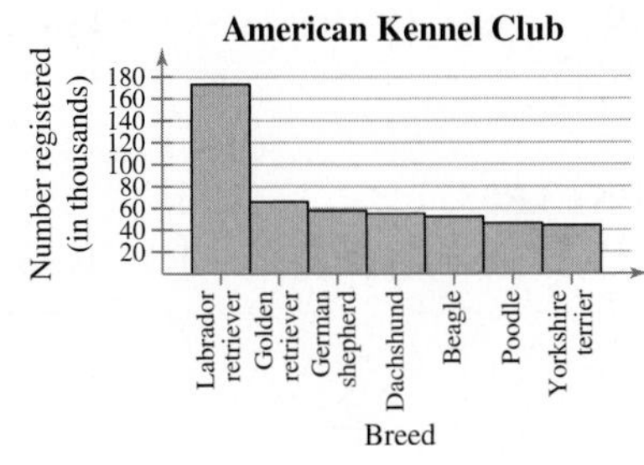

13. mean = 8.6
median = 9
mode = 9

15. 31.7

17. 79.6

19. skewed

21. skewed left

23. median

25. range = 2.8

27. population mean = 9
standard deviation ≈ 3.19

29. sample mean = 2453.4
standard deviation ≈ 306.10

31. Between \$21.50 and \$36.50

33. 30

35. sample mean ≈ 2.48
standard deviation ≈ 1.24

37. $Q_1 = 56$

39. IQR = 14

41. IQR = 4

43. 23% scored higher than 68.

45. $z \approx 2.46$, unusual

47. $z = 1.125$

Real Statistics–Real Decisions for Chapter 2

(page 106)

1. (a) Find the average price of gas for each city and do a comparison.

(b) Find the mean, range, and population standard deviation for each city.

2. (a) Construct a Pareto chart because the data in use are quantitative and a Pareto chart positions data in order of decreasing height, with the tallest bar positioned at the left.

(b)

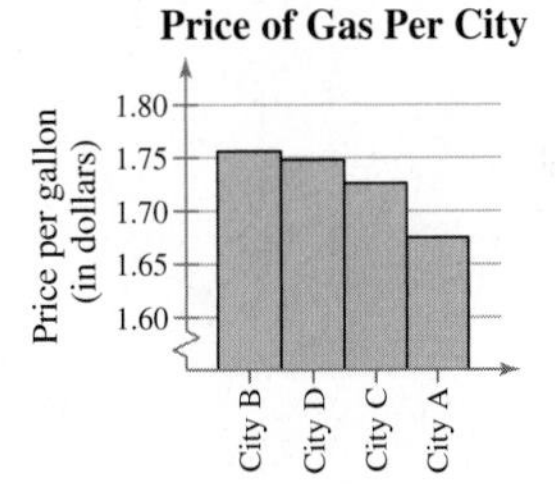

(c) Yes

3. (a) Find the mean, range, and population standard deviation for each city.

(b)

City A	*City B*
$\bar{x} = 1.675$	$\bar{x} = 1.756$
$\sigma \approx 0.03$	$\sigma \approx 0.02$
range = 0.09	range = 0.07
City C	*City D*
$\bar{x} = 1.726$	$\bar{x} = 1.748$
$\sigma \approx 0.03$	$\sigma \approx 0.03$
range = 0.08	range = 0.08

(c) Yes

4. (a) Tell your readers that on average, the cost for gas is higher in this city than in other cities.
   (b) Location, national inflation of gas prices, store's economic condition

Chapter Quiz for Chapter 2 *(page 107)*

1. (a)

Class limits	Midpoint	Class boundaries	Frequency, f	Rel freq	Cum freq
101–112	106.5	100.5–112.5	3	0.12	3
113–124	118.5	112.5–124.5	11	0.44	14
125–136	130.5	124.5–136.5	7	0.28	21
137–148	142.5	136.5–148.5	2	0.08	23
149–160	154.5	148.5–160.5	2	0.08	25

(b) Frequency histogram and polygon

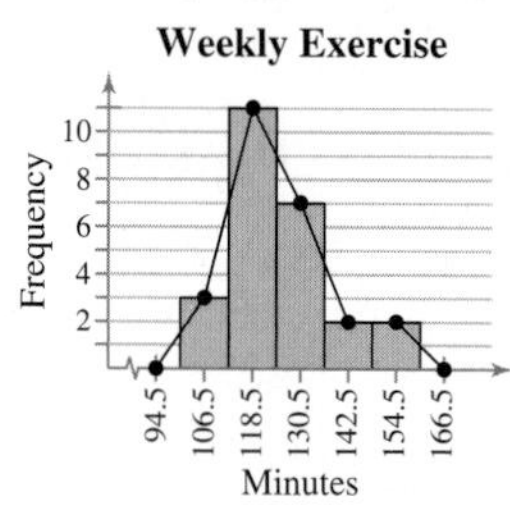

(c) Relative Frequency Histogram

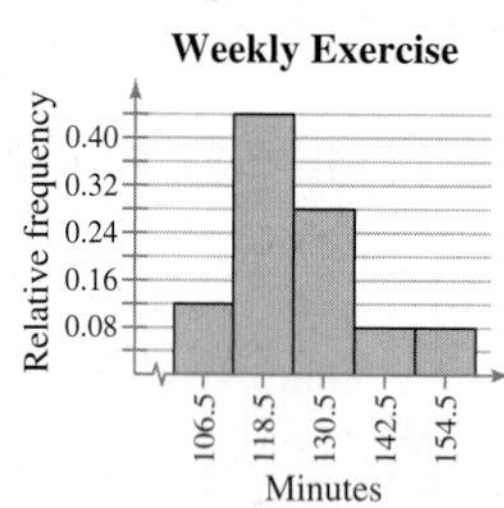

(d) Skewed

(e)

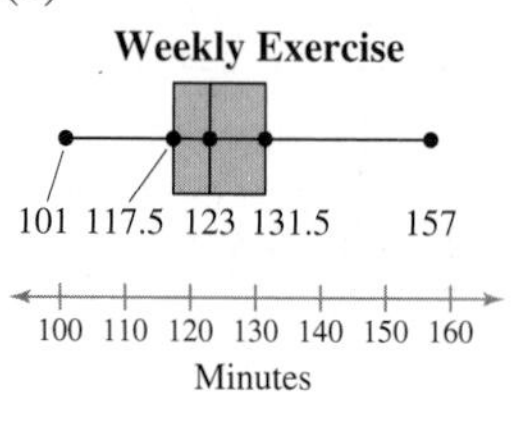

(f)

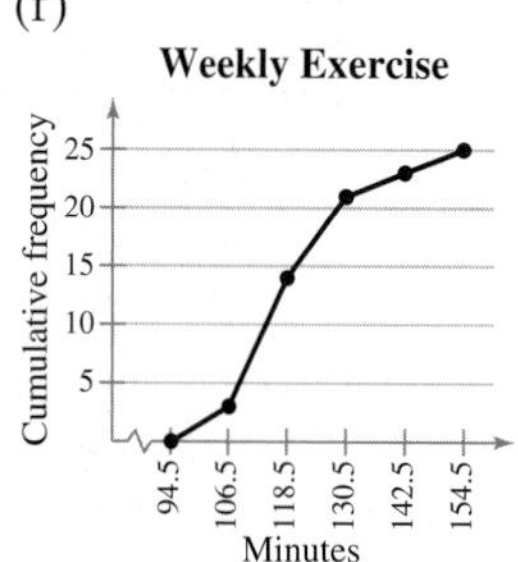

2. 125.22, 13.00

3. (a)

U.S. Sporting Goods

Footwear 13%
Recreational transport 34%
Clothing 22%
Equipment 31%

(b)

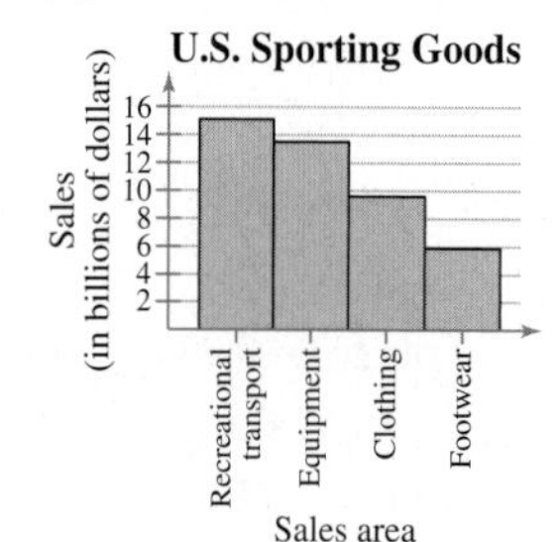

4. (a) 751.63, 784.5, none

   The mean best describes a typical salary because there are no outliers.

   (b) 575, 48135.13, 219.40

5. \$125,000 and \$185,000

6. (a) $z = 3.0$, unusual
   (b) $z \approx -6.67$, unusual
   (c) $z \approx 1.33$
   (d) $z = -2.2$, unusual

7. (a) $Q_1 = 72$, $Q_2 = 82$, $Q_3 = 90$
   (b) IQR $= 18$
   (c)

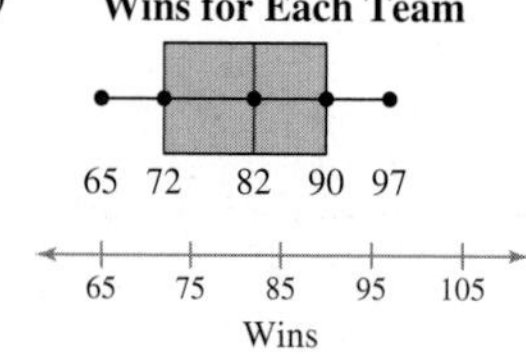

CHAPTER 3

Section 3.1 *(page 117)*

1. (a) Yes, the probability of an event occurring must be contained in the interval [0, 1] or [0%, 100%].
   (b) No, the probability of an event cannot be greater than 1.
   (c) No, the probability of an event cannot be less than 0.
   (d) Yes, the probability of an event occurring must be contained in the interval [0, 1] or [0%, 100%].
   (e) Yes, the probability of an event occurring must be contained in the interval [0, 1] or [0%, 100%].

3. $\{0, 1, 2, 3, 4, 5, 6, 7, 8, 9\}$

5. $\{(P, A), (P, B), (P, AB), (P, O), (N, A), (N, B), (N, AB), (N, O)\}$ where (P, A) represents positive Rh-factor with A-blood type and (N, A) represents negative Rh-factor with A-blood type.

7. Simple event because it is an event that consists of a single outcome.

9. Empirical probability since company records were used to calculate the frequency of a washing machine breaking down.

11. 0.52 **13.** 0.159 **15.** 0.000953 **17.** 0.072

19. 0.944 **21.** (a) 0.5 (b) 0.25 (c) 0.25

23. 0.747 **25.** 0.253

27. The probability of choosing a tea drinker who does not have a college degree.

29. (a) 0.506 (b) 0.117 (c) 0

31. (a) {(SSS), (SSR), (SRS), (SRR), (RSS), (RSR), (RRS), (RRR)}
    (b) {(RRR)}
    (c) {(SSR), (SRS), (RSS)}
    (d) {(SSR), (SRS), (SRR), (RSS), (RSR), (RRS), (RRR)}

33.

Supplier #1 — Not defective / Defective

Supplier #2 — Not defective / Defective

Supplier #3 — Not defective / Defective

35. (a) 0.444 (b) 0.556 **37.** 3:1

Section 3.2 *(page 125)*

1. Two events are independent if the occurrence of one of the events does not affect the probability of the occurrence of the other event.
   If $P(B|A) = P(B)$ or $P(A|B) = P(A)$, then Events A and B are independent.
3. False. If two events are independent, $P(A|B) = P(A)$.
5. Independent because the outcome of the 1st card drawn does not affect the outcome of the 2nd card drawn.
7. Dependent because the outcome of the 1st ball drawn affects the outcome of the 2nd ball drawn.
9. (a) 0.8 (b) 0.0032 (c) Dependent
11. (a) 0.0168 (b) 0.93
13. (a) 0.109 (b) 0.382 (c) 0.618
15. (a) 0.483 (b) 0.461 (c) 0.444
    (d) Dependent. Whether a person has at least one month's income saved depends on whether or not the person is male. $P(A') \approx 0.517 \neq 0.461 \approx P(A'|B)$, where $A = \{\text{have one month's income or more}\}$ and $B = \{\text{man}\}$.
17. (a) 0.0000000243 (b) 0.859 (c) 0.141
19. (a) 0.2 (b) 0.04 (c) 0.008 (d) 0.512 (e) 0.488
21. 0.012 **23.** 0.444
25. (a) 0.074 (b) 0.999 **27.** 0.954

Section 3.3 *(page 135)*

1. $P(A \text{ and } B) = 0$ because A and B cannot occur at the same time.
3. True **5.** Not mutually exclusive
7. Not mutually exclusive. The worker can be female and have a college degree.
9. Mutually exclusive. The person cannot be in both age classes.
11. (a) No, for five weeks the events overlap.
    (b) 0.423
13. (a) Not mutually exclusive, a carton can have a puncture and a smashed corner.
    (b) 0.126
15. (a) 0.066 (b) 0.873 (c) 0.226
17. (a) 0.11 (b) 0.72
19. (a) 0.014 (b) 0.226 (c) 0.774
    (d) b and c; the opposite of neither of the people being left handed is that at least one is.
21. Answers will vary.
    Conclusion: If two events, $\{A\}$ and $\{B\}$, are independent, $P(A \text{ and } B) = P(A) \cdot P(B)$. If two events are mutually exclusive, $P(A \text{ and } B) = 0$. The only scenario when two events can be independent and mutually exclusive is if $P(A) = 0$ or $P(B) = 0$.
23. $P(A \text{ or } B \text{ or } C) = 0.54$

Section 3.4 *(page 147)*

1. You are counting the number of ways two or more events can occur in sequence.
3. False; a permutation is an ordered arrangement of objects.
5. Permutation **7.** 6240 **9.** 4500 **11.** 40,320
13. 3,628,800 **15.** 720 **17.** 32,760
19. 9,189,180 **21.** 5,586,853,480 **23.** 56
25. (a) 70 (b) 16 (c) 0.086
27. (a) 120 (b) 12 (c) 12 (d) 0.1
29. 1.20×10^{-5} **31.** 6.00×10^{-20}
33. (a) 658,008 (b) 1.5197×10^{-6}
35. 1001; 1000
37.

Team (worst team first)	1	2	3	4	5	6	7
Probability	0.25	0.20	0.157	0.12	0.089	0.064	0.044

Team (worst team first)	8	9	10	11	12	13
Probability	0.029	0.018	0.011	0.007	0.006	0.005

39. 0.566

Uses and Abuses for Chapter 3 *(page 153)*

1. (a) 0.000001
   (b) 0.001
   (c) 0.001
2. Answers will vary.

Review Answers for Chapter 3 *(page 154)*

1. Sample space:
   {HHHH, HHHT, HHTH, HHTT, HTHH, HTHT, HTTH, HTTT, THHH, THHT, THTH, THTT, TTHH, TTHT, TTTH, TTTT}
   Event: Getting three heads
   {HHHT, HHTH, HTHH, THHH}
3. Empirical probability **5.** Subjective probability
7. Classical probability **9.** 0.71 **11.** 0.92
13. Independent **15.** 0.0417

17. Mutually exclusive

19. 0.60 **21.** 0.538 **23.** 144

25. 84 **27.** 2730 **29.** 2380

31. (a) 0.955 (b) 0.000000761 (c) 0.045
(d) 0.999999239

Real Statistics–Real Decisions for Chapter 3

(page 157)

1. (a) Answers will vary.

(b) Use the Multiplication Rule, Fundamental Counting Principle, and Combinations.

2. If you played only the red ball, the probability of matching it is $\frac{1}{42}$. However, because you must pick 5 white balls, you must get the white balls wrong. So, using the Multiplication Rule, we get

P(matching only the red ball and not matching any of the 5 white balls)

$$= \frac{1}{42} \cdot \frac{44}{49} \cdot \frac{43}{48} \cdot \frac{42}{47} \cdot \frac{41}{46} \cdot \frac{40}{45}$$

$$\approx 0.01356$$

$$\approx \frac{1}{74}.$$

3. The overall probability of winning a prize is determined by calculating the number of ways to win and dividing by the total number of outcomes.

To calculate the number of ways to win something, you must use combinations.

Chapter Quiz for Chapter 3 *(page 158)*

1. (a) 0.539 (b) 0.537 (c) 0.543 (d) 0.786
(e) 0.0292 (f) 0.657 (g) 0.097 (h) 0.568

2. Not mutually exclusive because both events can occur at the same time.

Dependent because one event can affect the occurrence of the second event.

3. (a) 518,665 (b) 1 (c) 551,299

4. (a) 0.94 (b) 0.00000181 (c) 0.999998

5. 4500 **6.** 303,600

Cumulative Test for Chapters 1–3 *(page 159)*

1. Quantitative, Ratio

2. Use the sampling method of data collection. The sampling technique should be a simple random sample because it would be difficult to collect this information from the entire population of students.

3.

Class Limits	Midpoint	Freq	Boundaries	Rel Freq	Cum Freq
90-111	100.5	10	89.5-111.5	0.33	10
112-133	122.5	8	111.5-133.5	0.27	18
134-155	144.5	3	133.5-155.5	0.10	21
156-177	166.5	4	155.5-177.5	0.13	25
178-199	188.5	5	177.5-199.5	0.17	30

4.

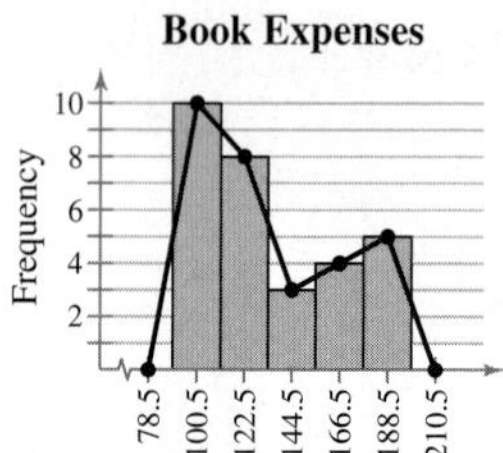

5.

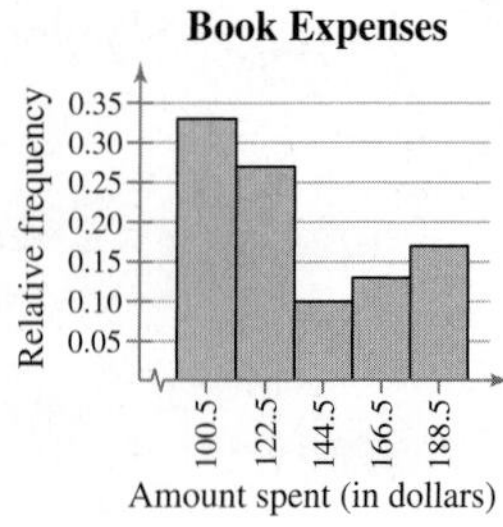

6. Key: 10 | 3 = 103

9	0138
10	349
11	0016789
12	037
13	26
14	
15	036
16	02
17	08
18	17
19	19

7. Skewed

8.

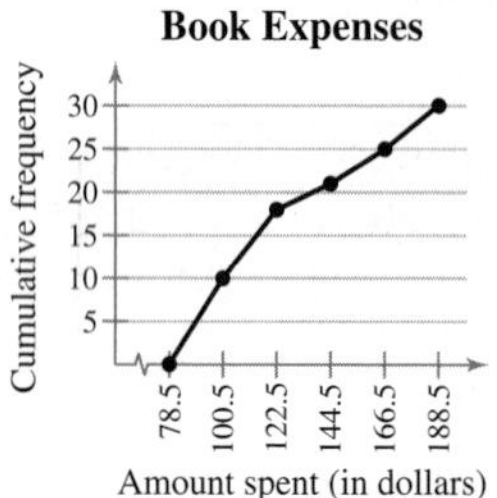

9. $\bar{x} = 133.8$

Median $= 121.5$

Mode $= 110$

Statistics, because they are characteristics of a sample.

10. Range $= 109$

$s^2 = 1057.959$

$s = 32.53$

11. $Q_1 = 110$

$Q_2 = 121.5$

$Q_3 = 160$

IQR $= 50$

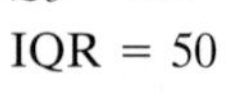

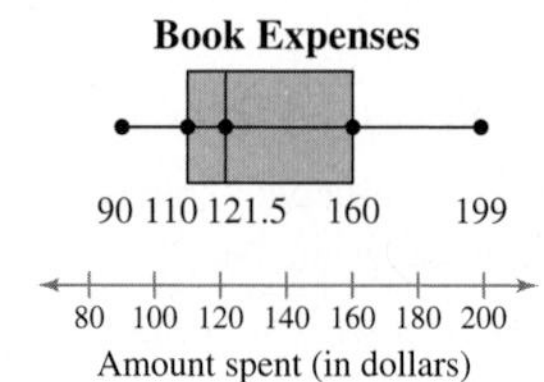

12. Min: $z = \dfrac{90 - 133.8}{32.53} \approx -1.35$

Max: $z = \dfrac{199 - 133.72}{33.099} \approx 2.00$

Neither of these values is unusual because each value is within 2σ away from the mean.

13. 0.467; 0.5 **14.** 0.7 **15.** 0.627 **16.** 142,506

CHAPTER 4

Section 4.1 *(page 169)*

1. A random variable represents a numerical value assigned to an outcome of a probability experiment.

   Examples: Answers will vary.

3. False. In most applications, discrete random variables represent counted data, while continuous random variables represent measured data.

5. True

7. Discrete because home attendance is a random variable that is countable.

9. Continuous because annual vehicle-miles driven is a random variable that cannot be counted.

11. Discrete because the random variable is countable.

13. Continuous because the random variable cannot be counted.

15. Discrete because the random variable is countable.

17. Continuous because the random variable cannot be counted.

19. (a) 0.35 (b) 0.90 **21.** 0.22 **23.** Yes

25. No, $\Sigma P(x) = 0.95$ and $P(5) < 0$

27. (a) 2.1 (b) 1.09 (c) 1.044

29. (a)

x	f	$P(x)$	$xP(x)$	$(x - \mu)^2 P(x)$
0	1491	0.6855	0	0.1723
1	425	0.1954	0.1954	0.0486
2	168	0.0772	0.1544	0.1734
3	48	0.0221	0.0663	0.1380
4	29	0.0133	0.0532	0.1628
5	14	0.0064	0.0320	0.1295
	2175	1	0.5013	0.8246

   (b) 0.5013 (c) 0.8246 (d) 0.9081

   (e) A household on average has 0.5013 dogs with a standard deviation of 0.9081.

31. (a)

x	f	$P(x)$	$xP(x)$	$(x - \mu)^2 P(x)$
0	300	0.432	0	0.252
1	280	0.403	0.403	0.022
2	95	0.137	0.274	0.209
3	20	0.029	0.087	0.145
	695	1	0.764	0.629

   (b) 0.764 (c) 0.629 (d) 0.793

33.

x	0	1	2	3
$P(x)$	0.216	0.432	0.288	0.064

35.

x	0	1	2	3
$P(x)$	0.125	0.375	0.375	0.125

37. (a) 18.375 (b) 41.734 (c) 6.460 (d) 18.375

39. (a) 2.151 (b) 1.146 (c) 1.071 (d) 2.151

41. (a) 2.55 (b) 1.90 (c) 1.38 (d) 2.55

43. (a) 0.8809 (b) 0.119 (c) 0.1126

45. −$0.05

Section 4.2 *(page 183)*

1. (a) $p = 0.50$ (b) $p = 0.20$ (c) $p = 0.80$

3. (a) $n = 12$ (b) $n = 4$ (c) $n = 8$

   As n increases, the distribution becomes more symmetric.

5. Is a binomial experiment.

   Success: baby recovers.

   $n = 5, p = 0.80, q = 0.20, x = 0, 1, 2, \ldots, 5$

7. Is not a binomial experiment because there are more than two possible outcomes for each trial.

9. (a) 0.088 (b) 0.104 (c) 0.896

11. (a) 0.069 (b) 0.089 (c) 0.911

13. (a) 0.301 (b) 0.654 (c) 0.346

15. (a) $n = 6, p = 0.36$ (b)

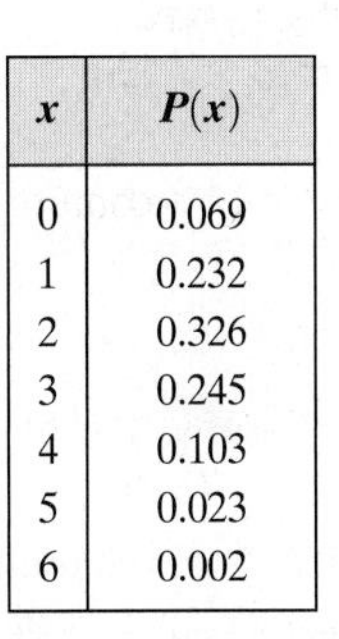

x	$P(x)$
0	0.069
1	0.232
2	0.326
3	0.245
4	0.103
5	0.023
6	0.002

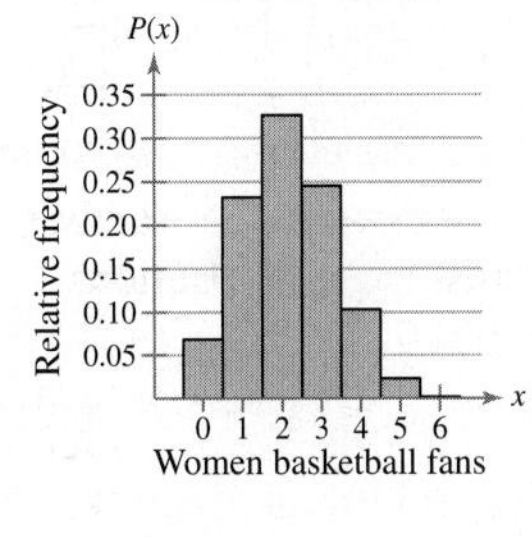

   (c) 2.16 (d) 1.382 (e) 1.176

   (f) On average 2.16, out of 6, women would consider themselves basketball fans. The standard deviation is 1.176 women.

   $x = 0$, 5, or 6 would be uncommon due to their low probabilities.

17. (a) $n = 4, p = 0.05$ (b)

x	$P(x)$
0	0.814506
1	0.171475
2	0.013537
3	0.000475
4	0.000006

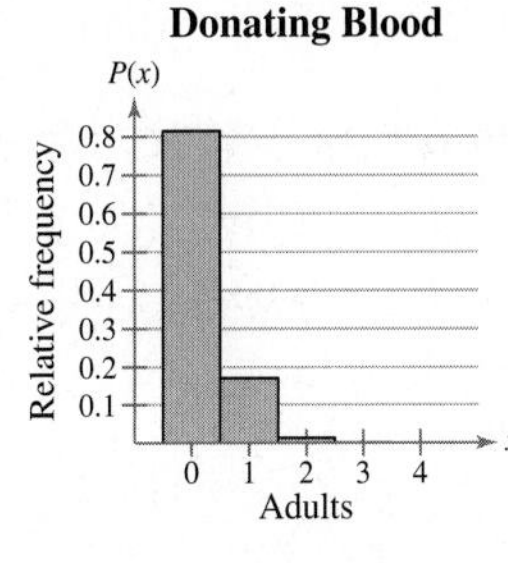

   (c) 0.2 (d) 0.19 (e) 0.436

   (f) On average 0.2 eligible adults, out of every 4, give blood. The standard deviation is 0.436 adults.

   $x = 2$, 3, or 4 would be uncommon due to their low probabilities.

19. (a) $n = 5, p = 0.28$ (b)

x	$P(x)$
0	0.193
1	0.376
2	0.293
3	0.114
4	0.0221
5	0.00172

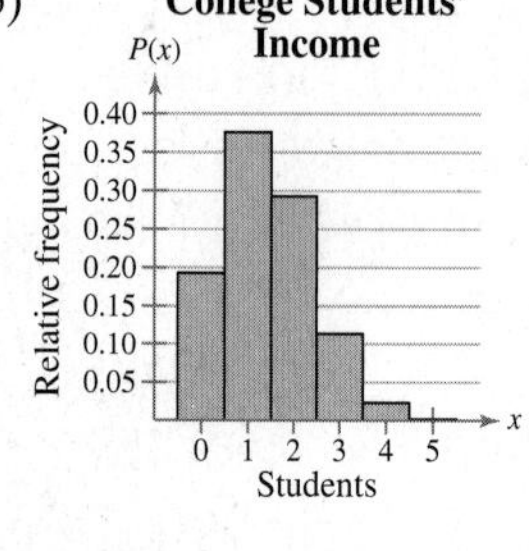

(c) 1.4 (d) 1.008 (e) 1.004

(f) On average 1.4 college students, out of every 5, earn at least $400 per month. The standard deviation is 1.004 college students.

$x = 4$ or 5 would be uncommon due to their low probabilities.

21.

x	$P(x)$
0	0.0277
1	0.1359
2	0.278
3	0.3032
4	0.1861
5	0.0609
6	0.0083

23. 0.033

Section 4.3 *(page 192)*

1. Geometric. We are interested in counting the number of trials until the first success.

3. Poisson. We are interested in counting the number of occurrences that take place within a given unit of space.

5. Binomial. We are interested in counting the number of successes out of n trials.

7. (a) 0.082 (b) 0.469 (c) 0.531

9. (a) 0.009 (b) 0.030 (c) 0.904

11. $\mu = 8$

(a) $P(4) = \dfrac{8^4 e^{-8}}{4!} \approx 0.057$

(b) $P(x \geq 4)$

$= 1 - (P(0) + P(1) + P(2) + P(3))$

$\approx 1 - (0.0003 + 0.0027 + 0.0107 + 0.0286)$

$= 0.9577$

(c) $P(x > 4)$

$= 1 - (P(0) + P(1) + P(2) + P(3) + P(4))$

$\approx 1 - (0.0003 + 0.0027 + 0.0107 + 0.0286 + 0.0573)$

$= 0.9004$

13. (a) 0.3293 (b) 0.8781 (c) 0.1219

15. (a) 0.1453996179

(b) 0.1453688667; The results are approximately the same.

17. (a) 1000; 999000; 999.50

On average you would have to play 1000 times until you won the lottery. The standard deviation is 999.50.

(b) 1000 times

Lose money. On average you would win $500 every 1000 times you play the lottery. Hence, the net gain would be −$500.

19. (a) $\sigma^2 = 3.9, \sigma \approx 2.0$

Standard deviation is 2.0 strokes

(b) 0.352

Uses and Abuses for Chapter 4 *(page 197)*

1. 40, 0.0812

2. 0.7386, Answers will vary.

3. The probability of finding 36 adults out of 100 who prefer Brand A is 0.059. So the manufacturer's claim is hard to believe.

4. The probability of finding 25 adults out of 100 who prefer Brand A is 0.000627. So the manufacturer's claim would not be believable.

Review Answers for Chapter 4 *(page 198)*

1. Discrete **3.** Continuous

5. No, $\Sigma P(x) \neq 1$ **7.** Yes

9. Yes

11. (a)

x	Frequency, f	$P(x)$
2	3	0.005
3	12	0.018
4	72	0.111
5	115	0.177
6	169	0.260
7	120	0.185
8	83	0.128
9	48	0.074
10	22	0.034
11	6	0.009
	650	1

(b)

Pages per Section

P(x)

Relative frequency

0.28 0.24 0.20 0.16 0.12 0.08 0.04

2 3 4 5 6 7 8 9 10 11 x

Pages

(c) 6.377, 2.858, 1.691

13. (a)

x	Frequency, f	$P(x)$
0	3	0.015
1	38	0.190
2	83	0.415
3	52	0.260
4	18	0.090
5	5	0.025
6	1	0.005
	200	1

(b)

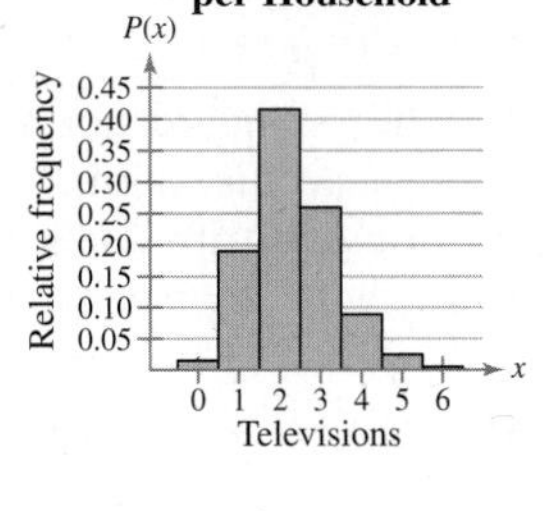

(c) 2.315, 1.076, 1.037

15. 3.37

17. Yes, $n = 12$, $p = 0.30$, $q = 0.70$, $x = 0, 1, \ldots, 12$

19. (a) 0.208
(b) 0.322
(c) 0.114

21. (a) 0.294
(b) 0.643
(c) 0.349

23. (a)

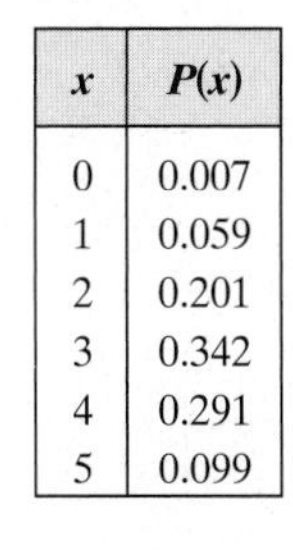

x	$P(x)$
0	0.007
1	0.059
2	0.201
3	0.342
4	0.291
5	0.099

(b)

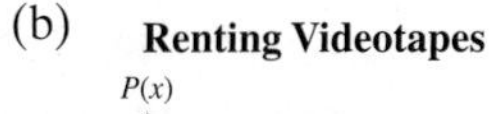

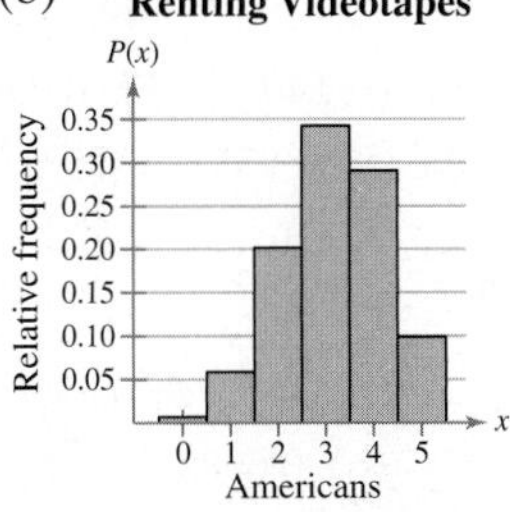

(c) 3.15, 1.1655, 1.080

25. (a)

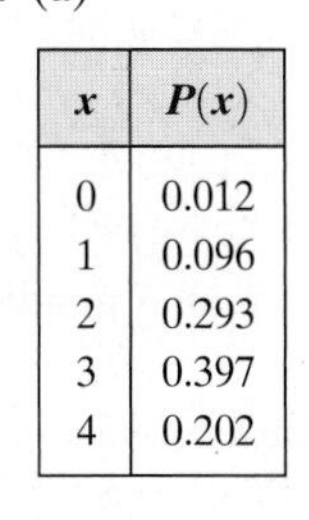

x	$P(x)$
0	0.012
1	0.096
2	0.293
3	0.397
4	0.202

(b)

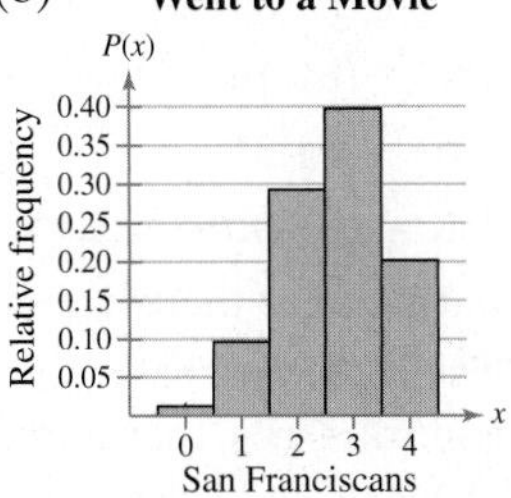

(c) 2.68, 0.8844, 0.940

27. (a) 0.096
(b) 0.518
(c) 0.579

29. (a) 0.782
(b) 0.192
(c) 0.026

Real Statistics–Real Decisions for Chapter 4

(page 202)

1. (a) Answers will vary. For example, calculate the probability of obtaining zero clinical pregnancies out of 10 randomly selected ART cycles.

(b) Binomial. Discrete because the number of clinical pregnancies are countable.

2. $n = 10$, $p = 0.305$
$P(0) = 0.0263$

x	$P(x)$
0	0.0263
1	0.1154
2	0.2279
3	0.2667
4	0.2048
5	0.1079
6	0.0394
7	0.0099
8	0.0016
9	0.0002
10	0.0000

3. (a) Suspicious, because the probability is very small.
(b) Not suspicious, because the probability is not that small.

Chapter Quiz for Chapter 4 *(page 203)*

1. (a) Discrete because the random variable is countable.
(b) Continuous because the random variable has an infinite number of possible outcomes and cannot be counted.

2. (a)

x	Freq, f	$P(x)$
1	57	0.361
2	37	0.234
3	47	0.297
4	15	0.095
5	2	0.013
	158	1

(b)

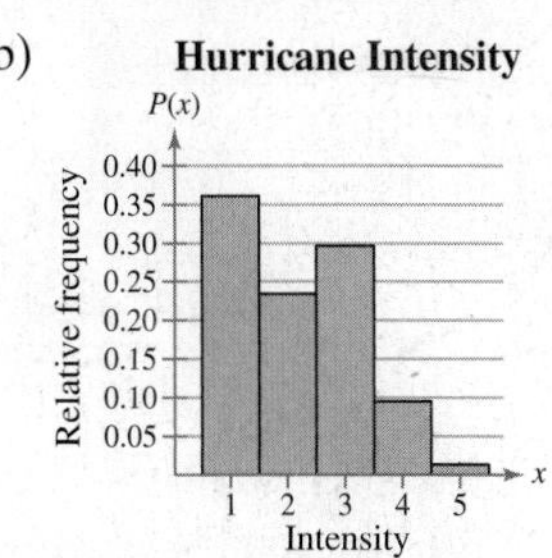

(c) 2.165; 1.125; 1.061
On average the intensity of a hurricane will be 2.165. The standard deviation is 1.061.

(d) 0.108

3. (a)

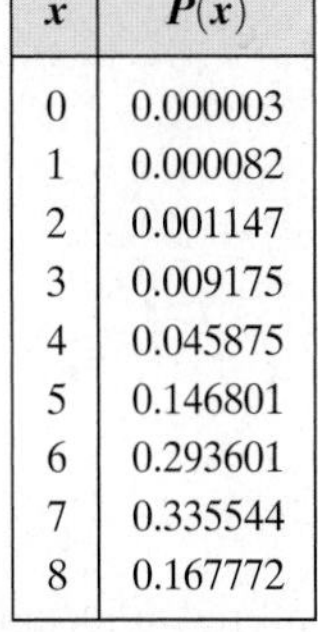

x	$P(x)$
0	0.000003
1	0.000082
2	0.001147
3	0.009175
4	0.045875
5	0.146801
6	0.293601
7	0.335544
8	0.167772

(b)

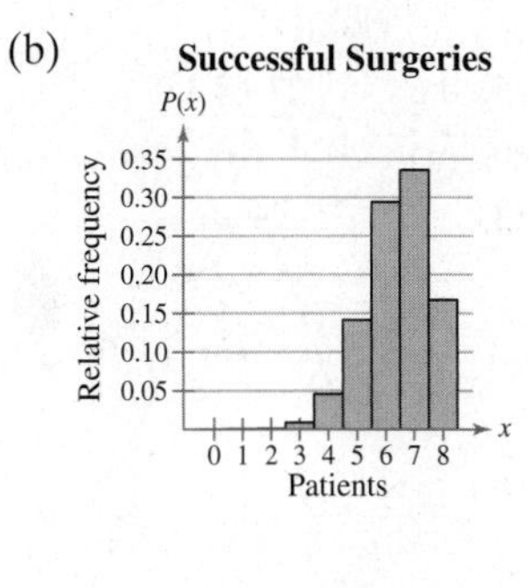

(c) 6.4, 1.28, 1.131 (d) 0.294 (e) 0.202

4. (a) 0.1755 (b) 0.4405 (c) 0.0067

CHAPTER 5

Section 5.1 *(page 210)*

1. Answers will vary.

3. Answers will vary.

Similarities: Both curves will have the same line of symmetry.

Differences: One curve will be more spread out than the other.

5. No, the graph crosses the x-axis.

7. Yes, the graph fulfills the properties of the normal distribution.

9. No, the graph is skewed to the right.

11. Mean and standard deviation.

13. $(9, 21)$ **15.** 0.68

17. (b) because the points of inflection are located at 2.98 and 3.02.

19. (a) 0.997 (b) $(2.96, 3.04)$

21. (a) $(19.86, 20.14)$ (b) $(19.79, 20.21)$

23. (a)

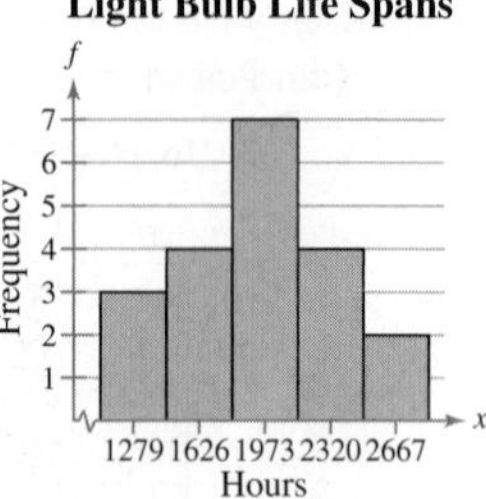

It is reasonable to assume that the life span is normally distributed since the histogram is nearly symmetric and bell shaped.

(b) 1941.35, 432.385

(c) The sample mean of 1941.35 hours is less than the claimed mean, so on the average the bulbs in the sample lasted for a shorter time. The sample standard deviation of 432 hours is greater than the claimed standard deviation, so the bulbs in the sample had a greater variation in life span than the manufacturer's claim.

25. 0.68 **27.** 0.475

29. (a) 320 (b) 1360 (c) 320

31.

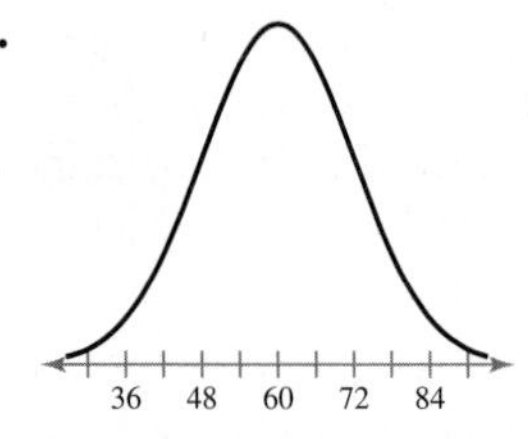

The normal distribution curve is centered at its mean (60) and has 2 points of inflection (48 and 72) representing $\mu \pm \sigma$.

33. (a) Area under curve = area of rectangle

$$= (1)(1) = 1$$

(b) 0.25 (c) 0.4

Section 5.2 *(page 219)*

1. $\mu = 0, \sigma = 1$

3. "The" standard normal distribution is used to describe one specific normal distribution $(\mu = 0, \sigma = 1)$. "A" normal distribution is used to describe a normal distribution with any mean and standard deviation.

5. 0.3849 **7.** 0.6247 **9.** 0.9382 **11.** 0.975

13. 0.8289 **15.** 0.1003 **17.** 0.005 **19.** 0.05

21. 0.475 **23.** 0.437 **25.** 0.95 **27.** 0.551

29. 0.2006 **31.** 0.05

33. (a) $A = 2.97$; $B = 2.98$; $C = 3.01$; $D = 3.05$

(b) 0.5; -1.5; -1; 2.5

(c) 2.5 is unusual.

35. (a) $A = 801$; $B = 950$; $C = 1250$; $D = 1467$

(b) -0.33; 1.12; 2.16; -1.05

(c) 2.16 is unusual.

37. 0.6915 **39.** 0.05 **41.** 0.9265 **43.** 0.9744

45. 0.2912 **47.** 0.1469 **49.** 0.4798 **51.** 0.3133

53. 0.901 **55.** 0.7540 **57.** 0.0098 **59.** 0.099

61. 0.9544, no because $95\% > 75\%$.

Section 5.3 *(page 225)*

1. 0.3055 **3.** 0.2573 **5.** 0.0566

7. (a) 0.1357 (b) 0.6983 (c) 0.1660

9. (a) 0.1587 (b) 0.7142 (c) 0.1271

11. (a) 0.0062 (b) 0.9876 (c) 0.0062

13. (a) 80.51% (b) 341

15. (a) 67.72% (b) 16

17. (a) 30.85% (b) 31

19. (a) 99.87% (b) 0.798

21. Out of control, since there is a point more than 3 standard deviations above the mean.

23. Out of control, since there are nine consecutive points below the mean and since two out of three consecutive points lie more than 2 standard deviations from the mean.

Section 5.4 *(page 234)*

1. -2.05 **3.** 0.85 **5.** -0.16 **7.** 2.39

9. -1.645 **11.** 0.84 **13.** -2.325 **15.** -0.25

17. 1.175 **19.** -0.675 **21.** 0.675 **23.** -0.385

25. -0.38 **27.** -1.645, 1.645 **29.** -1.96, 1.96

31. 0.325 **33.** 1.28 **35.** (a) 68.52 (b) 62.14

37. (a) 14.08 (b) 21.9 **39.** (a) 139.22 (b) 96.92

41. Tires which wear out by 26,800 miles will be replaced free of charge.

43. (a) 8.0155 (b) 7.6925

45. $A = 83.52$; $B = 76.68$; $C = 67.32$; $D = 60.48$

Section 5.5 *(page 246)*

1. {000, 002, 004, 006, 008, 020, 022, 024, 026, 028, 040, 042, 044, 046, 048, 060, 062, 064, 066, 068, 080, 082, 084, 086, 088, 200, 202, 204, 206, 208, 220, 222, 224, 226, 228, 240, 242, 244, 246, 248, 260, 262, 264, 266, 268, 280, 282, 284, 286, 288, 400, 402, 404, 406, 408, 420, 422, 424, 426, 428, 440, 442, 444, 446, 448, 460, 462, 464, 466, 468, 480, 482, 484, 486, 488, 600, 602, 604, 606, 608, 620, 622, 624, 626, 628, 640, 642, 644, 646, 648, 660, 662, 664, 666, 668, 680, 682, 684, 686, 688, 800, 802, 804, 806, 808, 820, 822, 824, 826, 828, 840, 842, 844, 846, 848, 860, 862, 864, 866, 868, 880, 882, 884, 886, 888}

   $\mu_{\bar{x}} = 4$, $\sigma_{\bar{x}} = 1.633$, $\mu = 4$, $\sigma = 2.828$

3. (c), since $\mu = 16.5$, $\sigma = 1.19$, and the graph approximates a normal curve.

5. 87.5, 1.804

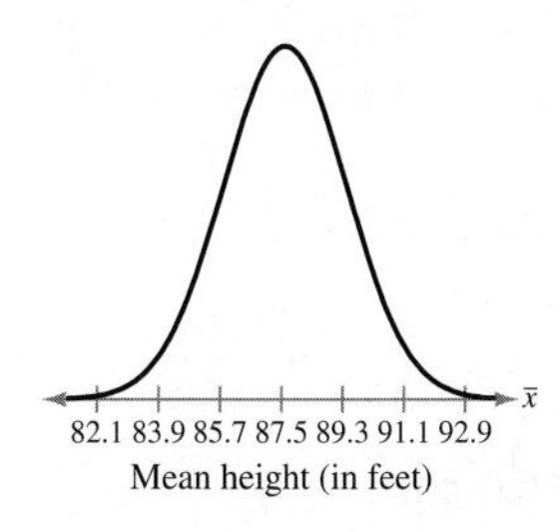

7. 115.6, 8.61

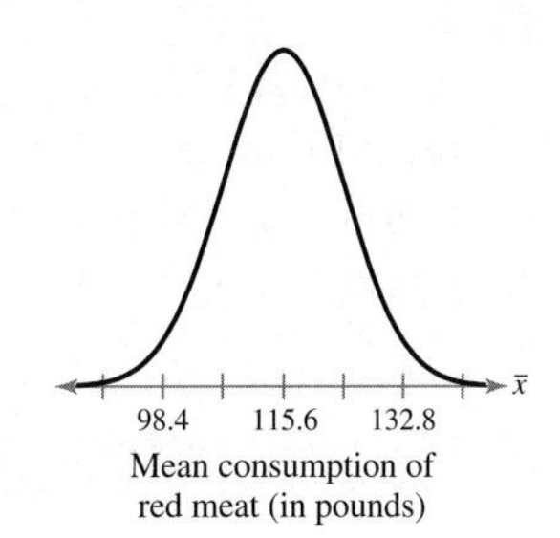

9. 87.5, 1.276; 87.5, 1.042

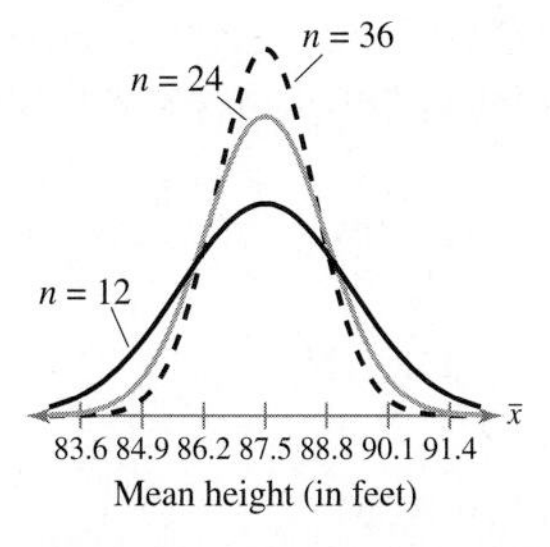

   As the sample size increases, the standard error decreases, while the mean of the sample means remains constant.

11. 0.0019 **13.** 0.6319 **15.** ≈ 0

17. It is more likely to select a sample of 20 women with a mean height less than 70 inches because the sample of 20 has a higher probability.

19. Yes, it is very unlikely that we would have randomly sampled 40 cans with a mean equal to 127.9 ounces.

21. (a) 0.0008 (b) Claim is inaccurate

    (c) No, assuming the manufacturer's claim is true, because 96.25 is within 1 standard deviation of the mean for an individual board.

23. (a) 0.0009 (b) Claim is inaccurate

    (c) Not unusual, assuming the manufacturer's claim is true, because 49,725 is within 1 standard deviation of the mean for an individual tire.

25. ≈ 0

Section 5.6 *(page 257)*

1. Cannot use normal distribution
3. Can use normal distribution
5. Cannot use normal distribution because $np \ngtr 5$.
7. Cannot use normal distribution because $nq \ngtr 5$.
9. (d) **11.** (a) **13.** (a) **15.** (c)
17. Binomial: 0.549; Normal: 0.5463
19. Cannot use normal distribution because $np \ngtr 5$.

    (a) 0.0000199 (b) 0.000023

    (c) 0.999977 (d) 0.1635

21. Can use normal distribution

    (a) 0.0465

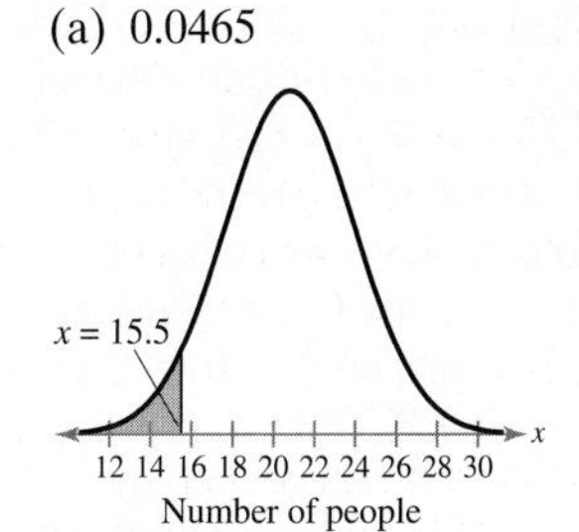

    (b) 0.9767

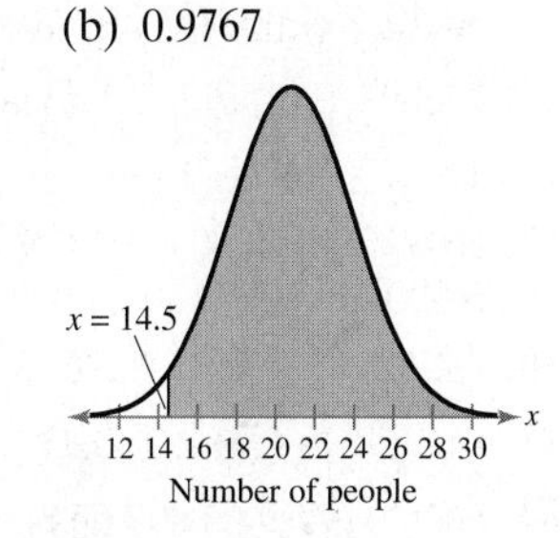

    (c) 0.9535

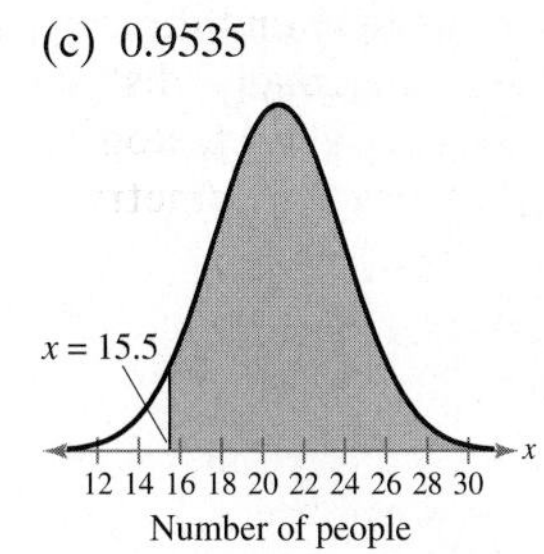

    (d) 0.1635

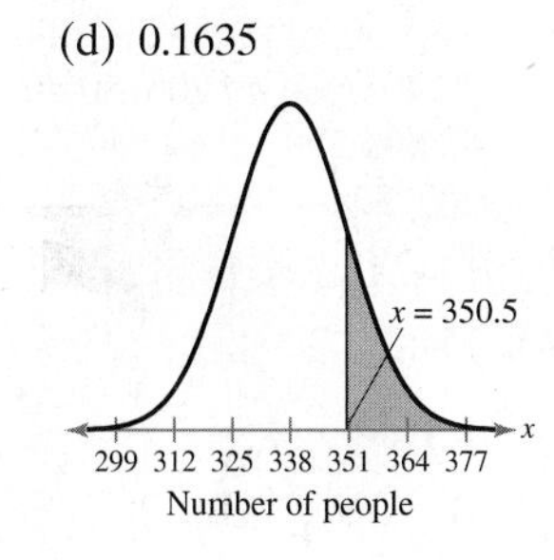

23. Highly unlikely. Answers will vary.
25. 0.1020

Uses and Abuses for Chapter 5 *(page 261)*

1. No, answers will vary.
2. It is more likely that all 10 people lie within 2 standard deviations of the mean. This can be shown by using the Empirical Rule and the Multiplication Rule.

   (a) By the Empirical Rule, the probability of lying within 2 standard deviations of the mean is 0.95. Let x = number of people selected who lie within 2 standard deviations of the mean.

   $P(x = 10) = (0.95)^{10} \approx 0.599$

   (b) P (at least one person does not lie within 2 standard deviations of the mean) $= 1 - P(x = 10) \approx 1 - 0.599 = 0.401$

Review Answers for Chapter 5 *(page 262)*

1. $\mu = 15, \sigma = 3$ **3.** (540, 800) **5.** 0.68
7. −2.25; 0.5; 2; 3.5 **9.** 0.2005

11. 0.3936 **13.** 0.0465 **15.** 0.4495 **17.** 0.3519

19. 0.1336 **21.** 0.8997 **23.** 0.9236 **25.** 0.0124

27. (a) 0.3156 (b) 0.3099 (c) 0.3446

29. -0.07 **31.** 1.13 **33.** 1.04 **35.** 43.9 meters

37. 45.9 meters **39.** 45.74 meters

41. {0 0 0, 0 0 200, 0 0 40, 0 0 600, 0 0 80, 0 200 0, 0 200 200, 0 200 40, 0 200 600, 0 200 80, 0 40 0, 0 40 200, 0 40 40, 0 40 600, 0 40 80, 0 600 0, 0 600 200, 0 600 40, 0 600 600, 0 600 80, 0 80 0, 0 80 200, 0 80 40, 0 80 600, 0 80 80, 200 0 0, 200 0 200, 200 0 40, 200 0 600, 200 0 80, 200 200 0, 200 200 200, 200 200 40, 200 200 600, 200 200 80, 200 40 0, 200 40 200, 200 40 40, 200 40 600, 200 40 80, 200 600 0, 200 600 200, 200 600 40, 200 600 600, 200 600 80, 200 80 0, 200 80 200, 200 80 40, 200 80 600, 200 80 80, 40 0 0, 40 0 200, 40 0 40, 40 0 600, 40 0 80, 40 200 0, 40 200 200, 40 200 40, 40 200 600, 40 200 80, 40 40 0, 40 40 200, 40 40 40, 40 40, 600, 40 40 80, 40 600 0, 40 600 200, 40 600 40, 40 600 600, 40 600 80, 40 80 0, 40 80 200, 40 80 40, 40 80 600, 40 80 80, 600 0 0, 600 0 200, 600 0 40, 600 0 600, 600 0 80, 600 200 0, 600 200 200, 600 200 40, 600 200 600, 600 200 80, 600 40 0, 600 40 200, 600 40 40, 600 40 600, 600 40 80, 600 600 0, 600 600 200, 600 600 40, 600 600 600, 600 600 80, 600 80 0, 600 80 200, 600 80 40, 600 80 600, 600 80 80, 80 0 0, 80 0 200, 80 0 40, 80 0 600, 80 0 80, 80 200 0, 80 200 200, 80 200 40, 80 200 600, 80 200 80, 80 40 0, 80 40 200, 80 40 40, 80 40 600, 80 40 80, 80 600 0, 80 600 200, 80 600 40, 80 600 600, 80 600 80, 80 80 0, 80 80 200, 80 80 40, 80 80 600, 80 80 80}

184,218.504, 184,126.153

43. 154.8, 8.72

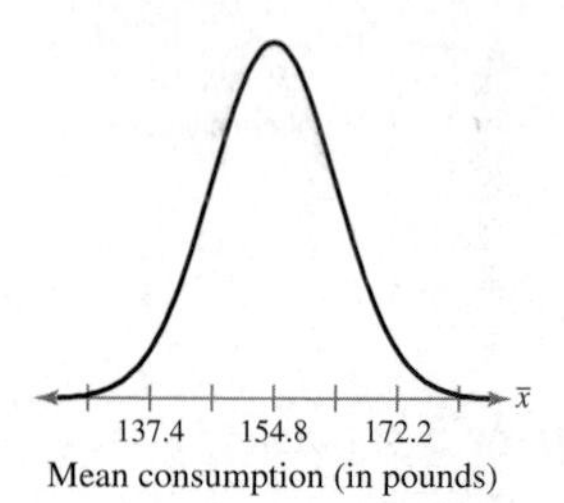

45. (a) 0.0485 (b) 0.8180 (c) 0.0823

(a) and (c) are smaller, (b) is larger. This is to be expected since the standard error of the sample means is smaller.

47. (a) ≈ 0 (b) ≈ 0 **49.** 0.0019

51. Cannot use normal distribution because $nq \ngtr 5$.

53. $P(x > 24.5)$

55. Use normal distribution.

0.0032

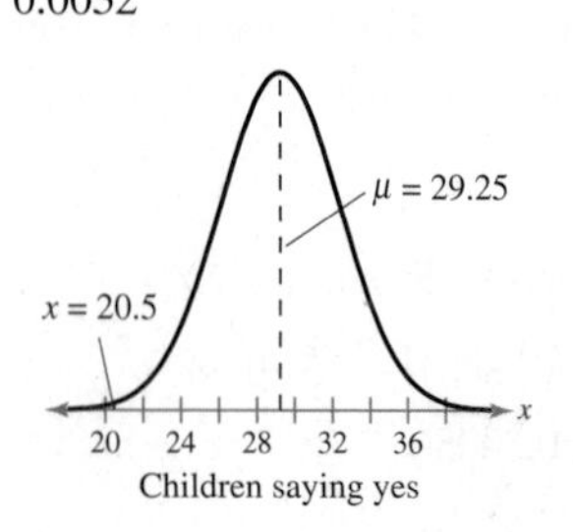

Real Statistics–Real Decisions for Chapter 5

(page 266)

1. (a) 0.8413

(b) 0.9999999997

2. (a) 0.9772

(b) 0.9999881476 (c) mean

3. Answers will vary.

Chapter Quiz for Chapter 5 *(page 267)*

1. (a) 0.9821 (b) 0.9994 (c) 0.9802 (d) 0.8135

2. (a) 0.9198 (b) 0.1940 (c) 0.0456

3. 0.1292 **4.** 0.5759 **5.** 77.64% **6.** 1509.8

7. 332.688 **8.** 253.052 **9.** 0

10. More likely to select one student with a test score greater than 300 since the standard error of the mean is less than the standard deviation.

11. Use normal distribution; 16.32, 2.285

12. 0.3594

CHAPTER 6

Section 6.1 *(page 277)*

1. You are more likely to be correct using an interval estimate since it is unlikely that a point estimate will equal the population mean exactly.

3. d; As the level of confidence increases, z_c increases therefore causing wider intervals.

5. 1.28 **7.** 1.15 **9.** 0.47 **11.** 1.76

13. 0.60 **15.** 0.685 **17.** 0.197

19. (14.775, 15.625) **21.** (4.179, 4.361)

23. (614.41, 647.39); (611.25, 650.55) 95% CI is wider.

25. (25.746, 27.854); (25.545, 28.055) 95% CI is wider.

27. (94.577, 105.423)

29. (96.165, 103.835)

$n = 40$ CI is wider because we have taken a smaller sample giving us less information about the population.

31. (9.719, 11.185)

33. (8.687, 12.217)

$s = 5.130$ CI is wider because of the increased variability within the population.

35. (a) An increase in the level of confidence will widen the confidence interval.

(b) An increase in the sample size will narrow the confidence interval.

(c) An increase in the standard deviation will widen the confidence interval.

37. (8.430, 9.704); (8.070, 10.064) 99% CI is wider.

39. 89

41. (a) 121 (b) 208

99% CI requires larger sample because more information is needed from the population to be 99% confident.

43. (a) 32 (b) 87

$E = 0.15$ requires a larger sample size. As the error size decreases, a larger sample must be taken to obtain enough information from the population to ensure desired accuracy.

45. (a) 16 (b) 62

$E = 0.0625$ requires a larger sample size. As the error size decreases, a larger sample must be taken to obtain enough information from the population to ensure desired accuracy.

47. (a) 42 (b) 60

$\sigma = 0.30$ requires a larger sample size. Due to the increased variability in the population, a larger sample size is needed to ensure the desired accuracy.

49. (a) An increase in the level of confidence will increase the minimum sample size required.

(b) An increase (larger E) in the error tolerance will decrease the minimum sample size required.

(c) An increase in the population standard deviation will increase the minimum sample size required.

51. (303.498, 311.252)

53. (13.680, 15.200)

55. (a) 0.707 (b) 0.949 (c) 0.962 (d) 0.975

(e) The finite population correction factor approaches 1 as the sample size decreases while the population size remains the same.

57. $E = \frac{z_c \sigma}{\sqrt{n}} \Rightarrow E\sqrt{n} = z_c \sigma \Rightarrow \sqrt{n} = \frac{z_c \sigma}{E} \Rightarrow n = \left(\frac{z_c \sigma}{E}\right)^2$

Section 6.2 *(page 289)*

1. 1.833 **3.** 2.947

5. (a) 2.450 (b) 2.664

7. (a) (10.855, 14.145)

(b) (11.157, 13.843); t-CI is wider.

9. (a) (4.059, 4.541) (b) (4.089, 4.511); t-CI is wider.

11. (59.482, 90.518); 15.518

13. (61.852, 88.148); 13.148; t-CI is wider.

15. (a) (3.604, 4.996) (b) (4.212, 4.388); t-CI is wider.

17. (a) 2174.75 (b) 100.341

(c) (2071.626, 2277.874)

19. (a) 909.083 (b) 305.266

(c) (635.374, 1182.792)

21. Use normal distribution because $n \geq 30$. (1.248, 1.252)

23. Use t-distribution because $n < 30$, the miles per gallon are normally distributed, and σ is unknown. (22.762, 25.238)

25. Cannot use normal or t-distribution because $n < 30$ and the times are not normally distributed.

27. $n = 25, \bar{x} = 56.0, s = 0.25$

$\pm t_{0.99} \Rightarrow 99\%$ t-CI

$\bar{x} \pm t_c \frac{s}{\sqrt{n}} = 56.0 \pm 2.797 \frac{0.25}{\sqrt{25}} \approx (55.860, 56.140)$

They are not making good tennis balls since desired bounce height of 55.5 inches is not contained between 55.850 and 56.140 inches.

Section 6.3 *(page 298)*

1. 0.080, 0.920 **3.** 0.120, 0.880 **5.** 0.066, 0.934

7. 0.721, 0.279

9. (0.064, 0.096); (0.058, 0.102) 99% CI is wider

11. (0.117, 0.123); (0.115, 0.125) 99% CI is wider

13. (0.053, 0.079); (0.049, 0.083) 99% CI is wider

15. (0.667, 0.775); (0.650, 0.792) 99% CI is wider

17. (a) 1068 (b) 822

(c) Having an estimate of the proportion reduces the minimum sample size needed.

19. (a) 1690 (b) 1267

(c) Having an estimate of the proportion reduces the minimum sample size needed.

21. (a) (0.554, 0.666)

(b) (0.383, 0.497) It is unlikely that the two proportions are equal because the confidence intervals estimating the proportions do not overlap.

23. (30.4%, 32.4%) is approximately a 95.2% CI.

25. If $n\hat{p} < 5$ or $n\hat{q} < 5$, the sampling distribution of $\hat{p}$ may not be normally distributed, therefore preventing the use of z_c when calculating the confidence interval.

27.

p	$q = 1 - p$	pq
0.0	1.0	0.00
0.1	0.9	0.09
0.2	0.8	0.16
0.3	0.7	0.21
0.4	0.6	0.24
0.5	0.5	0.25
0.6	0.4	0.24
0.7	0.3	0.21
0.8	0.2	0.16
0.9	0.1	0.09
1.0	0.0	0.00

p	$q = 1 - p$	pq
0.45	0.55	0.2475
0.46	0.54	0.2484
0.47	0.53	0.2491
0.48	0.52	0.2496
0.49	0.51	0.2499
0.50	0.50	0.2500
0.51	0.49	0.2499
0.52	0.48	0.2496
0.53	0.47	0.2491
0.54	0.46	0.2484
0.55	0.45	0.2475

$\hat{p} = 0.5$ gives the maximum value of $\hat{p}\hat{q}$.

Section 6.4 *(page 306)*

1. 16.919, 3.325 **3.** 35.479, 10.283

5. 52.336, 13.121

7. (a) (0.0000413, 0.000157) (b) (0.00643, 0.0125)

9. (a) (0.0305, 0.191) (b) (0.175, 0.437)

11. (a) (4.342, 44.636) (b) (2.084, 6.681)

13. (a) (359.596, 1829.774) (b) (18.963, 42.776)
15. (a) (6621.545, 24,422.477) (b) (81.373, 156.277)
17. Yes, because the confidence interval is below 0.015.

Uses and Abuses for Chapter 6 *(page 311)*

1. Answers will vary.
2. Answers will vary.

Review Answers for Chapter 6 *(page 312)*

1. (a) 103.5 (b) 9.016 **3.** (10.246, 10.354)
5. 47 **7.** 1.415 **9.** 10.255 **11.** (42.545, 63.055)
13. (73.634, 86.366) **15.** 0.420, 0.580
17. 0.292, 0.708 **19.** 0.402, 0.598
21. (0.387, 0.453) **23.** (0.240, 0.344)
25. (0.372, 0.432) **27.** 273
29. 23.337, 4.404 **31.** 14.067, 2.167
33. (0.003, 0.013); (0.055, 0.114)

Real Statistics–Real Decisions for Chapter 6
(page 315)

1. (a) No, there has not been a change in the mean concentration levels because the confidence interval for year 1 overlaps with the confidence interval for year 2.
(b) No, there has not been a change in the mean concentration levels because the confidence interval for year 2 overlaps with the confidence interval for year 3.
(c) Yes, there has been a change in the mean concentration levels because the confidence interval for year 1 does not overlap with the confidence interval for year 3.
2. The efforts to reduce the acetylene concentration are significant over the 3-year period.
3. (a) They used a point estimate because it is the most unbiased estimate of the population mean.
(b) No, because typically σ is unknown. They could have used the sample standard deviation.

Chapter Quiz for Chapter 6 *(page 316)*

1. (a) 98.110 (b) 8.847
(c) (89.263, 106.957); You are 95% confident that the mean repair cost is between \$89.26 and \$106.96.
2. 34
3. (a) 6.610 (b) 3.376 (c) (4.653, 8.567)
(d) (4.789, 8.431); The t-CI is wider since less information is available.
4. (3231.737, 4178.263)
5. (a) 0.660 (b) (0.643, 0.677) (c) 930
6. (a) (387.650, 1104.514) (b) (19.689, 33.234)

Cumulative Test for Chapters 4–6 *(page 317)*

1. 0.770; (0.732, 0.808) **2.** 2936 **3.** 0.455
4. 364.980; 83.945; 9.162
You would expect 364.98 women to say that the media have a negative effect on women's health. The standard deviation is 9.162.
5. Use normal distribution; 364.980; 9.162
6. (25.336, 25.864)
7. Normal distribution was used since $n \geq 30$.
8. You are more likely to select one woman with a BMI less than 20, because the standard deviation of the distribution of sample means is less than the standard deviation of the distribution of single values.
9. (a) (6.495, 18.506) (b) (2.549, 4.302)
10. (b) and (d) are unusual.

CHAPTER 7

Section 7.1 *(page 331)*

1. H_0: $\mu \leq 645$; H_a: $\mu > 645$ **3.** H_0: $\sigma = 5$; H_a: $\sigma \neq 5$
5. H_0: $p \geq 0.45$: H_a: $p < 0.45$
7. c H_a: $\mu < 3$ **9.** b H_a: $\mu \neq 3$
11. $\mu > 750$
H_0: $\mu \leq 750$; H_a: $\mu > 750$ (claim)
13. $\sigma \leq 1220$
H_0: $\sigma \leq 1220$ (claim); H_a: $\sigma > 1220$
15. $p = 0.44$
H_0: $p = 0.44$ (claim); H_a: $p \neq 0.44$
17. Type I: Rejecting H_0: $p \geq 0.24$ when actually $p \geq 0.24$
Type II: Not rejecting H_0: $p \geq 0.24$ when actually $p < 0.24$
19. Type I: Rejecting H_0: $\sigma \leq 23$ when actually $\sigma \leq 23$
Type II: Not rejecting H_0: $\sigma \leq 23$ when actually $\sigma > 23$
21. Type I: Rejecting H_0: $p = 0.60$ when actually $p = 0.60$
Type II: Not rejecting H_0: $p = 0.60$ when actually $p \neq 0.60$.
23. Left tailed because the alternative hypothesis contains $<$.
25. Two tailed because the alternative hypothesis contains $\neq$.
27. Two tailed because the alternative hypothesis contains $\neq$.
29. (a) There is enough evidence to reject the company's claim.
(b) There is not enough evidence to decide that the company's claim is false.
31. (a) There is enough evidence to support the Dept. of Labor's claim.
(b) There is not enough evidence to decide that the Dept. of Labor's claim is true.

33. (a) There is enough evidence to reject the manufacturer's claim.
(b) There is not enough evidence to reject the manufacturer's claim.

35. $\mu = 10$ **37.** (a) H_0: $\mu \leq 15$ (b) H_0: $\mu \geq 15$

39. If you decrease α, you are decreasing the probability that you reject H_0. Therefore, you are increasing the probability of failing to reject H_0. This could increase β, the probability of failing to reject H_0 when H_0 is false.

41. (a) Reject H_0
(b) Do not reject H_0
(c) Do not reject H_0

Section 7.2 *(page 344)*

1. $P = 0.1151$; fail to reject H_0

3. $P = 0.0096$; reject H_0

5. $P = 0.1188$; fail to reject H_0

7. c **9.** b

11. (a) fail to reject H_0
(b) reject H_0

13. 1.645 **15.** -1.88 **17.** ± 2.33

19. Right-tailed ($\alpha = 0.01$)

21. Two-tailed ($\alpha = 0.10$)

23. (a) Fail to reject H_0 because $-1.645 < z < 1.645$
(b) Reject H_0 because $z > 1.645$
(c) Fail to reject H_0 because $-1.645 < z < 1.645$
(d) Reject H_0 because $z < -1.645$

25. (a) Fail to reject H_0 because $z < 1.285$
(b) Fail to reject H_0 because $z < 1.285$
(c) Fail to reject H_0 because $z < 1.285$
(d) Reject H_0 because $z > 1.285$

27. Reject H_0 **29.** Reject H_0

31. (a) H_0: $\mu \leq 260$; H_a: $\mu > 260$ (claim)
(b) 0.838; 0.7995 (c) 0.2005
(d) Fail to reject H_0. There is insufficient evidence at the 4% level of significance to support the claim that the mean score for Illinois' eighth graders is more than 260.

33. (a) H_0: $\mu \leq 7$; H_a: $\mu > 7$ (claim)
(b) 2.996; 0.9987 (c) 0.0013
(d) Reject H_0. There is sufficient evidence at the 7% level to support the claim that the mean consumption of tea by a person in the U.S. is more than 7 gallons per year.

35. (a) H_0: $\mu = 15$ (claim); H_a: $\mu \neq 15$
(b) -0.219; 0.4129 (c) 0.8258
(d) Fail to reject H_0. There is insufficient evidence at the 5% level to reject the claim that the mean time it takes smokers to quit smoking permanently is 15 years.

37. (a) H_0: $\mu = 40$ (claim); H_a: $\mu \neq 40$
(b) $z_0 = \pm 2.575$; $z < -2.575$, $z > 2.575$
(c) -0.584
(d) Fail to reject H_0. There is insufficient evidence at the 1% level to reject the claim that the mean caffeine content per one twelve-ounce bottle of cola is 40 milligrams.

39. (a) H_0: $\mu \geq 750$ (claim); H_a: $\mu < 750$
(b) $z_0 = -2.05$; $z < -2.05$ (c) -0.500
(d) Fail to reject H_0. There is insufficient evidence at the 2% level to reject the claim that the mean life of the bulb is at least 750 hours.

41. (a) H_0: $\mu \leq 28$; H_a: $\mu > 28$ (claim)
(b) $z_0 = 1.55$; $z > 1.55$ (c) 1.318
(d) Fail to reject H_0. There is insufficient evidence at the 6% level to support the claim that the mean nitrogen dioxide level in West London is greater than 28 parts per billion.

43. Fail to reject H_0 because the standardized test statistic $z = -1.96$ is greater than the critical value $z_0 = -2.33$.

45. Using the classical z-test, the test statistic is compared to critical values. The z-test using a P-value compares the P-value to the level of significance α.

Section 7.3 *(page 356)*

1. Identify the level of significance α and the degrees of freedom, df $= n - 1$. Find the critical value(s) using the t-distribution table in the row with $n - 1$ df. If the hypothesis test is
(1) left tailed, use "One Tail α" column with a negative sign.
(2) right tailed, use "One Tail α" column with a positive sign.
(3) two tailed, use "Two Tail α" column with a negative and a positive sign.

3. 1.717 **5.** -2.101 **7.** ± 2.779

9. 1.328 **11.** -2.473 **13.** ± 3.747

15. (a) Fail to reject H_0 because $t > -2.086$
(b) Fail to reject H_0 because $t > -2.086$
(c) Fail to reject H_0 because $t > -2.086$
(d) Reject H_0 because $t < -2.086$

17. (a) Fail to reject H_0 because $-2.602 < t < 2.602$
(b) Fail to reject H_0 because $-2.602 < t < 2.602$
(c) Reject H_0 because $t > 2.602$
(d) Reject H_0 because $t < -2.602$

19. H_0: $\mu = 15$ (claim)
H_a: $\mu \neq 15$
$t_0 = \pm 4.032$
$t \approx -0.834$
Fail to reject H_0

21. H_0: $\mu \geq 8000$ (claim)
H_a: $\mu < 8000$
$t_0 = -2.492$
$t \approx -3.333$
Reject H_0

23. (a) H_0: $\mu \geq 100$; H_a: $\mu < 100$ (claim)
(b) $t_0 = -3.747$; $t < -3.747$ (c) $t \approx -4.472$
(d) Reject H_0. There is sufficient evidence at the 1% level to support the claim that the mean repair cost for damaged microwave ovens is less than \$100.

25. (a) H_0: $\mu \leq 1$; H_a: $\mu > 1$ (claim)
(b) $t_0 = 1.796$; $t > 1.796$ (c) $t \approx 2.309$
(d) Reject H_0. There is sufficient evidence at the 5% level to support the claim that the mean waste recycled by adults in the U.S. is more than 1 pound per person per day.

27. (a) H_0: $\mu = \$24{,}600$ (claim); H_a: $\mu \neq \$24{,}600$
(b) $t_0 = \pm 2.262$; $t < -2.262$ or $t > 2.262$
(c) $t \approx -0.572$
(d) Fail to reject H_0. There is insufficient evidence at the 5% level to reject the claim that the mean salary for full-time male workers over age 25 without high school diplomas is \$24,600.

29. (a) H_0: $\mu \geq 3.0$; H_a: $\mu < 3.0$ (claim)
(b) $P = 0.130$
(c) Fail to reject H_0. There is insufficient evidence at the 5% level to claim that teenage males drink less than three 12-ounce servings of soda per day.

31. (a) H_0: $\mu \geq 32$; H_a: $\mu < 32$ (claim)
(b) $P \approx 0.034$
(c) Fail to reject H_0. There is insufficient evidence at the 1% level to support the claim that the mean class size for full-time faculty is less than 32.

33. (a) H_0: $\mu = \$2116$ (claim); H_a: $\mu \neq \$2116$
(b) $P = 0.210$
(c) Fail to reject H_0. There is sufficient evidence at the 2% level to reject the claim that the typical household in the U.S. spends a mean amount of \$2116 per year on food away from home.

35. Because the P-value $= 0.096 > 0.01$, fail to reject H_0

37. Use the t-distribution because the population is normal, $n < 30$ and σ is unknown.
H_0: $\mu \geq 21$ (claim); H_a: $\mu < 21$
$t \approx -1.118$
P-value $= 0.163$
Fail to reject H_0. There is insufficient evidence at the 5% level to reject the claim that the mean gas mileage for the luxury sedan is at least 21 mpg.

39. Use the z-distribution because σ is known and the population is normally distributed.
H_0: $\mu \geq 21$ (claim); H_a: $\mu < 21$
$z \approx -0.894$
P-value $= 0.1867$
Fail to reject H_0. There is insufficient evidence at the 5% level to reject the claim that the mean gas mileage for the luxury sedan is at least 21 mpg.
We fail to reject the null hypothesis in both cases.

41. It is not necessary to do a hypothesis test to test the repairer's claim, because $\bar{x} \geq 50$. Note also that the test statistic will be positive and the critical value is negative.

Section 7.4 *(page 364)*

1. Verify that $np \geq 5$ and $nq \geq 5$. State H_0 and H_a. Specify the level of significance α. Determine the critical value(s) and rejection region(s). Find the standardized test statistic. Make a decision and interpret in the context of the original claim.

3. Use normal distribution
H_0: $p = 0.25$; H_a: $p \neq 0.25$ (claim)
$z_0 = \pm 1.96$
$z \approx -0.260$
Fail to reject H_0

5. Use normal distribution
H_0: $p \geq 0.60$; H_a: $p < 0.60$ (claim)
$z_0 = -2.33$
$z \approx -0.242$
Fail to reject H_0

7. Use normal distribution
H_0: $p \geq 0.48$ (claim); H_a: $p < 0.48$
$z_0 = -1.29$
$z \approx -1.34$
Reject H_0

9. (a) H_0: $p \geq 0.25$ (claim); H_a: $p < 0.25$
(b) $z_0 = -2.33$; $z < -2.33$ (c) $z \approx -0.163$
(d) Fail to reject H_0. There is insufficient evidence at the 1% level to reject the claim that at least 25% of U.S. adults are smokers.

11. (a) H_0: $p \leq 0.30$; H_a: $p > 0.30$ (claim)
(b) $z_0 = 1.88$; $z > 1.88$ (c) $z \approx 1.414$
(d) Fail to reject H_0. There is insufficient evidence at the 3% level to support the claim that more than 30% of U.S. consumers have stopped buying the product because the manufacturing of the product pollutes the environment.

13. (a) H_0: $p = 0.60$ (claim); H_a: $p \neq 0.60$
(b) $z_0 = \pm 2.33$; $z < -2.33$ or $z > 2.33$
(c) $z \approx -2.587$
(d) Reject H_0. There is sufficient evidence at the 2% level to reject the claim that 60% of people in the U.S. believe that government regulation of business does more harm than good.

15. H_0: $p \geq 0.52$ (claim); H_a: $p < 0.52$
$z_0 = -1.645$; rejection region: $z < -1.645$
$z \approx -0.566$
Fail to reject H_0

17. H_0: $p = 0.60$ (claim); H_a: $p \neq 0.60$

$z_0 = \pm 2.33$; $z \approx -2.587$

Reject H_0

The results are the same.

Section 7.5 *(page 373)*

1. Specify the level of significance α. Determine the degrees of freedom. Determine the critical values using the χ^2 distribution. If (a) right-tailed test, use the value that corresponds to df and α. (b) left-tailed test, use the value that corresponds to df and $1 - \alpha$. (c) two-tailed test, use the value that corresponds to df and $\frac{1}{2}\alpha$ and $1 - \frac{1}{2}\alpha$.

3. 38.885 **5.** 0.872 **7.** 7.261, 24.996

9. (a) Fail to reject H_0 (b) Fail to reject H_0 (c) Fail to reject H_0 (d) Reject H_0

11. (a) Fail to reject H_0 (b) Reject H_0 (c) Reject H_0 (d) Fail to reject H_0

13. H_0: $\sigma^2 = 0.52$ (claim); H_a: $\sigma^2 \neq 0.52$

$\chi_L^2 = 7.564$, $\chi_R^2 = 30.191$; $\chi^2 \approx 16.608$

Fail to reject H_0

15. H_0: $\sigma \geq 40$; H_a: $\sigma < 40$ (claim)

$\chi_0^2 = 3.053$; $\chi^2 \approx 11.444$

Fail to reject H_0

17. (a) H_0: $\sigma^2 = 3$ (claim); H_a: $\sigma^2 \neq 3$

(b) $\chi_L^2 = 13.844$, $\chi_R^2 = 41.923$; $\chi^2 > 41.923$ or $\chi^2 < 13.844$

(c) $\chi^2 \approx 24.267$

(d) Fail to reject H_0. There is insufficient evidence at the 5% level of significance to reject the claim that the variance of the life of the appliances is 3.

19. (a) H_0: $\sigma \geq 29$; H_a: $\sigma < 29$ (claim)

(b) $\chi_0^2 = 13.240$; $\chi^2 < 13.240$ (c) $\chi^2 \approx 19.159$

(d) Fail to reject H_0. There is insufficient evidence at the 10% level of significance to support the claim that the standard deviation for eighth graders on the examination is less than 29.

21. (a) H_0: $\sigma \leq 0.5$ (claim); H_a: $\sigma > 0.5$

(b) $\chi_0^2 = 33.196$; $\chi^2 > 33.196$ (c) $\chi^2 = 47.040$

(d) Reject H_0. There is sufficient evidence at the 10% level to reject the claim that the standard deviation of the waiting times is no more than 0.5 minute.

23. (a) H_0: $\sigma \geq \$18{,}000$; H_a: $\sigma < \$18{,}000$ (claim)

(b) $\chi_0^2 = 18.114$; $\chi^2 < 18.114$ (c) $\chi^2 \approx 28.521$

(d) Fail to reject H_0. There is insufficient evidence at the 10% level to support the claim that the standard deviation of the total charge for patients involved in a crash where the vehicle struck a wall is less than $18,000.

25. (a) H_0: $\sigma \leq 20{,}000$; H_a: $\sigma > 20{,}000$ (claim)

(b) $\chi_0^2 = 24.996$; $\chi^2 > 24.996$ (c) $\chi^2 \approx 16.011$

(d) Fail to reject H_0. There is insufficient evidence at the 5% level to support the claim that the standard deviation of the annual salaries for actuaries is more than $20,000.

27. P-value $= 0.385$

Fail to reject H_0

Uses and Abuses for Chapter 7 *(page 381)*

1. Answers will vary.

2. H_0: $p = 0.47$; Answers will vary.

3. Answers will vary.

4. Answers will vary.

Review Answers for Chapter 7 *(page 382)*

1. H_0: $\mu \leq 1593$ (claim); H_a: $\mu > 1593$

3. H_0: $p \geq 0.205$; H_a: $p < 0.205$ (claim)

5. H_0: $\sigma \leq 4.5$; H_a: $\sigma > 4.5$ (claim)

7. (a) H_0: $p = 0.85$ (claim); H_a: $p \neq 0.85$

(b) Type I error will occur if H_0 is rejected when the actual proportion of American adults who use nonprescription pain relievers is 0.85.

Type II error if H_0 is not rejected when the actual proportion of American adults who use nonprescription pain relievers is not 0.85.

(c) Two tailed

(d) There is enough evidence to reject the claim.

(e) There is not enough evidence to reject the claim.

9. (a) H_0: $\mu \leq 50$ (claim); H_a: $\mu > 50$

(b) Type I error will occur if H_0 is rejected when the actual standard deviation sodium content is no more than 50 mg.

Type II error if H_0 is not rejected when the actual standard deviation sodium content is more than 50 mg.

(c) Right tailed

(d) There is enough evidence to reject the claim.

(e) There is not enough evidence to reject the claim.

11. $z_0 = -2.05$ **13.** $z_0 = 1.96$

15. H_0: $\mu \leq 45$ (claim); H_a: $\mu > 45$

$z_0 = 1.645$; $z \approx 2.128$

Reject H_0

17. H_0: $\mu \geq 5.500$; H_a: $\mu < 5.500$ (claim)

$z_0 = -2.33$; $z \approx -1.636$

Fail to reject H_0

19. H_0: $\mu \leq 0.05$ (claim); H_a: $\mu > 0.05$

$z \approx 2.200$; P-value $= 0.0139$

$\alpha = 0.10 \Rightarrow$ Reject H_0

$\alpha = 0.05 \Rightarrow$ Reject H_0

$\alpha = 0.01 \Rightarrow$ Fail to reject H_0

21. Reject H_0. You have sufficient evidence to reject the claim.

23. $t_0 = \pm 2.093$ **25.** $t_0 = -1.345$

27. $H_0: \mu = 95$; $H_a: \mu \neq 95$ (claim)
$t_0 = \pm 2.201$; $t \approx -2.038$
Fail to reject H_0

29. $H_0: \mu \geq 0$ (claim); $H_a: \mu < 0$
$t_0 = -1.341$; $t \approx -1.304$
Fail to reject H_0

31. $H_0: \mu \leq 48$ (claim); $H_a: \mu > 48$
$t_0 = 3.143$; $t \approx 4.233$
Reject H_0

33. $H_0: \mu = \$25$ (claim); $H_a: \mu \neq \$25$
$t_0 = \pm 1.740$; $t \approx 1.642$
Fail to reject H_0

35. $H_0: \mu \geq 4$ (claim); $H_a: \mu < 4$
P-value ≈ 0.308; $t \approx -0.510$
Fail to reject H_0

37. $H_0: p = 0.15$ (claim); $H_a: p \neq 0.15$
$z_0 = \pm 1.96$; $z \approx -1.063$
Fail to reject H_0

39. Because $np = 3.6$ is less than 5, the normal distribution cannot be used to approximate the binomial distribution.

41. Because $np = 1.2$ is less than 5, the normal distribution cannot be used to approximate the binomial distribution.

43. $H_0: p = 0.20$; $H_a: p \neq 0.20$ (claim)
$z_0 = \pm 2.575$; $z \approx 0.561$
Fail to reject H_0

45. $H_0: p \leq 0.40$; $H_a: p > 0.40$ (claim)
$z_0 = 1.28$; $z \approx 2.632$
Reject H_0

47. $\chi_0^2 = 30.144$ **49.** $\chi_0^2 = 33.196$

51. $H_0: \sigma^2 \leq 2$; $H_a: \sigma^2 > 2$ (claim)
$\chi_0^2 = 24.769$; $\chi^2 = 20.230$
Fail to reject H_0

53. $H_0: \sigma^2 = 1.25$ (claim); $H_a: \sigma \neq 1.25$
$\chi_L^2 = 0.831$; $\chi_R^2 = 12.833$; $\chi^2 \approx 3.395$
Fail to reject H_0

55. $H_0: \sigma^2 \leq 0.01$ (claim); $H_a: \sigma^2 > 0.01$
$\chi_0^2 = 49.645$; $|\chi^2 = 172.800$
Reject H_0

Real Statistics–Real Decisions for Chapter 7
(page 386)

1. (a) Take a random sample to get a diverse group of people. Stratified
(b) Random sample
(c) Answers will vary.

2. $H_0: p \geq 0.40$; $H_a: p < 0.40$ (claim)
$p = \frac{15}{32} \approx 0.469$
$z_0 = -1.28$; $z = 0.794$
Fail to reject H_0; Answers will vary.

3. $H_0: \mu \geq 60$ (claim); $H_a: \mu < 60$
$t_0 = -1.345$; $t = -1.354$
Reject H_0; Answers will vary.

4. Answers will vary.

Chapter Quiz for Chapter 7 *(page 387)*

1. (a) $H_0: \mu \geq 94$ (claim); $H_a: \mu < 94$
(b) Type I error occurs if H_0 is rejected when actually the mean consumption is at least 94 pounds.
Type II error occurs if H_0 has not been rejected when actually the mean consumption is less than 94 pounds.
(c) Left tailed; z-test because $n \geq 30$.
(d) $z_0 = -2.05$; $z < -2.05$ (e) $z \approx -0.169$
(f) Fail to reject H_0. There is insufficient evidence at the 2% level to reject the claim that the mean consumption of fresh citrus fruits by people in the U.S. is at least 94 pounds per year.

2. (a) $H_0: \mu \geq 25$ (claim); $H_a: \mu < 25$
(b) Type I error occurs if H_0 is rejected when actually the mean mpg is at least 25.
Type II error occurs if H_0 has not been rejected when actually the mean mpg is less than 25.
(c) Left tailed; t-test because the population is normal, $n < 30$, and σ is unknown.
(d) $t_0 = -1.895$; $t < -1.895$ (e) $t \approx -1.131$
(f) Fail to reject H_0. There is insufficient evidence at the 5% level to reject the claim that the mean gas mileage is at least 25 miles per gallon.

3. (a) $H_0: p \leq 0.10$ (claim); $H_a: p > 0.10$
(b) Type I error occurs if H_0 is rejected when actually the proportion of microwaves needing repair is no more than 0.10.
Type II error occurs if H_0 has not been rejected when actually the proportion of microwaves needing repair is more than 0.10.
(c) Right tailed; z-test because np, $nq \geq 5$.
(d) $z_0 = 1.75$; $z > 1.75$ (e) $z \approx 0.755$
(f) Fail to reject H_0. There is insufficient evidence at the 4% level to reject the claim that no more than 10% of the microwaves need repair during the first five years of use.

4. (a) $H_0: \sigma = 105$ (claim); $H_a: \sigma \neq 105$
(b) Type I error occurs if H_0 is rejected when actually the standard deviation of the scores is 105.
Type II error occurs if H_0 has not been rejected when actually the standard deviation of the scores is not 105.

(c) Two tailed; χ^2 test because the population is normal.

(d) $\chi_L^2 = 3.565$, $\chi_R^2 = 29.819$; $\chi^2 < 3.565$ or $\chi^2 > 29.819$

(e) $\chi^2 \approx 15.056$

(f) Fail to reject H_0. There is insufficient evidence at the 1% level to reject the claim that the standard deviation of the SAT verbal scores for the state is 105.

5. (a) H_0: $\mu = \$53{,}102$ (claim); H_a: $\mu \neq \$53{,}102$

(b) Type I error occurs if H_0 is rejected when actually the mean salary is $53,102.

Type II error occurs if H_0 has not been rejected when actually the mean salary is not $53,102.

(c) Two tailed; t-test because the population is normal, $n < 30$, and σ is unknown.

(d) Not applicable.

(e) $t \approx -0.480$; P-value $= 0.640$

(f) Fail to reject H_0. There is insufficient evidence at the 5% level to reject the claim that the mean annual salary for full-time male workers over the age of 25 with a bachelor's degree is $53,102.

6. (a) H_0: $\mu = \$276$ (claim); H_a: $\mu \neq \$276$

(b) Type I error occurs if H_0 is rejected when actually the mean daily cost is $276.

Type II error occurs if H_0 has not been rejected when actually the mean daily cost is not $276.

(c) Two tailed; z-test because $n \geq 30$.

(d) Not applicable.

(e) $z \approx 1.775$; P-value $= 0.0759$

(f) Fail to reject H_0. There is insufficient evidence at the 5% level to reject the claim that the mean daily cost of meals and lodging for a family of four traveling in Massachusetts is $276.

CHAPTER 8

Section 8.1 *(page 395)*

1. State the hypotheses and identify the claim. Specify the level of significance and find the critical value(s). Find the standardized test statistic. Make a decision and interpret in the context of the claim.

3. (a) 2 (b) 7.603

(c) z is in the rejection region.

(d) Reject the claim

5. (a) 30 (b) $z \approx 1.84$

(c) z is not in the rejection region.

(d) Fail to reject H_0

7. Reject H_0

9. (a) H_0: $\mu_1 = \mu_2$ (claim); H_a: $\mu_1 \neq \mu_2$

(b) $z_0 = \pm 1.645$; $z < -1.645$ or $z > 1.645$

(c) $z \approx -2.786$

(d) Reject H_0. There is sufficient evidence at the 10% level to reject the claim that the mean braking distance is the same for both types of tires.

11. (a) H_0: $\mu_1 \geq \mu_2$; H_a: $\mu_1 < \mu_2$ (claim)

(b) $z_0 = -2.33$; $z < -2.33$ (c) $z \approx -1.536$

(d) Fail to reject H_0. There is insufficient evidence at the 1% level to conclude that the repair costs for Model A are lower than for Model B.

13. (a) H_0: $\mu_1 = \mu_2$; H_a: $\mu_1 \neq \mu_2$ (claim)

(b) $z_0 = \pm 2.575$; $z < -2.575$ or $z > 2.575$

(c) $z \approx 0.310$

(d) Fail to reject H_0. There is insufficient evidence at the 1% level to reject the claim that male and female high school students have equal ACT scores.

15. (a) H_0: $\mu_1 = \mu_2$ (claim); H_a: $\mu_1 \neq \mu_2$

(b) $z_0 = \pm 1.645$; $z < -1.645$ or $z > 1.645$

(c) $z \approx -0.606$

(d) Fail to reject H_0. There is insufficient evidence at the 10% level to reject the claim that the lodging cost for a family traveling in California is the same as Florida.

17. (a) H_0: $\mu_1 = \mu_2$ (claim); H_a: $\mu_1 \neq \mu_2$

(b) $z_0 = \pm 1.645$; $z < -1.645$ or $z > 1.645$

(c) $z \approx 0.645$

(d) Fail to reject H_0. There is insufficient evidence at the 10% level to reject the claim that the lodging cost for a family traveling in California is the same as Florida.

The new samples do not lead to a different conclusion.

19. (a) H_0: $\mu_1 \leq \mu_2$; H_a: $\mu_1 > \mu_2$ (claim)

(b) $z_0 = 1.96$; $z > 1.96$ (c) $z \approx 4.985$

(d) Reject H_0. At the 2.5% level of significance, there is sufficient evidence to support the claim.

21. (a) H_0: $\mu_1 = \mu_2$ (claim); H_a: $\mu_1 \neq \mu_2$

(b) $z_0 = \pm 2.575$; $z < -2.575$ or $z > 2.575$

(c) $z \approx 66.172$

(d) Reject H_0. At the 1% level of significance, there is sufficient evidence to reject the claim.

23. They are equivalent through algebraic manipulation of the equation.

$\mu_1 = \mu_2 \Rightarrow \mu_1 - \mu_2 = 0$

25. H_0: $\mu_1 - \mu_2 = -9$ (claim); H_a: $\mu_1 - \mu_2 \neq -9$

Fail to reject H_0. There is not enough evidence to reject the claim.

27. H_0: $\mu_1 - \mu_2 \leq 6000$; H_a: $\mu_1 - \mu_2 > 6000$ (claim)

Fail to reject H_0. There is not enough evidence to support the claim.

29. $-2.45 < \mu_1 - \mu_2 < 0.65$

31. There is not enough evidence to support the claim. I would not recommend using the herbal supplement with a high-fiber, low-calorie diet to lose weight because there was not a significant difference between the weight loss it produced and that of the placebo.

33. The 95% CI for $\mu_1 - \mu_2$ in Exercise 29 contained values less than or equal to zero and, as found in Exercise 31, there was not enough evidence at the 5% level of significance to support the claim.

If zero is contained in the CI for $\mu_1 - \mu_2$, you fail to reject H_0 because the null hypothesis states that $\mu_1 - \mu_2$ is less than or equal to zero.

Section 8.2 *(page 406)*

1. State hypotheses and identify the claim. Specify the level of significance. Determine the degrees of freedom. Find the critical value(s) and identify the rejection region(s). Find the standardized test statistic. Make a decision and interpret in the context of the original claim.

3. (a) $t_0 = \pm 1.725$ (b) $t_0 = \pm 1.833$

5. (a) $t_0 = -2.074$ (b) $t_0 = -2.306$

7. (a) $t_0 = 1.729$ (b) $t_0 = 1.895$

9. (a) $t_0 = \pm 2.771$ (b) $t_0 = \pm 3.106$

11. H_0: $\mu_1 = \mu_2$ (claim); H_a: $\mu_1 \neq \mu_2$

(a) -1.8 (b) $t \approx -1.199$

(c) t is not in the rejection region.

(d) Fail to reject H_0. There is not enough evidence to reject the claim.

13. There is no need to run the test since this is a right-tailed test and the test statistic is negative. It is obvious that the standardized test statistic will also be negative and fall outside of the rejection region. So, the decision is to fail to reject H_0.

15. (a) H_0: $\mu_1 = \mu_2$ (claim); H_a: $\mu_1 \neq \mu_2$

(b) $t_0 = \pm 1.645$; $t < -1.645$, $t > 1.645$

(c) $t \approx -0.833$

(d) Fail to reject H_0. There is not enough evidence to reject the claim.

17. (a) H_0: $\mu_1 \geq \mu_2$; H_a: $\mu_1 < \mu_2$ (claim)

(b) $t_0 = -1.350$; $t < -1.350$ (c) $t \approx -1.961$

(d) Reject H_0. There is enough evidence to support the claim.

19. (a) H_0: $\mu_1 \leq \mu_2$; H_a: $\mu_1 > \mu_2$ (claim)

(b) $t_0 = 1.345$; $t > 1.345$ (c) $t \approx 2.098$

(d) Reject H_0. There is enough evidence to support the claim.

21. (a) H_0: $\mu_1 = \mu_2$; H_a: $\mu_1 \neq \mu_2$ (claim)

(b) $t_0 = \pm 2.831$; $t < -2.831$, $t > 2.831$

(c) $t \approx -6.410$

(d) Reject H_0. There is enough evidence to support the claim.

23. (a) H_0: $\mu_1 \geq \mu_2$; H_a: $\mu_1 < \mu_2$ (claim)

(b) $t_0 = -1.282$; $t < -1.282$ (c) $t \approx -4.295$

(d) Reject H_0. There is enough evidence to support the claim and to recommend changing to the new method.

25. $-15.664 < \mu_1 - \mu_2 < -4.336$

27. $-0.580 < \mu_1 - \mu_2 < 2.580$

Section 8.3 *(page 417)*

1. Two samples are dependent if each member of one sample corresponds to a member of the other sample. Example: The weights of 22 people before starting an exercise program and the weights of the same 22 people six weeks after starting the exercise program. Two samples are independent if the sample selected from one population is not related to the sample selected from the second population. Example: The weights of 25 cats and the weights of 20 dogs.

3. Independent because the scores are from different students.

5. Dependent because the same adults were sampled.

7. Independent because different boats were sampled.

9. Dependent because the same tire sets were sampled.

11. H_0: $\mu_d \geq 0$; H_a: $\mu_d < 0$ (claim)

$\alpha = 0.05$, d.f. $= 9$

$t_0 = -1.833$ (left-tailed); $t \approx 21.082$

Fail to reject H_0

13. H_0: $\mu_d \leq 0$ (claim); H_a: $\mu_d > 0$

$\alpha = 0.10$, d.f. $= 15$

$t_0 = 1.341$ (right-tailed); $t \approx 67.778$

Reject H_0

15. H_0: $\mu_d \geq 0$ (claim); H_a: $\mu_d < 0$

$\alpha = 0.01$, d.f. $= 14$

$t_0 = -2.624$ (left-tailed); $t \approx -7.423$

Reject H_0

17. (a) H_0: $\mu_d \geq 0$; H_a: $\mu_d < 0$ (claim)

(b) $t_0 = -2.650$; $t < -2.650$

(c) $\bar{d} \approx -33.714$; $s_d \approx 42.034$ (d) $t \approx -3.001$

(e) Reject H_0. There is enough evidence to support the claim that the second SAT scores are improved.

19. (a) H_0: $\mu_d \geq 0$; H_a: $\mu_d < 0$ (claim)

(b) $t_0 = -1.415$; $t < -1.415$

(c) $\bar{d} \approx -1.125$; $s_d \approx 0.871$ (d) $t \approx -3.653$

(e) Reject H_0. There is enough evidence to support the claim that the fuel additive improved gas mileage.

21. (a) H_0: $\mu_d \leq 0$; H_a: $\mu_d > 0$ (claim)

(b) $t_0 = 1.761$; $t > 1.761$

(c) $\bar{d} = 17.6$; $s_d \approx 6.544$ (d) $t \approx 10.416$

(e) Reject H_0. There is enough evidence to support the claim that the new drug reduces systolic blood pressure.

23. (a) H_0: $\mu_d \leq 0$; H_a: $\mu_d > 0$ (claim)

(b) $t_0 = 2.764$; $t > 2.764$

(c) $\bar{d} \approx 1.255$; $s_d \approx 0.441$ (d) $t \approx 9.438$

(e) Reject H_0. There is enough evidence to support the claim that soft tissue therapy and spinal manipulation help reduce the length of time patients suffer from headaches.

25. $-1.763 < \mu_d < -1.287$

Section 8.4 *(page 426)*

1. State the hypotheses and identify the claim. Specify the level of significance. Find the critical value(s) and rejection region(s). Find $\bar{p}$ and $\bar{q}$. Find the standardized test statistic. Make a decision and interpret in the context of the claim.

3. H_0: $p_1 = p_2$; H_a: $p_1 \neq p_2$ (claim)

The test is a two-tailed test. Fail to reject H_0. There is not enough evidence to support the claim.

5. H_0: $p_1 \leq p_2$ (claim); H_a: $p_1 > p_2$

The test is a right-tailed test. Fail to reject H_0. There is not enough evidence to reject the claim.

7. (a) H_0: $p_1 = p_2$ (claim); H_a: $p_1 \neq p_2$

The test is a two-tailed test. Fail to reject H_0. There is not enough evidence to support the claim.

9. (a) H_0: $p_1 = p_2$ (claim); H_a: $p_1 \neq p_2$

(b) $z_0 = \pm 1.96$; $z < -1.96$ or $z > 1.96$

(c) $z \approx -5.060$

(d) Reject H_0. There is sufficient evidence at the 5% level to reject the claim that the proportion of adults using alternative medicines has not changed since 1990.

11. (a) H_0: $p_1 \geq p_2$; H_a: $p_1 < p_2$ (claim)

(b) $z_0 = -2.33$; $z < -2.33$ (c) $z \approx -2.859$

(d) Reject H_0. There is sufficient evidence at the 1% level to support the claim that the proportion of adults who are smokers is lower in Alabama than in Missouri.

13. (a) H_0: $p_1 \geq p_2$; H_a: $p_1 < p_2$ (claim)

(b) $z_0 = -2.33$; $z < -2.33$ (c) $z \approx -2.514$

(d) Reject H_0. There is sufficient evidence at the 1% level to support the claim that the proportion of college students who said they had smoked in the last 30 days in the private schools is less than the proportion in the public schools.

15. H_0: $p_1 \leq p_2$; H_a: $p_1 > p_2$ (claim)

$z_0 = 2.33$; $z \approx 0.881$

Fail to reject H_0. There is insufficient evidence at the 1% level to support the claim.

17. H_0: $p_1 = p_2$ (claim); H_a: $p_1 \neq p_2$

$z_0 = \pm 1.96$; $z \approx 3.797$

Reject H_0. There is sufficient evidence at the 5% level to reject the advocate's claim.

19. $0.0312 < p_1 - p_2 < 0.0328$

Uses and Abuses for Chapter 8 *(page 433)*

1. Answers will vary.

2. Answers will vary.

Review Answers for Chapter 8 *(page 434)*

1. H_0: $\mu_1 \geq \mu_2$ (claim); H_a: $\mu_1 < \mu_2$

$z_0 = -1.645$; $z \approx -1.862$

Reject H_0

3. H_0: $\mu_1 \geq \mu_2$; H_a: $\mu_1 < \mu_2$ (claim)

$z_0 = -1.282$; $z \approx -2.060$

Reject H_0

5. (a) H_0: $\mu_1 \geq \mu_2$; H_a: $\mu_1 < \mu_2$ (claim)

(b) $z < -1.645$ (c) $z \approx -2.713$

(d) Reject H_0

(e) There is sufficient evidence at the 5% level to conclude that the Arby's sandwich has fewer calories than the McDonald's sandwich.

7. H_0: $\mu_1 = \mu_2$ (claim); H_a: $\mu_1 \neq \mu_2$

$t_0 = \pm 1.96$; $t \approx 1.121$

Fail to reject H_0

9. H_0: $\mu_1 \leq \mu_2$ (claim); H_a: $\mu_1 > \mu_2$

$t_0 = 1.711$; $t \approx -1.460$

Fail to reject H_0

11. H_0: $\mu_1 = \mu_2$; H_a: $\mu_1 \neq \mu_2$ (claim)

$t_0 = \pm 3.169$; $t \approx 4.484$

Reject H_0

13. (a) H_0: $\mu_1 \leq \mu_2$; H_a: $\mu_1 > \mu_2$ (claim)

(b) $t_0 = 1.645$ (c) $t \approx 2.266$

(d) Reject H_0. There is sufficient evidence at the 5% level to support the claim that third graders taught with the directed reading activities scored higher than those taught without the activities.

15. Independent, because the two samples of laboratory mice are different groups.

17. H_0: $\mu_d = 0$ (claim); H_a: $\mu_d \neq 0$

$t_0 = \pm 1.96$ (two-tailed test); $t \approx 8.065$

Reject H_0

19. H_0: $\mu_d \leq 6$ (claim); H_a: $\mu_d > 6$

$t_0 = 1.282$ (right-tailed test); $t \approx 19.921$

Reject H_0

21. (a) H_0: $\mu_d \leq 0$; H_a: $\mu_d > 0$ (claim)

(b) $t_0 = 1.383$; $t > 1.383$ (c) $\bar{d} = 5$; $s_d \approx 8.743$

(d) $t \approx 1.808$ (e) Reject H_0

(f) There is enough evidence to support the claim.

23. The test is two-tailed. H_0: $p_1 = p_2$; H_a: $p_1 \neq p_2$ (claim)

$z_0 = \pm 1.96$; $z \approx -1.198$

Fail to reject H_0

25. The test is right-tailed. H_0: $p_1 \leq p_2$; H_a: $p_1 > p_2$ (claim)
$z_0 = 1.28$; $z \approx -1.970$
Fail to reject H_0

27. (a) H_0: $p_1 = p_2$ (claim); H_a: $p_1 \neq p_2$
(b) $z_0 = \pm 1.645$; $z < -1.645$ or $z > 1.645$
(c) $z \approx -0.776$
(d) Fail to reject H_0
(e) There is not enough evidence to reject the claim.

Real Statistics–Real Decisions for Chapter 8

(page 438)

1. (a) Take a simple random sample of records today and 10 years ago from a random sample of hospitals.
Use a cluster sample.
(b) Answers will vary.
(c) Answers will vary.

2. Use a t-test; independent; yes, need to know if normal or not; yes, need to know if population variances are equal or not.

3. H_0: $\mu_1 = \mu_2$; H_a: $\mu_1 \neq \mu_2$ (claim)
$t_0 = \pm 2.093$ $t \approx 2.974$
Reject H_0. There is enough evidence to support the claim.

Chapter Quiz for Chapter 8 *(page 439)*

1. (a) H_0: $\mu_1 \leq \mu_2$; H_a: $\mu_1 > \mu_2$ (claim)
(b) Right tailed z-test because n_1 and n_2 are each greater than 30 and the samples are independent.
(c) $z_0 = 1.645$ (d) $z \approx 28.387$
(e) Reject H_0. There is sufficient evidence at the 5% level to support the claim that the mean score on the science assessment for male high school students was higher than for the female high school students.

2. (a) H_0: $\mu_1 = \mu_2$ (claim); H_a: $\mu_1 \neq \mu_2$
(b) Two tailed t-test because n_1 and n_2 are less than 30, the samples are independent, and the populations are normally distributed.
(c) $t_0 = \pm 2.779$ (d) $t \approx 4.285$
(e) Reject H_0. There is sufficient evidence at the 1% level to reject the teacher's suggestion that the mean scores on the science assessment test are the same for nine-year old boys and girls.

3. (a) H_0: $p_1 \leq p_2$; H_a: $p_1 > p_2$ (claim)
(b) Right tailed z-test because we are testing proportions and $n_1\bar{p}$, $n_2\bar{p}$, $n_1\bar{q}$, and $n_2\bar{q} \geq 5$ and the samples are independent.
(c) $z_0 = 1.282$ (d) $z \approx 158.471$
(e) Reject H_0. There is sufficient evidence at the 10% level to support the claim that the proportion of accidents involving alcohol is higher for drivers in the 21 to 24 age group than for drivers aged 65 and over.

4. (a) H_0: $\mu_d \geq 0$; H_a: $\mu_d < 0$ (claim)
(b) Dependent t-test because both populations are normally distributed and the samples are dependent.
(c) $t_0 = -1.796$ (d) $t \approx -9.016$
(e) Reject H_0. There is sufficient evidence at the 5% level to conclude that the students' SAT scores improved on the second test.

CHAPTER 9

Section 9.1 *(page 453)*

1. Positive linear correlation

3. No linear correlation (but there is a nonlinear correlation between the variables)

5. c; You would expect a positive linear correlation between age and income.

7. b; You would expect a negative linear correlation between age and balance on student loans.

9. Explanatory variable: Amount of water consumed.
Response variable: Weight loss.

11. (a)

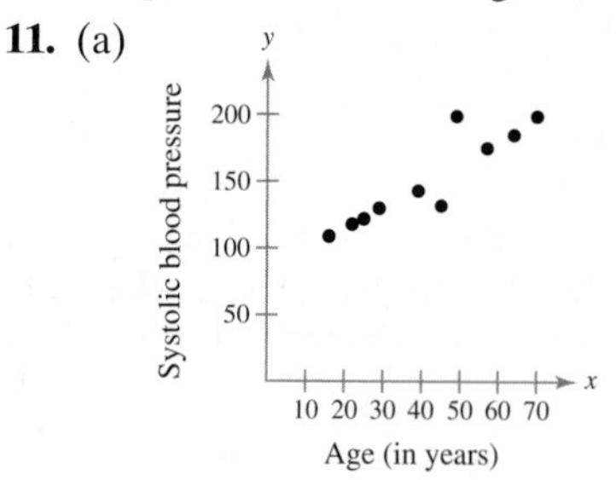

(b) 0.908 (c) Strong positive linear correlation

13. (a)

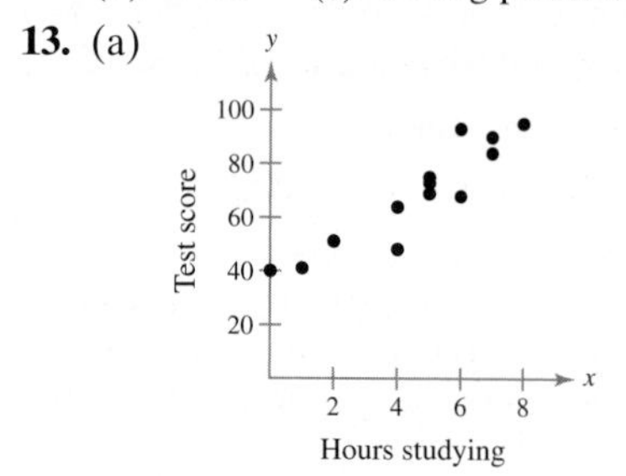

(b) 0.923 (c) Strong positive linear correlation

15. (a)

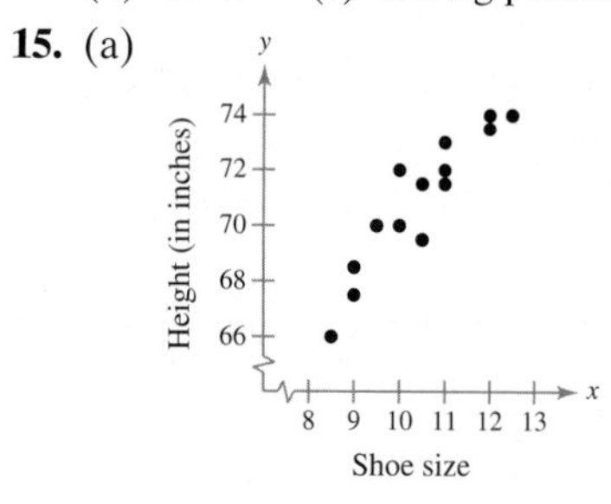

(b) 0.926 (c) Strong positive linear correlation

17. (a) 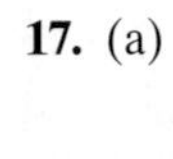

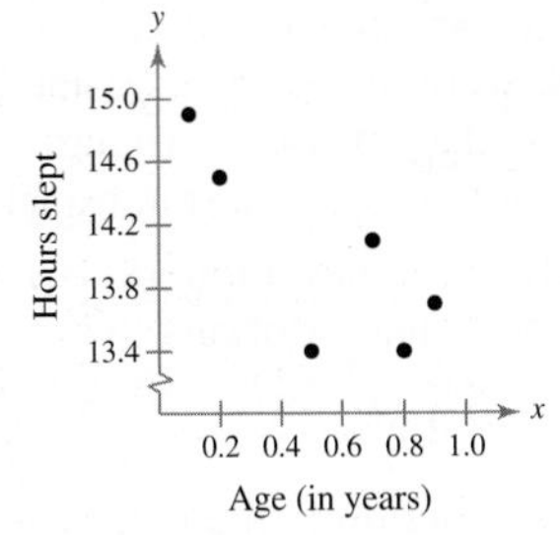

(b) −0.789

(c) Negative linear correlation

19. There is not a significant linear correlation between vehicle weight and the variability in braking distance.

21. There is a significant linear correlation between the data.

23. There is a significant linear correlation between the men's shoe size and height.

25. $r = -0.84$ represents a stronger correlation since it is closer to -1 than $r = 0.73$ is to $+1$.

27. State the null and alternative hypotheses. Specify the level of significance and determine the degrees of freedom. Identify the rejection regions and calculate the standardized test statistic. Make a decision and interpret in the context of the original claim.

29. The correlation coefficient remains unchanged when the x-values and y-values are switched.

31. Answers will vary.

Section 9.2 *(page 462)*

1. c **3.** d

5. g **7.** h

9. For $y = 0.4319x - 20.2970$, $\Sigma d^2 \approx 42.26$. For $y = 0.69x - 40$, $\Sigma d^2 \approx 54.04$. Because $42.26 < 54.04$, $y = 0.4319x - 20.2970$ is the line that fits the data better.

11. c **13.** a

15. $\hat{y} = 1.724x + 79.733$

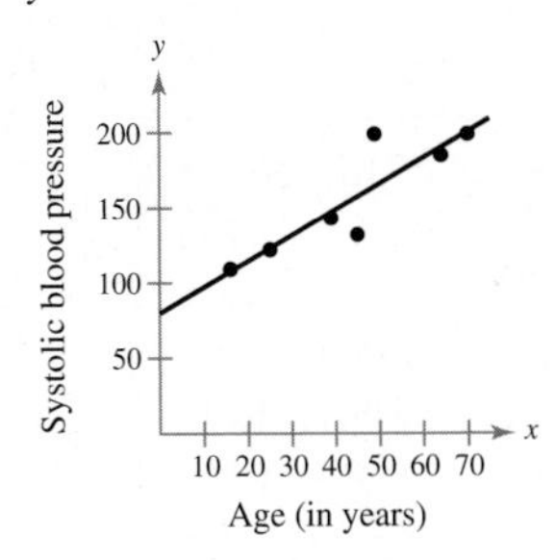

(a) 111 (b) 202

(c) 130 (d) 175

17. $\hat{y} = 7.350x + 34.617$

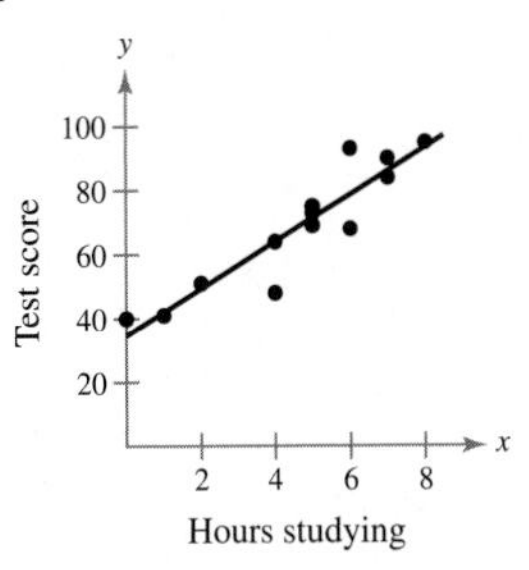

(a) 56.7

(b) 82.4

(c) It is not meaningful to predict the value of y for $x = 13$ because $x = 13$ is outside the range of the original data.

(d) 67.7

19. $\hat{y} = 1.870x + 51.360$

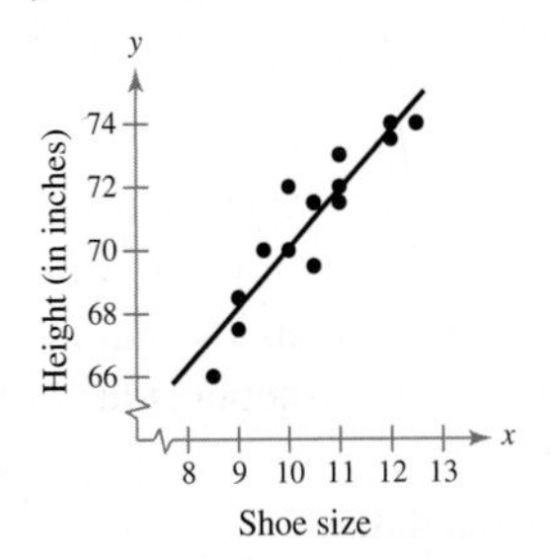

(a) 72.865

(b) 66.320

(c) It is not meaningful to predict the value of y for $x = 15.5$ because $x = 15.5$ is outside the range of the original data.

(d) 70.060

21. $\hat{y} = 0.000447x - 0.425$

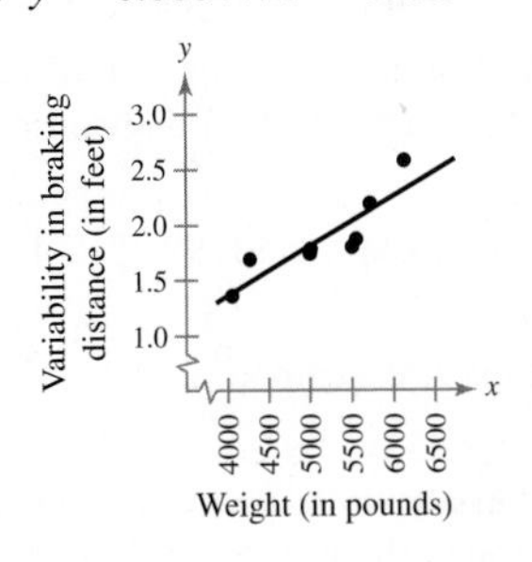

(a) 1.587

(b) 2.257

(c) It is not meaningful to predict the value of y for $x = 7500$ because $x = 7500$ is outside the range of the original data.

(d) 2.145

23. Substitute a value x into the equation of a regression line and solve for y.

25. (a) $\hat{y} = 1.724x + 79.733$

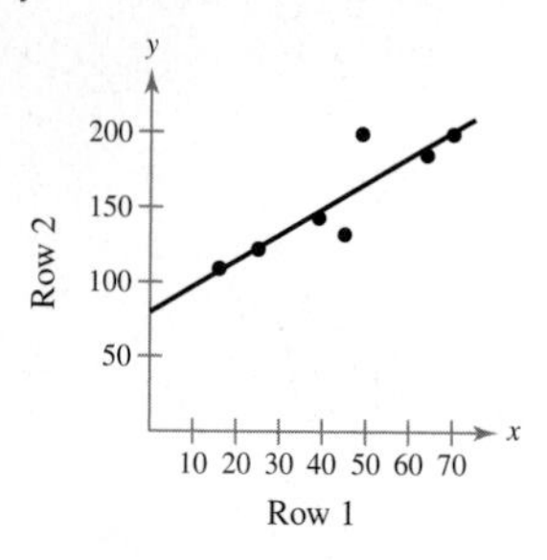

(b) $\hat{y} = 0.453x - 26.448$

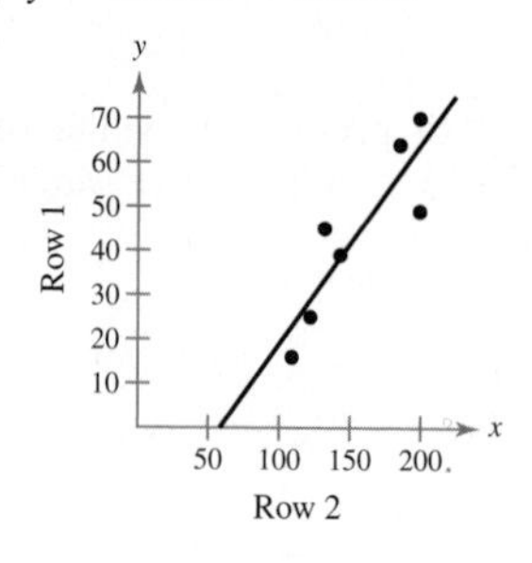

(c) The slope of the line keeps the same sign, but the values of m and b change.

27. Strong negative nonlinear correlation; as the number of subscribers increases, the average monthly bill decreases.

29. $-8.63; This is not a valid prediction because $x = 140$ is outside the range of the original data.

31. As the number of cellular phone subscribers increases, the average monthly cellular phone bill decreases. (Answers will vary.)

Section 9.3 *(page 474)*

1. $\Sigma(y_i - \bar{y})^2$; the sum of the squares of the differences between the y-values of each ordered pair and the mean of the y-values of the ordered pairs.

3. $\Sigma(y_i - \hat{y}_i)^2$; the sum of the squares of the differences between the observed y-values and the predicted y-values.

5. 0.063; 6.3% of the variation is explained. 93.7% of the variation is unexplained.

7. 0.794; 79.4% of the variation is explained. 20.6% of the variation is unexplained.

9. (a) 0.817; 81.7% of the variation in proceeds can be explained by the variation in the number of issues and 18.3% of the variation is unexplained.

(b) $s_e \approx 6029.907$; the standard deviation of the proceeds for a specific number of issues is about $6,029,907,000.

11. (a) 0.985; 98.5% of the variation in sales can be explained by the variation in the total square footage and 1.5% of the variation is unexplained.

(b) $s_e \approx 35.652$; the standard deviation of the sales for a specific total square footage is about 35,652,000,000.

13. (a) 0.998; 99.8% of the variation in the median weekly earnings of female workers can be explained by the variation in the median weekly earnings of male workers and 0.2% of the variation is unexplained.

(b) $s_e \approx 5.147$; the standard deviation of the median weekly earnings of female workers for a specific median weekly earnings of male workers is about $5.147.

15. (a) 0.992; 99.2% of the variation in the money spent can be explained by the variation in the money raised and 0.8% of the variation is unexplained.

(b) $s_e \approx 16.079$; the standard deviation of the money spent for a specified amount of money raised is about $16,079,000.

17. ($17,935,784,000, $47,264,966,000); you can be 95% confident that the proceeds will be between 17,935,784,000 and $47,264,966,000 when the number of initial offerings is 712.

19. ($679,861,000,000, $817,739,000,000); you can be 90% confident that the sales will be between $679,861,000,000 and $817,739,000,000 when the total square footage is 4.5 billion.

21. ($333.29, $400.13); you can be 99% confident that the median earnings of female workers will be between $333.29 and $400.13 when the median weekly earnings of male workers is $500.

23. ($723.415 million, $807.509 million); you can be 95% confident that the money spent in congressional campaigns will be between $723.415 million and $807.509 million when the money raised is $775.8 million.

25.

y
Trucks
$\bar{y} = 6.786$
$\bar{x} = 6.5$
5 6 7 8
x
Cars

27.

x_i	y_i	$\hat{y}_i$	$\hat{y}_i - \bar{y}$	$y_i - \hat{y}_i$	$y_i - \bar{y}$
8.1	7.8	7.893	1.107	−0.093	1.014
7.7	7.6	7.616	0.830	−0.016	0.814
6.5	6.5	6.785	−0.001	−0.285	−0.286
6.9	7.6	7.062	0.276	0.538	0.814
6.0	6.3	6.438	−0.348	−0.138	−0.486
5.4	5.8	6.022	−0.764	−0.222	−0.986
4.9	5.9	5.676	−1.11	0.224	−0.886

29. 0.887 **31.** (6.441, 8.237)

Section 9.4 *(page 481)*

1. (a) 2614.6 (b) 2298 (c) 2680 (d) 2233

3. (a) 7.5 (b) 16.8 (c) 51.9 (d) 62.1

5. $\hat{y} = -256.293 + 103.502x_1 + 14.649x_2$

(a) 34.16 (b) 0.988

(c) The standard deviation of the predicted sales given a specific total square footage and number of shopping centers is \$34.16 billion. The multiple regression model explains 98.8% of the variation in y.

7. 0.985

Uses and Abuses for Chapter 9 *(page 485)*

1. Answers will vary.

2. Answers will vary.

Review Answers for Chapter 9 *(page 486)*

1.

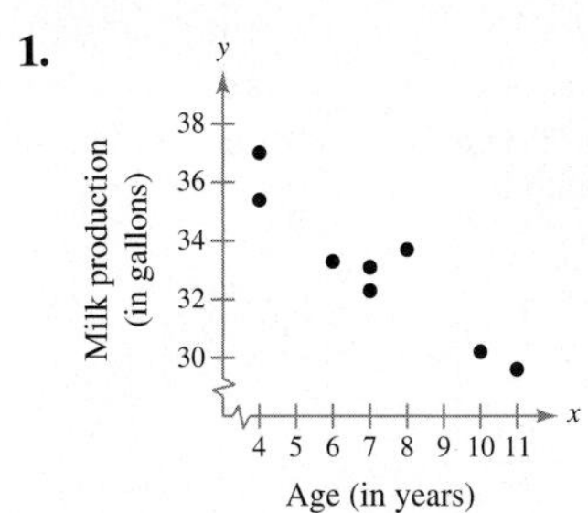

$r \approx -0.939$; negative linear correlation; milk production decreases with age.

3. $H_0: \rho = 0$; $H_a: \rho \neq 0$; $t_0 = \pm 2.797$; $t \approx 1.211$

Fail to reject H_0.

5. $H_0: \rho = 0$; $H_a: \rho \neq 0$; $t_0 = \pm 2.447$; $t \approx -6.688$

Reject H_0. There is enough evidence at the 5% level to conclude that there is a significant linear correlation between the age of the cow and its milk production.

7. $\hat{y} = 0.757x + 21.5$

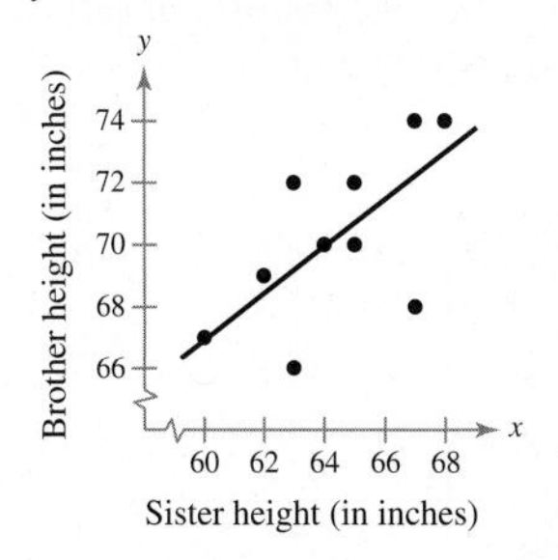

9. $r \approx 0.688$

(a) 68 (b) 71

(c) Not meaningful since $x = 71$ in. is outside the range of the data.

(d) Not meaningful since $x = 50$ in. is outside the range of the data.

11. 0.306; 30.6% of the variation in y is explained by the model.

13. 0.033; 3.3% of the variation in y is explained by the model.

15. (a) 0.897; 89.7% of the variation in y is explained by the model.

(b) 568.0; the standard error of the cooling capacity for a specific living area is 568.0 BTU/hr.

17. $65.71 < y < 74.19$ **19.** $8184.33 < y < 11{,}455.59$

21. $\hat{y} = 6.317 + 0.822x_1 + 0.031x_2 - 0.004x_3$

23. (a) 21.705 (b) 25.210 (c) 30.100 (d) 25.860

Real Statistics–Real Decisions for Chapter 9
(page 489)

1. (a)

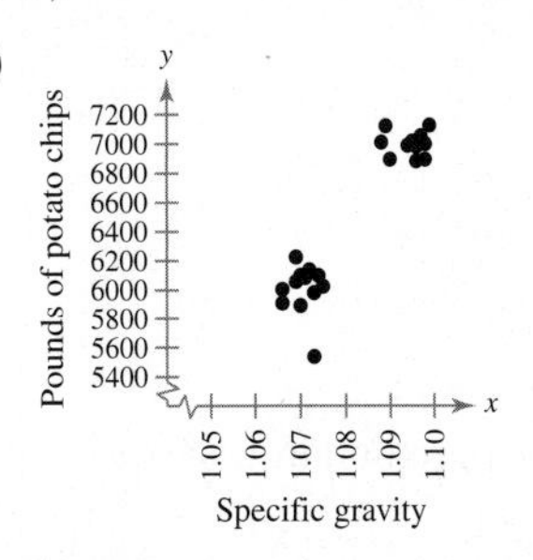

Positive correlation

(b) 0.931

(c) $H_0: \rho = 0$; $H_a: \rho \neq 0$

$t_0 = \pm 2.074$; $t \approx 11.96 > 2.074 \Rightarrow$ Reject H_0.

Or $n = 24$; critical value $= 0.404$;

$|r| = 0.931 > 0.404 \Rightarrow$ Reject H_0.

There is a significant linear correlation.

(d) $\hat{y} = 38535x - 35219$

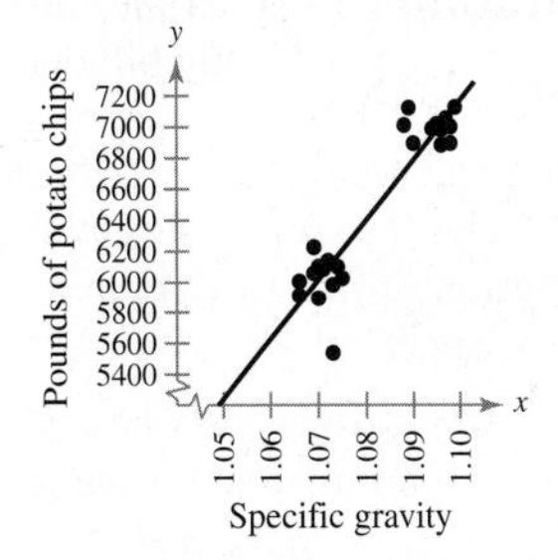

(e) Answers will vary.

(f) 0.867; 196.1; 86.7% of the variance in y is explained by the regression line; the standard deviation of pounds of potato chips yielded for a given specific gravity is 196.1 pounds.

2. Answers will vary.

Chapter Quiz for Chapter 9 *(page 490)*

1.

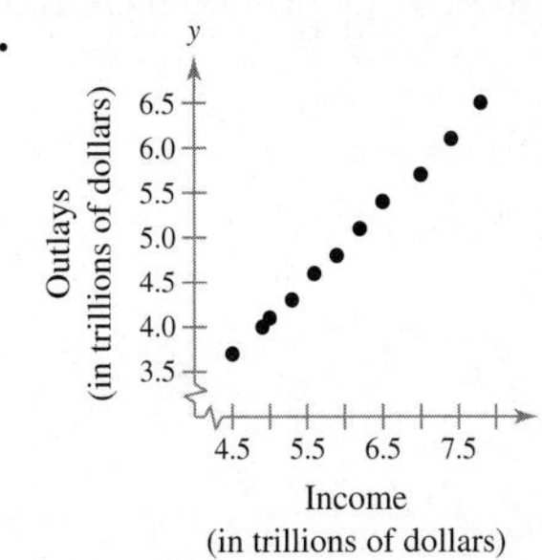

The data appear to have a positive correlation. The outlays increase as the incomes increase.

2. 0.999 → strong positive linear correlation

3. $H_0: \rho = 0$ and $H_a: \rho \neq 0$; $t_0 = \pm 2.262$; $t \approx 67.03$

Reject H_0. There is enough evidence to conclude that a significant correlation exists.

4. $\hat{y} = 0.843x - 0.129$

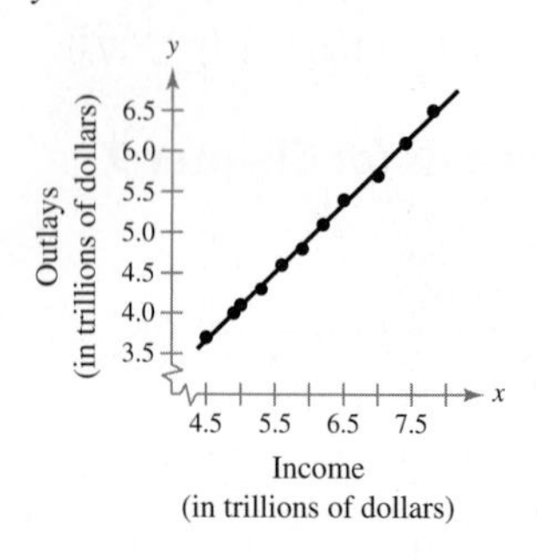

5. 4.339

6. $r^2 = 0.998$; 99.8% of the variation in y is explained by the regression model.

7. \$0.042 trillion; the standard deviation of personal outlays for a specified personal income is \$0.042 trillion.

8. (\$5.17 trillion, \$5.37 trillion); You can be 95% confident that the personal outlays will be between \$5.17 trillion and \$5.37 trillion when personal income is \$6.4 trillion.

9. (a) 1311.150 (b) 961.110 (c) 1120.900 (d) 1386.740

x_2 has the greatest influence on y.

Cumulative Test for Chapters 7–9 *(page 491)*

1. H_0: $\mu \geq 20.1$ (claim); H_a: $\mu < 20.1$

$z_0 = -2.33$; $z \approx -1.530$; Fail to reject H_0. There is not enough evidence at the 1% level to reject the claim that the mean per capita consumption of rice by people in the U.S. is at least 20.1 pounds per year.

2. Type I error will occur if H_0 is rejected when $\mu \geq 20.1$. Type II error will occur if H_0 is not rejected when $\mu < 20.1$.

3. H_0: $\mu_1 - \mu_2 = 0$ (claim); H_a: $\mu_1 - \mu_2 \neq 0$

$z_0 = \pm 1.96$; $z \approx -3.496$

Reject H_0

4. (a) Standard normal distribution because the sample size is greater than 30.

(b) Standard normal distribution because both sample sizes are greater than 30 and the samples are independent.

5. 0.989 → Positive linear correlation; the number of acres harvested increases as the number of acres planted increases.

6. $\hat{y} = 1.071x - 268.253$

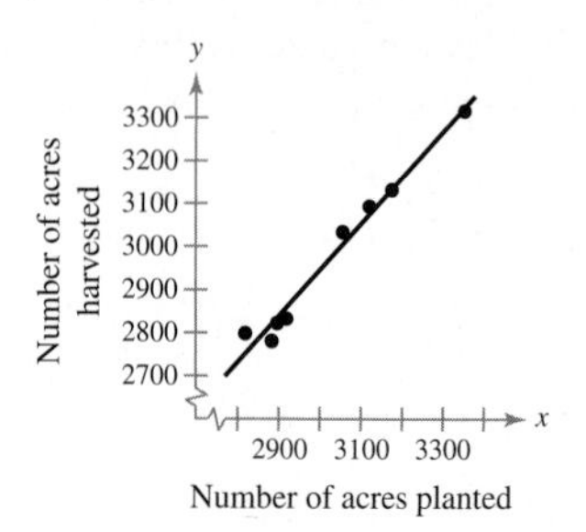

7. 0.978; $r^2 = 0.978 \rightarrow 97.8\%$ of the variation in y is explained by the model.

8. 31.46 thousand; The standard deviation of acres harvested for a specified number of acres planted is 31.46 thousand.

9. H_0: $\rho = 0$ (claim); H_a: $\rho \neq 0$

$t_0 = \pm 3.707$; $t \approx 16.378$

Reject H_0. There is enough evidence at the 1% level to reject the claim that there is no linear correlation between the number of acres of rice planted and the number of acres of rice harvested.

10. (2969.388 thousand, 3134.306 thousand)

CHAPTER 10

Section 10.1 *(page 500)*

1. (a) Claimed distribution:

Response	Distribution
Home	70%
Work	17%
Commuting	8%
Other	5%

H_0: Distribution of responses is as shown in table above.

H_a: Distribution of responses differs from the claimed distribution.

(b) $\chi_0^2 = 7.815$; $\chi^2 > 7.815$

(c) $\chi^2 \approx 3.754$

(d) Fail to reject H_0. There is not enough evidence at the 5% level to conclude that the distribution of responses differs from the claimed distribution.

3. (a) Claimed distribution:

Day	Distribution
Sunday	14.286%
Monday	14.286%
Tuesday	14.286%
Wednesday	14.286%
Thursday	14.286%
Friday	14.286%
Saturday	14.286%

H_0: The distribution of fatal bicycle accidents throughout the week is uniform as shown in the table above.

H_a: The distribution of fatal bicycle accidents throughout the week is not uniform.

(b) $\chi_0^2 = 10.645$; $\chi^2 > 10.645$

(c) $\chi^2 \approx 4.648$

(d) Fail to reject H_0. There is not enough evidence at the 10% level to conclude that the distribution of fatal bicycle accidents throughout the week is not uniform.

5. (a) Claimed distribution:

Object struck	Distribution
Tree	28%
Embankment	10%
Utility pole	10%
Guardrail	9%
Ditch	7%
Curb	6%
Culvert	5%
Sign/Post/Fence	10%
Other	15%

H_0: Distribution of objects struck is as shown in table above.

H_a: Distribution of objects struck differs from the claimed distribution.

(b) $\chi_0^2 = 20.090$; $\chi^2 > 20.090$ (c) $\chi^2 \approx 49.665$

(d) Reject H_0. There is enough evidence at the 1% level to conclude that the distribution of objects struck changed after warning signs were erected.

7. (a) Claimed distribution:

Response	Distribution
Not a HS grad	33.333%
HS graduate	33.333%
College (1yr+)	33.333%

H_0: Distribution of the responses is as shown in table above.

H_a: Distribution of the responses differs from the claimed distribution.

(b) $\chi_0^2 = 7.378$; $\chi^2 > 7.378$ (c) $\chi^2 \approx 5.637$

(d) Fail to reject H_0. There is not enough evidence at the 2.5% level to conclude that the distribution of responses is not uniform.

9. (a) Claimed distribution:

Cause	Distribution
Trans. accidents	41%
Assaults	20%
Objects/equipment	15%
Falls	10%
Exposure	10%
Other	4%

H_0: Distribution of the causes is as shown in table above.

H_a: Distribution of the causes differs from the claimed distribution.

(b) $\chi_0^2 = 11.071$; $\chi^2 > 11.071$

(c) $\chi^2 \approx 9.493$

(d) Fail to reject H_0. There is not enough evidence at the 5% level to conclude that the distribution of causes in the Western U.S. differs from the national distribution.

11. (a) Frequency distribution: $\mu = 69.435$; $\sigma \approx 8.337$

Lower boundary	Upper boundary	Lower z-score	Upper z-score	Area
49.5	58.5	−2.39	−1.31	0.0867
58.5	67.5	−1.31	−0.23	0.3139
67.5	76.5	−0.23	0.85	0.3933
76.5	85.5	0.85	1.93	0.1709
85.5	94.5	1.93	3.01	0.0255

Class boundaries	Distribution	Frequency f	Expected	$\frac{(O-E)^2}{E}$
49.5–58.5	8.67%	19	17	0.235
58.5–67.5	31.39%	61	63	0.063
67.5–76.5	39.33%	82	79	0.114
76.5–85.5	17.09%	34	34	0
85.5–94.5	2.55%	4	5	0.2
		200		0.612

H_0: Variable has a normal distribution.

H_a: Variable does not have a normal distribution.

(b) $\chi_0^2 = 13.277$; $\chi^2 > 13.277$ (c) $\chi^2 \approx 0.612$

(d) Fail to reject H_0. There is not enough evidence at the 1% level to conclude that the distribution of test scores is not normal.

Section 10.2 *(page 510)*

1. (a) H_0: Skill level in a subject is independent of location. (claim)

H_a: Skill level in a subject is dependent on location.

(b) 2; $\chi_0^2 = 9.210$; $\chi^2 > 9.210$ (c) $\chi^2 \approx 0.297$

(d) Fail to reject H_0. There is not enough evidence at the 1% level to conclude that skill level in a subject is dependent on location.

3. (a) H_0: Adults' ratings are independent of the type of school.

H_a: Adults' ratings are dependent on the type of school. (claim)

(b) 3; $\chi_0^2 = 7.815$; $\chi^2 > 7.815$ (c) $\chi^2 \approx 148.389$

(d) Reject H_0. There is enough evidence at the 5% level to conclude that adults' ratings are dependent on the type of school.

5. (a) H_0: Results are independent of the type of treatment.

H_a: Results are dependent on the type of treatment. (claim)

(b) 1; $\chi_0^2 = 2.706$; $\chi^2 > 2.706$ (c) $\chi^2 \approx 5.106$

(d) Reject H_0. There is enough evidence at the 10% level to conclude that results are dependent on the type of treatment. I would recommend using the drug.

7. (a) H_0: Reasons are independent of the type of worker.

H_a: Reasons are dependent on the type of worker. (claim)

(b) 2; $\chi_0^2 = 9.210$; $\chi^2 > 9.210$ (c) $\chi^2 \approx 7.326$

(d) Fail to reject H_0. There is not enough evidence at the 1% level to conclude that reason(s) for continuing education are dependent on the type of worker. Based on these results, marketing strategies should not differ between technical and non-technical audiences in regard to reason(s) for continuing education.

9. (a) H_0: Type of crash is independent of the type of vehicle.

H_a: Type of crash is dependent on the type of vehicle. (claim)

(b) 2; $\chi_0^2 = 5.991$; $\chi^2 > 5.991$ (c) $\chi^2 \approx 106.390$

(d) Reject H_0. There is enough evidence at the 5% level to conclude that the type of crash is dependent on the type of vehicle.

11. (a) H_0: Subject is independent of coauthorship.

H_a: Coauthorship and subject are dependent. (claim)

(b) d.f. = 4; $\chi_0^2 = 7.779$ (c) $\chi^2 \approx 5.610$

(d) Fail to reject H_0. There is not enough evidence at the 10% level to conclude that subject matter and coauthorship are related.

13. H_0: The proportions are equal. (claim)

H_a: At least one of the proportions is different from the others.

d.f. $= (r - 1)(c - 1) = 1$

$\chi_0^2 = 2.706 \rightarrow$ Reject H_0 if $\chi^2 > 2.706$

$\chi^2 \approx 5.106$

Reject H_0. There is enough evidence at the 10% level to conclude that at least one of the proportions is different from the others.

Section 10.3 *(page 521)*

1. Specify the level of significance α. Determine the degrees of freedom for the numerator and denominator. Use Table 7 to find the critical value F.

3. 2.93 **5.** 5.32 **7.** 2.06

9. H_0: $\sigma_1^2 \leq \sigma_2^2$; H_a: $\sigma_1^2 > \sigma_2^2$ (claim)

$F_0 = 3.52$; $F > 3.52$; $F \approx 1.010$

Fail to reject H_0

11. H_0: $\sigma_1^2 \leq \sigma_2^2$ (claim); H_a: $\sigma_1^2 > \sigma_2^2$

$F_0 = 5.26$; $F > 5.26$; $F \approx 1.007$

Fail to reject H_0

13. H_0: $\sigma_1^2 = \sigma_2^2$ (claim); H_a: $\sigma_1^2 \neq \sigma_2^2$

$F_0 = 4.65$; $F > 4.65$; $F \approx 1.083$

Fail to reject H_0

15. (a) Population 1: Company B

Population 2: Company A

H_0: $\sigma_1^2 \leq \sigma_2^2$; H_a: $\sigma_1^2 > \sigma_2^2$ (claim)

(b) $F_0 = 2.13$; $F > 2.13$ (c) $F \approx 1.077$

(d) Fail to reject H_0. There is not enough evidence at the 5% level to support Company A's claim that the variance of the life of its appliances is less than the variance of the life of Company B's appliances.

17. (a) Population 1: District 1

Population 2: District 2

H_0: $\sigma_1^2 = \sigma_2^2$ (claim); H_a: $\sigma_1^2 \neq \sigma_2^2$

(b) $F_0 = 2.63$; $F > 2.63$ (c) $F \approx 1.126$

(d) Fail to reject H_0. There is not enough evidence at the 10% level to reject the claim that the standard deviation of physical science assessment test scores for eighth-grade students is the same in Districts 1 and 2.

19. (a) Population 1: Before new admissions procedure

Population 2: After new admissions procedure

H_0: $\sigma_1^2 \leq \sigma_2^2$; H_a: $\sigma_1^2 > \sigma_2^2$ (claim)

(b) $F_0 = 1.77$; $F > 1.77$ (c) $F \approx 1.96$

(d) Reject H_0. There is enough evidence at the 10% level to support the claim that the standard deviation of waiting times has decreased.

21. (a) Population 1: California; Population 2: New York

H_0: $\sigma_1^2 \leq \sigma_2^2$; H_a: $\sigma_1^2 > \sigma_2^2$ (claim)

(b) $F_0 = 2.35$; $F > 2.35$ (c) $F \approx 2.495$

(d) Reject H_0. There is not enough evidence at the 5% level to conclude that the standard deviation of the annual salaries for actuaries is greater in California than in New York.

23. Right tailed: 8.94

Left tailed: 0.210

25. (0.366, 3.839)

Section 10.4 *(page 531)*

1. (a) H_0: $\mu_1 = \mu_2 = \mu_3$

H_a: At least one mean is different from the others. (claim)

(b) d.f.$_N = 2$; d.f.$_D = 27$; $F_0 = 3.35$; $F > 3.35$

(c) $F \approx 1.26$

(d) Fail to reject H_0. There is not enough evidence at the 5% level to conclude that the mean costs per month are different.

3. (a) H_0: $\mu_1 = \mu_2 = \mu_3$ (claim)

H_a: At least one mean is different from the others.

(b) d.f.$_N = 2$; d.f.$_D = 12$; $F_0 = 2.81$; $F > 2.81$

(c) $F \approx 1.77$

(d) Fail to reject H_0. There is not enough evidence at the 10% level to reject the claim that the mean prices are all the same for the three types of treatment.

5. (a) H_0: $\mu_1 = \mu_2 = \mu_3 = \mu_4$ (claim)

H_a: At least one mean is different from the others.

(b) d.f.$_N = 3$; d.f.$_D = 33$; $F_0 = 4.44$; $F > 4.44$

(c) $F \approx 5.21$

(d) Reject H_0. There is enough evidence at the 1% level to reject the claim that the mean number of days patients spend in the hospital is the same for all four regions.

7. (a) H_0: $\mu_1 = \mu_2 = \mu_3 = \mu_4$ (claim)

H_a: At least one mean is different from the others.

(b) $\text{d.f.}_N = 3$; $\text{d.f.}_D = 43$; $F_0 = 2.22$; $F > 2.22$

(c) $F \approx 3.04$

(d) Reject H_0. There is enough evidence at the 10% level to reject the claim that the mean price is the same for all four cities.

9. (a) H_0: $\mu_1 = \mu_2 = \mu_3 = \mu_4$

H_a: At least one mean is different from the others. (claim)

(b) $\text{d.f.}_N = 3$; $\text{d.f.}_D = 26$; $F_0 = 2.31$; $F > 2.31$

(c) $F \approx 6.36$

(d) Reject H_0. There is enough evidence at the 10% level to conclude that at least one of the mean prices is different.

11. (a) H_0: $\mu_1 = \mu_2 = \mu_3 = \mu_4$ (claim)

H_a: At least one mean is different from the others.

(b) $\text{d.f.}_N = 3$; $\text{d.f.}_D = 23$; $F_0 = 4.76$; $F > 4.76$

(c) $F \approx 0.89$

(d) Fail to reject H_0. There is not enough evidence at the 1% level to reject the claim that the mean amounts spent are equal for all regions.

13. (a) H_0: $\mu_1 = \mu_2 = \mu_3 = \mu_4$

H_a: At least one mean is different from the others. (claim)

(b) $\text{d.f.}_N = 3$; $\text{d.f.}_D = 18$; $F_0 = 5.09$; $F > 5.09$

(c) $F \approx 47.82$

(d) Reject H_0. There is enough evidence at the 1% level to support the claim that the mean credit card balance is different for at least one age group.

15. $\text{CV}_{\text{Scheffé}} = 6.660$

$(1, 2) \rightarrow 7.212 \rightarrow$ Significant difference

$(1, 3) \rightarrow 3.519 \rightarrow$ No difference

$(1, 4) \rightarrow 0.260 \rightarrow$ No difference

$(2, 3) \rightarrow 0.724 \rightarrow$ No difference

$(2, 4) \rightarrow 4.782 \rightarrow$ No difference

$(3, 4) \rightarrow 1.866 \rightarrow$ No difference

17. $\text{CV}_{\text{Scheffé}} = 6.930$

$(1, 2) \rightarrow 0.148 \rightarrow$ No difference

$(1, 3) \rightarrow 7.842 \rightarrow$ Significant difference

$(1, 4) \rightarrow 2.581 \rightarrow$ No difference

$(2, 3) \rightarrow 5.311 \rightarrow$ No difference

$(2, 4) \rightarrow 3.678 \rightarrow$ No difference

$(3, 4) \rightarrow 17.877 \rightarrow$ Significant difference

Uses and Abuses for Chapter 10 *(page 538)*

1. Answers will vary.

2. Answers will vary.

Review Answers for Chapter 10 *(page 539)*

1. Claimed distribution:

Category	Distribution
New patients	25%
Old/new	25%
Old/recurring	50%

H_0: Distribution of office visits is as shown in table above.
H_a: Distribution of office visits differs from the claimed distribution.

$\chi_0^2 = 5.991$; $\chi^2 \approx 74.101$

Reject H_0. There is enough evidence at the 5% level to conclude that the distribution of visits differs from the claimed distribution.

3. (a) Expected frequencies:

	HS—did not complete	HS complete	College 1–3 years	College 4+ years	Total
25–44	13,836.62	27,733.60	20,686.43	20,962.35	83,219
45+	15,047.38	30,160.40	22,496.57	22,796.65	90,501
Total	28,884	57,894	43,183	43,759	173,720

(b) H_0: Education is independent of age.

H_a: Education is dependent of age.

d.f. = 3

$\chi_0^2 = 6.251$; $\chi^2 \approx 2904.408$

Reject H_0

(c) There is enough evidence at the 10% level of significance to conclude that the education level of people in the U.S. and their age are dependent.

5. $F_0 \approx 2.295$ **7.** $F_0 = 2.39$

9. H_0: $\sigma_1^2 \le \sigma_2^2$ (claim); H_a: $\sigma_1^2 > \sigma_2^2$

$F_0 = 3.09$; $F \approx 2.419$

Fail to reject H_0

11. Population 1: Garfield County

Population 2: Kay County

H_0: $\sigma_1^2 \le \sigma_2^2$; H_a: $\sigma_1^2 > \sigma_2^2$ (claim)

$F_0 = 1.92$; $F \approx 1.717$

Fail to reject H_0. There is not enough evidence at the 10% level to support the claim that the variation in wheat production is greater in Garfield County than in Kay County.

13. Population 1: Male; $s_1^2 = 18{,}486.26$

Population 2: Female; $s_2^2 = 12{,}102.78$

H_0: $\sigma_1^2 = \sigma_2^2$; H_a: $\sigma_1^2 \neq \sigma_2^2$ (claim)

$F_0 = 6.94$; $F \approx 1.527$

Fail to reject H_0. There is not enough evidence at the 1% level to support the claim that the test score variance for females is different than that for males.

15. H_0: $\mu_1 = \mu_2 = \mu_3 = \mu_4$

H_a: At least one mean is different from the others. (claim)

$F_0 = 2.29$; $F \approx 6.60$

Reject H_0. There is enough evidence at the 10% level to conclude that the mean residential energy expenditures are not the same for all four regions.

Real Statistics–Real Decisions for Chapter 10

(page 542)

1. Claimed distribution:

Response	Distribution
Under 20	1%
20–29	13%
30–39	16%
40–49	19%
50–59	16%
60–69	13%
70+	22%

H_0: Distribution of response is as shown in table above. (claim)

H_a: Distribution of response differs from the claimed distribution.

$\chi_0^2 = 16.812$

$\chi^2 = 497.26$

Reject H_0. There is enough evidence at the 1% level to conclude that the distribution of the ages of telemarketing fraud victims differs from the survey.

2. (a)

Type of fraud	Age of victims				
	Under 20	20–29	30–39	40–49	50–59
Sweepstakes	10 (15)	60 (120)	70 (165)	130 (185)	90 (135)
Credit Card	20 (15)	180 (120)	260 (165)	240 (185)	180 (135)
Total	30	240	330	370	270

Type of fraud	Age of victims			
	60–69	70–79	80+	Total
Sweepstakes	160 (115)	280 (155)	200 (110)	1000
Credit Card	70 (115)	30 (155)	20 (110)	1000
Total	230	310	220	2000

(b) H_0: The type of fraud is independent of the victim's age.

H_a: The type of fraud is dependent on the victim's age.

$\chi_0^2 = 18.475$

$\chi^2 \approx 619.533$

Reject H_0. There is enough evidence at the 1% level to conclude that the victims' ages and the type of fraud are dependent.

Chapter Quiz for Chapter 10 *(page 543)*

1. (a) Population 1: San Jose; $s_1^2 = 429.984$

Population 2: Dallas; $s_2^2 = 112.779$

H_0: $\sigma_1^2 = \sigma_2^2$; H_a: $\sigma_1^2 \neq {}_2^2$ (claim)

(b) $\alpha = 0.01$ (c) $F_0 = 3.97$

(d) $F > 3.97$ (e) $F \approx 3.813$

(f) Fail to reject H_0

(g) There is not enough evidence at the 1% level to conclude that the variances in annual wages for San Jose, CA and Dallas, TX are different.

2. (a) H_0: $\mu_1 = \mu_2 = \mu_3$ (claim)

H_a: At least one mean is different from the others.

(b) $\alpha = 0.10$ (c) $F_0 = 2.43$

(d) $F > 2.43$ (e) $F \approx 7.39$

(f) Reject H_0

(g) There is enough evidence at the 10% level to conclude that the mean annual wages for the three cities are not all equal.

3. (a) Claimed distribution:

Education	25 & Over
Not a HS graduate	16.6%
HS graduate	33.3%
Some college, no degree	17.3%
Associate degree	7.5%
Bachelor degree	17.0%
Advanced degree	8.2%

H_0: Distribution of educational achievement is as shown in table above.

H_a: Distribution of educational achievement differs from the claimed distribution.

(b) $\alpha = 0.01$ (c) $\chi_0^2 = 15.086$

(d) Reject if $\chi^2 > 15.086$

(e) $\chi^2 \approx 5.189$

(f) Fail to reject H_0

(g) There is not enough evidence at the 1% level to conclude the distribution of educational achievement differs from the claimed distribution.

4. (a) Claimed distribution:

Education	25 & Over
Not a HS graduate	16.6%
HS graduate	33.6%
Some college, no degree	17.3%
Associate degree	7.5%
Bachelor degree	17.0%
Advanced degree	8.2%

H_0: Distribution of educational achievement is as shown in table above.

H_a: Distribution of educational achievement differs from the claimed distribution.

(b) $\alpha = 0.05$ (c) $\chi_0^2 = 11.071$

(d) Reject if $\chi^2 > 11.071$

(e) $\chi^2 \approx 76.299$ (f) Reject H_0

(g) There is enough evidence at the 5% level to conclude that the distribution of educational achievement differs from the claimed distribution.

CHAPTER 11

Section 11.1 *(page 552)*

1. A nonparametric test is a hypothesis test that does not require any specific conditions concerning the shape of populations or the value of any population parameters.

A nonparametric test is usually easier to perform than its corresponding parametric test, but the nonparametric test is usually less efficient.

3. (a) H_0: median $\leq$ 200; H_a: median $>$ 200 (claim)

(b) CV = 1 (c) $x = 5$

(d) Fail to reject H_0. There is not enough evidence at the 1% level to support the claim that the median amount of new credit card charges for the previous month was more than \$200.

5. (a) H_0: median $\leq$ 140,000 (claim); H_a: median $>$ 140,000

(b) CV = 1 (c) $x = 3$

(d) Fail to reject H_0. There is not enough evidence at the 5% level to reject the claim that the median sales price of new privately owned one-family homes sold in the past year is \$140,000 or less.

7. (a) H_0: median $\geq$ 1500 (claim); H_a: median $<$ 1500

(b) $z_0 = -2.055$ (c) $z \approx -3.040$

(d) Reject H_0. There is enough evidence at the 2% level to reject the claim that the median amount of credit card debt for families holding such debts is at least \$1500.

9. (a) H_0: median $\leq$ 36; H_a: median $>$ 36 (claim)

(b) CV = 3 (c) $x = 8$

(d) Fail to reject H_0. There is not enough evidence at the 1% level to support the claim that the median age of recipients of social science doctorates is greater than 36.

11. (a) H_0: median = 4 (claim); H_a: median $\neq$ 4

(b) $z_0 = -1.96$ (c) $z \approx -2.334$

(d) Reject H_0. There is enough evidence at the 5% level to reject the claim that the median number of rooms in renter-occupied units is 4.

13. (a) H_0: median = \$9.81 (claim); H_a: median $\neq$ \$9.81

(b) $z_0 = -2.575$ (c) $z \approx -0.961$

(d) Fail to reject H_0. There is not enough evidence at the 1% level to reject the claim that the median hourly earnings of male workers paid hourly rates is \$9.81.

15. (a) H_0: The headache hours have not decreased.

H_a: The headache hours have decreased. (claim)

(b) CV = 1 (c) $x = 3$

(d) Fail to reject H_0. There is not enough evidence at the 5% level to support the claim that the daily headache hours were reduced after the soft-tissue therapy and spinal manipulation.

17. (a) H_0: The SAT scores have not improved.

H_a: The SAT scores have improved. (claim)

(b) CV = 2 (c) $x = 4$

(d) Fail to reject H_0. There is not enough evidence at the 5% level to support the claim that verbal SAT scores improved.

19. (a) Fail to reject H_0.

(b) There is not enough evidence at the 5% level to reject the claim that the proportion of companies providing transportation home is equal to the proportion of companies that do not.

21. (a) H_0: median $\leq$ 418 (claim); H_a: median $>$ 418

(b) $z_0 = 2.33$ (c) $z \approx 1.459$

(d) Fail to reject H_0. There is not enough evidence at the 1% level to reject the claim that the median weekly earnings of female workers is less than or equal to \$418.

23. (a) H_0: median $\leq$ 24; H_a: median $>$ 24 (claim)

(b) $z_0 = 1.645$ (c) $z \approx 1.936$

(d) Reject H_0. There is enough evidence at the 5% level to support the claim that the median age of first-time brides is greater than 24 years.

Section 11.2 *(page 563)*

1. (a) H_0: There is no reduction in diastolic blood pressure. (claim)

H_a: There is a reduction in diastolic blood pressure.

(b) Wilcoxon signed-rank test

(c) CV = 2 (d) $w_s = 6$

(e) Fail to reject H_0. There is not enough evidence at the 1% level to reject the claim that there was no reduction in diastolic blood pressure.

3. (a) H_0: There is no difference in the earnings.
   H_a: There is a difference in the earnings. (claim)
   (b) Wilcoxon rank sum test
   (c) $z_0 = \pm 1.96$ (d) $z \approx -3.873$
   (e) Reject H_0. There is enough evidence at the 5% level to support the claim that there is a difference in the earnings.
5. (a) H_0: There is not a difference in salaries.
   H_a: There is a difference in salaries. (claim)
   (b) Wilcoxon rank sum test
   (c) $z_0 = \pm 1.96$ (d) $z \approx -1.819$
   (e) Fail to reject H_0. There is not enough evidence at the 5% level to support the claim that there is a difference in salaries.
7. Reject H_0. There is enough evidence at the 10% level to conclude that the gas mileage is improved.

Section 11.3 *(page 571)*

1. (a) H_0: There is no difference in the premiums.
   H_a: There is a difference in the premiums. (claim)
   (b) CV = 5.991 (c) $H \approx 14.05$
   (d) Reject H_0. There is enough evidence at the 5% level to support the claim that the distributions of the annual premiums in California, Florida, and Illinois are different.
3. (a) H_0: There is no difference in the salaries.
   H_a: There is a difference in the salaries. (claim)
   (b) CV = 6.251 (c) $H \approx 6.46$
   (d) Reject H_0. There is enough evidence at the 10% level to support the claim that the distributions of the annual salaries in the four states are different.
5. (a) Fail to reject H_0
   (b) Fail to reject H_0. There is not enough evidence to support the claim. This is the same decision found in part (a) using the Kruskal-Wallis test.

Section 11.4 *(page 576)*

1. (a) H_0: $\rho_s = 0$; H_a: $\rho_s \neq 0$ (claim)
   (b) CV = 0.929 (c) $r_s \approx 0.929$
   (d) Fail to reject H_0. There is not enough evidence at the 1% level to support the claim that there is a correlation between debt and income in the farming business.
3. (a) H_0: $\rho_s = 0$; H_a: $\rho_s \neq 0$ (claim)
   (b) CV = 0.497 (c) $r_s \approx 0.568$
   (d) Reject H_0. There is enough evidence at the 10% level to support the claim that there is a correlation between overall score and price.
5. Fail to reject H_0. There is not enough evidence at the 5% level to conclude that there is a correlation between science achievement scores and GNP.
7. Fail to reject H_0. There is not enough evidence at the 5% level to conclude that there is a correlation between science and mathematics achievement scores.
9. Fail to reject H_0.

Uses and Abuses for Chapter 11 *(page 581)*

1. Answers will vary.
2. Sign test → z-test
   Paired-sample sign test → t-test
   Wilcoxon signed-rank test → t-test
   Wilcoxon rank sum test → z or t-test
   Kruskal-Wallis test → one-way ANOVA
   Spearman rank correlation coefficient → Pearson correlation

Review Answers for Chapter 11 *(page 582)*

1. (a) H_0: median = \$13,500 (claim); H_a: median ≠ \$13,500
   (b) CV = 2 (c) $x = 7$
   (d) Fail to reject H_0. There is not enough evidence at the 1% level to reject the claim that the median value of stock among families that own stock is $13,500.
3. (a) H_0: median ≤ 6 (claim); H_a: median > 6
   (b) $z_0 \approx -1.28$ (c) $z \approx -2.032$
   (d) Reject H_0. There is enough evidence at the 10% level to reject the claim that the median turnover time is no more than 6 hours.
5. (a) H_0: There is no reduction in diastolic blood pressure. (claim)
   H_a: There is a reduction in diastolic blood pressure.
   (b) CV = 2 (c) $x = 3$
   (d) Fail to reject H_0. There is not enough evidence at the 5% level to reject the claim that there was no reduction in diastolic blood pressure.
7. (a) Dependent; Wilcoxon signed-rank test
   (b) H_0: Producers are not under reporting the caloric content of their foods.
   H_a: Producers are under reporting the caloric content of their foods. (claim)
   (c) CV = 8 (d) $w_s = 2$
   (e) Reject H_0. There is enough evidence at the 5% level to support the claim that the procedures are underreporting the caloric content of their foods.
9. (a) Independent; Wilcoxon rank sum test
   (b) H_0: There is no difference in the amount of time that it takes to earn a doctorate.
   H_a: There is a difference in the amount of time that it takes to earn a doctorate. (claim)
   (c) $z_0 = \pm 2.575$ (d) $z \approx -3.175$
   (e) Reject H_0. There is enough evidence at the 1% level to support the claim that there is a difference in the amount of time that it takes to earn a doctorate.

11. (a) H_0: There is no difference in salaries between the fields of study. (claim)

H_a: There is a difference in salaries between the fields of study.

(b) CV = 5.991 (c) $H \approx 22.98$

(d) Reject H_0. There is enough evidence at the 5% level to conclude that there is a difference in salaries between the fields of study.

13. (a) H_0: $\rho_s = 0$; H_a: $\rho_s \neq 0$ (claim)

(b) CV = 0.881 (c) $r_s \approx -0.429$

(d) Fail to reject H_0. There is not enough evidence at the 1% level to conclude that there is a correlation between overall score and price.

Real Statistics–Real Decisions for Chapter 11 *(page 586)*

1. (a) random sample

(b) Answers will vary.

(c) Answers will vary.

2. (a) Answers will vary.

(b) Sign test; You need to use the nonparametric test since nothing is known about the shape of the population.

(c) H_0: median ≥ 3.5; H_a: median < 3.5 (claim)

(d) $n = 19$; $\alpha = 0.05$; CV = 5

$x = 7$

Fail to reject H_0. There is not enough evidence at the 5% level to support the claim that the median tenure is less than 3.5 years.

3. (a) Wilcoxon rank sum test

(b) H_0: median male tenure $\leq$ median female tenure

H_a: median male tenure $>$ median female tenure

(c) $R = 177$; $\mu_R = 162$; $\sigma_R \approx 19.442$

$z_0 = \pm 1.645$; $z \approx 0.772$

Fail to reject H_0. There is not enough evidence at the 5% level to conclude that the median tenure for male workers is greater than the median tenure for female workers.

Chapter Quiz for Chapter 11 *(page 587)*

1. (a) H_0: There is no difference in the salaries between genders.

H_a: There is a difference in the salaries between genders. (claim)

(b) Wilcoxon rank sum test

(c) $z_0 = \pm 1.645$ (d) $z \approx -1.722$

(e) Reject H_0. There is enough evidence at the 10% level to support the claim that there is a difference in the salaries between genders.

2. (a) H_0: median $= 28$ (claim); H_a: median $\neq 28$

(b) Sign test (c) CV = 6 (d) $x = 10$

(e) Fail to reject H_0. There is not enough evidence at the 5% level to reject the claim that the median age in Puerto Rico is 28 years.

3. (a) H_0: $\rho_s = 0$; H_a: $\rho_s \neq 0$ (claim)

(b) Spearman rank correlation coefficient test

(c) CV = 0.881 (d) 0.095

(e) Fail to reject H_0. There is not enough evidence at the 1% level to conclude that there is a correlation between average hours worked and the number of on-the-job injuries.

4. (a) H_0: There is no difference in the annual premiums between the states.

H_a: There is a difference in the annual premiums between the states. (claim)

(b) Kruskal-Wallis test

(c) CV = 5.991 (d) $H \approx 1.43$

(e) Fail to reject H_0. There is not enough evidence at the 5% level to conclude that there is a difference in the annual premiums between the states.

Cumulative Test for Chapters 10–11 *(page 588)*

1. Fail to reject H_0 **2.** Fail to reject H_0

3. Reject H_0 **4.** Fail to reject H_0

5. Reject H_0 **6.** Reject H_0

Credits

TOC Ch. 1 Chicago skyline. Doug Segal. Panoramic Images; **Ch. 2** Akhiok child. Ray Corral Photography; **Ch. 3** Dam. Billie Johnson. U.S. Army Corps of Engineers, Washington; **Ch. 4** Storm Over Road. Rob Atkins, Getty Images. Inc.; **Ch. 5** Man on treadmill. Keith Brofsky, Getty Images/PhotoDisc. Inc; **Ch. 6** "Better Not Best" Chuck Kimmerle, Grand Forks Herald; **Ch. 7** 2002 Chevrolet Prizm. General Motors Media Archives. **Ch. 8** Vietnam Women's Memorial. Washington, D.C. Greg Staley, Glenna Goodacre Ltd.; **Ch. 11** Kennedy Space Center, Florida. NASA Headquarters.

CHAPTER 1 p. 0 Chicago skyline. Doug Segal, Panoramic Images; Los Angeles skyline. Larry Brownstein, PhotoDisc. Inc.; Houston, Texas. M. Mastrorillo. The Stock Market; Aerial view of Philadelphia, Pennsylvania. Lien/Nibauer, Liason Agency, Inc. **p. 1** Manhattan. Colin Patterson, PhotoDisc. Inc.; New York City skyline. PhotoDisc, Inc.; **p. 2** Household Income of Golf Fans. USA TODAY

CHAPTER 2 p. 30 Akhiok resident David Eluska; Akhiok resident with sea urchin harvest; Akhiok woman; Akhiok resident (smiling man); Akhiok children; Akhiok resident; young girl; Akhiok resident with drying salmon. **p. 31** Akhiok child; Town of Akhiok. **p. 58** Town of Akhiok; Fishing village-Akhiok. Alaska. All images Roy Corral Photography.

CHAPTER 3 p. 108 Dam. Billie Johnson, U.S. Army Corps of Engineers. Washington; Chinook or King Salmon migrating (Oncorhynchus tshawytschal). Tom and Pat Leeson Photography; David Samolik counts salmon smolts killed during the barging around Columbia River dams. Natalie B. Fobes Photography; Sockeye develop hooked jaws and humped backs when they are close to spawning, Alaska. Natalie B. Fobes Photography. **p. 109** Sunset in the Gorge. Jay Carroll Photo/Grafix.

CHAPTER 4 p. 160 National Center for Atmospheric Research. Boulder, Colorado; Storm over road. Rob Atkins, Getty Images, Inc.; Cumulonimbus cloud. Painted Desert, Arizona. Tom & Susan Bean. Inc.; View from mouth of ice cave past missive icicles. Michigan. G. Ryan & S. Beyer, Getty Images, Inc. **p. 161** Hay (Reeds). SuperStock, Inc. **p. 186** Desire for Information Doesn't Take a Vacation. USA TODAY.

CHAPTER 5 p. 204 Woman working out with dumbbells. close-up. David Madison, Getty Images, Inc.; Middle Age/Individual-Outdoor. SuperStock, Inc.; Swimming Laps. Getty Images, Inc./PhotoDisc, Inc.; Jogging. Sean Thompson, Getty Images. Inc./PhotoDisc, Inc. **p. 205** Man on Treadmill. Keith Brofsky, Getty Images, Inc/PhotoDisc, Inc. **p. 243** Speed Readers. USA TODAY. **p. 259** How Adults Get Physical. USA TODAY.

CHAPTER 6 p. 268 Better Not Best. Chuck Kimmerle, Grand Forks Herald; Wheat at Sunset. U. S. Wheat Associates; Close up of wheat. Chuck Kimmerle, Grand Forks Herald; Men standing by combines and machinery. U. S. Wheat Associates. **p. 269** Wheatfield with combines. U.S. Wheat Associates; Grain pouring into truck. U.S. Wheat Associates. **p. 283** Loggerhead sea turtle. Zig Leszczunski. Animals Animals/Earth Scenes. **p. 284** William Sealy Gosset. 1876–1937. English statistician, Brewer for Guinness beer. The Granger Collection. **p. 292** Karl Pearson. University College London; Sir Ronald Aylmer Fisher; Library of Congress; John W. Turkey. Princeton University Stanhope Hall.

CHAPTER 7 p. 318 Travis Walton. Michael H. Rogers; Cover of "Fire in the Sky. The Walton Experience" by Travis Walton. Published by Marlowe & Company. Michael H. Rogers; UFO hovering above London skyline, dusk (Digital Composite). Coneyl Jay, Getty Images, Inc.; A group of protesters. Joshua Roberts. Agence France-Presse; **p. 319** Close Encounters Film Still. The Museum of Modern Art; **p. 321** A man laying in bed taking medicine. R. B. Studio, The Stock Market; **p. 342** 2002 Chevrolet Geo Prizm. General Motors Media Archives. **p. 365** Freebies Work. USA TODAY.

CHAPTER 8 p. 388 USA. Washington DC. man touching Vietnam Veterans' Memorial. Paul Merideth. Getty Images. Inc.; Vets with a Mission doing volunteer work in Vietnam. George Stathos. Vets With A Mission; Film still of Born on the Fourth of July—Tom Cruise as Ron Kovic. Roland Neveu, Creative Artists Archive. **p. 389** Vietnam Women's Memorial. Washington. D.C. Greg Staley. Glenna Goodacre Ltd.; The Warriors. Deanna Gail Shlee Hopkins. **p. 403** Music sheets & notes. Masahiro Sano. Corbis/The Stock Market. **p. 407** Footwell intrusion. Insurance Institute for Highway Safety. **p. 423** Reflexology. Sauze, Francoise. Photo Researchers, Inc. **p. 428** Utility Satisfaction. USA TODAY.

CHAPTER 9 p. 440 Sprinter Marion Jones of the USA celebrates with an American flag after winning the gold medal in the 100 meters at the Olympics in Sydney. Australia. AP/World Wide Photos; Winter Olympic Games Happo'one. Hakuba. Japan; American Lanny Krayzelburg raises his hand after setting a championship record time during the heats of the men's 200-meter backstroke at the Pan Pacific Swimming Championships in Sydney. AP/Wide World Photos. **p. 441** Men rowing boats. Corbis/Sygma U.S. pole vaulter Stacy Dragila. AP/Wide World Photos.

CHAPTER 10 p. 492 Car crash-fronted impact. Insurance Institute for Highway Safety; 1997 Kia Sephia crash with dummy. Insurance Institute for Highway Safety; Pregnant crash test dummy being fitted with seatbelt. Andy Sacks, Getty Images, Inc. **p. 493** Crash test dummy/crash sled set-up. Brad Trent Photography; Crash-test dummies inside car with inflated airbags. close-up. Wayne Eastep, Getty Images, Inc. **p. 495** How often We See Red Lights Run. USA TODAY. **p. 500** Coffee in the Morning. USA TODAY. **p. 542** National Fraud Info Center.

CHAPTER 11 p. 544 Memorial of Space Shuttle *Challenger*. Smithsonian Institution Photo Services, Washington. D.C.; Kennedy Space Center Florida-The Space Shuttle *Challenger*. NASA Headquarters; Christa McAuliffe, teacher in space project. NASA/Lyndon B. Johnson Space Center; A 9′7″ × 16′ segment of Challenger's right wing is unloaded. NASA/Lyndon B. Johnson Space Center. **p. 545** Kennedy Space Center Florida-Crew members of Space Shuttle *Challenger*. NASA Headquarters; Kennedy Space Center Florida-Main engine exhaust. Solid Rocket Booster plume and expanding ball of gas. NASA Headquarters. **p. 555** After the Party. USA TODAY.

COVER Body Scan. ELSCINT. Photo Researchers. Inc.; Baseball. Baltimore, MD. Corbis Digital Stock; Newsstand. Brussels, Belgium. Craig Aurness. Corbis; Businesspeople Meeting in a Cafe. Jules Frazier, Getty Images. Inc./PhotoDisc. Inc.

Index

Table 6—Chi-Square Distribution

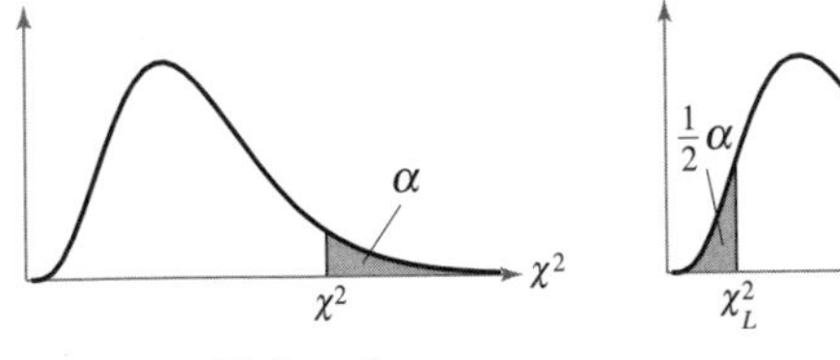

Right tail

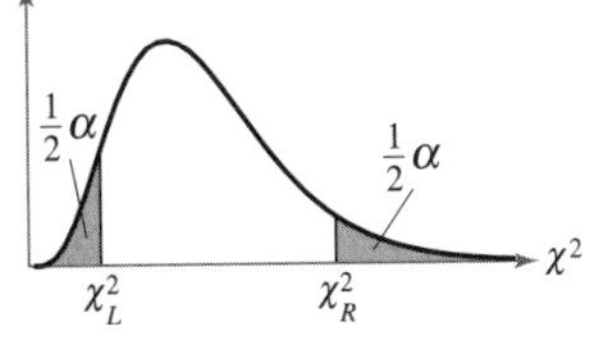

Two tails

Degrees of freedom	α									
	0.995	**0.99**	**0.975**	**0.95**	**0.90**	**0.10**	**0.05**	**0.025**	**0.01**	**0.005**
1	—	—	0.001	0.004	0.016	2.706	3.841	5.024	6.635	7.879
2	0.010	0.020	0.051	0.103	0.211	4.605	5.991	7.378	9.210	10.597
3	0.072	0.115	0.216	0.352	0.584	6.251	7.815	9.348	11.345	12.838
4	0.207	0.297	0.484	0.711	1.064	7.779	9.488	11.143	13.277	14.860
5	0.412	0.554	0.831	1.145	1.610	9.236	11.071	12.833	15.086	16.750
6	0.676	0.872	1.237	1.635	2.204	10.645	12.592	14.449	16.812	18.548
7	0.989	1.239	1.690	2.167	2.833	12.017	14.067	16.013	18.475	20.278
8	1.344	1.646	2.180	2.733	3.490	13.362	15.507	17.535	20.090	21.955
9	1.735	2.088	2.700	3.325	4.168	14.684	16.919	19.023	21.666	23.589
10	2.156	2.558	3.247	3.940	4.865	15.987	18.307	20.483	23.209	25.188
11	2.603	3.053	3.816	4.575	5.578	17.275	19.675	21.920	24.725	26.757
12	3.074	3.571	4.404	5.226	6.304	18.549	21.026	23.337	26.217	28.299
13	3.565	4.107	5.009	5.892	7.042	19.812	22.362	24.736	27.688	29.819
14	4.075	4.660	5.629	6.571	7.790	21.064	23.685	26.119	29.141	31.319
15	4.601	5.229	6.262	7.261	8.547	22.307	24.996	27.488	30.578	32.801
16	5.142	5.812	6.908	7.962	9.312	23.542	26.296	28.845	32.000	34.267
17	5.697	6.408	7.564	8.672	10.085	24.769	27.587	30.191	33.409	35.718
18	6.265	7.015	8.231	9.390	10.865	25.989	28.869	31.526	34.805	37.156
19	6.844	7.633	8.907	10.117	11.651	27.204	30.144	32.852	36.191	38.582
20	7.434	8.260	9.591	10.851	12.443	28.412	31.410	34.170	37.566	39.997
21	8.034	8.897	10.283	11.591	13.240	29.615	32.671	35.479	38.932	41.401
22	8.643	9.542	10.982	12.338	14.042	30.813	33.924	36.781	40.289	42.796
23	9.262	10.196	11.689	13.091	14.848	32.007	35.172	38.076	41.638	44.181
24	9.886	10.856	12.401	13.848	15.659	33.196	36.415	39.364	42.980	45.559
25	10.520	11.524	13.120	14.611	16.473	34.382	37.652	40.646	44.314	46.928
26	11.160	12.198	13.844	15.379	17.292	35.563	38.885	41.923	45.642	48.290
27	11.808	12.879	14.573	16.151	18.114	36.741	40.113	43.194	46.963	49.645
28	12.461	13.565	15.308	16.928	18.939	37.916	41.337	44.461	48.278	50.993
29	13.121	14.257	16.047	17.708	19.768	39.087	42.557	45.722	49.588	52.336
30	13.787	14.954	16.791	18.493	20.599	40.256	43.773	46.979	50.892	53.672
40	20.707	22.164	24.433	26.509	29.051	51.805	55.758	59.342	63.691	66.766
50	27.991	29.707	32.357	34.764	37.689	63.167	67.505	71.420	76.154	79.490
60	35.534	37.485	40.482	43.188	46.459	74.397	79.082	83.298	88.379	91.952
70	43.275	45.442	48.758	51.739	55.329	85.527	90.531	95.023	100.425	104.215
80	51.172	53.540	57.153	60.391	64.278	96.578	101.879	106.629	112.329	116.321
90	59.196	61.754	65.647	69.126	73.291	107.565	113.145	118.136	124.116	128.299
100	67.328	70.065	74.222	77.929	82.358	118.498	124.342	129.561	135.807	140.169

Key Formulas

From Larson/*Farber Elementary Statistics: Picturing the World,* Second Edition

CHAPTER 2

Class Width = (round up to next convenient number) of $\frac{\text{Maximum data entry} - \text{Minimum data entry}}{\text{Number of classes}}$

Midpoint = $\frac{(\text{Lower class limit}) + (\text{Upper class limit})}{2}$

Relative Frequency = $\frac{\text{Class frequency}}{\text{Sample size}} = \frac{f}{n}$

Population Mean: $\mu = \frac{\Sigma x}{N}$

Sample Mean: $\bar{x} = \frac{\Sigma x}{n}$

Weighted Mean: $\bar{x} = \frac{\Sigma(x \cdot w)}{\Sigma w}$

Mean for Grouped Data: $\bar{x} = \frac{\Sigma(x \cdot f)}{n}$

Range = (Maximum entry) − (Minimum entry)

Population Variance: $\sigma^2 = \frac{\Sigma(x - \mu)^2}{N}$

Population Standard Deviation:

$$\sigma = \sqrt{\sigma^2} = \sqrt{\frac{\Sigma(x - \mu)^2}{N}}$$

Sample Variance: $s^2 = \frac{\Sigma(x - \bar{x})^2}{n - 1}$

Sample Standard Deviation: $s = \sqrt{s^2} = \sqrt{\frac{\Sigma(x - \bar{x})^2}{n - 1}}$

Sample Standard Deviation for Grouped Data:

$$s = \sqrt{\frac{\Sigma(x - \bar{x})^2 f}{n - 1}}$$

Empirical Rule (or 68-95-99.7 Rule) For data with a (symmetric) bell-shaped distribution

1. About 68% of the data lies between $\mu - \sigma$ and $\mu + \sigma$.
2. About 95% of the data lies between $\mu - 2\sigma$ and $\mu + 2\sigma$.
3. About 99.7% of the data lies between $\mu - 3\sigma$ and $\mu + 3\sigma$.

Chebychev's Theorem The portion of any data set lying within k standard deviations $(k > 1)$ of the mean is at least $1 - \frac{1}{k^2}$.

Standard Score: $z = \frac{\text{value} - \text{mean}}{\text{standard deviation}} = \frac{x - \mu}{\sigma}$

CHAPTER 3

Classical (or Theoretical) Probability:

$$P(E) = \frac{\text{Number of outcomes in } E}{\text{Total number of outcomes in sample space}}$$

Empirical (or Statistical) Probability:

$$P(E) = \frac{\text{Frequency of event } E}{\text{Total frequency}} = \frac{f}{n}$$

Probability of a Complement: $P(E') = 1 - P(E)$

Probability of occurrence of both events A and B: $P(A \text{ and } B) = P(A) \cdot P(B|A)$

$P(A \text{ and } B) = P(A) \cdot P(B)$ if A and B are independent

Probability of occurrence of either A or B or both:

$P(A \text{ or } B) = P(A) + P(B) - P(A \text{ and } B)$

$P(A \text{ or } B) = P(A) + P(B)$ if events are mutually exclusive.

Permutations of n objects taken r at a time:

$${}_nP_r = \frac{n!}{(n - r)!}, \text{ where } r \leq n.$$

Distinguishable Permutations: n_1 alike, n_2 alike, ..., n_k alike:

$$\frac{n!}{n_1! \cdot n_2! \cdot n_2! \cdots n_k!}$$

where $n_1 + n_2 + n_3 + \cdots + n_k = n$.

Combination of n objects taken r at a time:

$${}_nC_r = \frac{n!}{(n - r)!r!}$$

Key Formulas

From Larson/*Farber Elementary Statistics: Picturing the World,* Second Edition

CHAPTER 4

Mean of a Discrete Random Variable: $\mu = \Sigma x P(x)$

Variance of a Discrete Random Variable:

$$\sigma^2 = \Sigma(x - \mu)^2 P(x)$$

Standard Deviation: $\sigma = \sqrt{\sigma^2}$

Expected Value: $E(x) = \mu = \Sigma x P(x)$

Binomial Probability of x successes in n trials:

$$P(x) = {}_nC_x p^x q^{n-x} = \frac{n!}{(n-x)!x!} p^x q^{n-x}$$

Population Parameters of a Binomial Distribution:

Mean: $\mu = np$ Variance: $\sigma^2 = npq$

Standard Deviation: $\sigma = \sqrt{npq}$

Geometric Distribution:
The probability that the first success will occur on trial number x is $P(x) = p(q)^{x-1}$, where $q = 1 - p$.

Poisson Distribution:
The probability of exactly x occurrences in an interval is $P(x) = \frac{\mu^x e^{-\mu}}{x!}$, where $e \approx 2.71828$.

CHAPTER 5

Standard Score, or z-Score:

$$z = \frac{\text{value} - \text{mean}}{\text{standard deviation}} = \frac{x - \mu}{\sigma}$$

Transforming a z-Score to an x-Value: $x = \mu + z\sigma$

$\mu_{\bar{x}} = \mu$ Central Limit Theorem (Mean of the Sample Means)

$\sigma_{\bar{x}} = \frac{\sigma}{\sqrt{n}}$ Central Limit Theorem (Standard Error)

$z = \frac{\bar{x} - \mu_{\bar{x}}}{\sigma_{\bar{x}}} = \frac{\bar{x} - \mu}{\sigma/\sqrt{n}}$ Central Limit Theorem

CHAPTER 6

c-Confidence Interval for μ: $\bar{x} - E < \mu < \bar{x} + E$ where $E = z_c \frac{\sigma}{\sqrt{n}}$ if σ is known and the population is normal or $n \geq 30$, and $E = t_c \frac{s}{\sqrt{n}}$ if the population is normal, σ is unknown, and $n < 30$.

Minimum Sample Size to Estimate μ: $n = \left(\frac{z_c \sigma}{E}\right)^2$

Point Estimate for p, the population proportion of successes: $\hat{p} = \frac{x}{n}$

c-Confidence Interval for Population Proportion p (when $np \geq 5$ and $nq \geq 5$): $\hat{p} - E < p < \hat{p} + E$, where

$$E = z_c \sqrt{\frac{\hat{p}\hat{q}}{n}}.$$

Minimum Sample Size to Estimate p: $n = \hat{p}\hat{q}\left(\frac{z_c}{E}\right)^2$

c-Confidence Interval for Population Variance σ^2:

$$\frac{(n-1)s^2}{\chi_R^2} < \sigma^2 < \frac{(n-1)s^2}{\chi_L^2}$$

c-Confidence Interval for Population Standard Deviation σ:

$$\sqrt{\frac{(n-1)s^2}{\chi_R^2}} < \sigma < \sqrt{\frac{(n-1)s^2}{\chi_L^2}}$$

Table 5—*t*-Distribution

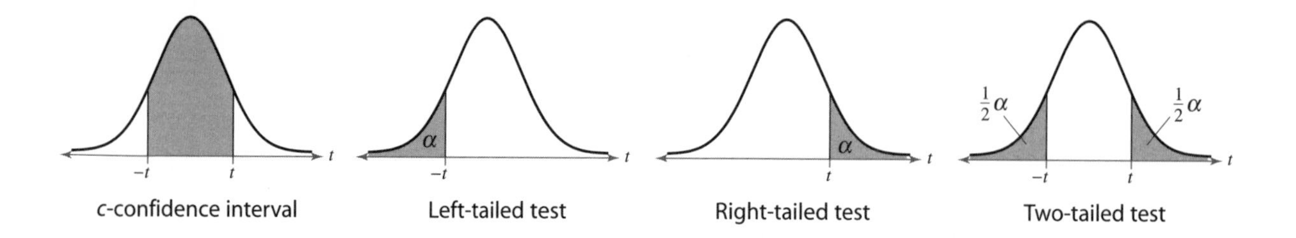

	Level of confidence, c	0.50	0.80	0.90	0.95	0.98	0.99
	One tail, α	0.25	0.10	0.05	0.025	0.01	0.005
d.f.	Two tails, α	0.50	0.20	0.10	0.05	0.02	0.01
1		1.000	3.078	6.314	12.706	31.821	63.657
2		.816	1.886	2.920	4.303	6.965	9.925
3		.765	1.638	2.353	3.182	4.541	5.841
4		.741	1.533	2.132	2.776	3.747	4.604
5		.727	1.476	2.015	2.571	3.365	4.032
6		.718	1.440	1.943	2.447	3.143	3.707
7		.711	1.415	1.895	2.365	2.998	3.499
8		.706	1.397	1.860	2.306	2.896	3.355
9		.703	1.383	1.833	2.262	2.821	3.250
10		.700	1.372	1.812	2.228	2.764	3.169
11		.697	1.363	1.796	2.201	2.718	3.106
12		.695	1.356	1.782	2.179	2.681	3.055
13		.694	1.350	1.771	2.160	2.650	3.012
14		.692	1.345	1.761	2.145	2.624	2.977
15		.691	1.341	1.753	2.131	2.602	2.947
16		.690	1.337	1.746	2.120	2.583	2.921
17		.689	1.333	1.740	2.110	2.567	2.898
18		.688	1.330	1.734	2.101	2.552	2.878
19		.688	1.328	1.729	2.093	2.539	2.861
20		.687	1.325	1.725	2.086	2.528	2.845
21		.686	1.323	1.721	2.080	2.518	2.831
22		.686	1.321	1.717	2.074	2.508	2.819
23		.685	1.319	1.714	2.069	2.500	2.807
24		.685	1.318	1.711	2.064	2.492	2.797
25		.684	1.316	1.708	2.060	2.485	2.787
26		.684	1.315	1.706	2.056	2.479	2.779
27		.684	1.314	1.703	2.052	2.473	2.771
28		.683	1.313	1.701	2.048	2.467	2.763
29		.683	1.311	1.699	2.045	2.462	2.756
∞		.674	1.282	1.645	1.960	2.326	2.576

Key Formulas

From Larson/*Farber Elementary Statistics: Picturing the World,* Second Edition

CHAPTER 7

z-Test for a Mean μ: $z = \dfrac{\overline{x} - \mu}{\sigma/\sqrt{n}}$, for σ known with a normal population, or for $n \geq 30$.

t-Test for a Mean μ: $t = \dfrac{\overline{x} - \mu}{s/\sqrt{n}}$, for σ unknown, x normally distributed, and $n < 30$. (d.f. $= n - 1$)

z-Test for a Proportion p (when $np \geq 5$ and $nq \geq 5$):

$$z = \frac{\hat{p} - p}{\sqrt{pq/n}}$$

Chi-Square Test for a Variance or Standard Deviation:

$$\chi^2 = \frac{(n-1)s^2}{\sigma^2} \quad (\text{d.f.} = n - 1)$$

CHAPTER 8

Two-Sample z-Test for the Difference Between Means:

(Independent samples; n_1 and $n_2 \geq 30$ or normally distributed populations)

$$z = \frac{(\overline{x}_1 - \overline{x}_2) - (\mu_1 - \mu_2)}{\sigma_{\overline{x}_1 - \overline{x}_2}}, \text{ where } \sigma_{\overline{x}_1 - \overline{x}_2} = \sqrt{\frac{\sigma_1^2}{n_1} + \frac{\sigma_2^2}{n_2}}$$

Two-Sample t-Test for the Difference Between Means: (Independent samples from normally distributed populations, n_1 or $n_2 < 30$)

$$t = \frac{(\overline{x}_1 - \overline{x}_2) - (\mu_1 - \mu_2)}{\sigma_{\overline{x}_1 - \overline{x}_2}}$$

If population variances are equal, d.f. $= n_1 + n_2 - 2$ and $\sigma_{\overline{x}_1 - \overline{x}_2} = \sqrt{\dfrac{(n_1 - 1)s_1^2 + (n_2 - 1)s_2^2}{n_1 + n_2 - 2}} \cdot \sqrt{\dfrac{1}{n_1} + \dfrac{1}{n_2}}$.

If population variances are not equal, d.f. is the smaller of $n_1 - 1$ or $n_2 - 1$ and $\sigma_{\overline{x}_1 - \overline{x}_2} = \sqrt{\dfrac{s_1^2}{n_1} + \dfrac{s_2^2}{n_2}}$.

t-Test for the Difference Between Means: (Dependent samples)

$$t = \frac{\overline{d} - \mu_d}{s_d/\sqrt{n}}, \text{ where } \overline{d} = \frac{\Sigma d}{n},\ s_d = \sqrt{\frac{n(\Sigma d^2) - (\Sigma d)^2}{n(n-1)}},$$

and d.f. $= n - 1$.

Two-Sample z-Test for the Difference Between Proportions:

Note: $n_1\overline{p}$, $n_1\overline{q}$, $n_2\overline{p}$, and $n_2\overline{q}$ must be at least 5.

$$z = \frac{(\hat{p}_1 - \hat{p}_2) - (p_1 - p_2)}{\sqrt{\overline{p}\,\overline{q}\left(\frac{1}{n_1} + \frac{1}{n_2}\right)}}, \text{ where } \overline{p} = \frac{x_1 + x_2}{n_1 + n_2}.$$

CHAPTER 9

Correlation Coefficient:

$$r = \frac{n\Sigma xy - (\Sigma x)(\Sigma y)}{\sqrt{n\Sigma x^2 - (\Sigma x)^2}\sqrt{n\Sigma y^2 - (\Sigma y)^2}}$$

t-Test for the Correlation Coefficient:

$$t = \frac{r}{\sqrt{\frac{1 - r^2}{n - 2}}} \quad (\text{d.f.} = n - 2)$$

Equation of a Regression Line: $\hat{y} = mx + b$

where $m = \dfrac{n\Sigma xy - (\Sigma x)(\Sigma y)}{n\Sigma x^2 - (\Sigma x)^2}$ and

$$b = \overline{y} - m\overline{x} = \frac{\Sigma y}{n} - m\frac{\Sigma x}{n}$$

Coefficient of Determination:

$$r^2 = \frac{\text{explained variation}}{\text{total variation}} = \frac{\Sigma(\hat{y}_i - \overline{y})^2}{\Sigma(y_i - \overline{y})^2}$$

Standard Error of Estimate: $s_e = \sqrt{\dfrac{\Sigma(y_i - \hat{y}_i)^2}{n - 2}}$

c-Prediction Interval for y is $\hat{y} - E < y < \hat{y} + E$, where

$$E = t_c s_e \sqrt{1 + \frac{1}{n} + \frac{n(x_0 - \overline{x})^2}{n\Sigma x^2 - (\Sigma x)^2}}. \quad (\text{d.f.} = n - 2)$$

Table 4—Standard Normal Distribution *(continued)*

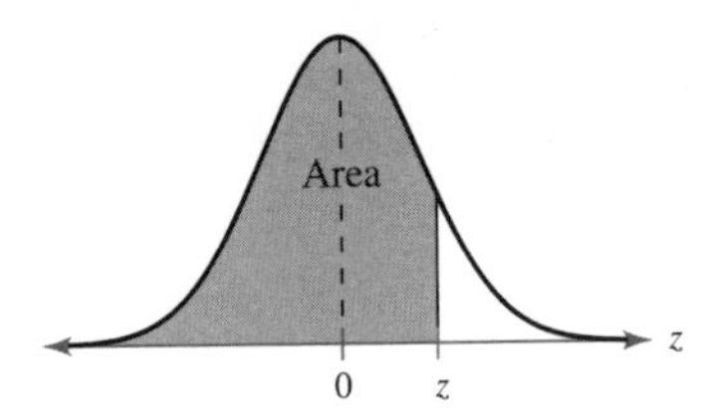

z	.00	.01	.02	.03	.04	.05	.06	.07	.08	.09
0.0	.5000	.5040	.5080	.5120	.5160	.5199	.5239	.5279	.5319	.5359
0.1	.5398	.5438	.5478	.5517	.5557	.5596	.5636	.5675	.5714	.5753
0.2	.5793	.5832	.5871	.5910	.5948	.5987	.6026	.6064	.6103	.6141
0.3	.6179	.6217	.6255	.6293	.6331	.6368	.6406	.6443	.6480	.6517
0.4	.6554	.6591	.6628	.6664	.6700	.6736	.6772	.6808	.6844	.6879
0.5	.6915	.6950	.6985	.7019	.7054	.7088	.7123	.7157	.7190	.7224
0.6	.7257	.7291	.7324	.7357	.7389	.7422	.7454	.7486	.7517	.7549
0.7	.7580	.7611	.7642	.7673	.7704	.7734	.7764	.7794	.7823	.7852
0.8	.7881	.7910	.7939	.7967	.7995	.8023	.8051	.8078	.8106	.8133
0.9	.8159	.8186	.8212	.8238	.8264	.8289	.8315	.8340	.8365	.8389
1.0	.8413	.8438	.8461	.8485	.8508	.8531	.8554	.8577	.8599	.8621
1.1	.8643	.8665	.8686	.8708	.8729	.8749	.8770	.8790	.8810	.8830
1.2	.8849	.8869	.8888	.8907	.8925	.8944	.8962	.8980	.8997	.9015
1.3	.9032	.9049	.9066	.9082	.9099	.9115	.9131	.9147	.9162	.9177
1.4	.9192	.9207	.9222	.9236	.9251	.9265	.9278	.9292	.9306	.9319
1.5	.9332	.9345	.9357	.9370	.9382	.9394	.9406	.9418	.9429	.9441
1.6	.9452	.9463	.9474	.9484	.9495	.9505	.9515	.9525	.9535	.9545
1.7	.9554	.9564	.9573	.9582	.9591	.9599	.9608	.9616	.9625	.9633
1.8	.9641	.9649	.9656	.9664	.9671	.9678	.9686	.9693	.9699	.9706
1.9	.9713	.9719	.9726	.9732	.9738	.9744	.9750	.9756	.9761	.9767
2.0	.9772	.9778	.9783	.9788	.9793	.9798	.9803	.9808	.9812	.9817
2.1	.9821	.9826	.9830	.9834	.9838	.9842	.9846	.9850	.9854	.9857
2.2	.9861	.9864	.9868	.9871	.9875	.9878	.9881	.9884	.9887	.9890
2.3	.9893	.9896	.9898	.9901	.9904	.9906	.9909	.9911	.9913	.9916
2.4	.9918	.9920	.9922	.9925	.9927	.9929	.9931	.9932	.9934	.9936
2.5	.9938	.9940	.9941	.9943	.9945	.9946	.9948	.9949	.9951	.9952
2.6	.9953	.9955	.9956	.9957	.9959	.9960	.9961	.9962	.9963	.9964
2.7	.9965	.9966	.9967	.9968	.9969	.9970	.9971	.9972	.9973	.9974
2.8	.9974	.9975	.9976	.9977	.9977	.9978	.9979	.9979	.9980	.9981
2.9	.9981	.9982	.9982	.9983	.9984	.9984	.9985	.9985	.9986	.9986
3.0	.9987	.9987	.9987	.9988	.9988	.9989	.9989	.9989	.9990	.9990
3.1	.9990	.9991	.9991	.9991	.9992	.9992	.9992	.9992	.9993	.9993
3.2	.9993	.9993	.9994	.9994	.9994	.9994	.9994	.9995	.9995	.9995
3.3	.9995	.9995	.9995	.9996	.9996	.9996	.9996	.9996	.9996	.9997
3.4	.9997	.9997	.9997	.9997	.9997	.9997	.9997	.9997	.9997	.9998

Key Formulas

From Larson/*Farber Elementary Statistics: Picturing the World,* Second Edition

Chapter 10

Chi-Square: $\chi^2 = \Sigma \dfrac{(O - E)^2}{E}$

Goodness-of-Fit Test: d.f. $= k - 1$

Test of Independence:

d.f. $=$ (no. of rows $- 1$)(no. of columns $- 1$)

Two-Sample F-Test for Variances: $(s_1^2 \geq s_2^2)$ $F = \dfrac{s_1^2}{s_2^2}$

(d.f.$_N = n_1 - 1$, and d.f.$_D = n_2 - 1$)

One-Way Analysis of Variance Test:

$$F = \frac{MS_B}{MS_W}, \text{ where } MS_B = \frac{SS_B}{k - 1} = \frac{\Sigma n_i\left(\bar{x}_i - \dfrac{\Sigma x}{N}\right)^2}{k - 1}$$

and $MS_W = \dfrac{SS_W}{N - k} = \dfrac{\Sigma(n_i - 1)s_i^2}{N - k}$.

(d.f.$_N = k - 1$, d.f.$_D = N - k$)

Chapter 11

Test Statistic for Sign Test: When $n > 25$,

$z = \dfrac{(x + 0.5) - 0.5n}{0.5\sqrt{n}}$, where x is the smaller number of $+$ or $-$ signs and n is the total number of $+$ and $-$ signs.

When $n \leq 25$, the test statistic is the smaller number of $+$ or $-$ signs.

Test Statistic for Wilcoxon Rank Sum Test: $z = \dfrac{R - \mu_R}{\sigma_R}$, where $R =$ sum of the ranks for the smaller sample,

$\mu_R = \dfrac{n_1(n_1 + n_2 + 1)}{2}$, $\sigma_R = \sqrt{\dfrac{n_1 n_2(n_1 + n_2 + 1)}{12}}$, and $n_1 \leq n_2$.

Test Statistic for the Kruskal-Wallis Test:

Given three or more independent samples, the test statistic for the Kruskal-Wallis test is

$$H = \frac{12}{N(N + 1)}\left(\frac{R_1^2}{n_1} + \frac{R_2^2}{n_2} + \cdots + \frac{R_k^2}{n_k}\right) - 3(N + 1).$$

(d.f. $= k - 1$)

The Spearman Rank Correlation Coefficient:

$$r_s = 1 - \frac{6\Sigma d^2}{n(n^2 - 1)}$$

Table 4—Standard Normal Distribution

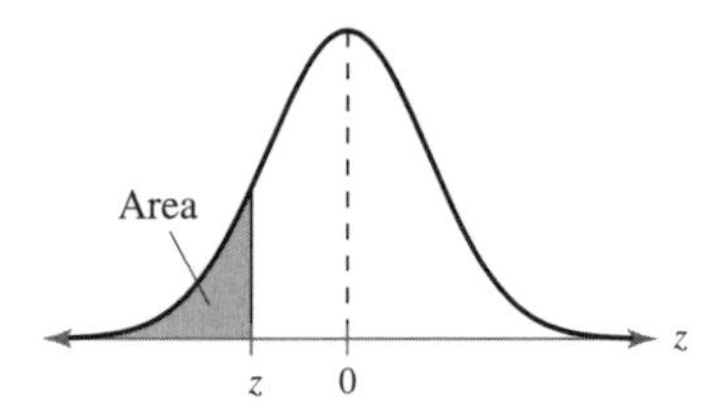

z	.09	.08	.07	.06	.05	.04	.03	.02	.01	.00
−3.4	.0002	.0003	.0003	.0003	.0003	.0003	.0003	.0003	.0003	.0003
−3.3	.0003	.0004	.0004	.0004	.0004	.0004	.0004	.0005	.0005	.0005
−3.2	.0005	.0005	.0005	.0006	.0006	.0006	.0006	.0006	.0007	.0007
−3.1	.0007	.0007	.0008	.0008	.0008	.0008	.0009	.0009	.0009	.0010
−3.0	.0010	.0010	.0011	.0011	.0011	.0012	.0012	.0013	.0013	.0013
−2.9	.0014	.0014	.0015	.0015	.0016	.0016	.0017	.0017	.0018	.0019
−2.8	.0019	.0020	.0021	.0021	.0022	.0023	.0023	.0024	.0025	.0026
−2.7	.0026	.0027	.0028	.0029	.0030	.0031	.0032	.0033	.0034	.0035
−2.6	.0036	.0037	.0038	.0039	.0040	.0041	.0043	.0044	.0045	.0047
−2.5	.0048	.0049	.0051	.0052	.0054	.0055	.0057	.0059	.0060	.0062
−2.4	.0064	.0066	.0068	.0069	.0071	.0073	.0075	.0078	.0080	.0082
−2.3	.0084	.0087	.0089	.0091	.0094	.0096	.0099	.0102	.0104	.0107
−2.2	.0110	.0113	.0116	.0119	.0122	.0125	.0129	.0132	.0136	.0139
−2.1	.0143	.0146	.0150	.0154	.0158	.0162	.0166	.0170	.0174	.0179
−2.0	.0183	.0188	.0192	.0197	.0202	.0207	.0212	.0217	.0222	.0228
−1.9	.0233	.0239	.0244	.0250	.0256	.0262	.0268	.0274	.0281	.0287
−1.8	.0294	.0301	.0307	.0314	.0322	.0329	.0336	.0344	.0352	.0359
−1.7	.0367	.0375	.0384	.0392	.0401	.0409	.0418	.0427	.0436	.0446
−1.6	.0455	.0465	.0475	.0485	.0495	.0505	.0516	.0526	.0537	.0548
−1.5	.0559	.0571	.0582	.0594	.0606	.0618	.0630	.0643	.0655	.0668
−1.4	.0681	.0694	.0708	.0722	.0735	.0749	.0764	.0778	.0793	.0808
−1.3	.0823	.0838	.0853	.0869	.0885	.0901	.0918	.0934	.0951	.0968
−1.2	.0985	.1003	.1020	.1038	.1056	.1075	.1093	.1112	.1131	.1151
−1.1	.1170	.1190	.1210	.1230	.1251	.1271	.1292	.1314	.1335	.1357
−1.0	.1379	.1401	.1423	.1446	.1469	.1492	.1515	.1539	.1562	.1587
−0.9	.1611	.1635	.1660	.1685	.1711	.1736	.1762	.1788	.1814	.1841
−0.8	.1867	.1894	.1922	.1949	.1977	.2005	.2033	.2061	.2090	.2119
−0.7	.2148	.2177	.2206	.2236	.2266	.2296	.2327	.2358	.2389	.2420
−0.6	.2451	.2483	.2514	.2546	.2578	.2611	.2643	.2676	.2709	.2743
−0.5	.2776	.2810	.2843	.2877	.2912	.2946	.2981	.3015	.3050	.3085
−0.4	.3121	.3156	.3192	.3228	.3264	.3300	.3336	.3372	.3409	.3446
−0.3	.3483	.3520	.3557	.3594	.3632	.3669	.3707	.3745	.3783	.3821
−0.2	.3859	.3897	.3936	.3974	.4013	.4052	.4090	.4129	.4168	.4207
−0.1	.4247	.4286	.4325	.4364	.4404	.4443	.4483	.4522	.4562	.4602
−0.0	.4641	.4681	.4721	.4761	.4801	.4840	.4880	.4920	.4960	.5000

Critical Values

Level of Confidence c	z_c
0.80	1.28
0.90	1.645
0.95	1.96
0.99	2.575

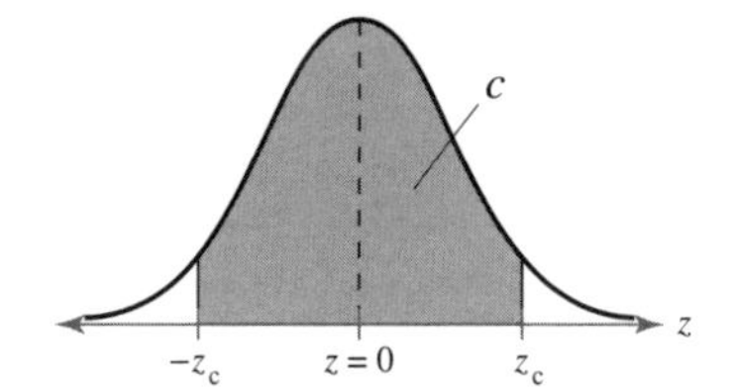